Trees and Shrubs
HARDY IN THE BRITISH ISLES

Volume IV

Ri – Z

Trees and Shrubs

HARDY IN THE BRITISH ISLES

W. J. BEAN
CVO, ISO, VMH

Eighth Edition Revised

CHIEF EDITOR

D. L. CLARKE
VMH

GENERAL EDITOR

SIR GEORGE TAYLOR
DSC, FRS, VMH

VOLUME IV

Ri–Z

JOHN MURRAY

First Edition (Vols I & II) 1914
First Edition (Vol. III) 1933
Seventh Edition 1950
Eighth Edition Revised (Vol. IV) 1980

*Published in collaboration with
The Royal Horticultural Society
Revisions and additions for Eighth Edition*
© *M. Bean and John Murray (Publishers) Ltd 1980*

First published 1980
by John Murray (Publishers) Ltd
50 Albemarle Street, London W1X 4BD

*Printed in Great Britain by
Butler and Tanner Ltd, Frome and London*

British Library Cataloguing in Publication Data

Bean, William Jackson
Trees and shrubs hardy in the British Isles
Vol. 4: Ri–Z.—8th ed. revised
1. Trees—Great Britain—Dictionaries
2. Shrubs—Great Britain—Dictionaries
I. Title
582'.16'0941   QK488

ISBN: 0-7195-2428-8

CONTENTS

PREFACE TO VOLUME IV

Publication of Volume IV brings to a close the major revision of the descriptive text of W. J. Bean's *Trees and Shrubs Hardy in the British Isles*. There is now only the *Supplement* to follow, and work on it is well under way.

That this point has been reached is undoubtedly due to the indefatigable industry and remarkable attention to detail of Mr Desmond Clarke, who has not only borne the brunt of the revision but has also been responsible in very large measure for its organisation. His achievements have twice been recognised by the Royal Horticultural Society, in the award of the Victoria Medal of Honour in 1970 followed by the award of the Veitch Memorial Medal in 1979.

I deem it correct to comment on the courage of John Murray in accepting the responsibility in difficult times of publishing a completely revised edition of Bean. Simon Young of John Murray has throughout been the administrator of this great project. He has been a stimulating colleague, a persistent stickler for detail and always at crucial times a driving force.

In this Edition for the first time since the work was originally published the hybrid garden roses are being treated. The formidable task of selection and concise description that this involved was willingly undertaken by Mr Graham Stuart Thomas, whose work as horticulturalist, author, botanical illustrator and photographer must surely be unique. His contribution is a major asset to Volume IV and we are deeply grateful for the care and skill he put into it. It is fitting that some of his photographs are also included.

Once more our thanks go to the Director of the Royal Botanic Gardens at Kew, Prof. J. P. M. Brenan, for generously providing facilities both in the form of assistance by his staff and in placing the Herbarium and Library at the Chief Editor's disposal; to Mr J. Robert Sealy, who has been with this revision all through, a triumphant fulfilment of his many years' association with W. J. Bean; to Mr P. S. Green, Mr C. Jeffrey and Mr N. Taylor, whose assistance with particular genera is acknowledged in the text; and to Mr G. E. Brown, Mr C. M. Erskine and Mr A. D. Schilling for material help; also to Mr R. D. Meikle for comments on the treatment of *Salix*.

We are grateful to the Keeper of Botany, Mr J. F. M. Cannon, for allowing access to the Herbarium of the British Museum of Natural History, and to Mrs Gillian Ballance for drawing to our attention the drawing of *Sorbus torminalis* by Miss B. O. Corfe that has been reproduced on the front of the dust jacket. Mr G. D. R. Bridson, the Librarian of the Linnean Society of London, has been unfailingly helpful to both Chief Editor and publisher and thanks are also accorded to the Regius Keeper, Mr D. M. Henderson, of the Royal Botanic Garden, Edinburgh, and his staff.

Our thanks also go to Miss Margaret Bean for again sharing the task of checking the proofs and for her enthusiastic encouragement throughout the project; to Mr H. G. Hillier for the many specimens collected from the arboretum he created at Romsey, Hants, and to Mr R. L. Lancaster for his continuing interest and assistance, and also to Mr P. H. B. Gardner for much help; to the late Mrs Madeline Spitta for making Dr W. Fox's papers and *Sorbus* photographs available; to Mr Alan Mitchell for giving measurements of outstanding specimen trees. Miss Mary Grierson's fine series of botanical drawings to supplement those retained from the Seventh Edition has now been completed. Some new photographers appear in the List of Illustration Sources, and especial mention must be made of Dr R. H. M. Robinson for the trouble he has gone to over photographing the garden roses.

Finally, renewed thanks are due to the Royal Horticultural Society for their continued support and a very warm word of thanks to our printers for their skilled copy preparation, and consistently high standard of proof-reading and presswork.

GEORGE TAYLOR

1980

APPROXIMATE METRIC EQUIVALENTS

INCHES TO MILLIMETRES

$\frac{1}{32}$ in	=	0·8 mm	$\frac{5}{8}$ in	= 15·9 mm	4 in	= 101 mm
$\frac{1}{16}$	=	1·6	$\frac{3}{4}$	= 19·1	5	= 127
$\frac{1}{8}$	=	3·2	$\frac{7}{8}$	= 22·2	6	= 152
$\frac{3}{16}$	=	4·8	1	= 25·4	7	= 178
$\frac{1}{4}$	=	6·4	$1\frac{1}{4}$	= 31·8	8	= 203
$\frac{5}{16}$	=	7·9	$1\frac{1}{2}$	= 38·1	9	= 229
$\frac{3}{8}$	=	9·5	$1\frac{3}{4}$	= 44·5	10	= 254
$\frac{5}{16}$	=	11.1	2	= 51	11	= 279
$\frac{1}{2}$	=	12·7	3	= 76	12	= 305

FEET TO METRES

2 ft	= 0·61 m	12 ft	= 3·66 m	60 ft	= 18·3 m
3	= 0·91	15	= 4·57	70	= 21·3
4	= 1·22	20	= 6·10	80	= 24·4
5	= 1·52	25	= 7·62	90	= 27·4
6	= 1·83	30	= 9·15	100	= 30·5
7	= 2·13	35	= 10·67	110	= 33·5
8	= 2·44	40	= 12·20	120	= 36·6
9	= 2·74	45	= 13·72	130	= 39·6
10	= 3·05	50	= 15·2	140	= 42·7
11	= 3·55			150	= 45·7

ALTITUDES TO METRES

500 ft	= 152 m	4,000 ft	= 1,219 m
1,000	= 305	5,000	= 1,524
1,500	= 457	10,000	= 3,048
3,000	= 914	15,000	= 4,572

TEMPERATURES: °F TO °C

0 °F	= −17·8 °C	45 °F	= 7·2 °C
10	= −12·2	50	= 10·0
20	= −6·7	55	= 12·8
32	= 0·0	60	= 15·6
40	= 4·4	65	= 18·3
		70	= 21·1

TREE MEASUREMENTS

All measurements of girth were taken at 5 ft (1·52 m) unless otherwise stated

LIST OF DRAWINGS IN THE TEXT

Those marked with an asterisk were drawn by Miss E. Goldring: the remainder are by Miss Mary Grierson.

SOURCES OF PLATES

B. Alfieri, 46, 61

J. E. Downward, 3, 5, 33, 38, 44, 45, 47, 58, 49, 66, 68, 79, 94, 95, 97, 100, 106, 108

W. Eng, 100

M. Hadfield, 89

L. W. Hammett, 83, 85

A. J. Huxley, 22, 34, 35, 41, 70, 107

E. F. Marten, 21, 51

W. Marten, 64

E. M. Megson, 1, 23, 43, 63, 72, 88, 105

B. O. Mulligan, 2, 10, 20, 53, 57, 67, 69, 80, 91

M. Nimmo, 74, 76

S. Orme, 4

R. H. M. Robinson, 8, 9, 12, 13, 14, 15, 16, 17, 18, 19, 25, 27, 28, 29, 30, 31, 77

H. Smith, 6, 26, 32, 37, 48, 52, 56, 60, 62, 65, 71, 75, 78, 82, 84, 86, 92, 93, 96, 99, 101, 102, 103

G. S. Thomas, 7, 11, 24, 81, 109

E. J. Wallis, 36, 39, 49, 54, 59, 73, 87, 90, 104, 110, 111

C. Wormald, 55

LIST OF PLATES

xiii

1 RIBES SANGUINEUM

2 RIBES ODORATUM

3 RICHEA SCOPARIA

4 ROBINIA PSEUDACACIA 'FRISIA'

5 ROMNEYA COULTERI
× R. TRICHOCALYX

6 Rosa banksiae

7 Rosa moschata

8 ROSA CENTIFOLIA 'MUSCOSA'

9 ROSA DAMASCENA 'TRIGINTIPETAL

10 Rosa ecae
11 Rosa filipes 'Kiftsgate'

12 ROSA GALLICA

13 ROSA GALLICA 'VERSICOLOR'

Rosa laevigata 'Cooperi'

15 Rosa roxburghii
f. normalis

16 Rose 'Alba Maxima'
17 Rose 'A Longues Pédoncules'

18 Rose 'Boule de Neige'

19 Rose 'Du Maître D'Ecole'

20 ROSE 'CANTABRIGIENSIS'

21 ROSE 'COMPLICATA'

22 ROSE 'FANTIN-LATOUR'

23 ROSE 'FRU DAGMAR HASTRUP' in fruit

24 ROSE 'FRÜHLINGSGOLD'

25 ROSE 'HANSA'

26 ROSE 'MOUSSU DU JAPON'

27 ROSE 'NIPHETOS'

28 ROSE 'OLD BLUSH CHINA'

29 ROSE 'PORTLANDICA'

30 ROSE 'QUATRE SAISONS BLANC MOUSSEUX'

31 ROSE 'REINE VICTORIA'

32 Rubus cockburnianus

33 Salix hastata 'Wehrhahnii'

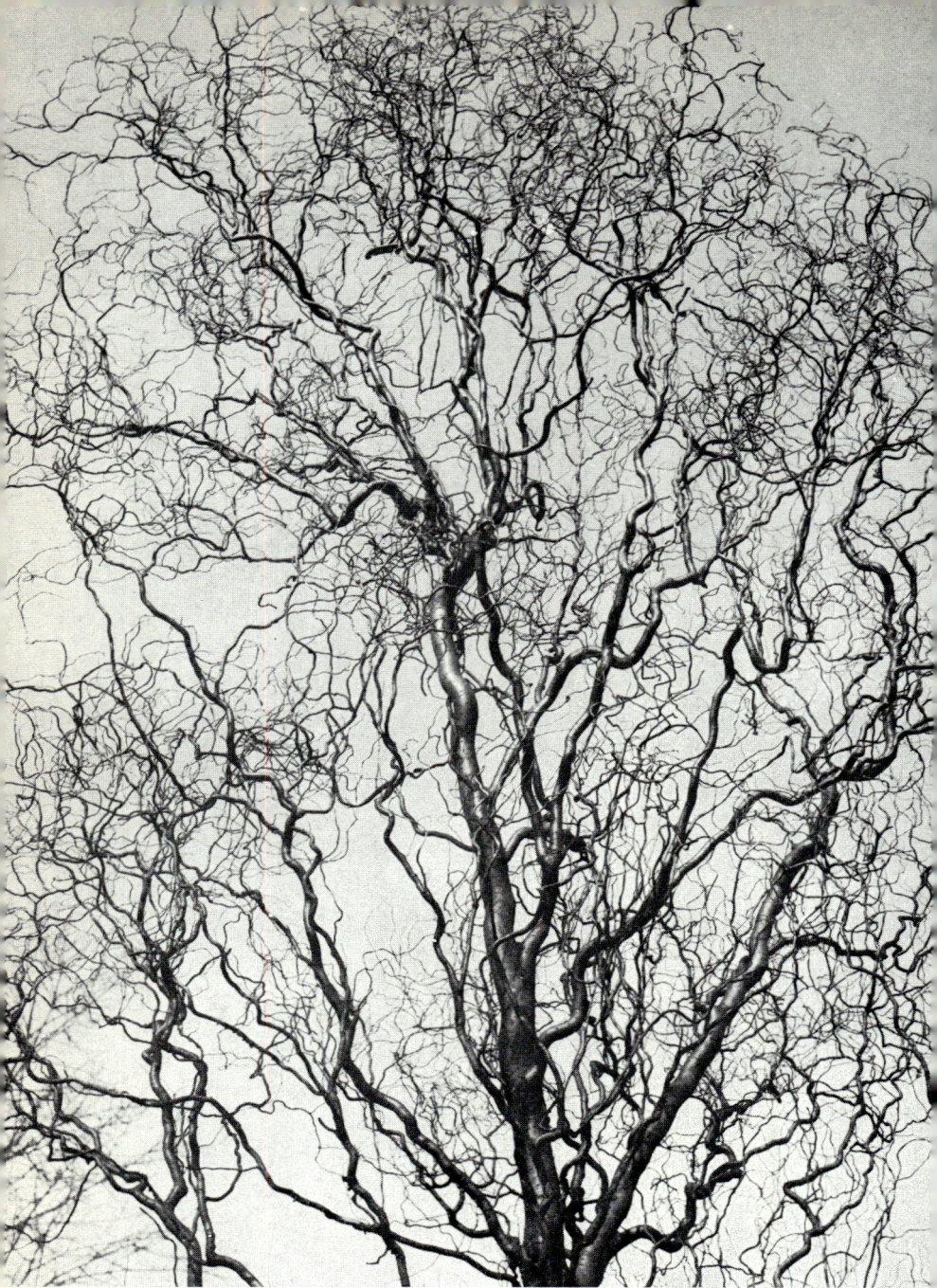

34 SALIX MATSUDANA 'TORTUOSA'

35 SALIX LANATA

36 SALIX × SEPULCRALIS 'SALAMONII'

37 Sambucus racemosa
'Plumosa Aurea'

38 Schima argentea

39 Sassafras albidum

40 SCHISANDRA RUBRIFLORA in fruit

41 SCHIZOPHRAGMA HYDRANGEOIDES

42 SCIADOPITYS VERTICILLATA

43 SENECIO (DUNEDIN HYBRID) 'SUNSHINE'

44 SEQUOIA SEMPERVIRENS 'CANTAB'

45 Sᴇǫᴜᴏɪᴀᴅᴇɴᴅʀᴏɴ ɢɪɢᴀɴᴛᴇᴜᴍ at Westonbirt

46 SEQUOIADENDRON
GIGANTEUM cones

47 SKIMMIA JAPONICA

48 SOLANUM JASMINOIDES

49 SOPHORA JAPONICA
at Kew

50 SORBARIA ARBOREA

51 SORBUS ALNIFOLIA

52 SORBUS ARIA 'MAJESTICA'

53 SORBUS CASHMIRIANA

54 Sorbus intermedia

55 Sorbus latifolia and S. torminalis

56 Spiraea 'Arguta'

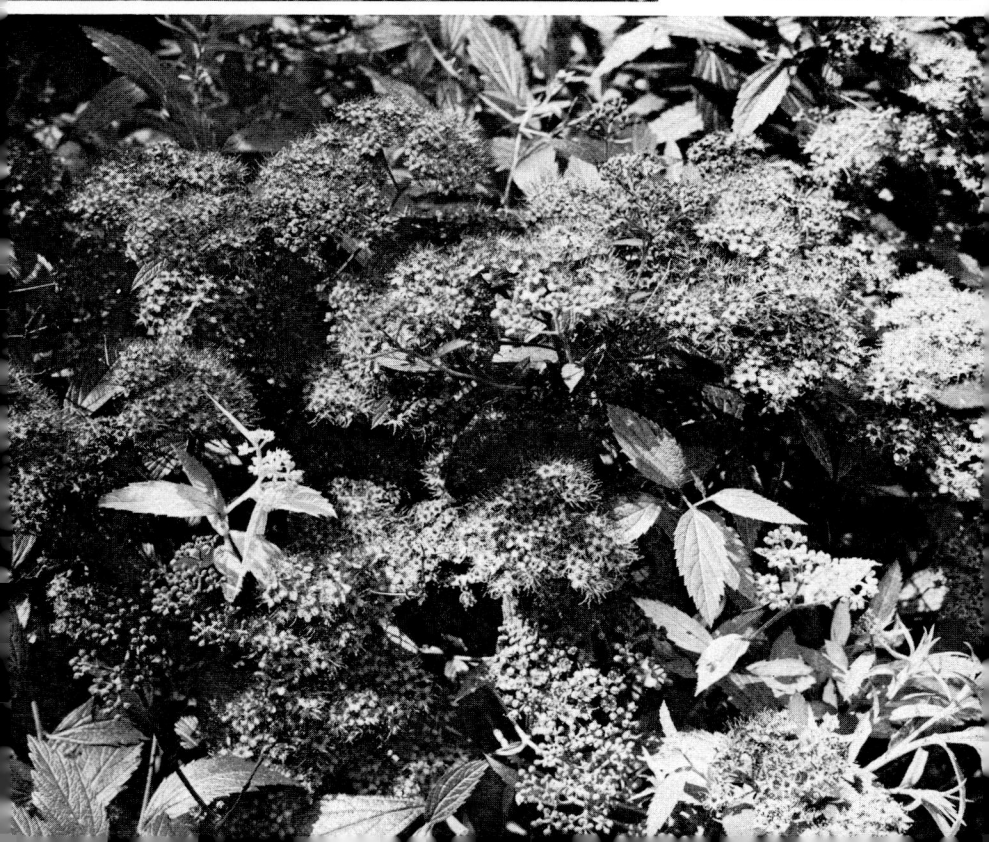

58 SPIRAEA JAPONICA
'ANTHONY WATERER'

59 Spiraea trichocarpa

60 Spiraea × vanhouttei

61 STACHYURUS CHINENSIS

62 STAPHYLEA 'COULOMBIERI'

63 STUARTIA SINENSIS bark

64 STUARTIA PSEUDOCAMELLIA

65 STRANVAESIA DAVIDIANA

66 Styrax hemsleyana

67 Sycopsis sinensis

68 SYMPHORICARPOS 'WHITE HEDGE'

69 SYMPLOCOS PANICULATA

70 Syringa 'Maud Notcutt'

71 Syringa reflexa

72 SYRINGA MEYERI 'PALIBIN'

73 Taiwania cryptomerioides

74 TAMARIX species, Poole Harbour, Dorset

75 Tamarix tetrandra

76 Taxodium distichum at F

77 Taxodium distichum

78 Taxus baccata 'Fastigiata'

79 TELOPEA TRUNCATA

80 TETRACENTRON SINENSE

81 Teucrium fruticans

82 Thuja occidentalis
'Ellwangeriana Aurea'

83 THUJA OCCIDENTALIS 'FILIFORMIS'

84 Tilia × europaea

85 TORREYA CALIFORNICA cones

86 TRACHYCARPUS FORTUNEI

87 Tsuga mertensiana

88 Tsuga heterophylla

89 Ulmus glabra

90 Ulmus 'Sarniensis'

91 Ulmus parvifolia

92 ULEX EUROPAEUS 'FLORE PLENO'

93 VIBURNUM × BODNANTENSE

94 VIBURNUM × CARLCEPHALUM

95 VIBURNUM DILATATUM

96 Viburnum plicatum

97 Viburnum tinus

98 Vitex agnus-castus

99 Vinca major

100 Vinca minor f. atropurpurea

101 VITIS COIGNETIAE

102 WEIGELA VARIEGATED
CULTIVAR

103 Weigela 'Rosea'

104 Wisteria floribunda
'Alba'

105 WISTERIA FLORIBUNDA

106 WISTERIA AND VIBURNUM MACROCEPHALUM

107 Xanthoceras sorbifolium

108 Yucca gloriosa

109 Yucca filamentosa,
3-yr-old from root cuttings

110 Yucca recurvifolia

111 Zelkova carpinifolia, many-stemmed form

RIBES   Currants and Gooseberries
GROSSULARIACEAE

A genus of about 150 species in the temperate parts of the northern hemisphere, extending southward in the New World to Chile and Argentina. They are deciduous, more rarely evergreen, shrubs with alternate, mostly three- or five-lobed leaves. Inflorescence racemose, sometimes reduced to a few-flowered cluster. The most conspicuous feature of the flower is the receptacle ('calyx-tube'), which varies in shape from tubular to widely cup-shaped or campanulate, its lower part adnate to the ovary; sepals normally five. Petals shorter than the sepals and usually inconspicuous. Stamens four or five. Styles two, more or less united. Fruit a berry, crowned by the persistent sepals.

The great majority of the species fall into two well-marked groups:

CURRANTS

Stems without spines at the nodes (except in R. *diacanthum* and some related species not treated here). Flower-stalks jointed. Inflorescence usually a raceme with more than five flowers (but see R. *fasciculatum*). In what might be termed the 'true' currants the flowers are perfect (subgen. *Ribes*, formerly *Ribesia*), but there is a group of species, constituting the subgenus *Berisia*, which are dioecious. To the latter belong R. *alpinum* and its allies, R. *orientale*, R. *tenue*, R. *maximowiczii*, R. *laurifolium*, R. *henryi*, R. *fasciculatum* and R. *gayanum*. In the two last-named species the flowers appear to be perfect, but the pollen in the female flowers and the ovules in the male flowers are sterile.

GOOSEBERRIES

Stems with spines at the joints. Flower-stalks not jointed. Inflorescence a racemose cluster with not more than four flowers, which are perfect. This is the subgenus *Grossularia*, by some authorities treated as a separate genus.

There is a small group of species—subgen. *Grossularioides*—which is intermediate between the currants and the gooseberries. They are armed like the latter, but the flower-stalks are articulated and the inflorescence is a many-flowered raceme, as in the currants. See R. *lacustre*.

The ribes present no difficulty in cultivation; they like a loamy soil of at least average quality, and the West N. American gooseberries need as sunny a spot as possible. They are propagated by seed or by cuttings. The latter will frequently form roots when made of leafless shoots in November and placed in the open air—as common gooseberries are—but they are more certain if placed under a handlight. A second method, better adapted to the currants, is to make cuttings of leafy shoots in July and August and place them in gentle bottom-heat, but most of them will strike root also in the open ground, of course much more slowly.

1

R. ALPINUM L.   MOUNTAIN CURRANT

A deciduous unarmed shrub, reaching in gardens 6 to 9 ft in height and as much or more in diameter, of dense, close habit; young twigs shining, and at first more or less glandular. Leaves broadly ovate or roundish, three- sometimes five-lobed, the lobes coarsely toothed, the base straight or heart-shaped, with five radiating veins; upper surface with scattered bristly hairs, the lower one usually shining and more or less hairy on the veins, $\frac{1}{2}$ to $1\frac{1}{2}$ in. long and wide; stalk glandular-downy, $\frac{1}{4}$ to $\frac{1}{2}$ in. long. Flowers unisexual, the sexes nearly always on separate plants, produced in the axils of bracts longer than the flower-stalk, greenish yellow; the males up to thirty together on small, erect, glandular racemes 1 to $1\frac{1}{2}$ in. long, the females fewer, and on racemes half as long. Fruits red, not palatable.

Native of Europe from N. England and Scandinavia to N. Spain, central Italy and Bulgaria, but rather patchily distributed and confined to the mountains

RIBES ALPINUM

in the southern part of its range; also of the Atlas and the Caucasus. In the British Isles it is a local species, found here and there in northern England and Wales, usually on limestone; it is abundant in woods near Fountains Abbey in Yorkshire. Although commonly under 6 ft high, there were tree-like plants 15 ft high of this species in an old hedge on the east front terrace of the old hall at Troutbeck, reputed to be 300 years old.

Although this currant has no special beauty of flower or fruit it makes a very neat and pleasing shrub, admirable for shady places. Occasionally plants with perfect flowers may be found.

cv. 'Aureum' ('Osborn's Dwarf Golden').—Leaves yellow throughout the summer, habit as in 'Pumilum', of which it may have been a branch-sport. First Class Certificate when shown by Osborn's nursery, Fulham, in 1881 (*Gard. Chron.*, Vol. 16 (1881), p. 333) as *R. alpinum* var. *pumilum* subvar. *aureum* or Osborn's Dwarf Golden Ribes; *R. alpinum* var. *aureum* Nichols., in *Dict. Gard.*, Vol. III (1887), p. 304). It colours best in full sun.

cv. 'Laciniatum'.—Leaves more deeply lobed and toothed than in the type.

cv. 'Pumilum'.—A dwarf variety with smaller leaves, 2 to 3 ft high but more in diameter; a very neat bush. It is female, though some flowers appear to have fertile stamens. It was described by Lindley in 1827; the original plant had grown in Miller's nursery, Bristol, for many years, but was of unknown origin.

cv. 'Pumilum Aureum'.—This appears to have been similar to 'Osborn's Dwarf Golden' and possibly represents the same clone; it was described in Belgium three years before Osborn's nursery exhibited their plant (Pynaert in *Rev. Hort. Belg.*, Vol. 4 (1878), p. 233), as *R. alpinum pumilum aureum; R. alpinum* var. *aureum* Rehd. (1902).

The so-called var. '*sterile*' appears to be merely the normal male-flowered plant. None of the forms of *R. alpinum* need a rich soil. They retain the neat, compact habit which is their greatest merit, in rather poor soil.

R. ambiguum Maxim.

A small, sparsely branched, stout-stemmed low shrub, usually growing as an epiphyte in the wild. Leaves orbicular, 1 to 2 in. wide, cordate at the base, shallowly three- or five-lobed, toothed, downy or glabrous above, downy and glandular beneath. Flowers bisexual, greenish, paired, on pedicels $\frac{3}{16}$ to $\frac{3}{8}$ in. long, each furnished with a pair of bracteoles near the apex; sepals oblong-elliptic, spreading to erect. Fruits glandular bristly, broadly ellipsoid, containing numerous small seeds.

Native of central and southern Japan, where it grows on trees in mountain forests. An interesting small shrub for a cool, sheltered position in leafy soil.

R. americanum Mill.   American Black Currant
R. *floridum* L'Hérit.; R. *missouriense* Hort.

This shrub is unarmed, and closely akin to the common black currant, which it resembles in having three- or five-lobed leaves with a coarse, irregular toothing and deeply heart-shaped base, and in possessing the same heavy odour, due to yellowish glands on the lower surface. The fruit also is black. The American species, however, is quite distinct in the flowers; these are nearly twice as long, more tapering and funnel-shaped, and yellow. Moreover, the bract from the axil of which each flower springs on the raceme is longer than the stalk. (In R. *nigrum* it is small and much shorter than the flower-stalk.)

Native of eastern N. America from Nova Scotia to Virginia, and west to the Rocky Mountains; introduced in 1729. As a garden shrub the only quality which recommends this currant is that its foliage becomes suffused with brilliant hues of crimson and yellow in autumn.

R. BRACTEOSUM Hook.   CALIFORNIAN BLACK CURRANT

An unarmed deciduous shrub 6 to 10 ft high; young shoots glabrous, except for a little loose down at first. Leaves handsomely five- or seven-lobed, 3 to 7 in. (sometimes more) wide, the lobes palmate, reaching half or more than half-way to the midrib, sharply and irregularly toothed, dotted with resin-glands beneath, bright green and soon quite glabrous above; stalk slender, often longer than the blade, glabrous except for a few bristles at the base. Racemes produced in May, erect, slender, up to 8 in. long. Flowers numerous, greenish yellow, erect, ⅓ in. across, each on a slender, slightly downy stalk about ¼ in. long. Fruits erect, resin-dotted, globose, ⅓ in. in diameter, black with a blue-white bloom. *Bot. Mag.*, t. 7419.

Native of western N. America; discovered by Douglas in 1826. An interesting species of the black currant group, very distinct in its large maple-like leaves (occasionally 10 in. across) and long, slender, erect racemes. Rarely seen, but quite hardy at Kew.

R. CEREUM Dougl.

A grey, deciduous unarmed shrub 3 to 6 ft high, of rounded, compact habit; young shoots downy and glandular. Leaves roundish or rather broader than long, three- or five-lobed, the lobes irregularly round-toothed, the base straight or heart-shaped, ½ to 1¼ in. wide, upper surface sown with white, resinous glands and slightly downy, lower one downy on the veins; stalk downy, glandular, nearly as long as the blade. Flowers tubular, ⅓ to ½ in. long, clustered two to five together at the end of a short, downy, glandular stalk, the individual flower almost stalkless, downy, white tinged with rose, produced in the axil of a comparatively large, toothed, glandular-downy, wedge-shaped bract. Style downy. Fruits bright red, ¼ in. in diameter. *Bot. Mag.*, t. 3008.

Native of western N. America; introduced in 1827 by David Douglas. It is a very pleasing shrub, conspicuous in the pale grey tint of its young leaves, and pretty in the delicate colouring of its abundant blooms. These appear with the young leaves in April.

R. INEBRIANS Lindl.   R. *cereum* var. *inebrians* (Lindl.) C. L. Hitchc.—Very similar to the above, and equally pleasing, this differs in having the bract at the base of each flower not toothed and pointed, the style glabrous, and the flowers deeper in colour. Introduced from western N. America in 1827.

R. CILIATUM Humb. & Bonpl.
R. *jorullense* Kunth

A deciduous unarmed shrub up to 10 ft high, of bushy rounded shape; young shoots slender, arching, downy and furnished with stalked glands. Leaves three- or sometimes five-lobed, doubly toothed, heart-shaped at the base, 1½ to 2 in. long and wide, dark dull green above and sprinkled with appressed bristles, downy beneath and furnished there with stalked glands, especially on the veins, margins glandular; stalk ¾ to 1¼ in. long, glandular. Flowers nearly ½ in. long, borne six to twelve together during April in nodding, stalked racemes about 2 in. long; greenish, downy outside, bell-shaped; ovary glabrous; main and secondary flower-stalks glandular and downy. Fruits globose, black, shining, as large as a red currant.

Native of Mexico. It is one of the currant section of the genus and is quite hardy at Kew, where it has attained 9 ft by 12 ft. It flowers freely every year. The stalked glands are black and shining.

R. CRUENTUM Greene
R. *roezlii* var. *cruentum* (Greene) Rehd.

A deciduous spiny shrub 3 to 6 ft high, more in diameter; young shoots minutely downy. Leaves roundish, ¾ to 1½ in. wide, three- or five-lobed, the lobes coarsely round-toothed, nearly or quite glabrous on both surfaces, stalk minutely downy, slender, ¼ to ½ in. long. Flowers ¾ in. wide, solitary, rarely in pairs, on a slender stalk ⅓ in. long, pendent. Receptacle ½ in. long, crimson, the tube narrowly bell-shaped, glabrous, the five sepals lanceolate, finally reflexed. Petals white, much shorter than the sepals; ovary covered with incipient spines. Fruits red, ⅔ in. across, with a hedgehog-like appearance due to their covering of numerous spines, each ⅙ to ¼ in. long. *Bot. Mag.*, t. 8105.

Native of California and S. Oregon; introduced in 1899. This interesting and remarkable gooseberry has flowers extremely pretty in their contrast of crimson and white, but they are not particularly abundant, usually one at each joint of the previous year's wood. The berries are remarkable in their prickliness. It is closely allied to, and perhaps only a variety of R. *roezlii*, but that species is distinctly downy on leaf and receptacle. Effective grown as a standard.

R. CURVATUM Small

A low, deciduous, bushy shrub 3 ft high; the shoots glabrous, purplish, armed with slender, simple or triple spines. Leaves roundish, usually 1 in. or less in diameter, three- to five-lobed, toothed, slightly downy; stalk slender, downy. Flowers produced singly or in pairs (rarely more) on pendent stalks, white. Receptacle bell-shaped with linear, much reflexed sepals ¼ in. long; petals very short, white; ovary covered with resinous glands; stamens ¼ in. long, erect, both they and the style downy. Fruits globose, glabrous, ⅓ in. across, purplish green.

Native of the south-eastern United States, hardy. I brought plants from the Arnold Arboretum to Kew in July 1910, which, so far as I am aware, were the first introduced to this country. R. *curvatum* is, however, no longer represented at Kew. It is closely allied to R. *niveum*, which it resembles in its white flowers and downy style and stamens, but the glandular ovary and often glabrous anthers are different. R. *curvatum* is also much dwarfer in habit, and comes from the opposite side of N. America.

R. DIACANTHUM Pall.

A deciduous shrub 4 to 6 ft high, armed with spines in pairs $\frac{1}{8}$ to $\frac{1}{5}$ in. long, or sometimes unarmed; young shoots not downy. Leaves obovate or rounded, often three-lobed, the lobes coarsely toothed, $\frac{3}{4}$ to 2 in. wide, the base ordinarily wedge-shaped but sometimes rounded, quite glabrous; stalk $\frac{1}{4}$ to $\frac{5}{8}$ in. long, more or less furnished with bristles. Flowers unisexual, the sexes on different plants. Males yellowish, in erect glandular racemes. Fruits roundish oval, about as big as a red currant, glabrous, scarlet-red.

Native of Siberia, Manchuria, etc.; introduced in 1781. This shrub, which has no particular merit, resembles R. *alpinum* in the plants being one-sexed, but differs in having prickles, and in the markedly wedge-shaped leaves. In spite of its prickles, it is undoubtedly a currant of the subgenus *Berisia* and does not resemble the gooseberries (subgenus *Grossularia*) in any other respect.

R. FASCICULATUM Sieb. & Zucc.
R. japonicum sens. Carr., not Maxim.

A deciduous unarmed shrub 3 to 5 ft high; young shoots finely downy. Leaves three- to five-lobed, the largest 2 in. long, $2\frac{1}{2}$ to 3 in. wide, the lobes coarsely toothed, usually more or less downy; stalk downy and with feathered bristles near the base. Flowers unisexual, the sexes on separate plants. Males clustered four to nine together in a stalkless umbel, each flower is on its own stalk without uniting on a common one, yellow, fragrant; females usually in pairs, sometimes three or four. Fruits erect on a stalk $\frac{1}{5}$ in. long, round, $\frac{1}{3}$ to $\frac{1}{2}$ in. in diameter, glabrous, bright scarlet.

Native of China, Japan, and Korea, and distinct from all other species in cultivation in having the flowers clustered in fascicles.

var. CHINENSE Maxim. R. *billiardii* Carr.—A taller shrub, partially evergreen, more downy than the type. The fruits of both are ornamental, and remain long on the branches.

R. GAYANUM (Spach) Steud.
Rebis (sic) *gayanum* Spach; *R. villosum* C. Gay, not Nutt.

An unarmed evergreen shrub 3 to 5 ft high; the young wood, leaf-stalks, flower-stalks, ovary, and calyx shaggy with soft hairs. Leaves stout, greyish, very broadly or roundish ovate, 1 to 2 in. long and broad, the three lobes rounded and toothed, the base usually straight, downy on both sides. Flowers bell-shaped, yellow, honey-scented, closely packed in erect, cylindrical racemes, 1 to 2 in. long, $\frac{1}{2}$ in. in diameter. Berries about the size of peas, purple-black, hairy. *Bot. Mag.*, t. 7611.

Native of central Chile. A handsome evergreen, and distinct in the shape and colour of its inflorescence, and the hairiness of its various parts. Some forms are less downy. Flowers in early June. It has been cultivated at Kew for many years, and is quite hardy.

R. GLANDULOSUM Grauer

R. prostratum L'Hérit.

A deciduous unarmed shrub with prostrate, rooting branches; young shoots glabrous. Leaves deeply five- to seven-lobed, 1½ to 4 in. wide, the lobes doubly toothed, bright green and smooth on both sides, except for occasional hairs on the veins beneath; stalk bristly at the base. Flowers greenish, produced eight to twelve together on erect, slender racemes, 2 to 3 in. long, stalks and ovary with gland-tipped hairs; sepals glabrous outside. Fruits red, glandular, ¼ in. in diameter.

Native of N. America, where it is widely spread over the cool moist regions on both east and west sides; introduced in 1812. It is distinct in its prostrate habit, nearly or quite glabrous, evil-smelling leaves, and red, glanded fruits. Nearly allied to this are:

R. LAXIFLORUM Pursh R. *affine* Dougl., not Kunth—This is also a prostrate shrub, but with leaves more downy beneath when young, the fan-shaped petals as broad as they are long (in R. *prostratum* they are much longer than broad), the sepals downy but not glandular outside, and the fruits black or dark purple, with glandular down and a glaucous bloom. Native of western N. America, whence it was introduced by Douglas in 1818.

R. COLORADENSIS Cov.—This is a third species belonging to the same group, being of prostrate habit; the young shoots are finely downy, the sepals with glandular hairs outside, the purplish petals twice as broad as long, the fruits black, not glaucous with bloom. Native of Colorado and New Mexico; introduced in 1905.

None of these three have much garden value, although their prostrate habit gives them interest.

R. × GORDONIANUM Lem.

R. beatonii Loud., *nom. nud.*

A hybrid between R. *odoratum* and R. *sanguineum*, raised in 1837 by Donald Beaton at Haffield, Herefordshire, and named after his employer William Gordon; it was put into commerce by Low of Clapton. Beaton, one of the most learned plantsmen of his time, later became head gardener to Sir William Middleton, Bt, at Shrublands Park, Ipswich, where he bred some of the forerunners of the present-day bedding pelargoniums. He made several contributions to Loudon's *Gardeners' Magazine*, and corresponded with Dean Herbert, the authority on the Amaryllidaceae, who named after him the genus *Beatonia*, now reduced to the rank of a subgenus of *Tigridia*.

Beaton's currant is intermediate in most respects between its parents—in habit, in the leaves being smaller and less hairy than those of R. *sanguineum*, and in the colour of the flowers, which are reddish outside, yellowish within, a curious blend. It is hardier than R. *sanguineum*, and can be grown in parts of the New England States where that species is too tender to thrive. It is interesting and not without beauty, but is inferior to both of its parents.

R. HENRYI Franch.

An evergreen shrub 3 to 4 ft high; young shoots glandular-bristly. Leaves tufted at the end of the shoots, obovate or diamond-shaped, tapering more or less equally towards both ends or more abruptly towards the apex, sharply pointed, finely and irregularly toothed, the teeth gland-tipped, with tiny bristles between them, 2 to 4 in. long, 1 to $2\frac{1}{4}$ in. wide, glabrous, yellowish green above with conspicuous sunken veins in about five pairs, lower surface pale, with short stiff hairs on the midrib and veins, between which the blade is thickly sprinkled with minute sticky glands; stalk $\frac{1}{4}$ in. or less long, glandular-bristly. Flowers unisexual, borne on separate plants, greenish yellow, produced early in the year along with new shoots in racemes 1 to 2 in. long, surrounded at the base by a cluster of pale green, membranous bracts; main and secondary flower-stalks clothed with glandular hairs, giving them a mossy appearance as seen under the lens. Fruits narrowly oval, $\frac{1}{2}$ in. long, glandular-hairy.

Native of Central China; introduced (apparently inadvertently) by Wilson in 1908. A plant which came up in a sowing of seeds of *Sinowilsonia henryi* at the Edinburgh Botanic Garden proved to be this species. It is closely akin to R. *laurifolium* and, like it, flowers in February and March, but the glabrous leaf-blades of that species easily distinguish it. The moss-like down of R. *henryi* is also much longer and more conspicuous. This currant is quite hardy.

R. LACUSTRE (Pers.) Poir.

R. *oxyacanthoides* β *lacustris* Pers.

A deciduous shrub 3 to 5 ft high, the stems thickly covered with slender prickles or stiff bristles; spines at the joints numerous, from three to nine arranged in a semicircle. Leaves 1 to $2\frac{1}{4}$ in. long and wide, handsomely and deeply three- or five-lobed, the lobes often again deeply cut; stalk and chief veins more or less bristly. Flowers from twelve to twenty in glandular-downy drooping racemes, 2 to 3 in. long, funnel-shaped, with short, spreading sepals brownish crimson inside, creamy white or pinkish outside. Fruits round, about the size of a black currant, covered with gland-tipped bristles, black.

Native of N. America on both sides of the continent, inhabiting cold damp localities; introduced in 1812. Although the general aspect of this shrub is that of a gooseberry, especially in the shape of its leaves and in its spines, it has the long racemes and flowers of the currants. Its multiple spines are also distinct. Although it has no lively colour to recommend it, it is pretty when its branches are strung with the graceful drooping racemes.

R. MONTIGENUM McClatchie R. *lentum* Cov. & Rose—This is another species which unites, as R. *lacustre* does, the two main subdivisions of the genus, but has shorter, fewer-flowered racemes (six to ten) and bright red fruits. Introduced from western N. America in 1905.

R. LAURIFOLIUM Jancz.

An unarmed evergreen dioecious shrub, rarely more than 3 ft high in cultivation, sparsely branched; branchlets sometimes reddish when young, at first glandular, then smooth and brown. Leaves leathery, ovate to oval, pointed, the largest 5 in. long, half as wide, coarsely toothed, dark dull green above, paler and brighter beneath, glabrous on both surfaces; stalk bristly. Flowers greenish yellow, produced in late winter or early spring. Male flowers $\frac{3}{8}$ to $\frac{1}{2}$ in.

RIBES LAURIFOLIUM

wide, in nodding racemes up to 2 in. or so long; receptacle cup-shaped, downy; abortive style bifid. Female flowers up to $\frac{3}{8}$ in. long, in racemes about 1 in. long, which are at first upright, later nodding; receptacle downy, flask-shaped or oblong; stamens reduced to staminodes. Bracts in both sexes about $\frac{1}{2}$ in. long, ciliate. Fruits ellipsoid, about $\frac{5}{8}$ in. long, reddish but said to be black when fully ripe. *Bot. Mag.*, t. 8543.

Native of W. China; discovered and introduced in 1908 by Wilson, according to whom it is rare in the wild. It is not a showy plant, but is interesting and welcome in flowering as early as February or March. Male plants are more ornamental than female, the racemes being longer and the flowers larger. Award of Merit 1912. It needs no pruning, apart from the occasional removal of the oldest stems.

R. LEPTANTHUM A. Gray

A deciduous spiny shrub 3 to 6 ft high, with slightly downy, occasionally glandular-bristly young branches; spines usually slender, solitary, up to $\frac{1}{2}$ in. long. Leaves roundish or somewhat kidney-shaped, $\frac{1}{4}$ to $\frac{3}{4}$ in. wide, deeply three- or five-lobed, toothed, the base mostly truncate; stalk as long as the blade, downy at the base. Flowers white tinged with pink, one to three on a short stalk; receptacle cylindrical, the sepals downy, ultimately reflexed. Fruits oval, shining, blackish red, slightly downy or glabrous.

Native of Colorado, New Mexico, etc.; one of the prettiest and daintiest of gooseberries, the branches being slender and densely clothed with tiny leaves. Introduced in 1893.

R. QUERCETORUM Greene R. *leptanthum* var. *quercetorum* (Greene) Jancz.— This has pale yellow flowers, fragrant, and produced two to four together. Native of California.

R. LOBBII A. Gray
R. subvestitum sens. Hook. in *Bot. Mag.*, t. 4931, not Hook. & Arn.

A deciduous spiny shrub 3 to 6 ft high; young shoots downy. Leaves roundish in the main, $\frac{3}{4}$ to 2 in. wide, three- to five-lobed, the lobes roundish toothed, downy or glabrous above, downy and glandular beneath and on the stalk. Flowers usually in pairs on a glandular-hairy stalk. Calyx-tube purplish red, downy, the sepals twice or thrice the length of the tube, recurved; petals white, erect, the stamens much protruded beyond them; anthers almost as broad as long; ovary covered with glands. Berry oblong, red brown, glandular. *Bot. Mag.*, t. 4931.

Native of western N. America from southern British Columbia to northern California; introduced about 1852 by W. Lobb for Messrs Veitch, but not often seen now, although, like the others of this group, very pretty when flowering in April. From the allied crimson-flowered gooseberries in cultivation, viz., R. *menziesii*, R. *roezlii*, and R. *cruentum*, this is very well distinguished in flower by the anthers being rounded at the top (in the others they are tapered like an arrowhead).

R. LONGERACEMOSUM Franch.

An unarmed shrub up to 12 ft high; young shoots glabrous. Leaves 3 to $5\frac{1}{2}$ in. long and wide, with three or five acute or acuminate lobes; petioles up to $4\frac{1}{2}$ in. long. Racemes pendulous, 12 to 18 in. long, sparsely set with greenish or reddish tubular-campanulate flowers on pedicels about $\frac{1}{4}$ in. long. Stamens and styles exserted. Fruits black, of about the size of a black currant, said to be of good flavour, but very thinly disposed along the main-stalk.

Wilson introduced this extraordinary currant in 1908 from W. China, where it had originally been discovered by the Abbé David. The one character that distinguishes it from its relatives is its remarkable tassel-like racemes more than 1 ft long. It is quite hardy, but uncommon in this country. A plant which grew originally in the collection of Janczewski, the monographer of *Ribes*, has attained a height of 12 ft in the Kornik Arboretum, Poland; it produces its racemes freely there, but does not fruit (*Arb. Kornickie*, Year Book No. 10 (1965), p. 60).

R. MAXIMOWICZII Batal.

A deciduous unarmed bush, ultimately 6 to 9 ft high, the young shoots clothed with pale hairs, some of which are glandular. Leaves of the black currant type, three- or sometimes five-lobed, 2 to 5 in. wide and about as long, glossy dark green and thinly downy above, clothed beneath with soft pale down, especially on the veins; stalk 1 to 2½ in. long, downy. Racemes slender, erect, 2 to 4 in. long, about ½ in. wide, main-stalk as well as individual flower-stalks very downy and glandular. Flowers ¼ in. wide, dull lurid red; receptacle funnel-shaped, glandular downy. Fruits globose, ⅜ in. wide, orange-coloured or red and covered with stalked glands.

Native of W. China, from Kansu to Szechwan; discovered by the Russian traveller, Potanin, in 1885; introduced by Wilson in 1904 and again in 1908 and 1910. It is a curious and remarkable currant on account of the racemes and very glandular fruit and the lurid hue of the blossoms which open in May.

var. FLORIBUNDUM Jesson   R. *jessoniae* Stapf—This variety differs from typical R. *maximowiczii* in the longer inflorescences (4 to 6 in. long), the more numerous flowers, and in the shorter and fewer glandular bristles on the berries. *Bot. Mag.*, t. 8840. The flowering material figured in the *Botanical Magazine* is from a plant at Borde Hill, Sussex, raised by F. D. Godman at South Lodge from seeds sent by Wilson from W. Szechwan during his first expedition for the Arnold Arboretum. The fruiting material figured was provided by Mrs Berkeley of Spetchley Park, Worcester (sister of the famous Miss Ellen Willmott), who reported to Kew that she had made a palatable jam from the fruits.

R. MENZIESII Pursh

A deciduous spiny shrub up to 6 ft high; young shoots downy and covered with long, slender bristles. Leaves roundish ovate in the main, 1 to 2 in. wide, deeply three- sometimes five-lobed, the lobes toothed, either glabrous or with gland-tipped hairs above, very downy and glandular beneath and on the stalk. Flowers in pairs or solitary on the slender, glandular, and downy stalk. Receptacle red-purple with a short, bell-shaped base, the sepals ⅓ in. long; petals white, sometimes rosy tinted. Fruits globose, covered with glandular bristles, the remains of the flower persisting at the top.

Native of western N. America from central California to Oregon; introduced in 1830. A vigorous gooseberry, which flowers freely in this country in May. From R. *lobbii*, with which it is much confused in gardens, it is distinguished by its bristly stems, the stalked glands on the ovary, and the tapered anthers. The contrast of purple and white in the flowers is pretty.

R. MOGOLLONICUM Greene

R. wolfii Rothrock in part only, *nom. confus.*

A sturdy unarmed shrub, said to become 9 to 11 ft high; young shoots glabrous or nearly so. Leaves three- or five-lobed, 2 to 3½ in. long and wide, heart-shaped at the base, glabrous above, downy only on the veins beneath, and with scattered glands which impart a somewhat disagreeable odour to the leaves when rubbed; stalk downy and glandular. Flowers greenish white, themselves short-stalked, but closely set on erect long-stalked racemes 1 to 1½ in. long; the stalks and ovary covered densely with stalked glands. Fruits ⅓ in. wide, roundish ovoid, glandular, purplish black. *Bot. Mag.*, t. 8120.

Native of Colorado, New Mexico, etc.; introduced to Kew in 1900, where it is very hardy and fruits freely. Its only interest for the garden is in the blue, ultimately black, glandular fruits arranged densely in more or less erect spikes. It belongs to the same group of currants as R. *sanguineum*, but has none of the flower beauty of that species.

R. NIGRUM L.   BLACK CURRANT

An unarmed shrub 5 or 6 ft high, distinguished by its peculiar odour, due to small yellowish glands sprinkled freely over the lower surface of the leaf, which is conspicuously three-lobed, deeply notched at the base, long stalked, coarsely toothed. Flowers bell-shaped, dull white, in racemes, each flower from the axil of a minute bract. Fruits black.

Native of Europe and Siberia, possibly of Britain. Several varieties of this species—so well known as the "black currant" of fruit gardens—have been distinguished. The two first mentioned are curious and interesting, but no others are worth cultivating as ornamental shrubs:

cv. 'DISSECTUM'.—Leaves very curiously cut, each of the three lobes reaching back to the stalk, and again bipinnately lobed (R. *nigrum* var. *dissectum* Bean, probably the same as R. *nigrum* f. *apiifolia* Kirchn.).

cv. 'LACINIATUM'.—The three primary lobes reaching nearly or quite to the stalk, and pinnately lobed.

cv. 'RETICULATUM AUREUM'.—Leaves mottled thickly with yellow. This or a similar variant was described by Mouillefert under the name R. *nigrum marmoratum*.

R. × CULVERWELLII Macfarlane   R. × *schneideri* Koehne—A hybrid between the black currant and the gooseberry. The typical form of the cross was raised by William Culverwell of Thorpe Perrow, Yorks, about 1880; it is a spineless shrub, and has flowers like the black currant, but the foliage and inflorescence are more suggestive of the gooseberry. An interesting curiosity, of no value either for fruit or for ornament. The Culverwell cross had the black currant as the seed-parent, but the type of R. × *schneideri*, of German origin and of the same parentage, was raised from the gooseberry pollinated by the black currant.

R. NIVEUM Lindl.

An armed deciduous shrub up to 9 ft high, the young shoots quite glabrous; spines solitary or in threes, about ½ in. long. Leaves between roundish and kidney-shaped, three- to five-lobed, 1 to 1½ in. across, usually truncate at the base, the lobes unequally and bluntly toothed. Flowers two to four together in slender-stalked, drooping clusters. Receptacle glabrous, white; sepals ultimately much reflexed, leaving the stamens exposed for ⅓ in. Ovary glabrous, stamens and style downy. Fruits globose, glabrous, black with a purplish bloom, about ½ in. in diameter. *Bot. Mag.*, t. 8849.

Native of western N. America; introduced in 1826. This gooseberry is rather pretty and distinct in its wholly white flowers, which open in April.

R. × ROBUSTUM Jancz.—A hybrid between R. *niveum* and either R. *oxyacanthoides* or R. *inerme* (the latter not treated here). The plant at Kew made a vigorous bush and was received in 1890 from Mr Nyeland, gardener to the King of Denmark.

R. ODORATUM Wendl.   BUFFALO CURRANT   [PLATE 2

R. *aureum sens.* Lindl. and other authors, not Pursh; R. *longiflorum* Nutt.; R. *fragrans* Lodd., not Pall.

A deciduous, lax-habited, spineless shrub 6 to 8 ft high, producing a crowded mass of stems which branch and arch outwards at the top; young shoots minutely downy. Leaves usually three-lobed, often broadly wedge-shaped or palmate, the lobes coarsely toothed, ¾ to 2 in. long, as much or more wide, pale green on both sides, and glabrous, or soon becoming so; stalks glabrous or downy, ½ to 2 in. long, very variable in length compared with the blade. Flowers spicily fragrant, bright golden yellow, appearing in April in semi-pendulous racemes 1 to 2 in. long, each flower with a cylindrical receptacle usually about ½ in. long, sepals slightly less than half as long as the receptacle, recurved after flowering. Fruits roundish, glabrous, dark purple or purplish black, about $\frac{7}{16}$ in. wide.

Native of the central USA from Arkansas and Minnesota westward to the Rocky Mountains; introduced in 1812. This species and R. *sanguineum* are by far the most attractive of the currants in their blossom. It is very distinct among them in its long, yellow receptacle ('calyx-tube').

cv. 'AURANTIACUM MINUS'.—Flowers fragrant, of a deeper, more orange shade than usual, and a more compact habit. The flowers are rather small, in this respect resembling those of R. *aureum*, but the young stems are downy, as in R. *odoratum*. Plants under this name were distributed by Booth's nursery at Flottbeck near Hamburg, shortly before 1864, but the plants at Kew came from Dr Dieck of Zöschen in 1889.

f. XANTHOCARPUM Rehd.—Fruits yellow.

R. AUREUM Pursh R. *tenuiflorum* Lindl. GOLDEN CURRANT—This species is very closely allied to R. *odoratum* and was named earlier. It differs in its glabrous or at most finely downy young stems, smaller flowers (receptacle usually not more than ⅔ in. long), with spreading sepals more than half as long as the receptacle. The flowers are fragrant; fruits palatable, red, yellow or black. R. *aureum* has a more western distribution than R. *odoratum*, extending across the Rocky Mountains to the eastern slopes of the Cascades, and has a variety—var. *gracillimum* (Cov. & Britt.) Jeps.—in the Coastal Range of California.

R. *aureum* was introduced to Britain in 1824 but is now rare in cultivation, the plants once grown under its name being usually R. *odoratum*.

R. ORIENTALE Desf.

R. *resinosum* Pursh; R. *punctatum* Lindl.

An unarmed deciduous shrub 5 or 6 ft high; young shoots and leaf-stalks covered with stiff, gland-tipped sticky hairs. Leaves of the red currant size and shape, but shining green and with bristly down on the nerves beneath; stalk ½ to 1 in. long. Flowers unisexual, the sexes on different plants, and produced on somewhat erect racemes 1 to 2 in. long; they are green suffused with red and covered with viscid hairs. Fruits red, downy.

Native mainly of S.W. Asia, but extending into S.E. Europe; in cultivation 1813. It is a species of little garden value, allied to R. *alpinum*, but in that species the fruits are glabrous and the stems and leaves almost so; also the leaf-buds are acute and elongate, broader and obtuse in R. *orientale*. The R. *resinosum* of Pursh, long thought to be a native of N. America, and figured as such in *Bot. Mag.*, t. 1583, is really this species.

R. PINETORUM Greene

A gooseberry growing 6 ft high, found in the Mogollon Mountains of New Mexico and in Arizona, often in pine-woods, an association from which it derives its name; introduced in 1902. It has the typically shaped leaf of the gooseberries, glabrous, blunt-toothed, and with long, slender stalks. The young shoots are quite glabrous; the spines solitary, in pairs, or in threes, rich brown, stout, slightly curved. Flowers solitary, orange-yellow, hairy outside; the sepals much reflexed, showing the erect petals. Fruits black-purple, globose, ½ in. in diameter, with numerous bristles. Although this species has some of the most brilliantly coloured blossoms among gooseberries, they are short-stalked and solitary (or very rarely in pairs) at each joint, and make no great display. They appear in May, when the leaves are one-third grown.

R. ROEZLII Reg.

R. *amictum* Greene

A deciduous armed shrub 3 to 6 ft high; young shoots downy. Leaves ½ to 1 in. wide, roundish or kidney-shaped in general outline, three- or five-lobed, the lobes with often sharp teeth, more or less downy on both surfaces, especially beneath; stalk ⅓ in. long, usually downy and sometimes glandular-hairy. Flowers solitary or in pairs on a short downy, often glandular stalk, pendent. Receptacle purplish crimson, downy, cylindrical, ¼ in. long, the sepals ⅓ in. long; petals rosy white, erect, shorter than the sepals. Fruits purple, ½ in. wide, covered with slender bristles.

Native of California. This pretty and curious gooseberry is not common in cultivation; the plant that was distributed for it from nurseries was as a rule either R. *lobbii* or R. *menziesii*. Its nearest ally is R. *cruentum* (q.v.). Greene's epithet *amictum* refers to the shape of the bract surrounding the base of each flower, which resembles the amice, or hood, worn by Roman Catholic clergy at mass.

R. RUBRUM L.   RED CURRANT

R. *sativum* (Reichenb.) Berger; R. *rubrum* subsp. *sativum* (Reichenb.) Syme; R. *rubrum* var. *sativum* Reichenb.; R. *vulgare* Lam., in part

Little need be said here about the red currant, so well known in its cultivated form in British fruit gardens. It is an unarmed, spreading shrub with three- or five-lobed leaves, 2 to 4 in. across, heart-shaped at the base, very downy beneath, and with scattered hairs above, at least when young; stalk from half to twice as long as the blade. Flowers saucer-shaped, flattish, greenish, produced in recurved racemes from the joints of last year's wood. Receptacle with a fleshy five-angled disk at the base, between the stamens and style. Fruits juicy, red and shining; white in a cultivated variety.

Native of western Europe, but probably not of Britain, where, however, it is widely naturalised, as it is elsewhere in the areas where it is cultivated. It is of little interest except in fruit gardens.

R. SPICATUM Robs.   R. *rubrum sensu* many authors, not L.—Near to R. *rubrum* but the receptacle of the flowers broadly funnel-shaped or bowl-shaped, without a disk at the centre. Racemes more or less upright in the flowering stage. It varies in the amount of down on the undersurface of the leaves and inflorescence. Native of N. Europe, including Britain (where it occurs in Scotland and N. England), and of Siberia, etc. In many works this species appears under the name R. *rubrum* L., but Wilmott pointed out in 1918 that this name belongs properly to the cultivated red currant of W. Europe and the wild plants agreeing with it (*Journ. Bot.*, Vol. 56, p. 22). This view is adopted here, as it is by D. A. Webb in *Flora Europaea*, Vol. 1 (1964), p. 383.

Perhaps not specifically distinct from R. *spicatum* is R. SCHLECHTENDALII Lange (R. *pubescens* (Swartz) Hedl.) with brownish flowers (greenish in R. *spicatum*), downy young stems, and leaves densely downy beneath. It is a native of W. Russia, Poland and parts of Scandinavia, and is cultivated for its fruits.

The taxonomic position of R. WARSZEWICZII Jancz. is uncertain. It was described in 1904 from a plant growing in the Botanic Garden at Krakow, Poland, said to have been raised forty years previously from seeds received from Siberia. It has larger flowers than in R. *spicatum*, pinkish, with a suggestion of a disk in the receptacle, borne in pendulous racemes, and large, acid fruits darker in colour than those of a Morello cherry.

R. MULTIFLORUM Roem. & Kit.—This is one of the red currant group and, as regards its flowers, the most attractive; they are yellowish green, crowded up to forty together on slender, cylindrical, pendulous racemes, sometimes 4 or 5 in. long. When well furnished with these the shrub is quite ornamental. For the rest, it is vigorous, up to 6 ft high, and has stout branches—stouter perhaps than those of any other currant; leaves of the red currant shape, grey with down beneath. Sepals reflexed. Stamens and style exserted. Fruits roundish, red when ripe, about ⅜ in. wide. It is a native of the Balkans and Italy, and the parent of some late-fruiting commercial varieties cultivated in central Europe. It was introduced to Britain in about 1818. *Bot. Mag.*, t. 2368.

R. TRISTE Pall.—This is a widely distributed relative of the red currant, occurring in the colder parts of N. America and Russia. It is a shrub of laxer habit than R. *rubrum*, the leaves white with down beneath when young; flowers purplish; fruit red, small and hard. It is said to be pretty and graceful in blossom in the United States and Canada, where it ranges from Newfoundland to Alaska, with extensions southward along the mountains, and inhabits bogs and wet woods. In Russia it ranges from the Pacific to E. Siberia.

R. PETRAEUM Wulfen—This is another of the red currant group, occurring in central Europe, the Pyrenees, etc., also in N. Africa. It has no value as an ornamental shrub, its flowers being green, suffused with purple, somewhat bell-shaped, in horizontal or slightly nodding racemes, 3 or 4 in. long. The leaves are more deeply lobed than in the common red currant, the lobes pointed. Fruits roundish, flattened somewhat at the end, red, very acid.

R. SANGUINEUM Pursh   FLOWERING CURRANT   [PLATE I

A deciduous, glandular, unarmed bush 7 or 8 ft high, usually considerably more in diameter; young shoots covered with a close, fine down. Leaves three- or five-lobed, palmately veined, the lobes broad and rounded, unequally toothed, the base conspicuously heart-shaped, 2 to 4 in. wide, less in length, glabrous or nearly so above, soft with pale down beneath; stalk ¾ to 2 in. long, covered with minute down like the young shoots, but with a few bristles near the base. Flowers deep rosy red, produced during April in drooping, finally ascending, racemes 2 to 4 in. long, 1 to 1½ in. wide, each flower ½ in. long and nearly as wide; the slender flower-stalk, ovary, and tubular receptacle dotted with glandular down. Fruits globose, ¼ in. in diameter, glandular, black covered with blue bloom. *Bot. Mag.*, t. 3335.

Native of western N. America; discovered by Menzies in 1793, and introduced by Douglas for the Horticultural Society in 1826. This currant is the finest of ribes and in the very front rank of all spring-flowering shrubs, being one of those that never fail to blossom well.

var. GLUTINOSUM (Benth.) Loud. *R. glutinosum* Benth.—Leaves glabrous or almost so beneath. Racemes drooping, the flowers sometimes very pale, almost white. Native of California.

cv. 'ALBESCENS'.—Flowers white with only a slight tinge of colour; leaves hairy beneath. Not so vigorous as the red-flowered sorts (R. *sanguineum* var. *albidum* Nichols., not Kirchn.).

cv. 'ALBIDUM'.—Flowers white, tinged with pink, in dense pendulous racemes. Leaves glabrous beneath. This is referable to the var. *glutinosum* (R. *albidum* Paxt.; R. *sanguineum* var. *albidum* (Paxt.) Kirchn.). It was raised around 1840 in Scotland.

cv. 'ATRORUBENS'.—Fruits deep red in the bud, open flower crimson, rather small, in short, dense racemes. In cultivation in the Horticultural Society's garden 1837. The name R. *s. atrosanguineum* has been used for this and for other forms of similar colouring; this name is not applicable to any one clone.

cv. 'BROCKLEBANKII'.—Leaves a good yellow, the colour lasting well. Of low, bushy habit. Flowers pale pink. A sort raised by the head gardener Thomas Winkworth at Haughton Hall, Tarporley, Cheshire, a few years before 1914 and named after his employer R. Brocklebank; put into commerce by Messrs Dickson of Chester. A.M. 1914. It is of value only for its foliage, which is brighter if the plant is pruned hard in late winter.

cv. 'CARNEUM GRANDIFLORUM'.—Flowers rosy on the outside, white or rose within, in compact trusses. Raised by Dauvesse of Orleans (*Rev. Hort.* 1874, p. 220).

cv. 'FLORE PLENO'.—Flowers double, the segments rosy red, neatly arranged in the form of a rosette. It was found around 1840 by David Dick, head gardener to the Earl of Selkirk, among seedlings raised by his predecessor (*Fl. d. Serres*, Vol. I (1845), pl. 50 and p. 106). 'ATROSANGUINEUM FLORE PLENO' was described as less vigorous than this, with more double but less regular flowers (*Rev. Hort.* 1913, p. 428).

cv. 'KING EDWARD VII'.—Flowers similar in colour to those of 'Pulborough Scarlet' but borne a week or two later on a more compact and spreading bush, to about 6 ft high and more in width, slower in growth. Put into commerce by Messrs Cannell. A.M. 1904.

cv. 'PULBOROUGH SCARLET'.—Flowers Cardinal Red in bud (RHS Colour Chart 53 B), the open flower (i.e., the interior of the sepals) near Bengal Rose (53 B-C), very freely produced in April in racemes 3 to 4 in. long. A vigorous, fast-growing shrub ultimately 10 ft high but less in width, flowering well from the start. Put into commerce by Messrs Cheal. A.M. 1959.

cv. 'SPLENDENS'.—This selection, put into commerce by Smith of Newry, Co. Down, around 1900, has now been largely superseded by 'Pulborough

Scarlet'. It grows as tall and flowers at the same time, but the flowers are not quite so richly coloured. It is still one of the finest varieties, however.

R. MALVACEUM Sm.   R. *sanguineum* var. *malvaceum* (Sm.) Loud.—Leaves bristly, rough, and finely wrinkled above, the lower surface and stalk covered with a grey felt with which are mixed glandular hairs. The flower-stalk and receptacle are also covered with bristly down. Flowers bright rose, smaller and not so beautiful as in R. *sanguineum*. From that species it is distinct in having the ovary and style covered with white hairs. Native of California.

R. SPECIOSUM Pursh

R. *fuchsioides* Berland.

A deciduous spiny shrub 6 to 12 ft high, the young shoots furnished with gland-tipped bristles. Leaves three- sometimes five-lobed, sparsely toothed, and from ¾ to 1¼ in. long and wide, with smaller ones often obovate and tapered at the base; usually quite glabrous; stalk slender, scarcely as long as the blade, with a few glandular bristles, especially at the base. Flowers rich red, usually two to five in pendulous clusters, the main-stalk longer and less glandular than the minor ones. Receptacle ½ in. long, glandular; sepals four, not reflexed;

RIBES SPECIOSUM

petals four, about as long as the sepals; stamens four, red, standing out ¾ in. beyond the receptacle. Fruits glandular-bristly, red, ½ in. long, sometimes seen in this country. *Bot. Mag.*, t. 3530.

Native of California; discovered by Menzies about 1793, and introduced from Monterey by a naval surgeon named Collie in 1828. As a flowering shrub it is the most beautiful of the gooseberries. Its young shoots are reddish, horizontal, or slightly dependent, and from their underside the richly coloured, fuchsia-like blossoms hang profusely in rows during April and May. It is very distinct in the parts of the flower being in fours (not the usual fives), and in the very long highly coloured stamens. It is one of the earliest shrubs to break into leaf—often in early February. It shows to best advantage perhaps against a wall, where it will grow 10 or 12 ft high, but is quite hardy in the open at Kew, where it has grown 6 or 7 ft high. It can be rooted from cuttings, but does not strike readily; layering is a more certain process.

R. TENUE Jancz.

A deciduous shrub up to 6 or 8 ft high of bushy habit and with slender unarmed, glabrous young shoots. Leaves broadly or roundish ovate in main outline, but deeply three- (sometimes five-) lobed as well as sharply and deeply toothed, each lobe pointed, the base cut straight across or slightly heart-shaped; 1 to 2¼ in. long, scarcely so much wide, sprinkled with appressed bristles; stalk slender, ⅓ to 1 in. long, reddish. Racemes of male flowers 1½ to 2 in. long, female ones shorter. Flowers brownish red, main flower-stalk slightly glandular. Fruits globose, red, ¼ in. wide.

Native of the Himalaya and W. China; introduced from the latter region by Wilson in 1900. It is closely related to our native R. *alpinum*, which differs in its greenish-yellow flowers. Beyond its neat habit and small handsomely cut leaves, it has no particular merit. It is one of the earliest of the currants to burst into leaf in spring and flowers in April.

R. UVA-CRISPI L.   COMMON GOOSEBERRY
R. *grossularia* L.

The common gooseberry is found wild in Britain, but is believed to be an escape from gardens. It is a genuine native of most parts of mountainous Europe, and on the Mount Atlas range in N. Africa. In a wild state it is distinguished by its bristly young wood, its downy calyx and hairy ovary, its style downy at the base, and its yellowish or red berry, more or less glandular-hairy. Some cultivated varieties have quite smooth berries.

There is a group of species closely allied to this, composed of deciduous, bushy, spiny shrubs with small green or purplish flowers which have no value in gardens apart from their use as fruit-bearers. They need not be given detailed notice, but may be included here.

R. OXYACANTHOIDES L.—This is widely spread over N. America. It has bristly branches, the leaves are downy, and more or less glandular, the stamens as long as the petals; the berries glabrous, red-purple. R. HIRTELLUM Michx.

is very near this species, but has glabrous shoots and stamens twice as long as the petals, which are purplish. Berry glabrous, purplish or black, ½ in. across—*Bot. Mag.*, t. 6892. It has borne very good fruit in the Isle of Wight, where it is known as "currant-gooseberry".

R. ROTUNDIFOLIUM Michx.—A native of the eastern United States, from Massachusetts to N. Carolina. Its solitary spines are small and inconspicuous; young wood and leaves downy, but not glandular or bristly; flowers greenish purple; receptacle and berry glabrous. The fruit is purple and of good flavour.

R. DIVARICATUM Dougl.—A native of the coast region of western N. America, of vigorous growth, and up to 10 ft high. Its young wood is armed with single or triple spines up to ⅔ in. long, and is sometimes bristly, usually glabrous. Leaves with appressed hairs above, almost or quite glabrous beneath. Receptacle downy, greenish purple, petals whitish, ovary and berry glabrous, the last globose, ¼ in. in diameter, black-purple. This species is nearly allied to R. *rotundifolium*, but is found wild on the opposite side of the continent, and is a bigger bush, well armed with long stout spines.

R. CYNOSBATI L. DOGBERRY.—Native of eastern N. America; introduced in 1759. Its stems are weakly armed or not at all; leaves and leaf-stalks downy; calyx green, bell-shaped with reflexed sepals; petals white; ovary bristly, the bristles not gland-tipped; style downy towards the base; fruit reddish purple, scarcely ½ in. in diameter, more or less covered with slender prickles. f. INERME Rehd. has its fruits smooth, not prickly.

R. VALDIVIANUM Phil.
R. *glandulosum sec.* C. Gay, not Ruiz & Pavon

A deciduous shrub 6 to 12 ft high, much branched; stems grey-hairy at first, sometimes producing sucker growths. Leaves ovate, 1 to 2½ in. long, often nearly as wide, the base varying from cordate to slightly tapered, mostly three-lobed and coarsely toothed, apex pointed to rounded, bright green above, paler beneath and downy on the veins; stalk ½ to 1¼ in. long. Flowers yellow, ⅛ in. wide, ¼ in. long, borne on downy, unisexual racemes, arching or pendulous, the males the larger and up to 3 in. long by ½ in. wide; each raceme comes from the axil of a small, pointed bract; main and individual stalks greyish-hairy. Fruits globose, ⅛ in. wide, purplish black, hairy, edible. *Bot. Mag.*, t. 9647.

Native of the forest region of Chile from 38° to 42° S., and of bordering parts of Argentina; in cultivation at Kew at the end of the last century; re-introduced by H. F. Comber in 1926 from San Martin de los Andes in Argentina. It blooms in late April and May. Except for its liability to injury by late spring frosts, it is hardy. Some of its forms are described as having green flowers. The yellow-flowered ones are quite attractive.

R. VIBURNIFOLIUM A. Gray

An evergreen unarmed shrub 7 or 8 ft high against a wall; young shoots slightly downy at first, with numerous resin-glands. Leaves ovate or oval,

$\frac{3}{4}$ to $1\frac{3}{4}$ in. long, $\frac{1}{2}$ to $1\frac{1}{4}$ in. wide, rounded at the base, blunt at the apex, coarsely toothed, glossy and glabrous above, almost or quite devoid of down beneath, but thickly sown with resin-dots which emit a very pleasant turpentine-like odour when rubbed; stalk downy, $\frac{1}{8}$ to $\frac{1}{6}$ in. long. Flowers $\frac{1}{3}$ in. across, produced in April in erect racemes about 1 in. long, terminating short, densely leafy shoots; dull rose-coloured, the sepals spreading. Fruits oval, red, $\frac{1}{3}$ in. long. *Bot. Mag.*, t. 8094.

Native of Lower California and Santa Catalina Island; introduced to Kew in 1897 but is no longer there. A remarkably distinct species, of little beauty, but interesting for its evergreen aromatically scented leaves. It is tender.

RICHEA EPACRIDACEAE

There are ten species in this Australian genus, all save one confined to Tasmania. It is the type of a small and remarkable group of the Epacris family in which not only are the leaves parallel-veined as in the grasses, but many species in habit and foliage resemble arborescent monocotyledons such as the cordylines and screw-pines (*Pandanus*). The resemblance to the latter is recalled in the name given to the tallest of the richeas, R. *pandanifolia*, which in places attains a height of 30 ft.

The leaves of *Richea* are rigid, with a sheathing base, crowded along the branches or tufted. Flowers in terminal clusters. Corolla closed at the apex except for a small aperture, not separated into lobes but splitting transversely near the base, the lower part persistent, the upper part deciduous (like the operculum of the eucalypts). Stamens usually five. Style one, with a small capitate stigma. Fruit a capsule. The mainly Australasian genus *Dracophyllum*, well represented in New Zealand, is allied to *Richea*.

The genus was named by Robert Brown in honour of the French naturalist C.-A.-G. Riche, who died in 1791 during Admiral d'Entrecasteaux's voyage in vain search of the French explorer La Pérouse.

R. SCOPARIA Hook. f. [PLATE 3

An evergreen shrub up to 10 ft high, but forming rounded, dense hummocks 2 to 3 ft high and up to 8 ft across in exposed places in the mountains, where it ascends to 4,500 ft. Leaves rigid, $\frac{3}{4}$ to 2 in. long, $\frac{1}{10}$ to $\frac{1}{5}$ in. wide, linear-lanceolate, tapering from the base to a slender point, sheathing and completely covering the stem by the broadened base, slightly decurved, glabrous. Flowers closely packed in a terminal, stiffly erect, spike-like raceme, 2 to 4 in. long, $1\frac{1}{4}$ in. wide. Corolla obovate to ovoid, $\frac{1}{4}$ to $\frac{3}{8}$ in. long; see further below and in introductory note. *Bot. Mag.*, t. 9632.

Native of Tasmania. It was introduced by Mr Overall, a Tasmanian nursery-man, but the plants now in cultivation derive from seed collected by H. F. Comber in 1930. According to him, R. *scoparia* varies greatly in colour—'white, pale or deep pink, red or maroon. One report credits it with yellow, but I hesitate to confirm this.' Four colour forms are figured in *Endemic Flora of Tasmania*, Part III, No. 111, including one with orange-coloured flowers tipped with red, collected by Lord Talbot de Malahide on Mount Wellington. Culti-vated plants have the flowers crimson, pink, or white tipped with pink.

R. *scoparia* is hardy, except perhaps in the coldest parts; it needs a moist, acid soil and a sunny position. It received an Award of Merit in 1942, when shown by Col. S. R. Clarke from Borde Hill, Sussex, where it still thrives. It is also cultivated on the rock garden at Kew, at Wakehurst Place in Sussex, and in the Edinburgh Botanic Garden. It is best propagated by seeds, which should be harvested while the capsules are still green, but particular colour-forms have to be perpetuated by cuttings, which can be struck under mist.

R. DRACOPHYLLA R. Br.—Allied to R. *scoparia* but with much larger leaves, 6 in. to 1 ft long, and producing its white flowers in much-branched panicles up to 10 in. long. It is a large shrub in the wild, found in the lower forests of Tasmania, and unlikely to be hardy here except in the mildest parts. It is figured in *Bot. Mag.*, n.s., t. 468, from a plant growing in the Australian House at Kew.

ROBINIA   LEGUMINOSAE

A genus of about twenty deciduous trees and shrubs confined to N. America, whose name commemorates Jean Robin, herbalist to Henry IV of France (*d.* 1629). They are amongst the most ornamental of all hardy trees both in leaf and flower. The leaves are pinnate, and the pea-shaped flowers are borne in pendulous racemes. Pods flat, many-seeded. Stipules often developing into spines.

All the species thrive well in a soil of moderate quality. If given very rich or manured soil they grow so coarse and rank that the danger of damage by wind, due to the brittleness of the branches, is increased. The best method of propagation is by seed, but in the case of R. *hispida* and R. *kelseyi*, and garden varieties, it is usual to graft them on roots or stems of R. *pseudacacia*. This should be done in spring with leafless scions, and the union is more quickly and surely effected if the plants can be kept in a warm greenhouse. Provided they are on their own roots, R. *hispida* and R. *kelseyi* can also be increased by suckers or root-cuttings and plants raised in this way are more desirable than grafted ones, which are short-lived and apt to sucker freely from the stock. The only hardy tree with which the robinias can be confounded is *Sophora japonica*, whose unarmed branches and autumnal flowering readily distinguish it.

The best study of the cultivated robinias is: The Hon. Vicary Gibbs, 'Robinias at Kew and Aldenham', *Journ. R.H.S.*, Vol. 54(1929), pp. 145–58.

R. BOYNTONII Ashe   BOYNTON ACACIA

R. *hispida* β *rosea* Pursh; R. *macrophylla* G. Don, not DC.

A deciduous shrub up to 10 ft high; young shoots unarmed, glabrous or very finely downy. Leaves 6 to 10 in. long, consisting of seven to thirteen leaflets which are oblong, blunt or pointed, 1 to 2 in. long, ½ to 1 in. wide, glabrous or soon becoming so. Flowers six to twelve together in loose racemes 2½ to 3½ in. long, produced from the lower leaf-axils of the young shoots in May and June, each flower barely 1 in. long; standard petal ¾ to 1 in. wide; calyx ¼ in. wide, bristly. The colour is described by Ashe as 'rose-purple, pink, or purple and pink on the outer portion, white or much paler at the base'. Pods glandular-bristly.

Native of the E. United States from N. Carolina and Tennessee to Georgia and Alabama. It may have been introduced long ago and grown under the name "R. *hispida rosea*", but the plant definitely named R. *boyntonii* by Ashe reached Kew in 1919. It is, however, not in cultivation at Kew at present. Ashe separates it from true R. *hispida* 'by its greater size, more oblong leaflets, many-flowered racemes, short calyx-lobes and smoothness'. The young shoots of that species are, of course, very bristly.

R. ELLIOTTII (Chapm.) Ashe

R. *hispida* var. *elliottii* Chapm.; R. *rosea* Ell., not Mill.

This is one of the dwarfest of the robinias, its maximum height being given as about 6 ft. Its stems are erect wands with a few short stout branches near the top; young shoots grey with down. There are eleven to fifteen leaflets to a leaf, which are oval, ½ to 1 in. long, grey-downy beneath. Flowers rose-purple or purple and white, nearly 1 in. long, produced in racemes five to ten together; flower-stalks and calyx grey-downy. Pods bristly. Blooms in May and June.

Native of the S.E. United States from N. Carolina to Georgia, chiefly near the coast. Like R. *boyntonii* it has been grown as "R. *hispida rosea*", but that species differs in not having the grey down on the young shoots, leaves, flower-stalks, and calyx which makes R. *elliottii* so distinct.

R. HARTWIGII Koehne

A deciduous shrub up to 12 ft high; young shoots, leaf-stalks, flower-stalks, and calyx all downy and furnished with stalked glands. Leaves up to 6 or 7 in. long, made up of eleven to twenty-three leaflets which are oval or inclined to ovate, ¾ to 1½ in. long, downy on both sides but more especially underneath, which is greyish. Flowers whitish to rosy purple, nearly 1 in. long including the stalk, densely borne in June and July twenty to thirty together on racemes up to 4½ in. long including an inch or more of bare stalk at the base. Calyx downy, ⅜ in. long, bell-shaped with triangular to awl-shaped lobes. Pods oblong, 2 to 3½ in. long, furnished with down and glandular bristles.

Native of the S.E. United States from N. Carolina to Alabama. This is one of the charming dwarf section of the robinias, its distinctive mark being the mixture of down and stalked glands on the young shoots and leaf-stalks. It was first named in 1913 but had been known in cultivation since 1904 at least. A bush originally obtained from the Arnold Arboretum flowered beautifully at Kew in July 1931. The species is worthy of wide cultivation.

R. HISPIDA L.   ROSE ACACIA

A deciduous unarmed shrub 6 to 8 ft high, of lax, rather gaunt habit, spreading by means of underground suckers, the branches covered with gland-tipped bristles ⅙ in. long. Leaves pinnate, 6 to 10 in. long; leaflets seven to thirteen, each 1½ to 2½ in. long, and from ¾ to 1½ in. wide, oval or ovate with a short bristle-like tip, very dark green; stalk hairy. Racemes 2 or 3 in. long, nearly as much wide, carrying five to ten flowers. The flowers are the largest and most showy among robinias, each about 1¼ in. long, with the rounded standard petal as much across, of a lovely deep rose; calyx ½ to ⅔ in. long, with long, slender, awl-shaped teeth, and bristly like the flower-stalk. Pods 1½ to 2½ in. long, ⅓ in. wide, thickly covered with gland-tipped bristles. Blossoms in May and June. *Bot. Mag.*, t. 311.

Native of the south-eastern United States; introduced in 1743. In the wild it spreads and renews itself by means of sucker-growths extending several feet in a single season, but in cultivation it is usually grafted as a standard on R. *pseudacacia* so as to form a low, bushy-headed tree. Undoubtedly one of the loveliest of all trees of that character, it is, unfortunately, very liable to lose its branches during storms, owing to the brittle nature of its wood. For this reason a secluded spot is desirable for it. A remarkable fact in connection with this tree is the rarity with which it produces seed. It has probably never borne pods in this country, and even in the wild they are very seldom seen. The pods in the Kew Herbarium are three contributed by T. Meehan of Philadelphia, to whom they had been sent in response to inquiries made in a public journal. He himself had made diligent search for seed-pods on the mountains of Tennessee, where the shrub grows in great abundance, but never found any. The defect seems to be in the male part of the flower, and due to the absence of pollen.

var. FERTILIS (Ashe) Clausen R. *fertilis* Ashe—Leaflets relatively narrower, elliptic or oblong-ovate, often acute, downy beneath. A more important difference is that this variety sets fertile seed. A parent of 'Monument', see below.

cv. 'MACROPHYLLA'.—This is distinguished chiefly by the branches and leaf-stalks being nearly or quite free of bristles. The stalks of the racemes and flowers are hairy, but by no means as markedly as in the type. The flowers are even larger and more brightly coloured, the leaflets rounder (R. *macrophylla* Hort.; R. *hispida* var. *macrophylla* DC.; R. *h.* f. *inermis* Kirchn.).

cv. 'MONUMENT'.—A compact shrub to 10 or 12 ft high, with rosy lilac flowers. Raised in the USA from var. *fertilis*, which it resembles in foliage. Also known as R. *fertilis* 'Monument'.

R. KELSEYI Cowell ex Hutch.

A lax-habited, deciduous shrub or small tree, with glabrous, slender branches. Leaves pinnate, 4 to 6 in. long; leaflets nine or eleven, oblong to ovate, 1 to 2 in. long, $\frac{1}{3}$ to $\frac{5}{8}$ in. wide, pointed, glabrous. Flowers brightly rose-coloured, in small clusters at the base of the young twigs; these clusters are sometimes simple racemes of three to eight flowers, but they are frequently forked or triplicate, the stalks always covered with glandular hairs. Each flower is $\frac{3}{4}$ to 1 in. long, with a rounded standard petal $\frac{3}{4}$ in. across; calyx $\frac{1}{4}$ in. long, glandular-hairy,

ROBINIA KELSEYI

teeth narrow, awl-shaped. Pods 2 in. long, $\frac{1}{3}$ in. wide, covered with reddish gland-tipped bristles $\frac{1}{6}$ in. long. *Bot. Mag.*, t. 8213.

The origin of this beautiful robinia is not definitely known. It was put into commerce about 1901, by Mr Harlan P. Kelsey, of Boston, USA, who informs me in a letter that it was 'discovered in our nursery apparently growing spontaneously. We thought at first it was a cross between R. *hispida* and R. *pseudacacia*, but now we think it is a true species that has crept into the collections from the southern Allegheny Mountains.' It was introduced to Kew in 1903, and is certainly one of the most beautiful shrubs added to gardens in recent years. The flowers appear in great profusion in June, and they are followed by handsome red pods. Its affinity with R. *hispida*, especially the smooth-branched form, is apparent, but it is abundantly distinct. Judging by its behaviour at Kew it can be made into a small tree, but it is very brittle. Increase is easily effected by grafting on roots of R. *pseudacacia* in spring.

R. × SLAVINII Rehd.—This hybrid originated from seed of R. *kelseyi* collected by B. H. Slavin in 1914 in the Durand-Eastman Park, Rochester, N.Y., the pollen parent being R. *pseudacacia*. The flowers are rosy pink, more numerous in the raceme than those of R. *kelseyi*; the hairs of the inflorescence are eglandular and the pod is roughened by small warts. The leaflets are relatively broader, up to ⅞ in. wide.

The same hybrid was raised by Messrs Hillier and named 'HILLIERI' in 1933. It makes a small, round-headed tree, with lilac-pink flowers in June, borne freely even on young plants. A.M. 1962.

R. LUXURIANS (Dieck) Schneid.

R. *neomexicana* var. *luxurians* Dieck; R. *neomexicana* sens. some authors, not A. Gray

A deciduous shrub or small tree 20 to 40 ft high, with a trunk 12 in. or more thick; branchlets downy. Leaves pinnate, 6 to 12 in. long, with downy stalks; leaflets fifteen to twenty-five, oval to slightly ovate, 1 to 1¾ in. long, ½ to ⅔ in. wide, with a bristle-like tip; stipules spiny, ultimately 1 in. long. Racemes 2 to 3 in. long, 2 in. wide, the stalk covered with brown shaggy hairs. Flowers ¾ to 1 in. long, pale rose, each on a hairy stalk ¼ in. long; the standard petal large, the calyx glandular, shaggy, with slender teeth. Pods 3 or 4 in. long, ⅓ in. wide, covered with gland-tipped bristles ⅛ in. or more long. *Bot. Mag.*, t. 7726.

Native of Colorado, New Mexico, Arizona, and S. Utah, in places at 7,000 ft above sea-level. First discovered by Dr Thurber in 1851; introduced to Kew in 1887. It flowers prettily every year in June, and frequently a second time in August. It differs from R. *pseudacacia* in its bristly pods, and from R. *viscosa* in the young twigs not being viscid. The larger of two examples at Kew measures 50 × 6 ft (1968).

R. *luxurians* was at one time confused with R. NEOMEXICANA A. Gray, a related species from New Mexico, which makes a shrub to about 6 ft high; its leaves have not more than fifteen leaflets and its pods are hairy but not glandular-bristly.

R. × HOLDTII Beissn.—A hybrid between R. *luxurians* and R. *pseudacacia* whose racemes are looser and longer than in R. *luxurians* and the flowers of a paler colour. The keel and wing-petals are almost white, the standard pale red with white markings. Habit and vigour of growth similar to those of R. *pseudacacia*. Pod rather glandular. Obtained by Mr Von Holdt, Alcott, Colorado, and put into commerce about 1902.

The same hybrid was raised in Späth's nurseries, Berlin, in 1893 from seeds of R. *luxurians*. This clone, named 'BRITZENSIS', is very near to R. *pseud-*

acacia in its flowers, but the pods are glandular. The tree named R. *coloradensis* by Dode was one of six raised by Vilmorin at Les Barres from seeds collected by E. L. Berthoud at Golden, Colorado, probably from a tree of R. × *holdtii*. Another of the seedlings is described in *Fruticetum Vilmorinianum* (1904), p. 54, as 'R. neomexicana var. ?'.

R. PSEUDACACIA L.   LOCUST, FALSE ACACIA   [PLATE 4

A deciduous tree 70 to 80 ft high, with a large, rounded head of branches, and a trunk 2 to 4 ft in diameter, covered with a rugged, deeply furrowed bark. Leaves pinnate, 6 to 12 in. long, leaflets in five to eleven pairs, oval or ovate, 1 to 2 in. (sometimes 2½) in. long in the typical form, covered with silvery hairs when quite young, eventually nearly glabrous. Stipules at first ½ in. long, downy, becoming stout, persistent spines 1 in. long; most conspicuous on young trees and suckers. Racemes 3 to 7 in. long, 1½ to 2 in. wide, pendulous. Flowers ¾ in. long, white, fragrant; each on a slender stalk, ⅓ in. long; standard petal blotched with yellow at the base; calyx ⅓ in. long, downy. Pods 2½ to 3½ in. long, ½ in. wide, upper seam winged, lower one thickened, containing four to ten seeds, glabrous.

Although now widely planted and naturalised in N. America, this species is native there only in the Appalachians from Pennsylvania southward, with a westward extension into Kentucky and Missouri and a secondary area west of the Mississippi in the Ozarks; it was in cultivation both in France and England by the 1630s. As an ornamental tree the robinia has much to recommend it. Its graceful feathery foliage is singularly effective in healthy trees, and when the tree is loaded with its white racemes in June the contrast of white and green is very effective. It grows with great rapidity when young, and its branches are apt to be broken off by wind in consequence. A judicious shortening back of the shoots in winter is helpful in inducing a sturdier growth. When old, the tree is apt to lose its large branches by their splitting off from the main trunk, or if the tree has been allowed to fork, nearly half of it may be lost at a time. The best way to prevent this is to keep the tree to a single leader until it is at least 25 ft high, so that one strong straight trunk is formed, and no branch allowed to develop sufficiently to rival it. But see 'Appalachia' and 'Bessoniana', neither of which should need this treatment. It is propagated by seeds or by the suckers the roots produce so plentifully, especially after the parent tree is felled.

Perhaps no American tree has made itself so thoroughly at home in Europe as this. The railway cuttings south of Paris are in places completely overrun with it, and I have noticed it thoroughly established in the Rhone Valley above Geneva, and on the hillsides between Trieste and the Castello di Miramare.

The locust produces a timber valuable on account of its peculiar quality of resisting decay in contact with the soil. On this account it is highly valued for making gate-posts and similar articles. Owing to the representations of William Cobbett, the famous Radical, who about 1825 to 1828 extolled the tree and its uses in his own peculiarly vigorous fashion, quite a mania for the tree was established. He himself set up as a dealer in seeds and plants, and to such purpose had he written up the tree and its virtues that he was, for a time, unable to meet the demand, although it is recorded that he imported seeds from America in tons. It did not prove a success as a forest tree, and is now rarely planted except for ornament. But every few years a controversy is started as to its value as a timber tree in Britain. There is no doubt about the value of the timber for certain minor purposes—it was once, and may be now, largely used for pins (treenails) to fasten timbers together in shipbuilding—but it is not produced in sufficient bulk, nor is it of a quality to render it of great value for constructive purposes.

Being so subject to wind-breakage, this species rarely makes a large specimen both in height and girth. Some of the largest recorded are: Frogmore, Berks, 75 × 16¾ ft (1964); Clare Court, Crondall, Hants, 50 × 15¾ ft (1973); Hampton Park, Seale, Surrey, 80 × 20½ ft (1969); Bayfordbury, Herts, 92 × 10½ ft, 25 ft bole (1973). The ancient tree at Kew, near the Main Gate, was planted in about 1762; it is 14¾ ft in girth.

An extraordinary number of seminal varieties of the locust have been raised in Europe. Between three and four dozen of them are cultivated at Kew, but many are not sufficiently distinct to require mention here. The following may be regarded as the most important:

cv. 'APPALACHIA'.—A tree of fairly broad, symmetrical habit, with a central stem running well up into the crown. It is a selection from the so-called Shipmast locust, a group of straight-stemmed clones cultivated near the seaboard in Massachusetts and on Long Island, New York (var. *rectissima* Raber). It is recommended as a street and landscape tree (*Dendroflora*, No. 8, pp. 51–2; *Morris Arb. Bull.*, Vol. 11 (1960), pp. 67–70).

cv. 'AUREA'.—Leaves golden at first, becoming lime green. In cultivation 1864. This has attained 80 × 5½ ft at Moor Park, Ludlow (1962). The example at Kew, by the Main Gate, is 46 × 5¼ ft (1968).

cv. 'BESSONIANA'.—A vigorous tree with an ovoid crown, a well-developed central stem, and almost unarmed branchlets. Shy-flowering, but a good street tree. Of continental origin, in cultivation in Britain by 1871.

cv. 'COLUTEOIDES'.—A small tree, exceptionally free-flowering, with closely set leaflets less than 1 in. long. Described in 1857 in a French publication and later distributed by Späth's nursery, Berlin. This robinia is not the same as the R. *p. sophoraefolia* of Loddiges's nursery, as is stated in many works.

cv. 'DECAISNEANA'.—See under R. × *ambigua*, p. 31.

cv. 'FRISIA'.—Leaves golden yellow throughout the summer; thorns on the young growths red. This striking robinia was found by W. Jansen around 1935 in a former nursery at Zwollerkerspel, Holland. A.M. 1964.   [PLATE 4

cv. 'PYRAMIDALIS'. ('Fastigiata').—A tree of narrow habit, resembling in form the Lombardy poplar; branchlets almost unarmed; flowers sparsely borne. It was put into commerce by Leroy of Angers, before 1843. It is not in the first rank of columnar trees, but might be of some use in dry soils, or where there is much reflected heat from paving or neighbouring buildings. The trees at Kew are 50 to 60 ft high, and a younger example in the R.H.S. Garden, Wisley, is as tall (1965–75).

cv. 'REHDERI'.—Resembling 'Umbraculifera' but more erect and less desirable. It was raised from seed at the Muskau Arboretum in 1859, and is named after the Director, who was the grandfather of Prof. Alfred Rehder of the Arnold Arboretum.

cv. 'ROZYNSKIANA'.—Leaves long, pendulous, with narrow, distant leaflets. Of open habit, with spreading branches. Put into commerce by Späth in 1903.

cv. 'SEMPERFLORENS'.—A large and vigorous tree which produces its flowers in two flushes, the first in June, the second around September, or sometimes flowers more or less continuously from midsummer onwards. It arose as a seedling in a French nursery and was put into commerce in 1874.

cv. 'TORTUOSA'.—Branches curiously twisted. Racemes small and thinly set with bloom. Known since early in the 19th century. Having many strong ascending limbs it is very subject to damage by wind. A tree at Kew, with a broken crown, measures 55 × 9 ft (1968).

cv. 'UMBRACULIFERA' ('*Inermis*') PARASOL OR MOP-HEAD ACACIA.— A round, compact unarmed bush, usually made into a small standard by grafting at 6 or 10 ft. It rarely or never flowers. Once common as a clipped specimen in villa gardens, it is now used to some extent as a street tree, but is not suitable for windy situations. Of uncertain but probably French origin, introduced to Britain around 1820. The botanical names originally given to it explain why it is sometimes wrongly known as 'Inermis': R. *inermis* Dum.-Cours., in part (1811), not Mirbel (1804); R. *umbraculifera* DC. (1813); R. *p.* var. *umbraculifera* (DC.) DC.; R. *p.* f. *inermis umbraculifera* Voss. The cultivar-name 'Inermis' properly belongs to an unarmed form with large leaflets raised in France early in the 19th century (R. *inermis* Mirbel; R. *p.* var. *inermis* DC.).

cv. 'UNIFOLIOLA' ('Monophylla').—Leaflets reduced in number to one, two or three, consisting either of the terminal leaflet alone (always much larger than in the type and often 4 in. long and 1½ in. wide), or sometimes with a few lateral leaflets of about the normal size. This remarkable variety arose in a French nursery about 1855. It flowers freely and is said to come about one-quarter true when raised from seeds. There are two examples at Kew in the *Robinia* collection: *pl.* 1889, 58 × 4¾ ft (broken) and 55 × 5 ft (1968), and one in Jephson Park, Leamington, 62 × 5¾ ft (1971).

There is also a fastigiate form with similar foliage, 'MONOPHYLLA FASTIGIATA' (R. *p.* f. *erecta* Rehd.) and one with slender semi-pendulous branches, 'MONOPHYLLA PENDULA', both distributed by Dieck of Zöschen. A broken tree of the former grows at Kew; *pl.* 1894, it measures 50 × 7¾ ft (1968), and there is a younger example in the Edinburgh Botanic Garden.

Information on other variants of R. *pseudacacia* will be found in the article cited in the generic introduction, pp. 149–57.

R. VISCOSA Vent.   CLAMMY LOCUST
R. *glutinosa* Sims

A deciduous tree 30 to 40 ft high, with a trunk 12 to 18 in. thick, and often furnished with large burrs; young branches covered with glands which exude a sticky substance that adheres to the fingers when touched. Leaves pinnate, 3 to 10 in. long, the main-stalk hairy and viscid like the young twigs. Leaflets eleven to twenty-one, oval or ovate, 1 to 2 in. long, $\frac{1}{3}$ to $\frac{3}{4}$ in. wide, dark green above, paler and at first slightly downy beneath, ultimately glabrous. Stipules at first $\frac{1}{4}$ in. long, becoming longer and spiny with age. Racemes 2 to $2\frac{1}{2}$ in. long, almost as wide, with a naked stalk half as long. Flowers $\frac{3}{4}$ in. long, ten to fifteen in a raceme, without fragrance; petals pale rose with a yellow blotch on the standard; calyx dark red, hairy. Pod 2 to $3\frac{1}{2}$ in. long, covered with viscid glands. *Bot. Mag.*, t. 560.

Native of the mountains of North Carolina, where it was originally discovered by William Bartram in 1776. It was introduced to France by Michaux in 1791, and six years later to England. It is a smaller tree than R. *pseudacacia* and of more stunted growth, but it flowers very freely and makes a bright picture towards the end of June. The viscid substance on the branches renders it easily distinguishable, although this appears to vary in amount, and is some-

ROBINIA VISCOSA

times not very discernible. According to Sargent it is one of the rarest of American trees, and from the time of Michaux to 1882 was never found in a genuinely wild state. But it is now naturalised in many parts of the eastern

United States. It is not common in English gardens, but there were several old trees at Kew, the largest of which had a trunk 5 ft 3 in. in girth at 5 ft from the ground, which is considerably more than the dimensions recorded of wild specimens. These old trees were almost unarmed, the spines only occurring on exceptionally vigorous shoots. A handsome tree in bloom.

The largest of the few trees recorded recently grows at Borde Hill, Sussex, and measures 42 × 4 ft (1968).

R. × AMBIGUA Poir. R. *dubia* Foucault, not Poir.; R. *intermedia* Soulange-Bodin—A small group of hybrids between R. *viscosa* and R. *pseudacacia*, showing the influence of the former in their moderate size, viscid young growths and pink flowers, and of the latter in the usually better developed spines. The type arose in France a few years before 1812 and was at one time cultivated under the name R. *dubia*, but seems to have become rare. The following members of the group are still available:

cv. 'BELLA-ROSEA'.—Branchlets viscid, spiny. Flowers deep rosy pink. It is said to have arisen in Holland around 1860.

cv. 'DECAISNEANA'.—This is retained by Rehder as a variant of R. *pseudacacia* but is now considered by most authorities to be a form of R. × *ambigua*. The branchlets are only very slightly viscid, but the influence of R. *viscosa* shows in the pink flowers and the reduced size of the stipular spines. It arose in Villevielle's nursery at Manosque in the department of Basse-Alpes, and was described in 1863. It is said to come more or less true from seed, and no doubt some of its seedlings, deeper in colour than the original, have been propagated; this would explain why there is some variation in colour in 'Decaisneana', despite the fact that it is normally propagated by grafting.

'Decaisneana' is perhaps the most handsome of the arborescent robinias, but it needs more than average summer heat if it is to grow and flower well. In cool gardens it flowers sparsely, and is unsuitable for exposed positions, owing to the brittleness of the wood.

ROMNEYA   PAPAVERACEAE

A genus of two semi-herbaceous shrubs, both Californian, one extending into Mexico. The leading characters of the genus are the pinnately divided leaves, the large white usually solitary flowers with four to six petals and three soon deciduous sepals, numerous stamens, sessile stigmas connate at the base, capsular appressed bristly fruits splitting by valves and containing numerous small seeds.

The generic name commemorates the Irish astromoner Dr F. Romney Robinson, a friend of Dr Coulter, who discovered the type-species (for further information on the history of the generic name see *Bot. Mag.*, n.s., t. 678).

R. COULTERI Harvey   CALIFORNIA TREE POPPY

A semi-shrubby plant, with succulent herbaceous stems 4 to 8 ft high, according to the mildness of the climate in which it grows, spreading by suckers. Leaves varying much in size according to the strength of the shoot which bears them, but averaging from 3 to 5 in. long, and of a very glaucous colour; they are obovate to pinnately lobed, the end lobe usually much the largest and itself more or less lobed, glabrous except for a few spine-like bristles on the stalk and midrib. Flowers fragrant, solitary or in pairs, terminating short twigs near the end of the stem, each one 4 to 5 in. across, with five or six overlapping, satiny-white, delicately textured petals surrounding a mass of golden-yellow stamens 1 to $1\frac{1}{2}$ in. across. Calyx glabrous. Flower-buds apiculate. *Bot. Mag.*, n.s., t. 648.

This beautiful plant has a limited range in S. California and is said to be abundant in the Sta Ana Mountains south-east of Los Angeles. It was discovered in 1833 by the Irish botanist Dr Thomas Coulter (*d.* 1843), but not introduced until 1875, when seeds were received by E. G. Henderson & Co. and by Thompson of Ipswich. The first recorded flowering took place, appropriately, in Ireland, where a small plant put out in the Glasnevin Botanic Garden in March 1876 opened one bud that autumn and flowered abundantly in the following summer, when 6 ft high.

R. *coulteri* is hardy over much of the British Isles, but needs a warm, sunny position and a deep, porous, nutritious soil; it is quite happy on chalk. The stems may be cut or destroyed in a hard winter, especially after a poor summer, but even then a crop of flowers will be produced on the new growths. The stems are anyway not of long duration and the oldest should be cut out each year. In cold gardens, with a shorter-than-average growing season, it is best given the extra summer heat provided by a south or west wall; grown in the open it may fail to open its flower-buds in most years.

R. *coulteri* is not easily raised from cuttings, but its thick fleshy roots afford a ready means of increase. They should be taken in late February, cut into 2 in. lengths, placed in pots or boxes and thinly covered with sandy soil. Mild bottom-heat is desirable. Suckers with roots attached can also be used, but should be established in pots before planting out.

R. TRICHOCALYX Eastw.   R. *coulteri* var. *trichocalyx* (Eastw.) Jeps.— Closely related to R. *coulteri* and with the same general distribution, but extending into the Mexican State of Baja California. It is commonest in Ventura and San Diego counties, and rare or absent in the Sta Ana Mountains in Los Angeles Co., where R. *coulteri* is abundant (C. B. Wolf, *Rancho Sta Ana Bot. Gard., California Plant Notes*, II (1938), p. 53). It differs in the following particulars: stems more slender; foliage of thinner texture, more finely cut, the basal lobes being usually ⅜ in. or less wide against ½ in. or more wide in R. *coulteri*; flowers more clustered; sepals, and hence flower-buds, covered on the outside with appressed bristly hairs, rounded at the apex (hence flower-buds not beaked); peduncles usually furnished with small but normal leaves, which sometimes extend almost to the base of the flower (in R. *coulteri* the peduncles are leafless or bear reduced, bract-like leaves); peduncles often with erect bristly hairs near the flower. *Bot. Mag.*, t. 8002.

R. *trichocalyx*, originally confused with R. *coulteri*, was separated from it by Alice Eastwood in 1898, after a study of plants grown in the Golden Gate Park, San Francisco. She sent seeds to Kew in 1902 from which plants were raised, but the species was then already in cultivation in the garden of Hiatt C. Baker of Almondsbury, near Bristol, and perhaps even in commerce (as R. *coulteri*).

In gardens, R. *trichocalyx* grows less tall than its relative and makes a more elegant plant, with its slenderer stems and more finely divided foliage. But, when well suited, it is even more invasive, sending up sucker shoots as much as 40 ft away from the parent plant. An interesting feature of at least some garden clones is that the flowers produce more abundant pollen than those of R. *coulteri*, whence perhaps the unpleasant odour attributed to them, for pollen in quantity is sometimes very rank-smelling.

R. *trichocalyx* needs the same conditions as R. *coulteri* and is propagated in the same way, though stem cuttings can also be used with fair success.

R. COULTERI × R. TRICHOCALYX   R. × *hybrida* Hort.; R. × *vandedenii* Correvon, *nom. nud.*—This hybrid was raised by W. B. Fletcher of Aldwick Manor, Sussex, R. *coulteri* being the seed-parent, and was shown by him to the Scientific Committee of the R.H.S. in 1916. The plants raised were intermediate between the parents in foliage; and in the plant shown the sepals had the apiculate apex of R. *coulteri* but with traces of bristly hairs inherited from the other parent. The same cross, according to Correvon, was made in Switzerland in 1909. But neither of the names cited above has been validated by a Latin description. [PLATE 5

The form distributed in Britain (as R. × *hybrida*) comes very near to R. *trichocalyx* but is more bushy and has fragrant flowers. It attains a height of 4 or 5 ft.

It is possible that the hybrid is also cultivated, or has been, under the name R. *coulteri*. Until 1914, this species was usually raised from imported seeds, which, if of garden origin, may well have produced hybrids as well as the true species.

A named hybrid between the two species is 'WHITE CLOUD', raised in the USA, with very large, fragrant flowers and intensely glaucous foliage. The flower-buds are smooth and apiculate as in R. *coulteri* but the peduncles are leafy as in R. *trichocalyx*. It has colonised an area of about fifty square yards in the Hillier Arboretum at Ampfield, Hants.

ROSA

ROSACEAE

CONTENTS

PREFACE

In the great family of Rosaceae, which gives to gardens more beautiful hardy deciduous trees and shrubs than any other, no genus stands out with greater distinctness than the one from which it derives its name. Writing at the end of the 16th century, L'Obel and Pena hailed *Rosa* as 'Orbis Deliciae', the Darling of the World. Yet even at that time the garden roses were not wildlings but the product of long centuries of selection by man. Many were hybrids of which even now we do not know the history or even the parentage. Today the situation is different and yet essentially the same. Most of the wild roses of the world have been brought into gardens at one time or another, yet few have gained a permanent footing and fewer still have ever been much grown outside collections. At the same time the store of garden roses has been enriched by sorts of a beauty and variety undreamed of even two centuries ago. There is no need to justify the omission from this work, as from previous editions, of the great race of Hybrid Teas and Cluster-flowered roses (Floribundas), a subject so well covered in many excellent and low-priced books that to intrude on it here would be presumptuous, even if they could be regarded as shrubs in the normal acceptation of the word. At the same time there is a large and now increasing number of roses available to gardeners, some very old and others of recent raising, that are undeniably shrubs or woody climbers and would have been treated in the present edition, albeit inadequately, even if Mr Graham Thomas had not so generously offered to undertake this task himself.

The section devoted to *Rosa* has therefore been divided into two parts. The first is botanical in the main, and is an expansion of the treatment contained in previous editions, with the main difference that proportionately more space has been given to the species that have contributed most to the formation of modern garden roses. The second part is entirely the work of Graham Thomas and, in addition to descriptions of some three hundred cultivars, it also contains sections on the history of the rose in gardens, as well as its cultivation and propagation. The two parts are complementary and linked by cross-referencing.

THE SPECIES

BOTANICAL CHARACTERS

At a rough estimate, something over 3,000 names of specific rank have been published in this genus, but the number of good species is comparatively small—probably no more than 150. An authoritative estimate is impossible, for no monograph on the genus has been published since Lindley's in 1820 and needless to say that is not of much relevance today.

The genus is mainly confined to the cooler parts of the N. Hemisphere and only three or four extend south of the Tropic of Cancer. As is the case with so many other woody genera, *Rosa* is most richly represented in E. Asia—especially in China, where the majority of the climbing species are to be found, as well as most of the most beautiful and individual of the shrubby species. But the rhododendron-rich, very rainy parts of the Sino-Himalayan region are poor in species of *Rosa* (most of the Chinese roses in cultivation were introduced by Wilson from Szechwan and Hupeh). Central Asia seems to be the main home of the groups centring around *R. xanthina* and *R. webbiana*. Europe and S.W. Asia can claim as their own the large and complex section *Caninae*, but is not otherwise at all rich in species. The majority of the N. American species belong to the wide-ranging section *Cassiorhodon* or to the closely related endemic section *Carolinae*; apart from these it has the endemic subgenus *Hesperhodos* and two outliers of mainly Old World groups—*R. gymnocarpa* and *R. setigera*.

No rose attains the dimensions of a tree, though a few have attained it in cultivation, e.g., the famous Hildesheim rose, an ancient specimen of the common brier, which is said to have attained a height of 45 ft and a girth of 2 ft. But that grew against a wall of the cathedral; and the 'tree' of *R. moschata* at Isfahan in Persia, named *R. arborea* by Persoon, probably had some support. The majority of the species are erect or bushy shrubs, and rarely attain a height of more than 10 ft. The climbing roses all belong to the section *Synstylae* or to the small E. Asiatic sections *Banksianae*, *Bracteatae*, *Chinenses*, and *Laevigatae*. No species is constantly prostrate, though some may become so in certain habitats, e.g., *R. wichuraiana* and *R. pimpinellifolia*.

All species of *Rosa* are armed, though in a few species, notably *R. blanda* and *R. banksiae*, the armature may be very sparse. The typical rose 'thorn' is the hooked or curved prickle of the common brier and most climbing species. But the prickles may be erect and straight, and with these as one extreme there is a series of types ranging through the awl-shaped prickle to the needle (acicle), the flexible bristle, and finally to the erect hair, usually gland-tipped. In cross-section the prickle may be terete or more or less flattened—even cartilaginous and translucent in the

37

pteracanthous form of *R. sericea*. The armature may be uniform, or a mixture of various types, and may vary from one part of the plant to another, strong growths from the base often having an armature markedly different from that of the branches. As a general rule the flower-bearing branchlets are the least armed part of the plant, and unfortunately many herbarium specimens consist only of such pieces. Even a cultivated plant, if it lacks vigorous growths, may give no idea of the true armature of the species to which it belongs.

The leaves of the roses are imparipinnate (except in subgenus *Hulthemia*, where they are simple). The number is rarely in excess of nine (four pairs of lateral leaflets) and only in a few species is it reduced to three, except immediately under the flowers (even *R. laevigata*, the most constantly trifoliolate species, sometimes has five leaflets, at least on cultivated plants). The shape of the leaflets varies even in the same species and is rarely of much use for identification. The marginal toothing may be simple or compound. Compound toothing is nearly always associated with the presence of glands on the back of each major tooth and may occur in species where the toothing is commonly simple. On the other hand, the presence of numerous aromatic glands on the undersurface of the leaflet is a constant feature of some species of the sections *Pimpinellifoliae* and *Caninae*.

Stipules are present in all species except *R. persica* (subgenus *Hulthemia*). In most groups they are adnate to the petiole for most of their length and persistent, but deciduous and largely free in the sections *Banksianae*, *Bracteatae* and *Laevigatae*. The actual shape of the stipules is not of much value in distinguishing related species.

The flowers are borne on laterals springing from wood of the previous season, or on extensions from existing laterals, though *R. rugosa* and some American species can produce flowers on the current season's growths. The ability of garden roses such as the Hybrid Teas to produce their flowers in flushes on the summer growths is not an inheritance from any wild rose but the result of mutations or new gene combinations which cause the extension growths to go over immediately to the flowering condition instead of producing flowers in the succeeding year as would normally be the case. The main source of remontancy in modern roses is the Chinese garden race descending from *R. gigantea* and the spontaneous form of *R. chinensis*.

Solitary flowers borne early in the season on short laterals are a feature of a few species, notably the *Pimpinellifoliae*. At the other extreme are the *Synstylae*, which mostly have elaborate inflorescences opening after the longest day. But most non-climbing species bloom around midsummer and bear their flowers singly or few in a cluster. Only in a few shrubby species is the standard number of flowers in excess of five. Many-flowered inflorescences are sometimes seen in species where three or five is the norm, but these are really the result of the merging of distinct, contiguous inflorescences. Bracts (reduced leaves) are usually present in the inflorescence except in some *Pimpinellifoliae*, and demarcate the pedicels from the lower part of the inflorescence, whose branches belong to the vegetative part of the plant. The pedicel is usually glabrous in the sense that it rarely has a coating of hairs (except in the *Bracteatae* and some *Synstylae*). But it frequently bears stalked glands or glandular bristles, and when these are present they usually extend to the receptacle. Indeed, pedicel and receptacle are intimately related, to such an extent that the upper part of the pedicel may become fleshy in the fruiting state.

In several genera of the Rose family, e.g., *Rubus* and *Potentilla*, the carpels are numerous and inserted on a convex torus. But in *Rosa*, the sole member of the tribe Roseae, the carpels line the inner walls (or the base only) of a concave receptacle (hypanthium, urceolus), on which the sepals, petals, and stamens are borne. This receptacle, which usually becomes fleshy in the fruiting stage, is variable in shape even within the same species and does not have the taxonomic value given to it by some of the earlier students of the genus. In the subgenus *Hesperhodos* the receptacle is open at the apex, but in other members of the genus is more or less closed by the development of a disk. In *R. pimpinellifolia* the disk is a mere rim. In most species it partly seals the opening of the receptacle, leaving an aperture through which the styles protrude. The disk may be flat, or built up into a mound or truncated cone.

The sepals equal the petals in number and alternate with them. They may be narrowly triangular in outline, i.e., broadening from the very base, as in the section *Chinenses* and most members of the section *Cassiorhodon* (*Cinnamomeae*), or be slightly constricted at the base, as in the section *Synstylae* and many species of the *Caninae*. After narrowing towards the apex the sepal often expands again into a prolonged 'tail'. The margins of the sepals may be quite entire (as those of the two inner sepals always are) or bear lateral appendages. These may be few, narrow, and inconspicuous, but are strongly and often elaborately developed in some of the *Caninae* and even more strikingly in the old cultivated hybrids— *R. alba*, *R. centifolia*, *R. damascena*, and *R. richardii*; also in some forms of *R. gallica*. This feature seems to be associated with polyploidy, though the correlation is not constant. Whether the sepals fall soon after the flowering stage or persist until the ripening of the fruit is generally agreed to be an important taxonomic character, and so too is the posture of sepals, if persistent. The *Caninae* are variable in both these respects.

The corolla in *Rosa* is on the average larger than in other members of the Rose family, a width of 2 to 3 in. being the norm, which is exceeded by some E. Asiatic climbing species such as *R. gigantea* and *R. laevigata*. Only in some species with many-flowered inflorescences is it as little as 1 in. wide. The normal number of sepals and petals is five; they are reduced to four only in one species—*R. sericea*—and the standard number is never exceeded in wild plants except as an abnormality. The predominant colouring is white, creamy white, blush, pink, or crimson; pure red is seen only in *R. chinensis* and *R. moyesii*. The more vivid shades of yellow in modern roses derive from *R. foetida* of the section *Pimpinellifoliae* and all the other truly yellow-flowered species belong to this section. The soft yellow and cream shades of the old Tea roses is the contribution of *R. gigantea*.

The petals themselves do not provide any diagnostic character of value, except that they are silky on the outside in R. *longicuspis* and R. *setipoda*.

The very numerous stamens add much to the beauty of all single and semi-double roses. Rarely are there fewer than thirty and the median number varies from one species to another. Boulenger names five in which the minimum number is 100 or more: R. *bracteata*, R. *palustris*, R. *roxburghii*, R. *rugosa*, and R. *setigera*.

The styles (one from each carpel) protrude through the aperture, their stigmas forming usually a sessile dome-shaped or conical head. But in the *Synstylae* the styles are agglutinated into a slender column which is exserted well above the disk and may be as long as the stamens. Exserted styles, though to a less marked degree, are also a feature of some *Caninae*, of the E. Asiatic R. *davidii*, and often of R. *gallica*.

After the petals have fallen the receptacle matures into the hep (or hip) so characteristic of the genus, enlarging and usually becoming fleshy but retaining the original shape and whatever outgrowths it had. Only in the anomalous subgenera *Platyrhodon*, *Hulthemia*, and *Hesperhodos* is the fruit (like the receptacle) always armed, but in a few species, such as R. *pimpinellifolia* and the R. *eglanteria* group, small prickles (as distinct from stalked glands) occur on the fruits. Only in the section *Bracteatae* is the fruit hairy. The size ranges from almost 1½ in. in width in R. *gigantea* and R. *rugosa* to barely ¼ in. in some species with many-flowered inflorescences such as R. *cymosa* or R. *multiflora*. Red, crimson, or scarlet are the predominant colours, but darker and more muted shades are common in the *Pimpinellifoliae*, and the subgenera *Hulthemia* and *Hesperhodos*, while R. *roxburghii* (subgen. *Platyrhodon*) is unique in its greenish fruits.

The sepals are deciduous from the ripe fruit, or earlier, in the *Synstylae* and other climbing species, in R. *gallica* and its derivatives, and in the *Carolinae*, usually persistent in the section *Cassiorhodon* (*Cinnamomeae*), while the section *Caninae* varies in this respect as in so many others.

The ripe carpel is a bony achene, containing a single seed.

CLASSIFICATION

subgen. ROSA

For characters, see generic introduction.

sect. CHINENSES.—Robust climbing shrubs in their wild state, with hooked prickles as their only armature. Stipules narrow, adnate to the petiole for most of their length, persistent. Leaflets glossy, glabrous, mostly five or seven, with more than twenty-five teeth on each margin. Flowers solitary, sometimes paired or in threes. Sepals entire or almost so, tapered from the base, deciduous in fruit. Styles free, slightly exserted. The two E. Asiatic species in this section—R. *chinensis* and R. *gigantea*—have contributed more to the formation of the modern garden roses than any other group.

sect. BANKSIANAE.—See R. *banksiae*. The leading characters of this small E. Asiatic group are the tall, lax, or climbing habit, sparse, simple armature of prickles, the slender free, deciduous stipules and corymbose inflorescence. Apart from R. *cymosa*, mentioned under R. *banksiae*, the only species is R. COLLETTII Crép., described from the Shan Hills of Burma; it is allied to R. *cymosa*.

sect. LAEVIGATAE.—See R. *laevigata*, the only species. It resembles the *Banksianae* in the stipules, which are free from the petiole for most of their length, the free part deciduous. But the flowers are solitary and larger, and the leaflets three only (sometimes five).

sect. BRACTEATAE.—See R. *bracteata*, which is the only species apart from the closely related R. *clinophylla*. The leading characters of this very distinct group are the almost free, deeply incised stipules, the large inflorescence bracts, and the woolly receptacle.

sect. SYNSTYLAE.—Climbing, sprawling or prostrate shrubs, with a simple armature of hooked prickles. Leaflets three to nine. Stipules adnate to the petiole for most of their length, persistent, entire, toothed, or (as in R. *multiflora*) laciniate. Inflorescence in most species a much-branched paniculate or corymbose cluster with numerous flowers, but sometimes few-flowered, as usually in R. *arvensis*. Petals usually white. Sepals with a few slender lateral appendages (larger and more numerous in R. *phoenicea*), deciduous from the ripe fruit. Styles more or less agglutinate, forming an exserted column.

Apart from the American R. *setigera* all the species are natives of the Old World, mostly in E. Asia. Those treated here, arranged geographically, are:

Europe: R. *arvensis*, R. *sempervirens*.

S.W. Asia: R. *moschata*, R. *phoenicea*.

Himalaya and N.E. India: R. *brunonii* (also in China), R. *longicuspis*.

China and continental N.E. Asia: R. *anemoniflora*, R. *cerasocarpa*, R. *filipes*, R. *gentiliana*, R. *helenae*, R. *henryi*, R. *lucens*, R. *maximowicziana*, R. *mulliganii*, R. *multiflora* (varieties), R. *rubus*, R. *soulieana*.

Japan: R. *luciae*, R. *multiflora*, R. *watsoniana* (garden origin), R. *wichuraiana*.

There is an interesting extension of the *Synstylae* into the northern tropical zone. R. LESCHENAULTIANA (Thory) Wight & Arn. is a native of the Nilgiris and other hills of S.W. India (Madras), while R. ABYSSINICA R. Br. inhabits the Horn of Africa and has two close relatives in the Yemen.

sect. PIMPINELLIFOLIAE.—Shrubs of diverse habit. Prickles straight or curved, sometimes flattened or even thin, cartilaginous and wing-like, often mixed with needles or bristles. Leaflets commonly more than seven (up to seventeen in R. *sericea*). Flowers rarely other than solitary, white or yellow. Petals normally the usual five, four in R. *sericea*. Sepals entire or with a few slender lateral appendages, erect and persistent in fruit. Styles free, included. A few species in the Old World.

R. ecae, R. foetida, R. hemisphaerica, R. hugonis, R. pimpinellifolia (the only European species), *R. primula, R. sericea, R. xanthina.*

sect. GYMNOCARPAE.—This small section is allied to the *Pimpinellifoliae*, but has the peculiarity that the top of the receptacle as well as the sepals is deciduous from the ripe fruit. The type-species, *R. gymnocarpa*, is W. American; the other species are Asiatic, those treated here being *R. albertiana, R. beggeriana,* and *R. willmottiae.*

sect. CASSIORHODON (*Cinnamomeae*).—Shrubs, often suckering. Prickles when present usually in pairs or clusters at the nodes (infrastipular); needles and bristles are often present, especially on strong growths (turions) and are the predominant form of armature in some species. Leaflets mostly five or seven (sometimes nine). Flowers in some shade of pink or red, solitary or few, sometimes fairly numerous in a corymbose cluster. Sepals entire or with a few slender lateral appendages. Styles included. Fruits crowned with the erect persistent sepals.

As understood by Crépin and those who have adopted his classification this is a large group, comprising most of the shrubby species of the subgenus *Rosa* apart from the 'cryptohybrid' *Caninae*. The following arrangement is mainly based on Boulenger's classification, in which only the first group belongs to the *Cinnamomeae* as he understood it:

1. Arranged geographically the species in this subgroup are:
Widely distributed in the N. Hemisphere: *R. acicularis* (N. America, N.E. Asia, Siberia, parts of European Russia, Sweden).

Europe: *R. glauca* (*rubrifolia*); *R. majalis* (*cinnamomea*), which ranges into Siberia.

E. Asia: Those treated here are *R. rugosa, R. setipoda,* and its two close allies *R. caudata* and *R. hemsleyana.*

N. America: *R. arkansana* and its allies, *R. blanda, R. californica, R. nutkana, R. pisocarpa, R. woodsii.*

2. The leading species in this subgroup is *R. webbiana,* to which *R. bella, R. fedtschenkoana,* and *R. sertata* are closely allied. The others treated here are *R. elegantula, R. forrestiana, R. latibracteata, R. multibracteata.* This group, which is near to the *Pimpinellifoliae,* is mainly confined to China but extends into the Himalaya and Central Asia.

3. This subgroup largely consists of the European *R. pendulina* and its allies—*R. oxyodon* (Caucasus), *R. macrophylla* (Himalaya and probably S.W. China), *R. wardii* (S.E. Tibet), and the Chinese *R. banksiopsis, R. davidii, R. moyesii, R. prattii, R. saturata, R. sweginzowii.*

sect. CAROLINAE.—A small group confined to eastern and southeastern N. America, near to the section *Cassiorhodon* (Group 1) but with the sepals deciduous in the fruiting stage. In setting up this group Crépin gave as the leading character that the achenes are inserted at the bottom of the receptacle only, but later workers have found this criterion unreliable. The species concerned are *R. carolina, R. foliolosa, R. nitida, R. palustris, R. virginiana.*

sect. CANINAE.—To this section belong all the roses of Europe with the exception of the two *Synstylae* (*R. arvensis* and *R. sempervirens*),

R. *pimpinellifolia*, R. *glauca* (*rubrifolia*), R. *pendulina*, R. *majalis* (uncommon outside Scandinavia), and R. *acicularis* (Scandinavia only). These are comparatively uncomplicated species, and it is the *Caninae*, a group of the most intricate variability, that was the chief preoccupation of the 19th-century rhodologists. A vast number of species was made out of the character-combinations that occur in this group, and even when the number of recognised species was reduced to a moderate number it remained the practice to inventorise and name every variation within it. For example, in Wolley-Dod's *Revision of the British Roses* (1931) five species are recognised in the subsection Caninae (R. *canina* aggregate), but these are subdivided into in all sixty-six varieties, many with two or more *formae*. In Boulenger's treatment of the same group, in *Roses d'Europe*, two species are recognised with thirty-three varieties between them, but the listing of the synonyms of this comparatively limited number of categories takes up sixty-three pages.

As is now well known, the species in this group are polyploids of a peculiar sort. The majority are pentaploids (35 chromosomes—five times the base-number of 7). In the reduction-division that leads to the formation of the sex-cells only 14 of these form pairs in the normal way, the other 21 remaining unpaired. In the pollen the unpaired chromosomes are lost, so that the pollen grains carry only seven chromosomes, as in a normal diploid. The ovules, on the other hand, contain all the unpaired chromosomes plus half the pairing set—28 in all. Thus the pollen contains only a small part of the gene-complement of the plant and its influence is weak. The bulk of the genes that determine the characters of the offspring are contained only in the ovules. Inheritance is therefore predominantly maternal, and minor variations are able to perpetuate themselves with little alteration from one generation to the next. The small number of species now recognised in this group are highly complex groups of microspecies and no attempt is made in modern floristic works to take account of every character-combination. It is likely, in any case, what with modern methods of agriculture, and afforestation, that many of the forms and hybrids that delighted the old rhodologists have long since departed with the rose-rich hedges and thickets in which they grew. Crépin's favourite hunting-grounds in Belgium had largely been destroyed by the time the centenary of his birth was celebrated in 1930.

It is now thought that the species of *Caninae* are what have been termed 'cryptohybrids', i.e., they are the product of ancient hybridisations between species that no longer exist, at least not in their primitive state. Dr Hurst's experimental crossings suggested that there may be a rose of the *Synstylae* in their ancestry, since hybrids of which a member of the *Caninae* was the pollen-parent showed *Synstylae* characters. For Boulenger's theories, see below.

The leading characters of the *Caninae* are: Stems erect or arching, armed with hooked or straight prickles. Leaflets mostly five or seven. Inflorescence a few-flowered corymb. Flowers white or light pink. Outer sepals with conspicuous lateral appendages. Styles free except in the *Stylosae*, sometimes slightly exserted. Sepals usually persistent until the fruit ripens, sometimes falling earlier.

subsect. CANINAE (*Eu-caninae*). DOG ROSES.—Prickles hooked or curved. Leaves glabrous or downy, not or only slightly glandular.

Species treated (see R. *canina*): R. *afzeliana*, R. *canina*, R. *coriifolia*, R. *corymbifera* (*dumetorum*), R. *obtusifolia*.

Boulenger considered that this group derives from R. *majalis* (*cinnamomea*) and R. *acicularis*. He therefore associated it with these species and their allies (see sect. *Cassiorhodon* Group 1) in his group CINNAMOMI-CANINAE.

subsect. VILLOSAE (*Vestitae*).—Prickles more slender than in the subsection *Caninae*, straight or curved. Leaflets with glandular-compound teeth, their undersurface usually densely woolly and often glandular. Pedicels bristly, as usually are the receptacles and fruits.

Species treated: R. *jundzillii*, R. *mollis* (very closely allied to R. *villosa*), R. *sherardii*, R. *tomentosa*.

Boulenger associated these species with R. *pendulina* (*alpina*) and its allies (see sect. *Cassiorhodon*, Group 3), holding that R. *pendulina* was ancestral to them. This group he named ALPINAE-VESTITAE.

subsect. RUBIGINOSAE. SWEET BRIER Group.—Prickles hooked, sometimes mixed with needles and bristles on the flowering branches. Underside of leaflets never woolly, often glandular.

Species treated: R. *biebersteinii* (*horrida*), R. *eglanteria* (*rubiginosa*), R. *micrantha*, R. *pulverulenta* (*glutinosa*), R. *serafinii*, R. *sicula*.

In Boulenger's classification these species are associated with the *Pimpinellifoliae*, considered to be ancestral to them, forming his group PIMPINELLI-SUAVIFOLIAE (the *Suavifoliae* of Crépin was a group of "species" all long since reduced to synonymy, mostly under R. *eglanteria*).

subsect. STYLOSAE.—One species. See R. *stylosa*, mentioned under R. *canina*.

A species usually included in the sect. *Caninae*, but out of place there, is R. *elymaitica* (q.v.).

sect. GALLICANAE (*Rosa*).—This section has a single species, R. *gallica*. Its other two components—R. *damascena* and R. *centifolia*—are of hybrid origin and not known in the wild state (the hybrid R. *alba*, sometimes placed in the *Gallicanae*, is really nearer to the *Caninae*).

The leading characters of R. *gallica* are the suckering habit, the mixed armature of needles and bristles, the reduced number of leaflets (three or five), and the reflexed, deciduous sepals.

R. *centifolia* has been proposed as the type-species of the genus *Rosa*. No worse choice could have been made, since this rose is certainly of hybrid origin but, if it is accepted, the proper title of the sect. *Gallicanae* would be sect. *Rosa*, according to the rule that a subdivision of a genus takes the name of the genus if it contains its type-species.

subgen. PLATYRHODON

Bark peeling. Receptacle with a torus (fleshy emergence) at the base, on which the ovules are inserted (these in *Rosa* are normally inserted at the bottom and on the walls of the receptacle). Only one species, R. *roxburghii*. This is sometimes included in the subgenus *Rosa* as the monotypic section *Microphyllae*.

subgen. HESPERHODOS

Receptacle without a disk. Fruits globose, prickly. According to some authorities the ovules are inserted on a torus as in the subgen. *Platyrhodon*, but others have found no trace of it. Two or three species in western N. America. See further under R. *stellata*.

subgen. HULTHEMIA

Leaves simple, without stipules. Flowers yellow with a maroon centre. Fruits globose, prickly. One or two species in S.W. and Central Asia. See further under R. *persica*.

SELECT BIBLIOGRAPHY

ANDREWS, H. C.—*Roses*. London, 1805–28.
   For dating of plates see *Journ. Arn. Arb.*, Vol. 18 (1937), or Stafleu, *Taxonomic Literature* (1967), pp. 7–8.
BOITARD, M.—*Manuel Complet de l'Amateur de Roses*. Paris, 1836.
BOULENGER, G. A.—'Les Roses d'Europe de l'Herbier Crépin' [Vol. I] in *Bull. Jard. Bot. Bruxelles*, Vol. 10, pp. 3–417 (1924–5); [Vol. II] in op. cit., Vol. 12, pp. 3–542 (1931–2).
   A study of the European species and natural hybrids, as represented in Belgium, Britain, France, Germany, Holland, and Switzerland. The hybrids are treated in Vol. II of this work (Vol. 12 of the *Bulletin*), pp. 257–506. A synoptic key to the species is provided in the same volume, pp. 507–11.
—— 'Revision des Roses d'Asie . . .', *Bull. Jard. Bot. Bruxelles*, Vol. 9, pp. 203–79 (1933) [Synstylae]; op. cit., Vol. 14, pp. 274–8 [Synstylae, supplement]; op. cit., Vol. 13, pp. 165–266 (1935) and Vol. 14, pp. 115–221 (1936) [Other species].
—— 'Introduction à l'étude du Genre *Rosa*', *Bull. Jard. Bot. Bruxelles*, Vol. 14, pp. 241–73 (1937).
BUNYARD, E. A.—*Old Garden Roses*. London, 1936.
BYHOUWER, J. T. P.—'An Enumeration of the Roses of Yunnan', *Journ. Arn. Arb.*, Vol. 10 (1929), pp. 84–107.
CLAPHAM, A. R., TUTIN, T. G., and WARBURG, E. F.—*Flora of the British Isles*, *Rosa*, Ed. 1 (1952), pp. 516–27; Ed. 2 (1962), pp. 405–13.

CRÉPIN, F.—*Primitiae Monographiae Rosarum.*
A series of papers on the taxonomy of *Rosa*, published in *Bulletin de la Société Royale de Botanique de Belgique*, 1869–82, later published separately and reprinted in *Plant Monograph Reprints*, ed. J. Cramer and H. K. Swann, 1972.

—— 'Sketch of a New Classification of Roses', *Journ. Roy. Hort. Soc.*, Vol. 11 (1889), pp. 217–28.

—— 'Nouvelle Classification des Roses', *Journal des Roses*, Nos 3–5, 1891.

—— 'La Question de la Priorité des noms spécifiques envisagés au point de vue du Genre Rosa', *Bull. Herb. Boiss.*, 3rd series, Vol. 5 (1897), pp. 129–63.

DE PRONVILLE, A.—*Nomenclature Raisonée des Espèces ... du Genre Rosa* Paris, 1818.
De Pronville's other work, *Sommaire d'une Monographie du genre Rosier* (1822), is based on Lindley's monograph.

DUHAMEL DU MONCEAU, H. L.—*Traite des Arbres et Arbustes ...*, Ed. 2 ('Nouveau Duhamel'), Vol. VII (ed. Loiseleur-Deslongchamps), pp. 17-60.

DUMONT DE COURSET, G. L. M.—*Le Botaniste Cultivateur*, Ed. 2, Vol. 5 (1811), pp. 465–89.

FALCONER, R. W.—'The Ancient History of the Rose'. A lecture read before the Botanical Society of Edinburgh in 1838 and printed in Loudon's *Gardener's Magazine*, Vol. 15 (1839), pp. 379–89.

FERGUSON, JESSIE.—'A Botanical Study of Rose Stocks', *Journ. Roy. Hort. Soc.*, Vol. 58 (1933), pp. 344-7.

FERNALD, M. L.—*Gray's Manual of Botany*, Ed. 8 (1950), pp. 868–74.

ERLANSON, EILEEN W.—'Cytological Conditions and Evidences for Hybridity in North American Wild Roses', *Botanical Gazette*, Vol. 87 (1929), pp. 443–506.

—— 'Experimental data for a Revision of the North American Roses', *Botanical Gazette*, Vol. 96 (1934), pp. 197–259.

GLEASON, H. A.—*Illustrated Flora of the Northeastern United States ...*, Vol. 2 (1952), pp. 322-7.

HERRMANN, J.—*Dissertatio Inauguralis Botanico-Medica de Rosa.* Strasbourg, 1762.

HITCHCOCK, C. L., *et al.*—*Vascular Plants of the Pacific Northwest*, Part 3 (1961), pp. 164-71.

HURST, C. C., and BREEZE, MABEL.—'Notes on the Origin of the Moss-Rose', *Journ. Roy. Hort. Soc.*, Vol. 47 (1922), pp. 26–42. Reprinted in G. S. Thomas, *The Old Shrub Roses*, pp. 57–94.
For a discussion of Dr Hurst's account, see the series of articles by C. Hartman Payne in *Gard. Chron.*, Vol. 72 (1922), indexed in that periodical under 'Rose, Moss'.

HURST, C. C.—'Origin and Evolution of Garden Roses', *Journ. Roy. Hort. Soc.*, Vol. 66 (1941), pp. 73–82, 242–50, 282–9. Reprinted in G. S. Thomas, *The Old Shrub Roses*, pp. 57–94.

IBN AL-'AWWAM.—*Kitab al-filahah*, translated as 'Le Livre d'Agriculture' by J. J. Clement-Mullet. Paris, 1864 (Roses in Vol. I, Chap. VI, Art. 28.61).

JORET, C.—*La Rose dans l'Antiquité et au Moyen Age.* Paris, 1892.

JUZEPCHUK, S. V.—*Rosa*, in *Fl. URSS* (Flora of the Soviet Union), Vol. 10

(1941), pp. 431–506.

KLASTERSKY, I.—*Rosa*, in *Fl. Europaea*, Vol. 2 (1968), pp. 25–32.

LINDLEY, J.—*Rosarum Monographia*. London, 1820.

LAWRANCE, MARY.—*A Collection of Roses from Nature*. London, 1799.

MANDENOVA, I. P.—'A Revision of Rosa in Turkey', *Notes Roy. Bot. Gard. Edin.*, Vol. 30 (1970), pp. 327–40.

MELVILLE, R.—*Rosa*, in C. A. Stace, ed., *Hybridisation and the Flora of the British Isles* (1975), pp. 212–27.

NATIONAL ROSE SOCIETY.—*See Royal National Rose Society*.

NILSSON, O.—*Rosa*, in P. H. Davis, ed., *Flora of Turkey*, Vol. 4 (1972), pp. 106–28.

PAUL, WILLIAM.—*The Rose Garden*. London, 1848 and later editions.

PREVOST fils.—*Catalogue Descriptif... des Espèces... du Genre Rosier*. Rouen, 1829.

REDOUTÉ, P. J., and THORY, C. A.—*Les Roses*. Paris, 1817–24 (Vol. I, 1817–19; Vol. II, 1819–21; Vol. III, 1821–4).
The plates are unnumbered; references are to the page facing the plate. For further details see Stafleu, *Taxonomic Literature* (1967), pp. 376–7 and 379. The third edition (1828–30) and the similar fourth edition of 1835 contain additional plates and text. A complete facsimile folio reprint of the original edition was published by De Schutter 1974–1977. An additional volume with commentaries and unpublished plates appeared in 1978.

REGEL, E.—'Tentamen Rosarum Monographiae', *Act. Hort. Petrop.*, Vol. 5 (1878), pp. 287–398.

REHDER, A.—*Rosa*, in *Pl. Wilsonianae*, Vol. II (1915), pp. 304–44.

RIVERS, T., jun.—*The Rose Amateur's Guide*... London, 1837 and later editions.

ROESSIG, C. G.—*Die Rosen nach der Natur*... (with a French translation by de Lahitte). Leipzig, 1802–20.

—— *Oekonomisch-botanische Beschreibung... der Arten der Rosen*. Leipzig, Part I, 1799, Part II, 1803.

ROYAL NATIONAL ROSE SOCIETY.—*The Rose Annual*, 1907–.

SERINGE, N. C.—*Rosa*, in De Candolle, *Prodromus*, Vol. II (1825), pp. 597–625.

—— 'Critiques des Roses déséchées de N. C. Seringe', *Mélanges Botaniques*, Vol. I (1819), pp. 3–63.

TÄCKHOLM, G.—'Zytologische Studien über die Gattung Rosa', *Act. Hort. Berg.*, Vol. 7 (1922), pp. 97–381.

THOMAS, G. S.—*The Old Shrub Roses*. 1955 and 1980.

—— *Shrub Roses of Today*. 1962. Revised edition 1980.

—— *Climbing Roses Old and New*. 1965. Revised edition 1978.

THORY, C. A.—*Prodrome de la Monographie des Espèces... du Genre Rosier*. Paris, 1820.

TRATTINICK, L.—*Rosacearum Monographia*. 4 vols. Vienna, 1823–4.

WYLIE, ANN P.—'The History of Garden Roses' (Masters Memorial Lecture, 1954), *Journ. Roy. Hort. Soc.*, Vol. 79 (1954), pp. 555–71, Vol. 80 (1955), pp. 8–24, 77–87.

WILLMOTT, ELLEN.—*The Genus Rosa*. London, 1910–14. 2 vols (Vol. I, 1910–11, Vol. II, 1911–14).

The technical descriptions and the introductory chapter were contributed by J. G. Baker, formerly Keeper of the Kew Herbarium (cf. R. A. Rolfe, *Gard. Chron.*, Vol. 62 (1920), p. 124).

WOLLEY-DOD, A. H.—'A Revision of the British Roses', *Journal of Botany*, Supplement, 1930–1 (also printed separately).

DESCRIPTIONS

R. ACICULARIS Lindl.

R. *alpina* Pall., not L.; R. *gmelinii* Bge.; R. *karelica* Fries; R. *sayi* Schweinitz *fide* Rehd.

A lax shrub to 8 ft high; stems glabrous, densely clad with slender, straight or slightly curved prickles and shorter bristly ones; flowering branches sometimes unarmed. Leaflets mostly five or seven, sometimes nine, ¾ to 2¾ in. long, elliptic, oblong-elliptic or ovate, usually acute at the apex, edged with coarse simple teeth, bluish green above, greyish and glabrous or sparsely downy beneath. Stipules narrow, glandular at the edge, the free part acute or acuminate. Flowers usually solitary, more rarely in twos or threes, 1½ to 2¾ in. wide, fragrant, rosy pink. Pedicels smooth or glandular-bristly. Sepals narrowly lanceolate, slightly expanded at the apex, more or less upright after flowering. Styles woolly, free. Fruits bright red, about 1 in. long, smooth, ellipsoid, globose or pear-shaped, often with a distinct neck, crowned by the persistent sepals.

Native of the Old World from European Russia and bordering parts of Scandinavia to the Pacific, south to N. China and Japan; also of N. America, though the var. *bourgeauiana* appears to be commoner there than the typical state. It was described by Lindley in 1820 from a garden plant introduced from Siberia. Although much cultivated in Siberia, and used there as a hedging plant, R. *acicularis* is uncommon in Britain outside scientific collections, though worth cultivating in semi-wild spots for its large pink flowers, borne in May, and abundant red fruits.

var. BOURGEAUIANA Crép. R. *engelmannii* S. Wats.; R. *acicularis* var. *engelmannii* (S. Wats.) Rehd.; R. *a.* subsp. *sayi* (Schweinitz) W. H. Lewis, in part—Undersurface of leaflets more or less glandular, their margins often edged with compound glandular teeth. Fruits variable in shape as in the type. Native of N. America, where, however, the species also occurs in its typical state. On the other hand, plants with glandular leaflets occur in Russia (R. *a.* var. *subalpina* (Bge.) Boulenger; R. *oxyacantha* Bieb.), but these are said to have commonly nine leaflets.

The synonymous name R. *engelmannii* S. Wats. is founded on a plant raised in the Arnold Arboretum from seeds collected by Engelmann in Colorado, but plants distributed in Britain under this name do not match the type, and some appear to have been R. *arkansana* (q.v.).

var. NIPPONENSIS (Crép.) Koehne R. *nipponensis* Crép., *nom. prov.*—Leaflets seven or nine, sometimes eleven on sterile shoots, elliptic, mostly rounded or subacute at the apex, not much over 1 in. long, finely toothed. Flowers, in the introduced plants, of a beautiful deep purplish red. *Bot. Mag.*, t. 7646.

This variety was, with hesitation, separated from R. *acicularis* by Crépin, whose description is based on specimens collected in Japan on Mt Fuji by Tschonoski and on others from plants grown in the Copenhagen Botanic Garden, whence it was introduced to Kew in 1894; the Copenhagen plants had been raised from seeds received from St Petersburg. It is doubtful whether this rose ever spread much beyond Kew and it is perhaps worthy of reintroduction. Typical R. *acicularis* also occurs in Japan.

R. × ALBA L.   WHITE ROSE
R. *alba* var. *vulgaris* Ser.

A spreading shrub 6 to 8 ft high, its branches green, with a pruinose bloom, armed with scattered slender or stout prickles of unequal size. Leaves with five or more rarely seven leaflets; rachis downy, prickly beneath. Leaflets ovate, broadly oblong-elliptic or roundish, up to 2 in. long, obtuse or often shortly acuminate at the apex, dull green and glabrous above, hairy beneath on the veins and slightly so on the blade, deeply and sharply toothed. Stipules broad, with narrow, spreading auricles. Flowers solitary or in threes, terminal or from the axils of the upper leaves or reduced leaves, semi-double or double, white, of medium size. Pedicels 1 to 1½ in. long, weakly glandular-bristly, the bristles more or less extending on to the ellipsoid receptacle. Sepals pinnated, with narrow, leafy tips, glandular-bristly on the back, reflexed and soon falling. Fruits said to ripen rarely.

An ancient hybrid of European gardens, long considered by botanists to be the double form of a wild species. However, in 1873 the Swiss rhodologist Christ described a wild plant which he identified as a hybrid between the dog rose R. *dumetorum* (*corymbifera*) and R. *gallica*, and remarked on its great similarity to R. *alba*. After this, with Crépin's endorsement, R. *alba* came to be accepted as a hybrid of the parentage suggested by Christ, but recently Klastersky has suggested that it may be a complex hybrid deriving from R. *gallica*, R. *arvensis*, and some white-flowered member of the Canina group (*Fl. Europaea*, Vol. 2 (1968), p. 26). Like R. *canina* and its allies, R. × *alba* is highly polyploid, all forms examined being hexaploid. The wild, single-flowered rose known as R. *alba* by early writers is of uncertain identity. Probably white-flowered dog roses were meant, or hybrids between the Canina complex and R. *arvensis*. The R *alba* var. *humilis* Thory in Redouté is probably R. × *polliniana*.

Typical R. × *alba* is little known today, and indeed the above description is drawn from old accounts, illustrations, and herbarium specimens. The rose 'Alba Maxima' (see page 170) is probably the R. *alba* of Gerard and Parkinson, though it does not seem to be quite the same as the R. *alba* of old continental works.

R. INCARNATA Mill.   R. *alba* var. *incarnata* (Mill.) Pers.; R. *carnea* Dum.-Cours.; R. *provincialis* var. *incarnata* (Mill.) Martyn—There has been much confusion over the name R. *incarnata*. As used by some pre-Linnaean botanists, it, or the plural *Rosae incarnatae*, meant what is now known as R. *damascena*, while some French botanists of the last century took Miller's R. *incarnata* to be a form of R. *gallica* with sparsely armed branches and glandular-compound leaflets. A comparison of authentic herbarium specimens shows, however, that Miller's R. *incarnata* is identical to the 'Cuisse de Nymphe' of French gardens. The English name for this—'Maiden's Blush'—is sometimes attributed to William Hanbury, who has a charming passage about it in his *Compleat Body of Gardening* (1770–1). In fact, Miller himself used it in the 1752 edition of his *Dictionary*.

The Maiden's Blush differs from R. *alba* in the colour of the flowers, the almost unarmed stems, the more numerous leaflets (mostly seven), and the

presence of numerous needle-like eglandular prickles on the flowering branchlets below the bracts. It is almost certainly the same as the R. *incarnata* of
Parkinson and a very old rose.

Miller was apparently unacquainted with the 'Great Maiden's Blush' (R. *alba*
var. *regalis* Thory), which is now commoner in gardens. But it is listed in
Weston's *Flora Anglicana* (1775) as 'Great Maiden's Blush Rose', with R. *incarnata major* as the Latin name.

The R. *incarnata* of *Bot. Mag.*, t. 7035, is not Miller's but a form of R. *gallica*.

R. CANINA aggregate × R. GALLICA.—As mentioned above, many authorities have considered that R. *alba* derives from a cross between R. *canina* or one
of its near allies and R. *gallica*. Spontaneous hybrids between the latter and
R. *canina sens. strict.* have been recorded from a few localities on the continent
of Europe. They show the influence of both parents in their armature, taking
large, hooked prickles from R. *canina* and bristles or needles from R. *gallica*,
which also contributes a suckering habit. The leaflets are usually five in number
against seven in R. *canina*. The flowers are larger than in R. *canina*, bright rose
or purplish rose, borne on pedicels which are shorter than in R. *gallica*, smooth
or glandular-bristly. Putative hybrids between R. *gallica* and R. *coriifolia* or
R. *afzeliana*, both members of the R. *canina* aggregate, have also been recorded,
but are less well authenticated. The wild rose found by Christ in Switzerland
and said to resemble R. *alba* (see above) may have been R. *gallica* × R. *coriifolia*
(Boulenger, *Bull. Jard. Bot. Brux.*, Vol. 12 (1932), pp. 462–76 and 459–61).

Altogether some twenty names of specific form have been given to forms or
putative forms of these hybrids. The oldest is R. *collina* Jacq., but Boulenger
remarks that there is nothing in Jacquin's description or figure to indicate the
influence of R. *gallica*. The name used by Rehder for R. *canina* × R. *gallica* is
R. × *waitziana*, but the rose so named by Trattinick is of uncertain identity.
For R. × *waitziana* var. *macrantha* (Desp.) Rehd., see R. 'Macrantha', in alphabetical order.

R. ALBERTII Reg.

A shrub 2 to 4 ft high, the stems armed with numerous, straight, needle-like
prickles. Leaves 1 to 3 in. long, composed of five to nine leaflets, which are
ovate, obovate, or roundish, ¼ to 1¼ in. long, sharply toothed, the teeth usually
compound-glandular, glabrous above, minutely downy beneath; rachis downy
and glandular. Flowers white, 1½ in. across, solitary. Pedicels glandular. Sepals
lanceolate, glabrous outside. Fruits about ¾ in. long, shortly stalked, slenderly
pear-shaped or ellipsoid, shedding when ripe both the sepals and the top of the
receptacle.

Native of Russia in the Altai Mountains and Central Asia; discovered by
Albert Regel in 1877, introduced by him to the St Petersburg Botanic Garden,
and named after him by his father. Although resembling R. *pimpinellifolia* in
some characters, it is nearer to R. *beggeriana*, with which it has in common the
peculiar fruits, devoid both of calyx and disk when ripe.

Plants at one time cultivated as R. *albertii* were wrongly named, and near to

R. ecae. The rose portrayed as *R. albertii* in Willmott, *The Genus Rosa*, p. 319, t., is also wrongly named.

R. ANEMONIFLORA Fort. ex Lindl.*

R. triphylla sens. Rehd. & Wils., not Roxb. ex Lindl.

A bush with spreading branches armed with scattered slender prickles. Leaves with three or five leaflets—the latter on the young barren shoots of the first year. Leaflets ovate or ovate-lanceolate, 1½ to 3 in. long, very finely and simply toothed, glabrous on both surfaces, dark green above, pale beneath. Stipules narrow, edged with glandular teeth or glandular ciliations. Flowers blush-white, 1 to 1½ in. across, in loose corymbs, double, the inner 'petals' (modified stamens) narrow and ragged. Pedicels slender, naked, or with a few glandular bristles. Styles united into a slender, hairy column.

This rose was introduced from China in 1844, by Fortune, who found it in a garden at Shanghai. Plants with single flowers have been found in Fokien province, but whether these were genuinely wild is uncertain. *R. anemoniflora* is thought by some authorities to be a hybrid between *R. multiflora* and *R. laevigata*, and also shows some resemblance to *R. banksiae*. But in its essential characters it is a member of the *Synstylae*. It is a curious and rather pretty rose, but not very hardy.

R. ARKANSANA Porter

R. blanda var. *arkansana* (Porter) Best

A suckering shrub up to 3 or 4 ft high, in the wild often a subshrub cut to the ground each winter; stems clad with slender, straight prickles and bristles, sometimes very densely so. Leaflets nine or eleven, more rarely seven, elliptic or sometimes obovate, 1 to 2 in. long, lustrous above, glabrous on both sides except for the sometimes downy veins beneath, edged with fairly deep, simple, eglandular teeth. Flowers pink, about 1½ in. across, borne around midsummer in lateral clusters, or later at the end of strong growths from the base. Pedicels and receptacle glabrous, sometimes slightly glandular. Sepals narrow, slenderly pointed, sometimes glandular on the back. Fruits globose to pear-shaped, about ½ in. wide, smooth or slightly glandular, crowned by the usually spreading sepals.

Native of the central USA from Wisconsin and Minnesota to Colorado and Kansas.

var. SUFFULTA (Greene) Cockerell *R. suffulta* Greene; *R. pratincola* Greene; *R. heliophila* Greene; *R. arkansoides* Schneid.—Rachis and underside of leaflets downy. More widely distributed than the typical state and extending into Canada.

R. arkansana has never been much cultivated in Britain, but it is an interesting rose owing to its ability to flower on strong shoots of the current season, and

* This name is illegitimate, being antedated by *R. anemoneflora* Andrews in *Roses*, Vol. I, t. 32 (1821), but no other name is at present available (1979).

is by all accounts a pretty one, especially its var. *suffulta*, of which there are forms with white and with deep pink flowers, and two named clones with double flowers—'WOODROW' and 'JOHN ALLEN'. Some hybrids have been raised from the var. *suffulta* in the USA.

R. ALCEA Greene is closely allied to R. *arkansana* var. *suffulta*, but more glandular. Another ally of R. *arkansana* is the Californian R. SPITHAMEA S. Wats.

R. ARVENSIS Huds.
R. *sylvestris* J. Herrm.

A deciduous trailing or climbing shrub, with long slender purplish branches no thicker than stout string, and armed with scattered, short, more or less curved prickles. Leaflets five or seven, varying in shape from orbicular to elliptic or narrowly to broadly ovate, $\frac{1}{2}$ to $1\frac{1}{2}$ in. long, glabrous on both sides or, more commonly, downy beneath at least on the veins, the underside often glaucous, teeth usually simple and eglandular. Stipules narrow, with spreading auricles. Flowers white, with little or no fragrance, solitary, or up to eight in a corymb. Pedicels slender, $\frac{1}{2}$ to 2 in. long, they and the ovoid to globose receptacle smooth or somewhat glandular. Sepals long-pointed, with a few slender appendages, usually glabrous and eglandular on the back. Styles exserted, united into a column, usually quite glabrous. Fruits $\frac{1}{4}$ to $\frac{3}{8}$ in. long, red, variable in shape, shedding the sepals when ripe.

Native of Europe and southern Anatolia; in the British Isles it is commonest in southern England, very rare in Scotland, widespread but local in Ireland. It is of interest as the only native member of the section *Synstylae*, with exserted styles united into a column, and is easily distinguished from other British roses by this character and by its slender shoots that often grow several yards in a season. The only other British species with joined styles is R. *stylosa* (q.v., p. 65), a sturdy bush with the dog rose habit.

As commonly found wild in Britain, R. *arvensis* is scarcely worthy of cultivation, though it has the ability to grow and flower in the shade of trees. There is, however, a more robust form found wild on the continent and occasionally in Britain, which has stouter, often pruinose stems, more numerous flowers in the inflorescence, and glossier leaflets. It is this form that is figured in Willmott, *The Genus Rosa*, p. 11, t.

R. 'CAPREOLATA' AYRSHIRE ROSE—A climber of extraordinary vigour, making shoots 30 ft long in a season. Leaves semi-persistent, with glossy leaflets green and glabrous beneath. Styles hairy. Fruits ovoid with a distinct neck, slightly glandular-bristly (R. *capreolata* D. Don ex Neil; R. *arvensis* var. *ayrshirea* Ser.; R. *arvensis* var. *capreolata* (Neil) Bean).

The original Ayrshire rose is now probably lost. Its history is given by Neil in *Edinb. Phil. Journ.*, Vol. 2 (1820), pp. 102–7, and by Sabine in *Trans. Hort. Soc.*, Vol. 4 (1822), pp. 456–67. In 1767 a consortium of Scottish gardeners including Dr John Hope, Regius Keeper of the Royal Botanic Garden, Edinburgh, sent a collector to Canada. One member was the Earl of Loudon of

Loudon Castle, and the original Ayrshire rose was raised from his share of the seeds. One of several seedlings, it was given by his gardener Douglas to a Mr Dalrymple of Orangefield, Ayrshire; planted against a wall by the roadside it soon attracted attention because of its great vigour and was put into commerce.

This, at any rate, is the story given to Neil and Sabine by Douglas, but Sabine adds a second version, according to which the Orangefield plant came from Germany by way of a garden in Yorkshire. If indeed the original parent grew in Canada, it must have been taken there by a settler, and this is not unlikely, considering that several European species are actually naturalised in N. America.

Although usually considered to be a cultivated variety of R. *arvensis*, there is a distinct possibility that the Ayrshire rose was a hybrid between that species and the related R. *sempervirens*. This is suggested not only by its botanical characters, but also by the fact that the group of Ayrshire ramblers raised from it by D. Martin of Dundee were all much alike yet presumably different from each other in some respects, since some fourteen were named. A pure species is unlikely to have given so much variation.

A few of the Ayrshire ramblers are still grown, though the only one of note is 'Splendens' (p. 201).

R. 'RUGA'—A presumed hybrid between R. *arvensis* and some form of R. *chinensis*, sent to the Horticultural Society from Italy before 1830 (R. *ruga* Lindl., *Bot. Reg.*, t. 1389).

R. × POLLINIANA (R. *arvensis* × R. *gallica*)—See page 128.

R. BANKSIAE R. Br. in Ait. f.   BANKSIAN ROSE   [PLATE 6

A climbing shrub up to 40 ft high, with slender, glabrous, unarmed shoots. Leaves with three or five leaflets, which are 1 to 2½ in. long, one-third to half as wide, oblong-lanceolate, pointed, simply toothed, glabrous on both surfaces except that the midrib beneath, and rachis, are sometimes slightly downy. Stipules very narrow, soon deciduous. Flowers white or yellow, 1¼ in. across, numerous in an umbel, each flower on a stalk about 1 in. long. Sepals ⅜ in. long, ovate, entire. Fruits globose, about the size of a pea, with the sepals fallen away.

Native of China, where it has long been cultivated.

cv. 'ALBA PLENA' ('BANKSIAE')—Flowers white, violet-scented, double. This is the typical form of R. *banksiae*, from which Robert Brown described the species in the second edition of Aiton's *Hortus Kewensis*. It had been introduced by the Kew collector William Kerr from Canton in 1807. *Bot. Mag.*, t. 1954.

cv. 'LUTEA'.—Flowers double, yellow, slightly fragrant. Introduced for the Horticultural Society by John Parks, who brought back several plants from China in 1824 (*Bot. Reg.*, t. 1105).

cv. 'LUTESCENS'.—Flowers single, yellow, fragrant. Of later introduction than the preceding. *Bot. Mag.*, t. 7171.

The Banksian rose is one of the most lovely of all, but unfortunately it is too fond of the sun to thrive in the cooler and rainier parts of the British Isles.

The double yellow Banksian ('Lutea') is the hardiest and most floriferous, consequently the one most commonly seen in British gardens, though unfortunately the least fragrant. It, and the others, need the protection of a sunny wall. Annual pruning is unnecessary but the older stems should be cut clean out periodically after flowering is over. The flowers are not borne on laterals from the previous year's growths, as in the common ramblers, but on the twigs produced by these laterals, so a stem will be two or three years old before it produces flowers. This should be borne in mind when pruning.

var. NORMALIS Reg.—This is the wild state of the species, with single, white, fragrant flowers. Of wide distribution in China, from Kansu and Shensi to Yunnan; first described from specimens collected by the Russian traveller Kirilov in Peking gardens. It was later found wild in Hupeh and Szechwan by Augustine Henry, and his interesting note on it will be found in *Gard. Chron.*, Vol. 31 (1902), pp. 438–9.

Curiously enough, this white, single form was already in cultivation in Britain when Kerr introduced the double, white-flowered type.

'Four years ago I found a rose growing on the wall of Megginch Castle, Strathtay, Scotland, which seemed to be a very slender-growing form of R. *Banksiae*. Captain Drummond of Megginch told me that it was a rose that his ancestor, Robert Drummond, had brought with other plants from China . . . in 1796. This old rose had been repeatedly cut to the ground by severe winters, and had rarely, if ever, flowered. The impression, however, was that it was white and very small. Captain Drummond kindly gave me cuttings, which I took to Nice, and this year they flowered, proving themselves to be the typical single white Banksian rose so long sought for and hidden away in this nook of Scotland for more than a hundred years.' (E. H. Woodall) *Journ. R.H.S.*, Vol. 35 (1909–10), p. 218).

R. CYMOSA Tratt. R. *indica* L., in part; R. *microcarpa* Lindl., not Retz. or Bess.; R. *sorbiflora* Focke; R. *bodinieri* Lévl.—An ally of R. *banksiae* widely distributed in central and southern China. It differs in its more prickly stems, much larger, compound inflorescences, resembling those of a *Sorbus*, and the longer sepals with lateral appendages. Wilson collected this species on several occasions, mostly in Hupeh, and probably sent seeds during his first expedition for the Arnold Arboretum, but it is not known to be in cultivation at present.

R. *cymosa* was made known to Western science by James Cunningham, who sent a fruiting specimen from Chusan in 1701. This is portrayed in Petiver's *Gazophylacii Naturae* (1704) with the phrase-name '*Rosa Chusan glabra Juniperi fructu*'. Fifty years later Linnaeus cited this name under his R. *indica*, but his description is of some other rose that cannot now be identified. For the R. *indica* of Lindley and other authors, see R. *chinensis*.

R. BEGGERIANA Schrenk ex Fisch. & Mey.
R. *anserinaefolia* Boiss.

A shrub 6 to 10 ft high; stems and branches armed with light-coloured, hooked spines. Leaflets usually seven or nine, $\frac{2}{3}$ to $1\frac{1}{4}$ in. long, oval to slightly obovate, grey-green and glabrous above, usually glandular and sometimes downy beneath, edged with ten to twenty simple or compound teeth. Flower-buds elongate, acute. Flowers white, 1 to 1½ in. across, in clusters of nine or more, produced from midsummer onwards at the ends of the new shoots. Pedicels slender, to about 1 in. long, glabrous or downy, sometimes glandular. Fruits globose, smooth, red at first, finally purplish, ¼ to ⅜ in. long, sepals at length falling away together with the top of the receptacle.

Native of Central and S.W. Asia (including Asiatic Turkey); introduced about 1881. It is a variable species, of which one botanist described or recognised fifty varieties. The cultivated form is of some value for its greyish leaves with a sweet brier fragrance, but the flowers are unpleasantly scented. For the fruits see further under R. *gymnocarpa*.

R. BIEBERSTEINII Lindl. in Loud.*

R. *provincialis* Bieb., not J. Herrm.; R. *ferox* Bieb., not Lawrance; R. *horrida* Fisch., nom. nud.; R. *horrida* Fisch. ex Crép. (1872), not Spreng. (1825); R. *turcica* Rouy

A dwarf, compact bush 1 to 2 ft high, of rounded form, armed with numerous strongly curved or hooked prickles usually intermixed with glandular bristles and needles. Leaves 1 to 2 in. long, with five or seven leaflets, which are elliptic or roundish, ¼ to ¾ in. long, coarsely but evenly compound-toothed, the teeth, rachis, stipules, and undersurface copiously glandular. Flowers white, solitary or two or three together. Pedicels very short, glandular. Sepals pinnately lobed, glandular-toothed and ciliate. Fruits roundish, red, about ½ in. wide, smooth or with a few short needles or bristles, devoid of sepals.

Native of S.W. Russia, Asia Minor, and the Balkans. This interesting and pretty little rose forms a dense mass of interlacing, very spiny twigs. It is allied to R. *pulverulenta* (*glutinosa*) but has coarser, mostly hooked prickles nearly always mixed with bristles and needles, white flowers and smaller fruits; also, although glandular it is not strongly aromatic like that species. Its armature distinguishes it from R. *sicula*, in which needles and bristles are lacking and the whole plant less glandular.

R. BLANDA Ait.

R. *fraxinifolia sens.* Lindl., not Borkh.; R. *solanderi* Tratt.; R. *subblanda* Rydb.

A shrub 4 to 6 ft high, usually quite unarmed except for a few slender scattered prickles near the base of vigorous stems; when, rarely, prickles occur on the branches they are sparse and straight, and do not occur in nodal pairs. Leaflets usually five or seven, elliptic or oblong-obovate, ¾ to 2¼ in. long,

* In Loudon, *Encyclopaedia of Plants* (1829), p. 444, the botanical part of which was the work of Lindley. This is a conventional renaming of the R. *ferox* of Bieberstein, which that author had confused with the earlier R. *ferox* of Miss Lawrance (for which see under R. *rugosa*). Lindley had listed R. *ferox* Lawrance on a previous page and had to find a new name for R. *ferox* Bieb. The name R. *biebersteinii* Lindl. is not invalidated by R. *biebersteiniana* Tratt. (1823), a synonym of R. *canina* sometimes wrongly cited as R. *biebersteinii*. The name R. *horrida* would be correct if Fischer had expressly intended it as a new name for R. *ferox* Bieb., but this is not the case.

dull green, glabrous on both sides or downy beneath, edged with eglandular, usually simple teeth. Stipules widening upwards, entire or somewhat toothed, downy or glabrous. Flowers solitary or in clusters of three to seven, 1¾ to 2½ in. wide, rosy pink, opening in late May or early June. Pedicels and receptacle glabrous. Sepals lanceolate, entire, ½ to 2 in. long, sometimes glandular. Fruits globose or broadest slightly above or below the middle, red, crowned by the erect sepals.

Native of eastern and central N. America; in cultivation 1773. A handsome rose, allied to the Old World R. *majalis* (*cinnamomea*), which flowers at about the same time. But in that species the prickles are more numerous and hooked, and occur in nodal pairs.

R. *blanda* has been used in the United States to breed thornless roses, either as ornamentals or for root-stocks. A presumed hybrid between R. *blanda* and R. *chinensis* was described in 1902 from the Forstgarten, Hannover-Münden (R. × ASCHERSONIANA Graebn.).

R. × MICHIGANENSIS Erlanson—A probable hybrid between R. *blanda* and R. *palustris*, showing the influence of the latter in the coarse, erect habit, the presence of prickles on the stems, the more finely toothed leaflets and spreading sepals. It was described from the region of the Great Lakes.

R. WOODSII Lindl. R. *fendleri* Crép.; R. *woodsii* var. *fendleri* Rydb.; R. *macounii* Greene—Allied to R. *blanda*, but with a more westerly distribution, from N.W. Canada south to Texas, east to around the 100th meridian. It is dwarfer, to about 4 ft, of stiff habit, rather more strongly armed, the prickles straight, paired at the nodes, often extending onto the flowering branchlets. Leaflets smaller than in R. *blanda*, to little more than 1 in. long, obovate to elliptic, the teeth often glandular. Sepals usually without glands on the back. Fruits smaller, to about ⅜ in. wide. The var. ULTRAMONTANA (S. Wats.) Jeps. is taller and laxer, with larger, always simply toothed leaflets.

R. *woodsii* was described by Lindley in 1820 from a cultivated plant supposed to have come from the Missouri river. As seen in cultivation, under the name R. *woodsii* var. *fendleri*, this is one of the prettiest of American roses, with foliage not unlike that of R. *pimpinellifolia* but grey green, bearing its flowers on short, closely set laterals.

R. *woodsii* and R. *blanda* overlap in distribution and produce fertile hybrids. For these the correct name would appear to be R. × DULCISSIMA Lunell *emend.* W. H. Lewis; see *Brittonia*, Vol. 14 (1962), pp. 65–71, where the differences between the two species are discussed.

R. BRACTEATA Wendl.   MACARTNEY ROSE
R. *lucida* Lawrance, not Ehrh.; R. *macartnea* Dum.-Cours.; *Ernestella bracteata* (Wendl.) Germain de St Pierre

An evergreen shrub of rambling habit, reaching on walls in favoured places a height of 20 ft. Branches very thick and sturdy, covered with brownish down, and armed with pairs of stout, hooked prickles and numerous scattered bristly ones. Leaflets five to eleven, obovate, often widely truncated at the end and

finely toothed, $\frac{3}{4}$ in. to 2 in. long, (in vigorous plants) $\frac{1}{2}$ to 1 in. wide, of a
very deep green and highly polished above, either glabrous or downy on the
midrib beneath; rachis glandular-downy. Stipules laciniated. Flowers 3 to 4 in.
across, white, borne singly on a very short stalk which is surrounded by several

ROSA BRACTEATA

large, laciniated, downy bracts. Receptacle and sepals (the latter $\frac{3}{4}$ in. long)
covered with a pale brown wool. Fruits globose, orange-red, woolly, about
$1\frac{1}{2}$ in. wide. *Bot. Mag.*, t. 1377.

Native of southeast China and Formosa; introduced in 1793 by Lord
Macartney. This distinct and remarkable rose is, unfortunately, not very hardy
except in the southwest counties and similar places, where its rich evergreen
foliage and large flowers make it one of the most striking of all the wild types.
Near London, even grown on a wall, it is occasionally damaged badly by frost.
Its flowers appear from June until late autumn, and have a delicate fruity
perfume.

The rose 'MARIE LEONIDA', with creamy white, double flowers, is a cross
between R. *bracteata* and R. *laevigata*. Its flowers do not open well in the average
British summer.

For 'Mermaid', a hybrid between R. *bracteata* and a Tea rose, see p. 191.

R. CLINOPHYLLA Thory R. *involucrata* Roxb. ex Lindl.—Closely allied to
R. *bracteata*, with the same laciniate bracts and stipules and tomentose fruits,
but the prickles not hooked, the leaflets narrower, acute, often downy beneath.
It is a native of India, where according to Hooker, 'it is the common rose of
the Bengal plains and foot of the Himalaya and the only really tropical species

of India'. It is usually found by riversides and other wet places. Farther east it occurs in Burma, where Kingdon-Ward found it on the Irrawaddy near Myitkyina, growing as a bush 10 to 15 ft high among rocks submerged for three months of the year, flowering in March and again in July (KW 6601, field note, and *Plant Hunting on the Edge of the World*, where it is mentioned on pp. 50, 138, as R. *bracteata*). It also occurs in Laos and possibly in S.E. China.

R. *clinophylla* was first collected in Nepal by Francis Buchanan-Hamilton, around 1803, and was described and named by Roxburgh from specimens sent by him to the Calcutta Botanic Garden. But by the time Lindley had published Roxburgh's name R. *involucrata* (1820), the species had been described by Thory, and figured, in the first volume of Redouté's *Les Roses* (p. 43, t. (1817)), under the name R. *clinophylla*. The type-plant grew in Boursault's garden, and had come from England, so the introduction to this country is earlier than 1818, the date given by Lindley in the *Botanical Register*. In 1834 Loudon saw a plant 11 ft high on a wall in Loddiges's nursery, which flowered magnificently, but the species seems to have dropped out of cultivation and should be reintroduced.

R. LYELLII Lindl.—This rose was described by Lindley in 1820 from a plant sent to Sir Joseph Banks from Nepal by Dr Wallich, though whether it was collected in the wild or in a garden is not recorded. It is beautifully figured in the frontispiece to his *Monograph* (plate 1). Although obviously near to R. *clinophylla*, and included in it by some botanists, it is distinct in its corymbose inflorescence with distant pairs of narrow bracts, and was judged by the Belgian rhodologist Crépin to be a hybrid of R. *clinophylla* with R. *moschata* (in which he included the Himalayan R. *brunonii*). According to Crépin, such a hybrid, in single or double forms, is cultivated in Indian gardens,* and he identified as R. × *lyellii* a rose distributed by the nurseryman William Paul under the name R. "*lucida duplex*" (figured in the frontispiece to Paul's *The Rose Garden*, ed. 1889). This came from the French rose-grower Jamain. A similar rose is figured in Willmott's *The Genus Rosa*, Vol. I, p. 129, t., wrongly as R. *involucrata*. The material portrayed was brought by Miss Willmott from France.

R. BRUNONII Lindl.   HIMALAYAN MUSK ROSE

R. *moschata sens.* Crép. and other authors, not J. Herrm.; R. *brunonis* Wall., *nom. nud.*; R. *brownii* Tratt.; R. *moschata* var. *nepalensis* Lindl.; ?R. *napaulensis* Andr.

A deciduous or semi-evergreen climber, reaching to the tops of lofty trees in the Himalaya, and to about 40 ft high in gardens. Leaves up to 7 or 8 in. long, with five to nine leaflets; rachis prickly and glandular. Leaflets ovate-lanceolate, elliptic or oblong-elliptic, acute or acuminate, mostly 2 to 3 in. long, simply and regularly toothed, upper surface dull or lustrous green, sometimes with a glaucous tinge, glabrous or thinly hairy, underside green or whitish, usually hairy, at least on the veins, sometimes slightly glandular. Stipules narrow, with spreading free tips, usually edged with glands or hairs.

* 'Nothing can be more ornamental than the double white rose of Northern India and the Deyra Doon, R. *Lyellii, kooza* of the natives . . .' (Royle, *Ill. Bot. Himal.* (1835), p. 203).

Flowers pale yellow in bud, opening white, 1½ to 1¾ in. across, borne in the second half of June or early July, arranged in corymbose clusters usually higher than wide, several clusters often united into a large compound inflorescence which on vigorous shoots of cultivated plants may be a foot across, though on wild plants the supplementary clusters from the upper leaf-axils do not usually reach to the same level as the terminal corymb, so that the compound inflorescence tends to be conical in outline. Pedicels hairy and usually more or less glandular, up to 1⅝ in. long. Receptacles ovoid or ellipsoid, with the same covering as the pedicels. Sepals with a few lateral appendages, slenderly pointed, longer than the rather narrowly ovoid flower-bud, hairy on the back, sharply reflexed at flowering-time and soon falling. Styles united into an exserted column. Fruits roundish to obovoid, ⅜ to ½ in. long. *Bot. Mag.*, t. 4030.

Native of the Himalayan region, where it ranges from 3,000 to 9,000 ft, climbing into alders, deodars, oaks, etc., extending to China (Yunnan and W. Szechwan); described in 1820 by Lindley, who named it after Robert Brown, the distinguished botanist; introduced from Nepal by Wallich in 1822. At Kew it was at first grown on a west wall, but was eventually found to be quite hardy save that the growths of young plants, sometimes 10 or 12 ft long, were cut back in hard winters. What was probably the Wallich introduction still grew at Kew in the Rose Dell until the 1960s, climbing 30 ft into a holly, whose dark glossy-green leaves were a perfect foil to the white flowers of its companion. A plant in the University Botanic Garden, Cambridge, climbing on a pine, is portrayed in *Garden*, Vol. 71 (1907), p. 251; at Arundel Castle, Sussex, planted about 1850, it was some 35 ft high in a yew in 1904 (*Gard. Chron.*, Vol. 36 (1904), p. 152 and figs. 62–3. In the Kew plant the leaves were dull green above; another form had grey-green leaves and was more tender, less vigorous, and with fewer flowers in the inflorescence.

A Chinese form of R. *brunonii* was introduced by Wilson in 1908 from the Wa-shan, W. Szechwan (W.1125). A plant under this number was cultivated in the Hanbury garden at La Mortola, Menton, and the fine form in commerce as 'La Mortola' may be of this provenance, though not certainly, since at least four forms of R. *brunonii* were grown there. Another reintroduction is Kingdon Ward 6309, from the Tsangpo Gorge at the eastern end of the Himalaya, collected 1924.

Until the early 1880s R. *brunonii* was grown under its correct name. But in 1879, the Belgian rhodologist Crépin, then the world authority on *Rosa*, published a paper in which R. *brunonii* was sunk in R. *moschata* and in this he was generally followed by other botanists. However, Crépin interpreted R. *moschata* in a very wide sense, including in it several other members of the *Synstylae* now treated as distinct species. R. *brunonii* is perhaps nearer to R. *moschata* than any of these, but not by much.

R. MOSCHATA J. Herrm.—Judging from old descriptions, portraits and herbarium specimens, this species, little known today, differs from R. *brunonii* in the following respects: It is a tall, lax shrub, scarcely a climber; leaves dark green and smooth above, whitish beneath, glabrous except for the downy midrib, up to no more than 2 in. long and ovate to lanceolate, relatively broader than in R. *brunonii*, very finely toothed; flowers larger, in lax corymbs, musk-scented, borne from August until the first frosts, the petals somewhat convex, acuminate at the apex (slightly retuse in R. *brunonii*); pedicels and receptacle covered with fine, appressed hairs, not or only slightly glandular (eglandular forms of R. *brunonii* have a much laxer and coarser indumentum on these parts); fruits not often described, but said by some authorities to be small and ovoid.

R. *moschata* is not known in the wild in its typical state. It was introduced to Britain in the reign of Henry VIII, from Italy. In Germany it was still a novelty in the 1580s, and not entirely hardy. From the fragrance of its flowers, likened to that of animal musk, it was even then called R. *moschata* or *muschata*. But the name *Rosa damascena* was also used for it, probably from the belief that it was the 'Nesrin' or 'Nefrin' of Arab medical works—a rose grown about Damascus whose flowers were used as a purgative. It was this property, and not the fragrance, that made the Musk rose of interest to the European medical botanists. 'The Musk Roses, called in Latin *Rosae Moschatae* and *Damascenae*, are the small, single, white roses, which blow not till autumn . . . the Best and most efficacious are those that grow in the hot countries, as *Languedoc* and *Provence*. . . . Three or four of these Musk Roses being bruised in a Conserve, or Infusion, purge briskly, so that sometimes they occasion blood; those of Paris do not work so strong, but are more purgative than the pale Roses.' (Lemery, *Traité des Drogues* (1698), an extract included in the English translation (1712) of Pomet's *Histoire Générale des Drogues* (1694).) By 'pale Roses' was meant the *Rosae Pallidae seu Incarnatae*, i.e., the R. *damascena* of modern authors.

Still a common garden rose in the early part of the 19th century, R. *moschata* has been displaced by its hybrids. Indeed, it was thought to be extinct in this country until Graham Thomas found it growing at Myddelton House, Enfield, once the home of E. A. Bowles, who records in *My Garden in Summer* (1914) that he had a young plant raised from a cutting brought from The Grange, Bitton (G. S. Thomas, *Climbing Roses*, pp. 37–8, 52–4 and fig. 3). In spite of its delicious fragrance, late flowering habit, and historical interest it remains rare.   [PLATE 7

Through 'Champney's Pink Cluster', its hybrid with the Pink China rose, R. *moschata* is an ancestor of the modern Hybrid Teas and the so-called Hybrid Musks. Crossed with R. *multiflora* it is a parent of the ramblers 'The Garland' and 'Madame d'Arblay'. See also 'Dupontii'.

var. NASTARANA Christ—Leaflets glabrous and green beneath, sometimes no more than 1 in. long. Inflorescence more glandular than in typical R. *moschata*. Described from Iran, where it is cultivated but said also to occur wild. The normal flowering-time has not been ascertained, but one of the two roses introduced by Paul's nursery around 1880 as R. *Pissardii* agrees quite well with the var. *nastarana* and this flowered into October. It had white, semi-double flowers, and may be the rose figured in Wilmott as var. *nastarana* (Vol. I, p. 39, t.).

Under var. *nastarana* Christ mentions the 'Gul e Rescht' or Rescht rose, a garden rose of Iran which is an obvious hybrid, with small, double, red flowers, strongly pinnated sepals and toothed stipules. It bears some resemblance to the Constantinople rose (R. *byzantina* Dieck), which Crépin judged to be a hybrid between R. *gallica* and R. *multiflora*. The second of the two roses introduced by Paul as R. *Pissardii* seems to have been similar to the Rescht rose.

R. 'PISSARDII'.—The rose named R. *Pissardii* by Carrière was found growing in Iran near Guilan on the Caspian by Pissard, Gardener to the Shah, and was brought to Teheran to ornament the gardens there (*Rev. Hort.*, 1880, p. 314 and plate; op. cit., 1888, p. 446).

This rose is usually considered to be synonymous with R. *moschata* var. *nastarana* (see above), but judging from the description and figure it was a hybrid. The broad stipules shown in the plate are quite unlike those of any form of R. *moschata*, and the scent of the flowers, according to Carrière, was intermediate between that of a Tea rose and R. *gallica*. They were single. Neither of the roses introduced by Paul's nursery as R. *Pissardii* agree with Carrière's description and figure (see above).

R. RUSCINONENSIS Gren. & Deségl. R. *sempervirens* var. *pilosula* Ser.; R. *sempervirens* var. *moschata* Gren. & Godr.—This rose, usually included in R. *moschata*, was described from plants naturalised in the Eastern Pyrenees (Le Roussillon), but similar forms are reported from Provence and Sicily. From typical R. *moschata* it differs in its less downy and more glandular pedicels. and glabrous receptacles, and in being summer-flowering. Also, judging from specimens collected in the Roussillon near Perpignan the flower-buds are more broadly ovoid than in either typical R. *moschata* or R. *brunonii*—a difference perhaps of some significance, since the shape of the flower-bud is of diagnostic importance in the *Synstylae*.

Roses naturalised in North Africa are usually referred to R. *moschata*, but from lack of material for study nothing useful can be said about them. They are reported to be summer-flowering.

R. CALIFORNICA Cham. & Schlecht.

A shrub 5 to 8 ft high, the stems armed with stout, flattened, usually recurved prickles with a broad base, paired at the nodes; strong shoots often bristly. Leaves 3 to 5 in. long; rachis downy. Leaflets mostly five or seven, oval or ovate, usually obtuse at the apex, finely downy or glabrous above, hairy beneath, especially on the midrib and nerves, edged with simple or compound teeth. Flowers about 1½ in. wide, pink, borne on long laterals, frequently over a dozen in a cluster, each subtended by a leafy bract. Pedicels glabrous or hairy, sometimes glandular. Sepals entire, finely tapered at the apex, glabrous or hairy on the back, sometimes glandular at the edge. Fruits globose or slightly elongated, ⅜ to ½ in. wide, contracted into a well-defined neck below the persisting erect sepals.

Native of western N. America from S. Oregon to Lower California, mainly represented in Britain by a double form. An interesting feature of this species, unusual in wild roses, is its ability to produce flowers at the end of strong seasonal shoots.

For a revision of the R. *californica* complex, see D. Cole in *Amer. Midl. Nat.*, Vol. 55 (1956), pp. 211–24.

cv. 'PLENA'.—The name R. *californica* f. *plena* Rehd. is based on the semi-double form portrayed as R. *californica* in Willmott, *The Genus Rosa*, Vol. I,

p. 223, t. The plant now grown as 'Plena', which may be a different clone, was introduced to this country by Mrs L. Fleischmann, who obtained her stock from the American firm Bobbink and Atkins of New Jersey (Graham Thomas, *Shrub Roses of Today*, p. 70 and photo 1). This is a very vigorous rose with semi-double flowers of a rich pink, borne in late June or July; a conspicuous feature of the buds is the long and slender tips to the sepals. Award of Merit 1958, when exhibited by Mrs Fleischmann and the John Innes Horticultural Institution.

R. CANINA L.   COMMON BRIER, DOG ROSE

A strong-growing shrub up to 12 ft high (sometimes much taller on the continent); stems armed with scattered prickles which are uniform, hooked, with no mixture of smaller, bristle-like ones. Leaflets usually five or seven, elliptic to broadly so or ovate, mostly 1 to $1\frac{1}{2}$ in. long, acute or acuminate, glabrous or almost so on both sides, sometimes glandular on the veins beneath, marginal teeth usually simple and eglandular, more rarely compound and glandular. Upper stipules broad. Flowers sweetly scented, $1\frac{1}{2}$ to 2 in. wide, white or pinkish, in clusters or solitary, opening around midsummer; pedicels glabrous, to about 1 in. long, they and the receptacle usually smooth, more rarely with some glandular bristles. Outermost sepals (which overlap their neighbours on both sides) pinnately lobed on both margins, the next similarly lobed on one side only (the overlapping side); the inner two, which are overlapped on both sides, have entire margins. Disk of receptacle broad. Styles

ROSA CANINA

glabrous or downy, not or slightly exserted from the very narrow aperture; stigmatic head usually conical, not concealing the entire surface of the disk. Fruits egg-shaped or roundish, bright red, with the sepals fallen away or remaining until the fruit changes colour.

The dog rose in one or other of its forms is spread over most of the cooler parts of Europe and S.W. Asia. It is naturalised in some parts of N. America. In the British Isles it is one of the commonest and most beautiful of wild shrubs, giving to the English country lanes one of their sweetest and most characteristic charms. For this reason the dog rose is out of place in the trim garden where so many other roses with a richer beauty compete for room. Yet in one sense the dog rose is the commonest of species in gardens, for in one form or another it provides the Canina stock on which most garden roses are budded.

The curiously diverse form of the sepals of R. *canina* (and of other members of the section to which it belongs) furnishes the answer to the Latin riddle*, which has been translated thus:

> Five brothers in one house are we,
> All in one little family,
> Two have beards and two have none,
> And only half a beard has one.

R. CORYMBIFERA Borkh. R. *dumetorum* Thuill.; R. *canina* var. *dumetorum* (Thuill.) Bak.; R. *canina* var. *corymbifera* (Borkh.) Rouy—Leaves simply serrate, downy on both sides, usually more densely so beneath, commonly rather broadly ovate or roundish. With the same distribution as R. *canina*, in which it is included by many authorities.

R. OBTUSIFOLIA Desv. R. *tomentella* Leman—Prickles usually more strongly hooked than in the preceding. Leaves relatively broad, often obtuse at the apex, downy beneath at least on the veins, edged with glandular-compound teeth. Flowers commonly white. Sepals rather short. Native of Europe, including England, where it is local.

The above two species are close to R. *canina*. Still part of the R. *canina* aggregate (sect. *Canina* subsect. *Canina*), but more distinct, are:

R. CORIIFOLIA Fr. (1814) ?R. *caesia* Sm. (1812); R. *canina* var. *coriifolia* Bak.; R. *dumalis* var. *coriifolia* (Fr.) Boulenger; R. *glauca* var. *coriifolia* (Fr.) Crép. —A shrub to about 6 ft high; prickles shorter than in R. *canina*. Leaves downy beneath, at least on the veins, usually obtuse. Flowers on very short pedicels. Receptacle with a wider stylar aperture than in R. *canina* and its close allies, the correspondingly rather narrow disk almost concealed by the broad, woolly stigmatic head. Sepals erect or spreading, usually persisting until the fruit ripens, or even later. This species is rare in S. England, common in Scotland and Scandinavia; it also occurs in the mountains of South and Central Europe and S.W. Asia. It was at one time considered to be a mountain race of R. *canina*.

---

* There are two versions of this riddle. What must be the older of the two is in Latin doggerel. In the second and more elegant version the half-bearded sepal asks the riddle:
> Quinque sumus fratres unus barbatus et alter
> Imberbesque duo, sum semiberbis ego.

cv. 'FROEBELII'. Stems sparsely armed. Leaves greyish above. Flowers white. (R. *laxa* Froebel, not Retz.; R. *canina* var. *froebelii* Christ ex Rehd.; R. *coriifolia* var. *froebelii* (Christ) Rehd.)

This is the well-known Laxa rose-stock, introduced from the Near East towards the end of the last century.

R. AFZELIANA Fr.   R. *glauca* Vill. ex Loisel., not Pourr.; R. *vosagiaca* Desportes; R. *dumalis sens.* Boulenger, ? not Borkh.—Similar to R. *coriifolia* in floral characters, but with glabrous leaflets, often glaucous above. It has much the same geographical distribution.

The following species of the section *Caninae* is usually placed in a subsection of its own (*Stylosae*):

R. STYLOSA Desv.   R. *systyla* Bastard—Prickles often large, with wide bases. Leaves (in the commonest British form) ovate-lanceolate, acuminate, to 2 in. long, dark and glossy above, simply toothed, more or less glabrous. Flowers white or pale pink, on rather long, usually glandular-bristly pedicels. Styles at flowering-time agglutinated into a shortly exserted glabrous column. Fruits ovoid, red, shedding the sepals when ripe.

A native mainly of western Europe, including the southern part of the British Isles. Crépin considered it to be a link between the *Caninae* and the *Synstylae*.

R. CAROLINA L.

R. *humilis* Marsh.; R. *pensylvanica* Wangenh.; R. *virginiana* Du Roi, not Mill.; R. *parviflora* Ehrh.; R. *lyonii* Pursh; R. *carolina* var. *villosa* (Best) Rehd.; R. *humilis* var. *villosa* Best

A small, suckering shrub, rarely more than 3 ft high in the wild; stems from the stolons densely clad with prickly bristles; upper stems with straight, slender prickles, which are usually paired at the nodes, scattered but fairly numerous between the nodes. Leaflets five or, more commonly, seven, elliptic or narrowly ovate, $\frac{5}{8}$ to $1\frac{1}{2}$ in. long, sharply toothed, the teeth rather spreading and averaging about twelve on each side, dull or slightly lustrous above, glabrous on both sides or softly downy beneath. Stipules narrow, the adnate halves of each pair more or less parallel and sometimes upfolded, forming a tube. Flowers in June or early July, solitary, or few in a corymb, $1\frac{1}{2}$ to 2 in. wide, light pink. Pedicels and receptacle usually clad with stalked glands. Fruits red, more or less globular, about $\frac{3}{8}$ in. wide.

Native of N. America from Maine to Florida, west to the Prairie States and Texas, inhabiting mainly dry and open habitats; cultivated by James Sherard at Eltham, in 1732, but uncommon in gardens, where it has been confused with R. *palustris*.

cv. 'ALBA'.—Leaflets hairy beneath. Flowers white (R. *virginiana* var. *alba* Baker in Willmott; R. *lyonii* f. *alba* Rehd.). Discovered at Cherryfields, Maine, in 1867.

cv. 'PLENA'.—Flowers double, the outer petals fading to white with age. Sepals with long, slender tips (R. *parviflora* var. *plena* Ehrh.). In the late 18th and early 19th centuries this seems to have been the commonest representative of the species in gardens, and the only one seen by Lindley, according to whom it 'does not yield in beauty to the most splendid varieties of *gallica*'. It was a weak grower and, according to Thory in Redouté's *Les Roses*, most gardeners lost it through weeding out the suckers that it needed to perpetuate itself. Probably for this reason it became extinct in Europe, but around 1955 it was rediscovered in the USA by Mr and Mrs Wilson Lynes, who sent plants to Graham Thomas (see his *Shrub Roses of Today*, p. 71 and fig. 2).

The rose figured in Willmott's *The Genus Rosa* as R. *humilis* var. *grandiflora* (Vol. 1, p. 207, t.) appears to be the same as one distributed by Smith of Newry towards the end of the last century as R. *lucida grandiflora*, with large flowers and obovate or broadly elliptic leaflets. This rose is the type of R. *carolina* var. *grandiflora* (Bak.) Rehd., said to occur wild in the USA. But, whatever the status of the wild plants, the type of this variety seems to be nearer to R. *virginiana* than to R. *carolina*.

R. PALUSTRIS Marsh. R. *carolina* L. (1762) and of many later authors, not L. (1753); R. *virginiana* Du Roi, not Herrm. nor Mill.; R. *corymbosa* Ehrh.; R. *pensylvanica* Michx., not Wangenh.; R. *hudsoniana* Thory—Taller than R. *carolina*, to about 6 ft. Stems with paired stout more or less curved prickles at the nodes, otherwise almost unarmed. A very marked difference between this species and R. *carolina* lies in the very fine, close-toothing of the leaflets, the average number of teeth on each side being twenty-six according to Mrs Erlanson, against an average of twelve in R. *carolina* and fourteen in R. *virginiana*. Stipules as in R. *carolina*, but more frequently inrolled into a tube. Flowers usually in corymbs, rarely solitary, with very numerous stamens, borne later than in R. *carolina* (July and August).

R. *palustris*, as its name implies, is usually found in swampy places, but has much the same geographical distribution as R. *carolina*. Like that species it spreads vigorously by suckers and does not need a wet soil in gardens.

cv. 'NUTTALLIANA'.—Flowers somewhat larger than average, borne as late as September. Received at Kew from William Paul's nursery 1894 (R. *carolina* var. *nuttalliana* Hort.).

A further difference between R. *carolina* and R. *palustris* is that the former is tetraploid and the latter diploid. Natural hybrids between them have been reported, and those studied by Mrs Erlanson were triploid and sterile, as might be expected. These hybrids had the habit of R. *carolina* but showed the influence of R. *palustris* in the more finely and more numerously toothed leaflets. Diploid and fertile plants apparently combining the characters of R. *carolina* and R. *palustris* are now thought to be the result of hybridisation between the latter and R. *blanda*.

R. CENTIFOLIA L.   HOLLAND OR PROVENCE ROSE

R. *provincialis* Mill., in part (1768), not J. Herrm. (1762)*; R. *gallica* var. *centifolia* (L.)
Reg.

A lax shrub to about 5 ft high, its stems armed with numerous prickles of various sizes, the larger ones hooked, the others almost straight and narrowly based. Leaves drooping, with five or seven leaflets; rachis rough with stalked glands but not prickly. Leaflets broadly ovate, dull green and glabrous above, downy beneath, edged with large glandular teeth. Flower-buds broadly ovoid. Flowers nodding, solitary or few in a cluster, borne in late June or July, clear pink, very double, goblet-shaped from the incurving of the petals, becoming more lax when fully blown, exposing the tightly packed petaloids. Sepals spreading, longer than the flower-buds and covered, like the pedicels and receptacles, with sticky aromatic glands. Fruits roundish or ellipsoid, with a pulpy flesh. Redouté, *Les Roses*, Vol. I, p. 25, t.

This rose belongs to a small group of garden hybrids whose history can be traced back to the late 16th century. What is believed to be the first of the Holland roses was described by Clusius (Charles d'Escluse) in his *Rariorum Plantarum Historia* (1601). In 1589 he received two plants from John van Hogheland of a rose cultivated in Holland as, reputedly, the R. *centifolia* of Pliny, of which one survived and flowered with him (probably at Leyden) in 1591. The flowers were very double, some with about one hundred petals, the outer ones much larger than the inner, which took the place of the stamens. The fragrance was like that of 'R. *praenestina*' (which, *sensu* Clusius, is R. *damascena*) but with a suggestion of the scent of R. *alba*; the colour of the flowers was not unlike that of 'R. *praenestina*'. He named it R. *centifolia batavica* (of Holland), though he himself doubted whether it really was the *Rosa centifolia* of Pliny, described as having scentless flowers. Clusius also mentions R. *centifolia batavica* II (altera). This he apparently never saw in flower, but he quotes van Hogheland, who sent him a plant in 1592, to the effect that it was the same as 'Number I' except in having smaller flowers. Linnaeus took 'Number II' as the type of R. *centifolia*, for no obvious reason but perhaps in the belief that 'Number I' was the very double 'cabbage' form and 'Number II' the normal form from which sprang the original Moss rose, which Linnaeus included in R. *centifolia* in the second edition of *Species Plantarum*.

The Holland rose had reached England by 1596, the date of Gerard's *Catalogus*, and in the following year he described it in his *Herball* under the name *Rosa Hollandica sive Batava*, adding that it 'is generally called the great Province Rose, which the Dutchmen cannot endure; for they say it came first out of Holland, and therefore to be called the Holland Rose; but by all likelihood it came from the Damask Rose, as a kind thereof, made better and fairer by art, which seemeth to agree with the truth.' Parkinson, in his *Paradisus* (1629), gives a fuller description agreeing with that of Clusius, and uses a name similar to Gerard's—R. *provincialis sive Hollandica Damascena*, 'The Great double Damask Province or Holland Rose'.

It is not known, and perhaps never will be, whether the Holland rose was raised in the Low Countries or imported from somewhere in southern Europe or

* See Note on the name R. *provincialis* on p. 70.

the eastern Mediterranean. Jean Bauhin, who saw it in flower at Pforzheim in Germany in July 1595, was told that it had been brought from 'the city of Delphi' and to have been purchased at a great price. But this sounds like a nurseryman's tale and should be regarded with scepticism, if only because Delphi had started to crumble into ruins well over a thousand years before the Holland rose emerged. R. *centifolia* is obviously quite near to R. *damascena*, differing in its glandular-toothed leaflets, broader receptacle and fruits, and the spreading, not reflexed, sepals. From R. *gallica*, with which botanists have compared or even united it, it differs in its much taller growth, more prickly stems and nodding flowers. The gland-edged leaflets are not a point of distinction from R. *gallica*, since they are a common feature of some wild forms of that species. It is perhaps significant that there seems to be nothing in the old literature to suggest a Dutch origin for the dwarf forms of R. *centifolia* (var. *pomponia*); all have French vernacular names and some might well be older than the Holland rose. Dr Hurst's view, however, was that it was bred from a cross between R. *damascena* and R. *alba*.

R. *centifolia*, now grown only by lovers of old-fashioned roses, has achieved immortality on the canvases of the Dutch and Flemish flower-painters and was first depicted by Jacques de Gheyn in a work dated 1603, not long after its introduction. Because of the complete doubleness of its flowers the Holland rose must have been sterile, and the few variants that existed up to the end of the 18th century must have descended from the original stock (of which there may have been more forms than those mentioned by Clusius) or have arisen later by sporting. By the middle of the 19th century 'Cabbage Rose' had become synonymous with 'Provence Rose', but earlier it had been recognised as a distinct variant, differing in its more fully double flowers, which had longer central petaloids and therefore lacked the hollow form of the classic R. *centifolia*. This form, which Andrews distinguished as R. *provincialis multiplex*, is beautifully portrayed in the plate reproduced in Blanche Henrey's *British Botanical and Horticultural Literature,* Vol. II, facing p. 49, which orginally appeared in Edwards's *A Collection of Flowers* (1783–95).

Early in the 19th century R. *centifolia* underwent a burst of evolution, following the appearance around 1800 of a single-flowered and fertile form, portrayed in Redouté, *Les Roses,* Vol. I, p. 77, t.

cv. 'ALBA'.—See 'Unique Blanche', p. 202, the name by which this rose is usually known at the present time; Andrews (*Roses,* t. 20) called it R. *provincialis alba*. Although discovered in Suffolk, it grew in a hedge bordering the garden of a Dutch merchant and in this connection it is interesting to note that Clusius mentions a rose which he called R. *centifolia batavica alba*.

f. ALBO-MUSCOSA (Bak.) Rehd.—This name is based on the rose portrayed in Willmott, *The Genus Rosa,* Vol. II, p. 349, t., as R. *centifolia albo-muscosa*. This rose is almost certainly 'Blanche Moreau' (see p. 173). R. *muscosa* var. *alba* Thory, which Rehder gives as a synonym, is a white-flowered sport of the Common Moss rose (see 'Muscosa' below), either 'Shailer's White' or the Bath Moss (Clifton Moss).

cv. 'BULLATA'. ROSE À FEUILLES DE LAITUE.—An exact counterpart of R. *centifolia*, apart from the luxuriant bullate (crinkled) foliage, of a brownish tint when young. It originated on the continent, before 1801. 5 ft. Very fragrant. (R. *centifolia* var. *bullata* Thory, in Redouté, *Les Roses*, Vol. I, p. 37, t.)

cv. 'CRISTATA'. CRESTED MOSS, CHAPEAU DE NAPOLÉON.—A sport which originated about 1820, identical to R. *centifolia* in every respect except the flowers. These are of the same good pink, but slightly less globular, and the sepals are extended into many divided wings or appendages; they frame the bud with a parsley-like frill of green. The French name recalls the cockade-like effect of these wings. 5 ft. Sweetly fragrant. (R. *centifolia cristata* Prévost; R. *centifolia muscosa cristata* Hook., *Bot. Mag.,* t. 3475; R. *centifolia* f. *cristata* (Prévost) Rehd.)

cv. 'MUSCOSA'. COMMON MOSS ROSE.—A sport from R. *centifolia*, characterised by the development of much-branched, moss-like aromatic glands on the calyx and pedicels, and the excessive glandularity of the leaf-rachis and branchlets. In size, foliage, and flower it is slightly less than R. *centifolia* itself, but both the rich pink colour and the fragrance are the same. This mutation first occurred on the continent before 1720, and was a great favourite in Victorian times (R. *muscosa* Ait.; R. *centifolia* f. *muscosa* (Ait.) Schneid.).   [PLATE 8

The Common Moss sported on at least two occasions to white blooms—Shailer's White Moss (1788) and the Bath or Clifton Moss shortly before 1818. What is grown today as 'Shailer's White' may not be the original clone but is a charming rose, its flowers faintly blush-pink at the centre on opening (G. S. Thomas, *The Old Shrub Roses*, pl. 19). 4 ft. Sweetly fragrant. It has reverted to the Common Pink Moss in recent years.

After the occurrence of a single-flowered sport of the Common Moss (R. *centifolia* f. *andrewsii* Rehd.; R. *muscosa simplex* Andr.) it became possible to raise seedling moss roses. The later hybrid moss roses derive from this, or from the moss form of the Autumn Damask (see R. *damascena*) crossed with Hybrid Chinas.

var. POMPONIA [Roess.] Lindl. R. *pomponia* Roess.; R. *dijoniensis* Roess.; R. *burgundiaca* Pers., not Roess.; ?R. *pulchella* Willd.—As usually defined, this variety differs from R. *centifolia* only in being smaller in all its parts. It seems to be really no more than an assemblage of miniature garden roses more or less agreeing with R. *centifolia* botanically but mostly disagreeing with it in floral style and probably of independent origin. The group, never a large one, is scarcely represented in gardens today. All the original forms originated in France, and several had been introduced to Britain by the 1770s.

The nearest to a true Centifolia was the Rose de Bordeaux, also known as the Gros Pompon de Bourgogne (R. *centifolia minor* Roess., *Die Rosen*, t. 20 and of Thory in Redouté, Vol. III, p. 33, t.). According to Thory this form was fertile and many subvarieties had been raised from it. The 'Petite de Hollande' (p. 195) belongs to this group.

The Bordeaux rose was scarcely a dwarf. But two miniatures were introduced to Britain from France late in the 18th century as the 'Rose de Meaux'—Meaux being a town to the east of Paris. Of these the one remaining in cultivation (see p. 178) appears to be the Lesser de Meaux (Lawrance, *Roses* (1799), t. 50, as R. *Pomponia*; Curtis, *Bot. Mag.*, t. 404 (1798), as R. *provincialis* var., and as Pompone rose or smaller de Meaux in text; Willmott, *The Genus Rosa*, Vol. II, p. 353, t., as R. *pomponia*). The Greater de Meaux is portrayed by Mary Lawrance (as the Rose de Meaux) in her plate 31 and seems very similar to the R. *Pomponia* of Roessig (*Die Rosen*, t. 37), which is the type of var. *pomponia*, and also to the R. *pomponia* of Redouté, Vol. I, p. 65, t. It is difficult to believe that this is the pompon of *Bot. Mag.*, t. 407, as Thory suggested. The Greater de Meaux, which may no longer be in cultivation, differed obviously from the Lesser in its longer more slender pedicels, less pinnated sepals and more Centifolia-like flower; also presumably in its larger stature.

Two other roses are placed under R. *provincialis* (i.e., R. *centifolia*) in the second edition of *Hortus Kewensis* (1810). One is the St Francis rose (Lawrance, t. 88), which was introduced to Britain in the 1770s. This had large flowers, but according to Trattinick it was dwarf in habit; he identified it with R. *gallica regalis* Thory in Redouté (Vol. II, p. 19, t.) and called it R. *pumila*. The other is the Rose de Rheims (Lawrance, t. 71), also with large flowers and not unlike the St Francis.

R. PARVIFOLIA Ehrh. (1791), not Pall. (1788)   R. *burgundiensis* West., in part only (1770); ?R. *burgundica* Durande (1782); R. *burgundiaca* Roess. (1802); R. *remensis* DC. (1805); R. *pomponia* var. *remensis* (DC.) Thory; R. *gallica* var. *parvifolia* (Ehrh.) Ser.; R. *centifolia* var. *parvifolia* (Ehrh.) Rehd.—This rose is described on p. 194 as 'Parvifolia' but, having botanical status, it is mentioned here. Strictly it should be called 'Burgundi-aca' and has generally been known as the Burgundy rose in Britain, but these or similar names have been used for the Centifolia pompons (see above) and would be more ambiguous than 'Parvifolia', which is founded on a well-established albeit illegitimate specific name.

This rose, an old garden variety perhaps deriving from a plant found originally in the wild, was considered by Lindley to deserve the specific rank given to it by Ehrhart. It is certainly out of place under R. *centifolia*, in which Rehder includes it as a variety, and the only question is whether it is a mutant of R. *gallica* or a hybrid of it.

Early accounts of this rose in French works are confusing. De Candolle, in *Flore Française* (1805), identified it with a rose grown in the Jardin des Plantes as R. *remensis* [of Reims], with the vernacular name Rose de Champagne; but the Rose de Rheims imported into Britain in the 1770s was different; as another vernacular name De Candolle adds 'Rose de Meaux', but neither of the two roses known in Britain under this name, and described earlier, were 'Parvifolia' (i.e., R. *remensis* as described by De Candolle). Other vernacular names for this rose given by Thory in Redouté, *Les Roses*, are also at variance with earlier accounts.

NOTE on the name R. *provincialis:* In the 1768 edition of his *Dictionary* Miller gave as the nomenclatural type of R. *provincialis* the R. *provincialis flore pleno ruberrimo* of the Leyden Catalogue of 1720 (the so-called *Index Alter*, attributed to Boerhaave). This is of uncertain identity, but a reading of Miller's account of R. *provincialis* and 'Provence' rose in the 1768 and earlier editions of the *Dictionary* makes it perfectly clear that his R. *provincialis* is R. *centifolia* as now understood, and not R. *gallica* var. *officinalis*—the name under which it appears in Rehder's *Bibliography*. The author of Miller's name was really Parkinson, who called the Holland rose R. *provincialis seu* [or] *Hollandica Damascena*. What Parkinson meant in this context by *provincialis* is not clear—perhaps simply 'provincial' or 'of the United Provinces', but, if so, he was departing from the nomenclature of some continental botanists of the 16th century, to whom R. *provincialis* meant the Damask rose.

The name R. *provincialis* in Miller's sense was taken up by Aiton in *Hortus Kewensis* (1789) and other works, but it dropped out of use after the publication of Lindley's *Monograph* (1820), in which R. *centifolia* is used in the modern sense. Lindley also cleared up the confusion over the name R. *centifolia*, which Miller had applied to the double form of R. *gallica* known as the 'Dutch Hundred-leaved'.

It has been widely believed that R. *centifolia* is called 'Provence rose' because it originated in Provence. In fact we owe the name to Miller, who was simply translating *provincialis* as 'of Provence', its normal meaning. His nomenclature was not universally followed, however. Andrews, for example, used the more non-committal name 'Province rose'. Lindley, too, evidently considered that Miller was wrong in associating

R. *centifolia* with Provence and instead called it 'Provins rose'—an extraordinary blunder, since the rose for which Provins was famous was a form of R. *gallica* (see p. 96). The similarity between the words 'Provins', 'province', and 'Provence' obviously invited confusion, especially as 'Provins' would have been pronounced 'province' in English. But the confusion seems to have existed only at the vernacular level. There is no evidence that the botanical name R. *provincialis* was ever used in the sense of 'Rose of Provins' until the student Herrmann did so in his doctoral thesis (1762).

The belief that R. *centifolia* originated in Provence has led some writers on the history of the Rose to ransack *The Romaunt of the Rose* (believed to be partly by Chaucer) for evidence of its presence there in the Middle Ages. The work is of course a translation of parts of *Le Roman de la Rose*, and the evidence should have been sought in the original. There is none to be found there—or indeed in the English translation. The 'Provincial roses' (i.e., rosettes) of *Hamlet* Act III, Scene ii, were in all probability fashioned after the Rose of Provins or rather after the common Apothecaries' or Officinal rose, for which the name 'Rose of Provins' was used by Shakespeare's contemporary Gerard (see further under R. *gallica* var. *officinalis*).

Although the name R. *provincialis*, without qualification, originally meant R. *damascena*, it was also used by some writers of the 16th century for other roses that were intermediate in flower colour between R. *gallica* and R. *alba*. Thus R. *provincialis praecox* was R. *majalis* and R. *provincialis minor* some dwarf rose with very double pink flowers.

The account of R. *provincialis* in Willmott, *The Genus Rosa* (Vol. II, p. 359) is confused. The rose actually portrayed is one of the forms or hybrids of R. *gallica* var. *holosericea*; it has nothing to do with the R. *provincialis* of Miller.

R. CERASOCARPA Rolfe

R. *gentiliana sens*. Rehd. & Wils., in part, not Lévl. & Van.

A deciduous climbing or semi-climbing shrub up to 15 ft high; young shoots somewhat glaucous, armed with a few scattered recurved spines. Leaves up to 7 or 8 in. long, consisting of three or (usually) five leaflets; rachis glandular, slightly prickly. Leaflets narrowly ovate or oval, long pointed, sharply and conspicuously toothed, 2 to 4 in. long, half as much wide, glabrous or nearly so, rather glaucous beneath. Flowers white, produced in June in fine corymbose clusters 6 in. wide, each flower 1½ in. across, borne on a glandular stalk ¾ to 1½ in. long. Receptacle obovoid, downy and glandular. Sepals linear, sometimes pinnately lobed, ½ in. long, downy and glandular. Fruits globose, downy, deep red, ½ in. wide, with the sepals fallen away. *Bot. Mag.*, t. 8688.

R. *cerasocarpa* was described in 1915 from a plant which flowered for the first time in June 1914 in the garden of Sir William Thistleton-Dyer in Gloucestershire; it had been obtained by him from Sir Thomas Hanbury of La Mortola, Italy, and had been raised from seeds collected in China. In the following year it was figured in the *Botanical Magazine*, but Sir William considered that even that excellent portrait did not do justice to the beauty of his plant. 'The solid trusses are unlike those of any rose I know, and suggest an Azalea' (*Gard. Chron.*, Vol. 74 (1923), p. 55). Whether this rose was ever propagated for general distribution it is impossible to say. It seems to be very near to R. *rubus* (q.v.)—nearer to that species than to R. *longicuspis*, with which Rolfe compared it.

The description of R. *cerasocarpa* in Boulenger's revision of the Asiatic *Synstylae* is inaccurate, and the species is wrongly placed in his key among those with rounded flower-buds.

R. HENRYI Boulenger R. *gentiliana sens.* Rehd. & Wils., in part, not Lévl. & Van.; R. *moschata* var. *densa* Vilm.—r. *henryi*, described in 1933, is essentially a renaming of the species treated in *Plantae Wilsonianae*, Vol. II, p. 312, under the name R. *gentiliana* (for the true species of that name, see under R. *multiflora*). As described, R. *henryi* does not seem to differ significantly from R. *cerasocarpa*, when allowance is made for the fact that the latter species was described from a single cultivated plant and R. *henryi* from a range of wild speciments. Boulenger would surely have noted the similarity if he had actually seen the specimens of Thistleton-Dyer's plant, or even the plate in the *Botanical Magazine*, but he seems to have been unaware of either. His conception of R. *cerasocarpa* was largely based on a fruiting specimen collected by A. Henry (7007) which Rolfe thought might belong to his species, and it is perhaps significant that another specimen of Henry 7007, in the Boissier Herbarium, is included by Boulenger in R. *henryi*.

R. *henryi* was introduced by Wilson in 1907 from W. Hupeh, China, under his numbers 609 and 609a. The nurseryman Paul of Cheshunt had plants, but whether he ever propagated and distributed this species is uncertain.

R. CHINENSIS Jacq.*

R. *sinica* L. ex Murr. (1774), form with an abnormal calyx; R. *indica sens.* Ait. f. and many other authors, not L.; R. *semperflorens* Curt.; R. *diversifolia* Vent., *nom. illegit.*; R. *bengalensis* Pers., *nom. illegit.*; R. *longifolia* Willd.

A shrub of variable habit, of moderate size or dwarf in its cultivated state, semi-scandent or tall and laxly branched in its putative wild states; branches armed with scattered, hooked and somewhat flattened prickles. Leaflets three or five, ovate to lanceolate, acuminate, 1 to 2¼ in. long, sometimes tinged with purple when young, serrated, glabrous and glossy above, underside glabrous except for the sometimes downy midrib. Petiole and rachis prickly and some-

* This is the earliest name for the species (1768). The R. *indica* of Linnaeus (1753) is a confused entity, and the only part of it that can be identified is R. *cymosa* (q.v., p. 55). The name R. *indica* was, however, widely used at one time for the species here described. In the narrow sense it was applied to the Pink or Blush China roses, and the Crimson Chinas were treated as a variety—var. *semperflorens* (Curt.) Ser. or even as a distinct species—R. *semperflorens* Curt. When the misused name R. *indica* was dropped in favour of R. *chinensis* no account was taken of the fact that the two names are differently typified. The type of R. *chinensis* was a Crimson China similar to the one named R. *semperflorens*, which should therefore have been treated as the typical variety, R. *chinensis* var. *chinensis*, but continued to be distinguished as R. *chinensis* var. *semperflorens*. The Pink China, which is the type of R. *indica sens.* Ait. f., Lindl. *et al.*, when transferred to R. *chinensis*, should have been given the varietal position as var. *pallida*, from R. *semperflorens* var. *pallida* Roessig. But botanically the difference between the Crimson and Pink Chinas is not great. The description given here comprises both, and the various old introductions are distinguished as cultivars.

what glandular-bristly. Stipules narrow, persistent. Flowers single or semi-double, varying in colour from light pink to crimson or crimson scarlet, solitary or in corymbs; pedicels variable in length, smooth or glandular, with a pair of narrow bracts at the base. Receptacle ellipsoid to globular, smooth or glandular; sepals with a few lateral appendages, reflexed at flowering-time and soon deciduous, glandular or smooth on the back. Disk thickened, more or less conical. Styles somewhat exserted, distinct, glabrous.

R. *chinensis* is essentially a race of garden roses. The type of the species, and all the forms that later found their way to Europe, are the result of more than a thousand years of mutation, intercrossing, and selection in the gardens of China. From the wild roses that are believed to have given rise to them they differ in their more modest, even dwarf, habit; and in their capacity, apparently not shared by their wild ancestors, of producing successive crops of flowers on the same stem—whence the Chinese name, which means Monthly Roses.

There can be no doubt that the China roses were bred from spontaneous forms that grow wild in the mountains of China. The first of these to become known to science was found by Augustine Henry in 1884 in the glens of Ichang, Hupeh (*Gard. Chron.*, Vol. 31 (1902), p. 438). This rose agrees in most of its essential characters with typical R. *chinensis*, but differs in its much more robust, scrambling habit and in the shorter, stouter pedicels. The flowers, according to Henry, vary from crimson to pink. They are solitary, and are borne on short laterals from the previous season's wood, as in R. *gigantea*. Henry's rose, and similar ones found by Wilson, were given botanical status by Rehder and Wilson as R. *chinensis* f. SPONTANEA. A similar rose, but with dark crimson flowers, was collected by Dr Rock in Kansu.

Of still greater interest are the roses of the section *Chinenses* found by Forrest in Yunnan, which seems to be the centre of variation of this group. Here R. *gigantea* occurs in its typical state, and also in forms with pale pink or pale yellow flowers. But other Forrest collections, from the north-western part of the province, are clearly referable to R. *chinensis* in a broad sense, differing from R. *gigantea* in such characters as the darker red flowers, glandular pedicels, appendaged sepals, and gland-edged stipules. A form of R. *chinensis* growing wild in thickets on the Salween–Kiuchiang divide has a compound inflorescence as in a Pink China, but grows 8 to 10 ft high and has flowers of a deep crimson-rose (F.21631). Other specimens are nearer to R. *gigantea,* while still showing some characters of R. *chinensis*. Forrest also found Crimson Chinas, near to the type of R. *chinensis*, but with one exception, annotated as 'semi-cultivated', all the specimens seen came from gardens. His F.24059, of which he also sent seeds, had dark crimson flowers; it was cultivated at the base of the Shweli–Salween divide and may well have derived from wild plants. It is evident from Forrest's collections—an important and neglected contribution to rhodology—that R. *chinensis* as well as R. *gigantea* is a native of Yunnan, and that both are cultivated in the province, together with various hybrids or intermediates between them. There is a strong probability that the China roses developed there, later finding their way to the gardens and nurseries of Canton, and thence to Europe. The cultivars of *Camellia reticulata* certainly had such a history.

The westward spread of the China roses must have begun at a comparatively early date, for they seem to have been common in the gardens of India when the sub-continent started to be botanised towards the end of the 18th century. But it is doubtful if the China roses were ever quite so much at home in Bengal as the French rosarian Boitard suggested in his *Manuel Complet* (1836): 'Le féroce tigre du Bengale, le hideux crocodile du Gange, se cachent quelquefois, pour attendre leur proie, dans les touffes épaisses du Rosier Toujours Fleuri.' By the time British forces seized Mauritius from the French in 1810 several sorts of China roses were established in the gardens there, and were probably introduced in the time of Pierre Poivre, who established a famous collection of Far Eastern plants on the island between 1767 and 1773.

The date of the first arrival of a China rose in Europe is not known. The type of *R. chinensis* is a specimen from the herbarium of the Dutch physician Gronovius of Leyden, dated 1733. This may have been brought from China in the dried state; or it may have been taken from a plant in the collection gathered together at Leyden by Boerhaave, which, according to Linnaeus, consisted of some seventy species 'brought' from all over the world' (*Hort. Cliff.* (1734), p. 191). Since the Boerhaave catalogue of 1720 lists only some twenty familiar sorts of roses the main acquisitions must have been after that date. The first recorded introduction of a Pink China was certainly to Holland, in about 1781. The introductions to Britain from 1789 onwards are described below.

cv. 'PALLIDA'. PINK OR BLUSH CHINA.—A shrub to about 3 ft high, taller on a wall. It agrees with typical *R. chinensis* in its glandular-bristly pedicels and pinnated sepals, but the stems and prickles are stouter and the flowers, semi-double and fragrant, are blush-pink and borne in clusters (*R. semperflorens* var. *pallida* Roessig in *Oek.-Bot. Beschr. Rosen* (1803); *R. semperflorens* var. *carnea* Roessig, *Die Rosen*, t. 19 (1802–20); *R. indica sens.* Andr., *Roses*, Vol. II, t. 66 (1805); *R. indica* var. *vulgaris* Thory in Redouté, *Les Roses*, Vol. I, p. 51, t. (1817)).

According to the younger Aiton, the Pink China was introduced to Britain by Sir Joseph Banks in 1789, but the first that is known to have entered commerce was seen growing in the garden of a Mr Parsons at Rickmansworth in 1793, and was propagated by the nurseryman Colvill. This 'Parsons' Pink' is the form portrayed by Andrews and later by Redouté. Roessig tells us nothing about the provenance of his *R. semperflorens pallida* (*carnea*), but his description tallies with that provided by Thory in Redouté.

Andrews called the Pink China 'one of the greatest ornaments ever introduced to this country'. No such praise was ever bestowed on 'Semperflorens', the Crimson China, which soon died out in Europe in its original form, while a Pink China, probably 'Parsons' Pink', is still grown in gardens; see 'Old Blush China', p. 193.

Through two lines of descent the Pink China is an ancestor of most modern garden roses. Crossed in South Carolina with *R. moschata*, it gave rise to 'Champney's Pink Cluster', parent of the Old Blush Noisette (R. × NOISETTIANA Thory), from which all the Noisettes and Tea roses descend. The second of its ancestral hybrids also arose outside Europe, on the Ile de Bourbon (Réunion), where sometime early in the 19th century it became hybridised with an Autumn Damask, giving rise to the race of Bourbon roses, from which, through the

Hybrid Perpetuals, most modern garden roses descend (see further on p. 164). It cannot be certain whether in either case it was 'Parsons' Pink' that was involved, as is usually assumed. A Pink China could have reached America at the same time as R. *laevigata*, by direct import from the Far East, while Réunion is likely to have had the same garden flora as Mauritius, which certainly did not owe its China roses to import from Britain (see above).

cv. 'PERSICIFOLIA' ('Salicifolia').—A curiosity merely, with abnormally narrow leaflets, figured in Redouté, *Les Roses* (Vol. II, p. 27, t.), as R. *indica* var. *longifolia* (Willd.) Thory. It is extremely improbable that this is really the R. *longifolia* of Willdenow, described from specimens of garden forms of R. *chinensis* received from India, which, judging from his description, and from Crépin's comments on the specimens in Willdenow's herbarium, seem to have had large leaves with rather long, lanceolate leaflets, in no way abnormal. The name R. *chinensis* var. *longifolia* (Willd.) Rehd. is based on the rose portrayed by Redouté.

cv. 'SEMPERFLORENS'. SLATER'S CRIMSON CHINA.—A small bush with slender branches, armed with scattered, slightly flattened prickles, obviously very near to typical R. *chinensis*, the flowers being semi-double, or paired and borne on long, slender glandular-bristly pedicels as described by Jacquin from the Gronovius specimen, deep crimson scarlet. Sepals pinnated, glandular on the back (R. *semperflorens* Curt., in *Bot. Mag.*, t. 284 (1794); R. *indica* var. *semperflorens* (Curt.) Ser.; R. *chinensis* var. *semperflorens* (Curt.) Koehne).

The rose figured in the *Botanical Magazine* as R. *semperflorens* was introduced from China by Gilbert Slater of Knots Green, near Leytonstone, Herts, about 1791. Like other original importations from China, Slater's Crimson was soon displaced by hybrids, but a rose agreeing well with it was found in Bermuda in 1953 and introduced to the John Innes Horticultural Institution in 1957 (G. R. Rowley, *Journ. R.H.S.*, Vol. 84 (1959), pp. 270–3, reprinted with additional illustrations in *The Rose Annual* (1960), pp. 31–4). Also in cultivation is 'Miss Willmott's Crimson China' (*The Genus Rosa*, Vol. I, p. 89, t.) and the differences between these two is given by Mr Rowley in his article. The rose referred to in garden literature as the Old Crimson China is, however, likely to be 'Cramoisi Supérieur', which is of later European origin.

A single-flowered Crimson China, cultivated by the French nurseryman Cels, is figured in Ventenat's *Hortus Celsianus*, t. 35 (1800–2) under the name R. *diversifolia*, and in Redouté's *Les Roses* (Vol. I, p. 49, t.) as R. *indica* simply. It was imported to this country by Lee and Kennedy and first flowered in their nursery in 1804 (J. Smith, *Exotic Botany*, Vol. II (1805), p. 91). At the present time there is a single-flowered Crimson China in cultivation as 'MISS LOWE' and another with darker flowers, the original name for which is uncertain.

Dr Hurst advanced the theory that Slater's Crimson was a parent of the original Portland rose ('Portlandica', see p. 196) and through it of the 'Rose du Roi', first of the Hybrid Perpetuals. This is impossible, since 'Portlandica' was in commerce well before 1791, and there is really nothing in its characters, except the vividness of its red flowers and its dwarf habit, that might suggest the influence of a Crimson China. But Slater's Crimson nevertheless probably played its part in the formation of the old Hybrid China and Portland groups, from which the modern Hybrid Teas partly descend. It is certainly a parent of the old bedding roses 'Cramoisi Supérieur' and 'Fabvier'.

cv. 'VIRIDIFLORA'.—A monstrous form, apparently derived from a Pink China, in which the petals are more or less green and the sexual parts are converted into narrow, toothed, leafy segments. This cultivar-name comes from R. *viridiflora*, given by A. Lavallée in 1856 to a rose grown by the French nurseryman Verdier, who had acquired it from America (*Hortic. Franc.* (1856), p. 218 and t. 19). What appears to have been the same clone had been exhibited at the Paris Exhibition in 1855 by Miellez, another French grower, and this too came from America, but via England (*Fl. des Serres*, p. 129 and t. 1136). The history of what is probably the American green rose in question is given in *Gard. Chron.*, Vol. 3 (1875), p. 20, by the nurseryman Robert Buist of the Rose-dale Nurseries, Philadelphia, a Scotsman trained at Edinburgh who had emigrated to the USA in 1828. The green rose, according to him, had been 'caught' at Charleston, South Carolina, about 1833. What became of the plants he sent to his friend Thomas Rivers in 1837 is apparently not recorded, but the American green rose is known to have been imported into England from Baltimore in 1853.

Another green rose was introduced to Kew around 1852 by Sir H. Barkly, the Governor of British Guiana, but this seems to have been a sport of some Bourbon rose.

The green rose distributed in this country by Paul's nursery came from Miellez (see above).

cv. 'MINIMA'. MISS LAWRANCE'S ROSE.—A miniature; flowers single, usually solitary, with flesh-coloured, acuminately tipped petals (R. *indica* var. *minima* Sims in *Bot. Mag.*, t. 1762, an inaccurate portrait showing bristles on the stems; R. *lawranciana* Sw.; R. *laurentiae* Andr., *Roses*, Vol. II, t. 75; ?R. *indica* var. *acuminata* Thory in Redouté, *Les Roses*, Vol. I, p. 53, t., almost certainly 'Minima', but Thory was misled by the inaccurate portrait of this into supposing that the two were different).

According to Sweet, this rose was introduced from Mauritius (Ile de France) in 1810, which, if true, means that it must have been brought home by a member of the invading force that captured Mauritius from the French in that year. The R. *pusilla* (i.e., dwarf rose) of the Mauritius Catalogue (1822), may be this rose. It was listed without description as a rose common in gardens on the island, and is perhaps the same as the R. *chinensis* of the 1816 catalogue, for which the vernacular name was 'Petit Rosier de l'Inde'. Sweet named the introduced rose in honour of Mary Lawrance, the botanical artist whose famous collection of rose portraits was published 1796-9.

Miss Lawrance's rose was the parent of a group of miniatures called Fairy Roses, of which Rivers had some sixteen varieties by 1837. These and other miniature derivatives of R. *chinensis* were sometimes known collectively in gardens as R. *lawranceana* or R. *laurentiae*. See also 'Rouletii'.

cv. 'PUMILA'. BENGALE POMPONE.—A miniature rose less than 1 ft high, with leaves, including petiole, 1½ in. long and mostly solitary, double, flesh-coloured flowers; petals acuminate at the apex, as in 'Minima' (R. *indica* var. *pumila* Thory in Redouté, *Les Roses*, Vol. I, p. 115, t.; R. *indica* var. *humilis* Ser.;

?R. *indica* var. *minor* Andrews, *Roses*, Vol. II, t. 68).

This rose was raised around 1806 at Colvill's nursery, Chelsea, and exported to France, where it soon became known as the 'Bengale Pompone'. According to Thory, it could be easily raised from cuttings, which flowered within three months of rooting when two or three inches high.

cv. 'ROULETII'.—An evergreen or semi-deciduous shrub from 4 to 9 in. high, of compact, bushy shape; young shoots glabrous, armed irregularly with spines $\frac{1}{16}$ to $\frac{1}{12}$ in. long. Leaflets oval-lanceolate, slenderly pointed, $\frac{1}{4}$ to $\frac{3}{4}$ in. long, about half as wide, purplish, glabrous, the main-stalk furnished with a few spines and glands. Flowers $\frac{3}{4}$ to 1 in. wide, very double, rosy pink, produced in erect clusters.

This pretty rose was first brought into notice by Henri Correvon of Geneva in 1922. He stated that a friend of his, Dr Roulet, after whom he named it, found it a few years previous to that date grown in pots as a window plant at Mauborget, near Grandson, in Switzerland. This village was afterwards completely destroyed by fire and the rose with it, but a single plant was subsequently found in a neighbouring village. From this all the plants now in cultivation have been raised (*Gard. Chron.*, Vol. 72 (1922), p. 342; H. Correvon, *Floraire* (n.d.), p. 120 and t. XIV).

'Rouletii' is quite hardy in the open ground. Grown in pots, it seems to flower more or less continuously. In the Swiss villages its height as a window plant was apparently 2 in. or so, but it grows 6 to 9 in. high in the open ground with us.

Nothing is known of the origin of 'Rouletii'. It is one of the pygmy forms of R. *chinensis*, and Boulenger suggested that it was the original 'Pumila' (*Bull. Jard. Bot. Brux.*, Vol. 14, pp. 367–71). Graham Thomas has noted that in fact it reverts to 'Pumila', of which it must therefore be a mutation, rather than the original form.

'Rouletii' is the parent of a new race of miniature roses, mostly bred in Holland, of which the best known are 'Oakington Ruby' and 'Tom Thumb' ('Peon').

R. × ODORATA (Andr.) Sw.   R. *indica odorata* Andr., *Roses*, Vol. II, t. 77 (1810); R. *indica fragrans* Thory in Redouté, *Les Roses*, Vol. I, p. 61, t. (1817); R. *indica* var. *odoratissima* Lindl.; R. *fragrans* (Thory) Thory—The type of this group—'Odorata' or 'Hume's Blush Tea-scented China'—was introduced from China a few years before 1810 by Sir Abraham Hume of Wormley Bury, Herts. Although perhaps no longer in cultivation, it is known from contemporary illustrations and herbarium specimens to have been near to R. *gigantea* in its essential characters—indeed, Rehder and Wilson considered them to be conspecific, and placed R. *gigantea* under R. *odorata* as a variety. This was, to say the least of it, a questionable decision, for 'Hume's Blush' clearly showed some characters of R. *chinensis*, notably the clustered flowers on glandular pedicels. Dr Hurst considered it to be a garden hybrid between R. *chinensis* and R. *gigantea*, which is likely, though there is always the possibility that it derived from some wild intermediate or hybrid between these two species.

'Hume's Blush' was tender and has probably disappeared from cultivation in Europe, leaving a large inheritance. Most modern Hybrid Teas have Hume's rose in their ancestry, through the old Pink Teas and Hybrid Chinas. Whose nose it was that first detected the fragrance of China tea in its flowers may never be known, but evidently a French one, for the epithet 'tea-scented' started in France.

cv. 'FORTUNE'S DOUBLE YELLOW' ('Beauty of Glazenwood', 'Gold of Ophir').—A tender climber, obviously very near to R. *gigantea*, as indeed Crépin pointed out when describing that species. Flowers semi-double, gamboge-yellow flushed with coppery red, borne singly or in small clusters at midsummer. *Bot. Mag.*, t. 4679. It was introduced in 1845 by Robert Fortune, who found it growing in the garden of a rich Mandarin at Ningpo. See further in: Graham Thomas, *Climbing Roses*, p. 105.

A rose similar to 'Fortune's Double Yellow' is widespread in the valleys of Yunnan, but according to Forrest it is always found near cultivation and is probably an escape from gardens. It is a climber to about 20 ft, with semi-double fragrant flowers, rose-coloured flushed with copper. It is near to R. *gigantea*, but with broader stipules and longer pedicels (F.21131 and F.21149).

'Fortune's Double Yellow' is included by Rehder in R. *chinensis* var. *pseudindica* (Lindl.) Rehd. This variety is founded on R. *pseudindica* Lindl., described from a Chinese painting in the library of A. B. Lambert (Lindley, *Monograph* (1820), p. 132). The plant portrayed had double, deep yellow flowers, but the botanical details shown in the portrait must have been fanciful, since the characters taken from it by Lindley do not accord with the section *Chinenses* or any other Chinese species.

cv. 'MUTABILIS'.—Young growths slender, wiry, with few prickles. Leaves small, purplish when young. Flowers single, fragrant; buds slender, pointed, flame-coloured, opening chamois-yellow, becoming after pollination coppery pink deepening to coppery carmine. It is sometimes harmed by severe weather, but will make growths 3 to 4 ft long if cut to the ground; in some sheltered gardens it makes a sheaf of branches 7 or 8 ft high; against a wall it will spread indefinitely. It is constantly in flower from late spring until autumn and makes a spectacular, bizarre display (R. *mutabilis* Correvon; R. *chinensis* f. *mutabilis* (Correvon) Rehd.). A.M. 1957.

The origin of this rose is uncertain. Correvon, who described it in *Revue Horticole*, 1934 (p. 60, with plate), had received it some forty years earlier from Prince Borromeo, in whose garden on the Isola Bella, Lake Maggiore, it had made a hedge some 7 ft high. It was distributed by Rovelli's nursery, Pallanza, on the same lake, and plants from that firm had reached Britain by 1916. It was apparently known horticulturally as "die Türkische Rose" or "R. *bicolor*", probably from confusion with R. *foetida* 'Bicolor', for which 'die Türkische Rose' is an old German name. This confusion in turn probably explains the name "R. *turkestanica*" once used for it.

Following the publication of Correvon's article, a correspondent wrote to the *Revue Horticole* from Madagascar, stating that a rose agreeing with 'Mutabilis' was grown in that country, to which it had been brought by French colonists from the island of Réunion. This, if correct, suggests a Chinese origin, since some China roses were in cultivation there and on Mauritius by the middle of the 18th century.

Rehder places this rose under R. *chinensis*, but it seems to be more at home under R. × *odorata*, showing as it does some characters of R. *gigantea*.

cv. 'OCHROLEUCA'. PARKS' YELLOW CHINA.—This rose, extinct in Europe, was brought from China in May 1824 by the Horticultural Society's collector A. D. Parks, and was described by Lindley in 1826 as R. *indica* var. *ochroleuca*. Crossed with an original Blush Noisette (R. *moschata* × Pink China) it is a parent of the old yellow Tea roses and yellow Noisettes, which, through it, derive their colouring from the yellow-flowered Yunnan form of R. *gigantea*.

cv. 'SULPHUREA'.—A seedling of 'Odorata' ('Hume's Blush'), figured by Andrews in *Roses*, t. 86 (1826); raised by Knight, the Chelsea nurseryman, it had pale yellow, semi-double flowers. Andrews called this rose R. *indica sulphurea*, and it is probable that the rose portrayed in the third edition of Redouté, *Les Roses* (1828–30), under the same name is the Knight seedling and not 'Parks' Yellow', as usually supposed.

R. CORYMBULOSA Rolfe

A deciduous shrub up to 6 ft high; young shoots not downy, becoming brown with age, usually unarmed or sometimes with a few scattered, mostly solitary prickles which are straight, slender, and ¼ in. or less long. Leaves 2 to 5 in. long, consisting of three or five leaflets; rachis downy, glandular, furnished with a few tiny prickles. Leaflets ovate-oblong (the side ones stalkless), pointed, often doubly toothed, ½ to 2 in. long, ¼ to 1 in. wide, dark green above, glaucous and downy beneath. Stipules edged with tiny glands. Flowers produced during July in corymbs of up to a dozen blossoms, each ¾ to 1 in. across and borne on a slender glandular stalk ¾ to 1½ in. long. Petals inversely heart-shaped, deep rose-pink, paling towards the base; anthers golden yellow. Receptacle obovoid, glandular. Sepals ¼ to ¾ in. long, downy, often widening at the apex. Fruits globose, ¼ to ⅓ in. wide, coral-red, crowned by the persistent sepals. *Bot. Mag.*, t. 8566.

Native of China, in the province of Hupeh, whence it was introduced to cultivation by Wilson in 1907, and of Shensi. It is a pretty rose both in flower and in fruit, distinct in its almost spineless branchlets and small flowers, which are normally borne in small separate clusters, not in a large compound inflorescence as shown in the *Botanical Magazine*. The leaves turn purplish red beneath in autumn.

R. DAMASCENA*   DAMASK ROSE

A shrub to about 7 ft high; stems and branches densely armed with curved prickles of various sizes, grading into stiff bristles. Leaves with five or seven leaflets; rachis hairy, prickly beneath. Leaflets oval or ovate, acute to obtuse at the apex, dull and glabrous above, greyish and hairy beneath, sharply and simply toothed. Flowers semi-double, fragrant, in shades of blush or pink, borne in lax clusters of up to a dozen, each on a long stalk which is densely covered with

* See Note on nomenclature overleaf.

glandular bristles and small prickles (but the inflorescence more compact in some of the Autumn Damasks). Sepals up to twice the length of the flower-bud, with slender, sometimes slightly expanded tails and with lateral appendages, glandular and hairy on the back, strongly reflexed at flowering-time, soon deciduous. Receptacle narrowly ellipsoid, or narrowly campanulate, sometimes (especially in some Autumn Damasks) funnel-shaped, with the same covering as the sepals.

R. *damascena* is not known in the wild state. Its affinity is with R. *gallica*, but its armature, although mixed as in that species, is denser and stronger, the prickles being more numerous and the bristles stiffer; the inflorescence is usually laxer, with more numerous flowers, the receptacles are more elongate, and the sepals longer and more pinnated (though strongly pinnated in some forms of R. *gallica*), and completely reflexed at flowering-time; it is also taller, and does not sucker. Showing little variation, it is probably a more or less fixed hybrid, with R. *moschata* as the other parent; Dr Hurst, however, suggested R. *phoenicea* for the Summer Damasks, which are the typical R. *damascena*, and R. *moschata* only for the Autumn Damasks (var. *semperflorens*).

R. *damascena* has been in cultivation in Europe at least since the early 16th century. The Spanish doctor Monardes, in a work written in 1551, called the Damasks *Rosae Alexandrinae* or *Rosae Persicae*, the former name indicating that they had reached Spain from Alexandria and the latter the place of their birth. In stating that the Italians, French, and Germans called this rose *Rosa Damascena*, from a belief that it came from Damascus, Monardes was confusing the Damask rose with the Musk rose, R. *moschata*, for it was to the latter that the name *Rosa damascena* was applied outside Britain, when used at all, the most frequent name for R. *damascena* being *Rosa incarnata*, or in the Low Countries and the Rhineland, *Rosa provincialis*; *Rosa pallida* was also used for it, especially by the apothecaries. But in British gardens it was called *Rosa damascena*, and appears under that name in all the editions of Miller's dictionary, as earlier in Gerard's *Herball* and Parkinson's *Paradisus* and *Theatrum Botanicum*.

The old medical botanists were concerned with R. *damascena* as the source of a purgative liquor, and make only passing reference to the fragrance of its flowers, for which, and as a source of rose-water, it was more commonly grown. It is, wrote Parkinson in the *Paradisus*, 'of the most excellent sweet pleasant sent, far surpassing all other Roses or Flowers, being neyther heady nor too strong, nor stuffing or unpleasant sweet, as many other flowers'. The distilled oil or spirit from its petals, he added in the *Theatrum*, 'serveth more for outward perfumes than inward Physicke . . . and yet there is by many times much more of them spent and used than of red roses, so much hath pleasure outstripped necessary use'. But even in Parkinson's time the Damask rose had a rival in R. *centifolia*, and by the 1830s had become rare. Today, the only pure Damasks still in gardens are the York and Lancaster rose and the Kazanlik ('Trigintipetala'), the other Summer Damasks being forms of comparatively recent introduction from Iran, or hybrids.

NOTE. The name R. *damascena* was first published by J. Herrmann in his *Dissertatio* (1762) and not, as has hitherto been assumed, by Miller in his *Dictionary* (1768). In the similar case of R. *virginiana* it is possible to get over the difficulty by making the convenient though not very convincing assumption that Herrmann's plant was an anomalous form of R. *virginiana* Mill. Such an expedient, in the present instance, is even less acceptable, since R. *damascena*, as usually understood, is included by Herrmann in R. *centifolia*. His R. *damascena* is a small shrub with sub-solitary medium-sized flowers, milky white with a red flush, borne on rather spiny pedicels; sepals pinnated; receptacle ovoid, spinose; leaflets five; stipules large, toothed; stem armed with incurved spines at the stipules. Herrmann cites as a synonym the R. *lacteola* of Jean Bauhin's *Historia* (1650), but Bauhin took the name and description from an earlier work, the *Hortus Medicus* of Camerarius (1588), from which we get the additional information that its flowers were very double and that it was cultivated in quantity around Bratislava. R. *lacteola* is figured in Besler's *Hortus Eystettensis* (1616), where it is shown as unarmed, and it is one of the five roses listed by Linnaeus in *Hortus Cliffortianus* (1737), where the extreme doubleness of the flowers was remarked on. Another pre-Linnaean name for R. *lacteola* was R. *alba, minor* of Caspar Bauhin's *Pinax* (1623). The only information in Herrmann's dissertation that suggests some connection between his R. *damascena* and Miller's is that, according to him, this rose was known in Germany as 'die Molcken-Rose' or 'Damascener-Rose', from which a purgative was made by infusing the flowers in whey (Molcke in German). The Damask rose as usually understood was certainly put to a similar use (as was R. *moschata*), but other roses may have the same property.

Despite its obscurity, R. *damascena* J. Herrm. invalidates the later R. *damascena* Mill. and another name is needed for the Damask rose. The next possible name in order of priority is R. *belgica* Mill., often made a synonym of R. *damascena*. But little is known today of the Belgic roses, and it is questionable whether they were of the same parentage as R. *damascena* Mill. A likelier candidate is R. *calendarum* Borkh. (1797), founded on an Autumn Damask; this name could be rejected as *nomen nudum* but was taken up by J. F. Gmelin in *Flora Badensis*, Vol. 2, p. 430 (1806) and probably still has priority over R. *bifera* Pers., also founded on an Autumn Damask, and published in November 1806.

The R. *damascena* of L'Obel (1581) is of uncertain identity. Later pre-Linnaean botanists gave the name as a synonym of R. *rubra*, i.e., R. *gallica*—surprisingly, since there is little or nothing in his description and figure to suggest that species. The rose was cultivated by Adrian van der Gracht at Ghent; it had stems armed with scattered, curved thorns, fragrant solitary flowers, which were semi-double, white with a flush of pink on some petals, roundish receptacles and entire sepals not exceeding the flower-bud in length. It was commonly known as R. *odoratissima* (L'Obel, *Stirp. Hist.* (1581), p. 618; *Icones* (1581), Vol. II, p. 206). Whatever this rose was, it was certainly not R. *damascena* in the modern sense, and in citing R. *damascena* L'Obel in the 1768 edition of his *Dictionary* Miller was guilty of carelessness, for the rose he actually describes is R. *damascena* as usually understood.

var. SEMPERFLORENS (Loisel.) Rowley   R. *bifera semperflorens* Loisel.; R. *centifolia* ζ R. *bifera* Poir.; R. *calendarum* Borkh.; R. *omnium calendarum* Roessig; R. *bifera* (Poir.) Pers.; R. *menstrua* Andr.   FOUR SEASONS or MONTHLY ROSE, AUTUMN DAMASK.—This group of old garden varieties has no constant botanical character to distinguish it from typical R. *damascena* Mill. and is probably of the same parentage (R. *gallica* × R. *moschata*). The names given to these roses all express the fact that, with suitable pruning, they had the ability to produce their flowers in two or three flushes during the growing season and could, with forcing, be flowered in the winter months. The epithet *bifera* was given by Poiret in the belief that 'le Rosier des Quatres Saisons' was the twice-bearing rose of Paestum often alluded to by the Roman poets; this had frequent flowers, but they were usually described as of a deep red colour.

The first reference in modern literature to a remontant Damask appears in

Ferrari's *Flora, seu de Florum Cultura,* a work published in Rome in 1633, where it is called *Rosa italica flore pleno perpetuo* and, in the Italian translation of 1638, the 'Rosa di ogni mese'. It was not known to Gerard or Parkinson, but is mentioned in the *Flora* of John Rea (1665). He calls it *Rosa mensalis* or the 'monethly' rose and remarks that it produced its flowers in three flushes (June, mid-August, and late September); it was 'in all the parts thereof very like unto the Damask Rose', but the flowers were 'something more double, and not all things so sweet'. According to Ferrari, the Italian monthly rose differed from the ordinary Damask only in being more prickly ('densioribus saevit aculeis').

Several sorts of Autumn Damask were grown, but during the first half of the 19th century they were displaced by the various hybrid remontant roses, which owe their 'perpetual-flowering' character partly to the Autumn Damasks and partly to the China roses. The Autumn Damask still in cultivation agrees very well with the botanical type of var. *semperflorens,* which was the common Quatre Saisons rose of the French (R. *bifera vulgaris* Thory)—a different rose from the old monthly Damask of British gardens. It is a bush to about 4 ft high, which if pruned in late winter will bloom from June until autumn. The semi-double pink flowers are borne in small clusters on short, stiff pedicels, and show what was, for Thory, the leading character of R. *bifera,* namely, its funnel-shaped and rather narrow receptacles (repeat-flowering Damasks with ellipsoid receptacles were placed by Thory under R. *damascena*). This peculiarity is not constant, however, judging from the cultivated plant, some of whose flowers have normal receptacles. It is perhaps the outward sign of partial infertility.

The Autumn Damasks are also represented in gardens by the old 'Quatre Saisons Blanc Mousseux' ('Perpetual White Moss'). This on at least two occasions has sported back to the pink-flowered moss-less Damask described above (see Graham Thomas, *The Old Shrub Roses,* p. 161 and plate IV).

cv. 'TRIGINTIPETALA' KAZANLIK ROSE.—In its botanical characters this is a typical summer-flowering Damask rose and needs no further description. Indeed, except in the absence of variegation in its flowers, there is little to distinguish it from the York and Lancaster.

This rose was named R. *gallica* var. *damascena* f. *trigintipetala* by Dr Dieck of Zöschen and was introduced by him about 1889 from the famous rose-fields of Bulgaria, situated on the southern side of the Balkan Mountains near Kazanlik, in the upper valley of the Tundzha, which have long been one of the principal sources of Attar of Roses. Dieck saw the same or a similar rose in Asia Minor and Cyprus, and took the epithet *trigintipetala* from the modern Greek name 'triandafil', or thirty-leaved (i.e., thirty-petalled). It is interesting that the same name, in the semi-italianised form 'Trentaphilla', is given as one of the names of the Damask rose in a commentary on the works of the Arab physician Mesuë, published in Venice in 1540. [PLATE 9

The identity of the Kazanlik rose seems to have been uncertain until specimens were received at Kew in 1874 and identified there by J. G. Baker as R. *damascena*. Another specimen in the Kew Herbarium, sent for identification by Messrs Dickson of Chester in 1886, is near to 'Trigintipetala'; it was received by them as R. 'Céleste'. R. *damascena,* as grown in Pakistan and Afghanistan, seems to be near to 'Trigintipetala'.

For accounts of the Kazanlik rose-fields see: *Gard. Chron.* (1859), p. 671; ibid. (1867), p. 606; ibid., Vol. 3 (1875), p. 202; ibid., Vol. 52 (1912), p. 425; *Journ. Hort.*, Vol. 33 (1877), p. 254, a reprint of a despatch to *The Times* from its Naval Correspondent, who was attached to the Turkish forces during the Balkan War of 1877; G. S. Thomas, *The Old Shrub Roses*, p. 156, and *Shrub Roses of Today*, pp. 210-12.

In earlier times there was a famous attar industry in the Fayyum Oasis, south-west of Cairo, while in India the largest fields were at Ghazipur, north-east of Benares (Hooker, *Himalayan Journals* (1854), p. 211; *Gard. Chron.* (1845), p. 211, quoting from Bishop Heber's *Indian Journal*). In India the principal centre is now at Jaunpur, north-west of Benares.

f. VERSICOLOR West.   YORK AND LANCASTER ROSE.—'This Rose in the forme and order of the growing, is neerest unto the ordinary damaske rose ... the difference consisting in this, that the flower (being of the same largenesse and doublenesse as the damask rose) hath the one halfe of it, some-times of a pale whitish colour, and the other halfe, of a paler damaske colour than the ordinary ... sometimes also the flower hath divers strips, and markes in it, as one leafe [petal] white, or striped with white, and the other halfe blush, or striped with blush, sometimes also all striped, or spotted over, and other times little or no stripes or markes at all, as nature listeth to play with varieties, in this and in other flowers ...' (Parkinson, *Paradisus* (1629), p. 414).

The York and Lancaster usually has the variegation that Parkinson mentions first ('party-coloured' as Rea termed it some years later) and is shown in Graham Thomas, *The Old Shrub Roses*, Plate IV, facing p. 104. A more irregularly coloured form was portrayed by Ehret in the painting reproduced in *The Rose Annual* 1977, facing p. 60. Except in the variegation of the flowers, the York and Lancaster is a typical representative of *R. damascena*; it makes a lax bush to about 7 ft high.

The variegated *R. damascena* was first described by Clusius in 1601, from information given to him by a Cologne gardener, and was named by him *R. versicolor*. It is sometimes stated that this is *R. gallica* 'Rosa Mundi', but Clusius distinctly stated that the flowers were like those of 'R. *praenestina*', the Plinian name used by him for *R. damascena*, and indeed his detailed description agrees very well with Parkinson's.

R. 'PORTLANDICA'   PORTLAND ROSE, SCARLET FOUR SEASONS.— A low-growing rose, spreading by suckers, its stems armed with fine prickles of various sizes. Flowers bright red, semi-double, in clusters of three or four, faintly scented, borne from midsummer into autumn (R. *Portlandica* West., *Fl. Angl.* (1775); R. *Portlandia* Andr.; R. *bifera Portlandica* Loisel.; R. *damascena coccinea* Thory in Redouté, *Les Roses*, Vol. I, p. 109, t.; R. *Paestana* Hort.).

The origin of this rose is not known, but according to Andrews it was named for the Duchess of Portland, who is said to have cultivated it in her garden at Bulstrode Park. It was in commerce in Britain by the 1770s. For the untenable theory that it was a hybrid between *R. damascena* and 'Slater's Crimson China', raised in Italy, see under *R. chinensis* 'Semperflorens'. A rose agreeing with the original Portland rose has been found in some English gardens, see p. 196.

The Portland rose is of historical interest as a parent of the 'Rose du Roi' (distributed in Britain as Lee's Crimson Perpetual), first of a small group known as the Damask Perpetuals or Portland Roses. This group played its part in the formation of the Hybrid Perpetuals, which had largely displaced it in gardens by the middle of the 19th century.

R. DAVIDII Crép.

A deciduous shrub of loose, spreading habit 6 to 12 ft high; young shoots glabrous, armed (often very strongly) with scattered, straight, or slightly curved spines. Leaves up to 6 in. or more long, composed of five to eleven (usually seven or nine) leaflets which are oval or ovate, pointed, simply toothed, $\frac{1}{2}$ to 2 in. long, $\frac{1}{4}$ to 1 in. wide, each pair increasing in size towards the end; dark green and glabrous above, rather glaucous and downy beneath, especially on the midrib and veins; rachis downy, sparsely spiny. Flowers bright rose-pink, $1\frac{1}{2}$ to 2 in. wide, produced during June and July in loose corymbs, each flower on a slender, downy, more or less glandular stalk 1 to $1\frac{1}{2}$ in. long. Sepals $\frac{1}{2}$ to $\frac{3}{4}$ in. long, downy, often glandular, ovate at the base but prolonged into an enlarged often spoon-shaped apex. Receptacle glandular and more or less downy. Styles exserted. Fruits pendulous, scarlet-red, bottle-shaped, $\frac{3}{4}$ in. long, narrowed at the top to a slender neck, above which are the persistent sepals. *Bot. Mag.*, t. 8679.

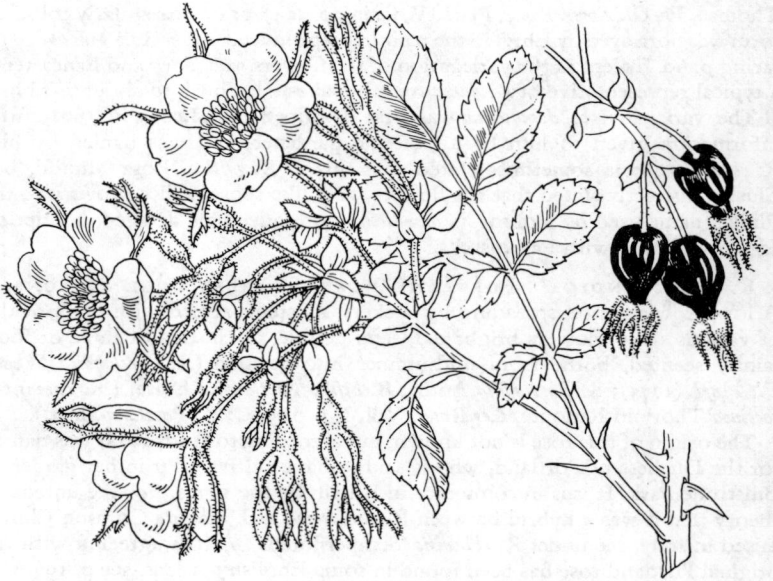

ROSA DAVIDII

Native of western and central China, and of S.E. Tibet; introduced by Wilson in 1903 and again in 1908. It is allied to R. *macrophylla*, but the flowers are usually more numerous in each inflorescence and smaller, the sepals are broad in the lower part and abruptly narrowed to a slender tip and, a difference given by Boulenger, the stylar aperture is narrower, being one-quarter to one-fifth the diameter of the disk, against one-third in R. *macrophylla*. R. *davidii* flowers rather later than most of the shrubby species. It is handsomest in autumn, when laden with its pendulous clusters of bright red fruits. As a vigorous shrub of spreading habit it is suitable for the semi-wild part of the garden.

cv. 'ACICULARIS'.—Stems densely set with needle-like prickles, which sometimes extend onto the flowering branchlets. Flowers about 1 in. wide, deep rose, borne in June (R. *macrophylla* var. *acicularis* Vilm.; R. *persetosa* Rolfe; R. *davidii* var. *persetosa* (Rolfe) Boulenger, *nom. illegit.*). Raised by Maurice de Vilmorin from seeds sent from Yunnan to his collection at Les Barres and introduced to Britain by Paul of Cheshunt, in whose nursery it flowered in 1912. It is a Chinese counterpart of R. *pendulina* 'Malyi' (q.v.).

var. ELONGATA Rehd. & Wils.—Leaflets longer (to 3 in. long). Flowers fewer in each corymb. This variety is very near to R. *macrophylla*. See that species for the Forrest introductions identified as R. *davidii* var. *elongata*.

R. BANKSIOPSIS Bak.—This was described from a plant raised by Miss Willmott from seeds collected by Wilson during his first expedition for the Arnold Arboretum (1907–8). According to Boulenger it represents a form of R. *davidii* with smooth pedicels. It is not known to be in cultivation.

R. MURIELIAE Rehd. & Wils.—Near to R. *davidii*, but with grey-green leaves and white flowers. A very elegant species, named after the daughter of E. H. Wilson, who introduced it from W. Szechwan, China, when collecting for Messrs Veitch. It is included in R. *davidii* by Boulenger.

R. 'DUPONTII'

R. *dupontii* Déségl., in part; R. *nivea* Dupont, not DC.; R. *moschata* var. *nivea* (Dupont) Lindl.; R. *damascena* var. *subalba* Thory, in part*.

A robust shrub of lax but not climbing habit, 6 to 8 ft high; branches unarmed, or bearing a few slender, short prickles. Leaflets usually five, sometimes three or seven, grey-green, ovate or oval, 1½ to 3 in. long, downy beneath, finely toothed. Flowers very fragrant, light creamy pink at first, fading to almost white, 2½

* Thory, in Redouté, gives R. *nivea* Hort. as one of several synonyms of his R. *damascena* var. *subalba*, but it is very improbable that the rose portrayed by Redouté under that name is the clone 'Dupontii' (*Les Roses*, Vol. I, p. 63, t.). Certainly the plant grown as R. *damascena subalba* in the Luxembourg garden in 1833 was not 'Dupontii', judging from the specimen in the Kew Herbarium. It seems that 'Dupontii' was also known as 'Henriette' or 'Belle Henriette'.

The name R. *dupontii* Déségl. is founded partly on a rose found in a hedge in Maine-et-Loire, which was transported to the Angers Botanic Garden. This may have been a spontaneous hybrid of R. *arvensis* or R. *sempervirens*.

to 3 in. across, single, borne around midsummer in clusters or singly. Pedicels long, downy and glandular, like the rather narrowly ellipsoid receptacle. Sepals reflexed at flowering-time, glandular, with several lateral appendages. Styles united in an exserted column. Fruits ellipsoid, orange. Willmott, *The Genus Rosa*, Vol. I, p. 43, t.

This beautiful rose was named *R. nivea* by the Paris nurseryman Dupont (*d.* 1817), who created the famous collection in the Luxembourg garden. De Pronville, who sent the plant described by Lindley in 1825, stated that it had been raised from *R. moschata*, though by whom he did not say; the largest plant known to him grew in Cels' nursery near Paris. The Belgian rhodologist Crépin was inclined to see in 'Dupontii' a variety of *R. moschata* (letter to Kew, 1888), but his later judgement was that it was a hybrid between that species and *R. gallica*. But a specimen sent by Kew to the Swiss authority Christ for naming was identified by him as *R. damascena*. 'Dupontii' is unlikely, in any event, to be the result of a direct cross between *R. moschata* and either *R. gallica* or *R. damascena*, so either de Pronville was wrong, or the seed-parent was itself a hybrid of *R. moschata*.

'Dupontii' is one of the least thorny of roses and one of the most beautiful. It received an Award of Merit in 1954.

R. 'FREUNDIANA'.—The name *R. freundiana* was given by Graebner to a rose cultivated at Dahlem, Germany, as *R. moschata alba hybrida*, and in the Berlin Botanic Garden as *R. gallica vittata* (*Gartenfl.* (1908), p. 470). This was supposed to be *R. moschata* × *R. gallica*, the parentage suggested by Rehder for 'Dupontii'.

R. ECAE Aitch.   [PLATE 10

R. xanthina sens. Hook. f. in *Bot. Mag.*, t. 7666, not Lindl.; *R. xanthina* var. *ecae* (Aitch.) Boulenger

A shrub up to about 4 ft high in the wild, of bushy habit, taller and laxer in cultivation; prickles crowded, up to ½ in. long, broad at the base, bristles none. Leaves 1 in. or less long, with usually seven, sometimes five or nine, leaflets, which are oval or almost round, ¼ in. or so long, glandular beneath, though sometimes very sparsely so, edged with proportionately large, simple, eglandular or slightly glandular teeth. Flowers solitary, about 1 in. across, rich buttercup yellow; pedicels and calyx glabrous. Fruits globose, about ⅜ in. wide, crowned with the deflexed, persistent sepals. *Bot. Mag.*, t. 7666, as *R. xanthina*.

Native of N.E. Afghanistan, N.W. Pakistan, and bordering parts of Russia; introduced to Britain by Dr Aitchison, who found it during the survey of the Kurram valley (now in Pakistan, near the border with Afghanistan). The name is an adaptation of Mrs Aitchison's initials—'E.C.A.' It is closely allied to *R. xanthina*.

R. ecae is an interesting and dainty rose, but now largely supplanted by other more robust members and hybrids of the Xanthina group. Although sometimes found in wet, shady places in the wild, it needs a sunny position in our climate. It does not grow well from cuttings and is best propagated by suckers or, failing that, by grafting.

R. *ecae* received an Award of Merit in 1933 when shown by the Knap Hill nursery. Their stock came from a garden in Rutland, where plants had been raised from seed presented to the owner by Dr Aitchison (*Journ. R.H.S.*, Vol. 72 (1947), p. 371).

For hybrids see 'Golden Chersonese' (p. 184) and 'Helen Knight' (p. 185).

R. KOKANICA (Reg.) Reg. ex Juzepchuk R. *platyacantha* var. *kokanica* Reg.; R. *xanthina* var. *kokanica* (Reg.) Boulenger; ?R. *turkestanica* Reg.; ?R. *primula* Boulenger—Leaflets somewhat larger than in R. *ecae*, edged with glandular-compound teeth and with larger glands on the undersurface. Otherwise scarcely differing from R. *ecae*. Native of Central Asia, probably extending into China. Both R. *kokanica* and R. *ecae* should probably be regarded as glandular sub-species of R. *xanthina*.

R. PRIMULA Boulenger (1936) R. *ecae* Hort., in part, not Aitch.; ?R. *platy-acantha* var. *kokanica* Reg.; ?R. *xanthina* var. *kokanica* (Reg.) Boulenger; ?R. *kokanica* (Reg.) Reg. ex Juzepchuk (1941)—A shrub to about 8 ft high in gardens, its branches armed with straight, somewhat compressed, broad-based prickles. Leaflets up to thirteen or even fifteen on sterile growths, but mostly nine to thirteen on flowering branchlets, obovate to oblanceolate or elliptic, up to ½ in. or slightly more long, densely coated beneath with large glands, edged with compound, more or less glandular teeth. Flowers 1¼ to 1¾ in. wide, primrose yellow.

R. *primula* has an interesting history. It was raised from seeds collected for the Arnold Arboretum by F. N. Meyer in 1910 near Samarkand in Russian Central Asia, and was at first grown as R. *ecae*, this being the name under which the seed or plants were distributed. However, the name was questioned by those who grew the true R. *ecae* as introduced by Aitchison, and in 1926 E. A. Bunyard sent a foliage specimen to Boulenger, then nearing the completion of his monumental treatment of *Rosa*. From this sterile piece Boulenger described R. *primula*, but he also cited the photograph of the Meyer "R. *ecae*" published in *New Flora and Sylva*, Vol. 8 (1936), to illustrate an article by the American rosarian Horace MacFarland, and took the epithet *primula* from the latter's description of the flowers as 'pale primrose' (p. 242 and fig.). Boulenger promised a more detailed consideration after he had studied a mature plant (one was sent to him by Bunyard in 1936/7). But illness, followed by his death in November 1937, prevented him from taking the matter further.

Although Boulenger treated R. *xanthina* in a wide sense, he nevertheless considered his R. *primula* to be distinct from it and its varieties (var. *ecae* and var. *kokanica*) in its more numerous leaflets with more numerous teeth, and more broadly based prickles. This conclusion, based on very limited material, might have been modified or withdrawn had he lived to study this rose in more detail. R. *primula* is very closely allied to R. *kokanica*, both differing from R. *ecae* mainly in having somewhat larger and much more glandular leaflets. Indeed, Boulenger identified as R. *xanthina* var. *kokanica* a specimen collected by Meyer near Samarkand. Plants very similar to R. *primula* occur in northern China.

R. *primula* received an Award of Merit in 1962. It is perhaps no more ornamental than other members and hybrids of the Xanthina complex, but the scent given off by its foliage, especially after rain, is more powerful than that of almost any other rose. The fragrance was likened by E. A. Bowles to that of Russia leather.

R. EGLANTERIA L.   SWEET BRIER, EGLANTINE
R. *rubiginosa* L.

An erect bush with arching branches, 6 to 8 ft high in gardens; stems and branches armed with numerous, scattered, hooked prickles, sometimes with the addition of stiff bristles on the flowering branchlets. Leaflets five, seven, or sometimes nine, ovate or roundish, compoundly toothed, nearly or quite glabrous above, covered beneath with sweet-smelling glands. Flowers pale pink, 1½ in. across, produced singly, in threes or sevens or even more together; pedicels and sepals bristly-glandular. Styles hairy. Fruits bright red, shining, egg-shaped, crowned with the persistent sepals.

Native of Europe and N. Africa, and with the dog rose one of the summer delights of English hedgerows, wherever the soil is calcareous. It is not so strong a grower as the dog rose, has smaller leaves, and is always distinguished by the sweet fragrance of the leaves. On this account, and unlike the dog rose, it may well be grown in gardens. It makes a charming low hedge clipped back annually in spring before growth recommences. The fragrance is most perceptible after a shower, and whenever the atmosphere is fresh and moist.

Double-flowered forms or hybrids of the sweet brier have been cultivated since the 17th century, and two raised before 1800 are still available; see 'Manning's Blush', p. 189.

R. *eglanteria* is one of the parents of a beautiful group of garden roses known as the Penzance Briers, which were raised by Lord Penzance from 1884 onwards by fertilising the flowers of this species with other species or with hybrid garden varieties. See further on p. 166.

The name R × *penzanceana* is not a collective name for the Penzance Briers. It is founded on 'Lady Penzance', a hybrid between R. *eglanteria* and R. *foetida* 'Persiana'.

From the Penzance brier 'Lucy Ashton' the German nurseryman Hesse raised 'Magnifica', which was used by Wilhelm Kordes to produce many shrub roses, including the beautiful 'Fritz Nobis'.

R. MICRANTHA Sm.—An ally of R. *eglanteria*, differing in its more arching branches, the absence of needle-like prickles on the branchlets, leaves less strongly scented, styles glabrous, stylar aperture narrower, sepals soon deciduous. A native of Europe, N. Africa, Asia Minor and the Caucasus; in Britain it is less committed to calcareous soils than the true sweet brier. Of little value in gardens.

Two other allies of R. *eglanteria* occurring locally in the British Isles are R. AGRESTIS Savi and R. ELLIPTICA Thuill. For these see Clapham, Tutin and Warburg, *Flora of the British Isles*.

R. ELEGANTULA Rolfe
R. *farreri* Stapf ex Stearn

A deciduous shrub a few feet high, the young sucker shoots copiously armed with short, slender prickles; branchlets sparingly armed with somewhat larger prickles. Leaves 2 to 4 in. long, composed of seven to eleven leaflets. Leaflets oval to ovate, ⅜ to 1 in. long, abruptly pointed or obtuse, sharply and mostly simply toothed, glaucous, especially beneath, downy on the midrib. Flowers solitary or a few together, 1 to 1½ in. wide, rich rose in the type, sometimes pale pink or white, opening in June; pedicels and receptacle smooth or sometimes glandular. Sepals abruptly narrowed to a slender tail, woolly at the margins and inside. Fruits bright red, top-shaped, ⅜ to ½ in. long, the sepals persistent.

Native of north-western and west-central China; described from plants raised by Veitch from seeds collected by Wilson (W.1280), which first flowered in 1908; re-introduced by Farrer from Kansu in 1915. It is allied to R. *sertata*, differing in the copiously prickly strong shoots, smaller flowers, and the sepals abruptly narrowed into a long, slender appendage. In the last-named character it resembles R. *davidii*, which differs in its leaflets with more numerous teeth, usually more numerous flowers in the inflorescence, and flagon-shaped fruits.

The stock of R. *elegantula* was sold at the winding-up sale of Messrs J. Veitch in 1913, but this species does not seem to have been much cultivated outside Kew. Indeed its very existence was overlooked when the plants from Farrer's re-introduction were being considered, for there is no doubt that R. *farreri* and R. *elegantula* are the same species. Even the following selection does not differ from the type in any significant botanical character, and is included by Rehder in R. *elegantula* without distinction (*Bibliog. Cult. Tr. & Shr.* (1949), p. 310).

cv. 'PERSETOSA'. FARRER'S THREEPENNY-BIT ROSE.—A bush up to 6 ft high, more in width. Young shoots with the same armature as typical R. *elegantula*; the upper twigs may be similarly armed or have a pair of spines at each joint. Leaflets seven or nine, up to about ⅝ in. long on the flowering branchlets, longer on the leaves of strong shoots. Flowers about ¾ in. wide, coral-red in bud, opening soft warm pink, solitary on smooth pedicels. Fruits about ⅔ in. long, ovoid, coral red. (R. *farreri* f. *persetosa* Stapf, *Bot. Mag.*, t. 8877.)

A very charming rose, beautiful when the flowers are in bud, when they are fully open, and again in autumn when the bush is hung with the brilliantly coloured fruits and specked with the purple and crimson of the changing leaves. This rose was selected by E. A. Bowles from plants raised from seed under Farrer's number 774, collected in Kansu in 1915. Other plants from this seed-collection were unremarkable. 'Persetosa' is quite hardy but succeeds best in part shade in south-eastern England. It is easily increased by cuttings.

R. ELYMAITICA Boiss.

A low, compact bush, whose stems are armed with stout, pale-coloured, very hooked prickles, ¼ to ⅓ in. long, some of which are arranged in pairs at the base of the leaf-stalks, some scattered. Leaves 1 to 2 in. long; leaflets mostly five, ⅛ to ½ in. long, oval or roundish, simply and coarsely toothed, downy above, felted beneath, of firm texture. Flowers rosy white, about 1 in. across, usually solitary on short, bristly stalks. Receptacle and sepals bristly. Fruits globose, ⅓ in. wide, dark red, glandular-bristly, crowned with the spreading sepals.

A little known species ranging from the mountains of Iran west to southern Transcaucasia, not closely allied to any other species. Crépin placed it in the *Caninae*, but all members of that group are polyploids of hybrid origin, whereas *R. elymaitica* is diploid. In the cultivated plant described above the leaflets are hairy on both sides, but they may be more or less glabrous. *R. elymaitica* was introduced to Kew in 1900 and proved hardy, but is no longer cultivated there.

R. FEDTSCHENKOANA Reg.

A shrub about 8 ft high, of vigorous suckering habit, armed with rather slender curved or straight prickles sometimes ½ in. long, pinkish when young, often reduced to bristles; year-old wood very dark. Leaflets five to nine, oval or obovate, more or less acute at the apex, rather coarsely toothed, glaucous green above, downy beneath. Flowers white, 2 in. across, produced singly or up to four on the peduncle, which is furnished with a downy, leaf-like glandular-margined bract at the base. Receptacle conspicuously covered with glandular bristles. Sepals entire, long-pointed, glanded like the receptacle, inner surface and margins downy. Fruits red, ¾ in. long, rather pear-shaped, with sepals attached. *Bot. Mag.*, t. 7770.

Native of Soviet Central Asia; discovered by the Russian traveller Fedt-schenko and described in 1878. The description given above is of the cultivated form, as figured in the *Botanical Magazine* from a plant received from the nurseryman Tom Smith of Newry, Co. Down, towards the end of the last century; the same form was also acquired by Kew from Fröbel of Zurich. It differs from *R. fedtschenkoana* as described in the *Flora* of the Soviet Union in having slender prickles passing into bristles and acute leaflets downy beneath (against an armature of uniform stoutish prickles and obtuse leaflets glabrous beneath). But it agrees well enough in other respects, and may represent some unrecorded variant.

In a border of wild roses the cultivated *R. fedtschenkoana* is at once marked by its pale, glaucous foliage, and this feature, combined with the fairly large, white flowers and the tall, suckering habit, render it unmistakable. The flowers are produced over a long period from midsummer onwards. This rose is highly praised by Graham Thomas and portrayed on Plate II of his *Shrub Roses of Today*.

R. FILIPES Rehd. & Wils.

A very large rambling shrub to about 30 ft high; shoots arching, glabrous, armed with hooked spines about ⅜ in. long. Leaves with five or seven leaflets, coppery when young; rachis and petiole glabrous, sparsely prickly and sometimes glandular. Leaflets elliptic or elliptic-lanceolate, $1\frac{1}{2}$ to $3\frac{1}{2}$ in. long, $\frac{3}{4}$ to 2 in. wide, shallowly toothed, glabrous above, typically glabrous and slightly glandular beneath, but with some hairs on the midrib and main-veins in the Kansu form. Stipules slender, fringed with glands. Flowers fragrant, white from cream-coloured buds, about 1 in. wide, borne in late June or July in several corymbose panicles near the ends of the laterals, forming together terminal masses of blossom often 12 in. or even more wide. Pedicels slender, glandular, not downy, 1 to $1\frac{1}{2}$ in. long. Sepals with slender, slightly expanded tips, glandular and slightly downy on the back, with a few slender lateral appendages. Petals obovate, not much overlapping. Stamens very numerous, with golden-yellow anthers. Styles exserted, united into a slightly hairy column. Fruits broadly ellipsoid to globose, orange, becoming crimson scarlet. *Bot. Mag.*, t. 8894.

Native of W. China; discovered by Wilson in N.W. Szechwan in 1908 and introduced by him from the type-locality near Wenchwan in that year and again in 1910. Plants from the second sending (W.4200) were flowering at the Sunningdale Nurseries and with Messrs Waterer of Bagshot by 1919. In *Plantae Wilsonianae* it is stated that *R. filipes* is confined to the area where Wilson first found it. But there is no doubt that the rose of which Farrer sent seed in 1914/5 under no. 291 is also *R. filipes*, though originally identified as *R. rubus* by Rolfe. Farrer (and his companion Purdom) first met this rose in southern Kansu, in a village near Siku that Farrer nicknamed 'Barley Bee' (Ban S'an). 'It is a huge rampageous bush, making shoots of 12 feet in the season, dark purple and smooth, set with smooth lucent Banksioid foliage of deep leathern green and particularly strong-minded thorns, ferocious though sparse. Next year that shoot, all along its length, is bowed with a burden of blossom in superb enormous lax clusters, opening of a nankeen buff, passing to pure snow-white, and diffusing upon the intoxicated air an intense sweetness that ripples for a hundred yards around in the end of June.' (*Journ. R.H.S.*, Vol. 42 (1916), p. 106; see also Farrer, *On the Eaves of the World*', Vol. II, p. 4).

Farrer's Barley Bee rose flowered with E. A. Bowles at Myddelton House, Enfield, in 1920, and it is a spray from his plant that is portrayed in the *Botanical Magazine*. *R. filipes* is now represented in the trade mainly by the fine clone 'KIFTSGATE'. The original plant at Kiftsgate Court in Gloucestershire was acquired from E. A. Bunyard's nursery in 1938, but nothing further is known of its origin. It is about 40 ft high and 60 by 30 ft in spread (Graham Thomas, *Climbing Roses*, p. 34). *R. filipes* is perhaps less tolerant of shade than its allies and grows slowly at first if planted under the branches of its intended host, but gains in vigour once its stems reach the sun. [PLATE 11

R. FOETIDA J. Herrm.

R. *lutea* Mill.; R. *eglanteria* L. (1760), not L. (1753); R. *eglanteria* var. *lutea* (Mill.) Ser.

A shrub 3 to 8 ft high, with erect or arching dark brown, later greyish stems, which are furnished with many slender, straight or slightly curved prickles abruptly widened at the base, up to $\frac{3}{8}$ in. long. Leaflets five, seven or nine, $\frac{1}{4}$ to $1\frac{1}{2}$ in. long, oval or obovate, rounded or broad cuneate at the base, edged with a few compound glandular teeth, brilliant parsley-green and glabrous or with scattered hairs above, glandular and more or less downy beneath, like the rachis and stipules. Flowers deep yellow, 2 to 3 in. across, usually solitary. Pedicels and receptacle glabrous and smooth in garden plants, sometimes hairy or glandular in Asiatic forms. Sepals ¾ to 1 in. long, lanceolate, with expanded leafy tips and sometimes with a few lateral appendages. Styles hairy. Fruits rarely seen on cultivated plants, but described as globose, red, and ½ in. wide. *Bot. Mag.*, t. 363.

The 'Austrian' brier has been known in gardens since the 16th century and was known simply as the yellow rose, or R. *lutea*. But this old name, used by Miller, has to give way to R. *foetida*, given to it in allusion to its unpleasantly scented flowers, with an odour once likened to that of bed-bugs.

R. *foetida* is most nearly allied to R. *hemisphaerica*, but in that species the main prickles are stouter, gradually narrowed from the base and curved or hooked, and the leaflets are eglandular, tapered at the base and grey beneath. Both species differ from R. *xanthina* and its allies in their tailed sepals with fairly numerous lateral appendages.

R. *foetida* has been reported from a wide area in south-west Asia and western Central Asia, but over much of that area it is only cultivated or naturalised, and some records of its growing wild are the result of confusion with glandular allies of R. *xanthina*. Boulenger's theory was that R. *foetida* is a hybrid between R. *hemisphaerica* and a glandular form of R. *pimpinellifolia*, taking from the latter its straight prickles, the shape of its leaflets, and their glands; it arose in Asia Minor and owes its dispersal entirely to the hand of man. Against that theory is the fact that it has been collected in Russian Central Asia in localities where the possibility of its being naturalised is almost out of the question. But, wherever its native home may lie, it does not occur wild in Europe, though it occurs as an escape in some parts of central and south-eastern Europe.

It has been stated that R. *foetida* rarely produces fertile pollen or seeds which, if universally true, would tend to support the hypothesis of a hybrid origin, but would also mean that every seemingly wild stand must mark the site of former cultivation. However, this may be true only of clones cultivated in Europe since the early 19th century, when this sterility was first mentioned.

In Britain R. *foetida* thrives and flowers best in the northern parts of the country, but succeeds well enough in the southern counties. Its reputation for being a 'bad Londoner' dates from the days when the atmosphere in and around the capital was more polluted than it is today.

cv. 'BICOLOR'. AUSTRIAN COPPER BRIER—This singularly and beautifully coloured rose has petals of a coppery red inside, yellow on the reverse. In other respects it is similar to ordinary R. *foetida*, in fact yellow flowers frequently appear on some of its branches (R. *lutea* var. *bicolor* (Jacq.) Sims; R. *bicolor* Jacq.; R. *punicea* Mill.; R. *eglanteria* var. *punicea* (Mill.) Thory).

The origin of 'Bicolor' is unknown. It has been cultivated since the late 16th century and probably reached Europe from Turkey through the Austrian dominions. It had been known in the Arab world since the 12th century, perhaps earlier.

cv. 'PERSIANA'. PERSIAN DOUBLE YELLOW.—Flowers very double, freely borne. According to William Paul it was introduced by Sir Henry Willock in 1837, but it must have remained scarce for many years, since the Kentish rose-grower Hooker of Lamberhurst had a small stock in 1843 for which he was asking 15/- a plant—an enormous price for a rose in those days. It was named R. *lutea persiana* by Lemaire in *Flore des Serres*, Vol. 4 (1848), t. 374.

'Persiana' is not to be confused with the typical double form of R. *hemisphaerica*, which in pre-Linnean times was regarded as the double form of R. *foetida*; nor with the R. *lutea flore pleno* of Sweet, which is the hybrid 'Williams's Double Yellow' (see p. 204).

R. *foetida* is a parent of the Pernetiana group of roses (see p. 165), now merged in the Hybrid Teas, and is also responsible for the yellow colouring of the Floribundas 'Goldilocks' and 'Allgold', and the shrub rose 'Agnes' (see p. 169). See also R. × *harisonii*.

R. FOLIOLOSA Torr. & Gr.

A shrub sometimes up to 5 ft high in gardens, though usually not much more than half that height, spreading by means of suckers; stems clustered, erect, either unarmed or with a few straight, slender prickles. Leaflets seven to eleven, narrowly oblong, ¾ to 2 in. long, glabrous and glossy above, downy on the midrib beneath, finely toothed. Stipules narrow. Flowers white or rosy pink in wild plants, but rich pink in the form cultivated in Britain, fragrant, solitary or few on short stalks, 2 to 2½ in. across, borne over a long period from midsummer onwards. Sepals ⅝ to 1 in. long, bristly outside. Fruits red, bristly, orange-shaped, ⅜ to ½ in. wide, sepals spreading. *Bot. Mag.*, t. 8513.

Native of the USA in western Arkansas, Oklahoma and parts of Texas; in cultivation at Kew by 1880, but little known in gardens until Canon Ellacombe's fine pink form was portrayed in the *Botanical Magazine* in 1913. It is distinct among American roses in its oblong, rather narrow, forward-pointing leaflets, closely set on the rachis. The foliage colours well in the autumn. The individual stems are short-lived and the oldest of them should be regularly removed.

R. × FORTUNIANA Lindl.

A climbing shrub, up to 30 or 40 ft high, introduced from China by Fortune about 1845. It has much the general character of the Banksian rose, having three or five leaflets to each leaf, glabrous and simply toothed. It is most probably a hybrid between that species and R. *laevigata*. The flowers are white and double as in typical R. *banksiae*, but larger, and with the bristly stalk and receptacle of R. *laevigata*, whose influence is further shown in the flowers being solitary, and in the large leaflets, which are downy only at the base of the midrib. It is a handsome and vigorous climber which thrives on sheltered sunny walls near London, but does not flower very freely.

R. × *fortuniana* was described in Paxton's *Flower Garden*, Vol. 2 (1851), p. 71. In the next volume of the same work the same name was used again, obviously owing to an editorial error, for 'Fortune's Double Yellow', for which see under R. × *odorata*, p. 77.

R. × FRANCOFURTANA Muenchh.

R. turbinata Ait.; *R. francfurtensis* Roessig; *R. campanulata* Ehrh.

This old garden rose is now as rare as it was once common, and is mentioned only for its historical interest. A shrub to about 6 ft high, its stems sparsely set with straight or curved prickles, the branches unarmed or with a few bristles. Leaflets five or seven, up to 2 in. long and 1¾ in. wide, broadly ovate to roundish, hairy on the veins beneath, coarsely toothed; rachis densely hairy. Stipules broad, hairy beneath. Flowers in corymbs of three to six, or solitary, terminating long growths, purplish pink, double, but with a central boss of stamens, subtended by very large bracts; pedicels glandular-bristly, the bristles extending to the base of the receptacle, which is top-shaped or bowl-shaped. Sepals entire or with a few lateral appendages, abruptly narrowed to a slender point, glandular at the edge and on the back, which is also hairy.

An old hybrid of European gardens, described by Charles d'Escluse (Clusius) in 1583; he saw it at Frankfurt, whence the name it bears in modern nomenclature. It is one of the least armed of roses and the original 'rose without a thorn' (*Rosa sine spinis*, so named by Tabernimontanus in 1590). According to Parkinson (1629) the flower is 'so strong swelling in the bud, that many of them break before they can be full blown.' Miller did not bother to describe the Frankfurt rose: 'it is of little value except for a stock to bud the more tender sorts of Roses upon, for the flowers seldom open fair and have no scent.' It was at one time a common hedging plant in central Germany, and was also planted on graves; possibly its thornlessness endowed it with religious significance, 'Rose without a thorn' being one of the epithets of the Virgin Mary.

Crépin suggested, with hesitation, that the Frankfurt rose was the result of a cross between *R. gallica* and *R. majalis* (*cinnamomea*). This cross is apparently unrecorded from the wild, no doubt because there is little overlap in the natural ranges of these species. But both, especially in their double forms, are old inhabitants of gardens.

The rose named *R. turbinata* var. *orbessanea* by Thory (Redouté, *Les Roses*, Vol. II, 21, t.) was considered by Lindley to having nothing in common with the Frankfurt rose except the top-shaped receptacle. It was more decorative and cultivated at one time in this country. A top-shaped or bell-shaped receptacle seems to be not infrequently associated with doubleness of flower (see *R. rapa* p. 147) and it may be that other roses with this character have been identified as the Frankfurt.

For the '*Rosa sine spinis*' of Parkinson, see under *R. pendulina*.

R. GALLICA L.*   RED ROSE   [PLATE 12

R. provincialis J. Herrm.; *R. centifolia sens.* Mill., not L.; *R. austriaca* Crantz; *R. pumila* Jacq.; *R. arvina* Krocker; *R. rubra* Lam.; *R. incarnata sens.* Boreau, not Mill.

A suckering shrub 1½ to 4 ft high; stems and branches clad with glandular bristles and a few slender prickles, the largest of which may be curved but are rarely hooked. Leaves composed of three or five leaflets, rarely seven except on sterile shoots; rachis downy or glabrous, glandular, usually more or less

prickly beneath. Leaflets elliptic, ovate or roundish, 1 to 3½ in. long, obtuse or
acute at the apex, rounded or slightly cordate at the base, dark green and
glabrous above, hairy beneath at least on the main veins, with a rather prominent
reticulation, teeth shallow, simple or glandular-compound. Stipules narrow,
with spreading, acute tips. Flowers solitary, but sometimes up to three or even
more in a cluster, bright rosy pink or crimson, sometimes dark velvety red
(though probably only in cultivars), usually sweetly scented, 1½ to 3 in. wide;
petals overlapping, with a white or yellowish claw. Pedicels up to 3 in. long,
clad with sessile or stalked glands, sometimes mixed with minute prickles.
Receptacle ellipsoid or ovoid, rarely globular, usually with the same covering
as the pedicels. Sepals of varying length, glandular on the back and at the edge,
with a few lateral appendages or sometimes strongly pinnated, spreading to
reflexed after flowering and soon falling. The glands in all parts of the inflorescence are aromatic, with a balsamic odour. Styles varying from glabrous to
woolly, ellipsoid to globular, about ½ in. long, slow to ripen, with a leathery
coat and acidulous flesh.

Native of southern and central Europe, east to the Ukraine, and of Asia
Minor, the Caucasus and Iraq; cultivated since time immemorial. In the western
part of its range R. *gallica* is patchily distributed and in the main confined to
calcareous soils. But so long has it been grown, both as an ornamental and
officinal plant, that it is difficult for botanists to distinguish the truly wild forms
from escapes and the relics of ancient cultivation. The wild European form is
sometimes distinguished as var. PUMILA (Jacq.) Ser.

Before the 19th century the red-flowered, semi-double officinal roses were the
commonest form of the species in cultivation; this and the other principal sorts
are mentioned below. From these an immense number of ornamental cultivars
were raised early in the 19th century (a few earlier). Paul listed almost 500 of
them in *The Rose Garden* (1848), mostly bred in France and many with variegated
flowers. See further under Gallica group, p. 163.

var. AEGYPTIACA Schweinfurth ex Täckholm—This variety is probably not
in cultivation but is mentioned for its historic interest. Judging from the
description it differs from R. *gallica* as known in Europe in its corymbose
inflorescence and in having sepals with a large leafy expansion at the apex
(as in the Abyssinian R. *richardii*). The leaves are thick, oval-elliptic, obtuse
at the apex, cordate at the base, with broad rounded teeth. It is common as a
hedge plant in Egypt, and known there as 'Ward Balladi'—the 'native' or 'country' rose—a name suggesting that it has been cultivated over a long period
(*Svensk. Bot. Tidskr.*, Vol. 26 (1932), pp. 346–64, an article dealing with Schweinfurth's interesting collections of *Rosa* in Egypt and Ethiopia).

var. HOLOSERICEA [Du Roi] Ser. R. *holosericea* Du Roi; R. *provincialis sens.*
Dum.-Cours., not Herrm. nor Mill. VELVET ROSES.—The first detailed

---

* The name R. *gallica* L. is discussed by Crépin in *Bull. Herb. Boiss.*, 2nd ser., Vol. 5,
pp. 139–41. Whatever may be the identity of the rose so named by Linnaeus in the first
edition of *Species Plantarum*, the R. *gallica* of the second edition is the species described
here. It has been stated that typical R. *gallica* has single flowers, but Linnaeus based
the name on R. *rubra multiplex* of C. Bauhin's *Pinax* and R. *rubra flore valde pleno et
semi-pleno, simplici fere* of J. Bauhin's *Historia;* it therefore covers the whole gamut from
double to almost single.

account of the Velvet roses was given by Roessig in his *Beschreibung* (1799, 1803) and some are portrayed in his *Die Rosen*. All had dark purplish red flowers, but there were two main sorts. In one the petals had a blackish sheen and the other, which he believed to be the original, a violet sheen. Of both there were single, semi-double and double forms, making six in all. He admitted that *R. holosericea* was near to *R. gallica* and indeed the differences he gives—the bristly but not prickly stems, pinnated sepals, round receptacles and non-exserted styles—are all part of the normal variation of *R. gallica*.

Velvet roses had been grown for some two centuries before Roessig's time and owed the name (and the epithet *holosericea*) to the velvety texture and lustre of their petals (the erroneous belief, current in the 19th century, that they were Moss roses, could have been dispelled by reading the descriptions in Gerard's *Herball* and Parkinson's *Paradisus*). The semi-double 'violet' form of the Velvet rose was named by Roessig *R. violacea* and the rose still cultivated as 'Violacea' (see p. 203) agrees well with his account. Unfortunately Roessig said nothing of the habit of var. *holosericea*, for 'Violacea' disagrees with *R. gallica* as usually defined, in its tallish, lax habit. A possible explanation is that the Velvet roses came from the Near East, where it may be that *R. gallica* is not always as dwarf as it is in Europe. Among the roses collected by Nancy Lindsay in Iran in the 1930s was one which she called 'the crested red cabbage'. This had leafy sepals and roundish leaflets as in 'Violacea', and very double, purple-red flowers.

The Velvet roses were certainly used in breeding the large race of Gallica roses that came into gardens towards the end of the 18th century. One still with us is 'Tuscany'. Another was the Maheka rose or 'Belle Sultane' of Redouté (Vol. III, p. 78, t.). It has been suggested that this is 'Violacea', but the Maheka rose, according to Dumont de Courset's description, had flowers which changed colour with age from rosy pink to 'rouge vif' and finally to dark brown.

var. OFFICINALIS Thory APOTHECARIES' or OFFICINAL ROSE.—Interpreted in a broad sense, this is really the typical variety of *R. gallica* (see footnote). In some works it is also called Rose of Provins (e.g., by Gerard) while in France the name 'Rosiers de Provins' eventually came to mean all the cultivated forms of *R. gallica*. The name derives from Provins, a town to the south-east of Paris, in the old County of Champagne, once famous for its officinal red roses. No doubt the local soil and climate, and the traditional skills of the growers, helped to give the Provins rose its pre-eminence. But Pomet, in his famous *Histoire des Drogues* (1694), insists that it is a distinct sort. To quote from the English translation: 'These *Provins* Roses are what are most esteemed of any Flowers in the whole World . . . . But since, of late years, these *Provins* Roses were dear, several Druggists and Apothecaries contented themselves with the common red roses that are cultivated about *Paris* and other parts . . . . Nevertheless, those who have made use of the other sort, have found, that they are not equal to the true *Provins* Roses, either in Beauty or Virtue; besides which, they will not keep so long, notwithstanding all their Pains to preserve them.'

According to tradition, the Provins rose was brought from the Near East by Thibaut IV, King of Navarre and Count of Champagne, who led the Crusade of 1239–40. This has the ring of truth, for there can be little doubt that the medicinal applications of the R. *gallica* were learnt from Arab works. The tannin-rich petals of the Provins rose were astringent and, either dried or converted into conserves or syrups were put to various uses, among them the treatment of tuberculosis, as laid down by the Arab physician Mesuë of Damascus, whose works became very influential in early European medicine (Linnaeus named after him the genus *Mesua*). But other forms of the officinal R. *gallica* must have had the same properties as the true Provins rose, even to a lesser degree. The dried petals of the Provins rose, according to Pomet, were exported in vast quantities to the West Indies and for this purpose their keeping qualities would have been of more importance than when they were used locally. The red rose was also much used to make rose-water and rose-vinegar, and was grown for that purpose in private gardens. There is also mention in the old literature of a very double form whose buds did not open properly; this was used for making sugar of roses.

Although the origin of the White Rose of York is uncertain, it is now held to be reasonably certain that the Red Rose of Lancaster was originally acquired as an emblem by Edmund, second son of Henry III and first Earl of Lancaster, through his marriage in 1275 to Blanche, widow of Henry I, King of Navarre and Count of Champagne (Capt. H. S. Lecky RN, 'The Rose in Heraldry', *Rose Annual* 1931, reprinted in the same publication in 1957, pp. 10–21).

The similarity of 'Provins' to 'province' and 'Provence' has been the source of much misunderstanding (see note, p. 70). Confusion has also been caused by the fact that in the nomenclature of the apothecaries the officinal red rose was known as *Rosa damascena* or the Red Damask. This usage passed into common speech—'. . . the vulgar idea is that a Damask Rose is a blowsy red Rose, like a rural lass at a country fair' (*Gard. Chron.* 1849, p. 596). But in garden nomenclature a Damask rose was R. *damascena* and 'damask' as an epithet had in earlier times meant flesh-coloured or light pink (incarnate).

cv. 'VERSICOLOR' ('ROSA MUNDI').—This differs from the common red rose 'only in the colour of the flowers, which in this case are for the most part of a pale blush colour, diversely spotted, marked and striped, throughout every leaf of the double flower, with the same red colour which is in the ordinary red rose.' Nothing is known of the origin of this rose nor by whom it was so delightfully named 'Rosa Mundi', 'Rose of the World'. Neither Gerard nor Parkinson were acquainted with it, and the first English description that can be found, quoted above, appears in Rea's *Flora* (1665), where the name 'Rosa Mundi' is used.   [PLATE 13

There was at one time a puzzling confusion between 'Rosa Mundi' and the York and Lancaster rose, which have quite different styles of variation and differ from each other botanically as R. *gallica* does from R. *damascena*. The confusion can be traced to Philip Miller's *Figures* (1770), Vol. II, t. 221, where the rose portrayed is 'Rosa Mundi' while the text refers throughout to the York and Lancaster and calls it by that name. Either there was an editorial blunder or Miller was not personally acquainted with either of these roses.

R. *gallica* enters into the parentage of the Hybrid Teas and their predecessors either directly, through the Hybrid Chinas or indirectly through R. *damascena* and the Portland rose ('Portlandica'), both of which derive from it.

R. GIGANTEA Collett ex Crép.

R. *odorata* var. *gigantea* (Crép.) Rehd. & Wils.; R. *macrocarpa* Watt ex Crép., not Mérat;
R. *xanthocarpa* Watt ex Baker in Willm.

A very vigorous deciduous or semi-evergreen climber attaining 40 ft or, in warm climates, 60 to 80 ft in height; young shoots growing 15 to 20 ft in length and as much as 1 in. in diameter in one season, armed with stout, uniform, hooked prickles. Leaves with usually seven leaflets; rachis and petiole glabrous, slightly prickly. Leaflets elliptic, oblong or ovate, rounded or tapered at the base, acuminate at the apex, finely and mostly simply toothed, 1½ to 3½ in. long, ½ to 1¼ in. wide, glabrous. Stipules narrow, with slender, spreading tips, not or scarcely glandular at the edge. Flowers solitary, rarely in twos or threes, fragrant, white or cream-coloured in the typical state, 4 to 5½ in. across; pedicels ½ to 1¼ in. long, they and the receptacles smooth and glabrous. Sepals entire, narrowly triangular, not constricted at the base, 1 to 1½ in. long, smooth on the back. Petals broadly wedge-shaped. Stamens white, with yellow anthers. Styles very downy. Fruits shedding the sepals when fully ripe, red or yellow flushed with red, 1 to 1⅜ in. wide, with a thick wall and small cavity. *Bot. Mag.*, t. 7972.

A native of N.E. India, Upper Burma and Yunnan; discovered by Sir George Watt in Manipur in 1882, but described in 1888 from specimens collected by Sir Henry Collett* in the Shan Hills of Upper Burma, growing at 3,500 to 5,000 ft, and introduced by him (the seeds were distributed from the Calcutta Botanic Garden in 1889). Judging from these specimens the flowers vary in size, in one instance being only about 3½ in. across. They were collected in March, and a fruiting specimen in April.

The largest flowered of all wild roses and one of the most rampant growers, this remarkable species has not proved very free-flowering with us. It first flowered with Lord Brougham at the Château Eléonore, Cannes, in 1898, and in England five years later, in March, at Albury, Surrey, where it was grown in a peach house. In the Temperate House at Kew it grew with excessive vigour but flowered regularly after 1912, when the borders were resoiled with a lighter mixture than had hitherto been used. With the protection of a wall, R. *gigantea* has survived and flowered for a time in southern England, and even in Suffolk, where a new and perhaps hardier introduction from India flowered on several occasions from 1912 onwards and withstood 25°F. of frost (*Gard. Chron.*, Vol. 85 (1929), p. 449). But it is most likely to be a permanency in a warm, sunny place in the milder parts. It grows well at Mount Stewart in Northern Ireland.

R. *gigantea* was reintroduced by Frank Kingdon Ward in 1948 from Manipur, where he found it flowering in early spring but still bearing ripe fruits. 'The chubby leaves, still soft and limp, were a deep red; the slim, pointed flower buds a pale daffodil yellow; but when the enormous flowers opened they were ivory white, borne singly all along the arching sprays, each petal faintly en-

---

* Henry Collett, of the Bengal Army, was a gifted amateur botanist and the author of *Flora Simlensis*. He found R. *gigantea* while serving as Colonel commanding a brigade in Burma and is said to have spotted it at a distance of two miles through his field-glass. Another of his discoveries at that time was *Lonicera hildebrandiana*. He was knighted on his retirement in 1893 and died in 1901.

graved with a network of veins like a watermark ... The globose hips look like crab apples. They are yellow with rosy cheeks when ripe, thick and iron hard ...' (*Plant Hunter in Manipur* (1952), pp. 45–6). The largest plants he saw had stems 'as thick as a man's forearm', but the original specimen at the Château Eléonore attained a girth of almost 5 ft at the base before it died.

R. *gigantea*, in its typical state, extends into the southern parts of the Chinese province of Yunnan. But of greater interest are the forms collected by Forrest in central and north-western Yunnan, which are of smaller stature than R. *gigantea* of Burma, even shrubs no more than 5 ft high, with fragrant flowers in shades of pale yellow or rose, yet agreeing essentially with R. *gigantea* and certainly not specifically distinct from it. It is perhaps these Yunnan forms, which Forrest found both wild and cultivated, that gave rise to the tea-scented roses of Chinese gardens (see further under R. × *odorata*, p. 77). For hybrids of R. *gigantea*, raised in recent years in Europe, America and Australia, see 'Belle Portugaise', p. 172. For "Cooper's R. *gigantea*" see under R. *laevigata*.

R. GLAUCA Pourr. (1788)*

R. *rubrifolia* Vill. (1789), *nom. illegit.*; R. *ferruginea* Vill. *sec.* Déségl., not Vill.; R. *rubicunda* Hall. f.; R. *glaucescens* Wulfen

A shrub of erect habit, 5 to 7 ft high, whose stems are covered with a purplish bloom, and armed with small decurved prickles; strong shoots clad with bristles and needles. Leaflets five or seven, ovate or elliptic, 1 to 1½ in. long, simply toothed, quite glabrous, of a beautiful coppery or purplish glaucous hue. Flowers clear pink, 1½ in. across, few in a cluster. Pedicels naked or with a few glandular bristles, which sometimes extend on to the receptacle. Sepals narrow, entire or occasionally with a few lateral appendages, 1 in. or more long, standing out beyond the petals. Fruits red, globose or nearly so, ½ in. or rather more long, usually quite smooth and with the sepals fallen away.

Native of the mountains of Europe from the Pyrenees to the Carpathians, south through western Yugoslavia to Albania. It was once grouped with R. *canina*, but is easily distinguished by the beautiful purplish grey colour of its leaves and its plum-coloured young stems, and also by the longer sepals. It is really more closely allied to R. *majalis* (*cinnamomea*) than to R. *canina*.

The leaf-colour makes R. *glauca* not only one of the most distinct of roses, but also the most ornamental in vegetative, as distinct from floral, characters. Planted in groups it makes a telling feature in the landscape all the summer

* The familiar name R. *rubrifolia* (1789) is antedated by R. *glauca* Pourret (1788). It is arguable that the latter name, although the earliest, should be discarded as a source of confusion, since the name R. *glauca* Vill. ex Loisel. (1809) has been widely used for another ally of R. *canina*. However, even if the name R. *glauca* Pourret for the present species were to be rejected it would be impossible to maintain the name R. *rubrifolia*, which Villars rendered illegitimate by citing his own earlier-named species R. *ferruginea* as a synonym. This was a taxonomic error, since his R. *ferruginea*, with leaflets hirsute on both sides, was clearly not the same as his R. *rubrifolia*, which always has perfectly glabrous leaflets. The illegitimacy of the name R. *rubrifolia* means that if the name R. *glauca* Pourret were to be left in abeyance it would be necessary to take up the next legitimate name in order of priority, which would appear to be R. *rubicunda* Hall. f.

through, and usually sets good crops of showy fruits.

For the hybrid 'CARMENETTA' see G. S. Thomas, *Shrub Roses of Today*, p. 51. The other parent is R. *rugosa*. (R. × *rubrosa* I. Preston).

R. × POKORNYANA Kmet—A rare natural hybrid between R. *glauca* and R. *canina*, occurring occasionally where the species grow together.

R. GLAUCA × R. PIMPINELLIFOLIA.—An evidently very ornamental hybrid of this parentage is portrayed by Redouté under the name R. *redutea glauca* Thory (*Les Roses*, Vol. I, p. 101, t.), and was cultivated early in the 19th century in the garden of the Horticultural Society. The origin of the cultivated plants is not recorded, but a colony of this hybrid was discovered in the Haute Savoie on La Croisette in 1898 (Boulenger, *Bull. Jard. Bot. Brux.*, Vol. 12 (1931), p. 325).

The R. *redutea rubescens* of Thory (Redouté, *Les Roses*, Vol. I, p. 103, t.) is not another form of this hybrid but R. *nitida*.

R. GLUTINOSA see R. PULVERULENTA

R. GYMNOCARPA Torr. & Gr.

A shrub usually 2 to 5 ft high in the wild, though said occasionally to attain 10 ft; stems slender, with straight, needle-like prickles often intermixed with bristles, sometimes almost unarmed. Leaflets five or seven, less commonly nine, ⅜ to ¾ in. long, elliptic to ovate or roundish, usually obtuse at the apex, sometimes slightly glandular beneath, teeth usually compound, glandular or not. Flowers rosy pink, 1 to 1½ in. across, solitary or two to four in a cluster. Pedicels glandular rosy pink, 1 to 1½ in. across, solitary or two to four in a cluster. Pedicels glandular-bristly, up to 1 in. long. Sepals glabrous and usually eglandular, ¼ to barely ½ in. long, triangular or ovate, with a slender prolongation. Fruits smooth, red, globose, pear-shaped or ellipsoid, to about ⅜ in. long; sepals, disk and styles deciduous from the ripe fruit.

Native of western North America; introduced about 1893. This pretty and graceful rose is the type of the mainly Asiatic section or subsection *Gymnocarpae*, characterised by the shedding of the entire top of the receptacle from the ripe fruit. It is closely allied to R. *willmottiae* and difficult to distinguish from R. *fargesiana*, a Chinese species not described here. Other species in this group are R. *beggeriana* and R. *albertii*.

This species is studied by G. A. Boulenger in *Bull. Jard. Bot. Brux.*, Vol. 14 (1937), pp. 279–88.

R. × HARISONII Rivers
R. *lutea* [*foetida*] var. *hoggii* D. Don in Sweet

A hybrid between R. *pimpinellifolia* and, almost certainly, R. *foetida*. The typical clone makes a gaunt shrub to about 6 ft high, which does not sucker freely. Leaflets five or seven on the flowering branchlets, resembling those of the burnet rose in size and shape, but with partly glandular teeth. Flowers solitary, loosely double, brilliant sulphur-yellow, borne in June. Pedicels finely prickly; receptacle widely bell-shaped, broader than high. Sepals glandular at the edge, slightly hairy on the back. Fruits almost black.

This rose was raised by George Folliott Harison (d. 1846), a New York lawyer, or possibly by his father Richard Harison, who served as US Attorney for the District of New York during the Presidency of Washington. Both were keen gardeners. The rose—whether a chance hybrid or the result of a deliberate cross is not known—was put into commerce by Thomas Hogg, nurseryman of New York, whence the name R. *lutea* var. *hoggii* used by David Don in Sweet's *British Flower Garden*, Vol. 4 (1838), t. 410; the plant figured was bought from Hogg by James M'Nab while on a tour of the eastern states for the Edinburgh Botanic Garden, but Hogg himself listed it as Harison's Yellow (Richardson Wright, *Amer. Rose Ann.* 1943, pp. 3–11).

Harison's rose is free-flowering, and a better garden plant than R. *foetida* 'Persiana'.

Under R. × *harisonii* Rehder places 'VORBERGII', possibly a seedling of the original clone, distributed by Späth's nursery early this century. The flowers are single, pale creamy yellow.

The following is probably of the same parentage as R. × *harisonii*:

R. 'LUTEA MAXIMA'.—Resembling R. *pimpinellifolia* in armature, but with buttercup-yellow flowers and further showing the influence of R. *foetida* in having sepals with occasional lateral appendages. The leaflets are broadly elliptic, about ¾ in. long or 1 in. long on strong shoots, with often double and slightly glandular teeth. The receptacles are vertically compressed as in typical R. × *harisonii*, but they and the pedicels are quite smooth (R. *spinosissima* var. *lutea* Bean).

Under the name "R. *ochroleuca*", but apparently also as R. *spinosissima lutea* there is, or was, in cultivation a rose very similar to the one described but with smaller leaflets, to about ½ in. long, and of denser habit. But the name R. *spinosissima* var. *lutea*, which starts in the first edition of this work, definitely belongs to the larger form—'Lutea Maxima'.

For another yellow-flowered rose of this group see 'Williams' Double Yellow', p. 204.

R. HELENAE Rehd. & Wils.

R. *moschata* var. *micrantha* Crép.; R. *floribunda* Baker, not Stev.

A deciduous rambling shrub up to 20 ft high; young shoots armed with short hooked prickles, becoming purplish brown. Leaves 3 to 7 in. long; main-stalk slightly downy and armed with small hooked prickles. Leaflets mostly five to nine, commonly seven, ovate, ovate-oblong or occasionally elliptic or obovate, sharply pointed, finely simple-toothed, ¾ to 2½ in. long, ⅜ to 1¼ in. wide, glabrous and rich green above, greyish beneath and glabrous except for some down on the midrib and main veins. Flowers fragrant, white, about 1½ in. wide, produced during June in many-flowered flattish corymbs

4 to 6 in. wide. Flower-buds roundish. Pedicels up to about 1 in. long, they and the receptacle densely glandular. Sepals awl-shaped, up to ½ in. long, glandular on the back, with a few lateral appendages. Styles exserted, united into a downy column. Fruits ellipsoid, egg-shaped or pear-shaped, about ½ in. long, orange-red or scarlet.

Native of China, where it is widespread in the mountains from Shensi south-wards through E. Szechwan and Hupeh and probably extends into the former Indochina. It was collected by Wilson in 1900, and earlier by Henry and others, but described from specimens sent by Wilson to the Arnold Arboretum in 1907 and introduced by him at the same time (W. 431b from W. Hupeh and W.666 from E. Szechwan*). According to Wilson it is very abundant in these regions, forming tangled masses often 20 ft high and as much through, and rambling over small trees in the margins of woods. It is a fine rose, producing great masses of flower and needs abundant space to show its full beauty. It is closely allied to the Himalayan R. *brunonii* and could be regarded as its main counterpart in China. It differs chiefly in the more glabrous leaflets, roundish flower-buds and more or less ellipsoid fruits. Also the inflorescence in R. *brunonii* tends to be more of a panicle than a corymb. The species is named after Helen, the wife of E. H. Wilson, who was killed in the accident in which he also lost his life on 15 October 1930.

Typically R. *helenae* is characterised by a roundish flower-bud, but there are plants in cultivation in which the buds are more tapered at the apex; in other respects they agree with R. *helenae*, but their provenance is unknown. Some of Forrest's collections in Yunnan differ from the typical state of R. *helenae* in their rounder fruits, but agree better with that species than with any other so far named.

R. *helenae* is a parent of the climbing rose 'LYKKEFUND', raised in Denmark and put into commerce in 1930. See G. S. Thomas, *Climbing Roses*, p. 69.

R. HEMISPHAERICA J. Herrm.

R. *sulphurea* Ait.; R. *glaucophylla* Ehrh.; R. *lutea* var. *sulphurea* (Ait.) Reg.; R. *rapinii* Boiss.
(wild state of species)

Growing on a wall, as this rose usually is in the British Isles, it will attain a height of 6 ft or more; branches slender, furnished with scattered, slender decurved prickles. Leaflets five to nine, obovate, ½ to 1½ in. long, rounded and coarsely toothed at the apex, cuneate at the base, glabrous and of a glaucous hue above, more glaucous beneath. Flowers solitary, drooping, delicate sulphur-yellow, 2 in. across, with numerous petals. Flower-stalks and receptacle glabrous or with glands. Sepals 1 in. long, the tips coarsely toothed, leaflike.

This beautiful yellow rose is known to have been in cultivation early in the 17th century, but owing to its difficult cultivation has always been very rare.

---

* Wilson's seed-numbers 431c and 666a were R. *rubus*, so it is understandable that this species and R. *helenae* were at first confused in gardens, including Kew. The rose exhibited by Messrs Paul of Cheshunt in 1915, allegedly raised from W.666 and identified as R. *floribunda* by Rolfe was not R. *floribunda* Baker (i.e., R. *helenae*), but R. *rubus*.

Near London especially it refuses to thrive, and in many places where it grows fairly well, its flowers do not expand properly. It is found in the gardens of Asia Minor, Persia, Armenia, etc. The English climate is too dull and damp to suit it, but one occasionally sees it doing well. In the garden of Bitton Vicarage, near Bath, Canon Ellacombe had it in splendid health for many years. The flowers do not open well in cold, wet summers.

The wild single-flowered state of the species (R. *rapinii* Boiss.) is a native of Asia Minor, the Caucasus, Russian Armenia and of northern and western Iran. It is near to R. *foetida*, differing in the following particulars: prickles more curved or even hooked, narrowing gradually from the base; leaflets obovate-cuneate, glaucous beneath, eglandular; stipules often distinctly toothed; flowers pale yellow; fruits orange or yellow.

R. × HIBERNICA Sm.

A shrub 6 to 9 ft high in gardens, with erect stems and arching branches armed with scattered prickles, the sucker shoots usually very freely furnished with prickles and bristles. Leaflets five to nine, oval or ovate, simply toothed, $\frac{3}{4}$ to 1 in. long, downy beneath, especially on the midrib. Flowers pink, usually one to three in a cluster (sometimes more), each on a glabrous stalk and $1\frac{1}{2}$ in. across. Sepals with an expanded leaflike tip, more or less pinnately lobed. Fruits globose, $\frac{1}{2}$ in. in diameter, red, crowned with the persisting sepals.

This interesting rose is a hybrid between R. *pimpinellifolia* (seed-parent) and the downy state of R. *canina* (sometimes separated from it under the name R. *dumetorum* Thuill. or R. *corymbifera* Borkh.). First discovered near Belfast, in 1802, by John Templeton (who thereby won a prize of five guineas for the discovery of a new Irish plant), this hybrid has since been found in a few other localities in the British Isles.

R. *pimpinellifolia* crosses with other members of the R. *canina* complex. The hybrid between it and R. *canina sens. strict.* was found in the last century growing near London on Ham Common and Barnes Common (R. × *hibernica* var. *grovesii* Baker). The rose named R. × *hibernica* var. *glabra* by Baker is now thought to be R. *pimpinellifolia* × R. *afzeliana*.

R. HUGONIS Hemsl.

R. *xanthina sens.* Boulenger, in part, not Lindl.; ? R. *pteragonis* W. Krause ex Kordes

A bush up to 8 ft high and more in diameter; branches slender, sometimes gracefully arching, armed with straight, flattened spines of varying length, which are associated on the barren shoots with numerous bristles; sometimes, too, the barren shoots may bear thin, flattened, triangular prickles, red and translucent, very like those of R. *sericea* var. *pteracantha* though not so large. Leaves 1 to 4 in. long. Leaflets five to eleven, elliptic or obovate, $\frac{1}{4}$ to $\frac{3}{4}$ in. long, finely toothed, deep grass green, perfectly glabrous on both sides. Flowers 2 in. across, cup-shaped, bright yellow, solitary on short lateral twigs. Pedicels

ROSA HUGONIS

slender, ¾ in. or less long, glabrous. Receptacle glabrous; sepals ½ in. long, entire, not prolonged at the apex. Fruits globose, blackish red when ripe, the calyx persisting at the top. *Bot. Mag.*, t. 8004.

A native of N.W. China; described in 1905 from a plant raised at Kew from seeds sent to England by Father Hugh Scallan (Pater Hugo). Little is known of him, but he collected in Shensi for his fellow-missionary Giraldi and no doubt the seed was collected in that province. Giraldi too sent seed of this species, to Germany, from which plants were raised. R. *hugonis* is closely allied to R. *xanthina*, in which it is included by Boulenger without distinction. Apart from the presence of bristles on the strong shoots, which typical R. *xanthina* lacks, it also differs in its more finely toothed, more numerous, perfectly glabrous leaflets.

R. *hugonis* is a most charming rose, and shares with R. *sericea* the distinction of being one of the earliest species to flower, usually by mid-May. It is beautiful even when not in flower, for its luxuriant, feathery masses of foliage. It is perfectly hardy and free when grown on its own roots, but does not thrive when grafted on Canina stocks. The group of plants at Kew near the old *Pinus pinea* are the originals from the seeds sent by Father Hugh Scallan and therefore more than three-quarters of a century old (1979).

The armature of R. *hugonis* is very variable. In the original introduction, bristles, at least permanent ones, seem to be confined to strong shoots, though the warts on the branches may be the swollen bases of transient bristles. In plants found by Wilson in the Min Valley of N.W. Szechwan bristles and needles extend even to the flowering wood. These wild plants, like those raised at Kew from the seed sent by Father Hugh, occasionally bore prickles on the strong shoots remarkably like those of R. *sericea* var. *pteracantha*. A yellow-flowered rose with similar wing-prickles was found by Farrer in Kansu, not far to the north of the Min Valley, and was introduced by him (Farrer 783). The existence of pteracanthous forms of R. *hugonis* suggests the possibility at least that some roses in cultivation which combine translucent wing-prickles with yellow flowers are forms of R. *hugonis* and not hybrids between it and R. *sericea* var. *pteracantha* as usually supposed. Such plants are usually grouped together under the heading R × *pteragonis*, this being the name given to a plant raised in Germany and supposed to be R. *sericea* var. *pteracantha* × R. *hugonis*. One of the roses in question is 'HIDCOTE GOLD', of which the original plant grew at Hidcote Manor. It is said to have been raised from seeds collected by Forrest, though, if indeed it is of wild provenance, it is more likely to have been introduced by Farrer (see above). Before Forrest attained his later renown the prefix 'F' before a seed-number indicated Farrer. 'Hidcote Gold' makes an arching bush to 6 ft high and agrees with the plants seen by Farrer in the large, flattened prickles of its stems and the bright yellow flowers, which open in early summer.

R. × INVOLUTA Sm.

R. × *gracilis* Woods; ? R. *rubella* Sm.; R. *wilsonii* Borrer, at least in part

A hybrid between R. *pimpinellifolia* and, it is now thought, R. *sherardii*, described from specimens collected in the 'Western Isles of Scotland'. Plants with R. *pimpinellifolia* as the seed-parent take after that species in armature. Leaflets usually seven broadly ovate to almost orbicular or elliptic, more or less double-toothed, downy beneath. Rachis downy to almost glabrous, with straight or slightly curved small prickles and a few glandular bristles. Flowers solitary, on bristly peduncles rarely more than ½ in. long. Fruits roundish, bristly; sepals persistent, erect or reflexed, glandular on the back. The reverse cross gives a more robust plant, more or less intermediate between the parents.

This hybrid occurs in Scotland, N. Wales and N. England, the form with the burnet rose as the seed-parent being the commoner.

R. × SABINII Woods—A natural hybrid between R. *pimpinellifolia* and R. *mollis*. The commoner form, with the burnet rose as the seed-parent, resembles the above but the leaf-rachis has only a few short, curved prickles, the peduncles are longer, to ¾ in. long, and the fruits are ovoid or urn-shaped, longer than wide, sparsely bristly, with erect sepals. It has more or less the same distribution in Britain as R. × *involuta* but is less common. The reverse cross is rare.

The hybrid R. *pimpinellifolia* × R. *tomentosa* is uncommon in Britain but has been collected in a few localities in S.E. England and the Midlands, the former being the seed-parent in all cases. The more indumented leaflets, hairy above and tomentose beneath serve to distinguish it from R. × *involuta* and R. × *sabinii*.

For further details, see R. Melville in C. A. Stace, *Hybridisation and the Flora of the British Isles* (1975), pp. 218–220, on which the above is largely based.

R. JUNDZILLII Besser

R. *marginata* of some authors, not Wallr.; R. *trachyphylla* Rau

An erect shrub to about 8 ft high, though usually not much more than half that height, of suckering habit; stems and branches armed with scattered straight or slightly decurved prickles, only very rarely mixed with needles and bristles. Leaflets of firm texture, five or seven, rarely nine, 1 to 1¾ in. long, elliptic to broadly so, usually acute or acuminate, glabrous above, more or less glandular and sometimes downy beneath, teeth compound-glandular, up to thirty on each side; rachis and petiole glandular, with or without down. Flowers in June or July, up to 3 in. across, light to rich rosy pink, produced singly or in twos or threes, sometimes in corymbs of up to eight. Pedicels up to 1½ in. long, clad with glandular bristles or needles which sometimes extend on to the receptacle. Sepals glandular on the back, the outer ones with up to six long lateral gland-edged appendages. Stigmas hairy, united into a large, rounded head. Fruits globose or slightly egg-shaped, bright red, shedding the sepals when fully ripe.

Native of Europe from central France eastward through Central Europe and N.W. Italy to the Balkans and S.W. European Russia; and of Asia Minor and the Caucasus. The plants of western Europe received many names before it was discovered that they belonged to this species, first described from the Ukraine early in the 19th century. *R. jundzillii* is one of the most distinctive and handsome of the European roses. Boulenger points out that it is really quite near to *R. gallica* and sometimes difficult to distinguish from hybrids between that species and *R. canina*. Although glandular, it is not aromatic, or at the most faintly turpentine-scented, and this character suffices to distinguish it from the sweet brier, to which it bears a superficial resemblance. It was at one time thought to be a native of Britain, but the plants so identified are a form of *R. tomentosa*.

R. LAEVIGATA Michx.   CHEROKEE ROSE
R. *sinica* Ait., not L.; R. *ternata* Poir.; R. *nivea* DC.; R. *hystrix* Lindl.

A climbing shrub, growing over the branches of trees in the wild state, its stems armed with hooked prickles, sometimes mixed with needles on the branchlets. Leaves trifoliolate, but sometimes with five leaflets on cultivated plants, brilliantly glossy green and quite glabrous. Leaflets shortly stalked, elliptic or ovate, simply toothed, 1½ to 4 in. long, about half as wide, of thick, firm texture. Stipules free from the petiole for most of their length, the free part soon deciduous. Flowers 3 to 4 in. across, pure white, fragrant, solitary, borne on a very bristly stalk, the bristles extending to the receptacle and usually to the backs of the sepals, which are 1 in. or more long, with leafy tips. Fruits red, ¾ in. wide, thickly set with bristles, the sepals persisting at the top for a long time. *Bot. Mag.*, t. 2847.

Native of S. China and Formosa, extending into the former Indochina and probably Burma; cultivated outside its natural range from an early date and naturalised in the southern USA. It was in fact first described in 1803 from specimens collected in Georgia, where it had been in cultivation since about 1780. According to Aiton, Philip Miller had it in the Chelsea Physic Garden in 1759 but, if so, it must have been lost. It is not known to have flowered in Britain until about 1825.

One of the most beautiful of all single roses when seen at its best, it is, unfortunately, too tender for the open air except in the warmer counties. Elsewhere it can only succeed in exceptionally sheltered sunny corners.

For hybrids of R. *laevigata* see 'Anemone', in the second part, p. 171; 'Marie Leonida', mentioned under R. *bracteata*, the other parent; and R. × *fortuniana*, in alphabetical order.

cv. 'COOPERI'. COOPER'S BURMA ROSE.—Although usually referred to as a form of R. *gigantea*, or as a possible hybrid between it and R. *laevigata*, this seems to belong wholly to the latter species. Some of its leaves have five leaflets instead of the normal three, but this is not infrequently the case in cultivated plants of R. *laevigata*. Although cut in severe winters, 'Cooperi' is probably hardier than the older form of R. *laevigata* and is said to flower more freely. A living plant has not been seen, but the flowers are said to acquire a pinkish tinge as they fade (R. *cooperi* Hort. ex A. T. Johnson, *Gard. Chron.*, Vol. 131 (1952), p. 80 and R. E. Cooper in Nat. Rose Soc., *Rose Annual* 1953, p. 139; R. 'Cooper Burmah' Hort. ex E. A. Bunyard, *New Fl. and Sylv.*, Vol. 10 (1938), p. 119; R. *gigantea* or R. *odorata* var. *gigantea* 'Cooper's Variety' Hort.).

From information kindly provided by the Regius Keeper of the Royal Botanic Garden, Edinburgh and the Director of the National Botanic Garden, Glasnevin, Eire, as well as from specimens preserved in the Kew Herbarium, it is possible to give what is probably the true history of this rose. It was introduced by Roland Cooper, who served 1921–7 as Superintendent of the Maymyo Botanic Garden in the Shan Hills of Burma and later became Curator of the Royal Botanic Garden at Edinburgh. He is known to have sent seeds of various Burmese plants to Lady Wheeler Cuffe in Ireland and he must have done so for the first time soon after his arrival, for in October 1921 Lady Cuffe gave to Glasnevin seeds of an unnamed rose received from Mr Cooper at Maymyo. In the same year Kew had from Glasnevin what must have been a share of this sending, since the plant there, flowering by the early 1930s, was recorded as having come originally from Cooper via Glasnevin. It was R. *laevigata*. The plant in the National Rose Society's Trial Grounds, of which Mr Courtney Page sent a specimen to Edinburgh in 1938 at Mr Cooper's request, has recently been re-examined by Mr Andrew Grierson and is also R. *laevigata*. Where the NRS plant came from is unfortunately not recorded, but Mr Page's recollection was that it had come from Kew as a small plant some seven years earlier, i.e., about 1931.   [PLATE 14

Where Roland Cooper collected the seeds is uncertain (he himself could not remember). But they were sent so soon after his arrival in Burma as to suggest that they were collected from a plant in the Maymyo Botanic Garden itself, and in view of the later confusion between Cooper's rose and R. *gigantea* it is even possible that the rose grown there as R. *gigantea,* and mentioned by Cooper in the article cited above, was really R. *laevigata*. It is almost certainly Cooper's rose that is mentioned by Boulenger, under R. *laevigata*, in *Bull. Jard. Bot. Brux.*, Vol. 14 (1936), p. 197, footnote, and p. 199.

R. LONGICUSPIS Bertol.
R. *moschata* var. *longicuspis* (Bertol.) Cardot

A large evergreen or semi-evergreen scrambling shrub, or a tall climber with a stout trunk, its branches armed with short, curved prickles. Leaves up to 8 in. long, dark green, they and the shoots reddish when young. Leaflets five or seven, of leathery texture, 2 to 4 in. long, elliptic or narrowly ovate, with slender acuminate tips (but more shortly acuminate on the flowering branchlets), rounded at the base, edged with numerous small teeth, glabrous on both sides except for occasional down on the midrib beneath, venation prominent on the undersurface. Stipules narrow, finely toothed at the edge. Flowers white, about 2 in. wide, up to fifteen or so in a lax panicle; flower-buds narrowly ovoid. Pedicels 1 in. or slightly more long, they and the receptacles glandular and often hairy, sometimes hairy and eglandular. Sepals hairy and glandular on the back, about 1 in. long, often slightly expanded at the apex, with a few lateral appendages. Petals densely silky on the back. Styles exserted, united into a hairy column. Fruits globular or broadly ellipsoid, up to ¾ in. or slightly more long; sepals deciduous from the fully ripe fruit.

R. *longicuspis* was described in 1861 from a specimen collected by Hooker and Thomson in the Khasi Hills of Assam, and occurs in all the lower ranges of north-eastern India (including the outer Himalayas), probably extending into Burma and with close relatives in China. It is allied to R. *brunonii*, differing in the leathery, more acuminate leaflets almost glabrous beneath, fewer-flowered inflorescence, and the silkiness of the underside of the petals. But it is possible that intermediates occur in the eastern Himalaya.

There are no specimens in the Kew Herbarium of cultivated R. *longicuspis*, except for one from Headfort, Eire, apparently from a private introduction. But seeds were sent by Kingdon Ward from the Naga Hills of N.E. India under his KW 7740, collected in 1927, and plants may have been raised from these in private gardens. He may also have collected seed from the plants he found in the Assam Himalaya near Shergaon in 1935. A rose has been distributed as R. *longicuspis,* and is described and figured under that name in Graham Thomas, *Climbing Roses*, p. 35 and fig. 2. The author now accepts that it is not that species; as he remarks, it is near to R. *mulliganii,* for which see under R. *rubus.*

R. LUCENS Rolfe.—The type of this species was a plant at Kew received from Vilmorin, but raised from Wilson's 1334 (not W. 1234 as stated in Rolfe's description); the seed was collected in W. Szechwan, China, in 1908. It is certainly very near to R. *longicuspis*, as Rolfe admitted, and is included in it by Rehder. The chief difference, not noted by Rolfe, is that the petals are glabrous on the back. The leaflets are shorter than in R. *longicuspis* and more shortly acuminate, but there is little or no difference in the inflorescence, nor in the size or shape of the fruits, which in Wilson's specimen 1334 are about as large as in R. *longicuspis*, though somewhat smaller on the cultivated plant that Rolfe took as the type of R. *lucens*. He also identified as R. *lucens* the plants raised from W. 4127, collected in W. Szechwan during Wilson's second expedition for the Arnold Arboretum. These might still be found in private gardens, probably under the name R. *longicuspis*, since W. 4127 was referred to that species by Rehder and Wilson. For a rose distributed commercially as R. *longicuspis* see under R. *mulliganii*, p. 133.

R. SINOWILSONII Hemsl. & Wils.—This species was discovered by Wilson on Mt Omei in W. Szechwan, China, in 1904 and introduced by him. It is very near to R. *longicuspis*, with the same large leaves richly coloured when young, and large flowers with silky-backed petals, but the pedicels, receptacles and sepals, though sometimes slightly glandular, are not or only slightly hairy, and the flower-buds, narrowly ovoid and tapered in R. *longicuspis*, are broadly ovoid in R. *sinowilsonii*.

At Kew, R. *sinowilsonii* grew for many years on the wall of the Herbaceous Ground. At Wakehurst Place in Sussex it has climbed 45 ft in a Chamaecyparis, and is also grown at Borde Hill in the same county. This species is really more remarkable for its fine foliage than for its flowers, which do not make much display. The leaves are sometimes 1 ft long.

R. 'MACRANTHA'

R. *macrantha* Hort., not Desportes; R. *waitziana* var. *macrantha* Rehd., in part

A vigorous shrub with spreading and arching branches attaining a width of about 10 ft and a height of 5 ft; stems armed with sparse, straight or slightly curved prickles mixed with minute straight prickles and stalked glands. Leaflets three or five, of firm texture, ovate or oblong-ovate, acute or acuminate, up to 2 in. long, edged with sharp mostly simple teeth, dull rich green above, glabrous on both surfaces but with glands and minute prickles on the midrib beneath; rachis prickly and glandular but almost devoid of hairs. Stipules parallel-sided, with ascending free tips, gland-fringed. Flowers single, pink in the bud, opening light pink, fading to white, about 3 in. wide, sweetly scented, borne in cymose clusters of three or five at the tips of the shoots and in the upper leaf-axils. Pedicels glandular. Receptacle broadly ellipsoid to globular, slightly glandular in the lower part. Sepals with numerous lateral appendages but not much prolonged at the apex, hairy at the edge and inside, slightly glandular on the outside. Stamens numerous and conspicuous. Disk mounded; styles slightly exserted. Fruits red, with persistent sepals, ripening late. Willmott, *Genus Rosa*, Vol. II, p. 403, t.; *New Fl. & Sylv.*, Vol. 12 (1940), fig. XLVII.

'Macrantha' shows the influence of R. *gallica* and could be a seedling of some garden hybrid with that species in its make-up. Canon Ellacombe was growing it in his garden at the Bitton Vicarage by 1888, and thirteen years later it was figured in *Revue Horticole* (1901, p. 549) with a description and discussion by Mottet, on which the account in Willmott's *The Genus Rosa* is largely based. Mottet assumed, like most later authors, that this rose was R. *macrantha* Desportes (for which see below), though in fact it is very different. He tells us nothing about the provenance of the plant he describes, but the plant listed as R. *macrantha* in the catalogue of the Roseraie de l'Haÿ for 1902 came from Messrs Paul of Cheshunt.

R. 'Macrantha' (of gardens) is one of the most beautiful of single roses. Being of lax habit it needs support if to be grown upright, but makes a useful ground-cover, especially on banks.

'MACRANTHA DAISY HILL' is similar, but the flowers have a few extra petals, and open better in wet weather than 'Macrantha'. It was raised by

Smith of Newry, Co. Down. This rose in turn was used by Kordes to produce some fine hybrids such as 'Raubritter', for which see p. 197.

R. MACRANTHA Desportes   R. *canina* var. *grandiflora* Thory in Redouté, *Les Roses*, Vol. III, p. 75, t.; R. *canina fulgens* Lemeunier ex Thory; R. *macrantha* var. *lemeunieri* Franch.—This rose, at least according to the received version, was found by the French rosarian Lemeunier growing in a hedge near La Flèche in the department of Sarthe. He sent a plant (probably a propagation) to the Luxembourg garden, where it flowered in 1822. It was portrayed by Redouté in the same year under the name R. *canina* var. *grandiflora* Thory, and given specific rank by Desportes in 1838 (*Fl. Sarthe*, p. 77). It has been a source of puzzlement that Desportes' own description, although agreeing with Thory's, was based partly on a specimen identified as 'Avessé, Martigné, (Goupil)'. The explanation appears to be that this specimen came from the garden of Lemeunier's friend and fellow rosarian Goupil, of the Château de Martigné, Avessé, and was one of the former's propagations from the original plant (Gentil, *Roses Indig. Sarthe* (1897), pp. 66–75). In this work Gentil voices the suspicion that the La Flèche plant was not in fact spontaneous but actually raised by Lemeunier himself, and accuses him and Desportes of inflating the flora of Sarthe with garden plants. Lemeunier is known to have raised roses from seed, one of his productions being the rose portrayed by Redouté as R. *muscosa anemoneflora* (Vol. III, p. 97, t.) In the Kew Herbarium there is a specimen of R. *canina grandiflora* Thory collected in the Luxembourg garden in 1829. From this, and from the original description and portrait, it is plain that this rose and the 'Macrantha' of gardens are not the same. The true R. *macrantha* of Desportes shows no obvious influence of R. *gallica*, having strong, uniform, hooked prickles. The leaves have the rachis hairy beneath; leaflets five or seven with whitish undersides; sepals with long expanded tips; and flowers of a vivid rose. The buds, according to Thory, were covered before expansion with a glaucous bloom, which can also be seen on the herbarium specimen.

Subsequently, Desportes himself found a rose near La Flèche which he identified as R. *macrantha* and still another, also referred to R. *macrantha*, was found near the neighbouring town of Angers and transported to the Botanic Garden there. Specimens from these roses were distributed to various herbaria as R. *macrantha* and are probably responsible for the belief that R. *macrantha* is a hybrid between R. *gallica* and R. *canina*, since, unlike the true R. *macrantha*, they showed the influence of both species in their armature. It is tempting to suppose that the 'Macrantha' of gardens descends from the Angers plant, which might have been propagated and distributed by one of the many local nurseries. But Boreau, the Director of the garden, gives a description of the Angers plant in his *Flore du Centre de la France*, ed. 3 (1857), Vol. II, p. 227, and further details are provided by Crépin in his *Primitiae* (*Bull. Soc. Bot. Belg.*, Vol. 8 (1869), p. 285). From these it is evident that the 'Macrantha' of gardens is an altogether different plant, whose origin remains obscure.

R. MACROPHYLLA Lindl.

R. *alpina* (*pendulina*) var. *macrophylla* Boulenger

A shrub 8 to 12 ft high, with erect stems and arching branches, sometimes unarmed but usually furnished with straight prickles up to $\frac{1}{2}$ in. long, more or less directed upwards, often paired at the nodes. Leaves up to 8 in. long, consisting of from five to eleven leaflets which are 1 to $2\frac{1}{2}$ in. long, elliptic, elliptic-oblong or elliptic-ovate, usually acute or acuminate at the apex, more rarely obtuse, glabrous above, usually downy and sometimes glandular beneath, toothed almost to the base, the teeth simple or compound, usually twenty or more on each side of the longest leaves. Stipules usually broad and often red-tinged. Flowers in June or July, 2 to 3 in. across, deep dog-rose pink or bluish pink, solitary, or more commonly, in bracted clusters of up to five. Pedicels and receptacles more or less glandular bristly. Sepals entire (very rarely with one or two lateral appendages), 1 to 1½ in. long, expanding into a leafy tip, more or less glandular on the back. Stigmas woolly. Fruits flagon-shaped to roundish, up to 1½ in. long, always with a more or less distinct neck at the apex, crowned by the persistent sepals.

Native of the Himalaya, with close relatives in China. This fine rose was introduced about 1818, and is among the handsomest of the genus in its fruits, which often hang in numerous clusters. In describing R. *macrophylla*, Lindley remarked on its similarity to the European R. *pendulina* (which he knew as R. *alpina*). In this century, Boulenger went so far as to make it no more than a variety of that species. The difference, according to him, is in average characters, R. *macrophylla* having the prickles usually twinned at the nodes (rarely so in R. *pendulina*), its leaflets somewhat longer, more frequently downy beneath, with more numerous teeth, and the inflorescence more frequently many-flowered.

Some of the roses of W. China are closely allied to R. *macrophylla* but in the present state of our knowledge it seems better to regard them as distinct species. See further under R. *moyesii*. The varieties of R. *macrophylla* named by Vilmorin from introductions by the French missionaries have been disposed of as follows: var. *acicularis* is transferred to R. *davidii*; var. *rubrostaminea* is, according to Rehder, a form of R. *moyesii*; var. *crasseaculeata* probably belongs to R. *setipoda*.

cv. 'KOROLKOWII'.—Stems stout, almost thornless, purplish red. Leaves mostly of nine leaflets up to 2½ in. long, half as wide, glandular and silky-hairy on the veins beneath. Flowers pink, 2½ in. wide. Receptacle very glandular. Fruits vermilion, 1½ to 2 in. long, tapering to the base (R. *korolkowii* Lav.; R. *macrophylla* var. *korolkowii* (Lav.) Vilm. & Bois). The original plant grew in the Segrez Arboretum and was propagated by Vilmorin, but its origin is uncertain. It is possibly the same as the R. *cinnamomea* var. *korolkowii* of Regel's *Tentamen* and, if so, it was found by Korolkow in a garden at Khiva in Uzbekistan. It is surprising, however, that even a cultivated form of R. *macrophylla* should have found its way to an area so far from the western limit of the species.

cv. 'MASTER HUGH'.—Fruits strikingly large, globose to ovoid. Raised from seeds collected by Stainton, Sykes and Williams in Nepal in 1954. Award of Merit 1966, when exhibited by Maurice Mason of Talbot Manor, Kings Lynn, Norfolk, and named by him. The original wild plant was 15 ft high, growing on a steep hillside among conifers at Kali Gandakhi.

R. 'RUBRICAULIS'.—Young stems bloomy, they and the rachis, stipules, bracts and receptacles dark purplish red. This beautiful but slightly tender rose was named by Messrs Hillier and raised by them from Forrest 15309, collected in the Yungpei mountains of N.W. Yunnan at 9,000 ft in 1917. The seeds were originally distributed as R. *macrophylla*, under which 'Rubricaulis' is listed by Messrs Hillier, but Forrest's corresponding field-specimen was later identified by Byhouwer as R. *davidii* var. *elongata*, a variety which could constitute a link between R. *davidii* and R. *macrophylla*. From F.14958, also now referred to this variety, Messrs Hillier raised 'GLAUCESCENS', with glaucous leaves and pruinose stems. The seeds were collected by Forrest on the Mekong-Salween divide, N.W. Yunnan.

For 'AUGUSTE ROUSSEL', a hybrid between R. *macrophylla* and a Tea rose, see p. 172. 'DONCASTERI' also probably derives from R. *macrophylla*; see p. 179.

R. SATURATA Bak.—A close ally of R. *macrophylla* found by Wilson in W. Hupeh and bordering Szechwan, and introduced by him in 1907. The leaflets are lanceolate, acute, almost 2 in. long, whitish beneath, and the large pink flowers are usually solitary. Evidently a handsome species, but perhaps no longer in cultivation in Britain. It was described from a plant in Miss Willmott's garden, raised from the Wilson seed.

R. MAJALIS J. Herrm.   CINNAMON OR MAY ROSE
R. *spinosissima* L. (1753), *nom. confus.*; R. *cinnamomea* L. (1759), not L. (1753)*; R. *fecundissima* Muenchh.

A strong-growing bush 6 to 9 ft high, spreading by suckers, stems erect, reddish brown, much branched near the top, with usually a pair of hooked prickles (sometimes a cluster) at the nodes, and others scattered on the stems, especially near the ground, where the armature may consist of needles and bristles. Leaflets usually five or seven, elliptic, oblong or obovate, obtuse, rounded or acuminate at the apex, usually narrowed to the base, 1 to 1⅜ in. long, simply toothed, but the lower third or quarter entire, green or grey-green and sometimes downy above, greyish and more or less downy beneath. Stipules broad, those on vigorous shoots often rolled inwards. Flowers in May on often unarmed branchlets, solitary or few in a cluster, in various shades of pink, double in the type and often double or with extra petals on naturalised plants, about 2 in. across. Pedicels short, usually not more than twice as long as the receptacle, glabrous and smooth. Sepals entire, slightly expanded at the apex, woolly at the edge and sometimes with hairs on the back. Fruits globose or slightly elongated, red, ½ in. wide, crowned by the erect sepals.

R. *majalis* has its main distribution in northern Europe and Siberia, but occurs locally in Central Europe, often in wet habitats. It is an old inhabitant of gardens, cultivated in Britain since the 16th century. Double forms have always been commoner in gardens than the single, and it was on one of these that Herrmann founded the species, taking the epithet from the old vernacular name 'May rose'. It was also known as R. *veneta*. The more familiar epithet *cinnamomea* probably refers to the colour of the stems, resembling that of stick-cinnamon, rather than to the fragrance of the flowers and certainly not to that of the leaves, for these are scentless.

* The R. *cinnamomea* of the first edition of Linnaeus' *Species Plantarum* (1753) is R. *pendulina*, and debars the use of this name for the species here described, to which Linnaeus transferred the name R. *cinnamomea* in the tenth edition of his *Systema Naturae* (1759).

R. *majalis* is the type of the large and amorphous section *Cassiorhodon* (*Cinnamomeae*); see further on page 42. Its closest counterpart in the New World is R. *blanda* (q.v.).

The following natives of N.E. Asia are closely allied to R. *majalis* and included in it by Boulenger: R. DAVURICA Pall.; R. AMBLYOTIS C. A. Mey.; R. MAR-RETII Lévl. They differ from R. *majalis* in minor characters such as the darker, purplish brown stems and the presence of glands on the undersides of the leaflets.

R. MAXIMOWICZIANA Reg.

R. *coreana* Keller, not R. *koreana* Komar.; R. *kelleri* Baker

This species, a native of the Russian Far East, Manchuria and Korea, is fairly closely allied to R. *multiflora*, differing in having the leaflets glabrous beneath except for down on the midrib, more shallowly toothed stipules (incisions not more than half the width of the limb), glabrous inflorescence and larger flowers ($1\frac{1}{2}$ to 2 in. long). The stems, at least the stronger ones, are armed with bristles as well as small hooked prickles. It may not be in cultivation in its typical state. In the cv. 'JACKII' the stems lack bristles, the stipules are scarcely toothed and the flowers are more numerous in each inflorescence than is usual in R. *maximowicziana*. It was described by Rehder (as R. *jackii*) from a plant raised at the Arnold Arboretum from seeds collected by J. G. Jack in 1905 near Seoul in Korea, and was subsequently placed under R. *maximowicziana* as var. *jackii* (Rehd.) Rehd. Although introduced to Kew in 1910 it was later lost and is not known to have spread into gardens.

R. MOYESII Hemsl. & Wils.

R. *holodonta* Stapf, in part*; R. *fargesii* Hort.; R. *moyesii* var. *fargesii* Rolfe; ? R. *macrophylla* var. *rubrostaminea* Vilm.; R. *alpina* (*pendulina*) var. *macrophylla* (Lindl.) Boulenger, in part, not R. *macrophylla* Lindl.

A shrub 6 to 10 ft high, of sturdy habit; stems erect, armed with stout, pale, scattered, broad-based spines, very abundant on the barren shoots, the lower part of which is also abundantly furnished with fine needle-like prickles; flowering shoots much less prickly. Leaves 3 to 6 in. long, with from seven to thirteen leaflets, which are ovate to roundish oval, $\frac{3}{4}$ to $1\frac{1}{2}$ in. long, simply or doubly toothed, glabrous except on the midrib beneath, which is downy and

---

* R. *moyesii* was described from two specimens, both from the vicinity of Tatsien-lu (Kangting)—Pratt 172 (collected 1893) and Wilson 3543 (collected 1903). Both had the deep red flowers characteristic of typical R. *moyesii*, but the Pratt specimen has broad-ovate, oblong-ovate or broad-elliptic leaflets, obtuse or at the most subacute at the apex and a serration of mostly compound or bicuspid teeth, while in the Wilson specimen the leaflets are longer, relatively narrower, acute, and simply toothed. In the article accompanying plate 9248 of the *Botanical Magazine* (1931) Stapf argued that the specimens represented two distinct species. For reasons explained in that article he

sometimes prickly, dark green above, pale or somewhat glaucous beneath; common stalk glandular and prickly. Flowers an intense blood-red, 2 to 2½ in. across, mostly solitary or in pairs; stalk and receptacle glandular-bristly. Sepals 1 in. or more long, with expanded tips and a few glands outside, downy inside. Fruits red, flagon-shaped, 1½ in. or more long, crowned by the erect, persisting sepals, glandular-hairy towards the base, or sometimes all over. *Bot. Mag.*, t. 8338.

Native of W. Szechwan, China; first found by A. E. Pratt in 1893 growing near Kangting (Tatsien-lu); Wilson found it in the same area in 1903 and introduced it, sending seed again during his second expedition for the Arnold Arboretum in 1911. The name commemorates the Rev. J. Moyes, a missionary in W. China. R. *moyesii* was first exhibited in flower by Messrs Veitch in June

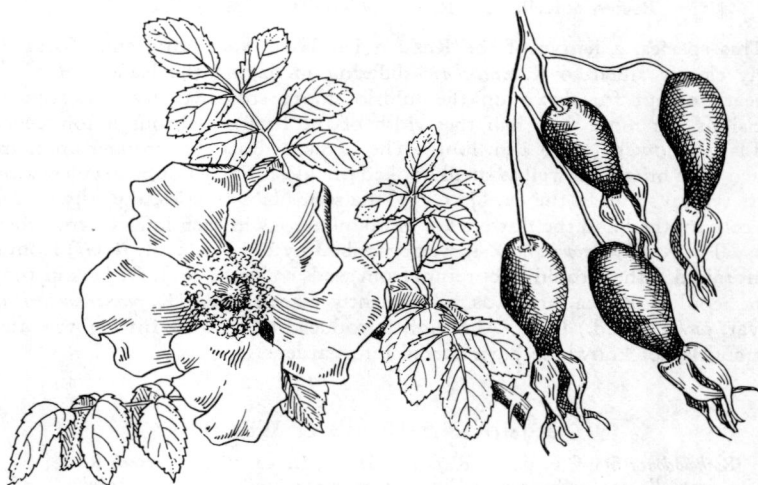

ROSA MOYESII

1908 and has since become one of the most admired and widely grown of rose species. It is interesting that the blood-red flowers that make it so distinct among wild roses are not a constant feature of the species in the wild, or even in the type-locality. Plants with pink flowers were raised by Veitch from the original Wilson seeds and predominate among garden seedlings, 'Geranium' (see below) being a fortunate exception. R. *moyesii* is a perfectly hardy shrub of rather gaunt habit, becoming almost a small tree in some gardens.

---

adopted the Pratt specimen as the type of R. *moyesii*, and made the Wilson specimen the type of a new species—R. *holodonta* Stapf. However, in *Plantae Wilsonianae* (1915), Rehder and Wilson took the Wilson specimen to be the type of R. *moyesii* and made no mention of the Pratt specimen. It is assumed here that the two specimens represent states of the same species; but cultivated plants agree better with the Pratt specimen.

Although founded on the Wilson co-type of R. *moyesii*, Stapf's R. *holodonta* also included R. *moyesii* f. *rosea* (q.v.) and specimens of R. *sweginzowii* (q.v.) with simply toothed leaflets.

cv. 'FARGESII'.—Plants cultivated by Messrs Veitch in their Coombe Wood nursery as "R. *Fargesii*" reached gardens, including Kew, through the winding-up sale of 1913-4, and were found to differ in no way from cultivated forms of R. *moyesii* except that the flowers were pink. However, they were given botanical status by Rolfe as R. *moyesii* var. *fargesii*. One plant of 'Fargesii' has been found to be tetraploid, which would make this variant more useful for breeding than the Wilson introduction, if it is really true that all the plants raised from the seeds he sent are hexaploid. 'Fargesii' received an Award of Merit in 1922 as a fruiting shrub.

No explanation has been found for the name R. *fargesii*. Presumably Veitch obtained his stock from Vilmorin, who certainly received seeds of many Chinese plants from Père Farges, but R. *moyesii* does not occur in the area where he collected (N.E. Szechwan and the Tapashan of bordering Shensi). However, the rose named R. *macrophylla* var. *rubrostaminea* by Vilmorin, which is thought to be R. *moyesii*, was raised from seed said to have come from 'Tibet', which suggests a collection in the area of Tatsien-lu, capital of a semi-independent principality then usually regarded as part of Tibet. The sender of the seed was not stated by Vilmorin, but is likely to have been Père Soulié, who collected in that area 1890-4 (the seeds of *rubrostaminea* germinated with Vilmorin in 1894).

In the current edition of Rehder's *Manual* (Ed. 2) R. *moyesii* var. *fargesii* Rolfe is recognised and is described as having obtuse, broad-oval to suborbicular leaflets 1 to 2 cm. long, against ovate or elliptic to oblong-ovate and acute, 1 to 4 cm. long in typical R. *moyesii*. By that reckoning all cultivated forms of R. *moyesii* would be referable to var. *fargesii*. See further in footnote to R. *moyesii*.

f. ROSEA Rehd. & Wils. R. *holodonta* Stapf, in part only—The type of this *forma*, collected by Wilson in the Mupin area of W. Szechwan (W. 1123) is doubtfully referable to R. *moyesii* and bears a strong resemblance to R. *davidii* var. *elongata* Rehd. & Wils., described from the same area. A plant raised at Kew from W. 1123 had numerous flowers in the inflorescence, which is not true of R. *moyesii*; Wilson's 3544 (Veitch expedition), placed under R. *moyesii* f. *rosea*, is also scarcely any form of R. *moyesii*. It must at any rate be accepted that f. *rosea* is not an appellation that can be attached to any pink-flowered seedling of R. *moyesii*, and still less can Dr Stapf's R. *holodonta* be used in this sense (see footnote). The plant figured in the *Botanical Magazine*, t. 9248 is R. *sweginzowii* (see below). A rose distributed commercially as R. *holodonta* is near to R. *davidii* var. *elongata*.

cv. 'GERANIUM'.—Of more compact and bushy habit than normal, with lighter green foliage. Flowers clear geranium-red. Fruits relatively wider than is usual in R. *moyesii*, with a shorter neck. Raised in the R.H.S. Garden, Wisley, shortly before 1937; selected and named by Brian Mulligan, then Assistant to the Director. If there is room only for a single plant of R. *moyesii* this should be chosen. It received an Award of Merit in 1950 for its fruits when exhibited by A. T. Johnson, who did much to popularise it in his writings.

cv. 'SEALING WAX'.—Another Wisley seedling (see 'Geranium'), selected for its large, bright-red hips. Flowers pink.

'Hillieri', described on p. 185, is a seedling of R. *moyesii*; it is sometimes placed under R. × *pruhoniciana* Schneid., the name given to a supposed hybrid between R. *moyesii* and R. *willmottiae*; but 'Hillieri' is unlikely to be of that parentage. Other seedlings of R. *moyesii* are 'Highdownensis' (p. 185) and 'WINTONIENSIS' (not treated in the second section), a shrub to about 12 ft high with flowers of a vivid pink, borne singly or in twos; it is near to R. *moyesii* but shows some characters of R. *setipoda*. It was raised by Messrs Hillier and named in 1928.

R. SWEGINZOWII Koehne R. *moyesii* sens. Stapf, in part, not Hemsl. & Wils.; R. *holodonta* Stapf, in part; R. *alpina* (*pendulina*) var. *macrophylla* (Lindl.) Boulenger, in part, not R. *macrophylla* Lindl.—This species comes very near to R. *moyesii*, but the branches are usually armed with flattened, triangular spines mixed with bristles, and even when they are unarmed, as is sometimes the case, the characteristic armature can be seen on the main stems. The foliage resembles that of R. *moyesii*, though the leaflets are on the average larger; in the type the leaflets are compoundly toothed, but they are simple in the Wilson specimens referred to R. *sweginzowii* in *Plantae Wilsonianae*. The flowers are solitary, or in twos or threes, sometimes in clusters of up to six or so, pink, 1½ to 2 in. wide. The fruits are glossier than in R. *moyesii*, and often relatively broader, with a shorter neck. *Bot. Mag.*, t. 9248, as R. *holodonta*.

A native of China from N.W. Szechwan northwards through Kansu and Shansi, east to Chihli; described in 1910 from a plant cultivated in an arboretum near Riga, which had probably been raised from Kansu seeds. It was introduced to Britain in 1903 by Wilson from N.W. Szechwan, near the Kansu border, but the plants raised by Veitch from his seeds appear to have been identified as a form of R. *macrophylla*. He reintroduced it from the same locality in 1910 (W. 4028); the plant portrayed in *Bot. Mag.*, t. 9248, as R. *holodonta*, was from this sending. Seeds were also sent by Farrer from S. Kansu in 1913, from which a plant in the R.H.S. Garden (Award of Merit 1922) was raised. Another provenance is Hers 625, sent in 1923.

R. *sweginzowii* is variable, but the best forms bear fine crops of fruit in August and September. The habit is elegant and the height up to 12 ft.

R. *sweginzowii* 'Macrocarpa', raised in Germany, is possibly a hybrid.

R. WARDII Mulligan R. *sweginzowii* var. *inermis* Marquand & Shaw—A shrub up to 6 ft high in the wild; branches reddish brown, with a sparse armature of slender, straight or slightly curved prickles, paired at the nodes. Leaflets five to nine, mostly ½ to ¾ in. long, edged with fine simple or compound-glandular teeth, glabrous above, downy on the main veins beneath and slightly glandular there. Flowers solitary or up to three, on short laterals, with white roundish petals. Sepals caudate or merely acuminate at the apex. Stigmas woolly. Disk maroon.

This species was discovered by Kingdon Ward in August 1924 in Kongbo, S.E. Tibet, and introduced by him; he found it in fruit on Lake Trasum (Pasum) and collected it in flower a few days later some twenty miles beyond the lake. It was described in 1940 from his herbarium specimens (*Journ. R.H.S.*, Vol. 65, pp. 57–9). In this article, Mr Mulligan notes that the plant raised at Wisley from Kingdon Ward's seeds (KW 6101) differed in certain minor respects from the wild material, the pedicels being glandular (against smooth and downy in the types), the sepals shorter and the flowers smaller (1¼ in. wide against 2 in.

in the wild plants). For this reason he distinguished the Wisley plant as var. *culta*.

R. *wardii* is uncommon in gardens and lacks vigour. Its most notable feature is the white flowers to which the coloured disk gives a maroon eye. Although once known as 'the white Moyesii' it seems to be nearer to R. *macrophylla*.

R. MULTIBRACTEATA Hemsl. & Wils.

R. *reducta* Baker

A deciduous bush up to 6 ft high; young shoots slender, glabrous, armed with slender, straight, pale spines usually in pairs, ⅓ to ½ in. long. Leaves 1½ to 2½ in. long, composed of five to nine leaflets; rachis slightly prickly and glandular. Leaflets obovate or oval to almost quite round, simply or doubly toothed, dark green and glabrous above, greyish beneath and downy on the midrib, ¼ to ⅜ in. long, from two-thirds to quite as much wide. Flowers sometimes numerous in terminal clusters, sometimes few or even solitary, subtended by a group of leaf-like bracts, each flower being 1 to 1½ in. wide, bright pink. Individual flower-stalks ¼ to ¾ in. long, downy and glandular. Receptacle ¼ in. long, slender, covered with gland-tipped bristles. Sepals ⅓ in. long, ovate, drawn out into slender tips, downy inside, glandular outside. Styles exserted, very downy. Fruits globose, ⅓ in. wide, orange-red with the sepals at the top and a few glandular bristles persisting.

Native of W. China; discovered by Wilson in the valley of the Min River, N.W. Szechwan and introduced by him in 1908. It is a very attractive and useful rose, flowering after most of the species are over and making in time a dense bush 12 ft or more wide. It is allied to R. *webbiana*, differing in its many-flowered inflorescence and exserted styles.

R. *multibracteata* is a parent of 'Cerise Bouquet', see p. 176.

R. FORRESTIANA Boulenger—This species resembles R. *multibracteata* in its foliage and in having flowers with exserted styles, but the flowers are borne in lateral clusters of up to five, on very short pedicels, and the bracts are almost orbicular, shortly acuminate. Fruits ovoid, with a distinct neck, bristly, up to ½ in. long. An uncommon species in gardens, with pink flowers about 1½ in. wide, opening in late June or July, of laxer habit than R. *multibracteata*. A native of N.W. Yunnan, discovered and introduced by Forrest.

The closely related R. LATIBRACTEATA Boulenger has the same unusually shaped bracts, but the leaflets are larger, to 1 in. long, and the styles are not exserted. The flowering type-specimens of these two species were collected in the same locality, and both were originally identified as R. *multibracteata*.

R. MULTIFLORA Thunb.

R. *polyantha* Sieb. & Zucc., not Roessig

A wide-spreading bush, ultimately 10 to 15 ft high, sending out each year from the main body of the plant long arching stems which are clothed with blossom the following June. Branches glabrous, armed with small decurved prickles. Leaves 3 to 6 in. long, more on vigorous shoots. Leaflets seven or nine, 1 to 2 in. long, obovate or elliptic, acute or acuminate at the apex, more or less cuneate at the base, commonly downy beneath, but sometimes almost glabrous, simply toothed. Stipules deeply laciniated, and with glandular teeth. Flowers white, about 1 in. across, numerously borne in branching panicles 4 to 6 in. across and as much high. Pedicels and receptacle hairy, the former sometimes with glands as well as hairs. Sepals shorter than the petals even in the bud, with up to three narrow lateral appendages. Petals narrow, not much overlapping. Stamens golden yellow. Styles exserted, united into a glabrous column. Fruits ovoid to round, ¼ in. long, red, with the sepals fallen away. *Bot. Mag.*, t. 7119.

Native of Japan and Korea, with varieties in China and Formosa; long known in gardens by its double and variously coloured forms; the single-flowered type with white petals was introduced to France about 1860 and to Britain shortly after 1875. The distinctive mark of R. *multiflora* is the conspicuously laciniated stipules and, in the typical state, the rather rigid, paniculate inflorescence.

R. *multiflora* is one of the most beautiful of wild roses; of a robust and very graceful habit, a single bush grows 10 ft or more high, still more in diameter, every branch wreathed with blossom during June. The lower branches take root if resting on loose soil, and for ordinary purposes afford a sufficient means of increase. When more are needed they can be obtained from cuttings with the greatest ease. R. *multiflora* is useful for clothing high fencing, for planting on banks, and in any place where its vigorous growths can have ample space to develop. In the USA, where R. *multiflora* is more appreciated than here, other uses have been found for it—for the prevention of soil erosion; as a crash-barrier on the central reservation of motorways; and to provide sanctuary and food for birds. Both in Europe and America it is much used as a stock for grafting garden roses, especially ramblers.

var. CALVA Fr. & Sav. R. *calva* (Fr. & Sav.) Boulenger—Undersurface of leaflets glabrous, except for occasional down on the midrib. Pedicels glabrous, sometimes glandular. This variety occurs in China as well as in Japan.

cv. 'CARNEA'.—Flowers double, flesh-pink. Introduced from China by Thomas Evans of the East India Company in 1804 (R. *multiflora* Sims in *Bot Mag.*, t. 1059, not Thunb.; R. *multiflora* var. *carnea* Thory). This, the first form of R. *multiflora* to be introduced to Europe, is perhaps no longer in cultivation but similar plants are widely cultivated in China. It evidently derives from var. *cathayensis*.

var. CATHAYENSIS Rehd. & Wils. R. *cathayensis* (Rehd. & Wils.) Bailey R. *calva* var. *cathayensis* (Rehd. & Wils.) Boulenger—Flowers larger than in the type, up to 1¾ in. wide, rosy pink, few or many in flattish corymbs; pedicels glabrous but often densely glandular. Leaflets, according to the original description, varying from glabrous to densely hairy beneath. Stipules often less deeply incised than in the type. This variety was described from W. Hupeh, China where it is said to be common on streamsides and is also cultivated in Chinese gardens, in single- and double-flowered forms.

cv. 'GREVILLEI' ('Platyphylla'). SEVEN SISTERS ROSE.—An exceedingly vigorous rose with large, rather rugose leaflets. Flowers up to fifty in a cluster, but usually around thirty, of varying shades between cerise-purple and white. 'The most astonishing curiosity is the variety of colours . . . white, light blush, deeper blush, light red, darker red, scarlet, and purple—all on the same clusters.' So wrote R. Donald, proprietor of the Goldsworth Nursery in a letter to Loudon (*Gard. Mag.*, Vol. 1 (1828), p. 467). The shades listed by Donald are seven in number—whence, according to Loudon, the popular name 'Seven Sisters Rose' (actually taken from the Chinese name for this variety, probably given for the same reason).

'Grevillei' is said to have been introduced between 1815 and 1817 from Japan, but if the introducer was the Hon. Charles Greville, one of the founders of the Horticultural Society, as is usually supposed, the introduction must have been earlier, as he died in 1809. The botanical name for the Seven Sisters rose is *R. multiflora* var. *platyphylla* Thory (Redouté, *Les Roses*, Vol. 2, p. 67, t.). The plant so named and figured is certainly very like 'Grevillei' but has larger and fewer flowers. It was introduced to France by the rose-grower Noisette, who saw it growing in the garden of a 'maraicher' (market-gardener) near London, during his visit to this country in 1817. He was given the whole plant, which had been raised from seeds received from Japan. *R. multiflora* var. *platyphylla* would be the collective name for the Seven Sisters Rose, of which there are no doubt several slightly differing forms, long cultivated both in China and Japan. But the plant figured under that name in *Bot. Reg.*, t. 1372 (1830) is the introduction to Britain, i.e., 'Grevillei'. This is still in cultivation. Its long growths, which bear the next season's flowers, are apt to be cut by frost, so a sheltered warm position should be given to it.

cv. 'WILSONII'.—At the winding-up sale of Veitch's Coombe Wood nursery in 1913 Kew purchased a rose under the unpublished name *R. multiflora Wilsonii*. A very similar rose, also acquired from Veitch, was exhibited by Messrs Paul of Cheshunt in 1915 under the name "*R. Wilsonii*". Both are a good match for flowering and fruiting specimens collected by Wilson in 1900 in W. Hupeh, China, under number W.178, referred by Rehder and Wilson to var. *cathayensis*, and it is highly probable that the cultivated plants derive from seeds gathered at the same time. The Paul plant, which may have been propagated and put in the trade, had rather finely toothed leaflets, obovate or narrowly so or elliptic, acuminately tapered or acute at the apex, cuneate at the base, downy beneath on the midrib; rachis very glandular; stipules fringed; flowers white, single, about 2 in. across, in a fine broadly pyramidal truss; pedicels slender, glandular-bristly, 1¾ to 2 in. long. Fruits orange-red, globular, about ¼ in. wide.

For another "*R. Wilsonii*", see below under R. *gentiliana*.

Although usually considered to be a hybrid, the once popular 'CRIMSON RAMBLER' is probably a Sino-Japanese garden variety of R. *multiflora*. Being scentless and subject to mildew, it has dropped out of cultivation, but deserves mention as a parent of hybrids. It was introduced by Thomas Jenner of Easter Duddingston, Midlothian, as part of a consignment of Japanese plants which he received in 1878 from Prof. R. Smith, then Professor of Engineering at Tokyo University. Jenner named it 'The Engineer' in his friend's honour, and in 1889 gave stock to John Gilbert, a small Lincoln nurseryman, who exhibited it under this name in 1890, when it received an Award of Merit. Lacking resources for large-scale propagation, he passed the stock to Turner of Slough, who renamed this rose 'Crimson Rambler'—whence 'Turner's Crimson Rambler', as this rose was often called (*Gard. Chron.*, Vol. 16 (1894), pp. 248–9).

In Willmott's *The Genus Rosa* it is suggested that 'Crimson Rambler' is R. *multiflora* × R. *chinensis*, but the shallow toothing of the stipules in this variety, which apparently led him to this conclusion, is also seen in R. *multiflora* 'Carnea' and often in var. *cathayensis*.

Through various lines of descent, R. *multiflora* is a parent or ancestor of many modern garden roses, some of which, such as the Floribundas, lie beyond the scope of this work. The following rough and ready classification of the hybrids of R. *multiflora* may help to explain its role as a parent.

R. MULTIFLORA × R. GALLICA (or hybrid of R. *gallica*)—The rose 'DE LA GRIFFERAIE', once much used as a stock, shows the influence of both the suggested parents; see further on p. 178. Very similar to this is 'BYZANTINA', the Constantinople rose, which Dieck found in Bulgaria and put into commerce (*Gartenflora*, 1889, p. 159). The Belgian authority Crépin identified the Constantinople rose as R. *multiflora* × R. *gallica*, a parentage that Dieck found hard to accept, on the grounds that the rose had reached Bulgaria too early—by the 1820s—for R. *multiflora* to be a possible parent. It is, however, by no means impossible that cultivars of R. *multiflora* had reached the gardens of S.W. Asia from China at an early date. The 'Gul e Rescht' of Persian gardens, which Christ placed under R. *moschata* var. *nastarana* might well be a hybrid of R. *multiflora*, judging from its laciniated stipules and small double flowers.

R. MULTIFLORA × R. MOSCHATA/BRUNONII—Hybrids believed to be of this parentage form a rather miscellaneous assemblage. The oldest to have survived are 'THE GARLAND' and 'MADAME D'ARBLAY' (p. 201). 'BLUSH RAMBLER', of more recent origin, is a cross between 'The Garland' and 'Crimson Rambler' (see above), thus having R. *multiflora* in its parentage on both sides. In 1886 the French nurseryman Bernaix put into commerce 'POLYANTHA GRANDIFLORA', a seedling of R. *multiflora* of which the other parent was either R. *moschata* or one of its hybrids. A similar cross, using R. *brunonii* as the seed-parent, was made by Paul of Cheshunt and the result given the preposterous name R. *himalaica alba magna*; the flowers were semi-double, blush fading to white, and the epithet *magna* must have been given in allusion to its extraordinary vigour (*Gard. Chron.*, Vol. 62 (1917), p. 23). This is probably the rose now in commerce as 'PAUL'S HIMALAYAN MUSK RAMBLER'. The rose called R. *moschata floribunda*, distributed by Smith of Newry, is of similar parentage.

R. MULTIFLORA × HYBRID TEAS, HYBRID PERPETUALS, NOISETTES, ETC.—Rambling roses raised from these crosses have been largely superseded by those deriving from R. *wichuraiana*, but a few survive, mostly with 'Crimson Rambler' in their parentage.

But the groups to which R. *multiflora* has made its most important contribution are not of climbing habit. The beautiful Hybrid Musks (see p. 163) owe their clustered flowers more to R. *multiflora* than to R. *moschata*, despite their name.

In the second and later generations the Multiflora ramblers have given rise to the group known as the Dwarf Polyanthas or Polyantha Pompons (see p. 167). The modern Floribundas derive partly from these, by crossing with Hybrid Teas, etc., and partly from the Hybrid Musks, so thə contribution of R. *multiflora* to the bedding roses of today is a large one.

R. GENTILIANA Lévl. & Van. R. *multiflora* var. *gentiliana* (Lévl. & Van.) Yu & Tsai—Little is known of R. *gentiliana*, inadequately described from a specimen collected by the French missionary d'Argy in the Chinese province of Kiangsu, probably from a garden plant or escape, since the flowers were semi-double. The type-specimen was sent to Miss Willmott and a drawing made from it appears in *The Genus Rosa* (Vol. II, p. 513, t.); it was subsequently lost or destroyed. For R. *gentiliana sens*. Rehd. & Wils., see R. *henryi*, p. 72.

Boulenger identified as R. *gentiliana* a rose cultivated at Kew, which had been received from the National Rose Society before 1934 as "R. *wilsonii*", and his amplified description of this (?) species was based on this (*Bull. Jard. Bot. Brux.*, Vol. 14, p. 277 (1937)). Why he should have reached this conclusion is not clear, for this rose does not agree at all well with what is known of R. *gentiliana*. In some respects it resembles R. *multiflora* 'Wilsonii' but the stems have a mixed armature of short, hooked prickles and others needle-like or tending to bristles and the leaflets are sometimes roundish and purple beneath when young. The flowers are pink-flushed at first, in a denser truss, and the fruits are larger, about ½ in. wide. The styles are only partly united—a character which, with the mixed armature, suggests hybridity.

R. NITIDA Willd.

A low bush, rarely more than 2 ft high, freely suckering, with erect, often reddish stems, densely furnished with prickly bristles. Leaves 2 to 3 in. long, very shining green, becoming purplish red in autumn; stipules with glandular-toothed margins. Leaflets five to nine, narrow oblong, tapering at both ends, from ½ in. to 1¼ in. long, one-quarter to one-third as wide, finely and sharply toothed, glabrous all over, and of firm texture. Flowers bright rosy red, 2 to 2½ in. across, usually solitary, occasionally two to three together; flower-stalks and sepals bristly or glandular, the latter entire, lanceolate, and reflexed. Fruits globose, ⅓ in. wide, scarlet, bristly, with the sepals fallen away.

Native of eastern N. America; introduced in 1807. A charming little rose, very distinct among dwarf kinds by its shining, narrow leaflets, its very prickly stems, and highly coloured flowers. The leaves turn bright red in autumn.

R. NUTKANA Presl

R. *fraxinifolia sens*. Borrer in Hook., *Fl. Bor. Amer.*, not Borkh.

A robust shrub 6 to 10 ft high, its prickles stout, usually straight, with a broad flattened base, paired at the nodes, sometimes ½ in. long on the young barren stems, often absent from the flowering branchlets. Leaves 3 to 5 in. long; rachis glandular. Leaflets five to nine, elliptic or ovate, ¾ to 2 in. long,

typically edged with compound glandular teeth (but see var. *hispida*), glandular
and sometimes downy beneath. Flowers solitary or in twos or threes, bright
red, 2 to 2½ in. across. Receptacles and pedicels smooth or glandular-bristly.
Sepals 1 to 1½ in. long, narrow, with an expanded leaf-like apex, smooth or
glandular-bristly and more or less downy. Fruits globose or orange-shaped,
red, ½ to ⅝ in. wide, crowned with the long, erect sepals.

Native of western N. America; described in 1851 from a specimen collected
on Nutka Sound in 1791 by Haenke, botanist on the Malaspina expedition;
specimens collected later by Menzies, Douglas and others were identified as
R. *fraxinifolia*, which, in Lindley's sense, is R. *blanda*. There appears to be no
record of its cultivation in Britain before 1884. It is a handsome wild rose,
perhaps the handsomest of W. American species, and flowers and fruits well in
this country.

var. HISPIDA Fern. R. *spaldingii* Crép.; R. *macdougalii* Holzinger—Leaflets
simply toothed and eglandular. Prickles weaker, not much enlarged or flattened
at the base. Of more easterly distribution than the typical state and in places
intergrading with it (Cronquist in Hitchcock *et al., Vasc. Pl. Pacif. Northwest*,
Part 3 (1961), p. 170). Introduced to Kew before 1905.

'Cantab' (p. 175) is a hybrid of R. *nutkana*.

R. PENDULINA L.

R. *alpina* L.; R. *cinnamomea* L. (1753) (see R. *majalis*, footnote); R. *pyrenaica* Gouan;
R. *pendulina* f. *pyrenaica* (Gouan) R. Keller

A suckering shrub of variable habit, averaging 2 to 4 ft high, but occasionally
reaching 10 ft and sometimes very dwarf; stems and branches usually reddish
brown, often almost unarmed, prickles when present slender; sometimes the
stems and even the branchlets are quite densely furnished with needles and/or
bristles (see cv. 'Malyi'). Leaves 2 to 6 in. long, with mostly seven or nine
leaflets; rachis glabrous or downy, usually glandular. Leaflets variable in shape,
mostly elliptic to broadly so, and acute at the apex, glabrous above, under-
surface glandular or not, sometimes downy, teeth simple or compound, often
glandular. Stipules more or less glandular, the adnate part commonly widening
upwards, more rarely parallel sided. Flowers solitary or in twos or threes, deep
pink or purplish pink, 1½ to 2½ in. wide, opening in May. Pedicels and receptacle
smooth or clad with turpentine-scented glandular bristles. Sepals usually entire,
more or less expanded at the apex, smooth or glandular on the back, variable
in length. Stigmas hairy. Fruits red, flagon-shaped to roundish, usually constricted at the apex, smooth or bristly, often pendulous, up to 1¼ in. long,
crowned by the persistent sepals. *Bot. Mag.*, t. 6724.

Native of southern and central Europe, ascending in the Alps to almost
8,000 ft; cultivated since the 17th century. It is a rose of great interest to many
because of its unarmed condition, and is sometimes known as the 'rose without
a thorn', though this name was given originally to R. × *francofurtana*. Often
the only prickles are a few weak ones at the base of the branchlets. It has fine
foliage, and is also very handsome in fruit, the heps often being large, highly

coloured, and flask-shaped, as in R. *moyesii* and often in R. *macrophylla*, to both
of which it is closely allied. Another point of similarity to R. *macrophylla* is
that the pedicels, receptacles and sepals are often tinged with vinous red.

R. *pendulina* is a very variable species, but the variations are quite uncorrelated
and several of the numerous varieties that have been named can be seen in a
single stand. Three were once found on a single plant. The plate in the *Botanical
Magazine* represents a dwarf plant with glandular pedicels, in this respect
resembling the type of R. *pyrenaica* Gouan from the E. Pyrenees, but forms with
glandular pedicels are not confined to the Pyrenees, nor necessarily dwarf.

cv. 'MALYI'.—A compact bush 3 to 6 ft high, the stems armed towards the
base with short spines and bristle-like prickles. Leaflets oval or roundish,
½ to 1¼ in. long, mostly doubly toothed, glabrous on both surfaces. Flowers
deep red, 1½ in. across, usually solitary, sometimes in threes. Pedicel, receptacle
and the narrow-lanceolate sepals glandular. Fruits ¾ to 1 in. long, bottle-shaped,
red, crowned with the sepals (R. *malyi* Kerner; R. *pendulina* var. *malyi* R. Keller).

This pleasing rose was introduced to Austria shortly before 1869 by Hof-
gartner Maly from the mountains behind the Dalmatian Coast; some of the
plants went to the Innsbruck Botanic Garden, whence this variant was distributed. It was at one time thought to be a hybrid between R. *pendulina* and
R. *pimpinellifolia*, recalling the latter in its armature and the shape of the leaflets,
but similar plants occur in parts of Switzerland where R. *pimpinellifolia* is
absent. For genuine hybrids between these two species, see R. × *reversa*.

cv. 'NANA'.—A freely suckering selection, growing to about 1 ft high and
suitable for the rock garden (R. *alpina* var. *nana* F. Barker, *Gard. Chron.* Vol. 107
(1940), p. 34). The origin of this is unrecorded, but a dwarf form from the
Pyrenees was in cultivation as early as 1885.

Probably referable to R. *pendulina* is the old double-flowered Virgin rose,
which is the *Rosa sine spinis* of Parkinson's *Paradisus*. This was not the same
as R. × *francofurtana* (q.v.), which Parkinson also knew and described.

R. × INERMIS Thory R. *alpina* var. *turbinata* Desv.—According to Crépin,
this double-flowered rose, figured in Redouté (Vol. II, p. 93, t.) is probably a
hybrid between R. *pendulina* and R. *gallica*; it was at one time common in the
gardens of Switzerland but had become rare by the end of the 19th century.

R. × SPINULIFOLIA Dematra R. *spinulifolia dematrana* Thory; R. *vestita*
Godet—A natural hybrid between R. *pendulina* and R. *tomentosa*, first described
in 1818 from the Fribourg Canton of Switzerland. It is of no value for gardens,
but is of interest as one of the roses figured by Redouté in his famous work
(Vol. III, p. 7, t.). For a discussion of this hybrid see *Bull. Jard. Bot. Brux.*,
Vol. 12 (1932), pp. 391–403).

R. OXYODON Boiss. R. *alpina* var. *oxyodon* (Boiss.) Boulenger—An endemic
of the Caucasus, very closely allied to R. *pendulina* and differing from it, according to Boulenger, only in average characters such as having the flowers less
frequently solitary and sometimes as many as seven in the inflorescence, and
the receptacle usually smooth even when the pedicel is glandular.

R. PERSICA Michx. ex Juss.

R. *berberifolia* Pall.; *Hulthemia berberifolia* (Pall.) Dumort.; *H. persica* (Michx.) Bornm.; *Lowea berberifolia* (Pall.) Lindl.

A thin, straggling bush, 2 or 3 ft high, with slender, wiry, downy stems furnished with hooked spines and slender prickles, spreading by means of underground suckers. Leaves glaucous, simple (consisting of one leaflet), stalkless, obovate or oval, $\frac{1}{2}$ to $1\frac{1}{4}$ in. long, toothed towards the apex, covered with fine down. Flowers about 1 in. across, solitary at the end of the shoot on a slender, spiny stalk, the petals deep yellow with a crimson spot at the base; calyx-tube thickly covered with pale prickles $\frac{1}{8}$ in. long. Sepals lanceolate, downy, more or less prickly. Fruits globose or oblate, very prickly, crowned by the persisting sepals. *Bot. Mag.*, t. 7096.

Native of Iran, Afghanistan and of Russia (Central Asia and the steppe region of S.W. Siberia)*, often found on saline soils; introduced from Iran in about 1790. This remarkable rose is distinguished from all others by the undivided leaf, the absence of stipules, and the bicoloured petals recalling those of some halimiums. It was separated from *Rosa* as early as 1824 under the generic name *Hulthemia*, though it is open to doubt whether its differential characters are really weighty enough to justify so drastic a step, especially as this species hybridises fairly readily with other roses of the same chromosome number (diploid). The rank of subgenus given to it by Focke surely suffices to give recognition to its distinctness, as in the case of other aberrant roses (see R. *stellata* and R. *roxburghii*), and dispenses with the necessity for a third genus to accommodate the hybrids between R. *persica* and other roses.

With regard to the cultivation of R. *persica*, Lindley wrote: 'Drought does not suit it, it does not thrive in wet, heat has no beneficial effect, cold no prejudicial influence, care does not improve it, neglect does not injure it.' Lindley was inclined to exaggerate for the sake of effect, but it is a fact that this species has not survived in the open for more than a few years in our climate. Coming from arid regions with hot, dry summers, its failure with us is no doubt due to inadequate ripening of the wood and excessive winter wet.

A plant in a cool unshaded house at Kew succeeded for more than twenty years, planted near the glass in loam mixed with lime rubble. Out-of-doors it would be most likely to succeed in some sun-trap on a mound of loam and rubble, and covered with a glass light in winter. Of various modes of propagation tried with this rose, the only one that has succeeded is to sever the suckers from the main plant, and then allow them to remain undisturbed for several months, to form roots of their own before taking them from the soil. But fertile fruits are produced if two or more clones are grown together.

R. *persica* has always been a rare plant in gardens. It was re-introduced around 1968 by the late Alec Cocker, who shared plants and seeds with J. L. Harkness. The latter's interesting account of R. *persica* and the hybrids he has raised from it will be found in *The Rose Annual* 1977, pp. 121–4.

* Most Russian plants have glabrous stems and leaflets, and are separated from R. *persica* in the *Flora* of the Soviet Union as *Hulthemia berberifolia*. The typical downy form of R. *persica* also occurs in Russia near the borders with Iran and Afghanistan.

R. × HARDII Cels × *Hulthemosa hardii* (Cels) Rowley—A hybrid between R. *persica* and R. *clinophylla*. Leaves composed of from one to seven narrowly obovate leaflets, toothed, glabrous on both surfaces, stipular. Flowers 2 in. across, petals yellow with an orange spot at the base of each. Receptacle downy, with a few prickles. This hybrid, of which R. *clinophylla* is said to have been the seed parent, was raised in the Jardin de Luxembourg, Paris, by its Director J. A. Hardy. Apparently Hardy made several crosses between R. *persica* and other roses, of which Rivers saw the seedlings when he visited the Luxembourg garden in the late summer of 1835 (*Gard. Mag.*, Vol. 12 (1836), p. 226). R. × *hardii*, apparently the only one to be distributed, was acquired by the Paris nurserymen Cels Frères, who described it with a coloured plate in 1836 and took the opportunity of announcing that plants were available at 25 francs each (*Ann. Flore et Pomone*, 1835–6, p. 372). R. × *hardii* wants much the same treatment as R. *persica* in regard to warmth and sunshine and perfect root-drainage, but is hardier and more amenable to cultivation.

Two hybrids between R. *persica* and other roses have been found growing wild in Russia. Since *Hulthemia* is recognised in the *Flora* of the Soviet Union these are regarded as intergeneric hybrids, and for them the name × *Hulthemosa* was published in that work. Apart from these two wildlings, and R. × *hardii*, no hybrids of R. *persica* were known until Mr Harkness started to breed from it in the late 1960s. Some of these have flowered and are portrayed in the article cited above.

For a hybrid of R. *persica* found wild in Iran by the late Edward Hyams, see *The Rose Annual* 1979, pp. 167–71.

R. PIMPINELLIFOLIA L. BURNET ROSE

R. *spinosissima* L. (1771), not L. (1753); R. *scotica* Mill.; R. *spinosissima* var. *pimpinellifolia* (L.) Hook. f.

A dwarf bush with creeping roots, rarely more than 3 or 4 ft high in the typical state, with erect, short-branched stems covered with slender spines and stout bristles intermixed. Leaves closely set on the branches, 1 to 2½ in. long, composed of five, seven or nine leaflets, which are round or oval or broadly obovate, ¼ to ½ in. long, simply toothed, deep green, and except for the some-times downy midrib quite glabrous. Flowers 1½ to 2 in. across, white, creamy white or pale pink, solitary, borne in May. Pedicels and receptacle sometimes smooth, sometimes bristly or even prickly. Sepals usually entire and acute, woolly at the edge, smooth and glabrous on the back. Fruits dark brown, finally blackish, globose, ½ to ¾ in. wide, crowned with the sepals.

A species widely spread in the Old World from Europe (very rare in Scandinavia) to Asia Minor, the Caucasus, W. Siberia and Central Asia. In Britain it is com-monest near the sea, on fixed dunes, shingles and cliffs, often forming large colonies, but also occurs inland on limestone heaths and other dry open places. In gardens the species gives way as a rule to its numerous and variable progeny, some of which are very beautiful in their single or double flowers, either deep rose, white striped with rose, or pale creamy yellow. See further on p.161.

Yellow-flowered cultivars once placed under R. *pimpinellifolia* almost certainly owe their colouring to R. *foetida* (see R. × *harisonii*).

NOTE. The name R. *spinosissima* formerly often used for this species appears in the first edition of Linnaeus' *Species Plantarum* (1753). The nomenclatural diagnosis is taken from his earlier work *Flora Suecica* (1745), p. 407, and refers to a rose common on field-margins and in field-hedges in the Swedish province of Uppland and known there as the butter-hip rose in allusion to its pulpous fruits. This rose is R. *majalis* (*cinnamomea*), which Linnaeus had seen earlier during his journey to Lapland (*Fl. Lapponica* (1737), p. 203). In all the works cited, Linnaeus wrongly identifies the rose of Uppland and Lapland with the R. *campestris spinosissima flore albo odorata* of Caspar Bauhin's *Pinax* (1623), p. 483, which is cited as a synonym of R. *spinosissima* in *Species Plantarum*, and with the R. *pumila spinosissima foliis pimpinellifoliis glabris flore albo* of Jean Bauhin's *Historia Plantarum*, Vol. II (1651), p. 40. The rose so named by the Bauhins is indeed the Burnet or Scotch rose, R. *pimpinellifolia*, but this species is exceedingly rare in Scandinavia and had still to be discovered there in 1753.

Thus from the start R. *spinosissima* was a confused name. But in cases such as this it is the element actually known to and described by the author that must be taken as the type of the species, which in this instance is the rose of Uppland and Lapland—R. *majalis* (*cinnamomea*). The objection that Linnaeus described R. *majalis* in the first edition of the *Species Plantarum* as R. *cinnamomea* is invalid, since the rose so named is R. *pendulina*. It is the R. *cinnamomea* of the second edition that is the May or Cinnamon rose, which must now be known as R. *majalis* J. Herrm.

The unambiguous name R. *pimpinellifolia* was first published by Linnaeus in 1759 and is used in the second edition of *Species Plantarum* (1762). The R. *spinosissima* of this edition, as redefined by Linnaeus in 1759, is R. *pimpinellifolia*, so far as the diagnosis is concerned, though elements of the old confusion remain. This revised R. *spinosissima* differed from R. *pimpinellifolia* in having prickly peduncles against smooth in the latter, and some later botanists continued to recognise both species, adding other supposed differences.

In one of his last works, the *Mantissa Altera* of 1771, Linnaeus gives an excellent description of the Burnet rose under the name R. *spinosissima* and comes near to making R. *pimpinellifolia* a synonym of it. It is therefore not without good reason that some botanists have upheld the name R. *spinosissima*, and Crépin was perhaps a little unjust in writing of them 'Ils n'avaient pas étudié les faits avec assez d'attention.' (*Bull. Herb. Boiss.*, 2nd ser., Vol. 5, p. 143). But the fact remains that the R. *spinosissima* L. of the first edition of *Species Plantarum*, where botanical nomenclature starts, is R. *majalis*. Pointing this out in 1820, the Swedish botanist Wikström proposed that the name R. *spinosissima* should either be applied to that species or rejected '*in aeternum*' as a source of confusion (*Kongl. Vet. Acad. Handl.*, 1820, p. 268).

var. ALTAICA (Willd.) Thory   R. *altaica* Willd.; R. *spinosissima* var. *altaica* (Willd.) Bean—R. *altaica* Willdenow (1809) is a renaming of the R. *pimpinellifolia* of Pallas in *Flora Rossica*, Vol. II (1789), p. 62 and t. 75, which is a Siberian form of the species resembling 'Hispida' in its bristly but not prickly stems and its yellowish white flowers. Thory rightly placed R. *altaica* under R. *pimpinellifolia* as a variety, but Seringe, in de Candolle's *Prodromus*, gives a diagnosis of var. *altaica* that seems to be based mainly on R. *grandiflora* Lindl., which he cites as a synonym. The R. *spinosissima* var. *altaica* of previous editions of the present work, and of Rehder's *Manual* (1940 edition) is wholly R. *grandiflora* Lindl. except for the citation R. *altaica* Willd. But Lindley's plant (see 'Grandiflora') is at the opposite pole to R. *altaica* Willd. so far as armature is concerned. See further under 'Hispida'.

cv. 'GRANDIFLORA' ("Altaica").—A suckering shrub up to 6 ft high, differing from R. *pimpinellifolia* as usually seen in Europe not only in its large size but also in the absence or comparative scarcity of bristles among the prickles of the stems. The flowers are 3 in. across, creamy white, and the leaflets up to 1 in. long. *Bot. Reg.*, t. 888. (R. *grandiflora* Lindl.; R. *spinosissima* var. *altaica* (Willd.) Bean, not R. *altaica* Willd.; see further under var. *altaica*).

A group of this rose when in full bloom at the end of May makes a very beautiful picture. It was introduced from Siberia before 1820.

cv. 'HISPIDA'.—A shrub up to 6 ft high, with sturdy, erect stems densely covered with slender, brown bristles. Leaflets five to eleven, ¾ to 1¼ in. long, ⅜ to ½ in. wide, glabrous. Flowers yellow at first, changing to creamy white, 2 to 3 in. wide, opening the third or fourth week in May. *Bot. Mag.*, t. 1570. (R. *hispida* Sims (1813), not Muenchh. (1770); R. *lutescens* Pursh; R. *spinosissima* var. *hispida* (Sims) Koehne).

There was some doubt as to the origin of this rose, which was at one time called the 'Yellow American rose' and was included by Pursh in his *Flora* of North America, but there is every reason to believe that it came from Siberia. It has been raised from seed at Kew and comes quite true, which would appear to show that it is not of hybrid origin. It is one of the most lovely of single roses, but uncommon in cultivation.

At the end of the 18th century plants were raised in Sweden from seeds received from Russia, and were named R. *ochroleuca** by Swartz, who further distributed plants and seeds. The first full description, accompanied by an excellent portrait, was published by Wikstrom in *Kongl. Vetensk. Acad. Handl.* 1820, p. 268 and t. III. Judging from these, R. *ochroleuca* was very similar to 'Hispida', as indeed Wikstrom pointed out.

var. LUTEA Bean (in R. *spinosissima*)—See 'Lutea Maxima', mentioned under R. × *harisonii*.

var. LUTEA PLENA Hort.—See 'Williams's Double Yellow', p. 204.

f. LUTEOLA (Andr.) Rehd. (in R. *spinosissima*) R. *spin. luteola* Andr., *Roses*, Vol. II, t. 128.—The rose portrayed by Andrews was flowering in Knight's nursery, Chelsea, in 1821, and was said to be a native of Scotland. The flowers were yellowish, and the armature not unlike that of 'Hispida', but the leaflets were uncharacteristically elongate for R. *pimpinellifolia*, and the flowers were borne in July with a repeat in the autumn, which rules out the possibility of its being any form of that species. It was probably a hybrid, raised in Scotland.

var. MYRIACANTHA (DC.) Ser. R. *myriacantha* DC.—A very distinct variety, with the habit and flowers of ordinary R. *pimpinellifolia*, but with longer and

---

* The name R. *ochroleuca* has been variously used in gardens. The rose grown by Canon Ellacombe under this name seems, judging from herbarium specimens, to have been a beautiful rose, with flowers in twos or threes up to almost 4 in. across, but is obviously a hybrid. Plants distributed commercially as R. *ochroleuca* are mentioned on p. 101 under 'Lutea Maxima'. R. *spinosissima* var. *ochroleuca* (Swartz) Baker, in Willmott, *The Genus Rosa*, Vol. II, p. 255, is a compound of four discrepant elements: R. *ochroleuca* Swartz; the R. *spinosissima luteola* of Andrews (see f. *luteola*); R. *xanthina* (to which the specimen from N. China cited by him belongs); and the rose cultivated by Miss Willmott as R. *ochroleuca*, which is portrayed in the facing plate and is of uncertain identity.

more numerous spines. The best distinction, however, is furnished by the numerous glands on the leaves beneath as well as on the rachis, stipules, pedicels and sepals, and by the compound toothing of the leaflets. Described from southern France. Glandular plants occur elsewhere in the southern part of the area of R. *pimpinellifolia*, but do not always have the formidable armature of the type.

f. MEGALACANTHA Borbas ex Dengler—Prickles long, much flattened, with an elongated base. A variant of this nature was mentioned in previous editions as var. *macracantha*, an unpublished name written on a specimen in the Kew Herbarium, collected near Gap in the Alpine region of S.E. France, which has flat, rigid spines ⅝ in. long and ¼ in. wide at the base.

Numerous natural hybrids of R. *pimpinellifolia* are known, of which the following are treated in this work:

R. *pimpinellifolia* × R. *canina* and related species. See R. × HIBERNICA.
   ,,   ,,   × R. *sherardii* and related species. See R. × INVOLUTA.
   ,,   ,,   × R. *glauca*. See under the latter species.
   ,,   ,,   × R. *pendulina*. See R. × REVERSA.
   ,,   ,,   × R. *foetida*. See R. × HARISONII.
Crossed with Hybrid Teas, R. *pimpinellifolia* 'Grandiflora' has given rise to some of the finest of the shrub roses. See in the second section: 'Frühlingsanfang', 'Frühlingsgold', 'Frühlingsmorgen', 'Karl Foerster' and 'Maigold'.

R. PISOCARPA A. Gray

A shrub 3 to 6 ft high in the wild, sometimes taller; branches slender, unarmed or with a few straight prickles, which occur in pairs at the nodes. Leaflets five or seven, sometimes nine, oval or ovate, ½ to 1⅝ in. long, simply and finely toothed, downy beneath. Flowers 1 in. or rather more across, with rounded, overlapping, bright rosy petals, occurring in clusters of as many as four or five, but sometimes solitary. Pedicels and receptacle glabrous. Sepals slender, slightly expanded at the apex, glandular-bristly on the back. Fruits purplish red when fully ripe, globose or ellipsoid, up to ½ in. wide; sepals persistent. *Bot. Mag.*, t. 6857.

Native of western N. America from British Columbia to N. California. An interesting and brightly coloured rose, with small, nearly always clustered flowers. In the form figured in the *Botanical Magazine* the fruits are about the size of a pea, as in the type (whence *pisocarpa*), but they can be ½ in. or even slightly more wide.

R. × POLLINIANA Spreng.

R. *hybrida* Schleich.; R. *gallica* var. *hybrida* (Schleich.) Ser.; R. *arvensis* var. *hybrida* (Schleich.) Lindl.; R. *geminata* Rau; R. *arvina sens.* Boreau, not Krocker; R. *gallicoides* Déségl.

A hybrid between R. *arvensis* and R. *gallica*, occurring occasionally in the wild; it was first recorded by the Italian botanist Pollini growing at the foot

of Monte Baldo in northern Italy. The stand most frequented by the French
rhodologists of the 19th century grew near Lyons, and many names were
given to the forms that occurred in it. It has also been found in Switzerland,
Germany, N. Yugoslavia, etc. It is a variable hybrid, partly no doubt because
the parents are variable, partly, according to Crépin, because the products of
the cross are fertile.

The date of introduction of R. × *polliniana* is uncertain, but it was cultivated
by Canon Ellacombe at the Bitton Vicarage in the 1880s as R. *hybrida*. The
clone now cultivated makes a spreading, mounded shrub or a climber, its stems
armed with a few short curved prickles. Leaflets three or five, to about 2 in.
long, elliptic, oblong-elliptic or slightly obovate, coarsely toothed. Flowers
appearing at midsummer, fragrant, blush-coloured from pink buds, about 2 in.
across, borne in terminal cymes supplemented by flowers from the upper leaf-
axils. Pedicels up to $1\frac{1}{2}$ in. long, glandular. Sepals smooth on the back, variable
in length, the longer ones expanded at the apex, with a few lateral appendages.
Styles long-exserted, the stigmas almost at the same level as the anthers.
Willmott, *The Genus Rosa*, p. 333, t.

R. PRATTII Hemsl.

A shrub 4 to 8 ft high in the wild; prickles narrow, straight, slightly dilated
at the base. Leaflets five to seven, elliptic or lanceolate, rarely obovate, to about
$\frac{5}{8}$ in. long, acute, downy beneath, obscurely toothed. Flowers up to seven or so
in a corymbose cluster, pink, about 1 in. wide; pedicels slender, $\frac{1}{2}$ to $\frac{5}{8}$ in. long,
clad like the receptacle with gland-tipped bristles. Sepals entire, abruptly
narrowed at the apex, contracted at the base. Fruits orange-red or scarlet.

Native of W. Szechwan, where it is common in thickets at altitudes of
7,000 to 11,000 ft; discovered by A. E. Pratt and introduced by Wilson in 1903.
A pretty rose, now uncommon in gardens, allied to R. *davidii*.

R. PULVERULENTA Bieb. (1808)

R. *glutinosa* Sibth. & Sm. (1809); R. *dalmatica* Kern.; R. *glutinosa* var. *dalmatica* (Kern.)
R. Keller

A shrub of dwarf, compact, bushy habit, whose stems are copiously furnished
with stiff, whitish, straight or decurved prickles up to $\frac{3}{8}$ in. long, intermixed
with which are numerous small needle-like prickles and glandular bristles.
Leaves $1\frac{1}{2}$ to 3 in. long, pine-scented and often sticky, owing to the dense
glandularity of the rachis and leaflets, which are mostly five or seven in number,
rarely nine, oval or obovate to roundish, $\frac{1}{4}$ to 1 in. long, glabrous or more or
less downy, glandular on both sides, edged with compound glandular teeth.
Stipules glandular, broad, with short, triangular tips. Flowers rosy pink,
1 to $1\frac{1}{2}$ in. across, usually in pairs or solitary; pedicels $\frac{1}{2}$ to $\frac{3}{4}$ in. long, usually
densely covered with stalked glands and sometimes downy. Sepals up to 1 in.
long, slightly expanded at the apex, with a few slender, gland-edged appendages.
Styles hairy. Fruits globose or ellipsoid, or broadest slightly above or below
the middle, dark red, up to 1 in. long, smooth or glandular-bristly; sepals

usually persistent. *Bot. Mag.*, t. 8826.

Native of S. Europe from Italy and Sicily eastwards through S.E. Europe and Crete to Asia Minor, the Caucasus, the Lebanon, Iran and Afghanistan; introduced early in the 19th century. It is remarkable for its excessive covering of glandular hairs or bristles, more marked even than in R biebersteinii (*horrida*), from which it differs also in its hairy styles, very wide stylar aperture, fruits with usually persistent sepals and pink flowers. R. *sicula* has persistent sepals, but its wood lacks the bristles and needles so characteristic of R. *pulverulenta*.

R. *pulverulenta* is a variable species in such characters as the length of its prickles, presence or absence of down on the leaflets, the presence or absence of glandular bristles on the fruits and the size and shape of these. The plants portrayed in the *Botanical Magazine* and in Willmott, *The Genus Rosa* (p. 467, t.) came from the Darmstadt Botanic Garden; they have large ellipsoid densely hispid fruits and may derive from an introduction from the mountains above Kotor in S. Dalmatia, shortly before 1870, to the Vienna Botanic Garden. These plants have been distinguished as var. *dalmatica*.

R. × RECLINATA Thory

R. × *reclinata* is the name given by Thory in 1824 to the Boursault rose, figured in Redouté, *Les Roses* (Vol. III, p. 80, t.); also portrayed, on the previous plate is another rose, received from the nurseryman Cugnot, from which the Boursault rose is said to have been raised. As described, the latter is a sparsely armed climber bearing semi-double purplish pink flowers in clusters at the ends of long branchlets, the individual flowers nodding in the bud-stage, whence the epithet *reclinata*. Receptacle globose. Leaves glabrous, glossy, with three to seven leaflets. The other rose, from which the Boursault was said to have been raised, was identical according to Thory, except in having single flowers. He supposed these two roses to be hybrids between R. *indica* (*chinensis*) and R. *alpina* (*pendulina*).

This parentage is also suggested by Thory for the rose he named R. LHERI-TIERANEA (for the French botanist L'Héritier), which is figured by Redouté in the same volume (p. 21, t.). This had been raised by Vilmorin about 1812 from seed of "R. *indica*", and seems to have been very similar.

The Boursault rose, and perhaps Vilmorin's hybrid, gave rise to a small group of climbing roses called the Boursault roses, most of which, like the original parents, have disappeared from cultivation. See further on p. 161.

Whether R. *pendulina* really enters into the parentage of the original Boursault, or of 'Lheritieranea' is rather doubtful. The first to question Thory's suggestion was his contemporary Seringe, who thought R. *indica* × R. *fraxinifolia* [*blanda*] as more likely for the former, and actually placed the latter under R. *fraxinifolia* as a variety.

R. × REVERSA Waldst. & Kit.
R. *rubella sens.* Godet, not Sm.

A naturally occurring hybrid between R. *pendulina* and R. *pimpinellifolia*, combining in various ways the characters of the two parents. It resembles

R. *pimpinellifolia* in habit and armature, but grows taller than the western form of that species (the type, introduced from the Matras Mountains of N. Hungary, attained 5 ft in cultivation, though the parent plants were of lower growth). The leaflets are sometimes narrower than in the Burnet rose, with more numerous, slightly compound teeth. Flowers nearly always solitary, in shades ranging from white to rosy pink. Fruits usually pendulous, as in R. *pendulina*, but darker and sometimes almost black.

There has been confusion between this hybrid and forms of R. *pendulina*, which sometimes has an armature resembling that of R. *pimpinellifolia* (see 'Malyi' under the former species).

R. *rubella* Sm. is not this hybrid but a form of R. × *involuta* or, according to some authorities, a colour variant of R. *pimpinellifolia*.

R. RICHARDII Rehd.
R. *sancta* Richard, not Andr.

As seen in cultivation this rose is a low, rather open bush, whose weakish stems have a few hooked, scattered prickles of unequal size. Leaflets three or five, ovate or oblong, 1 to 2 in. long, often blunt at the apex, simply toothed, bullate above, hairy beneath; rachis downy and more or less prickly. Stipules broad, edged with glands. Flowers 2 to 3 in. across, pale rose, produced several together in a loose cluster, each flower on a slender glabrous and smooth stalk 1 to 2 in. long. Sepals downy and glandular, very large and pinnately lobed, the largest being $1\frac{1}{2}$ in. long, and $\frac{5}{8}$ in. wide at the base, with broad, leafy points. Styles hairy, free, long exserted. Willmott, *The Genus Rosa*, p. 337, t.

This rose, known only from cultivation, was described in 1847 from Abyssinia, where it is grown in the vicinity of churches and in the courtyards of monasteries—whence the epithet *sancta* used by Richard. How and when it reached Europe is not certain, but the firm of Dammann at Naples were propagating it by 1896; the nurseryman George Paul of Cheshunt had it by 1897 and sent a flowering specimen to Kew in December of that year, to show its perpetual blooming.

According to the botanist Schweinfurth, the vernacular name used for this rose in Tigre province is of Greek origin. It is certainly a rose of great antiquity, already cultivated early in the Christian era in the Fayyum oasis, south-west of Cairo, to which it may have been introduced from Greece or the Near East via Alexandria in Ptolemaic times. The evidence for its presence there lies in the remains of roses forming the funerary chaplets of mummies found by the British archaeologist Flinders Petrie in the Pyramid of the Labyrinths at Hawara, dating from some time between the 2nd and 5th centuries A.D. The fragments were sent to the Belgian rhodologist Crépin and were identified by him as R. *sancta*. The Fayyum oasis was later a centre of Coptic christianity, with which the Abyssinian Church had close relations.

R. *richardii* (*sancta*) is near to R. *gallica*. Its armature, and the long-exserted styles, suggest that the other parent was a member of the *Synstylae*—either R. *moschata* or, as Hurst suggested, R. *phoenicea*. Such a parentage would make it a sort of Damask rose.

R. ROXBURGHII Tratt.   BURR ROSE   [PLATE 15

R. *microphylla* Roxb. ex Lindl. (1820), not Desf. (1798); R. *forrestii* Focke; R. *roxburghii* var. *hirtula* (Reg.) Rehd. & Wils.; R. *microphylla* var. *hirtula* Reg.; R. *hirtula* (Reg.) Nakai

A sturdy bush that, in its normal single-flowered form, grows up to 10 ft high and as much in width; bark grey or fawn, peeling; branches stiff, armed with a few rigid, straight prickles in pairs. Leaves 2 to 4 in. long, consisting of nine to seventeen or even nineteen leaflets; rachis downy and with a few prickles. Leaflets elliptic, ovate or oblong-ovate, up to 1 in. or slightly more long, obtuse or acute at the apex, rounded to cuneate at the base, glabrous on both sides or downy beneath, simply toothed. Flowers usually solitary, delicate rose, 2 to 2½ in. across, pleasantly fragrant (for the flowers of the type, which are double, see below); pedicels and receptacle prickly. Sepals broadly ovate, lobed, downy. Fruits flattened, tomato-shaped, 1½ in. across, very spiny, yellowish green, fragrant.

Native of China and Japan; the type of the species was a double-flowered garden variety introduced from China to the Calcutta Botanic Garden, where it had long been cultivated (see below). The next introduction was of the Japanese race sometimes distinguished as var. *hirtula*, to which most single-flowered plants in gardens probably belong (*Bot. Mag.*, t. 6548 (1881)). The wild single-flowered form of China (f. *normalis* Rehd. & Wils.) was introduced by Wilson, according to whom it is abundant by waysides and in semi-arid river-valleys throughout the warmer parts of W. Szechwan.

R. *roxburghii* is a most distinct rose, with its peeling bark, its small, numerous leaflets, and especially by its large, spiny, apple-like fruit with no hint of red in it even when fully ripe. In the leafless state its open habit, stiff branches and peeling bark scarcely suggest a rose. The flowers tend to be concealed by the foliage but are deliciously fragrant and much visited by bees.

cv. 'ROXBURGHII' ('Plena').—As mentioned above, the type of R. *roxburghii* is a double-flowered form of Chinese gardens, cultivated in the Calcutta Botanic Garden. This was introduced, probably direct from China, and first flowered in Colvill's nursery, Chelsea, in 1824. The flowers are fully double, the outer petals light pink, the inner darker (*Bot. Reg.*, t. 919; *Bot. Mag.*, t. 3490). It is much less robust than the common single-flowered form.

R. *roxburghii* has produced no notable hybrids, though it has been crossed to a limited extent. It is interesting that it has had some influence on the Floribunda race, through a cross with the Floribunda 'Baby Château', which produced 'Cinnabar', 'Floradora' and 'Kate Duvigneau'. For its hybrids 'Jardin de la Croix' and 'Triomphe de la Guillotière', not treated here, see G. S. Thomas, *Shrub Roses*, p. 180, and his *Climbing Roses*, p. 154, respectively. For 'Coryana', a seedling of R. *roxburghii*, see p. 178. For R. *roxburghii* × R. *rugosa* see R. × *micrugosa* under the latter species.

R. RUBIGINOSA see R. EGLANTERIA

R. RUBUS Lévl. & Van.*

R. *ernestii* Stapf ex Bean; R. *ernestii* f. *velutescens* & f. *nudescens* Stapf; R. *moschata* var.
hupehensis Pampan.

A vigorous shrub of spreading or semi-scandent habit, 8 to 15 ft high;
young shoots hairy or almost glabrous, armed with hooked spines, often
purplish. Leaves 4 to 9 in. long, composed usually of five leaflets, reduced to
three under the inflorescence; rachis prickly, slightly downy and glandular.
Leaflets elliptic-ovate to oblong-obovate, acute or shortly acuminate, $1\frac{1}{2}$ to $3\frac{1}{2}$ in.
long (longer on sterile shoots), simply and sometimes deeply serrate, greyish
and hairy beneath, rarely quite glabrous, sometimes strongly tinged with purple
when young. Flowers white, about $1\frac{1}{2}$ in. across, with broad, overlapping
petals, borne in late June, July or early August, in dense clusters. Pedicels
short (barely 1 in. long) and moderately thick, they and the receptacle clad
with hairs and stalked glands. Sepals not much longer than the ovoid flower-bud,
downy and glandular on the back, entire or with a few slender appendages.
Styles united in a shortly exserted downy column. Fruits globular, $\frac{3}{8}$ to $\frac{1}{2}$ in.
wide, red.

Native of western and central China; discovered by Henry about 1886, but
the type of the species was collected in Kweichow. Wilson introduced it from
Hupeh under two numbers: W.473c, with leaflets velvety beneath, as in the
type of R. *rubus*, and W.666a (R. *ernestii* f. *nudescens* Stapf), in which they are
nearly glabrous. It was also raised from seeds collected by Farrer in Kansu
(Farrer 786) but his no. 291, distributed as R. *rubus*, is R. *filipes*. Although
hardy and vigorous, with attractively tinted young foliage, R. *rubus* has always
been rare in gardens.

R. *cerasocarpa* Rolfe is very near to R. *rubus* and probably no more than a
glabrous form of it.

R. MULLIGANII Boulenger—Near to R. *rubus*, but the leaflets sometimes
seven and the flowers larger on more slender pedicels up to $1\frac{1}{2}$ in. long, in a
laxer inflorescence. It was raised in the R.H.S. Garden at Wisley from seeds
collected by Forrest during his 1917–9 expedition to Yunnan and was named
by Boulenger in honour of Brian Mulligan, then Assistant to the Director at
Wisley, who sent him specimens for identification.

A rose distributed as R. *longicuspis* is near to R. *mulliganii* and possibly from
the same batch of seed.

R. RUGOSA Thunb.

R. *ferox* Lawrance

A shrub 4 to 6 ft high, and one of the sturdiest of roses. Stems stout, densely
covered with prickles of unequal size, the largest $\frac{1}{3}$ to $\frac{1}{2}$ in. long, they, as well

* In the article accompanying *Bot. Mag.*, t. 8894, Dr Stapf argued that the name
R. *rubus* should be dropped, on the grounds that Léveillé had described the styles as
free, and proposed the name R. *ernestii* in its place. But the name R. *rubus* is accepted by
Rehder (*Journ. Arn. Arb.*, Vol. 13, p. 312). The type of R. *rubus* in the Léveillé herbarium
is R. *rubus* as understood here.

as the stem itself, downy. Leaves 3 to 7 in. long, with large downy stipules.
Leaflets five to nine, oblong, 1 to 2 in. long, shallowly toothed except towards
the base, downy beneath, the very conspicuous veins giving them the wrinkled
appearance to which the specific name refers; common stalk downy and armed
with hooked spines. Flowers very fragrant, 3½ in. across, purplish rose, produced singly, or a few in a cluster from early summer onwards; petals overlapping. Receptacle glabrous, but the flower-stalk and sepals downy, the latter
1 to 1¼ in. long. Fruits rich bright red, tomato-shaped, 1 in. or more in diameter,
crowned with the sepals.

Native of the Russian Far East, Korea, Japan and N. China, commonest in
sandy soils near the sea; described in 1784 by Thunberg, who gives the Japanese
name as 'Ramanas'—a probable slip of the pen or misprint for 'Hamanas',
the actual name used for this rose in Japan being 'Hama-nashi' or shore-pear.
R. rugosa was in cultivation in Britain at the end of the 18th century under the
name R. ferox, though how and whence it arrived is uncertain (see further
under var. ventenatiana). If ever common in gardens in the earlier 19th century,
it must have become rare by 1870, for when introduced about that time, as
R. regeliana, it was hailed as something new and soon became valued for its
handsome foliage, long flowering period, fine fruits and ease of cultivation. It
was even said to have stimulated a taste for single-flowered roses, then quite
thrust aside by the productions of the commercial breeder.

R. rugosa is said to have been cultivated since A.D. 1100 in China, where
the ladies of the Court long prepared a kind of potpourri from its petals mixed
with camphor and musk. But the forms introduced in the last century came
from Japan, where too R. rugosa has long been grown and many colour forms
selected, varying from crimson to pink and white. No rose hybridises more
readily with others, and if seed be sown from plants growing with or near other
roses, little of the progeny comes true. The consequence was that a worthless
lot of mongrels appeared, some of which were named, but ought never to have
been allowed to survive their first flowering. But, for the breeder, R. rugosa
has important qualities, notably its great hardiness, and disease-resistant foliage.
Many deliberate crosses have been made between it and garden roses, some of
which are described by Graham Thomas in his section starting on p. 169.
Most of these hybrids are highly sterile, but a notable exception is R. kordesii,
mentioned below.

cv. 'ALBA'.—Flowers blush in bud, opening pure white, single. Vigorous
and free-fruiting. The origin of this form is uncertain; it may descend from the
white-flowered seedling raised by the nurseryman Ware of Tottenham early in
the 1870s.

var. VENTENATIANA C. A. Meyer R. rugosa var. kamtschatica (Vent.) Reg.;
R. kamatchatica Vent.—This differs from typical R. rugosa in having the stipular
(nodal) prickles distinct from the more scattered bristly ones, in the leaves
being more obovate and rounded at the apex, and in the smaller fruits. It is a
distinct race, confined to S. Kamchatka, and is considered by some authorities
to be a natural hybrid between R. rugosa, which is found on the peninsula, and
R. amblyotis, a close relative of R. majalis. It was described in 1800 from a plant
cultivated by the French nurseryman Cels, and was probably introduced around

1770, though by what means is not known. Since R. *rugosa* itself also grows on Kamchatka it is possible that the first introduction of the species, known as R. *ferox*, came at the same time.

R. 'CALOCARPA'.—A handsome rose with branches less thick than in R. *rugosa*, clusters of bright red fragrant flowers borne over a long period, followed by globose, scarlet fruits ¾ in. wide, crowned with the sepals. Raised by Bruant of Poitiers and sent out about 1891. A hybrid of R. *rugosa* crossed with "R. *indica*", possibly some form of Crimson China. It was stated by the raiser to come true from seed, a statement of very doubtful validity (R. *rugosa calocarpa* André; R. × *calocarpa* Bak. in Willmott).

R. 'HOLLANDICA'.—A hybrid of R. *rugosa*, much used as a stock for garden roses, especially standards. It is a suckering shrub with mauvish pink, almost single flowers, of no value as an ornamental, though often seen in neglected gardens. It flowers from summer to autumn.

R. × IWARA Sieb. ex Reg. R. *yesoensis* (Franch. & Sav.) Makino; R. *iwara* var. *yesoensis* Franch. & Sav.—A natural hybrid between R. *rugosa* and R. *multiflora*, occurring occasionally in the wild. Of spreading habit, intermediate between the parents. The introduced form (as R. *yesoensis*) had small white flowers and could be described as two beautiful species spoilt. ·

R. × JACKSONII Baker in Willmott, *Gen. Rosa*, Vol. I, p. 63, t. (R. *rugosa* × R. *wichuraiana*)—This cross was made at the Arnold Arboretum by Jackson Dawson, towards the end of the 19th century. Here belong 'Lady Duncan', named by Dawson, and 'Max Graf', of independent origin (see further on p. 190 and also R. *kordesii* below).

R. × MICRUGOSA Henkel R. *vilmorinii* Bean; R. *wilsonii* Hort., not Borrer (R. *rugosa* × R. *roxburghii* (*microphylla*))—One of the best primary crosses, being intermediate in habit and foliage, and having large, single, pale pink flowers 4 or 5 in. across. The fruits are greenish and bristly. It was found as a self-sown seedling in the garden of the Strasbourg Botanical Institute and described by Henri de Vilmorin in *Revue Horticole* 1905, p. 144. From the original hybrid Dr Hurst raised R. × *micrugosa* 'ALBA' with white, fragrant flowers borne over a long period. The growth is more erect, and the foliage lighter green.

R. × PAULII Rehd. (R. *rugosa* × ? R. *arvensis*).—See 'Paulii', p. 194.

R. KORDESII H. D. Wulff—This rose, raised by the famous German breeder Wilhelm Kordes, is an interesting example—one of many now known—of how what is in effect a new species can arise in cultivation as a result of hybridisation followed by a doubling of the number of chromosomes. The parent of R. *kordesii* is 'Max Graf', a very vigorous and hardy hybrid between R. *rugosa* and, probably, R. *wichuraiana*, both diploid. Herr Kordes acquired a plant as soon as 'Max Graf' was put into commerce in 1919 and, seeing its potentialities as a parent, attempted to breed from it. But, like most hybrids of R. *rugosa*, it proved to be sterile both in crossing and when selfed. However, two fruits were eventually obtained, from which two seedlings were raised. One of these proved to be completely fertile and was found, on cytological examination, to be tetraploid. This was named R. *Kordesii* in 1951. Using R. *kordesii* as a parent, Herr Kordes raised a group of repeat-flowering climbers with glossy, disease-resistant foliage, of which several are described in the second section (W. Kordes, in *The Rose Annual* 1965, pp. 99–102).

R. SEMPERVIRENS L.

R. *scandens* Mill.

A scrambling or prostrate shrub; stems up to 20 ft long, armed with small, hooked, rarely straight, prickles, sometimes unarmed. Leaves persisting through the winter, with three to seven leaflets (commonly five on the flowering branchlets); these are rather thick, usually glossy on both sides, glabrous except for the sometimes hairy midrib beneath, elliptic to narrow-ovate, usually acuminate at the apex, 1 to 1½ in. long, the terminal leaflet usually distinctly longer than the upper laterals, edged with mostly twenty to thirty teeth, which are sometimes compound and glandular. Stipules glandular at the edge, not toothed. Flowers white, usually fragrant, borne in June (sometimes later) in round-topped or pyramidal panicles. Flower-buds broadly ovoid. Pedicels ½ to 3½ in. long, they and the globose or ovoid receptacle usually clad with stalked glands. Sepals elliptic, commonly short-pointed and entire, glandular on the back. Styles more or less united in an exserted column, hairy. Fruits smooth or glandular, globose or ovoid, ⅜ to ⅝ in. long, devoid of sepals.

A native of S. Europe, extending in France to the Loire valley; and of N. Africa and Asiatic Turkey; long cultivated, though never frequent in British gardens. This species and R. *arvensis* are the European representatives of the section *Synstylae* and are closely related, so much so that in southern Europe, where R. *arvensis* is more variable than in Britain, the two species are sometimes difficult to tell apart. The most reliable differences are, according to Boulenger, that in R. *sempervirens* the leaves are usually persistent, the leaflets thicker and more lustrous, never hairy at the edge, with more numerous teeth (mostly twenty to thirty on each side on the largest leaflets, against ten to nineteen in R. *arvensis*), sepals glandular on the back, and the stylar column hairy.

R. *sempervirens* is of interest as the European counterpart of the Japanese R. *wichuraiana*, but is unlikely to be of any greater value in gardens than that species or R. *multiflora*. It has, however, made its contribution as a parent of some old hybrid ramblers, such as 'Félicité et Perpétue' and 'Aimée Vibert'; see further in the second section. It is also a probable parent on one side of the Ayrshire roses, if the original Ayrshire rose was a hybrid between it and R. *arvensis*; see further under that species.

R. PHOENICEA Boiss.—This species is a native mainly of Asiatic Turkey, Cyprus and the Lebanon, but extends into N.E. Greece. It is therefore inter-mediate geographically between R. *sempervirens* and R. *brunonii*, but its affinity is more with the former and with R. *arvensis*. It is a lax shrub armed with short, curved or hooked prickles. Leaflets mostly five, elliptic to roundish, short-acuminate coarsely toothed, more or less downy on both sides. Flowers small, white, up to forty in a corymb or panicle. An unusual character for a member of the *Synstylae* is that the sepals have well-developed lateral appendages and widely expanded tips. Pedicels and receptacle glabrous or slightly hairy, the latter elongate-ovoid or ellipsoid. Stylar column glabrous.

R. *phoenicea* is tender and not of much value in gardens, but is mentioned because Hurst suggested it as a parent of some of the damask roses (see R. *damascena*) and of R. *richardii* (q.v.).

R. SERICEA Lindl.

R. *wallichii* Tratt.; R. *tetrapetala* Royle; R. *omeiensis* Rolfe; R. *polyphylla* Bak., *nom. inedit.*; R. *mairei* Lévl.; R. *omeiensis* var. *polyphylla* Geier

A shrub of variable habit in the wild, attaining a height of 20 ft under favourable conditions but stunted at high altitudes and found only 6 ft high by Kingdon Ward in Manipur, despite the warm climate and low altitude; in cultivation it will reach 12 ft or so in height and more in diameter, if well grown. The species has a varied and complex armature: the prickles are curved or straight, often directed upward, broadly based, usually arranged in pairs at the nodes, and are often mixed with slender needles or bristles; on the young stems especially the prickles may be broad and flattened and sometimes attain a very large size (see f. *pteracantha*). Leaves up to 4 or 5 in. long, composed of seven to seventeen leaflets (three to eight pairs); rachis glabrous or downy. Leaflets ¼ to 1¼ in. long, obovate, elliptic or oblong, rich green above, glabrous or hairy beneath (rarely hairy on both sides), simply toothed, often only in the upper part of the leaflet. Flowers solitary on short laterals in May, 1½ to 2½ in. wide, white, creamy white or sulphur yellow, usually with only four petals, arranged in the form of a Maltese cross. Pedicels and receptacle usually smooth and glabrous. Sepals glabrous or silky-hairy outside, entire, usually acute. Stamens comparatively few (40 to 65). Fruits globular or pear-shaped, dark crimson, scarlet, orange or yellow, sometimes with a fleshy foot-stalk.

Native of the Himalaya, the hills of N.E. India, S.E. Tibet, N. Burma and China (Kansu to Yunnan and parts of central China); described by Lindley in 1820 from a specimen in the Banks herbarium gathered by one of Wallich's collectors in Nepal. It was introduced from the Himalaya, probably from Nepal, about 1822 and flourished at Kew, where in the 1870s there was a specimen about 15 ft high on a wall. But this rose seems to have been largely ignored by gardeners until re-introduced from China. In 1890, Père Delavay, the French missionary, sent two lots of seed from Yunnan to Maurice de Vilmorin's collection at Les Barres, and another lot came from E. Szechwan, probably from Père Farges. By 1904 there was a great assemblage of these Chinese forms at Les Barres, varying in the armature and colouring of the stems, the degree of pubescence on the leaves, the number and shape of the leaflets, the colour and size of the spines and fruits. The colour of the young wood and spines was sometimes bright red, and the fruits, normally bright red, were yellow in the forms from Szechwan.

ROSA SERICEA f. PTERACANTHA

Wilson, too, sent seed from China between 1900 and 1910, first for Messrs Veitch and later for the Arnold Arboretum, mostly from W. Szechwan but also from W. Hupeh. In these regions, according to Wilson, R. *sericea* is abundant in upland thickets, on the margins of woods and in forest glades, and shows the same variations as the Vilmorin seedlings. Forrest, too, sent seed, but the only plants known to have been at all widely grown from his introduction were raised from Wisley number A.867, collected in the hills north of Tengyueh in 1917–19.

In two respects the plants from Hupeh and Szechwan differ from the more typical Himalayan ones—the leaflets are more numerous, occurring in up to eight pairs against usually not more than five in the typical state; and the foot-stalk of the fruit is always fleshy. It was on the basis of this difference that Rolfe established the species R. *omeiensis* in *Botanical Magazine*, t. 8471 (1912), taking the epithet from Mt Omei in W. Szechwan, where specimens had been collected by Wilson and earlier by Faber. However, the typical R. *omeiensis* and typical R. *sericea* are so closely linked by intermediate states, especially in Yunnan, that there is really no solid character by which to distinguish them, except that, so far as is known, the fruits of Himalayan plants do not have thickened pedicels in the fruiting stage. It was therefore with very good reason that Rowley united, or rather re-united, these species in 1959 (*Bull. Jard. Bot. Brux.*, Vol. 30, p. 210).

When well grown, R. *sericea* makes a beautiful specimen, with its abundant, soft green, ferny foliage. In having only four petals to the flower it is unique among roses, but the character is not altogether constant; towards the end of the flowering season odd flowers may be seen bearing five petals. In the colour plate accompanying his original description of R. *sericea* Lindley shows five petals, but he may have assumed that the missing petals had dropped, since five-petalled flowers are rare on wild plants. It is also unlikely that the type really had the flowers of a dog rose colour, as shown in the plate.

var. HOOKERI Reg.—Stem-bristles gland-tipped and clammy. Undersides of leaflets and sometimes the pedicels and receptacles more or less glandular. *Bot. Mag.*, t. 5200.

var. OMEIENSIS (Rolfe) Rowley   R. *omeiensis polyphylla* Geier—This combination is available to distinguish from typical R. *sericea* those plants that have fruits with fleshy foot-stalks and leaflets in excess of eleven. Most Wilson introductions belong here, but the variety does not comprise all plants of Chinese provenance. Some plants raised from the seed sent by Wilson have the numerous leaflets of this variety but fruits with non-fleshy pedicels. These could take the name var. POLYPHYLLA [Bak. in herb.] Rowley.

f. PTERACANTHA Franch.   R. *sericea* var. *pteracantha* (Franch.) Boulenger; R. *omeiensis* f. *pteracantha* (Franch.) Rehd. & Wils.—As originally introduced to cultivation, this is a shrub of open, slender habit, eventually as large as the type. Stems covered when young with blood red, translucent prickles which are sometimes 1½ in. wide at the base, ½ to ¾ in. deep, flat and thin, contracting abruptly to a sharp point, and forming together interrupted wings down the stem. The second year they become grey and woody.

This remarkable form was introduced to gardens by the French missionary Delavay, who sent seeds to Maurice de Vilmorin's collection at Les Barres in 1890. It was put into commerce in Britain by Messrs George Paul and Son of Cheshunt, and received a First Class Certificate when exhibited by them on behalf of Messrs Vilmorin in 1905. The leaflets in this introduction are glabrous and number nine to thirteen on each leaf. The fruits are red, with a fleshy foot-stalk. *Bot. Mag.*, t. 8218.

The production of these huge wing-prickles is a random character. In the type specimens of f. *pteracantha*, collected by Delavay in Yunnan, the leaflets are hairy on both sides, although the plant raised from the seeds he collected in the same area as these, or at least the one propagated, had them glabrous, as described above. Pteracanthous forms have also been found in Manipur and the E. Himalaya, and according to Wilson are common on the wind-swept mountain-sides of W. Szechwan. His W.4095 and W.4118, from both of which plants were raised in this country, were collected from wing-prickled plants, though not all the seedlings showed this character, judging from specimens in the Kew Herbarium.

R. *sericea* is variable in the colouring of its fruits. The commonest colour is bright red or scarlet (often with a yellow foot-stalk in var. *omeiensis*). But they may be wholly yellow [f. CHRYSOCARPA (Rehd.); R. *omeiensis* f. *chrysocarpa* Rehd.] or dark purplish red [f. ATROPURPUREA (A. Osborn); R. *omeiensis* var. *atropurpurea* Rolfe ex A. Osborn]. The fruits usually ripen within two months, so there was nothing unusual in this respect about the form which received an Award of Merit as a fruiting shrub when exhibited by Sir Frederick Stern on July 7, 1935, as R. *omeiensis praecox*.

R. SETIGERA Michx.   PRAIRIE ROSE
R. *rubifolia* R. Br.

A rambling shrub making slender stems several yards long in a season, armed with short, hooked prickles, not downy. Leaves trifoliolate, with a downy, glandular stalk and narrow stipules edged with glands. Leaflets among the largest in the genus, up to 3 in. long by over 2 in. wide, ovate, coarsely toothed, deep green and glabrous above, pale and downy beneath. Flowers 2 to 2½ in. across, variable in colour from nearly white to crimson, several in corymbs; the stalk glandular. Sepals ovate, pointed, ½ in. long, very downy. Fruits globose, about ⅓ in. in diameter, with the sepals fallen away.

Native of E. and Central North America, from Ontario to Florida, and west to Kansas and Texas. Introduced in 1800. This is the most distinct and, in its flowers, perhaps the most beautiful of N. American roses. It is the only one from that region belonging to the group whose styles are united in a column (*Synstylae*); the only one with normally three leaflets, and the only climbing species. It is an attractive plant, producing its large, rich rosy blossoms in clusters 6 in. or more across, but they have little or no fragrance. Flowering in July and August when few wild roses or shrubs of any kind are in flower, its value is increased. It may be trained up rough branches of oak, then left to form a tangle.

Many hybrids have been raised from R. *setigera* in the USA. Of these the oldest were bred by Samuel Feast of Baltimore, Maryland, who made a sowing of the prairie rose in 1836 and crossed it with Hybrid Perpetuals, Noisettes, etc. Some of the offspring, put into commerce in 1843, are still cultivated, e.g., 'BALTIMORE BELLE' and 'QUEEN OF THE PRAIRIES'.

R. SETIPODA Hemsl. & Wils.
R. *macrophylla* var. *crasseaculeata* Vilm., *fide* Rehd. & Wils.

An arching shrub to 8 or 10 ft high and as much wide; prickles sparse, short and straight, much thickened at the base and sometimes laterally compressed. Leaves 5 to 8 in. long; rachis usually glandular and prickly, slender. Leaflets seven or nine, elliptic or elliptic-ovate, acute to obtuse at the apex, usually not more than 1¼ in. long on the flowering laterals, but to twice that length on strong shoots (which may bear an inflorescence at the apex), medium green above, underside greyish, glabrous except for the hairy midrib and main veins, sometimes glandular on the blade. Flowers in June, up to twenty or even more in a lax cluster, deep purplish pink passing to white at the centre, about 2 in. wide; bracts large and leafy. Pedicels ½ to 2 in. long, more or less densely clad with glandular bristles, which may extend onto the rather narrowly ellipsoid receptacle. Sepals about as long as the petals, glandular at the margin, expanded at the apex and with a few lateral appendages. Petals downy on the back in the type but only very slightly so in authentic cultivated plants. Fruits flagon-shaped, 1 to 2 in. long, dark red, crowned by the sepals.

A species of limited distribution in China; discovered by Wilson in N.W. Hupeh in 1901 and introduced by him at the same time to Messrs Veitch's

Coombe Wood nursery, where it first flowered in 1909. It is one of the most handsome of the Chinese roses, recalling R. *moyesii* and R. *davidii* in its fruits and perhaps most nearly allied to the latter.

R. CAUDATA Bak.—Very near to R. *setipoda* and from the same area. It was accepted as a distinct species by Boulenger, on the ground that the inflorescence is shorter than the subtending leaves (longer in R. *setipoda*). It was described from a plant in Miss Willmott's garden, raised from the seeds collected by Wilson during his first journey for the Arnold Arboretum (1907–8).

R. HEMSLEYANA Täckholm   R. *setipoda* sens. Rolfe, (?) not Hemsl. & Wils.—This species, if such it be, was described from a plant at Kew, which had been raised from the same batch of seed as R. *setipoda* (W.1047). It differs primarily in its chromosome number (it is hexaploid, while R. *setipoda sens. strict.* is stated to be tetraploid), but also in having sepals with unusually well developed lateral appendages, as in the dog roses. A similar specimen was collected by Wilson in the wild in 1907 and is referred to R. *setipoda* in *Plantae Wilsonianae* (W.272). R. *hemsleyana* is figured in *Bot. Mag.*, t. 8569, as R. *setipoda* of which it is probably no more than a form.

R. SICULA Tratt.

A close-habited, densely branched shrub of rounded habit, 2 to 5 ft high, its branches set with slender, curved (not strongly hooked) prickles of varying lengths, the longest about ¼ in. long, rarely mixed with needles and glandular bristles, though more commonly so on the strong growths. Leaves 1½ to 2 in. long, composed of five to seven leaflets which are broadly ovate or round, ¼ to ¾ in. long, compound-toothed and with glands on the teeth, lower surface, rachis and stipules. Flowers 1 to 1¼ in. across, bright rose, usually solitary, sometimes two or three together. Pedicels smooth or glandular-bristly, sometimes downy, very short. Sepals lanceolate, with a few lateral appendages and glandular, ciliated margins. Styles downy. Fruits about the size of a large pea, red, finally black, crowned with the sepals. *Bot. Mag.*, t. 7761.

Native of the Mediterranean region, including N.W. Africa. A neat and pleasing little rose, seldom seen in gardens but quite hardy. It resembles R. *biebersteinii* (*horrida*) in its dwarf habit, small leaves and abundant spines, but differs in the particulars pointed out under that species. It is very similar to and confused with R. SERAFINII Viv., which differs from R. *sicula* in its hooked prickles and glabrous styles.

R. SOULIEANA Crép.

A very robust shrub, up to 10 or 12 ft high, forming an impenetrable tangle of branches wider than it is high. Shoots 10 to 12 ft long are made in a year on young vigorous plants; formidably armed with pale spines, which are compressed, decurved, scattered irregularly on the shoots. Leaves 2½ to 4 in. long, grey-green, composed of seven or nine leaflets, which are oval or obovate,

½ to 1 in. long, finely and simply toothed, perfectly glabrous on both surfaces except on the midrib, which, like the common stalk, is more or less downy. Flowers yellow in bud, opening white, 1½ in., produced abundantly in July on branching corymbs 4 to 6 in. across; stalk slender, and, like the calyx-tube, glandular; styles united; sepals attenuated, downy. Fruit orange-red, egg-shaped, ½ in. long, ⅓ in. wide, with the sepals fallen away. *Bot. Mag.*, t. 8158.

Native of W. China; discovered by the French missionary Soulié in W. Szechwan, and introduced by him to the Vilmorin collection at Les Barres about 1895, after he had been transferred to the Jesuit mission station at Tseku on the Mekong. A plant was sent to Kew by Maurice de Vilmorin in 1899, the year in which the seedlings first flowered. From the other cultivated Chinese *Synstylae* it is very distinct in its strongly armed stems, small, greyish leaflets, and broad stipules.

R. *soulieana* is one of the most robust of all roses, and well adapted to the wild garden, where it can have unlimited room and never be touched by the knife. In such a spot it is striking all the summer because of its luxuriant grey-green foliage, but especially in July when in flower, and in autumn when the fruits have coloured. It is hardy, though the long shoots sometimes die back in winter.

For 'Kew Rambler', a hybrid of this species, see p. 186.

R. STELLATA Wooton

Hesperhodos stellatus (Wooton) Boulenger; R. *vernonii* Greene; *Hesperhodos vernonii* (Greene) Boulenger

A deciduous shrub up to 2 ft high, with slender leafy shoots and of lax habit; young shoots thickly covered with starry down and armed with straight, pale, yellowish white, slender spines ¼ to ⅓ in. long, mixed with which are tiny prickles and stalked glands. Leaves ¾ to 1½ in. long, composed of usually three, sometimes five leaflets; rachis glandular-downy. Leaflets wedge-shaped or triangular, toothed mainly or only at the broad end; teeth comparatively large, blunt, ¼ to ½ in. long, glabrous and dullish green above, greyish and slightly downy beneath. Flowers solitary, 2 to 2½ in. wide, of a beautiful soft rose; petals inversely heart-shaped, deeply notched. Anthers yellow. Receptacle globose, covered with pale spines. Sepals ½ in. long, lance-shaped, two of them pinnately lobed, with a spoon-like tip, glandular and spiny outside, woolly on the margins. Fruits hemispherical, flat-topped, not fleshy, prickly, brownish red, ½ in. wide, the sepals persisting at the top.

Native of the south-western USA, where it ranges from W. Texas to Arizona. It was discovered in the Organ Mountains of New Mexico towards the end of the last century and was first grown in Britain by Dr Wallace of Broadstone, Dorset, who raised it from seeds collected in the type-locality by Prof. Cockerell of Colorado University and flowered it in 1912. A year later, in an article in *Nàture*, Prof. Cockerell pointed out that R. *stellata* and the related R. *minutifolia* were distinct among roses in having a non-fleshy fruit with a wide orifice and oval-elongate (not angled) achenes, and proposed for them a generic or sub-generic status under the name *Hesperhodos*. According to Boulenger, the most

ROSA STELLATA

significant character—fully justifying generic rank for this group—is that the receptacle entirely lacks the disk that in all other groups of *Rosa*, however aberrant, partly closes the mouth of the receptacle ('Monographie du Genre Hesperhodos', *Bull. Jard. Bot. Brux.*, Vol. 14 (1937), pp. 227–39). Cockerell's proposal of a possible generic rank for this group was also taken up by C. C. Hurst in *Rose Annual* 1929. However, other botanists consider that subgeneric rank in *Rosa* suffices to give recognition to the distinctness of this group.

For the cultivation of R. *stellata* see below.

var. M I R I F I C A (Greene) Cockerell R. *stellata* subsp. *mirifica* (Greene) W. H. Lewis; R. *mirifica* Greene; *Hesperhodos mirificus* (Greene) Boulenger—Young shoots without stellate down. Leaflets usually five, sometimes almost glabrous. It is more vigorous than the type, sometimes attaining 4 to 6 ft in the wild. A native of New Mexico, within the area of typical R. *stellata*. It was introduced to Kew in 1917 by means of seeds collected by Dr Alfred Rehder for the Arnold Arboretum in the Sacramento Mountains, where it is said to form patches acres in extent.

R. *stellata* and its variety, apart from their botanical interest, are remarkable in the garden for their gooseberry-like foliage and lilac-pink flowers (though the distinctive colouring is said to be lost if the plants are grown in a moist soil). The var. *mirifica* is quite at home in this country and spreads by underground suckers, which afford a simple means of propagation. It received an Award of Merit when exhibited by Kew in 1924. Typical R. *stellata*, as originally introduced, proved less amenable to cultivation, perhaps because it came from a drier region. Both, however, need abundant summer-heat and are most likely to grow in character in a well-drained soil and a warm, sunny position, though it has been found to do reasonably well on a practically sunless wall.

R. MINUTIFOLIA Engelm. *Hesperhodos minutifolia* (Engelm.) Hurst—This species, a native of Lower California, is the senior member of the group, discovered in 1882 and described in the same year. It was introduced to Kew in 1888 but did not long survive and is scarcely likely to be hardy. It differs from R. *stellata* in its minute leaflets ⅛ to ¼ in. long, densely downy beneath, few-toothed, borne on a pale, downy, threadlike rachis. Young shoots downy, with numerous slender spines up to ⅜ in. long. Flowers 1 in. wide, pink, or nearly white. Fruits globose, very spiny, ⅜ in. wide.

The most recent study of this group is: W. H. Lewis, 'The Subgenus Hesperhodos', *Ann. Miss. Bot. Gard.*, Vol. 52 (1965), pp. 99–113.

R. TOMENTOSA Sm.

R. *mollissima* Willd. *fide* Crép., *nom. confus.*; R. *scabriuscula* Sm.

A shrub to about 10 ft high, rather laxly branched; young stems green; prickles straight or slightly curved, usually rather stout, gradually narrowing from the base. Leaflets five or seven, elliptic to ovate or ovate-lanceolate, sometimes obovate, rounded to acute or acuminate at the apex, ⅝ to 1¾ in. long, the terminal leaflet not larger than the others, light green above, usually densely hairy on both sides, but the underside sometimes almost glabrous except on the veins and then rough to the touch, teeth simple or, more usually, compound and glandular. Stipules with short, triangular free tips. Flowers solitary, or few in a corymb, pink or white, fragrant, 1½ to 2¼ in. wide. Pedicels up to 1 in. long, clad with glandular bristles which sometimes extend onto the receptacle. Sepals up to 1 in. long, glandular or glandular-bristly outside, expanded at the apex, the outer ones with pinnately arranged appendages. Styles usually hairy. Stylar aperture narrow, ⅙ to ¼ the diameter of the disk. Fruits commonly globular, or broadest just above or below the middle; sepals erect or spreading, deciduous before the fruit is fully ripe.

Native of much of Europe, including the British Isles, and of Asia Minor and the Caucasus. It has a considerable resemblance to the common dog rose, which differs from it in its strongly hooked prickles and in having the leaflets glabrous on both sides or slightly downy beneath. It is suitable only for the wilder parts of the garden and produces very pleasant effects when laden with bright red fruits in autumn.

R. TOMENTOSA × R. GALLICA—This hybrid occurs occasionally in the wild, resembling R. *tomentosa* in most characters, but showing the influence of R. *gallica* in the mixed armature, leaves often with five leaflets, and larger, brighter pink flowers on longer pedicels. The name R. × *marcyana* is sometimes used for this cross, but the rose so named is probably a form of R. *tomentosa* (Boulenger, *Bull. Jard. Bot. Brux.*, Vol. 12 (1932), p. 445). The identity of the rose figured as R. × *marcyana* in Willmott, *The Genus Rosa*, Vol. II, p. 335, t., is uncertain.

R. SHERARDII Davies   R. *omissa* Déségl.—Of more compact habit than R. *tomentosa* but with similar prickles. Branches with a pruinose bloom when young (as in R. *mollis*). Leaflets often glaucous above, with usually compound teeth. Sepals ⅜ to ⅞ in. long, hence on the average shorter than in R. *tomentosa*, but with the same well developed lateral appendages. It resembles R. *villosa* and R. *mollis*, and differs from R. *tomentosa*, in the wide stylar aperture. Sepals persisting until the fruit is fully ripe.

R. *sherardii* was for a long time confused with R. *villosa*, R. *mollis* and R. *tomentosa* and, when finally recognised as a distinct species, was usually known as R. *omissa*. It is fairly widely distributed in Europe and occurs throughout Great Britain, though infrequently in the south and most commonly in Scotland. It is rare in gardens and deserves to be more widely grown, especially in its grey-leaved form.

R. VILLOSA L.*

R. *pomifera* J. Herrm., *nom. illegit.*; R. *villosa* var. *pomifera* (Herrm.) Desv.

A bush up to 8 ft in cultivation, of rather stiff habit, usually under 6 ft high in the wild, where it sometimes spreads by suckers and forms extensive stands; prickles scattered, slender, straight or slightly curved. Leaves 4 to 7 in. long; rachis glandular and downy. Leaflets five, seven or nine, 1¼ to 2½ in. long, ¾ to 1½ in. wide, oval to oblong-ovate, bluish green above, hairy on both sides and often very densely furnished beneath with resin-scented glands, edged with compound-glandular teeth. Flowers solitary or in clusters of three, rarely more numerous, deep rosy pink, 1½ to 2½ in. across. Pedicels short, about as long as the receptacle, which, like the pedicels, is densely covered with glandular bristles. Sepals fleshy at the base, glandular on the back, long-tailed, not constricted at the base, with a few lateral appendages. Stylar aperture wide. Fruits dark red, more or less bristly, globose to pear-shaped, 1 to 1½ in. long, surmounted by the erect sepals. *Bot. Mag.*, t. 7241.

Native of central and southern Europe, Asia Minor and the Caucasus; long cultivated in the British Isles and occasionally escaping or occurring as a relic of former cultivation. It is a remarkable rose, and, when well grown, one of the most striking, especially in the fruits, which are dark red and very large— whence the old name 'the apple-bearing rose' (*pomifera*). The fruits were at one time used for making preserves and it was even specially planted to supply them for that purpose.

* R. *villosa* L., in the first edition of *Species Plantarum* (1753) is founded on Haller's 'Rosa *foliis utrinque villosis, fructu spinoso*' (*Enum. Meth. Stirp. Helv.* (1742), p. 350). As a synonym Linnaeus gives 'Rosa *sylvestris pomifera major*', taken from C. Bauhin's *Pinax* (1623). The name R. *pomifera* J. Herrm. (*Dissertatio* (1762), p. 16) is superfluous and illegitimate, since he cites the *Species Plantarum* and Haller's phrase-name, but ignores Linnaeus' use of the epithet *villosa*, for which he substitutes *pomifera*. In his later works Linnaeus confused R. *villosa* with R. *tomentosa* and R. *mollis*, and for that reason some botanists have preferred the name R. *pomifera* J. Herrm., despite its illegitimacy.

ROSA VILLOSA

R. 'WOLLEY-DOD'.—Near to R. *villosa* but probably a hybrid. The large, semi-double flowers are clear pink, beautifully set off by the grey-green leaves. The fruits are smaller than in R. *villosa* and less freely borne. It is figured in Willmott, *The Genus Rosa*, from a plant in the garden of the Rev. Wolley-Dod at Edge Hall, Cheshire.

R. MOLLIS Sm.   R. *villosa* of some authors; R. *villosa* var. *mollis* (Sm.) Crép.; R. *villosa* subsp. *mollis* (Sm.) Keller & Gams—This species agrees with R. *villosa* in all essential characters, but the leaflets are smaller, to about 1½ in. long, the pedicels and receptacles less bristly and the fruits smaller (to about ⅞ in. long). Another point of distinction sometimes given is that the young stems have a pruinose bloom, which is said not to be the case in R. *villosa*.

R. *mollis* has much the same distribution as R. *villosa* except that it extends farther north, reaching the British Isles, where it is commonest in Scotland. So far as gardens are concerned it is an inferior substitute for R. *villosa*.

R. VIRGINIANA J. Herrm. (?)*

R. *virginiana* Mill.; R. *carolinensis* Marsh.; R. *lucida* Ehrh.; R. *pennsylvanica* Andr., not
Wangenh. nor Michx.; R. *carolina* var. *lucida* (Ehrh.) Farwell

A shrub to about 5 ft high in gardens, which, in the commonly cultivated form, suckers freely, and soon forms a dense mass of erect stems, dividing at the top into heads of stiff, brownish red branches and twigs; sucker-stems stout, densely clad with bristles and needles, in marked contrast to the branches, which bear prickles in pairs at the nodes but are otherwise unarmed (for the shape of the prickles see below). Leaves glossy green above, 3 to 5 in. long, composed of usually seven, sometimes nine leaflets, which are mostly obovate or oblong-obovate or oblong-elliptic, 1 to 2 in. long, rather coarsely toothed except towards the base, glabrous above, often the same below, but occasionally downy on the midrib as well as on the rachis. Stipules leafy, widening toward the apex. Flowers borne in July or early August in clusters of often three, sometimes solitary, 2 to 2½ in. across, pink. Pedicels and receptacle smooth or glandular. Sepals about 1 in. long, with long, slender points, glandular and downy, entire or with a few slender appendages. Fruits orange-shaped, ½ in. wide, red, crowned at first with spreading sepals which fall away when the fruit is ripe.

Native of eastern North America; in cultivation since early in the 18th century, perhaps earlier. It is a useful plant for forming thickets in the wild garden, but is too invasive for a choice position, which it would otherwise deserve, for its late-flowering and its glossy, always healthy foliage, turning red and purple in the autumn. It thrives in any soil but flowers better in one on the dry side. It is an excellent rose for windy positions.

The plant described above is probably an old clone of European gardens. It differs in two respects from R. *virginiana* as described in American works: it suckers freely, and its nodal prickles are straighter and more slender. In these two respects it resembles R. *carolina* but it is certainly not that species, though perhaps an intermediate or hybrid between them (such plants occur in the wild). It should be remarked that R. *lucida* Ehrh., once the established name for this species, was based on plants cultivated in Germany. Ehrhart did not mention the prickles or habit, but R. *lucida,* as cultivated in the Harbke arboretum, was suckering and had awl-shaped prickles, as in the common garden clone described above (Du Roi, *Harbk. Baumzucht,* ed. 2, Vol. 2 (1800), p. 564).

R. *virginiana* is recorded as an escape from gardens in some parts of Europe and was once used for fixing sand-dunes in the Loire estuary.

For R. *virginiana* var. *alba* and var. *grandiflora,* see R. *carolina.*

R. RAPA Bosc (?)R. *turgida* Pers.; R. *lucida fl. pleno* Savi; R. *virginiana* f. *plena* Rehd.—TURNIP ROSE.—This rose, often considered to be a double-flowered

* This species was once generally known as R. *lucida* Ehrh. (1789), but sometimes as R. *carolina*, with which it was confused. The earlier name R. *virginiana* Mill. (1768) was not in use, or at least not widely, until Baker took it up in Willmott's *The Genus Rosa* (Vol. 1 (1911), p. 197), and has since become established. Unfortunately it was overlooked that the name R. *virginiana* was first published by J. Herrmann in his *Dissertatio* (1762). This change of author would present no difficulties if Herrmann and Miller were describing the same species, but this is very doubtfully the case. Herrmann's R. *virginiana* was described from a herbarium specimen given to him by a friend in the Leyden Botanic Garden as 'Rosa *virginiana,* of all roses the smallest', and from his description it is difficult to believe that he was looking at R. *virginiana* Mill. (branch half-a-line (1/24 in.) thick, prickles flexible, flower apparently white, solitary, 1 in. wide, leaflets roundish, almost truncate at the apex, etc.). But R. *virginiana* does have very dwarf and anomalous forms, so it seems best to maintain the name R. *virginiana* with Herrmann as its author, until the matter has been dealt with by an American authority.

form of R. *virginiana*, has been known since the late 18th century, though now only from the many descriptions and portraits, since it may be lost to cultivation. It was a tall plant, sparsely armed, with leaves of a darker green than in R. *virginiana* as then cultivated. Flowers pink, shaded with violet, paling with age from the centre outwards (so Savi). Sepals pinnately lobed, with a spathulate apex, hence differing from those of R. *virginiana*. Receptacles turnip-shaped, whence the epithet *rapa*. It was most probably a hybrid, and the early flowering season—late May or June—supports this suspicion.

Another double-flowered form or hybrid of R. *virginiana* was figured by Andrews as R. *pennsylvanica flore pleno* (*Roses*, Vol. II, t. 102 (1806)). This was a dwarf, suckering rose flowering all the summer. A rose agreeing well with this was cultivated by Canon Ellacombe at Bitton in 1883; as in the Andrews plant the outer petals became bleached in hot weather (*Gard. Chron.*, Vol. 20 (1883), p. 41). For the double-flowered hybrid of R. *virginiana* still in gardens see 'Rose d'Amour', p. 197.

R. 'MARIAE-GRAEBNERIAE'.—A suckering shrub to 4 or 5 ft high, its stems sparsely armed with curved prickles. Leaves coarsely toothed, colouring in autumn. Flowers clear pink, borne in June and July, then intermittently until autumn. Fruits freely borne. A hybrid of R. *virginiana*, raised by Zabel at Hannover-Münden about 1880 (*Gartenflora* 1902, p. 564). The other parent was stated to be R. *carolina*, which at that time usually meant R. *palustris*, but possibly the true R. *carolina* was used. Ascherson, who described this rose, was the co-author with Graebner of a valuable work on the flora of Central Europe.

R. WATSONIANA Crép.

R. *multiflora* var. *watsoniana* (Crép.) Matsum.

A trailing shrub whose glabrous, slender stems are armed with small hooked prickles. Leaflets three or five, linear, 1 to $2\frac{1}{2}$ in. long, $\frac{1}{8}$ to $\frac{1}{4}$ in. wide, margins wavy, not toothed, downy beneath, mottled with yellow down the centre above; rachis downy, glandular and spiny. Stipules very narrow, not toothed. Flowers pale rose with narrow petals, $\frac{1}{2}$ in. wide, crowded in short, broad panicles. Sepals lanceolate, entire. Styles glabrous, united in an exserted column. Fruits globose, red, $\frac{1}{4}$ in. wide.

A garden rose of Japan, whence it was introduced to the USA before 1870 and thence to Britain. It shows some affinity with R. *multiflora*, but the long, narrow leaflets distinguish it, and in its essential characters it is quite different.

It is a rose of delicate constitution, but has been grown successfully at Kew and in other gardens, though the finest plants are in southern Europe. In previous editions it was added: 'Anywhere it must be regarded more as a curiosity than anything else', but some gardeners have a liking for it.

R. WEBBIANA Royle

A shrub up to 8 ft high in the wild, but often half that height or less; branches armed with straight, slender yellowish prickles abruptly widened at the base, up to $\frac{1}{2}$ in. long; on strong shoots the prickles may be very densely arranged and variable in size, though all of essentially the same form, the largest with enlarged, pad-like bases; flowering branchlets unarmed or with prickles like those of the branches, but shorter. Leaves 1 to 3 in. long, with five to nine leaflets; rachis glandular or not, usually unarmed. Leaflets broadly elliptic, obovate or roundish, ⅝ to 1 in. long, obtuse or truncate at the apex, cuneate to rounded at the base, simply toothed, usually entire in the lower third, somewhat glaucous above, pale and glabrous or slightly downy beneath. Flowers solitary on short laterals in May, 1½ to 2 in. wide, pale pink or white, or sometimes white in the centre shading to pink at the margin. Pedicels ⅜ to ½ in. long, sometimes slightly longer, they and the receptacles smooth or glandular-bristly. Sepals entire, glandular on the back, sometimes expanded at the apex. Fruits pitcher-shaped or globular, constricted at the apex, bright red, crowned by the persistent sepals.

Native of the central and western Himalaya, where it seems to be mainly confined to the drier inner valleys; also of Afghanistan, Tibet and parts of Russian Central Asia; the earliest recorded introduction to Kew was from Ladakh in 1879. The true species seems to be rare in cultivation (1979); see also R. *sertata* below.

var. MICROPHYLLA Crép. R. *nanothamnus* Boulenger—Of dwarf habit; prickles as long as in typical R. *webbiana* but the leaflets smaller, to ⅝ in. long, often shorter than the prickles. Flowers smaller, 1 to 1½ in. wide. Described from Kashmir, but extending throughout the N.W. Himalaya, mostly in the inner tracts, to Afghanistan and Russian Central Asia.

R. BELLA Rehd. & Wils.—This species, said to be widely distributed in N. China, is included in R. *webbiana* by Boulenger. It was described from plants raised at the Arnold Arboretum from seeds collected by William Purdom in Shansi, and introduced by him to the USA, but is scarcely known in this country, even from herbarium specimens. Plants grown as R. *bella* do not agree well with the original description and are of uncertain identity.

R. SERTATA Rolfe R. *webbiana sens.* Boulenger, in part—This species, widely distributed in China from Kansu to Yunnan, should perhaps rank as a subspecies of R. *webbiana*, differing chiefly in its laxer habit, much sparser prickles and longer leaves. It was introduced to Les Barres in 1897 by one of the French missionaries and distributed by Vilmorin as R. *webbiana*. But Rolfe described the species from plants at Kew raised by Veitch from seeds collected by Wilson. As seen in cultivation, usually as R. *webbiana*, it is a very graceful, free-flowering rose, growing to 8 ft high and as much in width, bearing rich rosy pink flowers about 2 in. wide, singly or sometimes up to five on each lateral, and bright red, pitcher-shaped fruits.

R. WICHURAIANA Crép.

R. *luciae* Crép. (1871), in part, not Crép. (1886); R. *luciae* var. *wichuraiana* (Crép). Koidz.

A procumbent shrub, evergreen in mild districts, rising only a few inches above the ground, and making shoots 10 or 12 ft long in a season; barren shoots unbranched, quite glabrous, armed at irregular intervals with solitary curved prickles ¼ in. long; flowering shoots branching and more slender. Leaves 2 to 4 in. long, consisting of five, seven or nine leaflets; rachis glabrous, armed beneath with small, hooked prickles. Leaflets oval, broadly ovate or almost orbicular, from ¼ to 1 in. long, coarsely toothed, deep polished green on both surfaces, glabrous except for occasional down on the midrib beneath. Stipules broad, ciliately toothed or laciniate, the segments equal to, or shorter than, the entire part of each wing. Flowers nearly 2 in. across, pure white, produced from July into September in panicles of six to ten blossoms rising out of the dense carpet of foliage. Pedicels glabrous. Sepals about ⅜ in. long, downy, entire or with a few slender lateral appendages. Styles united into a hairy exserted column. Fruits red, globose, ⅜ in. wide; sepals deciduous. *Bot. Mag.*, t. 7421 (as *R. luciae*).

Native of the coastal regions of Japan and of the Korean archipelago, with a close relative in Formosa; introduced to Kew from the USA in 1891, but probably cultivated earlier on the continent. It is part of *R. luciae* as Crépin conceived that species when he first described it in 1871, and it was at first known by that name in gardens. He separated it from *R. luciae* in 1886, naming it after the botanist Max Wichura of Breslau, who collected the type-specimens while attached to a Prussian diplomatic mission to the Far East.

Although somewhat eclipsed now by the large number of exquisite hybrids raised from it, it is well worth growing for its own sake. It flowers when nearly all other wild roses are past, and for making a low covering for a sunny bank few plants are better suited. In the USA it is known as the Memorial Rose, from its use in covering graves. Its flowers are very fragrant.

Rose breeders were quick to see the potentialities of *R. wichuraiana*, with its glossy, mildew-proof foliage and late flowering season and many of the most popular rambling roses of today were raised by crossing it with Teas, Hybrid Teas, Chinas and other garden roses. Most of these date from the first three decades of this century, but there is another group of more recent origin deriving from *R. wichuraiana* through *R. kordesii* (see under *R. rugosa*). Through its hybrids, *R. wichuraiana* has played a part—though compared to *R. multiflora* a minor one—in the breeding of the Dwarf and Hybrid Polyanthas.

R. × JACKSONII (*R. wichuraiana* × *R. rugosa*)—This hybrid has been mentioned under *R. rugosa*. Here belongs 'Max Graf', the parent of *R. kordesii*.

R. × BARBIERIANA Rehd.—This is a re-naming by Rehder of the hybrid between *R. wichuraiana* and 'Crimson Rambler', raised by Barbier of Orleans and misleadingly named *R. wichuraiana rubra* by André. The same cross, made in the USA by Walsh, was named 'Evangeline' (see p. 180).

R. LUCIAE Franch. & Rochebrunne ex Crép., *emend.* Crép.—This species, as originally described, included elements which Crépin later distinguished as a separate species—*R. wichuraiana* (see above). As amended, *R. luciae* differs from *R. wichuraiana* in the following characters: leaflets thinner and less glossy, fewer in each leaf (five or at the most seven), longer (up to 1¾ in. long), mostly acute or acuminate at the apex, the terminal markedly longer than the lateral ones, flowers smaller, about 1 in. wide. It is more widely spread in Japan than *R. wichuraiana* and has a more inland distribution. The specific epithet commemorates the wife of Dr Savatier, a French naval doctor who botanised in Japan during his service there 1866–75.

R. *luciae*, if ever introduced to Europe, cannot have been widely grown, since no reference to it (as a species distinct from R. *wichuraiana*) can be found in the literature. There are no grounds for supposing that it enters into the parentage of the Wichuraiana group of ramblers.

R. WILLMOTTIAE Hemsl.

A densely branched shrub 5 to 10 ft high and up to 10 ft wide, stems glaucous when young; branches slender, red-brown, armed with straight prickles ¼ to ⅜ in. long, mostly occurring in pairs. Leaves ¾ to 2 in. long, glabrous, composed of usually nine leaflets, which are oblong, obovate or nearly round, ¼ to ½ in. long, greyish above, toothed except towards the base. Flowers 1 to 1½ in. across, bright purplish rose, produced singly on short lateral twigs in May. Pedicels ⅜ to ½ in. long, smooth and glabrous. Sepals lanceolate, ½ in. long, entire, glabrous outside. Fruits roundish, bright orange-red; sepals deciduous, together with the top of the hypanthium. *Bot. Mag.*, t. 8186.

Native of N.W. China, fairly common, according to Wilson, in the more arid river valleys of N.W. Szechwan. He discovered and introduced it when collecting for Messrs Veitch. He re-introduced it in 1910 and on both occasions the seed came from near Sungpan Ting in the Min valley. R. *willmottiae* was at first considered to be near to R. *webbiana* or even to R. *xanthina*. But it was pointed out by Rehder in 1928 that in the fruiting stage the top of the receptacle is shed together with sepals—a character considered to be of taxonomic importance, and one that places R. *willmottiae* in the same group as the American R. *gymnocarpa*, to which it bears a quite strong resemblance in other characters also.

R. *willmottiae* is a charming rose with elegant foliage, attaining a height of about 6 ft in gardens and as much in width. The flowers are of an unusual shade of lilac pink, with creamy stamens. A selection is 'WISLEY', with deeper coloured flowers than the normal form.

R. XANTHINA Lindl.

A shrub up to 12 ft high, with a uniform armature of straight or slightly curved prickles, which are abruptly or gradually narrowed from a broad base, and sometimes much flattened on sterile growths; in some forms the branches are sparsely armed. Leaflets five to thirteen, obovate, oval or almost orbicular, edged with simple, rather coarse teeth, upper surface glabrous or sometimes hairy, especially along the midrib, underside hairy, more rarely glabrous. Stipules narrow. Flowers usually solitary, bright yellow, about 1½ in. wide, double in the typical form, borne in early June. Pedicels usually less than 1 in. long, smooth (more rarely glandular). Sepals glabrous or nearly so, to about ⅝ in. long. Styles hairy. Fruits globular to broadly ellipsoid, red or maroon, soon falling, crowned by the erect, spreading or reflexed sepals.

R. *xanthina* was described by Lindley in 1820 from a Chinese painting in Lambert's library, and his description is so short that it can be quoted here: 'A rose with all the appearance of R. *spinosissima* except in having no setae and double flowers the colour of R. *sulphurea*'. These words are quite enough to establish the species, since it is precisely the yellow flowers and the absence of bristles from the shoots that distinguishes R. *xanthina* from R. *pimpinellifolia*. There is the further difference that the fruits are black or dark purple in R. *pimpinellifolia*, red or brownish red in R. *xanthina*.

Typical R. *xanthina*, as portrayed in the Chinese painting, has double flowers and later proved to be a common garden plant in N. China and Korea. There is no record of its introduction to western gardens before 1907, when F. N. Meyer sent seeds to the Arnold Arboretum, collected near Peking, from which both double- and single-flowered plants were raised, which first flowered in 1915. It is uncommon in British gardens. The double-flowered rose distributed by Smith of Newry as early as 1915 under the name R. *xanthina* appears to have been a hybrid of R. *spinosissima* with double flowers.

f. SPONTANEA Rehd. R. *xanthina* f. *normalis* Rehd. & Wils., in part only; ? R. *platyacantha* Schrenk—Flowers single. This is the normal state of the species, widely distributed in N. China and also a native of Korea. R. *platyacantha* from Central Asia appears to differ in no essential character from spontaneous R. *xanthina*. The first introduction to be recorded was in 1907, by F. N. Meyer to the USA (see above) and seeds were later sent by Purdom from Shensi and by the Belgian engineer Hers from Honan and Shansi. The rose known as 'Canary Bird', although sparsely armed except on strong growths, clearly belongs to R. *xanthina* f. *spontanea* and is the commonest representative of the species in gardens. It makes a fine arching shrub to about 5 ft high, bearing canary-yellow flowers about 2 in. wide in late May or early June but, like R. *hugonis,* is subject to die-back if grafted. It received an Award of Merit when exhibited by Messrs L. R. Russell in 1945, and an Award of Garden Merit in 1966.

var. ECAE (Aitch.) Boulenger.—See R. *ecae*.

var. KOKANICA (Reg.) Boulenger.—See R. *kokanica* and R. *primula*, under R. *ecae*.

ROSES IN THE GARDEN

G R A H A M   S T U A R T   T H O M A S

HISTORICAL INTRODUCTION

The genus *Rosa* has had a greater effect on gardens—and their design—in this country than any other genus of hardy plants. Rhododendrologists may feel inclined to dispute this statement but roses are grown in far more gardens throughout the country than rhododendrons. Certain roses have been treasured for hundreds, even thousands, of years in the cradle of our Western civilisation, and for equally long other kinds have been treasured in China. In Europe and adjoining countries R. *gallica* and some of its derivatives were appreciated as much for the fragrance that lingers in the dried petals as for the beauty of the flowers, which is understandable since the uses of plants were valued before their beauty. Much later, in perhaps mediaeval times, their value was enhanced by the discovery of distilling flowers to create scented waters. Meanwhile as a garden plant, there is evidence that the compact types were valued as hedges to surround a garden; their prickly stems would deter intruders and also made a support for drying washed linens. By degrees their beauty captivated gardeners and chance hybrids or sports were carefully preserved; eventually, long after they had occurred and without record of the provenance, they were recorded and described in early books.

In this way the Gallica derivatives of the Old World were preserved until the great awakening to the individual beauties of plants in the 19th century. From mediaeval times through the 17th and 18th centuries roses and all other flowers, ornamental and otherwise, were grown in areas devoted to produce of various kinds, fruit, vegetables, herbs and simples, with an ever-increasing number of purely ornamental plants; none of these was given space in the great formal layouts followed by the informal landscape gardens of the wealthy trend-setters, in the 17th and 18th centuries.

Early in the 19th century however the growing of roses and ornamental plants generally was given a tremendous fillip. So many new plants had been brought to Europe—and specially to Holland and Britain—that there were enough delights to fill even a large garden. Napoleon's wife, the Empress Josephine, made a garden at La Malmaison near Paris where she grew every rose she could find, and many other plants as well. Her garden may not only be regarded as housing the first international collection of roses—and thus starting their present popularity—but as being one of the first gardens in the new genre, where the design was explicitly to demonstrate the beauties of the plants themselves. This style has since become known as the Gardenesque, a term invented by Loudon. P.J. Redouté painted over 150 of her roses for *Les Roses*, 1817–24.

153

As the century passed by, artificial aids to the display of roses in gardens were invented, treillage of various kinds, arches, pillars and swags, principally designed to show off the grace of lax growing roses and other climbing plants.

In a less decorative way roses were increasing too. Around 1800 four ancient Chinese garden hybrid roses had reached Europe. The European garden roses were mostly sturdy bushes flowering at midsummer only; the Chinese were less sturdy but flowered for so long as the weather was warm, from spring to autumn. During the century hybrids were raised at first by chance and later deliberately—combining in great part the special attributes of these two historic groups of roses, at first in the race known as Bourbon, and later in the Hybrid Perpetuals. By 1906 some 11,000 names of roses were recorded by Léon Simon and Pierre Cochet in their *Nomenclature de tous les Noms de Roses*.

In 1876 the National Rose Society was founded and had an immense influence on the rose. Its shows were greatly concerned with prize-winning blooms. For this reason, with the multitudes of varieties available from such noted firms as Thomas Rivers and Sons of Sawbridgeworth and William Paul at Cheshunt, gardens of specialists tended to become mere assemblies of rose bushes. In order to accommodate ever more plants in a given area, and to add height to the weaker growers, the art of budding on to stems of *Rosa canina*, dug out of the hedgerows, was invented. Thus was the society (now the Royal National Rose Society) responsible for the new conception of the rose as a florists' flower, rather than a garden shrub. Today the Society's widespread influence embraces all aspects of rose culture; its grounds are laid out in the gardenesque style of La Malmaison and contain examples of all types of roses, species and hybrids, old and new.

True to his general attitude to horticulture, and aided by the introduction at the turn of the century from China of two noted species, William Robinson (1839–1935) sought to call attention to the shrubby beauty of roses lost through the excessive desire for specimen blooms. *Rosa hugonis* and *R. moyesii* have proved two of the most popular of wild roses in the charm of their simple flowers, graceful growth and dainty foliage and of course the notable showy heps of the second species. These and other species mainly from the Far East arrived just at the time when the influx of foreign shrubs was at its peak, and with cotoneasters, berberises and the like caused gardeners to realize that the rose started life as a flowering shrub.

Gertrude Jekyll (1843–1932) in her book *Roses for English Gardens* (1902) had the same desire, but also was the principal devotee of the use of the graceful rambling roses, for use in covering arches, sheds and fences, and scrambling into trees. In addition her great artistic capabilities led us away from the dull repetition of roses in rows in beds and borders to their greater use in formal and informal garden design. The Victorian craze for bedding plants in serried display on the lawn—a direct descendant of the parterre of the 17th century—began to give way to beds of roses, since by

1920, and indeed earlier, the Hybrid Tea race had begun to lose the gawkiness of many of the Hybrid Perpetuals, and was developing into a race of plants 2–3 ft in height in many colours. Very often sunken rose gardens were made, reminiscent once again of looking down on the pattern of a parterre. On the other hand apart from the stereotyped breeding of bedding roses Robinson and Jekyll were the main influences which opened the eyes of gardeners to the very varied beauties and uses of the rose. Today the excessive vigour of many moderns makes them unsuitable for small formal beds.

The end of the Second World War was a time of reassessment of our heritage of roses. Certain devotees had preserved many of the old French 19th-century roses and it was soon realised that they had vigour, unique floral shape and colouring. Against this had to be balanced the fact that most of them flowered only at midsummer. A few of the better Hybrid Perpetuals and Bourbons had survived the decades of neglect also, and together with a number of hybrids raised by Wilhelm Kordes in Germany and coupled with some of the more distinctive species, shrub roses began to be used by garden designers with increasing impetus (many of the Kordes roses were by-products in his courageous attempts to breed the hardiness of some species into the Hybrid Teas and Floribundas, thus making them suitable for the rigours of a continental climate). With the increase of amenity planting, forms and hybrids of *Rosa rugosa* were found important, being tough and hardy, dense and prickly, and with a long period of beauty from flower and hep. It is these of which the greatest number are produced and sold annually in Europe for public display and use, in parks, on road islands, etc.

The bulk of the rambling roses extolled by Jekyll were of pale colours. Later hybrids improved this matter; even so with few exceptions ramblers flower only once at mid-summer. *Rosa moschata* and *R. wichuraiana* flower from July onwards, and a few hybrids such as 'Aimée Vibert', 'Phyllis Bide' and 'Paul Transon' point to the fact that ramblers could be induced to flower for more than a few weeks, and also be fragrant. The small size of the ramblers' flowers was repeated in excellent dwarf hybrids which, originally known as Dwarf Polyanthas or Poly-poms produce masses of bloom over a long period and have been merged with the Hybrid Teas to produce the Floribundas. The wider appreciation of the diverse beauties of the rose today has resulted in modern strains once more embracing flowers of only five petals or of old-style dense doubling, and colours which were achieved earlier in the century. Though there are so many species with first class attributes of growth, foliage and flower colour, heps and long flowering period, it takes many generations of breeding for the characters of a given species to be absorbed and eventually given out again through the modern involved strains of roses. Even *R. rugosa*, a rose introduced from the Far East nearly 200 years ago, and repeatedly producing hybrids, has not as yet made its influence really felt. Shrub roses of graceful habit but with recurrent flowering are a long way off, though there is a glimmer of encouragement to be found in 'Autumn

Fire', 'Gold Bush' and 'Cerise Bouquet'.

COLOURS, SCENTS AND SHAPES

The old French Gallica rose hybrids were of white, pink (both light and dark), mauve, lilac and maroon. The darkest tints owe their origin to the potentialities of R. *gallica*. The most prized varieties were those with fully double flowers of cupped shape, often with the central petals turned into the receptacle devoid of its floral parts, providing what is known as a 'button eye'. Sometimes the centre revealed its pale green carpels as a pointel, or again the petals might be arranged in groups, known as 'quartered'. Their fragrance was sweet and soft, inherited from R. *gallica*, R. *canina* and R. *moschata*; the latter has always been renowned throughout history for intense sweet fragrance. Thus were the Gallicas, the Damasks, the Albas and the Centifolias heavily endowed. Apart from the above soft colours, clear light yellow, or brilliant sulphur was also found in R. *foetida* and R. *hemisphaerica*.

The influence of the characters of the China rose hybrids during the 19th century gradually brought about a considerable change in shapes, colours and fragrance. From R. *gigantea* the China hybrids brought pale yellow, a long bud composed of petals rolled or scrolled, and a delicate fragrance of tea. R. *chinensis*, the China rose itself, brought true dark crimson, a colour unknown in the old European roses. These tints brought delicate coppery and salmon-pink tones into the early Tea roses and the Tea-Noisettes, which were the first yellowish or apricot-tinted roses to be seen in Europe. Thus R. *gigantea* made new history in Europe by being the progenitor of climbing roses with large yellow blooms.

Known and grown since before 1600, *Rosa foetida* was eventually used for hybridising by Pernet-Ducher in Lyon, France in 1883. The result of bringing this species into the strain of Hybrid Teas was that not only its brilliant yellow colouring was used, but that its sport R. *foetida* 'Bicolor' ('Austrian Copper') provided brilliant tomato-red. Coupled with these strong colours it brought an unpleasant fragrance and the progeny later became subject to 'black spot' fungus. To the deepening or fading, or even changing, colouring of the old European and Chinese roses was brought the potential of red on the upperside of the petals and yellow at the back, as in the 'Austrian Copper'. Thus, this one rose believed to be a natural sport of R. *foetida*, unique in its colouring in the rose world, has changed the garden strains of roses from whites, pinks, mauves, maroons, pale yellows and flesh tints to a range of brilliancy. The brilliant tones were yet further developed by the Dwarf Polyantha 'Gloria Mundi' (a sport of the dark red 'Superb') which suddenly acquired a new pigment, pelargonidin, in 1929, which is the reason for the brilliant roses of today. The heavy fragrance of *Rosa foetida* has been almost bred out and today's roses whether shrub or bedding varieties or climbers display as mixed a series of hybrid scents as they do colours.

Display of Flowers, Fruits and Foliage

There are no truly spring-flowering roses in the British climate but in sheltered gardens on sunny walls the Banksian roses and R. 'Anemone', may be expected to provide flowers by mid May followed quickly by the Burnet or Scots roses, R. *pimpinellifolia*, and R. *hugonis*, R. *foetida* and various hybrids such as 'Maigold'. The main mass of old French roses and the species flower in the second half of June: R. *multibracteata*, R. *foliolosa*, R. *fedtschenkoana*, R. *moschata* and R. *wichuraiana* extend the season for species into September. Before then the heps of several early flowering species, such as R. *moyesii* and R. *rugosa* and their relatives will be developing, the main autumn show being in September and October. Heps of 'Autumn Fire' and 'Scarlet Fire' are late in colouring and the latter are often brilliant even in January. A few roses excel in autumn colour, particularly brilliants tints occurring in R. *virginiana*, R. *foliolosa* and R. *nitida*; R. *pimpinellifolia* turns to purplish red tones, R. *rugosa* to bright yellow. During the summer months foliage other than the normal green is found on R. *glauca* (R. *rubrifolia*); the leaves are glaucous reddish purple in full sun, glaucous grey in shade. R. *fedtschenkoana*, R. *beggeriana* and R. *murieliae* have pale glaucous leaves. Forms of R. *sericea* are noted for the bright red colouring of the large, flattened, translucent prickles.

In shape and colour the fruits of roses vary considerably. Most are rounded or oval, orange red. Very rounded fruits, small and nearly black are found in R. *pimpinellifolia*; large, rounded, red in R. *rugosa*; flagon-shaped orange-red in R. *moyesii* and its relatives; plum-purple and bristly in R. *villosa*; green and prickly in R. *roxburghii*.

CULTIVATION

It will be seen from the above cursory appraisal and historical notes that the rose offers many attractions to gardeners and that even its prickles are at times an asset. Roses thrive in a variety of soils but heavy clay, so often recommended, is unnecessary and even undesirable; pure chalk will produce luxuriant growth in the more vigorous species when grown on their own roots; any good well-drained garden soil will prove effective; light sandy soils will suit specially the Burnet and Rugosa roses. The more that roses are hybridised, the more will they in general prefer a well nourished garden soil. Most thrive in full sunshine, but the best quality blooms usually come from the cooler north and west, or partly shaded gardens in the south-east of the country. While R. *pimpinellifolia*, R. *virginiana*, R. *rugosa*, to name but a few, will take full exposure to wind and sunshine, many of the Chinese species seem to prefer a little shade, and indeed they blend well with semi-woodland planting, on somewhat acid or limy soil.

The species of the *Synstylae* section, such as R. *arvensis*, R. *brunonii*, R. *helenae*, R. *filipes* and others like the Dog Rose of our hedgerows, indicate from their very nature that they prefer their roots shaded. In nature they hoist themselves up through bushes and trees by means of their hooked prickles, eventually getting their flowering branches into the sun. In the garden roses will always be found at their best when growing in a cool root-run. This is sometimes difficult to achieve when siting a somewhat tender rose such as R. *bracteata*, R. *banksiae*, or Tea hybrids, which need and deserve the shelter of a sunny wall. Watering in times of drought, or shading of the root-area with other bushes and dwarf plants or a mulch is helpful. One has only to go to an old untended garden and find there big luxuriant plants of unpruned roses thriving on neglect to realize that the typical rose beds of today, on an open sunny lawn, with bare soil and hard annual pruning, is a type of cultivation very far from the ideal for a rose.

There are roses to suit almost any type of garden, or position in a garden except a bog or darkest shade from dense overhanging trees—though R. *arvensis* and 'Ayrshire Splendens' will thrive and flower, hanging down in long trails under dense trees. This is a special use for roses of the *Synstylae* section and also the ramblers. It is best to plant well away from the trunk of the tree, and so that the prevailing wind will carry the long shoots into the branches. Once inside the crown of the tree shoots will grow through the canopy of foliage and the real beauty of the rambling or climbing rose will then be achieved by their shoots hanging down under the weight of blossom. In smaller gardens arches, fences and sheds make good supports for the less vigorous ramblers—not species of excessive vigour such as R. *filipes*. Here again the real beauty will not be revealed until the shoots of later seasons hang down and out from those which are initially trained up. Ramblers are suitable too for growing over rough hedges, or with the support of posts and wires, to create hedges themselves.

Certain lax-growing roses are suitable for making a dense canopy for covering ground, sloping banks, etc. R. *wichuraiana*, 'Max Graf', 'Paulii' are some of the most suitable. The low arching growth of 'Raubritter' makes it particularly suitable for hanging over a low wall.

Shrubby roses have for long been used for hedges. R. *gallica* itself was used in mediaeval times. The sweet brier or eglantine can be kept reasonably within bounds by pruning. 'Erfurt', 'Golden Wings', 'Schneezwerg' and the more sturdy Hybrid Musk varieties are popular because they give two crops of bloom if cut back after the first crop is over. For really large hedges 'Nevada' is unequalled, but for a hedge of average size 3–5 ft in height few roses can equal the closely allied forms or hybrids of R. *rugosa*, to which one can take secateurs or shears in winter.

In association with other flowering shrubs, all the species other than the ramblers of the *Synstylae* section are desirable, several carrying on the flowering period through the summer, and vying with the best of fruiting shrubs though not so successfully with those noted for autumn colours. They are particularly acceptable in informal planting schemes, where modern bedding roses would be out of place. Species of naturally suckering habit such as R. *rugosa*, R. *foliolosa*, R. *virginiana*, R. *pimpinellifolia*, R. *nitida*, R. *pulverulenta*, together with R. *gallica* and many of the 19th-century old Gallica varieties will colonise large areas of ground spreading 3 ft or so underground annually when well suited, and growing on their own roots. Many of these will thrive in short grass, or can be used to fix sand dunes. They are nevertheless to be avoided when grown on their own roots in small or average gardens.

PROPAGATION

SEEDS.—Roses are very liable to chance hybridising, and seeds are not a reliable means of reproducing a good form of a species, and anything may arise from named hybrids. If hand pollination is carried out, seeds need stratifying in sand for a year; germination will occur the following spring or in later months or years. Small quantities can be tended in frames, and the resulting seedlings of hardy roses will be quite winter hardy. The seeds will benefit from being exposed to frost, when stratified or sown.

DIVISION OF THE ROOT STOCK.—This forms an easy method with roses which have a suckering habit as mentioned above. Portions of underground stems, sometimes with pieces of root attached, will almost invariably succeed if the work is done in autumn, winter or early spring.

LAYERS.—If their growth is conveniently near to the ground, and handling the shoots is not too prickly a job, most rose stems will root if bent into the soil and the elbow covered securely with soil. After a year in such a position a gentle tug will ascertain whether they are rooted, when they may be severed from the parent plant and put into the required position.

CUTTINGS.—Hardwood cuttings, preferably with a heel, give considerable success in a cold frame or in the open ground, if taken in October and inserted with a little sand at the base of the cutting, which should be about 9 inches long, with 6–7 inches underground. Rugosas and any vigorous modern hybrids together with Gallicas and Albas can be propagated thus. Many wiry stemmed species are less easy. For these, or for any others, short cuttings of half-ripened wood, or even soft tips may be tried under glass in mist, or as a substitute, small fully enclosed glass containers.

BUDDING AND GRAFTING.—Much has been written and spoken about the iniquities of these methods of propagation.

Unless the scion is inserted below the stem (i.e., just above the root) suckers from the rootstock may arise and cause a nuisance annually. Suckers from the rootstock can also arise from an errant rootstock raised from seed with a proclivity for this method of self-propagation; from damage by hoeing, forking, etc.; and from excessive manuring or from an unsuitable rootstock being chosen for a particular species or cultivar.

On the other hand as we have seen above some roses when on their own roots can be a nuisance in small gardens from their invasive roots. Moreover except in a few instances the provision of one of the accepted rootstocks adds usually greater erect vigour to weak and freely suckering roses. Yet again from a nurseryman's point of view the putting of the scion onto a proved root stock enables him to produce a large quantity of uniform plants of one variety.

Grafting is not often the means of propagation. It is done in January or early February on seedling or other rootstocks, under glass. Some of the resulting plants may be expected to flower during the first summer.

Budding is done in July and early August on seedling or other root-stocks, planted the previous winter. The resulting plants usually flower in the following summer if they are recurrent varieties, or in two years if they are summer-flowering only.

Many different rootstocks have been used in the past, but today a selected strain of *Rosa canina* is usually the choice. Forms of R. *multiflora* are also used.

CLASSIFICATION OF THE GARDEN HYBRIDS

Well before the middle of the 19th century the garden-bred roses had become so numerous, and many were of such complex parentage, that to arrange them botanically under this or that species had become impracticable. Yet it was necessary to bring the garden roses into some sort of order if nursery catalogues were not to become a jumble of names. And so, as in other large groups of cultivars, the growers devised their own nomenclature. In Britain, the first practical arrangement of the commercial roses seems to have been that of Thomas Rivers in the *Rose Amateur's Guide*. First published in 1837, this was hailed by Loudon as an important advance and is set out in detail in an appendix to the treatment of *Rosa* in his *Arboretum et Fruticetum Britannicum*, Vol. II, pp. 779–83. A similar classification, departing even farther from a botanical framework, was published by William Paul in *The Rose Garden* (1848). The groups recognised by these authorities all still exist, though some, with changing fashion, have become shadows of their former selves, and other groups have emerged, unknown when these works were first published.

In the descriptions of the cultivars treated in this section the group to which each belongs is indicated after the name, wherever this is practicable, and the terminology used is explained below. To make the account more complete, groups have been included—notably the Hybrid Teas and Floribundas—which are not further treated in this work.

ALBA.—The prototypes of this group are the old 'Alba Maxima' and 'Maiden's Blush'. To these many were added in the early part of the 19th century: William Paul listed sixty-one sorts in *The Rose Garden* (1848), some certainly of hybrid origin but all more or less conforming to the style of the old White Rose. The group as a whole exhibits strong upright growth with large sparse prickles; leaves hard and greyish; flowers from white to deep pink, mostly fully double, opening wide from unpropitious buds. Heps oval. Pruning as for Centifolia.

BOURBON.—A rather mixed group which sprang originally from a chance hybrid found on the Ile de Bourbon in 1817 by the Parisian botanist Bréon who was in charge of the botanic garden at that time. The parents are believed to have been the Autumn Damask and a Pink China. Seeds of it received by Jacques, gardener to King Louis Philippe in Paris in 1819, gave rise to a new rose called 'Rosier de l'Ile Bourbon'. By 1825 it had reached England. Thory, in Redouté, called it R. *canina Burboniana*. As a group the Bourbons are mainly vigorous shrubs, some rather lax, with smooth, pointed foliage, and somewhat globular blooms, borne singly or in clusters in summer. Many of the cultivars flower intermittently until the autumn. To encourage later flower crops the old weak wood should be removed in winter or early spring.

BOURSAULT.—Hybrids of the Hybrid China Roses and a species of the *Cassiorhodon* (*Cinnamoneae*) section, usually considered to be R. *pendulina*, but the chromosome count indicates that a related species was more likely. The cultivars were popular because of their very early flowering habit and being unarmed, also because of their comparatively large flowers (as compared with Ramblers). They are of lax arching habit, needing support as a rule, though 'Morletii' will make a good bush, and attain some 10 ft when trained up supports. They inherit smooth foliage from R. *chinensis*. The colours are pink to maroon; slight fragrance. The best flowers are borne on short laterals on strong young wood; prune to encourage this by removing old weak wood after flowering, or in winter.

The Boursault roses treated here are 'Blush Boursault', 'Madame de Sancy de Parabère' and 'Morletii'. Another still in cultivation is 'Amadis' (Laffey, 1829), with semi-double, crimson-purple flowers. For the original Boursault rose, see R. × *reclinata*, p. 130.

BURNET OR SCOTCH ROSES.—Forms of R. *pimpinellifolia* in white, pink and maroon, double, semi-double or single. Sabine records in the *Transactions* of the Horticultural Society, 1822, how Robert Brown and his brother transplanted some wild plants of the Burnet rose from the Hill of Kinnoul into their neighbouring nursery near Perth. One bore flowers tinged with red and subsequent seedlings from this started the fashion in selecting colour variants of this rose in Scotland, and in England, some 200 cultivars being listed. 'ANDREWSII', clear rose-pink, double; 'DOUBLE WHITE'; 'FALKLAND', pale pink double, with somewhat glaucous leaves; 'MARY QUEEN OF SCOTS', purple-maroon and grey-lilac, semi-double; 'WILLIAM III', maroon-red, double, are a few recognised cultivars. They are, like R. *pimpinellifolia*, very thorny, suckering shrubs, from 1½–5 ft with tiny leaves. In autumn these frequently turn to reddish purple or maroon before falling. Many have small, shining rounded heps, almost black. A few have yellow flowers and these can be traced to R. *foetida*, a fact borne out by the heavy scent, as opposed to the fresh sweetness of R. *pimpinellifolia*.

CENTIFOLIA.—Believed to be derived from a fusion of the Autumn Damask and the Alba roses. It probably originated in Holland and is depicted in Flemish paintings from 1603. It inherits characters from all the Old World ancestral roses. The outcome is a lax bush, with small and large thorns and prickles, limp dark green leaves, rounded and coarsely toothed; flowers large, globular when half open—the form almost always favoured by painters—filled with short petals in the centre; later the outer petals reflex. Calyx tube oval. During the 19th century many seedlings were raised from open pollinated flowers, with the result that later cultivars often show characters of the allied groups derived from R. *gallica*, in particular the variable foliage and the compact flowers and purplish colouring of some Gallicas. Immediately after flowering, prune away shoots that have flowered and also old weak wood to encourage strong shoots from the base; these should be reduced to conform to the overall size of the shrub in winter, to prevent wind-damage at flowering time.

CHINA.—A term given to the four original Chinese hybrids of R. *chinensis* which reached Europe around 1800, and also to hybrids raised from them which show close affinity to them. They are of small rather delicate growth, with smooth stems and occasional reddish prickles; small pointed dark green leaves; small flowers with five or more rather thin papery petals and, originally, little fragrance—apart from those deriving from 'Hume's Blush' and 'Parks' Yellow' Chinas which are tea-scented. They thrive best in the warmer parts of Britain, and in mild weather flower from spring to late autumn. In order to encourage plenty of new growth during the summer months, it is advisable to prune old and weak growth away in early spring. The China roses cannot be described as shrubs, but a few are included in this book because of their original influence on hybridising and because of their unique values. In warmer climates they develop into shrubs.

CLIMBER.—A term that embraces large-flowered hybrid roses of lax habit, mostly repeat flowering. Some have been raised unexpectedly, others originated as climbing sports of Hybrid Tea Roses, Floribundas and the like. There is, of course no such thing as a 'climbing' rose except in so far as hooked prickles enable them to grow through the branches of a tree. 'Climber' is a term of convenience, therefore, grouping together shrubs of lax habit which, for the sake of managing a garden, are conveniently trained upon supports, such as wall, fence or treillage. The still more lax ramblers (q.v.) are suitable for other purposes. Pruning consists of encouraging the growth of strong young shoots by shortening or cutting away weak and old wood in winter or early spring.

DAMASK.—The Damask roses are lax, open bushes, with small and large thorns and prickles, but few may be called typical except those described under R. *damascena*. In general the leaves are softer, more rounded than those of R. *gallica*, sometimes downy. Heps long and narrow. In some 19th-century hybrids these characters become merged with other roses of Gallica derivation. Flowers are loosely double or semi-double from white to deep pink. Pruning as for Centifolia. The Autumn Damask is described under R. *damascena*.

FLORIBUNDA.—By combining the Dwarf Polyanthas with Hybrid Teas, the Danish raiser Svend Poulsen raised 'Else Poulsen' and 'Kirsten Poulsen' in the early 1920s which produced continuously, from plants double the size of the Poly-poms, clusters of nearly single blooms in pink and crimson respectively. They were originally called Hybrid Polyantha Roses. Subsequently these and other hybrids, by crossing again with Hybrid Teas, gave rise to today's Floribundas. Many of the most vigorous modern cultivars such as 'Frensham', 'Iceberg', 'Yesterday' and 'Chinatown' will make large bushes, particularly if given only a modicum of pruning; this treatment usually results in a less continuous production of bloom. Pruning consists of shortening or removal of old and weak growths, to encourage strong young growths from the base, which produce flowers later in the season—the ideal aimed at in Floribundas which are essentially bedding roses; they are therefore excluded from the list of cultivars in this book.

GALLICA.—Strictly these are merely forms, or ancient hybrids of R. *gallica*. Many so called Gallica varieties are of 19th-century origin from seed of open-pollinated flowers. Some show parentage inherited from allied groups, such as Damask or Centifolia, in which case they are usually of lax growth as opposed to the sturdy upright bushes of R. *gallica*. In the true Gallica varieties the leaves are of harsh texture, dark green, long-pointed. Flowers borne aloft, erect. The buds are unpropitious, the greatest beauty is revealed when the flowers are fully expanded, showing perhaps quartering of petals and a 'button-eye'. Most are fully double. Colours range from pale pink to carmine, more or less flushed with maroon; some verge towards lilac and purple. None is white. Heps rounded. Armature is limited to small bristly thorns. Pruning as for Centifolia.

HYBRID CHINA.—A term used in the 19th century to include the many open-pollinated hybrids between the four original China Roses introduced to Europe around 1800 and the European hybrids of R. *gallica* parentage. Some were repeat flowering, but it is not a group that is known in gardens today.

HYBRID MUSK.—The Reverend Joseph Pemberton, of Romford, Essex sought to raise roses of simple needs for cottage gardens, as opposed to the sophisticated blooms of the Hybrid Tea Roses. Sometime before 1912 he used 'Trier' as a parent; this was raised by Peter Lambert in Germany in 1904 between a seedling of 'Aglaia' and the Hybrid Tea 'Mrs R. G. Sharman Crawford'. ('Aglaia' was the result of a cross between R. *multiflora* and the Noisette 'Rêve d'Or' which in its distant pedigree includes R. *moschata*.) The title chosen for this group, therefore, is far-fetched; in France the group is included with the Noisettes. Pemberton's roses are of mixed quality, some have never been popular, others have become household names. They are shrubs from 3 to 8 ft; one is a climber ('Moonlight') and all are recurrent flowering, mostly making sturdy growth. The flowers are borne in clusters, large or small in summer but in large clusters on the strong late shoots. After Pemberton's death new varieties were added by J. A. Bentall of Romford, Essex and also by Wilhelm Kordes of Germany. All except those by Kordes are markedly fragrant inheriting the musk-like quality of scent from the *Synstylae* roses in the parentage.

A full account of Pemberton's work will be found in the Royal National Rose Society's Annual for 1968 by G. S. Thomas. They are useful shrubs and make good informal hedges. Pruning consists of shortening flowering shoots and cutting away old and weak growth in winter or early spring to encourage new growths.

HYBRID PERPETUAL.—These are descended from a seedling raised in 1816 from 'Portlandica', known as 'Rose du Roi', hybridised with Bourbon Roses. As a general rule they are vigorous and prickly, with dark green leaves. The first crop of bloom is prolific, followed by spasmodic production, particularly from the cluster of buds usually produced at the tip of the tall new shoots. To obtain the maximum quantity of summer flower these long growths should be pegged down, tied to the base of neighbouring bushes, or bent over hoops, when they will flower in their second year along their entire length. Pruning consists of shortening or removal of old and weak growth in winter or early spring. Pink colouring predominates in this group, and the dark crimson of some, inherited from hybrids of 'Slater's Crimson China', is nearly always clouded with purplish tones. Most of the hundreds of varieties that were raised have been lost. They represented a groping after the garden ideal—the Hybrid Tea. They are mostly too tall for use as bushes for beds, and insufficiently shrubby to be included among the true shrub roses. They are therefore omitted from the account of cultivars that follows, with the exception of 'Reine des Violettes', which is reliably bushy. The following are some of the more bushy and satisfactory of the Hybrid Perpetuals:

'BARON GIROD DE L'AIN'.—A sport from 'Eugène Fürst', which occurred in 1897. Few prickles, broad leaves; flowers globular, petals nicked at the edge where the purplish crimson fades to white. 'EUGÈNE FÜRST' (1875) is similar to this, but without the white edges.

'BARONNE PRÉVOST (1842).—One of the more bushy and compact varieties; sturdy, erect growth and a fairly constant succession of large, flat, fully double flowers of rich pink suffused with lilac.

'FISHER HOLMES' (1865).—Rich scarlet lake, heavily suffused with crimson. Flowers fully double, globular.

'EMPEREUR DU MAROC (1858).—Rather spindly growth. Dark red flowers, shaded maroon but without purple; flat, quartered.

'GÉNÉRAL JACQUEMINOT' (1853).—Brilliant crimson.

'HUGH DICKSON' (1905).—Brilliant crimson; lanky grower, for pegging down.

'PAUL NEYRON' (1869).—Copious large foliage; flowers among the largest of the genus, deep rosy pink, lilac flush.
'SOUVENIR DU DOCTEUR JAMAIN' (1865).—Vigorous, for pegging down or a wall. Velvety dark wine-crimson, shaded maroon and purple.
'ULRICH BRUNNER FILS' (1882).—Very large, full carmine-cerise blooms; long growths; best pegged down.

HYBRID TEA.—With the constant desire in this century for larger flowers with longer petals, giving a more shapely bud, brighter colours, and constant flowering from summer to autumn, breeders resorted again and again to bringing in the asset of the Tea Roses, despite the fact that they were less hardy than the Hybrid Perpetuals, their rivals in popularity. In the late 19th century the desire was more for success on the show bench than in the production of plants of garden value. 'La France' (1867) is generally looked upon as the first Hybrid Tea and from this the group is dated, though several others had appeared which were of similar parentage. To the end of the century many hybrids had been raised, embracing colours from white to pink and crimson—mostly purplish—many showing the additional coppery yellow tones of the Tea Roses as well. There is a good illustration of 'La France' in the *Amateur Gardeners' Rose Book* by Julius Hoffman, pl. 9.

About 1838 Sir Henry Willock introduced from Persia a double form of R. *foetida*, the so-called Austrian Brier, which was called the Persian Yellow or R. *f. persiana*. The species had in its being the propensity to produce a colour-sport in which the brilliant yellow of the petals is contrasted by a brilliant orange-red upper surface. The large double yellow flowers of the Persian Yellow—far brighter than any of the Tea Roses of that date and only equalled by the delicate R. *hemisphaerica*—fired the rose breeder Pernet-Ducher of Lyon, France to try crossing it with Hybrid Perpetuals. All attempts failed except one with 'Antoine Ducher' which gave him a few seeds in 1888. One of two resulting seedlings showed the influence of R. *foetida* in its wood, prickles and foliage and in the colour of its flowers. Crossing this with Hybrid Teas, Pernet produced 'Soleil d'Or' in 1900. From this rose all the brilliant yellow, orange and flame roses of today are descended; R. *foetida*, a rose all on its own in many characters thus changed the entire colour scheme of garden hybrids from soft tones to brilliance or even garishness. This was furthered by the new pigment 'pelargonidin' which cropped up spontaneously after 1929 in Poly-pom Roses.

Pruning is aimed at the constant production of new shoots by reducing or removing old and weak wood in winter or early spring. This treatment maintains them for bedding uses, which is beyond the scope of this book. A few are of exceptional vigour, such as 'PRESIDENT HERBERT HOOVER' and 'JOANNA HILL'; the latter has been used as a parent for several shrub roses.

Moss.—Roses whose calyx-lobes, hep and pedicel are more or less covered with a moss-like growth of glandular branched processes and whose other green parts are sometimes adorned by glandular bristles. They are descended from two originals: R. *centifolia muscosa*, a sport of R. *centifolia* that arose before 1720, and R. *damascena bifera alba muscosa*, a sport from the Autumn Damask, which was recorded in France in 1835. The former flowers at midsummer only, the latter is repeat-flowering. The moss of the former is soft to the touch, of the latter bristly. In other characters they follow their parents. The most beautiful in flower of the Mosses—among which none surpasses the original—are of R. *centifolia* derivation, but the desire for repeat-flowering prompted the hybridists to concentrate on the Perpetual Damask Moss. The two races became merged in the late 19th century. Hybridisation was much helped by the occurrence of single-flowered sports from R. *centifolia muscosa* in France in 1814, the 'Single Moss', and in England in 1852, the 'Single Red Moss'. The hybrids with strong stalky growth usually gain this from the Damask. Attempts have been made, and still continue, to introduce yellow colouring, such as 'Golden Moss' (1932) and 'Robert Leopold' (1941). Otherwise the colouring is from white through pink to crimson and purplish maroon. The leaves are mostly of dark green borne on rather lax and often very thorny wood; the darker coloured hybrids usually have dark brownish moss and twigs, though there are exceptions such as 'Blanche Moreau' whose dark tinting is not normally associated with white flowers. Pruning for once-flowering moss roses is as for R. *centifolia*; for repeat-flowering cultivars harder pruning in winter is required.

NOISETTE.—Hybrids derived from a cross between R. *moschata* and the Pink China in the first place. The first hybrid, 'Champney's Pink Cluster' is not grown in this country; it originated about 1802 through John Champneys of Charleston, S. Carolina, USA, having fertilised R. *moschata* with pollen from 'Parson's Pink'. It flowered at midsummer only. Seeds from it were raised by Philippe Noisette, a nurseryman also of Charleston; one which flowered repeatedly until autumn was called 'Blush Noisette' and is found in many a garden in Britain. Subsequently the group was formed by other seedlings and hybrids with 'Parks' Yellow China' and gave rise to a group known sometimes as Tea-Noisettes (q.v.).

PENZANCE HYBRID SWEET BRIERS.—A race developed by Lord Penzance during the last years of the 19th century, using pollen from Hybrid Perpetuals and Bourbons to cross R. *eglanteria*, the Sweet Brier, or seedlings therefrom. These crosses gave rise to a dozen or more roses with single or semi-double flowers from blush to crimson, all inheriting some of the fragrance of the foliage of R. *eglanteria*. Flowers 2–3 in. across; a few bear heps. Vigorous, prickly growth to 9 ft. Some of the best known, and still grown, are 'AMY ROBSART', 'ANNE OF GEIERSTEIN', 'FLORA MCIVOR,' 'MEG MERRILIES' and 'ROSE BRADWARDINE'.

'LORD PENZANCE' and 'LADY PENZANCE' are less vigorous and

less fragrant in foliage; their single flowers yellowish pink, the yellow tint inherited from an admixture of 'Harison's Yellow' in the first and R. foetida 'Bicolor' in the second.

POLYANTHA POMPONS or DWARF POLYANTHAS.—The first roses of this group, sometimes called Poly-poms, were raised at Lyons in the 1870s from first-generation hybrids of R. multiflora. Dr Hurst's theory was that the pollen-parent of these hybrids was one of the dwarf Lawranceana roses (see p. ooo), whose influence came to expression in the next generation, but this theory has been challenged (see 'The Earliest Polyantha Roses' by Baronne E. de la Roche in the Year Book of 1969 of the International Dendrology Society, pp. 78–96). Two of the earliest varieties—'Paquerette' (1875) and 'Mignonette' (1882)—have respectively white and light pink flowers, and show the influence of a China rose. But others, the result of deliberate crosses, have Tea Roses in their parentage, while others again derive on one side from 'Crimson Rambler', itself a cultivar or hybrid of R. multiflora. The roses of this group are dwarf bushes with the small leaves and wood of a typical flowering shoot of R. multiflora, bearing a large number of small flowers in a typical multiflora fashion, but achieving popularity by the selection of colours from white, pink to crimson, orange-red and coral. Many of them are scentless, but are as perpetually in flower as any rose. They are valuable for small beds and miniature gardens. Pruning consists of the removal of old weak growth from the base in winter or early spring to encourage new basal growths. Since they are not shrub roses, further reference to them is omitted from this book, except for 'Cécile Brunner' and 'Perle d'Or'; 'Little White Pet' has a different origin.

PORTLAND.—A group of roses raised initially from 'Portlandica'. They are of compact habit, up to 4 or 5 ft high and wide, and bear a considerable resemblance to the Gallica group, except for their narrow calyx tubes and recurrent habit, coupled with a propensity to bear their topmost leaves more or less in a rosette under the flower. Colours white, pink, carmine and purplish. The influence of this group may be observed in several Moss roses, and also in the Alba rose 'Pompon Blanc Parfait' as well as 'Reine des Violettes (q.v.). Production of later blooms in the Portland Roses is encouraged by thinning out or shortening old and weak wood in winter or early spring, and the removal of spent flowers in summer.

RAMBLER.—A term that covers strong growing roses of lax habit, needing support. Many if left to their own devices will make dense thickets of overlapping shoots. They all derive from species in the Synstylae Section, mainly from R. multiflora, R. wichuraiana, R. arvensis, R. sempervirens and R. moschata. They mostly bear large trusses of small blooms, produced at midsummer except in a very few exceptions, among which are R. moschata, R. wichuraiana, 'Aimée Vibert' and 'Phyllis Bide'. In view of their graceful growth ramblers are particularly suitable for use on arches and swags, for training into trees and over hedgerows and large shrubs; their full beauty is revealed when, having initially been trained onto their support, the subsequent branches are allowed to hang down. When used for more formal effect, once-flowering ramblers should have old stems removed after flowering; this should be delayed until winter or early spring for repeat-flowering kinds. Many are scentless but there are exceptions.

Apart from isolated cultivars such as 'Ayrshire Splendens' and others showing affinity to R. *sempervirens* and R. *arvensis*, Ramblers fall into two fairly distinct groups, one typified by 'Dorothy Perkins', 'Sanders' White' and 'American Pillar', and the other by 'Albéric Barbier'. In his book *Climbing Roses Old and New*, G.S. Thomas attributes these distinct groups to the influence in the parentage of R. *wichuraiana* and R. *luciae* respectively. It is doubtful, however, whether R. *wichuraiana* was exclusively used in the first group, and when R. *luciae* was mentioned in contemporary records as a parent of the second group, R. *wichuraiana* was probably meant, since that species was originally included in R. *luciae* and the old name may have been used in some publications. Without definite proof, it is probable that the difference between the 'Dorothy Perkins' group and 'Albéric Barbier' group is largely due to the influence of the other parent used in the original hybridisation. When a Tea or related rose was crossed with R. *wichuraiana* (as in the Barbier hybrids), the R. *gigantea* element in the Tea parent would combine with the latter to give glossy, disease-resistant foliage, besides contributing a larger flower. In 'Dorothy Perkins', however, and in at least some of the Walsh hybrids, the bush roses used in the crosses were Hybrid Perpetuals, which have little, if any, R. *gigantea* in their make-up.

SHRUB.—A term devised to cover all roses which are not classed as bush or bedding roses (Hybrid Teas and Floribundas) nor as ramblers and climbers. Most species fall into this group. The rambling species of the *Synstylae* are also eligible according to the amount of ground available; unless attached to supports they will all make dense thickets of overlapping shoots. Some of the more lanky in growth of garden hybrids, such as many Hybrid Perpetuals, cannot truthfully be called shrubs, and only when their shoots are pegged down can they be used satisfactorily for bedding. Repeat-flowering shrub roses are rendered the more prolific by hard pruning in winter or early spring, but the grace of the plant is thereby lost.

TEA.—Pink Tea roses were derived from the Bourbon Rose crossed with 'Hume's Blush China'; the latter is presumed to have R. *gigantea* as one parent, from which long buds and petals resulted. The tea-scent was also inherited in this way, but perhaps more so when 'Parks' Yellow Tea-scented China' was used to hybridise with 'Blush Noisette', resulting in Tea Roses of yellowish colouring and also, more directly, the Tea-Noisettes.

'Safrano' (1839) was the first yellow Tea Rose, from which others were raised. 'Lady Hillingdon' (1910) is one of the richest in colour, apricot yellow, and is noticeably tea-scented. The Tea Roses need a warm climate to ripen their soft wood in autumn, and a good soil. Smooth twigs with few large reddish prickles; the young shoots and leaves are often purplish, blending with the colours of the flowers which though basically pink are suffused with coppery yellow to varying degrees; some are creamy white, some coppery yellow, others may verge towards red. Since their rather soft wood is apt to be damaged in winter, pruning should be left until spring, so that old, thin, or damaged wood can be cut away or reduced. Branches of yew or laurel placed around them will help to protect them in cold weather.

TEA-NOISETTE.—Hybrids in which the vigour and hardiness of the Noisette Roses have been blended with the coppery yellow tones of the Teas. Though a very varied group they formed the first large-flowered hybrid roses of yellowish colouring and have not been surpassed for their combination of size, dense flat array of petals when expanded, colouring and tea-scent. They are mostly of remarkably recurrent habit. Old and weak wood should be shortened or removed in winter or early spring to encourage plentiful new growth. The greatest production and the best flowers are usually produced on sheltered sunny walls in our warmer counties.

DESCRIPTIONS

'ADÉLAIDE D'ORLÉANS' (Rambler).—Raised in 1826 by Jacques, head gardener to the Duc d'Orléans, this is a hybrid of R. *sempervirens*, and like 'Félicité et Perpétue', is partially evergreen. Though so old, nothing like them has been raised since. The long trailing shoots have reddish prickles, and small dark green leaves. Buds deep creamy rose-pink, this tint remaining on the outer petals, while those inside the flower are blush white. The flowers are in small trusses and hang more or less vertically, being thus particularly appreciated above one's head. Yellow stamens. 15 ft or more. Midsummer. Delicate scent.
Thomas, *Climbing Roses*, frontispiece.

'AGNES' (Shrub). Dr W. Saunders, Ottawa, 1922. R. *rugosa* × R. *foetida* 'Persiana'.—This hybrid, the result of an unusual cross, shows characters of both parents, and is a useful and interesting plant. Excessively prickly like R. *rugosa*, it has bright green leaves like the other parent; stout, freely branching shoots build into a wide-spraying shrub. Its first crop of blooms is very free and here the large globular double flowers of the 'Persian Yellow' are in evidence, but they are of soft butter yellow. Receptacle bristly. 5 ft. Recurrent. Fragrant. A.M. 1951.
McFarland, *Roses of the World*, p. 2.

'AIMÉE VIBERT' (Rambler). 1828. Also known as 'Bouquet de la Mariée'.
—A rose of mixed parentage, but particularly showing the influence of R.
moschata in its flowers, and R. *sempervirens* in its magnificent, dark green, glossy,
long leaves. From R. *moschata* also it derives its flowering period, from late
summer to autumn. A luxuriant rambler for large supports, trees, etc., where its
masses of white flowers will show to advantage. Borne in small clusters, followed
by very large clusters on late shoots, they are pink-tipped in bud, opening to
pure white, fairly full, showing stamens; about 2 in. across. A yellowish sport
occurred in 1905 but has apparently been lost. Apart from R. *moschata* itself,
and 'Phyllis Bide', it is the only rambler that is continuously in flower until
autumn. 18 ft. Slightly scented.
    Choix des Plus Belles Roses, pl. 5.

'ALAIN BLANCHARD' (Centifolia/Gallica). 1839.—The rounded leaves and
thorny growth indicate Centifolia influence, the colour of the flowers derives
from Gallica. Semi-double flowers, cupped, of strong crimson, quickly becoming
mottled with maroon in sunlight. This and 'La Plus Belle des Ponctuées' are two
of the many 19th-century spotted varieties still left in cultivation. 4–5 ft strong
arching growth. Midsummer. Sweet scent.
    Thomas, *Old Shrub Roses*, pl. v.

'ALBA MAXIMA' (Alba). JACOBITE or CHESHIRE ROSE.—A big coarse
shrub with stout stems and large prickles, well covered with large grey-green
leaves. The flowers are rather flat, well filled with petals, but usually showing
some stamens; the array of petals lacks form as it does in 'Maiden's Blush'.
Big clusters of flowers which repay thinning. Very fragrant. 7 ft. Midsummer.
This rose sometimes reverts to a semi-double form, 'ALBA SEMIPLENA',
which in turn may sport back to 'Alba Maxima'.   [PLATE 16

'ALBÉRIC BARBIER' (Rambler). Barbier, 1900. R. *wichuraiana* × 'Shirley
Hibberd'.—The type of Wichuraiana rambler by which others are judged.
Exceptionally luxuriant and glossy dark foliage. Shapely butter-yellow buds
opening to large, flat flowers of creamy white, fully double and often quartered.
It is almost dense enough for ground cover. After the main crop of flowers is
over, the plant is seldom without a few blooms until autumn. 20 ft. Midsummer,
recurrent. Very fragrant.
    Trechslin and Coggiatti, pl. 1.

'ALEXANDRE GIRAULT' (Rambler). Barbier, 1909. R. *wichuraiana* × 'Papa
Gontier'.—This has all the attributes of 'Albéric Barbier' except that it seldom
produces late blooms. When first open, the flowers are of brilliant carmine,
fading quickly to lilac carmine, but always brightened by the yellow base to the
petals. It covers the lengthy treillage at the Roseraie de l'Haÿ near Paris very
effectively. Few prickles. 20 ft. Midsummer. Fragrant.
    Thomas, *Climbing Roses*, pl. iv.

'ALISTER STELLA GRAY' (Tea-Noisette). Raised by A. H. Gray, Bath,
introduced by G. Paul, 1894; known as 'Golden Rambler' in the USA; see also
'Claire Jacquier'.—Although suitable for training on a wall, it will make a big
open shrub for the back of the border. Wood mostly unarmed, except on very
vigorous shoots. Foliage brownish when young, smooth and soft green later;

many small shoots give a covering of blossom from small clusters, followed by very large clusters on strong new shoots, until autumn. Buds very shapely, of warm yolk yellow, opening to flat blooms with many narrow petals, creamy yellow fading to nearly white, 2½ in. wide. It is a very remarkable rose in many ways. 15 ft. Perpetual. Very sweetly scented. A.M. 1893.

Thomas, *Climbing Roses*, pl. v.

'ALOHA' (Shrub). Boerner, New York, 1949.—Although often classed as a climbing Hybrid Tea, this is in reality a shrub rose of heavy calibre, of dense upright growth. Copious, glossy dark foliage inherited from one parent, 'New Dawn'. Each flower is held erect, a remarkable study in its full-petalled shape, and also in colour; rose-pink on the upper surface of the petals, carmine on the reverse, but where they are gathered most closely in the centre a warm apricot tint obtains. 7–8 ft. Perpetually in flower. Fragrant.

'À LONGUES PÉDONCULES' (Moss). 1854.—As its name implies the flowers are long-stalked, with green moss, and on opening are small, rounded, of soft mauve-pink. The leaves are of soft green, small and rounded. A charming, little known rose of true Centifolia charm. 5 ft. Midsummer. Sweetly scented.

[PLATE 17

'ANDERSONII' (Shrub). Before 1912. R. *canina* × ?—Vigorous arching stems with large prickles. Flowers in clusters, about 2½ in. across, single, of brilliant pink with centre paler around the stamens. Long-pointed leaves, downy beneath. Heps oval, showy. Growth wider than high. 6 ft. Midsummer. Fragrant.

Willmott, Vol. II, p. 379.

'ANEMONE' (Climber). Schmidt of Erfurt, 1895. R. *laevigata* × Tea Rose; also known as 'Anemonoides'.—The supposed resemblance to a Japanese anemone is not obvious, but Schmidt was a raiser of these plants and hence his biased thoughts, perhaps. It is a sparsely leafed plant, owing leaf-shape and dark colour and glossiness, also the armature, to R. *laevigata*; flowers single, 4 in. wide, composed of broad rounded petals of warm clear pink veined with a deeper shade, but paler on the reverse; yellow stamens. A.M. 1900 as 'Sinica Anemone'. A sport occurred in California in 1913, with petals of glowing cerise-crimson, but much paler, even greyish, on the reverse. It is called 'RAMONA' and sometimes reverts to the original. A.M. 1950. Both are of rather stiff open growth, best on a sunny wall; prickles and thorns red-brown. 12 ft. Early summer; a few blooms later. Scented.

Willmott, Vol. I, p. 121; Trechslin and Coggiatti, pl. 25.

'ARTHUR DE SANSAL' (Portland). 1855.—Sturdy, bushy, erect growth with dark leaves. Flowers densely filled with small petals of darkest maroon-crimson, fading to crimson purple. The late summer shoots produce clusters of buds, which need thinning. 5 ft. Summer to autumn. Sweetly scented. Like other Portland varieties it is an old style of rose and yet recurrent flowering.

'ASSEMBLAGE DES BEAUTÉS' (Gallica). 1823. Also known as 'Rouge Éblouissante'.—There are few brighter cultivars among the old French roses. The vivid cerise-carmine is intensified by the multitude of small petals which reflex into ball-like shape, often with button eye and green central pointel. Sturdy bushy growth. 4 ft. Midsummer. Sweetly scented.

'Auguste Roussel' (Shrub). Barbier of Orleans, 1913. R. *macrophylla* ×
'Papa Gontier'.—This unusual cross makes an open, graceful, wide bush, with
rounded leaflets reminiscent of the species parent. The second parent has given
it size of flower; borne in clusters the blooms are shapely, semi-double, showing
stamens, with recurving petals of clear light pink. 6 ft. Midsummer. Fragrant.

'Autumn Fire' (Shrub). W. Kordes, 1961; original name 'Herbstfeuer'.—
An arching shrub, prickly, with abundant medium-sized foliage borne on red-
brown shoots. The flowers are spectacular, dark blood-red touched with scarlet
and maroon, semi-double, borne in clusters; a further crop appears in autumn. In
late autumn and winter the heps, probably larger than those of any known rose,
achieve orange-red colouring; they are long, flagon-shaped. 7 ft. Recurrent.

'Belle de Crécy' (Gallica).—The lax bushes are scarcely armed, and bear
neat dark green leaves. In spite of lax growth, the flowers are always turned upwards, fully double, perfectly formed, opening to a flat array of petals with
button eye. On opening they are of bright cerise pink, quickly turning to soft
violet and then to lavender-grey. 3–4 ft. Midsummer. Very fragrant.
   Trechslin and Coggiatti, pl. 2.

'Belle Poitevine' (Rugosa). 1894.—Typical deeply veined, dark green,
pointed leaves of R. *rugosa*. Yellow autumn colour with clusters of rounded
orange-red heps. The wood is greyish, densely covered with prickles and sturdy.
The flowers are loosely double, opening flat, 4 in. across, showing cream
stamens; mallow pink or magenta. 5 ft. Recurrent.
   Park, *World of Roses*, pl. 189.
   Closely related to this are: 'Souvenir de Christophe Cochet'
(1894), in which the leaves are less rugose and the flowers are of a harsh magenta
(portrayed in *Rosenzeitung*, 1893); and 'Delicata' (1898), in cool, soft lilac-
pink. These three roses are all markedly fragrant.

'Belle Portugaise' (Climber). H. Cayeux, Lisbon Botanic Garden, 1903.
R. *gigantea* × 'Reine Marie Henriette'. The large, elegant drooping leaves,
nodding long-petalled blooms and the fragrance are typical of the Tea race
derived from both parents. Large, loosely double flowers of creamy salmon with
deeper reverse. It requires a warm wall. 'La Follette', raised at Cannes by
Lord Brougham's gardener, Busby, about 1910, is similar in most respects but
is much richer in colour. 'Sénateur Amic' (Nabonnand, 1924) is a cross
between R. *gigantea* and 'General McArthur'; it has nearly single flowers of
vivid cerise-crimson at midsummer, like the above two. In Australia similar
crosses by Alister Clark have been raised, but need greater warmth than is
provided by British summers; among them are 'Flying Colours', 'Kitty
Kinninmouth' and 'Lorraine Lee'. They date from the early 1920s.

'Blairii Number Two' (Bourbon).—A vigorous climber with mahogany-
tinted young foliage. Flowers large, fully double, retaining a rich pink centre
while paling at the edge. 12 ft. Midsummer. Sweetly scented. 'Blairii Number
One' is seldom seen and is more fully double. Both grow at Hidcote.

'Blanc Double de Coubert' (Rugosa). 1892.—A hybrid of R. *rugosa*,
but with a more open, tall growth; typical array of dense prickles, and the usual
deeply veined, dark green, pointed leaves of that species. Flowers double, of

pure cold white slightly tinted blush in the bud. A most useful rose imparting pure white colouring. Yellow autumn colour; large rounded orange-red heps are occasionally produced. 6 ft. Recurrent. Very fragrant. A.M. 1895; A.G.M. 1966. *Journ. Roses*, February 1897; Trechslin and Coggiatti, pl. 3.

A sport, 'SOUVENIR DE PHILÉMON COCHET' was named in 1899; it is in all respects similar apart from the flowers in which most of the petals are crowded into a dense, level mass, surrounded by large outer petals. It occasionally reverts to the original.

'BLANC DE VIBERT' (Portland). 1847.—In this group of old autumn Damask hybrids a pure white is valuable. The fully double flowers have a hint of lemon in the centre. A neat erect bush with pale green foliage. 4 ft. Fragrant and recurrent.

'BLANCHEFLEUR' (Centifolia). 1835.—A somewhat hybridised cultivar with good large white flowers opening from red-tipped buds, often quartered, with rolled edges to the petals, a faint blush in the centre. Habit vigorous, open, prickly with light green leaves. Sweetly scented. 4 ft. Midsummer.

'BLANCHE MOREAU' (Moss). 1880. 'Comtesse de Murinais' × 'Quatre Saisons Blanc'.—This is a noted pure white Moss rose, with remarkable brownish green moss and thorns; a vigorous but slender, arching bush. Flowers at first cupped, then opening flat, rather small. Willmott's *The Genus Rosa* depicts this rose under the erroneous name of R. *centifolia albo-muscosa*, (Vol. II, p. 349). The blooms lack the full, rounded shape of the Centifolia Moss Rose. 5–6 ft. Midsummer, with a few later blooms (the presence of the Autumn Damask in both parents accounts for this second crop). Sweetly scented; a good rose for cutting.

'BLUSH BOURSAULT' (Boursault Climber). Before 1848. Also known as 'Calypso'.—Both this and the similar 'Madame de Sancy de Parabère' are valuable on account of their early flowering habit, and unarmed wood. Light green smooth leaves. Pale pink flowers, deeper in the centre, loosely double, opening flat from shapely buds. 15 ft. Early summer. Faintly fragrant.

'BLUSH NOISETTE' (Noisette). Also known as 'Flesh-coloured Noisette'. —A valuable rose, in constant flower from early summer onwards, making a big lax bush, with smooth purplish young shoots, almost without prickles, bearing small dull green, smooth leaves. Flowers in small clusters in early summer, followed by large clusters on strong new shoots; small, cupped, rather more than semi-double but revealing stamens, deep old rose in bud fading to creamy lilac-pink. A rose common in old gardens, preserved for its continued flowering and fragrance. Raised from seed of 'Champney's Pink Cluster' (a first cross between *Rosa moschata* and Parsons' Pink China) by Philippe Noisette of Charleston, United States, and distributed by his brother Louis in Paris in 1819. 15 ft if trained on a support. Perpetual.

Redouté, Vol. II, p. 77 ('Le Rosier de Philippe Noisette').

'BLUSH RAMBLER' (Rambler). B. R. Cant, 1903. 'Crimson Rambler' × 'The Garland'.—Sometimes found in old gardens, this is treasured for its fragrance and because it is nearly thornless. Closely related to R. *multiflora*. Light green foliage. Flowers semi-double, light pink. 15 ft. Midsummer.

'BOBBIE JAMES' (Rambler). Introduced by Sunningdale Nurseries, 1960; named in honour of the Hon. Robert James of St Nicholas, Richmond, Yorks.— An extremely vigorous, luxuriant plant, with copious bright green handsome leafage, on strong shoots attaining 25 ft or more. Blooms have six or seven petals, cream in bud opening nearly white, with bright yellow stamens; they are borne in large clusters. One of the best of the very vigorous ramblers, related to R. *multiflora*. Midsummer. Extremely fragrant.

'BOTZARIS' (Damask). Possibly affiliated to *Rosa alba*, judging from its coarsely toothed leaves, prickly stems, and distinctive fragrance. Very double flowers, opening flat, creamy white with lemon-white centre, fading to pure white, quartered and with button eye. 4 ft. Midsummer. One of the most beautiful of white varieties.

'BOULE DE NEIGE' (Bourbon). 1867.—Bold, leathery, smooth leaves are held on a vigorous upright bush, when grown on good soil. On poor soil it is often disappointing, but good flowers are well worth striving for. From red-tipped buds they open to pure white, fully double, of camellia-like perfection, reflexing in maturity to a ball. 4 ft. Recurrent. Sweetly scented.

Nestel's Illustrierte Rosengarten, pl. 5.   [PLATE 18

'BOURBON QUEEN' (Bourbon). 1835. Also known as 'Reine de l'Ile Bourbon'.—A popular old plant, often encountered. Leathery, dark green leaves, markedly serrate. The loose array of crinkled petals makes a large flower, magenta and pink tones predominating. 10 ft, but can be pruned to keep it as a bush. Midsummer. Fragrant.

'BREEZE HILL' (Rambler or Shrub). Van Fleet, USA, 1926.—This proably has R. *wichuraiana* as one parent, but unlike most of its relatives the foliage is dull green, rounded. Growth somewhat more bushy than most ramblers, and thus it will mound itself satisfactorily into a bush if desired. Large double blooms, rounded and full of petals, creamy apricot-rose fading to creamy buff. 12 ft with support. Midsummer. Very fragrant.

'BUFF BEAUTY' (Hybrid Musk). 1939.—The nearest to yellow in this group. Lax, arching branches which build up into a good shrub with pruning, or may be encouraged to grow into small trees or over hedges. The coppery brown young foliage, turning to dark green later, contrasts well with the flowers, which appear in small clusters in summer, but in long panicles on the strong growth in autumn. Blooms large and shapely, rich apricot-yellow, tinted coral in bud, fading slightly with age. 6 ft. Recurrent. A.G.M. 1966. Sweetly scented.

Thomas, *Shrub Roses of Today*, pl. 6.

'CAMAIEUX' (Gallica). 1830.—Loosely double flowers with broad petals, blush-white heavily striped and splashed with light crimson on opening; the colour intensifies to rich purple and then to lavender grey. A somewhat weak grower to 3 ft. 'TRICOLORE DE FLANDRE', 1846, is very similar; the flowers are more double, the petals more reflexed and rolled at the edges and on the whole the blush-white is less in evidence. Midsummer. Both are very fragrant.

ROSA 175

'CANTAB' (Shrub). Dr C. C. Hurst, Cambridge, 1927. R. *nutkana* × 'Red Letter Day'.—A neglected rose of great beauty with very large single blooms, 4 in. wide, borne in clusters on strong shoots, among small leaden green leaves. colour deep pink, lilac tinted, stamens cream. Large oval heps persist until late winter. 8 ft. Midsummer. Fragrant. A.M. 1939.

'CANTABRIGIENSIS' (Shrub). 1931.—A chance seedling which cropped up in the University Botanic Garden, Cambridge. The erect stems are covered with bristly thorns and eventually arch outwards, well clad in small green leaves. The shapely flowers are somewhat cupped, with five well rounded petals, of clear light yellow. Small orange-red heps in autumn. A strong healthy plant, more satisfactory than R. *hugonis*, which may be one parent. 7 ft. Early summer. A.M. June 1931. A.G.M. 1966. [PLATE 20

'CAPITAINE JOHN INGRAM' (Moss). 1856.—This is one of the most satisfactory of the darker coloured Moss Roses, making a fine large bush, with dark leaves and dark moss. Flowers densely filled with petals, reflexing considerably; intense dark purplish crimson shaded and mottled with maroon. The button eye reveals the lighter colour of the undersides of the petals. 5–6 ft. Midsummer. Sweetly scented.

Thomas, *Old Shrub Roses*, pl. 6.

'CARDINAL DE RICHELIEU' (Gallica).—This probably has some China influence in its parentage, on account of the smooth leaves, and shiny green wood. When opening the flowers show the paler reverse of the petals; they are cupped and make a rounded flower. Later the petals reflex showing the intense velvety maroon-purple colouring and also the pale centre. 5 ft. Vigorous. Midsummer. Some fragrance.

Trechslin and Coggiatti, pl. 5.

'CÉCILE BRUNNER' (Poly-pom). 1881. R. *multiflora* seedling × Tea Rose. Also known as 'Madame Cécile Brunner', 'Mignon' and the 'Sweetheart Rose'.— This does not conform to the accepted idea of a Poly-pom, being of thin upright growth. Small, pointed leaves, dark green. Of all roses this has the most exquisitely fashioned buds, like a tiny Hybrid Tea of the best quality. Delicate pink flowers, deeper in the centre and on the reverse. Late summer basal shoots bear very large clusters. 3 ft. Perpetual. Fragrant.

Journ. Roses, February 1885; Trechslin and Coggiatti, pl. 6.

This rose has produced sports: 'WHITE CÉCILE BRUNNER' arose in the nursery of Fauque et Fils, Orleans, in 1909. The white flowers are lemon-tinted in the centre. 'CLIMBING CÉCILE BRUNNER' originated with Hosp in California in 1894. This is probably now better known than the original, since it is very vigorous and produces plenty of shoots for propagating, which are scarce on the original and the white form. The flowers are somewhat larger and produced in great quantity at midsummer, a few later. It is suitable for training on a wall, as at the University Botanic Garden, Oxford, or into trees. 'MADAME JULES THIBAUD' is probably another sport; flowers coral pink.

The plant known in British gardens as 'BLOOMFIELD ABUNDANCE' is possibly a sport of 'Cécile Brunner', but until an established bush of 'Cécile Brunner' sports again, or the accepted 'Bloomfield Abundance' reverts, there is no conclusive evidence available; the type 'Bloomfield Abundance' was raised in the USA in 1920, with a stated parentage of Hybrid Teas.

'CÉLESTE' (Alba). Early 19th or late 18th century.—This appears in Redouté as 'Rosa damascena Aurore', but its grey leaves, particularly clear pink flowers, and few large prickles indicate quite clearly the Alba group. It originated in Holland. The soft colours of flower and leaf make a happy combination. The buds are of exceptional beauty, opening into a loosely double flower. Sturdy erect bush. 6 ft. Midsummer. A.M. 1948. Sweetly scented.

Redouté, Vol. II, p. 41, as R. *damascena* Aurore.

'CÉLINE FORESTIER' (Tea-Noisette). Trouillard, Angers, 1842.—This is a renowned Tea-Noisette of superlative quality; it has the soft light green leaves of so many of its group. Though taking some years to gather vigour it is when established strong growing, leafy and floriferous, but is best on a warm sunny wall to encourage early ripening and free flowering. Early flowers on short side shoots, later ones at the extremities of new shoots. From shapely buds they open flat, somewhat cupped with numerous short petals in the centre, surrounded by larger reflexing petals, quartered and often with button eye, of clear, almost sulphur yellow. Each bloom is a perfect circle. 12 ft. Perpetual. Very fragrant. *Journ. Roses*, October 1880.

'CELSIANA' (Damask). Named by Thory in Redouté in memory of the famous nurseryman Cels of Paris (*d.* 1806), who introduced it to France from Holland.—An excellent, free-growing shrub, with light green leaves, smooth and greyish. Flowers in loose clusters, large, semi-double, revealing yellow stamens; warm light pink on opening fading to blush. 4 ft. Midsummer. Sweetly fragrant.

Redouté, Vol. II, pl. 53.

'CERISE BOUQUET' (Shrub). W. Kordes, Germany, 1958.—Several of Kordes' noted shrub roses are by-products of his efforts to breed into popular races some of the hardiness and stamina of species. In this R. *multibracteata* is united in a superlative shrub with the rich crimson Hybrid Tea, 'Crimson Glory'. When established 'Cerise Bouquet' will throw out shoots 10 ft long, arching, graceful, and bearing small rounded dull green leaves. Both these characters and the plentiful pale green bracts below the flowers are inherited from the species. Flowers are large, fully double, of brilliant cerise-crimson, appearing freely in clusters in summer all along the great shoots and a smaller crop is usually given in autumn. A highly ornamental shrub needing plenty of space, since it is wider than high. 12 ft, or more when climbing into low trees. Recurrent. Very fragrant. A.M. 1958.

Thomas, *Shrub Roses of Today*, pl. 1.

'CHARLES DE MILLS' (Gallica). Also known as 'Bizarre Triomphant'.—An erect, almost thornless shrub with typical dark green leaves. The flowers are remarkable in colour and shape but are only slightly fragrant. When half open the petals present a dense flat array, later reflexing considerably, and revealing an empty receptacle in the centre. The medley of tints varies from intense crimson on opening to a dull wine-purple tint on maturing. 4 ft. Midsummer.

Park, *World of Roses*, pl. 176.

'CLAIRE JACQUIER' (Tea-Noisette). 1888.—This is almost identical in bloom to 'Alister Stella Gray', but is far more vigorous and has more pointed leaves with coarser serrations. It flowers only at midsummer, but is a decorative rose, rare in its colour and vigour. 25 ft. Midsummer. A.M. 1889. Sweet scent.

'COMMANDANT BEAUREPAIRE' (Bourbon). 1874. Also known as 'Panachée d'Angers'.—A thorny, arching shrub, as wide as high, with light green very pointed leaves. The flowers are prolifically borne, loosely double, rounded and cupped, of light pink profusely striped and splashed with madder, carmine and purple. 6 ft. Midsummer. See also 'Honorine de Brabant'. Sweet scent.

Edwards, *Wild and Old Garden Roses*, facing p. 87.

'COMPLICATA' (Shrub/Climber). Probably a hybrid of R. 'Macrantha' of gardens, which originated on the Continent in this century.—It forms an arching bush, densely mounded, or can be trained over hedgerows and into small trees. The blooms are borne in clusters along the length of the branches; large, single, of brilliant pink with paler centre around the yellow stamens. A spectacular shrub, 5 ft high and wide unless trained upwards, when it will ascend to 10 ft. Midsummer. Heps oval. Sweetly scented. A.M. 1951. F.C.C. 1958. A.G.M. 1965.   [PLATE 21

'COMTE DE CHAMBORD' (Portland). 1860.—A good sturdy plant with light green leaves. Large flowers with many rolled petals, sometimes quartered, opening flat; they are clear pink with a faint lilac tone. Its perpetual flowering habit and Gallica flower-shape bring together the best of both worlds. It is undoubtedly the most beautiful of the Portland Roses remaining in cultivation. 4 ft. Summer to autumn. Sweetly fragrant.

'COMTESSE DE MURINAIS' (Moss). 1843.—Light green, ribbed leaves, on a vigorous bush. The moss is hard to the touch, denoting Damask influence. Flowers large, full petalled, blush-white when first open, turning to milk-white, with pronounced button-eye. The petals are narrow and flat, sometimes quilled, and quartered. 6 ft. Midsummer. Sweetly scented.

'COMTESSE DU CAYLA' (China). 1902.—Slender weak growth with neat small leaves, purplish when young. Loose, semi-double flowers of brilliant coral-flame, fading to coral pink, reverse yellowish. This is of the same type as 'MADAME LAURETTE MESSIMY', 1887, whose flowers are of coppery salmon-pink with yellow base, and which is figured in *The Garden*, October 24, 1891, p. 378. A pair of desirable bedding roses, though they will ascend to many feet on a warm wall; constantly in flower. Delicate fragrance.

'CONDITORUM' (Gallica).—A plant grown in the University Botanic Garden at Oxford, and reputedly used for the extraction of attar in Hungary. The unsophisticated flower shape indicates an ancient origin. Loosely double flowers, showing stamens of dark magenta-crimson flushed and veined with purple in hot weather. 3–4 ft. Midsummer. Sweet scent.

'CONSTANCE SPRY' (Shrub/Climber). David Austin, Wolverhampton, 1961. 'Belle Isis' (Gallica) × 'Dainty Maid'.—Principally because Constance Spry had done so much to preserve the old French roses, and because she was specially devoted to those with the shape of R. *centifolia*, this was named in her honour.

The globular, deep centred, soft pink blooms do indeed resemble those of that rose. They are borne singly and in clusters on stout stems from the lax, long branches of the previous year. Young shoots brownish, likewise the prickles; good large foliage which is apt to burn in hot weather. Though often grown as a bush, it needs the support of fence or wall, or can be allowed to climb through other shrubs, into small trees, etc. 7 ft. Midsummer. Rich fragrance reminiscent of myrrh. A.M. 1965. A.G.M. 1966.

'CORNELIA' (Hybrid Musk). 1925.—A vigorous spreading bush with glossy dark green foliage. Flowers carried in graceful sprays, a few together in summer; later, on the strong young growth the panicles may exceed one foot in length. Buds unprepossessing until half open, when they show soft coppery pink shaded with apricot, fading to a lighter tint. In autumn the colour is intensified. 5 ft, wider than high. Recurrent; sweet far-carrying fragrance.

'CORYANA' (Shrub). Dr C. C. Hurst, Cambridge, 1926. R. *roxburghii* (seed-parent) × ?R. *macrophylla*.—This is a large, substantial shrub, densely branched to the ground, with good foliage; the single deep pink flowers, of good size, are rather hidden by the copious foliage. 8 ft. Little scent.

'COUPE D'HÉBÉ'. 1840.—Though usually classed as a Bourbon, this is obviously of more mixed parentage. The strong growths are copiously clothed in bright green leaves. Flowers of rather modern outline, shapely, of soft pink, fading very little. 8 ft, best with some support. Midsummer. Sweetly fragrant.

'CRIMSON SHOWER' (Rambler). Norman, 1951.—Though almost scentless, this valuable variety has become popular because of its unfading colour and the fact that it flowers when other ramblers have finished, extending into September. Small, semi-double flowers in large graceful trusses, rich crimson. 15 ft.

'D'AGUESSEAU' (Gallica). 1823.—Typical Gallica foliage of good size. Intense crimson, fading to intense cerise-pink; flowers full-petalled, reflexing, often quartered and with button eye. One of the most brilliant Gallicas and a vigorous plant to 5 ft. Midsummer. Sweet scent.

'DE LA GRIFFERAIE' (Shrub). Vibert, 1846.—Fully double flowers borne in small clusters reminiscent of many of the old Gallica roses; magenta-cerise, fading to lilac-white and deliciously scented. It is a vigorous plant achieving 6 ft. Stems stout and almost unarmed; leaves broad, dark green and rounded. The stipules are much frayed, which suggests that R. *multiflora* was one parent, the other presumably being one of the old French roses. It has for long been known as R. *multiflora* 'De la Grifferaie' in the nursery trade and was at one time used as an understock; it is tough and long-lived and its presence in old gardens usually indicates that it has survived the rose budded onto it.

'DE MEAUX' (Centifolia). 18th century or earlier.—An erect little bush with fresh green foliage. The whole plant is in miniature proportions but will achieve 4 ft. Tiny flowers of bright pink, deeper in the well-filled centre. 'White de Meaux' is an albino sport, white with pink central petals. There was also a Moss form 'Mossy de Meaux' which seems to have been lost; it was introduced in 1814. These varieties are earlier in flower than R. *centifolia*. Midsummer. Sweet scent.

Willmott, p. 353, as R. *pomponia*.

'Desprez à Fleur Jaune' (Tea-Noisette). Desprez, Yebles, France, 1830. 'Blush Noisette' × 'Parks' Yellow Tea-scented China'.—A most historic hybrid, and well worthy of preservation, apart from its garden value. It is very vigorous, with few prickles and light green leaves. The early flowers are on short side shoots, but every late-growing tip ends with a cluster, until autumn. About 2 in. wide, they are silky-petalled, fairly full, opening flat, showing stamens; creamy apricot-pink fading to pale peach-yellow. It needs a hot wall and plenty of space, where its growths can be trained up and then hang down. 15 ft or more. Perpetual. Probably the first yellowish climbing rose to be raised. Rich scent.

Choix des Plus Belles Roses, pl. 35.

'Deuil de Paul Fontaine' (Moss). 1873.—Of dwarf growth with very bristly stems, dark leaves, hard reddish moss which like its recurrent habit indicates Damask parentage. The flowers are globular, very double with many dusky tints from dark pink through purple and maroon to near black, with brownish shadings. Quartered. 2–3 ft. Midsummer, with a few later blooms. Fragrant.

Trechslin and Coggiatti, pl. 8.

'Devoniensis' (Tea). G. Foster, Devonport, introduced by Lucombe and Pince, 1841; climbing sport raised by Pavitt of Bath and introduced by Henry Curtis, Torquay, 1858.—The original bush form is now rarely seen. 'Climbing Devoniensis' closely resembles 'Gloire de Dijon' (q.v.) in all but colour and in being more bushy. Flowers equally full of petals, creamy white with a flush of pale apricot when opening and in the centre. 12 ft. Recurrent. Rich fragrance.

'Doncasteri' (Shrub). Probably raised by Dr C. C. Hurst, introduced between the wars by Mr Doncaster of J. Burrell and Co., Cambridge.—This is near to R. *macrophylla*. Compared with that species the leaves and flowers are smaller and darker. In early autumn it is laden with large, red flagon-shaped heps, at which time it has few peers. 6 ft. Early summer. A.M. 1958.

'Dorothy Perkins' (Rambler). Jackson Perkins, 1901. R. *wichuraiana* (or hybrid of) × 'Madame Gabriel Luizet' (a Hybrid Perpetual).—Large and small trusses of small, double, clear pink flowers, nearly scentless. Glossy, dark green leaves. Subject to mildew.

Journ. Roses, June 1908; Darlington, *Roses*, pl. 5.

'Dorothy Perkins' has given rise to a paler coloured sport, 'Lady Godiva', of considerable charm. This originated in 1908 and produced in 1932 a dwarf, spreading sport, 'The Fairy'.

'Duc de Guiche' (Gallica).—Probably one of the later cultivars on account of its splendid flowers, opening cupped with the outer petals reflexing, and later the whole flower becomes a ball. Intense magenta-crimson quickly showing delicate purple veins and flush in hot weather. A good bush with good foliage. 5 ft. Midsummer. Sweet scent.

'Duchesse de Verneuil' (Moss). 1856.—Leaves pointed, light green, on an effective upright plant. Clear bright pink flowers; petals paler on reverse which shows distinctly in the pronounced button eye. 5 ft. Midsummer. Sweet scent.

'Du Maître d'École' (Gallica).—A mystery rose, purchased by Mrs Constance Spry from Messrs Pajotin-Chedane at La Maître École, France, but this firm disclaim its introduction. Good foliage on a compact, nearly thornless bush, whose stems splay outwards under the weight of the very large, double, flat blooms, quartered and with button eye. The soft old rose tint gives way to lilac-pink with delicate mauve and coppery tints. 3 ft. Midsummer. Very fragrant.

[PLATE 19

'Eos' (Shrub). Ruys, Dedemsvaart, Holland, 1950. R. *moyesii* × 'Magnifica' (a hybrid of R. *eglanteria*).—A tall, stiff, prickly shrub with dainty small foliage inherited from the former parent, of dark leaden green. The long wand-like growths produce short side shoots the next year, each bearing a small cluster of brilliant coral-red flowers, with six or seven petals. Stamens yellow. Named after the Goddess of the Dawn, its colouring is appropriate, but being sterile, it produces no heps. For background planting. 12 ft. Early summer. A.M. 1956.

'Erfurt' (Shrub). W. Kordes, 1939. 'Eva' × 'Reveil Dijonnais'.—A first-rate compact shrub, with beautiful glossy foliage, coppery purple when young. The flowers are in constant production from summer to autumn, large, 3 in. wide, semi-double, of bright glowing pink but with large creamy white centre around the stamens. One of the best recurrent roses for smaller gardens. 3–4 ft. Fragrant.

Thomas, *Shrub Roses of Today*, pl. VII.

'Evangeline' (Rambler). Walsh, 1909. R. *wichuraiana* × 'Crimson Rambler'.—The single flowers of pale pink, carried in medium-sized trusses over glossy foliage, have a further attraction among so many scentless ramblers, for they are deliciously fragrant. 18 ft. Midsummer.

'Excelsa' (Rambler). Walsh, 1909.—This resembles 'Dorothy Perkins' except in colour; the clear bright crimson flowers are long lasting, but are surpassed later in the season by 'Crimson Shower' (q.v.). 18 ft. Midsummer.

'Fantin-Latour' (Centifolia hybrid).—The name is a mystery, apparently without foundation, but the rose is well-known and is certainly of the type of flower so successfully painted by the artist. It bears some resemblance to R. *centifolia*, in its wood, armature and foliage, less so in the flowers, though these are large and shapely, of a bland pale pink richly tinted in the folds. Very vigorous and branching. 6 ft. Midsummer. Sweetly scented. A.M. 1959. A.G.M. 1966.

[PLATE 22

'Felicia' (Hybrid Musk). 1928. 'Trier' × 'Ophelia'.—One of the most satisfactory of the Hybrid Musks, making good rounded bushes, well branched and suitable for hedging. Foliage broad, rich green. Flowers in summer in small clusters, very free, loosely double, warm apricot-pink on opening, fading to two tones of silvery pink; larger clusters appear in autumn. 5 ft. Recurrent. Sweetly scented. A.G.M. 1966.

'Félicité et Perpétue'. Jacques, head gardener to the Duc d'Orléans, 1827.—This is one of the very few roses owing affinity to R. *sempervirens*, and like that species is practically evergreen; moreover it is extremely hardy and old plants may be seen in old gardens at high altitudes in Wales and Scotland and elsewhere. Dense, overlapping, prickly growths which flower most freely when not pruned. Perfectly double blooms in small clusters, with a neat regular array of petals, milk-white, emerging from crimson-touched buds. 15 ft. Midsummer. *Rev. Hort. Belg.*, 1890, pl. 18; Thomas, *Climbing Roses*, pl. 1.

A dwarf sport occurred in 1879 and is named 'LITTLE WHITE PET'. It produces flowers constantly from summer to autumn and exactly resembles the original except for its growth which does not exceed 2 ft. *Gartenzeitung*, 1880, pl. 28.

'FÉLICITÉ PARMENTIER' (Alba, with Damask affinity). 1836.—The light yellowish green, pleated leaves, and pale green twigs and dark thorns separate this from other old French roses. The buds are yellowish, borne in clusters, and open into very full-petalled flowers of clear flesh pink, at first cupped and then reflexing into a ball, when the petals fade to creamy white at the edges. A sturdy bush, arching with the weight of the flowers. 5 ft. Midsummer. One of the most exquisite old French roses and deliciously fragrant. Like some other Alba roses, it is advisable to thin some of the large clusters of buds to two or three.

'FELLENBERG' (Noisette). Before 1842.—Of obscure origin, this is a bushy, well-known rose, of great hardiness, often pruned hard as a bedding plant. For this purpose it is well suited and is constantly in flower. As a lightly pruned shrub it will get tall; of rather open, awkward habit; dark lead-green small leaves, purplish when young. Flowers carried in small clusters in summer and prolifically later on summer shoots, small, double, deep warm pink or light crimson, showing stamens; not more than 2 in. in width. Fragrant. 7 ft. A.G.M. 1929. Willmott, Vol. I, p. 97.

'FERDINAND PICHARD' (Bourbon). R. Tanne, France, 1921.—In its foliage, which is smooth and pointed, this very closely resembles several established Bourbons, such as 'Honorine de Brabant', and may be a sport, but it is more compact in growth than these vigorous cultivars. Flowers clear pink heavily dotted, splashed, striped and flaked with vivid crimson; on fading the darker tint turns to purple; cupped in shape, medium size and fairly double. It is one of the most productive of cultivars. 5 ft. Perpetual. Sweet scent.

'FIMBRIATA' (Shrub). 1891. R. *rugosa* × 'Madame Alfred Carrière'. Also known as 'Dianthiflora' and 'Phoebe's Frilled Pink'.—There is little to show of the second parent; R. *rugosa* is represented by the prickly stout stems which build up into a big arching bush, freely set with slightly rugose foliage. Flowers in clusters, of palest pink, each petal daintily fringed at the edge. It comes to greatest perfection in the cooler north and in Ireland. 6 ft. Recurrent. Sweet scent. A.M. 1896.
Journ. Roses, September 1896; Thomas, *Shrub Roses of Today*, Fig. 5 (pencil drawing).

'FLORA' (Rambler).—An old hybrid of R. *sempervirens*, with glossy dark green leaves, lasting well into the winter, borne on long, trailing shoots with prickles. Flowers double, nodding, in small clusters, full-petalled, old rose pink with darker, cupped centre. 12 ft. Midsummer. Fragrant.

'FRANCIS E. LESTER' (Rambler). Lester Rose Garden, California, 1946.— Although descended from 'Kathleen', a Hybrid Musk, this is a rambler, of dense, rather bushy growth. Very dark green leaves, with faint edging of maroon.

Large and small clusters of single flowers with yellow stamens; clear pink in bud
fading to white on opening, like apple blossom. 14 ft. Midsummer. Extremely
fragrant.

'FRANÇOIS JURANVILLE' (Rambler). Barbier, France, 1906. R. *wichuraiana*
× 'Madame Laurette Messimy'. This has most of the attributes of 'Albéric
Barbier'; small leaves, glossy, dark green on lengthy trailing shoots. Flowers
about 2½ in. wide, opening flat, fully double, often quartered, opening rich coral
rose, fading to light rose, with stamens and bases of petals yellow. Best away
from walls and fences where it is apt to suffer from mildew. 20 ft. Midsummer
only. Sweet fragrance. Wisley Rose Award 1926.

'FRITZ NOBIS' (Shrub). W. Kordes, Germany, 1940. 'Joanna Hill' ×
'Magnifica' (a hybrid of R. *eglanteria*).—One of the few shrub roses which has a
sophisticated, modern shape to its flowers, reminiscent of a Hybrid Tea, in clear
pink of two tones. They are borne in large and small clusters all over a beautiful
bush, with somewhat zig-zag stems, arching effectively, and well set with good
green foliage. Large prickles. An exceptionally fine shrub wider than high.
Sweet fragrance. 6 ft. Midsummer. A.M. 1959. A.G.M. 1966.

'FRU DAGMAR HASTRUP' (Rugosa). 1914.—A seedling of R. *rugosa* with
compact bushy growth, densely covered with prickles. Leaves typically deeply
veined, small, rounded, turning yellow in autumn. Flowers in clusters, single,
clear light pink, avoiding the magenta colouring of most Rugosa derivatives,
with cream stamens, followed by large rounded dark red heps. Of suitable
growth for a hedge. Occasionally branches die. 4 ft. Recurrent. Sweet scent.
A.M. 1958. A.G.M. 1966.   [PLATE 23
Thomas, *Shrub Roses of Today*, pl. IV (flowers) pl. III (heps).

'FRÜHLINGSANFANG' (Shrub). W. Kordes, Germany, 1950. 'Joanna Hill'
× R. *pimpinellifolia* 'Grandiflora'.—Large single ivory-tinted flowers, about 4 in.
across, cover the arching bushes, which are themselves a mass of good dark
foliage. This is a most effective shrub, inheriting the spiny stems of 'Grandiflora';
maroon-red heps are produced in autumn when the foliage usually turns to
bright colours. For a woodland glade few shrubs are more effective. 7 ft. Early
summer. Powerful fragrance. A.M. 1964.

'FRÜHLINGSGOLD' (Shrub). W. Kordes, Germany, 1937. 'Joanna Hill' ×
R. *pimpinellifolia* 'Grandiflora'.—This is the most famous and satisfactory of a
group of six cultivars named by the raiser from similar crosses. While 'Frühlingsanfang' makes the most shapely well-filled bush, this excels in the beauty of
its semi-double, warm butter-yellow flowers, 4 in. across, fading to creamy
white, but at all times beautiful, and revealing dark yellow stamens. Spiny,
wide-arching branches, good leaves of dull green. 6 ft, and much wider. Early
summer and occasionally later. Powerfully fragrant. A.M. 1950. F.C.C. 1955.
A.G.M. 1966.   [PLATE 24
Gault and Synge, *Dictionary of Roses*, pl. 165.

'FRÜHLINGSMORGEN' (Shrub). W. Kordes, Germany, 1941. ('E. G. Hill'
× 'Catherine Kordes') × R. *pimpinellifolia* 'Grandiflora'.—This has never
proved such a good shrub as the foregoing two, making rather awkward growth,
sparse and spindly at times, but there is no doubt that the leaden green leaves

and exquisite blooms leave little to be desired. Flowers single, 4 in. across, rich rose-pink, passing to clear pale yellow in the centre, with the incomparable addition of maroon-coloured stamens. It usually produces a second crop in autumn. 6 ft. Recurrent. Rich fragrance, and has maroon heps. A.M. 1951.

'GARDENIA' (Rambler). Manda, USA, 1899. R. *wichuraiana* × 'Perle des Jardins'.—This has most of the attributes of 'Albéric Barbier', but is a richer yellow in bud, fading to cream. Leaves moderately glossy, on a vigorous and healthy plant. 18 ft. Creates a spectacular display on the pergola at Bodnant, N. Wales. Sweet fragrance. Few later blooms.

'GÉNÉRAL KLÉBER' (Moss). 1856.—This bears a marked resemblance to 'Duchesse de Verneuil' in its growth, light green foliage and moss. Very shapely long buds open to wide, beautiful blooms of soft clear pink. 5 ft. Midsummer. Sweet scent.

'GÉNÉRAL SCHABLIKINE' (Tea). 1878.—A useful hardy Tea rose, showing considerable China Rose influence, which has three good bursts of bloom, alternating with the main crops of modern Hybrid Teas, the first crop being very early. Good growth, with dark green smooth leaves, purplish when young. Flowers singly or in small clusters, deep coppery pink or carmine; shapely buds developing into many-petalled flat flowers; this shape indicates some influence of the old French roses, but the smooth stalks and leaves do not. 5 ft, or higher on a wall. Perpetual. Delicate fragrance.

'GERBE ROSE' (Rambler). Fauque et Fils, Orléans, 1904. R. *wichuraiana* × 'Baronesse Rothschild'.—Less rampant than the others of the 'Albéric Barbier' group to which this approximates, and comparable to 'Breeze Hill' in being of bushy growth. Almost unarmed, reddish shoots, with dark glossy leaves; large loosely double flowers, 3 in. across, petals cupped, quartered and crinkled, soft pink with a hint of lilac and cream. 10 ft. Midsummer, seldom without a flower later. Sweet fragrance.

'GEORGES VIBERT' (Gallica). 1853. Also known as 'La Pintade', 'Gallica Meleagris'.—A neat little rose, with small pointed leaves and erect growth. Flowers of medium size opening flat with many narrow petals, of blush pink with pink and carmine stripes and green pointel. 3–4 ft. Midsummer. Sweet scent.

'GLOIRE DE DIJON' (Tea-Noisette). Jacotot, Dijon, 1853. ?'Souvenir de la Malmaison' × Tea Rose.—Those who cannot provide the warmth needed by 'Maréchal Niel' could not do better than to grow this or 'Céline Forestier'. Purplish young wood with few prickles; leaves also richly coloured when young turning to dark green. The stems are vigorous, but often are bare for 6 ft or so. Flowers open from globular deep-centred buds, to a flat array of quilted and quartered petals, sometimes with button eye, resembling those of 'Souvenir de la Malmaison'. Colour variable, deep buff-yellow more or less warmed by pink and apricot, particularly in warm weather. 15 ft. Recurrent, profuse, seldom out of flower after midsummer. Rich fragrance.
Hoffmann, pl. 5.

'GLOIRE DE FRANCE' (Gallica).—Large fully double flowers of warm lilac-pink, fading to lilac-white around the edges. The bush is well branched and has good leaves. 3–4 ft. Midsummer. Sweetly scented.

'GOLD BUSH' (Shrub). W. Kordes, Germany, 1954. 'Golden Glow' × R. *eglanteria* hybrid. Named 'Goldbusch' in Germany.—The wide-arching, prickly stems have leaves of yellowish green, inheriting some of the aroma of those of the species. Along the branches are borne clusters of shapely apricot buds, opening to semi-double peach-yellow blooms showing yellow stamens. Usually wider than high; 5 ft. Recurrent. Sweet fragrance. A.M. 1965.

'GOLDEN CHERSONESE' (Shrub). E. F. Allen, Copdock, Suffolk 1963. 'Canary Bird' × R. *ecae*.—The dainty foliage borne on wiry, thin sprays of prickly wood and the bright yellow small flowers incline to the second parent, but it inherits greater size of flower from 'Canary Bird'. 6 ft. Spectacular in early summer. A.M. 1966.

Hollis, *Roses*, facing p. 171.

'GOLDEN WINGS' (Shrub). 1956. Roy Shepherd, USA, 1956. Parentage complex, including 'Ormiston Roy' and 'Soeur Thérèse'.—This is a pointer to what may occur in other colours of shrub roses, being of manageable, upright bushy growth, with some prickles, and leaden green foliage. The clusters of blooms are produced freely and followed by fresh clusters at the end of every new shoot. Flowers large, 4 in. across, single, of clear yellow, deeper in the centre around the amber-brown stamens. 6 ft. Perpetual. Powerful sweet fragrance. A.M. 1965. A.G.M. 1973.

Thomas, *Shrub Roses of Today*, pl. VII.

'GOLDFINCH' (Rambler). George Paul, 1907.—Like 'Blush Rambler', this is closely related to R. *multiflora*. It is almost unarmed and has light green leaves. The flowers are very fragrant, semi-double, in large trusses; they are yolk-yellow in the bud fading to milk-white, with dark yellow anthers. 15 ft.

'GRÜSS AN TEPLITZ' (Shrub). P. Lambert, Germany, 1897.—This is a hybrid of complex parentage and does not fit conveniently into any particular group, but has most affinity with the Bourbons from a garden point of view. The flower colour, intense dark crimson, is to be prized. Flowers double, nodding on opening from erect buds (a trait of R. *moschata*), borne in small clusters in summer, but in panicles on the stronger, later shoots, continuing until autumn. Young foliage purplish red, turning to green, with a thin reddish margin. It can be treated as a climber, but can be kept as a large shrub by pruning. 6 ft. Perpetual. A.M. 1899. Sweetly scented.

Hoffmann, pl. 19.

'HANSA' (Shrub). 1905.—A hybrid of R. *rugosa*, with typical dense prickles on a stout bushy plant; the foliage also is typically deeply veined, small, and of dark green, turning yellow in the autumn. Flowers double, well formed, of deep crimson-purple in clusters. The colouring of the flowers is less appealing and decisive than that of 'Roseraie de l'Hay' and it is a little farther removed from typical R. *rugosa*. On the other hand its flowers are constantly produced and it is much used as an amenity plant on the Continent. 6 ft. Recurrent. Sweetly scented.   [PLATE 25

'HEADLEYENSIS' (Shrub). Sir Oscar Warburg, Headley, Surrey. R. *hugonis* (seed-parent) × ?R. *pimpinellifolia* 'Grandiflora'.—Of the many hybrids of R. *hugonis* this is perhaps the most bushy and elegant, covered with creamy yellow flowers, and multitudes of small clear green leaves. Wide, graceful growth. 8 ft. Early summer. Sweet fragrance.

'HELEN KNIGHT' (Shrub).—This was raised from open-pollinated seed of R. *ecae*, at Wisley by F. P. Knight; since R. *pimpinellifolia* 'Grandiflora' (R. *spinosissima altaica*) was growing nearby it is presumed this plant supplied the pollen. It inherits much of the dark wood, dainty foliage, and brilliant yellow of the flowers of R. *ecae*; they are slightly cupped, 1½ in. across, and bright clear yellow. Faintly fragrant. Early summer. 7 ft. Raised in 1966.

'HENRI MARTIN' (Moss). 1863.—The largest in flower of the more brilliant carmine tints among the old French roses, with full, beautiful blooms reflexing to a ball. Long stalked blooms with clear green moss, above mid-green leaves. This is a large handsome shrub, graceful and prolific. 6 ft. Midsummer. Fragrant. On good soil it can be a superlative rose.

'HIGHDOWNENSIS' (Shrub). Sir Frederick Stern, Highdown, Worthing, 1928. Seedling of R. *moyesii*.—Extremely vigorous, producing shoots 10 ft high in one season when established, freely set with dainty burnished leaves. Flowers single, over 2 in. across, borne in conspicuous clusters, of vivid cerise-crimson. Little fragrance. Later, when the long new summer shoots have grown, showing the young foliage and red thorns and prickles, the flowering branches bear bunches of long, flagon-shaped scarlet heps. It is one of the most valuable hybrids of the *Cinnamomeae* (section *Cassiorhodon*) with two good seasons of display. 10 ft. Early summer. A.M. 1958.

Gard. Chron., January 6, 1934 (supplementary illustration).

'HILLIERI' (Shrub). Hillier & Sons, Winchester, 1924. A seedling of R. *moyesii*.—Of very open growth, wide spreading, with small leaves. Flowers single, in clusters, of intense maroon-red; a wonderful sight when seen with the sun shining through them, against the green foliage and blue sky. A few of the flowers develop large flagon-shaped heps. 10 ft. Early summer.

'HONORINE DE BRABANT' (Bourbon).—This is closely related to 'Commandant Beaurepaire', but considerably more vigorous and leafy. Flowers pale pink, daintily spotted and striped with mauve and violet. The blooms are produced in abundance in summer, and intermittently later, at which time they are inclined to 'ball'. 8 ft. Midsummer onwards. Wider than high. Sweetly fragrant.

'HUME'S BLUSH TEA-SCENTED CHINA' (China). 1809.—This hybrid, considered to be R. *chinensis* × R. *gigantea*, together with the other three hybrids introduced from China between 1789 and 1824, has had a considerable influence on European garden roses, particularly from its fragrance. See under R. *chinensis*. Redouté, Vol. I, pl. 61 (as R. *indica fragrans*).

'ISPAHAN' (Damask). Also known as 'Rose d'Isfahan'.—One of the first Damask roses to start flowering and usually the last to finish. Clear brilliant pink flowers of good shape reflexing loosely later. A vigorous shrub with leaden green leaves and few prickles. 6 ft. Midsummer. Sweet scent. Reputedly used in Turkey for distilling rose water.

'JACQUES CARTIER' (Portland). 1868.—A vigorous but compact shrub with light green leaves, the terminal leaflet very long and narrow. Flowers of good size, similar to those of 'Comte de Chambord', very full, quartered, with button eye. 5 ft. Summer to autumn. Sweet scent.

'JANET'S PRIDE' (Shrub). W. Paul and Sons, 1892.—Possibly a hybrid of *R. eglanteria* and *R. damascena*, it inherits something of the aromatic foliage of the former. Flowers about 2½ in. across, semi-double, bright cherry-pink with nearly white centres around the stamens, the paler tint extending through the petals in veins and splashes. 6 ft. Midsummer. Sweet scent.

Willmott, Vol. II, pl. 449.

'JUNO' (Centifolia). 1847.—This is an arching shrub with soft, leaden green leaves, drooping beneath the weight of the large globular blooms opening flat and quartered, with marked button eye. A superlative bloom, richly fragrant. It probably has some affinity with the China Rose, since it was originally classed as a '*Hybride non-Remontant*'.

'KARL FOERSTER' (Shrub). W. Kordes, Germany, 1931. 'Frau Karl Druschki' × *R. pimpinellifolia* 'Grandiflora'.—A vigorous sturdy bush with large prickles and green leaves of no particular distinction. Flowers singly and in clusters, shapely, pure white. 5 ft. Perpetual. A beautiful and useful plant, seldom grown, probably because, like its first parent, it is practically scentless.

'KEW RAMBLER' (Rambler). Royal Botanic Gardens, Kew, 1912. *R. soulieana* × 'Hiawatha'.—The use of *R. soulieana* has provided the small greyish foliage, against which the single pink flowers—with white zone around the stamens—show to great advantage; it also has provided the small orange heps which colour in autumn. Free flowering. 18 ft. Midsummer. Sweetly scented. A.M. 1922.

'KOENIGIN VON DANEMARCK' (Alba, with Damask affinity). Introduced by James Booth, Flottbeck Nurseries, near Hamburg, in 1826. Also known as 'Queen of Denmark'.—This bears a marked resemblance in its growth and prickles and clusters of flowers to 'Félicité Parmentier', but its leaves are particularly dark blue-green, a perfect foil for the blooms which are of intense carmine in the centre, clear pink at the circumference. Densely packed with short petals in flat, quartered array, the blooms often have a button eye around a green central pointel; later the petals reflex. It is indeed a queen among roses. Open habit, to 5 ft. Midsummer. Sweet fragrance.

Thomas, *Old Shrub Roses*, pl. 11.

'LA BELLE DISTINGUÉE' (Shrub). Before 1790.—Presumed a hybrid of *R. eglanteria*, though the fragrance of the flowers and leaves of that species is scarcely evident. Leaves small, of very dark green, on a dense rather upright bush. Flowers double, opening flat, borne in clusters, small, deep cherry red. 4 ft. Midsummer.

'LADY CURZON' (Shrub). Turner of Slough, 1901. Reputedly *R. rugosa* × *R.* 'Macrantha' of gardens.—This is a useful trailing rose for ground-cover, or for climbing over hedges and into trees. Excessively prickly stems and dark green rough leaves. The beautiful single flowers, 3½ in. across, are light pink with white zone around the stamens and create a fine effect. 12–15 ft. Midsummer. Sweet scent.

'LADY HILLINGDON' (Tea). Lowe and Shawyer, 1910. 'Papa Gontier' ×

'Madame Hoste'.—The rich purplish tint of the young shoots and leaves together with the soft deep apricot yellow of the flowers set this rose apart from all others. Nothing like it has since been raised. Fairly strong growth, few prickles, leaves dark green when mature. Flowers develop from long shapely buds, when they may be touched with red, to fairly full, loose shape, usually nodding, borne singly or in small clusters. The nodding flower is a disadvantage on a bush that will only exceed 3–4 ft in very mild areas, but an advantage in the climbing sport which appeared in 1917; fortunately 'CLIMBING LADY HILLINGDON' is as constantly in flower as the original bush, and may reach 20 ft. A sunny wall is best. Perpetual. The bush type received an A.M. 1910. Rich tea or apricot scent.

Thomas, *Climbing Roses*, pl. VI; Trechslin and Coggiatti, pl. 13.

'LAMARQUE' (Tea-Noisette). 1830. Originally 'Thé Maréchal' having been raised by Maréchal of Angers, France, the only rose attributed to him—and subsequently named in honour of Général Lamarque.—This is of the same parentage as 'Desprez à Fleur Jaune', but shows less of the Noisette. Flowers large, 3–4 in. wide, with quilled and reflexing outer petals, central ones plentiful, cupped and quartered, white with lemon in centre, prettily nodding. Leaves fresh green, smooth, limp. It needs a warm sheltered wall to give of its best, and will attain or exceed 10 ft. Midsummer, with a few later blooms. Tea fragrance.

Choix des Plus Belles Roses, pl. 30.

'LA PLUS BELLE DES PONCTUÉES' (Gallica).—This very vigorous plant has long erect shoots and good foliage. The flowers are loosely double, of bright pink, daintily spotted with white. 6 ft. Midsummer. Sweet scent. 'Alain Blanchard' is another of these ancient, spotted varieties.

'LA VILLE DE BRUXELLES' (Damask). 1849.—Luxuriant, rich green, long leaves. The very large flowers cause the branches to bend to the ground; it therefore needs support. A constant rich pink tone lasts through the life of the flowers which are cupped to start with, later reflexing, massed with petals, with button eye. 5 ft. Midsummer. Sweet fragrance.

'LEDA' (Damask). Before 1838.—Very dark, luxuriant foliage, on a sturdy bush. This originated in England and has developed a pink sport on occasions, but the blooms are usually of glistening milky white, touched with red-brown at the tips, for they emerge from buds of this colouring. Marked button eye; reflexing into a ball. 3 ft. Midsummer. Sweet scent.

'MADAME ALFRED CARRIÈRE'. 1879.—This is grouped usually with the Noisettes, but conforms to the characters of more modern climbing roses. Very vigorous, enough to cover a large wall, or by some cutting back, can be made into a large bush. Large, smooth, light green leaves on green wood with few prickles. Fully double blooms from shapely buds, borne in small clusters or singly, creamy blush fading to nearly white, sometimes quartered. 18 ft or more. Midsummer, but is seldom completely out of flower later. Sweetly fragrant.

Journ. Roses, April 1886.

'MADAME ANTOINE MARIE' (Tea). 1901.—The rich purplish young growth, stems and prickles give a lovely quality to the bush. Long shapely buds with large outer petals flushed with deep pink, opening to creamy yellow with blush tinting; the centres are filled with many smaller petals. Against a warm wall in our warmer counties it can be a great success, but like the few other Tea roses that have lingered in cultivation it needs cosseting. Delicate Tea scent. Perpetual.

Journ. Roses, February 1904.

'MADAME DELAROCHE-LAMBERT' (Moss). 1851.—A treasured variety on account of its good growth, shapely buds and repeat-flowering habit. Long leafy sepals, green moss. Flowers of intense rich crimson-purple, with rolled and reflexing petals. Brownish moss on the stems. 5 ft. Midsummer. Sweet scent.

'MADAME DE SANCY DE PARABÈRE' (Boursault Climber). 1874.—One of the few truly unarmed roses. A luxuriant climber with green wood, smooth limp leaves and the habit of flowering early. The blooms are large, fully double in the cupped centre, with large reflexing outer petals, all of clear rose pink. 15 ft. Early summer. Almost scentless.

Thomas, *Climbing Roses*, pl. III.

'MADAME FALCOT' (Tea). 1858.—The full-petalled, rich deep buff-yellow flowers are large, often with quartered centres; they are well contrasted by purplish young foliage, turning to dark green but with purplish reverses. Can be successfully grown in our warmer counties, especially against a sunny wall. Perpetual. Sweetly scented. 4 ft.

Nietner, *Die Rose*, facing p. 48; *Journ. Roses*, June 1880.

'MADAME HARDY' (Damask/Centifolia). 1832.—The standard by which the best old French roses are judged. Large, full-petalled, cupped blooms which at one stage reveal reflexed outer petals as in R. *centifolia*, usually with green centre —the perfect contrast to the white of the petals which only on opening show a faint blush. Rich green leaves on a vigorous bush with prickles and thorns. J. A. Hardy, who was in charge of the *Jardin de Luxembourg* in the early 19th century, and his wife, are respectively honoured by R. *hardii* and this unsurpassed old French rose. Sweetly scented. 6 ft. Midsummer.

Bunyard, pl. 14 (as R. *centifolia alba*).

'MADAME ISAAC PEREIRE' (Bourbon). 1880.—Huge double flowers at first cupped and quartered, later reflexing, of rich madder-carmine, shaded with magenta. At its best it is one of the most perfect and large of all roses, from half open flower to expanded bloom. 'MADAME ERNST CALVAT' is a sport of 1888, in which there is deep purplish tint on the young foliage, contrasting well with the warm flesh-pink of the flowers with darker reverses. Both cultivars produce their best blooms in autumn, the earliest summer blooms often being misshapen. Both can be treated as large shrubs but are best with support. 9 ft. Recurrent. See also 'Mrs Paul'. Powerful fragrance.

Journ. Roses, April 1893; Edwards, *Wild and Old Garden Roses*, facing p. 84.

'MADAME LAURIOL DE BARNY' (Bourbon). 1868.—With the smooth leaves and stems of the Bourbon group, this has not inherited the propensity to flower more than once; the flowers are however of superlative quality, nodding, very full of petals, globular, of soft clear pink with a slight mauve tint. Arching growth. 9 ft. Midsummer. Sweet fragrance.

'MADAME LEGRAS DE SAINT-GERMAIN'. Before 1848.—A rose of such mixed parentage that it has been listed in several groups, Alba and Noisette included. Few prickles on long arching green stems, set with smooth leaves. The blooms are in small clusters or borne singly, with a dense array of small petals which make a flat flower, pure white with lemon tint in the centre. Midsummer. Excellent as an arching bush of overlapping branches, or trained into trees, in which it will ascend to 15 ft. Sweet scent.

'MADAME LOUIS LÉVÊQUE' (Moss). 1873.—Remarkable, vigorous, upright growth, just strong enough to bear the copious, bright green, long-pointed foliage and heavy blooms. They are the largest of any Moss variety, globular and full of petals, inclined to 'ball' in wet weather, of uniform pale pink. 5 ft. Midsummer, and occasionally later. Sweetly scented.

'MADAME PLANTIER'. 1835.—Like 'Madame Legras de Saint-Germain', this is grouped usually as an Alba, sometimes as a Noisette. Again smooth green wood and leaves, and a profuse display of nodding double milk-white flowers, with lemon flush in the centre when first open. It will ascend into trees and hang from them in a massed display from 20 ft or make a low mound of overlapping branches. Midsummer. Sweetly fragrant.

'MADAME ZOETMANS'(Damask).—This is very close to 'Botzaris' but differs mainly in its smooth flower stalks. Fresh green leaves. Faintly blush on opening, the flowers quickly pass to white with a blush centre, full petalled with button eye. Early flowering. 3 ft. Midsummer. Sweetly scented.

'MAIDEN'S BLUSH' (Alba).—With typical sparse prickles and otherwise smooth wood of the Alba roses, this also has greyish leaves, and the flowers are borne in typical large clusters, and benefit from thinning. From greenish creamy buds the flowers develop into a mass of short, muddled petals of a very clear light pink, fading paler. One of the great roses of all time. Sturdy but arching growth to 5–6 ft. Midsummer. Extra sweetly scented.
Redouté, p. 97, as *Rosa alba regalis*.

'MAIGOLD' (Shrub or Climber). W. Kordes, Germany, 1953. 'Frühlingsgold' × 'McGredy's Wonder'.—Excessively prickly stems and very vigorous growth indicate an unmanageable rose. It can be left to mound up as a bush or be trained on supports. Exceptionally beautiful glossy dark leaves. Flowers in clusters, deep buff-yellow, semi-double, showing dark yellow stamens; they emerge from orange-red buds. It is a wonderful bit of colour; many of the summer's long growths bear bunches of flowers later. 5–10 ft. Recurrent. Heavy scent.

'MANNING'S BLUSH' (Shrub). Before 1799. Hybrid of R. *eglanteria*.— Inheriting the aromatic foliage of its parent, this old variety is a valuable garden plant; it is compact, with arching branches, with small rich green leaves. The flowers are in small clusters, about 1½ in. across, fully double, blush white, emerging from a somewhat mossy or glandular calyx. 5 ft. Midsummer. Sweet fragrance.
Lawrance, pl. 41.

'MARÉCHAL DAVOUST' (Moss). 1853.—Attractive pointed leaves on a vigorous bush. The brownish mossy buds hold shapely flowers which later reflex and show button eye and green pointel. The colour is intense, from rich deep crimson-pink with paler reverses, becoming suffused with cerise and purple and soft mauve. 5 ft. Midsummer. Sweet scent.

'MARÉCHAL NIEL' (Tea-Noisette). Henri Pradel, Lyon.—This was the first large-flowered yellow climbing rose of any size and quality, but is not reliably hardy. It needs cultivation under glass to protect the thin petals, but will grow well on a warm sunny wall where in hot summers the size (5 in. wide) and quality of its blooms will astonish. They are composed of a mass of petals, quartered and sometimes button eyed, emerging from beautiful, long buds, characteristically nodding. The colour is a very soft, buttery, yet sulphur-tinted yellow. Leaves elegant, light green, smooth, coppery when young, borne on green wood, sparsely prickly. 12 ft. Midsummer, with several later. Rich Tea-scent. Introduced in 1864.
Hoffmann, pl. 8.

'MARIE DE BLOIS' (Moss). 1852.—Though this does not bear fine shapely flowers, they are borne from summer until autumn. Probably related to the Portland Roses. Vigorous and bushy plant, copiously clothed in fresh green leaves, young shoots and thorns reddish, flowers deep pink. 5 ft. Fragrant.

'MARIE LOUISE' (Damask). 1813. Raised at Malmaison.—Owing to rather lax growth and the weight of the extra large flowers the plant should be supported. A leafy plant; the flowers are very full, with large outer petals, reflexing into a ball, and pronounced button eye. Probably a Gallica hybrid. 4 ft. Midsummer. Sweetly fragrant.

'MARIE VAN HOUTTE' (Tea). 1871. 'Madame de Tartas' × 'Madame Falcot'. —This may be considered as one of the most satisfactory of Tea roses, especially in a sunny position in our warmer counties; wall protection will help to ripen the wood, which is purplish when young like the leaves, which turn to dark green later. Shapely buds. Petals light yellow, creamy tinted, suffused with rich pink round the edges. Large, quartered blooms of great beauty. 5 ft. Perpetual. Delicious Tea-scent.
Hoffmann, pl. 10.

'MARTIN FROBISHER' (Shrub). Canada Department of Agriculture, 1968; seedling of 'Schneezwerg', open-pollinated.—A good bush with few prickles, covered with light, brownish green, deeply veined leaves, which are distinctly toothed. The flowers are borne in small clusters, rose-pink paling towards the edges, fully double, opening flat, with occasional button-eyes. Rugosa scent. In constant flower. 8 ft × 6 ft.

'MAX GRAF' (Shrub). Bowditch, Connecticut, USA, 1919. R. *rugosa* × R. *wichuraiana*.—This lingered in cultivation for many years eventually achieving notoriety by doubling its chromosomes in the hands of Kordes and so bringing R. *rugosa* into the strain of modern climbers. Apart from this, the search for good ground covering shrubs has given it prominence at last. It forms a low mound, constantly spreading, dense, prickly and with glossy foliage. Clusters of small bright pink flowers are freely produced. It is weed proof, and a boon for banks and large areas or it can be trained over hedges and into small trees. 3 ft, spreading. Midsummer. Fragrant. A.M. 1964.

'MAY QUEEN' (Rambler). Manda, New Jersey, USA, 1898. R. *wichuraiana* ×
'Champion of the World, (Bourbon).—Prolific growth, somewhat more bushy
than others of the 'Albéric Barbier' group, making dense ground cover. Leaves
soft green. Flowers 3 in. across well filled with quartered petals, and often with
button eye; rose pink developing a lilac tint. Green wood with few reddish
prickles. 15 ft. Midsummer. Sweet fragrance.
NOTE: Van Fleet sent over a hybrid of similar parentage and under the same name,
but the plant described is almost certainly Manda's 'May Queen'.

'MERMAID' (Climber). W. Paul, Waltham Cross, 1917. R. *bracteata* × yellow
Tea Rose.—Like 'Nevada', an equally remarkable cultivar, this is sterile. Its
brown wood has large hooked prickles which make it an unpleasant neighbour.
But the luxuriant, long, dark green glossy leaves make a good setting for the
4 in.-wide single yellow blooms; the stamens last in beauty after the petals have
dropped. In constant production from midsummer until autumn. Though it
will thrive and flower well on a wall of any aspect—which must be large to
accommodate it—it is advisable to plant it facing south or west, because in
extremely severe winters it may be killed to the ground. 20 ft. Perpetual. A.M.
1917. A.G.M. 1933.

'MOONLIGHT' (Hybrid Musk). 1913. 'Trier' × 'Sulphurea'.—The reddish
brown wood, thorns and young foliage give way to dark green. Flowers milky
white, small, semi-double, opening from creamy buds; they appear in large and
small clusters in summer, but on the strong summer shoots may be a foot high
and wide. Although this can be kept to the outline of a shrub, it is extremely
useful for training into small trees, when its shoots, once anchored, progress in
every direction. 6 ft or more. Sweet fragrance. Perpetual. A.M.1913.

'MORLETII' (Boursault). Morlet, 1883. Also known as R. *inermis morletii*,
and sometimes grown as 'R. *pendulina plena*'.—This is a beautiful shrub, much
neglected, and has no prickles. The long, arching, reddish-purple shoots are
beautiful in winter; the emerging young foliage is glaucous purple; it excels in
long-lasting autumn colour. The flowers are small, but in clusters, semi-double,
magenta. 6 ft. Early summer. Practically scentless.

'MOUSSELINE' (Moss). 1855.—This appears to be synonymous with 'Alfred
de Dalmas'. Apart from being reliably repeat-flowering, its bushy growth is
clothed in a dense array of spoon-shaped leaves. The moss is not very evident.
Creamy blush blooms, small but with well-filled centres. Few thorns. Probably
closely related to the Portland Roses. 3–4 ft. Sweetly scented.

'MOUSSU DU JAPON' (Moss). Also known as 'JAPONICA'.—Of unknown
origin, but exhibiting the maximum amount of dense green moss on calyx and
stalk, stem and leaf stalks and leaf surfaces in a remarkable fashion. The young
foliage is often of metallic lilac tint. Flowers small, of poor shape, magenta
passing to lilac. A dense dwarf bush. 3 ft. Midsummer. Sweet scent. [PLATE 26

'MOYESII SUPERBA' (Shrub). Introduced by Van Rossem, Holland. Reportedly R. *moyesii* × Tea Rose.—Inherits the vigorous growth and stout wood
of R. *moyesii*, and achieves about 7 ft in height, eventually making a densely
branched shrub. The leaves are dark green, resembling but larger than those of

R. *moyesii*. Beautifully poised flowers, borne on plum-coloured pedicels, with calyx and receptacle of the same tint, large, semi-double, very dark maroon-crimson. It is unfortunately practically scentless and produces no heps. It makes however a considerable effect in the garden at midsummer.

'MRS ANTHONY WATERER' (Shrub). Waterer, Knaphill, 1898. 'Général Jacqueminot' × R. *rugosa* hybrid.—This makes a spreading bush 4 ft × 7 ft, with freely branching stems armed with reddish prickles, and carrying dull, dark green leaves. Bright crimson flowers, flushed with purple, loosely double and richly scented; they are borne at midsummer mainly but a few occur later. Sweetly scented.

'MRS COLVILLE' (Shrub).—A hybrid of R. *pimpinellifolia*, perhaps with R. *pendulina*, indicated by the long heps, as opposed to the rounded ones of R. *pimpinellifolia*. It inherits the freely suckering habit of the former parent, but its red-brown stems bear few thorns. Flowers single, intense crimson-purple with a white zone around the stamens. Foliage small, dark green. 3 ft. Early summer. Fragrant. A.M. 1956.

'MRS HERBERT STEVENS' (Tea). McGredy, 1910. 'Frau Karl Druschki' × 'Niphetos'.—Flowers nodding, with a delicate Tea fragrance, white with a lemon flush in the centre. The petals form a long shapely bud, opening loosely full, with a rolled and quilled shape. Their papery texture spoils the blooms in very wet weather and imparts a tendency to 'ball' under cool, damp conditions. But, for a Tea, this is a remarkably hardy rose. A.M. 1910. Wisley Rose Award 1925. The nodding flowers are better displayed in the sport 'CLIMBING MRS HERBERT STEVENS' (1922).

Thomas, *Climbing Roses*, pl. VI.

'MRS PAUL' (Bourbon). G. Paul, 1891. A seedling of 'Madame Isaac Pereire'. —This is remarkably like its parent except in colour, and once again the autumn blooms are best. The petals are pearly white with flesh-pink reverse. 8 ft. Recurrent. Very fragrant. A.M. 1890.

The Garden, 1890, page 484.

'NARROW WATER' (Shrub).—This originated in a garden at Narrow Water in Ireland, and was catalogued by the Daisy Hill Nurseries, Newry, Co. Down in 1883. It is fairly frequently found in old gardens without a name, or confused with R. 'Pissardii', and is probably a hybrid either with that rose or with some other combination of R. *moschata* and one of the China roses. It is valued because of its constant production of semi-double lilac-pink flowers, borne in clusters, from midsummer till autumn, on a wide, stout shrub. Smooth, pointed leaves. 6 to 8 ft. Sweet scent.

'NEVADA' (Shrub). Pedro Dot, Spain, 1927. 'La Giralda' (Hybrid Tea) × tetraploid form or hybrid of R. *moyesii*.—A magnificent arching shrub, creating a dense dome of light green leafage; plum-coloured stems and few prickles. Each branch in its second and later seasons is covered with large semi-double blooms, opening flat, over 3 in. across, for its whole length. No other shrub rose creates quite such a dramatic and splendid effect. For the rest of the season the bush is seldom without a flower or two; they are creamy white, sometimes flushed red in the bud and often flushed with pink in hot weather. 8 ft. F.C.C. 1954. A.G.M. 1966.

Thomas, *Shrub Roses of Today*, fig. 8 (pencil drawing).

Pink sports have occurred. One is 'MARGUERITE HILLING' (H. Sleet, Surrey, 1959), with deep flesh-pink flowers. Not so effective as its parent. 8 ft. Recurrent. A.M. 1960.

'NIPHETOS' (Tea). 1843.—It is probable that the plant in cultivation today is 'Climbing Niphetos', a sport which occurred in 1889. Owing to the loose formation of its thin petals which make long elegant buds, it is most suited to cultivation in a cold house in our warmer counties. Vigorous growth and good leaves. Pure white with lemon flush in the folded centre. Few prickles. 9 ft. Perpetual. Delicate Tea-scent. F.C.C. 1888 (climbing form).   [PLATE 27

Hariot, pl. 11.

'NUITS DE YOUNG' (Moss). 1845.—A slender but vigorous bush with small, dark leaves, purplish when young. Dark moss. Small flowers of intense, uniform maroon-purple, dark and velvety, lit by a few yellow stamens; few petalled, reflexing prettily. 5 ft. Midsummer. Fragrant.

'NYVELDT'S WHITE (Shrub). Nyveldt, Holland, 1955. A hybrid of R. *rugosa*, R. *majalis*, and R. *nitida*, raised as a hedging rose.—Like some other Rugosa hybrids, it has flowers of pure white, single, showing stamens, about 2½ in. across, borne at midsummer and intermittently later, with the orange-red heps. Fresh green slightly rugose leaves. Sturdy, growth to 5 ft. Recurrent. Sweet scent.

'OEILLET PANACHÉE' (Moss). 1888. Also known as 'Striped Moss'.— Though inferior in bloom and general quality to other striped roses, it is the only striped Moss. Small leaves, wiry growth. Flowers of blush pink, striped and particoloured with carmine, petals narrow, forming a flat rosette. 3 ft. Midsummer. Sweet scent.

'OLD BLUSH CHINA' (China).—It is believed this rose, known as 'Common Blush', 'Monthly Rose' or 'Blush China' is identical with 'Parsons' Pink China', introduced 1789 by Sir Joseph Banks. A rose of this kind has been grown for at least one thousand years in China. It was one of the four hybrids of R. *chinensis* and R. *gigantea* introduced at the turn of the century (see R. *chinensis*). Smooth wood with few reddish prickles, neat small dark green leaves and a constant succession of blooms so long as the weather is mild, from late spring until autumn. They are mostly borne in small clusters in summer, but in large clusters on the strong summer's shoots. Crimson in bud, opening to loosely double blooms of soft pink, deepening with age and daintily veined. Against a warm wall it may reach 9 ft but in the open is usually about 2–3 ft. Recurrent. Fragrant.

Redouté, Vol. I, p. 51; Willmott, Vol. I, p. 79.   [PLATE 28

'OMAR KHAYYAM' (Damask).—This is apparently an old type of Damask rose, markedly prickly with light green, oblong, widely spaced leaflets. Medium sized flowers of soft pale pink, quartered, with folded petals and a button eye. 3 ft. Midsummer.

'Omar Khayyam' was raised at Kew from a hep which had been collected in 1884 from a rose bush growing by the tomb of Omar Khayyam at Nishapur. A cutting planted in 1893 on the grave of Edward Fitzgerald at Boulge, Suffolk, was almost dead by 1948 when it was propagated by Messrs Notcutt (*Gard. Chron.*, 1894, p. 746; Frank Knight in *Journ.* R.H.S., Vol. 73 (1948), pp. 150–2 and Vol. 74, pp. 544–5, fig. 201).

A similar rose is 'GLOIRE DE GUILAN', introduced by Nancy Lindsay in 1949 from the Caspian region of Iran. She reported that it was used there for distilling rose-water and attar. Fresh green leaves and fresh clear pink flowers, with quartered and folded petals. 3 ft. Midsummer. Both these roses are sweetly fragrant.

'ORMISTON ROY' (Shrub). S. G. A. Doorenbos, The Hague. Second generation hybrid from R. *pimpinellifolia* × 'Allard'.—Small bright green leaves on a thorny bush. Flowers single, bright yellow, about 2 in. across, followed by large rounded dark maroon heps on similarly coloured pedicels. Its chief claim to fame is as parent to 'Golden Wings' (q.v.). 4 ft. Early summer. Heavy fragrance. A.M. 1955.

'PARKS' YELLOW TEA-SCENTED CHINA' (China). 1824.—The last of the four hybrids of R. *chinensis* to reach Europe by this date. (See R. *chinensis*.) This seems to be extinct but its yellow colouring and tea-scent had a lot of influence in breeding the Tea-Noisettes later in the century.

'PARVIFOLIA'. 18th century or earlier. For synonyms, etc., see under R. *centifolia*.—This gives no impression of R. *centifolia*, under which some botanists place it, and has not been observed reverting to it. It may be a 'witch's broom' or some other errant growth, propagated vegetatively since its discovery. The erect, clustered, nearly unarmed growths produce numerous very small dark green pointed leaves. The shoots are crowned with small clusters of miniature rosette-flowers, densely double, of dark pink suffused purple, with paler centres. Fragrant. 3–5 ft. There are two forms in gardens, one taller than the other, and larger in all its parts. Both have been cultivated in Britain since the latter part of the 18th century. Midsummer.

Willmott, p. 355.

'PAULII' (Shrub). G. Paul, Cheshunt, before 1903. R. *rugosa* × R. *arvensis*. Sometimes labelled R. *rugosa repens alba*.—Like 'Max Graf' this lingered in obscurity until reappraised as a ground-cover shrub; being very prickly and making a dense mound of interlacing branches, it is impenetrable for weeds or animals. Dark, somewhat rugose leaves. Flowers in clusters, freely disposed over the entire plant; the petals are narrow, giving a starry effect, pure white, with yellow stamens. 'PAULII ROSEA' is a counterpart in clear bright pink; the petals fade to white in the centre with good effect around the stamens. This sported back to 'Paulii' in the garden of A. T. Johnson, N. Wales. It does not grow quite so strongly as 'Paulii' but is an effective and beautiful low shrub. 4–5 ft high, width unlimited. Midsummer, occasional later blooms. Sweet fragrance.

Les Plus Belles Roses, p. 441 (as R. *gallica* × R. *rugosa*); Trechslin and Coggiatti, pl. 33 (both 'Paulii Rosea').

'PAUL RICAULT' (Centifolia hybrid). 1845.—Like 'Blanchefleur' this is a hybridised rose, and with similar stout prickly wood and open habit. Globular, but quartered blooms of rich old rose, with rolled petals. One of the most free-flowering of the Old French roses, and frequently found in old gardens, where it is treasured for its cabbagy blooms. 6 ft. Midsummer. Sweet fragrance.

'PAUL'S SINGLE WHITE'. G. Paul, Cheshunt, 1883. Probable hybrid of R. *moschata*.—A free-growing bush, with few thorns and prickles, and light green smooth, long leaves. The flowers are carried in small clusters from early summer until autumn, 2 in. wide, in clusters, white with blush tint in bud, showing stamens. 5 ft. Perpetual. Sweet scent. F.C.C. 1887.

The Garden, Vol. 29 (1886), pl. 526, facing page 28.

'PAUL TRANSON' (Rambler). Barbier, Orléans, 1900. R. *wichuraiana* × 'L'Idéal'.—Of the 'Albéric Barbier' group, and with equally dark glossy leaves but rather smaller. Young shoots and prickles purplish. Buds coppery orange, opening to salmon-coral, fading to creamy salmon with yellow tints in the centre. Fairly full double flowers, flat with pleated and often quartered petals. 15 ft. Midsummer, but seldom out of flower later. Sweet fragrance.

'PAX' (Hybrid Musk). 1918. 'Trier' × 'Sunburst'.—The most lax in growth of this group, but even so it can be kept as a bush by pruning; it is more effective when climbing into small trees or through other shrubs. Leaves good dark green, red-brown when young, contrasting well with the large creamy white blooms, opening from yellowish, pointed buds. They are loosely semi-double, cupped, and show yellow stamens. Long sprays of flowers are produced on the strong summer's growth after the summer crop. 6 ft. Recurrent. Sweet fragrance.

Thomas, *Shrub Roses of Today*, pl. v.

'PENELOPE' (Hybrid Musk). 1924. 'Ophelia' × seedling.—Perhaps the most popular of this group, it makes a sturdy bush with thick branches from the base when old; it is admirable as a hedge. Broad dark green leaves. The flowers are carried in small clusters in summer, considerably larger in autumn. The salmon-orange tint of the opening flowers tones well with the creamy pink of the open flowers, loosely semi-double. 6 ft high and as wide. Recurrent. Heps green and coral, long lasting. Sweetly fragrant. A.G.M. 1956.

'PERLE D'OR' (Poly-pom). Introduced by Dubreuil, Lyon, 1883. R. *multiflora* seedling × 'Madame Falcot'.—This resembles 'Cécile Brunner' in many ways and it may be added that no roses resemble these two in their exquisite buds and freedom of flowering. 'Perle d'Or' has buds of warm yolk-orange, opening to a light, creamy salmon-pink; petals numerous, narrow, often quartered, making a flat flower, and reflexing. Somewhat prickly bush with brownish young shoots and foliage, later dark green, neat and pointed. It will build up to a filled bush of 4 or more feet. Perpetual. Sweetly musk-scented.

Kingsley, *Roses*, p. 129; Trechslin and Coggiatti, pl. 6.

'PETITE DE HOLLANDE' (Centifolia). Before 1802.—While 'De Meaux' and 'Parvifolia' are complete miniatures this is just R. *centifolia* writ small, with otherwise typical growth, leaves and flowers. It is the best Provence Rose for small gardens. *New Flora and Silva*, Vol. 2, fig. 2 (monochrome photograph). Another miniature called 'SPONG', dating from 1805, is midway in size of flower and foliage between 'Petite de Hollande' and 'De Meaux'; it has the unfortunate habit of retaining its dead brown petals on the stems. 4 ft. Sweet fragrance.

'PETITE LISETTE' (Damask). 1817.—A small type of Damask rose with small, downy greyish foliage and small flowers of blush pink; the neat circular blooms are perfectly formed, well filled with folded petals radiating from a pronounced button eye. 4 ft. Midsummer. Sweetly fragrant.

'PHYLLIS BIDE' (Rambler). Bide and Sons, Farnham, 1923. Parentage stated to be 'Perle d'Or' × 'Gloire de Dijon'.—This is one of the very few ramblers which continue to flower until the autumn. Rather stiff yet slender habit with small leaves and prickles. Clear yellow, double, small flowers are flushed with pink. 12 ft. Perpetual. Fragrant. A.M. 1924.

'POMPON BLANC PARFAIT' (Alba). 1876.—A small, erect bush with small greyish leaves and small rounded blooms, packed with small petals, opening flat and reflexing, of pale uniform lilac-pink. Few thorns. Very long flowering period, sometimes into August. 5 ft. Summer. Sweetly scented.

'PORTLANDICA' (Portland). In cultivation 1775. Also known as 'Scarlet Four Seasons' and "R. *paestana*".—The first known hybrid of Gallica derivation, giving a brilliant carmine colouring coupled with the recurrent-flowering influence of the Autumn Damask. It makes a neat, suckering bush, with bright green leaves and semi-double flowers, produced so long as growth continues, in clusters. 2 ft.   [PLATE 29
Redouté, Vol. I, p. 109, as R. *damascena coccinea* or 'Le Rosier de Portland'.

'PRÉSIDENT DE SÈZE' (Gallica). Before 1836. Also known as 'Madame Hébert'.—A good sturdy bush with bold foliage. Flowers large, fully double, with petals rolled at the edges; on opening there is intense crimson-purple in the centre but the edges of the petals are lilac-white in dramatic contrast. A faded flower still retains the dark centre. 'JENNY DUVAL' is of similar colouring. 4–5 ft. Midsummer. Rich fragrance.

'PRINCESSE LOUISE' (Rambler). 1829.—Of the same origin as 'Adélaide d'Orléans' and 'Félicité et Perpétue', of similar vigour and foliage. The flowers are borne profusely, and are between the above two varieties in shape and fullness, they open creamy blush-white from pink buds. Delicate fragrance. 12 ft.

'PROSPERITY' (Hybrid Musk). 1919. 'Marie Jeanne' × 'Perle des Jardins'.— A splendid upright shrub with good dark foliage. Flowers are carried in small sprays in summer, but in wide heads on the late summer shoots, fully double, rounded, cream-pink in the bud opening to ivory white. 'PINK PROSPERITY' is warm, clear pink but is probably not a sport. These two cultivars and 'Wilhelm' and its sport are all of a type, upright and manageable. 6 ft. Recurrent. Sweet fragrance.

'QUATRE SAISONS BLANC MOUSSEUX' (Moss).—This was recorded by E. A. Carrière in *Production et Fixation des Variétés dans les Végétaux* (1865); it was a sport from R. *damascena* var. *semperflorens*. The present rose is an exact counterpart of the latter except for the flowers, which are white and less shapely, and for the mossy calyx and pedicel; normally it also has very short glandular hairs on the surface of the leaves. The moss on the calyx is stiff and bristly to the touch, quite different from that of R. *centifolia* ' Muscosa'. Because of its recurrent flowering habit, it was used in hybridising in an effort to produce recurrent flowering Moss Roses. It fairly frequently reverts to the pink non-mossy type. Erect vigorous growth; sweetly scented. Consistently in flower. 5 ft.   [PLATE 30

'RAUBRITTER' (Shrub). W. Kordes, Germany, 1936. 'Macrantha Daisy Hill' × 'Solarium'.—Thorny interlacing branches making a dense mound with small dark leaves. The blooms are in clusters, small, remaining in a unique globular shape, clear rose pink, semi-double. Each bloom lasts for about 7 days but bleaches in hot sun. Liable to mildew, but a rose for placing on banks, over low retaining walls, against steps. 3 ft, but much wider. Midsummer. Not fragrant.

'REINE DES VIOLETTES'. Mille-Mallet, 1860.—Usually grouped with the Hybrid Perpetuals, it has little to do with them, and leans more towards the Bourbon group, for it makes a good branching shrub, with plentiful greyish green smooth leaves, gathered into almost a rosette under the flowers, as in the Portland Roses. Dark, soft, grape-purple fading to soft lilac-grey, the blooms are 4 in. across, densely filled with short petals, opening flat, quartered, with button eye, showing pink reverse. It thus has the best of the Gallica flower shape coupled with abundant bloom, summer and autumn. 5 ft. Very fragrant.

Thomas, *Shrub Roses of Today*, pl. 11.

'REINE VICTORIA' (Bourbon). 1872.—Elegantly poised, pointed leaves on erect stems, from a slender erect bush. Flowers of soft old rose, having very curved petals, which results in an almost globular shape, retained until they fall. It has a more famous and popular sport, 'MADAME PIERRE OGER' 1878, identical except for its colouring, which is a cream-pink. A.M. 1951. In both these cultivars hot sunshine brings out intense colouring on the exposed portions of the petals. 8 ft. Recurrent. Both are sweetly scented.   [PLATE 31

Trechslin and Coggiatti, pl. 19 ('Madame Pierre Oger').

'RÉNÉ D'ANJOU' (Moss). 1853.—Like 'A Longues Pédoncules' and 'Nuits de Young' this shows Centifolia parentage in its graceful flower stems. Brownish green moss envelops the shapely buds; the soft warm pink petals fade to lilac-pink veined with a darker shade and somewhat crinkled. Bronzy young foliage. 5 ft. Midsummer. Sweetly scented.

'ROSE D'AMOUR'. Also distributed as R. *virginiana* 'Plena'.—The 'Rose d'Amour' is probably a hybrid of R. *virginiana*, and forms a lanky, open bush and is best trained on a support. It has ascended to about 10 ft on a wall at Wisley. Small, fairly glossy leaves, producing bright autumn tints on dark brown shoots. The exquisite scrolled buds open into perfect rosettes, fully double, deep pink, and like those of R. *virginiana* open mainly after midsummer, and continue into early autumn in cool districts.

The 'D'ORSAY ROSE', which has long grown at Hidcote, and was so named by Nancy Lindsay, is of shorter growth but of marked similarity of bloom. In this the leaves and stipules are narrower and a pair of prickles is borne beneath each leaf. The leaves are less glossy.

For a note on these two roses, see the Royal National Rose Society's *Annual* for 1977, pp. 27–28, and the drawings between pp. 60 and 61.

'ROSE-MARIE VIAUD' (Rambler). 1924.—A seedling from 'Veilchenblau' which confirms its general appearance of R. *multiflora*. It is practically unarmed, with smooth green wood, fresh green leaves with typically laciniate stipules. It has not inherited the fragrance of 'Veilchenblau', but has some of its colour. Flowers fully double, small, in large clusters, vivid lilac-cerise fading to uniform pale lavender. The flower stalks are prone to mildew, but otherwise it is a vigorous and healthy plant and highly effective late in its season. 15 ft. Midsummer. No scent.

Thomas, *Climbing Roses*, pl. 11.

'R O S E N W U N D E R' (Shrub). W. Kordes, 1934. W. E. Chaplin (Hybrid Tea) × 'Magnifica' (R. *eglanteria* hybrid).—The flowers are loosely double, cupped, of coppery light crimson, with a rich fragrance, appearing at midsummer. It is a lax grower, needing support; the reddish twigs and coppery brown spring foliage, coupled with abundant heps in autumn, make it a valuable garden plant. Sweetly fragrant.

'R O S E R A I E D E L' H A Ÿ' (Rugosa). J. Gravereaux, Roseraie de l'Haÿ, 1901.— The sturdy, bushy growth and the dense covering of prickles are typical of R. *rugosa*; the foliage, glossy, dark green, deeply veined, narrow and acutely pointed, turns yellow in autumn. Buds particularly long and shapely, opening vivid, rich crimson-purple, 4½ in. wide, with cream stamens; borne in clusters. 5 ft. Recurrent. Rich fragrance. A.G.M. 1966.

Journ. Roses, August 1906; Thomas, *Shrub Roses of Today*, pl. IV.

'R U S S E L L I A N A' (Rambler). Before 1826. R. *multiflora* hybrid. Formerly also known as 'Scarlet Grevillei'.—A very vigorous, thorny, climber with rough dark leaves. Today the flowers would not be described as scarlet, but intense cerise-crimson, fading to pale magenta; they contain many small petals, and are borne in large and small clusters at midsummer. 10 ft. Midsummer. Sweetly scented.

'S A L E T' (Moss). 1854.—A reliable bush with light green leaves, and slightly mossy buds. The flowers are rather poor in shape but are produced intermittently until autumn, of clear soft pink. Few thorns. 5 ft. Sweetly scented.

'S A N D E R S' W H I T E R A M B L E R' (Rambler). 1912.—A valuable plant on many counts. It is practically unarmed, makes a dense mass of trailing shoots if not trained up a support, and thus is an effective ground-cover. It flowers late in the rambler season. Unlike its close relative 'Dorothy Perkins' it is not subject to mildew. Small glossy leaves, abundant double, white, small blooms, with stamens, in large and small clusters. 18 ft. Late summer. Sweetly scented.

'S A R A H V A N F L E E T' (Shrub). Dr W. van Fleet, Glenn Dale, USA. R. *rugosa* × 'My Maryland'.—Excessively prickly, vigorous upright shoots, gaunt and leafless often for 3 ft but branching out above, where they are well covered with foliage, mid-green, bronzed when young. It is suitable for the backs of shrubberies, when its unceasing production of blooms can be enjoyed above other shrubs. Flowers in large and small clusters, semi-double, silvery pink, showing cream stamens. 9 ft. Recurrent. Sweetly scented. A.M. 1962.

Park, *World of Roses*, pl. 187.

'S C A B R O S A' (Rugosa). Before 1939. Origin unknown.—Very near to typical R. *rugosa* in quality and style of growth, foliage, flower and hep, but larger in all its parts. It makes a particularly leafy, lush bush, flowers single, in clusters, violaceous crimson with cream stamens, followed by bunches of rounded orange-red heps, over 1 in. across. Yellow autumn colour. 5 ft. Recurrent. Sweetly scented. A.M. for fruits, 1964.

Park, *World of Roses*, pl. 190.

'SCARLET FIRE' (Shrub). W. Kordes, Germany 1952. R. *gallica* form × 'Poinsettia'. Known in Germany as 'Scharlachglut'.—Smooth, brown wood and prickles, strong arching habit and good foliage make an impressive shrub. The flowers are borne in clusters over a long period, single, 3 in. across, velvety, blazing scarlet with yellow stamens. The resulting heps are bright red, pear-shaped, and last until late winter, when no other rose is so colourful. 8 ft. Slight fragrance. Midsummer. A.M. 1960, A.G.M. 1966.

Hollis, *Roses*, p. 173.

'SCHNEELICHT' (Shrub). 1896. Reported to be R. *rugosa* × R. *phoenicea*.— This certainly shows some Rugosa influence, and perhaps now that roses have come to be recognised as flowering shrubs of considerable value, it may yet become popular. Dull green deeply veined rugose foliage covering a big arching shrub. The flowers are of pure cold white, single and showing stamens, borne in clusters along the stems, and when in flower it is highly effective. 6 ft. Mid-summer. Sweet scent.

'SCHNEEZWERG' (Shrub). Lambert, Germany, 1912. Hybrid of R. *rugosa*.— Erect branching growth, with pairs of grey prickles below each leaf. Leaves rich dark green, somewhat rugose. Flowers in clusters, pure white, semi-double, 2 in. across, opening flat and showing bright yellow stamens. Later crops coincide with small orange heps. By clipping it would make a dense hedge. 6 ft. Perpetual. R. *rugosa* seems to excel in producing pure cold-white flowers devoid of cream or pink, witness 'Blanc Double de Coubert', R. *rugosa* 'Alba', 'Schnee-licht', 'Paulii'. A.M. 1948.

'SCINTILLATION' (Shrub). David Austin, Wolverhampton, introduced by Sunningdale Nurseries, 1966. R. 'Macrantha' of gardens × 'Vanity'.—This joins 'Macrantha' itself, 'Complicata', 'Lady Curzon', R. × *polliniana* and a few others of lax growth which slowly make mounds of interlacing branches, or may be trained over hedges and into trees, in which case the normal mound of 5 ft or so will be exceeded. The leaden green leaves make a good foil for the abundant clusters of semi-double blush-pink flowers at midsummer. Fragrant.

'SERRATIPETALA' (China). 1912. Origin uncertain, known in France as 'Oeillet de St Arquey'.—The smooth wood and small leaves, purplish when young, place it quite clearly within the China group; it is a sturdy wide-branching small shrub in the open but will attain great height against a warm wall. Flowers crimson, paler in the centre, double, with fringed petals, like those of a carnation. 4 ft. Midsummer, with occasional blooms later. Slight fragrance. The foliage of this rose closely approaches specimens of R. *chinensis*, collected in W. China.

'SILVER MOON' (Rambler). Dr van Fleet, 1910.—An extremely vigorous climbing rose with dark glossy leaves inherited from two species in its reputed parentage; R. *laevigata* and R. *wichuraiana*. It is possible that the Tea rose 'Devoniensis' was also in the parentage. The long-petalled buds are yellow, opening to creamy white, nearly single large blooms in clusters. Rich fragrance, but a short flowering season. The glaucous stems are set with a few purplish prickles. Sweet scent.

'SISSINGHURST CASTLE' (Gallica).—An obviously ancient cultivar owing to its flower-shape which is quite unsophisticated. It was found battling with the weeds at Sissinghurst Castle, Kent, when the garden was made by the Hon. V. Sackville-West. It has been seen also in a Devon garden, long established, but its name is lost. The rich plum-coloured petals are margined and flecked with magenta, showing yellow stamens. The name 'Rose des Maures' has no foundation. 3 ft. Midsummer. Sweetly scented.

'SLATER'S CRIMSON CHINA' (China). Introduced c. 1791, reintroduced 1953, see p. 75.—No China rose is vigorous or of coarse growth; this is particularly slender in habit, with neat, dark leaves, purplish when young, The flowers are borne singly or in small clusters, with short petals making a rosette, of dark pure crimson, lighter in the centre. Recurrent, throughout the growing season. Rather tender. 3 ft, higher with shelter.

'SOMBREUIL' (Tea). 1850.—This was raised from 'Gigantesque' a hybrid rose of 1845. The result is near to the Tea group and may be classed with them; though the buds are of long shape, the blooms open quite flat, filled with small petals and frequently quartered, with button eye. Ivory white with flesh tint in the centre. A good hardy plant with smooth dark green leaves. 'CLIMBING SOMBREUIL' seems to be the only form in cultivation and has been called, erroneously, 'Colonial White'; fortunately it is constantly in flower. 10–15 ft. Perpetual. Sweet Tea fragrance.

Hariot, pl. 12.

'SOUVENIR DE LA MALMAISON' (Bourbon). 1843. After the death of the Empress Josephine, the Grand Duke of Russia saw this rose at La Malmaison and asked that it be named after her.—The original bush form is one of the freest of recurrent blooming roses. The flowers are filled with petals, opening flat, quartered, of clear blush pink. It is best in hot weather. 4 ft. The climbing sport occurred in Australia in 1893; it flowers at midsummer and again in autumn, but is not continuous. 9 ft. The bush form has produced other sports, among them 'SOUVENIR DE ST ANNE'S' commemorating Lady Ardilaun's garden near Dublin; it is nearly single with a little more colour in the reverse of the petals. Both it and 'KRONPRINZESSIN VIKTORIA', which occurred in 1888 in Germany, are remarkably recurrent; the last is fully double, white with lemon tint in the centre. A rich pink sport has also occurred but seems to be lost. In all cultivars the later blooms achieve greatest perfection. 4–5 ft. Recurrent. Delicately fragrant in the doubles, more so in 'St Annes'.

'Souvenir de la Malmaison' is figured in: Paul, *The Rose Garden*, pl. 15; Hoffmann, pl. 4.

'SOUVENIR D'UN AMI (Tea). 1846.—Like 'Niphetos' this needs the protection of a cold house to develop its fragile flowers to perfection, though the plant is perfectly hardy. Small leaves, dark green, young wood and leaves purplish. Loosely double blooms, cupped and showing stamens, of light coppery pink. In warmer climates, for instance in New Zealand, it will achieve 10 ft but in the warmer counties of England seldom exceeds 4 ft. Recurrent. Delicate Tea scent.

'Choix des Plus Belle Roses', pl. 18.

'SPLENDENS' (Rambler). Also known as 'Ayrshire Splendens' or 'Myrrh-scented' Rose.—This is the most noted of the Ayrshire roses (see p. ooo), partly for its unusual fragrance. It has slender trailing shoots, which will ascend 25 ft or more into trees, and hang down festooned with flowers even in dense shade—a characteristic of R. *arvensis*. Loosely double flowers, borne singly or in small clusters, crimson in bud, opening to cream, with orange-yellow stamens. Midsummer.

'STANWELL PERPETUAL' (Shrub). Introduced by Lee and Kennedy of Hammersmith in 1838, having been found as a seedling in a garden at Stanwell, Middlesex.—It is pretty certain that one parent of this rose was R. *pimpinelli-folia*, the other, on account of its flower shape and continuous production, may have been the Autumn Damask. A lax, arching shrub with numerous sharp prickles and thorns and small greyish leaves, against which the clear pale pink of the flowers is particularly agreeable. Fairly full flowers but showing stamens, with quilled and quartered petals. 5 ft. Recurrent. Very sweetly scented.

Kingsley, *Roses*, pl. 45.

'ST NICHOLAS' (Damask hybrid). 1950. A seedling which cropped up in the garden of the Hon. Robert James at Richmond, Yorks, probably a hybrid with a Gallica rose.—Sturdy erect bush, dark green leaves and hooked prickles. Flowers semi-double, of warm rich pink, paler towards the yellow central stamens. 4–5 ft, bushy and vigorous. Midsummer. Red heps, long lasting. Fragrant.

'SURPASSE TOUT' (Gallica).—Cupped, reflexing, fully double flowers of very good shape, of brilliant crimson, fading to strong cerise, with delicate veining of a darker shade. This, 'D'Aguesseau' and 'Assemblage des Beautés' are the nearest to red in the Gallica roses. 4 ft. Midsummer. Sweet scent.

'THE GARLAND' (Rambler). Wells, 1835. Reputedly R. *moschata* × R. *multiflora*.—It has characteristic purplish brown prickles on green wood; also its flower stems and pedicels always assume a vertical position. The masses of small flowers, borne in large trusses, are creamy white, semi-double, opening from flesh-tinted buds, followed by small, oval, red heps. 15 ft. 'MADAME D'ARBLAY' raised by Wells at the same time is more vigorous and the blush of the buds remains in the open flowers. Both are sweetly scented.

'TOUR DE MALAKOFF' (Centifolia). 1856.—A hybrid of great vigour, making an ungainly, spindly shrub, but capable of climbing through other growths to 9 ft, when its large blooms nod downwards. Smooth rather small leaves. Flowers a rare mixture of tints, vivid magenta to intense parma-violet, veined and flushed with carmine and purple, eventually reflexing into a mass of cool lilac-grey. Fragrant.

'TREASURE TROVE' (Rambler). 1979.—Mr John Treasure at Burford House, Tenbury Wells, found this seedling near to his plant of R. *filipes* 'Kiftsgate', and since this rose is very free in all gardens with its progeny, 'Treasure Trove' is likely to be R. *filipes* 'Kiftsgate' crossed with a modern rose. The original plant has in 18 years achieved 33 ft in width and 20 ft in height and is still expanding. The young foliage is ruby red turning to mid green. The blooms average about 20 in a truss, semi-double, cup shaped on opening, of warm apricot showing yellow stamens. The colour fades first to delicate pink and later to blush. To date this appears to be the most flamboyant of 'Kiftsgate' seedlings. Delicious and pervasive scent. Midsummer.

Rose Annual 1979, p. 155.

'TUSCANY' (Gallica). Before 1800.—This rose has neat, folded leaves, and is no doubt a very old cultivar since, like 'Sissinghurst Castle', and 'Conditorum' its flowers are of an undeveloped shape. Semi-double, opening flat, of intense dark maroon purple, the flowers are of telling colour in the garden. The centre is lit by yellow stamens. Slightly fragrant. 4 ft. Midsummer.

Botanical Register, Vol. 6, pl. 448.

'TUSCANY SUPERB' (Gallica). Also known as 'Superb Tuscany'. Described by William Paul of Cheshunt in 1848.—This resembles 'Tuscany' but has broader, more rounded leaves and a more sturdy growth. The flowers are of similar colour, larger, fuller, but show fewer stamens. 4 ft. Midsummer. Slight fragrance.

'UNIQUE BLANCHE (Centifolia). 1778, discovered in Needham, Suffolk. Also known as 'White Provence', 'Unique' or 'Vierge de Cléry'.—This is probably a sport from R. *centifolia* and is typical in all its parts, though less vigorous. Flowers cupped, opening from red-flushed buds, creamy white, with glistening almost transparent, narrow petals; button eye. It flowers, like R. *centifolia*, late in the summer season. 4–5 ft. Summer. Sweetly scented.

Redouté, Vol. I, p. 111 as *Rosa centifolia mutabilis*.

'VANITY' (Hybrid Musk). 1920. 'Château de Clos Vougeot' × seedling.— A remarkable shrub for large gardens. Being of vigorous, open, thrusting growth it is best to plant 3–4 ft apart, in groups of three or four; these groups may well achieve 8–12 ft. Foliage rather sparse, soft green, on glaucous stems. In summer the flowers are produced in clusters and in hot weather are of vivid carmine-pink; in autumn, the colour is a clear light pink, when the clusters may be a foot or two across on branches 6 ft long. Flowers have 5 or 6 petals and are 2½ in. wide, showing yellow stamens. Sweetly fragrant. 8 ft. Recurrent. A.M. 1956, F.C.C. 1958, both given in autumn.

Thomas, *Shrub Roses of Today*, pl. v.

'VARIEGATA DI BOLOGNA' (Bourbon). 1909.—This bears a marked affinity to 'Honorine de Brabant' in its growth and pointed foliage. The flowers are fully double, of excellent, full, reflexed shape, white, conspicuously and neatly striped with vivid crimson-purple. Though it can be grown as a bush it is better with some support. 9 ft. Midsummer. Sweet scent. Subject to black spot.

Trechslin and Coggiatti, pl. 38.

'VEILCHENBLAU' (Rambler). Schmidt of Erfurt, 1909. 'Crimson Rambler' (q.v. under R. *multiflora*) × 'Erinnerung an Brod', a Hybrid Perpetual with (reputedly) R. *setigera* in its pedigree.—The green wood is almost unarmed, and bears glossy bright green leaves, long-pointed. Flowers in large and small clusters, the crimson-purple buds opening to violet and fading to lilac-grey, more or less streaked with white; semi-double, with good stamens. The colour is best in shade of a wall. It flowers early in the rambler season. 12 ft. Midsummer. Sweetly scented.

Thomas, *Climbing Roses*, pl. II.

'VICOMTESSE PIERRE DE FOU' (Tea-Noisette). Sauvageot, France, 1923.
'L'Idéal' (Noisette) × 'Joseph Hill' (Hybrid Tea).—The vigorous plant has good
broad foliage and large prickles. At first sight it might be taken for a Hybrid
Tea, but its parentage results in an unusual dusky coppery orange passing to
coppery pink colouring, and also the full, quartered bloom. 15 ft. Recurrent.
Rich fragrance.
Thomas, *Climbing Roses*, fig. 8 (pencil drawing).

'VIOLACEA' (Gallica). 18th century or earlier; see further on p. 000.—An
old cultivar with single dark crimson flowers rapidly assuming a purplish flush;
yellow stamens. The winged calyces suggest possible Centifolia inheritance,
particularly in view of its tall growth. Almost unarmed. Small rounded leaves.
6 ft. Midsummer. Sweet scent.
Sitwell and Russell, Pt I, frontispiece.

'VIOLETTE' (Rambler), France, 1921.—This should not be confused with
'Violetta', a less worthy rambler. Almost unarmed green wood, dark green
leaves. Large and small trusses of small flowers, semi-double with stamens, of
dark crimson purple, darkening to maroon and later fading to maroon-grey; occasional white streaks. 15 ft. Midsummer. No scent.
Thomas, *Climbing Roses*, pl. 11.

'WEDDING DAY' (Rambler). Sir Frederick Stern, Highdown, 1950. Hybrid
of R. *sinowilsonii*.—This is tremendously vigorous, achieving at least 25 ft, with
green wood and scattered prickles, and glossy rich green leaves. Large clusters
of flowers; buds yolk-yellow, petals broadly wedge-shaped with mucronate
apex giving a starry effect, creamy white with orange-yellow stamens. After
rain the petals become spotted with pink. Midsummer. Very fragrant. A.M.
1950.

'WILHELM' (Hybrid Musk). W. Kordes, Germany, 1934. Known as 'Sky-
rocket' in the United States.—A spectacular shrub, vigorous, with upright
branches, clothed with good leaves. Clusters of few flowers appear on small
shoots in summer, quickly overtaken by large clusters on strong shoots, which
continue to be produced until autumn. Very dark red on opening, changing to
crimson with purplish flush, paler around the yellow stamens; semi-double.
A sport occured in 1947 named 'WILL SCARLET'; it has flowers of a brighter,
clearer colour, described as hunting pink, or scarlet. Both cultivars will achieve
8 ft and are in constant production; their red heps last well into the winter.
'Will Scarlet' received an A.M. in 1954. Both are slightly fragrant.

'WILLIAM ALLEN RICHARDSON' (Tea-Noisette).—In 1878 this sport
occurred on 'Rêve d'Or' raised nine years previously in France; while 'Rêve
d'Or' is a fairly free grower (it is deep buff-yellow fading to butter-yellow) its
sport was an epoch-making rose of its time on account of its brilliant colouring—
orange fading nearly to white in hot weather. It was the first in this colour.
Young foliage mahogany brown; some prickles on stouter growths. Both roses
are slightly fragrant, semi-double, and achieve about 8 ft. Recurrent. Midsummer.
A.M. 1897.
Hoffmann, pl. 13.

'WILLIAM LOBB' (Moss). 1855. Also known as 'Duchesse d'Istrie', 'Old

Velvet Moss'.—Extra vigorous, gawky growth which makes the plant only tolerable behind others. Foliage small, dark leaden green, in contrast to the dense light green moss copiously spread down the stems. Large semi-double flowers at first dark crimson-purple, fading to lavender-grey, lightened by nearly white bases of the petals. 8 ft. Midsummer. Sweetly scented.

Edwards, *Wild and Old Garden Roses*, facing p. 87, erroneously captioned 'Reine des Violettes'.

'WILLIAMS' DOUBLE YELLOW' (Shrub). John Williams, Pitmaston, near Lancaster. R. *foetida* (seed-parent) × R. *pimpinellifolia*.—If the parentage is correct, this is closely allied to R. × *harisonii* (see page 100). But it has no stamens, only abortive carpels, in the flowers. Appearing in early summer, the semi-double flowers are of light bright yellow; the petals are apt to turn brown on fading and to stay on the plant. It forms a scattered mass of shoots, spreading by underground runners. It has been called 'Old Double Yellow Scots Rose' or 'Prince Charlie's Rose'. It inherits the heavy scent of R. *foetida*. 4 ft.

Sweet, *British Flower Garden*, 2nd series, Vol. 4, pl. 353.

'ZÉPHIRINE DROUHIN' (Bourbon). 1868.—A popular unarmed climber, with smooth leaves, coppery purple when young. Flowers of uniform cerise-pink, semi-double, loosely fashioned, some in clusters. It is seldom out of flower. The sport 'KATHLEEN HARROP' occurred in 1919 in Ireland, and is a pleasing pink, with darker reverse. Both achieve about 10 ft. Recurrent. Wisley Rose Award 1925. A.G.M. 1966. Sweetly scented.

Trechslin and Coggiatti, pl. 40.

'ZIGEUNER KNABE' (Shrub). Lambert, Germany, 1909. Seedling of 'Russelliana' × ?R. *rugosa* hybrid.—A large arching shrub, twice as wide as high, with rough dark green leaves on prickly stems. Clusters of intense crimson-purple flowers appear early in the season; they are flat and reflexing, white towards the centre. It ushers in the purple roses to good effect in the garden, but is rather lacking in quality. 5 ft. Midsummer. Heps orange-red. Slightly scented.

Park, *World of Roses*, pl. 177.

REFERENCES
FOR ILLUSTRATIONS

BUNYARD, E. A.—*Old Garden Roses*. 1936.

Choix des Plus Belles Roses. 1845–54.

CURTIS, H.—*Beauties of the Rose*. 1850–3.

DARLINGTON, H. R.—*Roses*. 1911.

EDWARDS, G.—*Wild and Old Garden Roses*. 1975.

GAULT, S. M., and SYNGE, P. M.—*Dictionary of Roses in Colour*. 1971.

HARIOT, P.—*Le Livre d'Or des Roses*. 1904.

HOFFMANN, J.—*The Amateur Gardener's Rose Book*. Translated from the German by J. Weathers. 1905.

HOLLIS, L.—*Roses*. Ed. 2, 1974.

KINGSLEY, ROSE.—*Roses and Rose Growing*. 1908.

LAWRANCE, MARY.—*A Collection of Roses from Nature*. 1799.

Les plus Belles Roses au debut du XXe siècle. Published by the Société Nationale de Horticulture de France. 1912.

MCFARLAND, J. H.—*Roses of the World in Colour*. 1937.

Nestel's Illustrierte Rosengarten. 1866–9.

NIETNER, Th.—*Die Rose*. 1880.

PARK, B.—*The World of Roses*. 1962.

PAUL, W.—*The Rose Garden*. 1848 and later editions.

REDOUTÉ, P. J., and THORY, C. A.—*Les Roses*. 1817–24. (References are to the page facing the plate.)

SITWELL, S., and RUSSELL, J.—*Old Garden Roses*, Part One. 1955. With reproductions of eight paintings by Charles Raymond. Only the first two parts of this work were published.

THOMAS, G. S.—See Bibliography, p. 47.

TRECHSLIN, A. M., and COGGIATTI, S.—*Old Garden Roses*. 1975. Published in Switzerland as *Roses d'Antan*; English-language editions translated by H. N. Raban. The work contains 40 plates reproduced from paintings by Anne Marie Trechslin.

WILLMOTT, ELLEN.—*The Genus Rosa*. 1910–14. (The reference is to the page facing the plate.)

Periodicals

Journal des Roses.—Published monthly 1877–1914, one plate in each issue. Abbreviated *Journ. Roses*.

Rosen-Zeitung.—Published by the Verein deutscher Rosenfreunde, 1886–1933.

COLLECTIONS

There are collections of species and shrub roses in the following gardens:

The Royal National Rose Society's Display Garden, Bone Hill, Chiswell Green Lane, nr St Albans, Herts.

The Royal Horticultural Society's Garden, Wisley, Ripley, Surrey.

Sissinghurst Castle, nr Cranbrook, Kent (National Trust).

Mottisfont Abbey, nr Romsey, Hants (National Trust).

The Hillier Arboretum, Ampfield, nr Romsey, Hants (Hampshire County Council).

Hidcote Manor, nr Chipping Campden, Glos. (National Trust).

Castle Howard, Yorks (Castle Howard Estates Ltd).

ROSMARINUS Rosemary LABIATAE

A genus of two or three species in the Mediterranean region (including N. Africa), Portugal and N.W. Spain. Leaves aromatic, strongly revolute, linear. Flowers in short axillary racemes from the previous season's growth. Calyx campanulate, two-lipped. Corolla with a retuse or two-lobed upper lip, lower lip three-lobed, the middle lobe concave. Stamens two, exserted. Style exserted.

The generic name is taken from the Latin word for rosemary, which was also rendered as *ros maris* or simply *ros* by some Roman authors. *Ros* in Latin can also mean 'dew', whence the popular belief that *Rosmarinus* means 'sea-dew'.

R. officinalis L. Rosemary

An evergreen shrub of dense, leafy habit, forming a bush 6 or 7 ft high and as much wide; young stems slender, downy. Leaves opposite, linear, ¾ to 2 in. long, ¹⁄₁₆ to ⅛ in. wide; not stalked, blunt at the apex; margins recurved; dark rather glossy green above, white-felted beneath, aromatically fragrant when crushed. Flowers produced during May in clusters of two or three in the leaf-axils of the previous year's shoots. Corolla two-lipped, pale violet-blue and white; calyx darker and purplish, very downy.

Native of the Mediterranean region, the western Iberian peninsula and. Morocco; cultivated in Britain for four hundred years, probably much longer. It is the only cultivated species, but there are some distinct forms. Nearly related to the lavender, this shrub is also much associated with it in gardens. Its aromatic odour suggests nutmeg. A fragrant oil is extracted from the plant. The rosemary, which likes a sunny spot and not too heavy a soil, is scarcely so hardy as the lavender, although it is rarely injured. During the peculiarly trying winter of 1908–9, however, most of the old plants at Kew were killed, whilst two-year-old plants were not injured. It is readily increased by cuttings placed in a cold frame. Old specimens form short, rugged trunks, and are very picturesque.

Both in S. Europe and in Britain the rosemary fills a notable place in folklore. At one time it was believed to possess a stimulating influence on the memory, and was even known as 'herb of memory', hence the well-known line of Ophelia, 'There's rosemary, that's for remembrance'. The same idea has also given it a significance in association with the dead. In the old chanson we find the lines:

A l'entour de sa tombe, romarin l'on planta,
Sur la plus haute branche, le rossignol chanta.

R. *officinalis* is very variable in habit, from erect to ground-hugging; in the length and relative width of its leaves and in their colouring, which varies from grass-green to deep sea-green; in the length of the calyx and the density of its indumentum, and in the size and colouring of its flowers. There is also a variation in the composition of its essential oil, and hence in the fragrance of the leaves, some garden clones of wild provenance having an aroma which, though pleasant, is not that of the common culinary rosemary. But this is itself

ROSMARINUS OFFICINALIS   ROSEMARY

a cultivar.

Five varieties of R. *officinalis* were recognised by Dr Turrill in his study of the genus (*Kew Bull.*, 1920, pp. 105–8), based mainly on vegetative characters. But for the garden plants clonal names suffice. The following is a selection of the varieties in commerce:

cv. 'BENENDEN BLUE'.—Flowers vivid blue, almost gentian-blue. Leaves very narrow. Habit semi-erect. This was introduced from Corsica by Colling-

208 ROSMARINUS

wood Ingram, and received an Award of Merit when exhibited by him in May 1933.

The Corsican narrow-leaved rosemaries usually have flowers of a good colour; they were the best in this respect, but also the most tender, among the rosemaries grown experimentally in the 1860s by the French botanist Jordan. They have been distinguished botanically as var. *angustissimus* Foucaud & Mandon. Other garden clones of Corsican provenance, apart from 'Benenden Blue', are 'CORSICAN BLUE' (A.M. when exhibited by Messrs Jackman in 1946 as "R. *corsicus*"; and 'CORSICUS PROSTRATUS', distributed by Messrs Treseder.

cv. 'BLUE SPIRE'.—Of erect, compact habit. Leaves light green, rather narrow. Flowers clear blue. This seems to be hardy.

cv. 'MISS JESSUP'S UPRIGHT'.—An erect and very robust form with rather broad, deep sea-green leaves. One of the hardiest. This and similar selections have been grown as R. *off. fastigiatus*.

cv. 'MAJORCA PINK'.—Erect, with rather short, relatively broad leaves. Flowers lilac-pink, starting to open early in the New Year. Moderately hardy despite its provenance. In Jordan's trials (see under 'Benenden Blue') plants from the hills of southern France, of similar flower-colour and habit, were found to be the hardiest of all.

cv. 'PROSTRATUS'.—Of trailing habit. Foliage fresh green, dense. Flowers light blue. Very tender, and usually grown at the edge of a dry wall, where it can cascade downwards. Although there may be more than one clone under this name, the plant described is probably the one put into commerce by Smith of Newry around 1900, which was said to have come from a garden near Nice. But there is also reference in the literature to an introduction of a prostrate form from Capri (*Gard. Chron.*, Vol. 39 (1906), p. 381; Vol. 74 (1923), p. 24; Vol. 82 (1927), p. 402; *Fl. and Sylv.*, Vol. 1 (1903), p. 200).

In the R.H.S. *Dictionary of Gardening*, Vol. IV (1956), p. 1831, R. *off. prostratus* Hort. is referred to R. *lavandulaceus* [de Noë ex Debeaux]. This species, a native of N. Africa and S. Spain, which should probably take the name R. ERIOCALIX Jordan & Fourreau, is easily distinguished from R. *officinalis* by the two-layered indumentum of the calyx, the lower layer resembling that of R. *officinalis*, but usually more dense, the upper consisting of long hairs. None of the plants seen under the name R. *off.* 'Prostratus' show this character.

cv. 'SEVERN SEA'.—A free-flowering, rather tender sort, with arching, spreading branches and flowers of a fine blue. Raised by Norman Hadden in his garden at West Porlock.

cv. 'TUSCAN BLUE'.—Of erect habit. Leaves light green, rather broad. Flowers large, clear blue. Introduced by W. Arnold-Forster, who wrote: '. . . in Tuscany, hedges of this plant are conspicuous from a distance owing to their ceanothus blue. The plant is hardy in the mild counties but a good deal tenderer than the common sort; it only flowers freely if the yard-long spikes are topped.' (*Shrubs for the Milder Counties* (1948), pp. 171–2.)

RUBUS ROSACEAE BLACKBERRIES, ETC

A genus of some 250 or 300 species of evergreen or deciduous shrubs and a few herbaceous perennials, widely distributed over the world but absent from dry regions and rare in the tropics, where it is confined to mountainous regions. In addition there are a thousand or more microspecies, all contributed by the blackberries and their American allies; for these see further in the synopsis. The woody species mostly have short-lived stems which are furnished with prickles of various forms, or bristles; only rarely are they quite unarmed. Leaves simple (then often lobed) or variously compound. Stipules present, free or adnate to the petiole in their lower part. Flowers hermaphrodite (rarely unisexual) united into a determinate raceme, panicle or corymb, but sometimes solitary or few in a cluster. Calyx-lobes, petals and stamens inserted on the rim of a receptacle. Calyx-lobes normally five, persistent, spreading to reflexed after flowering or sometimes remaining erect and embracing the fruit. Stamens numerous, with slender filaments. Carpels few to many, more or less free at flowering-time, inserted on the convex centre of the receptacle. The fruit, in its most characteristic form, is represented in the bramble and the raspberry. In both the seeds are embedded singly in juicy droplets, which are united so as to form a rounded or hemispherical or thimble-shaped cup, fitted on the cone-shaped receptacle. In the raspberries the fruit can be easily pulled off the receptacle, but in the brambles the two adhere. But in some other groups of Rubi the drupelets may fall off separately, and in some they are more or less dry.

In the garden of ornamental shrubs the Rubi do not occupy anything like so important a place as their number would seem to justify. Comparatively few of them are worth growing for beauty of flower, but a considerable number are elegant in habit or handsome in foliage. Many species have their stems more or less covered with blue-white or purple bloom, and a few of the most striking are cultivated on that account. Others are grown for the beauty or edible values of their fruits.

The cultivation of the hardy Rubi presents no problems. They all like a loamy soil of good quality, and those of semi-scandent habit need some sort of support. This may be a stout post, up which the main shoots may be loosely tied, leaving the lateral branches free; it may be three or more rough oak branches set up to form a sort of pyramid; or the longer-stemmed ones may be used for covering pergolas or other structures of a similar nature.

In the case of the biennial-stemmed species, it is necessary to cut away the two-year-old stems which flower, bear fruit, and then die. With those whose stems are of longer duration, it is also advisable to cut away the older, worn-out stems occasionally. Some of the Rubi, especially those with biennial stems, have a tendency to decrease in vigour after a few years. The base in time forms a large woody root-stock which does not send up such vigorous stems as younger ones. The remedy is, of course, to renew the stock by seed or other means.

PROPAGATION.—The mode of propagation depends largely on the character of the individual species. Those that form thickets (like R. *odoratus* and R. *parviflorus*) can be divided up into comparatively small pieces; this is best done in autumn just before the leaves fall, or in spring. Apart from any desire to increase the stock, the plants are benefited by undergoing this process occasionally. Where division affords no means of increase, recourse must be had to either cuttings or layers. R. *deliciosus* is best increased by layering; the double-flowered brambles strike root quite well from cuttings. Seed is scarcely ever used as a means of increase, except for newly introduced species. But Watson, in his work on the British Rubi (see below), remarks that a wild bramble that has been found to produce fruit of high quality can be brought into the garden by sowing its seeds, which quickly develop into fruiting plants. But it should be added that the seeds should be sown at once and kept cool and damp during the winter (or be stratified in the usual way and sown in early spring).

SYNOPSIS

The following classification is largely based on W. O. Focke, 'Species Ruborum' in *Biblioteca Botanica*, Heft 72 (1910–11) and Heft 83 (1914). Two of Focke's subgenera have been omitted, neither having species in the Northern Temperate Zone or, so far as is known, in gardens.

For the European species of *Rubus* see *Flora Europaea*, Vol. 2 (1968); and for the American species L. H. Bailey, '*Rubus* in North America' in *Gentes Herbarum*, Vol. 5 (1941–5), a work of over 900 pages, of which almost 800 are devoted to the American blackberries and dewberries.

subgen. DALIBARDA.—A small group of creeping, unarmed herbaceous perennials, of which the type species—R. DALIBARDA (L.) L.—is a native of the forests of eastern N. America. The showy white flowers are usually sterile, the fruits, which have dryish drupelets, being produced by inconspicuous apetalous flowers. It was originally described by Linnaeus as *Dalibarda repens* and is still kept separate from *Rubus* by some botanists.

subgen. CHAMAEMORUS.—A single herbaceous species R. CHAMAEMORUS, the cloudberry, which is widely distributed in the northern hemisphere in high latitudes, but with many southward extensions. In Britain it occurs in Scotland, N. England and N. Wales. The flowering shoots are annual, from creeping underground stems. Leaves palmately lobed. Flowers solitary, unisexual. Fruits amber-coloured, with a few large drupelets.

subgen. CYLACTIS.—Another herbaceous group, with the majority of its species in Arctic regions and N. America. The only native member is the interesting R. SAXATILIS L., the stone bramble, a stoloniferous perennial bearing its small white flowers in terminal cymes; the fruits consist of a few large translucent drupelets. The Chinese R. XANTHOCARPUS Bur. & Franch. is sometimes cultivated for its yellow fruits.

subgen. CHAMAEBATUS.—A small subgenus of prostrate herbs with simple, lobed leaves. Stipules free, persistent, ovate. Flowers solitary or sometimes twinned, terminating erect few-leaved shoots. Sepals finely prickly, longer than the petals. Fruits with a few fleshy carpels. Here belongs the Himalayan R. CALYCINUS D. Don, which is in cultivation and makes a useful carpeter. The Japanese and Formosan R. PECTINELLUS Maxim. is closely allied to this.

subgen. DALIBARDASTRUM.—Of the subgenera so far enumerated this is the first that contains truly woody species. It is a perhaps rather artificial group of prostrate shrubs and subshrubs in the Himalaya and E. Asia, their stems usually clad with soft bristles. Leaves simple or trifoliolate. Stipules broad, free. Calyx longer than the petals, bristly.
R. *fockeanus*; R. *nepalensis* (*nutans*); R. *tricolor*.

subgen. MALACHOBATUS.—A fairly large group of more or less prickly (rarely unarmed) deciduous or evergreen shrubs, some prostrate or climbing. Leaves simple or compound. Stipules broad, often toothed or incised, free (i.e., not united to the petiole), usually deciduous. Flowers not showy. Fruits separating from the receptacle as in the raspberries (subgen. *Idaeobatus*), but in that group the stems are usually biennial and the narrow stipules are adnate to the petiole.
This group is distributed from the Himalaya to China, S.E. Asia and Malaysia. All the species have handsome foliage and the prostrate ones make useful ground-coverers.
R. *calycinoides*, R. *flagelliflorus*, R. *henryi*, R. *hupehensis*, R. *ichangensis*, R. *irenaeus*, R. *lambertianus*, R. *lineatus*, R. *maliformis*, R. *parkeri*, R. *playfairianus*, R. *setchuenensis*.

subgen. ANOPLOBATUS.—A comparatively small group of unarmed shrubs, whose stems persist for several years and have a peeling bark. Leaves simple, palmately lobed. Flowers large, white or pink, with spreading petals. Fruits separating from the receptacle, as in the raspberries. All the species are natives of the New World (N. America and Mexico) except R. *trifidus* of Japan and Korea. Nearly all the Rubi grown primarily for their flowers belong to this group.
R. *deliciosus* (and the related R. *trilobus*), R. *odoratus*, R. *parviflorus*, R. *trifidus*.

subgen. IDAEOBATUS. RASPBERRIES, THIMBLEBERRIES.—Stems mostly biennial, upright or arching, sometimes prostrate, prickly or bristly, rarely quite unarmed. Leaves simple in a few species, but mostly ternate or pinnate. Stipules narrow, adnate to the petiole at its base. Flowers bisexual, large and showy only in a minority of species. Fruits red, more rarely black, separating from the receptacle when ripe, i.e., the compound fruit hollow, not with a core as in the blackberries.
A widely distributed group, but represented in Europe only by the common raspberry R. *idaeus* and even that has a variety in the New World. The headquarters of the subgenus is in E. and S.E. Asia, but it is also represented in Africa, S. America, Australasia, etc. Of the few N. American species R. *occidentalis*, the related R. *leucodermis*, and R. *spectabilis* are treated here. Some of the species are notable for their pruinose stems, notably the Asiatic R. *biflorus*, R. *cockburnia-*

nus, R. *coreanus*, R. *lasiostylus* and R. *thibetanus*. The other species treated, all Asiatic, are: R. *adenophorus*, R. *amabilis*, R. *corchorifolius*, R. *crataegifolius*, R. *flosculosus*, R. *illecebrosus* (semi-herbaceous), R. *koehneanus*, R. *kuntzeanus*, R. *mesogaeus*, R. *palmatus*, R. *parvifolius*. R. *phoenicolasius*, R. *trianthus*.

subgen. LAMPOBATUS.—Climbing, prickly, evergreen shrubs, with leathery usually compound leaves. Flowers small, often unisexual, in usually elongate inflorescences. A perhaps rather artificial group, confined to subtropical or warm temperate regions. The species treated here are endemic to New Zealand—R. *cissoides* (and those described under it) and R. *parvus*, the latter unusual in its unifoliate leaves. Placed in this group, but not treated here, is R. LUCENS Focke, a tall, stout-trunked forest climber from the eastern Himalaya and the hills of Assam.

subgen. RUBUS (*Eubatus*).—Evergreen or deciduous shrubs with usually biennial, angular (less commonly terete), prickly or bristly stems. Leaves compound, being trifoliolate (ternate), digitate (leaflets five or seven, springing from a single point) or pedate (each of the two basal leaflets springing from the stalk of the leaflet above it), rarely pinnate. Stipules narrow, adnate to the petiole basally. Flowers in racemes, panicles or corymbs, rarely solitary. Fruits black or dark-coloured, the drupelets adhering to the core.

Of the six sections recognised by Focke in this subgenus four are confined to S. and C. America. The fifth, sect. *Ursinus* contains a single species, R. URSINUS Cham. & Schlecht (R. *vitifolius* Cham. & Schlecht.; R. *macropetalus* Dougl.). A native of western North America, this is of interest as a parent of the loganberry; it is very distinct from the next section in its pinnate leaves and unisexual flowers. The sixth section is:

sect. RUBUS (*Moriferi*) BLACKBERRIES (BRAMBLES), DEWBERRY, AMERICAN DEWBERRIES.—One of Focke's six subsections contains a few Mexican species. The others between them are responsible for the great majority —over three-quarters—of all the described species of *Rubus*, and are the domain of the study that has come to be known as batology, from *batos*, the Greek word for bramble. But it is now accepted, except in a few instances, these are not species as usually understood but apomictic 'microspecies' that have arisen during and since the Ice Age as the result of hybridisation between a limited number of normal species. Since these batological species are 'facultative' apomicts, i.e., occasionally breed sexually, they are not wholly debarred from further hybridisation among themselves. The evolution of new 'species' is therefore a continuous process and indeed many of those that have been described may have arisen in historic times or even be of quite recent origin—only a few are at all widely distributed.

subsect. SUBERECTI.—Erect shrubs forming a stool or even suckering for some distance; stems arching at the apex, but not as a rule tip-rooting. Inflorescence usually taking the form of a raceme or corymb, terminating a short lateral, or the lateral flower-bearing throughout its length. The type-species of this group is the European R. NESSENSIS W. Hall (R. *suberectus* Sm.), but commoner in Britain is R. PLICATUS Weihe & Nees, easily distinguished from the true blackberries by its slender stems, thin leaflets, sparsely prickly shortly racemose inflorescence, green, white-edged concave sepals and long, spreading stamens. The European species are few and taxonomically simple. But the American species that Focke groups with them in the subsect. *Suberecti* are a very numerous and complex group, which take the place in N.E. North America of the Old World brambles and are usually called blackberries there, or 'high blackberries' to distinguish them from the species of the subsect. *Procumbentes.* Barely ten species had been described in the American group before 1900; of the eighty-six species treated by Fernald in *Manual of Botany* (ed. 8, 1950) the majority were described by L. H. Bailey in the 1940s. Unlike the European species, many of the Americans have fruits of high quality and have given rise to numerous commercial varieties, of which only 'KITTATINNY' (R. *bellobatus* Bailey) ever made its mark in Britain.

subsect. PROCUMBENTES. AMERICAN DEWBERRIES, LOW BLACK-BERRIES.—Usually prostrate or procumbent shrubs with tip-rooting stems, which are armed with terete prickles or are bristly. Flowers in short usually erect racemes or corymbs, sometimes solitary. A North American group, from which some commercial fruiting varieties derive. A little over 100 species are recognised by Fernald in the work cited above, most of them first described by L. H. Bailey in the early 1940s. R. *hispidus* is sometimes cultivated as an ornamental ground-cover and is described in alphabetical order.

subsect. SENTICOSI and subsect. GLANDULOSI (of Focke's classification; corresponding to the subsections *Silvatici, Discolores* and *Appendiculati* of *Flora Europaea* and sections *Silvatici, Discolores, Sprengeliani, Appendiculati* and *Glandulosi* of Watson's *Handbook*). BLACKBERRIES (BRAMBLES).—Stems at first erect or spreading, later arching down and rooting at the tip, usually angled, plane or concave between the angles, armed with more or less flattened hooked or straight long-based prickles, which in some subgroups are the sole armature and then confined to the angles; in other subgroups they may be scattered round the circumference of the stems and are mixed with pricklets, needles (acicles) and stalked glands. Inflorescence paniculate (branches with more than one flower) or sometimes more or less racemose, terminating a leafy lateral from the previous year's stem (primocane), the lowermost inflorescence-branches usually subtended by true leaves, the upper by reduced leaves or bracts, the rachis and branches variously armed. Flowers variable in many characters, the petals pink or white. Fruits black, sometimes of excellent quality.

This, in terms of the number of species described, is an immense group, daunting even to the professional taxonomist unless he happens to be batologically inclined. Yet it is of limited distribution, being largely confined to the climatically more oceanic parts of western Eurasia; it scarcely extends into Russia except in the Crimea and Caucasus; only one species reaches as far as the N.W. Himalaya; it is rare in the Mediterranean region, which only a few species penetrate; and even in the Alps the number of species is by no means large. A few species have been introduced to N. America and become naturalised there, and others have become noxious weeds in New Zealand and Chile, in areas where nothofagus forest has been cleared to make way for pasture.

The taxonomic complexity of the brambles stems from the fact that, by crossing and re-crossing over a long period, they have come to combine in so many different ways the characters of the putative ancestral species, coupled, as has already been mentioned, with the apomictic mode of reproduction, which permits every viable character-combination to multiply and spread. Two plants may agree in the majority of their characters and yet differ in others that are too weighty to permit their being lumped together as states of the same species. So, quite logically, the batologists gave specific rank to every new combination of characters that came to their notice. On the other hand, some botanists of the last century, impatient of these refinements, simply lumped together all the plants in this group under the name R. *fruticosus* L., which was really another way of saying 'a bramble is a bramble'.

In *Flora Europaea*, Vol. 2 (1968), fifty-eight species of bramble are fully described and keyed out (nearly all these occurring in Britain) and another 376 are listed, without description, under the species with which they have key-characters in common. The latest treatment on the British brambles is contained in: W. C. R. Watson, *Handbook of the Rubi of Great Britain and Ireland* (1958), in which 352 species are treated in the group under consideration, about one-fifth of them endemic and many of them not even listed in *Flora Europaea*. Of course many of the species recognised by Watson are local or rare, but the number that are both widely distributed and reasonably abundant is so large that to give even a selection of them would be pointless and even misleading. But some of the brambles are cultivated for ornament or for their fruits: see R. *laciniatus* and R. *ulmifolius*.

subsect. C A E S I I.—Here belongs R. *caesius*, the common dewberry (q.v.) and the hybrids between it and the brambles (*Rubi corylifolii*).

R. ADENOPHORUS Rolfe

A robust shrub, deciduous, 8 ft or more high; stems erect or arching towards the top, stout, armed with stiff, short, broad-based spines, densely clothed with bristles, and with stalked glands. Leaves of the first-year (or barren) shoots mostly pinnate, 8 to 12 in. long, with five leaflets, those of the flowering shoots shorter, with three leaflets or sometimes simple. Leaflets obliquely obovate or ovate, 2 to 5 in. long, 1¼ to 3½ in. wide, tapered, rounded, or heart-shaped at the base, slender-pointed, sharply and doubly toothed, dull and hairy on both sides; main-stalk bristly and furnished with stalked glands like the shoots. Flowers produced in July in terminal, cylindrical panicles 4 to 5 in. long, the petals pink, toothed, the flower-stalks and calyx densely clothed with bristles and stalked glands. Fruits black, about ½ in. wide, edible.

Native of Central China; introduced by Wilson in 1907. The most remarkable feature of this bramble are the conspicuous dark glands, resembling minute black-headed pins, stuck among the bristles on the stems and leaf-stalks, but extraordinarily abundant on the sepals and flower-stalks. The leaf next to the panicle is often simple.

R. AMABILIS Focke

A deciduous shrub up to 6 or 7 ft high; young shoots slightly downy and armed with small prickles. Leaves pinnate, 4 to 8 in. long, composed of seven to eleven leaflets; main-stalk prickly. Leaflets very shortly stalked, ovate, pointed, sharply and doubly toothed, $\frac{3}{4}$ to 2 in. long, $\frac{1}{2}$ to 1 in. wide (terminal one larger), usually downy on the veins and armed with a few prickles on the midrib. Flowers white, $1\frac{1}{2}$ to 2 in. wide, solitary at the end of short leafy twigs, petals overlapping. Fruits conical, red, edible, $\frac{5}{8}$ in. long.

Native of W. China; discovered and introduced by Wilson in 1908. This rubus is distinct in its graceful habit, its handsome, much laciniated leaves, and its large solitary flowers which open in June and July.

R. BIFLORUS Buch.-Ham. ex. Sm.

A deciduous shrub, with erect stems up to 10 ft high, and 1 in. thick at the base, covered with a thick, white, waxy coating, and armed with straight broad-based spines. Towards the top the stems branch freely, the branches also being white, and, like the leaf-stalks and often the midrib, spiny. Leaves 4 to 10 in. long, composed of three or five leaflets, which are dark green above, covered beneath with a close white felt, ovate, pointed, sharply and irregularly toothed, and from $1\frac{1}{2}$ to 4 in. long. Flowers terminal and axillary, white, $\frac{3}{4}$ in. across; fruits yellow, roundish, $\frac{3}{4}$ in. in diameter, edible. *Bot. Mag.*, t. 4678.

Native of the Himalaya up to 10,000 ft, eastward to China; introduced in 1818. Among the longer cultivated, white-stemmed raspberries this is by far the most effective although it is equalled by some of the newer Chinese species (see R. *cockburnianus* and R. *lasiostylus*). Its flowers are of little consequence, being small and of little beauty. It should be raised from seed (which ripens here), and planted in groups of not less than half a dozen. The soil should be a good loam, the aim being to produce stout thick stems, for the stouter they are, the whiter and more persistent is their waxy covering. After the previous year's stems have flowered and borne fruit, they should be cut away (usually about August) leaving only the virgin growths of the year. During autumn and winter a group of this rubus makes one of the most notable plant pictures in the open air.

var. QUINQUEFLORUS Focke—A vigorous Chinese form introduced by Wilson in 1908, with the terminal inflorescence composed most frequently of five (sometimes up to eight) flowers. In typical R. *biflorus* it usually has two or three flowers.

R. CAESIUS L.   DEWBERRY

A deciduous shrub, with slender creeping stems, prickly, and covered with a whitish bloom when young. Leaves usually composed of three leaflets which are green and slightly hairy on both sides. Flowers white, in small clusters. Fruit composed of a few large carpels, covered with a blue-white bloom when ripe.

This is one of the British brambles easily distinguished from all the forms of common blackberry by the few but large 'pips' composing the fruit, and by their being covered, like the young stems, with a white or bluish bloom. It is common in Britain and over Europe, extending into N. Asia. Of no value for gardens.

R. *caesius* hybridises with the true blackberries, and in that way numerous minor species have arisen of intermediate character. Twenty of these are recognised by W. Watson in his treatment of the British Rubi, of which the most interesting is R. BALFOURIANUS Bab., which probably arose originally from a cross between R. *caesius* and R. *gratus*. Not only are the pink or white flowers sometimes as much as 2 in. wide, but they are borne throughout the summer and early autumn, and the fruits, though often abortive, are sometimes large, with a mulberry flavour (Watson, *Handbook of the British Rubi* (1958), p. 54 and fig. 3). Apart from these microspecies of hybrid origin, normal hybrids between the dewberry and the blackberries occur; usually these are infertile, but a single individual can cover a wide area by tip-layering.

R. CALYCINOIDES Koidz.

R. *fockeanus* Hort., not Kurz

An evergreen, prostrate, spreading shrub, self-rooting freely from branches on the ground; young shoots, leaf-stalks, underside of leaves and flower-stalks all covered densely with pale down. Leaves ½ to 1½ in. long, broadly ovate to cordate, three-lobed, the lobes rounded, toothed, very much wrinkled, the dense network of veins sunken above, prominent beneath; stalks ¼ to 1⅛ in. long, sparsely prickly. Flowers ⅝ in. across, solitary or in pairs, terminal mostly on short, leafy, lateral shoots; petals white, roundish, minutely ciliate; sepals large, downy, their lobes toothed. Fruits scarlet, ⅝ in. long, style and ring of stamens persisting. *Bot. Mag.*, t. 9644.

Native of Formosa; introduced by the late Lord Headfort from seed collected by a Japanese. It is quite hardy and without being in any way showy it flowers and develops fruits every year and makes an interesting firmly matted ground-cover several feet across.

Until Messrs Hillier pointed out the error in their *Manual* R. *calycinoides* was grown as R. *fockeanus*, a quite different species, for which see under R. *nepalensis*.

NOTE. The name R. *calycinoides*, proposed by Hayata and published by Koidzumi in 1917, is unfortunately invalid, having been applied earlier by O. Kuntze to an Indian species. Furthermore, the Formosan R. *calycinoides* is probably no more than a small-leaved variety of R. PENTALOBUS Hayata, described, also from Formosa, in 1908.

R. CHROOSEPALUS Focke

A large, semi-evergreen, straggling shrub, with round, slender, glabrous stems armed with short, decurved prickles. Leaves simple, heart-shaped, with a long tapering apex, 3 to 7 in. long, more than half as wide, the margins very finely and sharply toothed, and often scalloped into a few broad, very shallow lobes, of firm texture, glabrous above, but conspicuously silvery beneath with a close felt; stalks glabrous, 1 to 2½ in. long, with one or two spines. Flowers borne in a terminal panicle, 6 to 9 in. long, each flower ½ in. across with no petals, but a coloured, downy calyx. Fruits black, small, and of poor flavour.

Native of Central China; originally discovered by Henry; introduced to cultivation by Wilson about 1900. Its leaves bear a striking resemblance to those of *Tilia tomentosa*. A remarkably distinct as well as rather handsome and effective shrub.

RUBUS CISSOIDES

R. CISSOIDES A. Cunn.   BUSH LAWYER

R. australia var. *cissoides* (A. Cunn.) Hook. f.; *R. australis* sens. some authors, in part, not Forst.

A climbing evergreen dioecious shrub, sometimes reaching to the top of lofty trees in the wild; main stems stout, unarmed; branchlets armed with hooked, reddish prickles. Leaves with three to five leaflets, which are glabrous and glossy above, of leathery texture, serrate, varying much in shape: in some forms (believed to be a juvenile phase) they are linear-lanceolate, 3 to 6 in. long but only ⅛ to ⅜ in. wide, in others ovate-lanceolate, 2½ to 6 in. long and about 1 in. wide, with a somewhat cordate base, in others again relatively broader, with a truncate or oblique base; main-stalk and the stalks of the leaflets armed with recurved prickles. Flowers unisexual, white, about ½ in. across, produced in often much-branched panicles up to 2 ft long. Fruits about ¼ in. wide, reddish orange, only produced when both sexes are grown.

Native of New Zealand. It is the most handsome of the New Zealand Rubi, at least in flower, but tender away from a wall outside the milder parts. It has been confused with R. AUSTRALIS Forst., which has smaller, thinner, long-stalked leaflets up to 2 in. long and 1¾ in. wide; shorter panicles, to 8 in. long, sometimes reduced to racemes; and yellowish fruits.

R. SQUARROSUS Fritsch R. *cissoides* var. *pauperatus* Kirk—In the forests this is a tall climber, its leaves composed of three or five leaflets, which are ovate or ovate-lanceolate, up to 2½ in. or slightly more long. Flowers yellowish, in panicles to 6 in. long. Prickles of branchlets yellow. In open places, however, it forms a low, intricately branched shrub with green stems and 'skeletonised' leaves: the blade of the leaflet is reduced almost to nothing, being ¼ to ½ in. long and ¹⁄₁₆ to ⅛ in. wide, while the main stalk and the stalks of the leaflets are up to 8 in. long and usually strongly armed with prickles. This form is well illustrated in Davies, *New Zealand Native Plant Studies* (1956), p. 71. According to Cockayne, such plants, growing in shade, develop larger leaflets and may bear flowers (which apparently the skeletonised form never does).

R. COCKBURNIANUS Hemsl.   [PLATE 32
R. *giraldianus* Focke

A vigorous deciduous shrub up to 8 or 10 ft high, its biennial stems much branched towards the summit, pendulous at the ends, covered with a vividly white, waxy covering, not downy, armed rather sparely with broad-based spines. Leaves pinnate, consisting of usually nine leaflets, and from 5 to 8 in. long, the main-stalk downy, and armed with hooked spines. Leaflets 1½ to 2½ in. long, ¾ to 1¼ in. wide, the terminal one the largest, ovate or rather diamond-shaped, lateral ones oval-lanceolate, all unequally and rather coarsely toothed, slender-pointed, glabrous above, white beneath with a close felt. Inflorescence a terminal panicle; the flowers small and of little beauty, purple. Fruits black.

Native of China; first found in W. Szechwan by A. E. Pratt; introduced by Wilson in 1907. Its claims to recognition in the garden are its remarkably white stems, which are as notable in this respect as those of R. *biflorus*, and its arching, pendulous branches, which give a remarkable fountain-like aspect to the shrub.

R. CORCHORIFOLIUS L.f.

A deciduous shrub of vigorous growth, spreading by underground suckers; stems erect, 6 to 8 ft high, branching towards the top, round, covered with an exceedingly fine down when young, and furnished with rather broad-based prickles. Leaves simple, ovate, with a heart-shaped base, 3 to 7 in. long, two-thirds as wide, those of the sterile sucker stems very deeply three-lobed, purplish when young, margins irregularly toothed, upper surface dull dark green, nearly glabrous, the lower one paler and downy about the veins, midrib spiny; leaf-stalk 1 to 1½ in. long, spiny. Flowers white, borne singly or a few together on short lateral twigs. Fruit large, bright red, and, Mr Wilson informed me, of 'delicious, vinous flavour'.

Introduced by Wilson in 1907 from Central China, but described and named
by the younger Linnaeus as long ago as 1781 from Japanese specimens. It
appears to be widely spread in China, Korea and Japan. It may prove useful in
the wild garden, judging by the way it spreads in borders.

R. COREANUS Miq.

A deciduous shrub, 8 to 10 ft high (it has been found 15 ft high in the wild),
with erect or arching, stout, biennial stems, branching towards the top; glabrous,
but covered with a blue-white bloom, and armed with stiff, broad-based spines,
up to ½ in. long. Leaves pinnate, 6 to 10 in. long, composed usually of seven
leaflets, which are ovate or broadly oval, from 1½ to 3 in. long, 1 to 2 in. wide, the
lateral ones stalkless or nearly so, tapering at the base and smaller than the terminal one, which is broader, rounded or heart-shaped at the base, and stalked; all
are parallel-veined, dark lustrous green, coarsely toothed, except towards the
base, and have silky hairs on the veins when young. Flowers borne in flattish
clusters 1 to 3 in. across, terminating short shoots from the wood of the previous
year. Fruits of various colours from red to nearly black, edible but small, and of
poor flavour.

Native of Korea and China; introduced from the latter country in 1907 by
Wilson, who found it at altitudes up to 6,000 ft. It is one of the handsomest of
all Rubi in its vigorous blue-white stems and beautiful pinnate foliage.

R. CRATAEGIFOLIUS Bunge

An erect, deciduous shrub of stiff habit, 6 to 8 ft high, with stout biennial
stems branched towards the top, grooved and armed with small scattered
prickles. Leaves on the barren shoots of the year, large, palmately three- or
five-lobed, 5 to 8 in. across, heart-shaped at the base, sharply and often doubly
toothed, downy beneath; stalks and midrib prickly; leaves of the flowering twigs
much smaller, usually three-lobed. Flowers ¾ to 1 in. across; produced in clusters
terminating short twigs; petals white, prettily crimped at the margins; calyx segments lanceolate, much decurved. Fruits the size of a small raspberry, red.

Native of China, Korea, and Japan. The epithet *crataegifolius* is only appropriate to the small leaves of the flowering twigs; on the barren, first-year stems
they are more like those of a vine or maple, and in good soil are sometimes of
very large size—8 to 12 in. across.

R. DELICIOSUS Torr.

A deciduous shrub of sturdy habit, reaching 6 to 10 ft in height, bark peeling;
branches often arching or pendulous, quite unarmed, downy when young.
Leaves like those of a blackcurrant in shape and size, being three- or five-lobed,
with jagged edges, the base truncate or heart-shaped, 1½ to 3 in. long, rather more
wide, downy on both sides when young, especially beneath; stalk 1 to 1½ in.

long. Flowers mostly solitary, pure white, 2 in. across, borne in May on short twigs from the previous year's branches; sepals downy, ovate, $\frac{1}{2}$ in. long. Fruits $\frac{1}{2}$ in. across, dry, and of no flavour. *Bot. Mag.*, t. 6062.

RUBUS DELICIOSUS

Native of the Rocky Mountains of Colorado, New Mexico and Arizona; discovered in 1820 by Dr James. The fruit is not delicious but no doubt the name refers to the delight the flowers gave to the eye, for in this respect it is the most lovely of all Rubi, the blossoms being as beautiful as single roses, and as profusely borne. It was introduced in 1870. It is not very easily increased by cuttings (especially the better of two forms in cultivation), but can be layered, although the layers will sometimes take a twelvemonth before they become sufficiently rooted to be removable. A good loamy soil, a sunny position, and an occasional pruning out of the old wood complete its requirements. It is one of the elite of hardy shrubs.

R. TRILOBUS Ser. R. *mexicanus* O. Kuntze—Closely allied to R. *deliciosus* and described earlier. Focke, the authority on *Rubus*, remarked that he could see no difference between the two species, except that the terminal lobe in R. *trilobus* is longer and the internodes longer, so that the branchlets are more widely spaced. Native of S. Mexico. It was in cultivation in the middle of the last century, probably from seeds collected by Hartweg, but the present stock was introduced by E. K. Balls and Dr W. Balfour Gourlay from the Pico de Orizaba, Veracruz, at 9–10,000 ft in 1938 (*Bot. Mag.*, n.s., t. 452). The differences from R. *deliciosus* given in the article cited are that the leaves are larger, darker green, more finely toothed, with more acute lobes; and in having the sepals red on the inside in the fruiting stage. R. *trilobus* attains about 12 ft in gardens and a plant from the 1938 introduction received an Award of Merit when exhibited by Collingwood Ingram on May 20, 1947, from his garden at Benenden, Kent.

R. TRILOBUS × R. DELICIOSUS ('*Tridel*') 'BENENDEN'.—From this cross, made about 1950 with R. *deliciosus* as the pollen parent, Collingwood Ingram raised three seedlings, of which the best was given the clonal name 'Benenden'. This had flowered by 1954 and was described by the raiser in that year (*Journ. R.H.S.*, Vol. 79 (1954), p. 540). It received an Award of Merit in 1958, and Award of Garden Merit in 1962 and a First Class Certificate in 1963. It is a vigorous shrub, producing annual stems 8 ft or more high, arching over in the following year and bearing in May or early June pure white flowers 2¼ to almost 3 in. wide. The foliage is variable in shape, but mostly takes after that of the pollen parent.

R. FLAGELLIFLORUS Focke

A climbing evergreen shrub, with slender, graceful stems growing 5 or 6 ft in length in one season; when young they are covered with a whitish felt, sprinkled amongst which are tiny decurved prickles. Leaves broadly ovate, long-pointed, the base heart-shaped, the largest are 6 or 7 in. long, and about two-thirds as wide, shallowly lobed on the margins as well as finely and sharply toothed, the upper surface has appressed hairs between the veins, the lower one is covered with a thick, yellowish felt; stalk 1½ to 2½ in. long, slightly spiny. Flowers white, borne in axillary clusters. Fruits shining black, ½ in. wide, edible.

Native of Central and W. China, up to 6,000 ft; introduced for Messrs Veitch by Wilson about 1901. In habit this is one of the most elegant of the Chinese Rubi, and one of the handsomest in its foliage. When trained up a post or other support, the slender, whip-like shoots push out in all directions. The leaves often put on a marbled appearance in the shade. The appropriate name of R. *flagelliformis* was rather commonly applied to this plant, but the one given above is correct.

R. FLOSCULOSUS Focke

A deciduous shrub up to 10 or 12 ft high, the stout stems erect, arching at the much-branched top, biennial, glabrous except for a few spines. Leaves pinnate, 4 to 7 in. long, composed of five or seven leaflets which are ovate, ¾ to 1½ in. long, the terminal one larger, often three-lobed, and 3 in. long, glabrous above or becoming so, covered beneath with a close white felt; coarsely, often doubly toothed. Flowers small, pink, $\frac{1}{4}$ in. wide produced in narrow, cylindrical racemes 2 to 4 in. long, terminating the shoot, and in shorter ones from the axils of the terminal leaves. Fruits small, very dark red, or black.

Native of Central and W. China; introduced by Wilson in 1907. A very vigorous, pinnate-leaved bramble, allied to R. *cockburnianus*, but with dark purplish brown stems.

R. HENRYI Hemsl. & Kuntze

An evergreen, elegant, scandent shrub, growing 20 ft high where support is available; stems slender, cord-like, armed with a few spines. Leaves three-lobed, 4 to 6 in. long, glabrous above, covered beneath with a close white felt; stalk 1 to 1$\frac{1}{2}$ in. long; lobes of varying depth but usually reaching about three-fourths down the blade, narrow (from $\frac{3}{4}$ to 1 in. wide at the base), tapering to a long fine point, finely toothed. Flowers pink, of little beauty, $\frac{3}{4}$ in. across, borne six to ten together in terminal and axillary racemes 3 in. or so long; petals and sepals of about equal length, the latter covered with glandular hairs, and ending in a tail-like point. Fruits shining black, $\frac{1}{2}$ in. wide.

Native of Central and W. China; first discovered near Ichang by Henry, in whose honour it is named. Introduced by Wilson in 1900. It is mainly represented in cultivation by:

var. BAMBUSARUM (Focke) Rehd. R. *bambusarum* Focke—Leaves composed of three distinct leaflets, which are narrowly lanceolate, 2$\frac{1}{2}$ to 5 in. long, $\frac{3}{8}$ to $\frac{3}{4}$ in. wide, on stalks $\frac{1}{8}$ in. or less long. Native of Central China and, like the typical state, discovered by Henry and introduced by Wilson in 1900. It is notable for its elegant and rapid growth. When trained up a pillar or similar support, its slender branches arch outwards in all directions. Growths 10 to 12 ft long are made in one season. The panicles of black fruits, 3 to 5 in. long, are also handsome. *Bot. Mag.*, n.s., t. 33. It is perfectly hardy and received a First Class Certificate when exhibited by Messrs Veitch in 1907.

R. HISPIDUS L.

A low semi-evergreen shrub, with mostly prostrate, very slender, wiry stems, armed with tiny decurved spines and more or less covered with bristles. Leaves trifoliolate, the common stalk longer than the leaflets, which are short-stalked, obovate, tapering to the base, sharply and coarsely toothed towards the apex, 1 to 1$\frac{3}{4}$ in. long, $\frac{1}{2}$ to 1 in. wide, glabrous or nearly so on both surfaces. Flowers white, $\frac{1}{2}$ to $\frac{3}{4}$ in. across, produced in few-flowered corymbs from the leaf-axils and the ends of erect shoots 6 to 12 in. high. Fruits at first turning red, nearly black when ripe, less than $\frac{1}{2}$ in. long, and composed of few carpels, sour.

Native of eastern N. America; introduced in 1768, but rarely seen nowadays. It flowers in June and July. Growing very quickly, it soon forms a low, dense tangle, and makes a pretty almost evergreen ground cover.

R. HUPEHENSIS Oliver

R. swinhoei of early editions, not Hance

A prostrate or climbing evergreen shrub, with round, slender, dark-coloured stems, thinly furnished with a cobweb-like down when young, and armed with a few small decurved spines. Leaves simple, oblong-lanceolate; 3 to 4½ in. long, by about 1½ in. wide; the base rounded, the apex long-pointed, margins finely toothed; veins in nine to twelve pairs; upper surface smooth except for tiny bristles along the veins, lower one covered with a close grey felt; leaf-stalk ¼ to ½ in. long. Flowers usually three to seven in short, terminal, very glandular racemes, of little or no beauty; calyx covered with grey felt like the leaves; petals soon falling. Fruits described as at first red, then black-purple, austere.

Native of Central China; originally described in 1899, but introduced to gardens by Wilson from Hupeh in 1907. The foliage is handsome, and distinct from that of any other cultivated species except R. *malifolius*; the inflorescence also is conspicuous in its glandular hairiness.

R. ICHANGENSIS Hemsl. & Kuntze

A deciduous shrub, with long, slender stems armed with small hooked spines, and furnished with numerous dark, glandular bristles. Leaves narrowly ovate-cordate (often with angular lobes towards the base), the sinus open and rounded, 3½ to 7 in. long, 1 to 2½ in. wide, glabrous on both surfaces, margins sparsely toothed. Flowers white, ¼ in. wide, produced in an elongated terminal panicle, supplemented below by short racemes in the axils of the uppermost leaves, the whole measuring 8 to 12 in. or even more in length; flower-stalks glandular-hairy, sepals erect, enclosing the small white petals. Fruits bright red, small, but of good flavour.

Native of Central and W. China; discovered by Henry, and introduced in 1900 by Wilson. He stated that it is one of the finest of Chinese Rubi in regard to its fruits—panicles of which he had often found over 2 ft in length.

R. IDAEUS L.   WILD RASPBERRY

A deciduous shrub, with erect biennial stems, 3 to 6 ft high, more or less downy; sometimes without prickles, but usually armed with weak ones. Leaves pinnate and composed of five leaflets on the lower part of the sterile (first year) stems, mostly of three leaflets at the upper part of the same, and on the flowering branches. Leaflets ovate, 1½ to 4 in. long, coarsely toothed, green and soon quite glabrous above, covered with a white felt beneath; the terminal one is the largest and broadest, and sometimes heart-shaped at the base. Flowers produced in a panicle at the end of short twigs springing from the year-old stems, small, pinkish. Fruits red and juicy.

This shrub, the source of the common raspberry of the fruit garden (where varieties with yellow and whitish fruits are grown), is found wild in British woods, and all through Europe and N. Asia to Japan. It is only of interest on this account, being of little value as an ornament.

var. ANOMALUS Arrhenius *R. leesii* (Bab). Bab.; *R. idaeus* var. *leesii* Bab.; *R. idaeus* f. *obtusifolius* (Willd.) Focke; *R. obtusifolius* Willd.—This differs in having much more rounded leaflets than common *R. idaeus*, the central one being rarely stalked. It is found wild in Devon and Somerset and elsewhere in England and Scotland, as well as on the continent. It rarely sets good seeds because of a defect in the ovary, but it has been raised from them and found to breed true. It is said to have been raised in cultivation by crossing a raspberry with pollen of a strawberry (see *Gard. Chron.*, Vol. 20 (1883), p. 12, fig. 3, and pp. 150, 214, 276, 342).

var. STRIGOSUS (Michx.) Maxim. *R. strigosus* Michx. AMERICAN RASP-BERRY.—Stems densely clad with bristles which are frequently gland-tipped. Inflorescence-axes also bristly and glandular. Native of N. America, where it ranges across the continent from Newfoundland to British Columbia, and south to Virginia. It is the source of several American varieties of raspberry, the European sorts being mostly unsuitable for the American climate except in the Pacific West. Other American varieties derive from *R. occidentalis* (q.v.) or from hybrids between it and *R. idaeus* var. *strigosus*.

R. LOGANOBACCUS Bailey *R. ursinus* var. *loganobaccus* (Bailey) Bailey LOGANBERRY.—A hexaploid 'hybrid species' of which one parent was an octoploid form of the W. American species *R. ursinus* (*vitifolius*) and the other the raspberry variety 'Red Antwerp'. See further in Crane and Lawrence, *Genetics of Garden Plants* (1952), pp. 239–41. It was raised in California in 1881.

Another blackberry-raspberry cross is the so called Veitchberry, raised for Messrs Veitch by their well known hybridiser John Seden and originally called 'The Mahdi'. Like the loganberry, this too behaves as a species, coming more or less true from seed when selfed. The blackberry parent was *R. ulmifolius* (op. cit., p. 241). 'BEDFORD GIANT' is a seedling of the Veitchberry.

R. ILLECEBROSUS Focke

R. rosaefolius var. *coronarius* f. *simpliciflorus* Mak.; *R. commersonii* var. *illecebrosus* (Focke) Mak.; *R. sorbifolius* Hort., not Maxim.

A subshrub with creeping, underground stems, sending up green annual flowering shoots 2 to 3 ft high, which are glabrous, angled, and armed with curved prickles. Leaves pinnately compound; leaflets mostly five or seven, lanceolate, 1 to 3 in. long, ½ to ⅞ in. wide, acuminate, glabrous or slightly downy above, usually downy on the veins beneath; rachis prickly. Flowers white, about 1¾ in. wide, borne in late summer in few-flowered bracted corymbs. Stamens numerous. Fruits red, round or broadly ellipsoid, about 1¼ in. wide, with numerous drupelets.

Native of Japan; introduced to the USA towards the end of the last century and thence to Europe. It is grown for its ornamental strawberry-like fruits, which are sweet but insipid, and is recorded as an escape from gardens on the continent and in N. America. Strictly it is not a shrub, as its woody stems creep underground, and the annual stems die back each winter.

R. *illecebrosus* is allied to the wide-ranging Asiatic R. ROSIFLORUS Sm., of which a double-flowered form 'Coronarius' is sometimes grown in greenhouses.

R. IRENAEUS Focke

An evergreen prostrate shrub; stems round, slender, covered with a dense grey down, amidst which are set numerous small decurved prickles. Leaves roundish with a heart-shaped base, and an abrupt, pointed apex, 6 in. or more across, margins toothed and bristly, sometimes obscurely lobed; upper surface glabrous, dark green, lower one covered with a pale brown felt, and more or less hairy on the yellow veins; stalks 1½ to 3 in. long. Flowers white, ¾ in. wide, produced singly or in pairs in the leaf-axils, and in a small terminal cluster. Fruits large, red.

Native of Central and W. China; introduced about 1900 by Wilson for Messrs Veitch. It is one of the most striking and remarkable of simple-leaved Rubi, the foliage being of a shape and size suggestive of a coltsfoot leaf, but having on the upper surface a curious metallic lustre. It has some value as a handsome covering for semi-shaded slopes, or wherever a low evergreen vegetation is desired.

R. KOEHNEANUS Focke

R. *incisus* var. *koehneanus* (Focke) Koidz.; R. *incisus* var. *subcrataegifolius* (Lévl. & Van.) Rehd.; R. *microphyllus* var. *subcrataegifolius* (Lévl. & Van.) Ohwi; R. *crataegifolius* var. *subcrataegifolius* Lévl. & Van.

A deciduous shrub of bushy, rounded habit, a few feet high, the erect, or nearly erect, biennial stems covered with purplish bloom, but with few or no prickles. Leaves simple, three- or five-lobed, or sometimes scarcely lobed at all, heart-shaped at the base, 1½ to 5 in. long, about the same wide, glabrous and green above, white but not downy beneath, margins sharply toothed; leaf-stalk often as long as the blade. Flowers ¾ in. across, produced usually three together; stalks glabrous, ¾ in. long; petals white, oblong, calyx downy within, the triangular lobes shorter than the petals. Fruits orange red, composed of comparatively few large carpels. *Bot. Mag.*, t. 8264.

Native of Japan; introduced by Späth of Berlin, and originally distributed as "R. *morifolius*". It is rather pretty in blossom, the flowers being abundant, and the purple-red anthers contrasting well with the white petals.

R. *koehneanus* is perhaps only varietally distinct from R. INCISUS Thunb. (1784), for which a possible earlier name is R. *microphyllus* L.f. (1781). This is a smaller plant than R. *koehneanus*, with more numerous prickles, leaves to only 2 in. or so long, and mostly solitary flowers. The plant in commerce as R. *microphyllus* 'VARIEGATUS' is probably referable to R. *incisus*, but its flowers have not been seen and indeed it may be sterile. The young leaves are red, becoming green with irregular patches of white suffused pink. It is a rather weak grower, its slender stems covered with a whitish bloom and armed with numerous small prickles.

R. KUNTZEANUS Hemsl.
R. *innominatus* var. *kuntzeanus* (Hemsl.) Bailey

A deciduous shrub, with erect, sturdy biennial stems, 6 to 10 ft high, branching towards the top, covered with soft, grey, velvety down, and armed with short broad-based, scattered prickles. Leaves from 6 to 12 in. long, composed of three or five (pinnately arranged) leaflets, the side ones of which are obliquely ovate, 2 to 4 in. long, 1 to 2½ in. wide, fine-pointed, rounded at the base, irregularly toothed and very shortly stalked, slightly hairy and dark glossy green above, covered beneath with a close white felt, interspersed with hairs on the veins, terminal leaflet larger, broader, longer-stalked, often three-lobed, and heart-shaped at the base. The main-stalk has hooked prickles and is covered with the same velvety down as the stem. Flowers small (⅓ to ½ in. wide), produced in large terminal panicles, 1 to 1½ ft long; petals pink and soon falling. Fruits orange-red, rounded, ½ to ¾ in. wide, of good flavour.

Native of Central and W. China; first introduced to Kew by Henry from Ichang in 1886, but most of the plants now in cultivation were introduced by Wilson between 1900 and 1907. The species is of some promise as a fruit-bearer, but has little to recommend it for ornament. It has been confused with R. INNOMINATUS S. Moore, a species very closely allied, but distinct in its glandular stems, leaf-stalks, inflorescence, and calyx.

R. LACINIATUS Willd.   CUT-LEAVED BRAMBLE
R. *fruticosus* var. *laciniatus* Weston

A deciduous shrub of rambling or scandent habit, the angled stems well armed with stout, recurved spines, and hairy. Leaves composed of five (sometimes three) leaflets, radially arranged; the common stalk 2 to 3 in. long, beset with hooedk spines. Leaflets stalked, and either pinnate, or deeply and pinnately lobed; final subdivisions of leaf coarsely and angularly toothed, spiny on the stalk and midrib, downy especially beneath. The leaves vary much in size, and on vigorous shoots will, including the stalk, reach 8 to 12 in. in length. Flowers in large terminal panicles; flower-stalk hairy and spiny; petals pinkish white; calyx with narrow, downy, reflexed segments spiny at the back, ½ to ¾ in. long, ending in a tail-like point. Fruits black, and both in size and flavour one of the finest of blackberries.

The origin of this handsome and useful bramble is not known. It was apparently grown in the Jardin des Plantes, Paris, in the 17th century for it was illustrated by Plukenet (*Phytografia*, t. 108, fig. 4) in 1691 from a specimen given him by William Sherard, who had himself collected it in the Jardin des Plantes. It was known to Philip Miller, who mentioned it in his *Dictionary* from the 6th edition (1752) onwards, but did not name it. It was named R. *fruticosus* var.

laciniatus by Weston in 1770, while in 1806 Willdenow illustrated and described it as R. *laciniatus* from a plant growing in the Berlin Botanic Garden. He did not mention Weston, so presumably did not adopt the epithet from him.

R. *laciniatus* comes more or less true from seed and wild plants, sprung no doubt from seed dropped by birds, may nearly always be found in the vicinity of cultivated plants. A selection is now extensively cultivated for its fruits in gardens, being perhaps the best of all blackberries for that purpose. The foliage is very handsomely divided, and the plant is sometimes grown on pergolas and trellises for its sake as well as for the fruit. It is useful also for growing on the boundary fences, fruiting freely there.

cv. 'ELEGANS'.—In this bramble, perhaps not really a form of R. *laciniatus*, the leaves are much smaller, and more hairy on the upper surface. It does not flower freely and fruits are rarely if ever borne (R. *laciniatus* var. *elegans* Bean; R. *laciniatus minor* Hort.; R. *quintlandii* Hort.). It was once grown by lovers of curiosities.

The commercial fruiting variety 'OREGON THORNLESS', raised in the USA, is thought to derive from the parsley-leaved blackberry. Having elegant foliage as well as excellent fruits, and not being excessively vigorous, it could be admitted into the ornamental part of a garden, trained up a pillar or trellis.

R. LAMBERTIANUS Ser.

A straggling sub-evergreen shrub, with slender, four-angled stems viscous when young, and armed with short decurved spines. Leaves glossy green on both surfaces, simple, sometimes three-, or obscurely five-lobed, sometimes merely wavy; broadly ovate or triangular, 3 to 5 in. long, nearly as much wide at the heart-shaped base, toothed, slightly downy on the veins above, more so beneath; stalk 1 to 2 in. long; stipules ⅓ in. long, with usually five linear lobes. Flowers white, ⅓ in. across, produced in a terminal panicle 3 to 5 in. long, calyx segments downy, ovate-lanceolate. Fruits red, small.

Native of Central China; introduced by Wilson in 1907. It is a luxuriant, very leafy, scandent shrub, suitable for planting as a rough group in thin woodland.

var. GLABER Hemsl. R. *hakonensis* Franch. & Sav.; R. *lambertianus* subsp. *hakonensis* (Franch. & Sav.) Focke; R. *lambertianus* var. *hakonensis* (Franch. & Sav.) Rehd.—Similar in habit to the above, stems round and like the leaves glabrous or nearly so. Fruits yellow. Native of Japan as well as China; introduced from the latter country by Wilson in 1907.

R. LASIOSTYLUS Focke

An erect-growing deciduous shrub, with biennial stems, 4 to 6 ft high, covered with a blue-white, waxy bloom, and closely set with bristle-like spines, ¼ in. or less in length, not downy. Leaves composed of three or five leaflets, and on young vigorous plants as much as 14 in. long, but usually some 6 or 8 in. long; side leaflets ovate, 2 to 4 in. long, coarsely and unevenly toothed, very sparsely hairy above, covered with a close white felt beneath, terminal leaflet much larger especially in the trifoliolate leaves, often lobed, heart-shaped at the base. Flowers small, with reddish purple petals which are shorter than the calyx segments, and soon fall. Fruits 1 in. across, roundish, red, and downy, with an agreeable acid taste. *Bot. Mag.*, t. 7426.

Native of Central China; originally discovered by Henry in Hupeh and introduced by him to Kew in 1889; it was later reintroduced by Wilson from the same province. It is one of the most effective of the white-stemmed raspberries.

R. LINEATUS Reinwardt

A deciduous or semi-evergreen rambling shrub up to 10 ft high; stems slender, downy, furnished with a few tiny prickles. Leaves made up of usually five, sometimes three leaflets radiating from the end of a downy main-stalk that is 1½ to 3 in. long. Leaflets oblanceolate to oblong, shortly and slenderly pointed, tapered at the base, scarcely stalked, evenly set all round with sharp triangular teeth; middle leaflet the largest and from 4 to 9 in. long by 1 to 2½ in. wide, lowest pair often about half the size, upper surface dark green with a line of white down on the midrib, under surface covered completely with shining silky down, veins parallel in thirty to fifty pairs. Flowers in short axillary clusters, white, sepals longer than the petals, downy. Fruits small, red or yellow.

RUBUS LINEATUS

Native of the Himalaya, S.W. China, and Malaysia. I first saw it cultivated out-of-doors at Caerhays, Cornwall, in 1916; it was then 10 ft high. It was also grown at that time by Harry White at the Sunningdale Nurseries. Amongst the

Rubi it is remarkably distinct in its five-foliolate leaves with the leaflets arranged as in the horse chestnut, in the singularly beautiful silvery sheen beneath them, and in their very numerous parallel veins, of which I have counted as many as fifty pairs on one leaflet. It is not hardy at Kew, but Messrs Hillier report that at Winchester it is injured only in hard winters. The plants in cultivation were raised from seed collected by Forrest, who found it in Yunnan as long ago as 1905. Henry had previously found it in the same province.

R. SPLENDIDISSIMUS Hara R. *andersonii* Hook. f. (1878), not Lefèvre (1877). —Near to R. *lineatus*, but its branches, petioles and pedicels are clad with a more woolly indumentum and are also furnished with long, gland-tipped bristles. The inflorescences are larger and more open and the leaflets up to 3 in. wide. Native of the Himalaya from E. Nepal to Bhutan; introduced in 1971 by the University of North Wales Expedition to Nepal.

R. MALIFOLIUS Focke

A deciduous shrub whose prostrate or climbing stems are sparingly armed with short recurved spines, otherwise glabrous. Leaves oval or ovate, 2 to 5 in. long, 1 to 2 in. wide, rounded at the base, glabrous above, downy on the veins beneath, the margins set with broad, shallow teeth, each tooth ending in a small abrupt point; veins in seven to ten pairs, parallel; stalk ¼ to ⅝ in. long. Flowers in terminal racemes, 2 to 4 in. long, each flower 1 in. across, the petals white, roundish, overlapping; anthers downy; sepals ovate, downy like the short flower-stalk. Fruits of goodly size, black.

Native of W. China, where it is common in thickets at 2,000 to 4,000 ft; also of Central China, but rare. It is an elegant species, and in regard to its flowers is one of the handsomest of Chinese Rubi, but according to Wilson the fruit has an unpleasant flavour. It differs from R. *hupehensis* in the inflorescence being without glands. The specific name refers to the apple-like foliage.

R. MESOGAEUS Focke

A strong-growing deciduous shrub with erect stems unbranched the first year, springing from the ground like raspberry canes and growing 10 ft long in a season, arched at the top; they are quite velvety with down which persists through the winter; prickles ⅛ in. or less long, curved. Leaves trifoliolate, with a velvety, prickly main-stalk, the side leaflets very shortly stalked. The leaflets vary much in size and are from 2 to 7 in. long and from half to nearly as much wide; they are ovate to roundish ovate, coarsely toothed, the terminal one the largest and more or less lobed, mostly slenderly pointed, rounded or slightly heart-shaped at the base, slightly downy above, grey-velvety beneath. Flowers small, pinkish white, borne in short axillary clusters; sepals ultimately reflexed; fruits black, round, ⅓ in. wide with a few hairs at the summit of each pip. Blossoms in June.

Native of Central and W. China, where it is widely spread; introduced by

Wilson in 1907. It is related to the common raspberry. Notable chiefly for its strong growth and large leaves. It has been confused in gardens with R. *pedunculosus*, which has wholly downy fruits.

R. NEPALENSIS (Hook. f.) O. Kuntze

R. *nutans* var. *nepalensis* Hook. f.; R. *nutans* Wall. ex Edgew., not Vest; R. *barbatus* Edgew., *nom. prov.*; R. *barbatus* Edgew. ex Rehd., not Fritsch; R. *nutantiflorus* Hara

An evergreen, prostrate shrub, rising only a few inches above the ground; the stems creeping, unarmed, but thickly covered with soft purplish bristles and rooting at almost every joint. Leaves trifoliolate, with bristly stalks 1½ to 2 in. long; leaflets glossy green above, bristly on the veins beneath, sharply toothed, the terminal one the largest and from 1 to 2½ in. long, rhomboidal, often rounded at the apex; the side ones half to two-thirds as large, all three very shortly stalked. Stipules toothed or laciniated. Flowers pure white, 1½ in. across, borne in the leaf-axils and at the top of erect, leafy shoots 6 or 8 in. high, each flower on a slender stalk 1½ to 2½ in. long, bristly like the reddish calyx. *Bot. Mag.*, t. 5023.

Native of the Himalaya; cultivated at Kew since the mid-19th century. The cheerful leaves and large flowers render this one of the most pleasing of dwarf Rubi. The plant is rare in gardens, but may be recommended as a low covering for sheltered semi-shaded slopes, etc.

R. *nepalensis* is a variable species. The plant described above agrees fairly well with the type of R. *nutans*—a name that has had to be discarded for nomenclatural reasons. Plants agreeing with the type of R. *nepalensis*, which is smaller in all its parts, are also in cultivation.

R. FOCKEANUS Kurz—Closely allied to the preceding, but with more slender stems which are hairy but not bristly, and entire stipules. It is of wider distribution than R. *nepalensis*, extending from the Himalaya to Western and Central China.

R. × NOBILIS Reg.

A hybrid between R. *odoratus* and R. *idaeus*, raised by C. de Vos, at Hazerswoude, near Boskoop, in Holland, about 1855. It is intermediate between the parents, having erect, sturdy stems peeling like those of R. *odoratus*, but less glandular-hairy and not so tall. Leaves trifoliolate, large, hairy on both surfaces. Flowers purplish red, produced in terminal corymbs in June and July. A handsome, vigorous shrub of about the same value for ornament as R. *odoratus*—the mother plant. The leaves resemble R. *idaeus* in being trifoliolate, but the flowers owe their colour and size largely to R. *odoratus*.

R. OCCIDENTALIS L.   BLACK RASPBERRY

A deciduous shrub with arching, biennial stems 6 to 10 ft long, very glaucous and armed with scattered short spines. Leaves dark green, composed of three (sometimes five, pinnately arranged) leaflets, which are ovate, 1½ to 4 in. long, pointed, coarsely and unequally toothed, covered with a close white felt beneath. Flowers white, ½ in. across, produced in terminal few-flowered corymbs in June; prickles in inflorescence straight and terete. Fruits purple-black, hemispherical.

Native of eastern and central North America and the parent of several commercial fruiting varieties grown there. In this country it is only worth growing for the long, arching, blue-white stems, and even in this respect it is not the equal of Asiatic species such as R. *biflorus* and R. *cockburnianus*. There is a variety with yellow fruits.

R. × NEGLECTUS Peck—A natural hybrid between the above and R. *idaeus* var. *strigosus*. The form introduced to Kew in 1893 had dark red fruits and prickly blue-white stems.

R. LEUCODERMIS Torr. & Gr. R. *occidentalis* var. *leucodermis* (Torr. & Gr.) Focke—Closely allied to R. *occidentalis*, which it replaces in western N. America. It differs in its lighter green leaves and in having hooked, flattened prickles in the inflorescence. Introduced by Douglas in 1829. It has been grown in gardens for its blue-white stems, but the name is better known than the plant, for what used to be grown as 'R. *leucodermis*' was the Himalayan R. *biflorus*, a finer species with much whiter stems. R. GLAUCIFOLIUS Kell., from the same area, is closely related to R. *leucodermis* but has procumbent or prostrate main branches and is less prickly.

R. ODORATUS L.

A vigorous, deciduous shrub, with stout, erect, very pale brown stems up to 8 ft high, bark peeling; young stems covered with glandular hairs. Leaves simple, amongst the largest of hardy Rubi, five-lobed, vine-like, 4 to 10 (or even 12) in. across; lobes pointed, sharply and irregularly toothed, hairy on both sides, but especially beneath, soft and velvety to the touch. Flowers fragrant, bright purple, 1½ to 2 in. across, borne in large, branching, corymbose clusters at the ends of the shoots; the stalks conspicuously furnished with dense glandular hairs, the calyx similarly covered, each of its five divisions narrowed to a tail-like point. Fruits flat and broad, red when ripe, but rarely seen in this country. *Bot. Mag.*, t. 323.

Native of eastern N. America introduced in 1770. Next to R. *deliciosus*, this is perhaps the most ornamental of Rubi, in regard to blossom. It flowers from July to September, and few shrubs at that time equal it in beauty and fragrance. It loves a semi-shaded spot, where its flowers are protected from the fierce midday and early afternoon sun; in such a place the blossoms last longer. It is a rampant grower, and soon forms a thicket; good soil should be provided and the plants are all the better if pulled apart every few years, and planted more thinly. The old stems should be removed every winter. It is very similar in growth to R. *parviflorus* (q.v.), but starts to flower a month later.

R. × FRASERI Rehd.   R. *robustus* Fraser, not Presl.—A hybrid between R. *odoratus* and R. *parviflorus* (seed-parent) raised in 1918 by George Fraser, Ucluelet, British Columbia. Described by Fraser as a vigorous, compact shrub to 8 ft high, resembling R. *odoratus* in its non-suckering habit but with acuminately lobed leaves inherited from the seed-parent. Flowers rosy at first, becoming pale purple.

R. PALMATUS Thunb.
R. *microphyllus sens.* Rehd., ? not L.f.

A deciduous shrub, 5 or 6 ft high in the open (thrice as much in a cool greenhouse); stems not downy, but armed with small, flattened prickles. Leaves usually palmately five-lobed, sometimes three-lobed, sometimes seven- or nine-lobed, 1 to 3 in. long, margins doubly toothed, green on both surfaces with silky hairs along the midrib and veins; stalk $\frac{3}{4}$ to $1\frac{1}{2}$ in. long, with hooked spines. Flowers white, $1\frac{1}{2}$ in. across, solitary, produced from the axils of terminal leaves on short shoots that spring from the previous year's growths. Petals of narrowly oval outline, their ends rounded; calyx downy outside, glabrous within, the lobes narrow, long-pointed, and toothed; stalk slender, $\frac{1}{2}$ to $\frac{3}{4}$ in. long. Fruits roundish, yellow and juicy, $\frac{3}{4}$ in. across. *Bot. Mag.*, t. 7801.

Native of China and Japan. In the Temperate House at Kew, trained on a pillar, this shrub was 20 ft or more high, but in the open and unprotected it is rather a low shrub. Although hardy enough, it apparently needs somewhat warmer conditions than the open air affords near London to bring out its best qualities.

R. PARKERI Hance

A deciduous shrub of climbing habit; stems biennial, round, slender, armed with short, scattered, decurved spines, and thickly covered with greyish hairs, many of them gland-tipped. Leaves simple, broadly lanceolate, long-pointed, heart-shaped at the base, 4 to 7 in. long, about half as wide, the margins wavy, and sharply and finely toothed; upper surface bristly, especially along the midrib and veins, the lower one covered with a dense brownish red down; leaf-stalk up to 1 in. long, hairy and prickly. Flowers borne on an elongated, lax panicle, the calyx being remarkable for its dense covering of reddish glandular hairs. Fruits black, ripening early.

Native of China, where it was originally discovered in the province of Szechwan by E. H. Parker, in 1881; introduced in 1907 by Wilson, who found it near Ichang. This bramble has distinct and striking foliage, and its habit is elegant.

R. PARVIFLORUS Nutt.   THIMBLEBERRY
R. *nutkanus* Moçino ex Ser.

A vigorous deciduous shrub, up to 8 ft high, with erect, unarmed stems, and peeling bark; young shoots downy and slightly glandular. Leaves simple, five-lobed, vine-like, 4 to 8 (or more) in. across, irregularly toothed, downy on both sides especially beneath; leaf-stalk 2 to 5 in. long, set with glandular hairs. Flowers pure white, 1½ to 2 in. across, borne three to ten in terminal clusters during June, and continuing for several weeks; the flower-stalk is glandular-hairy and the calyx is very downy, each lobe contracted at the apex into a short tail. Fruits large, hemispherical and flattened, red; said to be sometimes pleasantly flavoured in the wild. *Bot. Mag.*, t. 3453.

Native of N. America and N. Mexico; introduced by Douglas in 1827. Very similar in its growth and foliage to R. *odoratus*, but easily distinguished by its white flowers in smaller clusters; the shoots, too, are not so conspicuously downy and glandular, and are darker coloured. Like that species it forms, when left to itself in good soil, dense thickets, which should be overhauled every winter and the worn-out stems cut out. Easily increased by pulling old plants to pieces. Fruits ripen most seasons from the earliest flowers, but are insipid and worthless in this country.

R. PARVIFOLIUS L.
R. *triphyllus* Thunb.

A low, deciduous shrub, forming a tangle of slender, downy, prickly stems a few feet high. Leaves composed of usually three, but sometimes five, leaflets, borne on a common stalk 1½ to 2 in. long, downy, and covered with prickles. Leaflets of various shapes and sizes, usually roundish or widely obovate, ¾ to 2 in. long, coarsely toothed, dark green and glabrous above, clothed with a close white felt beneath, the terminal one the largest and longest stalked. Flowers produced from the leaf-axils near the end of the shoot in few-flowered corymbs on downy, prickly stalks; petals bright rose-coloured, erect. Fruits roundish, red, edible.

Native of Japan and China; according to Bentham some of its forms are also native of Australia. It was originally imported by the Horticultural Society in 1818. It is rather pretty in blossom, but not more so than many of our native brambles. Its relationship, however, is not with them but with the raspberries (subgen. *Idaeobatus*).

R. PARVUS J. Buchanan

An evergreen, unifoliolate, dioecious, prostrate shrub, the slender stems often partially buried in the soil and rooting from the joints; prickles few and small. The whole plant except the flower-stalks is glabrous. Leaves 1 to 3 in. long, ¼ to ⅝ in. wide, linear to linear-lanceolate, more or less cordate, pointed, the margins densely and regularly set with small, sharp teeth; stalk ½ to 1 in. long, midrib sparsely spiny beneath. Flowers solitary or in twos or threes, axillary or terminal, white, about 1 in. wide, petals ovate, spreading.

Native of the South Island of New Zealand; in cultivation at Leonardslee, Sussex, in 1916. It is uncommon in gardens but used to be grown by Fred Stoker, who had a low patch several feet across in his garden in Essex, flowering in May and June. In a sunny position the dark green leaves are tinged with bronze, especially in winter, but they may be badly burned by prolonged frost.

With even more vividly tinted foliage is the hybrid R. × BARKERI Ckn., discovered wild in Westland in 1898 and distributed by the New Zealand botanist Leonard Cockayne. The other parent is probably R. *schmideloides*, a New Zealand species not treated here. See further in *Gard. Chron.*, Vol. 82 (1927), p. 405.; also in that excellent work L. J. Metcalf, *The Cultivation of New Zealand Trees and Shrubs* (1972), pp. 244–5, where R. *parvus* and the hybrid are both figured.

R. PEDUNCULOSUS D. Don

R. *gracilis* Roxb., not Presl; R. *niveus* Wall., not Thunb.

A deciduous shrub, with very stout, erect, biennial stems, 1 to 1½ in. thick and in vigorous plants 4 to 6 yards long, covered with a soft, thick, velvety down, and sprinkled over with minute prickles. Leaves 6 to over 12 in. long, composed of three or five leaflets. Side leaflets about half the size of the terminal one, stalkless or nearly so, obliquely ovate, coarsely and doubly toothed, slightly hairy above, covered with a close white felt beneath, and with silvery hairs on the veins; terminal leaflets ovate to roundish heart-shaped, long-stalked, from 3 to 5 in. long and wide, in other respects the same as the side ones. Flowers white or pale pink, ½ in. across, the petals shorter than the sepals. Fruits blue-black, small.

Native of the Himalaya and of W. and Central China, whence it was introduced about 1901. The Chinese plants are chiefly remarkable for their vigour; Wilson stated that it is occasionally 20 ft high. It is the most robust of all cultivated Rubi; hardy in Britain.

R. PHOENICOLASIUS Maxim.   WINEBERRY

A deciduous shrub making spreading stems 8 to 10 ft long in favourable situations; the stems are biennial, round, and together with the branches and leaf-stalks are covered densely with reddish, gland-tipped bristles mixed with which are a few slender prickles. Leaves 5 to 7 in. long, composed of three leaflets. The terminal leaflet is stalked, 2 to 4 in. long, roundish or broadly ovate, the base rounded or heart-shaped, the margins coarsely toothed and lobed; the side leaflets differ only in being obliquely ovate, stalkless, and much smaller; all are sparsely hairy above, white-felted beneath. Flowers in terminal racemes, the chief feature being the calyx, which is covered with glandular hairs, and measures 1½ in. across, the five segments being very narrow and pointed; petals ¼ in. or less in length, pink. Fruits conical, ¾ in. long, bright red, sweet and juicy. *Bot. Mag.*, t. 6479.

Native of Japan, Korea and N. China; introduced about 1876. This raspberry is hardy at Kew, bearing fruit regularly in the open, but it would probably succeed better against a wall. The species is noteworthy, not only for its fruits, but also for its excessively bristly stems, its tiny petals, and large star-shaped calyx which persists and spreads out beneath the fruit. It flowers in June under glass, a few weeks later out-of-doors.

R. PLAYFAIRIANUS Focke

A rambling evergreen or semi-evergreen shrub; young stems dark coloured, round, very slender, string-like, armed with tiny hooked prickles, and covered with web-like down when young. Leaves composed of three or five leaflets radiating from the end of the stalk, which is 1½ to 2½ in. long, and prickly; leaflets lanceolate, sharply toothed, the terminal one the largest, and sometimes 6 in. long, the basal pair 1 to 3 in. long, dark glossy green above, covered beneath with a pale grey felt. Flowers about ½ in. across, produced in small terminal panicles, and in the leaf-axils near. Fruits resembling a raspberry, but black; ripe in July and August.

Native of Central and W. China; introduced in 1907 by Wilson, who states that it is found in thickets at 3,000 to 6,000 ft. It is a very graceful plant when trained up a support, and the shape of its leaves is very uncommon among hardy Rubi, being more suggestive of *Ampelopsis*.

R. SETCHUENENSIS Bur. & Franch.

R. omeiensis Rolfe; *R. clemens* Focke

A large straggling shrub, with round stems, unarmed, but furnished with small, stellate hairs. Leaves of maple-like form, five-, or obscurely seven-lobed, with a heart-shaped base, 3 to 7 in. long and as much wide, irregularly toothed, stellately downy beneath, less so above; stalk 2 to 3 in. long; stipules ½ to ¾ in. long, cut up into deep, narrow segments. Panicles many-flowered, terminal; flowers ½ in. across, with downy stalks; calyx downy, the lobes pointed, triangular; petals purple. Fruits black, well-flavoured, ripening late.

Native of W. China; discovered by Prince Henri d'Orléans near Tatsien-lu in 1890, later found on Mt Omei by Wilson, who introduced it for Messrs Veitch, with whom it flowered in August 1908. It grows up to 6,000 ft elevation, and is perfectly hardy. It makes growths 10 or 12 ft long in a season. The stipules are rather remarkable.

R. SPECTABILIS Pursh   SALMONBERRY

A deciduous shrub with erect stems, 4 to 6 ft high, glabrous, but armed with fine prickles. Leaves 4 to 6 in. long, composed of three leaflets which are ovate, from 1½ to 4 in. long, doubly toothed, almost or quite glabrous on both surfaces, the terminal one the largest and broadest. Flowers produced singly or a few together on short shoots springing from the older wood; purplish red, 1 in. or so across, fragrant; calyx downy, with broad pointed lobes not so long as the petals. Fruits orange-yellow, large, somewhat egg-shaped.

Native of western N. America; introduced by Douglas in 1827. It flowers freely towards the end of April, and is very pretty then. In this country the fruits do not ripen freely and indeed are not of much value even in the salmon-berry's native country. '. . . children like them, and grown-up people are not above trying a few. The Indians gather baskets full of the young shoots before they have become hard, and, when peeled, eat them with dried salmon roe.' (George Fraser, *Gard. Chron.*, Vol. 95 (1934), p. 93.) The plant spreads rapidly by means of sucker growths from the base, and soon forms a dense thicket. Plants should be overhauled annually, and the worn-out stems removed. Propagation is easily effected by dividing up the plants or removing offsets.

R. *spectabilis* was at one time used experimentally in Britain as game-covert, which may explain its occurrence here and there as an apparently naturalised plant.

R. THIBETANUS Franch.

R. *veitchii* Rolfe

An erect deciduous shrub, 6 ft or more high; stems biennial, glabrous, round, covered with a purplish bloom, and set irregularly with straight, slender prickles. Leaves pinnate, 4 to 9 in. long, composed of seven to thirteen leaflets, the main-stalk prickly; leaflets oval or ovate, more or less oblique, stalkless, coarsely and angularly toothed, dark lustrous green, and with minutely silky hairs above, whitish felted beneath; the lowest leaflets are 1 to 2 in. long, each successive pair diminishing in size towards the apex of the leaf which is termi-nated by a long, deep-lobed leaflet. Flowers ½ in. across, slender-stalked, solitary in the leaf-axils, or a few together in terminal flattish panicles; the calyx very downy, with triangular lobes; petals purple. Fruits roundish, ⅝ in. across, black with a bluish bloom.

Native of W. China; discovered and introduced by Wilson for Messrs Veitch, with whom it flowered in August 1908. Wilson found it in the Min River Valley at elevations of 4,000 to 6,000 ft, where it is rare. Of the Chinese Rubi introduced in this century it is one of the most distinct and attractive-looking, both for its blue-purple stems and very handsomely cut foliage.

R. TRIANTHUS Focke

R. *conduplicatus* Rolfe

A deciduous shrub of wide-spreading habit, the biennial stems erect, much branched, spiny, blue-white, 4 to 6 ft high. Leaves simple, 3 to 6 in. long, 1½ to 4½ in. wide, ovate to triangular, distinctly three-lobed on the barren stems, less markedly lobed on the flowering shoots, middle lobe long, taper-pointed, irregularly toothed, quite glabrous on both sides, whitish beneath, dark green above; there are hooked spines on the midrib and veins beneath; stalk ½ to 1½ in. long, similarly armed. Flowers pinkish white, insignificant, produced a few together on cymes that are terminal on short lateral twigs. Fruits dark red.

Native of Central China up to 4,000 ft; introduced for Messrs Veitch by Wilson in 1900. It is distinct from most Rubi in the absence of down or hairs, but has not much garden value.

R. TRICOLOR Focke

R. *polytrichus* Franch., not Progel

A quite prostrate, semi-evergreen shrub with round stems, devoid of prickles and spines, but densely clothed with yellow-brown bristles about ⅛ in. long. Leaves simple, heart-shaped, 3 or 4 in. long by two-thirds as much wide, irregularly toothed, pointed, dark green above, covered with a close whitish felt beneath; there are about seven pairs of parallel veins, which on the under-surface are furnished with bristles, but on the upper surface the bristles are confined in rows between the veins. Leaf-stalk 1 to 1½ in. long, bristly like the stems. Flowers white, 1 in. across, produced singly in the leaf-axils near the end of the shoot, and in a small terminal panicle. Fruits bright red, and of good size and flavour. *Bot. Mag.*, t. 9534.

RUBUS TRICOLOR

Native of W. China; first discovered by the French missionary Delavay, but introduced to cultivation by Wilson in 1908. It is remarkably distinct on account of the very bristly stems and leaf-stalks. Coming from elevations up to 10,000 ft it is quite hardy, and makes a good ground-cover in shady places.

R. TRIFIDUS Thunb.

An evergreen or sub-evergreen shrub, with erect stems, 4 to 7 ft high, zigzagged towards the top, not (or but little) branched the first year, beset with glandular hairs at first, but soon becoming glabrous. Leaves dark lustrous green, usually five- or seven-lobed, 4 to 10 in. across; the lobes reach half or two-thirds of the way to the stalk, are ovate, pointed, doubly toothed, slightly hairy on the chief veins above and below; stalks 1 to 2½ in. long. Flowers 1 to 1¼ in. wide, rosy-white, produced singly in the terminal leaf-axils and in a few-flowered, terminal corymb, each on a downy stalk ¾ to 1½ in. long. Calyx very downy outside. Fruits described as red and edible.

Native of Japan; introduced about 1888. This rubus is distinct from all other cultivated species in its large, handsome, deeply lobed leaves, which bear a great resemblance to those of *Fatsia japonica*, except that they are not so large. It is worth growing for their sake. The flowers, which appear in May are not freely borne, and I have never seen the fruit. The leaves of the flowering shoots are frequently three-lobed, and it is to them that the specific name refers.

R. ULMIFOLIUS Schott

R. *rusticanus* Mercier

A vigorous shrub whose arching stems are clothed with tufted down and armed with long, broad-based prickles (these mainly confined to the angles of the stems). Leaves composed of three or five leaflets radially arranged, which are slightly downy above, white-felted beneath, rather finely toothed. Inflorescence a long, usually rather narrow panicle, with spreading, armed branches, they and the long pedicels densely coated with appressed hairs. Flowers usually pink, sometimes bright rosy red or white, with roundish petals. Sepals reflexed, densely hairy on the outside. Fruits aromatic but with rather small, dryish drupelets.

A native mainly of W. Europe (including Britain) and the Mediterranean region. Unlike most blackberries it is diploid and reproduces itself sexually (i.e., not by apomixis). Although not in itself of any ornamental value and only second rate as a fruiting bramble, it has given rise to two garden varieties and is the parent, through its unarmed form, of two of the best fruiting sorts.

cv. 'BELLIDIFLORUS'.—This is the very handsome well known double flowered pink bramble, which in some gardens makes a gay display in July and August. Each flower produces an extraordinary number of narrow petals—whence *bellidiflorus*, from the resemblance of the flowers to those of a double daisy (R. *bellidiflorus* Kirchn.; R. *ulmifolius* f. *bellidiflorus* (Kirchn.) Voss). Being just as vigorous and prickly as the wild type it is only suitable for the wild garden, where it could be grown with the white double bramble—see below.

f. INERMIS (Wild.) Rehd. R. *inermis* Willd.; ?R. *fruticosus inermis* West.—A remarkable form absolutely devoid of spines or prickles. One may thrust one's hand into the middle of the bush without getting a scratch.

The origin of this form is unknown. It was described by Willdenow in 1809, but it may be the same as the thornless bramble of Miller's *Dictionary* (1768), named by Weston two years later. It was received at Kew in 1877 from the firm of R. Smith and Co. of Worcester.

Early in the 1920s this bramble was crossed at the John Innes Horticultural Institution with R. HASTIFORMIS W. C. R. Watson (R. *thyrsiger* Bab., not Banning) a rare British endemic noted for its excellent fruits. The latter is tetraploid and of the four seedlings raised three were triploid and sterile, as might be expected from the parentage, R. *ulmifolius* being diploid. The fourth, however, proved to be a fertile tetraploid and was put into commerce as Blackberry 'JOHN INNES' (Crane and Lawrence, *Genetics of Garden Plants* (1952), p. 249). This carried within it the genes for thornlessness, and from it was bred 'MERTON THORNLESS', one of the best fruiting varieties.

cv. 'VARIEGATUS'.—A handsome variegated form, the main part of the leaf being green, the midrib and veins picked out in bright yellow. It needs a more sunny place than the others.

R. *ulmifolius* belongs to the subsection *Discolores*, the type-species of which is R. DISCOLOR Weihe & Nees (R. *procerus* P. J. Mueller; R. *armeniacus* Focke). This is not a British native, but is of interest as the species to which belongs the well known commercial variety 'HIMALAYA GIANT', which came originally from Armenia or the Caucasus. It was misleadingly named "Himalaya Berry" in the United States, whence it reached Britain early this century. In Germany it is known as 'Theodor Reimers' (Hegi, *Fl. Mitteleuropa*, Vol. IV 2A (1961–6), p. 323).

The double-flowered form of R. *ulmifolius* has been mentioned above. Another double-flowered bramble of the subsection *Discolores*, of uncertain taxonomic position, is:

R. 'DOUBLE WHITE'.—A strong-growing bramble with arching, angled soon glabrous canes, which are armed with short prickles. Flowering branches and inflorescence-axis densely set with strongly recurved prickles and hairy throughout. Leaflets mostly three, but often five on the canes and mostly reduced to a simple, lobed leaf under the inflorescence, deeply double-toothed, glabrous above at flowering-time, finely felted beneath, terminal leaflet ovate, short-acuminate at the apex, truncate to rounded at the base, up to 2½ in. or slightly more long and two-thirds as wide, the lateral slightly shorter and often cuneate at the base; stalks of the lateral leaflets about ⅔ in. long on the canes, very short on the flowering branches; leaf-rachis prickly. Flowers white, double but with a central boss of stamens, borne in late July or August in erect panicles, the lowermost two or three branches subtended by leaves, the leafless upper portion dense, blunt-ended, the whole 7 or 8 in. long; pedicels densely armed with straight prickles. Calyx white-felted; sepals reflexed at flowering time (R. *thyrsoideus flore pleno* Hort., not R. *thyrsoideus* Wimm.; ?R. *linkianus* Ser.).

This handsome bramble, suitable only for the wild garden, is of uncertain origin; it is doubtfully the same as the double-flowered "R. *fruticosus*" of 18th- and early 19th-century writers, but was in cultivation in Britain by 1882. It is perhaps nearest to R. *candicans* Weihe ex Reichenb. (R. *thyrsoideus* Wimm.) but differs in the more prickly and less leafy inflorescence. It appears to be the same as the bramble that appears in some recent works as R. *linkianus* Ser., the name given by Seringe in 1825 to a double-flowered bramble cultivated in the Berlin Botanic Garden, which Link had shortly described in 1822 under the illegitimate name R. *paniculatus*. Whether 'Double White' is clonally the same as the Berlin plant it is impossible to say. Certainly R. *linkianus*, as understood by Focke, is near to R. *candicans*, which would have to be regarded as the normal wild form of R. *linkianus* if indeed the two are conspecific, the latter name having several years priority.

RUSCUS   LILIACEAE

Strictly speaking, the species of this genus should be regarded as shrub-like rather than as true shrubs, none having really woody stems. According to the older view the genus belongs to the Asparagus group of the Lily family, but Dr Hutchinson, in his *Families of Flowering Plants* (1934) removed it to a separate family, the Ruscaceae, together with *Danae* (q.v.) and *Semele* (with a single species in the Canary Islands). Like the asparagus they renew themselves by stems from the base, and the tender young shoots are eaten in some parts of Europe. They are evergreen, the 'leaves' mostly alternate, sometimes in whorls. These apparent leaves should properly be termed cladodes; they are modified stems, flattened and resembling leaves, and performing the same functions. The true leaves are scale-like, each bearing a cladode in its axil. Flowers with a perianth of six segments, small, inconspicuous, borne on the upper or lower surface of the cladode, usually at its centre, each inflorescence subtended by a bract that varies in size and texture according to the species. Mostly the flowers are unisexual and borne on different plants. Hence the rarity of fruits on the two commonly cultivated species. The filaments of the stamens are united into a hollow column; this column is present in flowers of both sexes, but only in male flowers does it bear anthers. Ovary in female flowers concealed at the base of the staminal column; style short with a globose or mushroom-shaped stigma. The fruits are red berries.

The genus ranges from Madeira to Iran and contains about six species. The most important work on it is: P. F. Yeo, 'A Contribution to the Taxonomy of the Genus Ruscus', in *Notes Roy. Bot. Gard. Edin.*, Vol. 28 (1968), pp. 237–64. This work deals principally with the series *Simplices* (stems unbranched; cladodes unarmed, bearing the inflorescences on the upper or lower side), to which all the species treated here belong, except R. *aculeatus*, which, with the little known R. *hyrcanus* Voronov, forms the series *Ramosae* (stems branched; cladodes armed; inflorescence always on the upper side of the cladode).

The commonly cultivated species thrive in almost any soil and are admirable in very shady places. The best time to break up the plants for propagation is spring.

R. ACULEATUS   BUTCHER'S BROOM

An evergreen, well-armed shrub, spreading and renewing itself by means of sucker growths springing from the base, 1½ to 3 ft high, the crowded erect stems having many rigid branches near the top; stems grooved. Cladodes ovate, stalkless, ¾ to 1½ in. long, ¼ to ¾ in. wide; slightly glossy on both sides, tapering at the apex to a slender, stiff spine. Flowers ¼ in. across, dull white, borne singly or in pairs (apparently stalkless) in the centre of the cladode, but really produced in the leaf-axil, the stalk being united to the midrib of the cladode. The flower-bud forms early in the year, and opens in spring. Fruit a globose or oblong, bright red berry, ½ to ⅝ in. in diameter, borne, like the flower of course, in the centre of the cladode.

Native of Europe, N. Africa and the Near East; widespread in S. England, becoming rarer northward, absent from N. England and Scotland. The butcher's broom is remarkable in being the only shrubby plant of the monocotyledonous type native of the British Isles. It is not, ordinarily, a showy plant, but is always interesting for the curious position of its flowers and fruit. When laden with the latter it is very ornamental indeed; but the plants are mostly unisexual, and the fruits are not commonly seen in gardens, because one of the sexes, but more especially the female, is wanting. There is a hermaphrodite form in commerce, however. It is especially useful for planting in dense shade where very few evergreens will thrive. It is said to have obtained its common name through being used in the shape of brooms by butchers to clean their blocks. In S. Italy I have seen it used as a garden besom, as birch and ling are used in this country. 'The bunches of this plant, with the ripe fruit upon them, are frequently cut, and put into basons of sand, mixing them with the stalks of ripe seeds of male Piony, and those of the wild Iris or Gladwyn, which together make a pretty appearance in rooms . . .' (Miller's *Dictionary* (1768)).

var. ANGUSTIFOLIUS Boiss.—Cladodes very narrow, at least four times as long as broad. Commonest in the eastern part of the range of the species.

var. PLATYPHYLLUS Rouy—R. *a.* var. *latifolius* Bean—Cladodes up to 2 in. long and ¾ to 1 in. wide. S.W. Europe.

R. HYRCANUS Voronov—This species replaces R. *aculeatus* in the Crimea, Transcaucasia and N. Iran. It differs chiefly in the more spreading main stems with the branches clustered in whorls at their extremities, and in the more numerous flowers in the inflorescence. It is still rare in cultivation in Britain.

R. HYPOGLOSSUM L.

An evergreen shrub, 8 to 18 in. high, forming compact tufts, and increasing by new sucker growths from the sides; stem somewhat arching, as thick as a lead pencil, scarcely woody, unbranched, green. Cladodes not spiny, the lower ones narrow-oval, the upper ones oblanceolate, tapered at both ends; glabrous and glossy on both sides, with prominent longitudinal veins, 3 to 4½ in. long, 1 to 1½ in. wide. On the upper side is borne a leaf-like bract, lanceolate, 1 to 1½

in. long, ¼ to ⅓ in. wide, in the axil of which a few small, yellowish flowers
appear in April and May. Berry red, globose, ¼ to ½ in. wide.

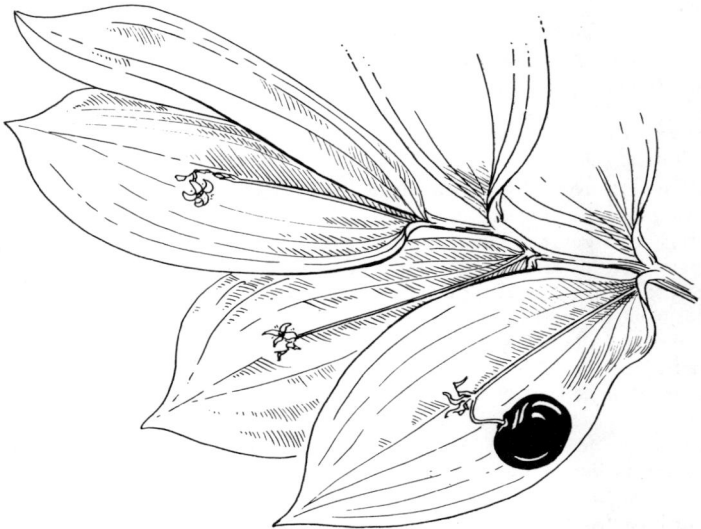

Native of S. Europe, the Danube region and Asiatic Turkey; cultivated in
Britain since the 16th century. No evergreen shrub thrives better than this in
shade and in competition with the roots of greedy trees; in this is its chief value in
gardens. It flowers in cultivation, but does not fruit unless both sexes are grown,
being completely dioecious. Because of its large bracts it was once known as the
'Bislingua' or 'Double Tongue', one 'tongue' being the bract, the other the
cladode.

R. COLCHICUS P. F. Yeo—Allied to the above, but with the inflorescence
borne on the underside of the cladode and the inflorescence bracts shorter (to
⅜ in. long), with fewer veins (three to six). Native of the eastern Black Sea region.
It was described in 1966 by Dr Yeo, who also procured plants from Russia for
the University Botanic Garden, Cambridge. It had previously been confused with
R. *hypophyllum*, although more closely related to R. *hypoglossum*.

R. MICROGLOSSUS Bertol.—This is intermediate between R. *hypoglossum* and
R. *hypophyllum*, resembling the former in foliage and habit, but differing in the
much smaller inflorescence bracts, about ¼ to ½ in. long and 1/10 in. wide, with
three or four pairs of veins, and in the flowers sometimes being borne on the
lower surface of the cladode. It is most probably a hybrid between the two

species, distributed by human agency, but possibly a relict species. It has long been cultivated and occurs apparently wild, though never far from habitations, in Italy and bordering parts of France and Yugoslavia. Only female plants are known, which may all belong to a single clone (Yeo, op. cit., pp. 254–6).

R. HYPOPHYLLUM L.

R. *trifoliatus* Mill.

A shrub with erect unbranched stems up to 2 ft high (at least in the wild). Cladodes oval to ovate, 2 to 3½ in. long, ½ to 2 in. wide, shortly stalked, abruptly pointed, dark green. Flowers produced on the upper or lower surface of the cladode, each on a slender stalk about ¼ in. long; bracts leafy or papery, much smaller than in R. *hypoglossum*, about ¼ in. long, linear or lanceolate, with up to three, rarely four, veins. Berries globose, red, ½ in. wide.

Native mainly of N. Africa as far east as Tunisia, but extending to the northern littoral of the Mediterranean; in cultivation 1768, perhaps earlier. It has a general resemblance to R. *hypoglossum*, but is readily distinguished by the shorter and broader cladodes and the tiny bracts. It is too tender to have much value in gardens near London.

The ruscus figured in *Botanical Magazine*, t. 2049 (1819) as R. *hypophyllum* is a distinct species, confined to Madeira, recently described by Dr P. F. Yeo under the name R. STREPTOPHYLLUS (*op. cit.*, pp. 250, 260).

RUTA   RUE   RUTACEAE

A small genus of aromatic subshrubs, ranging from the Canary Islands through the Mediterranean region to the Caucasus, with pinnately dissected leaves. Inflorescence cymose, terminal. Flowers yellowish, tetramerous (the terminal flower usually pentamerous). Petals spoon-shaped. Ovary inserted on a raised disk. Fruit a lobed capsule. The closely related *Haplophyllum* (not treated in this work) has more numerous species, mainly in the E. Mediterranean and Central Asia; it is often included in *Ruta* as a subgenus.

R. GRAVEOLENS L.   RUE

An evergreen shrub, with erect, half-woody branches, rarely seen more than 3 ft high. Leaves of a markedly glaucous hue, alternate, variable in length, but usually 3 to 5 in. long, pinnately decompound, the leaflets usually confined to the upper half, the ultimate subdivisions obovate, ⅛ to ½ in. long. Flowers ¾ in. wide, arranged in terminal corymbs, rather dull yellow; the sepals and petals usually four, sometimes five; the stamens twice as many. Petals scoop-shaped with jagged edges. Fruit a usually four-celled capsule.

Rue is known best, of course, as a garden herb with an acrid taste, once used in domestic and especially rustic medicine for colic, hysteria, promoting perspiration, etc. Applied locally it is a powerful irritant. These properties are due to a volatile oil which permeates the leaves and younger parts of the plant. Given in too large doses it is dangerous, and produces symptoms of acrid narcotic poisoning. The species is a native of S.E. Europe. Owing no doubt to its medicinal properties it has been grown in gardens from time immemorial and has become widely naturalised in Southern Europe generally. It finds frequent mention in Shakespeare as 'herb of grace':

I'll set a bank of rue, sour herb of grace.
—The gardener, in *King Richard II.*

It should find a place in all extensive shrub collections not only for its associations, but for its beauty also. When fully in flower the dark yellow blossoms contrast prettily with the glaucous foliage, and they continue to open from June onwards for some months. It is quite easily increased by cuttings, and will thrive all the better if lime or chalk be mixed with the soil where these are naturally absent.

cv. 'JACKMAN'S BLUE'.—Of compact, dwarf habit, with vividly blue-grey foliage. The original plant grew in the garden of Mr Greener, a part-time grower of nursery stock, who lived at Ottershaw, Surrey, and died in 1939. It was noticed by the late Rowland Jackman when he called to buy some plants, and from the cuttings given to him derives all the stock of this now much planted variety (*Gard. Chron.*, Vol. 171 (1972), p. 29). It was put into commerce by Messrs Jackman shortly after the second world war.

cv. 'VARIEGATA'.—Leaves bordered with white.

SABIA   SABIACEAE

An Asiatic genus of some twenty species of climbing shrubs with undivided, untoothed, alternate leaves and small flowers produced in the leaf-axils. The parts of the flowers are usually in fives. The only ally near to this genus amongst hardy trees and shrubs is *Meliosma*, whose species are very different in their tree or bush form, their toothed or pinnate leaves, and their usually terminal panicles of flowers.

S. LATIFOLIA Rehd. & Wils.

A deciduous scandent shrub up to 10 ft high; young shoots soon glabrous, yellowish green or purplish. Leaves alternate, oval, oval-oblong or slightly obovate, base rounded to tapered, often abruptly pointed, toothless, $1\frac{1}{4}$ to $5\frac{1}{2}$ in. long, $\frac{3}{4}$ to 3 in. wide, at first furnished with short hairs on both surfaces, eventually becoming glabrous except on the midrib and veins beneath; stalk $\frac{1}{4}$ to $\frac{5}{8}$ in. long, hairy. Flowers borne during May usually three together in axillary cymes $\frac{1}{4}$ to $\frac{1}{2}$ in. long, the stalks downy. Each flower is $\frac{1}{4}$ in. wide, somewhat globose through the incurving of the five oval petals which are greenish yellow at first, changing to reddish brown, and edged with minute hairs. Sepals five, minute, roundish ovate, minutely ciliate; stamens rather longer than the petals. Fruits bright blue, consisting of two compressed, kidney-shaped parts attached to a slender stalk $\frac{3}{4}$ to 1 in. long; each part of the fruit is $\frac{1}{3}$ in. long. *Bot. Mag.*, t. 8859.

Native of W. China; discovered by Pratt about 1888; introduced by Wilson in 1908. It was cultivated by Miss Willmott at Warley Place, Essex, where it became 10 ft high growing against a north wall. Here it flowered and ripened fruit in 1919. It is very rare.

S. SCHUMANNIANA Diels

A deciduous climber up to 10 ft high, with glabrous, slender young shoots. Leaves toothless, narrowly oblong to oval-lanceolate, slenderly pointed, shortly tapered to rounded at the base, 1 to 4 in. long, $\frac{1}{3}$ to $1\frac{1}{2}$ in. wide, glabrous; stalk slender, $\frac{1}{6}$ to $\frac{1}{3}$ in. long. Flowers $\frac{1}{4}$ in. long, cup-shaped, three to six in axillary cymes, with a slender main-stalk up to $1\frac{1}{2}$ in. long. Sepals small, rounded-triangular; petals greenish or dull purple, oval, blunt, $\frac{1}{5}$ in. long; stamens about as long as the petals. Fruits kidney-shaped, $\frac{1}{4}$ in. wide, blue-black, ripe in October, the sepals persisting at the base.

Native of W. China; introduced by Wilson in 1908. It differs from *S. latifolia* in its narrower, smaller, glabrous leaves and longer, more slender flower-stalks. It was introduced to Kew from the Arnold Arboretum in 1913 and has ripened fruit there. Quite hardy but of no great garden value. Flowers in May.

SAGERETIA   RHAMNACEAE

A genus of no garden merit with about thirty-five species in Asia (as far west as Asia Minor), N.E. Africa, and America, allied to *Rhamnus*. The generic name commemorates Auguste Sageret, a French botanist (1763–1851). The type of the genus, *S. theezans* (L.) Brogn., was described by Linnaeus in 1771 from a specimen collected in China by his pupil Osbeck. The odd specific epithet refers to the fact that the leaves of this species were used as a surrogate for tea.

S. PAUCICOSTATA Maxim.

A freely branching deciduous shrub up to 10 ft high; young shoots long and slender, glabrous, grooved, pale grey, armed with short, stiff, spine-tipped, axillary, leafy branches. Leaves alternate, elliptical to ovate, pointed, tapered at the base, minutely toothed, $\frac{1}{2}$ to 2 in. long, $\frac{1}{4}$ to $\frac{3}{4}$ in. wide, glabrous, veins prominent beneath, one to three each side the midrib; stalk $\frac{1}{8}$ to $\frac{1}{4}$ in. long, very slender. Flowers stalkless, very small and white, ranging from solitary to forming terminal leafy panicles 1 to 3 in. long, and opening in May. Fruit a subglobose, black drupe, $\frac{1}{6}$ in. wide.

Native of N. China and in cultivation at Kew since before 1919. It is very hardy. The specific name refers to the few veins of the leaf as compared with those of other species.

SALIX   WILLOW   SALICACEAE

Salix is a large genus, though the number of species in it depends very much on how narrowly or broadly these are defined. By some reckonings it has 500 species, by others only 300. They vary from stately timber trees such as *S. alba* or *S. nigra*, to tiny shrubs like *S. herbacea*, creeping along the ground and only rising an inch or two above it. But the great majority of the species are erect shrubs or small trees. The twigs of many species are very tough and flexible, and supply much the greater part of the material from which baskets and wickerwork are made in the temperate parts of the northern hemisphere. But the twigs of some, though tough, are easily snapped off in their entirety at the point of union with the older branchlet. These root readily in a favourable soil. This curious characteristic is best known in, and gives the popular name to, the crack willow *S. fragilis*, but there are several more that have it equally marked. The winter-buds of the willows are covered with a single scale; terminal winter-buds are not formed (except in a few dwarf alpine willows).

The leaves are deciduous (semi-persistent in a few mostly subtropical species), simple, usually short-stalked, toothed or entire, alternate or much more rarely, nearly or quite opposite, notably in *S. purpurea* and some of its allies. They are very variable in shape, usually lanceolate and slender-pointed in tree-willows, sometimes almost orbicular in dwarf species, and with every gradation in between. In the remarkable *S. magnifica* from China, the leaves are over 8 in. long and 5 in. wide; at the other extreme, the alpine shrublet *S. serpyllifolia* has them less than $\frac{1}{2}$ in. long. Stipules produced by willows are variable both in their size and persistence. They occur most markedly on the strongest shoots, and are frequently entirely absent from weaker ones. In some tree species glands are present near the junction between leaf-blade and petiole, and are sometimes converted into minute stipules.

The inflorescence is a catkin (ament) borne at the tip of a shoot that (in most species) springs from an axillary bud on the previous year's wood. This flowering shoot may be 2 in. or even more long, and bear leaves that are normal except in having no growth-buds in their axils, the whole shoot being in effect a leafy peduncle with a terminal inflorescence. However, the shoot is more commonly quite short and the leaves below the catkin are then small and bract-like; or, as in the sallows and many other species, the peduncle is so much reduced that the catkin is to all intents sessile in the scars of the previous year's leaves. In some alpine species of the subgenus *Chamaetia* there are no specialised flowering shoots: the catkin is terminal on a normal growth equipped with both leaves and axillary buds. See also *S. exigua* for the inflorescence of the *Longifoliae*. The willows are dioecious, i.e., the catkins are composed of either male or female flowers, and the individual plant bears catkins of one sex only. Rarely and abnormally (especially in hybrids) some flowers are bisexual, or some catkins bear both male and female flowers; see further under *S. aegyptiaca* and *S.* × *sepulcralis*. The catkins vary much in length, shape and relative width. They may be produced before the leaves (precocious) as in the sallows, or in such species as *S. daphnoides* and *S. viminalis*; with the unfolding leaves (coaetaneous); or during summer after the leaves are expanded (serotinous), as in *S. triandra* and *S. pentandra*.

Each flower is subtended by a scale (sometimes called a bract), which is entire or, very rarely, slightly toothed, usually persistent, long-hairy, sometimes merely downy or shortly ciliate, rarely quite glabrous. The scale may be uniformly coloured, or reddish or pinkish at the tip, or, as commonly in the subgenus *Vetrix* (*Caprisalix*), dark brown, dark purple, or almost black in the apical part. Between the flower and the axis of the catkin there is a minute fleshy body usually termed a nectary (but in older works called 'gland'). In many species there is a second nectary between the flower and the subtending scale, at least in the male flowers, and sometimes the two nectaries are united into an irregularly lobed disk (the nectaries are in fact homologous with the disk seen in the flowers of *Populus*, and those willows that have united nectaries usually show other 'poplar characters'). The nectaries are too minute to be of much use for identification, except to the specialists, and are omitted from the descriptions of most species in this work.

The flowers have no sepals or petals (though some botanists consider the nectaries to represent a much reduced perianth). The male flowers have mostly two stamens, but up to twelve in the mainly subtropical section *Humboldtianae*, and three to five in some other members of the subgenus *Salix*. In *S. purpurea* and a few other species (some closely related to it) the filaments are completely connate and the stamen apparently solitary. The female flower consists of a single ovary, which may be sessile or borne on a 'stalk' (strictly a pedicel or stipe). The ovary is variable in shape, and may be glabrous, silky or coated with matted hairs. The style is single, or occasionally forked in the upper part, and is very short or lacking in some groups. There are two stigmas, each entire or more often retuse or more or less deeply divided into two lobes or slender arms.

It is an interesting fact that the flowers are insect-pollinated (many willows are indeed useful bee-plants). In this respect *salix* differs not only from most other catkin-bearing genera (beech, oak, hazel, etc.) but even from the allied poplars, all of which rely on the wind to spread their pollen. An exception is *S. arbutifolia* (often separated from *Salix* as the genus *Chosenia*) and it is said that some dwarf willows of high latitudes are partly wind-pollinated. The fruit, which usually ripens very quickly, is a capsule splitting in two at the top, each half recurving. The seeds are minute, with a tuft of pale hairs at one end. They soon lose their vitality but germinate rapidly if they fall on favourable ground. It is said that seeds of the American black willow (*S. nigra*) grow 4 ft high by autumn.

The genus is widely distributed in the northern hemisphere. The section *Humboldtianae* is mainly tropical and subtropical and extends into the southern hemisphere. At the other extreme, dwarf willows extend farther towards the North Pole, and farther towards the eternal snow, than any other woody genus. In this respect the willows have proved themselves more adaptable and adventurous than the allied poplars.

Many natural hybrids have been described in *Salix*, but they occupy more space in the literature of the genus than they do in the wild, and with few exceptions (e.g., *S.* × *rubens*) are never so frequent under natural conditions as to blur the boundaries between species. They rarely occur in closed plant communities, even where, as in the Alpine valleys, four or five species may grow in the same locality, and are most likely to be found in such habitats as road- or railway-embankments, landslides, new scree or waste ground, where the more highly adapted parents may be at a disadvantage. Many hybrids that have been described, or identified in herbarium annotations, are certainly no more than forms of variable species. Either the writer had only a limited knowledge of the species; or, though knowing his subject well, he nevertheless defined the species arbitrarily, calling on hybridity to explain variations that lay outside the subjective boundaries he had laid down.

Some hybrids, though occurring rarely in the wild, or even uniquely, have been brought into gardens and multiplied there. Of greater importance are the hybrids of the various species used in basketry, many of which were widely planted in the osier-grounds and for estate use, thus acquiring so extensive a distribution as to be taken for species by early salicologists such as Sir James Smith, who denied that hybrids ever occurred in the genus. Hybrids have also been raised artificially by researchers and enthusiasts. It is interesting that female plants often far outnumber males in the progeny of these crosses.

For the natural hybrids occurring in Britain see: R. D. Meikle, *Salix*, in C. A. Stace ed., *Hybridization and the Flora of the British Isles* (1975), pp. 304–38.

Since prehistoric times the stems of the willows have been put to innumerable uses, from crude hampers and hurdles to the finest sort of basketware. Great osier-beds once stretched along the Thames from

Reading downwards to supply the London basket-makers, and today
there are still extensive beds in areas so damp or so subject to inundation
as to be unavailable for ordinary crops. But some of the finest quality
wicker-work is made from willows grown on ordinary farm land. Often
several forms of osiers used in basket-making are derived from a single
species. These clones, although they vary much in quality for their particular purpose, show no significant botanical differences, and are known in
the osier trade by colloquial names. Some of these clones were introduced
from the continent within the past few centuries, but some may be very
ancient, for the Celts were skilled basket-makers and taught their craft to
the Romans.

The value of several sorts of willow trees for producing wood from
which cricket-bats are made is alluded to under *S. alba* 'Caerulea', *S.
fragilis* and *S.* × *rubens*. But these, as well as *S. alba* itself, are useful also
for other purposes, especially where a non-splintering wood is required,
and where it is subject to rough friction as on cart or wheelbarrow bottoms. For wattling the wasting banks of rivers or other pieces of running
water nothing equals the branches of willows. Some of the tree species
or the taller sallow hybrids are useful for fast-growing shelter-belts. All
the willows are good bee-plants, the early-flowering kinds being the most
valuable. Altogether the willows are a useful and undemanding race. In
his edition of Miller's *Dictionary* (1807) Thomas Martyn summed up the
long service of the willows to man when he wrote: 'The antient Britons
made boats of wicker covered with skins (coracles) by which they passed
rivers and arms of the sea; so light as to be carried by a single man.
Modern Britons wield bats of Willow, in the game of Cricket.'

Most willows are propagated extremely easily by means of leafless
cuttings, which may be put in the open ground at any time between
November and early March. Pieces one to several years old may be used;
and of the tree sorts like *S. alba*, *S. fragilis* and *S.* × *rubens*, it is usual to
put in 'sets', i.e., naked rods, 8 to 12 ft long, and as thick as, or thicker,
than a broomstick. But cuttings of half-ripened wood, taken from June
onwards, will also root readily and this is the most convenient method of
increasing the dwarfer species. Only a few willows are hard to strike, one
notable example being male plants of pure *S. caprea*. Plants raised from
cuttings in the nursery should be put in their permanent places at not
more than two years old and should be planted a few inches deeper than
before; the best results will, indeed, be obtained by putting the cutting
into its destined place at the start if due protection and care can be given.
This is always done with the big 'sets' just mentioned.

The majority of willows abhor dryness at the root, but will thrive in
ordinary situations if the soil be deep. The sallows will grow well on
poor, unimproved soils and make useful pioneers and soil-fixers. *S.
purpurea* too is tolerant of dry soils, while *S. daphnoides* has been used on the
continent for the fixing of sand-dunes. A few species and hybrids of the
Alpine type are suitable for the rock garden and some of the more vigorous
ones make useful ground-cover.

The willows are subject to several fungal diseases with similar symptoms, all causing die-back of the young growths and lesions on the older stems. These have been studied mainly in relation to the clones grown in osier grounds, since the cankering and discoloration they cause render the rods unfit for basketry. In gardens the most serious damage is likely to be suffered by the various hybrids of *S. babylonica*, whose pendulous growths may be greatly shortened by a severe attack. Both Willow Anthracnose and Willow Scab can cause this damage, and the former is most prevalent in a wet spring.

The bacterial Watermark disease has caused severe damage in plantations of the Cricket-bat willow, *S. alba* 'Caerulea', but is mainly confined to areas where this is grown extensively. Even if not lethal, an attack causes staining in the wood of the trunk and makes it commercially valueless.

The attacks of a Gall Mite cause the catkins of some willows, especially the sallows and their allies, to become woody and persist on the branches as unsightly protuberances. The form or hybrid of *S. purpurea* once known as *S. helix* bore the vernacular name 'rose-willow' from the curious rosettes of short leaves sometimes formed at the tips of the branchlets, and the same abnormality has been observed in other willows. It is caused by the punctures of a species of Gall Midge.

CLASSIFICATION

It is hoped that the following synopsis may help to make this large and confusing genus a little more comprehensible. It is largely based on A. K. Skvortsov, *Ivi SSSR* (Willows of the USSR) (1968), a taxonomic and geographic survey, with keys, that covers all the Old World willows with the exception of some little studied groups largely confined to China and the Himalaya. Most of the American species have been fitted into this classification with the aid of some recent works, notably: Robert E. Dorn, 'A synopsis of American Salix', *Canadian Journal of Botany*, Vol. 54 (1976), pp. 2769–89; George W. Argus, *The Genus Salix in Alaska and the Yukon* (1973); Arthur Cronquist, *Salix*, in C. L. Hitchcock *et al.*, *Vascular Plants of the Pacific Northwest*, Part 2 (1964).

subgen. SALIX

Trees or large shrubs. Leaves elliptic or lanceolate, much longer than wide, acute or acuminate, toothed. Catkins slender, sometimes drooping; scales uniformly coloured, usually deciduous in the fruiting catkins. Nectaries in male flowers two (sometimes more numerous and then united into a disk); in female flowers one or two. Stamens free, two or more.

sect. HUMBOLDTIANAE (*Nigrae*)—Buds triangular, not appressed to the shoot, their scales free from each other on the inner (adaxial) side. Nectaries in male flowers sometimes more than two and then more or less connate at the base. Stamens three to ten, with small anthers. Ovary stalked; stigmas very short, sessile or on a short style. A group of about a dozen species, mostly in the tropics and subtropics, but in N. America extends in the cold temperate zone; see *S. nigra*. In western Eurasia its northernmost representative is S. ACMOPHYLLA Boiss., a native mainly of Iran and bordering parts, but extending as far west as Palestine and S.E. Turkey. S. HUMBOLDTIANA Willd., of Central and S. America, would probably be hardy in Britain; it has a fastigiate form cultivated in the warmer parts of America and also in New Zealand.

sect. AMYGDALINAE—A group of two or three species in the Old World. See *S. triandra.*

sect. URBANIANAE—The one (perhaps two or three) species in this E. Asiatic section are mentioned under *S. arbutifolia.*

sect. PENTANDRAE—Trees or large shrubs. Leaves glossy, glandular-serrate, often balsam-scented when young, usually with well-developed glands at the junction with the petiole. Catkins borne after the leaves unfold, on leafy twigs. Catkin-scales usually glandular at the apex. Nectaries two (sometimes more numerous in the male flowers and then connate into a disk enclosing the bases of the stamens). Stamens three to ten. Ovary stalked; style very short; stigmas bilobed. About eight species in Eurasia and N. America. Those treated here are *S. pentandra* and the American *S. lucida* and *S. lasiandra.* Apart from *S. caudata,* which is doubtfully distinct from *S. lasiandra,* the only other American species is S. SERISSIMA Fern., a close ally of *S. pentandra* remarkable for producing its catkins in autumn.

The E. Asiatic section GLANDULOSAE, probably not represented in cultivation in Britain, is intermediate between the Humboldtianae and Pentandrae. The type is S. CHAENOMELOIDES Kimura (*S. glandulosa* Seem., not Raf.) a native of Japan, Korea and China.

sect. SALIX (*Amerina, Albae*)—A group of three species in W. Eurasia. See *S. alba* (here taken to be the type-species of *Salix*) and *S. fragilis.* The third species is S. EXCELSA Gmel., a native of S.W. Asia. In this section the stamens are reduced to two

sect. LONGIFOLIAE—There are about five species in this very distinct section, natives of N. America and Mexico. See *S. exigua.*

sect. SUBALBAE—A group of a few species in E. Asia, often regarded as the counterpart of the more western *S. alba* and *S. fragilis.* They differ from most other species of the subgenus Salix in the very short catkins with persistent bud-scales and in having the catkin-buds of different form from the growth-buds. See further under *S. babylonica, S. matsudana* and *S. jessoensis.*

subgen. CHAMAETIA

Of creeping or procumbent habit or erect but dwarf. Leaves always relatively broad. Catkins on leafy laterals, the leaves on these being of the same size as those on the vegetative growths, and as numerous. Nectaries large, often more than one, sometimes lobed. Catkin-scales persistent. Stamens two or three, rarely solitary. All the sections are represented in the arctic and boreal regions of Eurasia and America, and many of the species occur at high altitudes in mountains farther south.

sect. CHAMAETIA (*Reticulatae*)—See *S. reticulata,* the only European representative, to which the American S. NIVALIS Hook. (*S. saximontana* Rydb.) is closely allied. Of the other two species S. VESTITA Pursh is American but also occurs in the Altai Mountains of Russia.

sect. RETUSAE (*Herbaceae*)—Cushion-forming or prostrate shrubs, sometimes with underground stems. Stipules inconspicuous or wanting. Catkins terminating normal leafy shoots. Filaments of stamens glabrous. Ovary narrow, tapered into the style; stigmas small. See *S. herbacea* and *S. retusa.* The Japanese S. PAUCIFLORA Koidz. belongs to this section.

sect. MYRTOSALIX—Small erect shrubs, more rarely prostrate. Leaves glossy above, often persisting withered on the branches. Stipules usually present, symmetrical.

Catkins terminating leafy shoots, their scales dark-coloured. Ovaries glabrous or clad with matted hairs; style well developed. See *S. myrsinites* and *S. uva-ursi*. The latter is an American species. So too is S. DODGEANA Rydb., a rare native of the Rocky Mountains, closely allied to S. ROTUNDIFOLIA Trautv. of the Aleutians, Alaska, etc.; it is a mat-forming species with minute leaves, often found in hollows where the snow lies long.

sect. GLAUCAE (*Ovalifoliae*)—Prostrate or small, erect shrubs, rarely medium-sized. Leaves entire or faintly toothed; stipules narrow or wanting. Catkins on leafy stalks, but these often markedly shorter than the vegetative shoots. Nectaries sometimes more than two in the male flowers, the adaxial nectary often large and lobed. Ovaries stalked, with a well-developed style. See *S. arctica*, *S. glaucosericea* and *S. pyrenaica*. The small section MYRTILLOIDES, see *S. myrtilloides*, is related to the *Glaucae*.

Also belonging to this subgenus, but scarcely represented in British gardens, is the section LINDLEYANAE, for which see *S. lindleyana*.

subgen. VETRIX (*Caprisalix*)

To this taxonomically difficult group belong about two-thirds of the species of *Salix*. They are shrubs or small trees. Catkins usually emerging before or with the leaves; scales dark, at least at the tip. Nectaries normally one in both sexes. Stamens two, the filaments sometimes wholly or partly united.

This group is exceedingly variable in foliage. In the great majority of the species the catkin-buds are markedly larger than the vegetative ones, and are produced near, but not extending to, the apex of the shoot, the uppermost buds, as well as those at the base of the shoot, being vegetative.

sect. ERIOSTACHYAE (incl. *Magnificae*)—A taxonomically rather isolated and primitive section, anomalous in this subgenus in having two nectaries in the male flowers. The species are shrubs with large or very large leaves. Catkins on leafly stalks, of a remarkable length in some species. Stamens two. Ovary sessile or shortly stalked, with a short style. The species, number uncertain, are natives of China and the E. Himalaya. See *S. fargesii*, *S. magnifica* and *S. moupinensis*.

sect. CORDATAE (*Hastatae*)—Mostly small or medium-sized shrubs. Leaves finely toothed or entire. Stipules well-developed, symmetrical or almost so, often a conspicuous feature. Catkins usually with or shortly after the leaves. Ovary glabrous, acute, with a definite style; stigmas small. A northern and alpine group, mainly American. See *S. cordata*, *S. hastata*, *S. rigida*. The American section BALSAMIFERAE is included in the *Cordatae* by some authorities; see *S. pyrifolia*, the only species in this section.

The new section GLABRELLA, set up in 1968, comprises some species previously grouped with *S. hastata*; see *S. glabra*.

sect. NIGRICANTES—A very small group, confined to the Old World. See *S. nigricans*.

sect. VETRIX (*Capreae*)—Shrubs or sometimes small trees; wood under the bark often striated (with raised ridges). Stipules well-developed in most species, asymmetrical. Leaves unequally toothed or entire, broad. Catkins almost always produced before the leaves. Stamens two. Ovary stalked, the stalk often elongating as the capsule matures; style very short. A large group in the temperate regions of the northern hemisphere, mostly in the Old World. See *S. aegyptiaca*, *S. aurita*, *S. caprea*, *S. cinerea*, *S. discolor*, *S. silesiaca*, *S. starkeana*.

sect. ARBUSCELLA (*Phylicifoliae*)—Shrubs, sometimes dwarf. Catkin-buds not markedly different from the vegetative. Leaves discolorous, being dark green above, white or grey and finely reticulate beneath. Catkins before, with or after the leaves. Ovary clad with appressed hairs, tapered into a definite style. A group of about eighteen species in the northern parts and mountains of the Old World, a few in N. America. See *S. arbuscula* and *S. phylicifolia*.

sect. VIMEN (*Viminales*)—Trees or large shrubs. Stipules present, of various shape. Petioles subtending the catkin-buds much expanded at the base. Leaves narrow, with numerous parallel lateral veins, entire or finely toothed, silky beneath at least when young. Catkins usually precocious. Stamens two. Ovary short-stalked or sessile, with a definite style; stigmas divided. A group of about ten species in the Old World. See *S. sachalinensis* and *S. viminalis*; also *S. dasyclados*, p. 306.

sect. SUBVIMINALES (*Gracilistylae*)—Two or three species, closely allied to the section *Vimen*, but with the filaments of the stamens partly or wholly united. See *S. gracilistyla*. Another close ally of the section *Vimen* is *S. elaeagnos* (q.v.), the only species of the section CANAE; in this species too the stamens are partly united.

sect. VILLOSAE—Shrubs (one species with a short trunk). Leaves entire or obscurely toothed, their undersides and the stems white-woolly. Catkins usually before the leaves, thick and very hairy. Ovary sessile or short-stalked; style elongate. About five species in the colder parts of the N. hemisphere. See *S. candida* and *S. lapponum*. This section has some affinity with *Vimen*, indeed *S. lapponum* is grouped with *S. viminalis* by some authorities.

sect. LANATAE (*Chrysanthae*)—This group of about five species bears much resemblance to the preceding, but the undersides of the leaves have a greyer, less matted indumentum and the stipules are generally larger and almost symmetrical. In the American species the uppermost buds of the shoot are usually all catkin-bearing, the uppermost catkin apparently terminating the shoot. See *S. lanata*. The N.W. American *S.* RICHARDSONII has recently been reduced to the rank of a subspecies of *S. lanata* by A. K. Skvortsov. *S.* BARRATTIANA Hook. ranges from Alaska to British Columbia, while the rare *S.* TWEEDYI (Bebb) Ball (*S. barratiana* var. *tweedyi* Bebb) has a more southern distribution.

sect. DAPHNELLA (*Pruinosae*)—See *S. daphnoides*, under which are mentioned the other three species in this section.

sect. INCUBACEAE—A group of probably only three species, these confined to the Old World. See *S. repens*.

sect. HELIX (*Purpureae*)—Shrubs or small trees, with slender, flexible branches. Leaves usually narrow, not revolute, venation not prominent. Catkins slender, mostly before the leaves in the cultivated species. Stamens two, their filaments partly or wholly connate. Styles short or wanting, with small stigmas. A group of nearly thirty species confined to the Old World, the majority in Central and E. Asia. See *S. caesia* and *S. purpurea*.

S. AEGYPTIACA L.   MUSK WILLOW

S. nitida S.G. Gmelin; *S. medemii* Boiss.; *S. phlomoides* Bieb., in part; *S. cinerea* var. *medemii* (Boiss.) Boiss.; *S. caprea* of many authors, in part, not L.

A shrub to about 15 ft, rarely a small tree to twice that height; branchlets downy until the second year, slowly becoming glabrous and brown or reddish brown prominently striated under the bark, the striations short and irregular; catkin-buds large, ovoid, reddish brown. Leaves obovate to elliptic, acute to subacute or rounded and acuminate at the apex, broad-cuneate to rounded at the base, 2 to $4\frac{3}{4}$ in. long, 1 to 2 in. or slightly more wide, upper surface hairy at first, soon glabrous, underside conspicuously net-veined, clad with a grey indumentum which may persist until leaf-fall, or largely disappear by late summer, margins shallowly and irregularly crenate or dentate; petiole stout, $\frac{3}{16}$ to $\frac{1}{2}$ in.

or slightly more long. Stipules obliquely semicircular, deeply toothed, deciduous. Catkins appearing in early spring (sometimes in late winter), almost sessile, with a few small silky leaves at the base; scales acute, pale at the base, darker towards the apex, densely silky-hairy. Male catkins about $1\frac{1}{4}$ to $1\frac{3}{4}$ in. long, $\frac{5}{8}$ to $\frac{7}{8}$ in. wide; stamens two, filaments hairy at the base, anthers smaller than in *S. caprea*. Female catkins and flowers similar to those of *S. cinerea*. *Bot. Mag.*, n.s., t. 91.

Native of S.E. Anatolia, S.E. Transcaucasia and N. Persia; introduced to the Botanic Garden at Innsbruck in 1874 by Dr Polak, doctor to the Shah of Persia, and in cultivation at Kew five years later. At one time a perfumed drink was made in Moslem lands from its male catkins, which were also sugared and eaten as a sweetmeat, and used for perfuming linen. For these it was cultivated from Egypt to Kashmir and central Asia, so the epithet *aegyptiaca* is not so inappropriate as it would otherwise seem to be. According to Brandis, in *Forest Flora of North-west India* (1874), the willow gardens at Lahore, in what is now Pakistan, consisted entirely of male trees of '*Salix caprea*', by which he undoubtedly meant *S. aegyptiaca*. This plantation was made during the reign of the Sikh Maharajah Ranjit Singh (*d.* 1839); the original cuttings came from Kashmir, to which *S. aegyptiaca* was probably introduced from Persia in Moghul times.

S. aegyptiaca is mainly represented in British gardens by the Persian garden clone introduced to Europe by Dr Polak. This is of great vigour, with large leaves and stout stems, and is abnormal in its flowers. Sometimes two or three catkins develop from a single bud; they are basically male, but often bear some female flowers, and the filaments of the stamens are often connate at the base. In a mild season the catkins may be in full flower by the middle of January. There is an example at Kew on the Palace Lawn about 12 ft high, more in width. This clone received an Award of Merit when shown from Kew in 1925, and again in 1957 when exhibited by Patrick Synge on January 22. It is easily increased by hardwood cuttings. More normal, less remarkable plants are also in cultivation.

S. aegyptiaca is closely related to *S. caprea*, but the wood is striated under the bark, as in *S. cinerea*.

S. ALBA L.   WHITE WILLOW

A tree up to 90 ft high in Britain (very rarely taller), of elegant habit, branches ascending at a fairly steep angle, pendulous at the ends; bark shallowly fissured; twigs at first grey with silky down, slowly becoming glabrous and brown; buds flattened, appressed. Leaves lanceolate, $1\frac{1}{2}$ to $3\frac{1}{2}$ in. long, $\frac{1}{4}$ to $\frac{5}{8}$ in. wide, much tapered at both ends, very finely toothed, permanently covered beneath with silky down, less so above; stalk $\frac{1}{8}$ to $\frac{1}{2}$ in. long. Catkins appearing with the leaves on short leafy laterals, dense, cylindric, more or less erect, about $1\frac{1}{2}$ in. long; scales yellowish, deciduous, downy at the base and on the margins; axis downy. Male flowers with two, rarely three stamens, filaments united at the base, hairy in the lower part; nectaries two. Female flowers with a single nectary; ovary glabrous, almost sessile; style short, the stigmas two-lobed or merely notched.

Native of Europe and W. Asia; widely distributed in the British Isles, though

SALIX

not genuinely indigenous throughout its range. It varies considerably in the colours of the leaves and young shoots, some being much more silvery than others. Its timber was at one time put to many uses. 'In the roofs of houses, rafters of this tree have been known to stand a hundred years. . . . The wood is also used in turnery, mill-work, coopery, weather-boarding, &c; and the stronger shoots and poles serve for making hoops, handles to hay-rakes, clothes-props. . . . The bark, which is thick, and full of cracks, is in nearly as great repute for tanning as that of the oak; and it is also used in medicine, in the cure of agues, as a substitute for cinchona. . . . The charcoal is excellent for use in the manufacture of gunpowder, and for crayons.' (Loudon, *Arb. et Frut. Brit.*, Vol. III, p. 1525 (1838)).

f. ARGENTEA Wimm. *S. alba* var. *sericea* Gaud.; *S. alba* var. *splendens* (Bray) Anderss.; *S. splendens* Bray ex Opiz; *S. alba* f. *splendens* (Bray) Schneid.; *S. alba* var. *leucophylla* Hartig; *S. argentea* Hort. ex K. Koch; *S. regalis* Hort. ex K. Koch SILVER WILLOW.—This is the most striking of all the forms of *S. alba* in the intense silvery hue of its leaves, conspicuous in their shining whiteness at long distances. It occurs occasionally in the wild, where it is usually of dwarf stature. Cultivated plants, too, are less robust than the common white willow, and probably belong to one or only a few clones, distinguished by such epithets as *argentea* and *regalis*.

cv. 'BRITZENSIS'.—First-year stems bright red in winter, likened in the original description to those of *'Cornus sibirica'*. Leaves about ⅞ in. wide. Raised from seed by the German nurseryman Späth at Britz near Berlin, and put into commerce in 1878. Male. The cultivar-name 'Chermesina', wrongly used for 'Britzensis', derives from *S. alba* var. *chermesina* Hartig; the willow so named had carmine-red twigs and was found near Braunschweig in Germany before 1851.

Like the var. *vitellina* this is usually grown for the winter-colour of its stems and therefore pruned hard each spring. Left to itself it makes a tree of narrow habit.

cv. 'CAERULEA' CRICKET BAT WILLOW.—This fine tree, sometimes called the 'blue willow', occasionally reaches a height of 100 ft and 15 to 18 ft in girth. It differs from the white willow in its pyramidal growth and erect branching, and by the leaves losing their silky down and becoming glabrous late in the summer, and blue-grey beneath. It is a female clone, which first came to notice at the end of the 18th century, at which time it was, and largely still is, confined to the eastern counties, though whether it originated there is not known. It was named *S. caerulea* by James Smith and described by him from a tree growing in the collection of Dr James Crowe at Lakenham in Norfolk (*Engl. Bot.*, t. 2431 (1812)). It has been suggested that 'Caerulea' is a hybrid between *S. alba* and *S. fragilis*, but such an origin would not explain its distinctive characters, and most authorities now accept it as a variant of the white willow.

The timber of 'Caerulea' is more prized by cricket-bat makers than any other. It grows with extraordinary rapidity in good situations (it likes a stiff, moist, but not waterlogged soil), and, raised from a cutting, will, in twelve or fourteen years, attain a girth of 4 to 5 ft.

Botanically, 'Caerulea' belongs to var. CALVA G. F. W. Meyer (*S. alba* subsp.

caerulea (Sm.) Rech. f.), which comprises all forms of *S. alba* in which the leaves are glabrescent and bluish grey beneath. These are said to occur occasionally with the type on the continent.

cv. 'CARDINAL'.—Stems coloured as in 'Britzensis', but the leaves narrower; it also differs in being a female clone. Cultivated since the 1880s as the Cardinal willow, it is probably older than 'Britzensis' (*S. alba* var. *britzensis sens.* J. Fraser in part, in *Rep. Bot. Exch. Club*, Vol. 9 (1930), p. 720; *S. cardinalis* Hort. ex A. B. Jacks., in *Tr. & Shr. at Westonbirt* (1927), p. 178).

cv. 'CHRYSOSTELA'.—This was described by the French dendrologist Dode as a male tree with the habit of a Lombardy poplar but broader; the name he chose for it, meaning 'column of gold', refers to the colouring of the young stems, which were as golden as in *S. alba* var. *vitellina* but red-orange at the tips (*Bull. Soc. Dendr. Fr.*, 1930, p. 93). Plants in commerce as "Chrysostella" may be of this clone, which Dode would certainly have distributed. These, if pruned hard in spring, produce orange-red stems maturing to yellow at the base.

cv. 'LIEMPDE'.—A tree with a well-developed central leader and ascending branches, forming a rather narrowly ovoid crown. Selected in Holland, where it is much planted, and available in Britain. This and other selections of *S. alba* are described in *Dendroflora*, No. 6 (1969), pp. 67–74, with a summary in English.

cv. 'VITELLINA TRISTIS'.—A semi-pendulous form of *S. alba* var. *vitellina* received by the botanist Seringe from Baumann's nursery in Alsace a few years before 1815. For the *S. alba* 'Tristis' of some continental works of dendrology, see 'Chrysocoma', under *S.* × *sepulcralis*.

var. VITELLINA (L.) Stokes *S. vitellina* L.; *S. alba* subsp. *vitellina* (L.) Arcangeli GOLDEN WILLOW—Twigs yellow or orange-yellow, the colour brightest in autumn and early spring. Leaves rather paler green than those of the white willow, and not so silky-hairy. This variant is not known in the wild (as early as 1623 it was referred to by Caspar Bauhin, the Swiss botanist, as 'the cultivated golden willow') but has been widely grown in Europe, possibly since Roman times, for its tough and flexible twigs, once much used for tying and bundling. Being of only second- or third-rate quality for basketry, the golden willow is now chiefly planted in gardens for the fine effect produced in winter by its yellow shoots. For this purpose it is pruned hard every spring so as to develop a low thicket of wands; several plants should be grouped together.

The var. *vitellina* is rather a group of clones than a proper botanical variety. In Britain the clone once commonly planted in osier-beds is female, and is distinguished by rather long and narrow catkin-scales—a character erroneously attributed by some writers to the var. *vitellina* as a whole. But some at least of the plants grown for ornament are male.

cv. 'VITELLINA PENDULA'.—See 'Chrysocoma' under *S.* × *sepulcralis*.

S. ARBUSCULA L. *emend.* Sm.
S. arbuscula var. γ L.; *S. formosa* Willd.

A dwarf shrub with spreading or ascending branches, rarely more than 2 ft high; mature shoots glossy and dark brown, glabrous, but the young shoots sometimes downy. Leaves ovate or elliptic, mostly $\frac{1}{2}$ to $\frac{3}{4}$ in. long, acute or obtuse at the apex, cuneate to rounded at the base, glabrous, lustrous and sometimes reticulate above, green or somewhat glaucous and sometimes appressed-hairy beneath, margins glandular-serrate or almost entire; petiole up to $\frac{1}{8}$ in. long. Stipules minute or wanting. Catkins appearing with the leaves, $\frac{1}{2}$ in. or less long, on short leafy peduncles; scales obtuse, clad with white hairs. Stamens with free, glabrous filaments, anthers yellow or tinged with red. Ovary sessile or shortly stalked, downy, style well-developed with two short, slender, notched or entire stigmas.

SALIX ARBUSCULA

Native of Scotland, where it is confined to a few localities in the mountains of Argyll and Perthshire, but with its main distribution in Scandinavia and N. Russia. It is of no ornamental value, and mentioned here only because it is of interest as a rare British native.

The following species are closely allied to *S. arbuscula* and were once included in it:

S. FOETIDA Schleich. ex Lam. *S. arbuscula* subsp. *foetida* (Schleich.) Braun-Blanquet; ?*S. arbuscula* var. *humilis* Anderss.—Leaves elliptic-lanceolate, up to 2 in. long, sharply serrated, the teeth tipped with large yellowish glands. Catkins relatively thicker than in *S. arbuscula*, about twice as long as wide. Native of the

Alps from France to the High Tauern, mainly on acid soils. It grows up to 6 ft high at low altitudes, becoming dwarf near the tree-line.

S. WALDSTEINIANA Willd. *S. arbuscula* var. *waldsteiniana* (Willd.) Koch; *S. a.* subsp. *waldsteiniana* (Willd.) Braun-Blanquet; *S. a.* var. *erecta* Anderss.— This is more distinct from *S. arbuscula* than is *S. foetida*. Leaves obovate to elliptic, ¾ to 2½ in. long, cuneate at the base, margins crenate-serrate, the teeth not glandular, or entire. Catkins with distinct peduncles, three to four times as long as wide. Native of the E. Alps, N. Balkans and Carpathians, on limestone formations. A medium-sized shrub.

S. ARBUTIFOLIA Pall.

Chosenia arbutifolia (Pall.) A. K. Skvortsov; *Chosenia bracteosa* (Turcz.) Nakai; *Salix bracteosa* Turcz.; *Chosenia macrolepis* (Turcz.) Komar.; *Salix macrolepis* Turcz., in part; *Chosenia eucalyptoides* (Schneid.) Nakai; *Salix eucalyptoides* F. N. Meyer ex Schneid.

A tree up to 100 ft high in the wild; bark of young trees covered with a white bloom, scaly on older trees; branchlets glabrous, often glaucous. Leaves lanceolate or oblong-lanceolate, 2 to 3½ in. long, acuminate at the apex, base cuneate, glabrous, entire or faintly toothed; petiole up to ¼ in. long. Catkins appearing with the leaves on short, leafy laterals, pendulous. Male catkins slender, ⅜ to 1 in. long; scales broad-obovate, obtuse, three- or five-veined. Stamens five, adnate to the scale. Female catkins about 2 in. long in fruit; scales as in male catkins, deciduous. Ovary glabrous, short-stalked. Style divided to the base, each arm with two linear stigmatic lobes, deciduous in the fruiting stage.

Native of N.E. Asia, including Japan (Hokkaido and one locality on the main island). It is an interesting willow, differing from all other species (except perhaps some dwarf species of high latitudes) in being wind-pollinated like the poplars, in the absence or virtual absence, of nectaries from the flowers; in the stamens being adnate to the subtending scale; and in the deciduous style. It has been separated from Salix as the genus CHOSENIA, but essentially the flower-structure is that of *Salix*, within which it could retain its place as a separate subgenus. It has been introduced, but is unlikely to grow to a large size in our climate.

Another anomalous species that may be mentioned here is:

S. URBANIANA Seem. *Toisusu urbaniana* (Seem.) Kimura; *Salix cardiophylla* Trautv. & Mey. subsp. *urbaniana* (Seem). A. K. Skvortsov; *Toisusu cardiophylla* var. *urbaniana* (Seem.) Kimura—A massive tree in the wild, up to 90 ft high and 7 ft in diameter, with a deeply furrowed bark. Leaves ovate or elliptic-ovate, to 6 in. long and 1⅞ in. wide, finely toothed, almost glabrous when mature, glaucous beneath; petiole up to 1¼ in. long. Catkins on leafy laterals. Male flowers with five to ten stamens, free from the scale, in catkins 2 to 3 in. long. Female flowers in longer catkins; style deeply cleft, with linear stigmas, deciduous. Native of Japan often by mountain streams, common around Lake Chuzenji. It is the type of the section URBANIANAE (subgenus *Salix*), consisting of a few species in N.E. Asia, probably reducible to one—*S. cardiophylla*. This section has been given generic rank by Kimura as *Toisusu*, and is grouped by him with *Chosenia* in the tribe Choseniae of the willow family. It resembles *S. arbutifolia* in the slender, deeply cleft, deciduous styles, but the flowers are insect-pollinated and nectaries are present in both sexes, those in the female flowers placed laterally, one on each side of the ovary instead of the normal 'fore and aft' position. *S. urbaniana* may not be in cultivation in Britain but is worthy of trial. Prof. Sargent of the Arnold Arboretum thought it the finest of all the Japanese willows.

S. ARCTICA Pall.

S. anglorum Cham.; *S. pallasii* Anderss.

This very variable species, widely distributed in arctic and subarctic regions, is represented in cultivation by the following more southern variety:

var. PETRAEA Anderss. *S. petrophila* Rydb.; *S. arctica* var. *antiplasta* (Schneid.) Fern.; *S. anglorum* var. *antiplasta* Schneid.; *S. arctica* var. *araioclada* (Schneid.) Fern.; *S. anglorum* var. *araioclada* Schneid.—A shrub with creeping, sometimes buried stems and more or less erect branchlets, forming mats a few inches deep; branchlets yellowish, often slightly downy when young, older wood brown or purplish brown. Leaves elliptic to broadly oblanceolate, 1 to 1⅞ in. long, ¼ to ⅝ in. wide, subacute to rounded at the apex, usually cuneate at the base, entire, glabrous when mature; petioles ⅛ to ½ in. long. Stipules small or wanting. Catkins borne on leafy laterals, many-flowered, stout, 1 to almost 2 in. long, erect; scales dark, clad with very long hairs. Stamens two, sometimes connate at the base. Ovary hairy, almost sessile; style well-developed, with slender bilobed stigmas.

Native of the mountains of western N. America as far south as Colorado and New Mexico; also of N.E. Canada, where it occurs in the Gaspé Peninsula, south of the mouth of the St Lawrence river. It is an interesting willow, distinct from most other cultivated dwarf species in its large, very hairy catkins.

S. AURITA L.   ROUND-EARED WILLOW

A much-branched shrub varying in height from 1 to 6 or 7 ft, according to soil and situation; young twigs slender, at first very downy, becoming glabrous the second year. Leaves obovate, blunt or pointed at the apex, tapered at the base, 1 to 2 in. long, ½ to 1¼ in. wide, rather indefinitely toothed; the upper surface dull dark green, wrinkled, and more or less woolly, lower surface covered with a permanent dull grey wool; stalk ⅙ to ⅓ in. long. Stipules conspicuous on vigorous shoots, and mostly persisting till the fall of the leaf. Catkins produced on the naked shoots in April, similar to those of *S. caprea* and *S. cinerea*, though rarely more than ¾ in. long. Flowers as in these two species, but the stigma almost sessile.

Native of Europe, common in the British Isles on acid soils. It differs from *S. caprea* in being a smaller, more bushy plant with smaller, wrinkled leaves and shorter catkins produced only just before the leaves. Also in having the wood striated under the bark, as in *S. cinerea*. From that species it is less easily distinguished, but it has the year-old twigs glabrous, whereas in typical *S. cinerea* they remain downy; *S. cinerea* var. *oleifolia* has glabrous year-old twigs, but its leaves are not wrinkled like those of *S. aurita*, nor are its stipules so large and conspicuous.

S. BABYLONICA L.

S. pendula Moench

A tree usually under 40 ft high, but up to 60 ft or even more if it lives long enough, the rugged trunk branching low and supporting a wide-spreading head of branches, the very slender terminal twigs of which hang down perpendicularly; these are glabrous except near the nodes. Leaves lance-shaped, with long, slender points, narrowed to the base, finely toothed, 3 to 4 in. long, ½ to ¾ in. wide, slightly silky when young, soon quite glabrous, light green above, glaucous beneath; petiole about $\frac{3}{16}$ in. long, downy in the groove on the upper side, otherwise glabrous. Stipules usually present on strong shoots, obliquely ovate or lanceolate, soon deciduous. Catkins always female on the commonly cultivated form, dense, slender, about 1 in. long, appearing in April with the young leaves, almost sessile, with one to three small leaves at the base; scales persistent, lanceolate-ovate, with a blunt or acuminate apex, yellowish, hairy towards the base and at the edge. Ovary almost sessile, ovoid, glabrous; style very short, stout; stigmas spreading, more or less two-lobed. Nectary one, posterior (between the ovary and the axis of the catkin), very short.

S. babylonica, at least in its typical state, is known only as a cultivated tree, introduced to Europe from the Near East in the first or second decade of the 18th century, perhaps some years earlier (see further below). The weeping forms by which it is chiefly known are doubtless mutants from some species of normal habit, but no such species is to be found wild anywhere in western or central Eurasia at the present time. The nearest ally of *S. babylonica* is *S. matsudana* (q.v.) of northern China, which indeed is part of *S. babylonica* as understood by botanists of the last century, and with good reason, for no one has yet advanced any convincing reason for regarding them as distinct species. The most likely hypothesis is that *S. babylonica* was brought from northern China along the ancient trade route to south-west Asia. In Tadjikstan, through which this route passed, there are at least three cultivated clones of *S. babylonica*, one of them male (Skvortsov, *Ivi SSSR*, pp. 114–5). In N.W. India and bordering Tibet male trees are by far the commoner and may well belong to a single clone, judging from the great similarity of herbarium specimens from that region. From the Near East, all the specimens seen (with the exception of some probable hybrids) are very like the female form of *S. babylonica* originally brought to Europe, which is the type of the species. But the centre of variation of the *S. babylonica-matsudana* complex, as might be expected, is in the Far East, where several cultivated forms have been given specific rank.

The first European to record the existence of *S. babylonica* was Sir George

Wheeler, Bt, who saw it near Bursa in western Anatolia in 1676; it has even been suggested that he introduced it to Britain, as he did *Hypericum calycinum* and *H. olympicum*. The earliest botanical mention of the weeping willow is to be found in Tournefort's *Corollarium*, in which that great French botanist recorded the plants he found during his journey in the Levant (1700–2); he too has been credited with its introduction. According to Peter Collinson, the original weeping willow was brought to England about 1730 by 'Mr Vernon, Turkey Merchant at Aleppo, from the river Euphrates, and planted at his seat at Twickenham Park, where I saw it in 1748'. It is, however, open to doubt whether this tree was the 'original' in the sense of being the only source of the plants cultivated in Europe. The London nurserymen were offering it in 1730, and may well have imported it on their own account. The most famous tree in the 18th century grew in the garden of the poet Alexander Pope at Twickenham. There is a story that he was one day in the company of Lady Suffolk, when she received a parcel from Spain tied up by willow twigs, and that, noticing one of the twigs was alive, he begged it and planted it at Twickenham, where it grew into the celebrated weeping willow of his villa garden. This was certainly not the first *Salix babylonica* to be grown in this country, and it is probable that the plant came from Pope's neighbour Vernon, or was simply bought from one of the London nurseries.

Linnaeus, author of the misleading name *Salix babylonica*, encountered the weeping willow in the Clifford garden at Hartecamp in the Netherlands, and described it in the *Hortus Cliffortianus* (1738). Unfortunately the tree there (and younger plants raised from it by layering) had not yet flowered, but there seems to be little doubt that Clifford's willow was the common female clone. Linnaeus evidently believed that the weeping willow came from the ancient Babylon and was the willow of the Psalmist: 'By the rivers of Babylon, there we sat down, yea, we wept, when we remembered Zion. We hanged our harps upon the willows in the midst thereof'. But, as is now well known, the 'willows' of Babylon, which was situated on the lower Euphrates, south of Baghdad, were *Populus euphratica*. The confusion arose because the name 'gharab', which in biblical times meant poplar, later came to signify willow, and was so translated when the vernacular versions of the Bible were compiled at the time of the Reformation.

S. babylonica is now cultivated in many parts of the world, but in Europe, at least in the colder parts, it has become a very rare tree, for it is by no means hardy, and even where it survives the average winter it may be damaged by spring-frost. As early as 1869 the German dendrologist Koch referred to it as a fast disappearing tree, and Elwes and Henry, writing early this century, could find only two examples of any size in Britain, both of them nearing the end of their lives. The various hybrids that came gradually to take the place of the original *S. babylonica* are discussed under *S.* × *sepulcralis*.

The Napoleon willow, of which so much has been written, was planted by the grave of Napoleon on St Helena after his death in 1821. There were several plants, which gradually died out. There is no doubt that they were typical *S. babylonica*, but the name "*Salix Napoleonis*" was used by unscrupulous nurserymen for other willows, to profit from the extraordinary magic of Napoleon's memory in the decades following his death. For further information see the interesting article by U. Forskhufvud, *Act. Hort. Gotoburg.*, Vol. 26, 5–18 (1963). The Napoleon willow that once grew at Kew was raised from a cutting brought home by the Kew collector Thomas Fraser in 1825. For an amusing account of the early history of this plant see the note by James Smith, a former Curator of Kew Gardens, in *Gard. Chron.* (1867), p. 105, or his book *Records Roy. Bot. Gard. Kew* (1880), pp. 261–2. This tree died in the drought of 1867, and no propagation from it has survived.

cv. 'CRISPA' ('ANNULARIS').—A curious form whose leaves are twisted into rings or spirally curved. It has little beauty (*S. babylonica* var. *crispa* Loud.; *S. annularis* Forbes). This sport has been known since early in the 19th century, and was said to be even more tender than the normal form. It is possible that some plants grown under the name at the present time are of independent origin and do not derive from *S. babylonica*.

S. 'LAVALLEI'.—Very near to *S. babylonica*, but of less pendulous habit. It is a male clone, with slender almost sessile catkins about 1 in. long (*S. japonica Lavallei* Hort.; *S. babylonica* var. *Lavallei* Dode). The original tree in Lavallée's Segrez Arboretum was said to have come from Japan and may have been the male clone now known there as *S. babylonica* var. *lavallei* f. *seiko* Kimura. 'Lavallei' was introduced to Kew shortly before 1898 but is no longer in the collection. Its catkins are portrayed in *Gard. Chron.*, Vol. 76 (1924), fig. 325.

S. BOCKII Seem.

A dwarf shrub of neat habit, 3 to 10 ft high; young shoots slender, covered with a dense grey down. Leaves oblong or obovate, tapered to a short stalk at the base, either rounded or pointed at the apex; margins entire or occasionally sparsely toothed, recurved, $\frac{1}{4}$ to $\frac{5}{8}$ in. long, $\frac{1}{8}$ to $\frac{1}{4}$ in. wide, dark bright green above, blue-white beneath and covered with silky hairs. Catkins produced in late summer and autumn from the leaf-axils of the current year's growth; females $1\frac{1}{2}$ in. long, $\frac{1}{2}$ in. broad; males shorter. Stamens two, but with their stalks united to the summit (as in *S. purpurea*); bracts of catkins narrowly lanceolate and pointed. *Bot. Mag.*, t. 9079.

A native of Western Szechwan, China, and abundant in river-beds up to 9,000 ft. It was introduced to the Arnold Arboretum by Wilson by means of cuttings in 1908 and a plant was obtained for Kew in 1910. This was female, and probably most of the plants originally grown in Britain descended from it; the more attractive but rarer male form is also in cultivation (both sexes were collected by Wilson in 1908). It is not unusual for normally spring-flowering species to produce the occasional catkin on summer shoots, but *S. bockii* is almost unique in regularly flowering late in the season on the growing shoots. It is then very ornamental.

S. bockii is probably not specifically distinct from S. VARIEGATA Franch. This species, apparently known in its typical state only from the gorges of the Yangtse Kiang, is prostrate, but in its essential characters scarcely differs from *S. bockii* and the two were united by K.-S. Hao in 1936 under the name *S. variegata*, which has priority. The two species were kept apart in *Plantae Wilson-*

ianae largely because of a biological difference in the plants seen by Wilson: the Yangtse plants are under water during the summer and flower on the ripened

SALIX BOCKII

wood in late autumn or early winter; *S. bockii* in W. Szechwan (as seen and introduced by Wilson) produces its catkins in late summer and autumn on young stems while these are still elongating. But the whole complex (which also includes *S. duclouxii* Lévl. from Yunnan) seems to vary in the rhythm of its growth and flowering.

S. 'BOYDII'

S. × *boydii* E. F. Linton

A shrub of erect, rigid growth, up to 3 ft high; shoots thinly downy at first; buds globose, soon glabrous. Leaves leathery, broadly obovate to nearly orbicular, up to 1 in. long, usually more or less heart-shaped at the base, at first grey with down above, becoming dark green, with the midrib and veins deeply impressed, covered permanently with a fine whitish wool beneath, the veins and midrib prominent; petiole very short. Catkins female, up to ¾ in. long, produced shortly before the leaves; scales obovate, rounded at the apex, silky; ovary ovoid, sessile, hairy.

'This unique plant is very near *S. reticulata* in the leaf-blades and very near *S. lapponum* in its inflorescence, and it might be that × *S. boydii* drew its origin from these two parents only; but there are two characters, the short petioles and the short nearly round glabrescent buds, which neither of these satisfactorily account for. These two characters suggest that a third species, viz. *S. herbacea*, may have entered into its composition, more or less remotely.' S. 'Boydii' was discovered towards the end of the last century by William Boyd (1831–1918), the raiser of many well-known alpine-garden plants, and the discoverer of the mysterious *Sagina boydii*. Brought into his garden at Upper Faldonside, Melrose, this willow was described in 1913 by the Rev. E. F. Linton, whose comments on its parentage are quoted above (from 'The British Willows', *Journ. Bot.*, Suppl. (1913), p. 38). The original plant, about fifty years old and some 3 ft high, is figured in *Qtly Bull. Alp. Gard. Soc.*, Vol. 8, p. 42 (1940). This plant, still 3 ft high, and another of similar size were moved by Mr Alex Duguid to Silverwells, Coldingham, Berwickshire, where they thrive and are used as stock plants for Edrom Nurseries.

Boyd's willow is now one of the most valued of the dwarfer kinds, though grown more for its picturesque habit than for its catkins, which are inconspicuous and not freely produced. It is of slow growth; plants at Faldonside propagated from the original were none of them more than 14 in. high when seven or eight years old (Linton, op. cit., p. 88). It received an Award of Merit in 1958.

S. CAESIA Vill.   BLUE WILLOW
S. myricaefolia Anderss.; *S. divergens* Anderss.

A shrub of low straggling habit, 2 to 4 ft high, very leafy, with glabrous dark brown young branches; buds glabrous, yellow. Leaves rather stiff, oval or obovate, tapered at the base, pointed, often abruptly, at the apex, sometimes wavy but not toothed at the margin, ¾ to 1½ in. long, ⅜ to ⅞ in. wide (somewhat larger on strong shoots), perfectly glabrous on both sides, dull green above, bluish beneath; stalks very short, usually ⅛ in. or less long. Stipules usually minute. Catkins ½ to ¾ in. long, produced in April or May on short leafy shoots; catkin-scales obtuse, thinly hairy. Filaments of stamens free or united for up to half their length. Ovary sessile or almost so, hairy; style about half as long as the ovary.

Native mainly of Siberia and Central Asia; also found in the Alps from the Dauphiné to the Vorarlberg and Tyrol, though nowhere occurring in quantity in that region; introduced in 1824. One of the most distinct of European willows, easily recognised by its glabrous, stiff, very short-stalked leaves, bluish beneath. It was at one time met with in gardens as "*S. zabellii pendula*", being grafted on standards and in that way transformed into a small weeping tree—pretty, but, treated in this way, usually short-lived.

S. CANDIDA Fluegge ex Willd.   SAGE WILLOW

S. incana Michx., not Schrank

A shrub up to 5 or 6 ft high of stiff, erect habit, the young shoots covered with a close, white wool. Leaves linear to narrow-oblong, tapered at both ends, 1½ to 4½ in. long, ⅛ to ⅞ in. wide, upper surface wrinkled, at first white with down which afterwards falls away, leaving it dull green, lower surface permanently covered with a thick tomentum of fine, matted hairs, margins decurved, obscurely toothed or entire; stalk ⅛ to ½ in. long. Stipules small and deciduous, except sometimes on strong shoots. Catkins produced in April as the leaves unfold, cylindric, densely flowered, almost sessile; scales pale brown. Male catkins about 1 in. long; stamens with free, glabrous filaments and purplish red anthers; nectary one. Female catkins 1½ to 2½ in. long in fruit; ovary downy, conic-ovoid, shortly stalked; style elongate, slender, reddish or purplish.

Native of North America from Labrador to British Columbia, south to Philadelphia, Iowa and Colorado; introduced in 1811. This distinct and hardy species is worth growing for the vivid whiteness of its young leaves.

S. CAPREA L.   GREAT SALLOW, GOAT WILLOW*

S. praecox Salisb.

A shrub or a small tree of bushy habit; young shoots at first grey with down, becoming smoother. Leaves varying in shape from roundish oval or oval lance-shaped to obovate, tapered, rounded, or heart-shaped at the base; pointed, sometimes blunt at the apex, toothed or entire, 2½ to 4 in. long, 1 to 2¼ in. wide, grey-green, wrinkled and slightly downy above, covered with a soft grey wool and prominently veined beneath; stalk ⅓ to ¾ in. long, woolly. Catkins produced on the naked shoots in March and April, stalkless; the males very silky, a little over 1 in. long, half as thick; anthers yellow. Female catkins ultimately 2 in. or more long; the seed-vessels white with down, and stalked; style very short.

Native of Europe and N.W. Asia, and common in Britain. Flowering branches of the male are often known in country places as 'palm', and are gathered by children the Sunday before Easter, when that day coincides with the opening of the flowers. This willow is one of those which bear seeds fairly freely in this country. It is often seen in hedgerows, where its yellow catkins make a cheerful display in early spring.

*So far as can be ascertained, the origin of the name 'goat willow' is to be found in the illustrated edition of the herbal of Jerome Bock (Hieronymus Tragus), first published in 1546 (Vol. III, p. 1078). The artist, David Kandel, enlivened his woodcuts of trees and shrubs with animal or human figures, some alluding to a property or association of the plant, others perhaps intended purely as decoration. In which category comes the he-goat that browses the catkins of the sallow it is impossible to say; there is no clue in the text, nor is there in the herbal of Tabernimontanus, published later in the century, where two sallows are figured, as *S. caprea, rotundifolia* and *S. caprea, latifolia*. Although Linnaeus did not cite these names when describing *S. caprea* he was certainly aware of them. The foliage of the sallows was at one time used as fodder for sheep and goats and much liked by them; this may be the explanation. Or the goat may have been a jocular allusion to the author, whose name means goat in German (a goat's head appears above his portrait in the frontispiece).

var. COAETANEA Hartm. *S. coaetanea* (Hartm.) Floderus: *S. caprea* var. *sericea* Anderss.; *S. caprea* var. β Wahlenb.; *?S. sphacelata* Sm., not Schleich.— Leaves smaller, obovate to elliptic, cuneate at the base, rather shortly stalked, sparsely silky-hairy above even when mature, densely white tomentose beneath. Native of Scandinavia and Finland. Plants found in the Highlands of Scotland are perhaps referable to this variety; they are usually of dwarf habit.

cv. 'KILMARNOCK' ('Pendula')—Branches stiffly pendulous. The Kilmarnock weeping willow was put into commerce by the nurseryman Thomas Lang of Kilmarnock, Ayrshire, and was first advertised in 1853, by which time he had propagated about 1,000 plants, most of them layers. The original plant came from James Smith of Monkwood Grove near Ayr, 'an old and enthusiastic botanist', who, according to an account published some fifty years after his death, found it on the banks of the Ayr. It is puzzling that there are two clones under the name *S. caprea* 'Pendula', one male and the other female. According to the advertisements in the *Gardeners' Chronicle* for 1853, the Kilmarnock willow bore 'gold-coloured catkins' and must therefore have been male. The history of the female is unknown, but it was in cultivation at Kew by 1880. It is recorded that Lang went back to Smith for more plants and it is possible that among these later acquisitions there was a pendulous female, also collected by Smith, or even that he had discovered a whole colony of pendulous plants. Or the female may have arisen later as a bud-mutation from the original Kilmarnock clone. The female clone has recently been named 'WEEPING SALLY' by Roy Lancaster in *The Garden* (*Journ. R.H.S.*), Vol. 101 (1976), p. 75.

S. × REICHARDTII A. Kern.—A natural hybrid between *S. caprea* and *S. cinerea*, widespread in the British Isles, where the second parent is usually the common sallow (*S. cinerea* subsp. *oleifolia*, syn. *S. atrocinerea*). Being fertile, and back-crossing in both directions, the hybrid is very variable and sometimes difficult to distinguish from the parents. Intermediate forms have relatively narrower leaves than in *S. caprea* and more sparsely indumented beneath, but with the characteristic prominent venation of that species. The wood under the bark may be smooth as in *S. caprea* or show the influence of *S. cinerea* in being slightly striated. The hybrid is uncommon in closed communities and mostly found in disturbed habitats, e.g., waste ground in semi-urban areas.

S. CINEREA L.

S. aquatica Sm.

A low, spreading bush in its typical state, rarely more than 15 ft high; buds and twigs densely grey-downy; second year wood prominently striated under the bark. Leaves obovate to oblanceolate or elliptic, 2 to 4 in. long, ¾ to 1¾ in. wide, tapered at the base, shortly pointed or rounded at the apex, entire or inconspicuously toothed, dull green and glabrous above, covered beneath with a permanent soft grey felt; petiole up to ½ in. long. Stipules usually conspicuous, up to ⅜ in. long. Catkins and flowers essentially as in *S. caprea*, but the former somewhat smaller and appearing a few weeks later, usually not before April.

Native of Europe, W. Asia and N. Africa, uncommon in Britain, where it is mainly confined to wet habitats in eastern England (but see var. *oleifolia*). It is allied to *S. caprea*, and best distinguished from it by the striated wood, the narrower leaves and persistently hairy twigs.

subsp. OLEIFOLIA (Sm.) Macreight *S. oleifolia* Sm.; *S. cinerea* var. *oleifolia* (Sm.) Gaud.; *S. atrocinerea* Brot. COMMON SALLOW.—Taller growing than the typical subspecies, sometimes a tree up to 40 ft high. Mature twigs and buds glabrous or almost so. Leaves smaller, to about 2½ in. long, the coating beneath rather thin, rough to the touch, and composed of grey hairs intermingled with brown hairs. Native of W. Europe, common in the British Isles in a wide variety of habitats. The thin indumentum of the leaf-undersides, composed of a mixture of grey and brown hairs, rough to the touch, serves to distinguish it from typical *S. cinerea* and from *S. caprea*. Also, in *S. caprea* the wood is smooth under the bark and the leaves are broadest about the middle, whereas in the common sallow, *S. cinerea* var. *oleifolia*, the wood under the bark is ridged and the leaves are broadest above the middle.

The flowering twigs of the common sallow are gathered like those of the goat willow on Palm Sunday, especially in years when Easter falls late.

cv. 'TRICOLOR'.—Leaves blotched and dotted with yellow and white. Male. Of little beauty.

S. CORDATA Michx.

S. adenophylla Hook.

A shrub up to 8 ft high, of loose habit, sparsely branched; twigs covered with a thick silky coat of hairs. Leaves arranged very closely on the branchlet (about four to the inch), ovate with a heart-shaped base, rather abruptly pointed, very finely, closely, and regularly toothed, many of the teeth glandular, especially at the base, 1½ to 2 in. long, ¾ to 1¼ in. wide, covered with long, whitish, silky hairs on both surfaces, not so thickly as on the twigs; stalk from ⅕ to ⅓ in. long. Stipules persistent, obliquely heart-shaped, glandular-toothed, ⅓ in. in diameter. The female plant only appears to be in cultivation; this has catkins 1½ to 3 in. long, borne on short leafy shoots.

Native of N. America from Labrador to Wisconsin, south to Philadelphia and Illinois. It is one of the most distinct of cultivated willows, especially in the extreme downiness of the younger parts, in the broad, closely set leaves, and large persistent stipules.

S. DAPHNOIDES Vill.   VIOLET WILLOW

S. pulchra Wimm. (not Cham.)

A tree of erect, vigorous habit up to 40 ft high; young shoots at first downy, becoming glabrous, and covered with a conspicuous plum-coloured bloom; twigs brittle. Leaves oval-lanceolate, tapered at both ends, but more gradually at the point, finely toothed (the teeth glandular); 1½ to 4½ in. long, ⅜ to 1 in. wide, somewhat leathery, glabrous dark green and glossy above, blue beneath; stalk ⅙ to ½ in. long. Catkins produced in March; males 1 to 2 in. long, ½ to ¾ in. wide, rather striking, and resembling those of the goat willow; females more slender; ovary almost sessile, glabrous; style about half as long as the ovary, usually yellow; stigmas linear, erect.

Native of Europe from Scandinavia to the Alps and N. Italy, east to the Urals. In Britain it has been found growing in a few out-of-the-way, damp localities in the north of England, but is not a native nor even naturalised. As a willow for the garden it is worth growing for the beautiful purple or violet-coloured bloom on the shoots, and for its handsome catkins. It received an Award of Merit in 1957. If the plants are cut back every second spring the crop of young wands makes a pleasing winter effect. In the osier basket trade it is known as 'Violets' and 'French Purple'.

cv. 'AGLAIA' ("*Latifolia*").—Leaves broader than normal; stems scarcely pruinose. In cultivation since the 1830s, at least on the continent. The plants now in commerce are male, but the one received by Kew in 1880 from the Belgian nurseryman van Houtte was female.

var. POMERANICA (Willd.) Koch   *S. pomeranica* Willd.—Leaves narrower, catkins more slender; usually of dwarf habit. It was described from the southern Baltic, but is said to occur elsewhere within the range of the species. A male clone in commerce on the continent has catkins about 3 in. long.

S. × CALLIANTHA J. Kerner   *S. daphnoides* × *S. purpurea*—An erect shrub with narrowly obovate leaves 4 to 5 in. long, silky on both sides when young, becoming glabrous, the upper surface dark green and glossy, margins crenate-serrate. Male catkins about 1 in. long, almost as wide when in flower; filaments united for about half their length; anthers at first red, golden-yellow when ripe. Described by J. Kerner from a plant growing near Vienna, at first thought by him to be *S. caprea* × *S. purpurea*.

S. × ERDINGERI J. Kerner   *S. daphnoides* × *S. caprea*—This hybrid was described from a cultivated female plant near to *S. daphnoides* in aspect but with shorter, broader leaves about 4 in. long and 1 in. wide, silky when young, soon becoming nearly glabrous. The catkins are cylindrical, 1¼ to 1¾ in. long, ½ in. wide, and rather effective when they appear in March. The flowers show the influence of *S. caprea* in their stalked, slightly hairy ovaries. The cultivated plants most probably derive from Kerner's type, but the hybrid is said to occur occasionally in the wild.

S. ACUTIFOLIA Willd.   *S. pruinosa* Bess.; *S. violacea* Andrews; *S. daphnoides* var. *angustifolia* Weinm.; *S. d.* var. *acutifolia* (Willd.) Doell; *S. d.* subsp. *acutifolia* (Willd.) Blytt & Dahl; *S. caspica* Hort., not Pall.—This is sometimes, and perhaps rightly, regarded as a variety or subspecies of *S. daphnoides*. It differs from that species in its more slender and often pendent shoots, its relatively narrower leaves more tapered at the apex, with fifteen or more pairs of lateral veins (eight to twelve in *S. daphnoides*). A further distinction is that in the female flowers the catkin-scales are about half as long as the ovary (about as long in *S. daphnoides*).

It is a native of Russia, where it is widely distributed from the west to eastern Siberia and Central Asia. It was introduced to Germany early in the 19th century and widely used there in coastal areas for fixing sand-dunes, under the name "*S. caspica*", which belongs properly to another species.

cv. 'PENDULIFOLIA'.—Branches arching; leaves up to 6 in. long, hanging vertically. A male clone, put into commerce by Späth in 1939 and later distributed by Dr Krüssmann from the Dortmund Botanic Garden. 'BLUE STREAK' is similar, but the leaves are shorter and the branches not so pendulous; also male.

S. KANGENSIS Nakai—This ally of *S. daphnoides* is in cultivation but is only of botanical interest, the young branches lacking the bloom of the western species. It is a native of Korea, the Ussuri region of Russia and of N.E. China. The fourth member of the section *Daphnella* is S. RORIDA Lakschewitz, with a wide distribution in N.E. Asia, including Japan. It is not known to be in cultivation.

S. DISCOLOR Muhl.

A shrub or low tree not more than 25 ft high; young shoots purplish brown, at first downy. Leaves oblong, oval, or obovate, tapered at both ends, toothed except towards the base, 2 to 5 in. long, $\frac{5}{8}$ to $1\frac{1}{4}$ in. wide, at first somewhat downy, soon becoming glabrous, bright green above, and blue-white beneath; stalk $\frac{1}{4}$ to 1 in. long. Catkins opening in March and April on the leafless shoots; males up to $1\frac{1}{2}$ in. long, cylindrical; stamens two, with glabrous stalks; female catkins up to 3 in. long in fruit. Ovary beaked, downy, with a distinct style.

Native of the eastern United States and Canada; introduced in 1811. It is rather striking in its deep brown branchlets and very glaucous under-surface of the leaves. This character serves to separate it from *S. caprea* and *S. cinerea*, and there is the further difference that in those species the stigma is almost sessile.

It is, however, allied to the Old World sallows. So too are:

S. HUMILIS Marsh.—A shrub to 10 ft high with downy or glabrous branchlets. Leaves up to 4 in. long, narrowly to broadly oblanceolate, dull green above, underside clad with persistent short, curled hairs or becoming almost glabrous, margins entire or slightly wavy-toothed. Stipules narrow or wanting. Catkins sessile, before the leaves, $\frac{1}{2}$ to 1 in. long; scales blackish, hairy. Anthers reddish or purplish. Ovary with a long beak, downy, stalked; stigmas almost sessile. A variable species, widely distributed in eastern and central N. America. Introduced to Kew from the Arnold Arboretum in 1889 and again in 1908 (a different form).

S. TRISTIS Ait. *S. humilis* var. *microphylla* (Anderss.) Fern.—Very near to *S. humilis* and from the same region, but dwarfer (to 3 or 4 ft) and smaller in all its parts. Leaves narrowly oblanceolate, up to 2 in. long and $\frac{1}{2}$ in. wide. Introduced 1765.

S. ELAEAGNOS Scop.

S. incana Schrank; *S. rosmarinifolia* Host, not L.; *S. riparia* Willd.

A shrub of dense, very leafy habit, bushy, up to 8 or 12 ft high, half as much more in diameter, rarely a small tree; young shoots clothed with a fine grey felt at first, becoming glabrous later; buds yellowish. Leaves linear, tapered at both ends, 2 to 5 in. long, $\frac{1}{8}$ to $\frac{7}{8}$ in. wide, made narrower by the decurved margins, dark green and glabrous above, covered with a blue-white felt beneath. Catkins erect, slender; females 1 to 1$\frac{1}{2}$ in. long, males shorter, appearing with the young leaves in April and May. Stamens two, connate for up to one-half their length.

Native of southern and central Europe, and of Asia Minor; introduced about 1820. In the Alps it occurs, usually on limestone, in the valleys of the larger rivers, ascending along streams into the subalpine zone, and is often associated with *Hippophae rhamnoides*. It varies in the width of its leaves, and the cultivated plants, which mostly have them no more than $\frac{3}{16}$ in. wide, should strictly be distinguished as subsp. *angustifolia* (Cariot) Rech. f.; no doubt the stock came originally from southern France or Spain, to which this narrow-leaved subspecies is said to be confined. It is one of the prettiest and most effective of the bush willows in foliage. Its leaves resemble those of *S. viminalis* only they are not so coarse, nor so glistening beneath. Very desirable for the banks of ponds, etc.

S. elaeagnos is the only member of the section *Canae*. It is quite closely allied to the section *Vimen* (*S. viminalis* and its allies) but with a different leaf-indumentum. The partial union of the stamens is also seen in *S. gracilistyla* (q.v.).

S. × SERINGEANA Gaud.—This natural hybrid between *S. elaeagnos* and *S. caprea* bears some resemblance to the osier-sallow hybrids (see *S.* × *sericans*). But the indumentum is more woolly, the leaves more parallel-sided, often more grey-hairy above when young, and the catkins more slender.

S. EXIGUA Nutt.

S. argophylla Nutt.; *S. longifolia* var. *argyrophylla* Anderss.

A thicket-forming shrub to about 12 ft high, or a small tree; young stems at first silky, becoming glabrous, or sometimes glabrous from the start. Leaves slender, 2 to 4 in. long, $\frac{1}{8}$ to $\frac{3}{8}$ in. wide (but larger and relatively broader on strong shoots), sharply acute to acuminate at the apex, tapered at the base to a very short petiole, clad on both sides when young with grey or silvery hairs, becoming grey-green and often glabrous above, more persistently hairy beneath, margins entire or inconspicuously toothed. Stipules wanting, except sometimes on vigorous shoots. Catkins 1$\frac{1}{4}$ to 2 in. long, appearing with the unfolding leaves on short leafy peduncles from the previous year's wood, or later on leafy laterals of the seaon's growths, terminal and solitary or supplemented (especially on male plants) by one or two axillary catkins; scales yellow, usually narrow and acute, deciduous. Stamens two; filaments clad with long hairs at the base. Ovary hairy or glabrous, sessile or almost so; stigmas divided, sessile.

Native of W. North America away from the coastal region, extending into Mexico. It is a variable species in itself, and is part of a taxonomically difficult and controversial group. The precise identification of the cultivated plants, which have not been seen to flower in this country, is uncertain, but they appear to belong to *S. exigua* as interpreted in recent American works. According to the older classifications they might have taken the name *S. argophylla* Nutt. They are, at any rate, very ornamental with their narrow, silvery foliage and would probably thrive best in a damp, grass-free, sandy soil.

S. INTERIOR Rowlee *S. longifolia* Muhl., not Lam.; *S. exigua* subsp. *interior* (Rowlee) Cronquist—Very closely allied to *S. exigua* but with green leaves and differing in various technical characters. Of wide distribution in America, from the Rocky Mountains to the Atlantic.

The section *Longifoliae* is confined to the New World. Unlike other willows, they spread widely by suckers from the roots. The catkins, at least the later ones, are borne on long, leafy laterals, often in clusters.

S. FARGESII Burkill

A deciduous shrub, up to 10 ft high, of wide-spreading habit; young shoots stout, quite glabrous, brownish green, changing by the second year to a dark shining brown; winter buds bright red. Leaves elliptic to elliptic-lanceolate, pointed, tapered at the base, finely toothed; 3 to 7 in. long, 1¼ to 3 in. wide, shining dark green above, wrinkled, and at first dull green beneath and silky hairy, especially on the midrib and veins; veins in fifteen to twenty-five pairs, deeply impressed above; stalk ½ to ¾ in. long and of the same colour as the young shoots. Catkins erect on short, leafy, silky stalks; the females cylindric, up to 6½ in. long, ¼ in. in diameter, the males up to 4½ in. long. Ovary glabrous, ovoid-cylindric; stigmas two, bilobed; bract oblong, rounded at the end, silky at the margin. Flowers in spring.

Native of central China; introduced by Wilson, who collected plants in W. Hupeh in November 1910 for the Arnold Arboretum, whence they (or cuttings from them) were distributed to Kew and other gardens. The plants came from woodlands near Fang Hsien at 6,000 to 8,000 ft, where *S. fargesii*, according to Wilson, makes a shrub 2 to 6 ft high and is often prostrate or procumbent (though in other localities he found it up to 10 ft high).

This fine, handsome willow is remarkable for its brightly coloured winter-buds, the dark glossiness of its younger bark, the large many-veined leaves, and slender, erect catkins. It is perfectly hardy, but the young growths are sometimes cut by frost. Most of the cultivated plants are female and probably belong to a clone originally distributed under the erroneous name *S. hypoleuca*. The species rightly so called (not treated here) was sent by Wilson under his number 4437, *S. fargesii* as number 4439.

S. fargesii is closely related to *S. moupinensis* (q.v.), and was described later. Schneider distinguished the two chiefly by two characters: in *S. fargesii* the rachis of the catkin and catkin-scales are densely long-hairy, glabrous or sparsely hairy in *S. moupinensis*; in *S. fargesii* the ovary is tapered into a long style, in *S. moupinensis* abruptly narrowed into a shorter style. Also the leaves of *S. fargesii* are more silky beneath; in *S. moupinensis* they are often quite glabrous. The young shoots of *S. fargesii* are stouter and the winter-buds larger, and, on the whole. the marginal toothing of the leaves is finer than in *S. moupinensis*. The two are also separated geographically, *S. moupinensis* west of the Red Basin, in the Sino-Himalayan floristic region, *S. fargesii* east of the Red Basin, in E. Szechwan and Hupeh.

S. FRAGILIS L.   CRACK WILLOW

A tree 80 to 90 ft high, with a rough corrugated trunk; branchlets growing at an angle of 60° to 90° to those from which they spring; young shoots glabrous. Leaves narrowly lanceolate to narrowly oblong, 2 to 7 in. long, $\frac{3}{8}$ to $1\frac{1}{4}$ in. wide, tapered at the base, the apex drawn out into a long slender point, distinctly and regularly toothed, usually somewhat silky at first, soon becoming glabrous; stalk $\frac{1}{4}$ to $\frac{3}{4}$ in. long. Stipules often present, semi-cordate or kidney-shaped, deeply toothed. Catkins 2 to $2\frac{1}{2}$ in. long, drooping, produced in April and May on short leafy shoots; scales pale straw-coloured, narrowly oblong or oblong-lanceolate, usually blunt at the apex, hairy except at the very tip, soon falling. Stamens two, hairy at the extreme base only. Ovary very shortly stalked (more distinctly so in fruit), flask-shaped, much tapered at the apex, shorter than the subtending scale; style slightly longer than the spreading stigmas.

Native of much of Europe, including Britain, extending in Russia as far east as the Altai and south to the Caucasus; also occurring in parts of S.W. Asia. It obtains its common name from the readiness with which the twigs snap off in their entirety at the joint when bent. Another peculiarity is that the rootlets which it sends into water are red. It is allied to *S. alba* with which it hybridises (see *S.* × *rubens*), differing in the wider angle of branching, its larger, glabrous, greener leaves, its larger catkins, and its stalked, more elongated ovaries. It produces a reddish timber, used for various purposes where a wood that is tough and capable of withstanding much friction is needed. It has been used for wheelbarrows and cart bottoms. Cheap cricket-bats are also made from it; manufacturers know it as the 'open-bark' willow.

S. × RUBENS Schrank *S. viridis* Fries—This willow occupies a place intermediate between *S. alba* and *S. fragilis*, and is considered to be a hybrid between them. It fills the gap between these two willows by an almost complete series of intermediate forms, sometimes approaching one of them in vegetative characters, whilst resembling the other in reproductive ones. What may be termed the central form is a tree branching at angles of about 60°, with leaves broader and larger than those of *S. alba*, and averaging 2 to 5 in. in length, $\frac{5}{8}$ to 1 in. in width, silky at first, but soon becoming glabrous, dark glossy green above, glaucous beneath. The male catkins are longer and more densely flowered than those of *S. alba* and the ovaries are more distinctly stalked and have more distinctly formed styles. This hybrid in one form or another is commoner in Britain than pure *S. fragilis*.

The timber of *S.* × *rubens* is of some value to cricket-bat makers, but ordinarily is much inferior to that of *S. alba* 'Caerulea', being heavier and coarser.

This refers to the central form of *S.* × *rubens*; as it approaches *S. alba* in relationship its value improves. By leaves alone it is sometimes difficult to distinguish between some of these forms and *S. alba* 'Caerulea', and the influence of *S. fragilis* is only to be seen in the stalked, more tapered seed-vessels. It is never pyramidal in growth like *S. alba* 'Caerulea'.

S. 'BASFORDIANA'—A vigorous tree with polished orange-yellow branchlets. Leaves bright green, finely serrated, glabrous, up to 6 in. long, ½ to ¾ in. wide. A male clone, with drooping catkins 2½ to 4 in. long, appearing with the leaves in April; scales acute, ciliate. It was raised by the willow-grower and basket-maker William Scaling of Basford, Notts, in the 1860s and named by him *S. Basfordiana*, under which name it was described by James Salter in 1882 (but see also 'Sanguinea' below). It is possibly a hybrid between *S. alba* var. *vitellina* and *S. fragilis* or *S.* × *rubens*. Pruned hard each spring it might be even more effective than *S. alba* var. *vitellina* in the winter-colour of its stems.

S. 'RUSSELLIANA'. DUKE OF BEDFORD'S WILLOW, LEICESTERSHIRE WILLOW—A large and vigorous tree with straight, slender branches. Twigs olive-brown. Leaves light green, deeply serrated, long-tapered at the apex, silky beneath when young, becoming glabrous. Only female trees are known: catkins as in *S. fragilis* though laxer; ovary slightly longer than the subtending scale (*S. russelliana* Sm.; *S. fragilis* var. *russelliana* (Sm.) Koch).

This willow first came to scientific notice when a Mr Bakewell sent it around 1800 from Leicestershire to the Duke of Bedford, after whom Smith named it in 1804 (Russell being the family name of the Dukes of Bedford). It is most probably a clone of *S.* × *rubens*, selected originally for its fast growth and excellent timber, and is still common in the north of England. For Dr Johnson's willow, supposed to have been 'Russelliana', see Loudon, *Arb. et Frut. Brit.*, Vol. III, p. 1517 and figs 1312–3.

S. 'SANGUINEA'.—Similar to 'Basfordiana' but with small, less tapered leaves up to 2½ in. long and ⅝ in. wide, and with redder twigs. A female clone; catkins 1¾ in. long, scales slightly ciliate or quite glabrous. Like 'Basfordiana' this willow was distributed by Scaling, who found it growing in the French Ardennes, probably as a cultivated tree, and later obtained cuttings. Salter considered 'Sanguinea' to be of the same "species" as 'Basfordiana' and his *S. basfordiana* is founded on both these willows.

S. DECIPIENS Hoffm. *S. fragilis* var. *decipiens* (Hoffm.) Koch—A smaller tree than *S. fragilis*; branchlets at first red on the exposed side, becoming clay-coloured and very lustrous, as if varnished. Leaves narrow-elliptic, rather short, up to 3½ in. long and 1 in. wide, coarsely serrated. It is not known in Europe as a truly wild tree, and in Britain is represented by a male clone, much planted for basket-work, with catkins not much over 1 in. long; according to Sir James Smith it was known in Norfolk and Cambridgeshire as the White Welsh willow, but more recent names for it are 'White Dutch' and 'Belgian Red'.

It has been suggested that *S. decipiens* is a hybrid of *S. fragilis* with either *S. triandra* or *S. alba*. According to another theory it represents the pure and original state of *S. fragilis*, which later became modified by hybridisation with *S. alba*.

S. GLABRA Scop.

A shrub to about 5 ft high, stoutly branched; young growths glabrous, becoming chestnut-brown when mature. Leaves short-stalked, obovate to broadly elliptic, to about $1\frac{7}{8}$ long and 1 in. wide (sometimes larger), acuminate at the apex, cuneate at the base, glabrous, rather thick, deep green and very lustrous above, paler green or glaucous with a waxy bloom beneath, distinctly toothed. Stipules rarely present. Catkins appearing with or after the leaves; scales yellow or brown, edged with long hairs. Male catkins on short leafy stalks; stamens much longer than the scales; anthers purple at first. Female catkins to about 3 in. long, on longer stalks. Ovary stalked, narrowly conic, glabrous, with a short style.

A native of the Alps (mainly in the eastern part) and of N.W. Yugoslavia. An interesting species, one of the most glabrous of the smaller willows, with very glossy leaves. It is usually placed in the section *Cordatae* (*Hastatae*), but has recently been separated, with three other species, as the section *Glabrella* A. K. Skvortsov, of which it is the type-species.

S. REINII Franch. & Sav. ex Seem. *S. glabra sens.* Franch. & Sav., not Scop.— This species differs from *S. glabra* only in a few minor characters—a remarkable fact considering that it is a native of Japan, with a few stations on the mainland of the Russian Far East and N.E. China. Franchet and Savatier, having proposed the name *S. reinii*, later withdrew it, having concluded that the Japanese plant was too near to *S. glabra* to merit separate naming.

S. GLAUCOSERICEA Floderus

S. sericea Vill., not Marsh.; *S. glauca* of many authors, in part, not L.

A shrub to 3 or 4 ft high, with yellowish angled branchlets; young twig densely hairy, becoming glabrous in the second year, except at the tip. Leaves oblanceolate, $2\frac{1}{4}$ to 3 in. long, $\frac{5}{8}$ to almost 1 in. wide, subacute at the apex, entire, pale green above, sea-green beneath, silky on both sides, lateral veins in seven to nine pairs; petiole about $\frac{1}{4}$ in. long, hairy. Stipules wanting except sometimes on strong shoots. Catkins with the leaves, erect, on short leafy peduncles; scale obovate, yellowish with a darker tip, hairy. Male catkins about $\frac{3}{4}$ in. long, half as wide; filaments of stamens hairy at the base; anthers purple. Female catkins about 2 in. long. Ovary narrowly ovoid-conic, hairy, very short-stalked; style about one-sixth as long as the ovary, divided at the apex; stigmas spreading, slender, bifid.

Native of the Alps from France to the High Tauern, mainly on the inner ranges above 6,000 ft and usually on acid soils. One of the most ornamental of alpine willows.

S. glaucosericea is really no more than one of the numerous states of the variable S. GLAUCA L., a native of the boreal regions and some high mountains of the northern hemisphere, absent from the British Isles.

S. GRACILISTYLA Miq.

S. *thunbergiana* Anderss.; *S. mutabilis* Hort.

A bush of a spreading habit, probably not more than 6 to 10 ft high; young shoots covered with grey down. Leaves oblong, oval, or narrowly ovate, tapered somewhat abruptly at both ends, 2 to 4 in. long, $\frac{1}{2}$ to $1\frac{1}{4}$ in. wide; indistinctly toothed except towards the base, grey-green above, and at first covered with appressed silky hairs which afterwards fall away except on the midrib; rather glaucous and persistently silky beneath; veins numerous, conspicuous, parallel; stalk $\frac{1}{6}$ to $\frac{1}{4}$ in. long; stipules up to $\frac{1}{3}$ in. long, persisting. Catkins produced on naked shoots in March and April; males grey suffused with red, 1 to $1\frac{1}{2}$ in. long; stamens in pairs, more or less connate, much longer than the scale. *Bot. Mag.*, t. 9122.

SALIX GRACILISTYLA

Native of Japan, Korea and N.E. China; introduced to Europe by Messrs Barbier of Orleans in 1895. It is an interesting willow, related, though not closely, to *S. viminalis* (see p. 253). The specific epithet refers to the long and slender style, not seen in cultivated plants, all of which are male. It is one of the most ornamental of willows, with pretty catkins and handsome many-ribbed leaves, the venation beautifully etched beneath.

cv. 'VARIEGATA'.—Leaves margined with white. A Japanese garden variety, perhaps worthy of introduction (*S. gracilistyla* var. *variegata* Kimura).

S. 'MELANOSTACHYS' ('Kurome', 'Kuro-yanagi').—Catkin-scales blackish red, almost glabrous; anthers brick-red, becoming yellow. Young stems glabrous. Leaves almost glabrous, rich green, thicker than in the cultivated clone of *S. gracilistyla*. (*S. gracilistyla* var. *melanostachys* (Makino) Schneid.; *S. melanostachys* Makino). A striking willow introduced to Europe by Messrs J. Spek of Holland in 1950.

S. 'THE HAGUE' ('Hagensis').—This hybrid was raised by S. G. A. Doorenbos at The Hague, reportedly by crossing *S. gracilistyla* with *S. caprea*. It is a vigorous, spreading shrub with rather thick, velvety stems. Leaves oblong, subacute or acuminate at the apex, rounded to truncate at the base, to about 4 in. long and 1½ in. wide, reticulate and slightly glossy above, whitish and permanently hairy beneath. Stipules ovate, about ⅜ in. long. A female clone, bearing abundant, closely set silky catkins about 1¾ in. long.

The name *S.* × LEUCOPITHECIA was given by Kimura to a male clone cultivated in Japan for its catkins. The parentage of this is similar to that of 'The Hague', namely *S. gracilistyla* × S. BAKKO Kimura, which is the Japanese counterpart of *S. caprea* and very closely allied to it.

S. HASTATA L.   HALBERD-LEAVED WILLOW

A prostrate or low, spreading shrub, sometimes erect and up to 5 ft high; young shoots hairy, purplish the second year. Leaves of hard texture, ovate, oval or obovate, tapered, rounded or sometimes (on vigorous shoots) heart-shaped at the base, always more or less acutely pointed, 1 to 4 in. long, ½ to 2¼ in. wide, ordinarily quite glabrous on both surfaces at maturity, but sometimes densely hairy when young, dull green above, glaucous beneath, veins in seven to ten pairs; stalk ⅛ to ⅜ in. long. Stipules usually present, often large and conspicuous, obliquely heart-shaped. Catkins appearing with or shortly before the leaves; scales lanceolate to obovate, blunt, usually dark at the tip, densely silky-hairy. Male catkins stout, up to 2 in. long, shortly stalked, with a few small silky hairs at the base. Stamens with yellow anthers, filaments glabrous. Female catkins longer-stalked than the male, with larger leaves on the peduncle. Ovary glabrous, shortly stalked (the stalk about one-third as long as the ovary); style up to one-half as long as the ovary.

Native of Eurasia, widely distributed, but mainly confined to the mountains in the southern part of its range, absent from Britain in the wild state; introduced in 1780.

cv. 'WEHRHAHNII'.—A very ornamental willow, bearing a profusion of silvery male catkins in April, beautifully shown up by the fresh-green young leaves. It was discovered around 1930 by Garteninspektor Bonstedt of Geismar, Germany, during a visit to the Engadine, and was named by him after his late friend H. R. Wehrhahn, author of a standard work on herbaceous plants. One of the first examples to be imported into this country was given by Frank Knight, then nursery manager to Messrs Notcutt, to Clarence Elliott, the well-known authority on alpine plants. Planted in 1953, this has attained a height of 6 ft and a spread of 12 ft in the garden of his son, Joe Elliott, at Broadwell, Moreton-in-Marsh. 'Wehrhahnii' received an Award of Merit in 1964, on April

21.   [PLATE 33

S. APODA Trautv.—This Caucasian species is allied to *S. hastata*, but is apparently always a low shrub, and there is the botanical distinction that the ovary is almost sessile, the style more developed, and the stigmas linear and entire. The female catkins are more shortly stalked than is usual in *S. hastata* and become up to 4 in. long in fruit.

The male clone of *S. apoda* introduced by Walter Ingwersen from the Caucasus in the 1930s is perhaps the most ornamental of the dwarf willows and received an Award of Merit when shown from his nursery in 1948. The stout, silky catkins are 1 to 1¾ in. long, produced just before the leaves in late March or April; anthers at first orange, becoming pale yellow; leaves light green, glabrous. It is a ground hugging shrub, best suited to the rock garden, where it will mould itself to the contours and eventually cover a wide area.

S. HERBACEA L.

A tiny shrub (the smallest of all British ones), reaching rarely more than 2 in. above the ground (3 or 4 in. in gardens); stems glabrous, or slightly silky when young, creeping and taking root and often buried in the soil. Leaves usually only two or three at the end of the twig; round, broadly oval or obovate, ¼ to ¾ in. long, finely round-toothed, often notched at the apex and indented at the base, glossy green on both sides and usually glabrous, sometimes slightly silky when young, prominently net-veined; shortly but distinctly stalked. Catkins ¼ to ¾ in. long, appearing in April on short stalks; scales yellow or brown. Stamens two. Ovary glabrous or nearly so, conic.

Native of the mountains of Europe, including the northern British Isles, west through Iceland to N. America, where it occurs in Arctic regions and extends south on the Atlantic side to the mountains of New England and the Adirondacks. In spite of its name it is a true shrub, and makes an interesting tuft for a damp spot in the rock garden.

S. × CERNUA E. F. Linton *S. herbacea* × *S. repens*—A prostrate shrub, the young growths at first densely coated with appressed hairs, becoming glabrous and brown. Leaves oblong or ovate, to about ¾ in. long, dark green and glossy, almost glabrous, above, prominently veined and usually coated with appressed hairs beneath, margins entire or finely toothed. Catkins with the leaves; scales sparsely hairy. Ovaries narrowly flask-shaped, varying from glabrous to densely downy, with a short style. This hybrid occurs occasionally in the mountains of Scotland.

S. × GRAHAMII Borrer ex Baker—A hybrid of *S. herbacea* discovered by Prof. Graham (d. 1845) at Frouvyn in Sutherland, introduced to cultivation by him, and described in 1867 from specimens in the Borrer herbarium. It is a procumbent shrub of which only female forms are known. The leaves are elliptic-oblong to rounded, up to about 1½ in., green and very sparsely hairy on both sides, with faintly toothed slightly wavy margins and a somewhat bent, acuminate tip. Stipules present, often persistent on strong shoots. Catkins ½ to in. long; scales roundish, edged with red. Ovary thinly hairy, the long style and stigma reddish.

cv. 'MOOREI'.—Very like the type of *S.* × *grahamii* in all respects, save that the catkin-scales are narrow-oblong (*S. moorei* F. B. White; *S.* × *Grahami* var. *Moorei* Lond. Cat. ed. 7).

'Moorei' was discovered by Dr David Moore on Muckish Mountain, Donegal shortly before 1870, and introduced by him to the Glasnevin Botanic Garden where it was propagated and further distributed. It is here assumed to be of the same parentage as *S.* × *grahamii*, but what species, crossed with *S. herbacea* could have produced these plants is uncertain and controversial. Five species have been suggested by one authority or another. It is possible that *S.* × *grahami* is a triple hybrid from *S. herbacea*, *S. repens* and *S. aurita* (R. D. Meikle in Stace *op. cit.*, p. 329).

S. × SADLERI Syme *S. herbacea* × *S. lanata*—A dwarf shrub taking after *S herbacea* in foliage, except that the leaves are larger and glaucous beneath Catkins on leafy peduncles, after the leaves, showing the influence of *S. lanata* in their silky scales. It was originally discovered in 1874 at the head of Glen Callater, Aberdeenshire, by John Sadler, who was then assistant to J. H. Balfour the Professor of Botany at Edinburgh, and later became curator of the Botanic Garden there. The same hybrid has been found elsewhere in Scotland, and in Scandinavia.

S. × SIMULATRIX F. B. White *S. herbacea* × *S. arbuscula*—A low, spreading shrub with glossy, reddish brown branches, its leaves broadly ovate to roundish soon glabrous, deep lustrous green above, paler and prominently veined beneath, to about ¾ in. long. It has been found in Argyll and Perthshire and also occurs in Scandinavia. Only female plants are known in Britain; ovaries almost sessile, densely woolly. A similar hybrid between *S. herbacea* and *S. foetida* (*S arbuscula* subsp. *foetida*) has been found in Switzerland and is cultivated in Germany in rock gardens.

S. POLARIS Wahlenb.—Very similar to *S. herbacea* in habit, but distinguished by its leaves being almost invariably entire, smaller, and not quite so rounded. The ovary is very hairy, not glabrous as in *S. herbacea* and the catkin-scales are dark brown or black. Native of Arctic regions. The true species may not be in cultivation.

S. HOOKERIANA Barratt

A shrub to about 6 ft high in cultivation and normally no taller in the wild though exceptionally it attains there the dimensions of a small tree; branches stout; young growths densely hairy and remaining so through the winter. Stipules very small, or lacking, except on strong shoots. Leaves oblong-elliptic oblong-ovate or obovate, acute to obtuse at the apex, cuneate at the base, mostly 1½ to 3 in. long and ¾ to 1½ in. wide, upper surface soon glabrous, glossy and reticulate, underside with an indumentum of variable persistence and density sometimes almost glabrous and glaucous, sometimes, as in the cultivated plant permanently coated with a loose, soft, whitish felt, margins obscurely toothed petiole ⅜ to ¾ in. long. Catkins produced before or with the leaves, sessile or short-stalked; bracts long-hairy, dark brown. Male catkins stout, 1 to 2 in long; stamens two, free, with glabrous filaments. Female catkins 1¾ to 4½ in. long; ovaries glabrous or hairy, short-stalked; style short, but longer than the slightly bilobed stigmas.

Native of the coastal regions of western N. America from Alaska to California. Although introduced to Kew towards the end of the last century it is scarcely known in gardens and deserves to be more widely grown, judging from the plants in the Hillier Arboretum, which make stiffly branched shrubs of picturesque habit, about 6 ft high (1979). They are male and we are told by the Canadian authority Dr George Argus, who kindly confirmed their identity, that they represent the tomentose form of the species. Never occurring far from the sea in its native habitats, *S. hookeriana* might succeed in exposed coastal gardens.

S. IRRORATA Anderss.

A shrub to about 15 ft high, rarely a small tree; young stems glabrous, purplish by winter and coated with a whitish bloom; winter-buds large, roundish, glabrous. Leaves 2½ to 4 in. long, oblong or narrowly lanceolate, acuminate, glabrous, bright green above, glaucous beneath, remotely toothed or entire; petiole ⅛ to ⅜ in. long. Catkins dense, almost sessile, appearing before the leaves, 1 in. or slightly more long; scales dark at the tips, obtuse, densely hairy. Stamens with glabrous filaments and reddish anthers. Ovary glabrous, very shortly stalked; style short, with stout, entire or bifid stigmas.

Native of the south-western USA (Colorado, S.E. Arizona and western New Mexico), common along mountain streams; introduced to Kew from the Arnold Arboretum in 1910 (a male clone). With its bloomy stems it is ornamental in winter, though no more so than *S. daphnoides*, from which it differs in its almost entire, very shortly stalked leaves and smaller catkins. Once established it should be pruned hard each spring. Award of Merit 1967. The specific epithet means 'dewy', in allusion to the pruinose stems.

S. LASIOLEPIS Benth.—Closely allied to *S. irrorata* and described one year earlier. The main difference appears to be that the mature stems are not bloomy (except in some areas where the species overlap) and that they and the leaves are usually downy when young. Its catkins are somewhat longer.

The affinity of these two species is uncertain and controversial. At any rate, *S. irrorata*, despite its bloomy stems, is in no way related to *S. daphnoides*. Even in winter it is distinguished by its appressed, beetle-shaped buds, very different from the diverging buds of *S. daphnoides* and its allies.

S. JAPONICA Thunb.

S. babylonica var. *japonica* (Thunb.) Anderss.

A shrub to about 6 ft high; branches slender, glabrous at maturity. Leaves narrowly to broadly lanceolate, 2 to 3 in. long and up to 1 in. wide, soon glabrous, the margins set with hook-like, finely pointed teeth except at the entire acuminately tapered apex, bright green above, glaucous or grey beneath; petiole up to ¼ in. long. Catkins appearing with the leaves; scales ovate, obtuse, sparsely hairy. The female catkins are slender and remarkable for their length, being up to 5 in. long even in the flowering stage, the male to 3 in. long. Stamens two, free. Ovary stalked, glabrous, with a short style. Both male and female flowers have a single nectary.

Native of Japan. A handsome species of uncertain affinity but certainly not related to *S. babylonica*, of which Andersson, the monographer of the genus, inexplicably considered it to be a 'minor modification'. It may not be in cultivation, but is worthy of introduction, for its distinctive foliage and long female catkins.

For the willow introduced as *S. japonica Lavallei* see under *S. babylonica*. Another willow grown in the last century as *S. japonica*, a female clone, may have been one of the allies of *S. jessoensis* (q.v.).

S. JESSOENSIS Seem.

A tree attaining a height of 80 ft in Japan, and a girth of 12 ft or more; bark shallowly fissured; winter growth-buds oblong, closely appressed to the shoot, the catkin-buds broadly ovoid; young growths densely downy. Leaves lanceolate to narrowly elliptic or oblong, usually finely tapered at the apex, to about 4 in. long and ½ in. wide, silky on both sides when young, almost glabrous when mature, dark green above, paler, often bluish beneath, finely toothed; petioles to about $\frac{3}{16}$ in. long. Stipules small, toothed, ovate or rhombic, slightly oblique. Catkins appearing with the leaves on short leafy stalks; scales truncate, hairy at the base. Male catkins 1 in. or slightly more long. Stamens two, with rather stout filaments. Female catkins to about 1¾ in. long, their scales persistent in the fruiting stage. Ovary silky, sessile; style very short; stigmas entire.

Native of Japan in Hokkaido and N. Honshu, known there as 'Shiro-yanagi' (white willow). It is one of a group of closely related species in N.E. Asia which have been regarded as the counterparts in that region of the western *S. fragilis* and *S. alba*, but their affinity is rather with *S. matsudana*.

S. KOREENSIS Anderss. (1868) ?*S. pierotii* Miq. (1867)—Allied to *S. jessoensis*, but the young growths and leaves more glabrous, the petioles longer, and the style longer, with bifid stigmas. Native of continental N.E. Asia and S. Japan.

S. pierotii Miq. was described from two specimens collected in the Japanese island of Kyushu, one by Pierot and the other by Siebold. The female specimen appears to have been *S. koreensis*, but the flowers of the male specimen were described, with a query, as having a single stamen and, unless defective, must have belonged to a different species.

S. LANATA L.   WOOLLY WILLOW   [PLATE 35

A low, sturdy bush, 2 to 4 ft high; branchlets stout, furnished when young with thick, soft, grey wool. Leaves silvery on both sides, with a rich coat of silky hairs, especially at first, oval to roundish or obovate, mostly abruptly pointed at the apex and tapered at the base, but sometimes rounded or heart-shaped, 1 to 2½ in. long, ¾ to 1½ in. wide, nearly always entire; stalk ⅛ to ¼ in. long; stipules up to ⅓ in. long, ovate, entire, prominently veined. Catkins produced in May, often solitary at the end of the previous season's growth, of a bright golden colour; males 1 to 2 in. long, ½ in. thick; females up to 3 in. long at the seeding stage. Ovaries glabrous, long-styled, almost sessile.

Native of high latitudes in Europe and Asia, extending south to Scotland, where it is found only in a few localities in Perthshire, Aberdeenshire and Angus, above 1,500 ft and ascending to 3,000 ft. It is one of the handsomest of dwarf willows, especially in spring, when the silver young foliage and golden catkins are in admirable contrast. It bears some resemblance to *S. lapponum*, but in that species the leaves are narrower, stipules are wanting or very small, the catkins are silvery and the ovary is hairy.

S. × BALFOURII E. F. Linton   *S. lanata* × *S. caprea*—This hybrid was described by Linton in 1913 from a foliage-specimen collected in 1837 by J. H. Balfour, Professor of Botany at Edinburgh University, in Glen Isla, Angus. But the male clone by which *S.* × *balfourii* is represented in cultivation almost certainly descends from the plant raised by the Rev. E. F. Linton himself by artificial cross-pollination in his garden at Edmondsham, near Bournemouth, and distributed as No. 88 in the Linton 'Set of British Willows'. This makes a shrub up to 10 ft high, perhaps taller. Leaves 1 to 3 in. long, broad-elliptic, roundish and often abruptly acuminate at the apex, grey with matted hairs beneath. Stipules conspicuous, at least on strong shoots, round or kidney-shaped. Catkins dense and very silky up to 2 in. long and ¾ to 1 in. wide, appearing in April just as the leaf-buds are breaking; stamens up to ½ in. long, with small anthers. The Linton clone is one of the finest of the willows grown for their early catkins, certainly a better garden plant than any male goat willow, and later flowering than *S. aegyptiaca*.

The hybrid between *S. caprea* and *S. lanata* also occurs in Norway and Sweden.

S. 'STUARTII'.—A probable hybrid between *S. lanata* and *S. lapponum*, nearer to the former but showing the influence of *S. lapponum* in the presence of hairs on the ovaries and their pedicels, the less copious, silvery hairs of the leaves (*S. lanata* × *S. lapponum* J. Fraser, *Gard. Chron.*, Vol. 85 (1929), p. 208; catkins, and the less prominent secondary veins on the undersides of the *S.* × *stuartii* Druce).

'Stuartii' came from the garden of Dr Charles Stuart of Chirnside, Berwick-shire (*d.* 1902). He was a plant breeder as well as a field naturalist, so it is im-possible to say whether the hybrid was raised by him or, like his discovery *Erica* × *stuartii*, collected in the wild. Some years after his death stock was acquired by the Craven Nursery Company, which had been set up by Reginald Farrer and his associates. They never propagated it commercially, but when the nursery was wound up after Farrer's death (1921), the stock was acquired by H. E. Mason of Alderley, who sent cuttings to some nurseries and private gardens (*Gard. Chron.*, Vol. 85 (1929), p. 159).

'Stuartii' is usually listed as a variety of *S. lanata* and could be regarded as a fine form of that species. So far as is known all the plants in general cultivation are female, though the suspicion has been voiced that more than one clone was distributed. The first private gardener to grow 'Stuartii' was A. T. Johnson, who was given cuttings by one of Farrer's associates in 1914. His plant was 3 ft high and 8 ft across in 1937 (*The Woodland Garden*, p. 117).

S. LANATA × S. REPENS—This hybrid was raised and distributed by the Rev. E. F. Linton. A low bush to 2 or 3 ft high; branchlets and buds dark-coloured, somewhat woolly. Leaves to about 2 in. long, obovate or broadly elliptic-oblong, thinly silky beneath. Catkins (female) about 1 in. long, lengthening in fruit, borne in May, with a few silky leaves at the base; ovaries narrow, thinly hairy, with long styles. This was Linton No. 99. In No. 100, also female, the leaves were narrower, more glabrous, and the ovaries were glabrous.

S. LAPPONUM L.

A shrub of spreading, much-branched habit, 2 to 4 ft high; young shoots dark brown, more or less downy. Leaves oval or somewhat obovate, occasionally lanceolate, tapered at both ends or sometimes rounded at the base, toothed only rarely, 1 to 3 in. long, $\frac{1}{3}$ to $1\frac{1}{4}$ in. wide, cottony above, becoming nearly or quite glabrous with age, lower surface permanently woolly beneath, silvery white at first, ultimately grey; stalk $\frac{1}{8}$ to $\frac{1}{3}$ in. long; stipules inconspicuous or absent. Catkins produced on the naked shoots in April and May, very silky; males about 1 in. long, stalkless; females longer, shortly stalked; ovaries hairy.

Native of Siberia, W. European Russia, Scandinavia, Scotland, the Lake District (one station on Helvellyn) and also occurring in some of the mountains of southern, central and southeast Europe, though not in the Alps. In Scotland it occurs above 1,000 ft, from Dumfriesshire and Angus north to Sutherland. For the distinctions between it and *S. lanata*, a much rarer species in Scotland, see under the latter. *S. lapponum* is very variable in the width of its leaves and in their indumentum. In some Scottish plants the leaves are almost linear; *S. stuartiana* Sm. (not to be confused with 'Stuartii', for which see under *S. lanata*) is a form with small, very woolly leaves, described from a specimen collected in the Breadalbane mountains by the Rev. John Stuart of Luss towards the end of the 18th century.

S. HELVETICA Vill. *S. lapponum* var. *helvetica* (Vill.) Anderss.—A native of the Alps, closely related to *S. lapponum*. The most obvious difference is in the colouring of the foliage: in *S. lapponum* the leaves are hairy, and more or less of the same colour, on both sides, while in *S. helvetica* they are green and lustrous above but contrastingly white-hairy beneath. Further differences are: catkins distinctly stalked, the female elongating considerably in fruit, and the less hairy ovary with a shorter style (up to half as long as the ovary). *S. helvetica* ranges from the French Alps and Maritime Alps to W. Austria and the Tatra Mountains, and is common in the Central Alps of Switzerland.

S. LASIANDRA Benth.

S. speciosa Nutt., not Host; *S. lancifolia* Anderss.; *S. lucida* var. *macrophylla* Anderss.

This fine willow belongs to the same group as *S. pentandra*, our native bay willow, and *S. lucida*, and is, according to Sargent, often a tree 60 ft high in western N. America, where it is native. It has the same dark green, shining leaves as its allies, the glandular teeth, the conspicuous stipules on strong shoots, the glandular leaf-stalks, yellow midrib, and the five or more stamens; but the leaf is, at first at any rate, pale or glaucous beneath and downy. In flower it is also distinguished by the scale, at the base of which the group of stamens or the ovary is attached, being toothed at the apex; it is entire in the other two. The leaves are 4 to 5 (sometimes 6 to 7) in. long, ½ to 1 (sometimes 1½) in. wide.

S. LINDLEYANA Anderss.

A low shrub with creeping stems, which are often buried in the upper surface of the soil, and with short, erect, glabrous branchlets. Leaves often very densely set, glabrous, distinctly stalked, mostly less than ½ in. long and often as small as ¼ in. long, $\frac{1}{16}$ in. wide, remotely toothed, the midrib usually deeply channelled above. Catkins very small (to about ½ in. long) ovoid to broadly elliptic, terminal on leafy branchlets; scales brown, uniformly covered, obtuse, glabrous; rachis sparsely hairy. Stamens two, filaments glabrous. Ovary glabrous, with a distinct, sometimes divided style.

Native of the Himalaya at high altitudes. It is of interest as the type of a section (*Lindleyanae*) which is closely allied to the *Retusae* of Europe and N. America. Some plants introduced in the 1970s from Nepal, and grown as *S. hylematica* (see below), probably belong to *S. lindleyana*, but flowering specimens have not been seen.

S. FURCATA Anderss. *S. fruticulosa* Anderss., in part (1860), not Lacroix (1859); *S. hylematica* Schneid., *nom. superfl.*—This Himalayan species is not related to *S. lindleyana* but the two are in some forms superficially similar. The best distinction between them appears to lie in the catkins, which in *S. lindleyana*, as in most other members of the subgenus *Chamaetia*, terminate leafy branchlets which are as long as, or not much shorter than, the sterile shoots. In *S. furcata* the catkins are shortly stalked or even sessile; usually they are longer (to 1 in. long), clustered; the rachis is commonly densely hairy. Normally *S. furcata* is a small, erect, rather stoutly branched shrub, but sometimes decumbent. Schneider placed this species in his new section *Denticulatae* (not treated here), but its affinity is uncertain. Indeed very little is known about the Himalayan species of *Salix*.

S. LUCIDA Muhl.

Usually a shrub, sometimes a tree up to 25 ft in height; young shoots glabrous, glossy; flowering twigs downy. Leaves lance-shaped, broadly wedge-shaped or rounded at the base, with long, slender, sometimes tail-like points; finely glandular-toothed, 3 to 5 in. long, $\frac{3}{4}$ to $1\frac{1}{4}$ in. wide, dark glossy green above, paler beneath; stalk $\frac{1}{4}$ to $\frac{1}{2}$ in. long, with several glands near the blade, downy in the groove on the upper side, and partially so up the midrib. Stipules large, roundish heart-shaped, glandular-toothed, often persistent. Catkins produced very abundantly on short, leafy twigs in April and May; males erect, $1\frac{1}{2}$ to $2\frac{1}{2}$ in. long, stamens five (sometimes three or four); females more slender, 2 to 3 in. long.

Native of N. America from Newfoundland to the eastern base of the Rocky Mountains. It is a handsome-leaved willow, and the only other with which it is likely to be confused is *S. pentandra*—its Old World representative. *S. lucida* differs in having a long drawn-out point to the narrower leaf, and the net-veining is not so prominent as in *S. pentandra*. (See also *S. lasiandra*.)

S. MAGNIFICA Hemsl.

A small tree or shrub, from 6 to 20 ft high, quite devoid of down in all its parts, the young shoots and conical buds purple, the former changing to red. Leaves oval or slightly obovate, entire, rounded or slightly heart-shaped at the base, the apex terminated by a short, abrupt, bluntish tip, 4 to 8 in. long, 3 to $5\frac{1}{4}$ in. wide; dull grey-green (with a bloom) above, pale and slightly glaucous beneath; stalk $\frac{1}{2}$ to $1\frac{1}{2}$ in. long, purplish. Male catkins 4 to 7 in. long; stamens two, four times as long as the scale. Female catkins longer, sometimes as much as 11 in.

Native of W. China; discovered in 1903 by Wilson in the mountains of Szechwan, at 9,000 ft altitude. It was not introduced at the time, and Wilson saw only two bushes then. In 1909 he found it again, and in abundance, 20 ft high. He sent cuttings to the Arnold Arboretum, where I saw it in 1910, and obtained it for Kew. This, I believe, was its first introduction to Europe. It is the most remarkable of all willows, and its leaves, in shape and colour, are more like those of *Arbutus menziesii* than a typical willow. Leaves have been borne on cultivated plants that measure 10 in. long, by $5\frac{1}{4}$ in. wide; the stalk 2 in. long. Wilson informed me that in a wild state the shoots change to red the first winter, and remain that colour for several years; also that the leaves die off a golden yellow. It has proved to be quite hardy.

S. MATSUDANA Koidz.   PEKING WILLOW

A deciduous tree 40 to 50 ft high; young shoots at first minutely downy, slender, yellowish, changing later to brownish grey and becoming glabrous. Leaves linear-lanceolate, slender-pointed, tapered at the base to a stalk $\frac{1}{12}$ to $\frac{1}{4}$ in. long, finely and regularly toothed (except those at the base of the shoot which are entire), 2 to 4 in. long, $\frac{1}{3}$ to $\frac{3}{5}$ in. wide, bright green above, glaucous beneath and soon quite glabrous. Female flowers in cylindrical spikes about 1 in. long with a few small entire leaves at the base, main-stalk downy. The flower is stalkless in the axil of an ovate bract two-thirds as long as the ovary which is glabrous or hairy, $\frac{1}{8}$ in. long, topped by a dark stigma. Male flowers on short cylindrical catkins about $\frac{3}{5}$ in. long, main-stalk villose; stamens two.

Native of the more arid parts of N. China (Inner Mongolia, Kansu, Shensi, Shansi, etc.) and widely cultivated in the north, growing excellently there and needing no water supply beyond the scanty summer rainfall (F. N. Meyer), but attaining its greatest dimensions at the foot of mountains and in the oases of Inner Mongolia. Its timber is, or was, put to many uses; the packing cases in which porcelain was sent to Europe were made from it, and it served for the construction of disposable boats, which were used to float the cotton crop down the rivers and then destroyed.

S. matsudana was introduced to Kew shortly after 1905 from the Arnold Arboretum, which had received cuttings in that year from a female tree, sent by J. G. Jack from either Korea or from near Peking. In 1913 another plant was received at Kew from the same source, also female, raised from cuttings sent by William Purdom in 1911 when collecting in N. China.

S. matsudana is very closely allied to *S. babylonica*, in which it has recently been included by the Russian authority A. K. Skvortsov—or rather reincluded, for *S. matsudana* is really part of *S. babylonica* as interpreted by earlier botanists, and was first separated from it by Koidzumi in 1915. The only difference between *S. matsudana* and typical *S. babylonica* that might be regarded as of specific value is that, according to Schneider, the female flowers of the former have two nectaries, against one (in the posterior position) in typical *S. babylonica*. However, this observation is not based on examination of a wide range of specimens. In *S. fragilis* the female flowers commonly have two nectaries, but the anterior nectary (between the ovary and the scale) is sometimes very small or altogether lacking, and it is very likely that *S. babylonica-matsudana* varies in the same way.

cv. 'PENDULA'.—Branches pendulous. Common in N. China as a cultivated tree. According to J. Hers, the Chinese name for it, *tao-tsai-liu*, means 'upside-down willow', from the belief that it can be raised by inserting cuttings the wrong way up. A pendulous male form of *S. matsudana* was sent to the Arnold Arboretum by F. N. Meyer in 1908, but the plants cultivated in this country are believed to derive from an introduction by J. Hers early in the 1920s, in the same consignment as 'Tortuosa' (see below). They are female, with silky ovaries. Young bark green.

cv. 'TORTUOSA' DRAGON'S CLAW WILLOW—Twigs and branches much contorted. Cultivated in N. China and introduced to France, probably to the firm of Vilmorin, by the Belgian engineer and plant collector Joseph Hers, early in the 1920s. It makes an interesting tree when young, but becomes less distinct with age.   [PLATE 34

cv. 'UMBRACULIFERA'.—Of bushy, rounded habit, without a central leader. Cultivated in N. China, where the common name for it is said to mean 'bread willow', from its loaf-like shape. It was introduced to the USA by F. N. Meyer in 1906; at the Arnold Arboretum, six years planted, it had made a specimen 20 ft high and 30 ft wide (*Journ. Arn. Arb.*, Vol. 6, p. 205).

S. × MEYERIANA Rostk.
S. cuspidata Schultz

This handsome willow is a hybrid between *S. pentandra* and *S. fragilis*, and has been found wild in Britain, as well as on the continent, in places inhabited by the parent species. In general appearance it very much resembles *S. pentandra*. The following distinctions, however, exist: the leaf is thinner, more slender, pointed, and sometimes glaucous beneath, and the tree is usually of larger size; the male flowers have fewer (three or four) stamens, and the scale is more hairy; the female catkins are more slender and more tapering, and the seed-vessels longer-stalked and more cylindrical. It is worth growing for its vigorous habit and its fine glossy foliage. Its leaves are oval inclined to ovate, or obovate, 1½ to 4½ in. long and ½ to 1½ in. wide, quite glabrous; the marginal teeth fine, regular, glandular.

The willow that received an A.M. in 1931 as ?*S.* × *meyeriana* was not this hybrid but most probably a form of *S. daphnoides*.

S. × EHRHARTIANA Sm.   *S. pentandra* × *S. alba*—As might be expected from the parentage, this hybrid is similar to *S.* × *meyeriana*, but the influence of *S. alba* shows in the hairiness of the young stems and leaves, and the almost sessile ovaries. It occurs very rarely in Britain and is by no means common on the continent. The British plants are all male, and probably planted.

S. × MOLLISSIMA Ehrh.

A group of hybrids between *S. triandra* and *S. viminalis*. The two parents are both important basket willows, with numerous commercial varieties, both male and female, and must over the centuries have crossed spontaneously on numerous occasions, giving rise to seedlings which have in turn proved useful for basket-work and been extensively propagated. The type was a female plant which seems to have been near to *S. viminalis*. It is not known in this country, where the group is represented by the following:

cv. 'LANCEOLATA'.—A large shrub with olive-brown branchlets; bark of the older stems flaking, as in *S. triandra*. Leaves 3 to 5 in. long, finely tapered at the apex, densely serrated, silky at first, soon glabrous, bright green. Stipules usually persistent. A female clone: catkins dense, narrowly cylindric, erect, up to 2½ in. long, appearing as the leaves unfold on short, leafy peduncles; scales strap-shaped, yellowish, thinly hairy. Ovary glabrous, more shortly stalked than in *S. triandra*, with a distinct style. This willow, fairly widely distributed in Britain, was described by Sir James Smith (as *S. lanceolata*) and belongs botanically to var. (nothomorph) *undulata* (Ehrh.) Wimmer (syn. *S. undulata* Ehrh.; *S. undulata* var. *lanceolata* (Sm.) Anderss.). It was at one time thought to be a hybrid between *S. alba* and *S. triandra*.

var. HIPPOPHAEIFOLIA (Thuill.) *S. hippophaefolia* Thuill.; *S. multiformis* var. *hippophaifolia* (Thuill.) Anderss.—Leaves relatively narrower than in 'Lanceolata', linear-lanceolate, also differing in being entire or with sparse glandular serrations. Stipules soon falling. Both male and female plants are known in Britain. Male catkins about 1 in. long, with two or three stamens. Female catkins with densely hairy scales; ovaries appressed-hairy at first and never completely glabrous (completely glabrous in 'Lanceolata').

S. MOUPINENSIS Franch.

A deciduous shrub or small tree which Wilson found 10 to 20 ft high; young shoots glabrous, becoming yellowish or reddish brown; winter buds slender, ¼ to ½ in. long and of a similar colour. Leaves oval or obovate, broadly tapered or almost rounded at the base, abruptly pointed at the apex, finely and regularly toothed, each tooth tipped with a gland, 2 to 5 in. long, 1 to 2¼ in. wide, upper surface bright green, glabrous, with a yellowish midrib; lower surface yellowish green, wrinkled with veins, usually more or less silky on the midrib if only when young; stalk ¼ to ⅝ in. long, glabrous or silky. Catkins very slender, the female up to 5 in. long; styles two, bifid; males shorter. Catkin-scales sparsely hairy or glabrous.

Native of W. Szechwan, China; discovered by the Abbé David in 1869, later by Henry; introduced by Wilson to the Arnold Arboretum in 1910 and thence to Kew in 1912. This willow is closely related to *S. fargesii* (q.v. for the points of difference). It is hardy, and handsome as willows go, but is not so outstanding as Farges' willow, nor so common in gardens.

S. MYRSINITES L.   WHORTLE WILLOW

A dwarf shrub 1 to 1½ ft high, of bushy habit, sometimes procumbent; young shoots slender, silky-hairy at first. Leaves ovate to elliptic (narrowly to broadly so) finely toothed, tapered at both ends, 1 to 2 in. long, ⅜ to 1 in. wide, bright green and, at least when dried, conspicuously net-veined on both sides, silky only when young; stalk ⅛ in. or less long. Stipules usually well-developed on strong shoots, about ¼ in. long. Catkins erect, borne on short leafy shoots in May; scales brownish purple, hairy. Male catkins cylindrical, up to 1¼ in. long; anthers usually reddish at first. Female catkins up to 2 in. long. Ovary downy, with a long style, the stigmas divided into narrow-lobes, usually purplish.

Native of northern Eurasia, from Scotland and the Orkneys eastward to Kamtschatka. It is suitable for the rock garden, but not so ornamental a species as *S. alpina* (see below). The withered leaves usually persist on the branchlets— a peculiarity that helps to distinguish it from *S. alpina* and *S. breviserrata* (see below). *S. arbuscula*, with which *S. myrsinites* might be confused, has smaller leaves, pale green and often hairy beneath, small, fugacious stipules, and yellow, often red-tinged anthers.

S. ALPINA Scop. *S. fusca* Jacq.; *S. jacquinii* Host; *S. jacquiniana* Willd.; *S. myrsinites* var. *jacquiniana* (Willd.) Koch—Closely related to *S. myrsinites* but always of procumbent habit and with entire, deciduous leaves. Native of the eastern Alps and the Carpathians. It is not unlike *S. retusa*, with which it occurs in the wild, but in that species the leaves are not conspicuously net-veined above, and the catkin-scales are yellow (in *S. alpina* they are brownish purple, as in *S. myrsinites*). A male plant of *S. alpina* received an Award of Merit in 1956.

S. BREVISERRATA Flod. *S. myrsinites* of many authors, not L. *sens. strict.*; *S. arbutifolia* Willd., not Pall.—Very near to *S. myrsinites*, from which it was separated by the Swedish botanist Floderus in 1940. Withered leaves deciduous. Catkins stouter, about half as wide as long. Native of the Alps.

S. MYRTILLOIDES L.

A shrub from a few inches to 2 ft or so high, with creeping underground stems and erect branches; young shoots and leaves soon glabrous. Leaves obovate, oblong-elliptic or sometimes ovate, obtuse and abruptly acuminate at the apex, $\frac{3}{8}$ to $1\frac{1}{2}$ in. long, $\frac{1}{6}$ to $\frac{7}{8}$ in. wide, entire and more or less decurved at the margin, dark dull green above, blue-green or purplish and prominently veined beneath. Catkins appearing in April or May. Male catkins $\frac{1}{2}$ to $\frac{3}{4}$ in. long, narrowly cylindrical, rather sparsely flowered, on leafy peduncles about $\frac{3}{8}$ in. long; anthers at first reddish, yellow when mature. Female catkins lax, on leafy peduncles about 1 in. long. Ovary glabrous, long-stalked; style very short, with short, purplish stigmas.

Native of N. Eurasia from E. Scandinavia to the Pacific; also of central Europe, where it occurs here and there on heaths and in bogs, nowhere common. The true species is rare in gardens.

S. × FINNMARCHICA Willd. *S. myrtilloides* × *S. repens*—A wide-spreading, vigorous shrub with short erect branchlets. The influence of *S. repens* shows in such characters as the longer duration of the silky hairs on the stems and leaves and the more hairy catkin-scales. One garden clone, originally distributed as *S. myrtilloides*, obviously belongs here and has been re-named accordingly. The branchlets are short, with oval leaves $\frac{3}{8}$ to $\frac{3}{4}$ in. long, medium green above, paler and silky beneath. There is, however, another clone in commerce as *S. myrtilloides* which is certainly near to that species, but possibly a form of *S. × finnmarchica*. The branchlets are taller than in the other clone, the leaves bluish and at first quite densely silky beneath, with occasional glandular teeth on the margins. Both make excellent ground-covers.

S. NAKAMURANA Koidz.

This species is represented in cultivation by:

var. YEZOALPINA (Koidz.) Kimura *S. yezoalpina* Koidz.; *S. cyclophylla* Seem., not Rydb.—A shrub with prostrate, rooting stems, glabrous and dark purplish brown when mature. Leaves leathery, broadly obovate, broadly oblong-elliptic or almost orbicular, to about 2 in. long, rounded at the apex, broad-cuneate at the base, almost entire, soon glabrous, glossy and reticulate above, the venation slightly raised beneath; petioles to about ⅜ in. long. Catkins many-flowered, erect, borne on leafy laterals, the males about 1 in. long, the females somewhat longer; scales dark, very hairy. Stamens two. Ovary glabrous, almost sessile; style about one-third as long as the ovary; stigmas slender.

Native of Japan in the mountains of Hokkaido. Both sexes are in cultivation, the female plants more vigorous and making wide mats.

S. nakamurana is closely allied to *S. arctica*. The typical state of the species occurs in the northern part of the main island of Japan.

S. NIGRA Marsh.
S. falcata Pursh; *S. nigra* f. *falcata* (Pursh) Rehd.

An elegant tree attaining in the wild, under optimum conditions, a height of 100 ft or more, but more commonly a medium-sized tree, often with several trunks; bark dark, deeply furrowed; branchlets yellowish at first, becoming reddish brown by autumn, soon glabrous; winter-buds small, with free inner margins. Leaves lanceolate or linear-lanceolate, sometimes falcate, tapered or rounded at the base, narrowing gradually to a long fine point, finely and regularly toothed, 3 to 5 in. long, ¼ to ¾ in. wide, palish green, and almost or quite glabrous on both sides except on the midrib; stalk ⅛ to ¼ in. long, downy. Stipules often large, semi-heartshaped and persistent. Catkins produced on short leafy shoots in April, 1 to 3 in. long, slender. Stamens three to five, the filaments hairy in the lower part. Ovary glabrous, distinctly stalked; style very short, with small bilobed stigmas.

Native of eastern and central N. America, extending into N.E. Mexico; introduced in 1811. It is the largest of the American willows and is lumbered in the Mississippi delta, where heights of up to 140 ft have been measured and diameters of 7 ft. But more frequently it is rather a huge bush than a tree. It is not so elegant a tree in this country as in the United States, although quite hardy. It has rather the aspect of a small, densely branched *S. alba*.

S. GOODDINGII Ball—Closely allied to *S. nigra*, but the branchlets remaining yellow through the winter, somewhat broader leaves, and often hairy ovaries. Native of California, the southwestern USA and northern Mexico. On pruned plants at Kew the young bark is olive-green.

S. NIGRICANS Sm.
S. phylicifolia var. β L.; *S. myrsinifolia* Salisb.*; *S. andersoniana* Sm.; *S. phylicifolia* var. *nigricans* (Sm.) F. B. White

A bushy shrub, 10 to 12 ft high, occasionally taller; young shoots and buds more or less downy. Leaves extremely variable in outline (roundish, elliptic,

* Considered by some authorities to be the correct name for the species; cf. Schneider, *Pl. Wils.*, Vol. III (1916), p. 123, footnote.

ovate, obovate or oblanceolate), shortly acute or acuminate at the apex, rounded or tapered at the base, toothed, 1½ to 4 in. long, ½ to 2 in. wide, upper surface dull, dark green at maturity, underside covered with a thin waxy bloom which disappears towards the apex of the blade, persistently downy, at least on the veins; stalk ¼ to ¾ in. long. Catkins appearing with the leaves, shortly stalked; scales hairy, dark at the tip. Male catkins ¾ to 1¼ in. long, ⅜ to ½ in. wide; stamens about twice as long as the scales, hairy towards the base. Female catkins up to 2 or 3 in. long in fruit; ovary distinctly stalked, usually glabrous, style about one quarter as long as the ovary or slightly longer, the stigmas divided, almost as long as the style.

Native of northern and central Europe, Siberia, etc., occurring in Britain from Yorkshire and Lancashire northwards, and in northern Ireland. It has been represented in gardens at different times by an extraordinary number of forms, varying in the shape and size of their leaves. Many of these are figured by Forbes in the *Salictum Woburnense*, but have little interest. The species is indeed one of the dullest and most uninteresting of hardy shrubs, and is not worth a place in the garden proper. It is seen in the seedling state oftener than most willows are. Some botanists have considered that *S. nigricans* and *S. phylicifolia* are not specifically distinct from each other. However, *S. nigricans* can usually be distinguished by its thinner, larger, duller green, more downy leaves, which, at least the younger ones, turn black on drying, and by the waxy bloom on the lower part of the underside.

S. × TETRAPLA Walker ex Sm.   *S. nigricans × S. phylicifolia*—A rather common natural hybrid between these two related species, and combining their characters in so many different ways that definition is impossible. Many of its forms have been described as species.

S. MIELICHOFERI Sauter   *S. glabra* var. *mielichoferi* (Sauter) Anderss.— Allied to *S. nigricans*, but more bushy, with thicker branchlets swollen at the nodes. Leaves green beneath, not blackening when dry. It could be confused with *S. glabra*, but in that species the leaves are very lustrous above and the undersides are coated throughout with a waxy bloom.

S. PENTANDRA L.   BAY WILLOW

A tree 29 to 60 ft high in gardens, often a shrub in the wild; twigs shining, brownish green, glabrous; buds yellow. Leaves aromatic, ovate to oval, rounded or slightly heart-shaped at the base, rather abruptly narrowed at the apex to a slender point, finely glandular toothed, 1½ to 4½ in. long, ¾ to 2 in. wide, glabrous, dark polished green above, dull and paler beneath; midrib yellow; stalk ¼ to ⅜ in. long, glandular near the blade. Male catkins cylindrical, about 1½ in. long; female catkins rather longer, both produced on leafy shoots in late May. Stamens five or more; seed-vessels glabrous, slightly stalked.

Native of much of Europe, except in the far north and the Mediterranean region, but in the British Isles not found wild south of Derbyshire, Yorkshire, N. Wales and N. Ireland; also widely distributed in temperate Asia. One of the

SALIX PENTANDRA

handsomest of all willows in the brilliant green of its large, broad leaves, resembling those of a bay laurel. In high latitudes it is a shrub, but in moist good soil it becomes a good-sized tree. There was one at Kew 50 ft high and 7 ft 8 in. in girth but it was destroyed by the great gale of 28 March 1916. There is a tree of about the same height at Wakehurst Place, Sussex (1978), and another, measuring 52 × 5½ ft (1975) in the University Parks, Oxford.

S. PETIOLARIS Sm.

S. gracilis Anderss.; *S. petiolaris* var. *gracilis* (Anderss.) Anderss.

A shrub of 6 ft or more high, with slender twigs which are slightly silky when young, soon becoming glabrous, and, later on, deep purple. Leaves narrowly lanceolate, tapered at both ends, finely and regularly toothed, 1½ to 4 in. long, ¼ to ⅜ in. wide, sometimes broader on strong shoots, silky only when quite young, soon glabrous, bluish beneath; petiole ¼ to ⅜ in. long. Stipules, when present, soon deciduous. Catkins appearing in spring on short leafy peduncles, at first under 1 in. long in both sexes, the female elongating in fruit, truncate at the apex; scales narrow, acute or truncate, brownish or yellow, sparsely hairy. Ovary conoid, stalked, long-tapered at the apex; stigmas sessile.

Native of N. America from Quebec to Alberta, south to New York, Wisconsin and Minnesota. A valuable basket-willow, cultivated as such in some parts of Europe. It was described by Sir James Smith in 1802 from a female plant in the Crowe collection, received from a Mr Dickson (who had forgotten its origin), and was for a short while believed to be a British native. Smith's description of the foliage must have been made from a strong shoot, since an authentic specimen, given by him to William Borrer of Henfield, has narrower leaves than those originally described. There seems to be no ground for following those authorities who have rejected the name *S. petiolaris* in favour of the later *S. gracilis*.

The species is variable in foliage; there are forms with very elegant short and narrow leaves.

S. PHYLICIFOLIA L.   TEA-LEAVED WILLOW

A bushy shrub, ¾ to 10 ft high; young shoots glabrous or slightly downy, shining, yellowish or brown. Leaves orbicular, oval, ovate, or obovate; slightly toothed or almost entire, ¾ to 3 in. long, ½ to 2 in. wide, shining green above, and either green or glaucous beneath, sometimes downy, sometimes glabrous; stalk ⅙ to ½ in. long. Catkins and flowers not much differing from those of *S. nigricans*, except that the ovaries are mostly silky or downy.

Native of N. Europe, including Britain (where its distribution is similar to that of *S. nigricans*), and of W. Siberia. Its affinity with *S. nigricans* has been mentioned under that species, but it is a brighter-looking, neater shrub, distinguished by the greater glossiness and smoothness of the young parts.

S. BICOLOR Willd.   ? *S. schraderiana* Willd., in part.—This very close ally of *S. phylicifolia* has a scattered distribution in southern and central Europe (absent from the Alps). According to the Swedish authority Floderus, it differs, among other characters, in the shorter and usually yellowish or orange buds; the leaves silky-hairy at first; and the more numerous and shorter catkins. But it has been questioned whether the differential characters are constant.

S. HEGETSCHWEILERI Heer   *S. rhaetica* Kern. ex Anderss.; *S. phylicifolia* var. *rhaetica* (Kern.) Anderss.; *S. phylicifolia* var. *hegetschweileri* (Heer) Anderss.; *S. bicolor* subsp. *rhaetica* (Kern.) Floderus—Often taller growing than *S. phylicifolia*, and said to attain 15 ft or more in cultivation. Leaves almost entire, mostly elliptic, acute or acuminate, about 1 in. wide. Perhaps not cultivated in Britain, but said to be a handsome species.

Eastwards *S. phylicifolia* gives way to S. PULCHRA Cham.; this ranges through Siberia into N. America, where the complex is also represented by S. PLANIFOLIA Pursh.

S. PURPUREA L.   PURPLE OSIER

S. helix of some authors, ? not L.

A shrub with thin, graceful branches forming a loose-habited, spreading bush, 10 to 18 ft high under cultivation, rarely a small tree; young shoots glabrous, glossy, usually purplish where exposed to the sun, but often yellowish. Leaves linear, or narrowly oblong, mostly broadening somewhat above the middle, pointed at the apex, rounded or abruptly tapered at the base, minutely toothed towards the apex, glabrous except when quite young, dark glossy green above, more or less blue or glaucous beneath; 1½ to 3 in. long, ⅛ to ⅓ in. wide; stalk about ¼ in. long. Catkins produced on the naked shoots in April, ½ to 1 in. long, slender; scales hairy, dark in the upper part. Stamens solitary, but with two anthers. Ovaries sessile, hairy; style very short.

Native of much of Europe (except Scandinavia), and of North Africa, ranging east through temperate Asia to N. China; in Britain it is fairly widespread in wet places, up to 1,500 ft, but is not genuinely wild in some localities where it occurs, and is rare or absent in some areas. It is a variable species, and is remarkable in having many of its leaves opposite as well as alternate. The bark is as bitter as quinine, and very rich in salicine. The twigs are very supple and tough, and much used in the manufacture of fine basketwork. The osiers known as 'Red-bud', 'Dicks', 'Kecks', and 'Welch' belong to this species. As a garden shrub it is worth growing for the sake of its loose, elegant growth and the vivid blue-white of the under-surface of the leaves. It thrives in dryish ground better than most willows.

f. GRACILIS (Gren. & Godr.) Schneid. *S. purpurea* var. *gracilis* Gren. & Godr.—This form is typified by Sir James Smith's description and figure of plants with slender catkins and of spreading decumbent habit that grew in the meadows opposite King Street, Norwich, and were considered by him to represent typical *S. purpurea* (*Engl. Bot.*, Vol. 20 (1800), t. 1388). Such plants are indeed part of the normal variation of *S. purpurea* and do not need a distinguishing name. Schneider considered that the horticultural clone called *S. purpurea nana* was referable to f. *gracilis*; see 'Nana' below.

var. LAMBERTIANA (Sm.) Koch   *S. lambertiana* Sm.; *S. purpurea* subsp. *lambertiana* (Sm.) A. Neumann ex Rech. f.—A variety distinguished by its larger leaves (up to 4 in. long and ¾ in. broad), distinctly wider above the middle; catkins also larger. Many botanists have considered *S. lambertiana* to be no more than a robust state of the species that merges into the typical state through intermediates. But, according to Dr Rechinger, who gives it the rank of a subspecies in *Flora Europaea*, this variant predominates in the lowlands of Europe, while the narrow-leaved typical state is commonest in mountainous regions.

The type of var. *lambertiana* was collected around 1800 by A. B. Lambert of Boyton, Wilts, according to whom the stand of this willow stretched for about sixteen miles along the River Wylye, in Wiltshire.

cv. 'EUGENEI'.—See under *S.* × *rubra*.

cv. 'NANA' ("*Gracilis*").—A roundish bush, usually less than 5 ft high and wide, which can be kept dense by annual clipping. Leaves slender, silvery grey. In recent years 'Nana' has been renamed 'Gracilis' in some catalogues and works of reference, owing to an unfortunate confusion between botanical and horticultural synonymy. It was originally distributed by Dr Dieck of Zoeschen towards the end of the last century, but its origin has not been ascertained.

cv. 'PENDULA'.—Grafted on standards 8 to 10 ft high, this makes a wide-spreading head with a tangle of more or less pendulous branches. It was at one time known erroneously as "American" weeping willow and also as "*S. Napoleonis*". 'SCHARFENBERGENSIS' is similar, but with shorter, more slender leaves.

cv. 'WOOLGARIANA'.—Leaves broadly oblanceolate, with a rather long cuneate base. Branchlets yellowish. (*S. woolgariana* Sm.) Named in honour of a Mr Woolgar, who had willow grounds near Lewes.

S. × DONIANA Sm. *S. purpurea* × *S. repens*—A shrub 3 to 6 ft high, with reddish brown or rust-coloured branchlets. The influence of *S. purpurea* shows in such characters as: leaves broadest above the middle, narrowed to the base, toothed, if at all, in the upper part, often opposite on the lower part of the shoot; catkins dense, cylindrical; filaments of stamens more or less connate. The presence of *S. repens* in the parentage shows in: leaves slightly revolute, silky beneath at least when young, sometimes entire; anthers yellow or brick-red; ovary definitely stalked, with a short style. An uncommon hybrid, described from a plant sent to Sir James Smith by George Don of Forfar (*d*. 1814), one of the first investigators of the flora of the Scottish Highlands and father of the botanists George and David Don. The original plant was female, as are most of the plants recorded from Britain, but a male was raised by the Rev. E. F. Linton and distributed by him.

S. × PONTEDERANA Willd. (excl. syn. *S. pontederae* Vill.) *S. sordida* Kern., not Schleich. ex Ser. *S. purpurea* × *S. cinerea*—A rather neat willow with downy twigs (often soon becoming glabrous.) Leaves narrowly obovate, oblong, or sometimes oval, tapered at both ends, most abruptly at the apex, varying from almost entire to rather prominently toothed, 1 to 2½ in. long, ⅓ to ¾ in. wide, dark glossy green above, conspicuously blue-white, and at first downy beneath. The influence of *S. purpurea* is seen in the glaucous under-surface of the leaf, and especially in the two stamens being more or less united by their stalks. Like nearly all hybrid willows, *S.* × *pontederana* varies in its approaches now to one parent now to another. It is at its best as a garden shrub when it most resembles *S. purpurea*.

This hybrid has been found on the River Tay near Perth, and also occurs on the continent.

S. × WIMMERIANA Gren. & Godr. *S. purpurea-caprea* Wimm.; *S.* × *pontederana* var. *grenierana* Anderss.—A rare hybrid in the wild, similar to *S.* × *pontederana* and difficult to distinguish from it. There is a male clone in commerce.

S. AMPLEXICAULIS Bory & Chaubard　*S. purpurea* subsp. *amplexicaulis*

(Bory & Chaubard) Schneid.—Leaves mostly opposite, almost sessile, truncate to cordate at the base, serrate almost throughout their length. Native of S.E. Europe and Asia Minor.

S. ELBURSENSIS Boiss.—Very near to *S. purpurea*, which it replaces in the Caucasus, Iran and E. Anatolia. Catkin-buds smaller. Catkins longer-stalked, the scales of the female catkins green or brownish, sparsely hairy (*Fl. Iranica*, Salicaceae (1969), p. 36). Introduced from N. Iran by Roy Lancaster in 1972.

S. INTEGRA Thunb. *S. multinervis* Franch. & Sav., not Doel; *S. purpurea* var. *multinervis* (Franch. & Sav.) Koidz.; *S. p.* subsp. *amplexicaulis* var. *multinervis* (Franch. & Sav.) Schneid.—Leaves mostly opposite, almost sessile, up to 2⅝ in. long and ⅞ in. wide, obtuse or subacute at the apex, cordate at the base. Catkins and flowers as in *S. purpurea*. Native of Japan, very near to the western *S. amplexicaulis*.

A hybrid of *S. integra* is 'GINME', a vigorous, spreading shrub, its green stems glabrous by late summer. Leaves closely set, oblong, obtuse to pointed at the apex, narrowed to a truncate base, up to 4½ in. long and 1¼ in. wide, finely serrate, rich green and slightly glossy above, blue-green, finely veined and sparsely hairy beneath. A female clone; catkins pinkish, shortly stalked, about ⅞ in. long, appearing in early spring. It is considered to belong to S. × TSUGALUENSIS Koidz., a natural hybrid between *S. integra* and the Japanese sallow *S. vulpina* (see p. 308). One of the best of the garden willows, free-flowering and with handsome foliage.

S. KORIYANAGI Kimura *S. purpurea* var. *japonica* Nakai—Near to *S. purpurea*, from which it differs in its leaves, which are narrowly oblong-lanceolate, tapered at the apex to a fine point. Other points of distinction given by Nakai are: the more hairy catkin-scales; longer stamens with darker anthers; ovary tapering into the style. Described by Nakai under the synonymous name from Korean plants; known in Japan only as a cultivated tree, grown for basketry and for ornament. The plants in commerce in Britain are a male clone of upright growth with green young stems.

S. MIYABEANA Seem. *S. dahurica* Turcz. ex Lakschewitz—This is the principal eastern counterpart of *S. purpurea*, with a wide distribution in N.E. Asia, including much of Japan and N. China. It differs from *S. purpurea* in having the leaves toothed to the base, and well-developed stipules, which are narrow and toothed. A graceful species, with narrow, finely tapered leaves. Both sexes are in cultivation. Some plants cultivated as *S. miyabeana* agree better with S. GILGIANA Seem., a closely related species confined to Japan and Korea, with stouter growths, downy when young.

S. PYRENAICA Gouan
S. ciliata DC.

A dense shrub with procumbent main stems and ascending branches, up to 3 ft high though commonly only half that height; branchlets soon glabrous, reddish brown. Leaves broadly oval, ovate-elliptic or obovate, $\frac{3}{4}$ to $1\frac{1}{4}$ in. long, upper surface green, fairly glossy, with scattered long hairs, underside at first hairy all over, becoming glabrous except on the main veins, margins entire, fringed with long hairs; petiole about $\frac{3}{16}$ in. long. Catkins lax, borne on leafy peduncles; scales obtuse, sparsely hairy. Stamens glabrous with purplish anthers. Ovary ovoid, woolly, almost sessile; style sometimes divided at the apex, stigmas bifid, the divisions slender.

An endemic of the Pyrenees and the Cantabrian mountains, allied to *S. glauca*. It is one of the first willows to have been recorded scientifically, for it was found by the Netherlands botanist Charles de l'Escluse (Clusius) during his journey to France, Spain and Portugal, 1561–5, and figured by him in a work of 1601. It was introduced to Britain in 1823 and reintroduced early this century.

S. PYRIFOLIA Anderss.   BALSAM WILLOW

S. balsamifera Barratt ex Bebb; *S. cordata* var. *balsamifera* Hook.

An erect, many-stemmed shrub, rarely a small tree; young branches reddish brown in their first winter; winter-buds bright red. Leaves very thin and purple-tinged when young, balsam-scented, ovate or oblong-lanceolate, 2 to 4 in. long, 1 to $1\frac{1}{2}$ in. wide, dark green above, paler and reticulate beneath, glabrous, margins finely glandular-serrate; petiole slender, up to $\frac{5}{8}$ in. long. Stipules wanting, or small and soon falling. Catkins on short leafy peduncles or almost sessile, male $\frac{7}{8}$ to $1\frac{3}{4}$ in. long, dense, female more slender, elongating to 3 in. in fruit; scales densely long-hairy, oblong, pale yellow. Ovary glabrous, slenderly stalked; stigmas almost sessile.

Native of N. America from Labrador to the Rocky Mountains, south to New York, Michigan and South Dakota. An interesting willow, allied to the section *Cordatae* but usually placed in a monotypic section—*Balsamiferae*. It is ornamental in winter, with its reddish young branches and bright red buds.

S. REPENS L.   CREEPING WILLOW

A low shrub of variable habit, often only 1 to $1\frac{1}{2}$ ft high in the wild, but sometimes 6 or 8 ft high in gardens; spreading by means of creeping or underground stems from which spring upright branches; young shoots silky. Leaves oblong or oval to lanceolate; normally $\frac{1}{4}$ to $\frac{3}{4}$ in. long and $\frac{1}{8}$ to $\frac{1}{3}$ in. wide, but in cultivation twice those dimensions; tapered about equally at both ends or more gradually towards the apex; glistening and silvery beneath, with a dense covering of silky hairs; dull or greyish green and more or less silky above, but sometimes glabrous, especially late in the season; stalk $\frac{1}{12}$ to $\frac{1}{6}$ in. long. Catkins produced on the naked shoots in April and May, sessile, ovoid to oblong, rarely more than $\frac{3}{4}$ in. long; scales usually dark at the apex, hairy. Anthers yellow. Ovary usually glabrous, long-stalked; style and stigmas variable.

S. repens in its typical state is fairly widespread in Europe (including the British Isles), especially on wet heaths and moorlands, and extends into Siberia

and Central Asia. Even in its typical state it is variable, and not clearly demarcated from the subsp. *argentea*. But it is easily distinguished among the smaller-
leaved willows by its creeping root-stock and the silvery under-surface of the
leaves. The subsp. *argentea* is more ornamental, and commoner in gardens, but
there is a dense-habited, sturdy form of typical *S. repens* with leaves about
⅔ in. long by ¼ in. broad, glabrous and rather glossy on the upper surface, very
glaucous beneath, that makes a neat bush for the rock garden.

subsp. ARGENTEA (Sm.) E. A. & G. Camus *S. argentea* Sm.; *S. arenaria* L.;
S. repens var. *argentea* (Sm.) Wimm. & Grab.; *S. repens* var. *nitida* Wenderoth—
A larger, more erect, stoutly branched shrub. Leaves relatively broader, tending
to obovate, those on strong shoots up to 1⅞ in. long and 1 in. wide, usually
permanently silky above. Ovary usually hairy.

In contrast to typical *S. repens*, this subspecies is in the main coastal in
distribution, inhabiting fixed sand-dunes from the Atlantic coasts of Europe
to the North Sea and the Baltic, but extends some way inland in Germany, and
is also found on some rocky heaths in N. Scotland.

var. FUSCA (L.) Wimm. & Grab. *S. fusca* L.—A minor variation of typical
S. repens with a short root-stock and erect stems. It occurs in fens.

cv. 'SERICEA PENDULA'.—Leaves silky above and beneath, closely set on
the shoot. Female. It is laxly branched and was at one time top-grafted, making
a very pretty weeping tree, but not long-lived. Some plants now grown as
S. repens argentea probably belong to this clone, which is evidently an old one,
judging from specimens in the Kew Herbarium.

cv. 'VOORTHUIZEN'.—A prostrate selection of neat and compact habit,
with small, greenish grey leaves. Of Dutch origin.

cv. 'WOLSEYANA'.—Although probably not naturally weeping, this was at
one time sold as a weeping tree. The leaves are much larger than in 'Sericea
Pendula', roundish, sparsely silky beneath. It was in cultivation as early as
1877 and used to be sold by several nurserymen.

S. × AMBIGUA Ehrh. *S. repens* × *S. aurita*.—A fairly frequent natural
hybrid, recorded from many parts of the British Isles, but of no horticultural
interest. The influence of *S. aurita* is usually to be seen in the more erect growth
and persistent stipules, and of *S. repens* in the silvery undersides of the leaves.

Hybrids of *S. repens* with *S. caprea* and *S. cinerea* also occur, though more
rarely. For another hybrid of *S. repens* see *S. × doniana*.

S. ROSMARINIFOLIA L. *S. sibirica* Pall.; *S. repens* var. *rosmarinifolia* (L.)
Wimm. & Grab.; *S. repens* subsp. *rosmarinifolia* (L.) Čelak.—Branchlets slender,
sparsely hairy. Leaves narrow-lanceolate, long-tapered to a sharp apex, more
abruptly to the base, silky beneath, entire or almost so, lateral veins in ten to
twelve pairs. Stipules small or wanting. Catkins globose. Ovary hairy. Native of
eastern Central Europe northwards to S. Sweden. Not to be confused with the
willow once commonly known as *S. rosmarinifolia* in gardens, which is
S. elaeagnos, a species which has the leaves white-woolly beneath, not silky, and
differs in many other characters. See also *S. × friesiana*.

S. rosmarinifolia has been confused with the following hybrid:

S. × FRIESIANA Anderss. *S. rosmarinifolia* of some authors, not L. *S. repens* × *S. viminalis*—This hybrid occurs occasionally in the wild and was raised artificially by the Rev. E. F. Linton. It differs from *S. rosmarinifolia* in the relatively broader, usually lanceolate, leaves, more persistently silky beneath, the ovoid, more silky male catkins and the linear-oblong nectaries (in *S. rosmarinifolia*, as in *S. repens*, the nectaries are very short, ovate, with an obtuse or truncate apex).

S. SUBOPPOSITA Miq. *S. repens* var. *subopposita* (Miq.) Seem.; *S. sibirica* var. *subopposita* (Miq.) Schneid.—A close ally of *S. rosmarinifolia*, but with the branchlets and buds more persistently hairy, and the leaves more densely silky when young; also in the more pronounced stipules. The specific epithet refers to the arrangement of the leaves, many of which are almost opposite. In the cultivated clone (male) the leaves are very closely arranged on the branchlet (five or even six to the inch); the stipules are ovate, with an oblique base, often stalked. Catkins ovoid, about ½ in. long, opening well before the leaves. Native mainly of Japan.

S. RETICULATA L.

A low or prostrate shrub, forming large patches on the ground, but only rising from it as a rule 5 or 6 in., rarely 12 in.; young branches somewhat angled, shining brown, and glabrous except at first, when they are more or less silky-hairy. Leaves mostly two to four on each twig, ½ to 1½ in. long; round, roundish oval or broadly obovate, not toothed, slightly indented, rounded, or sometimes tapered at the apex, deep green and much wrinkled above, glaucous white, prominently net-veined and sometimes silky beneath; stalk ¼ to ¾ in. long. Catkins cylindrical, ½ to 1 in. long, produced in May and June on slender stalks 1 in. long, at the end of the twig opposite the uppermost leaf.

Native of the arctic and subarctic regions of the northern hemisphere, extending southward in Eurasia to the mountains of central and southern Europe, eastern Siberia and the Russian Far East, and in America into the Rocky Mountains. In Scotland it is recorded from several localities in the Highlands above 2,000 ft, but is commonest in the Breadalbane Mountains of Perthshire, and is mainly confined to basic soils. Even in the Alps it is by no means common, and most likely to be found on the limestone ranges, in places where the snow lies long.

It is a very interesting dwarf willow, distinct from all others (except some allied American and Asiatic species) in its long-stalked leaves, long, naked peduncles, and above all in the large, deeply divided nectary, enclosing the base of the ovary in the female flowers.

S. RETUSA L.

A low, prostrate shrub, reaching only a few inches above the ground, the branches creeping and taking root; young shoots glabrous. Stipules wanting. Leaves obovate or lozenge-shaped, $\frac{1}{3}$ to $\frac{3}{4}$ in. long, $\frac{1}{6}$ to $\frac{1}{4}$ in. wide; tapered at the base, blunt or rounded, sometimes retuse, at the apex, not toothed, quite glabrous and green on both sides; stalk $\frac{1}{6}$ in. or less long; nerves in three to six pairs. Catkins erect, stalked, cylindrical, about $\frac{2}{3}$ in. long, produced at the end of short, leafy shoots in May and June. Catkin-scales oblong-obovate, obtuse, yellow or pale brown, glabrous except for occasional long hairs at the edge. Ovaries conical, glabrous, the stalk and style both short.

SALIX RETUSA

Native of the mountains of Europe (Pyrenees, Alps, Appenines), with a close ally in the Carpathians (*S. kitaibeliana* Willd.), not treated here; introduced in 1763. A neat little alpine shrub, forming close tufts in exposed places, but spreading more freely when planted in gardens. Suitable for the rock garden.

S. SERPYLLIFOLIA Scop.   *S. retusa* var. *serpyllifolia* (Scop.) Ser.—Closely allied to *S. retusa*, and is sometimes regarded as a variety of it. It differs chiefly in the smaller leaves, which in nature are only $\frac{1}{6}$ to $\frac{1}{3}$ in. long, forming with its stunted branches a close dense tuft. Under cultivation the plant becomes more creeping, and the leaves up to $\frac{2}{3}$ in. long. They are obovate, notched, rounded or pointed at the apex; nerves two to four each side. Native of the Alps of Europe, mostly at higher altitudes than *S. retusa*. Both species are, in the Alps, commonest on calcareous formations.

S. × COTTETII Lagger ex Kern. *S. retusa* × *S. nigricans*—A procumbent shrub with ascending branches; young stems and leaves at first hairy, becoming almost glabrous. Leaves elliptic to obovate, obtuse or slightly acute, up to 1⅝ in. long, about half as wide, finely toothed, equally green on both sides. Catkins ⅝ to ⅞ in. long on short leafy peduncles; scales oblong-obovate, darkened at the truncate or retuse apex. Originally described from a female plant found in Switzerland. A commercial clone, male, distributed as "*S.* × *gillotii*", has been identified as *S.* × *cottetii*.

S. RIGIDA Muhl.

S. cordata Muhl., not Michx.; *S. nicholsonii purpurascens* Dieck

A vigorous, richly leafy shrub or small tree, making long stiff shoots annually, and reaching 10 to 15 ft high; young shoots downy at first, getting glabrous by late summer. Leaves closely set on the branch, often furnished with a pair of large ear-shaped stipules, ovate-lanceolate, rounded or heart-shaped at the base, slender pointed, finely toothed, 3 to 6 in. long, ¾ to 1½ in. wide; green and glabrous on both sides except the midrib, which is slightly downy above; stalk ½ to 1 in. long. Catkins up to 2 in. long, produced on the naked wood in April, with one or a few tiny leaf-like bracts at the base of each. Stamens two. Ovaries long-stalked.

Native mainly of eastern and central N. America, but so closely related to the more western *S. lutea* and *S. mackenzieana*, into which it probably merges through intermediate forms, that its western limits are not clear. It was introduced to Britain in 1812. A very well-marked willow by reason of the large, long-stalked leaves usually cordate at the base, and the conspicuous persistent stipules. It varies in the shape and width of its leaves; in var. ANGUSTATA (Pursh) Fern. they are around ⅝ in. wide and tapered at the base.

S. rigida is represented in cultivation by a female clone with brownish red young leaves. This was, in fact, the characteristic of the clone distributed by Dieck in the last century under the name "*S. nicholsonii purpurascens*", but according to Fernald tinted young growths are a normal feature of this species.

S. MISSOURIENSIS Bebb   ?*S. eriocephala* Michx.; *S. rigida* var. *vestita* (Anderss.) C. K. Ball—Stems permanently downy or woolly. Leaves more tapered at the apex than in *S. rigida*, glaucous beneath. Stipules smaller. Central USA.

S. LUTEA Nutt. *S. cordata* var. *watsonii* Bebb; *S. rigida* var. *watsonii* (Bebb) Cronq.; *S. cordata* var. *lutea* (Nutt.) Bebb—Branchlets usually yellowish, they and the leaves quickly becoming glabrous. Leaves yellowish green above, glaucous beneath, narrowly to often broadly lanceolate, sometimes entire. Western N. America, as far west as W. Ontario, Montana, Nebraska and Colorado, south to California.

S. MACKENZIEANA Barratt ex Anderss. *S. cordata* var. *mackenzieana* Hook.; *S. rigida* var. *mackenzieana* (Hook.) Cronq.; *S. monochroma* Ball—Branchlets usually brown as in *S. rigida* from which it differs in its glossier, lighter green, mostly lanceolate leaves almost glabrous from the start, less conspicuous stipules, sparsely hairy catkin-scales and longer-stalked capsules (up to $\frac{1}{6}$ in. long). Like *S. lutea* a native of western N. America but with a more northern distribution. Described from specimens collected in N.W. Canada on the Mackenzie river and Great Slave Lake.

S. 'AMERICANA'.—A probable hybrid of *S. rigida*, cultivated on the continent as a basket willow. It was introduced from the USA to Germany in the last century by the basket-maker Ernst Hödt, and propagated by Otto Schön's willow-nursery, which distributed three million cuttings in 1914 (*S. americana* Hort. ex Schwerin, not Anderss.). Rehder suggested that it is a hybrid of *S. purpurea*, a species long cultivated in the USA, with *S. rigida* as the probable second parent, but it is now considered to be *S. rigida* × *S. gracilis*. It is a male clone.

<h3 style="text-align:center">S. × RUBRA Huds.</h3>
? S. helix L.

A hybrid between *S. purpurea* and *S. viminalis*, forming a shrub or small tree; young twigs slightly downy at first. Leaves linear-lanceolate, with long, tapered points, the base more abruptly tapered, distantly toothed except towards the base, 2 to $5\frac{1}{2}$ in. long, $\frac{1}{4}$ to $\frac{2}{3}$ in. wide, green and glabrous on both sides when mature, but grey and slightly downy beneath when young; stalk $\frac{1}{4}$ to $\frac{1}{2}$ in. long. Catkins produced on the naked shoots in April, 1 to $1\frac{1}{2}$ in. long. Stamens two, but with stalks united towards the base, or sometimes nearly to the anthers.

Native of Britain and Europe, and highly valued by basket-makers. The osiers known in the trade as 'Mawdesley's Long Skein' and 'Tulip Willow' belong to it.

S. 'EUGENEI'.—Of fastigiate habit, the branches and twigs all ascending at a steep angle; young bark pale green or greenish yellow. Leaves on strong shoots 2 to 4 in. long, linear-oblanceolate, tapered into the petiole, acute or short-acuminate at the apex, finely serrated in the upper half, sea-green above, the underside glaucous with a yellowish midrib, soon glabrous, lateral veins in about forty pairs on the longer leaves. A male clone, with slender, pinkish catkins $\frac{3}{4}$ to 1 in. long; anthers pale red (*S. purpurea* × *S. viminalis*, var. *eugenei* J. Fraser; *S. purpurea Eugenei* and *S. pyramidalis Josephinae* Hort. ex J. Fraser; *S. pyramidalis Eugenie* Hort. ex Dipp.; ? *S. pyramidalis Josephinae* Hort. ex K. Koch; ? *S. pyramidalis Josephine* Hort. ex Dipp.).

A vigorous but elegant willow, which James Fraser likened to a fine-leaved bamboo; it is also very pretty in spring, with its small but abundant catkins. It has been cultivated in Germany since the 1860s and has usually been placed under *S. purpurea* or *S. helix* (a name of uncertain application which has been used for forms of both *S. purpurea* and *S.* × *rubra*), but the Swedish authority Floderus saw a specimen from Messrs Hillier in the 1920s and identified it as

S. purpurea × *S. viminalis*. Although the rendering 'Eugenei' is established, the correct name should probably be 'Eugénie' or 'Eugeniae', for the Empress Eugénie, wife of Napoleon III. 'Josephine' may have been a different clone of similar habit.

S. 'FORBYANA'.—Near to *S.* × *rubra*, but now thought by some authorities to be a triple hybrid, the third parent being *S. cinerea* var. *oleifolia*. Twigs yellowish. Leaves broader than in *S.* × *rubra*, dark green and lustrous above. A female clone with catkins as in *S. purpurea*; occasionally male flowers are produced in the lower scales and these show the influence of *S. purpurea* in having the filaments connate. 'Forbyana' was described by Sir James Smith in 1804 from a plant in the Crowe collection at Lakenham, received from a Mr Forby. According to Smith it was known as the 'fine basket osier', but introduced later to the Thames osier-beds it was found to be too coarse for fine basket-work.

S. SACHALINENSIS Fr. Schmidt
S. opaca Anderss. ex Seem.

A tree to 30 ft high in the wild; branchlets reddish, purplish or yellow-brown, glabrous and lustrous; buds purplish red. Leaves lanceolate, acuminately tapered at the apex, cuneate at the base, up to 6 in. long and 1 in. wide, rich lustrous green above, paler beneath, glabrous on both sides, or the undersurface persistently downy, margins crenate; petiole up to 1 in. long. Catkins appearing before the leaves; scales black at the tip, obtuse, densely hairy. Male catkins almost sessile, 1½ in. long, ½ in. wide; filaments of stamens glabrous, anthers yellow flushed with red at the tip. Female catkins slender; ovary ovoid, silky, long-stalked, with a slender style branched at the apex, stigmas slender, bifid.

Native of Sakhalin, N. Japan, the Russian Far East and E. Siberia; introduced to Kew in the 1920s. It is mainly represented in gardens by the shrubby 'SEKKA' ('Setsuka'). This is valued by flower-arrangers for the curious fasci-ated and contorted growths produced on some of the stems, which are deep brownish red on plants grown in sun. It is a male clone, and the handsome catkins are borne freely both on normal and fasciated growths. 'Sekka' is very vigorous and needs a space 10 ft wide even when pruned hard every spring, as it should be; unpruned, it attains about 15 ft in height. It was introduced from Japan by Messrs Spek of Holland about 1950.

S. sachalinensis is closely allied to S. UDENSIS Trautv. & Mey., described earlier from the Okhotsk peninsula.

S. sachalinensis is allied to *S. viminalis*. Another east Asiatic representative of the same group is:

S. REHDERIANA Schneid.—A shrub or small tree to about 30 ft high; branchlets slender, dark brown or olive-green, glabrous or soon becoming so. Leaves lanceolate, slenderly acuminate, up to 5 in. long, glabrous above, silky beneath as in *S. viminalis* but sometimes becoming glabrous, margins shallowly glandular-crenate. Catkins before the leaves, the male about 1 in. long, half as wide. Ovary glabrous or sparsely silky, almost sessile; stigmas short, notched (hence very different from those of *S. viminalis*). Discovered by Wilson in W. Szechwan, China, and introduced by him by means of cuttings in 1908 and again in 1910.

S. × SEPULCRALIS Simonk. (1889)

S. × *salamonii* Carr. ex Henry (1913)

A group of hybrids between *S. alba* and *S. babylonica*, which have arisen spontaneously within the area where *S. alba* is native and *S. babylonica* still generally cultivated. The hybrid has been recorded from Central Asia, Trans-caucasia, Asiatic Turkey and Palestine. In western Europe it is probable that most of the cultivated trees belong to the clone 'Salamonii' (see below), but there is some evidence that other forms were distributed, either as *S. alba* or *S. babylonica*. The type of *S.* × *sepulcralis* was a tree cultivated in Hungary, but its origin is unknown. These hybrids (as grown in Europe) differ from *S. baby-lonica* in their greater height and robustness, thicker and shorter pendulous branchlets, the silkiness of the young stems and young leaves, and the stouter, more hairy, longer stalked catkins.

cv. 'SALAMONII' ("Sepulcralis").—This is one of the handsomest and most vigorous of all willows. It is not so weeping as *S. babylonica*, having inherited some of the firmer outlines of *S. alba*, but is still extremely graceful. It grows at least 60 ft high, forming a broad, shapely head of luxuriantly leafy branches; twigs silky when young. Leaves 2½ to 5 in. long, ½ to ⅞ in. wide, green above, blue-white beneath, silky beneath on first expanding and slightly so above, but less so than in *S. alba* and soon becoming as glabrous as those of *S. baby-lonica*, margins rather closely and finely toothed. They are slow to fall in autumn, some remaining on the tree as late as December. Catkins female (sometimes androgynous) about 1¾ in. long on peduncles about 1 in. long; scales ovate, obtuse or subacute, ciliate, eventually deciduous. Ovary ovoid, glabrous, sessile; style short. [PLATE 36

This willow first appeared on the property of Baron de Salamon at Manosque, Basses-Alpes, some years before 1864, and was put into commerce by Simon-Louis of Metz in 1869. It is fast-growing when young, but perhaps short-lived or subject to wind-breakage, for the large trees that once grew at Kew on the south side of the Lake are gone, and no old specimens have been recorded in recent years. But all the older weeping willows have become uncommon since 'Chrysocoma' began to spread into gardens at the end of the last century.

S. 'CHRYSOCOMA'. GOLDEN WEEPING WILLOW.—A large tree with stout branches ascending at an angle of 45° or 50°, the secondary branches steeply pendulous, usually reaching to the ground; young wood clear yellow by late summer, becoming greyish yellow in the second year, at first clad with silky appressed hairs, soon glabrous. Leaves narrow-lanceolate to narrow-elliptic, gradually tapered at the apex, shortly cuneate at the base, 3 to 4½ in. long, ⅜ to ⅝ in. wide, finely serrate, silky on both sides at first, soon glabrous above, becoming more slowly so on the glaucous underside; petioles downy, $\frac{1}{4}$ to $\frac{3}{8}$ in. long. Catkins in early spring with the leaves, slender, $1\frac{1}{2}$ to 2 in. long, on leafy stalks; they are either female or male, the two sexes occurring on the same or separate twigs, or are androgynous, the female flowers then occupying the apical part of the catkin; bisexual flowers also occur. Catkin-scales lanceolate to oblong, hairy towards the base, the hairs on the scales and rachis longer in the males (and on the male sector of a mixed catkin). Ovaries glabrous; style short (*S. vitellina pendula nova* Späth; *S. alba* var. *vitellina pendula* (Späth) Rehd.; *S. babylonica ramulis aureis* Hort.; *S. alba* 'Tristis' Hort., not *S. alba* var. *tristis* (Ser.) Koch; *S.* × *chrysocoma* Dode).

This fine weeping willow is of unrecorded origin and uncertain botanical status. It seems very likely that it derives from the old semi-pendulous form of *S. alba* var. *vitellina*, which was known to Loudon and described by Seringe in 1815 as *S. alba vitellina-tristis*, from a plant received from Baumann's nursery; it was also distributed as *S. aurea* and very possibly as *S. vitellina pendula*—whence the name *S. vitellina pendula nova* under which Späth put 'Chrysocoma' into commerce in 1888. Whether this tree owes its perfectly weeping habit to a mutation or to the influence of *S. babylonica* is a matter for dispute. The first to suggest that it was a hybrid of *S. babylonica* was the French dendrologist Dode, who re-named the Späth clone *S.* × *chrysocoma* in 1908; apart from the habit of the tree he adduced the androgynous catkins, so often a sign of hybridity. James Fraser, who made a careful study of the catkins and flowers of 'Chrysocoma' was convinced of its hybridity. If indeed it derives from *S. alba* and *S. babylonica* it should strictly be placed under *S.* × *sepulcralis*, but the name *S.* 'Chrysocoma' is unambiguous and leaves open the problem of its status. The origin of the name *S. babylonica ramulis aureis* has not been ascertained; the trees so named are usually supposed to be of the same clone as 'Chrysocoma', and the difference, if any, must be very slight (see *Journ. R.H.S.*, Vol. 58 (1933), pp. 401–2).

The golden weeping willow is the most frequently planted of all pendulous trees, and fine specimens are to be seen in suburban districts, where it can be something of a cuckoo-in-the-nest, often occupying virtually the whole garden. The largest specimen recorded grows at Trinity College, Cambridge, and measures 76 × 12$\frac{1}{4}$ ft (1969).

The only fault of 'Chrysocoma' is that with its heavy, ascending branches and rather brittle wood it becomes liable to wind-breakage when mature. All the old trees at Kew have lost large limbs in recent years.

The weeping willows so far discussed all derive from *S. babylonica* and *S. alba*. There are others, probably never much planted in Britain and certainly never much remarked on, of which the parentage is *S. babylonica* crossed with *S. fragilis* or, more likely, with the much commoner *S.* × *rubens* (*S. alba* × *S. fragilis*). These hybrids have been known in commerce as *S. babylonica* simply, as *S. babylonica mas* (also, rather oddly, as *S. babylonica foemina*), *S. sibirica pendula*, *S. Petzoldii*, etc., for none of which has a description been traced. The botanical names used for this group are *S.* × *blanda* Anderss. (1867) (see 'Blanda' below) and *S.* × *elegantissima* K. Koch (1871). The latter is, unfortunately, an ambiguous name. Koch's description was based on a tree known to him, said to be common in N.E. Germany and much hardier than *S. babylonica*.

But the horticultural synonyms that he cites may have belonged to different clones. This would explain why descriptions of *S.* × *elegantissima* by other botanists disagree with Koch's and indeed the supposition finds support in Koehne's *Deutsche Dendrologie* (1893), p. 91, where it is recorded that the old trees at Berlin labelled *S. elegantissima* by Koch himself were not uniform in their botanical characters.

There is record of "*S. babylonica mas*" being introduced to Britain in the 1870s, and it is possible that the trees studied by James Fraser in the 1920s, of which herbarium specimens still exist, belong to this clone. Branchlets dull green, terracotta or reddish. Leaves almost glabrous from the start, lanceolate, to 5½ in. long and 1 in. wide, dark green above, paler beneath. Catkins female or of mixed sex, to about 1¾ in. long on peduncles about ½ in. long; scales strap-shaped, acute or blunt. Ovary distinctly stalked, downy at the base. This agrees with *S.* × *elegantissima* as described by Rehder in the *Manual*, but not with Koch's original description, since the type of *S.* × *elegantissima* had glabrous, sessile ovaries. The tree distributed by Dieck towards the end of the last century as *S. elegantissima* seems to have been a different clone, with longer catkins on longer peduncles.

On the North side of the Lake at Kew there is an old tree resembling *S.* × *sepulcralis* in habit but belonging to this group and perhaps the true *S. elegantissima*. The catkins and flowers agree well with Koch's description and with a specimen, dated 1847, sent by Braun, Director of the Berlin Botanic Garden, where Karl Koch carried out his studies (Koch acted as Director for a year after Braun's death but, much to his disappointment, was not given the post).

S. 'BLANDA'.—This was described by Andersson in 1867 from a specimen collected in a garden at Hanau in Hessen, Germany. Dippel, who was Director of the Darmstadt Botanic Garden some thirty-five miles away, records that this willow was planted elsewhere in the Hanau district, and sent cuttings from a tree in the Castle Garden at Darmstadt to the nurseryman Späth of Berlin, who put this willow into commerce. 'Blanda' was judged by Andersson, from the herbarium specimen, to be *S. babylonica* × *S. fragilis*, but this clone remains an enigma. It is not truly a weeping tree at all, though the branchlets are somewhat pendulous. The leaves are dark green, glossy, and rather thick, quite unlike those of any other hybrid or putative hybrid of *S. babylonica*, rather bluntly toothed. The catkin-scales are narrowly triangular, acute and the ovaries (it is a female clone) glabrous, short-stalked. An authentic herbarium specimen at Kew, from the Hanau garden, was annotated by Linton, the British authority on willows, as *S. babylonica* × *S. pentandra*. Only young plants of 'Blanda' have been seen.

S. × SERICANS Tausch ex Kern.
S. × *smithiana* of some authors, ? not Willd.

A hybrid between *S. caprea* and *S. viminalis*, the common osier, making a tall shrub or small tree to 20 ft or so high; young stems sparsely hairy, soon becoming more or less glabrous. Leaves broadly oblong-elliptic or broadly lanceolate, to about 7 in. long, soon glabrous above, the undersides greenish, coated with a thin woolly indumentum, the lateral and cross-veins both prominent. Catkins produced before the leaves, $1\frac{1}{2}$ to 2 in. long. This hybrid is fairly widely distributed in the British Isles, in waste ground, hedgerows, etc., either spontaneously or planted. The very vigorous male clone in commerce as *S.* × *smithiana* probably belongs here and agrees quite well with the male that has been cultivated since early in the 19th century under the name "*S. rugosa*".

The hybrid between *S. viminalis* and *S. cinerea* also occurs. It differs in the more persistent down on the stems, the relatively narrower, lanceolate leaves, their undersides eventually almost glabrous and the cross-veins not prominent. This hybrid is less common than *S.* × *sericans* and male plants are rare.

S. × DASYCLADOS Wimm.—This controversial willow does not occur in Britain, either wild or naturalised. It is represented in commerce by the male clone 'GRANDIS', originally distributed as "*S. aquatica grandis*". It is a large shrub with stout stems, at first velvety, becoming glabrous or almost so. Leaves oblong-elliptic, acute, to 6 or 7 in. long, 1 in. or slightly more wide, sparsely silky beneath even when young; petioles 1 in. long, much expanded at the base when subtending the closely set, downy catkin-buds. Stipules conspicuous, broad based, more or less abruptly narrowed at the tip, withering before falling. The handsome catkins, borne before the leaves, are about 2 in. long. 'Grandis' is not the same as the male clone in commerce as *S. dasyclados* simply. This, a hybrid of uncertain identity, is of extraordinary vigour, making seasonal stems 10 ft or so long when young and pruned. The catkins appear in late winter or early spring, sometimes in autumn.

S. dasyclados was described by Wimmer in 1849 from a specimen collected near Troppau in Silesia. Many authorities, among them the Russian salicologist Skvortsov, consider it to be a good species, of wide distribution from Silesia and Brandenburg eastwards as far as the Lena river, reproducing itself by seed and occurring in the same sort of habitat as *S. viminalis*, to which it is allied. According to others, it is a hybrid or perhaps an aggregate of hybrids, deriving from *S. viminalis* and one or more of the sallows. Certainly, *S. dasyclados*, as often understood, includes such hybrids and it may be that the actual type of *S. dasyclados* was a hybrid, in which case the species would need another name.

The following appear to be allied to *S. dasyclados*:

S. × CALODENDRON Wimm. *S. acuminata* Sm., not Mill.; *S. dasyclados* of some authors, not Wimm.; *S.* × *smithiana* var. *acuminata* (Sm.) Anderss.—A shrub or small tree with stout, velvety young stems, the hairs persisting through the winter. Leaves oblong-elliptic or oblong-lanceolate, to about $5\frac{1}{4}$ in. long and $1\frac{1}{2}$ in. wide, tapered to an acute point, greyish and thinly downy beneath, with narrowly revolute margins. Stipules ovate-lanceolate, curved, retuse on the inner side at the base. Only female trees are known, with almost sessile, crowded, upward curved catkins 3 in. long. Of uncertain parentage, thought by Wimmer to be *S. dasyclados* × *S. caprea*. It occurs here and there in Britain, probably always planted, and is worthy of cultivation for its handsome catkins. Also known only as a female is S. × STIPULARIS Sm., with more slender leaves, whiter and velvety beneath. The specific epithet refers to the very large stipules of the strong shoots, which are lanceolate, tapered, and about 1 in. long. It was apparently widely distributed early in the last century as *S. viminalis*, under which name it was sent to the botanical artist Sowerby by several growers when the *English Botany* was being prepared. Further details will be found in an article by R. D. Meikle published in *Watsonia*, Vol. 2 (1952), pp. 243–8.

These osier-sallow hybrids, despite their handsome catkins, are too robust and too coarse in foliage to be acceptable in any but the largest garden. But, because of their vigorous and tall growth, they are useful for quickly providing shelter for choicer things, or as a temporary screen.

S. SERICEA Marsh.

S. petiolaris var. *sericea* (Marsh.) Anderss.; *S. pennsylvanica* Forbes

A shrub to about 15 ft high; branchlets glabrous or silky, brown, brittle at the base; winter-buds abruptly acute at the apex. Leaves lanceolate or oblong-lanceolate, up to 4 in. long, ½ to 1 in. wide, slenderly acuminate at the apex, finely serrated throughout, silky on both sides when young, usually permanently so beneath, becoming dark green and glossy above; petiole to ⅓ in. long. Catkins appearing in spring before the leaves, about 1 in. long, the females elongating in the fruiting stage; scales dark. Stamens two, the anthers red when young. Ovary short-stalked, ovoid, blunt at the apex, silky; stigmas almost sessile.

Native of eastern N. America; in cultivation 1829 and introduced on several occasions since. Although now uncommon in gardens it is one of the most ornamental of the medium-sized willows. It is grouped with the European *S. repens* by some authorities, but its taxonomic position is controversial.

S. SILESIACA Willd.

A shrub of the sallow group, growing to 6 ft or so high; young bark brown; growths loosely downy at first, becoming glabrous by the end of the season. Leaves rather thin, tinted brown when young, obovate or oval, 1½ to 4 in. long, about half as wide, downy when young, becoming glabrous above and nearly so on the greenish underside; stalk ¼ to ½ in. long. Catkins emerging in April, sometimes as the leaves unfold. Anthers red at first. Ovaries glabrous.

Native of the Balkans, Carpathians, Sudetenland, etc.; absent from the Alps and W. Europe. Of little garden interest.

A sallow very common in the Alps is:

S. APPENDICULATA Vill. *S. grandifolia* Ser.—A large shrub or small tree; young bark green or yellowish. Leaves obovate to oblanceolate, the venation deeply impressed above and raised beneath, becoming rather sparsely hairy beneath by autumn. Catkins less dense than in the native sallows, the females becoming very long and lax in fruit, with the ripening ovaries pointing out at a wide angle. It often occurs as a pioneer on landslides, etc., especially on calcareous formations.

The Japanese sallow S. VULPINA Anderss. (*S. daiseniensis* Seem., in part) is quite closely allied to *S. silesiaca* and like it has glabrous ovaries, an unusual feature in the section *Vetrix*. Leaves elliptic or oblong to slightly obovate, to about 3 in. long. Male catkins sometimes almost 2 in. long. On dried specimens at least the hairs on the catkin-scales are rusty brown, whence no doubt the epithet *vulpina*—foxy. It was offered by Mr Wada in the 1930s as an ornamental, but there is no record of its introduction to Britain.

Also belonging to this group is S. PEDICELLATA Desf., a native of the west Mediterranean region from Morocco and S. Spain to Sicily, with an outpost in the Near East. This may not be in cultivation, but the very closely allied S. CANARIENSIS Buch is growing at Kew. A native of Madeira and the Canary Islands it is of interest as the most tropical of the sallows. Young growths woolly at first, soon glabrous, sometimes coated with a pruinose bloom. Leaves up to 6 in. long and 1½ in. or slightly more wide, oblong to lanceolate, tapered or sometimes acute at the apex, finely reticulate and eventually almost glabrous beneath.

S. STARKEANA Willd.

S. arbuscula var. β L.; *S. depressa sens.* some authors, ? not L.; *S. livida* Wahl.

A small, erect slenderly branched shrub, usually less than 3 ft high; branchlets brown or yellow, glabrous at maturity; winter-buds yellowish, flattened and appressed. Leaves broadly lanceolate to roundish obovate, abruptly acuminate at the apex, the tip often bent sideways, 2 to 2⅜ in. long, ⅝ to $\frac{13}{16}$ in. wide, glabrous or sparsely hairy and reticulate above, glabrous and dull green or bluish green beneath, with the venation more or less prominent, margins glandular-serrate; petiole up to $\frac{3}{16}$ in. long. Stipules broad-elliptic to obliquely oblate, coarsely toothed, nearly always present. Catkins lax, to about 1¼ in. long, produced before the leaves on short leafy peduncles; scales lanceolate, greenish yellow, sparsely hairy on the surface, ciliate. Filaments of stamens glabrous. Ovary densely hairy, long-stalked, the stalk about equalling the ovary itself in length, ovoid-conic, developing a long beak in the fruiting stage; stigmas erect, ovoid, on a very short style.

Native of north-eastern Europe extending southward to the foothills of the Alps and the northern Ukraine; absent from the British Isles and from the Alps proper.

S. BEBBIANA Sarg. *S. rostrata* Richardson, not Thuill.; *S. livida* var. *rostrata* (Richards.) Dipp.; *S. depressa* subsp. *rostrata* (Richards.) Hiitonen—This species is very closely allied to *S. starkeana* but is generally much more robust, growing to 25 ft high and with leaves up to 4 in. long (at least in American plants). It is also more variable, some forms being very near in their botanical characters to *S. starkeana* as represented in Europe, others less markedly so. *S. bebbiana* ranges in N. America from Newfoundland to Alaska, south to California, then across the Bering Straits to the Russian Far East and Siberia, perhaps as far west as Lapland. It was introduced to Kew from the Arnold Arboretum in 1889.

S. × SUBALPINA Forbes

This, a hybrid between *S. elaeagnos* and *S. repens*, is said by Forbes (*Salictum Woburnense*, t. 93) to have been introduced from Switzerland. It is a low shrub of rather neat habit, branches ascending, downy, and retaining their down till the second year. Leaves oblong-lanceolate, usually tapered about equally at each end, 1 to 2½ in. long, ¼ to ⅝ in. wide, margins decurved; not or very slightly toothed towards the apex, bright green and downy (especially at first) above, permanently grey and woolly beneath; stalk ⅙ in. or less long. In his original description Forbes mentions having only seen the male. Catkins 1 to 1¼ in. long, slender, yellow.

As remarked in previous editions, the plants cultivated at Kew and Cambridge were also male; most probably they originated by vegetative propagation from the Woburn plant. But Forbes' name is accepted as the valid one for hybrids between *S. elaeagnos* and *S. repens*, which occur occasionally in the wild.

S. TRIANDRA L.

S. amygdalina L.

A shrub or small tree up to 30 ft high, of erect habit; bark flaking, brown when freshly exposed; young shoots glabrous, or downy and soon becoming glabrous, angled or furrowed. Leaves quite glabrous on both surfaces, lance-shaped, rounded or wedge-shaped at the base, tapered to a fine point, finely toothed, 2 to 4 in. long, ⅝ to 1 in. wide, dark green above, green or glaucous beneath; stalk ¼ to ½ in. long, with a few glands near the apex. Stipules well developed on sterile shoots, ³⁄₁₆ to ¾ in. wide, usually persistent. Catkins produced in April or May (or sporadically throughout the summer) on short leafy shoots; scales yellowish, thinly hairy, deciduous. Male catkins up to 2½ in. long; stamens three, anthers yellow, filaments hairy at the base. Female catkins shorter than the male; ovary flask-shaped, glabrous, stalked, with sessile stigmas.

S. triandra is of wide distribution in temperate Eurasia. It is one of the most valuable of the basket-willows, and has been so widely planted that its natural distribution in the British Isles is uncertain, but it is probably genuinely native in south-east and parts of central England in wet, low-lying places. The osiers known under the trade names of 'Black Hollander', 'Black Italian', 'Black Mauls', 'French', 'Jelstiver', 'Mottled Spaniards', 'Pomeranian', all belong to this species.

var. HOFFMANNIANA (Sm.) Bab. *S. hoffmanniana* Sm.—A shrubby tree to about 12 ft high with interlacing branches, forming a dense crown. Leaves narrow-ovate, rounded at the base, green beneath, 1½ to 2½ in. long. Stipules more persistent than is usual in *S. triandra*. Male catkins rather dense. This minor variant was discovered by William Borrer in Sussex; Sir James Smith, who described it in 1828, identified it with the willow figured as *S. triandra* in Hoffmann's *Salices* (1785), whence the specific epithet. The var. *hoffmanniana* is found in various parts of Britain, though not commonly. Nearly all the plants are male and may belong to a single clone, though there is no obvious reason why this willow should have been planted, as it is unsuitable for any practical purpose.

S. triandra varies in the size and relative width of its leaves, and in the colouring of their undersurface. Plants with the leaves green beneath are considered to represent the typical state of the species; those that have them glaucous beneath are sometimes distinguished as var. *discolor* Anderss. (or even as a subspecies). *S. medwedewii* Dode, shrubby, with very narrow leaves, intensely glaucous beneath, is included in *S. triandra* by A. K. Skvortsov in *Fl. Iranica*. It was described in 1908 from plants introduced to France from the Caucasus and could be regarded as a cultivar of *S. triandra*—cv. 'MEDWEDEWII'.

S. NIPPONICA Franch. & Sav.   ? *S. subfragilis* Anderss.; *S. triandra* var. *nipponica* (Franch. & Sav.) Seem.; *S. t.* subsp. *nipponica* (Franch. & Sav.) A. K. Skvortsov—This species, perhaps no more than a race of *S. triandra*, is a native of Japan and continental N.E. Asia, reaching almost as far west as the eastern limit of *S. triandra*. The differential characters given by von Seemen are: leaves more downy when young; leaves under the catkins almost entire; catkins denser-flowered. Skvortsov distinguishes it (as a subspecies) only by its branchlets being covered when mature with a pruinose bloom.

S. UVA-URSI Pursh   BEARBERRY WILLOW
S. cutleri Tuckerman

A creeping shrub with woody, branching stems, forming mats several feet across. Leaves narrowly to broadly elliptic or obovate, up to 1 in. long and ⅜ in. wide, acute or obtuse at the apex, cuneate at the base, finely glandular-serrate, upper surface glabrous or sometimes slightly hairy, glossy, underside paler green or glaucous, glabrous; petiole about ⅛ in. long. Stipules very small or wanting. Catkins appearing with or after the leaves on leafy peduncles, many-flowered, stout, about ⅜ in. long; scales obovate, silky, rosy-red at the tip. Stamen in effect solitary, the filaments being united throughout (more rarely are they free and the stamens two). Ovary distinctly stalked, glabrous, with a short style.

Native of Greenland and of Arctic eastern N. America, extending southward through Quebec and Newfoundland to some mountain-tops in New England. It is quite closely related to the European *S. retusa*, which differs from *S. uva-ursi* in its broader, usually truncate or retuse leaves equally green on both sides, almost glabrous catkin-scales and male flowers with always two stamens.

S. VIMINALIS L.   COMMON OSIER

An erect shrub or small tree, up to 20 ft high; young shoots grey with fine down at first, becoming glabrous and yellowish later. Leaves rather erect, linear or linear-lanceolate, tapering gradually to a fine point, not toothed; 4 to 10 in. long, ¼ to ½ in. wide; dull dark green and glabrous above, covered beneath with a shining, silvery grey, close down; stalk $\frac{1}{6}$ to $\frac{1}{2}$ in. long; midrib prominent. Catkins produced on the naked wood in March and April, up to 1 in. long, $\frac{3}{4}$ in. wide; scales narrow-oblong, obtuse, hairy, brown at the tip. Stamens two, with glabrous filaments. Ovary flask-shaped, sessile or almost so, with a long slender style and about equally long, slender stigmas.

SALIX VIMINALIS

Native of most of Europe outside the Mediterranean region, on the banks of streams, rivers and lakes, on flood-plains and in marshes, but so long cultivated that its distribution as a wild plant can no longer be known for certain. In the British Isles it is thought to be genuinely native at least in southern England. Outside Europe it occurs, wild or cultivated, in northern and central Asia, and has close relatives in the Far East.

S. viminalis is one of the most important of the basket-willows. The sorts known in the trade as 'Long Skein', 'Brown Merrin' and 'Yellow Osier' belong to this species.

S. KINUYANAGI Kimura—This willow is widely cultivated in Japan, to which it was probably introduced from Korea. It is closely allied to *S. viminalis*, differing in its stouter young stems and more densely arranged catkins with darker scales. Only male trees are known. This willow, as seen in the Hillier Arboretum, is very ornamental in spring when bearing its contiguous or overlapping catkins, which are ovoid or ellipsoid and about $\frac{3}{4}$ in. long. *S. kinuyanagi* is perhaps a cultivar of S. SCHWERINII E. Wolf, a native of the mainland of N.E. Asia.

S. WILHELMSIANA Bieb.

S. angustifolia Willd., not Wulf.; *S. dracunculifolia* Boiss.; ? *S. spaethii* Koopmann

A shrub up to 10 or 15 ft high; branchlets slender, arching, silky when young; winter-buds very densely set on the twigs. Leaves linear, crowded, often bent or sickle-shaped, 1 to 2 in. long, barely $\frac{3}{16}$ in. wide at the most, often half that width, silky on both sides when young, becoming more or less glabrous later, distantly and finely toothed, scarcely or very shortly stalked. Catkins slender, $\frac{3}{4}$ to 1 in. long, produced after the leaves on short leafy peduncles; scales pale yellow, obtuse, persistent, hairy at the base. Filaments of the two stamens usually connate throughout (hence each flower apparently with one stamen); anthers yellow; filaments glabrous. Ovary sessile, silky, developing into a very small capsule; style very short, with a minute stigma.

Native of S.W. Asia (E. Anatolia, Transcaucasia, Iran, Afghanistan) and of Russian Central Asia, of interest as a close relative of the better known *S. bockii* of China. A very elegant but slightly tender species, originally introduced under the name *S. microstachya*.

S. MICROSTACHYA Turcz. ex Trautv. *S. angustifolia* var. *leiocarpa* Ledeb.; *S. angustifolia* var. *microstachya* (Turcz.) Anderss.; *S. wilhelmsiana* var. *microstachya* (Turcz.) Herd.—Very closely allied to *S. wilhelmsiana*, differing in its glabrous ovaries and slightly broader leaves. Native of S.E. Siberia, Sinkiang and Mongolia, originally described from the region of Lake Baikal.

SALVIA   LABIATAE

A very large and wide-ranging genus of annuals and perennials, the latter mostly herbaceous or subshrubby; only a few species in Central and South America are true shrubs. The salvias are poorly represented in Europe outside the Mediterranean region, and only two are genuinely native in the British Isles.

The flowers in *Salvia* are borne in whorls, which are arranged in the form of spikes, racemes or panicles; sometimes the whorls are reduced to as few as two flowers. Calyx and corolla two-lipped. Tube of corolla variously shaped, sometimes with a ring of hairs inside; upper lip erect, hooded, the large spreading lower lip three-lobed, the centre lobe the largest, often toothed. Fertile stamens two instead of the four usual in the Labiates, the other two being either absent or reduced to staminodes. A remarkable and distinctive feature of *Salvia* is the structure of the stamens. Each has two arms (connectives) at the apex, one bearing a fertile cell, the other sterile; in some sections of the genus the latter is expanded and spoon-shaped and united by a sticky secretion to the corresponding arm of the other stamen.

Apart from *S. officinalis* and its allies very few salvias are both woody and hardy enough to be included in this work. Apart from those treated more fully below there are the following in cultivation, all on the border-line in both respects:

S. COERULEA Benth. *S. guaranitica* St Hil.; *S. ambigens* Briq.—A woody-based perennial to about 5 ft high, with ovate crenated leaves and deep blue flowers about 2 in. long, mostly four to six in each cluster, the inflorescence almost 1 ft long. *Bot. Mag.*, t. 9178. Native mainly of Brazil.

S. FULGENS Cav. CARDINAL SAGE.—A semi-woody much-branched perennial to about 4 ft. Leaves ovate, crenate, 1 to 3 in. long, hairy beneath. Flowers in summer and autumn, about 2 in. long, with a purplish red calyx and hairy bright crimson-scarlet corollas, arranged in rather distant clusters, about six flowers in each. Mexico.

S. INVOLUCRATA Cav.—A fairly woody perennial to about 3 ft high, sparsely branched, with glabrous, acuminate, long-stalked cordate leaves about 4 in. long. Although grouped by Bentham with *S. fulgens* and *S. microphylla* it is very distinct from both in its inflorescence, each whorl (of three to six flowers) being surrounded by pink floral leaves which drop off as the flowers expand. Corolla inflated, shortly lipped, crimson, bright rosy crimson in 'BETHELII' (the finest form raised by a Mr Bethell shortly before 1880). It is late-flowering, from about August and was once used for the winter decoration of conservatories. Mexico.

S. RUTILANS Carr.—This small subshrub is mainly winter-flowering and only suitable for a cool greenhouse or conservatory outside the mildest parts, though it may give a good display elsewhere, in a warm corner, if the autumn is mild. Flowers with bright red slender-tubed corollas; whorls arranged in branched spikes. Leaves pineapple-scented when crushed. Known only as a cultivated plant but near to *S. elegans* Vahl, a native of Mexico.

All the above are easily propagated by soft cuttings in summer. All need a very warm corner and a light soil low in nitrogenous matter, which encourages lush growth at the expense of flower.

S. GREGGII A. Gray

An evergreen shrub 3 to 4 ft high with square, slender, drooping, finely downy, very leafy branches. Leaves clustered, ¾ to 1½ in. long, narrowly oblong-lanceolate to oval, entire. Flowers mostly in pairs, borne from June onwards in slender, glandular-downy racemose inflorescences 2 to 6 in. long. Calyx ⅜ in. long, two-lipped, purplish, ribbed. Lower lip of corolla carmine-red, ¾ in. wide; the upper, hooded part more rosy and hairy.

Native of Mexico and Texas; discovered by Dr J. Gregg in 1848 but not described until 1870. It first flowered with Thompson of Ipswich in 1882. Perhaps scarcely as hardy as *S. microphylla*.

S. INTERRUPTA Schousboe

A plant with herbaceous shoots and a shrubby base, growing 2 to 3 ft high, of spreading habit; young shoots clothed with soft hairs. Leaves of variable shape and size, usually oblong-ovate in the main with a pair of lobes at the base, or pinnate with two pairs of leaflets; on weak or flowering shoots they are simply oblong-ovate; the larger leaves are up to 7 or 8 in. long, the smallest 1½ in. long, the stalks varying from 4 in. to 1 in. in length; both surfaces are wrinkled and hairy, the lower one especially; margins round-toothed. The inflorescence is a tall, open panicle 2 to 4 ft long, the flowers being borne on

SALVIA INTERRUPTA

a few branches towards the top. The whorls of flowers are 1 to 2 in. apart with six or nine flowers in each whorl, each on a downy stalk ⅛ to ¼ in. long. Corolla blue with a violet tinge, white on the throat and lower lip; opening dark, it pales with age, 1¼ in. long, 1 in. wide, downy; lower lip three-lobed, the middle lobe notched, ¾ in. wide; upper lip hooded, compressed. Calyx viscid, tubular, ½ in. long, ribbed, glandular-downy, two-lipped. Anthers attached to the stalk of the stamens by a secondary curved stalk. Style slender, exserted.

Native of Morocco; introduced to the Cambridge Botanic Garden in 1798. This handsome sage, which flowers from May to July, although it survives moderate winters unharmed, is not perfectly hardy. Probably for this reason it has at times disappeared from cultivation. It is sometimes grown in pots for conservatory decoration. It likes all the sunshine it can get.

Closely allied to *S. interrupta* is S. CANDELABRUM Boiss., an endemic of S. Spain.

S. MICROPHYLLA H.B.K.
S. grahamii Benth.

An evergreen shrub up to 3 or 4 ft high; young shoots soft, slender, square, downy, reddish purple. Leaves opposite, ovate, round-toothed, 1 to 1½ in. long (to 3 in. on sterile shoots); dull green and slightly downy on both surfaces; stalk ¼ to 1 in. long. Flowers produced mostly in pairs on terminal racemes up to 6 in. long, opening successively from June onwards. They are about 1 in. long, the lip of the corolla ½ to ¾ in. wide, rich red on first opening, changing with age to magenta-purple; the upper, hooded part of the corolla is rosy red. Calyx tubular, ⅓ in. long, ribbed, downy, reddish above.

Native of Mexico; introduced by G. J. Graham about 1830 to the Horticultural Society's garden at Chiswick and described by Bentham under the still familiar name of *S. grahamii*. But Bentham was aware that Graham's salvia was very near to the earlier-named *S. microphylla*, and there can be no doubt that the two represent states of the same species, the type of *S. microphylla* being dwarfer and with smaller leaves than the type of *S. grahamii* but otherwise essentially the same.

S. microphylla is quite hardy in the south and south-western counties and even in less favoured parts will live for several years in a sheltered nook or against a wall. There is no difficulty in keeping up a stock, as cuttings root most freely and will flower well in the following summer. A peculiarity of the plant is the strong odour given off by the leaves when they are crushed or rubbed. This almost exactly resembles the odour of black currant leaves.

var. NEUREPIA (Fern.) Epling *S. neurepia* Fern.; *S. grahamii* Hort., in part— Leaves on the flowering stems 1½ to 3 in. long, light green, rather thin. Flowers, at least on cultivated plants, with a clear bright red lower lip. This is commoner in gardens than the type and of the same order of hardiness.

S. OFFICINALIS COMMON SAGE

A sub-evergreen, aromatic shrub, usually 1 to 2 ft high, but said in favourable places to become three times as high; young stems square, and only half woody; the whole plant is covered with a short down which gives it a grey appearance. Leaves opposite, oblong, 2 to 3½ in. long, ½ to 1 in. wide; much wrinkled, round-toothed. Flowers arranged in whorls on terminal, erect racemes about 6 in. long. Corolla tubular, ¾ in. long, two-lipped, purple; calyx ribbed, funnel-shaped, two-lipped, about half as long as the corolla. Perfect stamens two. Blossoms from June onwards.

This species is considered to be genuinely wild at least in Spain and W. Yugoslavia, but is naturalised elsewhere in S. Europe; long cultivated as a medicinal and culinary herb and probably introduced to Britain in the early Middle Ages (or even in Roman times). It was highly valued in former times for making sage-beer—supposed to possess many healing virtues. It is still much used in the kitchen and indeed one of the most widely grown and sold of culinary herbs. The plant likes a sunny position, and is easily increased by cuttings placed under a handlight. Although rarely seen outside the kitchen garden, this plant is worth growing in a collection of old-fashioned fragrant plants for its crowd of erect racemes. But some of the clones of common sage sold for growing in the kitchen garden are selected for their leafiness and do not flower freely.

cv. 'Albiflora'.—Flowers white. Said to be the best culinary sage.

cv. 'Aurea'.—Leaves yellow; habit compact. This seems to have become rare.

cv. 'Icterina' ('Variegata').—Leaves variegated with yellow and light green (*S. off. icterina* Alefeld, possibly a misprint for *icterica*—jaundiced).

cv. 'Purpurascens' ('Purpurea').—Leaves purple when young. This is probably the red sage of Parkinson, cultivated since the early 17th century at least. There is more than one clone under this name.

cv. 'Rubriflora'.—Flowers reddish purple. Cultivated since the 16th century.

cv. 'Tricolor'.—Leaves grey-green, veined with yellowish white and pink, darkening to red or rosy-red. Raised in France towards the end of the last century. Probably a sport of the purple sage.

S. lavandulifolia Vahl *S. hispanorum* Lag.; *S. officinalis* var. *hispanorum* (Lag.) Benth.—Very closely allied to *S. officinalis* but with narrower oblong or linear-oblong leaves and the flowering stems not furnished with leaves. Native of Spain, cultivated in European gardens since the 16th century.

S. grandiflora Etlinger ? *S. tomentosa* Mill.—More robust than *S. officinalis*, with leaves 2 to 4 in. long and up to 2½ in. wide, rounded or slightly cordate at the base. Native of the Balkans, Asia Minor, etc., cultivated in Central and W. Europe since the 16th century and said to be more commonly grown in S. Germany and N. Switzerland than the common sage (Hegi, *Fl. Mitteleuropa*, Vol. V, p. 2483).

SAMBUCUS   Elder   caprifoliaceae

About forty species of elder are known, which are widely spread over the temperate and subtropical parts of the globe. Of these about nine shrubby ones are hardy in Britain. From the remainder of the hardy shrubs of the Caprifoliaceae, the elders are at once distinguished by their pinnate leaves, which always have an odd number (three to eleven) of toothed leaflets. There are, however, other differences of a technical character, considered by some botanists to be weighty enough for *Sambucus* to be given the rank of a monotypic family, the Sambucaceae. The flowers are borne in umbel-like inflorescences (sect. *Sambucus*) or in panicles (sect. *Botryosambucus*); to the latter section belong *S. melanocarpa*, *S. pubens* and *S. racemosa*. The flowers are very uniform in size and hue, being from $\frac{1}{8}$ to $\frac{3}{16}$ in. across, and of some shade of white. The various parts are normally in fives. Fruits $\frac{1}{4}$ in. or less in diameter, globose or nearly so, containing three to five one-seeded nutlets. All the cultivated species are deciduous, and have opposite leaves, the young shoots are soft and full of pith, but the wood of the trunk is hard and bony. A few species are herbaceous, of which the best known is *S. ebulus* L., which is naturalised in Britain but doubtfully a native.

The elders like a good, moist soil, and given this are not hard to accommodate. They can be propagated by cuttings either of leafless wood put in the open ground in early winter, or by half-ripened young wood with a heel in frames. The pruning of the sorts grown for their foliage should be done before growth commences.

S. CANADENSIS L.   AMERICAN ELDER

A deciduous shrub, up to 12 ft high, with white pith; young branches glabrous. Leaves pinnate, the leaflets mostly seven (but also five, nine, and eleven), oval, oblong, or roundish ovate; the largest $5\frac{1}{2}$ in. long, $2\frac{1}{2}$ in. wide, taper-pointed, sharply toothed, the lowest pair frequently two- or three-lobed, lower surface glabrous or slightly downy. Flowers in convex umbels, 4 to 8 in. across, white, produced in July. Fruits purple-black.

Native of eastern N. America from Canada to Florida; introduced in 1761. Nearly allied to *S. nigra*, it differs in the following respects: it never assumes a tree-like form or becomes half as high as *S. nigra*; the leaves have normally one more pair of leaflets; the flower clusters are more rounded and appear four weeks later; the fruit is not absolutely black. I have seen it making a very pleasing picture growing by the side of a stream in the Arnold Arboretum, Mass., flowering in July, but it is not so good in this country as *S. nigra*.

cv. 'ACUTILOBA'.—This is the counterpart of the cut-leaved form of the European elder, but is more graceful owing to the longer and more divided leaf, which is dark green. Put into commerce by the American firm of Ellwanger and Barry early this century.

cv. 'AUREA'.—Leaves golden yellow, fruits red.

cv. 'MAXIMA'.—The best and most remarkable form of the American elder in cultivation, originally sent out by Messrs Hesse of Germany under the erroneous name "*S. pubens maxima*". It is an extraordinarily robust variety, with leaves 12 to 18 in. long; the leaflets are often eleven to each leaf, and the enormous flower clusters are 10 to 18 in. across.

S. CERULEA Raf.   BLUE ELDERBERRY
S. glauca Nutt. ex Torr. & Gr.

A tree 13 to 30 ft (occasionally 50 ft) high in the wild, but a robust shrub in this country 5 to 10 ft high; young shoots glabrous. Leaves 6 to 10 in. long, glabrous; the leaflets usually five or seven, occasionally nine, ovate or oval, 2 to 6 in. long, ½ to 2 in. wide. Flowers yellowish white, produced during June in flat umbels up to 6 or 7 in. wide. Berries black, but covered densely with a pale blue bloom.

Native of western N. America. Its two most striking characteristics are its vigorous growth, which makes it even more tree-like in California than S. nigra is in Europe, its trunk being sometimes 18 in. in diameter; and the intensely glaucous hue of its berries.

S. MELANOCARPA A. Gray
S. racemosa var. melanocarpa (A. Gr.) McMinn

A deciduous shrub, 6 to 12 ft high, allied to S. pubens and S. racemosa, having its flowers and fruits in panicles as in those species, but the panicles are usually broader in proportion to their height. The berries, moreover, are not red but black, and without bloom. Leaflets five or seven (sometimes nine), their chief veins and midrib more or less downy beneath when young, but not so downy as S pubens. Native of western N. America; introduced to Kew in 1894.

S. NIGRA L.   COMMON ELDER

A deciduous shrub, 15 to 20 ft high, or a small tree 30 ft or more high; young branches glabrous. Leaves pinnate, 4 to 12 in. long, composed of three, five, or seven (usually five) leaflets, which are ovate, 1½ to 5 in. long, ¾ to 2 in. wide, sharply toothed, glabrous except for a few hairs beneath. Flowers yellowish or dull white, with a heavy odour, produced during June in flat umbels 5 to 8 in. across, each umbel composed of four or five main divisions which are again several times divided. Berries globose, shining black, ¼ in. wide, ripe in September.

Native of Europe, N. Africa and S.W. Asia. One of the best known of native shrubs, and to be regarded more often as a weed in gardens than anything else. Still, the elder, when made to assume the tree form by restricting it to one stem for 6 or 8 ft up, is not without a certain quaintness and charm. Its trunk is rough and crooked, and carries a large rounded head of richly leafy branches, laden with flower in June and with fruit in September. The seeds are spread by birds, and young elder plants spring up everywhere in woods, tall shrubberies, etc. In the neighbourhood of more important plants they must be rigorously pulled up. The species is chiefly represented in gardens by the numerous varieties that have sprung from it, some of which are mentioned below as worth cultivating. The type itself may be left to furnish out-of-the-way damp, dark corners, where little else will live.

No plant holds (or perhaps it is safer to say, used to hold) a more honoured place in domestic pharmacy than the elder. From its berries is prepared, by boiling with sugar, a wine or syrup which, diluted with hot water, is a favourite beverage in rural districts. It is usually taken just before bedtime and is considered a useful remedy for colds, chills, etc.

A large number of varieties have been obtained under cultivation, of which the following only need be mentioned as the most distinct:

cv. 'AUREA' GOLDEN ELDER—A good yellow-leaved shrub, useful for producing a broad patch of colour. It may be pruned back each spring. In cultivation 1883 and possibly the same as the elder once known as *S. n. aurea Dixonii*.

f. LACINIATA (L.) Zab. *S. nigra* var. *laciniata* L. PARSLEY-LEAVED ELDER.—The handsomest cut-leaved variety of common elder, the leaflets being pinnately divided into linear pointed lobes. Known since the 16th century and occurring occasionally in the wild.

cv. 'LINEARIS'.—In this form the blade of many leaflets is reduced to threadlike proportions, consisting of little more than the stalk and midrib. Others are ⅛ to ¾ in. wide, but distorted and shapeless. A curiosity only (*S. n. linearis* Kirchn.; *S. n. heterophylla* Schwer.).

cv. 'MARGINATA'.—A handsomely variegated shrub whose leaves are bordered with creamy white. The name *S. n.* var. *albo-variegata* was used for this in previous editions, but it belongs properly to an old white-splashed form. There is also a gold-margined variety, 'AUREO-MARGINATA'.

cv. 'PENDULA'.—A weeping form with stiff, pendulous branches.

cv. 'PYRAMIDALIS'.—A stiffly erect, inelegant form.

cv. 'ROSEO-PLENA'.—Flowers rosy coloured, with a double row of petals.

cv. 'ROTUNDIFOLIA'.—Leaves often with only three leaflets, proportionately broader, smaller and more rounded than in the type.

f. VIRIDIS Schwer. *S. n.* var. *viridis* West.; *S. virescens* Desf.; *S. n.* var. *virescens* (Desf.) DC.; *S. n.* var. *alba* West.; *S. n.* var. *leucocarpa* Sm.; *S. n.* f. *alba* (West.) Rehd.; *S. n.* var. *chlorocarpa* Hayne—A mutation, occurring occasionally in the wild, including Britain, in which the fruits lose their purple dye and become greenish or whitish and more or less translucent; known since the 16th century. The wine made from the berries is clear. The green-fruited and so-called white-fruited forms were united in previous editions of this work, as they were by Graf von Schwerin in his monograph on the genus. De Candolle, too, in the *Prodromus*, doubted if they merited separate naming.

There are two or three clones in cultivation in which the leaves are more or less flushed with purple. One of these, grown at Kew since 1957, derives from a plant found by Mr Robert Howat by a roadside in Yorkshire in 1954, and propagated by him. In this the flowers are pink-tinged, and the leaves bronze-purple, with or without streaks of bright green (*Gard. Chron.*, Vol. 156 (1964), p. 567). A clone cultivated by Messrs Hillier as 'Foliis Purpureis', and received from a garden in N. Ireland, received an Award of Merit in 1979.

S. PUBENS Michx.

S. racemosa var. *pubens* (Michx.) Koehne

This species, which occurs over a wide area of North America east of the Rocky Mountains is so closely allied to the Old World *S. racemosa* that some authorities do not separate them. The American shrub is distinguished by its young shoots, leaves and flower-stalks being downy, the brown pith, and the fruit-panicles not so densely packed with berries.

f. LEUCOCARPA (Torr. & Gr.) Rehd. *S. pubens* var. *leucocarpa* Torr. & Gr.— Fruits white.

f. XANTHOCARPA (Cockerell) Rehd. *S. racemosa* f. *xanthocarpa* Cockerell— Fruits yellow.

S. CALLICARPA Greene *S. racemosa* var. *callicarpa* (Greene) Jeps.;? *S. pubens* var. *arborescens* Torr. & Gr.—This replaces *S. pubens* in western N. America from Alaska to California. The mature leaves tend to be less downy beneath and the fruit-panicle flatter. Uncommon in cultivation, but evidently a fine shrub in the wild, attaining a height of 20 ft in places. Yellow-fruited plants are sometimes found.

S. RACEMOSA L.   RED-BERRIED ELDER

A deciduous shrub, 8 to 12 ft high, and as much through; young bark glabrous, pith white. Leaves pinnate, 6 to 9 in. long, composed of five leaflets, which are oval or ovate, 2 to 4 in. long, $\frac{3}{4}$ to $1\frac{3}{4}$ in. wide, taper-pointed, sharply and regularly toothed, glabrous on both surfaces. Flowers produced during April in terminal pyramidal panicles $1\frac{1}{2}$ to 3 in. high, scarcely so much wide; yellowish white. Berries scarlet; ripe in June and July; packed tightly in panicles.

Native of Europe, Asia Minor, Siberia, and W. Asia, cultivated in England since the 16th century. This very beautiful-fruited shrub is only occasionally seen in perfection in this country, although it grows well and flowers abundantly. It fruits admirably near Paris, and those who have visited the upland valleys of Switzerland in July will have marked its great beauty there. Whilst not a native of Britain it has established itself in a remarkable way in Scotland. On the slopes of the hills bordering the Tweed in one area, at least, above Peebles, it occurs in broad masses.

It would be an exaggeration to assert that *S. racemosa* never fruits in southern England, for it has done so in many places, but it cannot be regarded as a reliable fruiter in all areas. Why this should be is uncertain. Some gardeners have reported that fruits form but are taken by the birds before they have coloured. It has also been suggested that some seedlings and clones are self-incompatible, i.e., do not set fruit when self-pollinated, or pollinated by another plant of the same clone. But if we are denied too frequently its attractive fruits, it has on the other hand sported into a number of coloured and cut-leaved forms, which are amongst the best of their class, and thrive well.

f. LACINIATA (Koch) Zab.—Leaflets deeply and pinnately lobed, the lobes linear, pointed, not more than $\frac{1}{12}$ in. wide. This description is made from a cultivated plant of unknown origin, but laciniated plants occur occasionally in the wild, and indeed the botanical name is founded on a wild plant. See also 'Ornata', 'Plumosa' and 'Tenuifolia'.

cv. 'ORNATA'.—The lower leaves of the shoots cut more or less as in 'Plumosa', the uppermost more finely laciniated. Raised by Messrs Simon-Louis from 'Plumosa', described 1891.

cv. 'PLUMOSA'.—Leaflets up to 5 in. long and 1¼ in. wide, the teeth reaching half-way to the midrib. Put into commerce by Späth in 1886. He received it from Russia.

cv. 'PLUMOSA AUREA'.—A wholly golden form of 'Plumosa', and one of the most attractive of golden-leaved shrubs. Sent out by Messrs Wezelenburg in 1895, and given an Award of Merit in that year.   [PLATE 37

cv. 'PURPUREA'.—Petals rose-coloured on the back. Mentioned by Sweet in 1826.

cv. 'SPECTABILIS'.—Flowers nearly pure white.

cv. 'TENUIFOLIA'.—Leaflets divided quite to the midrib into long narrow segments, often doubly pinnate. A very handsome and graceful shrub with fern-like foliage, not a strong grower and therefore useful for small gardens. Of the same origin as 'Ornata'.

SANTOLINA*   COMPOSITAE

Two or three species of *Santolina* are not uncommon in cultivation. They are plants with semi-woody stems, strong-scented when crushed, and with yellow flower-heads composed of very numerous small florets, and without the ray florets common to so many plants of this family. They are of very easy cultivation, growing best in full sun in any soil that is well drained and not too rich. Cuttings taken about July, put in pots of sandy soil and placed in heat, root in a few days. All of them are seen at their best in a comparatively young state, and are apt to become shabby with age. *S. chamaecyparissus* is valuable for planting in masses on the front of a shrubbery, both for its whiteness and for its abundant blossom.

S. CHAMAECYPARISSUS L.   LAVENDER COTTON
S. incana Lam.

A white bush, 1 to 2 ft high in this country, forming a close, leafy mass; foliage persistent; stems semi-woody, covered the first season with a thick

* Revised by C. Jeffrey, of the Herbarium, The Royal Botanic Gardens, Kew.

white felt. Leaves alternate, very crowded on the shoots, the largest 1 to
1½ in. long, with clusters of shorter ones in their axils, all very narrow (⅛ in.
or less wide), and furnished with thick teeth or projections set in rows of about
four. The whole leaf is clothed with a white felt. Flower-heads bright yellow,
½ to ¾ in. across, hemispherical, solitary at the end of an erect, slender stalk
4 to 6 in. long, terminating short lateral twigs of the year. There are no ray-
florets.

Native of the Western and Central Mediterranean region; cultivated in
Britain since the middle of the 16th century. It is a beautiful and interesting
plant, probably the whitest of all hardy shrubs, and bears its showy flower-heads
in July so thickly that they almost touch. The plant has a rather agreeable
odour when lightly rubbed, but this becomes too strong and acrid to be
wholly pleasant when the leaves are crushed. Formerly used in medicine as
a vermifuge. The leaf has a curious structure suggestive of the stems of some
lycopods; it consists of a central axis on which are set, often in whorls, short,
thick, blunt projections.

var. NANA Hort. *S. chamaecyparissus* var. *corsica* Hort., not Fiori—A cultivated variant, 1 to 1½ ft high, with a lower, denser, more compact habit. Suitable
for the rock garden.

S. PINNATA Viv.

S. chamaecyparissus subsp. *tomentosa* (Pers.) Arcangeli

This species is mainly represented in cultivation by the following subspecies,
at present generally known in gardens as *S. neapolitana*:

subsp. NEAPOLITANA (Jord. & Fourr.) Guinea   *S. neapolitana* Jord. &
Fourr. in *Ic. Fl. Europ.* 2: 10, tab. 228 (1869); *S. italica* Hort.; *S. rosmarinifolia*
Hort., not L.—An evergreen, rather pleasantly scented shrub 2 to 2½ ft high,
producing a closely packed crowd of erect, slender branches covered with white
felt. Leaves 1 to 2 in. long, ⅛ to ¼ in. wide, either pinnate or with leaflets superposed in four rows, the leaflets (or perhaps better termed 'leaf-segments') are
$\frac{1}{12}$ to ¼ in. long, cylindrical, round-ended, $\frac{1}{36}$ in. in diameter, covered with
white down on the young non-flowering shoots, green and less downy on the
flowering ones. Flowers bright yellow, borne in a compact, circular, cushion-
shaped head, ¾ in. wide, in July, each head on a slender erect stalk 3 to 6 in.
long, from six to twelve heads being produced at or near the end of each shoot.

Native of S. Italy. This attractive shrub, often grown wrongly as "*S. rosmarinifolia*", differs from *S. chamaecyparissus* by the longer leaves and by the longer,
more slender segments of the leaf. It makes a bright display when in flower.
It should be grown in full sunshine and is better in rather poor soil than in
rich, the foliage being whiter and the growths sturdier and less liable to fall
apart, thus leaving the centre of the plant open and unsightly. Pruned plants
produce shoots which are at first quite green.

cv. 'EDWARD BOWLES'—Leaves greenish grey; flowers creamy white.

cv. 'SULPHUREA'—Leaves greyish green; flowers pale primrose-yellow.

subsp. ETRUSCA (Lacaita) Guinea *S. chamaecyparissus* var. *etrusca* Lacaita in *Nuov. Giorn. Bot. Ital.* 32: 215 (1925)—Leaves green, glabrous; flowers pale yellow to yellowish white. Native of Italy and Sicily, distinguished from the subsp. *neapolitana* by its shorter, denser habit, green leaves and creamy yellow flowers. Both the leaves and leaf-segments are shorter than in the typical subspecies (see below).

subsp. PINNATA—This, the typical subspecies of *S. pinnata*, has green, glabrous leaves 1 to 1½ in. long; leaf-segments ⅛ to 3/16 in. long, arranged in two or four rows. Flowers dull white. Native of Italy, flowering in July, distinguished by its white flowers and its green foliage with only a slight odour. Of little ornamental value.

S. ROSMARINIFOLIA L.   HOLY FLAX
S. viridis Willd.; *S. virens* Mill.

An evergreen bush 1½ to 2 ft high; stems glabrous, green. Leaves deep green, glabrous, the largest 1 to 2 in. long, about ⅛ in. wide, very similar in structure to those of the preceding species but with the teeth or projections more distant, forward-pointing or sometimes wanting. Flower-heads bright yellow, ¾ in. across, produced in July singly at the end of slender, erect, glabrous stalks 6 to 10 in. long.

SANTOLINA ROSMARINIFOLIA

Native of southern France, Spain and Portugal; in cultivation 1727. An attractive species, with foliage of a rich deep green. The plant is not quite as dense in growth, nor quite so hardy, as *S. chamaecyparissus*. The leaves emit an odour when rubbed, but it is neither so strong nor so pleasant as that of the latter species.

cv. 'PRIMROSE GEM'—Flowers pale primrose-yellow.

subsp. CANESCENS (Lag.) Nyman   *S. canescens* Lag.; *S. pectinata* Lag., not Benth.—Like the typical subspecies, but stems finely downy and the leaves pinnate, with linear, more crowded, laterally spreading lobes. Native of southern Spain.

SAPINDUS   SOAPBERRY   SAPINDACEAE

A genus of about thirteen species of deciduous or evergreen trees and shrubs in both the Old and New Worlds, mostly confined to the tropics and subtropics. Leaves alternate, usually even-pinnate, rarely simple, leaflets entire. Flowers small, unisexual or effectively so, in panicles. Sepals and petals four or five. Stamens eight or ten, inserted inside a fleshy ring-shaped disk. Ovary superior, three-celled, but usually only one carpel developing; style one, with a lobed stigma. Fruit drupe-like. The flesh of the fruits is rich in saponin, which causes it to lather in water, whence the generic name, from *Sapo indicus* or Indian soap, the fruits of the type-species, *S. saponaria* L, being used as soap in the West Indies; the specific epithet also refers to this property. The fruits of other species are put to the same use in some parts of the world.

The large family Sapindaceae is mostly tropical and subtropical. For other genera of the family treated in this work, see *Koelreuteria* and *Xanthoceras*. An economically important member of the family is *Litchi sinensis*, which is the source of the fruit known as lychee. The edible part, of grape-like taste and consistency, is an aril produced by the single small seed; the fruits themselves have a brittle, warty covering.

S. DRUMMONDII Hook. & Arn.   WILD CHINA-TREE
S. saponaria var. *drummondii* (Hook. & Arn.) Benson

A deciduous tree up to 40 or 50 ft high, with scaling bark; young shoots warted, slightly downy. Leaves alternate, pinnate, 10 to 15 in. long, made up of eight to eighteen leaflets which are lanceolate, slenderly pointed, tapered and often oblique at the base, not toothed, $1\frac{1}{2}$ to $3\frac{1}{2}$ in. long, $\frac{1}{2}$ to 1 in. wide, rather pale green and glabrous above, downy beneath when young especially at the base of the midrib. Flowers yellowish white, $\frac{1}{6}$ in. wide, produced during

June on pyramidal terminal panicles 6 or 8 in. long; sepals narrowly triangular; petals obovate, downy inside; stamens hairy, eight or ten. Fruits about $\frac{1}{2}$ in. wide, nearly globose, described as having a thin, dark, orange-coloured, semi-translucent flesh; they ultimately turn black.

Native of the southern United States, especially towards the west, and of N. Mexico; introduced to Britain in 1915. It is quite hardy but of little ornamental value. The popular name derives from a fancied resemblance between this species and *Melia azederach*, known as the China-tree or China-berry in the southern USA, where it is much cultivated.

SAPIUM EUPHORBIACEAE

A genus of about 100 species of trees and shrubs in the tropics and subtropics, a few extending into temperate regions. Leaves simple, pinnately veined. Flowers without petals, unisexual, borne in terminal bracteate spikes, racemes or panicles. Male flowers numerous, with two or three stamens. Female flowers few, at the base of the inflorescence; styles three. Fruits fleshy or dry, three-lobed and three-chambered, each chamber containing a single seed.

S. JAPONICUM (Sieb. & Zucc.) Pax & Hoffm.

Stillingia japonica Sieb. & Zucc.; *Excoecaria japonica* (Sieb. & Zucc.) Muell.-Arg.

A deciduous shrub or small tree, glabrous in all its parts. Leaves broad-ovate, 3 to 6 in. long, 2 to 4 in. wide, entire, dark green above, glaucous beneath, with a pair of glands at the base. Petioles about 1 in. long. Flowers borne in June in erect catkin-like racemes at the ends of leafy shoots, the numerous male flowers very shortly stalked, the few female flowers at the base on pedicels about $\frac{1}{2}$ in. long. Fruit a three-lobed roundish capsule flattened at both ends, about $\frac{7}{8}$ in. wide, containing three seeds.

Native of Japan, including the Ryukyus, and of China and Korea. Although not showy in flower it is of interest as one of the few woody members of the Spurge family hardy in this country. The leaves turn bright crimson in the autumn. It is in cultivation at the Hillier Arboretum, Romsey, Hants, and at Wakehurst Place, Sussex.

S. SEBIFERUM (L.) Roxb. *Croton sebiferus* L. CHINESE TALLOW TREE.—

A deciduous tree to 60 ft high in China, with long-stalked acuminate leaves resembling in shape those of a black poplar and becoming crimson in autumn. Inflorescence similar to that of *S. japonicum*, but the male flowers arranged in cymose clusters along the axis of the inflorescence. Native of China, but long cultivated in the warmer parts of the world. The seeds have a waxy layer under the epidermis, used for making candles and soap, and an oil extracted from the seeds themselves is used for the same purpose. It was in cultivation in Britain early in the 18th century and is sometimes grown in greenhouses for its handsome foliage and economic interest. It is not reliably hardy in this country, but is naturalised in the south-eastern USA, where, like *Rosa laevigata*, it was already established early in the 19th century.

SARCOBATUS   CHENOPODIACEAE

A genus of two species in western North America; for characters see below. It is the only cultivated member of the Goosefoot family that has spiny branches.

S. VERMICULATUS (Hook.) Torr.   GREASE WOOD
Batis (?) *vermiculatus* Hook.; *S. maximilianii* Nees

A deciduous shrub of lax habit, 6 to 9 ft high, more in diameter, making a dense thicket of stems, arching and spreading at the top; twigs angular, whitish, spine-tipped, usually glabrous. Leaves alternate, linear, $\frac{1}{2}$ to $1\frac{1}{2}$ in. long, $\frac{1}{16}$ to $\frac{1}{8}$ in. wide, grey, rather fleshy, stalkless. Flowers small, greenish, unisexual; males crowded in a spike $\frac{1}{2}$ to 1 in. long at the end of short lateral twigs, females appearing singly in the axils of the lower leaves of the same twig. Neither has any beauty, but they are interesting botanically. The male flower has neither calyx nor corolla, the stamens, about three in number, being arranged at the base of curious cup-like scales. The female flower is also without a corolla, but has a calyx which persists and enlarges and ultimately develops into a thin, papery disk, prominently veined, $\frac{1}{4}$ to $\frac{1}{3}$ in. across, with the seeds in the middle.

Native of the dry, alkaline, and saline regions of western N. America; introduced to Kew in 1896 but is no longer there. Like other shrubs from the same regions, it thrives quite well in ordinary garden soil. It flowers in July, but, as may be judged from the description, is of more botanical than horticultural interest.

SARCOCOCCA   SWEET BOX   BUXACEAE

A genus of evergreen shrubs in E. and S. Asia, the cultivated species coming from China and the Himalaya. They are allied to *Buxus*, but have

alternate leaves and female flowers below the male flowers in the inflorescence (in *Buxus* the leaves are opposite and the inflorescence
comprises several staminate flowers and a single terminal female flower).
Some species are rhizomatous, the rhizome and its branches each giving
rise at the apex to a single aerial stem which is usually unbranched for
some distance above the base. Two species, however, are densely bushy
with a much-branched main stem and a well-developed fibrous root-
system. The flowers have no petals; the males have four sepals and four
stamens, the females four to six sepals. Fruit a fleshy berry, either egg-
shaped or globose. The hardy species, all Chinese, are neat and pleasing
shrubs with only a modest beauty of flower, but healthy in appearance,
the flowers white, fragrant. Increased easily by summer cuttings. They
will thrive in any moist soil, and have a value in gardens on account of
their suitability for shaded spots.

For an account of the cultivated species, see: J. Robert Sealy, 'Species
of Sarcococca in Cultivation', *Journ. R.H.S.* Vol. 74 (1949), pp. 301–6.

S. CONFUSA Sealy

S. humilis Hort., in part, not Stapf; *S. ruscifolia* var. *chinensis* Hort., in part, not (Franch.)
Rehd. & Wils.

A densely branched evergreen shrub up to 6 ft high and as much through,
with a single main stem and fibrous root-system; branchlets downy. Leaves
elliptic or elliptic-lanceolate to elliptic-obovate, long-acute to acuminate, obtuse
to cuneate at the base, mostly $1\frac{1}{4}$ to 2 in. long and $\frac{1}{2}$ to $\frac{3}{4}$ in. wide, but up to
$2\frac{5}{8}$ in. long and 1 in. wide on vigorous shoots, upper surface dark green, lower
surface light green; petioles up to $\frac{1}{4}$ in. long. Flowers fragrant, crowded in
short clusters of one to three males and one to three females, but sometimes
reduced to one or two male or female flowers. Male flowers with four white
sepals about $\frac{1}{8}$ in. long, much exceeded by the filaments, which are two to
three times as long. Female flowers $\frac{3}{16}$ in. long, with four to six sepals. Ovary
with two or three stigmas. Fruits black. *Bot. Mag.*, t. 9449, as *S. humilis*.

A plant of unknown origin, in cultivation since 1916; it may have been
raised from seeds collected by Wilson in W. China, but is not matched by any
wild specimen. From all other species, *S. confusa* is distinguished by the female
flowers having either two or three stigmas, all other species having consistently
either two or three. Its bushy habit of growth distinguishes it from all other
species except *S. ruscifolia*, and from that species it differs in its leaves, in the
fruits becoming finally black, not red, and by the number of stigmas (constantly
three in *R. ruscifolia*). The flowers are sweetly scented, and are borne from late
December to February.

S. HOOKERIANA Baill.

An evergreen shrub up to 6 ft high, increasing by sucker growths from
the base; young shoots minutely downy. Leaves narrowly lanceolate to oblong-

lanceolate, slenderly pointed, wedge-shaped at the base, 2 to 3½ in. long, ½ to
¾ in. wide, bright green and quite glabrous; stalk about ¼ in. long. Flowers
small, fragrant, unisexual, white, crowded in the leaf-axils; styles three; fruit
nearly globose, ¼ in. wide, black. Flowers in late autumn.

A native of the eastern Himalaya, N. Assam and S.E.Tibet. Although
introduced in the last century, it is less common now than its Chinese variety
and rather less hardy. It received an Award of Merit when shown from Bodnant
in 1936.

var. DIGYNA Franch.—This is a variety from W. China, introduced in 1908
by Wilson. It is of dwarfer habit than the Himalayan type and is quite hardy.
Its chief botanical distinction is in having only two styles to each flower. From
S. ruscifolia it is distinct in its black fruits, that species having dark red ones.
Award of Garden Merit 1963.

cv. 'PURPLE STEM'.—A variant of the var. *digyna* in which the young
stems, petioles and midribs are dull purple.

S. HUMILIS Stapf
S. hookeriana var. *humilis* Rehd. & Wils.

An evergreen shrub of neat, tufted habit, 1 to 2 ft high, stems minutely
downy when young. Leaves narrowly oval, pointed, and somewhat more
tapered at the apex than at the base, 1 to 3 in. long, ⅓ to ¾ in. wide, glabrous
and glossy green above, paler beneath, with a prominent nerve parallel to
each margin; stalk ⅛ to ¼ in. long. Flowers in short, axillary racemes, white,
very fragrant, produced normally in early spring, sometimes in autumn;
stamens with flattened stalks, petal-like; anthers pink; styles two. Fruits round,
¼ in. in diameter, blue-black.

A native of western China, discovered by Augustine Henry and introduced
by Wilson in 1907. It is closely allied to R. *hookeriana* var. *digyna*, differing
in its dwarfer stature, its shorter, relatively much broader leaves and pink
anthers (cream-coloured in R. *hookeriana* var. *digyna*). It is a neat little shrub
sending up new stems from the ground like a butcher's broom, and normally
flowering in February.

For another plant once cultivated as *S. humilis*, see *S. confusa*.

S. RUSCIFOLIA Stapf

An evergreen shrub, 2 to 4 ft high; stems erect, branching towards the
top, minutely downy when young. Leaves 1 to 2½ in. long, half as wide;
ovate, rounded and triple-veined at the base, long and finely pointed, quite
glabrous, and of a very dark lustrous green above, paler beneath; stalk ⅛ to
¼ in. long. Flowers milk-white, fragrant, produced during the winter months
in the axils of the terminal leaves. Several flowers appear in each cluster,
which has a short stalk ⅛ in. or less long. Sepals four to six, about ¼ in. long;
stamens (of the male flowers) ¼ in. long. Fruit roundish, ¼ in. wide, crimson;
seeds black. *Bot. Mag.*, t. 9045.

Native of central and western China; discovered by Henry near Ichang in 1887, and introduced from the same neighbourhood by Wilson for Messrs Veitch in 1901. In var. CHINENSIS (Franch.) Rehd. & Wils. the leaves are narrower, up to ⅝ in. wide. This too was introduced by Wilson; it is commoner

SARCOCOCCA RUSCIFOLIA

in western China than the typical state and also more frequent in cultivation and more vigorous, but judging from wild specimens the two states are linked by intermediates.

S. *ruscifolia* and its variety, although their flowers possess only a very modest beauty, are, with their neat habit and dark, polished leaves, decidedly pleasing, and their cut stems are useful to the flower-arranger, as they last three weeks in water.

S. SALIGNA (D. Don) Muell.-Arg.

Buxus saligna D. Don; *S. pruniformis* Lindl., in part

An evergreen shrub, 2 to 3 ft high; stems erect, glabrous. Leaves 3 to 5 in. long, ½ to 1⅛ in. wide; narrow-lanceolate, with a long drawn-out point; base narrowly wedge-shaped; glabrous, glossy, with a marginal vein on each side extending all round the leaf; stalk ¼ to ⅜ in. long. Flowers greenish white, in short axillary racemes opening in winter and spring. Styles three. Berries egg-shaped, ⅓ to ½ in. long, purple.

Native of the western Himalaya from Afghanistan to Kumaon. The date of introduction is uncertain; it is usually given as 1820, but the species then introduced was *S. coriacea* (Hook.) Sw., which has a more easterly distribution in the Himalaya, and with which *S. saligna* was at that time confused. The true species was received at Kew in 1908 from Messrs Veitch, but the origin of these plants is uncertain.

SARGENTODOXA   SARGENTODOXACEAE

A genus of a single species, and itself the sole member of the family Sargentodoxaceae. It is allied to the Lardizabalaceae, in which it is placed by some botanists, But in the Lardizabalaceae the carpels in the female flowers are few, and arranged in whorls, whereas in *Sargentodoxa* the carpels are numerous and spirally arranged, as in the Schisandraceae. In foliage it bears some resemblance to *Sinofranchetia,* but the leaflets are virtually sessile and three-veined from the base, whereas those of the *Sinofranchetia* are distinctly stalked and pinnately veined. Also *Sinofranchetia* belongs to the Lardizabalaceae and has female flowers with only three carpels.

The genus is named in honour of Professor C. S. Sargent of the Arnold Arboretum.

S. CUNEATA (Oliver) Rehd. & Wils.
Holboellia cuneata Oliver

A unisexual, twining, deciduous climber 25 ft or more high; young shoots glabrous. Leaves dark glossy green, alternate, glabrous, composed of three leaflets borne on a common stalk 2 to 4 in. long. Side leaflets stalkless, obliquely ovate (or like a heart-shaped leaf halved lengthwise), pointed, up to 4½ in. long, half as much wide; middle leaflet obovate, oval, or lozenge-shaped, smaller than the side ones and on a stalk ½ in. long. Male flowers greenish yellow, fragrant, borne numerously in pendulous racemes 4 to 6 in. long; each flower has six petal-like, narrowly oblong sepals ½ in. long, ⅛ in. wide, and is borne on a stalk ½ to ¾ in. long; stamens six. Female flowers (borne on separate plants) also in pendulous racemes up to 4 in. long, with six similar greenish yellow, petal-like sepals, and the carpels crowded on a central cone ¼ in. high. When these carpels mature each develops into a roundish dark purplish blue 'berry' ¼ in. wide, carrying a single black seed and borne on a stalk from ¼ to ½ in. long. This stalk is the elongated base of the carpel. *Bot. Mag.,* t. 9111, 9112.

Native of Central China; discovered by Henry about 1887; introduced by Wilson in 1907. It first flowered in this country with C. J. Lucas at Warnham

Court, Horsham, in May 1922. It may need the protection of a wall in many places. In the shape of its leaflets it is distinct from any other hardy climbing shrub except *Sinofranchetia*, to which it has much resemblance.

SARGENTODOXA CUNEATA

SASA GRAMINEAE

A genus of dwarf or medium-sized bamboos, spreading by vigorous much-branched rootstocks, natives of Japan, Korea, Sakhalin and the Kuriles. Stems (culms) cylindrical, hollow, with persistent sheaths. Branches solitary at each node. Leaves relatively large, with numerous secondary veins and conspicuous cross-veins (not in fact very conspicuous to the naked eye, but easily seen under low magnification if the leaf is held up to the light). Inflorescence a panicle; flowers with three to six stamens.

The three species treated here are easily distinguished from *Arundinaria*, certainly from its dwarfer species, by the much broader leaves, rarely less than $1\frac{1}{2}$ in. wide. For dwarf bamboos included in *Sasa* by E. G. Camus, but not belonging to *Sasa* as now understood, see *Arundinaria*

chrysantha, A. disticha, A. humilis, A. pumila, A. variegata, A. viridistriata.
Arundinaria japonica, included in *Sasa* by some authorities, has broad leaves, but is far taller than any of the species treated here and does not run at the root.

S. PALMATA (F. W. Burbidge) E. G. Camus

Bambusa palmata Burbidge; *B. metallica* Mitf.; *Arundinaria palmata* (Burbidge) Bean

Stems 6 to 8 ft high, ¼ to ½ in. thick, more or less glaucous, with a few erect branches near the top, hollow; joints 5 or 6 in. apart. Stem-sheaths glabrous, terminated by a narrow, lanceolate tongue, which is strongly tessellated and edged with minute bristles, but soon falls away. Leaves bright green above, glaucous beneath, 6 to 13 in. long, 1½ to 3 in. wide, confined to the apex of the branches, broadly wedge-shaped at the base, with long, slender points; secondary veins seven to thirteen at each side the midrib, very strongly developed, and giving the leaf a ribbed appearance; tessellation minute; margins set with bristles, which fall away with age.

Native of Japan; introduced about 1889. This has the largest leaves of all hardy bamboos except *S. tessellata,* and is undoubtedly one of the noblest of them all. The stems and leaves are apt to get somewhat battered and shabby with age, and it is a good plan every few years to cut the plants back to the ground entirely. If this be done in May, taking care not to injure the young, pushing stems, the plant will soon be furnished with a perfectly fresh set of leaves. The only defect of this bamboo is its extraordinarily rampant habit. It is no uncommon thing for a young stem to push through the ground a yard or two away from the previous ones. It is not a suitable neighbour for other shrubs, but is very well adapted for the undergrowth of thin woodland.

S. palmata flowered for the first time in Europe in 1961–8.

S. TESSELLATA (Munro) Makino & Shibata

Bambusa tessellata Munro; *B. ragamowskii* Nichols.; *Arundinaria tessellata* (Munro)
E. G. Camus; *A. ragamowskii* (Nichols.) Pfitzer

Stems 4 to 6 ft high, ⅛ to ⅙ in. diameter, with a very small hollow up the centre; the joints 1 to 3 in. apart. Stem-sheath persistent, 8 to 10 in. long, clasping not only that part of the stem above the joint from which it springs, but also portions of the two or three stem-sheaths above it; it is fringed with hairs. Leaves somewhat ribbed, of varying size, the largest 18 in. long, and 3 to 4 in. wide in the middle, abruptly tapered at the base, very slenderly pointed, dark green above, glaucous beneath. The larger leaves have fifteen to eighteen secondary veins at each side the midrib, which is yellow; and tucked under one side of the midrib, especially towards the base, is a line of pale hairs.

Native of China; cultivated in England since 1845, probably before. It is the most striking of dwarf bamboos, with larger leaves than any other, tall or dwarf, and forms broad, rounded masses, the outer stems of which arch outwards to the ground, and out of which spring each summer the spike-like new growths. It has never been known to flower under cultivation. Very hardy. It differs from *S. palmata* in the dwarfer habit but larger leaves.

S. VEITCHII (Carr.) Rehd.

Bambusa veitchii Carr.; *Arundinaria veitchii* (Carr.) N. E. Brown; *Arundinaria albomarginata* Makino; *Sasa albomarginata* (Miq.) Makino & Shibata

Stems usually 1 to 1½, sometimes 3 to 4 ft high, with a single branch at each of the upper joints, green, round, ⅙ in. in width, joints 3 to 4½ in. apart, rather prominent. Stem-sheaths persistent, very downy at first; both they and the leaf-sheaths have at the apex a curious group of bristles (themselves minutely hairy), resembling in their tapering, twisted ends the arms of an octopus. Leaves narrow-oblong, 4 to 8 in. long, 1 to 2¼ in. wide, abruptly tapered at the base, and narrowed quickly also at the top to a short, slender point; at first dark green above, glaucous beneath, but afterwards turning yellow and finally pale brown at the margins; secondary veins 5 to 9 each side the midrib.

Native of Japan; introduced by Maries for Messrs Veitch about 1880. It forms dense, matted patches and spreads very rapidly. Pleasing in the summer and early autumn, the habit of decaying at the leaf-margins spoils its value. This character, which is equally apparent on plants wild in Japan, is not found, so far as I know, in any other hardy species.

This species is too invasive and shabby looking to be admitted into the garden, and does nothing to beautify woodland, where it will in time form vast thickets, difficult to eradicate.

SASSAFRAS   LAURACEAE

This small genus of deciduous trees demonstrates the affinity between the woody flora of eastern North America, which has one species, and eastern Asia, where there is one species in central China and one in Formosa—all three with the characteristic irregularly lobed foliage. Flowers mostly unisexual and borne on different trees. Inflorescence racemose, from few-scaled buds, developing before the leaves. Flowers without petals. Calyx composed of six segments. Male flowers with nine stamens in three whorls, the innermost each with a pair of glands at the base. Ovary one-celled, developing into a drupe inserted on a fleshy pedestal.

The generic name is probably a corruption of an Indian name for the American species.

S. ALBIDUM (Nutt.) Nees   SASSAFRAS   [PLATE 39

Laurus albida Nutt.; *S. officinale* var. *albidum* (Nutt.) Blake

A deciduous tree, occasionally 70 to 90 ft in the wild; young shoots glabrous or (in var. *molle*) downy. Leaves alternate, of variable shape, mostly oval, ovate or obovate, often with a conspicuous lobe on one or both sides, the sinus always rounded; they are 3 to 7 in. long, 2 to 4 in. wide, tapered at the base, prominently three-veined, glossy dark green above, pale and somewhat glaucous beneath, glabrous or almost so in the typical state, but downy on both sides when young and more or less permanently downy beneath in var. *molle*; stalk ½ to 1½ in. long. Flowers greenish yellow, produced in May in racemes 1 to 2 in. long, the sexes usually on separate trees. Corolla absent. Calyx about ⅜ in. long and wide, with six narrowly oblong lobes. Stamens nine in the male, perfect; six, and aborted, in the female. Fruits dark blue, roundish oval, about ⅜ in. long.

This species (including the var. *molle*, see below) is a native of the eastern and central United States, extending as far west as Kansas, Oklahoma and Texas, and from New England to Florida; also of Canada (Ontario).

var. MOLLE (Raf.) Fern. *Sassafras triloba* var. *mollis* Raf.; *Laurus sassafras* L.; *S. officinale* Nees & Ebermaier.—This variety, with the young branchlets and leaf-undersides downy, occurs almost throughout the range of the species, while the more glabrous, typical state is confined to the northern part of the area; but intermediates occur.

The sassafras is pleasingly aromatic, and many medicinal virtues were once imputed to it. Although it has no great beauty of flower, it is a striking and handsome tree in foliage, and is of ornamental value even when it remains shrubby, as it often does in this country. Although it has been in cultivation in Britain since the 1630s there are very few good specimens in the country. The best was at Claremont, near Esher, which was about 50 ft high in 1910, the trunk 7¼ ft in girth at 1 ft from the ground—a fine pyramidal specimen. No comparable tree has been recorded recently; one at Wakehurst Place, Sussex, is about 60 ft high but only 3¼ ft in girth (1974).

S. albidum is a perfectly hardy species except when quite young, though the unfolding leaves are sometimes crippled by late frost. It is usually raised from imported seed, but root-cuttings are another means of increase, and suckers are sometimes produced, which can be detached and established in pots. It needs a deep fertile soil.

S. TZUMU (Hemsl.) Hemsl.—This is the only other species in cultivation. It is a Chinese tree, introduced from Hupeh to the Coombe Wood nursery by Wilson in 1900; a tree which grew at Kew was the only one known to me. As compared with the American species, it has certain small differences in the structure of the flower, the shoots and leaves are glabrous, and it is remarkably distinct in growth. The original tree at Coombe Wood made enormous, succulent, erect growths every year, perhaps 6 or 7 ft long, with proportionately large leaves. These shoots were very much cut back in winter, but a woody trunk was gradually being formed, and a tree with age might become quite acclimatised. Leaves with the principal veins reddish; young wood purple-spotted. The specific name is founded on the native one ("tzu-mu"). According to Henry, who discovered it, the tree grows 50 ft high, and its timber is valued by the mountaineers where it is wild.

SATUREIA   LABIATAE

A genus of about thirty annual and perennial herbs, together with a few dwarf shrubs, natives of the warmer parts of the world, all aromatic. Flowers whorled or in lax cymes. Calyx tubular or campanulate, with five equal teeth. Corolla with a straight tube, two-lipped, the lower lip with three rounded lobes. Stamens four, shorter than the upper lip, curved. In *Hyssopus*, and usually in *Thymus*, the stamens are straight, exserted and spreading. *Satureia* is sometimes stated to be a large genus; it becomes so if *Micromeria*, *Calamintha* and some other related genera (none treated here) are included in it, as they are by some botanists.

S. MONTANA L.   WINTER SAVORY

An evergreen shrub of bushy shape, 1 to 1½ ft high, aromatically scented; young shoots slender, downy. Leaves opposite, stalkless, narrowly oblanceolate or linear-oblong, pointed, tapered gradually to the base, not toothed, ½ to 1¼ in. long, $\frac{1}{12}$ to $\frac{1}{6}$ in. wide, greyish green beneath and freely dotted with oil-glands, bristly hairy on the margins. Flowers produced during July and August in axillary racemes ½ to 1 in. long and in whorls towards the end of the shoot, the whole forming a leafy panicle 3 to 6 in. long. Corolla white to purplish, ⅜ in. long, two-lipped; upper lip slightly notched, lower lip three-lobed. Calyx tubular, ten- to thirteen-ribbed, downy, with five awl-shaped teeth; stamens four.

Native of S. Europe eastwards to the Caucasus, and of N. Africa; long cultivated for its pleasing odour and as a flavouring agent in cookery. It was also used by the old herbalists in affections of a flatulent nature and Gerard remarks that it 'doth prevail marvellously against winde.'

subsp. VARIEGATA (Host) P. W. Ball   *S. m.* var. *variegata* (Host) Vis.; *S. variegata* Host—Inflorescence laxer, owing to the slightly longer, more spreading flower stalks. Calyx campanulate. Flowers lilac or violet, the lower lip with darker speckling (whence presumably the epithet *variegata*), deeper coloured than the upper lip. N.E. Italy and Dalmatia. The most ornamental race—and one of several, since the species is very variable.

SAXEGOTHAEA   PODOCARPACEAE

A genus of a single monoecious species of conifer in temperate South America, belonging to the same subfamily of Podocarpaceae as *Podocarpus* itself and *Microcachrys*. Male inflorescences axillary, cylindric, borne near the tips of the branchlets. Pollen grains unwinged (winged in all other podocarps). Numerous fertile, pointed scales are borne at the end of short twigs, forming when ripe a small, roundish fleshy 'cone' containing up to about six seeds, each inserted in a groove at the base of the scale and surrounded by, but not united to, an epimatium (an excrescence from the subtending scale). The small shrub *Microcachrys tetragona* (q.v.) also has a cone-like fruit, but in most other podocarps only one or a few of the scales of the 'female' twigs are fertile.

The genus was named by Lindley in honour of Prince Albert, consort of Queen Victoria.

S. CONSPICUA Lindl.

An evergreen tree up to 80 ft high in the wild, with the aspect of a small-leaved yew; branches drooping; branchlets usually in whorls; bark of trunk peeling. Leaves linear or linear-lanceolate, $\frac{1}{2}$ to $1\frac{1}{8}$ in. long, $\frac{1}{10}$ in. wide, abruptly narrowed at the base to a short stalk; tapered more gradually at the apex to a very fine point, dull dark green above, with two comparatively broad, glaucous bands of stomata beneath. Male and female flowers on the same plant; the former in shortly stalked, cylindrical spikes $\frac{1}{4}$ in. long, produced in a cluster near the end of the shoot. The fruit is a small cone, solitary at the end of the twigs, globose in the main, $\frac{1}{2}$ in. diameter; the scales terminating in a broad, flattened, spine-like point. *Bot. Mag.*, t. 8664.

Native of Chile from $35°$ to $45°$ S, commonest in the Lake region, where it occurs in *Nothofagus dombeyi* forest and in the stands of *Fitzroya cupressoides*; it is also found in Argentina in the region of Lake Nahuel Huapi. It was introduced to Britain by William Lobb for Messrs Veitch, but has never become common. Even in the wild it is a slow-growing tree, so it is not surprising that no large specimens are known in this country. By far the finest is a tree at Woodhouse, Lyme Regis, measuring $55 \times 5\frac{1}{4}$ ft (1970); and there are two trees at Kilmacurragh in Eire, both about 40 ft high and $4\frac{3}{4}$ and $4\frac{1}{4}$ ft in girth (1966). The following smaller specimens have been measured recently in the south of England: Wakehurst Place, Sussex, *pl.* 1914, 29×2 ft (1969); Leonardslee, Sussex, $36 \times 2\frac{1}{4}$ ft (1969); National Pinetum, Bedgebury, Kent, *pl.* 1925, $23 \times 1\frac{1}{2}$ ft (1969); Killerton, Devon, $26 \times 1\frac{1}{2}$ ft (1970).

SCHIMA   THEACEAE

A genus of evergreen trees and shrubs, in which some fifteen species have been described, reduced to one in the revision referred to below. They are natives of the subtropical and warm temperate regions of S.E. Asia, but extending into the E. Himalaya and S.W. China (as far north as Mt Omei in W. Szechwan). Leaves alternate, entire or toothed. Flowers white, showy, borne singly on distinct pedicels in the upper leaf-axils;

pedicels with two opposite bracteoles. Sepals five, much smaller than the petals. Petals five, united at the base. Stamens numerous. Style one. Fruit a globose woody capsule flattened at the apex. Seeds flat, with a narrow wing at one side. Allied to *Gordonia*, but in that genus the capsule is oblong and the seeds are winged at one end.

A revision of *Schima* by S. Bloembergen was published in *Reinwardtia* Vol. 2 (1952), pp. 133–83.

The type-species of the genus is *S. noronhae* Reinwardt ex Blume of Java and S.E. Asia, but the first species to be described (in the genus *Gordonia*) was *S. wallichii* (DC.) Korthals, which, in the narrow sense, ranges from the eastern Himalaya to Yunnan. In the revision cited above, only this species is recognised, with two subspecies—subsp. *wallichii* and subsp. *noronhae* (Bl.) Bloembergen. Of the species treated here, *S. argentea* is in this revision sunk in *S. wallichii* subsp. *noronhae* var. *superba* (Gardner & Champion) Bloembergen, and *S. khasiana* becomes *S. wallichii* subsp. *wallichii* var. *khasiana* (Dyer) Bloembergen.

S. ARGENTEA Pritz.   [PLATE 38

An evergreen shrub or tree found in the wild state up to 60 ft or more high, but, as seen in cultivation hitherto, a spreading bush 6 ft or more high and as much in width; young shoots dark purplish, minutely downy. Leaves alternate, 3 to 5 in. long, ¾ to 2 in. wide, narrowly oval-lanceolate or oblanceolate, slender-pointed, tapering to a short stalk, entire, rather leathery, glabrous and deep shining green above, glaucous and minutely downy beneath. Flowers solitary or rarely in pairs, each on a silky-hairy stalk ½ to 1 in. long and produced as many as nine together from the terminal leaf-axils; they are 1 to 1½ in. across he five-petalled corolla is ivory-white, camellia-like. Stamens fifty to sixty., Calyx five-lobed, long ciliate, sometimes downy on the back. *Bot. Mag.*, t. 9558.

Species mainly of S. China; collected by Henry in Yunnan in 1898 and later by Forrest, who introduced it during his 1917–19 expedition, almost certainly under number F. 15029. The seed was collected in the Wei-Hsi valley in November 1917 at 8,000–9,000 ft, from trees 40 to 50 ft high, and distributed from the R.H.S. Garden at Wisley, where a plant was raised which was 8 ft high in 1936 but had died by 1953. The only plant known to have survived in S. England grows at Borde Hill in Sussex, 30 ft high; it flowers in most years but is not in the best of health. An example 36 ft high, its largest stem 2¾ ft in girth, grows at Trewithen in Cornwall (1971). It is evidently a species that needs the milder and rainier climate of the Atlantic zone. It received an Award of Merit when exhibited from Bodnant on October 4, 1955.

S. KHASIANA Dyer—This is a finer species than *S. argentea*, from which it differs in its larger and relatively broader leaves 5 to 7 in. long, 1⅜ to 2¾ in. wide, toothed at the margin, green beneath, and in the larger bracteoles on the pedicels, these being short and inconspicuous in *S. argentea* but ⅞ in. long and ⅜ in. wide in *S. khasiana*. The pedicels are also distinctly longer, being 1⅝ to 3⅝ in. long; in this respect the cultivated plant and many Chinese specimens differ also from typical *S. khasiana*. The flowers, at least on the cultivated plants, are larger than in *S. argentea. Bot. Mag.*, n.s., t. 143.

S. khasiana was described from the Khasi Hills of Assam, whence it extends through Burma into N.W. Yunnan and the former Indo-China. It was introduced by Forrest, probably under number F. 26026, collected November 1924 on the Shweli-Salween divide at 8,000 ft (the plant at Caerhays is under F. 24630, which belongs to a flowering specimen collected the same year in June in the same locality). The flowering material portrayed in the *Botanical Magazine* was received in September from Trewithen, where there are two fine specimens, the larger 52 × 4¾ ft (1971). Cut in hard winters they soon break into growth again and produce fertile seed. The tree at Caerhays measures 30 × 2¾ ft (1971), and there is another fine example at Trengwainton. *S. khasiana* received an Award of Merit when exhibited from Caerhays on October 20, 1953.

Both species can be propagated by cuttings.

SCHINUS   ANACARDIACEAE

Under cultivation in the open air, only one, or at most two, species of this genus are sufficiently hardy to thrive. These are evergreen shrubs, with the shoots often becoming spine-tipped, and the leaves alternate. Flowers very small and numerous on short racemes, yellowish or white. Fruit a round, one-seeded drupe. The genus is most nearly allied to *Pistacia* and *Rhus*. The species described below were long called *Duvaua*, being distinguished from *Schinus* proper by the simple leaves. S. MOLLE, the so-called "pepper tree," is very extensively cultivated in S. France, Italy, etc., where its much divided, pinnate leaves and drooping branches make it a singularly graceful tree, laden in autumn with beautiful clusters of red berries about the size of small peas. It is not hardy with us. Native of S. America.

The two following species do not require a rich soil, making shorter, hardier growth, and flowering better where it is rather poor. They do not transplant well. Propagated by cuttings made in August, and placed in gentle heat.

S. POLYGAMUS (Cav.) Cabrera

Amyris polygama Cav.; *S. dependens* Ortega, *nom. illegit.*; *S. bonplandianus* Marchand

An evergreen shrub up to about 15 ft high in this country; branches occasionally spine-tipped, green or purplish when young, glabrous. Leaves short-stalked, elliptic to broadly so, or oblong or lanceolate, more rarely obovate or ovate, cuneate at the base, obtuse to rounded at the apex, ⅝ to 2 in. long, mostly ³⁄₁₆ to ⁹⁄₁₆ in. wide, rarely to ¾ in. wide, entire or more or less toothed, glabrous on both sides; petioles to $\frac{3}{16}$ in. long. Flowers very small, about $\frac{1}{6}$ in. wide, greenish white or greenish yellow, borne in May in axillary racemose clusters $\frac{1}{2}$ to 1 in. long. Fruits in dense clusters, each one a dry, deep purple drupe about the size of a peppercorn.

Native of S. America, where it is widespread from Bolivia, Peru and Brazil south to Chile, Uruguay and Argentina; introduced in 1790. It is a very variable species in foliage, degree of spininess and in the size and shape of the inflorescence; the above description is based on the cultivated states, of which there have been many in British gardens, though the species is now rare. The form depicted in *Bot. Reg.*, t. 1573 (1833), as *Duvaua dependens*, probably represents the original introduction from Chile, with small obovate leaves obtuse or emarginate at the apex. This was perfectly hardy on a wall and set seed. In *Bot. Reg.*, t. 1568 is figured (as *Duvaua ovata*) a form with ovate, distinctly toothed leaves—even with a small lobe on each side near the base. This form is also figured in *Bot. Mag.*, t. 7406, as *S. dependens*. This form, also from Chile, is also hardy or almost so, and is probably what has been commonly cultivated as *S. dependens*.

Another introduction of *S. polygamus* was described in previous editions under the name *S. bonplandianus*. This had linear-oblong leaves $\frac{3}{4}$ to 2 in. long, $\frac{1}{4}$ in. wide, entire or sparsely toothed. It was introduced from Argentina by Low of Clapton about 1830 and was slightly tender.

S. LATIFOLIUS (Gillies ex Lindl.) Engl. *Duvaua latifolia* Gillies ex Lindl.— Closely allied to *S. polygamus*. Branchlets not spine-tipped. Branches glabrous or clad with spreading white hairs. Leaves ovate, narrowly ovate or broad-elliptic, rounded or truncate at the base, acute to rounded at the apex, mostly $1\frac{3}{4}$ to $2\frac{3}{8}$ in. long and $\frac{3}{4}$ to $1\frac{3}{8}$ in. wide, deep green above, light green below, downy or glabrous on both sides, prominently veined beneath, margins undulate and toothed. Introduced before 1829. It was hardy at Kew on the Temperate House Terrace.

The flowers of these two species have no bright colour to recommend them, but they are borne in such profusion that the shrub gives quite a pleasing effect. The leaves and other parts have a turpentine-scented sap, and from this derives what Loudon calls 'a pretty phenomenon'. It was described by Lindley as follows: the leaves or parts of leaves, when placed in water, 'after lying a short time, will be found to start and jump as if they were alive, while at the instant of each start a jet of oily matter is discharged into the water . . . Thus we have in every leaf a sort of vegetable battery, which will keep up its fire until the stock of ammunition is exhausted.'

SCHISANDRA   SCHISANDRACEAE

A small genus of more or less aromatic twining shrubs, one species in N. America, the others in E. and S.E. Asia. Leaves simple, without stipules. Flowers unisexual, the two sexes usually borne on different plants. Floral envelope not fully differentiated into calyx and corolla but consisting of up to about sixteen imbricated segments, the outermost of which are more or less sepaloid. Stamens five to fifteen, often partly connate, sometimes united by their filaments into a fleshy head. Carpels numerous, at first crowded but later becoming separated by the elongation of the floral axis and forming when ripe a spike-like infrutescence. In the related *Kadsura* the ripe carpels remain clustered in a globose or ellipsoid head.

The type-species of the genus is S. GLABRA (Brickell) Rehd. (*S. coccinea* Michx.), a native of the south-eastern USA. This was introduced to Britain in 1806 and found to be moderately hardy on a wall but seems to have dropped out of cultivation. Allied to it is S. REPANDA (Sieb. & Zucc.) Radlkofer (*S. nigra* Maxim.), a native of Japan and S. Korea, a probably quite hardy species with blue-black fruits.

The species mentioned below are all hardy except *S. propinqua*. They like a rich garden soil and can be increased by cuttings of half-ripened wood in mild bottom heat.

The genus *Schisandra* was revised by A. C. Smith in *Sargentia*, Vol. 7 (1947), pp. 86–156.

S. CHINENSIS (Turcz.) Baill.

Kadsura chinensis Turcz.; *S. japonica* (Sieb. & Zucc. ex A. Gr.) Hance;
Sphaerostema japonicum Sieb. & Zucc. ex A. Gr.; *Maximowiczia chinensis* (Turcz.) Rupr.

A deciduous, climbing shrub, growing 20 to 30 ft high; branchlets red, round, not downy, set with wart-like lenticels. Leaves 2 to 4 in. long, obovate or elliptical, tapering at the base to a slender stalk, remotely toothed; glabrous except on the principal veins beneath when young. Flowers produced during April and May, each on a slender stalk 1 in. long, two or three of them being borne in a cluster at the base of the young growths; they are pale rose-coloured, fragrant, ½ to ¾ in. across. After the female flowers are past, that portion bearing the carpels continues to lengthen until it is 2 to 6 in. long, and on it the berry-like, scarlet fruits are borne on a sort of pendulous spike. These remain on the plant during the winter. Male flowers with four or five stamens.

Native of China, Japan, Korea, Sakhalin and the Amur region; introduced in 1860. Although not showy in flower (the petals soon drop), its scarlet fruits are very handsome. The dried wood is charmingly fragrant.

S. HENRYI C. B. Cl.

A deciduous, glabrous climbing shrub with twining stems, triangular when young, each angle winged. Leaves leathery, shining, of variable shape, elliptical, ovate or cordate, pointed or rounded at the apex and sparsely toothed, 3 to 4 in. long; stalk 1 to 2 in. long. Flowers ½ in. across, unisexual, white, borne on a stout stalk 2 in. long. The column on which the carpels are borne elongates after the flowers are faded and becomes fleshy, and 2 to 3 in. long; on this the mucilaginous carpels are borne. They are eaten by the Chinese.

Introduced by Wilson for Messrs Veitch about 1900, from W. Hupeh and Szechwan, but discovered by Henry long previously. It is easily distinguished from *S. chinensis* by the lustrous, thicker leaves and triangular branchlets. Quite hardy at Kew.

S. GLAUCESCENS Diels—This is closely allied to *S. henryi*. It has not the angled young shoots of that species and the leaves are shorter stalked, more glaucous beneath, more tapered at the base and of thinner texture. Flowers orange-red; fruit scarlet. A deciduous climber growing 20 ft high which flowers in May and June. Introduced from Central China by Wilson in 1907. Another species of the same group is:

S. PUBESCENS Hemsl. & Wils.—This is easily recognised by the dense covering of short curled hairs beneath the leaf. Wilson says its 'attractive yellow flowers are succeeded by still more conspicuous orange-red fruits.' He is credited with introducing it in 1907, but it seems to be very uncommon.

S. PROPINQUA (Wall.) Baill.

Kadsura propinqua Wall.; *Sphaerostema propinquum* (Wall.) Bl.

A tall evergreen climber with glabrous, angled young stems. Leaves rather thin, ovate-lanceolate, narrowly oblong-ovate or almost elliptic, 2 to 5 in. long, ¾ to 2 in. wide, rounded or broad-cuneate at the base, narrowed at the apex,

glabrous, finely toothed or almost entire; petiole about 1 in. long. Flowers usually solitary, the outer segments greenish yellow, the inner orange, about ⅝ in. wide, borne on stalks not more than 1 in. long. Males with up to ten perianth segments, the stamens united into a more or less globose head. Female flowers with more numerous segments than in the male. Mature carpels scarlet, in spikes up to 6 in. long. *Bot. Mag.*, t. 4614.

Native of the Himalaya; in cultivation 1828. It is a tender species, at one time cultivated in greenhouses.

var. CHINENSIS Oliver—Leaves narrow-lanceolate or narrowly lanceolate-oblong, ¾ to 1 in. or slightly more wide, sometimes marbled with white. Flowers yellowish, borne in late summer; introduced by Wilson in 1907 from W. Hupeh, China. According to him it is a common species up to about 3,300 ft, growing in rocky places; he saw it always less than 10 ft high. It is hardier than the Himalayan plant and like it is easily distinguished from other species by its short-stalked flowers. Since they open so late, it would probably be necessary to grow both sexes of this variety if fruits are to be seen.

S. RUBRIFLORA (Franch.) Rehd. & Wils.   [PLATE 40

S. chinensis var. *rubriflora* Franch.; *S. grandiflora* var. *rubriflora* (Franch.) Schneid.;
S. grandiflora sens. Franch., not Hook. f.

A climbing deciduous shrub 10 to 20 ft high; young shoots slender, glabrous, at first reddish. Leaves mostly obovate, sometimes approaching oval, pointed, tapered at the base, toothed, 2½ to 5 in. long, 1½ to 2½ in. wide, quite glabrous; stalk ½ to 1½ in. long. Flowers unisexual, solitary in the leaf-axils at the base of the new shoots, 1 in. wide, deep crimson, each on a slender, pendulous, red stalk 1 to 1½ in. long. Sepals and petals five to seven in all, roundish. Fruit composed of roundish red carpels thickly disposed on the terminal half of a pendulous, red, slender stalk 3 to 6 in. long, each carpel about the size of a pea and containing two seeds. *Bot. Mag.*, t. 9146.

Native of W. Szechwan, China, thence ranging south and west as far as N.E. India; discovered on Mt Omei by E. Faber about 1887; introduced by Wilson in 1908. This is a handsome climber as regards both flowers and fruit, the former hanging downwards and in the bud state resembling ripe cherries. It is quite hardy and is now well established in cultivation; it can be trained on a wall or up a stout stake or pole. The fruit, which seems first to have been fully developed at Aldenham, Herts, in 1925, is ripe in September. The species was given an Award of Merit by the Royal Horticultural Society on September 22, 1925. Blooms in April and May.

S. rubriflora may also be in cultivation from seeds collected by Kingdon Ward in 1928 in the Mishni Hills, Assam.

S. GRANDIFLORA (Wall.) Hook. f. *Kadsura grandiflora* Wall.—*S. rubriflora* is perhaps no more than a geographical variant of this species, described in 1824, which ranges from the central Himalaya into China. Chiefly it differs from *R. rubriflora* in its white or sometimes light pink flowers.

S. INCARNATA Stapf *S. grandiflora* sens. Rehd. & Wils., not Hook. f.; *S. grandiflora* var. *cathayensis* Schneid., in part—Very near to *S. grandiflora* but recognised as a distinct species by A. C. Smith in his monograph. It differs in its always flesh-pink flowers (commonly white in *S. grandiflora*) on shorter pedicels and the male flowers with fewer stamens. From *R. rubriflora* it is obviously distinct in the colour of its flowers. Introduced by Wilson from W. Hupeh, China, first for Messrs Veitch, later, in 1907, for the Arnold Arboretum (W. 318, as *S. grandiflora*).

S. SPHAERANDRA Stapf *S. grandiflora* var. *cathayensis* Schneid., for the greater part—From *R. rubriflora* and other species mentioned above this differs in the narrower and shorter leaves, which are lanceolate, oblanceolate or narrow-oblong, and in the almost sessile anthers borne on a swollen receptacle. The flowers vary in colour, but are usually in some shade of red or magenta. Described in 1922 from specimens collected by Forrest in the Lichiang range of N.W. Yunnan, and by Schneider and Handel-Mazzetti in S.W. Szechwan. The *S. grandiflora* var. *cathayensis* of Schneider is mainly this species but partly *S. incarnata*.

S. SPHENANTHERA Rehd. & Wils.

A deciduous climber growing up to 16 ft high, devoid of down in all its parts; young shoots reddish brown. Leaves 2 to 4 in. long, obovate to oval or roundish, tapered at the base, more or less slenderly pointed, minutely and distantly toothed, pale green beneath; stalk slender, ¾ to 1½ in. long. Flowers unisexual, solitary in the lower leaf-axils of the new shoots, about ⅝ in. wide, greenish outside, orange-coloured within, pendulous on slender stalks 1 to 2 in. long; sepals and petals together about nine; male flowers with ten to fifteen stamens. Fruit clusters made up of numerous scarlet berries closely packed on the terminal two or three inches of a pendent stalk, which becomes thickened at the part bearing the berries. The entire length of the fruiting stalk may be up to 6 or 8 in. It flowers in April and May. *Bot. Mag.*, t. 8921.

Native of Hupeh and Szechwan, China; introduced by Wilson in 1907. It has flowered at Kew and Glasnevin and is quite hardy. It is easily distinguished from *S. rubriflora*, introduced by Wilson at the same time, by the colour of its flowers and by the minute, often scarcely noticeable, toothing of the leaves. *S. chinensis* is distinguished by having only five stamens.

S. LANCIFOLIA (Rehd. & Wils.) A. C. Smith *S. sphenanthera* var. *lancifolia* Rehd. & Wils.—Well distinguished by its angled young shoots and its narrow, lanceolate, slenderly pointed leaves which are remotely toothed, 2 or 3 in. long, 1 in. or less wide.

SCHIZOPHRAGMA   HYDRANGEACEAE

Two climbing deciduous shrubs, found in China and Japan, and very nearly allied to *Hydrangea*—especially the climbing section of that genus. Leaves opposite, long-stalked. Flowers in a large terminal cyme, the central flowers small and perfect, the outer ones sterile and reduced to one large showy bract borne at the end of a slender stalk. From *Hydrangea*, the only other genus with which it can be confused, *Schizophragma* differs in the sterile flowers consisting of but one bract instead of four, and in having the four or five styles united into one. The specialised function of the large bracts, is, no doubt, to attract insects to the inflorescence, and thereby bring about the fertilisation of the flowers. In the great majority of flowers each one does its own share in adversisement. Fruits top-shaped.

These two shrubs are easily cultivated. They like a good loamy soil and plenty of moisture, and can be increased by cuttings and layers. The only other necessity is something for them to climb over, and this may be wall, tree-trunk, or anything to which the aerial roots may attach themselves.

S. HYDRANGEOIDES Sieb. & Zucc.   [PLATE 41

A deciduous, climbing shrub, reaching 40 or more ft high in the wild; young stems glabrous, reddish, and furnished with aerial roots. Leaves broadly ovate, with a rounded, heart-shaped or tapering base; 4 to 6 in. long, 2½ to 4 in. wide, strongly veined, coarsely and angularly toothed, deep green and glabrous above, but paler, rather glaucous, and with silky hairs beneath; stalk 1 to 2 in. long. The leaves near the inflorescence are tapered at the base, those on sterile shoots heart-shaped. Flowers small, yellowish white, slightly scented, produced during July in a broad, flattish, cymose inflorescence 8 or 10 in. across. The chief feature of the inflorescence are the bracts, one of which terminates each main branch of the cyme, and is heart-shaped or ovate, pale yellow, 1 to 1½ in. long; flower-stalks furnished with a thin, loose down. *Bot. Mag.*, t. 8520.

Native of Japan, where, along with *Hydrangea petiolaris* it forms a conspicuous feature in the forests, often covering the trunks of large trees. In gardens it is rare, the plant introduced by Maries for Messrs Veitch under its name being in fact *Hydrangea petiolaris*, a species of similar habit but quite distinct in foliage and inflorescence. It was the latter, and not *S. hydrangeoides*, that received a First Class Certificate in 1885. The date of introduction of the present species is uncertain, but it appears to have first flowered in this country in 1905, in the garden of B.E. Chambers at Grayswood Hill, Haslemere.

The floral bracts of this species vary much in size and shape, but are usually smaller than those of the following species, from which it is also distinct in its coarsely toothed leaves.

cv. 'ROSEUM'.—Inflorescence bracts tinged with rose. A.M. 1939 when exhibited by Messrs Hillier, who raised it.

S. INTEGRIFOLIUM (Franch.) Oliver
S. hydrangeoides var. *integrifolium* Franch.

A deciduous, climbing shrub of robust growth, reaching probably 40 or more ft in height. It produces aerial roots from the branches by which it attaches itself like ivy to the object upon which it grows; young stems hairy or glabrous. Leaves ovate, with a heart-shaped or rounded base, tapering to a long fine point at the apex, 3 to 7 in. long, 1½ to 4½ in. wide, the margin entire or sparsely set with small thin teeth, hairy beneath on the midrib and veins; stalk 1 to 2½ in. long, more or less hairy when young. Flowers produced in a flat cyme up to 1 ft in diameter; the fertile flowers in the centre each ¼ in. across, and comparatively inconspicuous. But terminating each division of the inflorescence is the remarkable single sterile blossom, consisting only of one large white bract, narrowly ovate, up to 3½ in. long, 1¾ in. wide, and veined like a leaf with darker lines. *Bot. Mag.*, t. 8991.

Native of Central China, where it inhabits rocky cliffs; introduced by Wilson for Messrs Veitch in 1901. It grows well and is now well established in cultivation and is proving to be a free-flowering climber remarkable for the enormous white bracts accompanying the inflorescence. The stems of our young plants are very downy, and the down persists till the following year, but on Wilson's wild flowering specimens they are quite glabrous. Var. MOLLE Rehd., has the undersurface of the leaves and their stalks very downy.

S. integrifolium received a First Class Certificate in 1963 when exhibited by Sir Henry Price from Wakehurst Place, where there is a fine plant (*Journ. R.H.S.*, Vol. 88 (1963), fig. 169).

SCIADOPITYS   TAXODIACEAE

A genus of a single species in Japan, distinct from all other living conifers in its foliage. The leaves are of two kinds: those on the long shoots are small and scale-like, the upper ones crowded; the foliage leaves are flattened and 'double', each being apparently composed of two completely connate needles borne on an undeveloped spur springing from the axil of a scale-leaf. Owing to the crowding of the subtending scale-leaves, the foliage leaves form whorls at the ends of the shoots, like the ribs of an umbrella. The nature of the foliage leaves is a matter of controversy; according to some botanists they are what they appear to be, i.e., the result of the fusion of two distinct needles, while others consider them to be modified shoots (cladodes) and the occasional occurrence of branched needles is adduced to support this theory (*Rev. Hort.*, 1884, p. 16). Double needles have been found in Cretaceous deposits, so this form of photosynthetic organ is very ancient. Male and female flowers are borne on the same tree. The female cones are solitary at the ends of the branches; fertile scales in the flowering stage small, subtended by conspicuous bracts, but becoming much larger as the cone develops (it needs two years to ripen). Mature cones have numerous spirally arranged woody scales, each fertile one bearing five to nine narrowly winged seeds.

S. VERTICILLATA (Thunb.) Sieb. & Zucc.   UMBRELLA PINE
Taxus verticillata Thunb.   [PLATE 42

An evergreen tree attaining a height of 120 ft in its native forests, and with a girth there of up to 15 ft; bark reddish brown, shed in narrow vertical strips; branchlets brown, glabrous, bearing minute scale-leaves, scattered in the lower part, crowded at the apex of the shoot into two or three imbricating rows, where each subtends a leaf (by some interpreted as a cladode, see introductory note). Leaves in whorl-like clusters, persisting for about three years, 2 to 5 in. long, about ⅛ in. wide, slightly narrowed at top and bottom, minutely notched at the apex, dark green and slightly grooved above, lower surface paler, deeply grooved, with a white or yellowish line of stomata running along the groove. Male flowers in a raceme about 1 in. long. Cones 2 to 3 in. long, 1 to

2 in. wide, borne on a stout stalk; scales with broad reflexed margins. *Bot. Mag.*, t. 8050.

Native of Japan; introduced by J. G. Veitch in 1861 by means of seeds, but Fortune sent young plants to Standish's nursery in the same year, one of which was exhibited to the Royal Horticultural Society on June 5, a few days after

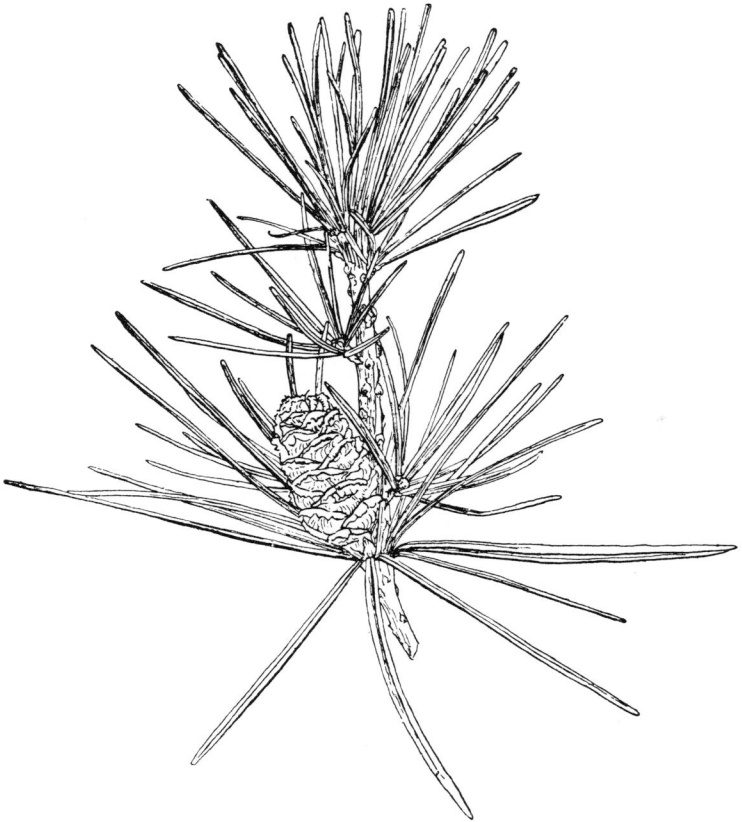

SCIADOPITYS VERTICILLATA

its arrival from Japan. In 1853 Thomas Lobb had sent a single plant to Veitch's Exeter nursery, which he had obtained from the Buitenzorg Botanic Garden in Java, but that was sickly on arrival and soon died.

The popular name for this conifer (and its generic name) refers to the arrangement of the leaves, which resemble the ribs of an umbrella. This remarkable and beautiful tree 'stands alone amongst Coniferae with no obvious affinities or immediate allies, and must be conjectured to have come down to us from a remote geological past which had obliterated all trace of its immediate ancestors.' (Thistleton-Dyer). Species of *Sciadopitys* occurred in Europe in the Tertiary and indeed were so common in Germany in the mid-Tertiary that the leaf-remains give a characteristic appearance to some layers of brown coal (Magde-frau, *Paläobiologie d. Pflanzen*, ed. 3 (1956), p. 356).

In gardens the umbrella pine is usually seen as a shrub or small tree, often many-stemmed from the base. What should perhaps be regarded as the forest form is less common, but any well-grown specimen is striking and ornamental. It should have an isolated position and thrives in a warm, lime-free soil. It is very hardy, but grows slowly when young.

cv. 'PENDULA'.—In *Forest Flora of Japan* (1894), p. 78, Sargent observes that there is a remarkable tree with pendulous branches in the Shiba Park, Tokyo. A tree of similar habit used to grow in Gunnersbury Park, London (*S. vert.* var. *pendula* Bean).

The finest example of *S. verticillata* known to Elwes and Henry grew at Hemsted, Kent, in the grounds of what is now Benenden School. In 1905 it measured 38 × 2 ft; it is now 74 × 4¾ ft (1970) and still as then a slender spire (cf. *Tr. Gt. Brit. and Irel.*, Vol. III, plate 159B). Others measured recently are: R.H.S. Garden, Wisley, 49 × 2¾ ft (1971); Tilgate Park, Crawley, Sussex, *pl.* 1905, 56 × 4¾ ft (1975); Leonardslee, Sussex, 55 × 3¼ ft (1966); Sheffield Park, Sussex, 50 × 4½ ft (1974); Bodnant, Denbighs., *pl.* 1901, 56 × 4½ ft (1974); Kilmory Castle, Argyll, 42 × 5½ ft (1976); Castlewellan, Co. Down, 37 × 4¾ ft (1976); Curraghmore, Co. Waterford, Eire, 46 × 5¾ ft (1975).

SECURINEGA   EUPHORBIACEAE

A genus of about twenty-five species in temperate and subtropical regions, allied to *Andrachne*. Flowers without petals, unisexual, axillary. Calyx with five sepals, persistent in fruit. Male flowers with five stamens which are longer than the sepals and inserted outside a lobed disk. Female flowers with an unlobed disk; styles three. Fruit a dehiscent capsule.

The generic name comes from the Latin *securis*, a hatchet, and *negare*, to refuse, in allusion to the hard wood of the type-species *S. durissima*, a native of the island of Réunion.

S. SUFFRUTICOSA (Pall.) Rehd.

Pharnaceum suffruticosum Pall.; *S. ramiflora* (Ait.) Muell.-Arg.;
Xylophylla ramiflora Ait.

A deciduous shrub, 3 to 5 ft high, with erect stems and long, graceful, slender, horizontal branches, all the parts devoid of down. Leaves alternate, oval, or slightly obovate, $\frac{3}{4}$ to 2 in. long, $\frac{1}{3}$ to 1 in. wide, mostly blunt or rounded at the apex, margin minutely undulated; dull green above, pale, rather glaucous beneath; stalk $\frac{1}{8}$ in. long. Plants unisexual; flowers greenish yellow, very small ($\frac{1}{10}$ in. across), produced during August and September in the leaf-axils of the current year's growth. The male flowers are densely packed a dozen or more together, opening successively; each flower on a stalk $\frac{1}{10}$ in. long; sepals and stamens five. The longer-stalked female flowers are borne singly in the leaf-axils. The seed-vessel, borne on a stalk $\frac{1}{4}$ to $\frac{1}{2}$ in. long, is about the size of a peppercorn, three-celled, the calyx adherent at the base.

Native of N.E. Asia, including China, Manchuria, and Siberia; introduced from the last named in 1783. It flowers very freely, but has little to recommend it except its graceful habit.

SENECIO *   COMPOSITAE

This is one of the largest genera of flowering plants and comprises over 1,000 species, including the common pest of our gardens, the groundsel.

The generic name, which was used by Pliny, is derived from the Latin *senex*, an old man, and refers to the usually grey, hair-like pappus of the seeds. The senecios have alternate leaves and the flowers (or 'florets' as they are usually called) are, as in all the Compositae, crowded in 'heads'. The florets are usually of two kinds: those in the centre of the head, of tubular shape and known as 'disk' florets; and those of the circumference, tongue-shaped, radiating, and known as 'ray' florets. (The daisy is the most familiar type.) But sometimes the florets are wholly discoid, as in *S. elaeagnifolius* and *S. reinoldii*. All those included in the following notes are natives of New Zealand, where there is the greatest concentration of species that are both woody and hardy enough for cultivation out-of-doors in the British Isles. S. ARGENTINUS Baker (syn. *S. leucostachys* Hort., not Baker), from Argentina and Uruguay, is grown for its much divided, silvery foliage, but it is scarcely shrubby enough to fall within the scope of this work.

Recently, B. Nordenstam in *Opera Bot.*, Vol. 44, pp. 25–33 (1978) has transferred the New Zealand species dealt with here to the genera *Urostemon* B. Nordenstam (*S. kirkii*) and *Brachyglottis* J. R. & G. Forst.

* Revised by C. Jeffrey, The Herbarium, The Royal Botanic Gardens, Kew.

(the rest). There is no doubt that the latter are more correctly placed in *Brachyglottis* than in *Senecio*, although the validity of the segregation of *Urostemon* from *Brachyglottis* is open to question. Since, however, work on the establishment of more rational generic limits within the *Senecio*-complex is still in progress, the species are here entered under their *Senecio* names.

Hybridisation amongst the New Zealand species is widespread, both in cultivation and in the wild, and the majority of plants in cultivation in the British Isles are of hybrid origin. Failure to recognise this has in the past caused much misidentification of cultivated material. Fortunately D. G. Drury, in *New Zeal. Journ. Bot.*, Vol. 11, pp. 731–784 (1973) has provided us with a reliable key to the New Zealand shrubby senecios and their hybrids, both cultivated and wild. His account gives full descriptions and illustrations of all the known hybrids, and the reader is referred to it for a fuller treatment. Most of the hybrids recorded as in cultivation in New Zealand have yet to be confirmed as in cultivation in the British Isles; some may eventually prove to be of garden merit.

Provided the climate is sufficiently mild for them, the senecios are of easy cultivation and succeed well in a light or sandy soil. *S. compactus*, *S. greyi*, *S. laxifolius* and the cultivated Dunedin hybrids succeed on lime or chalk, as probably do most of the others. The senecios are not sufficiently planted in sea-side gardens, for which places the toughness of the leathery leaves admirably fits them. Propagation is effected by late summer cuttings placed in very sandy soil and given if possible a mild bottom heat. (See *Olearia* for the distinction between that genus and *Senecio*.)

S. COMPACTUS Kirk

Brachyglottis compacta (Kirk) B. Nord.

An evergreen shrub of compact, much branched habit, 2 to 4 ft high, spreading twice as wide; young shoots, undersurface of leaves and flower-stalks all clothed with a pure white, dense felt. Leaves obovate or oval, rounded at the apex, tapered at the base, the lower larger ones mostly wavy at the margin, the small upper ones almost or quite entire; they are ½ to 2 in. long, ¼ to 1 in. wide, becoming smaller near the inflorescence, upper surface dark dull green, with persistent white margins when seen from above; stalks ¼ to ¾ in. long, felted. Flower-heads ¾ to 1 in. wide, three to ten of them forming a terminal raceme; ray-florets about twelve, bright yellow.

Native of the North Island, New Zealand. This handsome shrub is unfortunately not quite hardy near London, but is capable of surviving moderate winters if given a sheltered nook and a slight covering during cold spells. In the south and south-west it grows admirably without protection. Its compact habit, small leaves with crinkled margins, and clusters of racemes each terminating a slender shoot of the year, renders it distinct. It flowers in January and February in New Zealand and in the corresponding antipodal months (July and August) here. It occurs wild on limestone. Its nearesr relative is *S. monroi* which has conspicuously wavy margined, narrower leaves and glandular flower-stalks.

S. DUNEDIN HYBRIDS

S. × *crustii* Hort.; *S. laxifolius* Hort., not Buchan.; *S. greyi* Hort., not Hook. f.; *S. laxifolius* Buchan. × *S. compactus* Kirk; *S. greyi* Hook. f. × *S. compactus* Kirk; *S.* (*compactus* Kirk × *greyi* Hook. f.) × *S. greyi* Hook. f.; *S.* (*compactus* Kirk × *laxifolius* Buchan.) × *S. greyi* Hook. f.

Somewhat straggly evergreen shrubs 2 to 4 ft high, of bushy habit; young stems, undersurface of leaves and leaf-stalks all clothed in a dense white felt. Leaves obovate, broadly elliptic or ovate, rather abruptly narrowed or rounded at the base, blunt at the end, entire or obscurely wavy, 1 to 3 in. long, ½ to 1¾ in. wide, becoming smaller near the inflorescence, those at the first forks of the inflorescence narrowed to a stalk-like base, upper surface at first thinly grey-felted, becoming glabrous, dark green, without a persistent white margin when seen from above; stalks ⅜ to 1¼ in. long. Flower-heads about 1 in. wide, in loose terminal panicles; ray-florets 11–15, bright yellow.

The designation *Senecio* Dunedin Hybrids is given here to cover all cultivated senecios which derive apparently from hybridisation between *S. compactus*, *S. greyi* and *S. laxifolius,* including backcrosses. Though typical plants of the latter two species are distinct, more or less intermediate populations are found in both the North and South Islands of New Zealand, and the correctness of their maintenance as specifically distinct is open to doubt. From typical *S. laxifolius*, the hybrids may be distinguished by their relatively broader, more rounded leaves; from typical *S. greyi*, by the leaves at the first forks of the inflorescence having a narrowed stalk-like, not a sessile expanded, base; and from *S. compactus*, by their larger leaves, laxer habit and the absence of a conspicuous white margin to the upper surface of the leaves.

cv. 'SUNSHINE'.—This is the commonest variant in cultivation in the British Isles, widely and erroneously referred to in the horticultural literature as *S. greyi* or *S. laxifolius*. It appears to be of the origin *S. compactus* × *S. laxifolius* and has usually non-wavy leaves with gradually tapered bases, and eglandular, white-felted flower-stalks and involucral scales. However, variants that exhibit *S. greyi* characters, e.g., larger, more rounded leaves and glandular as well as white-felted flower-stalks and involucral scales, are also known, but have yet to be given names. 'Sunshine' is described in *The Garden* (*Journ. R.H.S.*), Vol. 102 (1977), pp. 161–3.   [PLATE 43

Senecio Dunedin Hybrids are the most frequently cultivated of the shrubby senecios in British gardens. Most past records of the cultivation of *S. laxifolius* and *S. greyi* refer in fact to these plants. The earliest known specimens date back to 1910–13 and originated from the Dunedin Botanic Garden, New Zealand, hence the name here given to them. The use of the horticultural name *S.* × *crustii* is avoided here, as it is now uncertain to what plant this name was originally applied, although probably it was to a plant of the parentage *S. laxifolius* × *S. compactus*.

S. GREYI Hook. f.

Brachyglottis greyi (Hook. f.) B. Nord.

An evergreen shrub found up to 8 ft high in the wild state, usually much less with us; young shoots stout and, like the undersurface of the leaves and leaf-stalks, densely clothed with a soft white felt, giving them a texture like that of chamois leather. Leaves oblong, sometimes inclined to ovate, rounded or tapered at the base, rounded to blunt at the apex, entire or obscurely wavy, 1½ to 4 in. long, ½ to 1¾ in. wide, dark green and glabrous above except for the felted margins; stalk ½ to 1½ in. long. Panicles terminal, 4 to 6 in. long, 2 to 5 in. wide, bearing numerous flower-heads each about 1 in. wide, the ray-florets twelve to fifteen and of a rich clear yellow; main and secondary flower-stalks glandular-downy, as are also the bracts they bear.

Native of New Zealand in the North Island. It is most closely akin to *S. laxifolius*, from which it is perhaps not specifically distinct, but typical *S. greyi* can be distinguished from typical *S. laxifolius* by the broader leaves less tapered at the base, by the glandular-hairy flower-stalks and involucral bracts, and by the leaves at the base of the inflorescence having a broad sessile base. In *S. laxifolius* and in *S.* Dunedin Hybrids the leaves at the bases of the main branches of the inflorescence are narrowed to a short stalk; by this feature, glandular variants of the latter may also be distinguished from *S. greyi* proper. In spite of this, the name *S. greyi* has been widely applied in horticultural circles to what are in reality *S.* Dunedin Hybrids, and true *S. greyi* is much less commonly met with in cultivation in the British Isles than perusal of the horticultural literature would lead one to believe. *S. greyi* is less hardy than both *S. laxifolius* and the hybrids, being a native of the northern, warmer part of New Zealand at altitudes not exceeding 1,500 ft, or only half those attained by *S. laxifolius* in the South Island. But *S. greyi* is perhaps the most beautiful of all the New Zealand senecios in cultivation. It gives a blaze of yellow about midsummer and there is often a second crop in autumn. It is admirable along the south coast, even in places exposed to the sea. There was a fine plant 15 ft in diameter in a garden at Monreith in Wigtownshire.

S. 'LEONARD COCKAYNE' *S. greyi* Hook. f. × *S. reinoldii* Endl.—An evergreen much-branched shrub 6 to 10 ft high; young shoots, leaf-stalks and undersides of leaves densely clad with soft white felt. Leaves broadly elliptic to ovate, more or less rounded at the base, shiny green above when mature, 3½ to 6 in. long, 2¼ to 3½ in. wide; stalks 1 to 2½ in. long. Flower-heads in large terminal corymbose panicles; flower-heads about ⅝ in. wide, with eight to ten recurving ray-florets about ¼ in. long.

A hybrid of garden origin, more or less intermediate between the parents, distinguished by its short, deeply toothed ray-florets. Not known for certain to be in cultivation in the British Isles, but grown in New Zealand gardens for its handsome foliage.

S. HECTORIS Buchan.

Brachyglottis hectoris (Buchan.) B. Nord.

An evergreen or semi-deciduous shrub of erect, sparingly branched, rather gaunt habit, growing 6 to 14 ft high in the wild state; young branches stout, covered with loose wool and bases of fallen leaves. Leaves oblanceolate or narrowly oval, tapered to a narrow, pinnately-lobed base, more abruptly pointed at the apex, conspicuously toothed (the teeth stand out at right angles from the leaf margin), 6 to 10 in. long, half as wide, thinly covered with grey cottony down beneath, smooth except for minute warts above; stalk stout, quite short. Inflorescence a terminal flattish or rounded corymb up to 10 in. wide and as much high, opening with us in July. Flower-heads 1½ to 2 in. in diameter; the ray-florets twelve to fourteen, pure white, linear, recurved; the florets of the centre (disk) yellow. Flower-stalks densely glandular-downy. *Bot. Mag.*, t. 8705.

Native of the South Island, New Zealand; introduced by Major A. Dorrien-Smith in 1910 and flowered by him three years later. It is a handsome species as regards blossom, although of rather ungainly habit. It is easily recognised by the size and shape of the leaves and especially by their curious lobing at the very base. It is not hardy at Kew, but Sir Herbert Maxwell had a plant 6 ft high by 5 ft wide in his garden at Monreith in Wigtownshire, and there is a large plant on a sheltered wall at Tyninghame House, East Lothian.

S. 'ALFRED ATKINSON' *S. hectoris* Buchan. × *S. perdicioides* Hook. f.— An evergreen, much-branched, very floriferous spreading shrub up to 8 or 9 ft high. Young branches purplish. Leaves elliptic-oblong, at first densely covered with buff-coloured hairs, when mature light green above, paler and thinly white-hairy beneath, 5 to 8 in. long, 1⅞ to 2¼ in. wide, margins somewhat wavy, finely toothed. Inflorescences terminal corymbs up to 12 in. or more wide. Flower-heads 1 to 1½ in. across; ray-florets about eight, white; disk yellow.

A garden hybrid, originating in the garden of a Mr Alfred Atkinson of York Bay, Wellington, New Zealand, and with a good horticultural reputation in that country, but not known for certain to be in cultivation in the British Isles.

S. HUNTII F. v. Muell.

Brachyglottis huntii (F. v. Muell.) B. Nord.

An evergreen shrub or small tree of rounded form 6 to 20 ft high; young shoots clothed with glandular down, viscid, stout. Leaves stalkless, narrowly oblong or narrowly obovate, pointed or blunt at the apex, tapered towards the base, entire, 2 to 4½ in. long, ¾ to 1 in. wide, downy above, felted beneath when young, becoming almost glabrous above with age and thinly rusty-downy beneath. Flower-heads very numerous and densely produced in terminal, rounded, or pyramidal panicles 3 to 5 in. wide. Each flower-head is ½ to ¾ in. wide, with fifteen to twenty yellow ray-florets; flower-stalks slender, glandular-hairy.

Native of the Chatham Islands, where it is described as forming a small tree handsome in its pale shining green leaves and bright yellow, copious blossoms. It is also of distinct appearance, with leaves rather resembling those of a shrubby spurge in their shape and close arrangement at the end of the branches. It is unfortunately not hardy in most parts of the country, but succeeds on the southern and western seaboards. It is at Monreith in Wigtownshire, where it commences to flower in June. Major A. A. Dorrien-Smith collected plants in the Chatham Islands in December 1909, when it was just coming into blossom.

S. HUNTII × S. Dunedin Hybrids—A garden hybrid, similar in appearance to S. *huntii* but with grey-green foliage, the leaves being grey-cottony above and densely grey- or white-felted beneath. Not known for certain to be in cultivation in the British Isles.

S. KIRKII Hook. f. ex Kirk

S. *glastifolius* Hook. f., not L. f.; *Urostemon kirkii* (Hook. f. ex Kirk) B. Nord.

An evergreen shrub described as of erect growth in its wild state and from 6 to 12 ft high, quite devoid of down in all its parts. Leaves very variable in shape; linear, oblanceolate, obovate, or ovate, 2 to 5 in. long, $\frac{1}{3}$ to $1\frac{1}{2}$ in. wide, always tapered at the base, but either pointed, blunt or rounded at the end and either entire or shallowly and widely toothed; stalk $\frac{1}{4}$ to $\frac{3}{4}$ in. long. Flower-heads produced numerously in a flattish, terminal, much branched corymb 4 to 12 in. wide. Each flower-head is $1\frac{1}{4}$ to 2 in. wide, with ten or less snow-white ray-florets; disk-florets forming a circular, central, yellow mass $\frac{1}{4}$ in. wide. *Bot. Mag.*, t. 8524.

Native of the North Island, New Zealand; common in wooded and hilly country up to 2,500 ft altitude. Cheeseman describes it as 'a very remarkable and beautiful species, the flowers being so abundantly borne as to conceal the leaves, the multitude of snow-white ray-florets rendering the shrub conspicuous from afar.' It was introduced to cultivation here by Major A. A. Dorrien-Smith and was flowered by him at Tresco Abbey in the Scilly Isles in April 1913. It is only in the warmest parts of our islands that it is likely to succeed.

S. LAXIFOLIUS Buchan.

S. *greyi* Hort., in part; S. *laxiflorus* Hort.; S. *latifolius* Hort.;
Brachyglottis laxifolia (Buchan.) B. Nord.

A laxly branched, evergreen shrub, 2 to 4 ft high, of bushy habit; young stems covered with grey down when young. Leaves alternate, 1 to 3 in. long, $\frac{1}{2}$ to $1\frac{1}{4}$ in. wide, elliptic to oblong-elliptic, mostly blunt at the apex, tapering at the base, obscurely crenate or entire, covered above when young with a grey, cobweb-like down, afterwards nearly glabrous, under-surface clothed with close white felt; stalk slender, $\frac{3}{8}$ to 1 in. long. Flower-heads 1 in. across, produced in summer in loose, terminal, broadly pyramidal panicles, 5 to 8 in. long,

3 to 5 in. wide. Ray-florets twelve to fifteen, golden yellow, fully spread; disk-florets very small and numerous, forming collectively a reddish brown centre ¼ in. across. *Bot. Mag.*, t. 7378.

Native of the mountains of the Nelson and Canterbury provinces of New Zealand, at 2,500 to 5,000 ft. It needs somewhat milder climatic conditions than those of east and middle England, and although several times tried in the open at Kew, it has never survived more than two or three winters except in specially sheltered nooks. In the slightly milder parts of the country it succeeds admirably.

S. laxifolius has been much confused with *S. greyi* and especially with *S.* Dunedin Hybrids, which latter have long been erroneously referred to as *S. laxifolius* in British gardens. The true *S. laxifolius* appears to be very rare in cultivation in the British Isles.

<h2 style="text-align:center">S. MONROI Hook. f.</h2>

Brachyglottis monroi (Hook. f.) B. Nord.

An evergreen, much branched shrub up to 4 ft high; young shoots, leaf-stalks, under-surface of leaves and flower-stalks all covered with a whitish felt.

SENECIO MONROI

Leaves oblong, oval, or rather obovate, conspicuously wrinkled or wavy at the margin, rounded at the apex, tapered at the base; ¾ to 2 in. long, ¼ to ⅝ in. wide, dull green, sticky and glabrous above; stalk ¼ to ⅜ in. long. Flower-heads in terminal compound corymbs, each section carrying three to five; the flower-head is ½ to ¾ in. wide, carrying ten to fifteen bright yellow ray-florets; flower-stalks long and slender, glandular-downy and sometimes also white-felted; involucral-scales glandular-downy, sometimes also white-felted. *Bot. Mag.*, t. 8698.

Native of the South Island of New Zealand, up to elevations of from 1,000 to 4,500 ft. The species is distinct in its wrinkled leaf-margins and glandular-downy flower-stalks, its nearest ally being *S. compactus* which has whitish, felted, but not glandular flower-stalks and usually only faintly wrinkled leaves. It is a handsome shrub whose inflorences may be 4 to 6 in. wide and are borne in July. No one succeeded better with it than the late Canon Boscawen at Ludgvan Rectory, near Penzance, or Lord Wakehurst in Sussex, who had a plant 4½ ft high and 9 ft in diameter. At Kew it just misses being hardy in a sheltered nook, but survives our milder winters.

S. PERDICIOIDES Hook. f.
Brachyglottis perdicioides (Hook. f.) B. Nord.

An evergreen shrub of bushy habit up to 6 ft or more high; young shoots slender, slightly ribbed. Leaves oblong or slightly obovate, rounded at the apex, tapered at the base; numerously toothed, 1 to 2 in. long, less than half as wide, dull green and glabrous; stalk slender, up to ½ in. long. Inflorescence a flattish or slightly rounded, terminal, erect, many branched corymb, 3 to 6 in. wide. Flower-heads small, ¾ in. long and as much wide, funnel-shaped; ray-florets two or three, bright yellow; disk-florets four to eight.

Native of North Island, New Zealand; discovered by Sir Joseph Banks and Solander when with Cook on his first voyage to New Zealand (1769–70), and not recorded again until found by Archdeacon Williams in 1870. It was grown by Major A. A. Dorrien-Smith at Tresco Abbey, Scilly; at Ludgvan Rectory in Cornwall; and Mrs Vereker, Sharpiton, S. Devon, sent it in flower to Kew in 1922. It is not likely to succeed out-of-doors except in these and similarly mild localities. It flowers during December in New Zealand, during June and July in England.

S. PERDICIOIDES Hook. f. × S. DUNEDIN HYBRIDS—An evergreen shrub of bushy habit up to 6 ft high and 9 ft across, much-branched; young shoots white-felted. Leaves ovate or elliptic, blunt at the apex and base, finely toothed, 1 to 3½ in. long, ½ to 1¾ in. wide, yellowish green above, white- or brownish-felted beneath. Inflorescences like those of *S. perdicioides* but flower-heads rather larger; flower-stalks and scales of involucre thinly felted, also glandular-hairy; ray-florets five to seven, bright yellow.

A garden hybrid, known from Tresco Abbey, Scilly; Glasnevin; and The Gardens, Ilnacullin, Glengariff, Co. Cork; it is also cultivated in New Zealand.

S. REINOLDII Endl.

Brachyglottis rotundifolia J. R. & G. Forst.; *Senecio rotundifolius* (J. R. & G. Forst.)
Hook. f., not Stokes nor Lapeyr.

An evergreen shrub or small tree varying from 6 to 30 ft high in the wild
state; young shoots grooved and, like the stalks and under-surface of the
leaves and the flower-stalks, clothed with a close, dense, white felt. Leaves
entire, orbicular, heart-shaped, or roundish ovate, blunt or rounded at the
apex, the largest 5 in. long by nearly as much wide, the smaller ones 2 to
3 in. long, glabrous and dark shining green above; stalks 1 to $3\frac{1}{2}$ in. long,
stout, grooved. Flower-heads in large terminal clusters of corymbs from 5 to
10 in. wide, each flower-head $\frac{1}{2}$ in. long, $\frac{1}{4}$ in. wide, white at the sides, yellowish
at the top without any ray-florets. The florets are erect and closely packed in
a faggot-like cluster. Flowers in June and July; rather unpleasantly scented.

Native of the South Island, New Zealand, where it extends from sea level,
often close to the sea, up to 3,500 ft altitude. Although it has little beauty of
blossom, it is a striking plant because of the size and roundness of its shining
leathery leaves. In September 1928 there was a plant 3 ft high and 4 ft in
diameter, growing in an exposed position on the rock garden at Edinburgh.
In Mr Cox's garden at Glendoick, in the valley of the Tay, Perthshire, a plant
was 5 ft high in June 1931. On all but our coldest shores it ought to be a
good seaside shrub. Kirk, in his *Forest Flora of New Zealand*, observes that
its power of withstanding the fiercest gales and dashing sea spray is marvellous,
and that he had never seen a leaf torn by the action of either. (See also
S. elaeagnifolius.)

S. ELAEAGNIFOLIUS Hook. f. *Brachyglottis elaeagnifolia* (Hook. f.) B. Nord.—
An evergreen shrub 4 to 10 ft high in the wild state; young shoots slightly
channelled, clothed like the under-surface of the leaves and the flower-stalks
with a pale, buff-coloured felt. Leaves, mostly oval or obovate, tapered at the
base, blunt or rounded at the end, 2 to 5 in. long, 1 to $3\frac{1}{2}$ in. wide, glabrous and
glossy above; stalk grooved, $\frac{1}{2}$ to $1\frac{3}{4}$ in. long. Inflorescence a terminal, pyramidal
panicle 3 to 6 in. high. Flower-heads $\frac{1}{3}$ to $\frac{1}{2}$ in. wide, with nine to twelve very
woolly scales surrounding the base of each of them.

Native of the North Island of New Zealand, where it ascends to 4,500 ft;
very abundant in some places. It is similar to *S. reinoldii*, especially in having
no ray-florets, but the leaves of that species are relatively broader in outline
and the inflorescence is shorter, often much broader and more rounded. Neither
has really any beauty of flower. *S. elaeagnifolius* is much the less common of the
two in cultivation in the British Isles, and some of the records of this species
refer to misidentified narrow-leaved variants of *S. reinoldii*.

cv. 'JOSEPH ARMSTRONG' *S. elaeagnifolius* Hook. f. var. *buchananii* Hort.,
not (Armstr.) Kirk—A low-growing variant 3 to 4 ft high. Leaves very thick,
2 to $3\frac{1}{2}$ in. long, 1 to $1\frac{3}{8}$ in. wide; stalks $\frac{3}{4}$ to $1\frac{1}{2}$ in. long; veins very distinct
above, the principal ones often whitened, especially at their confluence.

S. BUCHANANII Armstr., not Hort. *Brachyglottis buchananii* (Armstr.) B.
Nord.; *S. bennettii* Simps. & Thoms.—Like *S. elaeagnifolius*, but leaves thinner,

more flexible, with a thinner silvery-white felt beneath; inflorescence-branches rather more slender and involucral scales rather less blunt.

Native of the South Island of New Zealand, where it is rather common in montane forest and scrub, and formerly included within *S. elaeagnifolius*. Some of the plants in cultivation in the British Isles under *S. elaeagnifolius* perhaps belong here, but the classification of the rayless New Zealand species is yet to be fully worked out.

S. BIDWILLII Hook. f. *Brachyglottis bidwillii* (Hook. f.) B. Nord.—A compactly branched evergreen shrub 1 to 3½ ft tall, distinguished from the other rayless species by the smaller, very thick, elliptic-oblong or ovate-oblong leaves, ¾ to 1 in. long, ⅜ to ⅝ in. wide, on short stalks ⅙ to ¾ in. long; basal lateral veins relatively stronger than the others.

A native of New Zealand, not uncommon in montane and submontane scrub. Typical *S. bidwillii* is confined to the North Island; in the South Island it is replaced by var. VIRIDIS (Kirk) Cheesem., which tends to have a taller, rather more slender habit and larger leaves 1⅓ to 3 in. long, ⅝ to 1 in. wide.

SEQUOIA   TAXODIACEAE

A genus of a single species, distinguished within the *Taxodium* family by the following combination of characters: leaves persistent, of two forms, needle-like and apparently two-ranked on lateral branchlets, scale-like and loosely appressed on leading and fertile shoots; male cones minute, solitary, terminal or axillary on short twigs. Female cones ripening first year, with up to twenty peltate scales, each bearing up to seven seeds. See also *Sequoiadendron*. It was once thought that *Sequoia* was widely distributed in the northern hemisphere in pre-glacial times, but since the discovery of *Metasequoia* it has come to be realised that many sequoia-like fossil remains belong to that genus and not to *Sequoia*.

The genus was named by Endlicher in 1847 in honour of the talented Sequoiah (*d.* 1843), son of a British trader and a Cherokee Indian woman. He devised an alphabet for the Cherokee tongue, thanks to which the tribe quickly became literate and even had a newspaper in their own language. But neither he, nor his tribe, ever lived within the area of the redwoods.

S. SEMPERVIRENS (D. Don) Endl.   COASTAL REDWOOD
Taxodium sempervirens D. Don

An evergreen tree attaining a height of over 300 ft in the wild and already well over 100 ft high in this country, where healthy isolated trees form slender pyramids, furnished from base to summit with leafy branches. In their native forests crowded trees may have 75 ft or more of clean trunk. Bark of a rich brown-red, and of a fibrous nature, 6 to 12 in. thick in the giants of western N. America; young shoots and leaves not downy, arranged in two opposite rows. Leaves linear $\frac{1}{4}$ to $\frac{7}{8}$ in. long, $\frac{1}{20}$ to $\frac{1}{8}$ in. wide, terminated by a short abrupt point, very dark lustrous green above, with two broad stripes of white stomata beneath. On leading shoots the leaves are shorter and arranged all round the branchlet. Cones roundish oblong, $\frac{3}{4}$ to 1 in. long, about $\frac{1}{2}$ in. wide.

Native of California from Monterey northward, just reaching into Oregon, confined to a narrow belt near the coast where summer fogs off the Pacific are frequent and mitigate the summer heat and drought. The best development of the redwood is north of San Francisco, where the most extensive and purest stands occur, at altitudes of up to 3,000 ft. It was described by David Don in Lambert's *The Genus Pinus* from specimens collected by Menzies in 1796 and introduced to Britain via Russia shortly before 1843.

Like its ally *Sequoiadendron giganteum*, this is one of the vegetable wonders of the world, having been measured around 350 ft in height in its native stands and 25 ft in diameter, though only on the alluvial soils where it thrives best. The largest trees are probably near 2,000 years of age, but maturity is reached at 400–500 years and merchantable timber is produced by trees only 100 years old. Old natural stands have yielded 300,000 gross board-feet per acre, and even one million on very favourable sites. Young managed stands, grown on a 100 year cycle, are expected to produce 350,000 board-feet per acre (Fowells, *Silvics of Forest Trees of the United States*, Agr. Handb. 271 (1965), p. 666).

The germination rate of redwood seed is low, but in nature such vast quantities are produced that this is of little account. The redwood also has the remarkable ability to reproduce itself by root-sprouts, which, on rich soils, spring up round the stump only a few weeks after felling and quickly develop their own root-system.

Redwood timber is highly valued for joinery and building; it is reddish, free from resin, light and easily worked; being very durable, it is also used for shingles, sleepers, piles and posts.

Most conifers from the coastal regions of western North America thrive excellently in this country, and the coastal redwood is no exception, but the best growth is attained in the rainier and more equable conditions of the Atlantic zone. Some specimens planted soon after the original introduction still survive, as do many of those mentioned by Elwes and Henry in 1908 and nearly all those listed in the returns to the Conifer Conference of 1932. Thanks to Alan Mitchell's researches for the Forestry Commission we have abundant information on the rate of growth of the redwood in this country. The following is only a selection of the trees that he has measured recently; for a fuller list see his *Conifers in the British Isles* (1972).

Bowood, Wilts, *pl. c.* 1845, 105 × 20 ft (1975); Whiteways, Devon, 118 × 20¼ ft (1975); Leighton Park, Ackers Memorial Grove, nr Welshpool, 121 × 16¼ ft (1975); Bodnant, Denb., 133 × 14¼ ft (1974); Taymouth Castle, Perths., 130 × 22¼ ft and 120 × 21¾ ft (1974); Caledon Castle, Co. Armagh, 115 × 22¾ ft and 118 × 21¼ ft (1976); Curraghmore, Waterford, Eire, 102 × 21 ft

(1975); Coollattin, Carnew, Co. Wicklow, Eire, *pl.* 1851, 117 × 21 ft and 124 × 21 ft (1975).

Younger trees, to show rate of growth: Blackmoor, Hants, *pl.* 1915, 105 × 12 ft (1974); Forestry Commission, Rhinefield Drive, New Forest, *pl.* 1955, 80 × 3¾ ft and 75 × 5¼ ft (1976); Forestry Commission, Alice Holt, near Farnham, Surrey, *pl.* 1951, 56 × 5¾ ft (1975); Wakehurst Place, Sussex, *pl.* 1920, 94 × 12¾ ft (1975); National Pinetum, Bedgebury, Kent, *pl.* 1925, 92 × 11 ft (1975).

As will be seen from these measurements, the coastal redwood grows well even outside the Atlantic zone. It is, however, rather tender when young, and is not suitable for places exposed to cold winds, nor for shallow chalky soils.

cv. 'ADPRESSA' ('Albo-spica').—Leaves only ¼ to ⅜ in. long, loosely appressed (as on fertile shoots of the type); tips of young growths creamy white. Although sometimes referred to as dwarf, it is so only if leading shoots are cut out. Left to itself it makes a tree, e.g., 73 × 8½ ft (1971) at Grayswood Hill, Haslemere (*S. semp. adpressa* Carr.; *S. semp. albo-spica* Veitch).

cv. 'CANTAB' ('Prostrata').—Of semi-prostrate habit, with remarkably wide, glaucous, two-ranked leaves, not much over ½ in. long and about half as wide. It arose as a bud-mutation on a tree in the University Botanic Garden, Cambridge, and received an Award of Merit in 1951, when exhibited as *S. semp. nana pendula*—a name earlier used by Hornibrook for a different clone, not treated here and probably not in cultivation. It has also been grown as *S. semp.* 'Prostrata'. It is an unstable variant, which sends up strong erect stems, one of which has attained 20 ft in the Hillier Arboretum, but retains the broad leaves of the original mutant.   [PLATE 44

For propagation, see *Sequoiadendron giganteum.*

SEQUOIADENDRON   TAXODIACEAE

Although long included in *Sequoia*, *S. giganteum* differs from it in many respects and was given separate generic rank by Buchholz in 1939. The most important points of difference are the naked winter-buds, the acicular, juniperoid leaves, the cones requiring two years to ripen and remaining alive and green on the tree for some years, their scales with three to nine seeds arranged in two ranks, and the cotyledons of the seedlings usually four in number (two in *Sequoia*). *S. giganteum* was in fact originally described by Lindley in the genus *Wellingtonia*, named in honour of the 1st Duke of Wellington, but unfortunately a few years earlier Meissner thought fit to commemorate the Duke in the same manner, by naming after him an obscure genus allied to *Meliosma* and since sunk in it.

S. GIGANTEUM (Lindl.) Buchholz [PLATES 45, 46
BIG-TREE, SIERRA REDWOOD, WELLINGTONIA

Wellingtonia gigantea Lindl., *nom. illegit.; Sequoia gigantea* (Lindl.) Decne. (1854), not Endl. (1847); *Washingtonia californica* Winslow, *nom. prov.; Taxodium washingtonium* (sic) Winslow; *Sequoia wellingtonia* Seem.; *Sequoia washingtoniana* (Winslow) Sudworth

An evergreen tree, reaching ultimately from 250 to almost 300 ft in height, and forming a trunk 20 to 30 ft through at the enlarged, buttressed base. Bark 1 to 2 ft thick, rich brown-red, and of a fibrous spongy, texture. The head of branches in old trees commences at 100 to 150 ft from the ground, and consists of comparatively short, horizontal or drooping branches. The final ramifications of the branches are much divided, the ends forming a dense, bushy cluster of branchlets. Leaves varying in length from $\frac{1}{8}$ to $\frac{1}{2}$ in. long, but always more or less awl-shaped, triangular in section, tapering from the broad base (by which it is attached to, and extends down the branchlet) to a fine point. They are blue-green and always point forward, adhering for four or five years to the branchlet, which in the early stages they completely cover. Cones $1\frac{1}{2}$ to 3 in. long, $1\frac{1}{4}$ to 2 in. wide; seeds pale shining brown, $\frac{1}{8}$ to $\frac{1}{4}$ in. long, flattened. *Bot. Mag.*, tt. 4777–8.

Native of California on the western side of the Sierra Nevada from around 39° N to just below 36° N, mostly at altitudes between 4,500 and 7,000 ft, but ascending to over 8,000 ft in the southern part of its range. The northernmost stands are few and are widely separated by the sites of ancient glaciers which the big-tree never succeeded in colonising after the retreat of the ice. But south of the King's River (around 37° N) the range is more continuous and the finest and purest stands are to be found in this region, in Fresno and Tulare counties.

The first European to see the big-tree was probably John Bidwell, who passed through the Calaveras stand in autumn 1841, after he had become separated from the party with which he was crossing the Sierra from the United States into California, then still part of Mexico. He did not report his find at the time and more pressing matters prevented a second visit (*Garden and Forest*, Vol. 2 (1889), p. 614). There was probably more than one further sighting during the next ten years, but the existence of the big-trees first became known to the scientific world in 1852, when Dr Kellogg showed specimens to the newly formed California Academy of Sciences, which he had received from J. M. Hutchins, a settler in the Yosemite valley. These were also seen by William Lobb, then plant-hunting for Messrs Veitch, and in the autumn of 1853 he collected and brought home material from which Lindley described the new species, as *Wellingtonia gigantea*, in the *Gardeners' Chronicle* for December 24. Lobb's account of the Calaveras stand appeared at the same time: 'From 80 to 90 trees exist, all within the circuit of a mile, and these varying from 250 feet to 320 feet in height and from 10 to 20 feet in diameter ... A tree recently felled measured about 300 ft in length with a diameter, including bark, 29 feet 2 inches at 5 feet from the ground ... Of this vegetable monster, 21 feet of the bark, from the lower part of the trunk, have been put in the natural form in San Francisco for exhibition; it there forms a spacious carpeted room, and contains a piano, with seats for 40 persons. On one occasion 140 children were admitted without inconvenience.'

The felled tree mentioned by Lobb was reckoned to be about 3,000 years old. 'That is to say,' commented Lindley, 'it must have been a little plant when Sampson was slaying the Philistines, or Paris running away with Helen, or Aeneas carrying off good *pater Anchises* upon his filial shoulders.' The great age imputed to the big-trees has been questioned on the grounds that a tree may make more than one growth-ring in a single season. But it now seems to be generally accepted that some of the largest trees are indeed 3,000 years old, even 4,000.

The Calaveras stand, which lies in the Sierra east by north of San Francisco, was the first to be exploited as a tourist attraction—a chalet-style hotel had been built by 1855 and the butt of a felled tree converted into a dance-floor. It and the Mariposa Grove some fifty miles to the south were the stands most visited by 19th-century travellers. The latter, now in the Yosemite National Park, has as its largest tree the 'Grizzly Giant', though this cannot compare with others in the southern groves.

The finer and more extensive stands south of 37° N, discovered early in the 1860s, have suffered the most devastation. In an address to the Royal Institution in 1878 Sir Joseph Hooker said: 'The doom of these noble groves is sealed. No less than five saw-mills have recently been established in the most luxuriant of them, and one of these mills alone cut in 1875 two million feet of Big-tree lumber ... The devastation of the Californian forest is proceeding at a rate which is utterly incredible, except to an eye-witness ... before a century is out the two sequoias may be known only as herbarium specimens and garden ornaments; indeed, with regard to the Big-tree, the noblest of the noble coniferous race, the present generation, which actually witnessed its discovery, may live to say of it "The place which knew it shall know it no more." '

Fortunately, Hooker's prediction has not been fulfilled. By the end of the century the conservationists had prevailed, and most of the remaining stands of the big-tree are now under some form of public ownership or protection. The most active and influential of the conservationists was the splendid John Muir (1838–1914), born at Dunbar, East Lothian, who emigrated with his family to the USA in 1849. He had been agitating for a policy of forest conservation since the 1870s, and it was probably he who provided Hooker with some of the material for his lecture. The most extensive of the remaining southern stands is the Giant Forest in the Sequoia National Park, its most notable tree 'General Sherman', 272 × 101½ ft at 4½ ft (1955), with a branch at 130 ft which is 7 ft in diameter. 'General Grant', in the King's Canyon National Park a short way to the north is only slightly smaller. Despite many statements to the contrary it seems that no measured standing tree is as much as 300 ft in height, though heights of 275 ft are frequent in the southern groves.

The section of a big-tree in the British Museum of Natural History in South Kensington came from 'Mark Twain', which was the only good tree remaining in a ravaged stand south-east of Fresno, owned by the King's River Lumber Company. It was felled to provide the section, originally part of the Jesup

Collection of North American Woods, which is now displayed in the American Museum of Natural History in New York. This is the lowest section, and the South Kensington section immediately adjoined it, at about 18 ft from the ground (For the felling of 'Mark Twain' see *Garden and Forest*, Vol. 7 (1892), p. 541).

It is ironic that the timber of the big-tree, though virtually everlasting, is too brittle to be of much value, and brought little profit to the lumber companies. Much of it was simply used to make roof-shingles and fencing for the settlers in the foothills, whose grazing animals completed the destruction that axe and saw had begun.

There were two introductions of *S. giganteum* to Britain in 1853. The first was to Scotland, by John Mathew, who collected seeds in August, which he sent to his father Patrick Mathew of Gourdiehall, Perthshire. From this sending most of the oldest trees in Scotland probably derive. The second, and more important one, was by William Lobb, who, in addition to the botanical material, brought home a quantity of seed for Messrs Veitch. Sown at once, it germinated well, and Veitch started to send out seedlings in summer 1854, charging two guineas for one, six guineas for four and twelve guineas a dozen. On some properties the seedlings were put out in their final positions in the spring of 1855—only eighteen months after Lobb collected the seeds. Heights of up to 7 ft were reported in autumn 1857 and up to $9\frac{1}{2}$ ft in the following year.

Coming from altitudes where the temperature frequently drops below zero Fahrenheit, and the growing season is short, *S. giganteum* is perfectly hardy with us and grows well over much of the country in deep soil and a sheltered but not too shaded position (it is much less shade-tolerant than the coastal redwood). The following is a very short selection from the trees measured by Alan Mitchell: Rhinefield, New Forest, Hants, 155 × $24\frac{1}{2}$ ft (1975); Wellingtonia Avenue, near Wellington College, Berks, *pl.* 1869, 108 × $22\frac{3}{4}$ ft (1969)—this avenue was planted by John Walter III, owner of *The Times* newspaper and grandson of its founder; Stratfield Saye, Hants, *pl.* 1857, 118 × $24\frac{1}{4}$ ft (1968); Shirley Hall, Kent, 110 × $24\frac{3}{4}$ ft (1968); Cowdray Park, Sussex, *pl.* 1870, 118 × $28\frac{1}{2}$ ft (1973); Crichel House, Dorset, 112 × $29\frac{1}{2}$ ft (1971); Westonbirt, Glos., 145 × 21 ft (1971); Oakley Park, Shrops, 111 × 25 ft (1971); Whiteways House, Chudleigh, Devon, 108 × $26\frac{1}{2}$ ft (1975); Woodhouse, Lyme Regis, Devon, 158 × $22\frac{1}{4}$ ft (1970); Bodnant, Denb., *pl.* 1888, 132 × $15\frac{3}{4}$ ft (1966); Powis Castle, Montgom., 112 × $28\frac{1}{4}$ ft (1970); Taymouth Castle, Perths., *pl.* 1856, 157 × $23\frac{1}{4}$ ft (1970); Dunkeld House, Perths., *pl.* 1857, 149 × 22 ft (1970); Scone Palace, Perth, *pl.* 1866, 115 × $26\frac{3}{4}$ ft (1970); Dupplin (Gardens), Perths., 144 × $18\frac{1}{4}$ ft (1970); Castle Leod, Ross-shire, *pl.* 1854, 150 × 26 ft (1966); Balmacaan House, Invern., 95 × 27 ft (1970); Benmore, Argyll, in Avenue, 143 × $20\frac{3}{4}$ ft and 157 × $17\frac{3}{4}$ ft (1975); Castlewellan, Co. Down, 111 × 25 ft (1976); Powerscourt, Co. Wicklow, Eire, by river, 115 × 27 ft (1975).

The following younger trees show rate of growth in less than optimum conditions: National Pinetum, Bedgebury, Kent, *pl.* 1926, 77 × 15 ft (1974); Forest Research Station, Alice Holt, Hants, *pl.* 1951, 58 × 7 ft (1976).

cv. 'AUREUM'.—Young shoots deep yellow. The original plant was received as a seedling by the Lough Nurseries, Co. Cork, in 1856, and was later propagated by them.

cv. 'PENDULUM'.—An extraordinary tree with an erect leader and weeping branches hanging close to the stem, forming a narrow spire. It originated at Nantes in 1863. The best known example is in the Allard Arboretum near Angers, and the tallest in this country, at Bodnant, measures 100 × 7½ ft (1974). Sometimes the main stem leans or undulates and gives off some more or less vertical branches. These weird forms, Mr Hillier tells us, are the result of grafting on *Sequoia sempervirens*.

S. giganteum is usually propagated by means of seeds. They germinate readily when fresh or after stratification, but the seedlings are subject to attack by Grey Mould (*Botrytis cinerea*), especially if grown in too close conditions; many of the seedlings sent out by Veitch were lost in this way. Cuttings strike fairly readily, but are slow to form a leader if taken from side-branches. The cultivars are propagated by grafting on stocks of the species; *S. sempervirens* is sometimes used as the stock, but does not give satisfactory results (see above).

SHEPHERDIA ELAEAGNACEAE

Of the three genera constituting this family, *Shepherdia* differs from *Hippophae* and *Elaeagnus* in its opposite leaves, and in having eight stamens instead of four. It consists of three scaly N. American shrubs with male and female flowers separated on different plants, and both inconspicuous. There is no corolla, and the calyx is of four divisions. Fruit berry-like. Named in honour of John Shepherd, curator of the Liverpool Botanic Garden in the early part of the 19th century. The third species, not mentioned below, is S. ROTUNDIFOLIA Parry, an evergreen shrub not in cultivation.

S. ARGENTEA (Pursh) Nutt. BUFFALO BERRY
Hippophae argentea Pursh

A deciduous shrub, 3 to 12 ft high, with opposite, often spine-tipped branchlets, covered when young with silvery scales. Leaves opposite, oblong, with a rounded apex and wedge-shaped base, ¾ to 2 in. long, ⅛ to ⅝ in. wide, covered with silvery scales beneath, less so above. Flowers ¼ in. across, produced in small clusters during March from the joints of the previous year's growth; calyx of four oblong green segments, of little or no beauty. Fruit roundish egg-shaped, ⅙ to ¼ in. long, scarlet, acid but edible.

Native of the central United States and bordering parts of Canada; introduced in 1818. As with its ally, *Hippophae rhamnoides*, it is necessary to have plants of both sexes in order to obtain fruits, but these are rarely developed in this country. There was great confusion in gardens between this shrub and *Elaeagnus argentea*. The latter was often supplied for it, but is easily distinguished by its invariably alternate, much broader leaves; it is also more ornamental than the *Shepherdia*, which has not much to recommend it in this country. *S. argentea* differs from the following species in having narrower leaves with a silvery upper surface, and in the often thorn-tipped twigs.

S. CANADENSIS (L.) Nutt.
Hippophae canadensis L.

A deciduous, unarmed shrub, up to 6 or 8 ft high, of bushy habit; shoots covered with brownish scales. Leaves ovate or oval, $\frac{1}{2}$ to 2 in. long, $\frac{1}{4}$ to 1 in. wide, dull dark green above, and at first furnished with silvery starry tufts of hairs especially along the midrib and veins; the under-surface woolly and specked with numerous brownish scales; stalk $\frac{1}{8}$ to $\frac{1}{6}$ in. long. Fruits yellowish red, egg-shaped, $\frac{1}{4}$ in. long.

Native of N. America, where it is widely spread both in the United States and Canada; in cultivation 1759. I have seen this shrub growing wild on the cliffs of the Genesee River gorge in New York State, between Rochester and Lake Ontario, loaded with its beautiful reddish fruits in July; but in England they are rarely developed. The shrub is interesting, and the singular aspect of the under-surface of the leaf under the lens is worth notice, the thick basis of silvery hair-tufts being interspersed with brown scales, each scale with a dark, glistening, eye-like centre.

SHIBATAEA   GRAMINEAE

Two species of bamboos, natives of Eastern Asia, with zigzagged stems flattened between the joints and nearly solid, stem sheaths without bristles. The genus is named after K. Shibata, a Japanese botanist. The following is the only species in cultivation.

S. KUMASASA (Steud.) Makino
Bambusa kumasasa Steud.; *Phyllostachys ruscifolia* (Sieb.) Nichols.; *Bambusa ruscifolia* Sieb. ex Monro; *Phyllostachys kumasasa* (Steud.) Munro

Stems erect, but very zigzagged, $1\frac{1}{2}$ to $2\frac{1}{2}$ ft high, very much flattened between the joints, $\frac{1}{8}$ in. in diameter, the central hollow only large enough to admit a horse hair; joints 1 to $3\frac{1}{2}$ in. apart. Branches three or four at each joint, 1 to $2\frac{1}{2}$ in. long, bearing one to three leaves. Leaves narrowly ovate, broadly tapered at the base, slenderly at the apex, 3 to 4 in. long, ¾ to 1 in. wide, glossy dark green and glabrous above, slightly glaucous and downy beneath, both margins toothed; secondary veins, five to seven each side the midrib.

Native of Japan; cultivated by Messrs Veitch at Coombe Wood in the 1870s as "*Bambusa viminalis*", and probably introduced for them during the previous decade by John Gould Veitch. A pretty bamboo, being of a neat, tufted habit. It is one of the most distinct of all hardy bamboos, especially in its sturdy, zigzag stems, the great proportionate width of the leaves, their length of stalk, and the uniformly short branches.

SIBIRAEA ROSACEAE

A genus of a few species, scattered in the Old World from Yugoslavia to western China. It is closely allied to *Spiraea*, differing in being effectively dioecious and in having the carpels connate at the base.

S. LAEVIGATA (L.) Maxim.

Spiraea laevigata L.; *Sibiraea altaiensis* (Laxm.) Schneid.; *Spiraea altaiensis* Laxm.

A deciduous shrub of sturdy bushy habit, 2 to 5 ft high, with thickish, rather sparse, perfectly glabrous, brown branchlets. Leaves entire, narrowly obovate, 2 to 4½ in. long, ⅜ to ⅞ in. wide, sessile, tapering at the base, the apex with a short, abrupt point, glaucous green and quite glabrous. Flowers white, very small, produced from late April to early June in terminal, spreading, compound panicles 3 to 5 in. high. Petals obovate to roundish. Stamens about twenty-five in the effectively male flowers but shorter in the female flowers. Carpels five, present but minute and sterile in the male flowers. Follicles two-seeded.

Native of Russia in the Altai Mountains and Dzungarski Alatau; also occurring as a relict in Yugoslavia, where it is found in the Cabulja Planina above Mostar and in the Velebitska Planina above Karlobag on the Dalmatian coast. The Yugoslav plants are sometimes distinguished as var. CROATICA (Degen) Schneid., but do not differ from the Russian plants except in their dwarfer habit.

S. laevigata was introduced to Britain by Dr Solander in 1774, three years after it had been described. Although not particularly showy, this species is very distinct from the spiraeas and their other allies in its foliage, which in shape and colour is suggestive of a spurge. Plants 4 ft high are often as much as 7 ft across.

SIBIRAEA LAEVIGATA

SINOFRANCHETIA   LARDIZABALACEAE

A genus of a single species, allied to *Holboellia* and *Stauntonia*, but with
consistently trifoliolate leaves (in the other two genera the leaflets are
normally more than three in each leaf). The flowers are unisexual, smaller
than in *Holboellia* and *Stauntonia*. Carpels three, each developing into a
many-seeded berry. See also *Sargentodoxa*.

The genus was named in memory of Adrien René Franchet, once
attached to the Paris Museum of Natural History, and one of the most
capable botanists who ever worked at the Chinese flora. He died 14th
February 1900.

S. chinensis Hemsl.

A large deciduous glabrous climber, covering trees 40 ft or more high, and
with a main stem frequently 3 or 4 in. thick; young branches twining, covered
with purplish bloom. Leaves composed of three leaflets, borne at the end of a
slender purplish stalk 6 to 9 in. long. Side leaflets obliquely ovate-elliptic,

terminal one broadly obovate, longer-stalked, all glabrous, glaucous beneath, 3 to 6 in. long, short-pointed, entire. Flowers unisexual, dull white, small, inconspicuous, and of no beauty, produced in pendent racemes about 4 in. long, on short leafy shoots. Fruits about the size of a grape, blue-purple, and borne alternately at intervals on an elongated stalk 8 in. or more long. Seeds black, $\frac{1}{5}$ in. long.

Native of Central and W. China, up to 7,000 ft; introduced by Wilson for the Arnold Arboretum, and raised at Kew in 1908. Unlike its allies *Holboellia* and *Stauntonia*, it is quite hardy. Its value in gardens lies in its vigorous habit and fine glaucous foliage. Also in its fruits, which are borne by female plants even in the absence of a pollinator. It received an Award of Merit when shown by Sir Henry Price from Wakehurst Place in October 1948.

SINOJACKIA   STYRACACEAE

A genus of two species of small deciduous trees, natives of China. Flowers on slender pedicels, arranged in short racemes terminating leafy branchlets. Ovary inferior (superior in Styrax). Calyx top-shaped, shortly five- to seven-lobed. Petals five to seven, united at the base (in the allied *Halesia* the corolla is four-lobed). Style slender, longer than the stamens. Fruits woody, not winged or ribbed.

The genus is named after J. G. Jack of the Arnold Arboretum, a Canadian by birth, who was concerned with Chinese botany and introduced several species from N. China and Korea early this century.

S. XYLOCARPA H. H. Hu

A small deciduous tree to about 20 ft high. Leaves obovate, up to $2\frac{7}{8}$ in. long, $1\frac{3}{8}$ in. or slightly more wide, acuminate at the apex, cuneate to rounded at the base, finely toothed, glabrous on both sides except for some stellate down on the midrib beneath; petiole to about $\frac{1}{4}$ in. long. Flowers white, about 1 in. wide, arranged in cymes of three to five at the ends of lateral shoots which are leafy at the base; pedicels $\frac{1}{4}$ in. long, slender. Stamens twelve to fourteen, with yellow anthers. Fruits woody, indehiscent, obovoid but with a conical apex, about $\frac{5}{8}$ in. long, $\frac{1}{2}$ in. wide.

S. xylocarpa, the type of the genus *Sinojackia*, was discovered in 1925, and described by Dr Hu four years later. The type-specimens were taken from a single tree growing at a resort in the suburbs of Nanking known as the Sen-Tai-Tung cave. Shortly afterwards this tree, the only one found, was cut down during roadmaking, and the seeds that had been distributed to botanical institutions in Europe and America were immature, and failed to germinate. Thus, for a brief space, it seemed as if this new genus was extinct. Then, in 1933,

S. xylocarpa was found to be quite common in the hills of Pukow, on the other side of the Yangtse from Nanking, and was successfully introduced to cultivation later in the decade (H. H. Hu, in *Journ. R.H.S.*, Vol. 63 (1938), p. 383, and *New Fl.* & *Sylv.*, Vol. 13 (1940), p. 150).

Dr Hu considered *S. xylocarpa* to be a much finer species than *S. rehderiana*, described below, but it seems to be less common in gardens. It is available in commerce.

S. REHDERIANA H. H. Hu—A deciduous shrub or small tree, allied to *S. xylocarpa*, but with elliptic to elliptic-obovate leaves, laxer inflorescences, rather smaller flowers, and much narrower fruits, about 1 in. long and ¼ in. wide, beaked at the apex and tapered at the base. *Bot. Mag.*, n.s., t. 466.

This species was discovered by Dr Hu in 1930 in the hills above Nachang, the capital of Kiangsi province, and was introduced by him to Britain and other countries in the same year. It is perfectly hardy and flowers well, but is horticulturally too similar to the halesias to have become popular and is rarely seen outside collections.

SINOMENIUM   MENISPERMACEAE

The species described below is the only member of its genus, which is closely allied to *Cocculus*, differing in its larger, pendulous inflorescences, and male flowers with nine to twelve stamens, against three to six in *Cocculus*.

S. ACUTUM (Thunb.) Rehd. & Wils.

Menispermum acutum Thunb.; *S. diversifolium* (Miq.) Diels; *Cocculus divers*. Miq.; *C. heterophyllus* Hemsl. & Wils.

A deciduous climber, up to 20 ft high, with twining stems. Leaves very variable in shape, perhaps normally ovate-cordate and entire, but sometimes almost kidney-shaped, sometimes lobed like *Catalpa ovata*, sometimes deeply three- or five-lobed (with lanceolate lobes), sometimes shallowly so; often with a lobe on one side only; the base often truncate; 2 to 6 in. long, 1¼ to 4½ in. wide; deep bright green, glabrous, with three or five conspicuous veins radiating from the base; stalk slender, 2 to 6 in. long. Flowers small, yellow, unisexual, about ⅙ in. wide; sepals six, in two series of three each, downy beneath; petals very small; the flowers are borne in slenderly pyramidal panicles 6 to 12 in. long, the main and secondary flower-stalks downy. Fruit about the size of a small pea, globose, black, covered with blue bloom.

Native of E. Asia. Although described by Thunberg in 1784 from a Japanese plant, it does not appear to have been introduced to cultivation until Wilson sent seeds from China to Veitch's Coombe Wood nursery in 1901. It is perfectly hardy, and a vigorous grower.

var. CINEREUM (Diels) Rehd. & Wils. *Cocculus diversifolius* var. *cinereus* Diels—Leaves downy above, densely so beneath. Introduced by Wilson in 1907 from W. Hupeh, where, as in W. Szechwan, it is commoner than the glabrous type. Some of the plants raised from Wilson's 1901 seed may have belonged to this variety, which seems to be the commoner in gardens also.

SINOWILSONIA   HAMAMELIDACEAE

Like many genera of the Witch Hazel family, *Sinowilsonia* consists of a single species. Leaves alternate, deciduous, pinnately veined. Flowers without petals, unisexual, inconspicuous, the males in leafless catkin-like racemes, with five sepals and five stamens. Females with a stellately downy receptacle (calyx-tube) which encloses the ovary and is surmounted by five spreading sepals; they are arranged in racemes which are terminal on leafy shoots and at first about 1 in. long, greatly lengthening as the fruits ripen. The fruit is a capsule, partly enclosed by the receptacle and containing numerous black seeds.

The genus is named after the famous plant collector E. H. Wilson, who collected some of the specimens from which the genus was described, and introduced its one species.

S. HENRYI Hemsl.

A deciduous shrub or small tree, occasionally 25 ft high and upwards in China. Leaves broadly elliptic to obovate, 3 to 6 in. long, 2½ to 4½ in. wide, rather like those of a lime, but short-stalked, strongly veined beneath, and covered there with starry hairs (like the young shoots), the margins set with bristle-like teeth. Flowers greenish, in slender, terminal, pendulous racemes 9 in. long. Fruits very downy, egg-shaped capsules, ⅓ to ½ in. long, stalkless, and arranged on long slender spikes.

Native of central and western China, common in N.W. Hupeh, where it inhabits the banks of mountain streams at 3,000 to 4,000 ft; introduced by Wilson for the Arnold Arboretum in 1908. It is of botanical interest but of little ornamental value, the flowers having no great beauty.

SKIMMIA RUTACEAE

There are only half a dozen or so species in this genus, all evergreen shrubs or small trees, natives of the Himalaya and E. Asia. Leaves alternate, simple, entire, aromatic, dotted with pellucid glands. Flowers small, white or yellow, sometimes tinged with pink, normally unisexual and borne on different plants, arranged in short terminal compound panicles. Sepals, petals and stamens four or five. Carpels connate, two to five. Style short, its stigma obscurely two- to five-lobed. Fruit a drupe, red or black, with two to five stones.

The fruits are the most attractive feature of this genus, or rather of the red-fruited species grown in gardens, but to obtain these from *S. japonica* or *S. anquetilia* it is necessary to plant the two sexes together. But male plants have more numerous flowers in the inflorescence than the females, and are consequently more ornamental and fragrant in spring than the females. If there is room for only one plant it should be a male, or *S. reevesiana.*

The editor is greatly obliged to Mr Nigel Taylor of the Kew Herbarium for his assistance in the matter of *S. anquetilia, S. laureola* and *S. multinervia,* and for his helpful comments on the text as originally drafted.

The skimmias are of easy cultivation and like a good, moist soil; they will thrive and flower well in a moderately shady spot. Those who make a speciality of growing skimmias for their fruits assist fertilisation by artificial means (see under *S. japonica*).

The generic name derives from the Japanese 'Miyama-Shikimi' ('Hill or Wild Shikimi'—this word without qualification being the vernacular name for *Illicium religiosum*).

All the skimmias can be propagated by cuttings of half-ripe shoots, and this method is of course essential for the perpetuation of the named sorts. *S. reevesiana* may be propagated by seeds and this method can also be used for the other species, with the drawback that the sex of the seedlings will not be known until they flower.

S. ANQUETILIA N. P. Taylor & Airy Shaw

Limonia laureola sens. Wall. and other authors, not DC.; *Skimmia laureola* sens. Hook. f. and other authors in part, not (DC.) Walp.; *Anquetilia laureola* Decne.

A dioecious shrub usually under 4 ft high, sometimes prostrate. Leaves light green, up to almost 7 in. long, oblanceolate or oblong-elliptic, mostly tapering to the base for about two-thirds of their length, emitting when crushed a very heavy and, to some people, disagreable odour; petiole stout, up to ½ in. long. Inflorescence short and very compact. Sepals five, ovate, acute, or obtuse. Petals five, yellow, erect, with strongly incurved margins. Stamens about equalling the corolla, with orange anthers. Fruits dull red, up to almost ½ in. long, with two or three stones. *Bot. Mag.,* n.s., t. 789.

Native of the western Himalaya as far east as W. Nepal, extending into eastern Afghanistan; it appears to be common in the undergrowth of coniferous forest, up to 13,000 ft. The first recorded introduction was in 1841, when Dr Royle sent seeds to the garden of the Horticultural Society at Chiswick, though some nurseries may already have had it at that time. For the next decade it was the only species of *Skimmia* in gardens, but earned no praise and was soon overshadowed, first by *S. reevesiana* and later by *S. japonica*. By 1889 it was so little known in gardens that Dr Masters purposedly omitted it from his treatment of the cultivated skimmias, published in that year. It has, however, persisted in gardens. The male plant figured in the *Botanical Magazine* grows at Wakehurst Place in Sussex, and is an old one, of unrecorded origin. There is a female plant in the Hillier Arboretum near Romsey, Hants. Judging from these plants, *S. anquetilia* seems to be the least ornamental of the skimmias, the flower-trusses being small in relation to the size of the leaves, and the flowers unpleasantly scented.

S. anquetilia was first named in 1980 in the *Botanical Magazine*. Previously it had been known as *S. laureola*, a name that belongs properly to the very distinct black-fruited skimmia named *S. melanocarpa* by Rehder and Wilson. How this confusion arose is explained by Mr Taylor and Mr Airy Shaw in their article. From the other species of *Skimmia* it is easily distinguished by the combination of: leaves mostly tapering straight-sidedly from above the middle, yellow pentamerous flowers in small inflorescences, erect petals and red fruits.

S. JAPONICA Thunb.   [PLATE 47

S. oblata T. Moore

An evergreen bush of dense habit, usually 3 or 4 ft high, sometimes much taller. Leaves mostly in a cluster towards the end of the shoot, aromatic when crushed, usually 3 to 4 in. long, ¾ to 1¼ in. wide, pale or yellowish green, narrowly obovate or oval, thickly specked beneath with transparent glands; leaf-stalk short, stout. Flowers in terminal panicles 2 to 3 in. long, male and female flowers normally on different plants, fragrant, ⅛ in. across; petals usually four, sometimes five, dull white. Stamens four or sometimes five in the male plant, absent or very much aborted in the female. Fruits globular, or depressed at the top like an orange, bright red, ⅓ in. wide. *Bot. Mag.*, t. 8038.

Native of Japan, the Ryukyus and the Philippines, with variants as far North as Sakhalin and the southern Kuriles; perhaps also of China and Formosa. Cultivated at Kew as long ago as 1838, this species did not obtain any general attention from horticulturists until it was introduced from Japan by Fortune in 1861 to Standish's nursery. It received a First Class Certificate when exhibited in fruit in 1864*, and was then named *S. oblata* by Moore, under the impression that the plant now known as *S. reevesiana* was the true *S. japonica*. But the acclaim with which the Japanese skimmia was first greeted faded away when it was found that the plants sent out did not bear the expected beautiful fruits.

---

* It later emerged that the plant in question must have borne hermaphrodite flowers in that year, since no other skimmia was in flower with Standish at the same time. In later years it bore only female flowers, but in the meantime another plant of the original batch had flowered, which was male and was named *S. fragrantissima* in 1867. The

A few knowledgeable gardeners were aware of the dioeciousness of *S. japonica* but it was not until Dr Masters published his study of the skimmias in 1889 that it came to be generally realised. Even Charles Noble, Standish's former partner, confessed in a letter to Masters that he had never had fruits off his "*S. oblata*" and had never heard of male plants, though they had been available, as *S. fragrans* and *S. fragrantissima*, for many years before 1889.

cv. 'FOREMANII'.—See below, under *S.* × *foremanii*.

cv. 'FRUCTU-ALBO'.—Under this name Messrs Hillier offer an attractive white-fruited clone of dwarf habit.

cv. 'NYMANS'.—A free-fruiting female clone, first exhibited from Nymans in 1934. Leaves mostly oblanceolate, obtuse or bluntly pointed; petioles tinged with red.

var. REPENS (Nakai) Ohwi *S. repens* Nakai—A northern and montane variant of dwarf stature, sometimes prostrate. Cultivated plants perhaps referable to this variety are erect growing, but with small leaves and of dense habit.

cv. 'ROGERSII'.—See under *S.* × *foremanii*.

cv. 'RUBELLA'.—A bushy shrub to about 4 ft high. Leaves darkish green, elliptic to oblanceolate, mostly 3 to 4 in. long (to 5 in. in some seasons), acute or subacute at the apex (rarely obovate and obtuse); petioles dark red. A male clone. Flowers very fragrant, pink in the bud, borne in a fine panicle about 3 in. long and 2½ in. wide at the base; peduncles and rachis deep bronzy red, rendering the inflorescences conspicuous long before the flowers expand. Petals and stamens four (*S. rubella* Carr.; ? *S. intermedia* Carr.; *S. reevesiana* f. *rubella* (Carr.) Rehd.; *S. japonica* 'Rubella', *Journ. R.H.S.*, Vol. 87 (1962), p. 328).

This skimmia was introduced to France from China in 1865 by Eugène Simon and to Britain before the end of the century. Its taxonomic position is uncertain, but it seems to agree better with *S. japonica* than with *S. reevesiana*, under which it was placed by Rehder. In flower, 'Rubella' is the finest of the skimmias, and received an Award of Merit in 1962. It needs a half-shaded position and will flower from every terminal bud even under a dense canopy, provided the soil is not too rooty. There are other males in commerce. One, sent out as 'Fragrans', may not be the skimmia originally named *S. fragrans* but it is a fine ornamental, dwarfer than 'Rubella', with unusually glossy leaves and also differing from it in having no trace of a red infusion in any part.

cv. 'VEITCHII'.—As described by Carrière in 1874 this had very thick and leathery leaves, oval or obovate-elliptic, roundish at the apex, to about 4 in. long and 2 in. wide. Female. This may have been introduced by J. G. Veitch, who was in Japan at the same time as Fortune, but no reference has been found

---

origin of the male plant named *S. fragrans* by Carrière in 1869 was not stated, but some of the plants brought from Japan by Fortune were auctioned by Standish in autumn 1864 and it must have been one of these that the nurseryman William Bull exhibited as *S. fragrans* in the following year.

So far as is known, no plant of *S. japonica* is constantly hermaphrodite but it is likely that most females produce a few flowers in which the stamens are fertile. The mere presence of stamens in female flowers is of no significance, since close examination will show that these usually have abortive anthers.

to a Veitchian introduction. 'Veitchii' has been said to grow 4 ft high and 6 ft wide, but no authentic plant has been seen.

All these forms are easily increased by cuttings, and for the purpose of obtaining fruits one male need only be grown to, say, six females. In order to secure a crop of berries it is advisable to fertilise the flowers artificially. It is necessary, of course, to transfer the pollen from the male to the female, and this is usually done by taking some fluffy material (a rabbit's tail or camel-hair brush is often used) rubbing it over the male flowers as soon as the pollen is loose, and then dusting over the female flowers with it (the latter are easily recognised by the prominent ovary and stigma, and the abortive stamens). In some districts bees or other insects will do the business themselves, but it is safer to do it by hand. *S. japonica* grows well in the neighbourhood of towns, but does not, in my experience, fruit freely there, even with artificial fertilisation.

S. × FOREMANII Knight—Sometime in the 1870s the Scottish nurseryman Foreman made a cross between *S. japonica* and *S. reevesiana* at the Eskdale Nurseries, Dalkeith. Only one female was raised, and this he pollinated with the best of the male seedlings. The resulting second generation hybrid was exhibited at the Spring Meeting of the Caledonian Horticultural Society in 1881 and received a First Class Certificate (*Florist and Pomologist* 1881, p. 70).

There can be no doubt about the hybridity of the 1881 *S. foremanii*. But seven years later Foreman staged an exhibit at the R.H.S. Show, December 11, 1888, consisting of numerous pot-grown skimmias of his own raising, these too called *S. Foremanii* and awarded a First Class Certificate. But the parentage of these was given by the raiser as '*S. oblata* [*S. japonica*] × *S. fragrans*', which, if correct, meant that he had simply fertilised a female *S. japonica* with pollen of the same species. This prompted a reader of the *Gardeners' Chronicle* to ask 'Did you ever hear of a "hybrid" between a bull and a cow?' Dr Masters, who was investigating the skimmias at this time, suspected hybridity in the specimen sent to him by Foreman, largely on the grounds that some of the fruits were obovoid. He described this skimmia as having yellow-green oblanceolate or lanceolate leaves about 3 in. long and ¾ in. wide, tapered at each end, with reddish petioles; inflorescence many-flowered; berries scarlet, depressed-globose or pear-shaped (*Gard. Chron.*, Vol. 5 (1889), p. 553). Despite his sus-picions, Masters placed this skimmia under *S. japonica* as 'Foreman's Variety'. He seems to have been unaware of the earlier account of the undoubted hybrid mentioned above, otherwise he might have interrogated Foreman more closely. Whether any of the Foreman skimmias have been perpetuated vegetatively up to the present time it is impossible to say. Judging from cultivated plants and herbarium specimens, Foreman's name has been or is associated with three or four different skimmias, none of which agree with Masters' description. One characteristic of Foreman's variety (F.C.C. 1888) was that the flowers were as numerous in each panicle as in male forms and consequently just as well-scented as well as producing very large trusses of fruit (up to forty or so berries in each), when fully fertilised.

The account of Foreman's exhibit published in the horticultural press prompted W. H. Rogers of the Red Lodge Nurseries to inform Masters of a new skimmia that he had raised about 1877, when his "*S. oblata*", i.e. *S. japonica,* fruited for the first time. His letter, quoted by Masters, is somewhat obscure, but it appears that the plants he sent out as *S. Rogersii* were seedlings of the fruiting *S. japonica*, not a clone. The plant of which he sent a spray to Masters was, however, almost certainly a hybrid between *S. japonica* and *S. reevesiana* and accepted as such; the fruits were coloured as in the latter species, but squarish and depressed at the top (*Gard. Chron.*, Vol. 5 (1889), p. 553). The flowers were described by Masters as *structurally* hermaphrodite, which merely means that the organs of both sexes were apparently fully developed, not that the plant was actually self-fertile. As in the case of Foreman's skimmia, the plants under the name 'Rogersii' are not uniform. The name *S.* × *foremanii* var. *rogersii* (Mast.) Rehd. is based on the plant described by Masters.

S. LAUREOLA (DC.) Sieb. & Zucc. ex Walp.
Limonia laureola DC.

A low dioecious shrub of fairly compact habit, usually under 3 ft high. Leaves on flowering shoots narrowly oblong-obovate or oblong-elliptic, rarely quite elliptic, 2½ to 4½ in. long, 1 to 1½ in. wide, acute to slightly acuminate at the apex, tapering from the middle or slightly above it to a cuneate or roundish base, rich green above, aromatic when crushed; petiole green, ¼ to ⅜ in. long. Inflorescence of male plants fairly lax, pyramidal, up to 4 in. long and up to 3 in. wide at the base on the stronger shoots. Flowers sweetly scented, opening in spring. Sepals five, semi-ovate or deltoid, obtuse. Petals five, cream-coloured, semi-erect. Stamens five, with orange anthers. Fruits not seen, purplish black on wild plants.

Native of the Sino-Himalayan region from central Nepal to western Szechwan; described by de Candolle in 1824 from a specimen collected in Nepal. So far as is known it is represented in cultivation only by male plants, and it is from one of these, part of a large planting at Kew by the Victoria Gate, that the above description is mainly drawn. The origin of the Kew clone is unrecorded, nor is anything known of the provenance of very similar plants at Borde Hill in Sussex. Seeds of *S. laureola* were sent by Wilson from W. Szechwan and were distributed under the synonymous name *S. melanocarpa* (though some specimens originally identified as *S. melanocarpa* are *S. multinervia*, for which see below).

S. laureola, in its female form, is unlikely to be of much ornamental value. But the male plants are very handsome, with their large trusses of creamy fragrant flowers. They are somewhat dwarfer than the common male clone of *S. japonica*, from which they differ in their yellowish white pentamerous flowers, which do not open so widely as in that species. They stand full sun, but should also flower well in shade.

S. MULTINERVIA Huang *S. melanocarpa* Rehd. & Wils. in part (excluding type)—Allied to *S. laureola* and, like it, black-fruited, this species (described in 1958) differs most markedly in its size. In the wild it is usually 10 to 20 ft high, bnt somtimes a tree 50 ft high. The leaves are about twice as long as in *S laureola*, and the female inflorescences are as large and well-formed as those of male plants of that species or of *S. japonica*. There is an example 9 ft high in the Isabella Plantation, a woodland garden in Richmond Park near London, and younger propagations at Kew. *S. multinervia* ranges from eastern Nepal to western China, including Yunnan.

S. REEVESIANA (Fort.) Fort.

Ilex reevesiana Fort., *Gard. Chron.* 1851, p. 5; *S. fortunei* Mast.; *S. japonica sens.* Lindl., not Thunb.

A low evergreen shrub, usually not more than 2 ft high, with dark green, narrow elliptical leaves, averaging 2½ to 4 in. long, and from ¾ to 1 in. wide, tapering gradually and equally towards both ends. Flowers white, always bisexual in cultivated plants, ½ in. across, produced in terminal panicles 2 to 3 in. long. Fruits rich crimson, distinctly oval or pear-shaped, ⅓ in. long, persisting. Seeds pointed at both ends. *Bot. Mag.*, t. 4719.

SKIMMIA REEVESIANA

The plant from which this species was described was introduced by Fortune from a Shanghai nursery, where this skimmia was the rarest and most prized of the owner's possessions. The original had been brought from the Hwang Shan, a mountain some 250 miles to the south-west of Shanghai, and was called the 'Wang-shan-kwei', Kuei Hua being the Chinese name for *Osmanthus fragrans*. Fortune saw it in 1848 and it reached Standish and Noble's nursery in the following year; it was first exhibited in October 1852, but Lindley had seen a specimen in the previous year, which he identified as *S. japonica*. This is the name under which the species was grown until Dr Masters cleared up the confusion in 1889.

S. reevesiana dislikes a limy soil and needs one that is fairly rich and moist to be seen at its best. Being hermaphrodite and self-fertile, every plant is capable of bearing fruit without cross-pollination, but artificial fertilisation helps to secure a good set. The flowers are abundant and charmingly fragrant.

As represented in cultivation, *S. reevesiana* differs from *S. japonica* in its constantly hermaphrodite flowers with their parts in fives and in the obovoid or ellipsoid fruits; also in the narrower, darker green, more acuminate leaves. But the type-plant of *S. reevesiana* derived by seed or cuttings from the one individual of the Hwang Shan, which may have been a hermaphrodite or even apomictic variant of a species that is normally dioecious, as are the other skimmias. Of the Formosan plants referred to *S. reevesiana* by Li some are dioecious and some bear tetramerous flowers as in *S. japonica* (H.-L. Li, *Woody Flora of Taiwan* (1963), pp. 381–3). Some cultivated plants of *S. reevesiana* agree perfectly with the type. But others differ from it somewhat and may derive from a later, unrecorded introduction or be seedlings of the original clone.

cv. 'ARGENTEA VARIEGATA'.—Leaves margined with white. Introduced 1875 (*S. japonica argentea variegata* Nichols.; *S. fortunei* var. *argentea* Mast.).

cv. 'RUBELLA'.—See *S. japonica*.

SMILAX SMILACACEAE

A curious and interesting genus of mostly climbing plants, some with woody stems, some herbaceous, belonging to a small family allied to the Liliaceae and often included in it. Most of the 300 or so species are tropical or subtropical, and only few of those from temperate regions have become established in gardens. Of the woody-stemmed species there are both evergreen and deciduous as well as intermediate types. Leaves alternate, prominently three- to nine-ribbed, and net-veined between the ribs; from the stalks a pair of tendrils are developed by means of which the slender stems are supported. Stems round or angular, usually prickly, often springing from a fleshy or tuberous root-stock, the sexes usually but not always on separate plants: isolated plants having been known to produce fertile seed. The flowers have little beauty, and are always green or greenish. Fruit a black or red berry.

The chief value of the smilaxes in gardens is in producing rich, graceful masses of handsome foliage. They develop thickets of stems which are constantly being renewed from the base, and are happily placed when they can ramble over a tree-stump or some such support. Seed is rarely seen with us on many of the species, and propagation is best effected by dividing up the plants in spring.

The popular medicine, sarsaparilla, is a product from the root of various tropical American species.

Of the following sorts, *S. rotundifolia* and *S. hispida* are the most robust in my experience, but *S. excelsa* has also been known to make a vigorous tree climber in Surrey; and for the warmer counties, *S. aspera*, which has a very graceful inflorescence, is to be recommended.

S. ASPERA L.   ROUGH BINDWEED

An evergreen, semi-scandent plant with four- to six-angled stems and zigzag branches, armed with short, stout spines. Leaves very diverse in shape and size, but nearly always more or less heart-shaped at the base, and prickly on the margins, sometimes on the midrib also. As seen in cultivation the usual

SMILAX ASPERA

type of leaf is of a narrow, elongated, ovate shape, broadest almost or quite at the base, abruptly narrowed above the base, then tapering gradually to the point, five- to nine-nerved; sometimes the base is quite straight, and the leaf an elongated triangle; sometimes the leaf is heart-shaped and nearly as broad as long. They measure from $1\frac{1}{2}$ to 4 in. long, $\frac{3}{4}$ to 3 in. wide; stalk spiny or unarmed, $\frac{1}{4}$ to 1 in. long. Flowers pale green, fragrant, produced in terminal and axillary racemes, along which they are arranged in clusters of four to seven; the racemes vary from $1\frac{1}{2}$ to 4 in. long, and the flower-stalk of the individual flower is $\frac{1}{12}$ to $\frac{1}{3}$ in. long. Fruits about the size of a pea, red.

A species of wide distribution, from the Mediterranean region and the Canaries to India and southern Central Asia; cultivated in Britain since the mid-17th century. It is not reliably hardy in the colder parts of the country, and used to be commonly grown in cold greenhouses, where if planted out it makes a tangle of numerous stems eventually 8 to 10 ft high—a handsome cheerful evergreen, with graceful and fragrant, if not showy, flowers.

var. MACULATA (Roxb.) A. DC.—Leaves usually blotched with white. Native of N. India and other areas.

var. MAURITANICA (Desf.) Gren. & Godr.—Leaves larger, spines fewer.

S. BONA-NOX L.

A deciduous or partially evergreen climber, with angular or square branchlets, slightly armed with short, stout prickles. Leaves very variable, roundish, heart-shaped, fiddle-shaped, spear-shaped or three-lobed, 1½ to 4½ in. long, always pointed, green and glossy on both sides, often bristly or prickly at the thickened margins and on the nerves beneath, five- to nine-nerved; stalk ¼ to ½ in. long. Flowers deep green, produced in umbels, the main-stalk of which is ½ to 1 in. long. Berries black with a bluish bloom, round, ¼ in. across; six to twelve, sometimes more, in an umbel.

Native of N. America from Texas and N.E. Mexico to Florida, north to Kansas and Delaware; in cultivation 1739. A hardy species, allied to *S. rotundifolia*, differing in the thickened leaf-margins usually set with prickles or bristles and the longer-stalked umbels.

S. 'CANTAB'   CAMBRIDGE SMILAX
S. Cantab Lynch

An evergreen climber reaching 12 ft or more high; stems round, armed with sturdy unequal prickles, and furnished with curious minute tufts of bristles; branches square, often unarmed. Leaves of thin texture, triangular, with the base deeply heart-shaped, the apex pointed, the largest 5 in. long, rather more in width, five-nerved, green on both surfaces, with a few grey spots on the upper one, the margins slightly bristly. Flowers in umbels of eight to twelve, the main-stalk as long or rather longer than the leaf-stalk.

Probably a native of N. America, but of unknown origin, having been first described (*The Garden*, Vol. 56, p. 505) by Lynch from a plant growing against a wall in the Cambridge Botanic Garden, where it has stood for many years. It is a male plant, and may prove to be a hybrid or belong to a species previously named.

S. CHINA L.   CHINA ROOT

A deciduous rambling shrub, with round stems sparingly armed with slightly recurved prickles. Leaves 2 to 3 in. long, very variable, roundish ovate, or broadly oval, or sometimes broader than long, ending in a short abrupt point, the base tapered or truncate or slightly heart-shaped, five- or seven-veined; stalk $\frac{1}{3}$ to 1 in. long. Flowers yellowish green, often numerous in umbels, the main-stalk of which is about 1 in. long. There are often over twenty flowers in an umbel. Fruits $\frac{3}{8}$ in. in diameter, globose, bright red.

Native of China, Japan and Korea; introduced by Philip Miller from China shortly before 1759, and again by Wilson in 1907. It has a large, fleshy root-stock, said to be eaten by the Chinese. It also yields a drug known as 'China Root', once highly esteemed as a remedy against gout, though it is likely that other Chinese species were used for the same purpose (Norton, in *Pl. Wils.*, Vol. III, p. 4).

S. DISCOTIS Warb.

A deciduous climber growing 10 to 16 ft high; young shoots grooved or angled; stems armed with decurved spines $\frac{1}{6}$ in. long. Leaves ovate or narrowly oval, mostly heart-shaped at the base and pointed at the apex, $1\frac{1}{2}$ to $3\frac{1}{2}$ in. long, $\frac{3}{4}$ to $1\frac{3}{4}$ in. wide, glaucous beneath, three- or five-nerved; stalk $\frac{1}{12}$ to $\frac{1}{5}$ in. long. Flowers greenish yellow, in an umbel terminating a very slender main flower-stalk up to $1\frac{1}{2}$ in. long; each flower on a stalk $\frac{1}{4}$ to $\frac{1}{3}$ in. long. Fruits blue-black, $\frac{1}{4}$ in. or rather more wide.

Native of Central and Western China; introduced by Wilson about 1908. It is distinct on account of the comparatively large, semi-circular stipules up to $\frac{1}{5}$ in. long, combined with the black fruit, the heart-shaped base of the leaves and their glaucous under-surface.

var. CONCOLOR Norton—Leaves green on both sides and, at least in the type-specimens, larger and with longer petioles. Introduced by Wilson in 1907.

S. EXCELSA L.

A tall evergreen or late-deciduous climber, with squarish stems and branches armed with flat, stiff spines, $\frac{1}{4}$ to $\frac{1}{3}$ in. long. Leaves unarmed, broadly ovate, heart-shaped or truncate at the base, pointed; $1\frac{1}{2}$ to $3\frac{1}{2}$ in. long, often as broad or broader than long; five- or seven-nerved, green on both sides; stalk $\frac{1}{4}$ to $\frac{1}{2}$ in. long. Flowers six to twelve in an umbel, the main-stalk of which is $\frac{1}{2}$ to 1 in. long. Berries red, $\frac{1}{4}$ in. wide. *Bot. Mag.*, t. 9067.

'This noble liane is one of the most striking features in the lower sylvan belt of the southern littoral of the Caspian Sea and throughout the Caucasus and the Pontic ranges as far as the Sea of Marmora. Farther west it appears in isolated areas in Thrace and in southern and eastern Bulgaria, often accompanied, as in its Asiatic home, by the grape-vine. In Ghilan I have seen it climbing into the crowns of the tallest trees, garlanding their boughs and hanging down in long swaying festoons.' (Dr O. Stapf, in the article accompanying the plate in the *Botanical Magazine*).

S. excelsa was first recorded by the French botanist Tournefort, who saw it during his visit to the Levant in 1700–1. Philip Miller had it in cultivation at Chelsea by 1739. It is, for European gardeners, the most interesting of the taller species, extending as it does into Europe, where it was probably more widely distributed in earlier geological epochs than it is at present. Indeed, it has a very close ally in the Azores. It appears to be quite hardy, though the leaves may be burned by severe frost. It was reintroduced from N. Iran in 1972 by Mrs Ala and Roy Lancaster.

S. GLAUCA Walt.   SAW BRIER

A tall deciduous or partially evergreen climber, with round stems but angled branches, sparsely or not at all prickly. Leaves ovate with broadly tapering or rounded bases and fine points; 1½ to 3½ in. long, 1 to 2½ in. wide; green above, glaucous beneath, with usually three prominent nerves and two smaller ones at the margins; stalk ⅙ to ⅓ in. long. Flowers green, produced three to eight together in small axillary umbels, the main-stalk of which is ⅓ to 1 in. long. Berries black with a glaucous bloom, ¼ in. wide. *Bot. Mag.*, t. 1846.

Native of the eastern United States from Massachusetts southwards; introduced in 1815. It is hardy. The glaucous colour of the leaves beneath is its best distinguishing character.

S. HISPIDA Muhl.   HAG BRIER
S. tamnoides var. *hispida* (Muhl.) Fern.

A climbing deciduous shrub, with round stems furnished with slender bristles and straight prickles, densely so towards the base; branches almost without them. Leaves heart-shaped or broadly ovate, 2 to 6 in. long, 1½ to 4½ in. wide, five- or sometimes nine-nerved, finely pointed, green on both sides, margins often minutely jagged; stalk ¼ to ¾ in. long. Flowers greenish yellow, borne on an umbel with a main-stalk 1 to 2 in. long. Berries blue-black, globose, about ¼ in. wide.

Native of the eastern and central United States and Ontario; introduced early in the 18th century. This species thrives well in this country, and is well marked by its large leaves and very bristly stems.

S. LAURIFOLIA L.

A tall evergreen climber with thick, leathery leaves, 2 to 5 in. long, ⅝ to 2 in. wide, three-nerved, dark green and rather glossy above, oblong, tapered or rounded at the base, and with usually a short, abrupt, fine point; stalk ¼ to ½ in. long. Flowers greenish, in umbels, the main-stalk of which is about as long as the leaf-stalk. Fruits ¼ to ⅓ in. wide, one-seeded, black when ripe, but taking two seasons to become so.

Native of the south-eastern United States, west to Texas; in cultivation 1739. Very distinct in its three-nerved, leathery, evergreen, and comparatively narrow leaves, this is, unfortunately, rather tender. It succeeded well on a wall in the vicarage garden of Bitton, near Bath, and is suitable for the south and west generally, where it will make an interesting evergreen wall-covering.

S. MEGALANTHA C. H. Wright

An evergreen climber growing 15 to 20 ft high; young shoots grey, angled, armed with irregularly scattered, stout spines up to ½ in. long and furnished with tendrils proceeding from the much thickened leaf-stalk. Leaves very variable in size and shape, the largest broadly ovate, rounded at the base, 9 in. long by 6 in. wide, the smaller ones are lanceolate to narrowly oval, tapered at the base, some of them only 3½ in. long by 1 in. wide; all are of leathery texture, pointed, dark glossy green above, glaucous beneath, conspicuously three-nerved; leaf-stalk ½ to 1 in. long. Flowers (described by Wilson as greenish) produced from the leaf-axils of small young leaves on short branches in corymbose umbels; the perianth of the male flower has six slenderly pointed lobes; stamens six. Fruits globose, nearly ½ in. wide, coral-red, often one-seeded, borne in umbels over 2 in. across; main-stalk 1 in. long; individual stalks ½ to ¾ in. long.

Native of Hupeh, W. Szechwan, and Yunnan, China; originally collected by A. E. Pratt and E. Faber; the latter found it on Mt. Omei. It was introduced to cultivation by Wilson in 1907. This fine smilax—in foliage the most remarkable of all the species that succeed in our average climate—has grown vigorously without any protection at Kew since its introduction, in good loamy soil. It is perfectly hardy. Its unusually large fruits are very handsome, but even if it failed to produce them it is well worth cultivating as an evergreen foliage plant. Propagated, like the rest, by division.

S. ROTUNDIFOLIA L.   HORSE BRIER
S. caduca L.

A vigorous, deciduous or partially evergreen climber, with slender, round or more or less angled stems, armed with one or two short spines between each leaf (not at the nodes); the stems are sometimes 6 or 8 yards long; branches four-angled. Leaves ovate to broadly heart-shaped; 2 to 6 in. long, often broader than long, with a short abrupt point, prominently five-nerved, glabrous and glossy green on both sides; stalk ¼ to ½ in. long. Flowers greenish yellow, about ⅕ in. across, borne in umbels; main flower-stalk ¼ to ½ in. long, flattened. Berries roundish, black, ¼ in. in diameter, covered with glaucous bloom, usually three to six of them in one cluster.

Native of eastern N. America; introduced in 1760. This is the commonest, most vigorous and hardy of all the smilaxes in cultivation, making a dense thicket of stems. It is the common horse brier or green brier of the United States, where its stems are sometimes 30 to 40 ft long, stretching from tree

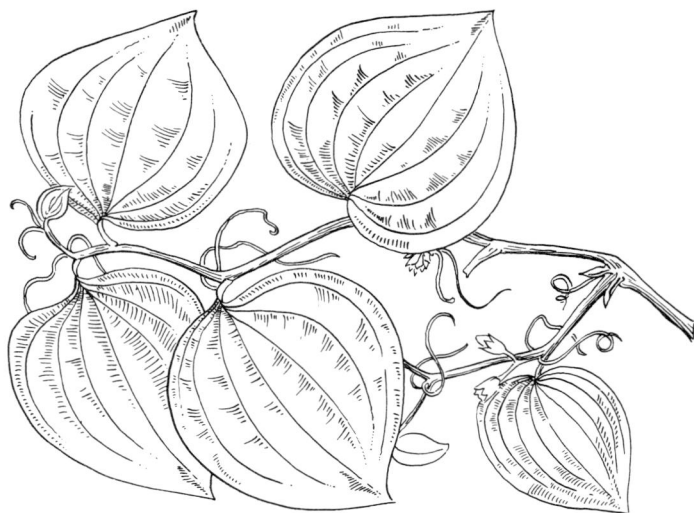

SMILAX ROTUNDIFOLIA

to tree. In gardens it may be trained up stout oak posts on which the stumps of the side branches have been left 2 or 3 ft long. Grown in this way it is very elegant.

S. SCOBINICAULIS C. H. Wright

A deciduous climber growing 15 ft or more high; branches slender, grooved, often very densely furnished with numerous black prickles, often unarmed. Leaves ovate to ovate-oblong, rounded or slightly heart-shaped at the base, finely pointed, five- or seven-veined, 3 to 5 in. long, 1 to 3 in. wide, glabrous and green on both surfaces; stalk $\frac{1}{4}$ to $\frac{3}{8}$ in. long, with tendrils attached to it near the blade. Flowers yellowish white, borne in axillary umbels $\frac{3}{4}$ in. wide, the main-stalk of which is $\frac{1}{3}$ in. long, the individual stalks $\frac{1}{8}$ to $\frac{1}{4}$ in. long. Fruits globose, black, $\frac{1}{4}$ in. wide, in clusters of nine to twelve, each containing one to three seeds.

Native of Central China; discovered by Henry; introduced by Wilson in 1907. The original type specimen collected in Hupeh by A. Henry about 1888, on which Wright founded the species, has the stem practically covered with the slender, bristle-like spines he described. Plants at Kew, however, raised from Wilson's seeds in 1908, had fairly stiff spines sprinkled freely over the stems but no bristle-like ones. The fruits, too, were much smaller. Wilson's No. 627 is certainly true but Nos. 455, 671, and 680, which have also been called *S. scobinicaulis*, seem to be distinct.

S. SIEBOLDII Miq.

A deciduous or semi-evergreen species, the stems round or somewhat ribbed, and more or less armed with slender prickles; branches distinctly angular. Leaves ovate with a heart-shaped base and a long fine point, five- or seven-nerved, margins minutely jagged; green both sides, 1½ to 3 in. long, two-thirds as wide. Flowers in small umbels of four to seven blossoms, green; main flower-stalk ½ to ⅔ in. long. Berries black, ¼ in. wide, often in threes.

Native of Japan, Korea, China and Formosa; introduced in 1908, perhaps before. It is a little known species in gardens, very distinct from *S. china*, which also occurs in Japan, in the smaller, black fruits, fewer flowered umbels, and triangular-ovate leaves.

S. WALTERI Pursh

A deciduous climber, with angled stems, armed only near the base; branches squarish, unarmed. Leaves ovate to ovate-lanceolate, 2 to 4½ in. long, ¾ to 2½ in. wide, broadly wedge-shaped to slightly heart-shaped at the base, ending in a short fine point, five- or seven-nerved, glabrous and green on both sides; stalk ¼ to ½ in. long. Flowers greenish, in short and flat-stalked umbels. Berries bright coral-red, globose, ⅓ in. wide.

Native of the eastern United States from New Jersey southwards; introduced early in the 19th century; rarely seen now. It is akin to *S. rotundifolia*, but has narrower, proportionately longer leaves, and is not so vigorous a grower. Also in *S. rotundifolia* the fruits are black.

SOLANUM   SOLANACEAE

A genus of around 1500 species, most of them herbaceous or semi-woody. Leaves simple or compound, sometimes solitary, more usually in cymose clusters arranged in the form of an umbel, raceme or panicle. Corolla star-shaped or rotate, five-lobed. Stamens five, the filaments short, inserted in the mouth of the corolla, the anthers long, leaning inwards and forming a cone. Fruit a many-seeded berry.

Apart from the species treated below, a few others with more or less woody stems are sometimes cultivated. Of the native bittersweet S. DULCAMARA L. there is a variegated form rather handsomely variegated with clear creamy white. The soft semiwoody shoots grow 6 to 8 ft high, but die back very much in winter, only the base being woody. The red berries are poisonous, so the plant is not desirable where there are children. The flowers are violet blue, and the leaves are frequently unequally lobed at the base. S. LACINIATUM Ait. f. of Australia (and of New Zealand according to some authorities) bears lanceolate or linear-lanceolate entire or pinnately lobed leaves up to 6 in. or so long, both forms appearing on the same plant. The bluish purple flowers are up to 2 in. wide, the fruits egg-shaped, orange-yellow, with conspicuous stone-cells in the flesh. It is a rather tender, soft-wooded plant attaining 10 ft or so (*Bot. Mag.*, t. 9154). S. AVICULARE Forst. f. of New Zealand (and probably Australia) is similar but less ornamental; the two are sometimes united under the name *S. aviculare*.

S. CRISPUM Ruiz & Pavon

A scandent, quick-growing, more or less evergreen shrub, with downy, scarcely woody young shoots. Leaves ovate, variable in size, usually 2½ to 5 in. long, mostly less than half as wide, taper-pointed, rounded or wedge-shaped, rarely heart-shaped at the base, minutely downy on both surfaces; stalk ½ to ¾ in. long. Flowers delicate bluish purple, fragrant, produced from June to September in long-stalked corymbs, 3 to 6 in. across. Each flower is 1 to 1¼ in. wide, the corolla with five ovate lobes, the yellow anthers closely packed in the centre. Fruits globose, ¼ to ⅓ in. wide. *Bot. Mag.*, t. 3795.

Native of Chile and Peru; introduced about 1830. This beautiful plant is only seen at its best in the milder counties of Great Britain, and few plants, even there, are more graceful and lovely. On a south wall at Kew it has grown and flowered for many years, but never with the vigour and profusion one sees in Devonshire and similar localities. It will grow 30 ft or more high if given support. The most beautiful effect I have seen produced by it was where it had been planted against the wall of a low shed, over the roof of which it had clambered. It may be pruned back in spring before growth commences. Where the climate is suitable it may be treated as a loose-habited, wide-spreading shrub, by pruning hard back annually. It will thrive in poor soil, even a chalky one.

cv. 'GLASNEVIN' ('AUTUMNALE').—A vigorous selection, flowering over a longer period than the older form, even in winter in a mild season. Distributed from the Glasnevin Botanic Garden before 1918. A.M. 1955 when exhibited from Kew, where it has long been cultivated.

S. JASMINOIDES Paxt.   [PLATE 48

A climbing evergreen shrub to about 20 ft high. Leaves simple, entire or with slightly undulate margins or sometimes with a few lobes or even free leaflets at the base, the entire leaves triangular-ovate, sometimes narrowly so, mostly 1 to 2 in. long and ½ to 1¼ in. wide, acuminate at the apex, truncate or slightly cordate at the base, bright green, smooth and glabrous above, glabrous beneath except for hairs in the axils of the principal veins; petioles elongating and twining round whatever support offers itself. Flowers up to twenty or so in broad lateral or terminal paniculate clusters; pedicels slender, often curved. Corolla pale blue or, in the commoner form, white (cv. 'ALBUM'), rotate, ¾ to 1 in. wide. Anthers yellow, united into a tube. Style hairy near the base, exserted beyond the anthers. Fruits not seen. *Bot. Mag.*, n.s., t. 568.

A native of S. Brazil, parts of Paraguay, also probably of Uruguay and bordering Argentina. Described by Paxton in 1840 from a cultivated plant, it was probably introduced by John Tweedie* from the Rio Grande do Sul in S. Brazil (see further in Dr Harley's article accompanying the plate in the *Botanical Magazine*. The type plant had pale blue flowers, but the commoner 'Album' was in cultivation by 1847 and was probably raised from the same batch of seed.

S. jasminoides is certainly more tender than *S. crispum* and may be cut to the ground in severe winters. But the loss of the younger growths from frost is of no account, some shortening back in spring being desirable in any case. Grown on a wall, it needs the support of wire or trellis, to which it will attach itself by its petioles.

S. VALDIVIENSE Dun.
S. euonymoides Rémy, not Sendtn.

A deciduous, sucker-producing shrub of loose, lax habit, growing up to 8 or 10 ft high and developing graceful arching shoots which when young are distinctly angled and downy at the angles. Leaves simple and entire, ovate-lanceolate, pointed, mostly tapered (but sometimes rounded) at the base; very variable in size, the largest 2 to 2½ in. long by ¾ to 1 in. wide, the smallest scarcely ½ in. long, downy on the margins; stalk $\frac{1}{12}$ to ¼ in. long. Flowers produced in short racemes ordinarily of two to seven blossoms, the main-stalk ⅛ to ½ in. long and downy, the individual flower-stalks up to 1 in. long, very slender and glabrous. Each flower is about ½ in. wide, pale mauve, white or lavender, with which the cluster of erect yellow stamens is in good contrast. Fruit globose, ¼ in. wide and of a "dullish slightly translucent olive green." (Comber.) *Bot. Mag.*, t. 9552.

Native of Chile and bordering Argentina from about 38; S. to 43; S.; known to botanists since about 1830, but probably not introduced until H. F. Comber sent seeds from near San Martin de los Andes, Argentina, in 1927. The leaves are not much developed at flowering time (May) and the long sprays, thickly wreathed with blossom, are then very handsome. According to Comber the aim in pruning should be to encourage the growth of long summer shoots which will flower the following spring. Worn out and weedy growths should be removed as soon as the flowering is over. Comber found it growing in 'moist sunny meadows.' He considered that unless the climate is sufficiently sunny and mild to ripen and preserve the long summer shoots through the winter

---

* John Tweedie worked as head gardener to several Scottish landowners before emigrating to Buenos Aires in 1825, at the age of fifty. During his explorations, which ranged from Brazil to Patagonia and were carried out at his own expense, he made important botanical collections and introduced many ornamental plants to the Glasnevin and Glasgow Botanic Gardens. He died in 1862.

it will scarcely be worth growing. Still, I have seen it flowering in several places in the south and thought it quite attractive. The sunniest possible spot should be chosen for it.

SOPHORA   LEGUMINOSAE

A genus of about eighty species of trees, shrubs, subshrubs and perennial herbs, scattered over the world. Leaves odd-pinnate, usually with numerous leaflets. Flowers in racemes or panicles, sometimes solitary. Corolla composed of standard, wings and keel, but not obviously of the pea-flower shape in *S. tetraptera* and its allies. Stamens free or almost so. Ovary stalked. Pods constricted between the seeds, hence necklace-shaped (moniliform), usually fleshy or woody and indehiscent. The cultivated species fall into two well-marked groups, the first represented by *S. japonica*, whose flowers are pea-shaped (papilionaceous) and the second by *S. tetraptera* and its allies (the Edwardsia group), in which the papilionaceous form of the corolla is obscured owing to the petals pointing forward. The latter group has a remarkable distribution: Lord Howe Island; New Zealand and Chatham Island; Hawaii; Easter Island; Juan Fernandez Islands; temperate S. America; Réunion (in the Mascarene group of islands). Thus the group is predominantly insular in distribution and only in New Zealand and Chile is it found more than a few miles from the ocean. See further in: Good, *Geography of Flowering Plants* (1953), p. 117 and fig. 34).

The generic name *Sophora* was originally applied by Linnaeus to *S. alopecuroides*, an herbaceous perennial mainly of W. Asia but extending into S.E. Russia and the Crimea, *sophora* or *sophera* being, according to him, an 'ancient' name for some similar plant. It is in fact an Arab name, probably for the plant that Linnaeus later named *Cassia sophera*. The rendering *Sophora* for the present genus allowed him to make a pun. For he remarked that the name could also signify 'of the wise men' (*sophorum*), since it brought 'knowledge and warning' that plants which combine papilionaceous flowers with free stamens, as in *Sophora*, could not be regarded as forming a distinct class (*Hort. Cliff.* (1737), p. 156). Because of its ten free stamens and single pistil he placed *Sophora* in his *Decandria Monogynia*, where it is associated with many genera quite unrelated to the *Leguminosae*, among them *Rhododendron*. The majority of pea-flowered genera of the Leguminosae have all the ten stamens, or nine of them, more or less connate, and those known to Linnaeus were placed in his *Diadelphia Decandria*.

S. AFFINIS Torr. & Gr.

A deciduous, round-headed tree up to 20 ft high, with a trunk 8 to 10 in. in diameter; young shoots slightly downy. Leaves pinnate, 6 to 9 in. long, with thirteen to nineteen leaflets which are oval, tapering about equally towards both ends, shortly stalked, 1 to 1½ in. long, about ½ in. wide, slightly downy beneath when young. Flowers in slender downy racemes 3 to 6 in. long, produced in June in the leaf-axils of the new growths, white tinged with rose, ½ in. or rather more long. Calyx downy, bell-shaped, with broad, shallow triangular teeth. Pods 1½ to 3 in. long, black, downy, persisting long on the tree. By the constrictions of the pod, each seed has its own oval or globose compartment, the inner wall of which is fleshy; the pod rather suggests a string of three or four beads.

Native of Texas and Arkansas, often on limestone hills; discovered in Texas in 1821; introduced to cultivation in 1890. Sargent observes that a domestic ink is sometimes made from the resinous exudations of the black fruits. In leaf this sophora resembles *S. japonica*, but that species is very distinct in having terminal, autumnal flower panicles. *S. affinis* would probably succeed best on the south or east coast, where it should make a pleasing small tree. After being probably extinct in British gardens it is again available in commerce (**1978**).

S. DAVIDII (Franch.) Skeels

S. moorcroftiana var. *davidii* Franch.; *S. viciifolia* Hance, not Salisb.

A deciduous shrub of rounded habit, from 6 to 10 ft high, and as much through, the young branchlets covered with greyish· down, the year-old branches more or less spiny. Leaves pinnate, 1½ to 2½ in. long, with seven

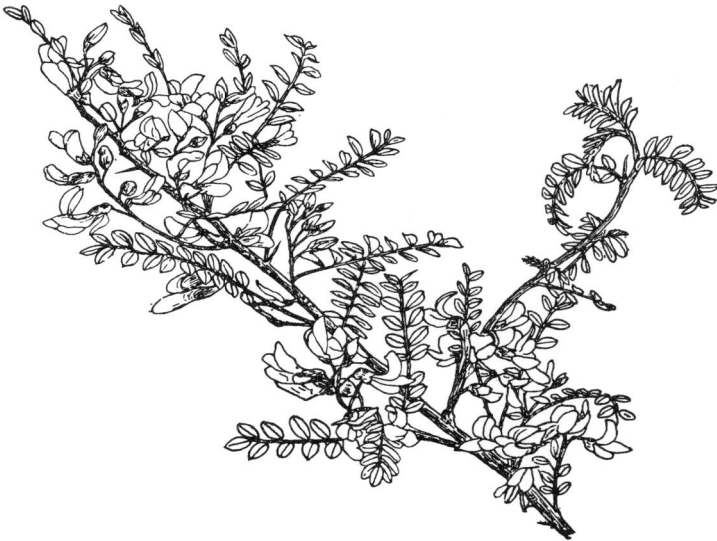

SOPHORA DAVIDII

to ten pairs of leaflets, which are $\frac{1}{4}$ to $\frac{3}{8}$ in. long, about $\frac{1}{8}$ in. wide, oval or obovate, with silky appressed hairs beneath. Racemes terminal on short twigs, produced from the buds of the previous year's shoots, 2 to $2\frac{1}{2}$ in. long. Flowers pea-flower-shaped; petals bluish white; calyx $\frac{1}{8}$ in. long, downy, short-toothed, violet-blue. Pod 2 to $2\frac{1}{2}$ in. long, about $\frac{1}{6}$ in. wide, downy, one- to four-seeded, constricted between the seeds. *Bot. Mag.*, t. 7883.

Native of China in the provinces of Yunnan, Szechwan, and Hupeh, up to 13,500 ft. It was introduced in 1897 to Kew, where it has grown well, and proved to be one of the most charming of Chinese shrubs, the branches being loaded with the blue and white racemes in June, their beauty greatly enhanced by the elegant fern-like foliage. It requires a good loamy soil, and a site exposed to full sunshine. According to Henry, in the elevated regions where it grows, it often covers large tracts of barren country, just as gorse does in Britain. It is propagated by cuttings made of young shoots with a heel of old wood, in July and August, and placed in a gently heated frame.

Nearly related to *S. davidii* is S. MOORCROFTIANA (Benth.) Baker, with similar foliage and habit, but which is more spiny, more downy, has smaller leaflets, yellow flowers, and a longer more slender calyx. Native of the drier parts of the Himalaya from Nepal westwards.

S. FLAVESCENS Ait.

Originally described by Aiton in the *Hortus Kewensis*, vol. ii., p. 43, in 1789, this species has since appeared at intervals, but has never obtained a secure footing in gardens. It is merely sub-shrubby with us, and thrusts up during the growing season shoots 3 ft or more long, which are, as a rule, cut back to ground-level in winter. The pinnate leaves are 6 to 9 in. long, with up to nineteen narrowly ovate leaflets $1\frac{1}{2}$ to $2\frac{1}{2}$ in. long. Flowers yellowish white, $\frac{1}{2}$ in. long, borne during July and August on a terminal cylindrical raceme or panicle up to 1 ft in length. The pods are 2 to $2\frac{1}{2}$ in. long and carry one to five seeds.

Native of China, where it has been collected by Delavay, Henry, Wilson, Forrest and others; also of Formosa. The shoots are really semi-herbaceous and too soft to survive the winters of our average climate, so that it forms in time a woody stool which produces gradually weaker shoots and finally succumbs. In W. Hupeh and E. Szechwan, where Wilson found it a shrub 6 ft high, he describes it as very common in sandy places.

S. JAPONICA L.   [PLATE 49

A deciduous tree, 50 to 80 ft high, of rounded habit and branching low down when growing in the open, but capable of forming a tall clean trunk when close planted. Bark downy when young, glabrous later and dark greenish brown; on old trunks it is grey, and corrugated rather like an ash. Leaves rich green, pinnate, 6 to 10 in. long, composed of nine to fifteen leaflets, which are ovate or oval, 1 to 2 in. long, half as wide, covered with small appressed hairs beneath. Flowers in terminal panicles 6 to 10 in. long and wide, creamy white, each about ½ in. long; calyx ⅛ in. long, bell-shaped, green, shallowly toothed. Pods 2 to 3½ in. long, glabrous, one- to six-seeded; rarely seen in Britain.

Native of China (not of Japan); introduced to France in 1747 and thence to England in 1753. It is one of the most beautiful of all leguminous trees, although it does not flower in a young state—not commencing until thirty to forty years of age. Old trees flower freely, especially after hot summers. The blossoms are not developed until September, and in wet cold summers do not develop at all. They do not fade on the tree, but drop off quite fresh, making the ground white beneath. On the continent of Europe, thanks to the warmer summers, it attains a larger size than with us, and frequently ripens seeds, which are the best means of increase. All parts of the plant, even the wood, contain a purgative principle said to be so potent that turners working on the green wood are immediately afflicted by colic, and that well-water becomes laxative if the leaves fall into it in autumn.

The largest specimen of *S. japonica* recorded recently grows at Syon Park, London; it measures 76 × 15¼ ft at 4 ft (1967). Others are: University Parks, Oxford, 59 × 9¼ ft (1975); Aldenham House, Herts, 60 × 6¼ ft (1976); Angelsey Abbey, Cambs., 48 × 5¾ ft (1973); Penrhyn Castle, Bangor, 44 × 6½ ft (1974). The decrepit tree at Kew in G.15 was planted soon after the introduction of the species and is kept for its historical interest.

cv. 'PENDULA'.—A very picturesque weeping tree with stiff, drooping branches. It should be grafted on stocks of the ordinary form 10 to 15 ft high. An admirable lawn tree, or for forming a natural arbour. It was distributed by Loddiges early in the 19th century, but whence he obtained it is unknown. Other pendulous forms are cultivated on the continent and may have originated there. Several were found among seedlings of a single sowing made at Paris in 1857 (*Rev. Hort.*, 1861, p. 85).

var. PUBESCENS (Tausch) Bosse *S. pubescens* Tausch; *S. korolkowii* Dieck ex Koehne; *S. japonica* var. *korolkowii* Zab. ex Henry—Once known in gardens as *S. Korolkowi*, this is no doubt a distinct variety of *S. japonica*. The most notable tree of the name grew in the famous Arboretum at Segrez, in France. I saw this tree in July 1904, and it was then covered with panicles of unexpanded flowers. When open they are described as dull white. The leaflets are longer than in *S. japonica* (fully 3 in.), but narrower in proportion; they are covered beneath with a very minute, close down. The young wood too is more downy, and of a lighter colour.

cv. 'VARIEGATA'.—Leaves margined with creamy white, but of little value.

cv. 'VIOLACEA'.—This was introduced from China to the Jardin des Plantes at Paris about 1858. Its flowers, which appear later than those of the normal form have the wing-petals and keel stained with rose-violet (*S. japonica violacea* Carr.; *S. violacea* Hort. ex Koehne). It has been cultivated at Kew for many years, but usually sets its flower-buds too late in the season to make a display. Except in colour it does not differ materially from the type. According to Henry the flowers of *S. japonica*, as he saw them in China, vary a good deal in colour, some forms being white, others yellow.

S. MACROCARPA Sm.
Edwardsia chilensis Miers ex Lindl.

An evergreen shrub or small tree; young wood covered with reddish brown down. Leaves pinnate, 3 to 5 in. long, composed of six and a half to twelve and a half pairs of leaflets, which are ¾ to 1½ in. long, ¼ to ⅜ in. wide, covered beneath and more or less above with reddish brown down. Flowers in short axillary racemes, yellow, 1 to 1¼ in. long. Calyx ¼ in. long, bell-shaped, shallow-toothed, covered with down like that of the leaves. Pods downy, 4 in. or more long, ½ in. thick where the seeds are enclosed, but much constricted between them; not winged, but thickened at the seams. *Bot. Mag.*, t. 8647.

Native of central Chile; introduced in 1822. This species is nearly akin to, and very much resembles the larger-leaved forms of *S. tetraptera*. It always differs from them, however, in having no wings to the pods, the flowers are not so large, and the leaves are pretty uniformly of the dimensions given above. It is not quite hardy at Kew, but thrives very well in more favoured localities.

S. SECUNDIFLORA (Ortega) Lag. ex DC.
Broussonetia secundiflora Ortega; *Virgilia secundiflora* (Ortega) Cav.

An evergreen tree 25 to 35 ft high with pinnate leaves 4 to 6 in. long, made up usually of seven or nine oblong or obovate leaflets which are 1 to 2 in. long, notched at the end. Flowers produced in racemes 2 or 3 in. long at the end of the leafy young shoots of the current year in spring. Each flower (of the normal pea-flower shape) is about 1 in. long, violet-blue and very fragrant (like violets).

Native of Texas, New Mexico, and North Mexico. It was figured by Ortega, the Spanish botanist in 1798 from a plant that flowered in Madrid. Decaisne, who figured it in the *Revue Horticole*, 1854, p. 201, observes that it withstands easily the climate of the *midi* of France. I do not know that it has been tried out-of-doors yet in this country but it is only likely to succeed in the sunniest, warmest localities and is better fitted no doubt for the south of Europe. It would be worthwhile trying it on a south wall, for Sargent describes it as one of the handsomest of small trees in the Texan forest, and hardy trees with fragrant violet-blue flowers are very rare.

S. TETRAPTERA J. S. Miller KOWHAI
Edwardsia grandiflora Salisb.; *S. tetraptera* var. *grandiflora* (Salisb.) Hook. f.

A shrub or small tree, varying from 15 to 40 ft high in the wild, the trunk 6 in. to 2 ft in diameter, usually shedding its leaves before the flowers expand in spring; young branchlets clad with a tawny down. Unlike *S. microphylla* it does not go through a juvenile phase. Leaves pinnate, $1\frac{1}{2}$ to 6 in. long, with ten to twenty pairs of leaflets, which are $\frac{1}{2}$ to $1\frac{3}{8}$ in. long, $\frac{3}{16}$ to $\frac{5}{16}$ in. wide, ovate or elliptic oblong, appressed silky-hairy on both sides. Flowers somewhat tubular, golden yellow, pendulous, clustered four to ten together in each raceme, opening in May. Calyx silky, obliquely bell-shaped, $\frac{1}{2}$ in. or more across, shallow-toothed. Standard forward pointing, about two-thirds as long as the keel; wings slightly longer than the standard. Pods 2 to 8 in. long, four-winged, with constrictions between the seeds. *Bot. Mag.*, t. 167.

Native of the North Island of New Zealand, in open places, often by stream-sides, or at the margins of forest; introduced in 1772 by Sir Joseph Banks, who accompanied Capt. Cook on his first exploratory voyage and brought back plants, two of which flowered and fruited in 1779. One of the most beautiful of New Zealand natives, the kowhai is fortunately hardy enough to thrive and flower against a sunny wall over much of the country and is unlikely to be killed outright by frost once it has built up a woody framework. But in cool and rainy localities it may not ripen its wood sufficiently to withstand severe winters. South of London it can be grown in the open in a sheltered spot but remains shrubby. In the milder parts it has attained tree dimensions, as at Mount Stewart in Northern Ireland, 23 × $2\frac{1}{4}$ ft (1976) and Mount Usher, Co. Wicklow, Eire, 26 × $2\frac{1}{4}$ ft. Even at Kew, in a south-facing bay on the Temperate House Terrace it attained a height of 18 ft.

S. tetraptera is well worth growing for its large, showy flowers and its foliage. The remarkable necklace-like pod, with four thin ridges traversing it lengthwise, is quite often seen in this country. *S. tetraptera* received an Award of Merit in 1943 when shown by the University Botanic Garden, Cambridge.

S. MICROPHYLLA Ait. *Edwardsia microphylla* (Ait.) Salisb.; *S. tetraptera* var. *microphylla* (Ait.) Hook. f.—A tree to about 30 ft high, usually going through a more or less prolonged juvenile state during which it forms a dense bush with slender intertwining branches. Leaflets smaller than in *S. tetraptera*, up to $\frac{3}{8}$ in. long at the most, and usually more numerous (up to forty pairs). Flowers as in that species but the standard petal relatively longer, about equalling the wings or a little longer. *Bot. Mag.*, t. 3735. This species is more widely spread in New Zealand than *S. tetraptera*, occurring on both the main islands and on Chatham Island. It was introduced at the same time. Both were planted in the Chelsea Physic Garden in 1774 and the plant of *S. microphylla* that still grows there on a wall of the main building is probably a descendant of the original Banks plant. It is hardy there and produces self-sown seedlings. This species received an Award of Merit in 1951.

var. LONGICARINATA (Simpson) Allan *S. longicarinata* Simpson; *S. tread-wellii* Hort. ex Allan—A small tree; leaflets only about $\frac{1}{6}$ in. long, in up to forty pairs. Flowers lemon-yellow, about 2 in. long. This was described in 1942 from plants found at Takaka in the northwestern part of South Island, growing on limestone rocks. It is perhaps not in cultivation in Britain but deserves to be introduced, being very ornamental (L. J. Metcalf, *Cult. N.Z. Tr. and Shr.* (1972), p. 256).

A Chilean sophora of the Edwardsia group is sometimes identified as *S. tetraptera*. It is not that species but near to, and perhaps not specifically distinct from, *S. microphylla*, from which it differs in its fewer leaflets (fifteen to seventeen pairs), in the slightly smaller flowers with the standard shorter than the wings, and in not going through a juvenile phase. It is a tree to about 25 ft high, occupying much the same habitats as its New Zealand relative, and is more widely distributed than the Chilean endemic *S. macrocarpa*, ranging from Maule province in central Chile through the forest region as far as Chiloe Island and the mainland opposite (Muñoz, *Flora Silvestre de Chile* (1966), pp. 191–2 and t. 38).

Early in the last century a sophora of unknown provenance grew in the Edinburgh Botanic Garden, where it was hardier than *S. tetraptera* or *S. microphylla*. It was named *Edwardsia macnabiana* by Graham in 1838 and portrayed in the *Botanical Magazine* two years later. Graham suspected that it might be no more than a seedling variation of *S. tetraptera*, but it seems to have agreed very well with the Chilean form or ally of *S. microphylla*. It is in cultivation from seeds collected by Comber in the Chilean Andes in 1927.

S. PROSTRATA Buchan. *S. tetraptera* var. *prostrata* (Buchan.) Kirk—A low intricately branched shrub, flat-topped and very dense in exposed places, sometimes prostrate. Leaves about 1 in. long, with up to eight pairs of leaflets, each leaflet about ⅙ in. long, oblong. Flowers to about three in a raceme, orange or brownish yellow. Pods less than 2 in. long, with up to five seeds. Native of South Island, New Zealand, in grassland and open rocky places, east of the Divide. *S. microphylla* sometimes bears flowers even in the juvenile state, and this species seems to be a modification that is fertile and permanently juvenile in habit.

SORBARIA   ROSACEAE

A genus of about seven species in the Himalaya and E. Asia, allied to *Spiraea*, but with pinnate leaves. Flowers small, white, in panicles. Carpels united at the base.

Bearing their panicles on the current season's growths, the sorbarias will flower more freely if the stems are shortened back in winter or early spring. On established plants old shoots should be cut clean out, to make room for replacement shoots from the base. The sorbarias thrive best in a good, moist soil. The commoner species may be increased by division or by taking rooted pieces off the plant; propagation by root-cuttings is also practised.

S. AITCHISONII (Hemsl.) Rehd.

Spiraea aitchisonii Hemsl.; *Sorbaria angustifolia* (Wenz.) Zab.;
Spiraea sorbifolia var. *angustifolia* Wenz.

A shrub of open, spreading habit, ultimately 10 ft high; branches red when young, perfectly glabrous, and very pithy. Leaves pinnate, 9 to 15 in. long, composed of eleven to twenty-three leaflets. Leaflets narrowly lance-shaped with long tapering points, 2 to 4 in. long, ¼ to ⅜ in. wide, evenly, sharply, and rather deeply toothed; green and quite glabrous on both surfaces, stalkless. Flowers white, ⅓ in. across, produced during July and August in pyramidal branching panicles from 1 to 1½ ft long, and 9 to 15 in. through; flower-stalks glabrous; seed-vessels red.

Native of Afghanistan, W. Pakistan and Kashmir; discovered by Dr Aitchison in the Kurram Valley, Afghanistan, in 1879; introduced to Kew from Kashmir by R. Ellis in 1880 (there was a second sending to Kew in 1895 by J. F. Duthie, also probably from Kashmir). It is closely allied to *S. tomentosa*, differing chiefly in the red young bark, the narrower leaves without down and with mostly simply (not doubly) toothed margins, and its larger flowers. On the whole it is superior to *S. tomentosa*, its foliage being more elegant and its flowers more effective. It is said to be hardy where *S. tomentosa* will not succeed.

S. ARBOREA Schneid. [PLATE 50

Spiraea arborea (Schneid.) Bean

A spreading deciduous shrub, usually 10 to 20 ft high, but said sometimes to be up to 30 ft; young shoots and leaf-stalks slightly downy. Leaves pinnate, with thirteen to seventeen leaflets which are oblong-ovate to lanceolate, long and slenderly pointed, 2 to 4 in. long, doubly toothed, stellately downy beneath but often only slightly so. Flowers white, ¼ in. wide, produced densely in fine, often pyramidal panicles up to and over 12 in. long, during July.

Native of Central and Western China, introduced by E. H. Wilson in 1908. It is closely akin to *S. tomentosa* but superior to that species which differs in the hairs beneath the leaves being simple (not clustered). *S. arborea* has also a shorter calyx-tube and longer stamens. It is the finest of the genus and may be pruned every winter. It likes a good loamy soil and gives finer panicles if top-dressed with manure occasionally.

var. GLABRATA Rehd. *Spiraea arborea* var. *glabrata* (Schneid.) Bean—This has glabrous shoots and leaves, often purplish, and narrower leaflets.

S. ASSURGENS Vilm. & Bois ex Rehd.

Spiraea assurgens (Vilm. & Bois) Bean

A deciduous shrub up to 8 or 10 ft high, with more or less erect stems, but making a shapely bush; young shoots round, not downy. Leaves pinnate, up to more than 12 in. long, consisting of eleven to seventeen leaflets which are 2 to 3½ in. long, ½ to 1 in. wide, lanceolate, with a long, slender, often curved point, doubly toothed, sometimes slightly downy on the veins beneath; veins in twenty-five or more pairs. Flower-panicles narrowly pyramidal, borne in July and August at the end of leafy shoots, 6 to 12 in. long, with their branches erect. Flowers white, ⅜ in. wide, with about twenty conspicuous stamens; calyx glabrous.

S. assurgens was raised by the firm of Vilmorin from seeds received from China in 1896, probably sent by one of the French missionaries. They first flowered it at Verrières, near Paris, in 1900 and sent it to Kew in 1903. It is allied to *S. sorbifolia*, which differs in having twice as many stamens in each flower but fewer (about twenty) veins to each leaflet. It is a worthy member of a hardy genus, but not the most striking.

S. SORBIFOLIA (L.) A. Braun
Spiraea sorbifolia L.

A shrub 3 to 6 ft high, which suckers freely; stems erect, very pithy, varying from nearly glabrous to downy. Leaves 8 to 12 in. long, composed of thirteen to twenty-five leaflets, which are lanceolate, 2 to 3½ in. long, ½ to 1 in. wide, sharply and conspicuously double-toothed, green on both sides; usually quite glabrous above and the same beneath. Flowers ⅓ in. across, white, produced during July and August in a stiff erect raceme 6 to 10 in. high; flower-stalks downy and glandular; ovaries glabrous or nearly so.

S. sorbifolia, besides being the type of the genus *Sorbaria*, is the most widely distributed of the species, extending from W. Siberia to the Pacific; how far the typical state extends outside Russia is uncertain, but probably to N. China and Korea. It is also the oldest representative of the genus in gardens, cultivated by Miller in the Chelsea Physic garden in 1759. It is distinguished from its near allies *S. aitchisonii* and *S. tomentosa* by its comparatively dwarf, stiff habit, and narrower, stiffer flower-panicles. Grown in rich soil it makes a handsome shrub.

var. STELLIPILA Maxim.   *S. stellipila* (Maxim.) Schneid.—This differs chiefly in the leaves being clothed with stellate hairs beneath, and in the downy ovaries. In habit and form of leaf it resembles the type. Native of Japan.

S. GRANDIFLORA (Sw.) Maxim.   *Spiraea grandiflora* Sw.; *Sp. sorbifolia* var. *alpina* Pall.; *Sp. pallasii* G. Don; *S. pallasii* Pojark.; *S. alpina* (Pall.) Dipp.—Previously treated as a variety of *S. sorbifolia*, this differs from it in its dwarf habit (less than 3 ft high in the wild), fewer and shorter leaflets with somewhat blunter teeth, and by the larger flowers, ½ to ⅝ in. across. Native of E. Siberia and the Russian Far East.

S. TOMENTOSA (Lindl.) Rehd.

Schizonotus tomentosus Lindl.; *Spiraea lindleyana* Wall. ex Loud.;
Sorbaria lindleyana (Wall.) Maxim.

A shrub of graceful spreading habit, up to 20 ft high; branches very pithy, green, glabrous. Leaves 10 to 18 in. long, pinnate, consisting of eleven to twenty-three leaflets, which are lanceolate or ovate-lanceolate, 2 to 4½ in. long, ½ to 1½ in. wide (the terminal one often larger and pinnately lobed); usually deeply and doubly toothed, glabrous above, furnished with loose, simple hairs beneath, especially about the midrib and veins. Flowers ivory white, scarcely ¼ in. wide, produced in terminal pyramidal, branching panicles 1 to 1½ ft long and 8 to 12 in. through; flower-stalks downy.

Native of the Himalaya from Nepal westward, and of Afghanistan and W. Pakistan; it was introduced to Britain by Dr Royle and was flowering in the Horticultural Society's garden at Chiswick by 1840. A very handsome, robust shrub, it is now less cultivated than its ally *S. aitchisonii*, from which it differs in its downy flower-stalks, and in the leaflets being broader, doubly toothed, and hairy beneath. From *S. sorbifolia* it differs in its strong spreading habit.

× SORBARONIA   ROSACEAE

A group of some half a dozen hybrids between *Aronia* and *Sorbus*—all shrubs or small trees. Where the whitebeam is the *Sorbus* parent the leaves are simple (× *S. alpina* and × *S. dippelii*). The other cultivated hybrids have *Sorbus aucuparia* or *S. americana* as the other parent, and the leaves are partly pinnate.

Although distinguished as early as 1789, *Aronia* was included in *Pyrus* or *Sorbus* until resurrected by Schneider in 1906; the hybrid genus × *Sorbaronia* was set up by him at the same time, but he never published a considered treatment of the group.

× S. ALPINA (Willd.) Schneid.

Pyrus alpina Willd.; *Aronia densiflora* Spach

A deciduous shrub whose young shoots are at first covered with loose white down, becoming dark with age. Leaves oval to obovate, tapering equally to both ends or more abruptly to the apex, finely toothed, 1½ to 3 in. long, about half as wide, at first soft and covered beneath with pale down; stalk ¼ to ½ in. long. Flowers in terminal corymbs 1 to 2 in. across, white, opening in May; styles three or four. Fruits ovoid to obovoid, dark brownish-red, ⅓ in. wide.

A hybrid between *Sorbus aria* and *Aronia arbutifolia*. It is very nearly allied to × *S. dippelii* but differs in having red fruits and in the leaves (upper surface especially) being less downy. Var. SUPERARIA Zab. is apparently a reversion towards *Sorbus aria* or perhaps a cross between that species and × *S. alpina*; the leaves are much larger and broader than in the latter, the flower clusters are also larger, the fruits deep red.

This shrub has been known in gardens since early in the 19th century. There is an example at Kew measuring 30 × 3½ ft (1971).

× S. DIPPELII (Zab.) Schneid.
Aronia dippelii Zab.

A bushy-headed shrub; young wood thickly covered with grey felt. Leaves 1½ to 3½ in. long, ⅝ to 1¼ in. wide, narrowly oval or oblanceolate, shallowly toothed, bright green and glabrous above, covered beneath with a close grey felt, tapering at the base to a stalk ¼ to ⅓ in. long. Flowers ⅓ in. across, white, with rose-coloured anthers, produced in small downy corymbs. Fruits top-shaped or roundish, ⅓ in. long, blue-black.

A hybrid between *S. aria* and *Aronia melanocarpa*, of unknown garden origin. Its affinity with *A. melanocarpa* is shown in the presence of glands along the upper surface of the midrib, and in the blackish fruits. It is an interesting and pretty round shrub, often made into a small tree by grafting on standards of mountain ash or hawthorn. It used to be called "*Pyrus alpina*" in gardens.

× S. HYBRIDA (Moench) Schneid.

? *Pyrus hybrida* Moench; ? *P. heterophylla* Pott; *Azarolus heterophylla* Boskh.; *Sorbus spuria* Pers.; *Pyrus spuria* (Pers.) Ser.; *P. spuria pendula* Loud.; *Sorbus hybrida pendula* Lodd. ex Loud.; *Sorbus heterophylla sens.* Dipp. (and Hedl.), ? not Reichenb.; × *Sorbaronia heterophylla* Boskh.) Schneid.; *Sorbus fallax* Schneid.; × *S. fallax* (Schneid.) Schneid.

A laxly branched shrub or small tree; young branchlets hairy. Leaves up to 3¼ in. long and 2 in. wide, variable in shape, some undivided, others pinnately cut or lobed at the base after the fashion of *Sorbus hybrida*, the terminal part of the leaf obtuse at the apex, sometimes acute or acuminate, especially on strong shoots, the lateral leaflets or lobes obtuse, glabrous above, hairy beneath at flowering-time. Flowers white, about ⅜ in. wide, borne in May or early June in small corymbs, the inflorescence-branches and receptacles densely white hairy. Fruits dark purple, globular or broad-ellipsoid, about ⅜ in. wide.

A hybrid of *Sorbus aucuparia*, of which the *Aronia* parent is uncertain, in commerce by the early 19th century and formerly known as '*Pyrus spuria pendula*'. It has never been common, however, and is now rare—deservedly so, for it has scant ornamental value. It gives some autumn colour but it is inferior to the aronias in that respect, and in its fruits, which are not freely produced. This is not the sorbaronia described in the 7th edition as × *S. hybrida* (and in earlier editions as *Pyrus spuria*); for this see under × *S. sorbifolia* below.

NOTE. The nomenclatural type of × *S. hybrida* is *Pyrus hybrida* Moench, described in 1785 from a plant that had been raised some six years earlier at Wilhelmshöhe (Lustschloss Weissenstein), near Cassel, Germany. The seed-parent was 'the arbutus-leaved pear' and the pollen-parent was considered by Moench to be *Sorbus aucuparia*. The seed-parent could have been either *Aronia arbutifolia* or *A. melanocarpa*, since the '*Pyrus arbutifolia*' of Moench's time covered both species. On the other hand he described the fruits as red when ripe, which is what one would expect from a cross between the red-fruited *A. arbutifolia* and the rowan. However, the cultivated plant as described above, which is × *S. hybrida sensu* Rehder and other authorities, had dark purple fruits—an unexpected though not impossible result from a cross between two red-fruited species. Dippel, who described the cultivated form under the name *S. heterophylla* Reichenb., suggested *A. melanocarpa* as the *Aronia* parent, but that is a very glabrous species and ought, on the face of it, to produce a sorbaronia a good deal less hairy than the cultivated × *S. hybrida*.

× S. SORBIFOLIA (Poir.) Schneid. *Mespilus sorbifolia* Poir.; *Pyrus sorbifolia* (Poir.) Wats.; *P. spuria sens.* Loud., not (Pers.) Ser.; *Sorbus* × *sargentii* Dipp.; *S.* × *sorbifolia* (Poir.) Hedl.; *Pyrus* × *mixta* Fern.—Similar to the preceding but almost glabrous in all its parts, the leaves and leaflets or lobes more sharply pointed, and the inflorescence-branches more slender. A probable hybrid between *Sorbus americana* and *Aronia melanocarpa*, described in 1816 from a plant cultivated in France but distributed earlier in Britain by the nurseryman Lee of Hammersmith. Towards the end of the last century the same hybrid was raised in Germany from seeds of *A. melanocarpa* received from Prof. Sargent of the Arnold Arboretum and was named *Sorbus* × *sargentii* by Dippel; this form was later distributed by Späth's nursery, Berlin. × *S. sorbifolia* gives quite good autumn colour but is very rare in gardens. Some of the leaves on strong shoots are very like those of *Sorbus* × *thuringiaca* in shape.

The sorbaronia described in the 7th edition of this work as × *S. hybrida* (and in earlier editions as *Pyrus spuria*) appears to be one received at Kew from Späth's nursery, Berlin, towards the end of the last century as *Pyrus hybrida*. Judging from herbarium specimens it was very near to × *S. sorbifolia* but with the young leaves cobwebbed beneath and the rachis of the inflorescence downy, especially near the base. A similar sorbaronia had, however, been in commerce in Britain and Germany by 1880, as *Pyrus spuria*, which seems to have been a catch-all name for the semi-pinnate members of this genus.

× SORBOCOTONEASTER   ROSACEAE

A genus uniting *Sorbus* and *Cotoneaster*, of which the hybrid described below is so far the only known representative.

× S. POZDNJAKOVII Pojark.

A shrub to about 10 ft high in the wild, sparsely branched; young growths woolly at first. Leaves pinnate, 1¼ to 2⅞ in. long, dull green above, hairy beneath, the terminal leaflet much longer and broader than the lateral ones, being 1 to 1⅞ in. long and up to 1 in. wide, elliptic or obovate, obtuse, entire or sometimes serrate in the upper part, lateral leaflets in one to three pairs, oblique at the base, the upper pair often partly connate with the terminal leaflet. Inflorescence few-flowered, terminal on a short leafy shoot; flowers white. Fruits globular, dark red or reddish black, ⅜ to   in. wide, with a juicy flesh, containing a few nutlets about ⅛ in. long.

A hybrid between *Sorbus sibirica* (a close ally of *S. aucuparia*) and *Cotoneaster melanocarpus*, collected shortly before 1951 in the upper valley of the Aldan, south of Yakutsk, in E. Siberia and described in 1953; named after the forester who discovered it. This interesting hybrid was introduced to Britain in 1958 by Messrs Hillier, who received scions from Siberia through the good offices of Dr D. K. Ogrin of the Faculty of Agriculture, Ljubljana, Yugoslavia.

× SORBOPYRUS   ROSACEAE

So far as is known this hybrid genus consists only of the Bollwyller pear and its variations, but the status of 'Malifolia' is somewhat controversial.

× S. AURICULARIS (Kroop) Schneid.
Pyrus auricularis Kroop; *P. pollveria* L.; *P. bollwylleriana* DC.

A deciduous tree, 20 to 40 ft high (sometimes 50 to 60 ft), forming a rounded bushy head; young branches more or less covered with loose down. Leaves ovate or oval, 3 to 4 in. long, 2 to 2½ in. wide, pointed, irregularly and coarsely, sometimes doubly toothed, rounded or rather heart-shaped at the base, upper surface covered at first with loose down which falls away as the season advances, lower surface permanently grey-felted; stalk 1 to 1½ in. long, woolly. Flowers white, ¾ to 1 in. across, produced about mid-May in many-flowered corymbs 2 to 3 in. across; anthers rosy red; calyx with its triangular lobes covered with a conspicuous pure white wool. Fruit pear-shaped, 1 to 1¼ in. long and wide, red, each on a stalk 1 to 1½ in. long, with sweet, yellowish flesh.

This interesting and remarkable tree is a hybrid between the common whitebeam (*Sorbus aria*) and the pear (*Pyrus communis*). It is said to have originated at Bollwyller, in Alsace, and is first mentioned by J. Bauhin in 1619 and figured by him in 1650. For three hundred years it has been propagated by grafts, for it produces very few fertile seeds, and these do not come true. The finest tree recorded in Britain grew at Bramford Hall, Ipswich, which, according to information received from Lady Loraine in 1904, was then over 60 ft high.

cv. 'BULBIFORMIS'.—A seedling of the Bollywyller pear, raised at Prague before 1878. Both in its foliage and in the larger fruits it was nearer to *Pyrus* than is the parent. (× *S. auricularis* var. *bulbiformis* (Tatar) Schneid.; *P. bollwylleriana* var. *bulbiformis* Tatar).

cv. 'MALIFOLIA'.—Leaves shorter and comparatively broader than in the type, often roundish oval, not so coarsely toothed, nearly always heart-shaped at the base, not so much felted beneath; flowers larger, 1 to 1½ in. across, fewer on the corymb and with stouter stalks, produced in late April and May. Fruits broadly top-shaped, about 2 in. long and wide, deep yellow when ripe. This interesting and handsome tree, although not so common as the Bollwyller pear, is on the whole more attractive. (*Pyrus malifolia* Spach; × *Sorbopyrus malifolia* (Spach) Schneid. *sec.* Bean; *Bollwilleria malifolia* (Spach) Zab.).

Spach, who named and described *Pyrus malifolia* in 1834, said that the original specimen at that time grew in the Ménagerie of the Jardin du Roi at Paris, and was 30 ft or more high. He suggested that it might be a hybrid between × *S. auricularis* and a garden pear (hence a back-cross) but it is more likely to have been a seedling variation of the Bollwyller pear, as 'Bulbiformis' is supposed to be. It should be added that the description given above is made from a tree distributed by the firm of Simon-Louis; their stock may have been of independent origin, but the description agrees in all essentials with that of Spach.

Zabel raised a seedling of the Bollwyller pear at Hannover-Münden which he was convinced was a hybrid between it and an apple. He suggested the same parentage for 'Malifolia', but the theory is rather far-fetched.

SORBUS   ROSACEAE

A genus of about 120 species (nearer 200 if European microspecies are included in the total). They are deciduous trees or shrubs, natives of the mountains and moister regions of the northern hemisphere, mostly in the Old World and with a centre in the Sino-Himalayan region.

The leaves are alternate, simple or pinnate(semi-pinnate in some species of hybrid origin); if simple, then toothed or sometimes lobed. Stipules present, usually deciduous, but persistent and forming a conspicuous feature of the flowering shoots in a few species of the section *Sorbus*. Inflorescence corymbose, more rarely paniculate, terminating a short leafy shoot, the (usually) two uppermost leaves subtending the outer branches of the inflorescence. Flowers up to 1 in. wide in section *Aria* but more commonly less than ½ in. wide. The body of the flower consists of a concave cone-shaped or urn-shaped receptacle (hypanthium, calyx-tube). Calyx-lobes (sepals) five, triangular or sometimes lanceolate. Petals five. Stamens most frequently twenty, in three whorls. Carpels two to five, partly or wholly enclosed within the tissues of the receptacle and more or less connate, each with two ovules. Styles as many as the carpels, free or partly connate. Fruit composed of the thin-walled carpels, each with usually only one seed, embedded in a dryish or juicy flesh developed from the wall of the receptacle. In the section *Sorbus (Aucuparia)* both the free upper part of the receptacle and the calyx-lobes persist in the fruiting stage and thicken, becoming in effect an integral part of the fruit; at the other extreme, in section *Micromeles*, the rim of the receptacle with its lobes is shed. In the other sections the calyx-lobes usually persist but do not become fleshy.

The genus is a polymorphic one, held together mainly by technical characters of the flower and fruit, but extraordinarily diverse in outward appearance. Quite an extensive area could be devoted exclusively to *Sorbus* without the planting becoming monotonous—except at flowering time, the genus having no great claim to floral beauty. But by summer its ally *Malus*, which vastly excels it in its flowers, has nothing to offer, while *Sorbus* comes into its own, with its endless variations of habit and foliage. Not many of the simple-leaved species are outstanding either in their fruits, which are no more ornamental than those of the hawthorns, or in autumn colour, and from August onwards the section *Sorbus* becomes the standard-bearer of the genus. Only a few in this section do not give autumn colour, and in some, such as *S. sargentiana, S. commixta* and 'Joseph Rock' it is outstanding; all have ornamental fruits, ranging in colour from scarlet as in the common rowan to yellow, orange, pink, crimson or pure white.

The cultivars and hybrids are propagated by grafting. Seeds are used by commercial growers only to raise the necessary stocks, but many species are apomictic and come true when raised in this way, while the others should do so unless the seeds are gathered in large and mixed collections. The seed is best sown in pans as soon as ripe, and overwintered in a frame or cold greenhouse.

Most members of the genus are undemanding, needing only a reasonably good soil and an open position, but the rowan group as a whole are sensitive to drought and excessive heat and may suffer in dry summers if planted in full sun and in a shallow, light soil. Only *S. insignis* is tender. The worst enemy of *Sorbus*, as of all members of the sub-family Pomoideae, is Fire Blight, for which see Vol. I, p. 730.

CLASSIFICATION

sect. S O R B U S *(Aucuparia)*.—This section, which comprises all the fully pinnate-leaved species except *S. domestica*, is the only one to occur in North America (see *S. americana, S. decora* and *S. sitchensis*) and in western Eurasia is represented only by the variable *S. aucuparia*. Although the type-species of the section, it is not really representative of the group as a whole. Its nearest allies in eastern Asia would seem to be *S. esserteauiana, S. pohuashanensis* and *S. amurensis*, which resemble *S. aucuparia* in their red fruits and in having white hairs on the bud-scales, on the undersurface of the leaflets and in the inflorescence. Despite the white hairs of its leaflets and its red fruits in broad trusses, the Chinese *S. scalaris* is not closely related to *S. aucuparia* and belongs to the same group as the Himalayan *S. foliolosa (wallichii)*. *S. americana* and *S. commixta* are both distinct from *S. aucuparia* in being almost glabrous, and the hairs on the latter, when present, are brown. In their very different ways *S. sargentiana* of Szechwan and *S. gracilis* of Japan are unique in the genus.

In the Sino-Himalayan region, where the section *Sorbus* is at its most varied, red-fruited species are uncommon, and very rare or even absent from Yunnan, their place being taken by what might be called the 'pernettya-fruited rowans', whose fruits ripen from green to porcelain white or from dark brownish red to crimson, light pink or rose-tinged white. The flowers of these are mostly borne in laxer and less branched inflorescences than in the red-fruited species, and are often pink or even crimson. Forrest made extensive collections of these interesting rowans and sent seeds on many occasions, but few of his introductions have survived. He himself did not care for them: 'In all my years in Yunnan, I haven't seen a member of the group to equal our own mountain ash, when well grown' (field-note under F.26547). Yet *S. hupehensis*, whose presence in gardens we largely owe to Forrest, is considered by some judges to be one of the most ornamental species. It is the tallest of the white- or pink-fruited rowans in cultivation. Mostly they are small trees or shrubs, or even dwarf shrublets (*S. reducta* and *S. poteriifolia*). Although introduced early this century, *S. vilmorinii* is rarely seen outside collections, though it is a delightful species, with its fern-like leaves and handsome fruits, and the same is true of *S. prattii*. A latecomer to gardens is *S. cashmiriana*, less ornamental in foliage than the others mentioned, but with large rosy flowers and white fruits the size of marbles.

sect. CORMUS.—The one species in this very distinct section, *S. domestica*, is confined to southwestern Eurasia. Its leading characters are the pinnate foliage, the five completely united carpels, and the large fruits with numerous stone-cells in the flesh.

sect. ARIA.—The members of this section range in size from large trees to shrubs, with simple, toothed or sometimes lobed leaves, which in most species have a thin to dense felt or a woolly tomentum beneath. The flowers range from $\frac{1}{2}$ to 1 in. in width, white. Styles two to five, free, more rarely partly connate. Fruits mostly $\frac{1}{2}$ to 1 in. long, sparsely to densely dotted with lenticels, red, brown or yellowish. Calyx-lobes usually persistent (if deciduous, then falling late), not becoming fleshy in fruit.

This section is mainly represented in cultivation by *S. aria*, the European whitebeam, and its cultivars. This has numerous relatives in Europe and southwest Asia, some of them microspecies of often very limited distribution, others, mainly in southwest Asia, of wider range. But these have scarcely been tested in gardens.

The Himalayan and Chinese members of the section *Aria* differ from most of the European species in having fruits that never become an honest red when ripe, but mostly they have fine foliage. See *S. cuspidata, S. mega-locarpa, S. pallescens* and *S. thibetica.*

sect. CHAMAEMESPILUS.—This section has one species, *S. chamae-mespilus,* endemic to the mountains of Europe. The flowers are pink, with erect sepals and petals.

sect. TORMINARIA.—Like the preceding, and the section *Cormus,* this is monotypic, with one species, *S. torminalis,* in Europe and south-west Asia. Leaves sharply lobed or lobulate. Styles, two, united for more than half their length. Fruits brownish, densely lenticellate, with two chambers, which are surrounded by a dense layer of stone-cells.

sect. SORBUS/sect. ARIA.—Intermediate between these two sections there is a group of species, which are believed to derive from hybridisation between *S. aucuparia* and tetraploid members of the section *Aria* such as *S. rupicola.* Some are known to be tetraploid or triploid, and to reproduce themselves apomictically. *S. hybrida,* the first of this group to be described, occupies a central position between *S. aucuparia* and *S. rupicola* and shows the influence of the former in the frequent occurrence of free leaflets at the base of the leaves, while in *S. meinichii,* as yet scarcely known in British gardens, the influence of *S. aucuparia* predominates. In the most ornamental of the group, the British endemics *S. anglica, S. arranensis* and *S. minima,* the influence of *S. aucuparia* is less obvious, and the leaves are merely deeply lobed. The status of *S. intermedia* is uncertain, some botanists holding that the deep lobing of the leaves derives from *S. torminalis* and not, as others believe, from *S. aucuparia.*

These species of hybrid origin should not be confused with the casual hybrids between *S. aucuparia* and *S. aria,* for which see *S.* × *thuringiaca.*

sect. ARIA/sect. TORMINARIA.—This group is a counterpart to the preceding, but the genes of the section Aria are here united with those of *S. torminalis.* The senior species of this group is *S. latifolia,* which makes a good specimen tree and has interesting though not showy brown fruits. But in other species, notably the British endemic *S. bristoliensis,* the influence of *S. torminalis* has brightened the colour of the fruits to orange-red.

Casual hybrids between *S. aria* and *S. torminalis* occur; see *S.* × *vagensis* under *S. latifolia.*

sect. MICROMELES.—In this polymorphic Asiatic section the leaves are simple, toothed, the lateral veins either straight and running out to double teeth or lobules, or curving. Flowers in most species very small, to about ⅜ in. wide, in mostly rather small often paniculate inflorescences. Carpels two to five, connate and enclosed in the receptacle. Styles two to five, free or connate in the lower part. Fruits rarely much more than ½ in. long, lenticellate or smooth. Calyx-lobes and top of receptacle usually deciduous, leaving a conspicuous scar on the top of the fruit.

This section has its headquarters in the rainier parts of the Sino-Hima-layan region, with eight species in the eastern Himalaya and northeast India, one of which extends as far southeast as Sumatra. The species best known in gardens are northern outliers of the group and are unusual in having red fruits. These are *S. alnifolia*, *S. folgneri* (the two most ornamental species) and *S. japonica*. Resembling these in the straight lateral leaf-veins, but with fruits of a nondescript colour, are *S. caloneura* and *S. meliosmifolia*. The species with curved lateral veins are represented in cultivation, though very rarely, by *S. keissleri* and *S. epidendron*.

SELECT BIBLIOGRAPHY

Fox, W.—Unpublished notes on *Sorbus*.
   At the time of his death in 1962 Dr Wilfrid Fox had brought near to completion a horticultural monograph on *Sorbus*, a genus of which he had planted a comprehensive collection in his arboretum at Winkworth near Godalming in Surrey, now the property of the National Trust. Dr Fox had been assisted in his work by the late Mrs Madeline Spitta, to whom these notes passed on his death. During the closing months of her life Mrs Spitta put the material in order, and entrusted it to the chief editor for use in preparing the present revised account of the genus, and it has been frequently consulted.

Gabrielian, E.—*The Genus Sorbus in Eastern Asia and the Himalayas.* Erevan, 1978 (in Russian).
—— 'The Genus *Sorbus* in Turkey', *Notes Roy. Bot. Gard. Edin.*, Vol. 23 (1961), pp. 483–96.
—— *Sorbus*, in Davis, P. H., ed., *Flora of Turkey*, Vol. 4 (1972), pp. 147–46.

Gilham, C. M., and McAllister, H. A.—'Tree Genera—6. *Sorbus* sect. *Aucuparia*', *Arboricultural Journal*, Vol. 3 (1977), pp. 85–95.

Handel–Mazzetti, H.—*Symbolae Sinicae*, Part VII 3 (1933), pp. 465–75.

Hitchcock, C. L., Cronquist, A., et al., *Vascular Plants of the Pacific Northwest*, Part 3 (1961), pp. 188–90.

Hedlund, T.—'Monographie der Gattung *Sorbus*', *Kongl. Svensk. Vet.-Akad. Handl.*, Vol. 35, no. 1 (1901), pp. 1–147.

Hensen, K. J. W.—'The Sorbus-collection in the Botanical Gardens and Belmonte Arboretum of the Agricultural University at Wageningen'. Parts I–III in *Jaarb. Nederl. Dendr. Ver.*, 20, pp. 121–39; 21, pp. 180–7; 22, pp. 48–56 (1956–62). Part IV in *Dendroflora* No. 3 (1966), pp. 60–72 (in Dutch).
—— —— 'Intermediate taxa between *Sorbus aria* (L.) Crantz and *Sorbus aucuparia* L. cultivated in the Netherlands', *Jaarb. Nederl. Dendr. Ver.* 21, pp. 189–204 (1958) (in Dutch).
—— —— 'Het Sorbus latifolia-Complex', *Belmontia*, fasc. 13 (1970), pp. 181–94.

Hutchinson, J.—Unpublished Notes on *Sorbus*.
   Early in 1943, Dr. John Hutchinson, of the Royal Botanic Gardens, Kew, undertook at the request of the Royal Horticultural Society to write a botanical account of the genus *Sorbus*. After drafting descriptions of most species of the section *Sorbus* (*Aucuparia*) he abandoned the task, which he is said to have found uncongenial, and handed over

the completed material to Dr Wilfrid Fox (q.v.), of whose papers on *Sorbus* it forms a part.

JONES, G. N.—'A Synopsis of the North American Species of Sorbus', *Journ. Arn. Arb.*, Vol. 20 (1939), pp. 1–43.

KARPATI, Z. E.—'Die Sorbus-Arten Ungarns und der angrenzenden Gebiete', *Fedd. Repert.*, Vol. 62 (1963), pp. 71–334.

KOEHNE, E.—*Sorbus*, in *Plantae Wilsonianae*, Vol. I (1913), pp. 457–83 (treatment of sect. *Sorbus* (*Aucuparia*)).

KOVANDA, M.—'Flower and Fruit Morphology of *Sorbus* . . .', *Preslia*, Vol. 33 (1961), pp. 1–16.

—— —— 'Taxonomic Studies in *Sorbus* subgen. *Aria*', *Act. Dendr. Čech.*, Vol. 3 (1961), pp. 41–83.

—— —— 'Spontaneous hybrids of Sorbus in Czechslovakia', *Act. Univ. Carol.*, 1961, pp. 41–83 (in English).

LILJEFORS, A.—'Cytological Studies in *Sorbus*', *Act. Hort. Berg.*, Vol. 17 (1955), pp. 47–113.

MCALLISTER, H. A., and WILLIAMS, C. M. '*Sorbus aucuparia* in Town and Country' (notes on the section *Sorbus*, privately circulated, 1975).

REHDER, A.—*Sorbus*, in *Plantae Wilsonianae*, Vol. II (1915), pp. 266–79 (sections *Aria* and *Micromeles*).

REGELINGSCOMMISSIE SIERBOMEN N.A.K.-B.—*Sorbus*, in *Dendroflora* No. 2 (1965), pp. 28–44.

RICHARDS, A. J., *Sorbus*, in Stace, C. A., ed. *Hybridisation and the Flora of the British Isles* (1975), pp. 233–8.

SCHNEIDER, C. K.—*Illustrierte Handbuch der Laubgehölze*, Vol. I (1906), pp. 667–98 and pp. 700–4.

YÜ, T. T.—*Sorbus*, in *Flora Reipublicae Popularis Sinicae*, Vol. 36 (1974), pp. 283–344.

—— and KUAN, K. C.—'Taxa Nova Rosacearum Sinicarum', with a classification of the Chinese species of *Sorbus* by Dr T. T. Yü, *Act. Phytotax. Sin.*, Vol. 8 (1963), pp. 207–210, 221–225.

WARBURG, E. F.—*Sorbus*, in Clapham, Tutin and Warburg, *Flora of the British Isles*, ed. 2 (1962), pp. 423–37.

—— and KARPATI, Z. E.—*Sorbus*, in *Flora Europaea*, Vol. 2 (1968), pp. 67–71.

S. ALNIFOLIA (Sieb. & Zucc.) K. Koch　　[PLATE 51

Crataegus alnifolia Sieb. & Zucc.; *Micromeles alnifolia* (Sieb. & Zucc.) Koehne; *Pyrus alnifolia* (Sieb. & Zucc.) Franch. & Sav., not Lindl.; *Pyrus miyabei* Sarg.

A tree growing to 70 ft high in cultivation, but the tallest in Britain are 40–50 ft high; bark smooth, grey and beech-like on mature trees; twigs slender grey-brown, with numerous lenticels; buds copper-brown, slender, glabrous, to $\frac{1}{4}$ in. or slightly more long. Leaves of thin texture, $1\frac{1}{2}$ to 4 in. long, $\frac{3}{4}$ to $1\frac{1}{2}$ in. wide, the apex pointed, the base rounded, the margins double-toothed and some times slightly lobulate in the upper part, lateral veins parallel, in six to ten, rarely to twelve pairs, silky hairy beneath when young, becoming glabrous later; petiole $\frac{1}{2}$ to $\frac{3}{4}$ in. long. Flowers pure white, $\frac{3}{8}$ to $\frac{1}{2}$ in. wide, produced in late spring in dense corymbs 2 to 3 in. across; receptacle and pedicels glabrous or slightly silky; petals rounded; anthers cream-coloured; styles two to four. Fruits bright red or sometimes deep pink, dotted with dark lenticels, roundish or obovoid, $\frac{3}{8}$ to $\frac{5}{8}$ in. long, ripening in late September or October; calyx deciduous. *Bot. Mag.*, t. 7773.

Native of Japan, Korea, and of northern and central China, as far south as Anhwei, Shantung and Kiangsi; put into commerce by the German nursery-man Späth in 1892, but perhaps cultivated earlier. It is a member of the section *Micromeles*, and perhaps its finest representative in gardens; indeed few other simple-leaved species surpass it as an ornamental. 'This species can fairly easily be identified by its slim, crowded buds and twigs, and many slender lateral branches which grow along the boughs often to within a foot of the main trunk. The foliage, which is very dense in consequence, is composed of thin leaves borne in crisp-looking, overlapping clusters reminiscent of those of the native Beech or Hornbeam. In its full summer foliage this is one of the most massive in appearance of all Sorbus but the tracery of its branches in winter is noticeably delicate . . . . This species does not bloom conspicuously every year, but at its best . . . the clusters are packed 2–2$\frac{1}{2}$ in. apart and the blossom lasts for a fortnight. With most Sorbus species the clusters are more widely spaced and much shorter lived.' (Fox MS).

In October the chief beauty of this species is its autumn colour of apricot-pink or orange-scarlet; the fruits are usually few in each cluster and in many gardens are soon taken by birds.

S. alnifolia is an uncommon tree in Britain, despite its merits. A tree planted by Dr Fox in the Winkworth Arboretum, Godalming, in 1938 measures 40 × 3$\frac{3}{4}$ ft at 3 ft (1975). A slightly older tree at Borde Hill in Sussex is 50 × 4 ft (1975). At the Arnold Arboretum, Boston, USA, Dr Fox saw a tree planted towards the end of the 19th century which was 70 ft high and 45 ft in spread (1955).

var. SUBMOLLIS Rehd.—This not very distinct variety was described as having the leaves densely downy beneath. Dr Fox records that even on a single tree there is variation from year to year in the density of hairs on the leaf-undersides. For *S. zahlbruckneri* Hort. (not Schneid.), sometimes placed under this variety, see Wilson 628 below.

S. alnifolia is also represented in cultivation by plants raised from seeds sent by Wilson from Hupeh, China, in 1901 (W. 628). This introduction was at first identified as *S. zahlbruckneri* and it was under this name that it received an Award of Merit when exhibited by Kew in 1924. Botanically, Wilson's form differs from the original introduction in the slightly larger leaves, and the larger, cherry-like fruits. But there is the further difference that all the recorded plants have a fastigiate habit. It is probable that they all belong to a single clone, named 'SKYLINE' by Messrs Hillier.

It is possible that *S. alnifolia* is also cultivated in this country from seeds collected by Wilson in Korea.

S. AMERICANA Marsh.

S. microcarpa Pursh; *Pyrus microcarpa* (Pursh) DC.;
Pyrus americana of some authors, not DC.

A tree 15 to 30 ft high in the wild, forming a rounded head, but often scarcely more than a shrub; branchlets stout, glabrous or at first downy, greyish brown by winter; winter-buds conical, dark purplish red, up to ¾ in. long and about ¼ in. wide, glabrous except for a tuft of brownish hairs at the apex, and some scattered white or brown hairs on the scales. Leaves pinnate, 6 to 12 in. long; lateral leaflets in five to eight pairs, 1½ to 4 in. long, ½ to ¾ in. wide, more or less gradually tapered to an acute apex, evenly serrate or doubly serrate, the teeth finely pointed, not spreading, light green and glabrous above, lower surface paler, slightly hairy at first or glabrous from the start. Flowers creamy white, about 3/16 in. across, produced about the beginning of June in flattish corymbs 3 to 5 in. wide; inflorescence-axes lenticellate, they and the receptacles glabrous. Stamens shorter than the petals. Fruits roundish, bright red, ¼ to ⅜ in. wide, ripening in September or October.

Native of eastern North America; date of introduction to Britain uncertain (the rowan known to Loudon as *Pyrus* (*Sorbus*) *americana* was *S. decora*). The nearest ally of *S. americana* is *S. commixta*, which has non-glutinous buds, redder branchlets, usually more acuminately tapered leaflets, and a taller less bushy habit. In their typical states *S. americana* and *S. decora* are distinct enough, but hybrids and intermediates occur; see further under the latter species.

S. americana fruits freely in this country, and its bunches of berries are as handsome as those of the rowan, but it does not make so large a tree. It is quite hardy and grows rapidly when young. It received an Award of Merit on 24 October 1950 when exhibited by the Crown Estate Commissioners from Windsor Great Park. The plants there were part of a gift from the Government of Canada. The leaves of these trees turn apricot or red in autumn, but on some soils *S. americana* does not colour so brightly.

cv. 'BELMONTE'.—Branches ascending, forming an ovoid crown. Selected at the Belmonte Arboretum, Holland, and named in 1964. It appears that when this cultivar was first released, some of the bud-wood distributed was taken from the wrong tree, which no doubt accounts for the fact that young trees in this country (1979) grown as *S. americana* 'Belmonte' are *S. commixta*.

S. ANGLICA Hedl.

S. mougeotii subsp. *anglica* Hedl., *nom. altern.*; *S. mougeotii* var. *anglica* (Hedl.) Butcher

A shrub to about 8 ft high in the wild, taller in cultivation (at least when grafted); branchlets grey in their second year, with pale lenticels; winter-buds small, grey-brown or greenish brown, white-hairy at the tip and at the edge of the scales. Leaves simple, broadly elliptic to slightly ovate, 3 to 4 in. long, 2 to 3 in. wide, obtuse at the apex, broad-cuneate at the base, with ten to twelve pairs of lateral veins, clad beneath with a close greyish white tomentum, sharply and irregularly serrate, lobed around the middle, the lobes not extending more than a quarter of the way towards the midrib; petiole ½ to ⅞ in. long. Inflorescence thinly woolly, variable in size. Flowers white, about ⅜ in. wide, with creamy filaments and anthers. Styles two or three. Fruits more or less globose, about ⅜ in. wide, very sparsely to fairly densely dotted with lenticels; calyx-lobes forming a cone, slightly fleshy at the base.

S. anglica is an interesting minor species, endemic to the British Isles, where it is found on carboniferous limestone in parts of southwest England, Wales and Ireland (Co. Kerry, Eire). It appears to be a complex species, with a number of slightly differing local races, all tetraploid and reproducing themselves apomictically. It is of hybrid origin, probably deriving three sets of chromosomes from the Aria group and one from *S. aucuparia*. When first noticed, it was considered to be the British form of *S. intermedia* and was first distinguished by Hedlund in 1914. It does not make a substantial tree as *S. intermedia* does; the leaves are not so deeply lobed; the lateral veins are in more numerous pairs and the lower ones do not arch so much; and the lobes are usually entire on the inner margin or have only one tooth there, while in *S. intermedia* the inner side has usually more than one tooth; and the fruits are more or less globular, longer than wide in *S. intermedia*. The difference in the colour of the leaf-indumentum, greyish white in *S. anglica* and yellowish in *S. intermedia*, applies only to herbarium specimens of some age.

S. anglica has the merit of being of smaller stature than *S. aria* and with more interesting foliage, though the leaves are not so white beneath. There are some five examples in the Hillier Arboretum, raised from seeds collected by Roy Lancaster above the Cheddar Gorge. These have bright scarlet fruits. On the plant in the Winkworth Arboretum they are darker, more crimson scarlet.

S. MOUGEOTII Soy.-Willem. & Godr.—Similar to *S. anglica*, but a tree up to 60 ft high, with relatively narrower leaves and, in the cultivated form, with more numerous lobes. Fruits usually without lenticels, about ⅜ in. wide, globular. *S. mougeotii* was discovered in the Vosges about 1850 by Dr Mougeot the French bryologist and brought into cultivation in the Botanic Garden at Nancy, where it was described in 1858. It is found as far west as the Pyrenees and in the east ranges as far as Austria. In the subsp. AUSTRIACA (G. Beck) Hayek, which appears to overlap with the typical state, the leaves are relatively broader than in *S. mougeotii* and more like those of *S. anglica*.

Some plants distributed as *S. mougeotii* are 'Theophrasta', q.v. under *S. latifolia*.

S. ARIA (L.) Crantz   WHITEBEAM
Crataegus aria L.; *Pyrus aria* (L.) Ehrh.; *Aria nivea* Host

A tree usually 30 to 50 ft high in gardens, but occasionally met with 60 to 80 ft high; main branches more or less erect; young branchlets clothed with loose white hairs, becoming nearly glabrous and lustrous dark brown by winter and furnished with pale lenticels; winter-buds ovoid, the scales greenish, brown and hairy at the edge, glabrous or slightly woolly on the back. Leaves simple, with eight to thirteen pairs of parallel ribs, elliptic to broadly so or ovate, rarely obovate, 2 to 4 in. long, half to two-thirds as wide, obtuse or acute at the apex, narrowed or rounded at the base, margins double-toothed except in the basal part which is simple-toothed or entire, upper surface lustrous, glabrous except when quite young, coated beneath with a close white felt; petiole $\frac{1}{2}$ to 1 in. long. Flowers dull white, about $\frac{1}{2}$ in. across, produced towards the end of May in corymbs 2 to 3 in. across; stalks and receptacles covered with white, matted hairs. Fruits ovoid or roundish, $\frac{3}{8}$ to $\frac{1}{2}$ in. long, scarlet-red, specked with brownish lenticels.

SORBUS ARIA

Native of southern England as far west as Dorset, but naturalised outside this area; in Ireland only in Galway; on the continent it is fairly widely distributed but is absent from Scandinavia and overlaps in southeast and parts of central Europe with minor related species; it does not occur in Asia.

There is no tree more characteristic of the chalk hills of England, or more beautiful in fruit. It is very effective in the breeze when the wind, by lifting the leaves, reveals the pure white undersurface in kaleidoscopic glimpses. A tree well laden with the bright red fruits is also one of the most beautiful of autumn pictures, but one that owing to the depredations of birds is often of short duration. It is best propagated by seeds, but the young plants grow very slowly at first. The timber is hard and heavy but too scarce to count for much in the timber trade. The largest tree recorded by Elwes and Henry grew at Camp Wood, near Henley-on-Thames, and was 75 × 4¾ ft in 1905. One of the largest recorded recently grows, appropriately, at Colesbourne in Gloucestershire, where it was planted by Elwes in 1904. It measures 50 × 6 ft (1975). Others are: Trawscoed, Aberystwyth, 58 × 6 ft (1969); Marble Hill, Twickenham, 50 × 4½ ft (1968).

The whitebeam has many varieties, some wild, others of garden origin. The most distinct are:

cv 'AUREA'.—See under 'Chrysophylla'.

cv. 'CHRYSOPHYLLA'.—Leaves yellow throughout the season, rather narrower than normal, elliptic tending to obovate, to about 4 in. long and 2 in. wide. Put into commerce by Messrs Hesse of Weener, Hanover, 'AUREA', from the same firm, does not retain its colour so well, and has still narrower leaves.

f. CYCLOPHYLLA (Beck) Javorka *Aria nivea* f. *cyclophylla* Beck—Laeves very broadly ovate or almost orbicular, up to 6 in. or slightly more long. Occasional in the wild; originally described from a tree noticed in Bosnia.

cv. 'EDULIS'.—Leaves relatively narrower than in the type. Fruits larger and juicier (*Pyrus edulis* Willd.; *S. aria* var. *edulis* (Willd.) Wenzig). Known since the early 19th century but perhaps no longer in cultivation. By some authorities it is included in f. *longifolia* because of the similarity in leaf-shape, but there are no grounds for supposing that narrowness of leaf is correlated with large fruits. A tree received by Dr Fox at the Winkworth Arboretum had narrowish leaves but normal fruits; it was of compact habit.

var. FLABELLIFOLIA.—See under *S. umbellata*.

cv. 'GIGANTEA'.—Possibly deriving from 'Majestica', differing in its more lobulate, larger leaves and longer fruits (about ¾ in. long). Raised by Messrs Lombarts of Holland and introduced in 1953. There is a similar tree at Kew, received as 'Majestica' early this century.

var. GRAECA.—See under *S. umbellata*.

f. INCISA (Reichenb.) Javorka *S. aria* var. *incisa* Reichenb.; *S. incisa* (Reichenb.) Hedl.—Leaves more definitely lobed than normal, but otherwise not differing from the species. Such plants occur in Britain, e.g., on the North Downs, and some were at one time confused with *S. intermedia*.

f. LONGIFOLIA (Pers.) Rehd. *S. aria* var. *longifolia* Pers.; *S. longifolia* (Pers.) Hedl., in part—Leaves relatively narrower than in the type, and often longer, obtuse at the apex. Hedlund's *S. longifolia* is based on specimens from Carinthia and neighbouring parts, in which the larger leaves have twelve pairs of lateral veins, and the fruits are unusually large, besides being somewhat longer than wide.

cv. 'Lutescens'.—A tree of conical habit. Leaves elliptic to obovate, cuneate to rounded at the base, to 4 in. or slightly more long, densely coated with silvery hairs on both sides when young and then very striking. Fruits in rather small trusses. It was put into commerce by Messrs Simon-Louis with its present cultivar-name before 1885, and is doubtfully the same as the *S. aria* var. *lutescens* of Hartwig (1892). The inappropriateness of the epithet *lutescens* for the present tree was remarked on soon after it came into commerce, and it has been suggested that it refers to the sulphur tinge of the leaves on the late-summer shoots of vigorously growing trees. 'Lutescens' has been widely used as a street-tree.

cv. 'Magnifica'.—A vigorous tree, of narrow habit when young, more spreading with age. Leaves up to 5 in. long and 3 in. wide, rounded at the apex, darkish green and glossy above, white-felted beneath, turning yellow in autumn and persisting on the tree until well after those of *S. aria* and other cultivars have fallen. Flowers about ⅝ in. wide in dense trusses about 4 in. wide. Fruits ovoid, about ½ in. long, bright red, lenticellate, not freely borne.

This whitebeam was raised from seed by Messrs Hesse of Weener, Hanover, and described by Herr Hesse in 1916. Beyond that, nothing is known of its history and it may be that it is not pure *S. aria*. However, it is not so distinct from *S. aria* as might be supposed from some catalogues, where the fruits are described as brown. This error is the result of confusion between *S. aria* 'Magnifica' and the form of *S. aria* × *S. torminalis* distributed by Messrs Hillier as *S.* × *magnifica*, for which see p. 443.

Herr Hesse's choice of name was unfortunate, for the epithet *magnifica* had been used earlier for a whitebeam similar to 'Majestica'.

cv. 'Majestica' ('Decaisneana').—This is the finest of all the varieties of whitebeam, having larger leaves and fruits than any. It is of unknown origin, but is recorded as having existed in the Segrez Arboretum in 1858. It was at one time known in the nurseries near Paris as the 'Sorbier du Nepaul,' but there is no evidence to show that it ever came from Nepal or any part of N. India. We are thus led to believe that, like many other fine varieties of trees, it originated as a chance and unrecorded seedling under cultivation. Its leaves are sometimes 7 in. long and 3 to 4 in. wide, the fruits ⅝ in. long. In other respects it does not differ from *S. aria*. *Bot. Mag.*, t. 8184 (*Aria majestica* Lav.; *Aria decaisneana* Lav.; *Hahnia aria* var. *majestica* Dipp.).   [Plate 52

'Majestica' was put into commerce in 1879 by Messrs Simon-Louis, who received their original stock from the Segrez Arboretum. There can be no doubt that the two names used by Lavallée—*Aria majestica* and *A. decaisneana*—refer to the same tree. In publishing the first name, without description, he remarked that a plate portraying it had been prepared; but when the plate appeared four years later in *Arboretum Segrezianum* it bore the name *Aria Decaisneana*. That only one tree was involved was confirmed by M. Jouin, the Simon-Louis nursery manager, in a letter preserved at Kew.

The name 'Majestica' is applicable only to the original clone, but it may be that seedlings have also been distributed under the name. For the false 'Majestica' at Kew, see under 'Gigantea'.

'Majestica' has attained 68 × 7 ft at 3 ft at Westonbirt, in Willesley Drive;

another example, *pl.* 1935, is 56 × 4 ft (1975–6).

cv. 'PENDULA'.—Habit pendulous. Leaves smaller than normal.

cv. 'QUERCOIDES'.—Of dwarf habit with leaves evenly lobed like those of the common oak, and upturned at the edges.

var. RUPICOLA.—See *S. rupicola*, under *S. umbellata*.

S. AUCUPARIA L. COMMON ROWAN, MOUNTAIN ASH
Pyrus aucuparia (L.) Gaertn.

A deciduous tree, 30 to 60 ft high, of erect growth when young, becoming more spreading and graceful with age; trunk smooth and grey; branchlets downy when young, becoming glabrous later; terminal bud very downy throughout the winter, not gummy. Leaves pinnate, 5 to 9 in. long, with mostly six or seven pairs of leaflets, which are narrowly ovate-oblong, 1 to 2½ in. long, smallest towards the apex, pointed, sharply toothed, downy beneath when young, becoming almost or quite glabrous by the autumn. Flowers white, unpleasantly scented, ⅓ in. across, produced very numerously in terminal, flattish corymbs 3 to 5 in. across; calyx and flower-stalks clothed with grey wool. Fruits in large showy clusters, ¼ in. to ⅜ in. across, round or slightly oval, bright red.

The mountain ash is widely spread over the cool, temperate parts of Europe and Asia, and is abundant in most parts of the British Isles. It is one of the most beautiful of our native trees alike in leaf, flower, and fruit. Its beauty no doubt is greatest when the branches are laden with the large nodding clusters of ripe fruits in September, but where bird life is abundant that beauty soon passes. Of neat habit and never of large size, it is a useful tree in small gardens, for which, however, some of the varieties mentioned below might be selected, leaving the typical mountain ash for the larger spaces and woodland. It is easily raised from seed, and grows quickly when young. On this account young trees are much used as stocks for grafting varieties of this and allied species on. It likes a cool, moist situation and is by nature adapted to a growing season shorter than that of southern England, ripening its fruits well before autumn has set in and dropping its blackened leaves before the foliage of most other trees has started to turn colour. Where the winters are longer the leaves persist into autumn and colour yellow or red. It is said that in Scotland the colour depends on the soil (Dr H. Tod, *Journ. R.H.S.*, Vol. 78 (1953), p. 105).

cv. 'ASPLENIFOLIA'.—Leaflets more than usually downy, the marginal teeth twice as deep as in the normal form and mostly themselves toothed. It is perhaps a fixed juvenile form, since such leaflets are usual on seedlings.

cv. 'BEISSNERI' ('Dulcis Laciniata').—Bark of branches and trunk light coppery brown. Leaves with a reddish rachis, the leaflets yellowish green, deeply and coarsely incised. Autumn colour yellow. Fruits as in 'Edulis' (*S. aucuparia* f. *beissneri* Rehd.; *S. a.* var. *dulcis laciniata* Beissn.). Discovered in the Erzgebirge (Krušne Hory) on the borders between Germany and what is now Czechoslovakia; described in 1899. A very distinct and ornamental rowan.

cv. 'DIRKENII'.—Leaves clear yellow when young. Raised by the Dutch nurseryman Dirken about 1880.

cv. 'EDULIS' ('Dulcis', 'Moravica').—A strong-growing tree with ascending branches. Leaflets widely spaced, rather narrow, tapered to an acute point, sparsely hairy beneath, margins toothed only in the upper half or even less. Fruits somewhat larger than normal, pleasant-tasting, sour but not bitter. Put into commerce by Dieck in 1887. It derives from the mountains of northwest Czechoslovakia, where similar trees were first recorded early in the last century. Botanically it is near to the subspecies *glabrata* (*S. aucuparia* var. *edulis* Dieck; *S. a.* var. *dulcis* Kraetzl; *S. a.* var. *moravica* Dipp.).

Like those of the Rossica group (see 'Rossica Major') the fruits of 'Edulis' are more suitable for making jam and jelly than those of the common rowan. It is recorded that in Germany 450 lb of fruit were harvested from fourteen thirty-year-old trees and turned into 230 lb of preserve.

cv. 'FASTIGIATA'.—More than one fastigiately branched form of the common rowan may have been distributed. The tree originally described by Loudon in 1838 had rather rigid branches and was raised by the Irish nurseryman Hodgins of Dunganstown, Co. Wicklow. For the fastigiate clone grown under this and many other names see the note below, after subsp. *sibirica*.

cv. 'FRUCTU LUTEO' ('Xanthocarpa').—Fruits orange-yellow. This makes a small tree with a rather open crown. Origin unknown (*Pyrus aucuparia* var. *fructu luteo* Loud.; *S. aucuparia* var. *xanthocarpa* Hartwig & Rumpler; *S. a.* var. *Fifeana* Hort. ex Hartwig). Award of Merit 1895.

subsp. GLABRATA (Wimm. & Grab.) Cajander	*P. aucuparia* var. *glabrata* Wimm. & Grab.; *S. glabrata* (Wimm. & Grab.) Hedl., *nom. illegit.*—A shrub to about 8 ft high. Leaflets less hairy beneath than in the typical subspecies, tapered to an acute point; inflorescence-axes glabrous or almost so; sepals rounded at the apex, hairy; fruits longer than wide. Scandinavia, the Baltic region and northwest Russia; also said to occur in parts of Central Europe.

f. INTEGERRIMA (Hartm.)	*S. aucuparia* var. *integerrima* Hartm.—Leaflets entire or almost so. Originally described in 1832 from Scandinavian plants. The name *S. aucuparia* var. *integerrima* was applied by Koehne in 1901 to a tree seen in the Jena Botanic Garden, with almost entire leaflets, the uppermost pair often connate with the terminal leaflet, and with decurrent bases. Although Koehne insisted that this was a mutant of *S. aucuparia*, it was thought by Schneider to be a form of *S. aucuparia* × *S. aria* (see *S.* × *thuringiaca*).

var. LANUGINOSA Beck	*S. lanuginosa* Kit. ex DC. *sec.* Hedl.—Undersides of leaflets, and inflorescence, more persistently hairy than in the typical state, leaflets often more coarsely toothed. Said to be the common state of the species in southern central and southeast Europe, on limestone.

For *Sorbus* (*Pyrus*) *lanuginosa* Hort. see *S.* × *thuringiaca* 'Decurrens'.

subsp. PRAEMORSA (Guss.) Nym.	*Pyrus praemorsa* Guss.; ?*S. maderensis* (Lowe) Decne; ?*Pyrus aucuparia* var. *maderensis* Lowe—Leaves with a short petiole, about ¾ in. long; leaflets obtuse at the apex, parallel-sided. Native of Calabria, Sicily and Corsica. Hedlund identified with this the interesting Madeira rowan (see synonyms), discovered in 1826 and known only from a few localities in the mountains, where it grows with *Vaccinium maderense*. According to Lowe it makes a stiffly branched shrub, bearing pleasantly scented flowers in short, dense corymbs. On herbarium specimens the hairs on the buds are rusty and the fruits covered with a bloom—characters also attributed to subsp. *praemorsa*. Young plants of the Madeira rowan are in cultivation (1979) raised from seeds collected by Brian Halliwell.

cv. 'ROSSICA MAJOR'.—Leaflets longer and relatively broader than normal, to 3⅜ in. long and 1 in. or slightly more wide, obtuse, remaining woolly beneath until late summer; rachis and petiole reddish. Fruits somewhat larger than normal, dark red; like those of 'Edulis' they lack the usual bitterness. This is a tree of excellent habit with ascending branches, forming an ovoid crown. It was introduced by Späth in 1903 from the Kiev region of Russia, where the fruits were used for making jam.

In 1898, Späth sent out 'ROSSICA', imported from the same region of Russia. It was less distinct from typical *S. aucuparia* in foliage and in the size of its fruits, and therefore of less value as an ornamental, though the fruits were used in Russia for the same purpose. Rehder included both clones in *S. aucuparia* var. *rossica* Koehne, with the result that 'Rossica Major' has come to be known as 'Rossica'.

cv. 'SHEERWATER SEEDLING'.—Branches ascending, forming a narrow crown, which opens in time and becomes about 15 ft wide. Fruits freely borne, in large trusses. The original plant grew by the Sheerwater stream near Woking, and was moved to a private garden, where Rowland Jackman noticed it and obtained graft-wood. It was put into commerce by his firm and is now used as a street tree.

subsp. SIBIRICA (Hedl.) Krylov *S. sibirica* Hedl.—Leaflets oblong-lanceolate, acute, 1½ to 2 in. long, up to ⅝ in. wide, glabrous on both sides. Inflorescence up to 5 in. wide, almost glabrous. Native mainly of Siberia.

The tree referred to above under *S. aucuparia* 'Fastigiata' is of fastigiate habit, closely so at first, but widening with age. It agrees with *S. aucuparia* in most essentials, but has larger, more conical winter-buds, with some brown hairs among the white; very stout annual growths; leaflets sometimes in nine pairs, though mostly six to eight; oblate, deep crimson scarlet fruits almost ½ in. wide, the calyx-lobes in the fruiting stage separated by sinuses and not meeting in the centre. The trusses are remarkably heavy, the branches often sagging under their weight.

This puzzling rowan is of unknown origin, but is possibly the plant mentioned in the present work for the first time in 1921 under its horticultural name "*S. americana nana*", with the suggestion that it might be *S. scopulina*. It is not that species, though it has been sold as such, and also as "*S. decora nana*", but is not *S. decora*; neither, for that matter, is it dwarf. What appears to be the same clone is in commerce in Holland, where it is accepted as the true *S. aucuparia* 'Fastigiata' and appears under this name in the Hillier Manual. In some respects it resembles the subsp. *praemorsa* and even the Madeira rowan, but disagrees in others, especially the more numerous leaflets. The large open-topped fruits shown by this tree are also a feature of *S. splendida*, described by

Hedlund in 1901 from a tree growing in the Uppsala Botanic Garden, which had been received from Hamburg in 1850. He judged this tree to be a hybrid between *S. aucuparia* and *S. americana,* and such a parentage (or *S. aucuparia* × *S. decora*) is possible for the tree discussed here. But the possibility remains that it is nothing but an aberrant form of *S. aucuparia.*

This many-named rowan is of some value for planting in confined places in towns, but is not a good garden tree, being ornamental only when in fruit. The habit is graceless and the foliage dull and coarse.

S. CALONEURA (Stapf) Rehd.
Micromeles caloneura Stapf; *Pyrus caloneura* (Stapf) Bean

A slender tree to about 35 ft in the wild, with glabrous young shoots and large, ovoid, glabrous winter-buds ¼ in. long and slightly more wide. Leaves elliptic to oblong, narrowed at both ends, 2 to 3½ in. long, half as wide, irregularly double-toothed, clothed when very young with a loose floss which soon falls away, leaving them glabrous above but with a few hairs beneath on the veins and in the vein-axils; lateral veins in nine to sixteen pairs, impressed above and prominent beneath; petiole ⅜ to ½ in. long, at first hairy. Flowers white, about ½ in. wide, produced in rounded, dense corymbs 2 to 3 in. across; flower-stalks and receptacle downy. Anthers pink. Styles five, sometimes three. Fruits somewhat pear-shaped, ⅜ in. long, brown, lenticellate; the calyx falls away from the apex completely, leaving a small pit there. *Bot. Mag.,* t. 8335.

A native of western, central and southern China, extending into the former Indochina and Malaysia; discovered by Augustine Henry; introduced by Wilson in 1904 for Messrs Veitch, in whose Coombe Wood nursery it flowered and fruited only five years later. It is a member of the section *Micromeles,* less common in cultivation than the closely allied *S. meliosmifolia.*

S. CASHMIRIANA Hedl.   [PLATE 53

A small, rather laxly branched tree; young stems soon glabrous and reddish brown, about ⅛ in. thick, sparsely lenticellate; winter buds dark purplish red, conoid, about ½ in. long, outer scales glabrous, the inner densely brown-hairy. Leaves with mostly six to nine pairs of leaflets, 3½ to 6½ in. long, including petiole ½ to 1 in. long. Leaflets 1 to almost 2 in. long, ⅜ to ⅝ in. wide, lanceolate to oblong, acute or subacute at the apex, sharply toothed almost to the base, upper surface rich green, glabrous, with the lateral veins slightly impressed, underside grey green, glabrous on the blade, the midrib at first coated with straight brown hairs but almost glabrous by flowering-time; rachis grooved, glabrous above, with scattered whitish hairs beneath. Flowers opening in May. Inflorescence almost glabrous, much branched, up to 7 in. wide, with forty to sixty flowers, which are ½ in. or slightly more wide. Receptacle glabrous; sepals sparsely ciliate. Petals pink in the bud, in the expanded flower white flushed with pink and with a rosy purple stain at the base. Stamens shorter than the petals, with rigid, flattened filaments and rosy purple anthers. Styles four or five. Top of ovary flattish, slightly hairy. Fruits at first green, becoming pure white except for the pink, fleshy sepals, globular-ovoid, $\frac{1}{2}$ to $\frac{5}{8}$ in. wide, filled with watery, bitter pulp.

Native of the western Himalaya, described in 1901 from a flowering specimen in the Falconer herbarium, collected in Kashmir some fifty or sixty years earlier. In Kashmir it appears to be confined to the inner ranges, e.g., in Baltistan and the area southeast of Nanga Parbat. Most herbarium specimens are from the Himalaya southeast of Kashmir, in the subalpine zone reaching from the River Chenab to the River Sutlej. Specimens collected on or near the borders between Pakistan and Afghanistan have been identified as *S. cashmiriana*, but their identity is uncertain.

S. cashmiriana was introduced to Britain early in the 1930s, though whence and by whom is not recorded. In 1934 Mr Ben Wells had some six or eight young plants in his nursery at Merstham, one of which was bought by Mr Butt of Chalford in the Cotswolds, with whom it fruited in 1940; a specimen was sent to Kew, where it was identified as *S. cashmiriana*. Mr Butt gave his plant to Wisley on giving up his garden. It died there, probably from drought, but it had been propagated by Messrs Marchant (Brian Mulligan, *Gardening Illustrated*, Vol. 70 (1953), p. 44). *S. cashmiriana* was exhibited by Messrs Marchant in 1952, and received an Award of Merit, since when it has become widely distributed in gardens.

S. cashmiriana has two seasons of beauty—in May when bearing its unusually large, pink-tinted flowers, and in October when the marble-sized white fruits hang from the branches, which by then are bare, for the leaves drop early in most seasons without colouring. The fruits are not liked by the birds and usually remain on the tree until they decay. The propagation of this sorbus presents no difficulty, since it comes true from seed.

It has been suggested that *S. cashmiriana* is no more than a white-fruited race of *S. tianschanica*. But the flowers of that species are even larger than in *S. cashmiriana* and an important difference given by Eleanora Gabrielian is that in *S. tianschanica* the carpels are free from each other to the base (an unusual feature in *Sorbus*), whereas in the present species they are partly united. A further difference is that in *S. cashmiriana* the leaflets are more numerous and that the buds are glabrous except for brown hairs at the tip.

The above description of *S. cashmiriana* is made from the cultivated form, which matches many wild-collected specimens, including the type-specimen. But on some wild plants the midribs of the leaflets and the underside of the rachis are quite densely white-tomentose and may represent the form named f. *jaeschkei* by Koehne. Their taxonomic position is uncertain.

S. CHAMAEMESPILUS (L.) Crantz

Mespilus chamaemespilus L.; *Crataegus alpina* Mill.; *Pyrus chamaemespilus* (L.) Ehrh.

A shrub of dwarf, compact habit, becoming eventually 5 or 6 ft high; branches short, stiff; young twigs covered at first with a whitish, cobweb-like substance. Leaves $1\frac{1}{4}$ to 3 in. long, $\frac{5}{8}$ to $1\frac{1}{2}$ in. wide, ovate, oval or slightly obovate, green and glabrous on both surfaces, apex rounded or pointed, margins finely toothed; stalk ⅛ to ⅓ in. long. Flowers rosy, crowded in umbels which together form a small terminal corymb. Receptacle woolly at the base like the flower-stalk, the teeth pointed, erect, almost glabrous outside, but covered with a thick white wool inside; petals erect, never spreading. Fruits ⅓ to ½ in. long, scarcely so wide, scarlet-red.

Native of the mountains of central and southern Europe, to which it is endemic; in cultivation 1683. It is not frequent in gardens, but is very ornamental when in fruit and well worth growing as one of the few truly shrubby species, especially for planting in some sunny spot where a slow-growing shrub is desirable. It is distinct from the Aria group in its quite glabrous leaves and upright petals, and is now usually placed in a separate section of *Sorbus*, of which it is the sole member.

Although usually very dwarf in the wild, *S. chamaemespilus* is said to have attained a height of 12 ft in a garden in Perthshire.

S. COMMIXTA Koehne

S. aucuparia var. *japonica* Maxim.; *S. japonica* (Maxim.) Koehne, not Hedl.; *S. serotina* Koehne (see cv. 'Serotina'); *S. pruinosa* Koehne; *S. reflexipetala* Koehne; *S.* or *Pyrus discolor* Hort., not *S. discolor* (Maxim.) Maxim.

A tree to about 35 ft high in Japan, with ascending branches; twigs glabrous, reddish brown, roughened with numerous prominent lenticels; winter-buds about ¾ in. long, acute, glabrous, green or reddish brown, usually not glutinous on trees cultivated in this country. Leaves up to 10 in. long, including petiole up to 1¾ in. long, with mostly six to eight pairs of leaflets; rachis grooved, not winged, glabrous, it and the petiole usually stained with red. Lateral leaflets narrowly to broadly lanceolate or oblong-lanceolate, 1½ to 2¾ in. long, ⅜ to 1 in. wide, acuminately narrowed at the apex to a slender tip, sharply serrated almost to the base, the teeth finely pointed, frequently double, upper surface dark or medium green, finely reticulated and sublustrous, paler beneath, glabrous on both sides, or with light brown hairs along the midrib beneath. Flowers opening in late spring. Inflorescence 4 to 6 in. across, flat, glabrous or sparsely furnished with pale brown hairs, densely lenticellate. Flowers ¼ to ⅜ in. wide, white; receptacle glabrous; petals roundish, about $\frac{3}{16}$ in. long. Stamens almost as long as the petals, with purplish anthers. Styles three or four. Fruits globose, bright red, lustrous, ripening in autumn, about ⅜ in. wide, with a dry flesh. *Bot. Mag.*, n.s., t. 166, as *S. serotina*.

A native of Japan, Sakhalin and Korea (for the closely related *S. randaiensis* of Formosa, see below). The species is a somewhat variable one in the length and relative breadth of the leaflets, in the degree of taper at the apex, and in their toothing, as well as in the presence or absence of hairs on the undersides of the leaflets and in the inflorescence (see further under var. *rufo-ferruginea*). But the variations are certainly not enough to justify the extraordinarily large number of species that Koehne, a notorious splitter, made out of them. There have been three recorded introductions of *S. commixta*, and their nomenclature is very confused:

1. *S. commixta* first reached this country from Späth's nursery towards the end of the last century and was grown as *Sorbus* or *Pyrus aucuparia* var. *japonica*. But it seems to have attracted little attention.

2. Towards the end of the 19th century, seed reached Europe, apparently from the Arnold Arboretum, under the erroneous name of *Pyrus* or *Sorbus discolor*, and in Britain this introduction continued to bear this name in gardens until the middle of the present century or even later. On the continent the nomenclature became more involved. Seedlings in Späth's nursery were identified by Koehne as *S. matsumarana*, a quite different species. Later he identified them as *S. commixta* but described one of Späth's trees as a new species, *S. serotina* (see cv. 'Serotina').

3. In the early 1930s Messrs Marchant received seed from the Chugai Nursery Company of Japan as *S. matsumarana*, and plants were distributed under this name. The leaflets in this form of *S. commixta* are brown-hairy along the midrib and rather broad and shortly pointed. A tree at Borde Hill in Sussex received as *S. matsumarana* from Messrs Marchant was identified by A. B. Jackson as *S. commixta* var. *rufo-ferruginea* (see below), and it is really intermediate between that variety and the typical state. Jackson's identification of the false *S. matsumarana* was not publicised, and it was under its erroneous name that this form of *S. commixta* received an Award of Merit on 28 October 1958 for its autumn colour and fruits. This form of *S. commixta* retains a narrow habit even in age, and grows rather taller than the maximum height given for wild trees; the tree at Borde Hill mentioned above, planted about 1930, measures 50 × 3½ ft (1971).

cv. 'EMBLEY'.—A selection from the former "*S. discolor*" of gardens, distributed by Messrs Hillier. It has steeply ascending branches, and colours rather later than trees grown as typical *S. commixta*.

Dr Wilfrid Fox had a tree of *S. commixta* at Winkworth received under the name *S. discolor* (see introduction 2 above). He was aware that this tree had nothing to do with the true *S. discolor*, and considered it to be an intermediate species or hybrid between *S. commixta* and *S. rufo-ferruginea* or *S. serotina*, both of which are here included in *S. commixta*. He had intended to describe it in his treatise on *Sorbus* under the name *S. embleyensis*, in honour of J. J. Crossfield of Embley Park, Romsey, in whose garden he first saw an example of the "*S. discolor*" of gardens. The original tree in the Winkworth Arboretum no longer exists, but a specimen from it is preserved at Kew. It was evidently not quite the same as the tree named 'Embley' by Messrs Hillier, which received an Award of Merit in 1971 for its autumn colour.

var. RUFO-FERRUGINEA Shirai ex Schneid.   *S. rufo-ferruginea* (Schneid.) Schneid.—Although *S. commixta* is often described as completely glabrous, light brown hairs often occur on the undersides of the leaflets, on the rachis, and in the inflorescence, but are not a conspicuous feature. In the var. *rufo-ferruginea*, however, the hairs are darker and denser; in one specimen in the Kew Herbarium, collected in the wild in late July, the whole undersurface of the leaflets is coated with brown wool. There is a fine group of the true variety in the Winkworth Arboretum. The plants are of spreading habit and colour beautifully in autumn; the fruits last until the end of November in some seasons. These plants match the type of var. *rufo-ferruginea* in having narrow leaflets ⅝ to ¾ in. wide.

cv. 'SEROTINA'.—This cultivar of *S. commixta* was described by Koehne, as *S. serotina*, from a plant raised in Späth's nursery, Berlin, from seeds received under the name *Pyrus discolor* (see introduction 2 above). A plant was acquired from Späth about 1930 for the Royal Horticultural Society's Garden at Wisley, and scions from this were given to Messrs Marchant in 1935. 'Serotina', as the name implies, is a late grower—Koehne noted that it leafed and flowered two weeks later than the plants that he considered to represent the true *S. commixta* (these too grew in Späth's collection). The tree in the Winkworth Arboretum also ripens its fruits and colours later in the autumn than most forms of *S. commixta*. The leaves of 'Serotina' have mostly six pairs of rather short, elegantly toothed leaflets; there are some brown hairs in the inflorescence and on the undersides of the leaflets, but these have mostly disappeared by late summer. Its fruits are brownish red.

The *S. serotina* of *Bot. Mag.*, n.s., t. 166 is not 'Serotina', but some other form of *S. commixta*. The material portrayed came from a tree at Dawyck, which had been bought from Messrs Veitch.

S. RANDAIENSIS (Hayata) Koidz. *S. aucuparia* var. *randaiensis* Hayata— This very close ally of *S. commixta* grows in the mountains of Formosa at altitudes of 10,000 to 12,500 ft. The differences between the two species are elusive, but the leaflets of *S. randaiensis* are narrower, being rarely more than ½ in. wide, and may be in as many as ten pairs. The trees under this name were originally distributed by Messrs Marchant, who received their stock from the Chugai Nursery Company of Kobe, Japan, in the late 1930s, and there is some doubt as to their authenticity. But whatever the identity of these trees, they are in effect a fine selection of *S. commixta*, very vigorous and of exceptionally narrow habit. They colour just as well in autumn and, unlike some forms of *S. commixta*, they are very free-fruiting, bearing the clusters on long spurs well clear of the foliage.

S. CUSPIDATA (Spach) Hedl.

Crataegus cuspidata Spach; *Pyrus vestita* Wall. ex Loud.; *Sorbus nepalensis* Hort. ex Hedl.; *Aria lanata* Decne., not *S. lanata* (D. Don) Schauer

A deciduous tree, of large size in the wild, but rarely seen more than 35 ft in cultivation. The habit is rather gaunt; branches few, thick, covered when young with a white wool, which afterwards falls away, leaving the shoots a smooth purplish brown; winter-buds stout, usually obtuse, almost glabrous. Leaves varying in shape even on the same spur, oblong-ovate or oblong-elliptic to broadly so, or broad-elliptic, obtuse to acute or shortly acuminate at the apex, rounded to cuneate at the base, sometimes decurrent at the base on to the petiole, 5 to 7 (sometimes to 9 in. long) by 2½ to 5 in. wide, the margins irregularly and shortly toothed, sometimes doubly so or slightly lobulate, upper surface at first coated with a white cobweb-like down, but soon becoming glabrous, lower surface covered with a persistent close felt, which is at first white or yellowish white, becoming grey later; nerves more or less parallel, in usually no more than twelve pairs; petiole ½ to 1 in. long. Flowers white, ⅝ in. to almost 1 in. across, borne in substantial corymbs 2 to 3 in. across; stalks and receptacle very woolly. Petals woolly within. Styles four to five, sometimes three, free or almost so. Fruits globose to obovoid or oblate, about ⅝ in. wide (sometimes larger), russet or dull yellow flushed with red, specked with round lenticels, ripening in early winter; sepals dry, erect or spreading, inserted round a fairly wide cavity filled by the protuberant top of the ovary. *Bot. Mag.*, t. 8259.

Native of the Himalaya from Garhwal eastward, and of north Burma; introduced in 1820, it is the most striking in its foliage of all the whitebeam group, but no large specimens have been recorded—at least none that belong unequivocally to this species. Although hardy, it will sometimes grow well for some years and without apparent reason, even in the middle of the summer, will droop and die.

S. cuspidata is variable in its foliage. An interesting form in the Hillier Arboretum makes a large shrub and has rather narrow, acuminate leaves of thinnish texture, bright green above, and unusually large fruits. It was raised from seeds received from the Indian seedsman Ghose of Darjeeling.

S. 'WILFRID FOX'.—A tree to about 40 ft high, narrowly fastigiate when young, widening with age. Leaves on mature trees mostly elliptic, obtuse, up to about 5 in. long and 2¾ in. wide, larger and broader on young trees, dark green and glossy above at maturity, undersurface with a persistent intensely white felt; petiole slender, to about 1½ in. long. Fruits roundish, lenticellate, about ½ in. wide, light russet-yellow flushed with red, ripe in early October.

This whitebeam, named by Harold Hillier in honour of Dr Wilfrid Fox, founder of the Winkworth Arboretum, was previously cultivated as "*S. nepalensis*". This is a horticultural synonym of *S. cuspidata*, but 'Wilfrid Fox' differs from that species in its thinner longer-stalked leaves and its early-ripening fruits. It is most probably, as Mr Hillier believes, a hybrid of garden origin between *S. cuspidata* and *S. aria*. A tree in the Copenhagen Botanic Garden was judged by Hedlund, the monographer of *Sorbus*, to be a hybrid of this parentage.

The parentage *S. cuspidata* × *S. aria* was suggested by A. B. Jackson for a tree at Borde Hill in Sussex which came from Veitch in 1907 as *Pyrus vestita*, i.e., *S. cuspidata*. The leaves are thinner than in that species, predominantly roundish, up to 4¾ in. long, tomentose beneath, distinctly and sharply lobulate in the upper part. Flowers almost 1 in. wide, in trusses 3 in. wide. Styles two or three, free. Fruits similar to those of *S. cuspidata*. Another possibility is that the second parent is not *S. aria* but *S. lanata* (see below), which would explain the sharp lobules, though the petioles are longer than in either species. Twenty-five years after planting this tree was 47 × 3¼ ft and is now 58 × 8 ft (1975).

S. HEDLUNDII Schneid. *S. thomsonii* Hort., not (Hook. f.) Rehd.—An ally of *S. cuspidata*, differing most obviously in having the lateral ribs coated with brown or orange hairs, contrasting with the white indumentum of the blade, but this character is not shown by young trees. Other differences are the rather finer toothing of the leaves and the more united styles. *S. hedlundii* was described by Schneider from a specimen collected by the younger Hooker on Tonglo, west of Darjeeling, where it grows with *Magnolia campbellii*; it is also known from eastern Nepal and western Bhutan.

The largest examples of *S. hedlundii* in cultivation grow at Mount Usher in Co. Wicklow, Eire, and were raised in the 1920s from a tree at Kilmacurragh in the same county, now dead, which is believed to have come from the Glasnevin Botanic Garden at the end of the last century. The Kilmacurragh tree was inexplicably identified by Sir Frederick Moore as *Pyrus thomsonii*, a sorbus of the *Micromeles* group to which *S. hedlundii* bears not the remotest resemblance, and it was under the erroneous name of *S. thomsonii* that this sorbus was put into commerce by the Slieve Donard Nursery Company, who received propagation material from Mount Usher.

Seed of *S. hedlundii* was collected by the University of North Wales expedition to E. Nepal in 1971 (B.L. & M. 20), from trees 60–70 ft high, growing on a northwest facing hillside above Pokhari.

S. LANATA (D. Don) Schauer *Pyrus lanata* D. Don—This species is of wide range, from eastern Afghanistan and bordering Pakistan through the Himalaya as far as western Nepal. Surprisingly there is no record of its having been grown in this country until quite recently, though it was introduced to Germany in 1868 by seeds received from Berthold Ribbentrop, who was at one time Chief Conservator of Forests in India. It differs from *S. cuspidata* in the sharply lobulate leaves less white beneath and sometimes almost glabrous by autumn. The fruits are apple-shaped or pear-shaped, and may be over 1 in. wide.

S. DECORA (Sarg.) Schneid.

Pyrus americana var. *decora* Sarg.; *Sorbus americana* Pursh, not Marsh.; *Pyrus americana* (Pursh) DC.; *P. sambucifolia* of some authors, not Cham. & Schlecht.; *Sorbus sambucifolia* sens. Dipp., not Roem.; *S. scopulina* of some authors, not Greene

A shrub or a small tree up to 35 ft high; branchlets fairly densely lenticellate, more or less downy when young; winter-buds dark crimson to almost black, glutinous, outer scales glabrous except for a few marginal hairs. Leaves with five to seven pairs of leaflets; rachis broadly grooved above, downy, or glabrous except for some glands and short hairs near the insertions of the leaflets. Lateral leaflets oblong or oblong-lanceolate, the longest about $2\frac{1}{4}$ in. long and $\frac{7}{8}$ in. wide, rather abruptly pointed, serrate, sometimes rather jaggedly toothed, glabrous and bluish green above, whitish and glabrous or slightly downy beneath (more densely so on sterile shoots). Inflorescence flat-topped, much branched, up to 6 in. wide, its branches and the pedicels more or less hairy. Flowers white, about $\frac{3}{8}$ in. across. Receptacle nearly glabrous outside, sepals ciliate. Petals rounded. Styles three or four. Fruits bright scarlet or vermilion, about $\frac{7}{16}$ in. wide.

Native of eastern N. America from Newfoundland to New York, perhaps farther south, west to Wisconsin and Iowa (for var. *groenlandica*, see below). It is fairly closely allied to *S. americana*, and where the two occur in the same region, *S. decora* grows at somewhat higher altitudes. It is said that intermediates or hybrids occur in the wild, but *S. decora* can usually be distinguished by its

bluish green relatively broader leaflets (length:breadth ratio 2–3:1 against
3½–5:1 in *S. americana*, according to Jones), by being acute or shortly and
abruptly acuminate at the apex, and by its larger flowers and fruits. Although
the two species have long been considered distinct, the fact that the epithet
americana has been applied by botanists to both species has inevitably caused
confusion in gardens, where *S. decora* has generally been known as *S.* (or *Pyrus*)
americana, and indeed often still is (1979).

S. decora makes in gardens a small, bushy-crowned tree, valued more for its
broad trusses of large fruits than for its autumn-colour, though the leaves
do turn to orange and russet on some soils. The fruits ripen in August and are
soon taken by birds in country gardens, especially in a dry season.

var. GROENLANDICA (Schneid.) G. N. Jones *S. americana* var. *groenlandica*
Schneid.—This interesting rowan is in some respects intermediate between
S. americana and *S. decora*. It is a native of Greenland and Labrador, extending
farther south at high altitudes, and is not known to be in cultivation in Britain.

A shrubby form of *S. decora* is in cultivation, originally distributed by Harry
White of the Sunningdale Nurseries as *S. sambucifolia*.

S. SCOPULINA Greene *S. sambucifolia sens.* Rydb., not Roem.—A shrub to
about 15 ft high, often many-stemmed from the base; young shoots coated
with appressed hairs; winter buds as in *S. decora*. Leaflets in five to seven pairs,
narrowly oblong or oblong-lanceolate, the longest nearly 3 in. long and about
⅞ in. wide, of rather thin texture, glabrous and dark green above, glabrous
or slightly downy beneath, closely and simply toothed. Inflorescence about
3½ in. wide, rather densely white-hairy. Fruits as in *S. decora*, but sometimes
bright orange.

A native of N. America from South Dakota to Alberta across the Rocky
Mountains to British Columbia, south to New Mexico; described from Colorado.
It was probably not in cultivation here until Dr Brian Mulligan, then Director
of the University of Washington Arboretum, Seattle, sent propagation material
to Messrs Marchant, who distributed plants in the 1950s. But it remains rare.
For the sorbus distributed under the name *S. scopulina* see under *S. aucuparia*.

S. DISCOLOR (Maxim.) Maxim.
Pyrus discolor Maxim.

A tree to about 35 ft high in the wild; branchlets slightly downy when young,
soon glabrous, purplish brown by winter, scarcely lenticellate; winter-buds
ovoid, reddish brown, about ⅜ in. long, almost glabrous, the outer scales sharply
keeled. Leaves about 6 in. long, with five to seven pairs of widely spaced
leaflets, which spread at almost a right-angle to the rachis, which is broadly
grooved but not winged, almost glabrous. Lateral leaflets narrowly lanceolate,
1¾ to 2 in. long and about ⅝ in. wide, tapered to a very acute apex, but the
lowest pair small and blunter, dull matt-green above, pale glaucous green
beneath, at first thinly hairy, later quite glabrous, serrated in the upper half or
three-quarters. Stipules persistent, leafy, up to ½ in. or slightly more long,

digitately lobed, with sharp-pointed, gland-tipped lobes, glabrous. Inflorescence laxly branched, at most about 4 in. wide, its branches and the pedicels sparsely hairy at first. Flowers white, about ½ in. wide. Receptacle slightly hairy at flowering-time. Petals orbicular, broadly clawed. Styles three or four. Fruits globose, white, about ⅜ in. wide.

A native of north China, where it grows, among other localities, in the mountains west of Peking. It was described by Maximowicz in 1859 from what must have been a very inadequate specimen, collected by the Russian doctor P. Y. Kirilov. The plants raised in Europe from the seeds sent by Dr Bretschneider, identified as *S. discolor*, appear to have been *S. pohuashanensis*, and other plants raised in Europe not long after, received as *Pyrus* or *S. discolor*, proved to be *S. commixta*. But the true species was introduced to Kew from Späth's nursery, Germany, in 1903 and again in 1924 from the Arnold Arboretum, raised from seeds collected by Hers in China. The true *S. discolor* appears to be no longer in cultivation in Britain, though it is grown in the Arnold Arboretum, and in the Russian Far East, and no doubt still exists in the mountains of north China.

S. ?× PEKINENSIS Koehne ?*S.* × *arnoldiana* Rehd.—Although *S. pekinensis* is usually regarded as synonymous with *S. discolor*, there is some evidence that the plants described by Koehne were natural hybrids between *S. discolor* and *S. pohuashanensis*, both of which grow in the mountains west of Peking. The plants from which Koehne drew his description had been raised by Herr Gebbers of Wiesenburg, from seeds sent by Bretschneider or by Brandt from Peking, and others by the German nurseryman Späth. They were certainly near to *S. discolor* in most characters, but the fruits were originally described by Koehne as 'rosy white, some bright yellowish red' and in a later note as 'dull reddish yellow or partly suffused with white, more or less translucent when ripe.' A tree obtained from Gebbers for the Arnold Arboretum bore white fruits, but a seedling raised from it had rosy fruits and the leaflets and inflorescences were more hairy. Rehder concluded that this was the result of pollination by *S. aucuparia* and called the plant *S.* × *arnoldiana*. The alternative hypothesis is that the parent plant was genetically impure in the first place.

Trees cultivated as *S. pekinensis*, and sometimes as *S. discolor*, bear dull yellow or apricot-pink fruits (Majolica Yellow in plants oddly named "*S. tetriensis*"). In one instance the flowers are known to be mostly sterile, which may explain the shy-fruiting of these plants, none of which show any influence of *S. aucuparia*. In Holland plants have been raised from *S. pekinensis* in which the fruits are orange-yellow and in these too they are sparsely borne. (*Dendroflora*, No. 2 (1965), p. 33). One of these has been named 'SCHOUTEN'.

Apart from 'Tetriensis', mentioned above, there are other derivatives of *S. pekinensis* in gardens, raised by the Dutch nurseryman Lombarts. Of these 'CARPET OF GOLD' is a slenderly branched tree of fastigiate habit, with elegant foliage. The fruits are orange-yellow, with red flecks, borne in rather sparse trusses. In 'GOLDEN WONDER' too the fruits tend to be few in each truss; they are pure orange-yellow, larger than in 'Carpet of Gold', but the foliage is coarser.

S. DOMESTICA L.   SERVICE TREE

Pyrus sorbus Gaertn.; *Cormus domestica* (L.) Spach

A deciduous tree, usually 30 to 50 ft (occasionally 60 to 70 ft high); trunk covered with a scaly, rough bark; shoots furnished with loose, silky hairs when quite young, which soon fall away; winter buds glutinous and shining. Leaves pinnate, 5 to 9 in. long, composed of thirteen to twenty-one leaflets, which are narrowly oblong, usually pointed, but sometimes rounded at the tip, 1¼ to 2½ in. long, ⅜ to ½ in. wide, margin set with slender teeth except towards the base, which is entire, glabrous above, more or less downy beneath, but becoming glabrous or nearly so by autumn. Flowers white, about ½ in. across, produced in May in panicles at the end of short branches and from the leaf-axils, the whole forming a rounded or rather pyramidal cluster 2½ to 4 in. wide. Receptacle and flower-stalks downy. Styles five. Fruits pear-shaped or apple-shaped, 1 to 1¼ in. long, green or brown tinged with red on the sunny side.

Native of southern and parts of eastern Europe, thence eastward to the northern Caucasus; also of North Africa. As an ornamental tree this is perhaps scarcely equal to its ally, the mountain ash, but is well worth growing for the beauty of its foliage, and for its flowers, which are larger than usual in *Sorbus*. It also attains to greater dimensions than any of its immediate allies.

The most famous of all British service trees was one which grew for some hundreds of years in Wyre Forest, in Worcestershire. The story of this tree was told by Robert Woodward in the *Gardeners' Chronicle* of 13 April 1907. It was first noted by one Edmund Pitt, in 1678, and was mentioned and discussed by various writers up to 1862, when it was set on fire and killed by a vagrant. This tree was considered to be an old one by Pitt in 1678, and there appears to be little doubt that the species lives for five or six hundred years. The Wyre Forest tree is the only one which gives the species any claim to rank as a British tree, for it has never been found truly wild elsewhere, and it is now generally accepted that even this tree was planted.

The fruit of the service tree is sometimes eaten in a state of incipient decay, especially in France, although Loudon observes that it is not highly prized, and is more frequently eaten by the poor than by the rich. On the other hand Mr E. Burrell, then gardener to H.R.H. the Duchess of Albany at Claremont, in a letter dated 11th Nov. 1883, observes that 'we are sending good fruits of the pear-shaped service for dessert at the present time.' This Claremont tree was blown down in 1902, and was then close upon 70 ft high. The timber is of fine quality, being very hard and heavy, but too scarce to count for much.

The form with pear-shaped fruit, which appears to make the finest tree, is distinguished as f. PYRIFERA (Hayne) Rehd.; the other, with apple-shaped fruit, as f. POMIFERA (Hayne) Rehd. Both are easily distinguished at any time from the mountain ash by the rough scaling bark; in autumn by the big fruits; and in winter by the glutinous, not very downy buds. The service tree should be raised from seed.

Early in the last century Lord Mountnorris of Arley Castle, Worcestershire, raised grafts of the Wyre Forest tree, and there is still a tree there about 65 ft high. The most famous examples of the service tree grow in the Oxford Botanic Garden, and are believed to have been planted in the time of Dr Sibthorp, Professor of Botany at Oxford 1784–95. They measure 75 × 7¼ ft (maliform) and 54 × 6½ ft (pyriform) (1970). Other large examples measured recently are: Park Place, Henley, 65 × 8 ft (1969), and Colesbourne, Glos., *pl. c.* 1905, 60 × 6 ft (1975).

S. EPIDENDRON Hand.–Mazz.

A tree up to 50 ft high in the wild, but said to be sometimes an epiphytic shrub (whence the epithet *epidendron*, living on trees); branchlets at first hairy, soon glabrous; winter-buds ovoid, about ¼ in. or slightly more long, glabrous. Leaves narrowly obovate to elliptic, up to 6 or 7 in. long on wild plants, rounded or abruptly acuminate at the apex, cuneate to narrowly so at the base, glabrous above, coated more or less densely beneath when young with rusty or sometimes whitish hairs, most of which are soon shed but a few persisting, especially on the veins, finely toothed almost to the base; lateral veins prominent beneath, in mostly ten to twelve pairs, the lowermost sending off numerous branches to the teeth; petiole varying on the same shoot from almost absent to about ½ in. long (but much longer in some herbarium specimens accepted as this species by Dr Yü). Inflorescence with a definite central axis, sometimes dense and apparently corymbose, sometimes taking the form of a conical panicle as long or even longer than wide; branches of inflorescence clad at flowering-time with rusty or fawn-coloured wool. Flowers white, about ¼ in. across. Petals hairy on the inside. Receptacle constricted at the apex, its free part and the sepals soon deciduous. Styles two or three, free, glabrous like the top of the receptacle. Fruits slightly lenticellate, globose, not seen in the ripe state on cultivated plants but about ¼ in. wide on wild trees, probably not colouring.

A native of S.W. China and north Burma; discovered by the Austrian botanist Handel-Mazzetti in 1915 and described by him. In his description he also cited two specimens collected by Forrest, who introduced this species in 1924–5; seed was also distributed after his death under number F.29021, from his 1930–1 collections. This species is very rare in cultivation. There are two shrubby examples in the Winkworth Arboretum, of unrecorded provenance, planted in 1952 and probably bought from Messrs Hillier, who for some years propagated a plant from Forrest's 1924–5 introduction. The healthier of the two is a fastigiate, slenderly branched bush about 20 ft high (1979), which bears inconspicuous flowers and fruits in some years. The young foliage is copper-tinted; the leaves persist until late autumn and sometimes turn pink or apricot before falling. Evidently from a different Forrest sending is a tree in the Edinburgh Botanic Garden planted in 1926 and measuring 20 × 1 ft (1967). This has much paler hairs than in the Winkworth plant or in any wild-collected specimen examined, and less prominent leaf-ribs, but there is no other species with which it agrees better. The flowers, at least in some years, are borne in quite showy lax panicles about 3 in. long and 4 in. wide.

S. ESSERTEAUIANA Koehne

Pyrus esserteauiana (Koehne) Bean; *Sorbus conradinae* Koehne

A shrub or small tree in the wild, up to about 40 ft high and 1 ft in diameter of trunk; young shoots grey with down; winter-buds ovoid, silky, ½ in. or slightly more long. Leaves pinnate, 6 to 10 in. long; leaflets mostly in five or six pairs, well-spaced on the rachis, which is tinted with crimson or purple. Lateral leaflets up to 3½ in. long and 1 to 1½ in. wide, lanceolate or lanceolate-oblong, gradually pointed, dark green and slightly glossy above, coated beneath with white hairs which have been partly shed by autumn, margins finely to rather jaggedly serrate; lateral veins in twelve to sixteen pairs, impressed above and prominent beneath. Flowers opening in May or early June, about ½ in. wide, white, borne in flat or slightly convex corymbs up to 5 in. wide, the inflorescence-branches, pedicels and receptacles clad in grey down; stipules of the leaves immediately below the inflorescence large and leafy, fan-shaped, toothed at the apex, persistent in the fruiting stage. Petals roundish, with a short claw. Styles three or four. Fruits ripening in October, bright red or orange-yellow in the wild, sometimes deep yellow on cultivated trees, about ¼ in. wide. *Bot. Mag.*, t. 9403.

Native of W. Szechwan, China; discovered and introduced by Wilson. He first sent seed in 1908 during his first expedition for the Arnold Arboretum, collected in the Mupin area, but plants from this introduction were not widely distributed in Britain (at least not as *S. esserteauiana*, see below). During his second expedition Wilson sent two batches of seed from farther north (W.4156 and 4321), and from the herbarium specimens he collected at the same time Koehne described *S. conradinae*, which is the name under which plants from the second introduction were distributed. Koehne admitted that *S. conradinae* and *S. esserteauiana* (also described by him) were allied, and distinguished the former by the whiter and denser down beneath the leaflets, by their greater relative width, and their more deeply impressed veins. However, J. R. Sealy, writing in the *Botanical Magazine* in 1935, concluded that the characters relied on by Koehne were not of specific value, and united the two species under the name *S. esserteauiana*, a judgement that has not been seriously challenged since.

S. esserteauiana, as usually seen in gardens, has a short trunk dividing into a few stems which are set with long spurs throughout their length but sparsely branched. The fruits ripen late and may persist until January; the yellow-fruited forms are the most desirable. The leaves turn red before falling, though not on dry soils. Award of Merit 1934 (yellow-fruited) and 1958 (red-fruited).

Although there is no doubt that *S. conradinae* is synonymous with *S. esserteauiana*, the decision to unite them was challenged by Dr Fox on the grounds that a plant grown by him at Winkworth under the name *S. conradinae* differed from *S. esserteauiana* in a number of respects, notably that it branched more freely, had broader leaflets less hairy beneath and ripened its fruits as early as August, bearing them in more abundant trusses (*Journ. R.H.S.*, Vol. 79 (1954), p. 92 and Fox MS). But if this sorbus is not *S. esserteauiana*, neither is it *S. conradinae*. In fact, it bears a strong resemblance to *S. pohuashanensis* but is more likely to be a hybrid of garden origin between *S. esserteauiana* and *S. aucuparia*. The tree in the Winkworth Arboretum that Dr Fox had in mind cannot be traced, but there is a specimen from it in the Kew Herbarium, and a similar tree at Borde Hill in Sussex, which was received as *S. conradinae* from the nursery attached to Vicary Gibbs' collection at Aldenham.

There is a sorbus in cultivation which may derive from the original introduction of *S. esserteauiana* by Wilson in 1908, differing from the common form (previously grown as *S. conradinae*) in its relatively narrower leaflets less hairy beneath and less impressed veins. It has been grown as "*S. wilsoniana*" and has also been misidentified as *S. pohuashanensis*.

S. 'KEW HYBRID'.—This fine rowan was put into commerce by Messrs J. O. Sherrard and Son of Newbury, Berks. It appears to be a hybrid of *S. esserteauiana*, of which the other parent could be either *S. pohuashanensis* or *S. aucuparia*. It is very free-flowering and bears heavy trusses of orange-red fruits $\frac{5}{16}$ to $\frac{3}{8}$ in. wide, ripening in September; the autumn colour is orange and purple. Mr Sherrard tells us that the original tree in the nursery was planted by his father in 1947 and is now about 23 ft high (1977). The provenance is unknown, but it is tempting to suppose that it was propagated from the tree at Kew that received an Award of Merit in 1948 (see 'Kewensis', under *S. pohuashanensis*).

S. 'WARLEYENSIS'.—This sorbus received an Award of Merit as *Pyrus* sp. when shown by Miss Ellen Willmott on October 6, 1931, and was later named *S. sargentiana* var. *warleyensis* by Marquand (*Gard. Chron.*, Vol. 94 (1933), p. 177). It had been raised at Warley Court from seeds collected in China by one of the French missionaries and presented to Miss Willmott's mother by Henri de Vilmorin. 'Warleyensis' scarcely belongs to *S. sargentiana* and is really nearer to *S. esserteauiana*, though the leaflets are more finely toothed than is usual in that species. It is probably no longer in cultivation, though 'Kew Hybrid' (see above) is near to it. Plants distributed in the 1930s under the name "*S. aucuparia Warleyensis*" are not *S.* 'Warleyensis' and are near to *S. pohuashanensis*; they were presumably propagated from some other plant at Warley Court.

S. WILSONIANA Schneid. *S. expansa* Koehne—Although the name *S. wilsoniana* used to occur frequently in horticultural literature it is doubtful if the species to which the name properly belongs was ever introduced. It bears some resemblance to *S. esserteauiana*, but the leaflets are at first clad beneath with a dense white wool, later becoming almost or wholly glabrous.

S. FOLGNERI (Schneid.) Rehd.

Micromeles folgneri Schneid.; *Pyrus folgneri* (Schneid.) Bean;
S. nubium Hand.-Mazz., *sec.* Yü; ? *S. aria* var. *chinensis* Henry

A tree up to 30 ft high, with slender, often semi-pendulous branches; young shoots at first covered with whitish felt, becoming glabrous by autumn; winter buds pointed, slender, glabrous. Leaves lanceolate or narrowly ovate, tapering to both ends, 2 to $3\frac{1}{2}$ in. long, $\frac{3}{4}$ to $1\frac{1}{4}$ in. wide, long-pointed, dark green and glabrous above, covered beneath with a close, beautifully silvery white felt; nerves parallel in eight to ten pairs; stalk about $\frac{1}{2}$ in. long. Corymbs 3 to 4 in. across, sometimes rather elongated, carrying numerous rather densely arranged flowers $\frac{1}{3}$ to $\frac{1}{2}$ in. in diameter and white; calyx and flower-stalk woolly.

SORBUS FOLGNERI

Fruits oval or obovate, $\frac{1}{2}$ in. across, ripening in late autumn or early winter, becoming red at least on the exposed side, devoid of calyx-teeth and with a small pit at the apex.

Native of Central and S.W. China; introduced by Wilson for Messrs Veitch about 1901. It varies somewhat in the more or less pendent character of its branches, and in one form, named f. PENDULA by Rehder, the habit is beautifully elegant, the branches arching outwards and drooping at the ends, and the leaves vividly white beneath.

In Dr Fox's manuscript notes there are interesting remarks about this species, here quoted in full:

The seasonal displays of *S. folgneri* can be among the finest exhibited by a wild Sorbus species introduced into this country, but under conditions of cultivation they are provokingly variable. The foliage, though quite distinct in character, may alter in tinge from year to year. The tree is therefore better suited to be a companion to other Sorbus, rather than to be the only representative of its kind in the gardens of southern England, where it is perfectly hardy.

Its unusually soft leaves droop in loose clusters, a habit which distinguishes it from those of the Whitebeam for which they might be mistaken, being simple in form, and sometimes beautifully white beneath. On windy days a tree of *S. folgneri* can usually be detected by the characteristic shimmering of its half-silvery leaves.

In the rare but unforgettable seasons when they turn golden pink on top in autumn, the foliage becomes brilliantly incandescent, and with its two tones of colour, is unmistakable.

cv. 'LEMON DROP'.—Fruits yellow. Semi-pendulous habit. Raised by Messrs Hillier.

There is a sorbus in cultivation which in general appearance strongly resembles *S. folgneri*, but differs in three respects: the leaves have only a thin veil of hairs beneath, the teeth ending the main lateral ribs are much longer than the others and on a few leaves could almost count as small lobes, and the calyx is persistent in fruit. A tree at Borde Hill of this character was identified by Bruce Jackson as S. ZAHLBRUCKNERI Schneid. It is strange that a tree so like *S. folgneri* should have fruits with a persistent calyx, which remove it from section *Micromeles* into section *Aria*, but that is equally true of *S. zahlbruckneri* itself, for Schneider, in describing it from a specimen collected by Henry in Szechwan, wrote that he would have taken it to be a *Micromeles* but for this character. It may be that its proper place is indeed in section *Micromeles*, as an anomalous species. The French botanist Cardot noted that on some wild specimens of *S. zahlbruckneri* the calyx was in fact deciduous, and equally that it was sometimes persistent in *S. folgneri*.

There is another example of *S. zahlbruckneri* in the Winkworth Arboretum, with smaller leaves than those of the Borde Hill tree—a difference probably to be accounted for by the drier soil in which it grows.

The sorbus mentioned above should not be confused with the Chinese form of *S. alnifolia* (q.v.) originally distributed as *S. zahlbruckneri*. This was raised from Wilson's seed-number 628, which was wrongly mentioned under *S. zahlbruckneri* in *Plantae Wilsonianae*.

S. FOLIOLOSA (Wall.) Spach
Pyrus foliolosa Wall.; *P. wallichii* Hook. f.; *Sorbus wallichii* (Hook. f.) Yü

A small lax shrub, often epiphytic, perhaps sometimes a small tree; branchlets at first densely coated with a white cobwebby indumentum, soon glabrous; winter-buds glabrous or almost so. Leaves to about 6 in. long, with five to nine pairs of leaflets, which are usually not much more than 1 in. long, oblong-lanceolate, acute, glabrous above, at first coated beneath with an indumentum like that of the branchlets, later glabrous, margins reflexed, toothed only near the apex. Inflorescence many-flowered, 2 to 4 in. wide, cobwebbed with white hairs at first. Styles three or four. Fruits pink or crimson, globose, about ¼ in. wide.

A native of the central and eastern Himalaya, possibly extending into Burma and China; described by Wallich in 1831, largely from a specimen which he had himself collected on Sheopore near Kathmandu in Nepal. This was in flower, and the description of the fruits was taken from another specimen, which is not the present species but *S. ursina*. It has also been a source of confusion and misunderstanding that he described the colour of the hairs on the undersides of the leaflets as 'subferrugineous'. They are now fawn-coloured on the Sheopore specimen in the Wallich Herbarium and may have originally had a brownish glint originally. The younger Hooker concluded that Wallich was describing *S. ursina* and applied Wallich's name *P. foliolosa* to that species, describing the present species, or rather redescribing it, under the name *P. wallichii*. Many botanists followed Hooker's nomenclature, but others, in the present century, have taken the same view as Hedlund, the Swedish monographer of *Sorbus*, that Wallich's *P. foliolosa* is the species here described and not *S. ursina*. The fact that the name *Sorbus* or *Pyrus foliolosa* has been used for two quite distinct species has been a source of confusion, especially in herbaria.

S. foliolosa is not in cultivation so far as is known (1979), but is likely to be introduced before long, as it occurs in easily accessible parts of Nepal.

In gardens the name *S. foliolosa* has been used for *S. vilmorinii* and also for a relative of that species (see 'Pearly King'). *S. foliolosa* var. *pluripinnata* Schneid. is *S. scalaris*, but plants distributed under this name are near to *S. vilmorinii*.

S. WATTII Koehne *Pyrus wallichii sens.* Hook. f., in part—This species was described from a specimen collected by Sir George Watt on Mt Japvo in the Naga Hills, Assam. It is certainly very near to *S. foliolosa*, but so far as is known, the young branchlets, undersides of leaves and inflorescence do not have the cobwebby tomentum of *S. foliolosa*. *S. wattii* was introduced by Kingdon-Ward in 1935 from the type-locality (KW 12571) as *Pyrus foliolosa*; according to the field-note the autumnal leaves are scarlet and the fruits rosy red, hanging in large bunches. A plant from the seeds collected by Kingdon Ward was given by the Edinburgh Botanic Garden to Col. S. R. Clarke and was grown by him for some years in his garden in the Isle of Wight and was propagated by Mr Hillier. By dint of re-grafting the species has been kept in existence in the Hillier Arboretum but evidently does not thrive on rowan root-stock. From herbarium specimens of the Isle of Wight plant and earlier propagations it would seem that *S. wattii* is indeed a very glabrous species, and that the hairs on the youngest leaves are brown. The leaflets are very variable, being obtuse and sparsely toothed near the apex on short shoots, acute and toothed to the base on vigorous shoots. The fruits ripened very late and were not freely borne.

S. GLOMERULATA Koehne
Pyrus glomerulata (Koehne) Bean

A deciduous shrub or small tree, with glabrous young shoots and winter-buds. Leaves pinnate, $2\frac{1}{2}$ to 6 in. long, with ten to fourteen pairs of leaflets which are attached to the grooved, slightly winged, glabrous common stalk at intervals of $\frac{1}{6}$ to $\frac{1}{2}$ in. Leaflets mostly oblong, sometimes ovate or narrowly oval, toothed towards the apex, pointed, $\frac{1}{4}$ to $1\frac{1}{8}$ in. long, $\frac{1}{8}$ to $\frac{1}{3}$ in. wide, glabrous or with a little down on the midrib beneath. Corymbs 2 to 3 in. long and wide, with glabrous stalks, the white flowers, each about $\frac{1}{3}$ in. wide, crowded at the top in a rounded cluster. Receptacle and styles glabrous. Fruit pearly white, $\frac{1}{4}$ to $\frac{3}{8}$ in. wide, globose.

S. glomerulata was described from specimens collected by Wilson near Changyang Hsien in Hupeh in 1907, but he had sent seed earlier to Messrs Veitch from the same locality. It is most likely to be confused with *S. koehneana*, which differs in its slightly downy inflorescence, more strongly toothed leaflets, often hairy buds, and in the style being downy at the base. It was praised in previous editions for its pretty foliage and pearl-like fruits, but is apparently no longer in cultivation.

S. GRACILIS (Sieb. & Zucc.) K. Koch
Pyrus gracilis Sieb. & Zucc.; *S. schwerinii* Schneid.

A small, deciduous tree or shrub of compact, bushy habit, up to 12 ft, young shoots slender, downy. Leaves pinnate, with a main-stalk 3 to 6 in. long, carrying three to five pairs of leaflets which are roundish oblong to oblong, 1 to 2½ in. long, toothed towards the rounded or pointed apex, slightly downy and pale beneath. Inflorescence few-flowered, 1 to 2 in. across, with persistent, green, fan-shaped stipules, which are toothed at the apex and about ⅞ in. wide. Flowers cream-coloured, opening in late May, about ¼ in. wide, with orbicular petals. Fruits erect, about ¾ in. long, yellowish red, pear-shaped in cultivated plants but usually globose or ellipsoid in the wild.

A native of the mountains of Japan, not introduced until the 1930s. It is a distinct and attractive species, the foliage being bronzy green at first, changing to rich red in the autumn. This character, combined with its relatively small size and slow growth, render it well adapted for small gardens. Out of flower it could be taken for a rose, and the likeness is greatest in autumn, when the scattered fruits stand out above the foliage like rose heps. A plant in Dr Fox's collection was 12 ft high and 7 ft in spread when eighteen years planted, but there are older plants in gardens only 8 ft high.

S. × HOSTII (Jacq.) K. Koch
Aria hostii Jacq.; *Pyrus hostii* (Jacq.) K. Koch *pro syn.*

A shrub to about 12 ft high, sometimes taller, of compact habit; twigs smooth and glossy, brown, sparsely lenticellate; winter-buds conical, ¾ to ⅝ in. long, the scales hairy at the edge, otherwise glabrous. Leaves simple, elliptic, 3½ to 4¾ in. long, about 2 in. wide, glossy dark green above, matted with grey hairs beneath, sharply toothed and also mostly shallowly lobed in the upper half, the lobes diminishing in size towards the apex; petiole about ⅝ in. long. Flowers borne in May in dense rounded clusters 1½ to 2½ in. wide, held erect on stiff white-hairy stalks. Petals trowel-shaped, ⅜ in. long, white on the inside, edged with deep pink, flushed with pink on the outside. Stamen-filaments cream-coloured, anthers pink at first, later cream flushed with pink. Styles two. Fruits ovoid, six to eight in each truss, held erect, slightly over ½ in. long, lustrous red, with small straw-coloured lenticels.

A hybrid between *S. chamaemespilus* and *S. mougeotii*, varying according to whether the second parent is the western, typical subspecies or the more eastern *S. mougeotii* subsp. *austriaca*. It was originally described in 1826 from a plant growing in the Vienna Botanic Garden. A form of this hybrid, of unrecorded provenance, was distributed in the 1870s by Messrs Simon-Louis of Metz, but the description given above is adapted from Dr Fox's account of the plant that once grew in the Winkworth Arboretum. This was 9 ft high when eight years old and about 4 ft in spread; it no longer exists but was undoubtedly authentic, which is more than can be said about some plants distributed as *S.* × *hostii*. The true hybrid is one of rare beauty both in flower and fruit, and valuable for the small garden, or where space is limited.

S. HUPEHENSIS Schneid.

Pyrus mesogaea Cardot; *P. hupehensis* (Schneid.) Bean, not Pampan.; *S. laxiflora* Koehne; *S. aperta* Koehne; *S. hupehensis* var. *aperta* (Koehne) Schneid.; *S. oligodonta* (Card.) Hand.-Mazz.; *Pyrus oligodonta* Card.; *S. hupehensis* var. *obtusa* Schneid.; *Pyrus glabrescens* Cardot; *S. glabrescens* (Cardot) Hand.-Mazz.; *Pyrus wilsoniana sens.* W.W. Sm., not Schneid.

A tree to 50 ft high; branchlets slender, grey-brown or purplish brown, soon glabrous; winter-buds reddish or purplish brown, glabrous except for some hairs

SORBUS HUPEHENSIS

on the margins of the scales and at the apex. Stipules on both sterile and flowering shoots small, soon falling. Leaves (including petiole 1 to 2 in. long) 4 to 7 in. long, with four to eight pairs of leaflets spaced ½ to ⅞ in. apart; rachis grooved, not or scarcely winged. Leaflets oblong or oblong-elliptic, sometimes oblong-obovate, 1 to 2¼ in. long, ⅝ to ⅞ in. wide, apex acute to obtuse or rounded in general outline, but often prolonged into a slender cusp or a thread-like mucro, glabrous and (in most cultivated plants) sea-green above, underside papillose, glabrous except for sparse white hairs at the base of the midrib, margins serrated, sometimes almost throughout, but in most cultivated plants only in the upper one-third or one-quarter. Flowers white, about $\frac{1}{4}$ in. wide, produced in May or early June in rather loose trusses; inflorescence-axes slightly hairy or glabrous, sparsely lenticellate, often becoming brilliant red. Stamens with cream-coloured filaments and pink or purple anthers. Styles four or five. Fruits globose, about $\frac{5}{16}$ in. wide, porcelain white except for a pink flush at the tip, or sometimes crimson-pink throughout. *Bot. Mag.*, n.s., t. 96.

Native of China, from Shensi to Yunnan, east into the central provinces, possibly extending into upper Burma and the Mishmi Hills of Assam. It was described from a specimen collected by Wilson in 1901 in W. Hupeh but he did not send seed until 1910, under W.4155, from the Min valley of W. Szechwan; the corresponding field-specimen is the type of *S. aperta*, here included in *S. hupehensis*. The greater part of the cultivated stock derives, however, from Forrest's sendings from N.W. Yunnan. The first of these was in 1910 (F.5550 from the Lichiang range), but the common white-fruited form of *S. hupehensis* derives from seeds collected during his 1917–19 expedition. The task of packaging and distributing the harvest of these two years was undertaken by the Royal Horticultural Society, who sent out the seed of this sorbus under the provisional number 'A.1812'. Recipients of this seed were later told, owing to a mis-identification by Sir William Wright-Smith, that A.1812 was '*S. wilsoniana* f. *glaberrima*' and it was under this name that Forrest's white-fruited *S. hupehensis* was grown in gardens for some years. The seed most probably came from the Lichiang range. This form of the species grows vigorously, and has attained a height of 50 ft in gardens. The white fruits, on brilliant red stalks, persist for some months after the leaves have fallen, and the autumn colour on some soils is red and yellow, or orange-yellow. It was this form that received an Award of Merit in 1930.

There is another form of *S. hupehensis* in cultivation, also introduced by Forrest, possibly under F.19645, collected in Yunnan on the Mekong-Salween divide. It was originally grown as "*S. wilsoniana*". The most obvious difference between this and 'A.1812' is that the fruits are crimson-pink or at least strongly tinged with that colour. Dr Fox observed in his notes that in the Winkworth Arboretum this form expands its leaves a month earlier than in the white-fruited form, that the flower-buds and pedicels are tinged with pink, and the anthers deeper pink or red. The leaves on this tree are very glaucous and in autumn assume a beautiful crimson-red, harmonising with the colour of the fruits. There is no valid distinguishing name for the crimson fruited form. A selection from it has been named 'NOVEMBER PINK' in Holland and another received an Award of Merit on 23 October 1962 when exhibited by the Sunningdale Nurseries as *S. hupehensis* "Rosea".* A seedling with crimson fruits and good autumn colour, named 'RUFUS', was raised at Windsor and received an Award of Merit on 21 November 1967 when shown by the Crown Estate Commissioners.

The practice has grown up in gardens of distinguishing the pink-fruited form as *S. hupehensis* var. *obtusa* or as *S. oligodonta*. These names are perfectly valid,

* The tree mentioned in the 7th edition of this work under the name *S. hupehensis* var. *rosea* (A.M. 1938) was *Malus hupehensis*.

but would be equally applicable to the white-fruited form. The variations on which some botanists have made distinct species, or varieties, out of the numerous variations of *S. hupehensis*, are based on foliage characters, and not on the colour of the fruits.

S. HYBRIDA L.

Pyrus pinnatifida Ehrh.; *Sorbus fennica* (Kalm) Fries; *Crataegus fennica* Kalm; *Pyrus fennica* (Kalm) Bab.; *Sorbus meinichii* Hort., in part, not (Hartm.) Hedl.; *Pyrus firma* Gibbs ex Osborn; *S. hybrida* 'Gibbsii' Hort.; *S. pinnatifida* Hort.

A small tree with spreading branches, forming a roundish irregular crown; branchlets grey-woolly when young, brown and more or less glabrous by autumn, sparsely lenticellate; winter buds ovoid, blunt, $\frac{1}{4}$ in. or slightly more long, basal scales brown, almost glabrous, the upper ones woolly. Leaves lobed and partly pinnate, broadly elliptic or ovate-elliptic in general outline, mostly broadly obtuse at the apex, 3 to $3\frac{1}{2}$ in. long and 2 to $2\frac{1}{2}$ in. wide under the inflorescence but larger on sterile shoots. They are lobed at least in the upper part, the lobes increasing in length downwards towards the base of the leaf, where there is usually one pair of free leaflets on those of the stronger flowering shoots, but sometimes two on strong shoots, while the leaves on weaker shoots may be merely lobed; lobes and free leaflets edged with sharply pointed teeth. The leaves are darkish dull green above, whitish tomentose beneath; petiole $\frac{5}{8}$ to $1\frac{1}{8}$ in. long. Flowers in May in clusters $2\frac{1}{2}$ to $4\frac{1}{2}$ in. wide, white, about $\frac{5}{8}$ in. across. Petals hairy inside. Anthers yellow. Styles usually three. Carpels free at the apex. Fruits globose, ripe in October, $\frac{1}{2}$ to $\frac{5}{8}$ in. wide, deep scarlet crimson, sparsely dotted with lenticels; calyx lobes inflexed, narrowly triangular, becoming fleshy only when the fruit is fully ripe, and even then horny at the tips.

A native of southern and southwest Scandinavia, rarely far from the coast. Linnaeus, who described this sorbus in 1762, considered it to be a hybrid that behaved like a species, and in this he has been proved right, for it is now known that *S. hybrida* normally reproduces itself apomictically, i.e., produces fertile fruits without fertilisation, and then breeds true, although of undoubted hybrid origin. Linnaeus suggested *S. aucuparia* and *S. intermedia*, but the objection to this hypothesis, which was long accepted, is that hybrids of this parentage have been raised artificially and are quite different from *S. hybrida*. It is now thought more likely that *S. hybrida* derives from hybridisation between *S. aucuparia* and *S. rupicola* (q.v. under *S. umbellata*).

S. hybrida was probably introduced in the last century, but the first recorded introduction was by the Earl of Ducie to his garden at Tortworth in Gloucestershire early this century; it seems to have been little known in gardens until the Hon. Vicary Gibbs exhibited it in 1920 under the strange name of "*Pyrus firma*", the specific epithet being perhaps the result of misreading the word *fennica* on a label. It was later correctly identified as *Pyrus pinnatifida*, but was given by Arthur Osborn the varietal name 'var. Gibbsii', no doubt to distinguish it from the sorbus then commonly grown as *Pyrus pinnatifida*, which was wrongly named and is in fact a form of *S.* × *thuringiaca* (q.v.). 'Gibbsii' is certainly very distinct from this sorbus, but does not differ from the normal wild *S. hybrida* except perhaps in having a trifle larger fruits. *S. hybrida* has also been grown wrongly as *S. meinichii*.

S. hybrida received an Award of Merit in 1920 as "Pyrus firma" and in 1953 as "*S. pinnatifida*" 'Gibbsii'.

S. PSEUDOFENNICA E. F. Warburg—Very similar to *S. hybrida* but a smaller tree with shorter leaves, yellowish green above, rather thinly tomentose beneath, and with smaller flowers and fruits. It is known only from Glen Catacol in the Isle of Arran, Scotland, and is thought to be the result of hybridisation between the triploid *S. arranensis* (q.v. under *S. minima*) and *S. aucuparia*. It was described in 1957.

S. INSIGNIS (Hook. f.) Hedl.

Pyrus insignis Hook. f.; *Sorbus harrowiana* (Balf. f. & W.W. Sm.) Rehd.; *P. harrowianus* Balf. f. & W.W. Sm.

A shrub, or more commonly a tree 25 to 40 ft high (said by Kingdon Ward to grow taller than this in the forests of N. Burma); young shoots stout, at first coated with brown and whitish hairs intermixed, later glabrous; winter-buds ovoid, ½ to ⅞ in. long, the outer scales glabrous except at the tip, the inner densely hairy. Stipules on sterile growths up to 1 in. wide and often persistent, toothed. Leaves up to 10 in. long with mostly two to five pairs of leaflets (but sometimes with a single pair and with up to seven or eight in KW 7746); rachis grooved, not winged; petiole with a wide, sheathing base, almost concealing the subtended bud. Leaflets leathery, oblong, mostly 2 to 5 in. long and ¾ to 1¼ in. wide (but even larger on some Forrest specimens), more or less equal, but the basal pair sometimes smaller and more elliptic, apex subacute to obtuse or even retuse at the apex, oblique at the base, dark green and soon glabrous above, undersides (and rachis) at first coated with rusty and some whitish hairs, becoming glabrous except for a few hairs on the midrib, glaucous and papillose, margins often revolute and then apparently entire, but fine serrations evident when the leaflet is flat. Flowers small, creamy white, in convex clusters up to 6 in. wide; inflorescence-branches at first densely hairy, becoming glabrous or retaining a few white, spreading hairs, lenticellate. Petals roundish, about ⅛ in. wide. Styles two or three. Fruits globose, white, pink or purplish red, about ⅜ in. wide, in rather sparse clusters.

Native of the eastern Himalaya, east through upper Burma to S.W. Yunnan and bordering S.E. Tibet; described by Hooker in 1878 from Sikkim specimens but not introduced until Forrest sent seeds in 1912 from the Shweli-Salween divide near the border between Yunnan and Burma (F.9040); raised from this seed it flowered at Caerhays, Cornwall, in May 1924. Forrest later sent seed from other localities in the Burma-China borderland, and there was one sending by Farrer and Cox from the same area in 1921. Kingdon Ward sent seed in 1926 from the Seinghku valley, Burma, near the Assam frontier (KW 6851).

In its foliage, *S. insignis* is the most remarkable of the pinnate-leaved group of *Sorbus*, remote in aspect from the common rowan and suggesting some evergreen species from a subtropical rain-forest. The original Forrest introduction is very tender but a plant at Nymans in Sussex, probably from KW 6851, has survived there for about half-a-century.

The Yunnan-Burma forms of *S. insignis* have generally been known by the name *S. harrowiana*, described from a fruiting specimen collected by Forrest in 1912. Here, following Dr Hutchinson, this species is included in *S. insignis*. It is true that most specimens from Yunnan have fewer and larger leaflets than those of Sikkim (on some Forrest specimens they are 6 in. long and almost 2 in. wide), but the difference is not one on which to base two separate species, and is not constant.

More distinct from typical *S. insignis* (and from *S. harrowiana*) is the sorbus introduced by Kingdon Ward from Mt Japvo in the Naga Hills of Assam in 1927–8. This has long been grown as Sorbus sp. KW7746 (its field-number), but Dr Hutchinson, when working on *Sorbus* in the 1940s, identified it as *S. insignis* and his judgement has been upheld by other botanists. It agrees with *S. insignis* in most of its essential characters, but is certainly untypical in having up to seven or eight pairs of leaflets not more than 3 in. long, with brown hairs persisting on the midrib beneath. The deep pink fruits are borne in clusters of up to 100. It is hardier than the Forrest introductions.

S. 'BELLONA'.—So far a shrub of pyramidal habit about 10 ft high (1979). Winter-buds with rusty-hairy inner scales. Leaflets in three to five pairs, palish matt-green above, grey beneath, glabrous on both sides at maturity, narrowly oblong-obovate, toothed only in the upper one-third to one-half, $1\frac{1}{4}$ to $3\frac{1}{2}$ in. long and up to 1 in. wide. Fruits small, pale pinkish red, in trusses about 6 in. wide. Raised at Kew in 1956 from seeds received as *S. harrowiana*. Award of Merit September 7, 1971. It is evidently a hybrid of *S. insignis* but the other parent is uncertain.

S. 'GHOSE'.—A tree, ultimate height unknown. Young parts clad with a whitish or pale brown woolly indumentum that persists more sparsely on the undersides of the leaflets and in the inflorescence until the autumn. Leaflets in six to eight pairs, oblong or narrowly oblong-lanceolate, up to $2\frac{1}{2}$ in. long, serrate in the upper half, acute at the apex, oblique at the base, light matt-green above; petioles swollen at the base, but not sheathing the buds. Inflorescence about 6 in. wide, many-flowered, with stout main branches; stipules leafy, fan-shaped, toothed, persistent in some seasons. Fruits when nearly ripe dull crimson, eventually redder and glossy, $\frac{1}{4}$ to $\frac{3}{8}$ in. wide.

This interesting sorbus was raised by Messrs Hillier from seeds received from the Indian seedsman Ghose of Darjeeling. It seems to be intermediate between *S. insignis* and *S. foliolosa (wallichii)*. It agrees in many respects with the description of S. ARACHNOIDEA Koehne, described from a specimen collected in Sikkim, and another from Chumbi, just across the border with Tibet, but these collections are not represented in the Kew Herbarium. Koehne likened his species to *S. scalaris*, with which 'Ghose' too has some affinity.

S. INTERMEDIA (Ehrh.) Pers. SWEDISH WHITEBEAM

Pyrus intermedia Ehrh.; *Sorbus suecica* (L.) Hartm.; *Crataegus aria* var. *suecica* L.;
S. scandica (L.) Fries; *Crataegus aria* var. *scandica* L.

A tree 20 to 40 ft, occasionally more, high, sometimes a shrub in the wild; shoots very woolly when young, becoming glabrous by winter. Leaves 2 to $4\frac{1}{2}$ in. long, 1 to 3 in. wide, broadly oval or ovate, tapering or rounded at the base, rounded or pointed at the apex, margins lobed towards the base, the lobes becoming reduced to double or jagged teeth near the apex, ribs in six to nine pairs, upper surface glabrous, polished green when mature, lower surface covered with a close grey felt. Flowers dull white, $\frac{3}{4}$ in. across, produced during May in large corymbs up to 5 in. across; calyx and flower-stalk very woolly. Fruits oval, $\frac{1}{2}$ in. long, red, surmounted by reflexed calyx teeth.

S. intermedia has its principal distribution on the mainland of southern Sweden, with outposts on the islands, in Denmark, and on the southern shores of the Baltic. Long cultivated in Britain, it is sometimes bird-sown in wild habitats, and is said to have become naturalised in a few places. But genuinely wild plants previously identified as *S. intermedia* now rank as distinct species and may in fact be unrelated to *S. intermedia*; see *S. arranensis*, *S. anglica*, and *S. minima*. [PLATE 54

Like *S. hybrida*, the Swedish whitebeam is a tetraploid apomict, coming true from seed and yet most probably of hybrid origin. The Swedish authority Hedlund suspected that it had *S. torminalis* in its make-up, and more recently Liljefors has suggested that it derives wholly from *S. torminalis* and some tetraploid member of the Aria complex, and that *S. aucuparia* consequently does not enter into its parentage either directly or indirectly. Other authorities, however, consider that the deep lobing of the leaves derives from *S. aucuparia*.

Although quite distinct in its fruits, the Swedish whitebeam bears a certain similarity to *S. latifolia* in its foliage, but the lobing is deeper, the lobes are rounded in outline at the apex, and are separated by a narrow sinus which is almost closed at the base. The leaves do not really suggest the influence of *S. torminalis* as those of *S. latifolia* do, but it is interesting that both in *S. latifolia* and *S. intermedia* the hairs on the undersides of the leaves become yellowish or brownish in herbarium specimens.

The Swedish whitebeam makes a broad-crowned, sturdy tree, branched to the base and usually with secondary branches drooping to the ground (though some trees lack these). It is of more value as a picturesque specimen, and for its flowers, than it is for its fruits, which are soon taken by the birds in most seasons. It is a good tree for industrial areas, as it withstands atmospheric pollution very well.

S. intermedia has attained 45 × $7\frac{3}{4}$ ft at Studley Royal, Yorks (1973), and there is another example almost as large at Swallowfield, Berks (1974). A tree in the Edinburgh Botanic Garden, *pl.* 1898, measures 33 × $5\frac{1}{2}$ ft (1967).

S. JAPONICA (Decne.) Hedl.

Aria japonica Decne.; *Micromeles japonica* (Decne.) Koehne *Pyrus commutata* Cardot

A tree to about 70 ft high in the wild, but barely half as high in Britain; branchlets dark brown, lenticellate, at first densely white-woolly; winter-buds conic, glossy, brown or greenish brown. Leaves simple, broadly ovate to obovate, cuneate at the base, 2½ to 5 in. long, lateral veins in eight to eleven pairs, straight and parallel, margins toothed and shallowly lobulate, upper surface dull green, glabrous at maturity, lower surface sparsely to densely clad with white wool. Inflorescence many-flowered, the flowers white and about ⅜ in. across; branches of inflorescence and receptacle white-woolly; calyx lobes acuminate, soon deciduous with the free part of the receptacle. Styles two. Fruits ellipsoid to pear-shaped, red, lenticellate, almost ½ in. long (but orange-yellow or golden yellow, somewhat larger and without lenticels in var. CALOCARPA Rehd.).

Native of Japan and Korea. The plants in cultivation, mostly small and not thriving, probably derive from seed of the var. *calocarpa*, sent by Wilson from Japan.

S. 'JOSEPH ROCK'

A tree so far about 40 ft high in cultivation; branchlets soon glabrous, grey-brown by autumn, with numerous warty lenticels; winter-buds narrow-ovoid, acute, with scattered brown hairs, especially at the edge of the scales and at the tip. Leaves 4½ to 7 in. long, including petiole; leaflets in seven to ten pairs, closely set; rachis grooved above, not winged, soon glabrous. Lateral leaflets narrowly oblong, 1 to 1¾ in. long, ⅜ to ⅝ in. wide, subacute and narrowly apiculate at the apex, obliquely rounded to truncate at the base, incise-serrate almost to the base, glabrous and fairly glossy above, undersurface glabrous by June but earlier with a few brown hairs on the midrib. Flowers in a rather open and irregular inflorescence about 4 in. wide and almost as long, opening in late May or early June; inflorescence-branches lenticellate, at first sparsely clad with brown hairs, glabrous by autumn, by which time they are bright red. Flowers about ⅜ in. wide; receptacle funnel-campanulate, with acute lobes. Stamens with creamy filaments, anthers pink, darkening to purple and finally to black. Fruits globular, about ⅜ in. wide, up to fifty or slightly more in each cluster, at first green, becoming white and finally amber yellow by October but remaining white on the shaded side and always with a pink tinge at the calyx. *Bot. Mag.*, n.s., t. 554.

This rowan, of uncertain origin, is one of the most beautiful of its group and indeed one of the first small trees of any genus to be considered when only a few can be grown. Given a good soil it should attain 25 ft in height and 10 ft in spread in fifteen years. Pretty in flower, its primrose-yellow fruits combined with the splendid autumn colour of crimson, purple and scarlet makes a unique combination, and even in midsummer it is a pleasant tree, graceful in habit and with fresh green foliage enhanced by the lighter green of the young fruits. It received a First Class Certificate in 1962.

The history of 'Joseph Rock' will be found in the article by C. D. Brickell in *Journ. R.H.S.*, Vol. 89 (1964), pp. 19–22. In brief, the original in the R.H.S. Garden at Wisley is believed to have come from the Royal Botanic Garden, Edinburgh, under Dr Rock's collector's number 23657. But the corresponding herbarium specimen is *S. hupehensis (oligodonta)*, and it is doubtful if 'Joseph Rock' could even be a hybrid of that species. J. R. Sealy, who went into the problem of its identity, remarked on its resemblance to *S. ursina* (q.v.), and to the Japanese and Korean *S. commixta* (q.v.).

'Joseph Rock' does not come true from seed, and the name applies only to plants propagated vegetatively and belonging to the original Wisley clone. Caution is needed in deducing its parentage from its seedlings, at least when it grows in large collections, but it may be of significance that of two raised by Messrs Hillier 'TUNDRA' has white fruits and resembles both *S. vilmorinii* and *S. koehneana*, while 'SUNSHINE', apart from its bright yellow fruits, bears a strong resemblance to *S. rehderiana*, a Chinese ally of *S. ursina* not known to be in cultivation. All the evidence, such as it is, suggests that 'Joseph Rock' is a chance natural hybrid, raised from seeds collected by Dr Rock in Yunnan in 1932.

S. KEISSLERI (Schneid.) Rehd.

Micromeles keissleri Schneid.; *Micromeles decaisneana* Schneid., not *Sorbus decaisneana* Zab.; *Pyrus keissleri* (Schneid.) Lévl.

A deciduous tree up to 40 ft high; young shoots felted at first, the lenticels numerous and distinct. Leaves obovate to oval, pointed, tapered at the base, finely toothed, $2\frac{1}{2}$ to 5 in. long, $1\frac{1}{4}$ to $2\frac{1}{2}$ in. wide, of firm rather leathery texture, dark glossy green above, paler beneath, at first shaggy, afterwards quite glabrous on both surfaces, veins in seven to ten pairs; leaf-stalk $\frac{1}{4}$ in. or less long. Flowers white, $\frac{3}{8}$ in. wide, produced in rounded or pyramidal clusters 2 in. across at the end of short leafy twigs; stalks woolly; petals broadly oval; stamens twenty; styles two or three, united near the base. Fruits flattish orange-shaped, $\frac{1}{2}$ to $\frac{3}{4}$ in. wide, dull green, spotted with pale dots; the calyx falls away from the top and leaves a broad circular pit or scar there. Blooms in April.

Native of central, western and southwest China, northern Burma and south-east Tibet; introduced by Wilson in 1904 for Messrs Veitch from W. Szechwan, and reintroduced by Forrest from Yunnan in 1931. This fine species has become rare in gardens. It '. . . is one of the first of the Sorbus to open its buds, in mid-April. The glossy young leaves, of an unusual yellowish green in the early spring, retain as they mature a fresh green colour so that by mid-November the tree looks like an evergreen.' (Fox MS). The plant in the Winkworth Arboretum, to which Dr Fox referred, attained a height of about 20 ft and bore fruits, but died in the late 1960s. The flowers are very fragrant, an unusual feature in this genus, but drop their petals after barely more than a day.

S. keissleri belongs to the section *Micromeles*, but is not closely related to any other cultivated member of that group.

S. KOEHNEANA Schneid.

Pyrus koehneana (Schneid.) Cardot

A deciduous shrub or small tree to about 15 ft high in the wild; young shoots glabrous, turning very dark; scales of bud often hairy. Leaves pinnate, 2 to 4 in. long on the flowering twigs, up to 6 in. long on the barren ones, with mostly eight to twelve pairs of leaflets; rachis slightly hairy at first, grooved and narrowly winged in the apical part. Leaflets oblong to narrowly oblong-ovate, parallel-sided, up to 1¼ in. long, $\frac{3}{16}$ to $\frac{5}{16}$ in. wide, obtuse to subacute and often finely mucronate at the apex, glabrous on both sides except for some hairs on the midrib beneath, sharply and slenderly toothed. Flowers about ⅜ in. wide, borne in lax sparsely hairy clusters 2 to 3 in. long and wide. Calyx-lobes triangular, glabrous outside, downy within. Fruits globose, porcelain white, about ¼ in. wide, the stalks becoming reddish.

Native of western China, extending to western Hupeh, where it was discovered by Henry. Wilson introduced it, probably from near the Hupeh-Shensi border, when collecting for Messrs Veitch (W.1098). Purdom sent seeds from Shensi in 1910, and plants cultivated on the continent probably derive from Giraldi's collections in the same province. A plant of *S. koehneana* was bought by Kew at the winding-up sale of Messrs Veitch's Coombe Wood nursery, but the true species has never been widely cultivated in this country. It is allied to *S. ursina*, but the leaflets are smaller and more glabrous, the hairs when present are white, and the fruits are purer white.

Plants seen under the name *S. koehneana* in some British collections are not that species and probably to be called S. SETSCHWANENSIS Koehne. This is a very glabrous species, with leaves 2 to 4 in. long, the leaflets in up to fifteen pairs, not more than ½ in. long except on strong shoots. Inflorescences very sparsely branched and few-flowered. Fruits white, about $\frac{5}{16}$ in. wide. The provenance of these plants is unknown. They are quite pretty in autumn when the leaves turn russet, but are of ugly, sparsely branched habit, the majority of the leaves being borne on clustered spurs at the ends of the branches. *S. prattii* is in similar style and more ornamental.

Some plants distributed to collections as *S. koehneana*, raised from Forrest seed, are *S. vilmorinii*.

S. LATIFOLIA (Lam.) Pers.   FONTAINEBLEAU SERVICE TREE

Crataegus latifolia Lam.; *Pyrus latifolia* (Lam.) Lindl.;
P. intermedia var. *latifolia* (Lam.) DC.   [PLATE 55

A tree 30 to 45 ft, sometimes over 60 ft high; branchlets downy when young, becoming by winter shining and quite glabrous. Leaves roundish ovate, 2 to 4 in. long, often nearly as wide at the base as they are long, the apex pointed, the base either truncate or broadly wedge-shaped, margin cut into triangular, pointed lobes which are sharply toothed, glabrous, dark lustrous green above, covered beneath with a greyish felt; ribs six to ten on each side; stalk downy, ½ to 1 in. long. Flowers white, ⅝ in. across, borne in corymbs

3 in. wide during May; stalks and calyx very woolly. Fruits globular, $\frac{1}{2}$ in. in diameter, dull brownish red, dotted with large pale lenticels.

This interesting tree is, at least in its typical state, confined to a small area with the Forest of Fontainebleau as its centre, and has been known there since early in the 18th century. Its origin and status has given rise to considerable difference of opinion, but it is now generally considered to be a species of hybrid origin, and is said to come true from seed. One parent is certainly *S. torminalis*. The other must have been some member of the *Aria* complex, but not necessarily *S. aria* itself, since other species of this group may have existed in northern France at the time when it arose. Trees similar to *S. latifolia* have been found in other parts of France and may be of similar though independent origin. *S. latifolia* was at one time confused with *S. intermedia* (q.v.), though the two are really quite distinct.

S. latifolia makes a handsome specimen, with a short trunk and spreading crown, but is uncommon.

S. latifolia is the type of a large group of minor and mostly very local species believed to be apomictic, that derive, as it does, from *S. torminalis* and some members of the *S. aria* group. While some of these are near to typical *S. latifolia*, others differ from it markedly in leaf or fruit or both. Forty-one of these are recognised (but not described) in *Flora Europaea*, the majority from Central Europe. Descriptions are given below of the three British species; of 'Theophrasta', apparently known only in cultivation; and two Hungarian species of recent introduction, the descriptions of which are taken from the work by Z. Karpaty listed in the Bibliography.

S. BRISTOLIENSIS Wilmott—A shrub or small tree to about 30 ft high; branchlets slender, glabrous by autumn, red-brown, sparsely lenticellate; winter-buds small, the scales glabrous except at the margins and the tips, brownish green with darker edges. Leaves of firm texture, oval, rhombic-elliptic or slightly obovate, obtuse at the apex, broad-cuneate at the base, $2\frac{1}{2}$ to 4 in. long, $1\frac{1}{2}$ to 3 in. wide (but on short sterile spurs narrowly obovate, tapered at the base), light green, glabrous, scarcely glossy above, clad beneath with a thin tomentum through which the undersurface shows green, lateral veins straight, in eight to ten pairs, the lower three or four ending in distinct but shallow lobes, which intergrade upward with less distinct lobes and finally with double or simple teeth. Flowers about $\frac{1}{2}$ in. wide, white, in May or early June; inflorescence branches woolly, almost glabrous by autumn and then light brown. Anthers pink. Fruits broadly ellipsoid, the largest almost $\frac{1}{2}$ in. long, bright reddish orange, glossy, sparsely dotted with small lenticels.

A very ornamental native species, known only from the Avon Gorge near Bristol (Leigh Woods and Clifton Down). It is triploid.

S. DEVONIENSIS E. F. Warburg FRENCH HALES.—This was once regarded as the Devon form of *S. latifolia*, which it resembles in its brown, heavily speckled fruits. It is, however, quite distinct in its leaves, which are about one-and-a-half times as long as wide (about as long as wide in *S. latifolia*), acute at the apex, rounded at the base, shallowly lobed. The influence of *S. torminalis* shows in the large, sharp projecting teeth terminating the laterals and the sharpness of the minor teeth. *S. devoniensis* was first recorded in 1797 by Polwhele, who gave its name as 'French Hail', where the second word, like the commoner 'Hales' is undoubtedly a corruption of the French 'alise' for the

SORBUS DEVONIENSIS

fruits of the service trees (*S. latifolia* is 'L'Alisier de Fontainebleau'). The word is translated, or rather taken over, by Chaucer in *The Romaunt of the Rose*— 'notes, aleys and bollas' (nuts, hales and bullace).

S. devoniensis is commonest in north Devon and its fruits were once sold in Barnstaple Market. It also occurs in south Devon, east Cornwall and southeast Ireland. It is a genuine species in the sense that it breeds true from seed, a fact first established in 1888 by T. R. Archer Briggs, author of *The Flora of Plymouth*. See also 'Theophrasta'.

S. KARPATII Boros—A shrub to about 15 ft high. Leaves broadly ovate, to about 3 in. long and 2⅜ in. wide, acute at the apex, rounded at the base, with seven to nine pairs of lateral veins, sharply and irregularly lobed, densely grey woolly beneath. Fruits globose, cinnabar-red, dotted with small lenticels. A native of Hungary, where it is confined to a small area in the southern part of the Vertes mountains, west of Budapest. Introduced late in the 1970s. M. Robert de Belder has this species in his collection at Kalmthout in Belgium, and praises it highly. Whether it is an improvement on our own *S. bristoliensis* remains to be seen.

The second of the Hungarian group to have been introduced, from the same area as *S. karpatii*, is S. PSEUDOVERTESENSIS Boros. This has similar fruits, but the leaves are ovate-oblong, to about 3½ in. long, 2 in. wide, acute or acuminate at the apex, cuneate or slightly cordate at the base, with nine to eleven pairs of lateral veins, obscurely lobed, finely and bluntly toothed, greenish and thinly tomentose beneath.

S. SUBCUNEATA Wilmott—A small tree. Leaves elliptic to rhombic-elliptic, 3 to 4½ in. long, acute at the apex, narrowly to broadly cuneate at the base, sharply toothed, lobed in the upper two-thirds, the lobes triangular, extending one-eighth to one-quarter of the way to the midrib, light green above, greyish white-tomentose beneath. Fruits brownish orange, becoming brown when fully ripe, lenticellate. This sorbus was described by Wilmott in 1934 from the Greenaleigh Woods, Minehead, and has a limited distribution along the coast from there to the wood above Watersmeet near Lynton. It is near to *S. devoniensis* but the leaves are always more or less cuneate at the base (never rounded) and are relatively narrower. A tree at Watersmeet itself, usually identified as 'No Parking', is anomalous and has been identified as both *S. devoniensis* and *S. subcuneata*, but is nearer to the latter.

S. 'THEOPHRASTA'.—A tree growing as large in height and girth as *S. latifolia* and indistinguishable from it in fruit. It is similar to *S. devoniensis*, and quite distinct from *S. latifolia*, in its leaves, but these are not lobed as in *S. devoniensis* but merely double-toothed. They are ovate or elliptic-ovate, 3 to 4½ in. long, ⅞ to 3 in. wide, acute or subacute at the apex, rounded or broad-cuneate at the base, with eight to eleven pairs of veins, dark green and glossy above, greenish grey and closely woolly beneath (*S. devoniensis* 'Theophrasta' Hensen; *S. aria* var. *Theophrasta* Hort.; ?*Pyrus Theophrastii* Hort.; *S. latifolia* Hort., in part, not (Lam.) DC.; *S. mougeotii* Hort., not Soyer-Willemet & Godr.).

This handsome relative of *S. latifolia* was first distinguished as such and fully described by Dr K. J. W. Hensen in 1966 (*Dendroflora* No. 3, p. 62 and fig. 1). The type is a plant received from the Dutch nurseryman P. Lombarts, who listed this sorbus in his catalogue for 1947/8 as *S. Theophrasta*. Mr Lombarts' stock came from a tree at Kew (no. 695), which was received from the Edinburgh Botanic Garden in 1922 as *S. aria* var. *theophrasta*. The tree at Edinburgh still exists, as does its offspring at Kew, but is of unknown provenance. It is perhaps of significance that the Lawson Company of Edinburgh were offering '*Pyrus theophrastii*' as early as 1874, but 'theophrasta'—'food of the gods', in allusion to the edible fruits—is almost certainly more correct. Another plant at Kew, agreeing perfectly with 'Theophrasta', was received before 1913 under the name *Pyrus* (or *Sorbus*) *mougeotii*, though it does not in the least resemble the sorbus to which the name *S. mougeotii* properly belongs.

In describing 'Theophrasta' Dr Hensen suggested that it was the same as the unnamed, cultivated sorbus shortly described by E. F. Warburg in *Flora of the British Isles*, under *S. devoniensis*. Certainly it is a good match for a set of specimens in the Kew Herbarium, from Burgh Heath, Surrey, annotated by Warburg as 'the most commonly cultivated form of *S. latifolia* s.l. [in the broad sense]. I have not yet found a name for it. It is pretty near to S. devoniensis.' There are many other specimens from southern England in the Kew Herbarium, from cultivated or perhaps naturalised trees, which too agree well with 'Theophrasta'.

S. 'Theophrasta' makes a handsome tree, similar at a distance to *S. latifolia*, and with identical fruits ripening at the same time. At Borde Hill in Sussex, where the two grow near together, 'Theophrasta' has made the finer specimen and comes into leaf about two weeks later.

S. × VAGENSIS Wilmott ?*S. rotundifolia* (Bechstein) Hedl.; ?*Pyrus rotundifolia* Bechstein, *nom. illegit.*; ?*S. confusa* Gremli, *nom. ambig.*; *S.* × *confusa* Gremli ex Rouy & Camus, *nom. illegit.*; *Pyrus decipiens* of some authors, not Bechstein— A natural hybrid between *S. aria* and *S. torminalis*, occurring occasionally with the parents, and at one time much confused with the various microspecies of the *S. latifolia* aggregate. Indeed it is impossible to determine to which category an intermediate between *S. aria* and *S. torminalis* belongs unless it has been studied in the field, or unless its breeding behaviour is known. These casual hybrids are often of low fertility, but in compensation for that they sucker freely when coppiced, so that a single plant may eventually give rise to a thicket of considerable extent.

The correct name for *S. aria* × *S. torminalis* is uncertain, for all the older names that have to be considered are either of uncertain application or illegitimate. *S.* × *vagensis*, although recent (1934), is probably the correct name and certainly the best founded, since the type-tree, at Symonds Yat in the Wye Valley, is well known and has been studied. Wilmott, who described it, took it to be a species, believing that the identical younger plants near it were seedlings. These, however, proved to be suckers, and plants raised by Edmund Warburg from seeds of the original tree were not uniform.

S. × *vagensis* is confined in Britain to the Wye Valley, no doubt because *S. aria* and *S. torminalis* so rarely occur together in this country. But this hybrid (called *S.* × *confusa* by French and some other botanists) is reported to be fairly common on the calcareous plateaux of Burgundy and Lorraine, and near Lake Annecy.

S. × *vagensis* is variable in the shape of its leaves, and of no greater ornamental value than *S. devoniensis* or *S. subcuneata*. Of more interest as an ornamental is the sorbus distributed by Messrs Hillier in the 1930s as *Sorbus* × *magnifica* Hesse, from an apparent confusion between this plant and the *S. aria magnifica* of Hesse. Its fruits are not unlike those of *S. torminalis*, which it also resembles in giving good red or yellow autumn colour, but the leaves are scarcely lobed and are loosely woolly beneath. What appears to have been a very similar plant grew in the Münden Botanic Garden in Germany; it was bred from and found to be a hybrid. A plant in the Winkworth Arboretum, Godalming, is probably this "*S.* × *magnifica*" but it seems to be rare in gardens. The original tree, which may have come from Messrs Hesse under some other name, grows in the Sarum Road nursery of Messrs Hillier.

S. MATSUMARANA (Mak.) Koehne
Pyrus matsumarana Mak.

A shrub or small tree; branchlets glabrous; winter-buds about ½ in. long, glabrous except for a few hairs at the edge of the scales. Leaves with four to six pairs of leaflets, which are oblong, ⅞ to 2½ in. long, up to ⅞ in. wide, obtuse and acuminately tipped or subacute at the apex, glabrous, glaucous green above, glaucous beneath, toothed mostly only in the upper half. Flowers in small, glabrous clusters 2 to 3 in. wide. Calyx-lobes with a few rusty hairs at the edge. Petals much longer than the stamens. Styles five. Fruits red, ellipsoid to globose, about ⅜ in. or slightly more long.

Native of the mountains of Japan (Hokkaido and northern part of the main island). It is probably not in cultivation (1979), and judging from herbarium specimens it is unlikely to be of much ornamental value. In previous editions it was stated to have been introduced in 1912, but the plant received by Kew in that year from Messrs Lemoine proved to be not *S. matsumarana* but *S. commixta*. The confusion may have arisen from the fact that plants raised by Späth from seed sent from the Arnold Arboretum as *Pyrus discolor* (or *P. aucuparia* var. *discolor*) were pronounced by Koehne in 1901 to be *S. matsumarana*, at a time when he knew them only as young seedlings. This introduction was in fact *S. commixta*. Another sorbus fairly common in collections, distributed originally by Messrs Marchant as *S. matsumarana*, is also a form of *S. commixta* and is mentioned under that species.

S. matsumarana finds its nearest ally in *S. sitchensis* of N. America.

S. MEGALOCARPA Rehd.
Pyrus megalocarpa (Rehd.) Bean

A small tree up to 25 ft high or a large shrub; young shoots stout, glabrous, reddish, becoming later dull purple, freely marked with lenticels; winter-buds very large, ovoid, ½ to ¾ in. long, viscous and shining. Leaves narrowly oval, sometimes obovate or ovate, broadly wedge-shaped or sometimes rounded at the base, finely and closely toothed, 5 to 9 in. long, 2 to 4½ in. wide, veins parallel, in fourteen to twenty pairs, glabrous on both surfaces except that when young there are tufts of down in the vein-axils; stalk ½ to 1 in. long. Corymbs appearing with or even before the unfolding leaves 4 to 6 in. wide, 3 to 4 in. high, carrying numerous flowers each ¾ in. wide. Petals dull white, round, ¼ in. wide. Sepals broadly triangular, pointed, $\frac{1}{12}$ in. long, glabrous inside, woolly outside, persisting at the top of the fruit. Styles three or four, united below the middle. Flower-stalks woolly when young. Fruits egg-shaped, ¾ to 1¼ in. long, up to ⅞ in. wide, russet-brown, minutely wrinkled. *Bot. Mag.*, n.s., t. 259.

Native of W. Szechwan, China; introduced by Wilson for Messrs Veitch in 1903. It has larger fruit than any of the *Aria* section except *S. lanata* and in foliage is one of the finest. The bark of the branchlets is very dark and its winter buds are remarkably large. It is quite hardy but, on account of its early growth, is liable to injury by spring frosts. The fruits have no beauty but the foliage occasionally turns a good red.

var. CUNEATA Rehd.—Leaves narrowed at the base, very shortly stalked. Fruits smaller, about ½ in. long. Introduced by Wilson in 1910 for the Arnold Arboretum, and commoner in gardens than the typical state. As seen in cultivation the inflorescence is somewhat laxer than in the Veitchian introduction.

S. MEINICHII (Hartm.) Hedl.

S. aucuparia var. *meinichii* Lindeb. ex Hartm.; *S. hybrida* var. *meinichii* (Hartm.) Rehd.

A small tree in the wild, which in gardens has been confused with *S. hybrida*, from which it is obviously distinct in having leaves with five (more rarely four) pairs of free leaflets, which are subacute to obtuse at the apex, narrowed and adnate to the rachis at the base; the upper part of the leaf, which is in effect a terminal leaflet, is lobed, the lobes decreasing in depth upwards, and more or less rhombic in outline. Flowers about ½ in. across. Fruits orbicular, about ½ in. wide, with a bitter flesh.

A native of southern and western Norway. It is a tetraploid apomict, considered by Liljefors to have three sets of chromosomes from *S. aucuparia* and one from the *Aria* group. It was introduced by Messrs Hillier in the late 1970s and so far as is known had not previously been in cultivation in Britain. A sorbus once in commerce under the name is *S. hybrida*, and there has also been confusion between *S. meinichii* and seedling forms of *S.* × *thuringiaca* (*S.* × *semipinnata*).

Similar to *S. meinichii* is S. TEODORI Liljefors, described in 1953 and named in honour of Teodor Hedlund, the authority on *Sorbus*. It has mostly four pairs of free leaflets with broader bases, acute at the apex, a terminal 'leaflet' narrowly cuneate at the base, and sweet fruits narrowed towards the apex. It is a triploid apomict with two sets of chromosomes from *S. aucuparia*, one from the Aria group, and is confined to Faro Island, Götland, Sweden. It is in cultivation in the Edinburgh Botanic Garden.

S. MELIOSMIFOLIA Rehd.

Pyrus meliosmifolia (Rehd.) Bean

A deciduous tree 25 to 35 ft high; young shoots glabrous, purplish brown; winter buds glabrous. Leaves ovate-elliptical, slenderly pointed, tapered at the base, more or less doubly toothed, 4 to 7 in. long, half as much wide, green on both surfaces, woolly on the midrib and veins beneath; veins parallel, in eighteen to twenty-four pairs; stalk ¼ in. or less long. Flowers white, hawthorn-scented, borne in April in dense corymbs 2 to 4 in. wide; inflorescence-branches sparsely hairy at flowering-time. Anthers pinkish brown. Fruits nearly globose, about ⅜ in. long, bronze-coloured and dotted with lenticels, ripening late; calyx deciduous, leaving a rounded depression at the apex.

S. meliosmifolia was discovered by Wilson in October 1910 growing in woodland near Mupin in W. Szechwan, and is cultivated, though not widely, from the seeds he collected on that occasion (W.4221). It is very closely allied to *S. caloneura*, described three years earlier, and a monographer of the genus might well decide to unite them. The differences are that in *S. meliosmifolia* the leaves are somewhat broader, less narrowed at the base, shorter-stalked and with more numerous pairs of veins. Differences between the cultivated plants of the two species given by Dr Fox are that *S. meliosmifolia* has much larger buds—¾ to 1 in. long against half that length in *S. caloneura*—and comes into leaf earlier.

The specific epithet refers to the resemblance of this species to leaves of many species of *Meliosma*, with their regular pattern of ridges and furrows.

S. MICROPHYLLA Wenzig *emend.* Hedl.
S. rufopilosa Schneid.

A tree usually under 20 ft in the wild, or a shrub; branchlets glabrous by autumn, purplish brown; winter-buds ovoid, about ¼ in. long, glabrous except for a tuft of brown hairs at the tip. Leaves with mostly ten to fifteen pairs of leaflets; rachis narrowly grooved, winged in the apical part, clad beneath with pale brown or whitish hairs, or the two intermixed. Leaflets oblong, up to ⅝ in. long and about half as wide, acute, deeply and sharply toothed throughout, glabrous or often somewhat hairy above, more densely so beneath, especially on the midrib. Flowers pink or even red, few in a narrow, lax cluster; inflorescence-branches brownish red, clad with brown or whitish hairs, which are rather denser on the pedicels, which, like the main branches of the inflorescence, are conspicuously lenticellate. Fruits white or pinkish, globular, ⅜ in. or slightly more wide.

Native of the rainier parts of the Sino-Himalayan region from Nepal to Yunnan; described by Wenzig mainly or wholly from specimens collected by Hooker and Thomson in the interior of Sikkim in 1849. Ludlow and Sherriff collected seeds of this species on at least one occasion, but the few plants in cultivation under the name *S. microphylla* were raised from seeds collected in Nepal in the late 1960s and early 1970s; these have not yet been seen in flower. This species gives good autumn colour in the wild.

In the western Himalaya (Kashmir, Simla region and perhaps Nepal) there is a small-leafleted species of *Sorbus* which may be specifically distinct from *S. microphylla*. It is more glabrous in all its parts; the leaflets are fewer; and the pedicels are not lenticellate. The colour of the flowers is not stated on the specimens seen. It is represented in the Wallich Herbarium by a specimen collected by Webb southeast of Simla early in the last century, to which the name *Pyrus microphylla* is attached in the catalogue (No. 676). This name is of no validity unless published with a description. It has been assumed that the name *S. microphylla* Wenzig is based on it, but he did not cite it and may not even have seen a specimen of No. 676; certainly his description is not made from the Webb specimen, but agrees with the Hooker specimens from Sikkim, two of which he did cite. These are *S. rufopilosa* Schneid., which therefore becomes a synonym of *S. microphylla* Wenzig.

S. MINIMA (Ley) Hedl.

Pyrus minima Ley; *Sorbus intermedia* var. *minima* (Ley) Bean

A shrub to about 10 ft high in the wild, with slender, spreading branches. Leaves simple, elliptic or oblong-elliptic, 2⅜ to 3¼ in. long, about half as wide, narrowed at the apex, cuneate to rounded at the base, glabrous above at maturity and dull green, grey-tomentose beneath, lateral veins in mostly eight or nine pairs, margins sharply serrated, lobed, the lobes extending about one-quarter of the way to the midrib; petiole about ½ in. long. Flowers white, barely ⅜ in. wide, in narrow, round-topped clusters; the whole inflorescence white-tomentose, but the receptacles rather more densely so than the peduncles and pedicels. Anthers cream-coloured. Fruits scarlet, freely borne, globose to oblate, ⅜ in. or slightly less wide, sparsely dotted with lenticels; calyx-lobes still dry when the fruit is ripe.

S. minima is a British endemic, known only from limestone cliffs above Crickhowell in Breconshire and one locality about two miles farther west; described by Ley in 1895. It is an apomictic triploid species, deriving from *S. aucuparia* and some member of the Aria group, probably *S. rupicola*.

S. ARRANENSIS Hedl. *S. intermedia* var. *arranensis* (Hedl.) Rehd.—Very like the preceding in foliage, but the leaves less straight-sided in outline, elliptic or rhombic-elliptic, more deeply lobed, the lobes extending one-half to three-quarters the way to the midrib and on the average with one more pair of lateral veins. There is little difference in the flowers or fruits. *S. arranensis* is an endemic of the Isle of Arran known since the early 19th century but confused with *S. intermedia* and *S. hybrida* until Hedlund distinguished it. It too is a triploid apomict probably of the same parentage as *S. minima*. Although now extinct there *S. rupicola* is known to have grown on the island in the last century.

Although not ranking high as an ornamental these two species are of graceful habit and attractive both in flower and fruit besides being of interest as endemic British species with no counterpart on the continent except in Norway. Of the Norwegian species S. LANCIFOLIA Hedl. is in cultivation in Britain and has about the same garden value as our species.

Another British endemic in this group is S. LEYANA Wilmott known only from one locality in Breconshire. For this see Clapham Tutin and Warburg *Flora of the British Isles*.

S. PALLESCENS Rehd.

S. ochrocarpa Rehd.

A tree to about 40 ft high in the wild or a large shrub; branchlets woolly at first, glabrous and brown by autumn, sparsely lenticellate; winter-buds ovoid, acute, about ¼ in. long, glabrous except for white hairs at the tip, with a few, rather lax scales. Leaves simple, of moderately thin texture, elliptic or elliptic-obovate, acute or acuminate at the apex, 3 to 4¾ in. long, 1¼ to 1¾ in. wide, glabrous, dark green and slightly glossy above, undersides clad with a white felt which is at first fairly dense, becoming thinner, the leaves being silvery green beneath by autumn, lateral veins straight, in ten to thirteen pairs, margins serrate, the teeth ending the main veins slightly longer than the others; petiole ½ to ⅝ in. long. Flowers white, about ⅜ in. wide, opening in late April or early May, borne ten to twenty in clusters about 3 in. wide; inflorescence-branches slender, sparsely lenticellate, at first hairy, glabrous by autumn. Styles two to five. Fruits globular, about ⅜ in. wide, with scattered lenticels, ripening late in the autumn, turning yellow before they decay; calyx-lobes persistent, reflexed.

A native of China, where it is widespread; discovered by Wilson in 1908. It is said to be very variable in the wild. The above description is made from a tree at Borde Hill in Sussex, raised from Forrest's seed-number 29044, collected in Yunnan shortly before his death. It measures 55 × 4 ft (1976) and there is an example at Westonbirt from the same batch of seed, which is 38 × 4¼ ft (1976).

S. pallescens makes a vigorous, leafy and freely branched tree, which is more than can be said of other East Asiatic members of the section *Aria*. The leaves are light bronze in spring and remain green and fresh until late autumn.

S. POHUASHANENSIS (Hance) Hedl.

Pyrus pohuashanensis Hance; *Pyrus aucuparia* var. *discolor* Hort. ex Rehd.;
?*S.* × *kewensis* Hensen.

A small tree in the wild, under 30 ft high, or a shrub; young shoots hairy, becoming reddish brown and glabrous; winter buds ovoid, densely to sparsely coated with white silky hairs, ¼ to ⅜ in. long. Leaves up to about 7 in. long including petiole 1¼ in. long, with five to seven pairs of leaflets; rachis broadly grooved above, white-hairy. Lateral leaflets oblong-elliptic or oblong-lanceolate, mostly 1¾ to 2¼ in. long, about ⅝ in. wide, acute at the apex, more or less oblique at the base, but the basal pair often much shorter and more obtuse than the others, sharply toothed in the upper half or two-thirds (but with coarse, spreading and partly double teeth in a sterile specimen collected by Purdom), glabrous above, undersurface whitish with darker reticulations, sparsely hairy, more densely so on the midrib, becoming nearly glabrous by autumn. Stipules under the inflorescence fan-shaped, toothed, up to about ½ in. long, sometimes much smaller or even lacking. Inflorescence dense, much branched, up to 5 in. wide, its branches slender, not lenticellate, sparsely to densely white-hairy. Flowers white, opening in May; receptacle sparsely hairy to almost glabrous. Petals broadly ovate, hairy on the inner surface. Stamens about as long as the petals. Fruits red, globose, about ⅜ in. wide.

A native of the mountains of northern China; discovered by Dr Bretschneider on the Pohuashan, a mountain about 60 miles west of Peking, in 1874 and introduced by him in 1882. It was reintroduced by Frank Meyer from the Hsiao-Wutai-shan, another of the mountains west of Peking, in 1913.

It is uncertain whether the true species is now in cultivation in Britain; at least there seem to be none that can be traced back to a direct introduction from the wild. Some plants under the name are a form of *S. esserteauiana*. Others really agree quite well with *S. pohuashanensis* except in being more robust, but they were apparently first distributed as "*S. conradinae*", properly a synonym of *S. esserteauiana*, which suggests that they were raised from seed of that species, the pollen-parent being probably *S. aucuparia*. The true *S. pohuashanensis* is related to both these species, and something very like it might arise from garden crosses between them.

S. 'KEWENSIS'.—A tree under the name *S.* × *kewensis* received an Award of Merit when exhibited by Kew in 1948 as a probable hybrid between *S. esserteauiana* and *S. pohuashanensis*. No description was published and no specimen was deposited in the Kew herbarium; nor, so far as is known, does this tree exist any longer. For the plants distributed by Messrs Hillier as 'Kewensis', see 'Chinese Lace' below.

S. × **KEWENSIS** Hensen—A tree received from Kew by the Wageningen Arboretum in 1948 was considered by Dr Hensen to be a hybrid between *S. pohuashanensis* and *S. aucuparia*, differing from the former in the small stipules under the inflorescence. But judging from wild material in the Kew Herbarium, the stipules in *S. pohuashanensis* are often very small, or even lacking; Roy Lancaster took the trouble during a visit to China in 1979 to examine the specimens of *S. pohuashanensis* in the Peking Herbarium, and was able to confirm this conclusion. For the *S. pohuashanensis* of the *Botanical Magazine*, identified by Dr Hensen as his *S.* × *kewensis*, see 'Pagoda Red' below.

S. 'PAGODA RED'.—A very small and slow-growing shrubby tree, agreeing with *S. pohuashanensis* except that the leaflets have coarse, spreading teeth and that there appear to be no stipules at all under the inflorescence. It agrees rather better with *S. amurensis* Koehne of the Amur region and Korea, but this species is allied to *S. pohuashanensis* and included in it by Dr Yü. The provenance of this plant is uncertain, but it is a good match for Wilson 9154, collected by Wilson during his expedition to Korea. 'Pagoda Red' received an Award of Merit when shown from Kew in 1971, and is portrayed in *Bot. Mag.*, n.s., t. 133, as *S. pohuashanensis*.

S. 'CHINESE LACE'.—A seedling of either 'Pagoda Red' or of another similar plant once grown at Kew, raised by Messrs Hillier and originally distributed as *S.* × *kewensis*. It is an interesting large shrub with abnormally narrow, jaggedly toothed leaflets. Inflorescences with fan-shaped stipules to about ¼ in. long and ⅜ in. wide. Fruits orange-red, glossy, very many in trusses to 6 in. wide.

S. POTERIIFOLIA Hand.-Mazz. *emend.* Hand.-Mazz.

S. reducta Hort., in part, not Diels; *S. pygmaea* Hutch. ex Bean in R.H.S. *Dict. Gard.*, Vol. IV, p. 1988 (1951), *anglice*

A dwarf or prostrate shrublet 6 to 12 in. high, its stems rooting here and there, covered with a smooth, grey bark; winter buds ovoid, about $\frac{3}{16}$ in. long, crimson, hairy towards the tip; young shoots very short, thinly hairy at first.

Leaves $1\frac{5}{8}$ to $3\frac{1}{4}$ in. long, with four to seven pairs of leaflets; rachis scarcely winged, broadly grooved above. Leaflets oval, broadly oblong-elliptic to ovate-elliptic, $\frac{1}{4}$ to $\frac{1}{2}$ in. long and $\frac{3}{16}$ to $\frac{5}{16}$ in. wide, very sharply serrate, glabrous. Inflorescence with three to five pink flowers, crowded at the top of a hairy peduncle, the flowers sessile or on short stalks. Receptacle glabrous, purplish, obconical, about $\frac{1}{8}$ in. long, its lobes ovate-triangular, also glabrous. Petals orbicular, about $\frac{1}{12}$ in. wide. Stamens ten. Styles three to five. Fruits about $\frac{3}{8}$ in. wide, glabrous.

A native of northwest Yunnan and northern Burma at altitudes of 10,000 to 13,500 ft. The type-specimen of *S. poteriifolia* was collected by its describer, the Austrian botanist, in 1916, on the Irrawaddy—Salween divide, at 12,500 to 13,000 ft, growing on micaceous schist, but Forrest had in fact collected it two years earlier, and did so on several occasions later. It was apparently not introduced until Kingdon Ward collected seeds in the Seingkhu valley, northwest Burma, in 1926 at 11–12,000 ft (KW 6968). He saw it in flower towards the end of June, and when he returned to collect seeds 'its numerous clusters of reddened berries presently turned snow white, beading the crinkly black stems like moonstones.' He sent seeds five years later from the neighbouring Adung valley, and also collected it during his last expedition to Burma in 1953.

Plants from KW 6968 were raised at Kew and one of these, six inches high, was shown to the Scientific and Floral B Committees of the Royal Horticultural Society on July 13, 1943, already in ripe fruit. At that time Dr Hutchinson considered the plant to be *S. reducta* (q.v.) and it was exhibited under that name. He later concluded that it was a new species, differing from *S. reducta* in its almost glabrous calyx-lobes, pink flowers and white fruits. It is also dwarfer, and semi-herbaceous. He had intended to publish a description of this new acquisition under the name *S. pygmaea* but he abandoned his studies of *Sorbus* later in 1943 and never did so, though the name was imparted to gardeners and came into use for the few plants then in cultivation. It has since proved, however, that *S. pygmaea* would have been a superfluous name for *S. poteriifolia*.

S. poteriifolia might have become extinct in gardens had not Harold Hillier kept some plants on his rock garden at Jermyns House, regrafting them whenever they had shown signs of weakening. Dr Hugh McAllister has raised seedlings from them in the Liverpool Botanic Garden, which agree perfectly with the parent, so this interesting species is becoming more widely available. The seedlings grow much better than the grafted parents.

As an authority on the classification and phylogeny of flowering plants, Dr Hutchinson was interested in *S. poteriifolia* and *S. reducta* as recently evolved and advanced species of the genus, approaching the herbaceous habit and with half the normal number of stamens.

NOTE. As first described by Handel–Mazzetti, *S. poteriifolia* included the element later separated by him under the name *S. filipes*, not treated here. The description given here also excludes specimen Forrest 20266, which Handel–Mazzetti continued to include in *S. poteriifolia* (see text). For the *S. poteriifolia* of gardens, see S. 'McLaren D.87', mentioned below.

S. 'MᴄLᴀʀᴇɴ D. 84' *S. poteriifolia* Hort., not Hand.-Mazz.—A tree so far about 30 ft high in gardens; edges of bud-scales, young growths, rachis and midribs of leaflets beneath all sparsely furnished with white hairs, the foliage becoming glabrous. Leaflets in mostly seven or eight pairs, 1 to 1⅜ in. long, about ½ in. wide, oblong, sharply toothed to the base. Inflorescences broad, many-flowered, rather lax, the main branches up to 2 in. long, borne on clustered spurs. Stipules under the inflorescence awl-shaped to semi-ovate, up to ¼ in. long, deciduous. Flowers pink. Fruits globose, deep rosy pink, about ⅜ in. wide.

This rowan was raised from seeds gathered by one of Forrest's collectors, some of whom continued to work for some time after his death at Lord Aberconway's expense. The seed was distributed as *S. poteriifolia*, but it is not that species. There are matching specimens among Forrest's collections in Yunnan, especially from the Mekong-Salween divide, taken from trees up to 30 ft high, some with more numerous leaflets than in the cultivated plants. This rowan received an Award of Merit for its fruits when shown by the Crown Estate Commissioners, Windsor Great Park, on September 21, 1951.

The correct name for this species is uncertain, but it may be *S. monbeigii* (Cardot) Hand.-Mazz.

S. PRATTII Koehne

S. pogonopetala Koehne; *S. munda* Koehne; *S. munda* f. *tatsienensis* and f. *subarachnoidea* Koehne; *S. pratti* var. *tatsienensis* (Koehne) Schneid.; *S. prattii* f. *subarachnoidea* (Koehne) Rehd.

A tree up to 20 ft high; winter-buds ¼ to ⅜ in. long, they and the young shoots clad with long rusty hairs. Leaves 3¾ to 5½ in. long, including petiole, with ten to fourteen pairs of leaflets; rachis grooved, slightly winged towards the apex of the leaf, covered beneath with a loose tomentum of pale brown, cottony hairs. Leaflets mostly ⅝ to ¾ in. long, ¼ to 5/16 in. wide, the lateral leaflets rounded to subacute at the apex, rounded at the base, medium green above, underside rather glaucous, cobwebby on the blade and with a tomentose midrib, later nearly glabrous, margins rather deeply toothed. Inflorescence long-pedunculate, lax, sparsely hairy to almost glabrous. Flowers white, about ⅜ in. wide; receptacle campanulate, the calyx-lobes broadly deltoid, short, rather thick, with an apiculate apex. Petals obovate to orbicular, slightly hairy inside at the base. Styles five. Fruits globose, about ⅜ in. wide, green when immature, becoming pearly white. *Bot. Mag.*, t. 9460.

Native of W. Szechwan and E. Kansu, perhaps also of N.W. Yunnan. It varies slightly in the size and number of its leaflets and the degree of their hairiness, but the varieties and independent species made out of these variations are not marked enough to merit recognition. *S. prattii* was introduced by Wilson in 1910 during his second expedition for the Arnold Arboretum (W. 4323). F.R.S. Balfour of Dawyck, who organised the British contribution to the financing of this expedition, gave most of his share of the seeds to Harry White of the Sunningdale Nurseries, and plants were also distributed by Vicary Gibbs of Aldenham, where *S. prattii* fruited in 1922. The description above is based on material from a plant at Dawyck, figured in the *Botanical Magazine*.

S. prattii makes an attractive small tree or large shrub, differing most obviously from *S. vilmorinii* in its pure white fruits, which do not go through a

pink phase. The flowers open in May and the fruits are usually white by
September. The cultivar-name 'Aeolus' has been given to a plant at Kew, which
received an Award of Merit for its fruits in 1971. It does not differ significantly
from the plant described above.

S. REDUCTA Diels

A dwarf shrub 6 in. to 2 ft high in the wild, spreading by underground runners
and forming an underscrub in thickets or amongst rocks in open alpine meadows;
annual growths short, thinly hairy at first, becoming glabrous and grey or
grey-brown, with a few large lenticels; winter buds narrow-ovoid, acute,
brownish red, almost glabrous except at the tips, about ¼ in. long. Leaves to
about 4 in. long including petiole, with four to seven pairs of leaflets; rachis
deeply grooved, it and the petiole tinged with brownish red. Lateral leaflets
oblong-elliptic to oblong-ovate, obtuse to abruptly acute at the apex, cuneate
at the base, ¾ to 1¼ in. long (on cultivated plants), about ⅜ in. wide, sharply and
rather deeply toothed, upper surface dark green, reticulate, glossy, glabrous
except for scattered whitish hairs at first, undersurface almost glabrous but
with scattered brownish hairs when unfolding, some of which persist for a time.
Stipules awl-shaped. Flowers white, about ⅜ in. wide, borne in May, few together
in lax corymbs; inflorescence branches glabrous or sparsely hairy; pedicels very
short. Receptacles glabrous, calyx-lobes ovate-triangular, acute, hairy inside near
the reddish apex. Stamens reduced in number, mostly about ten; styles and
carpels three to five. Fruits globose, pink, about ¼ in. wide.

S. reducta was discovered by Forrest in 1906 on the eastern flank of the
Lichiang range of Yunnan, China, but the cultivated plants derive from seeds
collected by Dr T. T. Yü in 1937 (Yü 14439). The Kingdon Ward introduction
originally named S. reducta is S. poteriifolia (pygmaea), q.v. It is easily grown
in any good, moist soil, soon forming a clump 3 ft or more across, but on poor
soils it is inclined to wander, sending up a shoot here and there. The leaves
usually turn bronze before falling.

S. SAMBUCIFOLIA (Cham. & Schlecht.) Roem.
Pyrus sambucifolia Cham. & Schlecht.

A shrub to about 10 ft high, sometimes reduced to a low, creeping shrublet;
branchlets prominently and densely lenticellate; winter-buds narrow-ovoid,
glabrous and glossy outside, the inner scales ciliate. Leaves with five to six pairs
of leaflets (fewer in some Russian forms). Leaflets 2 to 3½ in. long, up to 1 in.
wide, broadly to narrowly lanceolate, tapered to an acute apex, glabrous except
for hairs at their edge and sometimes on the midrib beneath, coarsely toothed.
Inflorescence few-flowered, sparsely brown-hairy or glabrous. Flowers large,
to about $\frac{9}{16}$ in. wide, white or pink-tinged. Styles mostly three. Fruits globose,
about ½ in. wide, red, with a sweet flesh, crowned with persistent, erect, leathery
calyx-lobes.

Native of the Aleutians, thence west and south through Kamchatka and the shores of the Sea of Okhotsk, Anadir, Sakhalin, the Kuriles to central Japan. It is absent from North America proper, but its name has been applied to many American species, especially to *S. decora* and *S. scopulina*. The true species is not known to be in cultivation in Britain, and is unlikely to be of much ornamental value, even if it thrives, as species from the region it inhabits rarely do in our climate. Among species of the section *Sorbus* (*Aucuparia*) it is apparently unique in having leathery and erect calyx-lobes in the fruiting state.

S. SARGENTIANA Koehne

A tree said to grow to 50 ft high in the wild; young growths stout, grey-brown or fawn by winter, about ½ in. thick, downy at first; winter-buds not unlike those of a horse-chestnut, ovoid, up to ⅞ in. long and ½ in. wide, often so densely varnished that the scales are scarcely distinguishable. Leaves pinnate, up to 12 in. or slightly more long (including petiole 2 to 3 in. long), with four to six pairs of leaflets; rachis red on the exposed side, slightly woolly at first beneath, narrowly grooved. Leaflets oblong-lanceolate, acuminately tapered to a sharp apex, up to 5 in. long and 1¾ in. broad, with very numerous closely set pairs of lateral veins (mostly twenty to twenty-five pairs), impressed above, prominent beneath, upper surface light green, the lower whitish and thinly downy, margins finely serrated except in the lower quarter. Stipules very large and leafy, persistent, broadly obovate-kidney-shaped, coarsely toothed, present both on extension growths and in the inflorescence. Flowers borne in early June, individually small (about ¼ in. wide) but crowded in flattish, white-hairy inflorescences up to 10 in. wide. Petals fugitive, orbicular. Styles three or four. Fruits flattened-globose, about 5/16 in. wide, matt orange-scarlet, in flattish more or less circular clusters, up to 500 or even more in each.

A native of W. Szechwan, China, described from specimens collected by Wilson in 1908 and 1910 for the Arnold Arboretum, after whose then Director it is named. It was also introduced by Wilson, and the plants cultivated in Britain probably derive from his 1910 collection in the Mupin area (W.4207). All things considered, *S. sargentiana* is the finest of the rowan group and was awarded a First Class Certificate in 1956. The handsome leaves, with larger leaflets than in any other species except *S. insignis*, are mahogany-coloured when they unfold in late May, and colour brilliantly—usually orange-red—in early November. The display of fruits is truly spectacular, but cannot be fully appreciated unless the tree is given a prominent position in nearly full sun. Even in winter it is a striking sight, with its large, richly coloured buds, some so glutinous that they seem to be encased in aspic. Its only fault is the stiff, open and sparsely branched habit. The heavily scarred spurs are a remarkable feature; they may remain on the tree for twenty years or more, gaining a few inches in length each season. Young grafted plants are sometimes so well spurred that they look like well pruned cordon-apples and may fruit to excess without making extension growths. These can be induced, however, by removing the terminal buds from some of the spurs to encourage the lateral buds to grow out.

Perhaps the finest examples of *S. sargentiana* in Britain are the fine pair at

Westonbirt in Gloucestershire, in Broad Drive. They measure 34 × 4¾ ft and 39 × 3¼ ft (1974).

S. SCALARIS Koehne

Pyrus scalaris (Koehne) Bean; *S. pluripinnata* (Schneid.) Koehne; *S. foliolosa* var. *pluripinnata* Schneid.

A tree with a spreading crown, up to 35 ft high in the wild; young shoots coated at first with white or greyish hairs, becoming glabrous, dark grey in their second year; winter-buds ovoid, about ⅜ in. long, the outer scales glabrous, the inner white-hairy. Leaves up to 8 in. long including petiole, with ten to sixteen closely set pairs of leaflets; rachis deeply grooved, winged in the apical part of its length, at first rosy purple, later green tinged with red, grey-hairy. Lateral leaflets narrowly oblong, sharply or bluntly acute at the apex, ½ to 1½ in.

SORBUS SCALARIS

long, ¼ to ⅜ in. wide, toothed in the upper half or one-third, sometimes almost entire, dark green and soon glabrous above, covered beneath with a whitish, cobwebby indumentum. Stipules large, toothed, up to ½ in. wide, present both on strong growths and under the inflorescence. Flowers dull white, opening in late May or early June, about ¼ in. across, in clusters up to 7 in. wide, all the inflorescence branches grey-woolly. Anthers cream-coloured. Styles usually three. Fruits globose, about ¼ in. wide, bright red, ripening in October, up to 200 or so in each cluster. *Bot. Mag.*, n.s., t. 69.

Native of W. Szechwan, China; discovered and introduced by Wilson in 1904. It is a species of great character but not commonly met with outside collections. The fern-like foliage is brownish crimson or bronze when it expands in April and renders the tree unmistakable when mature, with its narrow, dark green, closely set leaflets. The autumn colour is unreliable but the leaves turn orange-yellow on some soils before falling late in the autumn. The fruits are borne on spurs all along the branches, and are usually among the last to be taken by birds. As usually seen, *S. scalaris* is a short-trunked tree with spreading branches. It received an Award of Merit in 1934.

S. SITCHENSIS Roem.
Pyrus sambucifolia of some authors, not Cham. & Schlecht.

An erect many-stemmed shrub 5 to 15 ft high; young growths rusty-hairy, becoming glabrous and purplish, sparsely lenticellate. Leaves up to 8 in. long, including petiole; rachis sometimes crimson, obscurely grooved, usually glabrous except for hairs and glands at the base of the leaflets, which are in three to five pairs, oblong-elliptic, up to 2 in. long and $\frac{7}{8}$ in. wide, abruptly acute or rounded at the apex, sharply and rather deeply toothed in the upper half or two-thirds, glabrous on both sides except for some rusty hairs on the midrib beneath. Inflorescence rounded, with up to eighty flowers, its branches rusty-hairy. Flowers white with more or less orbicular petals; receptacle glabrous or slightly downy. Fruits described as red with a glaucous bloom; they are ellipsoid to globular, about $\frac{3}{8}$ in. wide.

Native of western N. America from Alaska and south Yukon east to Montana and Idaho; described from specimens collected by Mertens in south Alaska. It differs from its geographical neighbour *S. scopulina* in having brown rather than white hairs, and in the fewer leaflets, blunt or even almost truncate at the apex and less fully toothed. The inflorescence too is smaller and rounder. Its nearest ally appears to be the Japanese *S. matsumarana*.

var. GRAYI (Wenzig) C. L. Hitchc. *S. sambucifolia* var. *grayi* Wenzig; *S. occidentalis* S. Wats.—Leaflets toothed only at the apex. Native mainly of British Columbia, Washington and Oregon.

S. THIBETICA (Cardot) Hand.-Mazz.
Pyrus thibetica Cardot

A tree to 50 ft high in the wild, but sometimes a shrub. It is allied to *S. cuspidata* but with often more slender branches and with smaller leaves, to about 5 in. long, variable in shape as are those of *S. cuspidata* but more often almost orbicular, and clad beneath with a looser tomentum (not almost plastered as in *S. cuspidata*); in some specimens identified as *S. thibetica* they are almost glabrous beneath. The flowers are somewhat smaller, and the styles are two or three in number (rarely fewer than four in *S. cuspidata*). The fruits, according to collectors' notes are yellow with a red flush, or 'green-orange'.

S. thibetica was described from a specimen collected by the French missionary Soulié in 1895 in northwest Yunnan near the Tibetan border, and a photograph of the type-specimen is reproduced by Eleanora Gabrielian in her work on *Sorbus*, plate 42. Cardot described the fruits as dark red, a faulty inference as they were obviously immature (green, unripe fruits in this group become very dark when dry). *S. thibetica* appears to be fairly common in Yunnan, and to extend through Burma and the eastern Himalaya as far as central Bhutan (but other specimens from Bhutan seem to be *S. cuspidata*, which for its part extends as far as Burma at least).

Plants are in cultivation under the name *S. thibetica* KW 21175, raised from seeds collected by Kingdon Ward during his expedition in 1953 to The Triangle (the mountains between the two upper branches of the Irrawaddy). This has light brown young branchlets, they and both sides of the leaves thinly filmed with cobwebby hairs. The leaves are elliptic to broadly so or roundish, cuneate to almost truncate at the base, to about 5 in. long, shortly stalked, lateral veins in nine to ten pairs, straight and parallel, not much branched. Flowers very small and few in the inflorescence. Styles one to three, free. Fruits globular or broad-ellipsoid, up to ¾ in. long, yellow or orange when ripe in late October, but quickly decaying to toffee-coloured, lenticellate. This tree really agrees better with *S. wardii* Merr., described from a flowering specimen collected by Kingdon Ward on an earlier expedition to Burma, but this species has been submerged in *S. thibetica* by Yü.

With a rather better claim to be called *S. thibetica* is a sorbus in the Westonbirt Arboretum, named *S. 'Mitchellii'* ('JOHN MITCHELL')*. The young stems are soon glabrous, very dark in their second year. Leaves broadly elliptic, ovate-elliptic or roundish, 4 to 5 in. long, 3⅜ to 4¼ in. wide, truncate to rounded at the base, dark green and at first woolly above, silvery tomentose beneath, the felt looser than in *S. cuspidata*. Inflorescences 3 to 4 in. wide, loosely woolly. Flowers about ¾ in. wide. Styles mostly two. Fruits brown when fully ripe in October, globular, about ⅝ in. or slightly more wide, lenticellate.

The original plant at Westonbirt was planted in 1938 and measures 64 × 4¾ ft (1975). It was raised there from seeds, which may well have been collected in Yunnan, since 'John Mitchell' closely resembles specimens collected there by Forrest and others. It is believed that the seeds were received as *Pyrus vestita* (*S. cuspidata*), but Forrest's collections of *S. thibetica* were originally identified as that species.

This fine tree is named in honour of William John Mitchell, VMH, who was for many years curator of the Westonbirt Arboretum. There is an avenue in the Royal Horticultural Society's Garden at Wisley, and also a specimen there on Weather Hill.

* The name 'Mitchellii', being in Latin form and published after January 1, 1959, is not in accordance with the *International Code of Nomenclature of Cultivated Plants*. The rendering 'John Mitchell' is proposed here.

S. × THURINGIACA (Ilse) Fritsch

Pyrus thuringiaca Ilse; *S. semipinnata* (Roth) Hedl., not *Pyrus semipinnata* Roth (1827) nor *P. semipinnata* Bechst. (1821); *S. hybrida* of some authors, in part, not L.; *Pyrus pinnatifida* of some authors, in part, not Ehrh.

In areas where they grow together, *S. aria* and *S. aucuparia* occasionally hybridise, producing plants that in the first generation are intermediate and mixed in their foliage, having some leaves that are deeply lobed in the basal part or even with a pair of free leaflets at the base, and are also intermediate in leaf indumentum, fruits and winter-buds. Such hybrids occur as rare individuals in the company of the parents; like these they are diploid; and do not breed true when raised from seed. Such hybrids have been confused with *S. hybrida* L., which is a tetraploid apomictic species with a restricted distribution in Scandinavia (for the differences between it and the cultivated form of *S.* × *thuringiaca*, see below). The difference between *S. hybrida* and the trees found wild in Britain and on the continent was pointed out by Boswell-Syme in 1875 (*Journ. Bot.*, Vol. 13, pp. 286–7), but these natural hybrids between *S. aria* and *S. aucuparia* continued to be confused with *S. hybrida* until well into the present century.

The form of *S.* × *thuringiaca* commonly planted in Britain lacks a distinctive name, but is probably the same as the tree cultivated on the continent for which Hedlund took up the horticultural name '*S. quercifolia*'. It is a tree to 50 ft high, with very numerous ascending branches, forming in age a dense, rhombic crown; branchlets covered at first with a loose floss, soon glabrous, lustrous brown; winter buds ovoid, densely woolly. Leaves oblong-lanceolate or narrowly elliptic in general outline, 3 to 6 in. long, twice to two-and-a-half times as long as wide, narrowed to a mostly acute apex, dark green and rather rugose above (light green when young), covered beneath with a dull grey persistent tomentum. On flowering shoots and the short growths of old trees the leaves are deeply lobed in the lower part, the basal lobes extending to the midrib, the upper ones decreasing in depth and merging into double teeth and these into simple teeth near the apex; but on strong shoots there is a free pair of leaflets at the base, or even two pairs, the lowermost pair separated by about ⅜ in. from the pair of lobes or leaflets above it. The main lateral veins are in ten to twelve pairs, excluding the veins running out to the sinuses, which are sometimes as conspicuous as those running to the tips of the lobes or main teeth. The minor teeth are shortly acute, with callous tips. Flowers white, about ½ in. wide, produced in May in corymbs 3 to 5 in. wide, which are woolly as in both parents. Fruits bright red, globose or ellipsoid, about ⅜ in. wide, with a few lenticels; calyx-lobes erect or infolded, eventually more or less fleshy.

This tree has been in cultivation since late in the 18th century, but is of unknown origin, and may be the result of a cross between *S. aucuparia* and some long-leaved continental form of *S. aria*. It is surprisingly uniform in its foliage, the only essential difference between one leaf and another being in the presence or absence of free leaflets at the base. In wild presumed first generation hybrids between the two parents the foliage is more variable, some leaves being unlobed, and leaves with free leaflets at the base occurring only at the tips of strong shoots. From *S. hybrida* this tree differs obviously in its habit, but also in its longer, more pointed leaves with shortly pointed teeth, and quite different fruits.

This cultivated form of *S.* × *thuringiaca* makes a handsome specimen of rather sombre aspect. It has attained 52 × 4¼ ft in the Savill Garden, Windsor (1973), and 40 × 2¾ ft in the Winkworth Arboretum, Godalming, Surrey (1967).

cv. 'DECURRENS ('Lanuginosa').—Leaves with up to five or even seven pairs of leaflets but fewer on short shoots, the uppermost pair broadly decurrent onto the rachis; upper part of the leaf simulating a terminal leaflet, obovate, 1 to 2 in. long, lobed in the lower part and sometimes confluent with the upper pair of leaflets. Fruits not freely borne (*S. decurrens* (Koehne) Hedl.; *S. aucuparia* × (*S. aria nivea* × *S. aucuparia*) f. *decurrens* Koehne; *Pyrus lanuginosa* Hort., not Kit. ex DC.).

This sorbus, now uncommon, was long grown in gardens as *Pyrus lanuginosa*, and is excellently described and figured in Loudon, *Arb. et Frut. Brit.* (1838), Vol. II, p. 924, and Vol. VI, p. 146). Its origin is unknown, but central forms of *S.* × *thuringiaca* are known to produce seedlings that approach *S. aucuparia* in their numerous leaflets and no doubt 'Decurrens' arose in this way. Similar second generation plants have been found growing wild in Hungary.

Garden clones similar to 'Decurrens' are 'NEUILLYENSIS', raised in France in the last century and distributed by Messrs Simon-Louis. It is not known to be any longer in commerce in Britain but was probably ornamental in fruit, as the inflorescence was remarkably large, judging from herbarium specimens. Of more recent introduction is 'LEONARD SPRINGER', put into commerce by the Dutch nurseryman Lombarts in 1938, with four or five pairs of leaflets and a terminal 'leaflet' 2½ to 2¾ in. long. It produces abundant orange-red ovoid fruits about ⅝ in. long and is of more spreading habit than other cultivated forms of *S.* × *thuringiaca*. Its name is perhaps best placed directly under *Sorbus*, since it is by no means certain that it a form of *S.* × *thuringiaca* and not a cross between *S. aucuparia* and *S. hybrida* or *S. intermedia*.

Under the name *S. decurrens* a hybrid was distributed in the 1930s which is not 'Decurrens'. It is similar in foliage to the common garden form of *S.* × *thuringiaca*, but the segments of the leaves are more acute and the fruits are more freely borne in denser clusters on closely set spurs.

Seedlings of *S.* × *thuringiaca* with numerous leaflets have been confused with *S. meinichii* (q.v., under *S. hybrida*).

cv. 'FASTIGIATA'.—This appears to differ from the usual garden form of *S.* × *thuringiaca* only in its more narrowly fastigiate habit. It was put into commerce by Messrs Backhouse of York early this century.

S. TIANSCHANICA Rupr.

Pyrus tianschanica (Rupr.) Franch.

A tree to about 25 ft high in the wild, or a shrub; branchlets glabrous or almost so, olive-brown or reddish brown, purplish brown in the second season;

winter buds about ⅜ in. long, narrowly conical, acute, the scales white-hairy on the back or only at the edge. Leaves 4 to 6 in. long including petiole; rachis glabrous, or sometimes with woolly tufts at the insertions of the leaflets, which are in five to seven pairs, lanceolate, 1½ to 2 in. long, ⅜ to ⅝ in. wide, tapered to an acute apex, toothed in the upper half or two-thirds, almost glabrous beneath when mature. Inflorescence lax, 4 to 6 in. wide, more or less glabrous. Flowers white or sometimes tinged with pink, unusually large, about ⅝ in. wide on the introduced plants and apparently even wider on some wild plants; receptacle glabrous. Carpels free almost to the base. Fruits globose, about ⅜ in. wide, glabrous. *Bot. Mag.*, t.7755.

Native mainly of Soviet Central Asia, but extending southwest into Afghanistan, northwest Pakistan and the inner parts of Kashmir, and eastward as far as the Chinese province of Kansu; introduced from Russia in 1895. So far, this species has not been a success in this country, though with so wide a range in Asia it may yet provide a form adapted to our climate. A plant that once grew at Borde Hill in Sussex, planted in 1907, had not produced a fruit by 1932, when it was 5 ft high. Dr Fox's plant in the Winkworth Arboretum grew to be 9 ft high in nineteen years and in that time flowered once, without bearing fruits. In Scotland it succeeds better.

For a species supposedly related to *S. tianschanica*, see *S. cashmiriana*.

S. TORMINALIS (L.) Crantz   WILD SERVICE TREE

Crataegus torminalis L.; *Pyrus torminalis* (L.) Ehrh.   [PLATE 55

A tree from 30 to 40 ft high as a rule, but occasionally 60 to 70 ft, with a trunk girthing over 5 ft, branchlets covered at first with a loose floss, but soon quite glabrous and shining. Leaves 2½ to 5 in. long, nearly or quite as wide, of a broadly ovate or triangular outline, divided half-way to the midrib into three to five pointed lobes on each side, margins doubly-toothed, upper surface glabrous and lustrous dark green, lower surface paler and at first downy, afterwards glabrous; stalk 1 to 2 in long. Flowers white, ½ in. across; produced during June in rather lax corymbs 3 or 4 in. across; calyx and flower-stalks very woolly. Styles two, united in the lower part. Fruit oval or roundish, ½ in. long, brownish.

A native of Europe, North Africa and southwest Asia, occurring in England from Westmorland southward, but locally and never in large stands. It is rare in cultivation, yet few trees of its size are more striking, its leaves being large, boldly cut, and of a healthy polished green, turning crimson or yellow in autumn. The flowers are not very pure white, but attractive when seen in the mass. The fruits when bletted after the fashion of medlars have a similar flavour, but are not to be recommended. In southeast England they are known as 'chequers'. *S. torminalis* should be raised from seeds, and will grow in any soil that is not too poor or acid, but attains its largest dimensions on clay or loam.

The largest examples recorded in recent years are: White Beeches Wood, Chiddingfold, Surrey, 75 × 6¾ ft and 65 × 8¾ ft (1968); Yattenden Court, Berks, 63 × 11¼ ft (1968).

In the southeastern part of its range *S. torminalis* is more variable than in Britain. The leaves may be more deeply cut than in our trees, sometimes almost to the midrib, or at the other extreme, only shallowly lobed. They may also be more hairy on the undersides.

S. UMBELLATA (Desf.) Fritsch

Crataegus umbellata Desf.; *S. flabellifolia* (Spach) Hedl.; *Crataegus flabellifolia* Spach; *S. aria* var. *flabellifolia* (Spach) Wenzig

A small tree in the wild, with a rounded crown, or a shrub; twigs reddish brown, lenticellate, glabrous; winter-buds roundish, about ¼ in. long, glabrous or downy. Leaves simple, 1 to 2¾ in. long, varying on the same plant from obovate with a cuneate base—almost fan-shaped—to a more rounded shape, glabrous and slightly glossy above, densely and closely covered beneath with a vividly white tomentum, deeply and jaggedly toothed or double-toothed on the rounded or almost truncate upper part of the leaf, lateral veins in five to eight pairs; petiole ¼ to ¾ in. long. Flowers in late May or early June, pure white, about ⅜ in. wide, in loose clusters about 3 in. wide. Anthers pink. Styles two. Fruits ovoid to globose, about ½ in. long, said to be red on wild plants.

Native of Asia Minor, the Caucasus, the Crimea and parts of the Balkans; cultivated since early in the 19th century under various names. Although uncommon in gardens, it is one of the most ornamental of the Aria group in its jagged leaves intensely white beneath. An example in the Winkworth Arboretum, planted by Dr Fox in 1938, is an intricately branched shrubby tree about 20 ft high (1979). In cultivation *S. umbellata* ripens its fruits very late—in early November—and perhaps not at all in some seasons. Dr Fox describes the fruits of the Winkworth tree as greenish yellow except on the side exposed to the sun which is red. But plants in their native habitat may become fully red.

S. GRAECA (Spach) Kotschy *Crataegus graeca* Spach; *Pyrus aria* var. *cretica* Lindl.; *S. cretica* (Lindl.) Fritsch; *S. umbellata* var. *cretica* (Lindl.) Schneid.— This species, often considered to be no more than a variety of *S. umbellata*, differs in the more numerously toothed leaves with up to nine pairs of veins, more thinly and grey-hairy beneath. The leaves in the cultivated form are obovate, rounded at the apex, cuneate at the base, of rather thick texture, about 2½ in. long. Fruits dark red, with very few scattered lenticels, globose or broadly oblong-ellipsoid, about ½ in. long, ripening in September.

Despite its name, *S. graeca* is of wide distribution, from Iraq, the Lebanon and the Caucasus through Asia Minor and S.E. Europe to eastern Central Europe, west to Sicily and N. Africa. It was in cultivation in Britain as early as 1830 but the present stock is believed to derive from an introduction by E. K. Balls from Anatolia in the 1930s. The description of the foliage given above is taken from the cultivated form; in wild plants there is variation in the shape and relative width of the leaves.

In the Winkworth Arboretum this species and *S. umbellata* were planted by Dr Fox side by side and make an interesting and instructive pair. At least in these two trees there is a marked difference in habit as well as in the other characters mentioned, *S. graeca* being of rather fastigiate habit in contrast to the bushy habit of *S. umbellata*.

Considered as a variety of *S. umbellata* this sorbus would take the name *S. umbellata* var. *cretica*, but *S. graeca* is its correct name at the specific level.

S. RUPICOLA (Syme) Hedl. *Pyrus rupicola* Syme; *S. aria* var. *salicifolia* Myrin ex Hartman; *S. salicifolia* (Myrin) Hedl.—This species is not cultivated for ornament to any extent but deserves mention as a native of the British Isles (it also occurs in Scandinavia and Estonia). It is similar to *S. graeca*, but the fruits are more lenticellate and the teeth of the leaves are curved on their outer margin, pointing forward towards the apex of the leaf, whereas in *S. graeca* they point more or less in the same line as the lateral ribs (Warburg). Also *S. rupicola* is apparently always tetraploid and apomictic, while *S. graeca* is said to be mainly diploid and to reproduce sexually. In the British Isles *S. rupicola* occurs locally in rocky and often inaccessible places, usually on limestone, in Devon, Wales, the Pennines, Scotland and parts of Ireland.

Some minor and very local species of this complex were described by E. F. Warburg in 1957, and descriptions will be found in Clapham, Tutin and Warburg, *Flora of the British Isles*: S. EMINENS (Wye Valley and Avon Gorge, type-locality Offa's Dyke, Tidenham); S. HIBERNICA (local in central Ireland); S. LANCASTRIENSIS (Lancashire and Westmorland, type-locality Humphrey Head); S. PORRIGENTIFORMIS (N. Devon, the Mendips, Wye Valley, S. Wales, type-locality Offa's Dyke, Tidenham); S. VEXANS (Lynmouth to Culbone, type-locality a wood between Lynmouth and Watersmeet). Apart from the last named, all are confined to limestone, and all are known or presumed to be tetraploid or triploid apomicts. Described by Warburg later is: S. WILMOT-TIANA (type-locality above the cliffs of the Avon Gorge near Clifton).

S. URSINA (G. Don) Schauer

Pyrus ursina Wall. ex G. Don; *Sorbus foliolosa* (Wall.) Spach, in part; *Pyrus foliolosa* Wall., in small part only; *S. foliolosa* var. *ursina* Wenzig; *S. ursina* var. *wenzigiana* and *S. wenzigiana* (Schneid.) Koehne, in part

A tree to about 20 ft high in the wild, or sometimes a shrub; young growths at first sparsely hairy, becoming light brown or grey-brown, furnished with a few large, pale lenticels; winter buds ovoid, ¼ to ⅜ in. long, reddish brown, glabrous except for brown hairs at the tip. Leaves pinnate, with eight to eleven pairs of closely set leaflets; rachis deeply grooved, with the same indumentum beneath as the midribs of the leaflets. Lateral leaflets oblong, mostly ⅞ to 1¼ in. long, the middle pairs the longest, obtuse and finely aristate at the apex, rounded to truncate at the base, finely and closely serrated in the upper half, dark green, glabrous and finely reticulate above, glabrous beneath on the blade, but the midrib at first densely coated with a mixture of brown, spreading hairs and shorter more appressed white ones and retaining some hairs mostly of the latter kind, until autumn. Stipules deltoid, laciniated, seen only on strong branches. Flowers white, opening late May, about ⅜ in. wide. Inflorescence 4 to 5 in. wide, branched in the upper part, rusty-hairy at flowering time, almost glabrous by autumn, sparsely lenticellate. Receptacle soon glabrous. Styles four or five. Fruits at first green tinged with purple, pure white when mature except for a pink flush at the tip, about ⅜ in. wide.

Native of the Himalaya from the Simla area eastward and of southeast Tibet (Pome, Kongbo), with related but untypical forms in north Burma and Yunnan. The name *Pyrus ursina* was given in the catalogue to a set of specimens in the Wallich Herbarium, and was first validated by George Don in his *General History of Dichlamydeous Plants* in 1832. His description is probably based on the specimen collected by Wallich himself in Nepal, which was in unripe fruit, whence no doubt Don's innaccurate statement that the fruits are red; the other specimens in the set were collected in Kumaon. It was apparently not introduced to Britain until Col. Donald Lowndes collected seeds in Nepal in 1950, from which plants were raised by Messrs Hillier and propagated by grafting (distributed as *Sorbus* sp. Lowndes). The above description is made from these trees, which are quite typical, but some wild plants have a denser and more persistent indumentum on the undersides of the leaves and more acute leaflets. More distinct is var. WENZIGIANA Schneid., in which the leaflets are not only more densely indumented with brown hairs but are unusually narrow and acutely tapered, toothed only near the tip.

Col. Lowndes' introduction of *S. ursina* is perfectly hardy and vigorous, though far from being so ornamental as *S. vilmorinii* or *S. hupehensis*. The leaves do not colour in autumn, though they do so on some wild trees.

S. HIMALAICA Gabrielian—This interesting and probably very ornamental species was described by the Russian authority on *Sorbus* Eleanora Gabrielian in 1971, the type being a specimen collected in Nepal by Stainton, Sykes and Williams in 1954. She groups it with *S. ursina*, from which it differs most obviously in its larger, pink-tinged flowers almost ½ in. wide, the relatively narrower leaflets and the red fruits, as well as in other more technical characters. The herbarium specimens cited by the author for this species range from just east of Kashmir, through Nepal, Sikkim and Bhutan to southeast Tibet. It is not known to be in cultivation (1979).

S. REHDERIANA Koehne—This species was described from specimens collected by Wilson in W. Szechwan in the vicinity of Kangting (Tatsien-lu). It is evidently closely allied to *S. ursina*, but more glabrous, with more pointed somewhat longer leaflets and relatively longer and narrower inflorescences. A remarkably constant character is the thick, dark young wood. It ranges as far south as Yunnan, perhaps to north Burma, and west into Tibet, where it was collected by Ludlow and Sherriff sixty miles north of Lhasa. In var. CUPREONITENS Hand.-Mazz., described from Yunnan, there are light brown hairs on the midribs of the leaflets, the underside of the rachis and in the inflorescence. It appears to be intermediate between *S. rehderiana* and *S. ursina*.

Wilson sent seeds of *S. rehderiana* in 1908 and again in 1910, but it is not known to be in cultivation. Plants distributed in the 1930s under this name do not even remotely resemble the true species, and appear to be a hybrid of *S. aucuparia*.

S. VILMORINII Schneid.

Cormus foliolosa sens. Vilm., not (Wall.) Franch.; *S. foliolosa* Hort., in part;
Pyrus vilmorinii (Schneid.) Aschers. & Graebn.

A small tree to about 25 ft high in cultivation, of elegant, spreading habit; branchlets slender, at first rusty-hairy, later glabrous and grey-brown, with oval lenticels; winter buds ovoid, about $\frac{3}{8}$ in. long, hairy at the tip. Leaves mostly 4 to 6 in. long including petiole, with nine to fourteen pairs of leaflets, the longest ones near or slightly below the middle of the leaf; rachis grooved, slightly winged, sparsely furnished beneath at first with pale brown hairs. Leaflets oblong-elliptic to oblong-lanceolate, $\frac{1}{2}$ to $\frac{3}{4}$ in. long and $\frac{1}{4}$ in. or slightly less wide, obtuse at the apex but the midrib running out into a thread-like tip, broadly cuneate at the base, slit-toothed in the upper one-third to one-half,

SORBUS VILMORINII

dark green and glabrous above, underside paler, glabrous except for some pale brown hairs on the midrib, lateral veins beneath inconspicuous and not very clearly differentiated from the minor reticulations. Flowers white, about $\frac{1}{4}$ in. wide, borne in late May or early June, in rather lax, sparsely branched clusters; inflorescence branches and receptacle at first fairly densely but inconspicuously coated with pale brown hairs; calyx-lobes obtuse. Petals obovate, shortly clawed. Stamens shorter than the petals, with reddish purple anthers, darkening as they wither. Styles mostly three. Fruits globose or broadly ovoid, about $\frac{3}{8}$ in. wide, deep pink at first, paling to almost pure white. *Bot. Mag.*, t.8241.

Native of northwest Yunnan, China, possibly extending into Szechwan; introduced by Père Delavay, who sent seeds to Maurice de Vilmorin in 1889.

Where the seeds were collected is not recorded, but Delavay's collections in 1888–9 were from the mountains that stretch northward from Lake Tali to the Lichiang Range. It was described by Schneider in 1906 from the original plant at Les Barres. This was a graft of one of the two original seedlings. both of which had come to an untimely end, and all the original stock of *S. vilmorinii* descends from this one individual. It set fertile fruit for the first time in 1903, but the tree received from M. de Vilmorin at Kew in autumn 1905 must have been a graft, as it flowered and fruited well in 1907. In the previous year seeds were provided by Vilmorin to British nurseries, and the first seedlings were sent out in 1909. The oldest extant tree, growing at Borde Hill in Sussex, came from Messrs Veitch in 1914. It is still in perfect health and vigour, and measured 25 × 3¾ ft at 1 ft in 1968.

S. vilmorinii was reintroduced by Forrest during his last expedition (1930–1), but even now is rarely seen outside collections, though it makes a charming tree of moderate size, with elegant foliage, fern-like when it first expands and colouring bronze or orange late in the autumn. The chief attraction is the pernettya-like fruits, which are fairly numerous in each cluster and all the more conspicuous from the individual clusters being so closely grouped on contiguous spurs. It received an Award of Merit in 1916.

This species is closely allied to *S. ursina*, but with smaller and more numerous leaflets.

S. 'PEARLY KING'.—Near to *S. vilmorinii* but with larger leaflets, to 1 in. or slightly more long, in fewer pairs (mostly six to eight) and wider, more branched inflorescences. It is also of narrower habit, with straight, ascending main branches, but in flower and fruit it resembles *S. vilmorinii*. It is of unknown but probably wild origin. Grown under the erroneous name "*S. pluripinnata*", properly a synonym of *S. scalaris*, it was given its present name by Messrs Hillier. An identical plant has been in commerce as *S. vilmorinii robusta* and what appears to have been a very similar one was distributed by Messrs Hesse of Germany as *S. foliolosa*.

SPARTIUM   LEGUMINOSAE

A genus of a single unarmed species allied to *Genista*, with terete almost leafless branches and a one-lipped calyx.

S. JUNCEUM L.   SPANISH BROOM

A tall shrub of rather gaunt habit, with erect, cylindrical, rush-like stems, glabrous and dark green, which, in the almost entire absence of foliage, fulfil the functions of leaves. It grows 8 to 12 ft high. Leaves very few and deciduous, simple, linear, $\frac{1}{2}$ to $\frac{3}{4}$ in. long, with silky hairs beneath. Flowers fragrant, disposed in terminal racemes 12 or even 18 in. long, on the current season's growth. Each flower is about 1 in. long, pea-shaped (papilionaceous), shortly stalked, rich glowing yellow, with a showy, roundish standard petal nearly 1 in. across. The upper edge of the keel towards the base is sensitive. If it be touched by a pencil point (or the proboscis of an insect) the stamens spring out from the keel, ejecting the pollen in a little cloud. Pods $1\frac{1}{2}$ to 3 in. long, $\frac{1}{4}$ in. wide, hairy, five- to twelve-seeded.

Native of S. Europe, N. Africa, Anatolia, the Crimea and W. Syria; if the '*Spartum frutex*' of William Turner's *Names of Herbes* was this species, it was already established in gardens as early as 1548.

S. junceum is a useful shrub whose value is enhanced by its coming into bloom in June and lasting until September. In July, when it is at its best, it is very showy. It is admirable for planting on hot dry banks, especially if it be associated with a dwarfer shrub (like double-flowered gorse), which will hide its gaunt and naked base. But in the ordinary reaches of the garden also it makes very effective groups, and gives masses of welcome colour when shrubs generally have gone out of flower. It must be raised from seeds (which ripen in abundance), and kept in pots until planted out in permanence, for it dislikes disturbance at the root. Sometimes it is grown as a formal bush, being clipped over with shears in early spring before growth starts; shoots then spring out all over the bush, which blossom in their due season a few months later.

The shrub has some economic value in the south of Europe, yielding a fibre which is obtained from the branchlets by maceration, and is worked up into thread, cordage, and a coarse fabric.

cv. 'PLENUM'.—A double-flowered form propagated by grafting on young seedlings of the type. It was introduced by Peter Collinson from Nuremberg in 1746. He says: 'It cost me a golden ducat; came from thence down the Rhine, and was brought by the first ship to London in good order. I inarched it on the single-flowered broom and gave it to Gray and Gordon [two famous nurserymen].' It was, till recently, cultivated at Kew, but unless it had deteriorated, it was scarcely worth Collinson's trouble.

SPHAERALCEA   MALVACEAE

A genus of about sixty species, some of them herbaceous, others more or less shrubby, the majority natives of the New World, in both hemispheres, a few in S. Africa. They belong to the same group of the Mallow family as *Lavatera* and *Abutilon*.

S. FENDLERI A. Gray

A perennial with a woody base, its annual stems, leaves, inflorescence-axes and calyx all clad with stellate hairs. Leaves oblong-ovate to broadly ovate, up to 2 in. long, acute at the apex, cuneate to rounded at the base, sparsely toothed and mostly shallowly three-lobed; petiole up to 1 in. long. Flowers produced in summer and autumn in the upper leaf-axils, solitary or up to six together in a cymose cluster, on stalks up to 1 in. long; bracteoles linear, in whorls of three beneath each flower. Calyx with narrowly triangular lobes. Corolla about 1 in. wide, with five reddish orange obovate emarginate petals. Staminal column downy, about ¼ in. long, bearing numerous anthers. *Bot. Mag.*, n.s., t. 140.

Native of the south-western USA and N. Mexico. A very striking species, useful for its long period of bloom from mid-summer into autumn. It is one of those many plants that in our climate come near to being herbaceous perennials, yet are too tender for the conventional herbaceous border and need a favoured position such as the foot of a sunny wall. Even if the stems survive the winter they should be shortened to near the base in spring. Propagated by cuttings of the young shoots.

SPIRAEA   ROSACEAE

A genus of around seventy species of shrubs in the temperate regions of the northern hemisphere. Leaves deciduous, without stipules, simple, toothed or more rarely entire, sometimes lobed. The flowers are very uniform in size, being rarely more than ⅜ in. wide, but about ½ in. across in the double-flowered forms of *S. prunifolia* and *S. cantoniensis*. They are either white or in some shade of pink or crimson (all the early flowering sorts have white flowers), and in most species they are hermaphrodite. The inflorescence is a condensed raceme or rounded corymb produced from buds on the previous season's wood, or it terminates the growths of the season, then taking the form of a panicle or a sometimes complex corymb (see further under classification). Receptacle cup-shaped, campanulate or top-shaped, bearing on its rim five short sepals, five petals and numerous stamens, which in some species are longer than the petals. The receptacle is usually edged with a many-lobed, nectar-secreting disk. Gynoecium composed of five free or almost free carpels, inserted at the base of the receptacle, each bearing a style at its apex (or just below the apex on the outer side). The carpels usually enlarge as they mature, each becoming a dry follicle which opens along the inner suture, releasing a few minute seeds.

Spiraea, as once understood, was a much larger genus than now. In previous editions, following the older classification, the following were included in it (the most obvious differential characters are given in brackets): SORBARIA (leaves pinnate); CHAMAEBATIARIA (leaves doubly pinnate and fern-like); PETROPHYTUM (creeping or tufted shrubs; flowers in long-pedunculate racemes or heads); LUETKEA(dwarf shrub with much divided leaves); HOLODISCUS (flowers very small in plume-like panicles; carpels indehiscent in fruit); SIBIRAEA (near to *Spiraea* but carpels more united; easily recognised by its stout branches and the combination of slender panicles of white flowers and narrow entire leaves).

The old *Spiraea* also contained herbaceous plants, e.g., *Aruncus dioicus* (Walt.) Fern. (*Spiraea aruncus* L.); *Filipendula ulmaria* (L.) Maxim. (*Spiraea ulmaria* L.), the native Meadow Sweet; *Filipendula purpurea* Maxim. (*Spiraea palmata* Hort., not Thunb.); *Filipendula rubra* (Hill) Robins. (*Spiraea lobata* Jacq.; *S. venusta* Hort.).

As ornaments in the garden, the best of the spiraeas fill an important place. They flower with great freedom, are often very graceful, and except that some of the earlier flowering kinds are liable to injury by late frost, they are perfectly at home under cultivation. All like a good loamy soil, abundant moisture, and full sunlight. 'Arguta', 'Grefsheim' and *S* × *vanhouttei* make excellent low hedges.

PROPAGATION.—Some of the spiraeas spread by means of sucker growths from the base, and such are easily increased by dividing the plants into small pieces. The rest can nearly all be propagated easily by means of cuttings made of moderately firm wood placed in light sandy soil in gentle bottom heat in July and August. If this be not available, cuttings made of harder wood in September may be placed under bell-glasses out-of-doors in a sheltered spot.

The spiraeas produce fertile seed in abundance, but they cross-breed with such facility that seed can only be depended on to come true when the plants are fairly isolated from other species. Some of the very best spiraeas are hybrids, as may be gathered from the following descriptive notes, but they have become so numerous that they make the genus, as represented in gardens, excessively difficult to study and classify. Zabel of Münden devoted a volume* of one hundred and twenty-eight pages exclusively to their elucidation, but many are so similar to each other that their differentiation on proper is no longer possible within convenient limits.

PRUNING.—Few shrubs repay careful pruning better than the spiraeas, and in this matter they may be divided into two groups, viz. (1) those that

---

* H. Zabel, *Die strauchigen Spiräen der deutschen Garten* (1893). The bulk of Zabel's collection at the Forstakademie, Hannover-Münden, was propagated by Messrs Hesse of Weener, Hannover, and distributed by them from about 1894 onwards, each plant being sent out with the name and number under which it appears in Zabel's work. His collection consisted mainly of hybrids raised by open pollination, some by himself, others in nurseries or gardens. A generation or so later such hybrids would have received 'fancy names', but Zabel followed the then usual practice of giving botanical status to these garden productions. In the present revision this status is recognised only if the name is the valid designation for a simple interspecific cross.

flower early and from the buds of shoots made the previous year, such as *S.* 'Arguta', *S.* 'Grefsheim', *S. thunbergii, S.* × *vanhouttei, S. veitchii;* and (2) those that flower later in the year at the ends of the shoots of the current season, such as *S. japonica, S. douglasii, S. salicifolia,* etc. This matter is fully discussed in the introductory chapter on pruning, and from what is there stated it will be evident that the first group must only be pruned by thinning out the older and weaker wood; any shortening back of the shoots must mean a reduction in the next crop of blossom. The second group, on the other hand, is benefited by the shoots being shortened back. This should, of course, be done in later winter or early spring, and at the same time superfluous old shoots should be cut clean out. Unless pruning of either kind is done, many of the spiraeas get into a weedy, thin condition, and their blossoms will not bear comparison either in quantity or quality with that of properly pruned plants.

The group including *S. douglasii, S. tomentosa, S. salicifolia,* and their hybrids form dense thickets, and spread rapidly by means of underground suckers. These should be pruned as in group (2) (being late flowering), and it is also advisable at intervals of a few years to dig them up, divide them into smaller pieces, and after enriching the ground, replant them more thinly. This, of course, applies to ordinary cultivated shrubberies and borders, but they also make admirable masses for the wilder portions of the demesne, where they can safely be left to take care of themselves. In such places the reddish or rich brown stems of many spiraeas make a cheerful feature in winter.

CLASSIFICATION

It has been customary to group the species of *Spiraea* into three sections, according to the nature of the inflorescence, but the classification is to a large extent artificial, as Schneider pointed out early this century. The species treated here, and their hybrids, can be informally arranged as follows:

SALICIFOLIA Group.—This is the well marked section *Spiraea* (*Spiraria*), in which the inflorescence is a panicle terminating a long shoot of the season's growth. Here belong the Old World *S. salicifolia* (the type-species of *Spiraea*), and the American *S. alba, S. douglasii, S. latifolia* and *S. tomentosa.* There are numerous garden hybrids in this group, of which *S.* × *billiardii* is treated here.

JAPONICA Group (sect. *Calospira,* in part).—Inflorescence a flat or slightly convex, much branched corymb, usually borne at the end of a long shoot, as in the first group, more rarely on a short lateral (*S. bella*); the peduncles are usually furnished with slightly reduced leaves. This group occurs in both the the New and the Old World. The Asiatic species treated here are *S. japonica, S. bella,* its close ally *S. amoena,* and *S. betulifolia.* The last named also occurs in N. America where the other representative (in the western part of the continent) is *S. densiflora.* The geographically isolated European species *S. decumbens* and *S. hacquetii* should perhaps also be placed here. Among the hybrids that have arisen in this group are 'Margaritae' and those mentioned under it.

There are some fine hybrids between the Salicifolia and Japonica groups, for which see *S.* × *sanssouciana.*

CANESCENS Group (sect. *Calospira,* in part).—Inflorescence corymbose as in the preceding group, but of simpler form, not leafy, and borne at the end of a short leafy lateral. *S. canescens*, *S. henryi*, *S. wilsonii*, *S. veitchii*, *S. longigemmis* and *S. rosthornii* may also belong here.

S. × *brachybotrys* and *S.* × *fontenaysii* are hybrids between *S. canescens* and members of the Salicifolia group.

The remaining species, all natives of Europe and Asia, are usually grouped in the section *Chamaedryon*, a rather artificial aggregate. The inflorescence is usually simple, i.e., the peduncles each bear a single flower and are arranged in the form of a condensed raceme or umbel. But in several species the inflorescence is partly compound, the lower peduncles being branched and bearing several flowers. There is also great variation in foliage and in the form of the leaf-buds. The flower-clusters are borne in spring or early summer on the branches of the previous season on short leafy laterals, or are sessile on them. Grouped according to probable affinity the species treated here are:

S. prunifolia, *S. thunbergii*
S. crenata, *S. hypericifolia*
S. cana, *S. media*
S. chamaedryfolia, *S. flexuosa*
S. blumei, *S. chinensis*, *S. trilobata*, *S. yunnanensis*
S. alpina, *S. myrtilloides*
S. calcicola
S. arcuata, *S. gemmata*, *S. mollifolia*, *S. nipponica*, *S. trichocarpa*

Hybrids within this group are: *S.* 'Arguta', *S.* × *cinerea*, *S.* × *multiflora*, *S.* × *schinabeckii*, *S.* × *vanhouttei*. See also *S.* × *nudiflora*, p. 476.

S. ALPINA Pall.

A shrub 3 to 5 ft high, with erect stems; young shoots angled, finely downy, bright brown. Leaves ¼ to 1 in. long, ⅓ in. or less wide, narrowly oblong, or obovate, entire, glabrous, with feathered veins beneath. Flowers yellowish white, small, produced during May and June in small umbels; flower-stalks glabrous.

Native of Siberia from the Altai eastward, Mongolia and N.W. China; probably introduced to Britain shortly before 1824 (the plant portrayed by Loudon as *S. alpina*, said to have been introduced in 1806, is clearly not the present species). It is of no garden value. Wilson found what appears to have been a dwarf, more ornamental form in the uplands of W. Szechwan, but so far as is known this is not in cultivation.

S. ARCUATA Hook. f.

A medium-sized shrub with arching, strongly ribbed reddish brown stems, which are finely woolly when young, more or less glabrous by autumn. Buds hairy, slender, acute, longer than the petioles. Leaves dark green, shortly stalked, those on the strong shoots obovate or obovate-elliptic, to about ½ in. long, mostly crenately toothed or lobulate at the apex, on weaker shoots smaller and mostly entire; they are glabrous with somewhat impressed main veins above, undersides glabrous except for hairs on the prominent veins, margins with a woolly ciliation. Flowers typically pink but sometimes white, borne in early summer in dense or lax umbel-like or compound clusters up to about 1½ in. wide, terminating short leafy laterals. Peduncles finely woolly, each furnished with a bract, which is leaf-like in the outer part of the inflorescence. Receptacle shallow, glabrous. Follicles well exserted, glabrous and lustrous; styles inserted below the apex, outward-pointing.

Native of the Himalaya from Nepal (perhaps farther west) to S.W. China, and of S. Tibet. An ornamental and unusual spiraea, perhaps allied to *S. gemmata*, though differing from it in foliage. It is still rare in gardens.

S. 'ARGUTA' [PLATE 56
S. × *arguta* Zab.

A shrub of rounded, bushy habit, 6 to 8 ft high, and as much wide; branches graceful, slender, twiggy, and covered with down. Leaves oblanceolate, ¾ to 1½ in. long, ¼ to ½ in. wide, entire, or with a few teeth towards the apex; of a lively green and glabrous above, slightly downy and rather prominently nerved beneath. Flowers ⅓ in. across, pure white, produced during April and May in fascicles of four to eight, each flower on a slender glabrous stalk ½ in. or so long.

A seedling of *S.* × *multiflora* (*S. hypericifolia* × *S. crenata*), raised some years before 1884; the other parent is thought to be *S. thunbergii*. It is the most beautiful of the spring-flowering spiraeas, being quite hardy and never failing to produce a wealth of blossom. The flower-clusters are crowded on the upper side of shoots made the previous year, forming snowy white wreaths from 6 in. to 12 in. long. It is most conveniently increased by means of layers, its slender lissom branches adapting themselves admirably to this method.

S. 'GREFSHEIM'.—A shrub in the style of 'Arguta' but dwarfer, to about 4 or 5 ft and flowering somewhat earlier. Leaves on the season's growths narrowly elliptic or lanceolate, narrowed at both ends, mostly quite entire, 1½ to 1¾ in. long, about ⅜ in. wide, soft sea-green and at first downy above, permanently clad beneath with short appressed silky hairs. Upper flower-clusters sessile, the lower on definite leafy branchlets; pedicels downy. This spiraea arose in the Grefsheim nursery at Nes, Norway, as a self-sown seedling and was put into commerce in 1954. It is probably a hybrid between *S. hypericifolia* and *S. cana* (*S.* × *cinerea* Zab.). The spiraea grown as *S. arguta nana* or *compacta* is probably of the same parentage.

S. BELLA Sims

A shrub 4 to 6 ft high, with angular, slightly hairy young branches. Leaves thin, broadly ovate, pointed, doubly toothed towards the apex, 1 to 2¼ in. long on the barren shoots, shorter, relatively narrower, and simply toothed on the flowering ones, glabrous above, glaucous or whitish, and more or less downy beneath; stalks up to ¼ in. long. Flowers unisexual, about ¼ in. across, produced around midsummer in corymbs ¾ to 1½ in. across, at the ends of lateral shoots. Stamens longer than the petals (small and abortive in female flowers). Follicles glabrous except for down on the inner suture. *Bot. Mag.*, t. 2426.

SPIRAEA BELLA

Native of the Himalaya from Kashmir eastward, extending into China; introduced from Nepal about 1820. In spite of its name this shrub is not one of the best of the spiraeas, but was widely grown in the last century until displaced by *S. japonica*. Even with the exclusion of *S. amoena* (see below) the species is variable. The branches may be erect or arching and the flower-bearing laterals vary considerable in length, as does the width of the inflorescence, which on some wild plants may be more than 1½ in. across. *S. bella* is allied to *S. japonica* but differs most obviously in its dioeciousness and the shorter flowering branches.

S. AMOENA Spae   *S. expansa* K. Koch; *S. fastigiata* Schneid.; *S. bella sens.* Hook. f., in part, not Sims—A shrub up to 6 ft high, with slender, round, downy stems, erect and not much branched; buds hairy. Leaves lanceolate to ovate up to 4 in. long by 1¼ in. wide, coarsely and sharply toothed (both simply and doubly) except at the base, dark green above, hairy on the veins and rather glaucous beneath; stalks ¼ to ⅜ in. long. Flowers white with a flush of red, borne in flat compound corymbs from 2 to 8 in. across; calyx and flower-stalks downy. It blossoms on the shoots of the year in July and August.

Native of the Himalaya; in cultivation by the 1840s and perhaps introduced at the same time as *S. bella*. It is closely allied to that variable and little studied species and included in it by the younger Hooker as 'the *fastigiata* form of *S. bella*'. The two are difficult to separate in the herbarium, but in cultivation they are really quite distinct and it is convenient to keep them apart nomenclaturally. *S. amoena* has white flowers in large flat corymbs borne in late summer on long terete branches and in general aspect bears a greater likeness to *S. micrantha* (see below) than to *S. bella*, with pink flowers on short or shortish laterals in smaller corymbs opening earlier in the season.

Plants now in commerce under the name "*S. japonica fastigiata*" (1978) appear to be *S. amoena*, though a singularly fine form of it, with a very large central corymb supplemented on strong shoots by clusters at the ends of axillary growths, opening later. The flowers are male, pink in the bud, opening white, with a conspicuous ring of disk-glands, which are purplish pink at first, later yellowish; abortive carpels pink. Distributed by the Sunningdale Nurseries, the stock came from a Cotswold garden whose owner had acquired the original plant from a Scandinavian nursery. It is therefore of interest that this clone agrees in every discernible detail with specimens of a spiraea cultivated in Denmark as *S. expansa*, which were collected by the botanist Lange in August 1869 and August 1881 and are now preserved in the Kew Herbarium. It also agrees with Lange's description of *S. expansa*, obviously made from a cultivated plant, in *Bot. Tidskr.*, Vol. 13 (1882), p. 28. The possibility that this spiraea is a hybrid has been considered, but seems unlikely.

NOTE. It has been suggested that the name *S. amoena* Spae (1846) is illegitimate on the grounds that it is a later homonym of *S. amena* Raf. (1838). This would indeed be the case if 'amena' is to be regarded as an orthographic variant of 'amoena'. But Rafinesque did not explain his choice of epithet, which he may have taken from the Greek 'amenes', meaning weak. If the name *S. amoena* had to be discarded, the correct name for the species, if kept separate from *S. bella*, would be *S. expansa* K. Koch, not *S. fastigiata* Schneid., the name used by Rehder in the *Bibliography*.

S. MICRANTHA Hook. f.   *S. japonica* var. *himalaica* Kitamura—This comes near to *S. amoena* but the leaves are usually thinner, more acuminate at the apex, the flowers are mostly hermaphrodite, the inflorescence is more open, with woollier branches, and the follicles are densely hairy. It is also closely allied to *S. japonica*, but the densely indumented inflorescence, white or pale pink flowers and hairy follicles distinguish it. Native of the Himalaya from Nepal eastward, and of N.E. India. It was cultivated at Kew in the last century from seeds collected by J. D. Hooker in Sikkim but the plants now at Kew and in other collections were raised from B.L. & M. 292, collected by the University of North Wales Expedition to E. Nepal in 1971.

S. BETULIFOLIA Pall.

A shrub to about 2 ft high, of rounded habit, with zig-zagged terete or slightly angled, reddish brown stems; buds ovoid, glabrous. Leaves pinnately veined, mostly elliptic or oblong-elliptic, rounded to cuneate at the base, or sometimes obovate and tapered to a narrow base, obtuse at the apex, crenately toothed or double-toothed in the upper part or throughout, glabrous on both sides or downy beneath; petiole to $\frac{1}{4}$ in. long. Flowers white or rose, densely arranged in flat or slightly convex downy or glabrous corymbs up to $3\frac{1}{2}$ in. wide. Petals much shorter than the stamens. Follicles glabrous; sepals reflexed in the fruiting stage.

Native of N.E. Asia, south to Japan; introduced in 1812. The true species is rare in cultivation.

var. CORYMBOSA (Raf.) Maxim.   *S. corymbosa* Raf., not Muhl.; *S. betulifolia sens.* some authors, not Pall.—A shrub to about 3 ft high, with terete, sparsely branched stems. Leaves to about 3 in. long, coarsely and often doubly toothed in the upper part, glabrous, glaucous beneath. Flowers white, about $\frac{1}{6}$ in.

across in rounded corymbs 2 to 4 in. across. The chief difference from typical
S. betulifolia is that the sepals are upright in fruit. Native of the eastern USA;
introduced 1819. A handsome shrub, which, like typical *S. betulifolia* renews
itself by stems pushed up from the base annually; these should be encouraged
by pruning out the older wood. Both flower from about midsummer on the
shoots of the year.

In western North America *S. betulifolia* is represented by var. LUCIDA
(Dougl.) C. L. Hitchc. (*S. lucida* Dougl. ex Greene; *S. corymbosa* var. *lucida*
(Dougl.) Zab.), which is very near to the typical state of *S. betulifolia* but is
more constantly glabrous.

S. DENSIFLORA Nutt.—A deciduous shrub up to 2 ft high, with glabrous,
round, rich brown young shoots. Leaves oval or ovate, rounded at both ends,
rather coarsely toothed towards the apex; $\frac{2}{3}$ to $1\frac{3}{4}$ in. long, glabrous on both
surfaces, deep green above, paler beneath; stalk $\frac{1}{12}$ in. or less long. Flowers
rose-coloured, densely packed in dome-shaped corymbs 1 to $1\frac{1}{2}$ in. wide,
opening in June.   [PLATE 57
Native of western N. America, from British Columbia to Oregon. It was
no doubt gathered by early collectors like Douglas and Lobb, and, as it was
growing in Veitch's nursery at Exeter in 1861, was probably introduced by the
latter. It is useful as a very hardy, low-growing shrub that will keep dwarf
without pruning, and its rose-coloured flowers are pretty and attractive.
S. densiflora is very near to *S. betulifolia* var. *lucida*, differing in its rosy flowers.
The true species is still in commerce, but some plants in the trade as *S. densiflora*
are a dwarf form of *S. japonica* (see its cv. 'Nana').

var. SPLENDENS (Baumann) Abrams *S. splendens* Baumann ex K. Koch—
Stems and inflorescence-axes finely downy. This extends into California.
Described from plants distributed by Baumann's nursery and introduced to
Britain shortly before 1883.

S. VIRGINIANA Britt.—Near to *S. betulifolia* var. *corymbosa* but with the
leaves entire or almost so, oblong-lanceolate, glaucous beneath, $\frac{1}{2}$ in. or slightly
more wide. Native of the south-eastern USA.

S. × BRACHYBOTRYS Lange

S. luxurians Lav. ex Zab.; *S. pruinosa* Hort. ex Lange; ? *S. pruinosa* Hort. ex K. Koch

A vigorous shrub, up to 8 ft high, branches gracefully arching; young
wood downy, ribbed. Leaves oblong or ovate, $\frac{3}{4}$ to $1\frac{1}{4}$ in. long, $\frac{1}{3}$ to $\frac{3}{4}$ in. wide,
with a few teeth at the apex only; upper surface dull dark green, and slightly
downy, lower one pale and felted with fine grey down. Flowers rosy pink,
small, and crowded densely in stout panicles $1\frac{1}{2}$ to 3 in. long and about the
same wide; they are borne at the end of leafy twigs, 3 to 12 in. long, that
spring from the branches of the preceding year, expanding in June and July;
flower-stalks and calyx hairy.

A hybrid between *S. canescens* and probably *S. douglasii*, inheriting much of
the grace and vigour of the former. This is, indeed, one of the best of the

taller summer-flowering kinds, the long shoots made one year branching copiously towards the top the following one, when each twig carries its terminal panicle, the whole forming a fine sheaf of delicately coloured blossom.

This hybrid is of unknown origin. It was described in 1882 from a plant grown under the name *S. pruinosa*, which is the name under which it was received at Kew around 1880 from the nurseryman Booth of Hamburg; it, or a similar clone, came to Kew around the same time from Lavallée as *S. luxurians*. Zabel considered that the plants under these two names were identical.

S. × FONTENAYSII Lebas *S. × fontenaysiensis* Dipp.—Similar to the preceding but with the leaves almost glabrous beneath. The flowers are white in the typical form, pink in 'ROSEA'. Raised by Billiard of Fontenay-aux-Roses before 1866, from *S. canescens* crossed with *S. salicifolia*.

S. CALCICOLA W. W. Sm.

A shrub 2 to 5 ft high in the wild, its arching branches ribbed, dark reddish brown, slightly downy when young. Leaves on strong extension shoots semi-circular in outline, deeply three-lobed, the lobes coarsely and doubly crenate-toothed, truncate at the base, but on weaker shoots obovate-cuneate and deeply toothed at the apex; the leaves are up to ½ in. long on the stronger shoots, rather more in width, and are glabrous on both sides, pruinose beneath; petiole of the largest leaves about ¼ in. long. Flowers small, white, flushed with pink on the outside, borne few together in reduced corymbs on the previous season's wood, with a few minute leaves at the base; sometimes short spurs occur among the flowers, on which the leaves are up to about ⅜ in. long. Receptacle shallowly cup-shaped, it and calyx-lobes glabrous. Disk not lobed. Carpels glabrous, with terminal, erect styles.

This little-known species was found by Forrest in the Lichiang range, Yunnan, in June 1910, and described three years later, by which time it was already grown by Messrs Bees from seeds sent by Forrest. The type-specimen, collected in flower, did not bear fully developed leaves, and the description of these is taken from the plant in the Hillier Arboretum, which came from Kew and agrees well in other respects with the type. The species is of uncertain affinity, and perhaps not closely related to *S. hypericifolia*, under which Rehder places it in the *Manual*. It is a pretty spiraea, probably not attaining more than 3 ft in height in gardens. The specific epithet refers to its habitat in the crevices of dry limestone cliffs.

S. CANA Waldst. & Kit.

A shrub 3 ft or more high and as much through, of dense, twiggy habit, the young shoots round and covered with a thick grey down. Leaves narrowly oval or ovate, tapering at both ends, ⅓ to 1 in. long, about half as wide, nearly always entire; covered on both sides, but especially beneath, with a grey silky down. Flowers dull white, ¼ in. across, produced during May at the end of short leafy twigs in dense umbel-like racemes ¾ to 1 in. wide.

Native of N.W. Yugoslavia, just reaching into Italy; introduced in 1825. The leaf is very like that of *Salix repens* in its dense grey down, not, however, so silvery. One of the least attractive of the spiraeas, but of neat habit and quite hardy; also distinct in the character of its leaves.

S. CANESCENS D. Don

S. flagelliformis Hort.; *S. rotundifolia* Hort.

A shrub varying considerably in height, at its tallest 12 to 15 ft high, more often 6 to 8 ft high, the main stems erect, but producing towards the top slender, arching or pendulous branches growing 3 ft or more long in one season; the young branches ribbed and downy. Leaves ½ to 1 in. long, ¼ to ⅝ in. wide, oval or obovate, usually blunt and toothed at the apex, and always more or less tapering to the very short stalk at the base, dull green and with some down above, grey and more or less thickly downy beneath. Flowers white, or dull creamy white, small, produced during June and July in corymbs 1 to 2 in. across, at the end of short leafy twigs; flower-stalks and calyx grey-downy or even felted.

Native of the Himalaya; introduced in 1837. The chief distinguishing characteristic of this spiraea, and one which gives it a leading place in the genus, is its habit of producing in one season the long, thong-like shoots to which the former popular name "flagelliformis" refers. When, the following year, there springs from every bud a short erect twig, each crowned with its dense cluster of flowers, there are few more strikingly beautiful shrubs, especially at the date when it blossoms. The species is somewhat variable in the shape and size of the leaf, in the amount of down it bears, also to some extent in habit. It has, in consequence, received many names. The small greyish leaves tapering at the base, and the abundant clusters of white flowers set on the upper side of long arching branches, generally distinguish it.

The above account of *S. canescens* is taken from previous editions. This species has now been largely displaced in gardens by its Chinese allies, notably *S. veitchii*.

cv. 'MYRTIFOLIA'.—Leaves more elongate, almost glabrous beneath; flowers smaller.

S. CANTONIENSIS Lour.

S. reevesiana Lindl.

A deciduous or partly evergreen shrub, 4 to 6 ft high, of wide-spreading, graceful habit, producing a thicket of erect and outwardly arching stems; young stems glabrous. Leaves lozenge-shaped, 1 to 2½ in. long, ½ to ¾ in. wide, deeply and irregularly toothed (sometimes almost lobed) on the upper part, green and quite glabrous on both sides, with a glaucous tinge especially beneath; stalk slender, ¼ to ⅓ in. long. Flowers white, ⅓ in. across, produced during June in hemispherical corymbs 1 to 2 in. across, each corymb on a leafy stalk 1 to 2 in. long.

Native of China. This shrub is scarcely known in gardens except in its double-flowered state, 'FLORE PLENO' ('*Lanceata*'), in which the many-petalled blossoms are nearly ½ in. across; when freely borne they make a charming display. In the gardens of the south of France, Italy, and Dalmatia this double-flowered form is perhaps the most beautiful white-flowered shrub in April, its long sprays arching in every direction and laden with blossom. But in the Thames Valley it is rarely seen to perfection owing to injury by spring frosts. It can be got in better condition on a wall. Nearly allied to this species is *S. chinensis* (*q.v.*), also with fragrant white flowers in corymbs, but readily recognised by the yellowish felted under-surface of its leaves. It is spring-tender.

S. CHAMAEDRYFOLIA L.

An erect shrub, up to 6 ft high, the young shoots yellowish, glabrous, angular, zigzag. Leaves ovate or ovate-lanceolate, 1½ to 3 in. long, ¾ to 1½ in. wide, coarsely, irregularly, often doubly toothed, dark green and glabrous above, somewhat glaucous and slightly downy beneath. Flowers ⅓ in. across, white, produced in a corymb or corymbose raceme 1½ in. across; flower-stalks glabrous, slender, the lower ones ¾ in. long, becoming shorter towards the summit. Stamens conspicuously long.

S. chamaedryfolia is a rather variable species with a wide natural distribution from the eastern Alps, Carpathians and Balkans eastward to Siberia and central Asia; farther east its place is taken by *S. flexuosa* and other closely related species.

var. ULMIFOLIA (Scop.) Maxim. *S. ulmifolia* Scop.—Leaves ovate, the upper two-thirds coarsely toothed. The inflorescence is more of a raceme than a corymb, and from 1½ to 2 in. long; flowers white, about ½ in. across in cultivated plants. This is the handsomest form of *S. chamaedryfolia*, distinct in its broader leaves and more elongated inflorescence. It was described from Europe and is said to be the predominant state of the species in the western part of its range.

S. chamaedryfolia is in all its variations an attractive and quite reliable shrub, usually escaping spring frosts and flowering in May (var. *ulmifolia* towards the end of the month). It renews itself by sending up every year strong, erect sucker-growths from the ground, which produce flowers on short twigs in the following year; and, to give these their best chance, sufficient of the old shoots should be pruned out after flowering to enable them to develop strongly and freely.

S. × NUDIFLORA Zab. *S. hookeri* Hort. ex Zab.—A hybrid of *S. chamaedryfolia* var. *ulmifolia* and *S. bella*. Leaves almost glabrous, ovate, doubly toothed or incised, up to 2½ in. long on strong shoots, to 1½ in. on the flowering laterals, which are produced all along the previous year's growths. Flowers rose-coloured, borne in racemose umbels which are sometimes supplemented by clusters from the axils of bracts or reduced leaves, the whole inflorescence then up to 3 in. wide. An interesting hybrid which unites sect. *Chamaedryon* with the Japonica group of sect. *Calospira*. It was distributed on the continent by the Berlin nurseryman Lorberg, who obtained it from England under the name '*S. hookeri*', already in use for some other hybrid, and also a horticultural synonym of *S. bella*. It was given further circulation in the 1890s by Messrs Hesse under the name given to it by Zabel.

S. × SCHINABECKII Zab.—A handsome hybrid between *S. ch.* var. *ulmifolia* and *S. trilobata*. It is a twiggy bush to about 6 ft high with white flowers in stalked umbels, at their best in June. Lower leaves on the shoots roundish, the upper ovate, doubly toothed, about 2 in. long. The flowers are larger than is usual in *S. chamaedryfolia*, longer than the stamens (shorter in the species).

S. FLEXUOSA Fisch. *S. ch.* var. *flexuosa* (Fisch.) Maxim.—Closely related to *S. chamaedryfolia* and probably no more than a variety of it. It is distinguished by the more conspicuously angled (or winged) stems, the dwarfer habit, the smaller narrower leaves simply-toothed on the upper third or half only, sometimes almost entire, and by the flowers being fewer in the cluster. Native of Siberia.

S. CHINENSIS Maxim.

S. pubescens Lindl., not Turcz.

A shrub 3 to 5 ft high, of dense very leafy habit; young shoots downy. Leaves 1 to 1¾ in. long, ½ to 1¼ in. wide, varying from rhomboidal and tapering at both ends, to broadly ovate with a nearly truncate base, sometimes obscurely three-lobed, the upper part sharply and coarsely toothed, the teeth gland-tipped, upper surface furnished with scattered hairs, under-surface clothed with yellowish felt; stalk ¼ to ⅓ in. long. Flowers white, nearly ½ in. across, produced during June in stalked umbels or corymbs 1 to 2 in. wide; flower-stalks and calyx downy. The leaves remain very late on the branches.

Native of N. China and allied to *S. cantoniensis*, but readily distinguished by its downy shoots, flower-stalks, and yellowish felted leaves, the last named being considerably broader in proportion to their length than those of *S. cantoniensis*. It is not very hardy, and is killed to ground level in hard winters.

S. chinensis is very near to S. NERVOSA Franch. & Sav. of Japan, which was described earlier.

S. CRENATA L.

A shrub 3 to 5 ft high, bushy, with slightly angular stems; young twigs at first more or less downy, becoming glabrous later. Leaves narrowly to broadly obovate, ½ to 1⅓ in. long, ¼ to 1 in. wide; toothed only at the apex, slightly downy or glabrous beneath, with three distinct veins running lengthwise. Flowers white, small, produced during May in small hemispherical umbels at the end of short, leafy twigs.

Native of S.E. Europe, north to E. Czechoslovakia, east to central Russia, Asia Minor, the Caucasus and N. Iran; long in cultivation. It is probably most closely allied to *S. hypericifolia* which too has leaves three-veined from the base, but differs from *S. crenata* in its almost sessile inflorescence.

S. DECUMBENS W. Koch
S. procumbens Hort.

A dwarf shrub, 3 to 8 in. high, with slender, glabrous, often prostrate stems, from which the thin, wiry flowering branches ascend. Leaves obovate or oval, tapered at both ends, sharply, angularly and rather coarsely toothed towards the apex, ½ to 1½ in. long, ¼ to ½ in. wide, quite glabrous on both surfaces; stalk about ⅛ in. long. Flowers white, ¼ in. across, in corymbs 2 in. wide; seed-vessels glabrous, with the sepals deflexed.

An endemic species of the Carnic Alps and bordering mountains (Italy in Udine province, N.E. Slovenia, one station in Austrian Carinthia). It is a pleasing little shrub, one of the dwarfest of the spiraeas, and very suitable for the rock garden. In the wild it occurs only on limestone.

S. HACQUETII Fenzl & K. Koch ? *S. lancifolia* Hoffmanns.; *S. decumbens* var. *tomentosa* Poech; *S. d.* subsp. *tomentosa* (Poech) Dostal—Of similar habit to *S. decumbens*, but differing in the young stems, leaves, flower-stalks and calyx being downy; in the more prominent nerves beneath the leaf; also by the sepals being more erect in fruit. Like the preceding, this too has a very limited range, but farther west, in the Alpe Venezie of Italy. Of the same garden value as *S. decumbens*, of which it is now usually treated as a subspecies.

These two interesting species are the only European natives that bear their flowers in corymbs on leafy shoots, all the others, except the obviously distinct *S. salicifolia*, having their flowers in umbel-like racemes. Also, they are mainly or wholly dioecious, as is the Himalayan *S. bella*.

S. DOUGLASII Hook.

A shrub 4 to 6 ft high, forming a thicket of erect stems, reddish, covered when young with a very fine felt. Leaves narrow oblong, 1½ to 4 in. long, ½ to 1 in. broad, coarsely and unequally toothed on the terminal part only, dark green above, covered with a fine grey felt beneath. Flowers purplish rose, produced in an erect terminal panicle 4 to 8 in. high, very closely packed; flower-stalks and calyx grey-downy; stamens pink, standing out well beyond the petals; ovaries glabrous. *Bot. Mag.*, t. 5151.

Native of western N. America from British Columbia to N. California; discovered by David Douglas in British Columbia about 1827, and first raised in the Glasgow Botanic Garden from his seed. It flowers from the end of June to the end of July, and a patch several feet across makes a rather striking display. The shoots that flowered the previous summer should be pruned back in February or early March, and the plants are all the better if broken up every few years as advised in the introductory notes to this genus, and the soil enriched. It thrives especially well near water. It is allied to *S. tomentosa*, but differs in its glabrous ovaries, its longer more oblong leaves, and in flowering earlier.

var. MENZIESII (Hook.) Presl *S. menziesii* Hook.—Young growths and the undersides of the leaves glabrous or downy, not felted, and the inflorescence parts more finely and less densely hairy. This variety has a more northern distribution than the typical state, from N. Oregon to Alaska, and extends farther inland. Introduced in 1838. It has the same garden value as typical *S. douglasii* and is pruned in the same way. For the spiraeas once treated as varieties of *S. menziesii*, see below.

S. × BILLIARDII Herincq—A group of hybrids between *S. douglasii*, or its var. *menziesii*, and the Old World *S. salicifolia*, which seem to have occurred spontaneously in many gardens from about 1850 onwards, the two species being closely akin and flowering at the same time. Here belong:

cv. 'BILLIARDII'.—This, the original clone of the group, was raised by the nurseryman and breeder Billiard (*sic*, not Billard) of Fontenay-aux-Roses and was described in 1855.

cv. 'EXIMIA'.—Leaves oval or obovate, 3 in. long, ½ to 1 in. wide, toothed in the upper half or two-thirds, underside more or less grey-felted. Panicles broadly pyramidal and much branched, the lower branches 3 or 4 in. long, leafy. Some of the finer panicles measure 8 in. long by 6 in. wide (*S. menziesii* f. *eximia* Hort. ex Zab.).

cv. 'TRIUMPHANS'.—Leaves oval-lanceolate, 1½ to 2½ in. long, ½ to ¾ in. wide, toothed nearly to the base, green beneath and slightly downy, especially on the veins. Panicles broadly pyramidal, branching at the base, up to 8 in. high and 4 in. wide. Flowers bright purplish rose. This is perhaps the finest of the late-flowering spiraeas, making a splendid display from mid-July onwards (*S. menziesii triumphans* Hort. ex Zab.).

Other hybrids of *S. douglasii*, of similar type to the above, have been named but are no longer in general cultivation.

S. PYRAMIDATA Greene—A shrub to about 4 ft high, spreading by suckers. Leaves usually downy beneath, oblong-elliptic to ovate-lanceolate or elliptic, coarsely toothed. Flowers white, tinged with pink in the bud, arranged in dense, broadly pyramidal to rounded panicles. This species, or perhaps natural hybrid, is intermediate between *S. betulifolia* and *S. douglasii* var. *menziesii*, having the habit of the former and an inflorescence like that of the latter, but shorter and relatively broader. It occurs wild in southern British Columbia, Oregon and Idaho (Hitchcock *et al.*, *Vasc. Pl. Pacific Northwest*, Part 3, p. 193 and p. 195 (fig.)).

S. GEMMATA Zab.

S. mongolica sensu. Koehne, not Maxim.

A shrub 4 to 8 ft high, with slender, arching, more or less angular stems, quite glabrous; buds slender, pointed, longer than the leaf-stalk. Leaves narrowly oblong, from ½ to 1 in. long, ⅛ to ⅓ in. wide, often entire with a short abrupt tip, but sometimes blunt and with about three teeth at the end; green and quite glabrous on both surfaces. Flowers white, small, produced during May in umbels about 1 in. across. Pedicels glabrous, slender, each furnished with a bract, which in the lower part of the inflorescence is leaf-like and inserted near the base of the pedicel, becoming progressively narrower and moving nearer to the flower in the upper part. Follicles glabrous.

Native of N.W. China. It was introduced to St Petersburg (? by Potanin from Kansu) in 1886 and the seeds distributed from there under the erroneous name "*S. mongolica*", whence Koehne's error. It is a pretty, white-flowered shrub in the style of *S. nipponica*, which it resembles in its bracteate inflorescence but from which it differs in its narrower, more oblong, mostly entire leaves and its long slender buds (in *S. nipponica* the buds are flattened-ovoid, shorter than the petiole). It is by no means so attractive a species, however, and now uncommon.

S. HENRYI Hemsl.

A shrub of lax, spreading habit, 6 to 9 ft high, more in diameter; branches reddish brown, slightly hairy when young. Leaves of the barren shoots $2\frac{1}{2}$ to $3\frac{1}{2}$ in. long, $\frac{3}{4}$ to $1\frac{1}{4}$ in. wide, narrowly oblong or oblanceolate, coarsely toothed near the apex, those of the flowering twigs much smaller, $\frac{3}{4}$ to $1\frac{1}{2}$ in. long, oblong or obovate, more shallowly toothed at the apex than the others, sometimes entire; all covered more or less with loose, greyish down beneath. Flowers white, $\frac{1}{4}$ in. across, produced in June on rounded corymbs, 2 in. across, which terminate short, leafy twigs; flower-stalk and ovary downy. *Bot. Mag.*, t. 8270.

SPIRAEA HENRYI

Native of central and western China; named in honour of Dr A. Henry, who first discovered it near Ichang in 1885; introduced for Messrs Veitch by Wilson in 1900. It is a fine shrub, and stands in the front rank of spiraeas, but on account of its wide-spreading habit needs plenty of space for lateral development; it is better as an isolated plant than grouped in a shrubbery. Shoots 5 ft or more long are made in a season. It is allied to the Himalayan *S. canescens*.

S. WILSONII Duthie—This is closely allied to, perhaps only a variety of *S. henryi*. It is distinguished among other points by its glabrous ovary, and glabrous or slightly silky flower-stalks. Leaves of flowering shoots entire, downy above, duller green. Introduced in 1900 from Hupeh. *Bot. Mag.*, t. 8399. For another ally of *S. henryi*, see *S. veitchii*.

S. HYPERICIFOLIA L.

This species is represented in cultivation by the following subspecies: subsp. OBOVATA (Waldst. & Kit.) H. Huber; *S. obovata* Waldst. & Kit. ex Willd.; *S. hypericifolia* var. *obovata* (Waldst. & Kit.) Maxim.—A bushy shrub, 5 or 6 ft high, with graceful, arching, twiggy branches, which, when young, are brown and usually covered with fine down, becoming grey with age. Leaves obovate, with a tapering base, and about three teeth at the apex, or none at all, ¾ to 1¼ in. long, ¼ to ½ in. wide, of a greyish green, slightly downy beneath, three or five nerves running lengthwise; stalk very short. Flowers pure white, ¼ in. across, produced during early May in clusters from the buds of the previous summer's shoots; each flower on a usually downy stalk about ⅜ in. long.

Native of N. Spain and S. France. At its best it is a pretty shrub, although not in the very first rank of spiraeas; starting later into growth than several other of its white-flowered allies, it escapes the damaging influence of late spring frosts.

The typical subspecies differs from the above in its narrower, oblanceolate leaves, usually acute at the apex and more or less entire, and in having the petals slightly longer than the stamens (about equalling them in subsp. *obovata*). It is much more widely distributed than subsp. *obovata* and does not occur in western and central Europe. It is now uncommon in cultivation, but the form of the species grown in 17th and 18th century gardens as '*Hypericum frutex*' was probably typical *S. hypericifolia*. It was apparently so called from the resemblance of its leaves to those of the common St John's Wort, *H. perforatum*, and was supposed to have come from Canada.

S. hypericifolia is a parent of the hybrids 'Arguta' and 'Grefsheim'.

S. JAPONICA L. f.
S. callosa Thunb.

A shrub to about 5 ft high, its branches usually terete, sometimes flattened and angled, glabrous or downy. Leaves short-stalked, lanceolate to ovate (sometimes broadly so), up to 3 in. or so long, acute at the apex, cuneate at the base, sharply toothed or incised, glabrous on both surfaces or downy beneath, the underside green or somewhat glaucous. Flowers about ⅜ in. across, rosy pink, produced in July and August at the ends of the current season's shoots in an often large and complex arrangement of flat corymbs. The whole inflorescence, in some garden clones, may be anything up to 12 in. wide, and consists of a terminal corymb of varying size, around which are arranged supplementary corymbs at the ends of laterals from the upper leaf-axils, which open later; but even on one and the same plant there is great variation in the size and complexity of the inflorescence, depending on the vigour of the shoots; separate and small corymbs are sometimes borne on short laterals lower down the shoot. Inflorescence axes glabrous or downy. Calyx-lobes erect at flowering-time, later spreading. Stamens longer than the petals and a conspicuous feature of the flower. Follicles glabrous except along the suture, erect or slightly spreading.

Native of Japan, with varieties in China, where it is widespread, except in the north. It is represented in the gardens of the western world by cultivars, of which the first to be introduced was the Chinese 'Fortunei' (c. 1850), followed by 'Albiflora' (to France before 1864) and 'Bullata' (before 1881). The others were probably either imported from Japan at a later date or raised in European gardens.

The garden varieties of *S. japonica* are among the most valuable and trouble-free of late-summer shrubs. They should be pruned in spring by cutting clean out sufficient of the older wood to prevent crowding, and then shortening back those selected to remain.

cv. 'ALBIFLORA' ('Alba').—An erect shrub of dwarfer and weaker growth than the other garden clones; young shoots downy, distinctly flattened and angled. Leaves light green, lanceolate, slenderly pointed, usually under 2 in. long, glabrous. Flowers white (*S. callosa* var. *albiflora* Miq.; *S. fortunei alba* Hort.; *S. callosa alba* Clemenceau; *S. japonica* var. *alba* Nichols.; *S. albiflora* (Miq.) Zab.; *S. leucantha* Lange; *S. japonica* var. *albiflora* (Miq.) Koidz.). A pretty shrub, growing to not much more than 2 ft high; introduced to Europe before 1864, probably from Japan.

It is unfortunate that most authorities (outside Japan) have followed Zabel in giving this clone the rank of a distinct species and treating other smaller growing clones with slightly flattened branches as hybrids between this "species" and *S. japonica* under the name *S.* × *bumalda* Burven. *sens.* Rehd. (*S.* × *pumila* Zab.). However, Japanese botanists, followed here, include it in *S. japonica*, and Dr Hara has pointed out that *S. japonica* is a variable species and that in Kyushu the branches are usually flattened (*Flora of Eastern Himalaya* (1966), p. 637).

'Albiflora' has given rise to a mutation in which some trusses bear pink flowers.

var. ALPINA Maxim.—See under 'Nana'.

cv. 'ALPINA'.—See 'Nana'.

cv. 'ATROSANGUINEA'.—Described by Zabel as having strongly villous inflorescences, dark rose-coloured flowers and persistently hairy stems.

cv. 'BULLATA'.—A dwarf shrub of very compact, rounded habit, rarely more than 12 or 15 in. high; young shoots erect, covered with rusty coloured down. Leaves $\frac{1}{2}$ to 1 in. long, almost or quite as much wide, broadly ovate, often recurved, coarsely and irregularly toothed, and nearly or quite glabrous on both surfaces except for a few hairs at the base and on the stalk; nerves prominent beneath. Flowers scarlet-rose, small, produced towards the end of July in great numbers in flat branching corymbs 3 in. wide at the end of the current season's growths; flower-stalks downy (*S. bullata* Maxim.; *S. japonica* var. *bullata* (Maxim.) Mak.; *S. crispifolia* Hort.).

A Japanese garden variety in cultivation in Britain by 1881. It is one of the dwarfest of spiraeas and one of the prettiest; very suitable for the rock garden or wherever small shrubs can be accommodated, and protected from stronger growing neighbours. The plant is almost hidden by its flowers in July.

Although often treated as a distinct species, it is now generally accepted that it is no more than an aberrant garden variety of *S. japonica*. The varietal epithet refers to the puckering of the leaf-blade, often noticeable between the veins.

cv. 'BUMALDA'.—Branches downy when young, slightly ridged. Leaves broad-lanceolate, at first brownish red, sharply and deeply toothed, glabrous. Inflorescence downy. Flowers carmine-pink (*S. bumalda* Burvenich, *Rev. Hort.*, 1891, p. 12; *S. × pumila* f. *bumalda* (Burv.) Zab.; *S. japonica* var. *bumalda* (Burv.) Bean).

'Bumalda', described in 1891, is of unknown origin; Kew received it in 1885 from Froebel of Zürich, but whether he raised it or imported it from Japan has not been ascertained. In foliage and habit it bears a certain likeness to 'Albiflora' (q.v.) but the notion that it is a hybrid between it and typical *S. japonica* is unproven.

As originally introduced, 'Bumalda' was a rather dwarf plant with a high proportion of its leaves edged or marked with yellow, or even wholly yellow, but it appears to be an unstable clone.

A few years before 1890 a branch-sport appeared on 'Bumalda' in the Knap Hill Nursery, with flowers of a much more beautiful shade of carmine. This is *S. japonica* 'ANTHONY WATERER' which is now the most frequently planted of the garden varieties. It received a First Class Certificate in 1891. It is itself somewhat sportive and there may be several sub-clones in commerce. A particularly fine form, growing to 4 or 5 ft high, shows no trace of the variegated leaves which the original 'Anthony Waterer' inherited from 'Bumalda', which, too, may have sported into taller and laxer forms. [PLATE 58

cv. "FASTIGIATA".—See *S. amoena*, treated under *S. bella*.

cv. 'FORTUNEI'.—One of the tallest of the garden varieties, growing to about 5 ft high; branches terete. Leaves reddish when young, oblong-lanceolate, 3 to 4 in. long, acuminate at the apex, sharply toothed, glaucous beneath, glabrous on both sides. Flowers deep pink, in large compound, downy corymbs. *Bot. Mag.*, t. 5164 (*S. callosa* sens. Lindl., not Thunb., *sens. strict.*; *S. fortunei* Planch.). This, the first representative of *S. japonica* in western gardens, was sent by Fortune to Standish and Noble in 1849 or the following year during his second visit to China; they were already offering plants at 10/6 each by the time he returned home in the autumn of 1851. It was the commonest form of the species in gardens for the rest of the century and was considered to represent the typical *S. callosa* of Thunberg, i.e., *S. japonica*. Indeed the description of *S. japonica* in previous editions appears to have been largely based on the Chinese 'Fortunei'. By the 1930s it had become so rare that a specimen was sent to Kew from an English nursery as a possible new species.

The Fortune introduction is the type of var. FORTUNEI (Planch.) Rehd., a native of E. and C. China, agreeing essentially with the original introduction.

cv. 'FROEBELII'.—Leaves brownish red when young, oblong-ovate, to about 3½ in. long, glaucous beneath, glabrous. Flowers bright crimson, in broad, compound corymbs; pedicels hairy but main inflorescence-axes almost glabrous. Distributed by Froebel of Zürich, before 1894 (*S.* × *pumila* f. *froebelii* Zab.; *S.* × *bumalda* f. *froebelii* (Zab.) Rehd.). A shrub of neat, erect habit, flowering in late summer. The similar 'COCCINEA' was introduced from Japan by Messrs Gauntlett of Chiddingfold, at an uncertain date.

cv. 'GLABRATA' ('Glabra').—A strong-growing form with corymbs, including the supplementaries, over 1 ft in diameter; flowers rosy pink. Leaves broadly ovate, the largest 4 to 5 in. long and 2½ to 3 in. wide, rounded at the base, and, like the young wood and flower-stalks, glabrous (*S. glabrata* Lange; *S. japonica* var. *glabrata* (Lange) Nichols.).

Of unknown origin, described by the Danish botanist Lange as *S. glabrata* and introduced to Kew from Denmark; the *S. pumila* var. *glabra* of Zabel, sent out by Messrs Hesse, is probably the same clone.

The botanical name var. GLABRA (Reg.) Koidz (*S. callosa* var. *glabra* Reg.) is of general application to glabrous forms of the species. The type was a cultivated plant, but bore no resemblance to 'Glabrata', judging from the authentic specimen in the Kew Herbarium.

cv. 'GOLDFLAME'.—Very effective in spring, the unfolding leaves being rich bronze-red, becoming light russet-orange, but rather jaundiced in appearance by the time the flowers appear; these are deep rosy red, but in rather small widely spaced corymbs. A few leaves are partly deep green and reversion shoots with leaves wholly of that colour sometimes grow out from the base of the plant and should be removed at once. Of American origin, introduced in the early 1970s.

cv. 'LEUCANTHA'.—A seedling of 'Albiflora', with rather larger and relatively broader more coarsely toothed leaves and a larger inflorescence. Flowers white, as in the parent. Raised by Zabel and later distributed by Messrs Hesse (*S.* × *pumila* var. *leucantha* Zab.).

cv. 'LITTLE PRINCESS'.—See under 'Nana'.

cv. 'MACROPHYLLA'.—Leaves as large as in 'Glabrata' or larger, but curiously inflated (bullate); inflorescence poor and small. It is really only worth growing as a curiosity, though a redeeming feature is that the leaves colour quite well in the autumn. Put into commerce by Messrs Simon-Louis, with whom it arose as a seedling.

cv. 'NANA' ('Alpina').—A very dense shrub, attaining after fifteen years or so a height of about 2½ ft and a spread of 6 ft; young stems and inflorescence-axes finely woolly. Leaves glabrous except at the margin, ovate or elliptic-ovate, mostly ½ in. or slightly more long, but ¼ in. long on the weakest shoots and around 1 in. long on the strong growths that emerge here and there, rounded to broad-cuneate at the base, with about five pairs of veins. Flowers lilac-pink, starting to open rather earlier than in most garden varieties of

S. japonica, mostly arranged in simple corymbs (but compound on the strongest shoots).

Although not outstanding in flower-colour, this makes a neat formal specimen. It needs no pruning beyond a light shearing in spring.

The origin of this variety is unknown. It has also been grown as *S. j. alpina* and erroneously as *S. densiflora*. 'Nyewoods' is probably the same clone, but 'LITTLE PRINCESS', selected in Holland from plants grown as *S. j. alpina*, is somewhat more robust, with larger leaves.

S. japonica var. ALPINA Maxim. was described in 1879 from a plant found wild on Mt Hakone in the central island of Japan, with procumbent or erect stems, growing a 'handbreadth' high. This variety, according to Maximowicz, was also cultivated in Tokyo gardens. But 'Nana' is perhaps of garden origin, since it does not come true from seed as a natural mountain variety might be expected to do, its self-sown seedlings being laxer and with larger leaves, though the characteristic rather woolly indumentum is retained.

cv. 'RUBERRIMA'.—Described by Zabel as having flowers coloured dark rose in downy corymbs and soon glabrous stems.

The spiraea once grown at Kew as *S. japonica* var. *ruberrima* was a hybrid of *S. japonica* with either *S. bella* or *S. amoena*; it came from St Petersburg. 'BUMALDA RUBERRIMA' was raised by Messrs Lemoine by crossing 'Bumalda' and 'Bullata' (*S.* × *lemoinei* Zab.).

S. LONGIGEMMIS Maxim.

A shrub 4 or 5 ft high, with glabrous, erect, angular stems, and curiously flat, leaf-like winter buds often ⅓ in. long. Leaves ovate-lanceolate,. wedge-shaped at the base, sharply and deeply, often doubly, toothed, 1½ to 3 in. long, ¾ to 1¼ in. wide, bright green and glabrous above, rather glaucous and hairy on the veins beneath when young; stalk ⅛ to ¼ in. long. Flowers white, ¼ in. across, produced towards the end of May in broad, rounded, corymbose panicles 2 to 3½ in. across and 1 to 2 in. long; stamens prominent; flower-stalk and calyx downy.

Native of W. China; described from specimens collected in Kansu but reported from as far south as Yunnan. It is a very pretty white-flowered spiraea, blossoming late enough to escape injury by frost, but now uncommon in cultivation. It was introduced towards the end of the last century and used to grow well at Grayswood Hill, Haslemere—the upper side of the branches wreathed with corymbs terminating long leafy twigs (*Gardening Illustrated*, 13 July, 1912). It is in cultivation at Kew.

S. 'MARGARITAE'

S. × *margaritae* Zab.

A shrub 4 or 5 ft high, its stems erect, reddish brown, and downy. Leaves 2 to 3½ in. long, ¾ to 1½ in. wide, narrowly oval or oblong, coarsely, sharply, and irregularly toothed at the terminal part, entire and narrowly wedge-shaped at the base. Flowers bright pink, $\frac{1}{8}$ in. wide, produced from July onwards in large, flat corymbs 3 to 6 in. across, terminating the growths of the year.

This spiraea was raised by Zabel from *S.* 'Superba' (see below), supposed to be a hybrid between *S. japonica* 'Albiflora' and *S. betulifolia* var. *corymbosa*; the pollen parent was probably *S. japonica*. However that may be, it is certainly one of the very best of the late summer-flowering group. A large mass of it makes a very striking effect from July to September. It should be pruned every winter or early spring in the same way as recommended for the Japonica group, i.e., to cut out entirely the older shoots and prune the younger ones back to within 1 ft of the ground, leaving only sufficient—say one every 6 in. or so—to furnish the plant during the ensuing summer. Treated in this way the shrub does not get to be more than 3 ft high, and becomes a sheet of blossom. If the corymbs are cut off as they fade, a succession of flowers may be obtained until September.

This hybrid was named by Zabel after his daughter.

S. 'SUPERBA'.—Mentioned above as the seed-parent of 'Margaritae', this was put into commerce by Froebel of Zürich in the 1870s as *S. callosa (japonica)* var. *superba*. It is not in general cultivation in Britain.

S. × FOXII K. Koch ex Zab.—A hybrid between *S. japonica* and *S. betulifolia* var. *corymbosa*, resembling 'Margaritae' but with blush-pink or white smaller flowers in June or July and glabrous leaves, green beneath. It was described in 1870 by K. Koch, under *S. japonica* but with the suggestion that it might be a hybrid between that species and *S. betulifolia*. Strictly both 'Margaritae' and 'Superba' should be placed under *S. × foxii*.

S. MEDIA F. Schmidt

S. confusa Reg. & Koernicke; *S. oblongifolia* Waldst. & Kit.

An erect shrub, up to 4 or 6 ft high, with glabrous, round stems sometimes downy when young. Leaves ovate or oblong with a wedge-shaped base, 1 to 2 in. long, $\frac{1}{8}$ to $\frac{3}{4}$ in. wide, the terminal part sharply toothed or with a few large teeth only near the apex, sometimes entire, upper surface glabrous, lower one more or less hairy or sometimes glabrous; stalk $\frac{1}{6}$ in. or less long. Flowers white, $\frac{1}{8}$ in. across, produced during late April and early May in long-stalked racemes 1 to $1\frac{1}{2}$ in. across each terminating a short leafy twig.

Native mainly of Russia, where it extends from the European parts almost to the Pacific, but occurring also in eastern Central Europe and the northern part of the Balkans; its western limit is in Austria (Steiermark) and the mountains north of Trieste. It is a pretty species, but liable to be injured by late spring frosts.

f. GLABRESCENS (Simonkai) Zab.—Leaves quite glabrous or soon becoming so.

var. MOLLIS (Koch & Bouché) Schneid. *S. mollis* Koch & Bouché—Leaves hairy on both sides, more densely so beneath. Possibly a natural hybrid between *S. media* and *S. cana*. S.E. Europe.

SPIRAEA MEDIA

var. SERICEA (Turcz.) Maxim. *S. sericea* Turcz.—Leaves entire, or few-toothed at the apex, downy above at first, later glabrous, usually permanently silky-hairy beneath. Native of the Russian Far East, Sakhalin, N. China, Korea and Japan.

S. media bears some resemblance to *S. chamaedryfolia*, which has more toothed leaves, angled stems, and longer petals (about $\frac{3}{16}$ to almost $\frac{1}{4}$ in. long, against $\frac{1}{8}$ in. long in *S. media*). Also the styles are terminal in *S. chamaedryfolia* while in *S. media* they are inserted on the outer side of the carpels, just below the apex.

S. MOLLIFOLIA Rehd.

A deciduous shrub up to 6 ft high, with arching branches; young shoots very hairy at first, becoming purple, nearly glabrous and very distinctly angled the second year; buds up to $\frac{1}{6}$ in. long, brownish purple. Leaves oval, oblong or obovate, tapered at both ends, usually more abruptly so at the apex, mostly entire, sometimes three-toothed at the apex, $\frac{1}{2}$ to $\frac{3}{4}$ in. long, half as wide, silky all over. Flowers white, $\frac{1}{3}$ in. in diameter, borne during June and July in corymbs about 1 in. across, terminating short leafy twigs that spring from the growths of the previous year; stamens twenty.

Native of W. Szechwan, China; discovered by Wilson in 1904; introduced in 1909. It is quite distinct from all other cultivated spiraeas in the combination of its silky leaves, and long, slender winter buds. The closely allied *S. gemmata* is distinguished 'by its narrower, glabrous foliage, glabrous inflorescence and slenderer branches.' (*Pl. Wils.*, Vol. I (1913), p. 442).

S. MYRTILLOIDES Rehd.

S. virgata Franch., not Raf.

A deciduous shrub 4 to 8 ft high, with angular downy shoots, becoming glabrous and dark coloured the second year. Leaves oval, mostly rounded at the apex, broadly tapered at the base, entire, or obscurely toothed towards the apex, $\frac{1}{4}$ to $\frac{5}{8}$ in. long, about half as much wide, glabrous on both surfaces, glaucous beneath; stalk $\frac{1}{20}$ in. long. Flowers white, scarcely $\frac{1}{4}$ in. wide, produced in early June in rounded clusters $\frac{1}{2}$ to 1 in. wide at the end of short leafy shoots springing from the growths of the previous year; flower-stalks downy. Calyx glabrous except near the flower-stalk, its lobes triangular. Petals roundish obovate. Ovary glabrous.

Native of western and central China; described by Franchet under the illegitimate name *S. virgata* from specimens collected by Delavay in Yunnan. It was introduced to the Arnold Arboretum in 1908 from W. Szechwan by Wilson, who also collected the type specimen of *S. myrtilloides*, but probably did not reach this country until some years later, when Forrest sent seeds from Yunnan.

S. myrtilloides has no outstanding qualities as a garden plant and is not in general cultivation. It appears to be nearest in most characters to *S. alpina* and *S. media*.

S. NIPPONICA Maxim.

S. bracteata Zab.

A deciduous shrub of rounded, bushy habit, growing 4 to 8 ft high, the branches, leaves, and flower-stalks quite glabrous; young wood reddish. Leaves very broadly obovate or oval, sometimes nearly round, $\frac{1}{2}$ to 1 in. long, sometimes entire, but usually with a few broad teeth at the rounded apex; stalk $\frac{1}{6}$ to $\frac{1}{8}$ in. long. Flowers borne in June, pure white, $\frac{1}{3}$ in. across, crowded densely in rounded or conical clusters, 1 to 1½ in. wide. Each cluster is borne at the end of a leafy twig, 1½ to 3 in. long, springing from the wood of the previous year; petals overlapping. The synonym refers to the bracts on the flower-stalks, which are leafy and conspicuous on the outside of the inflorescence, but present throughout. *Bot. Mag.*, t. 7429.

SPIRAEA NIPPONICA

Native of Japan (for introduction see 'Rotundifolia'). It is sometimes injured by late frosts, but when these are escaped, few June-flowering shrubs are more lovely. The individual flowers are beautifully formed, and the clusters are all borne on the upper side of the horizontal or arching branches. It is perfectly hardy, but needs liberal conditions at the root, even more than the majority of spiraeas do. The great thing is to get a comparatively few long shoots rather than a crowd of small twiggy ones. Old shoots should be removed as soon as they produce nothing but twiggy shoots; vigorous plants produce so much new wood that all flowered stems can be cut out.

cv. 'ROTUNDIFOLIA'.—This is the typical state of the species, as described above. It was introduced by Siebold to his nursery at Leyden but was not generally available in commerce until Messrs Lemoine acquired stock and distributed it shortly before 1882 (*S. rotundifolia alba* Hort.; *S. media* var. *rotundifolia* Nichols.). *S. nipponica* Maxim. is simply a renaming of *S. media* var. *rotundifolia* of Nicholson, i.e., the Siebold clone; Maximowicz also cited a wild-collected specimen, which too had rather broad leaves. It is this round-leaved clone that is figured in the *Botanical Magazine* and received an Award of Merit in 1955.

cv. 'SNOWMOUND'.—Leaves narrower than in the typical round-leaved form, being narrowly oblong-obovate, 1 to 1½ in. long and ⅜ to ½ in. or slightly more wide on the strongest shoots, around ⅝ by ¼ in. under the flower-clusters. Everything said in praise of *S. nipponica* above applies equally or perhaps in greater measure to this spiraea, which in good, moist soil will grow to 4 or 5 ft high and more in width in three years. It has been confused with var. *tosaensis*.

var. TOSAENSIS (Yatabe) Mak. *S. tosaensis* Yatabe—Leaves linear-oblong, tapered at the base, entire or with two or three teeth at the apex, ⅜ to 1 in. long, ⅛ to ¼ in. wide. Native of Japan in the island of Shikoku, where it was discovered on the banks of the river Watarigawa in 1891; introduced to Kew in 1923. It is distinct in its narrow leaves which occur on the slender twigs seven to ten to the inch. It is perfectly hardy but is too modest in flower to attract much notice. For var. *tosaensis* Hort. see 'Snowmound'.

S. PRUNIFOLIA Sieb. & Zucc.

This species is scarcely known in cultivation except by the double-flowered form, to which the following description refers: A shrub 4 to 6 ft high, the branches gracefully arching and forming a dense bush as much in diameter as it is high; young shoots downy. Leaves ovate, 1 to 1¾ in. long, ½ to ¾ in. wide, downy beneath (especially when young), finely and evenly toothed, stalk ⅛ in. or less long. Flowers produced during late April and May in fascicles three to six together, each flower on a glabrous, slender stalk, ½ to ¾ in. long; petals pure white and so numerous as to form a flower like a small 'bachelor's button', ½ in. across.

Native of China, and much cultivated there; it was found by Wilson in its double-flowered state in W. Hupeh. This form was originally introduced from Japan by Siebold about 1845. In the *Gardeners' Chronicle* for 20th February 1847, an advertisement sets forth that 'the stock of this magnificent novelty bought at Dr Siebold's sale, is now in the possession of Louis van Houtte, florist at Ghent,' and plants to be delivered the following April are offered at one guinea each. It is still one of the most beautiful of hardy shrubs, producing during the summer slender shoots, 1 to 2 ft long which, the following May, are wreathed from end to end with blossom. For the needs of most gardens it can be increased sufficiently quickly by taking off the side suckers from old plants and potting them, then placing them in a mild bottom heat; but if such conveniences are not available they can be planted in the open ground—a slower, less certain process.

The single-flowered plant is in cultivation, and is distinguished as *S. prunifolia* f. *simpliciflora* Nakai. In my experience it is an absolutely worthless shrub because of its extraordinary sterility. A plant was obtained from the Continent for Kew in 1887, but although I have known this and others raised from it for twenty years, I have never yet seen it in flower. But this, of course, is more likely to be an individual than a racial characteristic, seeing the floriferousness of the double-flowered form.

S. ROSTHORNII Pritz.

A deciduous shrub up to 6 ft high, of spreading habit; young shoots slightly downy. Leaves ovate to lanceolate, slenderly pointed, broadly wedge-shaped or almost rounded at the base, sharply, jaggedly, unevenly toothed; 1½ to 3 in. long, ¾ to 1¼ in. wide, bright green above, more or less downy on both surfaces but especially beneath; stalk ⅛ to ¼ in. long. Flowers white, about ¼ in. wide, produced in early June in flattish corymbs 2 to 3½ in. across that terminate leafy shoots; flower-stalks all downy.

Native of W. Szechwan, China; introduced under Wilson's No. 965 in 1909. Nearly related to *S. longigemmis* and having the same curious elongated winter buds, but that is a glabrous shrub as regards the young shoots and nearly so as regards the leaves.

S. PRATTII Schneid.—Made synonymous with this species in *Plantae Wilsonianae*, vol. I, this seems to differ in its sturdier growth; more compact, densely flowered, rounded corymbs; smaller, more abundant and more densely pubescent leaves. It was found by A. E. Pratt in W. Szechwan (No. 190).

S. SALICIFOLIA L.   BRIDEWORT

A shrub 3 to 6 ft high, with running roots and forming ultimately a dense thicket of erect stems, which are soon quite glabrous. Leaves lanceolate or narrowly oval, but sometimes broadest above the middle, 1½ to 3 in. long, ½ to 1 in. wide, pointed, sharply and often doubly toothed, glabrous and green on both surfaces. Flowers rose-tinted white, crowded on erect, terminal, slightly downy panicles about 4 in. high, and 2 in. wide at the base.

Native of the Old World, where it is widely distributed from Central Europe through Russia and Mongolia to N. China, Korea and Japan; cultivated since the 16th century and widely naturalised in Europe, including Britain, to the west and north of its natural range. When once it obtains a footing, it appears to be able to hold its own against any other vegetation, spreading by its creeping suckers and forming an almost impassable thicket. Left to itself in this way, its inflorescences become poor; but cultivated in good garden soil and occasionally divided, it makes a handsome show in June and July.

S. ALBA Du Roi   *S. salicifolia* var. *paniculata* Ait.—Often regarded as a variety of the preceding, this differs chiefly in its much larger compound panicles 8 to 12 in. long and as much wide, and the usually white flowers. The leaves are mostly oblong-lanceolate, 2 to 3 in. long, ⅜ to ¾ in. wide. Native of N. America, where it is widely distributed in the north-eastern parts of the USA and extends into Canada.

S. LATIFOLIA (Ait.) Borkh.   *S. salicifolia* var. *latifolia* Ait.—This N. American species is closely allied to the preceding, differing in the broadly oval to obovate leaves and glabrous inflorescence (downy in *S. alba* as in *S. salicifolia*). The flowers are often pink or flesh-coloured. It does not extend so far west as *S. alba*, but reaches farther north, to Newfoundland.

S. × SANSSOUCIANA K. Koch

A hybrid between *S. japonica* and *S. douglasii* originally named and described from a form raised in the government nursery at Sans Souci, near Potsdam, Germany, shortly before 1857. However, the form of the cross cultivated in this country is of independent origin. This is:

cv. 'NOBLEANA'.—A shrub 4 or 5 ft high, with erect, brown stems covered with a close grey felt. Leaves oblong to narrowly oval, 2 to 4 in. long, ¾ to 1¼ in. wide; mostly tapering, sometimes rounded at the base, irregularly and rather jaggedly toothed except near the base, green, downy on the veins above, covered with a dull greyish, close down beneath. Flowers bright rose, produced during July, and densely crowded in broad, corymbose panicles which form an inflorescence 3 to 10 in. across, terminating the shoot of the year; flower-stalks and receptacle grey-felted. (*S. nobleana* Hook., *Bot. Mag.*, t. 5169).

This hybrid was sent to Kew in 1859 by Charles Noble of Sunningdale and was named *S. nobleana* by Sir William Hooker. In an accompanying letter, still preserved at Kew, Noble stated that it had been raised from a plant of *S. douglasii* growing by the side of one of *S. japonica*. Hooker nevertheless identified it with a spiraea that had been collected by William Lobb in California, and for many years 'Nobleana' was regarded as a native of that State. There is no doubt, however, that Lobb's plant is different, and that Noble's plant had the origin he indicated.

Also coming under *S. × sanssouciana* is 'INTERMEDIA', a hybrid between *S. japonica* 'Albiflora' and *S. douglasii*, raised by Lemoine. It is similar to *S. × semperflorens* 'Syringiflora' (see below), but the leaves are finely woolly beneath.

S. × SEMPERFLORENS Zab.—A group of hybrids between *S. japonica* and the Old World *S. salicifolia*. It is represented in commerce by 'SYRINGIFLORA', a spreading shrub to about 4 ft high. Leaves lanceolate or lanceolate-oblong, up to 3 in. long and ⅞ in. wide, acuminate, almost glabrous, toothed above the middle. Flowers light pink, borne over a long period in late summer in broadly conical trusses; stamens longer than the petals. The parent on the *S. japonica* side is 'Albiflora'. A fine hybrid, distributed, and perhaps raised, by Messrs Lemoine.

S. × WATSONIANA Zab. (1907) *S. × nobleana sens.* Zab. (1893), not Hook. f. —Similar to *S. × sanssouciana* and of about the same garden value as 'Nobleana'. Like that hybrid it unites *S. douglasii* with the Japonica group, but the second parent instead of *S. japonica* is its American ally, *S. densiflora* var. *splendens*. It was raised by Zabel, the latter being the seed-parent. The two parents both occur wild in western N. America and it has been suggested that *S. subvillosa* Rydb., described in 1908 from a specimen collected wild in Oregon, may belong to *S. × watsoniana*, but other authorities consider it to be a variation of *S. douglasii* var. *menziesii*.

S. SARGENTIANA Rehd.

A deciduous shrub 4 to 6 ft high, with long, slender, arching, round young shoots, at first downy, soon glabrous. Leaves narrowly oval to narrowly obovate, wedge-shaped and entire at the base, more or less toothed towards the tip, $\frac{1}{2}$ to 1 in. long, $\frac{1}{8}$ to $\frac{1}{2}$ in. wide, dull green above, paler and downy beneath, the few veins running lengthwise; stalk $\frac{1}{16}$ to $\frac{1}{8}$ in. long. Flowers creamy white, $\frac{1}{4}$ in. wide, produced during June in rounded clusters 1 to $1\frac{3}{4}$ in. wide that terminate short leafy twigs springing from the virgin shoots of the previous year; main and secondary flower-stalks downy.

Native of W. China; discovered and introduced by Wilson in 1908–9. In its graceful habit, small leaves and mode of flowering this resembles the well-known *S. canescens* which can be distinguished by its ribbed stems. It is also closely related to *S. henryi*, which has much larger but similarly toothed leaves. *S. sargentiana* is a distinctly pretty shrub, perfectly hardy and well worth cultivation. It received an Award of Merit in 1913.

S. THUNBERGII Blume

A shrub 3 to 5 ft high, often more in diameter, of very twiggy, bushy habit; branchlets slender, angled, downy. Leaves linear-lanceolate, 1 to $1\frac{1}{2}$ in. long, $\frac{1}{8}$ to $\frac{1}{4}$ in. wide, taper-pointed, the margins set with a few incurved teeth, smooth and pale green on both sides. Flowers pure white, $\frac{1}{4}$ in. across, produced on the leafless, wiry twigs during March and April in clusters of two to five, each flower on a glabrous, slender stalk, $\frac{1}{4}$ to $\frac{1}{3}$ in. long. Calyx shallow, smooth.

Native of China; but first introduced from Japan, of which country, however, it is not native. This is the earliest of all the spiraeas to flower in the open, and in ordinary seasons is at its best by the middle of April. The fascicles of blossom spring directly from the shoots made the previous summer, and if the season has been sufficiently sunny and hot to have thoroughly ripened the wood, the plants will be almost hidden by the profusion of flowers. The habit of the plant is graceful owing to the arching form of the slender branches, and altogether there are few more attractive shrubs in bloom in early April. The leaves are slow to fall in autumn and sometimes remain on the bush throughout the winter.

S. TOMENTOSA L.   STEEPLEBUSH

A shrub 3 to 5 ft high, with spreading underground roots, ultimately forming a thicket of erect angled stems which when young are covered with brownish felt. Leaves ovate, $1\frac{1}{2}$ to 3 in. long, $\frac{3}{4}$ to $1\frac{1}{2}$ in. wide, coarsely and irregularly toothed almost to the base, dark green and nearly glabrous above, covered with a close, yellowish grey felt beneath. Flowers purplish rose, densely produced in erect, terminal, branching panicles 4 to 7 in. long, $1\frac{1}{2}$ to $2\frac{1}{2}$ in. wide during late summer.

Native of the eastern United States; introduced, according to Aiton, in 1736. It is allied to the western *S. douglasii*, and is often confused with it;

it is, however, distinguished by the thicker, browner (or yellowish) felt beneath the leaves, which are toothed much nearer the base; by flowering some weeks later, and by the ovaries being woolly (glabrous in *S. douglasii*).

f. ALBA (Weston) Rehd.—A pretty white-flowered form.

The cultivation of these handsome spiraeas is the same as for *S. douglasii*.

S. TRICHOCARPA Nakai   [PLATE 59

A deciduous shrub 4 to 6 ft high, with rigid, spreading branches; young shoots glabrous, distinctly angled. Leaves oblanceolate to oblong, abruptly pointed, tapered at the base, with a few teeth near the apex or none at all; 1 to 2½ in. long, ½ to 1 in. wide, vivid green above, rather glaucous beneath, quite glabrous; stalk ¼ in. or less long. Flowers white, ⅓ in. wide, produced during June on rounded corymbs 1 to 2 in. wide, growing upwards, and terminating short leafy twigs that are clustered on the preceding year's shoots, the whole forming a handsome arching spray sometimes over 1 ft long. Petals roundish, notched; flower-stalks and seed-vessels downy.

Native of Korea; discovered in 1902; introduced to cultivation in 1917 from the Diamond Mountains by Wilson, who considered it one of the best of all spiraeas. It is distinct in its glabrous, nearly entire leaves, very pale beneath. It is considered to be nearest to *S. nipponica* and has similar leafy bracts on the inflorescence, but *S. nipponica* has much shorter, more oval leaves and a more compact, less spreading habit. *S. trichocarpa* thrives in the garden of the late E. H. M. Cox at Glendoick in Perthshire, where it has grown since Wilson introduced it, and flowers well every year even where densely shaded in summer (E. H. M. and P. A. Cox, *Modern Shrubs* (1958), p. 169; see also *New Fl. and Sylva*, Vol. I (1928), pp. 12–13 and fig. iii; *Gard. Chron.* Vol. 74 (1923), p. 87 and fig. 32).

This species received an Award of Merit in 1942.

S. TRILOBATA L.

A twiggy shrub, 3 to 4 ft high, of broad but compact habit, young shoots and leaves glabrous; stems round, often zigzagged in growth. Leaves roundish, ½ to 1 in. (rarely 1½ in.) long, and about as much wide, coarsely toothed, sometimes obscurely three- or five-lobed, the base rounded or sometimes slightly heart-shaped, rather glaucous green. Flowers white, small, produced during June, packed very numerously in umbels ¾ to 1½ in. across; each umbel terminating a short leafy twig, springing from the previous year's growth, every flower having a slender, glabrous stalk ⅓ to ¾ in. long.

Native of N. Asia, from Korea and N. China to Siberia and Turkestan; introduced in 1801. Although its flower-buds are sometimes injured by frosts, this is a very pretty shrub of neat habit.

S. BLUMEI G. Don   *S. chamaedryfolia sens.* Bl., not L.—A native of Korea and Japan, this is nearly allied to *S. trilobata*, but differs in the shape of the leaf, which is ovate or lozenge-shaped, longer than it is wide, the base wedge-shaped. Flowers white, crowded in umbels 1 in. wide. A shrub 3 to 6 ft high.

S. × VANHOUTTEI (Briot) Zab.   [PLATE 60

S. aquilegifolia vanhouttei Briot

A shrub 6 ft high, with gracefully arching, glabrous brown stems. Leaves rhomboidal or obovate, sometimes distinctly three-lobed, more or less broadly tapering and entire at the base, coarsely toothed on the upper half; ¾ to 1¾ in. long, ½ to 1¼ in. wide; dark green above, rather glaucous beneath, glabrous on both sides. Flowers white, ⅓ in. across, produced during June in umbel-like clusters 1 to 2 in. across; calyx-lobes erect.

A hybrid between *S. trilobata* and *S. cantoniensis*; raised by Billiard, a nursery-man at Fontenay-aux-Roses, near Paris, about 1862. At its best it is probably the finest of all the white-flowered spiraeas, except perhaps S. 'Arguta'; in low-lying situations it is subject to injury by late spring frosts. In more elevated gardens, or where the plant is not forced into premature activity by unseasonable warmth, there is no more desirable shrub, for it is very hardy. Its stems, at first erect, afterwards arching outwards at the top, bear the extraordinarily profuse blossoms on the upper side of the branches. It is one of the spiraeas which should have the older wood thinned out after flowering to allow light and air to enter and help in the development of the younger growths. It is very valuable for forcing early into bloom for indoor decoration, and used to be exhibited in this state at the spring shows under the erroneous name of *S. confusa*, a synonym of *S. media*—a less vigorous shrub with longer stamens and the calyx-lobes ultimately reflexed.

S. VEITCHII Hemsl.

A strong-growing shrub, 10 or 12 ft high eventually, producing gracefully arching shoots, 2 to 3 ft long in a season; young branches reddish, slightly downy. Leaves ¾ to 2 in. long, ⅓ to ¾ in. wide, oblong or obovate, not toothed, glabrous on both surfaces or very slightly downy beneath. Flowers white, small, crowded in dense corymbs, 1½ to 2½ in. across. Calyx and flower-stalks covered with a fine down. Fruits glabrous when ripe. *Bot. Mag.*, t. 8383.

Native of western and central China; discovered by Wilson in W. Hupeh in 1900, and introduced by him for Messrs Veitch. It is a fine species (Mr Wilson told me he considered it the best of Chinese spiraeas), somewhat similar in general aspect and in producing its flowers on short leafy twigs from the growths of the previous summer, to the well-known *S. canescens*. It is readily distinguished from that species, however, by its glabrous, entire leaves and glabrous fruit. Its entire leaves also distinguish it from two other allies—*S. henryi* and *S. wilsonii*. I saw the plants first introduced in their young state in the Coombe Wood nursery, when they were making shoots as much as 8 ft long in a season; when these, the following June, were wreathed from end to end with clusters of pure white blossom, they made a picture of remarkable beauty.

S. YUNNANENSIS Franch.

S. sinobrahuica W. W. Sm.

A deciduous shrub 4 to 7 ft high; young shoots covered with a thick tawny down; buds white-woolly. Leaves broadly ovate or obovate, lobed and toothed at the upper part, entire and tapered towards the stalk, $\frac{1}{2}$ to $1\frac{1}{8}$ in. long, from two-thirds to nearly as much wide, dull green and downy above, grey-tawny and velvety beneath; stalk $\frac{1}{16}$ to $\frac{1}{8}$ in. long. Flowers creamy white, $\frac{1}{3}$ in. wide, produced during June in rounded clusters 1 in. across at the end of short leafy twigs. There are ten to twenty flowers in a cluster, each borne on a slender downy stalk. Petals roundish. Calyx downy like the flower-stalk. Stamens twenty.

Native of Yunnan, China; discovered by Père Delavay and introduced by Forrest, who found it at upwards of 10,000 ft altitude, varying apparently in height from 3 to 8 ft. It is very distinct in the tawny down that covers the young shoots, under-surface of the leaves, flower-stalks and calyx; also in the goodly size (for a spiraea) of the blossoms. It is evidently nearly akin to *S. chinensis*.

STACHYURUS   STACHYURACEAE

This genus, once included in the Theaceae (Ternstroemiaceae), now ranks as a separate family, of which it is the only member. Leaves deciduous; stipules present, but minute and soon falling. Flowers bisexual in axillary racemes, formed in autumn and opening in spring before the leaves. Sepals and petals four, imbricated. Stamens eight, in two whorls. Ovary four-chambered with numerous axile ovules (interpreted by some botanists as one-chambered, with the ovules arranged on parietal placentae which grow out from the wall of the ovary and meet in the centre). Style short, simple, with a capitate stigma. Fruit a leathery many-seeded berry. It is a small genus, with about ten species in E. Asia.

Of the species treated here, only *S. praecox* and *S. chinensis* are frequently planted, and both are hardy. They prefer an acid, humus-rich soil, but are said to be lime-tolerant; it is always advisable to add peat or leaf-mould to the soil when planting, especially if it is on the heavy side. A sunny or half-shaded position is necessary. Propagation is by cuttings made of fairly firm wood taken in July with a heel attached, and placed in gentle heat.

The genus is reviewed by H.-L. Li in *Bull. Torr. Bot. Club*, Vol. 70 (1943), pp. 615–28.

S. PRAECOX Sieb. & Zucc.

S. japonicus Steud.

A deciduous shrub, said to become as much as 10 ft high in Japan, but rarely more than half as high in England. Leaves ovate-lanceolate, glabrous, 3 to 7 in. long, with a long slender apex, toothed at the margin. Flowers twelve to twenty together, in stiff drooping racemes 2 to 3 in. long, each flower ⅓ in. across, pale yellow. *Bot. Mag.*, t. 6631.

Native of Japan, and quite hardy. Its greatest merit in the garden is its early-flowering nature. In favourable years it will be in full flower by the middle of February, and ordinarily, not more than a month later. The flower-spikes are formed in the axils of the leaves and attain their full length in autumn, and, although exposed to whatever inclemencies the winter may bring, remain unscathed. Unseasonable warmth in the early part of the year, followed by a rough cold spell, will sometimes injure the flowers. But on the whole they are very hardy, and when the reddish leafless branches are hung with yellow racemes 1 in. or less apart there are few things in the garden more pleasing at that early season.

var. MATSUZAKII (Nakai) Makino *S. matsuzakii* Nakai; *S. lancifolius* Koidz. —A more robust variant of the species, with stouter branches and larger leaves and fruits, confined to maritime localities in southern Japan and the Ryukyus.

S. CHINENSIS Franch.—This Chinese species, closely allied to *S. praecox*, was introduced by Wilson in 1908. Although, as seen growing side by side, they appear distinct, there is really very little on which one can seize to differentiate them. In habit *S. chinensis* is the stronger and more vigorous, sending up strong arching shoots, varying from green to dark brown, sometimes red as in *S. praecox*. The leaves are relatively broader, ovate or oblong-ovate, and more abruptly acuminate; the style is about equal in length to the petals or slightly exserted, and the fruits smaller. It flowers at Kew about a fortnight later than *S. praecox*, and is as hardy and, if anything, more attractive.   [PLATE 61

S. HIMALAICUS Benth.—A tall straggling shrub in the wild. Leaves more slender than in *S. praecox*, oblong to lanceolate, 3 to 5 in. long, 1½ to 2 in. wide, acuminately narrowed to a long, slender tip, very finely and densely toothed, short-stalked. Racemes 2 to 4 in. long. Flowers yellow or pinkish; style included. Of wide distribution from the Himalaya and N.E. India to central and southern China and Formosa. A rather tender species that needs the protection of a wall except in the milder parts.

STAPHYLEA   BLADDER-NUT   STAPHYLEACEAE

A genus of about ten species in the northern hemisphere, all shrubs but sometimes large enough to be considered small trees. Leaves opposite, trifoliolate in most species but pinnate with five to seven leaflets in two. Flowers in terminal racemes or panicles, white or white flushed with pink. Sepals five, petaloid. Petals five, upright. Carpels two or three, free, or connate at the base (the styles sometimes connate at the apex), developing into a two- or three-lobed inflated capsule, each chamber of which contains a few lustrous seeds.

The bladder-nuts have a rather remarkable distribution over the North Temperate Zone. They spread all round the world, but most of the species have each their own separate area. Thus, starting at home, we have *S. pinnata*, which extends through Europe to Asia Minor, then come *S. colchica* in the Caucasus, and *S. emodi* in the Himalaya and Afghanistan. Crossing into China, there is *S. holocarpa* (and perhaps one or two more species); then *S. bumalda* carries the genus to the western shores of the Pacific. Across that ocean the roll is taken up on the western side of N. America by *S. bolanderi* and some central American species, and on the Atlantic side by *S. trifolia*.

The bladder-nuts are planted in gardens for the beauty of the foliage and flowers, and for their interesting fruits. All those given separate mention below are hardy. Their needs are simple—a good, loamy, moist soil and a fairly sunny spot. They can be increased by cuttings.

S. BUMALDA DC.
Bumalda trifolia Thunb.

A deciduous shrub 3 to 6 ft high, of neat habit. Leaves of three leaflets which are 1½ to 3 in. long, ovate-lanceolate, sharply toothed, downy on the midrib and veins. Flowers greenish white, ⅓ in. long, borne in a terminal cymose cluster 1½ to 3 in. long. Fruit a membranous inflated capsule about 1 in. long and wide, in two flattened obovate parts, each terminated by the bristle-like, persistent style; seeds yellowish.

Native of Japan, where it inhabits mountainous regions; also of Central and W. China. The Japanese form has not proved of much value in gardens, being rather tender and having few attractions. It flowers in May and June.

S. COLCHICA Stev.

A deciduous shrub, 6 to 10 ft high, with stiff, erect branches. Leaves composed of three or five leaflets, which are ovate-oblong, 2½ to 3½ in. long, nearly or quite glabrous, shining beneath, the margins set with fine, rather bristle-like teeth, the terminal leaflet is stalked, the lateral ones stalkless. Flowers in erect panicles terminating the young shoots and lateral twigs, the largest up to 5 in. long, and as much wide, each flower ¾ in. long and wide, the sepals spreading, narrow oblong, very pale green; petals white, erect, narrow, recurved at the tips. Fruit a two- or three-celled inflated capsule, 3 to 4 in. long, 2 in. wide, the apex of each division ending in a long, fine point. Seeds ⅓ in. long, pale brown. Flowers in May. *Bot. Mag.*, t. 7383.

A native of the S.W. Caucasus; introduced to Britain shortly before 1879. It is the handsomest of the staphyleas, allied to the more western and more widely distributed *S. pinnata*, but distinguishable from it in leaf by the shining lower surface, and in fruit by the much larger capsules. It received a First Class Certificate when shown by Messrs Veitch in 1879.

var. KOCHIANA Medved.—Filaments of stamens hairy (said to be glabrous in the type).

S. 'COULOMBIERI'.—A deciduous shrub of vigorous habit, considered by some to be a hybrid between *S. colchica* and *S. pinnata*, by others a variety of the former. The leaves are composed of three or five leaflets, which are larger than in either of the reputed parents, the terminal one often 5 to 6 in. long; they are

STAPHYLEA 'COULOMBIERI'

ovate-oblong, toothed, dark green on both sides, and very lustrous beneath. Flowers white, and intermediate in size between those of the parents; the panicles are not so large as in *S. colchica*, the blossoms more compact, and the sepals and petals wider and shorter. The fruit is intermediate in size, being a two- or three-celled capsule, 1½ to 2 in. long, the seeds rather larger. (*S.* × *coulombieri* André; *S. colchica* var. *coulombieri* (André) Zab.).   [PLATE 62

This handsome shrub was first noticed as showing hybrid characters in the nursery of Coulombier, at Vitry, in France, in 1887. It had been obtained by him from the famous arboretum of Segrez in 1872, beyond which date its history is unknown. But it may well have originated there as a chance hybrid. It is most closely related to *S. colchica*, especially in the shining green under-surface of the leaves, but the much smaller fruits and the differences in the sepals and petals distinguish it.

S. 'GRANDIFLORA'.—A very distinct form with much longer, laxer panicles, and larger individual flowers; the leaflets are rather longer than in ordinary *S.* 'Coulombieri', but proportionately narrower (*S. colchica* f. *grandiflora* Zab.; *S. coulombieri* var. *grandiflora* (Zab.) Bean; *S.* × *elegans* Hort. ex A. T. Johnson, not Zab.).

S. 'HESSEI'.—Leaflets mostly five, sometimes three, slenderly pointed. Flowers in more or less pendulous panicles. Sepals rose-coloured, especially at the apex. Petals stained deep pink at the base. Very free-flowering. Raised by Messrs Hesse of Weener, Hannover, some years before 1898. It was considered by Zabel to be a hybrid between *S. colchica* var. *coulombieri* (see 'Coulombieri') and *S. pinnata* and named by him *S.* × *elegans* var. *Hessei*. The typical S. × ELEGANS of Zabel was not a plant raised by himself, as has been erroneously stated. He received it in 1871 from the Flottbeck Nurseries, Hamburg, as *S. colchica*, and described it as a rather tender shrub with the habit of *S. pinnata*, mostly five leaflets, large pink-flushed flowers in ovate panicles, not freely borne and rarely developing fruits. He judged it to be a hybrid between *S. colchica* and *S. pinnata* (the earlier named *S.* × *coulombieri* André was con-sidered by Zabel not to be a hybrid between these two species but a variety of the former; see above). Whether the supposedly hybrid *S. elegans* is in cultiva-tion it is impossible to say; the plant sometimes grown under the name appears to be *S.* 'Grandiflora'.

S. HOLOCARPA Hemsl.

A deciduous shrub, 20 to 30 ft high in the wild state, and often tree-like; young shoots glabrous. Leaves of three leaflets, which are oblong-lanceolate, abruptly acuminate, the terminal one stalked and 2 to 4 in. long, the side ones almost stalkless; all finely toothed, and downy at the base beneath. Flowers white or rose-coloured, about ½ in. long, borne in April or May in slender panicles up to 4 in. long, which spring direct from axillary buds on the wood of the previous year. Fruit a three-celled, pear-shaped, inflated capsule, 2 in. long, 1 in. wide, tapering gradually at the base but terminating in a short, sharp point. Seeds shining grey, about the size of large shot. *Bot. Mag.*, t. 9074.

Native of central and, more rarely, of western China; discovered by Henry in Hupeh; introduced by Wilson in 1908 for the Arnold Arboretum. Wilson called it the most beautiful species of the genus, but it has never become common outside collections. It is hardy enough except when it is young and making long sappy growths. But the panicles emerge from the buds on the naked wood quite early in the spring and may be destroyed by frost before the flowers expand. Both in the wild and in cultivation it grows to about 30 ft high and may even form a small tree with a single trunk. But the flowers make a finer display if it is grown as a shrub.

The pink-flowered form, usually called 'Rosea', is more admired than the white; the young leaves are chocolate-bronze in colour, later sea-green, and add much to the beauty of the plant. It received an Award of Merit in 1953 (the white-flowered form was accorded the same award in 1924).

var. ROSEA Rehd. & Wils.—The flowers in typical *S. holocarpa* are often pink. In this misleadingly named variety the leading character is not the colour of the flowers, though these are pink, but their larger size and leaves being woolly all over on the underside when young, later woolly on the lower part of the midrib. But the 'Rosea' of gardens, at least as distributed commercially, has the youngest leaves quite glabrous beneath, except on the midrib.

S. EMODI Brandis—This species, a native of the N.W. Himalaya, Pakistan and Afghanistan, is nearer to *S. holocarpa* botanically than it is to the geographically more adjacent *S. colchica* and *S. pinnata*. The leaves are trifoliolate, each leaflet 3 to 6 in. long, elliptic or ovate-elliptic, rounded at the base, downy or glabrous beneath, the terminal leaflet long-stalked, the lateral almost sessile, finely and evenly toothed. Flowers white, borne with or after the leaves in pendulous panicles. Fruits three-lobed at the apex. It has been introduced but is uncommon. The specific epithet derives from Emodus, the classical name for the western Himalaya.

S. PINNATA L.

A deciduous shrub up to 12 or 15 ft high. Leaves pinnate, composed of usually five leaflets, occasionally three, rarely seven; they are 2 to 4 in. long, ovate or ovate-oblong, toothed, dull green above, pale and dull beneath, with down near the base of the midrib. Flowers in terminal drooping panicles 2 to 4 in. long, white, each flower about $\frac{1}{2}$ in. long, the sepals as well as petals erect. Fruit a two-celled, bladder-like capsule 1 to 1$\frac{1}{2}$ in. long, about the same wide, each cell containing one or more seeds about the size of a large pea, brownish yellow.

Native of Europe from S.E. France and Italy eastward to the western Ukraine and the Balkans; and of parts of S.W. Asia (Anatolia, Transcaucasia and bordering parts of Syria). It is the best known of the bladder-nuts, and although not a native of Britain is now naturalised in the hedgerows and copses of some parts. In 1596, according to Gerard, it grew in the Strand 'by the Lord Treasurer's House.' It is not so handsome and notable a shrub as *S. colchica*, from which, as well as from 'Coulombieri', it is distinguished by the dull under-surface of the leaves and erect sepals, and from *S. colchica* in particular by the much smaller fruits, containing seeds twice as large. It merits a place in the garden for its curious and interesting fruits as well as its flowers and foliage.

S. TRIFOLIA L.

A deciduous shrub up to 10 or 15 ft high. Leaves of always three leaflets, which are broadly ovate, 2 to 4 in. long, occasionally doubly toothed, pale and downy all over the lower surface, dark green and less downy above, the middle leaflet is long-stalked, the side ones very shortly so. Panicles short, drooping, 1½ to 2 in. long, either terminating the leading shoot or small side twigs; flowers bell-shaped, dingy white. Fruit a usually three-celled capsule, 1¼ to 1½ in. long, less in width; seeds yellowish, $\frac{3}{16}$ in. long.

Native of the eastern United States; cultivated in England in 1640, but not ornamental enough ever to have been extensively grown. It is easily distinguished from the other species in cultivation by the very downy leaves. There appears to be rarely more than one seed to each cell, often none.

STAUNTONIA   LARDIZABALACEAE

A genus of about fifteen species in E. and S.E. Asia, closely allied to *Holboellia*, differing in the stamens being united instead of free. The name commemorates Sir George Staunton, who accompanied Lord Macartney on his famous embassy to China in 1792.

S. HEXAPHYLLA (Thunb.) Decne.

Rajania hexaphylla Thunb.

An evergreen climbing shrub, whose main stem is sometimes 4 or 5 in. thick near the base. Leaves long-stalked, compound, consisting of three to seven leaflets radiating from a common centre. Leaflets ovate to elliptical, acutely pointed, the side ones usually oblique, of leathery texture, glabrous, 2 to 5 in. long; stalks 1 to 2 in. long. Flowers fragrant, unisexual, produced three to seven together in a raceme, white tinged with violet, ¾ in. across; they have six fleshy sepals, but no petals; the males with six stamens, the females with three ovaries. Fruits of the size of a walnut, purple, sweet and watery, eaten by the Japanese.

Native of S. Korea, Japan and the Ryukyus; introduced in 1874. In foliage it is one of the most handsome of climbers, hardy on a south or west wall, but growing most luxuriantly in the milder parts. Fruits have been borne by female plants in the absence of a male, which suggests that this species is not completely dioecious. It received an Award of Merit when exhibited in flower from Wakehurst Place, Sussex, in 1960.

STEPHANANDRA ROSACEAE

An Asiatic genus of deciduous shrubs, comprising four species. They are allied to *Spiraea*, but differ in having leaves with stipules and flowers with a single carpel, developing into a capsule which splits only at the base, and contains one or two black, lustrous seeds. Leaves alternate. Flowers dull white or greenish, in terminal corymbs or panicles. The generic name derives from the Greek *stephanos*, crown, and *andros*, man, from the stamens being persistent and forming a wreath round the capsule.

The stephanandras like a good, moist soil in sun or half-shade. They are easily propagated by cuttings or by division of the plants; *S. incisa* also by root-cuttings put in heat in March.

STEPHANANDRA INCISA

S. INCISA (Thunb.) Zab.

Spiraea incisa Thunb.; *Stephanandra flexuosa* Sieb. & Zucc.

A deciduous shrub of graceful habit, with glabrous, wiry, zigzag branches, forming a dense, rounded bush ultimately 3 to 5 ft high (rarely taller), sending up sucker growths freely from the base; sometimes a spreading, mounded bush only 2 ft or so high. Leaves triangular in the main, truncate or heart-shaped at the base, tapering to a slender apex, 1½ to 3 in. long, somewhat less in width at the base, the margins cut into deep lobes, the lobes toothed; stipules linear, toothed, $\frac{1}{4}$ in. long. Flowers greenish white, $\frac{1}{5}$ in. wide, crowded on panicles 1 to 3 in. long and terminating short side-twigs from the previous year's shoots; stamens ten.

Native of Japan and Korea; introduced to Kew, in 1872, by way of St Petersburg. It has proved quite hardy, and is now generally cultivated for the beauty of its handsomely cut, fern-like foliage, and for the brown of its naked stems and branches in winter. The finest specimen I have seen is in Lord Annesley's garden at Castlewellan, which some years ago was 8 ft high and more in diameter—an exceedingly elegant bush. The flowers appear in June, but have little beauty.

cv. 'CRISPA'.—Of procumbent habit, rooting at the nodes and spreading widely, eventually about 2 ft high. Leaves rather more deeply incised than normal, crisped when young. Raised in Denmark and introduced to this country in the late 1950s. A useful ground-cover, but not suitable for very dry soils. The leaves turn orange in the autumn. Forms of *S. incisa* no taller than this have long been cultivated.

S. TANAKAE (Franch. & Sav.) Franch. & Sav.

Neillia tanakae Franch. & Sav.

A deciduous shrub of twiggy habit, up to 6 ft or perhaps more high, with glabrous, slender branches. Leaves broadly ovate or triangular, 2 to 5 in. long and from two-thirds to quite as much wide, the point long and slender, the base rounded or heart-shaped, the margins double-toothed, and frequently with one or two pairs of angular lobes more or less developed near the base, veins hairy when quite young, becoming glabrous; stipules heart-shaped, toothed, $\frac{1}{4}$ in. long. Flowers yellowish white, produced in June and July in a lax, branching panicle 2 to 4 in. long; each flower $\frac{1}{5}$ in. across on a stalk about as long; stamens fifteen to twenty. *Bot. Mag.*, t. 7593.

Native of Japan, introduced to Kew in 1893. It differs from *S. incisa* in the larger, less deeply cut leaves (which turn an orange colour in autumn), in the broader stipules, the more numerous stamens, and in the longer, more slender, and less densely flowered panicles. The flowers become a purer white under sunnier skies than ours. The species is hardier than *S. incisa* and its stems are brighter brown.

STEWARTIA *see* STUARTIA

STRANVAESIA   ROSACEAE

A genus of about five species of evergreen shrubs and small trees, ranging from the Himalaya and N.E. India to China, S.E. Asia, the Philippines and Borneo. It is closely allied to *Photinia* and there is no constant character of foliage or flower to distinguish them. The differential character that has been mainly relied on is that in *Stranvaesia* the closely united carpels form, in the fruiting stage, a sort of stone, which is enclosed by the receptacle at the base and is loosely covered by the free upper part of the receptacle. This 'stone' is said to dehisce when the fruit is fully mature, but this character has not been observed on living plants, and the splitting or breaking of the upper part of the carpels seen in some herbarium specimens may be the result of pressing and drying. *Stranvaesia* was submerged in *Photinia* by C. Kalkmann in *Blumea*, Vol. 21 (1973), pp. 416–17; *S. davidiana* had been transferred to that genus earlier. *Stranvaesia* is, however, recognised by Vidal, whose monograph on the genus was published in *Adansonia*, Vol. 5 (1965), pp. 229–35.

The genus was named by Lindley in honour of the Hon. William Fox Strangways, later Earl of Ilchester (*d.* 1865), 'a learned and indefatigible investigator of the flora of Europe.'

S. DAVIDIANA Decne.   [PLATE 65

Photinia davidiana (Decne.) Cardot

An evergreen shrub or small tree to about 35 ft high; young shoots slender, clad rather densely with whitish hairs, later more or less glabrous. Leaves entire, sometimes undulate, green on both sides, variable in length and shape, lanceolate, oblong, elliptic or oblanceolate, 2 to 5 in. long, ⅜ to about 1 in. wide as usually seen in gardens, but up to 1¼ in. wide in the more robust forms, narrowed at the apex to a slender acute or acuminate tip, sometimes more broadly acute; petioles clad with the same type of hair as the young growths, ½ to ¾ in. long. Stipules awl-shaped, soon falling. Flowers white, about ¼ in. across, opening around midsummer, arranged in rather lax hairy corymbs about 3 in. wide. Petals five, roundish, concave, soon falling. Stamens about twenty with red or pinkish anthers. Fruits bright red, globose, ¼ to ⅜ in. wide. *Bot. Mag* , tt. 8862, 9008.

A native of western, central and southern China, extending into Vietnam; discovered by Père David near Mupin in W. Szechwan in 1869; introduced by Wilson from Mt Omei, not far from the type-locality, in 1903, when collecting for Messrs Veitch, and again four years later during his first expedition for the Arnold Arboretum (W. 1064 from the Panlanshan, W. Szechwan). A plant, almost certainly from W. 1064, received at Kew from the Arnold Arboretum in 1910, was described by Dr Hutchinson in 1920 as a new species, *S. salicifolia* (*Bot. Mag.*, t. 8862), characterised by rather narrow, more or less parallel-sided leaves; a broader-leaved plant from the batch of seed was grown at Aldenham. This species, which Dr Hutchinson compared with *S. undulata*, not mentioning *S. davidiana* at all, is a very minor variation of the species. Some plants cultivated as typical *S. davidiana* derive from seeds collected by Forrest in N.W. Yunnan on the Mekong-Salween divide during his 1917–19 expedition; the plant portrayed in *Bot. Mag.*, t. 9008, raised from this sending, had broader and more acuminate leaves than in "*S. salicifolia*".

S. davidiana, as usually seen in gardens, makes a rather gaunt, sparsely branched and not very leafy shrub, often tall enough to be reckoned as a tree, and looks better when surrounded or fronted by other shrubs than when grown as a specimen. But some plants offered as *S. salicifolia* are of more compact habit. It is not particularly ornamental in flower—in this respect it is inferior to the pyracanthas—but few shrubs give a more reliable display of fruits, which are too dry to be of much interest to birds. The leaves persist for more than a year and usually turn bright red before falling. It is perfectly hardy, crops well even in a semi-shaded position, and will grow in any soil that is not too dry or excessively chalky. No pruning is needed.

var. UNDULATA (Decne.) Rehd. & Wils. *S. undulata* Decne.—This differs from the typical state of the species in its shorter leaves, to about 3½ in. long, more undulate at the margin, and in the sometimes almost glabrous inflorescence. But intermediates occur, and Rehder and Wilson, in reducing *S. undulata* to the rank of variety, questioned whether it really deserved to be recognised even at that level. It was originally described from a specimen collected in Kweichow, and was introduced by Wilson in 1901 from W. Hupeh, where it is said to be common and very variable. *Bot. Mag.*, t. 8418.

Whatever may be the botanical standing of this variety, the plants from the Wilson introduction are distinct enough in being of lower and more spreading growth, often broader than high, and in having scarlet or orange-scarlet fruits. There is a yellow-fruited form—'FRUCTU LUTEO'—which was apparently first raised, before 1920, at the Donard nursery in Co. Down. It has also given rise to 'PROSTRATA', which has prostrate stems but sends up erect branchlets here and there; it is vigorous and eventually takes up an excessively large space. The original plant was raised at Hidcote Manor, probably from seeds collected by Forrest in Yunnan.

S. NUSSIA (D. Don) Decne.

Pyrus nussia D. Don; *Stranvaesia glaucescens* Lindl.; *Photinia nussia* (D. Don) Kalkmann

A small evergreen tree, the branchlets covered when young with a loose, whitish down, ultimately glabrous. Leaves leathery, lanceolate to obovate, 2½ to 4 in. long, ¾ to 2 in. wide, dark shining green and glabrous above, paler, glossy and slightly downy on the midrib beneath, finely toothed. Flowers white, about ½ in. across, produced in July in flattish, terminal, hairy-stalked corymbs 2 to 4 in. across; flower-stalk and receptacle woolly. Fruits hoary with down when young, becoming pale red and glabrous, ¼ in. long, pear-shaped.

Including its varieties, not treated here, *S. nussia* is of wide distribution from the central and eastern Himalaya to S. China, the Indonesian region and the Philippines; it was introduced from the Himalaya in 1828. It is not reliably hardy near London unless grown on a wall, but flowers and fruits in the open ground in the milder parts and attained a height of 20 ft at Binstead in the Isle of Wight.

STUARTIA THEACEAE

A small genus of evergreen or deciduous trees and shrubs in the S.E. United States and E. Asia, belonging to the same subdivision of the Tea family as *Camellia*. Leaves simple, alternate, on short often winged petioles. Flowers solitary, more rarely in twos or threes, shortly stalked, mostly 2 to 3 in. wide. Calyx with five or six imbricated sepals connate at the base, subtended by one or two bracts. Petals five (rarely more numerous), white, connate at the very base. Stamens numerous, their filaments connate at the base and forming a tube. Fruit a woody ovoid or roundish often beaked loculicidal capsule, each of its normally five chambers containing up to four seeds.

Stuartias have been too much neglected in gardens; they have great beauty, and flower in July and August, when few shrubs remain in blossom. They are evidently not among the most rubust, for the American species, although first introduced more than a hundred years ago, must still be classed with the rarest inhabitants of our gardens. A sheltered sunny position should be selected for them, and care should be taken that they do not suffer from excessive drought. Whilst a peaty soil is not essential for them, they are undoubtedly benefited by having some of it, as well as leaf-soil, mixed with the ordinary loam of the garden, especially when young. Still a warm sandy loam free from lime suits them well. I find that the root shelter they obtain by being planted in a bed of Erica is very grateful to them, and the soil which suits heaths suits them also. Stuartias are not easy to propagate except by seeds which are occasionally borne by good-sized plants. They should, like most of those of the family, be sown as soon as obtained. Failing them, cuttings may be used. These should be taken from ripened wood in late summer and inserted in very sandy soil under a cloche in a cool frame, or even in pure sand. It is wise to put plants in their permanent sites as early as possible.

The genus was named in honour of John Stuart, Earl of Bute (1713–92), a gifted amateur botanist who was chief adviser to Augusta, Princess Dowager of Wales, when she founded the Botanic Garden at Kew, in 1759–60, and served as Prime Minister 1762–3. As explained under *S. malachodendron*, Linnaeus was misled into spelling the generic name "*Stewartia*". This unintentional mis-spelling was emended to *Stuartia* by L'Héritier in 1785, and the generic name was almost universally so spelt throughout the 19th century.

An excellent account of the hardy species of *Stuartia*, by Stephen Spongberg, will be found in *Journal of the Arnold Arboretum*, Vol. 55 (1974), pp. 182–214.

S. MALACODENDRON L.

S. virginica Cav.

A deciduous small tree or shrub, 15 ft or more high. Leaves 2 to 4 in. long, ovate, oval, or obovate, more tapered at the base, and less distinctly stalked than in *S. ovata*; the apex is pointed or blunt, the margins toothed, the lower surface more or less hairy. Flowers solitary in the leaf-axils, $2\frac{1}{2}$ to $3\frac{1}{2}$ in. across; calyx $\frac{3}{4}$ in. across, with five broad, pointed, hairy divisions; petals white, silky on the back; stamens purple, with bluish anthers, forming a conspicuous and beautiful centre to the flower. Fruit woody, egg-shaped, $\frac{1}{2}$ in. in diameter. *Bot. Mag.*, t. 8145.

Native of the south-eastern USA from Virginia to Florida, west to Louisiana and Arkansas, easily distinguished from the other American species, *S. ovata*, by the united styles and by the smaller and differently shaped leaves. It was introduced to Britain by John Clayton, probably from Virginia, and flowered in 1742 in the garden of Mark Catesby, the famous author of *The Natural History of Carolina*. A year or so later it was portrayed by the great botanical artist Ehret from a plant growing in the garden of the young John Stuart, Earl of Bute, at Caenwood House, London (now Kenwood House). The Ehret portrait was sent to Linnaeus by Isaac Lawson, with a dedication to the Earl in which his family name was mis-spelt "Stewart", and it was from this portrait, and from a Clayton specimen passed on to him by Gronovius that Linnaeus described the genus in 1746.

By the 1830s *S. malacodendron* had become a rather uncommon plant in gardens, the reason being, according to Loudon, that it was slower-growing than its fellow-American *S. ovata*. The only large plants he mentions grew at the Mile End nursery, where they had probably been planted by James Gordon the Elder. It is now one of the rarest of American shrubs and should be planted more often. Starting to flower in July, it continues into August, being at that time one of the handsomest and most distinguished shrubs in flower. But needing more than average summer-heat, it is not a species for the cooler and rainier parts of the country. The largest plants recorded grew at Syon House, London; at the Knap Hill Nursery near Woking, and at Stoke Poges in Buckinghamshire—all in or near the Thames Valley.

STUARTIA MALACODENDRON

S. MONADELPHA Sieb. & Zucc.

A deciduous tree up to 70 ft high in the wild; bark smooth, peeling in small, thin flakes; young shoots hairy at first, slender. Leaves ovate to ovate-lanceolate, $1\frac{1}{2}$ to 3 in. long, $\frac{5}{8}$ to $1\frac{1}{4}$ in. wide, rounded to widely cuneate at the base, the apex narrowed to a slender point, finely serrate, appressed downy beneath at first; stalk $\frac{1}{2}$ in. or less long. Flowers white, 1 to $1\frac{3}{8}$ in. wide. Bracts two, persistent, oblong, longer than the calyx-lobes, which are ovate or deltoid and about $\frac{1}{4}$ in. long. Petals spreading, silk-downy on the back. Stamens united at the base, with violet anthers. Fruits ovoid, beaked, about $\frac{3}{8}$ in. long, covered with yellowish appressed hairs.

Native of S. Japan and of Quelpaert Island; introduced about 1903. Wilson found it abundant on the Island of Yakushima with a smooth pale trunk sometimes 3 ft in diameter, though in Japan proper, where it grows in the zone of beech forests, it is usually more slender. In gardens, where it is rare outside the larger collections, it makes a small tree with no outstanding qualities. The bark does not peel so freely as in *S. sinensis* and the flowers and fruits are the smallest in the genus.

S. OVATA (Cav.) Weatherby

Malacodendron ovatum Cav.; *S. pentagyna* L'Hérit.

A deciduous shrub, up to 15 ft in height, with erect branches but a bushy habit; young shoots, leaf-stalks, and often the leaves tinged with red. Leaves ovate, $2\frac{1}{2}$ to 5 in. long, about half as much wide, rounded at the base, pointed, toothed more or less distinctly on the margin, or entire, hairy beneath, more especially when young; petioles winged. Flowers produced singly in the leaf-axils, on hairy stalks, about $\frac{1}{4}$ in. long, each bearing a single bract. Sepals five, about $\frac{1}{2}$ in. long, broadly strap-shaped or ovate, densely hairy; petals five or six, creamy white, prettily crenulated, one of them often deformed. A conspicuous feature of the flower is the cluster of normally whitish, but sometimes purple stamens; styles three to five, not united; the finest flowers are over 4 in. across, others under 3 in. *Bot. Mag.*, t. 3918.

Native of the south-eastern USA from Virginia and Kentucky south to Alabama and Georgia; introduced in 1795. Loudon records that around 1837 there were specimens at Dropmore 10–12 ft high and others almost as tall at White Knights, both gardens in the Thames Valley, making a splendid display every year in July and August. Like the other American species it is now very rare in gardens and worthy of further trial on acid soils in south-eastern England.

var. GRANDIFLORA (Bean) Weatherby *S. pentagyna* var. *grandiflora* Bean— 'I distinguish by this name [see synonym] the beautiful form with purple stamens, which give a much more striking character to the flower than the ordinary whitish ones, especially as it measures 4 to $4\frac{1}{2}$ in. across the petals. This form is found along with the white-stamened one in the woods of Georgia; there appears to be no other character to differentiate them, but the stamens are always purple.' (W. J. Bean, in the present work, Ed. 1, Vol. II, p. 555 (1914)).

S. PSEUDOCAMELLIA Maxim.   [PLATE 64

A deciduous tree to about 60 ft high in Japan, with a flaking bark; branchlets and often the leaves quite glabrous, the latter sometimes silky beneath. Leaves 2 to 3½ in. long, ovate or obovate, tapering at the base to a short stalk, finely toothed. Flowers produced singly in the leaf-axils on a short stalk, ½ in. or less in length; each flower 2 to 2½ in. across, white and cupped. Bracts below the flower shorter than the sepals. Petals five, roundish, concave, covered with silky hairs on the back, the margins irregularly jagged. Sepals densely hairy. Stamens numerous, incurved, orange-yellow. Ovary conical, surmounted by five united styles, the stigmas only spreading. Fruit a broadly ovoid, hairy capsule, 1 in. long. *Bot. Mag.*, t. 7045.

Native of Japan; introduced to Britain around 1880, but cultivated previously in the United States and in France, where it first bore fruit in the nursery of Messrs Thibaut and Keteleer at Sceaux, near Paris, in 1878. It is not, perhaps, quite so striking as either of the American species, but is still a beautiful tree, and is evidently more at home in English gardens. When the seasons are suitable the leaves turn brilliant yellow and red before falling. The ugly specific name refers to the resemblance of the flowers to those of a single camellia.

Apart from *S. sinensis*, this species is the only stuartia to have attained the stature of a tree. The largest recorded are: Leonardslee, Sussex, 56 × 2¾ ft (1968); Westonbirt, Glos., 46 × 1¾ ft (1971); Killerton, Devon, 50 × 3¾ ft (1970).

S. pseudocamellia received a First Class Certificate when exhibited by Messrs Veitch in 1888.

var. KOREANA (Rehd.) Sealy *S. koreana* Rehd.—The Korean race of *S. pseudocamellia* is perhaps not even varietally distinct from the Japanese. But, as represented in cultivation, it differs from trees of Japanese provenance in having flowers which, instead of being cup-shaped, are rather widely open and about 2¾ in. across; the leaves are broader, 1¼ to 4 in. long and ¾ to 3 in. wide, broadly tapered or even rounded at the base, and less silky-hairy beneath. *Bot. Mag.*, n.s., t. 20.

The var. *koreana* was introduced by Wilson in 1917 but seems to have been unknown in this country until Arthur Osborn of Kew saw small trees in the Arnold Arboretum flowering beautifully in 1930. It was introduced to Britain in the following year, and Osborn's high opinion of this tree was amply confirmed when flowering branches from Exbury were shown at Vincent Square on July 3, 1945. It is now well established and perhaps the best of the genus for small gardens, flowering well in July in a sunny position and turning reddish brown or orange in the autumn. Its ultimate height in this country is uncertain, but in the wild it is said to attain 45 ft.

S. PTEROPETIOLATA W.-C. Cheng

Hartia sinensis Dunn

An evergreen shrub or tree from 20 to 50 ft high, its young shoots silky-hairy at first, becoming glabrous and afterwards brown or greyish. Leaves alternate, 3 to 5 in. long, 1 to 2 in. wide, elliptical inclined to ovate or obovate, pointed, more or less rounded at the base, toothed, each tooth tipped with a dark gland, dark glossy green and glabrous above, paler, conspicuously veined and at first silky-hairy beneath (chiefly on the midrib), becoming nearly or quite glabrous; stalk up to ¾ in. long, hairy, winged. Flowers white, 1 to 1½ in. wide, produced singly in early summer from axillary buds on short leafy shoots, each on a stout stalk ¼ to ½ in. long. Petals five, roundish ovate with jagged margins. Calyx-lobes ovate, silky. Fruit conical, woody, ¾ in. long. Stamens very numerous, united at the base to form a short tube; anthers golden yellow. *Bot. Mag.*, n.s., t. 510.

Native of S. China; discovered by Augustine Henry in S.E. Yunnan, and originally described in a new genus, *Hartia*, united with *Stuartia* in 1934. Forrest collected this species in 1912, west of Tengyueh, on the border between Yunnan and Burma, and probably introduced it in that year. The material portrayed in the *Botanical Magazine* came from a plant grown under glass in the Edinburgh Botanic Garden, which had been raised from F. 24406, collected in 1924 on the Shweli-Salween divide at 8–9,000 ft, where it grew as a shrub 10 to 20 ft high; the leaves, Forrest remarked, were used locally for making tea.

S. pteropetiolata is a very tender species, only suitable for the mildest parts. It has succeeded remarkably well at Caerhays Castle in Cornwall, where there are five specimens, three of them about 58 ft high and 2½ to 3¼ ft in girth (1966).

S. SERRATA Maxim.

A deciduous tree probably 30 ft and upwards high; young shoots often reddish and slightly hairy at first, becoming glabrous. Leaves of rather leathery texture, oval or obovate, tapered towards the base, more abruptly so towards the pointed apex, margins set with incurved teeth; 1½ to 3 in. long, half as much wide; dull green and glabrous above; paler, downy in the vein-axils and usually hairy on the midrib beneath; stalk ⅛ to ½ in. long. Flowers solitary in the leaf-axils of the young shoots, cup-shaped, 2 to 2½ in. wide, opening in June; flower-stalk ⅙ to ¼ in. long, hairy. Bracts leaf-like, longer than the sepals. Petals five, creamy white, stained with red outside, scoop-shaped, 1 in. wide, the margins jagged. Stamens numerous, free, their stalks silky at the base; anthers yellow. Sepals five (occasionally six), ovate-oblong, ½ to ¾ in. long, pointed, glabrous except for minute hairs on the margin; persisting to the fruiting stage and then much reflexed. Seed-vessel woody, ovoid, ¾ in. long, quite glabrous, each of the five divisions tapered to a beak at the top. Seeds winged like those of an elm. *Bot. Mag.*, t. 8771.

Native of S. Japan, with its main distribution in Kyushu and Shikoku. It was scarcely known in gardens until portrayed in the *Botanical Magazine* in 1918 from flowering specimens sent to Kew by Sir Edmund Loder in the previous year from his garden at Leonardslee, Sussex; it had also been in cultivation for some time at Nymans in the same county. Like all its kind, it is an attractive tree, and flowers freely. Its nearest allies are *S. sinensis* and *S. monadelpha*, in both of which, as in the present species, the bracts on the pedicels equal or exceed the sepals in length. But in those two species the ovaries and capsules are hairy, glabrous in *S. serrata*. *S. sinensis* is further distinguished by its unmistakeable bark, and *S. monadelpha* by its very small flowers and capsules.

This species received an Award of Merit when shown from Borde Hill, Sussex, in 1932.

S. SINENSIS Rehd. & Wils. [PLATE 63
S. gemmata Chien & Cheng

A deciduous small or medium-sized tree; bark smooth, peeling (see further below); young shoots clothed at first with fine hairs. Leaves oval or ovate-oblong, 1½ to 4 in. long, ⅝ to 1¾ in. wide, tapered at the apex, cuneate at the base, toothed, at first hairy on both surfaces (more densely so above) and at the margin, becoming almost glabrous, bright green on both sides; stalk hairy, ⅛ to ¼ in. long. Flowers solitary in the leaf-axils, white, fragrant, 1½ to 2 in. across on pedicels up to ⅜ in. long; bracts ovate or ovate-oblong, about as long as the sepals. Stamens united in the lower third. Ovary downy. Capsule ovoid, five-angled and five-beaked, about ¾ in. wide. *Bot. Mag.*, t. 8778.

Native of central and eastern China; introduced by Wilson about 1901 from W. Hupeh, when collecting for Messrs Veitch. It is not remarkable for its flowers, but few hardy trees have a more beautiful bark, which, in its summer condition, is as smooth as alabaster and the colour of weathered sandstone; in autumn it turns purple, later brown, and peels away in translucent scrolls, exposing the fresh inner coating. It is perfectly hardy at Kew, where there are two trees showing the characteristic bark. But the finest examples recorded all grow in woodland conditions in Sussex: Tilgate Park, Crawley, 46 × 3¼ ft, a superb specimen (1974); Borde Hill, Sussex, *pl.* 1912, 38 × 2¾ ft (1973); Wakehurst Place, 36 × 3¼ ft at 3 ft (1969).

S. sinensis is nearly allied to *S. monadelpha*, which is, however, a native of Japan only. Rehder and Wilson observe, moreover, that the "capsule of *S. sinensis* is the largest in the genus (⅘ in. in diameter), that of *S. monadelpha* is the smallest" (⅛ in. in diameter). *Plantae Wilsonianae*, Vol. II, p. 396.

S. ROSTRATA Spongberg—Formerly confused with *S. sinensis*, this species differs in its shallowly furrowed bark; smaller winter-buds; subglobose capsules downy at the base only, abruptly beaked and with four seeds in each chamber (against ovoid with a tapered beak and downy throughout, with two-seeded chambers in *S. sinensis*). Native of E. China; described in 1974 (*Journ. Arn. Arb.*, Vol. 55, pp. 198–202). It was introduced to the USA from the Lushan Botanic Garden in 1936, as *S. sinensis*, but is not known to be in cultivation in Britain.

STYRAX STYRACACEAE

A genus of about 130 species of trees and shrubs in America, E. Asia (including the Himalaya), and Malaysia, only a few occurring in cool-temperate regions and only one (*S. officinalis*) a native of Europe. The cultivated species are deciduous. Leaves alternate, without stipules, they and the branchlets, pedicels, etc., often clad with an indumentum of stellate hairs. Flowers in racemes, which are sometimes branched, or in clusters, or solitary. Calyx shortly lobed or entire. Corolla deeply divided into five or sometimes up to eight segments. Stamens twice the number of the corolla-lobes. Ovary superior (inferior in *Halesia* and the other cultivated Styracaceae). Fruit a dry or fleshy drupe, dehiscing irregularly and containing one or two large seeds.

The generic name comes from the classical Greek name for *S. officinalis*; this is feminine when applied to the tree, neuter only when used for the resin extracted from it. But, as will be seen from the synonyms of the species treated below, the genus has often been treated as of neuter gender, and is masculine in Miss Perkins' monograph on the genus, in Engler, *Pflanzenreich* Heft 30 (1907).

All the species of *Styrax* need a sheltered spot and should be given careful attention when young. When first planted out they should be given a light soil to which some fine peat and, if available, some decayed leaves have been added. Once established, they will root into the surrounding soil. Propagation is by cuttings of half-ripened wood taken about July, and rooted in bottom-heat. Many of the cultivated species produce good seed—often too much for their health. This should germinate in the following spring if sown as soon as ripe or stratified; otherwise it may need a year or even two to germinate.

S. AMERICANA Lam.   AMERICAN STORAX

S. laevigata Ait.

A deciduous shrub, 3 to 8 ft high; young shoots nearly glabrous. Leaves narrowly oval, or obovate, $1\frac{1}{2}$ to $3\frac{1}{2}$ in. long, $\frac{1}{2}$ to $1\frac{1}{4}$ in. wide, the base wedge-shaped, the apex mostly pointed, minutely toothed, dark green above, paler beneath, almost or quite glabrous on both sides; stalk $\frac{1}{6}$ in. or less long. Flowers white, pendulous, $\frac{3}{4}$ to $1\frac{1}{4}$ in. across, produced in June and July one to four near and at the end of short leafy twigs, each flower on a slender stalk $\frac{1}{4}$ to $\frac{1}{2}$ in. long. Petals $\frac{1}{8}$ to $\frac{3}{16}$ in. wide, pointed; calyx triangular-lobed; stamens erect, $\frac{1}{2}$ in. long. Fruit roundish oval, $\frac{1}{4}$ in. wide, covered with fine grey down, and supported at the base by the persistent five-lobed calyx.

Native of the south-eastern United States; introduced in 1765. This shrub has long been cultivated at Kew, but grows slowly, really needing a warmer climate. It was killed to the ground by the frosts of February 1895, but sprang up again later. It is better adapted for our south-west counties, where it is a pretty shrub; yet neither as hardy nor as beautiful as *S. japonica*, to which in its pendulous blossoms it bears some resemblance but is easily distinguished by its narrower petals.

S. DASYANTHA Perkins

A deciduous shrub or small tree, the young branchlets furnished at first with reddish brown down, becoming glabrous. Leaves almost sessile, obovate to broadly oval, 2 to 4 in. long, 1½ to 3 in. wide, tapered more or less at the base, pointed, the upper part minutely toothed, the lower surface at first covered with tufted hairs but almost glabrous by autumn. Flowers white, ½ to ¾ in. long, produced in July in slender terminal racemes 2 to 4 in. long, augmented by clusters of two to four flowers in the uppermost leaf-axils; pedicels ¼ to ½ in. long, felted. Calyx cup-shaped, ¼ in. long, felted outside, with several short but unequal, pointed teeth. Corolla-segments lanceolate, covered outside with tufted, yellowish white down.

Native of Central and Western China; discovered by Augustine Henry in Hupeh and introduced by Wilson from W. Szechwan in 1900. Although an attractive species, it is not in the same class as *S. hemsleyana*, from which it differs, among other characters, in its glossier, shorter leaves. There is some doubt as to the identity of this cultivated form, which differs from the type of *S. dasyantha* (the Henry specimen from Hupeh) in its larger flowers, fewer in each inflorescence, and comes near to the Himalayan S. SERRULATA Hook. f., which is probably not in cultivation.

There has been some confusion in gardens between *S. dasyantha* and *S. japonica*, but the latter has fewer flowers in a laxer inflorescence and its pedicels and calyx are glabrous.

var. CINERASCENS Rehd.—Branchlets and undersurface of leaves more densely and persistently downy. Introduced from the Lushan Botanic Garden in the 1930s as *S. philadelphoides*, but not the species described under that name by Miss Perkins. Some plants under this erroneous name, and possibly from the same source, are more glabrous and better called *S. dasyantha* simply.

S. HEMSLEYANA Diels   [PLATE 66

A deciduous tree, 20 ft or more high, young shoots covered at first with tufted down. Leaves obovate or unequally ovate, 3 to 5½ in. long, 2 to 3½ in. wide, usually more gradually tapered at the base than at the apex; finely and rather distantly toothed, prominently veined, glabrous and pale green above, sparsely furnished beneath with tufted (stellate) down; stalk ¼ to ⅝ in. long. Flowers pure white, produced in June on terminal downy racemes or few-branched panicles, 4 to 6 in. long, each flower on a stalk ⅙ in. long. Corolla ¾ in. long, about 1 in. wide, the five lobes narrowly oval, downy outside, joined at the base into a tube ¼ in. long. Calyx bell-shaped, ¼ in. long, slender-toothed, covered with reddish brown tufted down. *Bot. Mag.*, t. 8339.

Native of Central and Western China; introduced by Wilson in 1900. It is of remarkable beauty in flower, and striking too in size of leaf, but not an easy species to suit. Although hardy, it needs a deep moist soil and a sheltered position, well clear of the root-run of forest trees. Perhaps the finest specimen in the country, or at least the best displayed, grows in one of the walled en-

STYRAX HEMSLEYANA

closures at Trengwainton in Cornwall. Measured examples are: East Bergholt Place, Suffolk, 28 × 3¼ ft (1972); Hollycombe, nr Liphook, Hants, 46 × 2¼ ft (1974); Caerhays, Cornwall, 40 × 2¼ ft (1971).

S. hemsleyana bears a certain resemblance to *S. obassia*, but the leaves of that species are rounder, more coarsely toothed, and much more downy beneath, and the bud is enclosed by the base of the leaf-stalk.

S. JAPONICA Sieb. & Zucc.
S. serrulatum sens. Hook. f. in *Bot. Mag.,* not Roxb.

A small deciduous tree, 10 to 25 ft high, rarely taller, of very graceful habit; the branches slender, sometimes drooping; young shoots at first furnished with scattered tufts of down, which soon fall away. Leaves usually oval, tapering about equally at both ends, 1 to 3½ in. long, ½ to 1½ in. wide, but occasionally obovate or even roundish, margins set with minute, shallow, distant glandular teeth, dark glossy green above, glabrous on both surfaces except for tufts of down in the vein-axils; stalk ⅓ in. or less long. Flowers pure white, perfectly pendulous, ¾ in. in diameter, borne on short lateral shoots carrying about three leaves and three to six blossoms; each flower on a glabrous slender stalk, 1 to 1½ in. long. Corolla of five pointed divisions, which are united near the base, ⅝ in. long, downy outside. Calyx glabrous, funnel-shaped, ⅙ in. long, persisting at the base of the roundish, egg-shaped fruit, which is ½ in. long. Flowers in June. *Bot. Mag.*, t. 5950.

Native of Japan, China and Korea; introduced to Kew by Richard Oldham in 1862. It is a small tree of singular grace and beauty, very hardy, but preferring a sheltered spot and one, if possible, shaded from morning sun, for the flower-buds and the young shoots are liable to injury by late spring frosts. It should be given a light loamy soil to which either peat or leaf-soil, or both, have been added. Apart from its susceptibility to late frost, especially in low-lying situations, it is one of the most desirable of all hardy trees of its type, and amply repays the trouble of preparing a suitable medium for the roots, if that does not already exist.

The first introduction to Britain from China was by Wilson, who sent seeds to Messrs Veitch early this century; plants from this source were distributed when the nursery stock at Coombe Wood was auctioned after the winding-up of Messrs James Veitch and Son, which took place in the autumns of 1913 and 1914 (many of Wilson's introductions first reached gardens through this sale). But the plants known as "*S. fargesii*" or *S. japonica* "var. *fargesii*" derive from seeds sent by Père Farges in 1898, probably from N.E. Szechwan, to the Vilmorin collection (Les Barres seed-number 1901). The plant acquired by Kew from Vilmorin in 1924, as *S. japonica fargesii*, was presumably a seedling of one of the originals at Les Barres; another, from the same source, flowered at Nymans in Sussex in 1934 (J. Comber in *New Fl. and Sylv.*, Vol. 6 (1934), p. 121); a third tree from Vilmorin grows in Collingwood Ingram's garden at Benenden, Kent, and this received an Award of Merit when exhibited by him on May 29, 1945. The description of *S. japonica* var. *fargesii* in the R.H.S. *Dictionary of Gardening* appears to have been based on the Benenden tree. Finally, it should be mentioned that two trees received at Kew as *S. japonica fargesii* came from Messrs Lemoine in 1926.

It is likely that many of the taller and larger-leaved trees of *S. japonica* grown in British gardens are of Chinese provenance, but even in Japan the species is variable, and it would be impossible to deduce the origin of a tree from its foliage or flowers.

S. OBASSIA Sieb. & Zucc.

A small deciduous tree, 20 to 30 ft high, of rather narrow proportions; young wood covered at first with tufted hairs, soon glabrous. Leaves broadly oval or almost round, 3 to 8 in. long, and from two-thirds to as much wide;

distantly toothed, except near the base; upper surface deep green and glabrous except on the veins, the lower surface densely clothed with velvety tufted down; stalk $\frac{1}{4}$ to 1 in. long, the base enclosing the bud. Flowers fragrant, about 1 in. long, pure white, drooping, produced in June on terminal racemes 6 to 8 in. long, each flower on a downy stalk $\frac{1}{3}$ in. long; the common stalk is almost glabrous. Corolla deeply five-lobed, the lobes about $\frac{3}{4}$ in. long, $\frac{1}{4}$ in. wide, minutely downy. Calyx between funnel- and bell-shaped, from five- to ten-lobed, downy, $\frac{1}{4}$ in. long, persistent and enlarging with the fruit, which is egg-shaped, about $\frac{3}{4}$ in. long, velvety. *Bot. Mag.*, t. 7039.

Native of Japan, Korea and N. China; introduced from Japan by Maries for Messrs Veitch in 1879. This is one of the most beautiful and striking of flowering trees and an older inhabitant of British gardens than the equally fine *S. hemsleyana*. It needs the same conditions as that species, but is perhaps rather slower-growing and slower to flower. It attained a height of 20 ft in Veitch's Coombe Wood nursery near Kingston-on-Thames. Examples measured recently are: Grayswood Hill, Surrey, 28 × 3$\frac{1}{4}$ ft at 1 ft (1976); Ladhams House, Goudhurst, Kent, 52 × 3$\frac{1}{2}$ ft (1975); Endsleigh, Devon, 26 × 3$\frac{1}{4}$ ft (1974); Caerhays, Cornwall, 36 × 3 ft (1971).

S. OFFICINALIS L.   STORAX

A shrub or small tree to about 20 ft high in the wild, the young shoots, the undersides of the leaves and the inflorescence parts all covered with a whitish stellate down. Leaves up to 3$\frac{1}{2}$ in. long, ovate, often broadly so, and sometimes heart-shaped at the base, entire. Flowers white, borne in June in short terminal clusters of three to eight; they are fragrant, about 1$\frac{1}{4}$ in. wide, recalling those of an orange-tree; segments of the corolla six to eight, narrow-oblong, very downy. Calyx with very short teeth. Fruits globose, $\frac{3}{4}$ in. wide, with the remains of the style at the top and the persistent woolly calyx beneath. *Bot. Mag.*, t. 9653.

Native of the E. Mediterranean and the Near East, with a western limit in S. Italy; naturalised in Italy around Bologna and in France in the department of Var. This beautiful and interesting species has been cultivated in Britain since the end of the 16th century, when Gerard had two plants in his garden 'the which I have recovered of the seed'. It is not a tender species, but coming from a region of hot and dry summers it needs the extra heat of a south or west wall to ripen its wood thoroughly; in areas where summer temperatures and sunshine are above the average it should succeed in a sheltered spot in the open ground, provided it is not bloated by overfeeding.

The fragrant resin known as 'storax' is obtained from this shrub by wounding the stem and used to be imported in small pieces, 'storax in the tear', or in larger, masses 'storax in the lump'. It was used medicinally as an expectorant, and there were also made from it 'sundry excellent perfumes, pomanders, sweet waters, sweetbags, sweet washing-balls, and divers others sweet chains and bracelets.' Storax was also used as incense, and it is recorded that the dust cast up by the boring of an insect which attacks the wood was used in Greece for the same purpose.

var. CALIFORNICA (Torr.) Rehd. *S. californica* Torr.—It is remarkable that this Californian styrax differs so little from *S. officinalis* that some botanists refuse to give it even varietal rank. See further in the article with *Bot. Mag.*, t. 9653.

S. SHIRAIANA Makino

A small deciduous tree; young shoots covered with minute stellate down. Leaves mostly obovate and tapered at the base, sometimes roundish; coarsely and unevenly toothed (or almost lobed) above the middle, entire towards the base; 1½ to 4 in. long, 1 to 3 in. wide, stellately downy beneath, more especially on the midrib, thinly so elsewhere; veins in six to eight pairs; stalk ⅛ to ½ in. long. Flowers white, very shortly stalked, produced during June in racemes of eight or ten that terminate young, leafy, lateral shoots and are about 2½ in. long. Corolla ⅝ to ¾ in. long, funnel-shaped, five-lobed, the lobes ovate, pointed and about half as long as the tubular part, downy. Calyx bell-shaped, with five triangular, sometimes bifid, lobes, thickly covered with tawny down. Fruit globose to egg-shaped, ½ in. long.

Native of Japan, where it is said to be sparingly distributed. This is a very distinct species, firstly in the tube of the corolla being twice as long as the lobes (usually it is shorter) and, secondly, in the shape of the leaves with their broad ends and deep toothing. They bear a considerable resemblance to those of *Hamamelis japonica*. *S. shiraiana* was introduced to the USA in 1915 but did not reach Britain until some thirty years later and is still too rare in gardens for its merits to be assessed.

S. SHWELIENSIS W. W. Sm.

A deciduous shrub up to 25 ft high in the wild state, with reddish brown, stellately downy young shoots. Leaves elliptical, pointed, usually tapered but sometimes rounded at the base, shallowly toothed; 2 to 4 in. long, ¾ to 2 in. wide; dark dull green above, greyish beneath, both sides soft and velvety to the touch; chief veins in five to seven pairs; stalk ⅙ in. or less long. Flowers white, about 1 in. wide, produced on short leafy shoots as in *S. langkongensis*. Corolla lobes downy outside and at the margins. Calyx bell-shaped, ¼ in. long, very downy, minutely toothed or almost entire; flower-stalk about ¼ in. long, often thickening towards the calyx and giving it a funnel-like shape.

Native of Yunnan and probably of bordering Burma; discovered by Forrest in 1913 and introduced by him in 1919 from the Mekong-Salween divide (F. 18249, shrub 10–12 ft high, growing in open thickets). It is now rare.

S. LANGKONGENSIS W.W. Sm.—Another of Forrest's discoveries and introductions, allied to the preceding and also shrubby, differing in its smaller leaves whiter beneath and less velvety, and its less toothed calyx. Forrest found it in flower in 1910 as a shrub 1 to 4 ft high, but later collections were slightly taller. He introduced it during his 1917–19 expedition (F. 16929).

Neither of these species is superior to *S. wilsonii* and both are probably more tender or at least more demanding of summer heat.

S. VEITCHIORUM Hemsl. & Wils.

A small tree, 12 to 30 ft high; young shoots, leaf-stalks, and calyx covered with a close, grey, starry down. Leaves lanceolate, with a long tapered point, and a wedge-shaped or slightly rounded base, remotely and shallowly toothed, 3 to 5 in. long, 1 to 1¾ in. wide, of thin texture, downy on both surfaces, but especially on the midrib and veins beneath; stalk ¼ to ⅓ in. long. Flowers white, nearly 1 in. across, produced at the end of the shoots and in the upper-most leaf-axils on the current season's growth, forming a group of slender panicles each 4 to 8 in. long. Calyx minutely five-toothed.

This species was discovered by Wilson and introduced by him for Messrs Veitch from W. Hupeh, China, in 1900; all his collections (save one doubtful one) were from a single locality near Fang Hsien. It is a hardy species, allied to *S. hemsleyana*, but less common in gardens.

S. WILSONII Rehd.

A deciduous shrub, sometimes of tree-like form, 6 to 10 ft high, of much-branched, twiggy habit; young shoots furnished with starry down. Leaves ovate, ½ to 1 in. long, half to two-thirds as wide, rounded or broadly tapered

STYRAX WILSONII

at the base, often bluntish at the apex, the lower half not toothed, the terminal part either three-lobed or sparsely toothed; green and minutely downy on both sides when young. Flowers nodding, pure glistening white, $\frac{5}{8}$ to $\frac{3}{4}$ in. across, produced one to four in the leaf-axils, and at the end of short lateral twigs in June. Corolla lobes ovate-oblong, $\frac{1}{3}$ in. long, pointed, covered with minute starry down outside; calyx green, scurfy, with lance-shaped lobes $\frac{1}{12}$ in. long. Stamens clustered in an erect columnar group, $\frac{2}{3}$ in. high, their stalks white, the anthers yellow. *Bot. Mag.*, t. 8444.

Native of W. China; introduced by Wilson when collecting for the Arnold Arboretum in 1908. It is a very pretty shrub, remarkable in flowering when a few inches high and when only two or three years old. In June, 1913, I saw in Mr Chenault's nursery at Orleans, a plant 6 ft high in full blossom. It was one of the most beautiful objects I have ever seen.

S. wilsonii is hardy enough, but flowers and fruits so excessively that it is short-lived and has now become rather rare. Dead-heading might have saved the many plants that have been lost.

SUAEDA   CHENOPODIACEAE

A genus of about 100 species of glabrous herbs or small shrubs, distributed throughout most of the world on saline soils. Leaves small, fleshy, alternate. Flowers axillary, solitary or in cymes, inconspicuous. Perianth segments five, somewhat fleshy. Stamens five. Ovary with three to five stigmas, developing into an achene.

S. VERA Forskål ex J. F. Gmel.

S. fruticosa of many authors, not Forskål

A sub-evergreen shrub, 3 or 4 ft high, with glabrous, erect branches. Leaves alternate, linear, nearly cylindrical, fleshy, $\frac{1}{4}$ to $\frac{3}{5}$ in. long, blue-green, borne at very close intervals on the stem. Flowers small, green, stalkless, one-third as long as the leaves, produced during July in the leaf-axils of the current year's shoots, either singly or two or three together, insignificant.

Native of the maritime districts of S. and W. Europe, including some parts of the east and south coasts of Britain, and of N. Africa. It has rather a heath-like aspect, with its slender, erect stems and closely set, short leaves—but the latter are, of course, much more thick and fleshy. The shrub has no beauty of flower, but the habit and foliage are sufficiently interesting and graceful for it to be planted in brackish places, or in positions exposed to salt spray where comparatively few shrubs will thrive. It succeeds well in sandy soil, and can be increased by cuttings placed under a handlight. If it gets too ungainly in form it should be pruned back in spring, but the semi-woody shoots are frequently cut back by winter frost.

SYCOPSIS   HAMAMELIDACEAE

A genus of about seven species of evergreen shrubs or trees, ranging from N.E. India to China and Malaysia. It is the only evergreen genus of the Hamamelis family represented in British gardens except for *Distylium*. The flowers lack petals and are either male or hermaphrodite, both types sometimes occurring in the same cluster. The hermaphrodite flowers are strongly protandrous, i.e., the stamens develop before the ovary and are shed before it reaches its full size. For a discussion of the flowers of this genus and the related *Distylium*, see *Journ. Arn. Arb.*, Vol. 51 (1970), pp. 316–24 and 324–37.

An intergeneric hybrid between *Sycopsis* and *Parrotia* has been raised in Switzerland and was named × SYCOPARROTIA SEMIDECIDUA by Endress and Anliker in 1968.

S. SINENSIS Oliver   [PLATE 67

An evergreen bushy shrub or a small tree, up to 20 ft high in China; young shoots at first scaly. Leaves rather leathery, strongly nerved, entire or slightly toothed towards the apex, ovate or ovate-lanceolate, 2 to 4½ in. long, one-third to half as wide, glabrous and dark green above, paler and quite glabrous beneath; leaf-stalk and young wood slightly warted. Flowers in short dense clusters less than 1 in. long, the chief features of which are the six to ten stamens and the reddish brown bracts that enclose the inflorescence. Anthers yellow, tinged with orange. Fruit a dry, woolly, egg-shaped capsule, ⅓ in. long. *Bot. Mag.*, n.s., t. 655.

Native of Central China; introduced by Wilson for Messrs Veitch in 1901. It is perfectly hardy at Kew, and can be increased by means of cuttings made of fairly ripened wood and placed in heat. Its neat habit and distinct appearance, combined with its evergreen nature, make it welcome in gardens, and when well flowered it is quite handsome. It blossoms usually in March, sometimes in February.

S. TUTCHERI Hemsl.—So far as is known, this species is not in cultivation. Plants distributed under the name *S. tutcheri* derive from one at Kew, originally so labelled, which appears to be no species of *Sycopsis* but a form or relative of *Distylium racemosum*.

SYMPHORICARPOS   CAPRIFOLIACEAE

An unimportant genus of shrubs, of which only two have been widely cultivated in Britain, though several others have been introduced. Leaves opposite, not toothed but sometimes with a wavy lobing. Flowers of no beauty, but nectar-rich and much visited by bees, arranged in dense terminal or axillary racemes, sometimes solitary. The genus is allied to *Lonicera*, but the corolla is more or less regular and the berry is two-seeded. The generic name was coined by J. J. Dillen (Dillenius) and published by him in the *Hortus Elthamensis* (1732), the catalogue of the Sherard garden at Eltham in Kent. It refers to the arrangement of the fruits, which are so clustered as to resemble a compound fruit; they are commonly white, but red in *S. orbiculatus*, partly pink in *S. mexicanus* and dark blue in *S. sinensis*.

The genus has about fifteen species in North and Central America, and was thought to be wholly confined to the New World until Wilson found S. SINENSIS Rehd. in W. China. This differs from all the others in its dark blue fruits, which are ovoid, about $\frac{5}{16}$ in. long and covered with a plum-like bloom. It was introduced to Britain in 1912 from the Arnold Arboretum but has never spread into gardens, probably because the fruits are too near to black.

A monograph of the genus by G. N. Jones was published in *Journ. Arn. Arb.*, Vol. 21 (1940), pp. 203-52.

All the species grow well in any moist soil, and are easily propagated by cuttings, several by division.

S. ALBUS (L.) Blake   SNOWBERRY
Vaccinium album L.; *S. racemosus* Michx.

A deciduous shrub 3 to 4 ft high; young shoots slender, usually somewhat downy. Leaves oval to oval-oblong, roundish at the base, sometimes lobed, downy beneath, $\frac{3}{4}$ to 2 in. long, apex blunt. Flowers of little beauty, $\frac{1}{4}$ in. long, produced at the end of the twigs during June and July in spikes or clusters up to $1\frac{1}{2}$ in. long; corolla pink, downy inside; stamens and style rather longer. Fruits globose or ovoid, $\frac{1}{2}$ in. wide, snow-white, pulpy when ripe.

Native of North America, where it is widely distributed from the Atlantic westward, giving way in the Pacific region to the following variety, by which the species is almost wholly represented in cultivation:

var. LAEVIGATUS (Fern.) Blake *S. racemosus* var. *laevigatus* Fern.; *S. rivularis* Suksdorf—A taller shrub and up to 3 to 6 ft high, forming dense thickets of erect, many-branched, glabrous stems. Leaves up to 3 in. long, glabrous; fruit rather larger than in the type. This variety is a native of the western side of N. America and is the one most commonly grown in Britain and known generally as 'snowberry'. Introduced in 1817. This well-known shrub ripens its fruit in October, and having apparently no attraction for birds, they remain on the twigs up to New Year or later, interesting for their pure whiteness. Whilst the plant repays good cultivation by the greater size and abundance of the fruit (which often weigh down the branches in graceful arches), there are few shrubs more useful for filling up dark out-of-the-way corners where its invasive habit will not cause problems. Although deciduous, its stems and twigs are dense enough to make an effective screen.

S. 'Doorenbos Hybrids'

These hybrids, which are likely to displace the older sorts in gardens, were raised in Holland by G. A. Doorenbos, who made a mixed planting of *S. albus* var. *laevigatus*, *S. orbiculatus* and *S.* × *chenaultii*, and selected the best of the seedlings raised from it by open pollination. The first of these hybrids was named in 1940, but British gardeners did not make their acquaintance until 1955, when Mr Doorenbos sent fruiting sprays by air to the R.H.S. Show of October 4–5. The most admired of these hybrids is 'Mother of Pearl' (*S.* × *doorenbosii* Krüssmann) which is a vigorous, semi-pendulous shrub 5 or 6 ft high with broad-elliptic or obovate leaves and dense clusters of berries up to ½ in. wide, white with a pink cheek. This is considered by Dr Krüssmann to be a hybrid between *S. albus* var. *laevigatus* and *S.* × *chenaultii*. It makes a good hedging plant and received an Award of Merit when exhibited by Messrs Notcutt in 1971. Also a good hedger, but needing closer planting, is 'White Hedge', which is of stiffly erect habit to about 5 ft high and bears large white berries in upright mostly terminal clusters well above the foliage. Although near to the common snowberry it does not sucker. In 'Erect', of similar habit, the fruits are purplish pink. 'Magic Berry' has rosy lilac berries, colouring earlier, and is of low, bushy habit. [Plate 68

S. mollis Nutt.

S. ciliatus Nutt.

A low or prostrate shrub; young shoots covered with soft down. Leaves roundish to oval, sometimes shallowly lobed, ½ to 1 in. long, velvety and grey with down, especially beneath. Flowers few and small, produced singly or in short clusters at and near the end of the twigs; corolla widely bell-shaped, about ⅛ in. long and broad, pinkish white, enclosing the glabrous style. Fruit white, globose, about ¼ in. wide.

Native of California; distinct and interesting for its decumbent habit and densely downy leaves.

S. occidentalis Hook. Wolfberry

A deciduous shrub up to 6 ft high. Leaves oval or oblong, stout, up to 2 in. or more long, glabrous or more or less downy beneath. Flowers in dense spikes or racemes, both in the leaf-axils near the end of the shoot and at the end itself. Corolla open funnel-shaped, deeply five-lobed, densely hairy inside, ¼ in. long, pinkish; style and stamens slightly protruded, the former glabrous. Fruit dullish white, globose, about ⅓ in. wide.

Native mainly of the Rocky Mountains but extending eastwards to Michigan. It has been confused with *S. albus*, but is an inferior shrub with smaller, duller fruits; it differs also in the deeper-lobed corolla and in the protruded style and stamens. Of little garden value.

S. ORBICULATUS Moench CORALBERRY, INDIAN CURRANT

Lonicera symphoricarpos L.; *S. vulgaris* Michx.; *Symphoria glomerata* Pursh

A deciduous shrub, 3 to 7 ft high, of dense, bushy habit; branches thin, densely leafy, spreading, very downy. Leaves oval or ovate, with a rounded base, ½ to 1¼ in. long, ¼ to ¾ in. wide, dark dull green above, hairy and somewhat glaucous beneath; stalk 1/12 in. long. Flowers produced in August and September in short, dense clusters in all the leaf-axils from the lower side of the twigs. Corolla bell-shaped, ⅛ in. long, dull white, the style hairy. Berries purplish red, between egg-shaped and globose, ⅙ in. long.

Native of the USA, where it is widely distributed from the Atlantic States to S. Dakota, Kansas and Texas; also of Mexico; in cultivation 1732. A neat bush with the leaves arranged in opposite rows on the branches, but with little beauty of flower. The fruits are pretty, and when freely borne make the shrub extremely ornamental in autumn and winter, but it does not bear fruit so freely in this country as in its native one, except after a hot summer.

f. LEUCOCARPUS (D. M. Andrews) Rehd.—Fruits white.

cv. 'FOLIIS VARIEGATIS' ('VARIEGATUS').—Leaves smaller than in the type, bordered unevenly with yellow. A good variegated shrub, raised before 1837 (*S. glomerata fol. varieg.* Loddiges ex Loud.). It is inclined to revert if grown in too much shade.

S. orbiculatus differs from all the rest of the species here mentioned in having a downy style and red berries. These characters and the long array of short flower-spikes beneath the branches make it the most distinct of the cultivated members of this genus.

S. × CHENAULTII Rehd.—An upright shrub 5 or 6 ft high, much branched, the young shoots downy. Leaves ovate, to about ⅞ in. long, dark green above, glaucous and densely hairy beneath. Flowers in short spikes; corolla funnel-campanulate, the tube about twice as long as the lobes. Fruits globose, ripe in late autumn, stippled with red on the exposed side. A hybrid between *S. orbiculatus* and the Mexican *S. microphyllus* (see under *S. rotundifolius*), raised by Messrs Chenault of Orleans early this century but of comparatively recent introduction to Britain. It is a parent of some of the Doorenbos hybrids.

S. × *chenaultii* 'HANCOCK', raised in Canada about 1940, is of procumbent habit and attains in time a considerable width by self-layering. It has been recommended as a ground-cover.

S. ROTUNDIFOLIUS A. Gray

A deciduous shrub, 2 to 3 ft high; branches very leafy, covered at first with minute down. Leaves roundish to oval or ovate, ⅓ to 1 in. long, pointed or blunt at the apex, more or less downy beneath, sometimes with sinuous margins, but otherwise entire. Flowers stalkless, produced in June and July in two- to five-flowered spikes in the upper leaf-axils, and the end of the shoot. Corolla pinkish white, ¼ to ⅓ in. long, between funnel and bell-shaped, shallowly five-lobed; hairy towards the base inside; style glabrous and, like the stamens,

enclosed within the corolla. Fruit white, oval or nearly globose, ¼ in. wide.
Native of the southern Rocky Mountains, of little garden value. It belongs
to the subgenus *Anisanthus*, in which the corolla is funnel-shaped or rotate,
the tube twice or three times as long as the lobes.

S. MICROPHYLLUS H.B.K.   *S. montanus* H.B.K.—A native of Mexico,
introduced in 1829 and figured in *Bot. Mag.*, t. 4975, but probably not now in
cultivation. It is a shrub to about 10 ft high, with acute leaves not much over
1 in. long, glaucous and downy beneath. Flowers mostly axillary, solitary or in
pairs. Fruits white or white flushed with pink. It is of interest as a parent
of *S.* × *chenaultii* (see under *S. orbiculatus*) and through it possibly of some
of the Doorenbos hybrids. It is the type-species of the subgenus *Anisanthus*.

S. OREOPHILUS A. Gr.—Another close ally of *S. rotundifolius*, native of the
southern Rocky Mountains; introduced in 1898. It differs from that species in
minor characters and is only of botanical interest.

SYMPLOCOS   SYMPLOCACEAE

A genus of about 350 species of trees and shrubs in the tropics and subtropics of America, Asia, Australia and Polynesia, a few extending into
temperate regions. Leaves alternate, simple, without stipules. Flowers
in panicles, racemes or clusters. Calyx four- or five lobed. Corolla acti-
nomorphic, more or less deeply divided into mostly five to ten lobes.
Stamens four to many, usually arranged in fascicles and more or less
adnate to the corolla. Ovary inferior or semi-inferior with two to five
chambers. Style one, with up to five stigmas. Fruit a drupe or a berry.
Symplocos is the largest genus in the small family Symplocaceae and in
some interpretations its sole member. It was placed in Styracaceae by
Bentham and Hooker and is still considered to be allied to that family
by some authorities. According to others it is near to the Theaceae.

S. PANICULATA (Thunb.) Miq.   [PLATE 69]

Prunus paniculata Thunb.; *S. crataegifolia* D. Don; *S. chinensis* (Lour.) Druce

A deciduous shrub or small tree, of light and elegant aspect; young shoots
hairy. Leaves oval, ovate, or somewhat obovate; 1½ to 3½ in. long, ¾ to 1¾ in.
wide; tapering at both ends, finely toothed, slightly hairy above, more so
on the veins beneath; stalk ⅛ to ⅓ in. long, hairy. Flowers fragrant, white,
⅓ in. across, produced during late May and early June in terminal hairy panicles,
and in the leaf-axils on small lateral twigs; the whole inflorescence is 1½ to
2½ in. long. The stalk of the axillary inflorescence appears to spring from

the stem some distance above the leaf-axil itself, which seems to be due to its union to the branchlet. Petals five, united only at the base; stamens about thirty in five clusters, one cluster attached to the base of each petal. Fruits roundish oval, mostly one-seeded, becoming bright blue in autumn. *Bot. Mag.*, n.s., t. 149.

S. paniculata, taken in a broad sense, is a variable species in several of its characters, including the colour of the fruits, and is of wide distribution from the Himalaya to Japan and Formosa. The plant described above, which is the form now cultivated, belongs to a blue-fruited race occurring in Japan, Korea and parts of China. It was introduced to the Parsons Nursery at Flushing, New York, by Thomas Hogg in 1875, and to Britain in the 1890s (for the Himalayan race, introduced earlier, see below).

S. paniculata received an Award of Merit as a flowering shrub when exhibited by Dame Alice Godman of South Lodge, Sussex, in 1938. But, pretty though the flowers are, it is for its turquoise-blue fruits that this species is grown. For a full crop of these it is necessary to grow several plants in a group, and these must be seedlings or belong to at least two different clones, since the species is usually or always self-incompatible and produces fruits only if pollinated by a different individual (all members of a clone count, of course, as a single individual). For this reason the fruits had rarely been seen in their full beauty until loaded branches were exhibited in 1947 at the Autumn Show of the R.H.S., from the Savill Garden, Windsor Great Park. It then received an Award of Merit for its fruits and a First Class Certificate in 1954, again from Windsor, whence too came the material portrayed in the *Botanical Magazine*. Mr T. H. Findlay tells us that the original plants in the Savill Garden came from Harry White of the Sunningdale Nurseries, and from the fruits produced by these, seedlings were raised and added to the plantings.

S. paniculata is hardy, and indeed in the Arnold Arboretum it has withstood the cold winters of New England since 1880. But it is not a plant for the cooler and rainier parts of the country, needing more than average summer heat to fruit well. The only qualification to its hardiness is that the fruits are sometimes spoilt by early autumn frost. It is best propagated by seeds, which may be slow to germinate unless sown at once or stratified.

The Himalayan race of *S. paniculata* named *S. crataegifolia* by Don in 1825, occurs from the Indus eastward into China and perhaps Japan. It was introduced in 1824 but is not of any horticultural importance, the fruits being black. In the genus *Symplocos* the name *S. crataegifolia* has priority over *S. paniculata* and was at one time used for the whole species, but under modern rules it must take the name *S. paniculata*, having been described by Thunberg (in *Prunus*) in 1784:

S. TINCTORIA (L.) L'Hérit.   HORSE SUGAR, SWEETLEAF
Hopea tinctoria L.

This native of the south-eastern United States was introduced to Britain in 1780 and several times later, but is not hardy. It is a small semi-evergreen tree, whose sweet-tasting leaves are greedily eaten by sheep and cattle. The fragrant flowers are borne in spring in almost sessile axillary clusters on the previous season's wood, which is leafless at flowering-time in the northern forms, and are followed by oval, yellowish brown fruits. It might prove hardier if re-introduced from the northern end of its range or from a high altitude (according to Sargent it ascends to 4,000 ft on the Blue Ridge of North and South Carolina).

SYRINGA*   LILAC   OLEACEAE

A group of small trees and shrubs, consisting of about two dozen species, confined to the Old World. Two are found in Europe, one in the Himalaya, the others in E. Asia as far south as the Chinese province of Yunnan. The cultivated species are deciduous, and have opposite leaves, usually neither toothed nor lobed; but in one species they are pinnate (*S. pinnatifolia*), and in another pinnately lobed (*S. laciniata*). The flowers appear in panicles, often pyramidal, but sometimes of indeterminate shape. Corolla tubular, with four lobes; calyx bell-shaped, unevenly toothed; stamens two. Seed-vessel a capsule of flattened or spindle shape, composed of two valves, which split from the top downwards when ripe.

There are two distinct subgenera:

subgen. LIGUSTRINA.—Corolla-tube short (about as long as the calyx), white or cream-coloured; stamens clearly exserted. This group, which in some respects resembles the privets (*Ligustrum*), is composed of only two species: *S. pekinensis* and *S. reticulata*.

subgen. SYRINGA.—Corolla tube long, enclosing the stamens. Flowers usually in some shade of purple, sometimes white.

This subgenus is subdivided into four series by Rehder:

ser. VILLOSAE.—The species of this well-marked group form a terminal bud and flower on the leafy shoots of the year.

ser. SYRINGA (*Vulgares*).—In this group, of which the common lilac is the type, the flowers are produced in leafless panicles from upper axillary buds on the previous season's growth and the terminal bud is usually lacking. Leaves simple or pinnately divided, glabrous or nearly so.

ser. PINNATIFOLIAE.—The one species in this series is closely allied to the preceding, but has completely pinnate foliage and terminal growth-buds.

ser. PUBESCENTES.—This group resembles the series *Syringa* in the

* Botanical section revised by P. S. Green of The Herbarium, The Royal Botanic Gardens, Kew.

manner in which the flowers are borne and in the absence of terminal buds, but in other respects approaches the series *Villosae*. The type of the group is *S. pubescens*, and the other species recognised are all closely allied to it.

The standard work on the genus is: Susan D. McKelvey, *The Lilac* (1928). It is profusely illustrated and treats the garden varieties and hybrids as well as the species. The author was for many years a Research Associate at the Arnold Arboretum, Massachusetts.

The cultivation of the common lilac is dealt with under the heading of *S. vulgaris*, and the soil and general treatment are the same for the rest of the genus. All of them can be propagated by layers, most of them by cuttings. Cuttings should be made of mature shoots in August, and placed in a sheltered position under handlights. Softer cuttings taken earlier will often take root in gentle heat.

The lilacs are subject to a bacterial disease, which causes the young shoots and inflorescences to wilt and blacken, and is often taken to be the result of frost damage. The affected parts should be cut out immediately the attack is noticed.

S. × CHINENSIS Willd. ROUEN LILAC

S. laciniata × *S. vulgaris*; *S. dubia* Pers.; *S. rothomagensis* Mordant de Launay; *S. varina* Dum.-Cours.

A deciduous bush of dense rounded habit, 10 to 15 ft high. Leaves ovate, 1½ to 2½ in. long, ⅝ to 1¼ in. wide, rounded or broadly wedge-shaped at the base, taper-pointed, glabrous; stalk ⅓ to ½ in. long. Flowers of the common lilac shade, intermediate in size between those of the common and Persian lilacs, somewhat loose; corolla tube ⅓ in. long, lobes ¼ in. long.

A hybrid between *S. laciniata* and *S. vulgaris*, raised at Rouen in the last quarter of the 18th century by M. Varin, director of the local Botanic Garden. It was introduced to Britain about 1795. Émile Lemoine, who later repeated the cross and raised many seedlings from it, suggested that *S. vulgaris* was probably the seed-parent. The epithet *chinensis* relates to an incorrect guess at the country of origin and is quite misleading. It has, however, been cultivated in Peking for a number of years and was collected there by Joseph Hers in 1921.

The Rouen lilac is a bush of great beauty when in flower, the growths made during the summer producing the following May a pair of flower-trusses 3 to 6 in. long at each joint towards the end, so that the whole makes a heavy, arching, compound panicle. It sometimes produces fertile seed, at least on the continent.

cv. 'ALBA'.—Flowers light pink, almost white. Probably a branch-sport from the original Rouen lilac, which may have occurred in several gardens. See also *S.* + *correlata*.

cv. 'BICOLOR'.—A branch-sport of the Rouen lilac noticed by Victor Lemoine in a private garden in 1850 and put into commerce by his firm. Flowers described as slaty-grey, with a bluish violet throat.

cv. 'DUPLEX'.—Flowers double. Raised by Émile Lemoine from a cross between *S. laciniata* (seed-parent) and a double-flowered common lilac. Put into commerce about 1897.

cv. 'METENSIS'.—Flowers rosy lilac. A branch-sport of 'Saugeana', which occurred on a plant growing in the Place d'Esplanade at Metz. It was propagated by Messrs Simon-Louis and distributed in 1871–2.

cv. 'SAUGEANA'.—Flowers rather darker and redder than in the original form. This is said to have been named for the nurseryman Saugé, who was son-in-law of Varin, raiser of the original Rouen lilac, and to have been raised from seed, around 1809. The same, or similar, lilac was distributed by Loddiges, the Hackney nurseryman, as *S. chinensis rubra*.

S. + CORRELATA A. Braun—This was described in 1873 by A. Braun from a plant growing in the Berlin Botanic Garden. He observed that the white flowers resembled those of *S. vulgaris* while the leaves were those of the Rouen lilac, and suggested that the plant might be the result of a back-cross between the latter and a white-flowered common lilac. It is considered by Hjelmqvist to be a periclinal chimera (graft-hybrid), whose outer tissues derive from the common lilac, the inner from the Rouen. It is not certain if the true plant is in cultivation in Britain.

S. EMODI Wall. ex Royle   HIMALAYAN LILAC

A large robust shrub, 10 to 15 ft high, the branchlets dark olive green or brownish, but freely spotted with long, narrow, pale excrescences. Leaves 3 to 8 in. long, and about half as wide, oval or sometimes ovate or obovate, tapering at the base, dark dull green above, pale, or almost white beneath. Panicles mostly columnar, 3 to 6 in. long, one or three of which terminate the young shoots. Flowers not pleasantly scented, expanding in June. Corolla ⅜ in. long, scarcely as much wide across the lobes, white or slightly purple tinted. Calyx bell-shaped, very shallowly lobed. Seed-vessels ¾ in. long, each half ending in a slender, almost tail-like point.

Native of the western Himalaya; long known in gardens, but not common. It is useful in flowering rather late. Closely allied to *S. villosa*, it is scarcely as good a shrub, and differs in its leaves being whiter beneath and downy only on the midrib, or glabrous. The seed-vessel also differs in being rather longer and in having the more attenuated apices mentioned above. *S. emodi* never seems to have the magnificent inflorescences characteristic of vigorous specimens of *S. villosa*, nor are the flowers ever so richly coloured. Series *Villosae*.

cv. 'AUREO-VARIEGATA'.—Leaves broadly, irregularly, and rather effectively margined with yellow.

SYRINGA EMODI

S. × HYACINTHIFLORA (Lemoine) Rehd.
S. oblata Lindl. × S. vulgaris L.

A race of early flowering lilacs, resembling both parents, first produced by Lemoine of Nancy, France, by crossing *S. oblata* with *S. vulgaris azurea plena*. The first clone bore double flowers but later crosses by Emile Lemoine between *S. oblata* var. *giraldii* and forms of the common lilac, *S. vulgaris*, produced a range of cultivars including 'Lamartine', 'Buffon' and 'Necker' (single-flowered), 'Vauban' and 'Claude Bernard' (double). More recently the cross has been repeated by breeders in the USA and Canada; of these newer cultivars the best known in Britain were raised by W. B. Clarke in California; see further on p. 545.

S. JOSIKAEA Jacq. f. ex Reichenb.

Belonging to the same group of lilacs as *S. villosa* and *S. emodi*, this is inferior in many respects to both. Its flowers, however, are of a deeper lilac than either. The leaves are whitish beneath, as in *S. emodi*, and of the same shape, 2 to 5 in. long. Panicle slender, 4 to 8 in. long, 2 to 4 in. wide. Corolla $\frac{1}{2}$ in. long, $\frac{1}{4}$ in. or less across the lobes. Seed-vessel $\frac{5}{8}$ in. long, bluntish at the end. Blossoms in early June. Series *Villosae*. *Bot. Mag.*, t. 3278.

This lilac was first noticed about 1830 in what was then Hungary, having been sent by the Baroness von Josika to J. F. von Jacquin, the botanist of Vienna, who named it in compliment to her. Wild in the mountains of Transylvania and the Ukrainian Carpathians, it and *S. vulgaris* are the only native European species of lilac. It is distinguished from *S. villosa* in flower by the denser arrangement of the flowers in whorls and the slightly more funnel-shaped and purple-coloured corolla.

Syringa josikaea has proved a good species for crossing, with the development of a number of late flowering hybrids. S. × HENRYI Schneid. is the collective name for crosses with *S. villosa*, first produced by L. Henry in Paris towards the end of the last century, to which was later given the name 'LUTÈCE'. *S.* × *henryi* has also been used in producing crosses with *S. sweginzowii* to which the name S. × NANCEANA McKelvey has been given and of which 'Floréal' (see p. 550) is perhaps the best known cultivar. S. × JOSIFLEXA Preston ex J. S. Pringle covers crosses between *S. josikaea* and *S. reflexa*, the first having been raised by Dr Isabella Preston of Ottawa in the 1920s and given the name 'Guinevere', but other cultivars have been raised, outstanding amongst these being 'Bellicent'. See further on p. 550.

S. JULIANAE Schneid.

A deciduous spreading shrub of stiff, bushy habit, about 6 ft high; young shoots slender, very downy, the down persisting for two years. Leaves 1 to 2 in. long, $\frac{1}{2}$ to 1 in. wide, oval (sometimes inclined to ovate or obovate), tapered at the base, finely pointed, dull dark green, with appressed hairs above; grey and very hairy beneath; stalk $\frac{1}{6}$ to $\frac{1}{3}$ in. long, hairy. Panicles 2 to 4 in. long, usually in pairs from the terminal buds of the previous year's shoots, sometimes from the two or three uppermost pairs, hairy like the shoots. Flowers fragrant, $\frac{1}{4}$ to $\frac{1}{3}$ in. long, $\frac{1}{8}$ to $\frac{1}{6}$ in. across the lobes. Calyx violet-coloured, glabrous, with short pointed lobes. The hairy flower-stalks (about $\frac{1}{8}$ in. long) carry one to three blossoms. Corolla deep lilac outside, pale inside the lobes. Series *Pubescentes*. *Bot. Mag.*, t. 8423.

Native of W. China; introduced by Wilson for Messrs Veitch about 1900. It is allied to *S. pubescens* and *S. patula*, but is more downy than the first and its flowers are more deeply coloured. The second species has a downy calyx. *S. julianae* flowers in May and June, and is both distinct and pretty, but not equal to the best lilacs.

SYRINGA JULIANAE

S. KOMAROWII Schneid.

S. sargentiana Schneid.

A deciduous shrub up to 15 ft high; young shoots pale brown, distinctly warted. Leaves mostly oval, but sometimes obovate or ovate-lanceolate, tapered sometimes slenderly and equally towards both ends, sometimes more abruptly towards the apex, 3 to 7 in. long, 1 to $2\frac{3}{4}$ in. wide, dark green above and downy on the sunken midrib when young, yellowish green beneath and downy more or less all over; stalk $\frac{1}{3}$ to $\frac{3}{4}$ in. long. Inflorescence borne on a leafy shoot, nodding, 4 to 6 in. long, 2 in. wide, of cylindric shape, made up of whorls of densely packed flowers, the main-stalk strongly warted and sparingly downy. Flowers deep rose, pink, or lilac-coloured, about $\frac{1}{2}$ in. long, the four lobes of the corolla $\frac{1}{12}$ in. long, erect or rather spreading; calyx cup-shaped, $\frac{1}{12}$ in. long, with shallow triangular lobes or truncate, rather downy like the short flower-stalk. Seed-vessel $\frac{1}{2}$ in. long, nearly glabrous. Flowers in June. Series *Villosae*.

Native of W. Szechwan, China; introduced by Wilson in 1908, but apparently represented in the St Petersburg Herbarium in 1893. It is a handsome lilac closely related to *S. reflexa*, which differs in its more slender, longer and more recurved panicles and usually more warted fruits.

S. LACINIATA Mill.

S. persica var. β L.; *S. persica* var. *laciniata* (Mill.) West.

A handsome small shrub most easily distinguished by its variable pinnatisect leaves with from three to nine parallel oblong lobes. It produces clusters of violet-purple coloured flowers towards the ends of the branches and is one of the most attractive species. Series *Syringa*.

A native of Kansu in western China it was presumably taken to Persia as a cultivated plant and from there made its way into western gardens in the 17th century. It is cultivated even today in gardens in Kabul and introductions from there have been incorrectly called *S. afghanica*, a species so far only known from a few collections in the wild with small entire privet-like leaves.

S. MEYERI Schneid.

This lilac was introduced by F. N. Meyer from Chihli, N. China, in 1908 to the United States Dept. of Agriculture by means of cuttings. It is a deciduous shrub of dense, compact habit growing up to 5 or 6 ft high, with slightly downy, squarish young shoots. Leaves oval, sometimes inclined to obovate, $\frac{3}{4}$ to $1\frac{3}{4}$ in. long, not quite so wide, glabrous except occasionally for down on the veins beneath. Two pairs of veins run from the base of the leaf to the apex parallel with the margins. The violet-purple flowers are produced in May and June, densely packed in panicles up to 4 in. long and $2\frac{1}{2}$ in. wide. Corolla $\frac{1}{2}$ in. long, with spreading lobes giving it a diameter of over $\frac{1}{4}$ in. Calyx and flower-stalks either glabrous or slightly downy. Seed-vessel $\frac{1}{2}$ to $\frac{3}{4}$ in. long, warted. Series *Pubescentes*.

Meyer's lilac is only known as a cultivated plant in N. China. It is most closely related to *S. pubescens*, whose leaves are not generally so tapered at the base, more downy beneath, and have three or more pairs of veins. Mrs McKelvey considered that it may eventually prove to be a selected form of that species.

cv. 'PALIBIN'.—A dwarf, slow-growing, compact selection of *S. meyeri*, which has become very popular as a rock garden shrub. It was introduced under the incorrect name of *S. palibiniana* and more recently has been listed as *S. velutina*. Both these names are properly synonyms of *S. patula* (q.v.). This lilac has been given the clonal name 'Palibin' (*Bot. Mag.*, n.s., t. 778). [PLATE 72

S. MICROPHYLLA Diels

A deciduous shrub 5 ft or more high, of spreading habit, with slender, downy young shoots. Leaves roundish ovate, pointed or rounded at the apex; $\frac{1}{2}$ to 2 in. long, $\frac{1}{3}$ to $1\frac{1}{4}$ in. wide, dark green above, greyish green beneath, slightly downy on both surfaces, ciliate; stalk $\frac{1}{6}$ to $\frac{1}{3}$ in. long. Panicles 2 to

4 in. long and 1½ to 2 in. wide, produced in pairs at the end of the shoot and often supplemented by lateral ones. Flowers very fragrant, pinkish-lilac, the corolla slender-tubed, ⅜ in. long, with the spreading lobes oblong, round-ended, ⅛ in. long; calyx downy, helmet-shaped, 1/16 in. long, with very short triangular lobes. Seed-vessel spindle-shaped, ½ in. long, warted. Series *Pubescentes*.

A native of N. and W. China; discovered by the Italian missionary Giraldi in 1893; introduced to the Coombe Wood nursery of Messrs Veitch by Purdom in 1910. It is variable in the downiness of the young shoots, leaves, etc., and separate species have been made of its more glabrous forms, which are, however, linked to the typical state by intermediates. It is a very distinct lilac on account of its small size and the often nearly orbicular shape of its leaves. It flowers in June but a second crop of flowers is often produced in autumn on the current year's leafy shoots. A particularly floriferous form has been named *S. microphylla* 'SUPERBA'.

S. OBLATA Lindl.

A deciduous shrub, 10 to 12 ft high, or a small tree, similar in habit to the common lilac; young shoots glabrous, round; buds purplish. Leaves very broadly heart-shaped to reniform, often considerably wider than long, being 1½ to 4 in. wide, 1½ to 3 in. long, short-pointed, glabrous on both surfaces, stalk ¾ to 1 in. long. Flowers pale lilac, produced at the beginning of May in short broad panicles, usually in pairs from the uppermost joints of the previous year's wood. Corolla-tube ½ in. long, about ⅔ in. across the lobes; calyx slightly glandular, with pointed lance-shaped lobes. Seed-vessel ⅝ in. long, slender-pointed. Series *Syringa*. *Bot. Mag.*, t. 7806.

Native of N. China; introduced by Robert Fortune from a garden in Shanghai in 1856. It is very closely allied to *S. vulgaris*, but is easily distinguished by the wider leaves and by flowering about a fortnight earlier. My experience of it is that it is the most unsatisfactory of all the lilacs except *S. reticulata* var. *mandshurica*. It is excited into growth by mild weather in early spring, only to have its young leaves and flowers destroyed by later frost. Probably in higher localities it may succeed better, for the shrub itself is perfectly hardy, and in climates with a much harder but more settled winter than ours flowers abundantly. The leaves turn red in autumn.

This is the species which is one parent of the early hybrids *S.* × *hyacinthiflora*. In addition to the original introduction, which is widely cultivated in N. China, other varieties have been introduced.

var. ALBA Hort. ex Rehd. *S. affinis* L. Henry.—A white-flowered form cultivated in N. China.

var. DILATATA (Nakai) Rehd.—Native to Korea and introduced by E. H. Wilson to the Arnold Arboretum in 1917. A spreading shrub with an open spreading inflorescence and flowers with a corolla tube of about ⅘ in., long and slender, and lobes which reflex on opening.

var. GIRALDII (Lemoine) Rehd.—A native of northwestern China introduced by Rev. Giuseppe Giraldi as seed from Shensi in the 1890s. Of taller habit, the flower clusters are open and spreading as in var. *dilatata* and the corolla long and slender, although at most ¾ in. long, but also with lobes that curl back.

S. × DIVERSIFOLIA Rehd.—A hybrid which arose at the Arnold Arboretum in 1929 from open pollinated *S. pinnatifolia* and nearby *S. oblata giraldii*. The first clone produced, 'WILLIAM H. JUDD', is noteworthy mainly because of the novelty of its characteristic entire, pinnatifid or three- to five-lobed leaves. Its white flowers often open as early as mid-April at Kew, a fortnight or so before the start of the main lilac season.

S. PATULA (Palibin) Nakai

Ligustrum patulum Palibin; *S. velutina* Komar. (but see below);
S. palibiniana Nakai (but *not* of gardens; see below)

A deciduous shrub up to 10 ft high, with slightly downy or glabrous, sometimes glandular, purplish young shoots. Leaves oval, ovate or rhomboidal to lanceolate, long- to short-pointed, tapered at the base, 2 to 2½ in. long, ½ to 2 in. wide; dark dull green and glabrous or slightly downy above, paler and more or less downy beneath; leaf-stalk ¼ to ½ in. long. Panicles often in pairs from the terminal pair of buds of the previous season's shoots, each 4 to 6 in. long, rather thinly set with blossom. Flowers opening in late May and June, fragrant. Corolla very slender, ⅓ in. long, less in diameter, of various shades of lilac outside, white within; anthers purple, near to but not reaching the mouth. Flower-stalks and calyx often purplish, varying from downy to glabrous. Seed-vessel ½ in. long, pointed, warty. Series *Pubescentes*.

Native of Korea and N. China; discovered in the former country in 1897; introduced to St Petersburg soon after. Mrs McKelvey, in her monograph, sank several names under *S. velutina* which were previously regarded as specific. Amongst them is *S. palibiniana*, which had been cultivated under that name since 1917, but only differs in the amount of down carried by the various parts (and which should not be confused with 'Palibin', the clone of *S. meyeri* which was introduced some years later, incorrectly under the name *S. palibiniana* and more recently listed, also incorrectly, as *S. velutina*). There appears, indeed, to be in this species every gradation between glabrousness and the velvetiness its synonym implies.

S. PEKINENSIS Rupr.

Ligustrina pekinensis (Rupr.) Dieck

A deciduous small tree of spreading, graceful habit, up to 20 ft high eventually; young shoots glabrous. Leaves ovate, oval, or ovate-lanceolate, 2 to 4 in. long, 1 to 2 in. wide, mostly tapering, sometimes rounded at the base, long and tapering at the apex, quite glabrous on both surfaces; stalk slender, ½ to ¾ in. long. Flowers cream-coloured, very densely clustered in numerous loose panicles 3 to 6 in. long, produced in pairs. Seed-vessel ⅝ to ¾ in. long, glabrous, pointed at the end.

Native of the mountains of N. China, where it was discovered by the Abbé David. It was raised at Kew in 1881 from seed sent from Peking by Dr Bretschneider. Botanically allied to *S. reticulata*, it is very distinct as seen growing. It has much more slender branches, the leaves are smaller, the inflorescence instead of being sturdy, pyramidal, and erect, is smaller and is a loose, rather shapeless panicle; the seed vessel, too, differs in the more pointed apex. It is perfectly hardy, and has grown more quickly at Kew than *S. reticulata*. It flowers freely towards the end of June.

cv. 'PENDULA'.—Raised from Chinese seed in the Arnold Arboretum, Mass., and a very graceful, pendulous tree.

S. × PERSICA L.   PERSIAN LILAC
S. ?afghanica × S. laciniata

A deciduous shrub, 4 to 6 ft high, of dense, bushy, rounded habit; young shoots slender, glabrous. Leaves lance-shaped or ovate lance-shaped (rarely three-lobed), with a long tapering apex and a more abruptly tapered base, green and glabrous on both sides, 1 to 2½ in. long, ⅓ to ½ in. wide; stalk ⅓ in. long. Flowers of the common lilac shade and fragrance, produced in May from the uppermost buds of the preceding summer's growth in small, sometimes branching panicles, 2 to 3 in. long and as much wide. Corolla-tube about ¼ in. long, the four spreading lobes rather shorter. Calyx funnel-shaped with four short, pointed lobes. Seed-vessels ½ in. long, cylindrical. Series *Syringa. Bot. Mag.*, t. 486.

A hybrid of unproved but ancient origin, which has been cultivated from time immemorial in Persia and India and had reached Europe by the early 17th century. It has been suggested that one of its parents may be the rare and little known *S. afghanica* Schneid., only known from a few collections in the wild and not yet introduced to cultivation.

The Persian lilac is a delightful shrub, both in its neat habit and its fragrant blossom. There is also a white-flowered form of it, cv. 'ALBA', in cultivation since the 18th century. Both are increased by cuttings of nearly ripe wood.

S. PINNATIFOLIA Hemsl.

A deciduous shrub, 8 to 12 ft high, of elegant bushy habit, the young shoots and every other part of the plant free from down. Leaves pinnate, 1½ to 3½ in. long, composed of seven, nine, or eleven leaflets, which are dull green, stalkless, ovate-lanceolate, ¾ to 1¼ in. long, ¼ to ⅜ in. wide, pointed, the base rounded or in the case of the terminal leaflets frequently attached to the common stalk by a portion of the blade. Flowers white, with a slight lilac tint, produced in late April or early May in panicles 1½ to 3 in. long, which spring usually in opposite pairs from the joints of the previous year's wood. Corolla-tube ½ in. long, the lobes at the mouth spreading and giving the flower a diameter of ¼ in.; calyx-lobes rounded.

Native of W. China; discovered by Wilson in 1904 in W. Szechwan, where it grows at altitudes of 7–9,000 ft. The pinnate leaves of this species at once suggest an affinity with *S. laciniata*, but they are divided (except sometimes near the apex) into quite distinct leaflets, and not merely lobed as in the other. It has been placed by Rehder in a series of its own, ser. *Pinnatifoliae*. It has interest as a distinct and hardy lilac, but its garden value is not equal to that of *S. laciniata* or *S.* × *persica*. It is one parent of *S.* × *diversifolia* (see under *S. oblata*).

S. POTANINII Schneid.

A deciduous shrub of graceful habit ultimately 9 to 12 ft high; young shoots minutely downy. Leaves mostly oval, slenderly pointed, tapered at the base, 1½ to 3 in. long, ¾ to 1½ in. wide, dark green and minutely but densely downy above, thickly covered with soft down beneath; stalk ⅛ to ¼ in. long. Inflorescence loosely pyramidal, erect, 3 to 6 in. long, 2 to 3 in. wide; main and secondary flower-stalks downy. Flowers fragrant, ⅓ to ½ in. long, white to pale rosy purple, the tube very slender, the four lobes ⅛ in. long, narrowly oblong; anthers yellow. Calyx downy, cup-shaped, shallowly toothed or nearly truncate. Seed-vessel ⅝ in. long, pointed, glossy, smooth or minutely and sparsely warted. Flowers in June. Series *Pubescentes*. *Bot. Mag.*, t. 9060.

Native of W. China; discovered in Kansu in 1885 by the Russian traveller Potanin. Wilson found it in 1904 and again in 1908 near Kangting (Tatsien-lu) in W. Szechwan, but apparently it did not reach Britain until Farrer sent seeds from S. Kansu in 1914. A beautiful form from this sending, raised at Highdown, near Worthing, by the late Sir Frederick Stern, is portrayed in the *Botanical Magazine*. The flowers of the Highdown plant are almost pure white, delightfully fragrant, with yellow anthers. This species is related to *S. julianae*, but that species has smaller leaves, dark violet anthers and a glabrous calyx. Nearly related to these two species is:

S. PINETORUM W.W. Sm.—According to Forrest, who collected it in June 1914, in the Lichiang range of Yunnan, this is a shrub 4 to 8 ft high with pale lavender-rose flowers. They have yellow anthers, and the leaves, 1 to 1½ in. long, are hairy on the midrib and veins beneath. Forrest introduced it, but it is uncertain if the true species is now in cultivation (1979).

S. × PRESTONIAE McKelvey

In 1920 Dr Isabella Preston of the Central Experimental Farm, Ottawa, crossed *S. reflexa* with *S. villosa* and produced a wide range of most desirable late-flowering lilac hybrids, combining in different degrees the various characteristics of the parent species. The cross has been repeated and many cultivated varieties are now available. See further on pp. 550–1. For the history of this group see *Ornamental Flowering Trees and Shrubs*, R.H.S. Conference Report (1940), pp. 135–40.

S. PUBESCENS Turcz.

S. villosa sens. Bot. Mag., t. 7064, not Vahl

A deciduous shrub or small tree, 12 to 15 ft high, forming a rounded head of branches; young shoots glabrous. Leaves 1 to 2½ in. long, ¾ to 1½ in. wide, broadly ovate, sometimes roundish, tapered abruptly at the apex to a short point, rounded or broadly wedge-shaped at the base, dull green and glabrous above, pale and with a little scattered down beneath, most abundant on the midrib; stalk ¼ to ½ in. long. Flowers fragrant, pale lilac or nearly white, produced along with the young leaves during early May in leafless panicles from one or both of the terminal buds of last year's shoots. The panicles are 3 to 5 in. long, 2 to 3 in. wide, the corolla-tube slender, ½ in. long; lobes ⅛ in. long, the incurving of the margins making them cupped. Calyx very short, with triangular lobes. Series *Pubescentes. Bot. Mag.*, t. 7064.

Native of N. China; introduced by Dr Bretschneider in 1881. It is only a second-rate lilac in this country, owing to the frequent injury of the young growths and panicles by late frost. In the United States, where the summer heat is greater, and the seasons better defined, it is very beautiful. The confusion in the naming of this shrub and *S. villosa* is alluded to under that species.

S. REFLEXA Schneid. [PLATE 71

A deciduous shrub up to 12 ft or perhaps more high; young shoots somewhat angular, stout, warty, becoming grey the second season. Leaves oval-oblong, sometimes obovate or ovate-lanceolate, pointed, mostly tapered at the base, 3 to 8 in. long, nearly half as much wide, dark green above, paler beneath; there are many short hairs on the midrib and chief veins beneath, otherwise they become nearly glabrous before falling. Flowers densely packed in a series of whorls on a terminal, leafy, arching or pendulous, cylindrical or narrowly pyramidal panicle, 4 to 10 in. long and 1½ to 4 in. wide, opening in June, not fragrant. Each flower has a narrow funnel-shaped tube about ⅓ in. long, rich pink or purplish pink outside, whitish within; and four ovate, pointed lobes inflexed at the tip which give the flower a diameter of ⅜ in. Calyx cup-shaped, with small erect teeth; glabrous or slightly downy. Seed-vessel cylindrical, ¾ in. long, warted. Series *Villosae. Bot. Mag.*, t. 8869.

Native of Hupeh, Central China; discovered by Henry in 1889; introduced in 1910 by Wilson, who found it at elevations of 8,000 to 9,000 ft. It is undoubtedly one of the handsomest of the Chinese lilacs and perfectly hardy. The most distinctive character is afforded by the shape and pose of the inflorescence which, in being densely packed with blossom, in being of cylindrical shape and more or less pendulous, differs from all other cultivated lilacs except *S. komarowii* (q.v.).

Syringa reflexa has been successfully used in various interspecific crosses. Hybrids with *S. sweginzowii* were first raised in the early 1930s by Messrs Hesse of Germany, who coined the collective name *S. × swegiflexa* for plants of this parentage. (See also *S. × josiflexa*, under *S. josikaea*, and *S. × prestoniae*).

S. RETICULATA (Bl.) Hara

Ligustrum reticulatum Bl.; *S. amurensis* var. *japonica* (Maxim.) Fr. & Sav.;
Ligustrina amurensis var. *japonica* Maxim.; *S. japonica* Decne.

A deciduous tree up to 30 ft high, of erect habit, often a shrub; young shoots not downy, but marked with small, round, pale dots. Leaves ovate with a long tapering point, rounded or broadly wedge-shaped at the base; 3 to 8 in. long, about half as wide, and either glabrous or slightly downy beneath, glabrous above; stalk ½ to 1 in. long. Flowers cream-coloured, somewhat privet-scented, produced at the end of the branch, usually in a pair of broad pyramidal panicles, 8 to 12 in. long, 6 to 8 in. through. Corolla ¼ in. across, the short tube almost hidden in the calyx, which is bell-shaped and scarcely lobed. Seed-vessel ¾ in. long, scimitar-shaped, glabrous, blunt at the end. *Bot. Mag.*, t. 7534.

Native of Japan; introduced to the Arnold Arboretum in 1876 and thence to Kew in 1886. Professor Sargent, who saw it wild on the hills of central Hokkaido (Yezo), says that there it is an ungainly, straggling tree, 25 to 30 ft high, with a trunk rarely 12 to 18 in. thick. I saw it flowering in June, 1910, in the Arnold Arboretum and other places near Boston, Mass., and it was the most striking tree then in flower, some being specimens over 30 ft high, of shapely, rather columnar habit, and laden with blossom. In Britain it does not succeed so well and remains more a shrub than a tree, but even here it is very attractive at the end of June.

var. MANDSHURICA (Maxim.) Hara *Ligustrina amurensis* var. *mandshurica* Maxim.; *Syringa amurensis* Rupr.; *Ligustrina amurensis* (Rupr.) Rupr.—Of the lilacs now in cultivation which represent the subgenus Ligustrina (or privet-like species), this is the least satisfactory in my experience. It was discovered in Manchuria by Radde, a Russian botanist, in 1857, and like many other shrubs from the same region, its flower-buds are easily excited into premature growth by warm January and February days, and are almost invariably cut off by late frosts. I have never seen a perfect panicle at Kew, although the flowers set freely enough. The species is a sturdy bushy shrub, 6 to 8 ft high, or a small tree. Leaves 2 to 4 in. long, 1 to 2 in. wide, ovate or oval, usually with a drawn-out apex, the base more or less tapered; stalk about ½ in. long. Flowers dull white, not very pleasantly scented, produced during June in panicles which, when perfectly developed, are 4 to 6 in. long, 3 to 4 in. wide; tube of corolla very short.

S. SWEGINZOWII Koehne & Lingelsh.

A deciduous shrub up to 12 ft high, with slender, glabrous, purplish-grey young shoots. Leaves of thin texture, ovate-lanceolate or oval-lanceolate, often long and slenderly pointed, sometimes more abruptly pointed, the base tapered or almost rounded, 2 to 4 in. long, 1 to 2 in. wide, dark green and glabrous above, paler and hairy on the chief veins beneath, ciliate; stalk ¼ to ½ in. long. Panicles terminal, erect, usually 6 to 8 in. long, opening in

June, sometimes supported by lateral ones; flower-stalks mostly glabrous, purplish. Flowers fragrant, pale rosy lilac, white inside the corolla lobes. Corolla ⅓ in. long, slender-tubed; lobes spreading; anthers yellow, inserted near the top. Calyx truncate or with shallowly triangular teeth, glabrous, purplish. Seed-vessel ½ in. long, pointed, smooth, shining. Series *Villosae*.

Native of China; first named in 1910 from a cultivated plant in a private garden near Riga; it must, therefore, have been introduced to Europe early in the century, if not before. It was collected in Szechwan by Wilson for Messrs Veitch in 1904, and he found it again in August 1910 at Sungpan, in the northern part of the same province. It is a very hardy, charming lilac of graceful habit which first came into notice in this country through being exhibited by the Hon. Vicary Gibbs at Westminster on 8 June 1915, when it received an Award of Merit. Messrs Lemoine sent out a lilac which they called 'S. *Swegin-zowii superba.*' Except that it has panicles somewhat larger than ordinary, such as might be developed under very good cultivation, it does not seem to differ from the type. The species is akin to *S. villosa*, but has thinner, smaller leaves and a more slender corolla tube, from which the anthers do not protrude.

S. TIGERSTEDTII H. Sm.—A medium-sized shrub closely related to *S. sweginzowii* and still relatively little known. It was introduced by Prof. Harry Smith of Uppsala in 1934 from W. Szechwan, China, and bears erect panicles of very pale pink flowers with a fragrance likened by the collector to that of carnations, but more spicy (H. Smith in *Lustgården*, Vol. 28–9, pp. 105–10 (1948)). It was introduced to Britain about 1954 and is available in commerce. Its name commemorates C. G. Tigerstedt, owner of the Mustila Arboretum in Finland.

S. TOMENTELLA, Bur. & Franch.

S. wilsonii Schneid.; *S. alborosea* N.E. Br.; *S. adamiana* Balf. f. & W.W. Sm.

A deciduous shrub up to 10 or 15 ft high; young shoots usually without down but sprinkled with pale warts. Leaves oval to ovate, pointed, more or less wedge-shaped at the base, 2 to 6 in. long, half as much wide, dark green, glabrous or slightly downy above, pale and downy (sometimes very much so) beneath, ciliate; stalk ¼ to ½ in. long. Panicles erect, terminal, up to 8 in. long and 5 in. wide, rather loose; flower-stalks reddish, more or less downy, some-times glabrous. Flowers pale lilac-pink, white inside, with a fragrance resembling, but not so strong as, that of common lilac. Corolla-tube about ½ in. long with four lobes spreading sufficiently to give the flower a diameter of ⅓ in. Calyx cup-shaped, reddish, 1/16 in. long, truncate or slightly toothed, glabrous or downy. Seed-vessel shining, spindle-shaped, ⅔ in. long, not downy. Series *Villosae. Bot. Mag.*, t. 8739.

Native of W. China; first described in 1891 from material collected by Prince Henri d'Orléans the previous year. It had been collected a year previously by A. E. Pratt, but did not reach cultivation until 1904, when Wilson introduced it from W. Szechwan to Veitch's nursery at Coombe Wood. This sending was named *S. alborosea* by N. E. Brown. In 1908 Wilson again collected it in W.

Szechwan and of part of his material Schneider made a new species, viz., *S. wilsonii*, and identified the remainder as *S. tomentella*; they are now regarded as one species, the differences consisting chiefly in the degree of pubescence.

In all its forms *S. tomentella* is very handsome and, like others of the *Villosae* group, valuable in coming into flower in June after the common lilac and its varieties are past. The often very densely downy undersurface of the leaves is distinctive.

S. VILLOSA Vahl

S. bretschneideri Lemoine; *S. emodi* var. *rosea* Cornu

A deciduous shrub, 10 ft or more high, of robust habit; branches erect, stout, stiff, nearly or quite glabrous when young, marked with a few pale dots. Leaves oval or oval lance-shaped, pointed, rounded or wedge-shaped at the base, 2 to 6 in. long, 1 to 2½ in. wide, glabrous and dark green above, glaucous and thinly furnished with bristle-like hairs or nearly glabrous beneath; stalk ¼ to 1¼ in. long. Panicles terminal and axillary, often three at the end of a leafy shoot; they are usually 6 to 10 in. long (but I have measured exceptionally fine ones 18 in. long), half to two-thirds as wide. Corolla lilac-rose, ½ in. long, the lobes ⅛ in. long, rounded, spreading. Calyx bell-shaped with four short, pointed lobes; slightly hairy or glabrous. Seed-vessel about ½ in. long. Series *Villosae*. *Bot. Mag.*, t. 9284.

Native of N. China; discovered early in the 18th century by Père d'Incarville, the Jesuit missionary; introduced around 1880 by Dr E. Bretschneider from the mountains west of Peking, where it sometimes attains the dimensions of a tree. Much confusion existed for a time as to the correct name for this lilac but eventually it was shown to be the true *S. villosa* and that the plants then grown under that name were really *S. pubescens*.

This beautiful lilac, perhaps the most robust of its section of the genus, flowers at the end of May and early in June, after the flowers of the common lilac and its varieties have faded. It is one of the most desirable of hardy shrubs, vigorous in constitution, and free flowering. It differs from *S. vulgaris* and its allies in forming a true terminal bud, and in flowering on the current year's shoots. As will be noticed by the synonyms recorded above, it has been referred to as *S. emodi*, to which it is allied, but from which it differs in its larger, more open inflorescence.

S. VULGARIS L.   COMMON LILAC

A deciduous shrub or small tree, up to 20 ft high, usually producing a crowd of erect stems, but occasionally a single trunk over 2 ft in girth, clothed with spirally arranged flakes of bark; shoots and leaves quite glabrous. Leaves heart-shaped or ovate, 2 to 6 in. long, from three-fourths to almost as much wide near the base; stalk ¾ to 1¼ in. long. Panicles pyramidal, 6 to 8 in. long, usually in pairs from the terminal buds. On cultivated improved varieties, panicles 12 to 18 in. long are produced. Flowers 'lilac', delightfully fragrant;

corolla-tube $\frac{1}{3}$ to $\frac{1}{2}$ in. long, the lobes concave; calyx and flower-stalks more or less furnished with minute gland-tipped down. Seed-vessels smooth, $\frac{2}{3}$ in. long, beaked. Series *Syringa*.

Native of the mountainous regions of E. Europe. Introduced to W. Europe in the 16th century. It has been cultivated in England for over three hundred years, and is now as characteristic a feature of village scenery as almost any native shrub.

For a selection of garden varieties of *S. vulgaris* (and of *S.* × *hyacinthiflora*, its hybrid with *S. oblata*), and a note on their cultivation and pruning, see pp. 544–50.

S. WOLFII Schneid.

S. formosissima Nakai

A deciduous shrub up to 20 ft high; young shoots glabrous or nearly so, turning grey. Leaves oval-lanceolate, slenderly pointed, more or less abruptly tapered at the base, glabrous except for down beneath (especially on the veins), 3 to 7 in. long, about half as wide, dark green above, pale beneath; stalk $\frac{1}{2}$ to 1 in. long. Panicle terminal, up to 12 in. long, opening in June; flower-stalks downy. Flowers lilac, fragrant; corolla $\frac{1}{2}$ in. long, the lobes erect; anthers inserted half-way down the tube, primrose yellow. Calyx usually more or less downy, sometimes glabrous; cup-shaped, scarcely or not at all lobed. Seed-vessel $\frac{1}{2}$ in. long, glabrous, blunt ended. Series *Villosae*.

Native of Korea and Manchuria; introduced about 1909. This handsome and very hardy lilac is related to *S. villosa*, which differs, however, in the corolla lobes being spreading and in the anthers being near the mouth. It is still more closely allied to *S. josikaea*, the Hungarian lilac, which has the same erect corolla lobes but has smaller anthers inserted even lower than those of *S. wolfii*. *S. josikaea* is not so vigorous, has smaller leaves paler beneath, and is not so handsome a shrub.

S. YUNNANENSIS Franch.   YUNNAN LILAC

A deciduous shrub up to 12 ft high, of rather erect, slender habit; young shoots minutely downy or glabrous, reddish, thinly but conspicuously warty. Leaves oval, oblong-lanceolate or narrowly obovate, tapered at the base, pointed, $1\frac{1}{2}$ to $3\frac{1}{2}$ in. long, $\frac{3}{4}$ to $1\frac{1}{2}$ in. wide, dull green above, glaucous beneath, glabrous on both surfaces but margined with minute hairs; stalk reddish, $\frac{1}{2}$ to $\frac{3}{4}$ in. long. Panicle terminal, up to 8 in. long and 6 in. wide, flower-stalks minutely downy. Flowers fragrant, opening in June. Corolla pale pink, becoming almost white, $\frac{1}{4}$ to $\frac{3}{8}$ in. long, $\frac{1}{4}$ in. across the spreading lobes; anthers yellow, reaching the mouth of the tube. Calyx glabrous, cup-shaped, with very small teeth, reddish. Seed-vessel smooth or with a few warts, $\frac{2}{3}$ in. long. Series *Villosae*.

Native of Yunnan, China; first named in 1891 from specimens collected by Delavay; first introduced to this country by Forrest when collecting for

A. K. Bulley, 1904–7. It is a pleasing lilac and is most closely related to the Himalayan *S. emodi*. It has, especially, the same smooth, pale, almost glaucous under-surface of the leaves. The Himalayan lilac is distinct in its much larger leaves and in the anthers being more protruded. Plants grown at Kew and Aldenham as *S. yunnanensis* appear to fit the type in every way except that the panicles are quite glabrous.

THE GARDEN LILACS

In this section a brief account is given of the principal cultivars of lilac, which, as in so many other shrubby genera, have largely displaced the original wild forms in gardens. They fall into two groups. The first and by far the largest and most important, comprises the garden-bred sorts of *S. vulgaris*, to which have been added in this century various hybrids between it and the closely related *S. oblata*. The second group is made up of hybrids between members of the series *Villosae*. The two groups are as distinct horticulturally as they are botanically, and no viable hybrid between them has been raised.

The International Registration Authority for cultivars of the genus *Syringa* is The Royal Botanic Gardens, Hamilton, Ontario, Canada. A tentative *International Register of Cultivar Names in the Genus Syringa*, compiled by Dr Owen M. Rogers, was published in 1976 by the Agricultural Experiment Station of the University of New Hampshire, Durham, New Hampshire, USA. To this we are indebted for details of raiser and date of introduction, when not given in Susan McKelvey's *The Lilac*, in which a detailed account is given of all cultivars raised before the mid-1920s. This work is one of the great classics of horticultural literature, besides being the authoritative monograph on the genus *Syringa*. For an excellent short account see: F. P. Knight, 'Lilacs', *Journ. R.H.S.*, Vol. 84 (1959), pp. 486–99.

The Kew collection of lilacs has been greatly extended in recent years. The most complete collection is to be seen in Withdean Park, Brighton. It was established in the early 1960s by Mr J. R. B. Evison, then in charge of the Department of Parks and Gardens at Brighton, under whose care it is.

I Lilacs of the Vulgaris Group (Series Syringa)

At the beginning of the last century the gardener had few sorts of garden lilac to chose from. Apart from the common blue there was only the white and the purple or Scotch lilac, and a few minor variations of these two, but by the middle of the century new varieties were coming into commerce in increasing numbers. The first to receive a cultivar-name of modern form was 'Charles X' (1831–2), but the use of names of this kind, usually the sign that a horticultural group has come of age, did not become general until the 1870s. Few of the old lilacs have remained in commerce, but most were acquired by the Arnold Arboretum in the 1870s and 1880s when they were still comparatively new, and still grow there, in a collection that must be the largest in the world— about 400 different varieties. Most of the sorts still available have been raised since the 1880s and a remarkably high proportion were the creation of the great firm Lemoine, of Nancy, which with its countless philadelphus, deutzias, weigelas and lilacs has done so much to beautify the gardens of the temperate world. The first double lilacs were raised by Victor Lemoine in the 1870s, from crosses between single-flowered varieties and a double, almost sterile mutant called *S. vulgaris azurea plena*. The history of this cross, as told by Victor Lemoine to the American lilac enthusiast T. A. Havemeyer, is quoted in McKelvey, *The Lilac*, p. 264.

Some lilacs raised this century are the result of crossing *S. vulgaris* with the closely related *S. oblata*; see *S. × hyacinthiflora*, p. 531. It would be inconvenient and indeed pointless to list these hybrids separately, and they are identified below by placing Hyacinthiflora in brackets after the name. The distinctive character of these hybrids lies not in the flowers themselves as in the fact that they are borne earlier than in the varieties deriving from *S. vulgaris* purely—in the first half of May or even in late April—and thus help to lengthen the season. They show hybrid vigour and tend to grow rather tall and lanky, a character that makes them suitable more for growing at the back of a mixed planting than as specimens. There must, incidentally, be a suspicion that some of Lemoine's later productions have *S. oblata* in their parentage even though they are classified as cultivars of *S. vulgaris*. 'Maréchal Foch' is an example.

CULTIVATION AND PRUNING.—The lilacs of this group need full sun or the lightest shade, and thrive best in areas where the summers are warmer than the average. To be obtained at their best they must be given generous treatment. Any soil is suitable except a very acid one, and a deep, stiff but well-drainèd loam is best of all. A chalky soil is very much to their liking, provided it is deeply dug and well fertilised. A mulch of compost or well-rotted manure helps them, and so too does a dressing of bone meal. They need a minimum spacing of 10 ft, and 15 ft is not too much. An important item in the cultivation of the finer lilacs is the removal of the flower-trusses as soon as they fade, so as to prevent the formation of seed, thereby concentrating the energies of the plant in the new growth and the succeeding crop of flower. This dead-heading may not be practicable once a plant has reached its full height, but should always be done when the plant is young. The lilacs need no systematic pruning, but in order to obtain fine trusses the weaker and superfluous shoots may be cut out at the same time as the old inflorescences are removed. But some modern sorts, especially those of the Hyacinthiflora group, are of vigorous and rather lax habit. These should have their branches shortened by about one-third after flowering, to stop them becoming too lanky. Isolated bushes— and a fine, shapely lilac is an admirable ornament for a lawn—should be trained to a single stem by removing all the lower buds and subsequently the lower branches. However, in North America, where borers are an insect pest, they are better grown on several stems, each arising from ground-level—the vigour and size of the plant being maintained by cutting out the oldest shoots in sequence. As the lilacs of this group form no terminal bud, and naturally fork their branches every year, some pruning and training is at first needed to get a tree-like example.

Young lilacs should be treated with patience, as some years may pass before they produce trusses of the characteristic size, shape and colour. The flower-buds of newly planted lilacs should always be removed. Old plants that have grown out of shape or lost vigour respond very well to hard cutting, which should always be done in winter or early spring.

PROPAGATION.—Most lilacs of this group can be propagated by means of half-ripe cuttings, or by soft-cuttings in a mist unit, but the young plants develop slowly. It is of more interest for the private gardener to know whether the plants he buys are on their own roots or grafted. Propagation by layering has long been practised by British growers and is gaining favour on the continent and in the United States. The advantage of having the plant on its own roots is obvious, but its price is inevitably higher than one that is grafted or budded. Plants on common lilac stock grow well and are said to give better blooms than 'own-root' plants, but the disadvantage is that, unless a watch is kept, the variety in time becomes overwhelmed by suckers; if it is not known in advance that the plant is grafted then the case is worse. With privet as the stock the scion does not grow so well nor last so long, if it has to rely for the privet's roots for its sustenance, but plants produced by modern methods are supposed to be planted with the graft union well below the surface of the soil, the purpose of the privet's roots being simply to sustain the plant until the scion has struck its own roots.

The following selection comprises most of the common lilacs available in Britain, and all those that have received awards from the Royal Horticultural Society. The colours of the lilacs are notoriously difficult to describe, partly because they are impure, and none the worse for that, and partly because they change so much in hue and tone as the flower ages. Susan McKelvey, in her splendid treatment of this group, often devotes five lines to describing the colour of a single cultivar, using the Ridgway colour chart. But she admitted that these detailed specifications were intended only for verification, and stressed that tone, i.e., the degree of saturation, is more important than hue. In the following summary account, the cultivars, apart from the whites, are arranged in two colour-groups—those whose flowers are in the paler shades of near-blue, lavender and pinkish mauve, and those in which the colour is more saturated. The lilacs in the first group assort together amicably. The dark lilacs need more careful placing.

SINGLE WHITE FLOWERS

'JAN VAN TOL'. Van Tol, Boskoop, c. 1916.—Flowers snow-white, about 1¼ in. wide, in long, drooping panicles. Corolla-segments narrow, widely spaced. The result of a cross between 'Marie Legraye' and 'Mme Lemoine'.

'MME FELIX'. Felix and Dykhuis, 1924.—Flowers in shapely, erect panicles. Raised from 'Marie Legraye'. A good lilac for forcing.

'MARIE LEGRAYE'. Mme Legraye (?), before 1879.—This old variety was once much used for forcing, and is the parent of many other whites. Flowers ivory-white from yellowish buds. Anthers conspicuous (considered to be a defect in a forcing lilac). Panicles rather narrow and lax.

'MAUD NOTCUTT'. Notcutt, 1956.—Flowers flat, with broad segments, about 1 in. wide, well displayed in shapely trusses about 1 ft high, carried well above the foliage. Habit fairly upright. According to the raisers the cut sprays last well in water. A.M. 1957. [PLATE 70

'MONT BLANC'. Lemoine, 1915.—Flowers from greenish buds, very large, in unusually tall and broad conical trusses.

'PRIMROSE'. G. Maarse, Aalsmeer, Holland, 1915.—A sport from 'Marie Legraye', with primrose-yellow flowers. A weak grower. A.M. 1950.

'VESTALE'. Lemoine, 1910.—A fine, vigorous lilac, bearing its flowers in long, well-filled rigidly erect trusses. A.M. 1931.

DOUBLE WHITE FLOWERS

'EDITH CAVELL'. Lemoine, 1916.—Flowers large, from cream-coloured buds. Panicles large and rather lax.

'MME LEMOINE'. Lemoine, 1890.—Flowers large, more or less hose-in-hose, in erect, compact trusses. Very free-flowering, even when young. A seedling of 'Marie Legraye', crossed with a double lilac. F.C.C. 1897; A.G.M. 1937.

'MISS ELLEN WILLMOTT'. Lemoine, 1903.—Flowers opening from greenish buds, pure white, hose-in-hose, with broad lobes. Trusses long, open, pyramidal. A.M. 1917. It is usually listed in catalogues as 'Ellen Willmott'.

'MONIQUE LEMOINE'. Lemoine, 1939.—Flowers hose-in-hose, in broad trusses. A.M. 1958.

'SOUVENIR D'ALICE HARDING'. Lemoine, 1938.—Flowers warm white, hose-in-hose, in tall trusses. Useful for its late flowering, at the end of May or in early June.

SINGLE PALE FLOWERS

'AMBASSADEUR'. Lemoine, 1930.—Flowers pale near-blue from pink buds, flattish, well-shaped, with a white eye. Trusses broad.

'BLUE HYACINTH' (Hyacinthiflora). Clarke, 1942.—Flowers hyacinth-shaped, with long, slender tubes, well spaced, pale blue from mauve buds.

'BUFFON' (Hyacinthiflora). Lemoine, 1921.—Flowers mauve-pink, about 1 in. wide, with cupped lobes, borne in dense, broad trusses. Lax habit.

'CAPITAINE BALTET'. Lemoine, 1919.—Flowers large, violet-purple, fading to a bluer tone. Trusses large, not dense.

'CLARKE'S GIANT' (Hyacinthiflora). Clarke, 1948.—Flowers mauvish pink on the inside, pale lavender outside, about 1 in. wide, in very large, loose trusses. Erect, vigorous habit. A.M. 1958.

'DIPLOMATE'. Lemoine, 1930.—Put into commerce at the same time as the similar 'Ambassadeur', this seems to be uncommon in the trade. The small flowers are slaty blue from deep rosy pink buds, freely borne in open trusses.

'ESTHER STALEY' (Hyacinthiflora). Clarke, 1948.—Flowers vivid carmine-red in the bud, opening a pleasant shade of bright pink. Vigorous, erect habit. A.M. April 25, 1961. A.G.M. 1969.

'FIRMAMENT'. Lemoine, 1932.—Flowers near-blue, from pink buds, medium-sized. A charming lilac, in the same style as 'Ambassadeur' and 'Diplomate'.

'LAMARTINE' (Hyacinthiflora). Lemoine, 1911.—Flowers about 1 in. across, pinkish mauve, opening towards the end of April. A tall erect bush. This was one of a batch of hybrids between *S. oblata* var. *giraldii* and various cultivars of *S. vulgaris* raised by Emile Lemoine and put into commerce in 1911 (his father Victor Lemoine died in that year at the age of almost ninety). 'Lamartine' was planted at Kew in 1912, and received an Award of Merit when shown from there on April 26, 1927. According to the raiser himself, these hybrids did not prove popular with gardeners. Perhaps their lanky habit and early flowering told against them.

'LUCIE BALTET'. Baltet, before 1888.—Flowers flesh-pink in open trusses. A charming lilac, now uncommon but represented in the Kew collection. Mrs McKelvey considered it to be 'one of the loveliest of the single lilacs'.

'MME CHARLES SOUCHET'. Lemoine, 1949.—Flowers large, in broad trusses, near-blue, borne very freely. Possibly a derivative of the Hyacinthiflora group, as it flowers in early May.

'MARÉCHAL FOCH'. Lemoine, 1924.—Carmine-pink buds, opening pinkish mauve and ageing to lilac pink. Large, well-formed trusses. Its tall and narrow habit, as well as its early flowering suggests some Hyacinthiflora influence in its makeup, though, like the preceding, it is classified as a pure *S. vulgaris*.

DOUBLE PALE FLOWERS

'ALPHONSE LAVALLÉE'. Lemoine, 1885.—Large flowers, lavender blue shaded with pink, from purple buds. One of Victor Lemoine's first double lilacs, now uncommon.

'BELLE DE NANCY'. Lemoine, 1891.—Flowers deep purplish pink in the bud, opening lilac-pink and fading to bluish mauve, in rather long, narrow clusters.

'CAPITAINE PERRAULT'. Lemoine, 1925.—Flowers large, in wide trusses, rosy mauve. Useful for prolonging the season as it is one of the last to flower.

'EDOUARD ANDRÉ'. Lemoine, 1900.—Carmine pink in the bud, opening rosy pink, in broad, open trusses.

'EDWARD J. GARDNER'. Gardner, Wisconsin, USA.—Flowers semi-double, light pink, in long trusses; buds rich purple. A fine new variety, introduced to this country by Messrs Notcutt.

'GENERAL PERSHING'. Lemoine, 1924.—Flowers of variegated appearance, lilac-pink mixed with paler pink and light violet.

'KATHERINE HAVEMEYER'. Lemoine, 1922.—Flowers large, at first lavender-purple, fading to a pinker tone. Trusses dense, broadly pyramidal. One of the finest and most widely grown of lilacs. A.M. 1933. A.G.M. 1969.

'MME ANTOINE BUCHNER'. Lemoine, 1909.—Flowers hose-in-hose, deep vinaceous purple in the bud, opening soft carmine-rose tinged with mauve. Trusses long and open. A lovely, softly multi-coloured lilac. A.G.M. 1969.

'MICHEL BUCHNER'. Lemoine, 1885.—Flowers pale rosy lilac, in tall trusses.

'OLIVIER DE SERRES'. Lemoine, 1909.—Flowers light violet shading to a pinker tone, in large trusses. Strong growing, but not flowering freely when young.

'PRESIDENT GRÉVY'. Lemoine, 1886.—Flowers deep pink in the bud, opening lavender blue with a paler margin. Clusters large and pyramidal. A.M. 1892.

SINGLE DARK FLOWERS

'ANDENKEN AN LUDWIG SPÄTH' ('Souvenir de Louis Spaeth'). Späth, Berlin, 1883.—Flowers deep wine-red, holding their colour well, borne in slender trusses about 1 ft long. Vigorous, spreading habit. Although nearing its centenary, this lilac has maintained itself against all rivals, and remains one of the most popular of all. A.G.M. 1930.

The use in Britain of the French name for this lilac is inappropriate, as was pointed out by the Royal Horticultural Society in 1930. It was suggested that the name might be anglicised to 'Souvenir of L. Spaeth' (*Journ. R.H.S.*, Vol. 55, p. 279).

'CHARLES X'.—This cultivar, put into commerce in the 1830s, probably derived from the old purple lilac, and was raised in France. What is believed to be the true clone is still offered by nurserymen; it has magenta-purple flowers in conical trusses.

'CONGO'. Lemoine, 1896.—Flowers deep purplish red in the bud, paling somewhat when open to a rich lilac-purple, in large, open clusters. Compact habit.

'ETNA'. Lemoine, 1927.—Buds dark claret-red, opening rich purplish red. Late.

'MME F. MOREL'. Morel, 1892.—Flowers reddish purple in the bud, opening light violet-purple, in very large clusters.

'MASSENA'.—Flowers with very concave lobes, deep reddish purple. Spreading habit. A.G.M. 1930.

'RÉAUMUR'. Lemoine, 1904.—Flowers deep reddish purple, shaded with violet. Late flowering. Vigorous, bushy habit.

'SENSATION'. De Maarse, Boskoop, 1938.—A remarkable lilac, quite unlike any other. The flowers a rich vinous red, edged with white, in tall trusses. A plant at Kew produces a branch-sport with white flowers.

DOUBLE DARK FLOWERS

'CHARLES JOLY'. Lemoine, 1896.—Flowers dark reddish purple paling to a whitened plum-purple, the recurved segments showing the paler undersides. Erect habit.

'CONDORCET'. Lemoine, 1888.—Flowers purplish red in bud, opening violet purple, in dense trusses.

'MARÉCHAL LANNES'. Lemoine, 1910.—Flowers large, more or less double, in broad-based clusters, violet with an infusion of pink.

'MRS EDWARD HARDING'. Lemoine, 1922.—Flowers semi-double, purplish red, rather quickly fading to pink. Usually considered to be the best double red. Tall growing. A.G.M. 1969.

'PAUL THIRION'. Lemoine, 1915.—Flowers likened by Mrs McKelvey to large double violets in form, carmine in bud, opening rosy red, fading to lilac pink, in dense, broad trusses. Late flowering (after 'Mrs Edward Harding'). A.G.M. 1969.

Iĩ HYBRIDS OF THE VILLOSAE SECTION

The first crosses between members of the section *Villosae* were made in France late in the 19th century, and a few later by Lemoine (see under *S. josikaea*), but none of those that were put into commerce attained much popularity and only 'Floréal' is mentioned below. Those best known in Britain were raised in Canada and belong for the most part to the Prestoniae group (see *S.* × *prestoniae*) or to the similar Josiflexa group (*S.* × *josiflexa*, see under *S. josikaea*). The Prestoniae group is the larger by far of the two, with about 100 named cultivars, but the amount of variation in this group does not justify such a multiplication of names. Although the parentage of most is known, some of the later cultivars were raised from open-pollinated plants, and could therefore have more than two species in their parentage. The flowers of these lilacs, with the long, slender tube and narrow limb characteristic of the section, render the group as a whole very distinct from those deriving from the common lilac. This, and their privet-like scent, may account for the fact that they have not yet won general popularity. They are of easy cultivation in any good soil and attain mostly a height of 10 or 12 ft and as much in width. Unless otherwise stated they flower in June. They are easily propagated by cuttings.

'AUDREY' (Prestoniae). Preston, 1927.—Trusses conical, dense, about 9 in. long and wide. Flowers light magenta pink outside, almost white within. A.M. 1939.

'BELLICENT' (Josiflexa). Preston, before 1948.—Flowers clear pink (pale Rose Bengal), fragrant, in trusses about 9 in. long. The most admired of the Ottawa hybrids, with flowers of an unusually clean shade of pink. *Journ. R.H.S.*, Vol. 84 (1959), fig. 163. It flowers towards the end of May. F.C.C. 1946; A.G.M. 1969.

'ELINOR' (Prestoniae). Preston, 1928.—Flowers in dense panicles about 7 in. long, Magenta Rose in the bud, opening pale lilac-pink outside, pale violet inside. A.M. 1951; A.G.M. 1969.

'ETHEL M. WEBSTER' (?Prestoniae). Preston, before 1948.—Flowers flesh pink, in lax panicles.

'FLORÉAL' (Nanceana). Lemoine, 1925.—Flowers in rather dense trusses about 10 in. long, light lavender-purple. A spreading shrub, flowering freely in May. Lemoine distributed it originally as '*S. henryi* Floréal'. Mrs McKelvey pointed out to him that as a hybrid between *S.* × *henryi* and *S. sweginzowii* it could not be regarded as a variety of the former, to which he replied: 'Les catalogues d'horticulteurs n'ont pas la prétention d'etre des documents botaniques, et je ne désire pas changer ce nom.' (McKelvey, *The Lilac*, p. 108).

'FOUNTAIN' (?Swegiflexa). Preston, before 1948.—This is believed to be a cross between *S. reflexa* and the original clone of *S.* × *swegiflexa*, sent out by Messrs Hesse in 1935. If so, it is a back-cross and certainly seems to be near to *S. reflexa*. It makes a large bush, with drooping panicles of pink flowers in May.

'GUINEVERE' (Josiflexa). Preston, before 1938.—Trusses about 9 in. long and wide, conical, drooping at the tip. Flowers purple in the bud, opening lilac-purple and fading to a pinker tone. This is the type of *S.* × *josiflexa*.

'HIAWATHA' (Prestoniae). Skinner, 1934.—Flowers in rather short, broad trusses, deep reddish purple in the bud, paler on opening. Low, spreading habit. It is earlier flowering than the Prestoniae hybrids raised by Miss Preston herself. It is recorded that in selecting seedlings Miss Preston chose those in lighter shades, while Mr Skinner favoured the darker-coloured sorts.

'ISABELLA' (Prestoniae). Preston, 1927.—Flowers pale lilac-purple outside, almost white within, in trusses about 1 ft long, dense. This is the type-clone of *S.* × *prestoniae*, described by Mrs McKelvey in her monograph and named by her in honour of Miss Preston. A.M. 1941.

'KIM' (parentage uncertain). Preston, before 1948.—Flowers mallow-purple, in broad panicles about 10 in. long. A.M. 1958.

TAIWANIA   TAXODIACEAE

A genus of one or two species; if two, one (the type) in Formosa and the other, closely related to it, in the Chinese province of Yunnan and bordering Burma. Leaves of young trees strongly resembling those of *Cryptomeria*, but on coning trees thick and appressed. Cones terminal, ½ in. or slightly more long, with twelve to twenty scales; seeds winged, usually two on each fertile scale.

T. CRYPTOMERIOIDES Hayata   [PLATE 73

An evergreen tree averaging in its adult state 150 to 180 ft, but occasionally over 200 ft in height, the trunk (sometimes 30 ft in girth) clothed with reddish brown bark separating from it in loose strips. As in so many conifers, the leaves of juvenile trees are quite distinct from those of adult ones. The former are arranged equally all round the stem, about twenty to the inch, awl-shaped, ¼ to ⅜ in. long, 1/12 in. wide at the base, spine-tipped and of hard texture. Leaves and twigs of adult trees very like those of *Athrotaxis laxifolia*, the former being ⅛ to ⅙ in. long, triangular in cross section, incurved at the shortly pointed apex, with stomata on all three surfaces; they are close enough together completely to hide the stems. Cones terminal, ovoid-cylindrical, ½ in. long, ¼ to

⅜ in. wide; scales numerous, rounded, overlapping.

Native of Formosa, especially on Mt Morrison; first introduced to England in 1920 by E. H. Wilson, who had raised plants in the Arnold Arboretum from seed he had himself collected in Formosa. This tree is one of the tallest conifers in the Old World and is only surpassed in height by some of those native in the Himalaya. The shoots of young trees are very like those of *Cryptomeria japonica* and the adult tree is in general aspect also similar. In youth it is extremely elegant, the branches curved gracefully upwards, the branchlets slender and more or less drooping.

T. cryptomerioides is rare outside collections. The tallest example grows at Killerton in Devon, 41 × 1¾ ft (1973). At Wakehurst Place in Sussex it is 24 × 1 ft (1970).

The mainland race, or relative, of the Formosan species was first collected in Burma by J. H. Lace in 1912, near the Pyehpat bungalow, on the track between Myitkyina and Hpimaw, on the Burma-Yunnan frontier. Frank Kingdon Ward met with it in March 1939 near Htawgaw (just below 26° N). Here, and for some way north it used to be common, but most accessible trees had been felled and the planks exported to China to make coffins (Kingdon Ward, *Burma's Icy Mountains* (1949), pp. 222, 241, 267–71, plate facing p. 272; *Pilgrimage for Plants* (1960), Chap. 12; and *Blackwood's Magazine*, No. 1490 (1939), pp. 769–84). In China the mainland taiwania was found by Handel-Mazzetti, the Austrian botanist, growing in western side-valleys of the Salween around 28° N, near the Burma frontier. From his coning specimens Gaussen described T. FLOUSIANA, but it is controversial whether the Sino-Burman trees are really sufficiently distinct from the Formosan type to merit specific rank.

NOTE.—In Vol. II of the present work, p. 491, it is stated that the wood of *Juniperus recurva* var. *coxii* was used for coffin-making by the Chinese, and the name 'Coffin Juniper' is given to this tree in most works devoted to the conifers. But C. W. D. Kermode, writing in *Indian Forester*, Vol. 65 (1939), pp. 204 et seq., points out that trees in the Maymyo Botanic Garden, Burma, introduced from the northeast frontier as *J. recurva*, were in fact the taiwania. Before the discovery and identification of the taiwania on the Burma-China frontier it seems to have been generally believed that the coffin tree was some species of giant juniper. According to Kermode's enquiries *J. recurva* was not used for making coffins, and in view of his statement the name 'Coffin Juniper' for *J. recurva* var. *coxii* should perhaps be replaced by 'Cox's Juniper' if a vernacular name is needed. But an element of doubt remains, since several species of tree with durable, aromatic timber were used by the Chinese for making coffins, and if *J. recurva* var. *coxii* was not so used in 1939 the reason may be that all suitable trees had been felled.

TAMARIX   TAMARISK   TAMARICACEAE

A group of shrubs or small trees, natives of the Old World, and often inhabiting maritime situations or places where the soil is permeated with saline substances. Some half a dozen species are grown in British gardens, all distinguished by the feathery character of their branches, the minute scale-like leaves resembling those of some junipers, and the small flowers crowded on short racemes. There are few genera of shrubs whose nomenclature is more obscure and involved, many of the species needing microscopical examination for their identification. The flowers have four or five sepals and the same number of petals. Stamens, in the cultivated species, four or five, opposite the sepals, inserted on or between the lobes of a disk, which surrounds the base of the ovary. Styles short, three or four. Fruit a capsule; seeds numerous, with a tuft of hairs at one end. The latest revision of the genus is: B. R. Baum, *A Monographic Revision of the Genus Tamarix*, Jerusalem, 1966. See also, by the same author, *Tamarix* in *Flora Europaea*, Vol. 2 (1968), pp. 292–4; and in *Baileya*, Vol. 15 (1967), pp. 19–25.

The tamarisks are easily cultivated, and none of them appears to find the peculiar conditions under which they occur wild essential, although perhaps they do not thrive so well in their absence. Although some of them come from hot, dry regions, the saline substances which are absorbed by the plant in such places prevent excessive transpiration. But when these are absent from the soil, and nature's safeguard against too great a loss of moisture no longer exists, a more regular supply of moisture at the root becomes necessary. This simply means that in inland districts they need a fairly good, deep loam. No shrubs are more easily propagated than these. It is only necessary to make cuttings of the previous summer's wood about the thickness of a lead pencil and, say, 8 in. long, and place them in the open ground in early winter, burying about two-thirds of the cutting. On the south coast of England, where hedges are often made of *T. gallica* or *T. africana*, the process consists of simply cutting out pieces the length and thickness of a stout walking-stick, sharpening them at one end, and driving them in the ground where the hedge is to be.

For exposed seaside places there are few shrubs so beautiful and so conveniently managed as the tamarisks. In gardens the late summer or autumn flowering species may be cut back every February if it be desirable to keep them low.

T. GALLICA L.
T. anglica Webb

A shrub or small tree up to 25 ft high in the wild; bark of the one-year-old branchlets brown or purple. Leaves minute, narrow at the base, glabrous. Flowers in slender racemes 1 to 2 in. long, white tinged with pink outside, each flower produced in the axil of a narrow bract. Petals deciduous, elliptic or elliptic-ovate, rounded at the apex, about one-twelfth of an inch long. Stamens five, each inserted on a lobe of the disk and widening into it (the disk therefore not showing rounded teeth between the points of attachment as in *T. ramosissima* and *T. chinensis*).

Native of western Europe, ranging to northwest France and N. Africa; long cultivated in the maritime parts of Britain and naturalised in some localities, or at least apparently wild. But some plants grown as *T. gallica* are *T. chinensis* (q.v. under *T. ramosissima*), while others are the following:

T. AFRICANA Poir.   *T. hispanica* Boiss.—A native of southwest Europe and N. Africa, closely allied to *T. gallica* and with a similar flower-structure. Although this species normally flowers on the previous year's wood, it also has 'aestival' forms, i.e., bearing their flowers on the season's wood, and these are not easy to distinguish from *T. gallica*. The most reliable differences would appear to be that in *T. africana* part at least of the petals are persistent and that they are subacute to acuminate at the apex. The petals are also somewhat longer, to ⅛ in. long, but this difference only holds good for the vernal forms of *T. africana*. A specimen collected near Weymouth in Dorset has been identified as *T. africana* by Baum, and a plant from the Hampshire coast appears also to belong to this species. See also *T. chinensis* under *T. ramosissima*.

T. HISPIDA Willd.
T. kaschgarica Lemoine

A deciduous shrub, up to 3 or 4 ft high, distinct from all other cultivated tamarisks in the downiness of its young branches and leaves. It has a rather erect, compact habit, and the leaves are very glaucous, the largest less than ⅛ in. long, sharply pointed, but comparatively broad at the base, the smallest only one-third or one-fourth the size. Flowers bright pink, opening in late August and September, and borne in erect racemes 2 or 3 in. long terminating the branchlets.

This handsome tamarisk, easily distinguished from the others here mentioned by its hairy twigs and leaves, was introduced to cultivation by the Russian traveller Roborowsky, who collected seeds near Kashgar in W. Asia and sent them to Messrs Lemoine of Nancy. It was put on the market in 1893. It has also been found in the deserts east of the Caspian Sea. Whether it is not quite hardy, or whether (as is more likely) it does not get enough sun in England to ripen its wood properly, this species has not proved long-lived at Kew. Its glaucous white colour, its handsome flowers, and the fact that it blooms during the whole of September, make it a charming acquisition wherever it thrives, but it is evidently better suited for a continental climate than for ours.

T. RAMOSISSIMA Ledeb.
T. pentandra Pall., in part, *nom. illegit.*; *T. pallasii* Desv., in part, *nom. confus.*; *T. odessana* Stev. ex Bunge

T. ramosissima, better known in gardens as *T. pentandra*, is a species of wide distribution, from southern Russia and Asia Minor through much of southwest and central Asia to China, usually on saline soils. From other cultivated species, except the closely related *T. chinensis*, it is distinguished by the following combination of characters: Flowers in racemes usually 2 in. or more long, arranged in panicles and usually borne in late summer on the new growths; sepals, petals and stamens five; disk five-lobed, but the lobes usually indented and thus apparently ten-lobed; stamens with slender filaments, inserted in the sinuses between the lobes (in *T. africana* and *T. gallica* each stamen is inserted on a lobe and separated from its neighbours by a sinus). *T. ramosissima* was at one time fairly common in gardens under the name *T. odessana*, but in this country at least its main representative is the following cultivar, better known as *T. pentandra*:

cv. 'ROSEA' ('Aestivalis').—A shrub or small tree, ultimately from 12 to 15 ft high, or upwards, with long, slender, plumose branches. Leaves very small, pointed, the largest ⅛ in. long, arranged at intervals along the flowering shoots; the smallest one-fifth as large, and crowded fifty or more to the inch. Flowers arranged densely in slender, sometimes branching racemes, 1 to 5 in. long, each tiny blossom ⅛ in. across, rosy pink; they cover the whole terminal part of the current year's shoot, which is thus transformed during August into a huge plume-like panicle of blossom as much as 3 ft long. Bracts longer than the pedicels. Sepals microscopically toothed. Petals elliptic. *Bot. Mag.*, t. 8138.

This beautiful tamarisk is quite hardy, and one of the most pleasing of late-flowering shrubs. It should be planted in groups large enough for its soft rosy plumes to produce an effect in the distance. To obtain it at its best, it is necessary to cut it back every winter almost to the old wood. It then sends up the long slender branches which flower for six weeks or so in August and September. It is propagated with the greatest ease by making cuttings, 6 to 9 in. long, in early winter of the stoutest part of the season's growth, and putting them in the ground out-of-doors like willows.

This tamarisk was raised towards the end of the last century by Messrs Chenault of Orleans. It occurred in a seed-bed of *T. hispida*, and was originally sent out by them as "*T. hispida aestivalis*". Dr Stapf of Kew identified it as the present species, but used for it first the confused name *T. pallasii* Desvaux, and later, in the *Botanical Magazine*, the equally confused name *T. pentandra*. He gave the Chenault plant the distinguishing epithet *rosea*, which is preferable to *aestivalis* as a cultivar-name, and was accepted by the raisers at the time.

'RUBRA' is a sport of 'Rosea', with darker pink flowers, and Messrs Jackman have raised and put into commerce 'PINK CASCADE', a shrub of great vigour, bearing flowers of a slightly richer pink than in 'Rosea'.

T. CHINENSIS Lour. *T. gallica* of some authors, not L.; *T. juniperina* Bge.; *T. elegans* Spach; *T. indica* Hort. ex Spach—Very closely allied to *T. ramosissima*, differing, according to Baum, in the 'smaller, entire sepals, ovate petals and shorter bracts.' (*Fl. Europ.*, Vol. 2 (1968), p. 293). Also, whereas *T. ramosissima* normally bears its inflorescences on the season's growths, *T. chinensis* has forms that produce them earlier in the season on the previous year's wood, as in *T. parviflora*. These 'vernal' forms have generally been distinguished as a separate species—*T. juniperina*—while the forms flowering in late summer on the green wood have been confused with *T. gallica*. *T. chinensis* is a native of eastern and central Asia, introduced to Europe at an unknown date, but probably in the 18th century. It was certainly in European gardens by the 1830s. According to B. R. Baum, in his monograph on the genus, *T. chinensis* is widely naturalised in N. America, and it seems probable that some plants cultivated in Britain in coastal localities and usually identified as *T. gallica* are really this species (though some are no doubt the true *T. gallica*, and others its relative *T. africana*).

The following vernal and almost sterile form of *T. chinensis* has been cultivated in Europe since the 1870s:

cv. 'PLUMOSA'.—A shrub or small tree, becoming in time gaunt in habit, the very distinct plumose branches covered with pale green foliage. In their final subdivisions the branchlets are the thinnest of cultivated tamarisks, scarcely thicker than threads, but through its close branching, this species is the densest in habit. The larger leaves scattered on the thicker branchlets are $\frac{3}{16}$ in. long, pointed, and ultimately decurved; they become smaller on each subdivision until, on the final ramifications they are about $\frac{1}{32}$ in. long. Flowers bright pink in the bud state, paler after opening; produced in May on the twigs of the preceding season; racemes $1\frac{1}{2}$ to 2 in. long.

It is the most graceful of hardy tamarisks, and is worth growing for the fine plumose effect of its branches, which stand out very prominently when associated with other shrubs, not only for their elegance but also for the peculiar freshness of their pale green. It has lived outside for many years at Kew, and forms a rugged trunk, but rarely flowers. It is cut back in hard winters.

T. PARVIFLORA DC.

A shrub or small tree to about 15 ft high, with usually arching branches; bark of second-year wood brown or purplish brown. Leaves sessile, acute, about $\frac{1}{10}$ in. long, with colourless margins, not much widened at the base. Racemes 1 to 2 in. long and a little less than $\frac{1}{4}$ in. wide, borne on the wood of the previous season; bracts slightly longer than the pedicels, green at the base, the upper part translucent and more or less stained with purple. Petals light rosy pink, spreading, about one-twelfth of an inch long. Stamens four, inserted on the lobes of the disk and separated by shallow sinuses; anthers pink.

Native of the Aegean and the Balkans, possibly of N. Africa. From all other cultivated species except the rare *T. tetrandra* it is distinct in flowering from the old wood and in having flowers with four stamens. It is a shrub of great beauty and grace, admirable in masses.

In the commonly cultivated form of *T. parviflora* the racemes are densely flowered and also densely arranged on the branches; they are mostly up to $1\frac{1}{2}$ in. long, a few to 2 in. In another form, grown as *T. tetrandra*, they are longer, laxer and rather more widely spaced.

T. parviflora does not inhabit saline soils in the wild and has become naturalised in some parts of S. Europe outside its natural range, and in N. America.

T. TETRANDRA Pall.—Like the preceding, this flowers on the previous year's wood and the flowers have normally four petals and stamens (sometimes five petals and five antesepalous stamens and sometimes with a few extra stamens opposite the petals). The bark is black. The racemes are somewhat longer and broader than in *T. parviflora* and the bracts are green except at the apex. Petals ⅛ in. long, so somewhat longer than in *T. parviflora*. Antesepalous stamens inserted between the lobes of the disk, which is fleshier and more conspicuous than in *T. parviflora* (Baum, *Monographic Revision* and *Flora Europaea*). Native of the eastern Balkans, Asia Minor and southern Russia (including the Crimea). It is uncertain whether the true species is in cultivation.

TAPISCIA ? STAPHYLEACEAE

The generic name *Tapiscia* is an anagram of *Pistacia*, to which this species, the only one of its genus, bears some resemblance. Its taxonomic position is uncertain, and it should perhaps rank, with one other genus, as a distinct family, near to the Staphyleaceae, in which it is usually placed.

T. SINENSIS Oliver

A deciduous tree, usually about 30 ft high in the wild (very rarely as much as 80 ft, with a trunk 12 ft in girth). Leaves pinnate, 12 to 18 in. long, composed of five to nine leaflets, which are ovate, heart-shaped at the base, pointed, toothed, 3 to 5 in. long, greyish beneath. Flowers honey-scented, male or bisexual, in axillary panicles, those bearing the male flowers composed of many slender spikes, the fertile panicles shorter and more stoutly branched, with larger flowers. Fruits egg-shaped, black, about ⅜ in. long.

Native of Central China, where it occurs at comparatively low elevations; introduced by Wilson in 1908, when collecting for the Arnold Arboretum. Wilson introductions from the 1907–8 expedition are uncommon in this country, and this species, which is also tender, has never spread into gardens, or even collections.

TAXODIUM TAXODIACEAE

A genus in which three species are at present recognised, though these are perhaps reducible to one, *T. distichum*, q.v. for the generic characters.

T. DISTICHUM (L.) Rich.   SWAMP CYPRESS, BALD CYPRESS
Cupressus disticha L.   [PLATES 76, 77

A deciduous, usually pyramidal tree, 100 to 150 ft high, the tapered trunk erect, buttressed at the base, and measuring above it 4 to 6 ft in diameter. In damp situations the roots produce curious woody protuberances, which occasionally stand up several feet out of the ground, being several inches thick, and hollow. The young shoots are of two kinds: (1) the leading ones, which are persistent and have the leaves spirally arranged; (2) the others, very slender, annual, and falling away in autumn along with the leaves; this latter kind of shoot has no buds. Leaves spirally attached, but spreading (except in the leading shoots) in two opposite horizontal rows; linear, pointed, $\frac{3}{8}$ to $\frac{5}{8}$ in. long, $\frac{1}{16}$ to $\frac{1}{12}$ in. wide, of a soft yellowish green. Male and female flowers separate, but on the same tree; the former in slender panicles 4 or 5 in. long. Cones globular, $\frac{3}{4}$ to $1\frac{1}{2}$ in. wide.

Native of the United States, from S. Delaware to Florida, west to Texas, and extending northwards in the Mississippi valley to about 37° N. Most of the stands occur at less than 100 ft above sea-level, where the base of the bole is submerged during a portion of the year. Here it is at an advantage over competing trees that lack its ability to grow under such conditions, though experiment has shown that it actually grows best on soils that are merely moist and deep.

The swamp cypress was introduced to Britain by John Tradescant (*d.* 1638), as recorded by his friend John Parkinson in the *Theatrum Botanicum* (1640), but the oldest trees of known date are those that grow at Syon House, planted a century or so later.

This tree is one of the most beautiful and interesting that can be grown in wet places, but it thrives well in ordinary soil. Its fine feathery foliage, of the tenderest green in spring, and dying off a rich brown in autumn, has nothing similar to it in the whole range of hardy trees. It is perfectly hardy, and thrives well in Britain, despite the coolness of our summers. The largest specimens are nearly all south of the Severn-Wash line and mostly grow near water or on alluvial soils. It is also very accommodating. A dry hollow at Kew, in which a deciduous cypress was growing, was turned into a lily pond in 1896 by puddling over with clay. The tree was left standing, and its trunk became permanently immersed in 2 or 3 ft of water. The tree showed no ill effects from this sudden and drastic change in its root conditions, but on the contrary grew much better for it. The only visible effect was that the immersed part of the trunk became soft and spongy.

The 'knees' produced by the roots of the deciduous cypress are often known as pneumatophores—a term coined at a time when it was believed that they serve to supply air to the roots when these are immersed in water. But research in the United States has shown that their removal makes no difference to the health of the tree. In cultivation they are produced only by trees growing by still or running water, and then not always.

Seed is the usual means of increase when commercial quantities are needed, but cuttings taken from extension growths in late summer root quite easily.

The following are some of the notable specimens recorded in recent years: Syon House, London, three trees *pl. c.* 1750, Duke's Walk, 105 × 13 ft (1976), south side of Lake, 96 × 14½ ft at 4 ft (1967), Park, 90 × 14½ ft (1967); Dulwich Art Gallery, London, 85 × 10¼ ft (1976), very fine; Dulwich, Ash Cottage, 98 × 11 ft (1976); Knap Hill Nursery, Surrey, 97 × 11¼ ft (1974); Burwood Park, Walton-on-Thames, Surrey, 115 × 18¼ ft (1973); Brockett Park, Herts, eighteen trees by the River Lee, the best 105 × 12½ ft and 77 × 14 ft (1976), cf. 86 × 9 ft and 80 × 10 ft, the best measured by Elwes in about 1904; Broadlands, Romsey, Hants, 118 × 15½ ft (1976); Bowood, Wilts, 82 × 11 ft (1975); Elvaston Park, near Derby, 82 × 12¾ ft (1977); Orton Hall, Hunts, *pl.* 1830, 87 × 9½ ft (1974).

T. ASCENDENS Brong. *Cupressus disticha* var. *imbricaria* Nutt.; *Taxodium distichum* var. *imbricarium* Croom; *T. ascendens* var. *erectifrons* Beissn. POND or UPLAND CYPRESS.—A smaller tree than *T. distichum*; bark deeply furrowed; 'knees' rarely produced. Deciduous branchlets upcurved. Leaves about ⅜ in. long, awl-shaped, appressed to the branchlet. Cones as in *T. distichum*. This species does not extend so far west as *T. distichum*, and is absent from the Mississippi valley. It may occur with its relative, of which it is often considered a variety, but is usually found at a somewhat higher altitude, around ponds. It was originally described from a tree cultivated at Paris under the name *"Cupressus sinensis"*; the epithet *ascendens* refers to the posture of the deciduous branchlets, not to the habit of the tree.

The pond cypress was introduced to Britain late in the 18th century. Examples are: Syon House, London, 78 × 8½ ft (1967); Sheffield Park, Sussex, 53 × 3½ ft (1968); Nymans, Sussex, 46 × 3¼ ft (1970); Knap Hill Nursery, Surrey, 56 × 6 ft and 56 × 7¼ ft (1974).

f. NUTANS (Ait.) Rehd. ?*Cupressus disticha* var. *nutans* Ait.; *Glyptostrobus pendulus* Endl.; *T. distichum* var. *pendulum* Carr.—This differs from *T. ascendens* only in the slenderer more or less pendulous deciduous branchlets. It has been suggested that the posture of the branchlets is determined by the conditions in which the tree grows, and even that it varies seasonally. However, the trees commonly grown under the name *T. ascendens nutans* are of narrow habit, with some of the main branches steeply ascending, the others short and horizontal; the branchlets are spreading and unusually long; the leaves are very short, closely set, appressed at first, later spreading and two-ranked. Examples are: Kew, 52 × 4 ft; Grayswood Hill, Haslemere, 44 × 3½ ft (1976).

T. MUCRONATUM Ten.   MEXICAN CYPRESS, AHUEHUETE

T. distichum var. *mexicanum* (Carr.) Gord.; *T. mexicanum* Carr.;
T. distichum var. *mucronatum* (Ten.) Henry

Various botanical differences have been put forward to justify a specific rank for this tree but, according to the Mexican authority Martinez, none is really reliable, though the leaves are on the average shorter than in *T. distichum*. It is often said to differ from its N. American ally in being evergreen, but even that is not altogether reliable as a specific character, since trees may be leafless for a time in the cooler parts of the country. Differences of a physiological nature are that knees are rarely produced and the different flowering time— towards the end of the year after the rainy season is over.

Distinct species or not, the Mexican cypress is a splendid tree, found, wild or planted, throughout the more elevated parts of Mexico and extending into Guatemala. Its main habitat is along the banks of rivers and mountain streams, but it also occurs away from visible water. Historically it is the senior tree, cultivated since pre-Colombian times in the Vale of Mexico. Legend has it that the fine trees, some 500 in all, that ornament the Chapultepec Park in Mexico City were planted during the reign of the last Aztec ruler, Montezuma II (*d.* 1520), though they may be older. Remnants of a formal planting at El Contador, northeast of Mexico City, near Texcoco, formed part of the garden of the local Aztec ruler Nezahualcoyotl (Martinez, *Las Pinaceas Mexicanas* (1963), pp. 196–9). He died about 1472, so the trees must be some 500 years old. The most famous ahuehuete, and perhaps the most frequently measured tree in the world, grows at Sta Maria del Tule, a few miles south of Oaxaca. This has a tight-tape girth of almost 40 yards, and vast ages have been assigned to it. It is not unlikely that it is really the result of the fusion of three distinct trees; even if it grew from a single seedling it is not necessarily more than 1,000 years old. For a note on this tree, with two portraits, see *Journ. R.H.S.*, Vol. 93 (1968), pp. 478–80 and fig. 240–1.

TAXUS   YEW   TAXACEAE

A genus of eight or nine closely allied species, distinguished one from another mainly by vegetative characters. They are natives of the more humid regions of the northern temperate zone, though one species (*T. celebica*) extends to just below the equator in Indonesia. They are evergreen shade-bearing trees or large shrubs, their leaves spirally arranged but appearing more or less two-ranked on all except vertical shoots. Male and female flowers usually borne on separate plants, in the leaf-axils of the previous season's wood. Male flower a globose cluster of stamens. Female flower consisting of a solitary terminal ovule protected by a few whorls of green bracts, which resemble a growth-bud. Fruit a bony-shelled seed, enclosed in a fleshy aril, ripening the first year.

Taxus is the type of the small family Taxaceae, to which *Torreya* (q.v.) also belongs. The other members of the family, not treated here, are *Pseudotaxus* (one species in China, probably hardy), *Amentotaxus* (four species in China, one at least probably hardy); and *Austrotaxus* (one species in New Caledonia). This family was once grouped with the conifers but is now considered by most authorities to constitute a distinct Order, the Taxales. The Cephalotaxaceae and Podocarpaceae resemble the Yew family in their solitary seeds not borne in cones, but it is now held that these groups are true conifers in which the ovule-bearing scales have become much reduced or modified. Thus the woody arborescent gymnosperms cultivated out-of-doors in this country really represent three Orders of the plant kingdom: the Ginkgoales, with one family and genus; the Taxales, with one family, the Taxaceae; and the Coniferales, with five families—Pinaceae, Cupressaceae, Araucariaceae, Taxodiaceae, as well as the anomalous Podocarpaceae and Cephalotaxaceae.

T. BACCATA L.   COMMON YEW

A tree 30 to 40, rarely 50 or 60 ft high, forming in age a short, enormously thick trunk, clothed with red-brown peeling bark, and crowned with a rounded or wide-spreading head of branches. Leaves spirally attached to the twigs, but by the twisting of the stalks brought more or less into two opposed ranks; they are of a dark, glossy, almost black-green above, grey, pale green or sometimes yellowish beneath, the stomatic lines indistinct, linear, $\frac{1}{2}$ to $1\frac{1}{4}$ in. long, $\frac{1}{16}$ to $\frac{1}{12}$ in. wide, more gradually tapered to a fine point than any other of the species here mentioned. Flowers unisexual, with the sexes almost invariably on separate trees, produced in spring from the leaf-axils of the preceding summer's twigs. Male a globose cluster of stamens; female borne close to the end of the shoot, and consisting of an ovule surrounded by small bracts. What is usually termed the fruit is a fleshy cup developed from a disk in which the ovule is set. This cup is bright red (sometimes yellow), juicy, and encloses the nut-like seed except at the top.

Native of Europe (including Britain), N. Africa, and W. Asia. In Europe the yew has been on the retreat ever since man acquired the ability to fell it. Many ancient stands on the continent have been reduced to stunted remnants or exterminated, place-names often indicating its earlier presence in areas from which it has long since disappeared. In the British Isles too, where the climate is more favourable to it than in continental Europe, it is now far less common than in former times. Evelyn wrote in 1662: 'He that in *winter* should behold some of our highest *Hills* in *Surrey*, clad with whole *Woods* of these two last sort of *Trees* [yew and box] . . . might without the least violence to his Imagination, easily phansie himself transported into some new, or enchanted Country . . .' The most complete surviving stand is in Kingley Vale on the South Downs near Chichester. There are smaller pure stands on the chalks of southern England and the limestones of the northwest, and the yew is still fairly frequent as an understorey in beechwoods.

No tree has become more woven into the history and folklore of Great Britain than the yew. All through the Middle Ages and until gunpowder came into general use, yew wood was more valued than any other for the manufacture of bows, long the national weapon of offence; but Spanish-grown wood was considered the best. In earlier ages still, before the conversion of this country to Christianity, yews were, no doubt, sacred trees, and the Druids erected their emples near them. The early Christians made a practice of building their churches on sites previously held sacred by the Britons, and thus perpetuated that association of the yew with religious sites which has lasted until now.

The yew is the longest lived of all native trees, retaining its health and verdure even when the trunk has been hollow for many centuries, and clinging to life even when much of that has been destroyed. The actual age of the many large trees in the British Isles is impossible to estimate with any certainty, since the earlier measurements (in some cases going back to the 17th century) are mostly unreliable. What is certain is that the age once often attributed in local guide-books and histories to yews of large girth and venerable appearance was greatly exaggerated. A tree 5 ft in diameter at 5 ft is unlikely to have started its life much earlier than the reign of Elizabeth I. Swanton, in the work cited below, gives the following estimates for the rate of growth of yews in terms of the number of years needed to gain 1 ft in diameter:

Youth (to 12 ft in girth): 65 years or less.
Maturity (12 to 24 ft in girth): 75 years.
Old Age (24 to 30 ft in girth): 80 years.
Extreme old age (over 30 ft): 100 years.

Converted into gain in girth over a ten-year period these estimates become for Youth: 5½ in. or more; for Maturity: 5 in. (c) 4½ in.; for Old Age: 3½ in. For trees 30 ft in girth he refused an age of more than 800 years, but was prepared to accept a pre-Conquest date for the two famous Surrey churchyard yews (Tandridge and Crowhurst), and those at South Hayling and Mamhilad (see below). But recent measurements by Alan Mitchell show that a very slow girth-increase—about 2 in. in ten years—may set in well before a girth of 20 ft is reached. If an average of 5 in. in ten years is assumed up to 15 ft in girth, the age of trees 30 ft in girth must then be well over 1,000 years. The following works have been devoted to the British yews:

LOWE, JOHN.—*The Yew-trees of Great Britain and Ireland*. 1897.
CORNISH, VAUGHAN.—*The Churchyard Yew and Immortality*. 1944.
SWANTON, E. H.—*The Yew Trees of England*. 1958.

The following measurements (except of the Darley Dale tree) are by Alan Mitchell. All the trees in the first list grow in churchyards, except the Keffolds yew (which is not in Lowe's work and was scarcely known until E. H. Swanton called attention to it in 1901 in *The Surrey Magazine*). Loose, Maidstone, Kent, male, 45 × 30¾ ft (1961); Cudham, Kent, 35 × 27½ ft at 3 ft (1965); Ulcombe, Kent, 40 × 27 ft at 3 ft and 35 × 32 ft at ground level (1965); Keffolds, Haslemere, Surrey, 48 × 29 ft (1961), a splendid female tree with a bole of 9 ft; Hambledon, Surrey, 40 × 31 ft (1959), hollow; Crowhurst, Surrey, 35 × 30¼ ft (1959), hollow; Tandridge, Surrey, 45 × 33¼ ft (1959), a female tree with three stems; Crowhurst, Sussex, 38 × 27¾ ft (1965), a fine female tree; Selborne, Hants, male, 38 × 27¾ ft (1965); South Hayling, Hants, 32 × 33½ ft at 3 ft (1961); Tisbury, Wilts, 35 × 31 ft (1959); Woolland, Dorset, 55 × 29 ft (1963), a fine tree, but one limb lost; Church Preen, Salop, 45 × 22½ ft (1962), bole 8 ft; Mamhilad, Monm., 30 × 31 ft at 6 in. (1959); Gresford, Denb., 35 × 29½ ft (1961), a fine male tree; Darley Dale, female, 26½ ft at ground level (Swanton, 1950).

The preceding have been mentioned because of their large girth. The following, none of them in churchyards, are remarkable for their height: Close Walks, Midhurst, Sussex, 85 × 8¼ ft and 72 × 10½ ft (1967); Fairlawne, Tonbridge, Kent, 74 × 8 ft (1965); Pusey House, Farringdon, Berks, 73 × 8½ ft (1955); Walcombe, Wells, Som., 75 × 8¾ ft (1964); Lowther Castle, Penrith, Cumb., 71 × 10¾ ft and 71 × 13 ft (1967).

Only fragments remain of the famous Fortingall yew, at the entrance to Glen Lyon in Perthshire. An interesting account of it by the late Euan Cox will be found in *New Flora and Sylva*, Vol. 3 (1931), pp. 135–8; see also Loudon, *Arb. et Frut. Brit.*, Vol. IV, p. 2079. Although of no great girth, a yew at Whittinghame in East Lothian is remarkable for its wide, low crown of drooping branches. It is portrayed in Elwes and Henry, *Trees of Great Britain*, Vol. 1 (1906), pl. 36.

A peculiar mystery is attached to the poisonous quality the yew is known to possess, owing to its uncertain and apparently capricious effects. One may go into parks where yews are standing, and see them eaten off by cattle up to the grazing line as other trees are, and yet no case of poisoning heard of; on the other hand, deaths of horses, cattle, and calves turned into fresh fields where they were able to get at yew bushes have occurred so often as to leave no doubt that the yew is poisonous. It appears as if the poison acts only on certain states of the stomach. In my opinion it is more virulent when the stomach is empty, perhaps only then. It also appears that semi-dried twigs and foliage are more dangerous than green ones, and it has been surmised that the male tree is more poisonous than the female. The poison does not appear to be of an acrid or irritant nature, but brings about death rather by interference with the heart's action. Neither the Canadian nor the Himalayan yew is known to be poisonous. The red fleshy cup that surrounds the seeds is frequently eaten by children without ill effects, but the seeds themselves contain the alkaloid known as taxine that is found in the leaves, and may be the principle that has caused so many fatalities.

Yew timber possesses remarkable strength and durability, and was once highly valued in this country, especially for indoor use (furniture, etc.); it is also very resistant to decay from wet out-of-doors.

The yew bears clipping exceptionally well, and on that account makes excellent evergreen hedges. It is also the best, frequently the only tree used for topiary work, *i.e.*, training and clipping into formal and fantastic shapes. The most remarkable examples in this country are at Levens Castle, in Westmorland (Cumbria). William Barron's Yew Garden at Elvaston Castle near Derby, planted for the 4th Earl of Harrington in the middle of the last century, is portrayed in *Gard. Chron.*, Vol. 9 (1891), fig. 103. Some of the topiary work and yew specimens are preserved in the Elvaston Castle Country Park. There are topiary features in many other gardens open to the public, of which those at Packwood House (National Trust) and Compton Wynyates, both in Warwickshire, are well known. For the history of the former, see G. S. Thomas, *The Gardens of the National Trust* (1979), p. 192. The Levens trees were planted in 1692, and have been annually clipped ever since—a remarkable testimony to the adaptability and vitality of the yew. Hedges and topiary yews are clipped in August or September, but if drastic cutting is necessary, as when a hedge has become overgrown, or a tree is to be shaped, this should be carried out in May.

The tree is an extremely hardy one, and is adapted to almost any soil, but like most trees is best suited on a good loam. It is one of the best evergreens for calcareous soils. Common yew is mostly raised from seed which, collected when ripe in autumn, should be kept a year before sowing, mixed with sand or soil and turned occasionally. Named varieties are easily raised from cuttings of small shoots placed under a cloche in late July or August.

Yew hedges and topiary may be severely damaged by attacks of the aphid known as Yew Scale. The honeydew exuded by the insects nurtures a sooty mould, which is usually the first sign of an attack. The pest can be checked by regular spraying with a systemic insecticide.

There is now a great number of varieties of yew in cultivation, mostly of seedling origin, and the most notable are described below. The most comprehensive account in English is to be found in: den Ouden and Boom, *Manual of Cultivated Conifers* (1965). Still of interest is: Vicary Gibbs, 'Taxaceae at Aldenham and Kew', *Journ. R.H.S.*, Vol. 51 (1926), pp. 189–210 (other species of *Taxus* are also treated in this article).

cv. 'ADPRESSA'.—A very striking and handsome form that would be considered a distinct species if its origin were not known. It is a wide-spreading shrub (female) of dense habit, with leaves only $\frac{1}{4}$ to $\frac{1}{2}$ in. long, $\frac{1}{10}$ in. or slightly more wide, abruptly pointed at the apex. It was found, either in 1828 or 1838, as a seedling in a bed of hawthorn at the nurseries of F. and J. Dickson at Chester, who slowly built up a stock, about half-a-dozen of which were inadvertently sold to a representative of the Chelsea nurseryman Knight by a member of the staff who was unaware of their true value and charged the price of ordinary yew. Francis Dickson, who was away at the time, wrote to Knight demanding their return but was met with a blank refusal. It was Knight who named this yew *T. adpressa*, later going over to '*T. tardiva*'. But Francis Dickson called it *T. brevifolia* and stuck to this name, which properly belongs to the Pacific yew.

'Adpressa' eventually makes a large shrub or even a small tree. Planted in 1840, it is 30 × 3$\frac{1}{2}$ ft (1974) at Orton Hall, Hunts (cf. 15 ft high in 1931). Similar variations have since been recorded in seed-beds and in the wild.

cv. 'ADPRESSA STRICTA'.—Of more ascending habit than 'Adpressa'. Put into commerce by Standish and Co. of Ascot and given a First Class Certificate when exhibited by them in 1886. Female.

cv. 'ADPRESSA VARIEGATA'.—Leaves at first silvery yellow, later margined with yellow. Female (*T. adpressa variegata* in *Journ. R.H.S.*, Vol. 11 (1889), p. cx; *T. baccata adpressa aureo-variegata* Beissn.). Raised (or distributed) by the Handsworth nursery of Fisher and Holmes. First Class Certificate 1889. Sometimes known as 'Adpressa Aurea'.

cv. 'AMERSFOORT'.—An even more remarkable departure from the norm than 'Adpressa', the leaves being oblong or oblong-elliptic, up to $\frac{3}{8}$ in. long and about half as wide, roundish and mucronate at the apex, radially arranged. Of stiff, upright habit, slow-growing. The original plant grew at the Psychiatric Hospital at Amersfoort in the Netherlands, and propagations from it were at first identified as belonging to a species of *Podocarpus* (*Baileya*, Vol. 9 (1961), p. 133 and fig. 46).

cv. 'ARGENTEA MINOR' ('Dwarf White').—A very slow-growing dwarf with drooping branchlets, its leaves with a narrow white edge.

f. AUREA Pilger—Yews with leaves tinted or edged with yellow have been known since the 17th century, but the first commercial distribution recorded in the literature appears to have been the *T. baccata foliis variegatis* of Loddiges nursery, described by Loudon as having the leaves variegated with whitish yellow and being seldom more than a large shrub. This may be the yew known as the 'Old Gold-striped', which was male and of low, spreading habit. But there is also record of a form which, except in leaf-colour, resembled the common yew.

The old golden yews seem to have attracted little attention until the energetic William Barron started to collect plants for his employer the 4th Earl of Harrington, buying every one he could lay his hands on, regardless of size, for he was a pioneer of tree-transplanting and an example of the machine he invented and marketed for this purpose is still in existence. By 1849 there were some one thousand golden yews at Elvaston Castle. Among them was a male tree bought from Lee of Hammersmith for forty guineas. This sported a female branch, and from its fruits Barron raised several improved golden yews.

cv. 'BARRONII FOEMINA'.—Of dense pyramidal habit. Leaves variegated with bright golden yellow. Raised by William Barron around 1860 from the monoecious golden yew mentioned above. He also distributed a male clone of the same origin.

cv. 'CHESHUNTENSIS'.—A form intermediate, and probably a hybrid, between the common and Irish yews; it has a wider habit than the latter, but the leaves are similarly arranged all round the twig. Raised by Messrs Paul of Cheshunt.

cv. 'DECORA'.—As now grown in Britain a dwarf, pendulously branched form with dark green, glossy leaves about ½ in. long (H. J. Welch, *Dwarf Conifers*, fig. 242). This is not the *T. b.* var. *decora* of Hornibrook.

cv. 'DOVASTONIANA'. WEST FELTON YEW.—One of the most distinct and handsome forms. It is a short-trunked tree, sometimes a shrub, with numerous stems, its branches widely spreading, the branchlets pendulous. Leaves very dark green.

The original tree was planted at West Felton near Shrewsbury about 1776 by John Dovaston, who bought it from a cobbler for sixpence, hoping that the fibrous roots of the yew would fix the soil at the mouth of a well he had just dug. 'They did so; and . . . the yew grew into a tree of the most extraordinary and striking beauty . . . . Though a male tree, it has one entire branch self-productive, and exuberantly profuse in female berries . . .' (J. F. M. Dovaston, the planter's son, in a letter to Loudon printed in *Arb. Frut. Brit.* (1838), Vol. IV, p. 2082). In 1836 the tree was 56 ft in spread and about 5 ft in girth at 5 ft. Its present dimensions are 56 × 12 ft, with a clear bole of 6–7 ft (1979). For its portrait in 1900 see *Gard. Chron.*, Vol. 27 (1900), p. 147. The tree was named *T. baccata Dovastoniana* by Leighton in *Fl. Shrops.* (1841), p. 497.

cv. 'DOVASTONII AUREA'.—Of similar habit to the preceding but with golden yellow branchlets and yellow-margined leaves. Raised in France and, named in 1868.

cv. 'ELVASTONENSIS'. Leaves golden at first, becoming bright orange in winter. Male. A branch-sport from a common yew found by William Barron at Elvaston Castle and distributed by him after he set up his own nursery.

cv. 'ELEGANTISSIMA'.—The yew originally so named was raised in the Wetley Rocks nursery of a Mr Fox, in the 1840s. 'He had an Irish and a Golden Yew growing near each other; the former got fertilised by pollen of the latter, and a 'wee striped plant' made its appearance between them. This was offered to me for seven guineas, when only a few inches high, which offer I declined. It then fell into the hands of Messrs Fisher and Holmes, who made the most of it, and it obtained a wide circulation' (William Barron, *Gard. Chron.* 1868, p. 921). This 'Elegantissima' was a male of erect, uniform growth and a straw-coloured variegation. It is probably now lost to cultivation, but many improved golden yews were raised from it by backcrossing it with the Irish yew. See further under 'Fastigiata Aurea'.

What is now grown as 'Elegantissima' is a vigorous female clone of dense, ascending habit, with leaves golden at first, later straw-coloured. It becomes green if grown in shade. It is possibly the 'Elegantissima Foemina' put out by Smith of Worcester, but they also had an 'Elegantissima Superba' and there is also mention in the literature of 'Elegantissima Nova'.

cv. 'ERICOIDES'.—A slow-growing form with erect or spreading stems and narrow heath-like leaves standing out around the shoot. A plant at Kew, acquired about 1876, was 7½ ft high and 21 ft in circumference of spread in 1926 (V. Gibbs, op. cit., p. 202). It is now rare.

cv. 'FASTIGIATA'. IRISH YEW.—Of columnar habit, with numerous spires, its leaves standing out all round the twig, dark, dull green. Two young plants were originally found by a Mr Willis on a rock in the mountain above Florence Court in Co. Fermanagh called Carricknamaddow or 'The Rock of the Dog'. One, which the finder planted in his own garden, died in 1865. The second he took to his landlord the Earl of Eniskillen at Florence Court, and it is from this plant that all the true Irish yews are descended by vegetative propagation. Cuttings were given to Lee of Hammersmith at an unrecorded date but probably around the turn of the century, since the Irish yew was available at a low price from both British and continental nurseries by 1838, under the name 'T. *hibernica*'.   [PLATE 78

The Irish yew has attained a height of 40 to 55 ft in the western parts of the British Isles, and there the individual spires are long and slender; elsewhere it is squatter, with a broader top, and grows slowly (A. F. Mitchell, op. cit., p. 286).

Being propagated by cuttings, the Irish yew is female, like the original parent. When its flowers are fertilised by pollen of a normal yew, as would normally be the case, the seedlings usually take after the male parent. But branches bearing male flowers have been observed on several occasions on the Irish yew, and the fastigiate seedlings that are occasionally produced are possibly the result of self-fertilisation. The name *T. baccata erecta* has been applied to more than one of such seedlings. The only one to rival the Irish yew is 'OVEREYNDERI', raised at Boskoop around 1860. This was introduced to Britain some forty years later, but has never become established here. It is not naturally of so fastigiate habit as the Irish yew and makes a broad bush unless pruned. 'FASTIGIATA ROBUSTA', of columnar habit, was found in a Swiss garden and put into commerce in 1950. Female. The leaves are lighter green than in the Irish yew (Van Ouden and Boom, *Man. Cult. Conif.* (1965), pp. 398, 402).

cv. 'FASTIGIATA AUREA'. GOLDEN IRISH YEW.—This is really a collective name for plants more or less agreeing with the Irish yew in habit but with golden or gold-variegated foliage. Dallimore remarked: 'It is a very conspicuous variety, and varies somewhat in colour according to the nursery from which it is procured.' (*Holly, Yew and Box* (1908), p. 201). Such plants were raised by at least two nurserymen—Waterer of Knap Hill and F. and J. Dickson of Chester—by raising seedlings from the Irish yew, planted near 'Elegantissima' (q.v.), which was itself a hybrid of the Irish yew.

cv. 'FASTIGIATA AUREOMARGINATA'.—A selection of the golden Irish yew, raised at the Handsworth Nursery around 1880. Leaves with a margin of bright yellow, becoming duller later. Male.

cv. 'FRUCTU-LUTEO' ('Lutea').—Differing from the common yew only in the orange-red arils of its fruits. This very striking form appears to have originated in Ireland. It was first noted about 1817 on the demesne of the Bishop of Kildare at Glasnevin near Dublin, but the immediate parent of the cultivated stock was apparently a tree at Clontarf Castle, mentioned by Mackay in his *Flora Hibernica* (1836).

cv. 'GLAUCA'.—The plant cultivated under this name, or as 'Blue Jack', is of bushy habit, eventually 20 ft or so high and as much in width, with leaves of a dark glossy green, bluish beneath. Male (?*T. baccata glauca* Carr.).

cv. 'NANA'.—Three cultivars have been distributed as *T. b. nana*. According to Hornibrook, it belongs properly to the dwarf otherwise known as *T. b. Foxii*, which has very short dark green leaves and grows slowly to a height of about 3 ft, making a laxly branched, pyramidal bush. The *T. b.* var. *nana* of previous editions was renamed *T. b. parvula* by Vicary Gibbs; this is a semi-procumbent form, perhaps no longer in commerce. The third "nana" was raised and named by William Paul and was renamed *T. b. Paulina* by Gibbs. This is of compact, dome-shaped or conical habit, to about 4 ft high.

cv. 'NEIDPATHENSIS'.—Differing from the type only in its steeply ascending branches and short, rather dense leafage. It originated at Neidpath Castle, Peebles, but appears to have been first distributed by continental nurseries.

cv. 'NUTANS'.—A slow-growing, flat-topped, irregularly branched bush to about 3 ft high and wide; branches nodding at the tips. Leaves radially arranged, less than ½ in. long, some reduced to appressed scales. Raised in Belgium and first distributed by a Dutch nursery in 1910 (den Ouden and Boom, *Man. Cult. Conif.*, p. 410).

cv. 'PUMILA AUREA'.—A very dwarf and compact bush with golden leaves retaining their colour throughout the winter. Grown and named by Vicary Gibbs at Aldenham, but of unrecorded origin. It was propagated and probably most of the stock was dispersed at the sale of the Aldenham collection in 1932. Mr Welch believes that two old plants in the National Pinetum at Bedgebury belong to this clone (*Dwarf Conifers*, p. 290).

cv. 'PYGMAEA'.—The dwarfest of the yews, eventually about 15 in. high and wide, of dense habit, with radially arranged leaves less than ½ in. long. (den Ouden and Boom, op. cit., p. 403). It has the same history as 'Nutans'.

cv. 'REPANDENS'.—A semi-prostrate shrub forming an undulating mass eventually 2 or 3 ft high and 10 to 15 ft across. Female. Distributed by Parsons of Flushing, New York, before 1887. An excellent ground cover.

cv. 'SEMPERAUREA'.—A large shrub, wider than high, with ascending branches and short branchlets. Leaves golden, holding their colour through the winter. Of unrecorded origin, before 1900. A plant at Kew was 9 ft high and 15 ft wide when thirty years planted.

cv. 'STANDISHII'.—Resembling the Irish yew in habit but of much smaller stature and slower growth. Leaves crowded, rich golden yellow. The best golden yew for small gardens and with the finest colouring of all. Described in 1908 but distributed earlier by Standish and Co. of Ascot.

cv. 'WASHINGTONII'.—A large shrub of spreading habit. Leaves golden at first, later yellowish green. Female. Distributed by Smith of Worcester in 1874 as a variety of *T. canadensis*, and probably raised in N. America. A plant at Kew, acquired in 1876, was 12 ft high and 69 ft in circumference of spread in 1926.

T. BREVIFOLIA Nutt.   PACIFIC YEW

A small tree 20 to 30, rarely 50 to 70 ft high, the trunk clothed with thin reddish brown bark; branchlets slender, winter buds clothed with loose yellowish, pointed scales. Leaves ¼ to ⅔ in. long, $\frac{1}{16}$ in. wide, linear, rather abruptly narrowed at the apex to a fine point; dark green above, paler green beneath, arranged in two opposite horizontally spreading rows and persisting four or five years. Fruit as in *T. baccata*.

Native of western N. America from S. Alaska to California, nowhere frequent; introduced in 1854. This yew is rare in cultivation, the form so-called being usually a form of *T. baccata*. On the other hand, the yews differ so little from each other in essential points that it may easily be lost among the numerous forms of common yew.

T. CANADENSIS Marsh.   CANADIAN YEW

A shrub of spreading habit, often low and straggling, sometimes 4 to 6 ft high; winter buds small, roundish, the scales loose, roundish at the apex, ridged at the back. Leaves ½ to ¾ in. long, $\frac{1}{16}$ to $\frac{1}{12}$ in. wide; linear, terminated by a fine rather abrupt point, shortly stalked; dark glossy green above, paler green beneath. Fruit red, as in *T. baccata*.

Native of eastern N. America, from Newfoundland to Virginia; introduced in 1800. The Canadian yew is distinguishable from the English yew by the invariably shrubby habit, by the more abruptly pointed leaves, and by the leaf-buds, but can scarcely be said to differ from it more than the varieties of common yew do among themselves. It has little to recommend it beyond its botanical interest, except that it is the hardiest of the yews and can be grown where it is too cold for *T. baccata*.

T. CELEBICA (Warb.) Li

Cephalotaxus celebica Warb.; *T. chinensis* (Pilg.) Rehd., in part; *T. mairei* (Lemée & Lévl.) S. Y. Hu; *Tsuga mairei* Lemée & Lévl.; *Taxus speciosa* Florin

A tree to about 50 ft high in the wild, a spreading shrub as seen in Britain. The leaves are mostly rather longer than in the common yew and relatively wider, slightly falcate, pectinately arranged, and in cultivated plants are rather distantly arranged on the branchlet. The stomatal bands are yellowish green beneath, and are not papillose as in *T. wallichiana*.

T. celebica is the most widely distributed yew in China, except in the north, and also occurs in Formosa, the Philippines and Celebes. It was introduced by Wilson in 1908 from Szechwan, and seeds may later have been sent by Forrest, who collected it several times in fruit in N.W. Yunnan. It was at first grown as *T. chinensis* but, as pointed out in previous editions, it is quite different from the collections on which the name *T. chinensis* is founded (see *T. wallichiana*). *T. celebica* is of no ornamental value in this country and is represented only in a few collections. Plants in the National Pinetum at Bedgebury are 15 to 20 ft tall. Fruits do not ripen in this country.

T. CUSPIDATA Sieb. & Zucc.   JAPANESE YEW

A tree 40 to 50 ft high in Japan, with a trunk girthing about 6 ft; in cultivation a low tree or spreading shrub; older bark reddish brown. Leaves $\frac{1}{2}$ to 1 in. long, $\frac{1}{12}$ to $\frac{1}{8}$ in. wide, linear, tapered rather abruptly at the apex to a fine point, rounded, and with a distinct stalk at the base $\frac{1}{12}$ in. long, dark green above, with a broad, tawny yellow strip composed of ten to twelve stomatic lines on each side of the green midrib beneath. The leaves are arranged approximately in two ranks, and stand more or less erect from the twig, often forming a narrow V-shaped trough. Fruits as in *T. baccata*, but more profusely borne, often clustered.

Native mainly of Japan but also occurring in continental N.E. Asia; introduced by Fortune in 1855 by means of plants which had probably come from Japan, though he sent them from China. Several distinct forms of this yew are grown in the USA, some with a central leader, others many-stemmed from the base. The following seems to be commonest in Britain:

var. NANA Rehd.   *T. cuspidata brevifolia* Sieb., *nom. nud.*; *T. c.* var. *compacta*

Bean—Dense and shrubby, with spreading branches and radially arranged leaves. Although originally described from plants in the USA this occurs in the wild and is recognised in Ohwi's *Flora of Japan* (1965), p. 110. It is inappropriately named, plants in the Arnold Arboretum, forty years old, being 15 ft high and 20 ft across (D. Wyman, *Shrubs and Vines* (1969), pp. 458, 460). This variety was probably introduced to Europe by Siebold.

f. THAYERAE (Wils.) Rehd. *T. cuspidata* var. *thayerae* Wils.—A group of seedlings raised on the Thayer Estate, Mass., in 1916–7, selected plants of which, growing in the Arnold Arboretum, were named by Wilson in 1925. All are of wide-spreading habit, up to 8 ft high and twice as wide. By some authorities they are referred to *T.* × *media* (see below).

T. cuspidata thrives extremely well in the trying New England climate, and is apparently one of the best evergreens introduced there. Whilst the general aspect is the same as that of the English yew, it can be distinguished by the marked yellow tinge of the under-surface of the leaves, and by the longer, more oblong winter buds with looser, more pointed scales.

T. × MEDIA Rehd.—A group of hybrids between *T. cuspidata* and *T. baccata*, the first of which were raised at the Hunnewell Arboretum, Mass., about 1900. Of the many clones now available the oldest in British gardens is 'HICKSII', a free-fruiting variety resembling the Irish yew in habit but with larger, more glossy, spine-tipped leaves. Other columnar forms are now in commerce. Also available is 'HATFIELDII' (male), which makes a dense pyramid about 10 ft high and as much across; it is one of the original Hunnewell seedlings and is named after the then superintendent of the estate, T. D. Hatfield.

T. WALLICHIANA Zucc.

T. baccata has its eastern limit in N. Iran. From Afghanistan eastwards through the Himalaya and N.E. India to S.W. China it is replaced partly by *T. wallichiana*, which may not be in cultivation. This differs from *T. baccata* in its usually sickle-shaped, relatively narrower and longer leaves, recurved at the margin, tapered to a usually spiny point, minutely papillose over the whole surface beneath (Florin, *Act. Hort. Berg.*, Vol. 14 (1948), p. 382).

var. CHINENSIS (Pilger) Florin *T. baccata* subsp. *cuspidata* var. *chinensis* Pilger; *T. chinensis* (Pilger) Rehd., in part only—Leaves with the same papillose undersurface as in the typical variety, but shorter and more abruptly narrowed at the apex. Described from specimens collected in Central China by Henry and by Farges. Confused with *T. celebica* by Rehder.

TELOPEA PROTEACEAE

A genus of three or four species in E. Australia and Tasmania (which has the endemic *T. truncata* described below). It belongs to the same subdivision of the Proteaceae as *Embothrium*, from which it differs in having the flowers clustered in condensed racemes, surrounded by an involucre of coloured bracts.

The generic name derives from the Greek 'telopos' and alludes to the conspicuous flowers, which can be seen from afar.

T. TRUNCATA (Labill.) R. Br. TASMANIAN WARATAH
Embothrium truncatum Labill. [PLATE 79]

An evergreen, varying from a shrub of low, spreading habit to a small tree up to 25 ft high; young shoots stout, round, covered with brownish down. Leaves alternate, of stiff leathery texture, usually closely set on the branchlets where some persist for two years. They are of oblanceolate shape, tapered gradually to the base, more abruptly to the apex which may be rounded or pointed, sometimes having two or three large teeth, or sometimes (especially on young cultivated plants) being deeply bilobed, 2 to 5 in. long, ½ to 1¾ in. wide, dullish green above, rather glaucous beneath. Flowers rich crimson, crowded in a terminal head 2 to 3 in. wide, each blossom of much the same shape as those of *Embothrium coccineum*; the slender curved perianth is about 1 in. long, splitting and showing the conspicuously exposed, curving style 1 to 1¼ in. long, with the large knob-like stigma at the end. The flowers are filled with honey and open in June. Seed-pod woody, cylindrical, curved, 2 to 3 in. long, terminated by a 'tail' ½ to ¾ in. long. *Bot. Mag.*, t. 9660.

Native of Tasmania at 2,000 to 4,000 ft, occasionally descending to lower altitudes in the rainiest parts of the country. One writer remarks: 'the vivid scarlet colouring of the flowers shining out amongst the sombre blue-greens of the gum forests is certainly one of the most beautiful sights the Tasmanian bush affords.' R. C. Gunn wrote to Kew from Tasmania in 1844: 'I really think this plant will do well in the open air in Britain; it is only found in the cool mountainous parts of the island and I have tried in vain to coax it to grow in my garden, but the summer heats have always destroyed it.' This forecast has proved correct, for the Tasmanian waratah does indeed thrive in the British Isles in localities where the rainfall is above average and the soil acid. At Wakehurst Place in Sussex a plant now 18 ft high has lived in a sheltered position for over half a century and flowers and fruits regularly. Somewhat taller examples grow in Ireland. *T. truncata* is nearly hardy and should survive severe winters in a protected position. It is not shade-loving, and is best placed where its roots are shaded and its head in the sun. It is propagated by seeds or cuttings. For further details see the interesting note by the late Lord Talbot de Malahide in Curtis, *Endemic Flora of Tasmania*, Part IV (1973), pp. 296–7. According to him, plants with yellow and with creamy yellow flowers have been brought into cultivation in Tasmania.

Seeds of *T. truncata* were collected by Harold Comber during his expedition to Tasmania in 1929–30, and his interesting account of this species will be found in *Gard. Chron.*, Vol. 93 (1933), p. 27. Contrary to what has been stated, he did not introduce it, nor did he claim to have done so. It received an Award of Merit when exhibited by Mrs Sebag-Montefiore from her garden near Plymouth on May 29, 1934, and a First Class Certificate in 1938, when shown by Lionel de Rothschild of Exbury.

TERNSTROEMIA   THEACEAE

A genus of almost 100 species of evergreen trees and shrubs, mostly in the tropics and subtropics of Asia and America. Flowers bisexual or some functionally male, solitary or clustered. Sepals five, imbricated. Petals five, imbricated, connate at the base. Stamens numerous, with glabrous anthers (hairy in the allied *Cleyera*). Fruit a leathery berry with a few large seeds suspended from the top of the central column.

T. GYMNANTHERA (Wight & Arn.) Sprague
Cleyera gymnanthera Wight & Arn.; *T. japonica* Thunb., in part

An evergreen shrub or small tree with a much-branched head, and warted, not downy branchlets. Leaves alternate, crowded at the apex of the shoot, obovate or oblanceolate, 1½ to 3 in. long, ½ to 1½ in. wide, tapered gradually at the base to a short, stout, purplish stalk, more abruptly tapered to a rounded or bluntish apex; they are dark varnished green, thick and leathery, and quite glabrous. Flowers fragrant, solitary on stalks about ¾ in. long, nodding, of short duration, produced in July and August from the leaf-axils, and from the axils of fallen scales on the lower naked part of the shoot. Corolla yellowish white, about ⅜ in. across; petals five. Fruit globose, and about the size of a cherry, yellow, tinged with rose on the sunny side, the rounded sepals persisting at the base.

A species of wide distribution in E. Asia, from India eastwards, extending north as far as S. Japan and S. Korea; introduced in the early part of the 19th century but probably lost until reintroduced early in the present century. It has withstood 20° F of frost at Kew but is not reliably hardy outside the mildest parts and not of much ornamental value. It has been confused with *Eurya japonica*, but this has broader-ended, more distinctly obovate leaves and is dioecious. The Japanese plant named *T. japonica* by Thunberg is *Cleyera japonica*.

cv. 'VARIEGATA'.—Leaves with a margin of creamy white, later stained with rosy pink; centre of leaf marbled with grey.

TETRACENTRON TETRACENTRACEAE

Tetracentron is the only genus of its family, and itself contains a single species. The family is a very ancient one, most closely allied to the Trochodendraceae, also consisting of a single genus and species (see *Trochodendron*). Although unalike in foliage and inflorescence, the two families have several essential characters in common, one of the most remarkable being the structure of the water-conducting tissues (xylem), which are made up, not of vessels, as in most flowering plants, but of tracheids, resembling those of the conifers but rather more primitive in form. For the foliage of Tetracentron, see below. The flowers are grouped in clusters of four on pendulous spikes and are bisexual. Sepals four, imbricated. Petals none. Stamens four. Ovary superior, composed of four united carpels. Styles four, with a minute stigma, at first terminal and erect but becoming recurved and then apparently borne at the base of the ovary. Fruit a loculicidal capsule, containing a few oily seeds.

See further in: Smith, A. C., 'A taxonomic review of *Trochodendron* and *Tetracentron*', *Journ. Arn. Arb.*, Vol. 26 (1945), pp. 123–42; and Bailey, I. W., and Nast, C. G., 'Morphology and relationships of *Trochodendron* and *Tetracentron*', ibid., pp. 143–54 and 267–76; Takhtajan, A., *Flowering Plants. Origin and Dispersal* (1969), translated from the Russian by C. Jeffrey, pp. 52, 94.

T. SINENSE Oliver [PLATE 80

A deciduous tree 50 to 90 ft high in the wild; young branches dark, glabrous. Leaves, except on extension growths, borne singly below the apex of slow-growing and long-lived spurs, each year's growth marked by the scar left by the fallen leaf. The leaves are ovate or heart-shaped, long-pointed, with five or seven prominent nerves radiating from the base, 3 to $4\frac{1}{2}$ in. long, 2 in. wide, the margins evenly set with blunt teeth. Inflorescence a pendulous, catkin-like spike springing from the end of a spur and bearing numerous very small yellowish flowers. See further in generic introduction.

A native of central and western China, upper Burma, S.E. Tibet and (var. *himalense*) of the Himalaya as far west as E. Nepal; discovered by Augustine Henry in Hupeh; described in 1889 and introduced by Wilson in 1901 when collecting for Messrs Veitch. Like its cousin *Trochodendron aralioides*, it is more than just a botanical curiosity, being handsome in foliage and a picture of great elegance and beauty when bearing its slender catkins around midsummer. It is much hardier and less demanding than was once supposed. The largest example grows at Caerhays in Cornwall and measures 40 ft in height, with several stems, the stoutest $4\frac{1}{4}$ ft in girth (1975). But the other sizeable specimens are well distributed: The High Beeches, Handcross, Sussex, 39 × $2\frac{1}{2}$ ft (1974); University Botanic Garden, Cambridge, 37 × $2\frac{3}{4}$ ft + 2 ft (1976); Edinburgh Botanic Garden, *pl.* 1905, 30 × $3\frac{1}{4}$ ft (1970).

var. HIMALENSE Hara & Kanai—It was believed until recently that

T. sinense had its western limit in upper Burma. But in 1963 the Japanese botanical expedition to E. Nepal discovered there a tetracentron that differs from *T. sinense* only in the mostly rather larger, longer-tipped leaves, with finer tips, and the laxer fruiting catkins. It is possible that this variety extends into upper Burma and Yunnan (Hara in *Fl. East. Himalaya* (1966), pp. 85–6). The variety was again found in 1967 by the Japanese expedition to Bhutan and is also now known to occur in the Assam Himalaya.

TEUCRIUM   GERMANDER   LABIATAE

A large genus of herbs, some woody at the base, and of dwarf shrubs. Leaves opposite, entire, toothed or pinnately lobed. Flowers in whorls, which are arranged in the form of often one-sided racemes, spikes or (sect. *Polium*) in heads. As always in the Labiatae the corolla is two-lipped, but the upper lip is split down as far as the calyx so that the bottom half of it appears to belong to the lower lip, which is thus apparently five-lipped. Stamens four, protruding through the slit of the upper lip. The generic name derives from the Greek *teukrion* and is associated in popular etymology with Teucer, Prince of Troy.

T. CHAMAEDRYS L.   WALL GERMANDER

An evergreen, semi-shrubby plant, herbaceous at the top, woody at the base, with a creeping root-stock and ascending, very downy branches, 8 to 12 in. high. Leaves opposite, mainly ovate, but conspicuously toothed or almost lobed, tapering at the base to a winged stalk, $\frac{1}{2}$ to $1\frac{1}{4}$ in. long, almost half as much wide, bright green above, hairy on both sides. Flowers arranged on a terminal raceme 2 to 5 in. long, two or three together in the axils of the leaves or bracts. Corolla rose-coloured, $\frac{1}{2}$ to $\frac{5}{8}$ in. long, two-lipped, the lower lip veined with darker rose. Calyx $\frac{1}{4}$ in. long, tubular, with five sharp lobes, hairy and, like the floral bracts, purplish.

Native of much of Europe (though not of Britain nor of Scandinavia), N. Africa and parts of S.W. Asia; recorded in Britain from the 17th century onwards as an escape growing on the walls of gardens, ruins, etc. It is very hardy and easily grown, flowering from July to September (earlier in warm gardens). At one time it was considered to be a valuable specific for gout and was an important ingredient in the popular medicine known as 'Portland powder', a name it acquired through its having (reputedly) cured an 18th-century Duke of Portland of that complaint. Even higher up the social scale, the gout of the Emperor Charles V was cured 'by a vinous decoction of it, with some other herbs, taken for sixty successive days' (Martyn, in his edition of Miller's *Dictionary*).

The specific epithet is taken from the old Greek name, meaning dwarf oak—whence *Quercula minor*, as it was called by some of the herbalists. The English name germander is a corruption of *chamaedrys*, through the French 'germandrée'.

T. MARUM L. CAT THYME.—A small subshrub with woolly stems, usually entire linear-lanceolate or rhombic leaves to about ⅜ in. long, hairy beneath, and purple flowers in dense racemose clusters. Native mainly of the islands of the western Mediterranean, once common in gardens. It is as attractive to cats as nepeta and is sometimes grown as a room plant on the continent, for its aromatic foliage. T. SUBSPINOSUM Willd. is closely related to this, differing chiefly in having the lateral branches converted into spines. Native of Mallorca.

T. FRUTICANS   SHRUBBY GERMANDER   [PLATE 81

An evergreen shrub of diffuse habit, naturally 7 or 8 ft high, stems square, and covered with a close white felt. Leaves opposite, ovate, ½ to 1½ in. long,

TEUCRIUM FRUTICANS

about half as wide, broadly wedge-shaped or rounded at the base, bluntish at the apex; dark, rather bright green and glabrous above, white, with a close felt beneath, fragrant when crushed; stalk ¼ in. or less long. Flowers produced during the summer and sometimes in autumn singly in the axils of the small uppermost leaves or bracts—the whole forming a raceme 3 or 4 in. long. Calyx ¼ in. long, with five ovate, pointed, leaf-like lobes, white beneath. Corolla pale purple or lavender-coloured, forming a short tube at the base, to which the four long stamens are attached, then developing into a large five-lobed lip 1 in. long (like the lip of an orchid flower in shape), the basal pair of lobes the smallest and palest; flower-stalk white, ¼ in. or less long.

Native of Portugal, the W. Mediterranean region and the Adriatic; introduced early in the 18th century. It is very pretty, and its curiously shaped labiate flower makes it one of the most easily recognised of shrubs. It can be grown in the open in the milder parts, but elsewhere it needs the protection of a wall and is hardiest in a light, well-drained soil. Easily increased in summer by cuttings of half-ripened wood. It needs no regular pruning apart from a light trimming after the main flush of flowers is over and the removal of winter-killed wood in spring.

cv. 'AZUREUM'.—Flowers darker blue than in the form usually cultivated. Introduced by Collingwood Ingram from the High Atlas of Morocco. A.M. May 16, 1936. Capt. Ingram also gave cuttings to friends in the South of France, where it makes a wonderful display. But in this country it is definitely tender and needs a very warm corner.

T. MONTANUM L.
Polium montanum Mill.

An evergreen, semi-shrubby plant prostrate and woody at the base, 6 to 12 in. high, often reduced to tufts one-third those heights in dry, barren, rocky places; young shoots slender, wiry, erect, downy. Leaves opposite, crowded, linear to narrowly oblong, tapered towards both ends, ¼ to 1 in. long, $\frac{1}{12}$ to $\frac{1}{6}$ in. wide, bright green and thinly downy above, grey-white with thick down beneath; margins decurved. Flowers densely packed in terminal, roundish clusters 1 in. across. Corolla ½ in. long, yellow, upper lip veined with purple; calyx tubular, downy, ¼ in. long, with five teeth.

Native of the mountains of S. and S.E. Europe, north to parts of the Rhineland and the Carpathians, etc., east to Asia Minor; in cultivation early in the 18th century. It is easily distinguished from *T. chamaedrys* by its clustered yellow flowers and narrow, usually toothless leaves. It flowers from July to September. In the Alps it is common on calcareous formations.

T. montanum belongs to the section *Polium*, of which the type is T. POLIUM L., a variable species widely distributed in the Mediterranean region and once cultivated for its medicinal properties. The flowers are white or purplish red and the leaves always more or less crenated.

THERORHODION   ERICACEAE

A genus of two species in N.E. Asia, the Aleutians and W. Alaska. Although often included in *Rhododendron* it differs from it in the inflorescence, which is a raceme bearing in its lower part persistent, leafy, sterile bracts, the few flowers being confined to the apex of the raceme; the bracteoles on the pedicels, and the bracts subtending the flowers, are also persistent. The inflorescence in *Rhododendron* is also basically racemose, but does not have persistent bracts below the flowers.

T. CAMTSCHATICUM (Pall.) Small
Rhododendron camtschaticum Pall.

A deciduous shrub growing in low dense tufts 4 to 10 in. high, producing its flowers on stems up to 6 in. high. It spreads by means of underground suckers. Young shoots furnished with scattered bristles. Leaves stalkless, obovate, ¾ to 2 in. long, ⅓ to ¾ in. wide, thin in texture, glabrous above, slightly bristly beneath, and conspicuously so on the margin. Flowers solitary or in pairs (rarely in threes) on an erect, slender, glandular-bristly stem, the lateral flower or flowers produced on stalks ¾ to 1½ in. long; corolla 1½ to 1¾ in. across, rosy crimson, with five open, spreading, oblong lobes, the three upper ones spotted. Calyx green, 1 in. across, the lobes narrowly oblong, bristly; stamens ten, very downy at the bottom. *Bot. Mag.*, t. 8210.

Native of Japan in Hokkaido (and of one locality in Honshu), thence north through Sakhalin and the Kuriles to the Aleutians and W. Alaska; also on the mainland of Russia from the Ussuri region to Kamchatka; introduced to Britain a few years before 1802. This remarkable and pretty species thrives and flowers well on the rock garden at Edinburgh, but is difficult to suit in southern England, where, for no obvious reason, it succeeds in one garden and fails in another. It needs a light, acid, humus-rich soil, but special mixtures are unlikely to help; it is unsuitable for gardens with poor air drainage, where the expanding flower-shoots may be killed by late frost. Beyond that, no useful advice can be given, except that it should not be coddled.

THUJA   THUYA   CUPRESSACEAE

A genus of six species (but see *T. orientalis*), all of them evergreen trees with thin, scaling or shreddy bark. They resemble *Chamaecyparis* in their flat, pinnately divided branchlets and scale-like leaves, but are very distinct from it in the cones, which are egg-shaped or oblong, and have flat, oblong and (except in *T. orientalis*) thin scales—very different from the peltate or top-shaped scales of *Chamaecyparis* or the true cypresses. A closer ally in the Cupressaceae is *Libocedrus*.

All the thuyas like a good, moist soil, and though best raised from seeds can be increased by cuttings.

The generic name derives from a Greek word used by Theophrastus for an African cypress-like tree with aromatic wood,—probably *Tetraclinis articulata* (not treated in this work), a tender conifer mainly confined to Morocco and belonging to the same sub-family of the Cupressaceae as *Thuja*. The word was taken into classical Latin as 'thya' or 'thyia' and was later applied by some pre-Linnaean botanists to the American *T. occidentalis*, which had been introduced to France from Canada in the 1530s. They, however, rendered the name as 'thuya', as did Linnaeus in *Genera Plantarum* (1754). But in *Species Plantarum* (1753), where the name officially starts, he spelt it 'thuja', and this rendering is correct under the rules of botanical nomenclature, according to which the letter 'j' must be preserved in cases such as this, where it represents consonantal 'i' (like the English 'y' of 'yes'). No doubt the main purpose of this rule is to preserve originally mediaeval spellings such as 'Juglans' and 'Juncus', which in classical Latin were written 'iuglans' and 'iuncus'. The present work is in breach of the *Code* in using the familiar spellings 'Buddleia' and 'Satureia' in place of the technically correct but misleading 'Buddleja' and 'Satureja'. The uncouth but correct 'Thuja' has, however, become so well established in horticultural literature that it would be pointless now to compound the offence by calling the genus *Thuya* or *Thuia*, as did most botanists until recently.

For reasons that can only be guessed at, *T. occidentalis* became known in France as 'L'Arbre de Vie'—a name that in its Latin form 'Arbor Vitae' came to be widely used in pre-Linnaean times for this conifer and was later extended first to *T. orientalis*, called 'the Chinese Arbor Vitae' by Miller, and then to other members of the genus. But like so many so-called popular names, it is more common on the printed page than in the parlance of tree-growers or foresters.

T. OCCIDENTALIS L.   WHITE CEDAR, ARBOR VITAE

An evergreen tree, 50 to 60 ft high, with a trunk 2 to 3 ft in diameter; in cultivation a pyramidal shrub or tree rarely more than half as high; branches usually upturned towards the end; branchlets three or four times pinnate, the ultimate subdivisions much flattened, $\frac{1}{16}$ to $\frac{1}{12}$ in. wide. Leaves scale-like, about $\frac{1}{12}$ in. long, the lateral ones pointed, prominently keeled and overlapping the middle ones; they are a dull yellowish green above, paler and grey green beneath (not with whitish patches, as in *T. plicata*), the middle ones beneath are furnished with a raised roundish gland in the centre. Cones about $\frac{1}{3}$ in. long, oblong, with eight or ten scales.

Native of N. America from the Gulf of St Lawrence to Manitoba, south to New York State and the region of the Great Lakes and with a scattered distribution farther south in the Appalachians; it is often found in swampy locations, but the best stands are on well drained soils. A young plant brought back from the St Lawrence by Jacques Cartier, probably in 1536, lived for many years in the royal garden at Fontainebleau and must have been one of the very first species of N. America to reach the Old World. By the time the first permanent settlements were established there it was already widely cultivated in Europe, including Britain, as the Arbor Vitae (see introductory note). According to Parkinson (*Paradisus*, 1629), all these trees were propagations from the original.

T. occidentalis is not in the first rank of conifers, being often thin in habit (especially on dry soils) and dull in colour, frequently putting on a yellowish brown appearance in winter. It often grows slowly, and as an ornamental conifer is much interior to its western ally *T. plicata*. Only three really healthy-looking specimens have been recorded recently: Little Hall, Canterbury, *pl.* 1906, 62 × 4½ ft (1973); Carey House, Wareham, Dorset, 68 × 6 ft (1968); Trawscoed, nr Aberystwyth, 63 × 7¼ ft (1969). This thuya does, however, have some value for forming evergreen shelter hedges, especially in very cold areas where *T. plicata* does not thrive.

In previous editions it was remarked that very numerous forms of garden origin have been named, but the number available early this century was nothing compared to those raised and distributed in N. America in recent years. For a comprehensive account of these cultivars specialist works must be consulted, especially Van Ouden & Boom, *Manual of Cultivated Conifers* (1965), where sixteen pages are devoted to the cultivars of this species. Only the best known can be mentioned here:

cv. 'AUREA'.—A large, broadly conical shrub with golden leaves, turning gold-bronze in winter.

cv. 'BEAUFORT'.—Similar in habit to the type, but the leaves variegated with white. Raised in Holland.

cv. 'BUCHANANII'.—Branchlets very slender, with the subdivisions far apart. It makes a small, narrow tree. Raised in the USA before 1887.

cv. 'CAESPITOSA'.—A low, cushion-shaped shrublet with short and narrow sprays. It was noticed by Hornibrook growing in the Glasnevin Botanic Garden and described by him in 1923.

cv. 'CRISTATA'.—This is very distinct in the penultimate subdivisions being curiously curved like a cock's comb, the ultimate divisions often developed on one side only. Of lax habit, eventually 6 to 10 ft tall.

cv. 'ELLWANGERIANA'.—An inelegant lanky shrub with slender, curving branches, some of which bear typical leaves, others the needle-like leaves seen on seedlings, others again with both types of leaf. Shrubs over forty years old retain this dimorphic character. It attains a height of 10 ft in time. Raised by the American firm of Ellwanger and Barry, before 1868. 'ELLWANGERIANA AUREA' is a golden-leaved, slow-growing sport from this, raised by Späth and put into commerce in 1895. [PLATE 82

cv. 'ERICOIDES'.—In this the whole foliage is needle-like and of the juvenile or seedling type. The plant is moderately dwarf and at first compact; later its shape is often ruined by heavy falls of snow. It was raised in the USA before 1867 and at first often called "Retinispora dubia"—*Retinispora* being the pseudo-genus founded on the juvenile forms of *Chamaecyparis pisifera*. But by 1888 it had coned in the USA and revealed itself to be a form of *T. occidentalis*.

cv. 'FASTIGIATA'.—Of conical habit, with ascending branches. Foliage-sprays mostly held obliquely, light green. 'COLUMNA', put into commerce by Späth in 1904, is said to be an improvement. See also 'Malonyana'.

cv. 'FILIFORMIS'.—A mound-like shrub, eventually attaining a large size, with thread-like stems, drooping at the tip; ultimate branchlets sparse and short. [PLATE 83

cv. 'GLOBOSA'.—Of compact, globular habit, attaining 4 or 5 ft in height and width; foliage grey-green. Raised by Smith of Worcester before 1874. Two other compact forms of moderate size are 'HOVEYI' with bright green leaves, raised in the USA around the middle of the last century, and 'WOODWARDII', also of American origin, with dark green leaves. In both these the sprays tend to be held vertically as in *T. orientalis*. 'LITTLE CHAMPION', raised in Canada, is reported to be an improvement on 'Woodwardii' (*Dendroflora*, No. 8 (1971), pp. 58–9).

cv. 'HOLMSTRUP'.—Of dense, columnar habit, growing slowly to about 10 ft. Raised at Holmstrup in Denmark shortly before 1951. It has been recommended for dwarf hedges.

cv. 'LUTEA' ('George Peabody').—Leaves golden throughout the year. Of narrow habit, to 40 ft or so high. Raised in the USA and introduced to Britain around 1870. One of the finest golden conifers. There is an example in the National Pinetum, Bedgebury, *pl.* 1926, 34 × 3¼ ft (1969). 'LUTEA NANA' is of similar colour but dwarfer. It is distinguished from 'Ellwangeriana Aurea' by the absence of juvenile foliage.

cv. 'MALONYANA'.—Of very slender, columnar habit; foliage dark green, not browning in winter. Fast-growing when young and eventually 65 ft high. The original plants, dating from the end of the last century, grow in the arboretum formed by Count Ambrozy at Mlynany (formerly Malonya) in Czechslovakia (J. P. Krouman, *Gard. Chron.*, Vol. 154 (1963), p. 457). It is available in commerce.

f. MASTERSII Rehd. *T. occ.* var. *plicata* Mast., not Hoopes; *T. plicata* Hort., not D. Don—A small tree of dense habit, with short branches. Foliage usually brownish green, the sprays curved out of the horizontal plane or erect. This is the old "*T. plicata*" of gardens and was probably a seed-strain descending from some 18th-century introduction (Masters, *Gard. Chron.*, Vol. 21 (1897), p. 258; Kent, *Veitch's Manual of the Coniferae*, Ed. 2 (1900), p. 247). See also 'Wareana'. The established though incorrect use of the name *T. plicata* for this variant of *T. occidentalis* explains the prejudice in the trade against the (correct) use of the name for the western red cedar, which continued to be sold under the name *T. lobbii* for half-a-century or more after Dr Masters elucidated the nomenclatural confusion in 1897.

cv. 'OHLENDORFII' ('Spaethii').—Leaves mainly of the juvenile kind, being awl-shaped, spreading, but incurved at the tips, about ½ in. long, borne on stiff, erect stems; from some of these leaves spring thick shoots bearing leaves which are of the adult kind, though abnormally small. The size of the plant and the proportion of juvenile foliage, depends on the type of cutting from which it is propagated. Hornibrook found that plants raised from juvenile growths taken from the base of the plant remained dwarf and compact and retained the juvenile type of foliage. Plants from cuttings taken higher on the shoots produced mainly adult foliage.

This variety was distributed by Späth towards the end of the last century, but was raised by Ohlendorf of Hamburg.

cv. 'RHEINGOLD'.—Foliage mostly juvenile, bronzy yellow; young growths rosy. An ovoid shrub to about 4 ft high, much used as an associate to winter-flowering heaths. It was raised by the nurseryman Vollert of Lubeck and described in 1904 by Beissner, who emphasised that it was quite distinct from 'Ellwangeriana Aurea'. It appears, however, that at one time some plants sent out as 'Rheingold' were not the true clone but propagations from juvenile foliage of 'Ellwangeriana Aurea'. The conifer that received an Award of Merit in 1902 when shown by Turner of Slough as *T. occ. ellwangeriana pygmaea aurea* is unlikely to have been the true 'Rheingold', as has been suggested.

cv. 'SPIRALIS'.—A narrow tree to about 50 ft high, with short, densely set branchlets. Foliage dark green, arranged at the ends of the branchlets in fan-like spirals. First distinguished in the USA in 1923, but of unknown origin.

cv. 'VERVAENEANA'.—A small, dense tree with slender branchlets. Leaves light yellow or greenish yellow in summer, brownish orange in winter. Raised by Vervaene in Belgium before 1862. There is an example 30 ft high at Wakehurst Place, Sussex.

cv. 'UMBRACULIFERA'.—A bun-shaped shrub to about 3 ft high and 5 ft wide. Leaves glaucous. Raised in Germany before 1892.

cv. 'WAREANA'.—Of dense habit, with deep vivid green foliage never tinged with brown. Branches stout, horizontal; branchlets short, clustered. Very like the old "*T. plicata*" (see f. *mastersii* above), except in the colouring of its foliage (*T. occidentalis Wareana* Hort., not *T. Wareana* Booth; *T. occ. robusta* Carr.). Probably raised, as Van Ouden and Boom suggest, by the nurseryman Weare of Coventry, not Ware. According to Loudon, the nursery was in decay when he saw it in the 1830s, following a change of owner, but had some fine things in its American department. Not one to mince his words, he called the excellent foreman 'a pearl cast before swine.' The well-known nursery of T. S. Ware was at Tottenham, near London, and so far as is known did not distribute conifers.

T. ORIENTALIS L.

Platycladus orientalis (L.) Franco; *Platycladus stricta* Spach;
Biota orientalis (L.) Endl., *nom. illegit.*

A shrub or small tree, 30 to 40 ft high, very distinct among the thuyas and cypresses by reason of the more or less erect or upward-curving branches bearing the spray or branchlets in the vertical plane, and in being of the same colour on both sides. There are two distinct types in cultivation; the one tall, somewhat columnar, and comparatively thin-branched, sometimes called var. *pyramidalis*; the other a dense, rounded or broadly pyramidal shrub with numerous branches springing from near the ground. The latter is the more effective for gardens. Ultimate subdivisions of the branchlets $\frac{1}{16}$ in. wide, flattened; green on both sides. Lateral leaves with their edges overlapping the middle ones, about $\frac{1}{12}$ in. long, scale-like; middle ones grooved; all marked with numerous white stomata. Cones roundish egg-shaped, $\frac{3}{4}$ in. long, erect, purplish; scales six, rarely eight, thick and woody, with a hooked, horn-like boss near the apex. Seeds wingless.

Native of China, where it has long been cultivated on graves and in temple gardens; not indigenous in Japan, though widely grown there. As a truly wild tree it seems to be rare and its natural distribution uncertain. Fortune saw trees in the mountains west of Peking that he took to be wild, and there seems to be no doubt about the wild status of the stands in N.W. Yunnan, first recorded by Forrest and later by Handel-Mazzetti and Joseph Rock. It grows, for example, in some of the side valleys of the Salween-Mekong divide, on steep dry slopes or even on vertical cliff-faces. Rock also saw it in a remote part of W. Szechwan not visited by Wilson. A point of great phytogeographical interest is that *T. orientalis* may be genuinely wild in N.E. Iran (*Fl. Iran.*, Cupressaceae (1968), p. 3).

T. orientalis was in cultivation in Holland early in the 18th century. Later the French missionaries sent seeds from Peking to Paris, whence it was introduced by Philip Miller to the Chelsea Physic Garden around 1740. It was reintroduced by Fortune in 1861, and no doubt there were other acquisitions of seed from China earlier in the century.

T. orientalis is now rare in this country in its normal arborescent form, and does not really thrive in our cool climate. Alan Mitchell has found that it is commoner in towns and cities than it is in large gardens and collections, and is healthier there (*Conif. Brit. Isles* (1972), pp. 290–1). The conditions under which it grows in S.W. China suggest that a rubbly, well-drained soil suits it best. A self-sown plant grows at Kew in the boundary wall of the garden of the Wood Museum and is probably the very one mentioned by Elwes and Henry in 1906. If so, it has survived there for seventy years. The largest examples of *T. orientalis* in Britain are 35–50 ft high and up to 6½ ft in girth.

Although *T. orientalis* agrees with other species of the genus in having only two seeds to each scale (three to five in *Thujopsis*), it differs from them in the thicker cone-scales and wingless seeds. If removed from *Thuja* it would become the sole member of the genus *Platycladus*, taking the name *P. orientalis* (L.) Franco. The name *Biota orientalis*, once widely used for it, is illegitimate.

cv. 'AUREA NANA'.—A dwarf roundish to ovoid bush; foliage golden yellow when young, light green later; branchlets arranged in definite vertical planes. One of the most frequently planted of dwarf conifers. Several variants of this character have been raised and the origin of the present stock is uncertain. It is unlikely to be the '*T. aurea*' raised by Messrs Waterer of Knap Hill, since they had plants up to 5 ft high and 15 ft in circumference in 1859.

cv. 'CONSPICUA'.—Of narrower, more conical habit than 'Aurea Nana' and less regularly arranged sprays. The colouring is similar but more persistent. Raised in the USA towards the end of the last century.

cv. 'ELEGANTISSIMA'.—Of conical habit, to about 20 ft high. Foliage greenish yellow, browning in winter. Rollison's nursery, 1858.

f. FLAGELLIFORMIS (Jacques) Rehd.   *T. orientalis flagelliformis* Jacques; *T. pendula* D. Don in Lamb.; *T. filiformis* Lodd. ex Lindl.; *T. orientalis* var. *pendula* (D. Don) Mast.—The most remarkable but least ornamental of all the forms. The branchlets are not arranged in sprays but long and cord-like, produced in crowded clusters. Leaves slender, awl-shaped, about ⅜ in. long, produced in pairs or threes. Although the foliage is of a juvenile type, this form occasionally produces fertile cones, the seed of which mostly produces ordinary *T. orientalis*. Introduced by Loddiges about 1800, probably from China, where, as in Japan, it is cultivated in gardens. According to Siebold, there are several sub-forms in Japan, propagated by cuttings or grafting.

In 'FILIFORMIS ERECTA' the branches are ascending and the plant remains moderately dwarf.

cv. 'HILLIERI'.—Foliage yellow at first, becoming green. Compact and slow-growing, but not dwarf. Raised by Messrs Hillier, before 1924.

cv. 'JUNIPEROIDES' ('Decussata').—A compact shrub to about 4 ft high. Leaves of the juvenile type, rigid, grey green in summer, purplish in winter, spreading and awl-shaped, arranged in decussate pairs (*Retinospora juniperoides* Carr.; *T. orientalis* var. *decussata* (Beissn.) Mast.; *Biota orientalis* var. *decussata* Beissn.). Of uncertain origin. It is uncommon in gardens and almost lost until Hornibrook distributed cuttings from a plant at Rostrevor in Co. Down.

A better known juvenile form, and considered by Hornibrook to be perhaps the finest of all conifers of this type, is 'ROSEDALIS' ('Rosedalis Compacta') with slender branches and soft leaves that are creamy yellow in spring, becoming light green and finally, in winter, plum-purple. 'MINIMA GLAUCA' is very dwarf, with sea-green, semi-juvenile foliage. All these juvenile sorts need a sheltered position, especially 'Rosedalis'. See also 'Meldensis'.

cv. 'MELDENSIS'.—As now cultivated this is an oval dwarf bush, rather sparsely branched, with foliage of the juvenile kind, but shorter than in 'Juniperoides', dull green in summer, purplish in winter. Many if not all of these plants descend from one noticed by Graham Thomas in a garden at Wadhurst in Sussex late in the 1930s. This had been propagated from another plant in the same garden which attained a height of about 4½ ft and a spread of 9 ft before it died. This plant in turn was a cutting taken in 1904 from one growing at 'The Gardens', a nursery at Chudleigh in Devon, which was 18 to 20 ft high. Murray Hornibrook identified the Wadhurst plant as 'Meldensis' and suggested that the Chudleigh tree was a reverted 'Meldensis' which still bore the occasional juvenile shoot. (G. S. Thomas, *Gardening Illustrated*, June 17, 1941, p. 272). The original 'Meldensis' was raised at Meaux near Paris about 1852, and was originally believed to be a hybrid between *T. orientalis* and a juniper. Judging from early descriptions 'Meldensis' was not dwarf, and it may be that the present form is a sub-clone meriting a distinctive name. It is reasonably stable, while its immediate parent at Wadhurst tended to revert.

cv. 'SEMPERAUREA'.—A compact shrub of rounded habit, to about 10 ft high, with golden foliage holding its colour in winter. Raised by Messrs Lemoine before 1871.

cv. 'SIEBOLDII'.—A compact roundish bush to about 6 ft high with bright green leaves borne on fine, densely arranged vertical sprays. A variety of Japanese gardens, introduced by Siebold. It is said to come true when raised from seed and that plants once grown as *T.* (or *Biota*) *orientalis nana* or *compacta* are the same.

T. PLICATA D. Don WESTERN RED CEDAR

T. gigantea Nutt.; *T. lobbii* Hort. ex Gord.; *T. menziesii* Dougl. ex Endl.

A tree up to 200 ft high in the wild, with a trunk sometimes 15 ft in diameter at the buttressed base; in cultivation a slender, pyramidal tree in some places already approaching 100 ft in height. Unless close-planted it retains its branches to the ground, but is inclined to become thin at the top. Branches curving upwards at the ends, branchlets drooping, strong-smelling and slightly aromatic when crushed; ultimate subdivisions $\frac{1}{16}$ to $\frac{1}{12}$ in. wide, flattened. Leaves dark glossy green, scale-like, $\frac{1}{12}$ to $\frac{1}{8}$ in. long; the lateral ones the longer, with their edges infolded and overlapping the flatter ones above and below the twig; they are all sharply pointed and have glaucous patches beneath. Cones egg-shaped, $\frac{1}{2}$ in. long; scales about ten, with a small triangular boss just beneath the apex.

Native of the coastal ranges of western N. America from S.E. Alaska to northernmost California, and with a second area in the northwestern Rocky Mountains. It was discovered at the end of the 18th century during the Malaspina expedition and was described by David Don in Lambert's *The Genus Pinus* from a specimen collected by Taddaeus Haenkel, assistant to Née, chief botanist to the expedition, on Nootka Sound (Née himself is usually credited with the discovery though he in fact remained in Mexico while the expedition explored northwards). It was introduced by Thomas Lobb in 1853 (possibly two years earlier to Scotland by John Jeffrey, for the Oregon Association), and the horticultural name *T. lobbii* has clung to it ever since.

T. plicata is distinguished from the E. American *T. occidentalis* by the glaucousness of the underside of the branchlet, the absence or obscurity of gland-pits on the upper surface of the leaves, and its much more rapid, cleaner growth. From *T. standishii* it differs in its denser habit, finer spray, and different odour.

Although not strong, the timber of the western thuya is easily worked, remarkably light, does not warp, and is remarkably resistant to decay. Imported into Britain, western red cedar is much used for garden buildings, outdoor furniture, cladding, etc., and in its native country for making roof shingles. It is also the wood from which most pencils are made.

In Britain *T. plicata* has proved by far the handsomest and best growing of the thuyas. Although commonest as a specimen tree, it is also useful for hedges and shelter belts and has been planted to some extent as a forest tree. In its native forests the western red cedar attains maturity at an age of about 100 years, but home-grown timber from stands half that age is of good quality and is used mainly for fencing, gates and ladder poles. The principal drawback of this species is that seedlings are often attacked by the fungus *Didymascella* (*Keithia*) *thujina*.

T. plicata grows well over much of the country, on all except very dry soils, but the best specimens are in the Atlantic zone, as is usually the case with the conifers of the American Pacific region. The most remarkable English specimens are: Stourhead, Wilts, behind the Lake, *pl. c.* 1854, 115 × 16¼ ft (1976), by the entrance, 105 × 14¼ ft (1976); Fonthill, Wilts, from 1860 seed, 120 × 15 ft (1963) and 115 × 14 ft (1976); Bicton, Devon, 128 × 14¼ ft (1968) and 120 × 15 ft (1963); Rhinefield Drive, New Forest, *pl.* 1861, 132 × 12 ft (1976); and, to indicate growth outside the optimum area, National Pinetum, Bedgebury, Kent, *pl.* 1916, 63 × 5¼ ft (1970).

Among the largest examples in Scotland are: Strone, Argyll, *pl.* 1875, 129 × 16¼ ft (1976); Inveraray, Argyll, Lime Kilns, 117 × 15¼ ft (1969); Belladrum, Inverness-shire, 115 × 16¼ ft (1970).

T. plicata has given rise to fewer cultivars than *T. occidentalis*:

cv. 'ATROVIRENS'.—Foliage dark green and glossy, in broad sprays. Smith of Worcester, 1874.

cv. 'EXCELSA'.—A fast growing, columnar tree with short horizontal branches and dark green glossy foliage. Put into commerce by J. Timm and Co. of Elmshorn, Germany. It is an excellent hedging plant.

cv. 'FASTIGIATA'.—Branches steeply ascending; foliage dense. As a hedging plant it needs less clipping than the normal form and also makes a good specimen (*T. menziesii fastigiata* Carr.; *T. plicata* var. *pyramidalis* Bean).

cv. 'GRACILIS'.—Sprays finer and with smaller leaves. Rare in cultivation and of no value.

cv. 'HILLIERI'.—A slow-growing bush with stout, crowded branches and branchlets, the leaf-sprays borne in irregular whorls. Eventually 8 or 10 ft high. 'A very irregular bush, with foliage so dense that one wonders how light or moisture can ever penetrate its centre.' (Hornibrook). Raised by Messrs Hillier about 1900.

cv. 'ROGERSII' ('Rogersii Aurea').—A dwarf, very slow-growing golden bush of ellipsoid or broadly conical habit, densely foliaged. Raised by George Gardiner for Messrs Rogers about 1914 but not described until 1929, when the mother plant was only 2¾ ft high (*Gard. Chron.*, Vol. 85 (1929), p. 50; ibid., Vol. 149 (1961), p. 502).

Another of Mr Gardiner's raising is 'STONEHAM GOLD', put into commerce by Messrs Rogers in 1948 but raised much earlier. It is more vigorous and less dense, the older leaves dark green, the younger bright gold. Both were raised from a tree grown under the name *T. plicata aurea*.

cv. 'ZEBRINA'.—Spray with curious patches of yellow interspersed with the ordinary green of the current year's shoots. It is a more attractive conifer than this description would suggest, the tree being shapely and dense, and from a short distance appearing golden-green—a colour that is constant throughout the year. Its origin is uncertain, but it is probably a renaming of one of several clones cultivated in the last century as *T. gigantea aureovariegata*. A tree under the present name was introduced to Kew about 1900. The oldest dated examples were planted soon after that and are now 60 to 70 ft or slightly more high (measured 1968–76). Trees of this size grow at Bicton, Devon; Stourhead, Wilts; Nymans and Tilgate, Sussex; Grayswood Church, near Haslemere, Surrey; Hardwicke, Suffolk.

cv. 'SEMPERAURESCENS'.—Leaves yellowish green. Rare. There is an example at Westonbirt, Glos.

T. STANDISHII (Gord.) Carr.

Thujopsis standishii Gord.; *Thuja japonica* Maxim.

A tree up to 100 ft high in Japan, with a slender trunk and a shaggy, deep red bark; branches horizontal, curved upwards at the ends; leaf-sprays arching downwards, their ultimate subdivisions about $\frac{1}{10}$ in. wide. Leaves scale-like, about $\frac{1}{8}$ in. long, the lateral pairs with their edges turned inwards and clasping the flatter ones above and below the twig, blunt, thickened and incurved towards the apex, rather pale yellowish green on the upper side of the twig, glaucous on the lower side, except at the points. Cones oblong, $\frac{3}{8}$ in. long, composed of about ten broadly oval, overlapping scales, only two pairs of which bear seeds.

Native of Japan, in the mountains of Honshu and Shikoku; introduced by Fortune for Standish of Bagshot in 1860–1. Fortune only saw it as a cultivated tree about Tokyo, and one of the first Europeans to see it growing wild was the Veitchian collector Maries, who gathered specimens in 1878 in the mountains of Honshu. This species has the most open branching of the cultivated species, and in its coarse branchlets it resembles *Thujopsis dolabrata*.

Examples measured recently are: Kew, *pl.* 1895, 38 × 3¼ ft (1965); National Pinetum, Bedgebury, *pl.* 1926, 30 × 3 ft (1970); Leonardslee, Sussex, one of three, 51 × 4¾ ft (1969); Linton Park, Kent, *pl.* 1866, 67 × 7½ ft (1970); Little Hall, Kent, *pl.* 1906, 51 × 5 ft (1973); Longleat, Wilts, 54 × 6¼ ft (1971); Westonbirt, Glos., *pl.* 1875, 63 × 4 ft (1971); Trentham Park, Notts, 60 × 4¼ ft (1968); Powerscourt, Co. Wicklow, Eire, 56 × 8½ ft (1975); Inistioge, Co. Kilkenny, Eire, 55 × 8½ ft (1975).

T. KORAIENSIS Nakai   *T. kongoensis* Nakai—According to Wilson, this species, as he saw it in Korea, is 'remarkable in its variations of habit from a sprawling shrub of nondescript shape to a slender, graceful, narrowly pyramidal tree.' (*Journ. Arn. Arb.*, Vol. 1 (1919), p. 186). He gives its maximum height as 25 to 30 ft, but this it rarely attains and usually it forms an impenetrable tangle 1 to 2 yards high.

T. *koraiensis* is closely allied to T. *standishii*, but the leaf sprays are flatter and the undersides of the leaves are unusually glaucous for this genus, almost silvery. The aroma of the crushed leaves has been likened by Alan Mitchell to that of rich fruit-cake 'with plenty of almonds', against that of oil of citronella or lemon-scented verbena in T. *standishii*.

Wilson introduced this species in 1917 from the Diamond Mountains of northern Korea, and probably all the largest plants in collections are from his seeds (W. 9244); these, or plants raised from them, were distributed by F. R. S. Balfour of Dawyck, who organised the British share of the financing of Wilson's expedition. These plants are mostly 20 to 33 ft high and 1 to 2 ft in girth.

THUJOPSIS CUPRESSACEAE

A genus of a single species, closely allied to *Thuja* but differing in the cones, which are almost globular, with woody scales, the fertile scales (usually six) bearing three to five seeds (two or three in *Thuja*).

T. DOLABRATA (L. f.) Sieb. & Zucc.
Thuja dolabrata L.f.

A tree up to 40 or 50 ft high, or a shrub of pyramidal form; branchlets arranged in opposite rows (distichous), the ultimate subdivisions much flattened, about ¼ in. wide, dark glossy green above, with conspicuous glaucous patches beneath. Leaves hard and rigid, borne in four ranks; those of the lateral ranks strongly keeled, ⅙ to ¼ in. long, incurved at the point, their edges overlapping the leaves of the middle ranks, which are appressed and rounded at the apex. Cones ½ to ¾ in. long, subglobose; the six or eight scales thick, woody, ending in a horn-shaped boss; seeds winged.

Native of Japan. The typical race extends from the southern islands to central Honshu, but the finest stands are on two peninsulas at the northernmost end of Honshu. These belong to the northern race var. HONDAE Makino, which extends into southern Hokkaido and has as its distinguishing characters: leaves smaller, foliage denser, cones more globose, cone-scales with an indistinct mucro. The first introduction of *T. dolabrata* was in 1853, when Thomas Lobb sent a plant to Messrs Veitch of Exeter from the Buitenzorg Botanic Garden in Java, which soon died, leaving no issue. Later in the 1850s Captain Fortescue brought a plant from Japan which was grown at Castle Hill, Devon, and propagated by cuttings. But the first commercial introductions were by John Gould Veitch and Robert Fortune in 1860–1. The former collected seed of what must have been var. *hondae* during his short stay at Hakodate, the southern port of Hokkaido, in autumn 1860; Fortune's sending (to Standish's nursery) was of seed collected in a cemetery near Tokyo.

Seen at its best, this is a striking and beautiful conifer, needing a sheltered position and tolerant of shade. Even in favoured localities it grows slowly and when young is very dense in habit at the base, but as it increases in height the upper growth is apt to become thin and attenuated. Some plants eventually become trees with a single stem, but many of the largest specimens recorded are many-stemmed from the base. Examples from Cornwall are: Scorrier House, *pl.* 1868, 52 × 5 ft (1965); Lamellen, St Tudy, 53 × 5¾ ft (1963); Pencarrow, 66 × 3 ft + 2¾ ft (1975); Tregrehan, 66 × 7¾ ft, short bole (1971); Boconnoc, 67 × 6¼ ft (1970); Penjerrick, 69 ft, many-stemmed (1965). Others are: Bicton, Devon, 57 ft, many-stemmed (1968); Killerton, Devon, 57 ft and almost 50 ft in spread, with seventeen stems; Melbury, Dorset, 67 × 4¼ ft (main stem) (1971); Lydhurst, Warninglid, Sussex, 59 × 3¼ ft (1965); Sheffield Park, Sussex, 51 × 3¼ ft (1968); Stonefield, Argyll, two trees, the taller 56 × 5 ft (1969); Castlewellan, Co. Down, 59 ft, many-stemmed (1976).

cv. 'LAETEVIRENS' ('Nana').—A dwarf shrub to about 3 ft high, with finer, lighter green foliage than normal. Introduced by J. G. Veitch in 1860–1 (*T. laetevirens* Lindl.; *T. dolabrata* var. *laetevirens* (Lindl.) Masters; *T. d. f. nana* (Endl.) Beissn., in part.

cv. 'V A R I E G A T A'.—Leaf-sprays splashed with white. Introduced by Fortune in 1861. A valueless variety, soon reverting almost wholly to the normal green.

Imported seed of *T. dolabrata* does not germinate well, but cuttings strike readily and are said to be the usual means of propagation in Japan, even for forest trees.

THYMELAEA   T H Y M E L A E A C E A E

A genus of about twenty herbs (some annual) and evergreen shrubs, confined to the Old World and mainly natives of the Mediterranean region and Asia. It is allied to *Daphne*, differing chiefly in the fruit, which is dry and usually enclosed in the persistent receptacle. The 'thymelaia' of the ancient Greek writers was probably *Daphne gnidium*, and the generic name *Thymelaea*, as originally used by Miller, included both the present genus and the daphnes.

T. N I V A L I S (Ramond) Meissn.
Passerina nivalis Ramond

A semi-prostrate evergreen shrub, 4 to 8 in. high, with half woody, slightly hairy, unbranched shoots. Leaves densely arranged in whorls of threes (about seven whorls to the inch), stalkless, linear, $\frac{1}{3}$ to $\frac{1}{2}$ in. long, about $\frac{1}{10}$ in. wide, bluntish pointed, slightly hairy about the margins, dull greyish green, rather fleshy. Flowers solitary in each leaf-axil, stalkless, $\frac{1}{4}$ in. across, scarcely so long, yellow. Calyx tubular at the base, dividing at the top into four ovate lobes, two of which are conspicuously broader than the other two. Stamens yellow, eight, in two series of four, inserted near the apex of the calyx-tube; very shortly stalked.

Native of the Pyrenees. A pleasing little evergreen for the rock garden, flowering abundantly in March, and quite hardy. It is scarcely distinct specifically from T. T I N C T O R I A (Pourr.) Endl., also a native of the Pyrenees but extending into N.E. Spain and with two stations in France. Another close ally is T. D I O I C A (Gouan) All., of wider distribution from Spain to N.W. Italy (*Fl. Europ.*, Vol. 2, p. 260).

THYMUS   THYME   LABIATAE

A genus of small evergreen aromatic plants, woody at least at the base. Leaves small, opposite, usually dotted with oil glands. Flowers in axillary whorls, which are sometimes crowded into terminal clusters. Calyx cylindric or bell-shaped, more or less two-lipped. Corolla two-lipped or almost symmetrical, the upper lip not hooded; tube straight. Stamens four, usually exserted.

Thymus is confined to the Old World. Most of the species that have been described belong to the very complex section *Serpyllum*, which ranges across Eurasia to the Russian Far East and Japan. It is largely this section, with its numerous local races, that accounts for the varying estimates of the number of species to be recognised in this genus, which ranges from thirty-five to four hundred. The section *Thymus*, the type of which is *T. vulgaris*, is also a large one, and extends into Central Asia. All but fifteen of the sixty-five species recognised in *Flora Europaea* belong to these two sections.

The most ornamental of the thymes belong to the section *Pseudothymbra* but these are intolerant of winter wet or even tender, and are better suited to the alpine house than to the rock garden. The distinctive features of this group are the long, slender corolla tube and the conspicuous, often coloured inflorescence-bracts. Those known to have been introduced or reintroduced in recent years are:

T. CILICICUS Boiss. & Bal.—A.M. 1962 to a form introduced by Peter Davis from S.W. Anatolia *c.* 1958.

T. INTEGER Griseb.—A native of Cyprus introduced by Peter Davis in 1941. A.M. 1947.

T. LONGIFLORUS Boiss.—A native of S.E. Spain introduced by Peter Davis and Vernon Heywood in 1947. A.M. 1951. They also introduced *T. murcicus* Porta, which is included in *T. longiflorus* in *Flora Europaea*.

T. MEMBRANACEUS Boiss.—A close ally of *T. longiflorus* introduced from S.E. Spain by T. A. Lofthouse in the 1920s. *Bot. Mag.*, n.s., t. 58. A.M. 1930.

For descriptions of these species see the excellent treatment of *Thymus* by Dr W. T. Stearn in the R.H.S. *Dictionary of Gardening*.

The species treated below are all very hardy and thrive in light, sandy, or calcareous soils. Propagation is by cuttings, but the mat-forming sorts can also be increased by division.

T. CAESPITITIUS Brot.
T. micans Sol. ex Lowe; *T. azoricus* Lodd.

A procumbent subshrub of compact habit. Leaves up to $\frac{3}{8}$ in. long, and $\frac{1}{16}$ in. wide, dark green, fleshy, glabrous except for some marginal hairs near the base. Flowering stems erect, to about $2\frac{1}{2}$ in. long. Inflorescence oblong, lax, few flowered. Calyx much shorter than the corolla, campanulate; upper teeth very small, lower almost as broad as long, fleshy, glabrous except at the margin. Corolla ¼ to ½ in. long, lilac-pink.

Native of the western Iberian peninsula, Madeira, and of the Azores, whence it was introduced early in the 19th century. Once confused with *T. serpyllum* it is distinct in fleshy leaves, the longer corolla and especially in the broad lower teeth of the calyx. It flowers in late summer and forms small hummocks.

T. SERPYLLUM L.

An evergreen subshrub a few inches high, with trailing, rooting stems, woody at the base; younger stems wiry, hairy all round. Leaves firm, elliptic or elliptic-ovate, mostly ⅛ to ⅜ in. long and about half as wide, blunt, narrowed at the base, hairy at least on the margin, sometimes on both surfaces, prominently veined beneath, dotted with oil glands. Flowering stems erect, bearing the rosy purple flowers in summer and early autumn in dense rounded heads ½ in. wide. Corolla scarcely ¼ in. long.

Native of Europe, including Britain, where it is rare, however. Our common thyme is:

T. DRUCEI Ronniger *T. britannicus* Ronniger; *T. pseudolanuginosus* Ronniger; *T. praecox* subsp. *articus* (E. Durand) Jalas, in part.—This is very near to *T. serpyllum*, from which it was not distinguished until 1924. It is a variable polyploid species, with the habit of *T. serpyllum*, differing in the obscurely four-angled stems, two of the opposite faces being hairy, the other two glabrous or almost so. Less constant differences are the slightly larger, obovate leaves and broader flower-heads. In *Flora Europaea* this species is included in *T. praecox* subsp. *arcticus*, which, according to that work, differs from *T. serpyllum* in leaf-venation: the lateral veins curve along the margin, anastomosing at the apex of the leaf, while in *T. serpyllum* they fade away before reaching the apex. However, in the British Isles *T. drucei* (which is founded on British specimens) is not so clearly differentiated from *T. serpyllum* as its continental counterparts.

T. drucei, the British wild thyme, is much commoner in the British Isles than *T. serpyllum* in the narrow sense, and is the thyme of the chalk downs, always loved for its sweet scent, the 'wild thyme' of Shakespeare, and the 'close cropped thyme' of Kipling,

'. . . that smells
Like dawn in Paradise.'

T. drucei is much more variable than *T. serpyllum* and most of the creeping ornamental thymes treated as varieties of the latter belong to *T. drucei* or to other members of the *T. serpyllum* aggregate. But the treatment of these is beyond the scope of this work.

T. DOERFLERI Ronniger *T. hirsutus* Bieb. var. *doerfleri* (Ronniger) Ronniger —Of uncertain taxonomic status, this thyme belongs to the same group as *T. drucei*, and is mentioned in *Flora Europaea* under *T. praecox* (q.v. above). Stems hairy all round. Leaves ⅜ to ½ in. long, densely covered with long and short hairs. Flowers purplish pink. Apparently confined to the Koritnik massif on the border between Albania and Yugoslavia; discovered by J. D. Doerfler in 1916 and introduced by him to the Vienna Botanic Garden, thence to Kew. *Bot. Mag.*, n.s., t. 381.

T. PULEGIOIDES L. *T. chamaedrys* Fr.; *T. serpyllum* subsp. *chamaedrys* (Fr.) Vollmann—The leading characters of this species are the absence of runners, the usually larger leaves (to about ½ in. long) with less prominent veins, and the longer, four-angled flowering stems, hairy mainly at the four corners. It is widespread in Europe, including Britain, where it is mainly confined to calcareous soils in S. and E. England.

T. VULGARIS L.   GARDEN THYME

An evergreen shrub, much branched, 6 to 12 in. high, with a woody base and slender, semi-herbaceous, greyish, downy young shoots. Leaves stalkless, ¼ to ½ in. long, $\frac{1}{16}$ to ⅛ in. wide, linear to ovate, toothless, grey downy, dotted with numerous oil-glands; margins recurved. Flowers lilac-coloured or pale purple, opening from May to July in axillary whorls, the whole forming a terminal spike 1 to 2 in. long and ½ in. wide; bracts no wider than the leaves. Corolla about ¼ in. long; calyx hairy, about as long as the corolla-tube, cylindric, with three very short triangular teeth and two longer, awl-shaped ones.

Native of S. Europe from Portugal to Greece, especially in the Mediterranean region; also of Corsica and the Balearic Isles. It has been grown in Britain from ancient times chiefly as a flavouring herb and for its pleasant aromatic odour. Oil of thyme, produced from the plant by distillation, chiefly in the south of France, is used for scenting soaps and as a local external stimulant. According to Gerard, the herbalist, thyme taken internally has many virtues, some curiously diverse, such as being 'good against winde in the belly' and 'profitable for such as are fearfull melancholicke and troubled in minde.' It is scarcely employed at all in English medicine to-day.

T. × CITRIODORUS (Pers.) Schreber   *T. serpyllum* var. *citriodorum* Pers.; *T. serpyllum* var. *citratus* West. LEMON THYME.—This old garden plant is the '*Serpyllum citratum*' of Parkinson—'The wilde Tyme that smelleth like unto a Pomecitron or Lemon . . .' (*Paradisus*). Linnaeus recorded it in his *Species Plantarum* (1753) as a variety of *T. serpyllum* under the Greek letter γ, without naming it, but citing *Serpyllum foliis citri odore* of Caspar Bauhin's *Pinax* (1623). Since then it has received many names, either as a species or as a variety of *T. serpyllum*. But it is intermediate in its characters between *T. vulgaris* and *T. pulegioides* and Ronniger's view, adopted here, was that it is a hybrid between them.

Miller found that it did not come true from seed and insisted that it must be increased by slips or cuttings. Unfortunately, later gardeners have been less scrupulous, and the true variety has had to be reselected from stocks of seedling origin. The lemon thyme makes a spreading bush up to 1 ft or slightly more high, differing from the common garden thyme in its somewhat broader, ovate or lanceolate, more or less glabrous leaves.

T. CARNOSUS Boiss. *T. nitidus* Hort., not Guss.—Related to *T. vulgaris*, but with fleshy leaves, wider inflorescence-bracts and whitish flowers. It is more erectly branched than the cultivated forms of *T. vulgaris*, almost fastigiate, to about 1 ft high. Native of southern Portugal.

TIEGHEMOPANAX   ARALIACEAE

A genus of trees and shrubs with about seven species in Australia (others in New Caledonia). Leaves pinnate, alternate. Flowers in umbels, arranged in the form of a panicle. Pedicels jointed. Petals four or five, valvate in the bud. Stamens four or five. Styles two, free almost to the base. Fruit a drupe.

The species treated below appeared in previous editions under the name *Panax sambucifolium*, but *Panax*, as now understood, consists entirely of herbs. The genus *Tieghemopanax* is closely allied to *Polyscias*, from which it was separated in 1905; it is not recognised by all authorities, and the species described appears in some works under the name *Polyscias sambucifolius*.

T. SAMBUCIFOLIUS (Sieber) Viguier

Panax sambucifolium Sieber ex DC.; *Polyscias sambucifolius* (Sieber) Harms; *Nothopanax sambucifolium* (Sieber) K. Koch

A shrub or small tree free from down in all its parts. Leaves pinnate, 3 to 10 in. long, usually made up of nine to eleven leaflets. Leaflets stalkless; lance-shaped or narrowly oval, finely pointed, tapered at the base, quite toothless, 1 to 3 in. long, ¼ to ¾ in. wide, dark green above, glaucous beneath, the midrib raised to a knife-like edge above. Flowers greenish, borne in panicles made up of spherical umbels ½ in. wide, of no beauty. Fruits produced in a group of globose clusters 1 in. or so wide, each fruit globose, ⅓ in. wide, of a watery, translucent, faint blue colour, very handsome and persisting a long time. Except in colour they resemble white currants. *Bot. Mag.*, t. 6093.

A native of temperate E. Australia; introduced to Kew from the Melbourne Botanic Garden shortly before 1873. In 1913 three plants were put out in the open garden at Nymans in Sussex and were 8 ft high in 1917 (Miss Muriel Messel in *A Garden Flora* (1918), p. 129). These were killed in the severe winter of 1928–9 but lived long enough to show that this handsome species should be hardy enough to succeed in the more maritime counties. The trees at Nymans agreed with the figure in the *Botanical Magazine* in having simply pinnate leaves and entire leaflets, but in some plants referred to this species the leaves are doubly pinnate and the leaflets toothed. The Nymans plants flowered in late summer but did not produce fruit.

TILIA LIME OR LINDEN TILIACEAE

A genus of about thirty species of large or medium-sized, deciduous trees, with more or less zigzagged young shoots; winter buds prominent. The inner bark is tough and fibrous, and that of some species is used for making rough ropes and mats. Leaves alternate, but set in two opposite rows on the branches, toothed, usually heart-shaped at the base. Flowers produced in summer on the shoots of the current year, in axillary, slender, long-stalked cymes. One of the most characteristic features of the genus is the large membranous bract, several inches long, to whose midrib the lower part (sometimes more than half) of the main flower-stalk is united, thus giving it the appearance of rising directly from the centre of the bract. The bract may reach to the base of the peduncle, and is then said to be sessile, or it may end some way above the base, and is then said to be stalked. In some Chinese species, not yet introduced, the bract is almost free from the peduncle. The flowers are very uniform in the limes, and help little to differentiate species. They are fragrant, $\frac{5}{8}$ to $\frac{3}{4}$ in. across, dull white or yellowish white. Sepals and petals five. In most species petal-like scales (staminodes) are present in the flowers, opposite the petals, but these are lacking in *T. platyphyllos* and its allies, and usually in *T. cordata*. The fruits are nut-like, with up to three seeds, with a thick or thinnish, downy, ribbed or smooth shell.

The limes belong to the temperate latitudes of the northern hemisphere, but do not occur in western N. America nor in the Himalaya. Mostly they thrive well in gardens, preferring a rich moist soil. The American species are not of much account with us, and the Asiatic species are uncommon in Britain, even in collections, which is surprising, for some, such as *T. insularis*, *T. oliveri* and *T. tuan* are beautiful trees of moderate size.

The limes provide excellent pasture for hive-bees, and a mixed planting would provide it over a period of some two months. Further investigation is needed concerning the toxicity to bees of the flowers of *T. tomentosa* and *T. 'Petiolaris'*, but it seems that bumble-bees rather than hive-bees are affected. For limes as bee-pasture see the interesting article by Dr R. Melville of Kew in *Kew Bulletin* 1949, pp. 147–51.

There is no modern monograph on the genus *Tilia* but the American species have been treated by George N. Jones in *Taxonomy of American Species of Linden (Tilia)*, Illinois Biological Monographs 39 (1968). This work reduces to three the fifteen species recognised by Sargent in the second edition of his *Manual of the Trees of North America*.

As with all forest trees, the limes should, if possible, be raised from seed. Failing that, they may be raised from layers, or, in the case of named varieties, by grafts. Grafted plants, however, frequently make very unshapely trees. The graft is taken, as a rule, from side branches, with the distichous (or two-ranked) arrangement alluded to above. The leading shoot often retains this character for many years, and shows a tendency to grow horizontally rather than erect. Often, too, the stock grows in thickness less quickly than the scion, or *vice versa*, with the result that there is formed an unsightly break in the trunk.

There has been considerable confusion in gardens over the nomenclature of the limes, largely due to a great number of hybrid or intermediate types. They interbreed with great facility under brighter skies than ours.

T. AMERICANA L.   AMERICAN LIME, AMERICAN BASSWOOD
T. glabra Vent.; *T. canadensis* Michx.; *T. nigra* Borkh.

A tree up to 70 or 90 ft in the wild (rarely to 130 ft) and 6 to 9 ft in girth (rarely to 15 ft); bark smooth and grey on young trees, becoming thick and furrowed; twigs glabrous and glossy, becoming dark brown; winter-buds glabrous, ovoid or broadly ellipsoid, with two or three outer scales. Leaves roundish ovate, abruptly acuminate at the apex, heart-shaped or sometimes truncate at the base, 2¼ to 6 in. long, almost as wide, coarsely toothed, dark green and glabrous above, usually quite glabrous beneath even when young, except for minute tufts of down in the axils (but the leaves of basal sprouts are downy beneath, at least on the veins); petioles glabrous, 1½ to 3 in. long. Flowers in pendulous cymes, yellowish white, ½ in. or slightly more wide, opening around midsummer; pedicels glabrous. Floral bracts spathulate, 3 to 4 in. long. Sepals lanceolate, acuminate, slightly downy on the outside, more densely so within. Petals ¼ to ⅜ in. long, about one-third as wide. Staminodes present, slightly shorter than the petals. Fruits woody, ⅜ in. or slightly less wide, downy.

Native of northeastern and east-central N. America, from New Brunswick and Maine west to southern Manitoba and North Dakota, south to New Jersey, North Virginia, Kentucky and Missouri. In the United States it is a valuable timber tree and attains a large size on deep, moist soils. Like many limes, it has the ability to regenerate by sprouts, and often these spring up around the base of aged standing trees. It has been cultivated in Britain since the middle of the 18th century, but it is not one of the first-rate limes in this country, being apt to die back when young and not flowering so freely with us as on the continent. In its normal state it is easily distinguished from *T. heterophylla* by the glabrous leaves, glossy beneath.

Examples of *T. americana* measured recently are: Grayswood Hill, Surrey, 75 × 8 ft (1976); Westonbirt, Glos., in Willesley Drive, 77 × 5½ ft (1974); Kinfauns Castle, Perthshire, 66 × 6 ft (1970).

cv. 'DENTATA'.—A vigorous tree said to attain 100 ft in Germany, with a rounded crown when old. Leaves up to 8 in. long, deeply toothed (*T. longifolia dentata* Hort. ex Kirchn.; *T. americana* f. *dentata* (Kirchn.) Rehd.). A similar form is offered by some Dutch and German nurseries as *T. americana* 'NOVA'.

cv. 'FASTIGIATA'.—A fastigiate form selected in the Rochester Parks, New York, about 1927.

var. VESTITA (Döll) V. Engler *T. nigra* var. *vestita* Döll; ?*T. neglecta* Spach; *T. pubescens sens.* Hook. and other authors, not Ait.; *T. americana* var. *pubescens*

Dipp., not Loud.; *T. michauxii sens.* Sarg. in *Manual*, Ed. 1, not Nutt.—Typically, *T. americana* has the leaves glabrous beneath except on strong sterile shoots. But some trees, especially towards the northern end of its range, have the leaves on flowering shoots sparsely clad beneath with simple or few-branched hairs.

The name *T. neglecta* occurs frequently in the literature of American limes. According to G. N. Jones, some herbarium specimens so identified are this downy-leaved state of *T. americana* or are sprout-shoots of the normal state. *T. neglecta*, as understood by Sargent in the second edition of his *Manual* is, however, *T. caroliniana*. The name has also been used for the sparsely indumented state of *T. heterophylla*. The confused use of the name *T. neglecta* arises from the fact that Spach described this species from a tree growing in the garden of the Paris Museum in 1834, and that there is no type-specimen at Paris. A probably authentic specimen of *T. neglecta* preserved in the Kew Hervarium suggests that the last-mentioned judgement is correct.

The name *T. michauxii* Nutt. has also been used in differing senses. The lime so named by Nuttall was a sparsely indumented form of *T. heterophylla*. But *T. michauxii* of the first edition of Sargent's *Manual*, which was the one current when the present work was originally published, was *T. americana* var. *vestita*.

T. CAROLINIANA Mill.

T. pubescens Ait.; *T. americana* var. *pubescens* (Ait.) Loud.; *T. floridana* Small; *T. neglecta sens.* Sarg., not Spach; *T. ashei* Bush; for other synonyms, see below

A tree up to 70 ft high in the wild; branchlets glabrous when mature. Leaves broadly ovate, short-acuminate at the apex, oblique and truncate or cordate at the base, 2½ to 6 in. long, 2 to 4¾ in. wide, coarsely toothed, upper surface of the mature leaf glabrous, underside at first coated with a dense rusty or pale tomentum of stellate hairs which has been largely shed by flowering time, exposing the pale green or glaucous blade and the small axillary tufts; petioles slender, glabrous, usually shorter than the blade. Flowers pale yellow, slightly smaller than in *T. americana*, borne around midsummer, up to thirty or so in each inflorescence; pedicels stellate-downy. Sepals densely downy on the outside. Fruits felted, about ¼ in. wide.

Native of the southeastern USA from North Carolina to central Florida, west through the Gulf States to the eastern parts of Texas, Okhlahoma, Texas and Missouri; introduced by Mark Catesby from the Carolinas in 1726. It is probably tender, and no sizeable trees have been recorded.

The following species, described by Sargent and featuring in the second edition of his *Manual* (1922) are included in *T. caroliniana* by G. N. Jones: *T. cocksii, T. crenoserrata, T. georgiana, T. littoralis, T. nuda, T. phanera, T. texana, T. venulosa.*

T. CORDATA Mill. SMALL-LEAVED LIME

T. parvifolia Ehrh.; *T. microphylla* Vent.; *T. ulmifolia* Scop.

A tree sometimes 80 to 90 ft high on the continent, usually much smaller in Britain; young shoots glabrous or nearly so. Leaves rounded, heart-shaped, 1½ to 3 in. long, nearly or quite as much wide, with a short tapered apex, sharply and rather finely toothed, dark green and glabrous above; pale, sometimes whitish beneath, with tufts of red-brown hairs in the axils of the veins; stalks slender, glabrous, 1 to 1½ in. long. Flowers yellowish white, fragrant, produced in the latter part of July in ascending or horizontally poised, slender-stalked cymes 2 or 3 in. long. Floral bract 1½ to 3½ in. long, ⅜ to ¾ in. wide, glabrous. Fruits globose, covered (especially at first) with a loose greyish felt, not ribbed, thin-shelled.

Native of most of Europe, and of the Caucasus, extending to 63° N in Sweden, and northwest Russia, but in the British Isles confined to England and Wales as far north as the Lake District and Yorkshire, usually on limestone formations. It is less common in cultivation than *T.* × *europaea*, its hybrid with *T.* *platyphyllos*, but is really more deserving of a place in the landscape than either. As usually seen it is a neat, small tree, but it is long-lived, and in time attains a large size, at least in areas where the summers are warmer than average. The largest specimens recorded in recent years are: Oakley Park, Shropshire, 97 × 17¾ ft and 115 × 11¾ ft (1971); Whitfield House, Heref., 98 × 11 ft (1973); Tottenham House, Savernake, Wilts, 105 × 10¾ ft (1967); Westonbirt Arboretum, Glos., 100 × 7¼ ft (1976); The Vyne, Basingstoke, Hants, 105 × 7¾ ft (1972); Westleton Church, E. Suffolk, 65 × 14 ft (1968).

For the Shrawley Wood stand of *T. cordata* near Worcester, see the article by Miles Hadfield in *Qtly. Journ. For.*, Vol. 57 (1963), pp. 35–43.

cv. 'SWEDISH UPRIGHT'.—Of very slender habit, though with horizontal or drooping branches. It was collected in Sweden in 1906 by Dr Alfred Rehder for the Arnold Arboretum, where the original tree was 35 ft high and 12 ft in spread in 1964 (Wyman, *Trees for American Gardens* (1965), p. 451).

Another selection of narrow habit is 'ERECTA', said to make a good street-tree even in industrial areas, with rather small, almost orbicular leaves, colouring yellow in the autumn and remaining on the tree until November (*Dendroflora*, No. 7 (1970), p. 72).

T. × FLAVESCENS Döll—A putative hybrid between *T. cordata* and *T. americana*, described in 1843 from a tree growing at Karlsruhe. The trees distributed by the Späth and Simon-Louis nurseries are of uncertain origin. They are near to *T. cordata* but with larger leaves.

T. × EUCHLORA K. Koch

T. dasystyla Hort., in part, not Stev.; *T. rubra* var. *euchlora* (K. Koch) Dipp.

A tree as yet about 50 ft high in this country, but probably considerably higher naturally, of graceful, often rather pendulous growth; young shoots glabrous. Leaves roundish ovate, oblique and heart-shaped at the base, with short, tapered points, 2 to 4 in. long, often more in young trees, and as much or more wide; rich glossy green and glabrous above, pale green beneath and glabrous, except for tufts of hairs in the axils of the veins; marginal teeth small, regular and slender; stalk glabrous, 1 to 2 in. long. Flowers produced in the latter half of July, three to seven together in cymes 2 to 4 in. long, yellowish white. Floral bract linear-oblong, or narrowly lance-shaped, 2 to 3 in. long, ¼ to ⅝ in. wide, glabrous, shortly stalked. Fruits distinctly ovoid, tapered to a point, shaggy, with pale brown wool, ¼ to ⅓ in. long.

Of doubtful origin; introduced about 1860. In some respects this is the most beautiful of the limes on account of its bright green large leaves and pleasing form. It is remarkably free from insect pests. In the summer of 1909, when not only limes but nearly every other tree and shrub was infested with aphides and other pests, I examined specimens of this lime at intervals during the summer, and never found a single parasite on the leaves. Yet it is quite uncommon in this country. On the continent, however, its qualities are better appreciated, and it is being much planted in streets. Its brilliantly glossy, rounded, nearly glabrous leaves and pendulous branches very well distinguish it.

To the above account, taken unchanged from previous editions, it should be added that the merits of *T.* × *euchlora* are now as much appreciated in Britain as on the continent. Being largely resistant to aphis infestation it does not drip honey-dew and is therefore suitable for street-planting. Its only disadvantage as a street tree is that the lower branches tend to droop with age, and give off pendulous secondary branches. It is an undemanding tree, holding its foliage well in hot, dry summers and tolerant of smoke pollution. Its beautiful, almost yellow flowers afford excellent forage for bees.

The history of *T.* × *euchlora* remains obscure, but it was apparently first distributed by Booth's Flottbeck Nursery near Hamburg as *T. dasystyla*. It is considered by some authorities to be a hybrid between that species (or *T. caucasica*) and *T. cordata*, by others as a variant of *T. caucasica*. The seed from which it was raised almost certainly came from the Crimea, but it is not a native of that region in the sense of having a wide distribution there.

The trees at Kew, planted in 1871–2, measure 60 × 5 ft and 60 × 8 ft (1974). Others measured recently are: Westonbirt, Glos., Willesley Drive, 62 × 5 ft (1971), and The Downs, 48 × 6 ft (1975); Tortworth, Glos., 50 × 6¼ ft (1973); Queens Road, Cambridge, 60 × 5 ft (1976).

cv. 'REDMOND'.—As yet scarcely known in Britain, this cultivar is said to differ from the usual form of *T.* × *euchlora* in its dense, pyramidal habit. It was selected in the USA and put into commerce there in 1927.

T. × EUROPAEA L. *emend.* Sm. COMMON LIME
T. vulgaris Hayne; *T. officinarum* Crantz, in part; *T. intermedia* DC.

A tree reaching well over 100 ft, sometimes 150 ft high; young branches glabrous. Leaves 2½ to 4 in. long, nearly as wide, obliquely heart-shaped at the base, with a short, tapered apex, sharply toothed, dark green and smooth above, pale green beneath, with tufts of hairs in the main axils; stalk slender, 1 to 2 in. long, glabrous. Flowers yellowish white, fragrant, produced in pendent, slender-stalked cymes, 3 or 4 in. long, during early July. Floral bracts 3 to 4½ in. long, ½ to ⅞ in. wide; slightly downy on the midrib at the back. Fruits roundish oval, the shell thick and tough with ribs only faintly showing. [PLATE 84

This tree is of uncertain origin, but is now generally believed to be a hybrid between *T. cordata* and *T. platyphyllos*, between which it is intermediate. From the former it is obviously distinct in its greater vigour, larger leaves resembling those of *T. platyphyllos* in size and venation, and the pendulous inflorescence. From *T. platyphyllos*, to which it is nearer, it differs in the glabrous branchlets, in having the leaves glabrous beneath except for axil-tufts, the more numerous flowers in the inflorescence, and the only faintly ribbed fruits. It is, however, less demanding than *T. platyphyllos*, growing well on soils too poor and dry for that species. The hybrid has been observed in the wild, but is not common, probably because the difference in flowering time between the two parent species is a bar to crossing, even where they grow together.

T. × *europaea* is the common lime of the British Isles, and was one of the most popular of all trees for avenues, streets, gardens and parks. It has the objectionable habit of dropping its leaves early, especially in dry summers, and is very subject to attacks by aphides, whose excrement is very sticky, turns black on the leaves rendering them very unsightly in late summer, and drops onto anything below.

Rarely producing fertile seed in economic quantities, the common lime has always been propagated by layering from stools, and the ease with which it could be produced in this way explains its wide use as a tree for formal avenues, where uniformity of growth is essential. Unfortunately, the clone used in Britain for some two centuries has an objectionable propensity to form huge burrs on the trunk, that sprout into dense thickets of succulent shoots, which if not removed ultimately completely hide the trunk.

The common lime reaches to great age, and has the faculty of staying alive for many years after the centre of the trunk has decayed. In consequence, many famous and historical trees and avenues exist. But most of the large limes of Central Europe are *T. platyphyllos* or *T. cordata*. The following are some of the largest recorded specimens: Bramshill, Hants, 80 × 21¼ ft (1965); Bowood, Wilts, 125 × 14 ft (1975); Welford Park, Berks, 135 × 11 ft (1966); Holme Lacey, Heref., 124 × 21¾ ft at 3 ft (1973); Bicton, Devon, 130 × 15 ft (1964); Duncombe Park, Yorks, 150 × 12½ ft and 138 × 12¼ ft (1972).

Several gall-producing insects infest the leaves, the commonest being one which produces the curious 'nail-gall', a conical, pointed growth on the surface of the leaf, ¼ to ⅜ in. long. None of these pests appear to do much permanent damage.

The following clones, long propagated in Holland, are described by H. J. Grootendorst in *Dendroflora* No. 7 (1970), pp. 74–6:

'PALLIDA'.—Of broadly conical habit. Branchlets reddish brown. Leaves yellowish green beneath (*T. vulgaris pallida* Hort., not *T. europaea* var. *pallida* Reichenb., which was a wild form of *T.* × *europaea* near to *T. cordata*). This lime is known in Holland as 'Koningslinde' (Royal Lime) and in Germany as 'Kaiserlinde'. Similar to this, and perhaps a sport of it, is 'WRATISLAVIENSIS', in which the young leaves are yellow, becoming yellowish green. It was raised at Breslau towards the end of the last century.

'ZWARTE LINDE'.—This, the Dutch 'Black Lime' has very dark twigs and a broadly ovoid crown rounded at the summit, with almost horizontal lower branches.

For further information on *T. × europaea* and the two parental species, see Elwes and Henry, *Trees of Great Britain and Ireland*, Vol. VII (1913), pp. 1656–73; and Miles Hadfield, 'Notes on Lime Trees in Britain', *Qtly. Journ. For.*, Vol. 55 (1961), pp. 303–12 and Vol. 56 (1962), pp. 41–8.

T. HENRYANA Szysz.

A tree 30 to 80 ft high in the wild, the branchlets at first stellately downy, ultimately glabrous. Leaves obliquely and broadly ovate, heart-shaped or cut

TILIA HENRYANA

off straight at the base, shortly taper-pointed, 2 to 5 in. long, 1½ to 3 in. wide, the margin set with bristle-like teeth $\frac{1}{10}$ in. long, the midrib and veins downy above, the whole under-surface covered with dull brownish stellate down; there are tufts of down in the vein-axils; stalks 1 to 1½ in. long. Flowers whitish, numerous (twenty or more), on cymes 4 to 6 in. long; floral bracts of similar length, ½ to ¾ in. wide, stellately downy, especially behind.

Native of Central China; discovered by Henry in western Hupeh. According to Wilson it is an uncommon species in the wild, but grows taller than any other lime of Central and Western China. The trees he saw in 1907 in western Hupeh, at 4,000 ft, were up to 80 ft high and 28 ft in girth. As a general rule, trees from that part of China grow well in this country, but *T. henryana* is among the exceptions. Wilson is believed to have introduced it in 1901 for Messrs Veitch, but there is no record of a plant from this sending. The largest known examples in the British Isles, at Birr Castle in Eire, were raised by Lord Rosse from seeds received from the Lushan Botanic Garden in 1938; they measure 26 × 2½ ft and 33 × 2 ft (1975). The small plant at Kew was raised from seeds received from Nanking in 1934.

T. henryana flowers in late summer or early autumn. It is remarkable for the almost hair-like teeth of its leaves.

T. HETEROPHYLLA Vent.   WHITE BASSWOOD

T. alba Ait., in part, *nom. ambig. propos.*; *T. michauxii* Nutt. (but see below); *T. heterophylla* var. *michauxii* (Nutt.) Sarg.; ?*T. neglecta* Spach; *T. monticola* Sarg.

This species has a more restricted distribution in the wild than *T. americana* and is at its best in the Appallachians. In its typical state it is easily distinguished from *T. americana* by having the undersurface of the leaves clad with a close white felt of stellate hairs, which conceals the tufts of light brown hairs in the axils of the nerves.

Forms of this species exist in which the indumentum of the leaf-undersides is sparse or even almost lacking, but even these forms can be distinguished from *T. americana* by the closer smaller, shorter-pointed, less curved teeth. These forms have been distinguished as *T. heterophylla* var. *michauxii* (*T. michauxii* Nutt.), but it should be noted that the name *T. michauxii* Nutt. as interpreted by Sargent in the first edition of his *Manual* (1905) is *T. americana* var. *vestita*, and this appears to be also true of the *T. michauxii* of previous editions of the present work. The sparsely indumented forms of *T. heterophylla* are by some authorities considered to be the result of hybridisation between that species and *T. americana*, but are included by G. N. Jones in *T. heterophylla* without distinction.

No sizeable examples of *T. heterophylla* have been recorded but there seems to be no reason why its Appallachian forms should not succeed with us.

T. INSULARIS Nakai

A deciduous tree 40 to 80 ft high in the wild, with a grey-barked trunk 1 to 3 ft in diameter; young shoots glabrous or sparsely hairy. Leaves of firm texture, roundish ovate, sometimes almost kidney-shaped, coarsely toothed, rounded or shortly pointed, heart-shaped at the base, 2 to 3½ in. long, usually as wide or even wider than long, glabrous except for tufts of pale down in the vein-axils; stalk 1 to 1½ in. long. Flowers ½ in. wide, numerous on the cyme, borne on slender stalks, the bract 1½ to 2½ in. long, ½ in. or more wide, not downy. The whole inflorescence is 3 to 4 in. long; fruits obovoid, $\frac{3}{16}$ in. long.

Native of Korea on Daghelet Island (Cheju Do), whence it was introduced by Wilson in 1919. It is allied to *T. cordata* and flowers at about the same time, differing in the coarser acuminately tipped teeth, sometimes enlarged into a lobule when terminating a main lateral nerve, in the regular presence of staminodes in the flowers, and in the long-stalked pendulous inflorescences 2 to 3 in. across, with up to thirty-six flowers in each. A charming small tree, which deserves to be more widely planted. It is, however, uncommon in cultivation, the largest examples being: Kew, *pl.* 1928, 38 × 2¼ ft (1967); Edinburgh Botanic Garden, 33 × 2½ ft (1970); Westonbirt, Glos., Pool, 50 × 2½ ft (1970); Thorp Perrow, Yorks, 42 × 2½ ft (1974).

T. insularis is scarcely more than a local race of the following, described earlier:

T. JAPONICA (Miq.) Simonkai   *T. cordata* var. *japonica* Miq.—As in the preceding, the inflorescences are many-flowered and pendulous, and the flowers have staminodes; the bracts are usually long-stalked. Leaves more or less orbicular in outline, not lobulate, up to 3 in. long, cordate. It is uncertain if the true species is in cultivation. A tree at Borde Hill, planted as *T. japonica*, has ovate leaves, mostly with a narrow lobule on one side, inflorescences with up to only ten flowers, not pendulous; bracts sessile or short-stalked. It is perhaps referable to *T. amurensis* Rupr., a native of Amurland, N. Korea, etc., and thus a near neighbour of *T. japonica*, which extends from Japan to E. China. The Borde Hill tree, whatever its correct name, is one of the most ornamental of the limes, with a conical, densely leafy crown, and flowering from almost every node in some years. The bright green young leaves appear early in the spring but are not damaged by frost. The tree was planted in about 1910 and measures 43 × 4½ ft (1968).

T. INTONSA Wils. ex Rehd. & Wils.
T. tonsura Hort. Veitch.

A deciduous tree up to 65 ft high in the wild, with a trunk up to 9 ft in girth; young shoots densely clothed with short, yellowish brown hairs, glabrous the second year. Leaves roundish heart-shaped, often obliquely so, terminated by a short abrupt point, margins evenly and distinctly toothed, 1½ to 4 in. long, rather less wide, glabrous above except for some minute hairs on the chief veins, grey-green and densely clothed with starry down beneath, especially on the midrib and the seven to nine pairs of lateral veins; there are also tufts of down in the vein-axils; leaf-stalk 1 to 2 in. long, downy. Flowers solitary or three on the cyme, yellowish white, the bract narrowly oblong, clothed with starry down, especially at the back, the flower-stalk united with it as far as the middle. Petals $\frac{1}{8}$ in. long, $\frac{1}{10}$ in. wide. Fruits egg-shaped, $\frac{3}{8}$ in. long, distinctly five-angled, covered with a pale close felt.

Native of W. Szechwan, China, where Wilson discovered it in 1903 and introduced it at the same time. It is quite hardy and grows well at Kew, where it has flowered and borne fruit. Its most distinctive character perhaps is the woolliness of its shoots and general downiness, to which the specific name ('unshorn') refers. Examples recorded are: Wakehurst Place, Sussex, 51 × 2$\frac{1}{4}$ ft (1974); Colesbourne, Glos., 67 × 4$\frac{1}{4}$ ft and 66 × 5$\frac{3}{4}$ ft (1971).

As admitted in *Plantae Wilsonianae*, this species may be no more than a local variant of T. CHINENSIS Maxim., a species not differing much from the above description, except that the young shoots are glabrous and the fruits more strongly ribbed. It is of wide range, from Kansu through Szechwan to Yunnan. The lime described by Schneider under the name T. *chinensis* is not the true species but T. *tuan* var. *chinensis*.

T. KIUSIANA Makino & Shirasawa

A tree said to grow to about 35 ft high in Japan; branchlets slender, clad with stellate hairs when young. Leaves of firm texture, oblong-ovate or ovate-elliptic, mostly 1 to 2$\frac{1}{4}$ in. long, $\frac{5}{8}$ to 1$\frac{1}{4}$ in. wide (sometimes to 3 in. long and 2 in. wide), acuminate at the apex, oblique at the base and sometimes slightly cordate on one side, darkish green above, undersides paler, downy, and with yellowish brown tufts of hairs in the axils, finely toothed; petioles very short, to about $\frac{1}{2}$ in. or slightly more long. Inflorescence with up to thirty-six flowers, pendulous. Bracts sessile, 1 to 2 in. long. Flowers about $\frac{1}{4}$ in. wide. Fruits globose, not ribbed, thinly hairy.

A native of Japan in Kyushu, Shikoku and the southern part of the main island; described in 1900 but not introduced until the 1930s. With its small very shortly stalked leaves it is one of the most distinct of limes, but rare in gardens. There is an example at Westonbirt, Glos., about 16 ft high. It belongs to the same group as T. *japonica*.

T. MAXIMOWICZIANA Shirasawa

A tree 70 to 100 ft high; young shoots downy. Leaves roundish ovate, 3 to 6 in. long, scarcely as wide; contracted at the apex to a short point, heart-shaped at the base, coarsely toothed, dark green and slightly downy above, covered beneath with grey stellate down, and furnished with conspicuous tufts in the axils of the veins; stalk 1$\frac{1}{2}$ to 3 in. long. Flowers not seen in this country but described as being produced in clusters of ten to eighteen, the floral bracts 3 to 4 in. long, downy. Fruits $\frac{3}{8}$ in. long, ribbed.

Native of Japan in Hokkaido and the northern part of the main island. It was introduced to Kew from the Arnold Arboretum in 1890, but this tree did not thrive and had not flowered by 1913. The present tree, planted in 1910, flowers freely in July and measures 55 × 4$\frac{1}{4}$ ft (1967). The species is uncommon in Britain, and the only other sizeable example grows in the Thorp Perrow collection.

From other Asiatic species with the leaves tomentose beneath this differs in having tufts of brownish hairs in the leaf-axils.

T. MIQUELIANA Maxim.

T. mandshurica sens. Miq., not Rupr. & Maxim.; *S. franchetiana* Schneid.

A tree 40 ft high, the young shoots, leaf-stalks, and especially the under-surface of the leaves covered with a dull grey felt. Leaves ovate, 2 to 5 in. long, 1½ to 3½ in. wide, heart-shaped at the base, taper-pointed, coarsely toothed (sometimes lobed), dark glossy green above, without tufts in the vein-axils beneath. Flowers numerous, sometimes over twenty on the cyme; floral bracts 3 to 4½ in. long, ⅝ to ¾ in. wide, with scattered starry down. Fruits globose, felted, ⅜ in. long.

Native of E. China (Kiangsu). It is sacred to Buddhists, and has long been cultivated about temples both in China and in Japan, from which it was first described. The description given above is of the form cultivated at Kew, but there are others in which the leaves are relatively much broader. It is a slow-growing tree with us, but flowers freely in the second half of July. The Kew tree, though planted in 1904, measures only 20 × 1¼ ft (1974).

T. 'MOLTKEI'

T. × moltkei Späth ex Schneid.

A vigorous tree to about 80 ft high, resembling *T. americana* in its glabrous branchlets and buds, and in the shape, size and toothing of the leaves, but differing in the undersurface of the leaves, which is sparsely coated with stellate down and lacks axillary tufts. It was put into commerce by the German nursery-man Späth in 1883, and named by him in honour of the famous General Moltke, who planted a tree outside Späth's house in 1888, next to a specimen of *T. tomentosa*, planted by Bismarck. Späth considered it to be a hybrid between *T. americana* and *T. tomentosa*, but the rather pendulous habit suggests that the second parent was *T.* 'Petiolaris', to which it is inferior as an ornamental. It is no longer much planted either here or in Germany.

Recorded examples are: Kew, 66 × 6 ft (1971); Stratfield Saye, Hants, 77 × 8 ft (1968); National Botanic Garden, Glasnevin, *pl.* 1888, 72 × 6 ft (1975).

T. 'SPECTABILIS'.—Similar to the preceding, but with some stellate pubescence on the branchlets and buds. Distributed by Dieck of Zöschen and described in 1893. (*T. × spectabilis* Dipp.; not Host; *T. blechiana* Dieck ex Dipp.). This is probably a hybrid between *T. americana* and *T. tomentosa*. Similar trees, of unknown origin, were in commerce several decades before 1893, and there is an example at Kew, received from Booth of Hamburg in 1872, which measures 74 × 6¾ ft (1974).

T. MONGOLICA Maxim.   MONGOLIAN LIME

A tree up to 60 ft high in cultivation; young branchlets glabrous, reddish by autumn. Leaves tinted red when young, 1½ to 3 in. long, ovate to broadly so, acuminate, truncate to cordate at the base, coarsely toothed, the teeth triangular, with slender points, sometimes (especially on young trees) three- or five-lobed, dark green and glossy above, pale beneath and glabrous except for tufts of down in the vein-axils; petiole about 1 in. long, reddish. Flowers produced in late July, often numerous (sometimes thirty or more) in the cyme; floral bract stalked, 2 to 3 in. long, ½ in. wide. Fruits obovoid to roundish, downy, thick-shelled.

Native of N. China and the Russian Far East; discovered by Father David in 1864 in the Pohuashan, west of Peking, but described from specimens collected by Przewalski in Outer Mongolia seven years later; introduced to the Jardin des Plantes at Paris in 1880 by Dr Bretschneider, and thence to Kew in 1904.

Unlike so many trees of continental northeast Asia, this lime grows and flowers well in this country, and is one of the most distinct in its leaves, recalling when unlobed those of a silver birch. It is one of the numerous Asiatic allies of T. cordata and flowers at about the same time.

The following examples have been recorded: Kew, pl. 1904, 47 × 4¼ ft (1967); Edinburgh Botanic Garden, pl. 1936, 41 × 3 ft (1970); Thorp Perrow, Yorks, 50 × 3½ ft (1974); Wakehurst Place, Sussex, 62 × 4 ft + 3¼ ft and 60 × 3 ft (1974). The tree at Kew derives from the Bretschneider introduction, probably from the mountains west of Peking. But there was another introduction by William Purdom for Messrs Veitch in 1913, also from the Peking region.

T. OLIVERI Szysz.

T. pendula V. Engler, not Rupr. & Maxim.

A tree up to 80 ft high in the wild; young shoots glabrous. Leaves 3 to 5 in. long, mostly somewhat less in width, abruptly taper-pointed, heart-shaped to almost truncate at the base, dark green and glabrous above, pure white beneath with a close white felt, axillary tufts absent, margins edged with rather fine and distant teeth; petiole 1 to 2 in. long. Flowers small, up to about twenty in each pendulous cyme; floral-bract sessile. Fruits thickshelled, downy, more or less warted, globose to ellipsoid or obovoid, slightly ribbed.

A native of Central China, common, according to Wilson, in the moist woods of northwestern Hupeh; discovered by Henry in 1888 and introduced by Wilson in 1900 for Messrs Veitch. It is one of the Chinese white limes, allied to T. tomentosa, which it resembles in its flowers, but they open earlier, at about the same time as T. cordata, the leaves are more finely and more distantly toothed, and the young growths are glabrous. It is an attractive tree, smaller than T. tomentosa, but still needing plenty of room owing to its spreading branches. There are two examples at Kew; one, on the lawn west of the Iris Garden (H. 16) measures 36 × 4¾ ft (1972); the other, in the Lime collection, is about the same height and 3¼ ft in girth. A crowded specimen at Westonbirt measures 80 × 5¼ ft (1977).

An excellent account of this species by Nigel Muir will be found in *Gard. Chron.*, Vol. 183 (1978), pp. 21–2.

T. PAUCICOSTATA Maxim.

A small deciduous tree, the young shoots glabrous. Leaves very obliquely ovate, the base cut straight across in a slanting direction or slightly heart-shaped, the apex acuminate, margins conspicuously and fairly regularly toothed except at the apex and the base, 2 to 3½ in. long and 1½ to 2½ in. wide in adult trees, much larger (up to 5 or 6 in. long) in young, cultivated ones, dull dark green and glabrous above, green beneath, and with tufts of rusty brown down in the axils of the veins, but not at the base, where the main veins join the leaf-stalk; stalk glabrous, ¾ to 1½ in. long. The cymes carry seven to fifteen flowers and the bract is glabrous, 2 to 3 in. long. Fruits roundish or slightly obovoid and ribbed.

Native of western and central China, occurring as far south as Yunnan; discovered in Kansu by Potanin in 1875. A plant in the Coombe Wood nursery of Messrs Veitch, introduced by Wilson from northwest Hupeh in 1901, was probably this species; some grafts from it were made, but there is no present record of the Wilson introduction in cultivation. It was collected in Yunnan by Forrest, who may also have sent seeds. There is a small example in the Edinburgh Botanic Garden, planted in 1934.

T. paucicostata belongs to the same group as *T. cordata* and *T. japonica*, but differs in the ovate leaves, obliquely truncate at the base.

T. 'PETIOLARIS' PENDENT WHITE LIME

T. tomentosa var. *petiolaris* (DC.) Kirchn.; *T. petiolaris* DC. *sensu* Kirchn., and *sensu* Hook. f. in *Bot. Mag.*, t. 6737, not DC.; *T. americana pendula* Hort.; *T. alba pendula* Hort.

A round-topped tree, already over 100 ft high in Britain, with pendulous branches and a singularly graceful habit; young shoots downy. Leaves roundish ovate, heart-shaped or nearly straight across at the base, mostly oblique, pointed, regularly and sharply toothed, 2 to 4¼ in. long, about three-fourths as wide, dark green and slightly downy above, white with a close felt beneath; stalk downy, up to 2½ in. long. Flowers dull white, three to ten together in drooping cymes 2 to 3 in. long. Floral bract as long as the cymes, narrowly obovate, sprinkled with minute tufted down. Fruits globose to orange-shaped, grooved, warty, ⅓ in. wide.

A tree of unknown origin. It first came to botanical notice in the descriptive catalogue of the Muskau Arboretum (1864), where it had been received (possibly from Booth of Hamburg) under the erroneous name of *T. americana pendula*. In renaming it *T. tomentosa* var. *petiolaris*, Kirchner confused it with *T. petiolaris* DC., described from a specimen taken from a tree growing in the Botanic Garden at Odessa. This tree, as it later proved, was simply a form of *T. tomentosa* with longer-stalked leaves than normal, but Hooker, in the *Botanical Magazine*, took up the name *T. petiolaris* for the present tree, by which it has been known

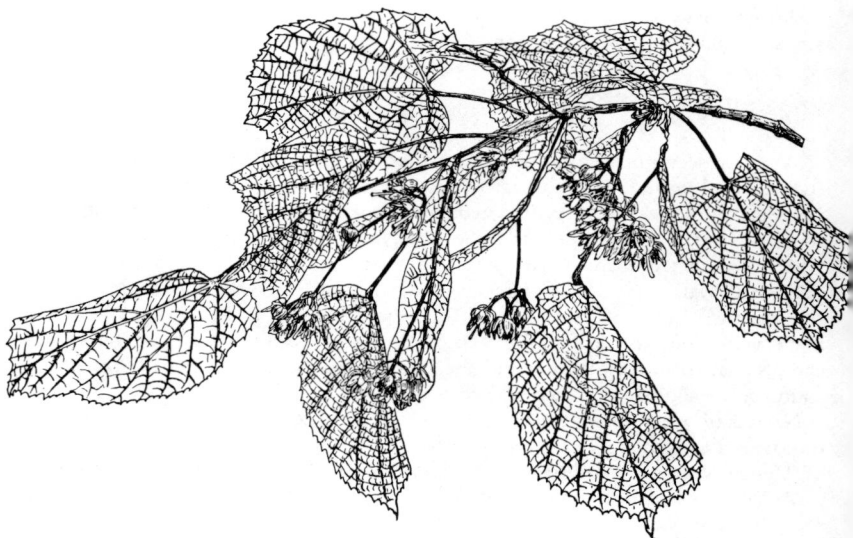

TILIA 'PETIOLARIS'

ever since. It is doubtful whether 'Petiolaris' is anything more than a pendulous
form of *T. tomentosa*, and considered as such could take the name *T. tomentosa*
'Pendula'. Its pendulous habit is obviously not a specific character. Neither is
the length of the petioles of much significance, since it is barely more than on
some wild trees. The wartiness of the fruits, sometimes advanced as a mark of
difference from *T. tomentosa* is in fact normal for that species. We are left with
the depressed-globose shape of the fruits, but this might be no more than an
outward sign of sterility, for the pendent white lime rarely produces fertile seed.

'Petiolaris' was in cultivation in Britain by the 1840s, since Elwes and
Henry mention a tree in the University Botanic Garden, Cambridge, planted
about 1842. But it appears to have been unknown to Loudon. It is one of the
most beautiful of limes, and its fragrance is perceptible many yards away when it
flowers in late July and August, at the end of the lime season. Bees find something narcotic in the flowers, as they may be seen in the evenings lying in scores
beneath the tree, and many do not recover.

Among the largest specimens of the pendent white lime are: University Parks,
Oxford, 102 × 9¾ ft (1975); Magdalen Bridge, Oxford, 90 × 11¾ ft (1968);
Egrove House, Kennington, Oxfords., 100 × 12½ ft (1968); Stratfield Saye,
Hants, 107 × 10¼ ft (1968); Westonbirt, Glos., in Willesley Drive, 102 × 9½ ft
(1976); Botanic Garden, Bath, 108 × 12 ft (1975); Warwick Castle, 98 × 10 ft
(1975).

T. 'ORBICULARIS'.—A tree to about 80 ft high, with the habit of 'Petiolaris', though somewhat more pendulous and with a more conical crown. The

leaves have shorter stalks, are very glossy above, and the felt of the undersurface is grey rather than silvery. They remain on the tree until late October. (*T. argentea orbicularis* Carr.; *T.* × *orbicularis* (Carr.) Jouin).

This tree arose in the nursery of Messrs Simon-Louis, near Metz, about 1868. The seed-parent was *T.* 'Petiolaris', and E. Jouin, the tree-nursery manager, considered that the pollen-parent was *T.* × *euchlora*, which grew nearby. Two examples were planted at Kew in 1900; they are 72 ft and 82 ft high, 5¾ ft in girth (1974).

An illustrated account of this hybrid by Nigel Muir was published in *Gard. Chron.*, Vol. 182 (1977), pp. 28–30.

T. PLATYPHYLLOS Scop.

T. europaea var. β L.; *T. officinarum* Crantz, in part

A tree of the largest size, 100 ft or more high, with a straight, clean trunk, and a shapely, rounded head of branches; young shoots downy. Leaves roundish heart-shaped, occasionally oblique, 2 to 5 in. long, nearly or quite as much wide; shortly taper-pointed, sharply toothed, dark green and slightly downy above, densely so beneath, especially on the veins and midrib; stalk downy, ½ to 2 in. long. Flowers yellowish white, produced in late June, usually in three- but sometimes six-flowered, lax, pendent cymes, 3 or 4 in. long. Floral bracts 2 to 5 in. long, ½ to 1¼ in. wide, downy, especially on the midrib and back. Fruits somewhat pear-shaped, ⅓ to ½ in. long, prominently five-ribbed, downy.

A native of Europe, extending into S.W. Asia, but less common in that region than the related *T. caucasica*. In Britain it is probably indigenous in a few localities such as the Wye Valley, around Sheffield, and near Richmond in Swaledale (see further in the article by C. D. Piggott in *Journ. Ecol.*, Vol. 57 (1969), pp. 491–504). In Central Europe it usually occurs at somewhat higher altitudes than *T. cordata*, preferring cooler and moister conditions. Although both live to a great age, most of the ancient trees in Germany and Central Europe are said to be *T. platyphyllos*.

Although not so commonly grown in England as *T.* × *europaea*, it is a more shapely and cleaner grown tree. The trunk does not produce the numerous swollen burrs covered with adventitious buds that are so characteristic of *T.* × *europaea*. From *T.* × *europaea* and *T. cordata* this lime is easily distinguished by its larger downy leaves, the downy shoots, and the larger five-ribbed fruit.

T. platyphyllos is a variable species, especially in the degree of hairiness of the young twigs and leaves. In the more northern race, which extends into southwestern European Russia, the twigs, petioles and leaf-undersides are densely to moderately downy, and there may be some down on the upper surface of the leaves. This is subsp. CORDIFOLIA (Bess.) Schneid. (*T. cordifolia* Bess.). The typical subspecies (subsp. PLATYPHYLLOS), with a more southern distribution, is only sparsely hairy, and the hairs on the undersides of the leaves are mostly confined to the main veins. The subsp. PSEUDORUBRA Schneid. is almost wholly glabrous; this is mainly confined to southeastern Europe, but extends into the southern Alps. But in horticultural nomenclature it is simpler to ignore these geographical variations, and place all cultivars and minor variants directly under the name *T. platyphyllos*.

cv. 'AUREA'.—Twigs golden-yellow in winter (*T. europaea* var. *aurea* Loud.; *T. p. aurantia* Hort.). An example at Kew measures 71 × 6 ft (1967). There may be more than one clone under this name.

In 'HANDSWORTH', of which there is a small tree at Kew, the twigs are greenish yellow in winter.

cv. 'COMPACTA'.—An interesting variation of rounded, compact habit. The original plant in the Wageningen Arboretum in Holland is only about 15 ft high after some forty years of growth.

cv. 'FASTIGIATA'.—Branches steeply ascending (*T. p.* f. *fastigiata* Rehd.; *T. p. pyramidalis* Hort., not Schneid.).

cv. 'LACINIATA'.—Leaves mostly smaller than normal, jaggedly and irregularly toothed, often with deep sinuses near the base; some are almost of the normal length, but very narrow and incised (*T. europaea* var. *laciniata* Loud.).

This seems to be the commonest and oldest of the cut-leaved clones. It flowers abundantly, often rather earlier than the normal form. The largest examples are mostly 50 to 60 ft high and 5 to 5¾ ft in girth. It was mentioned in previous editions as *T. p.* var. *asplenifolia*, but there seems to be little doubt that it is the form first named var. *laciniata* by Loudon and figured in *Arb. Frut. Brit.* (1838), Vol. V, p. 21.

Two other cut-leaved forms have been distributed commercially. In 'FILICI-FOLIA NOVA' the leaves are narrowly triangular, often irregularly lobed as in 'Laciniata' but not so deeply; they are also more glabrous. In the other, sent out by Baumann's nursery, probably as *asplenifolia nova*, the leaves were very narrow and deeply cut. But some trees in gardens may be seedlings of 'Laciniata'.

var. OBLIQUA (Host) Simonkai *T. obliqua* Host—Leaves obliquely truncate at the base, larger than normal, almost glabrous beneath. Floral bracts stalked. This is said to be common in gardens in Central Europe, and may be a cultivar.

cv. 'RUBRA' ('Corallina'). RED-TWIGGED LIME.—Twigs red in winter. Leaves downy beneath (*T. europaea* var. *rubra* West.; *T. e.* var. *corallina* Ait.; *T. rubra* DC., in *Cat. Pl. Monspel.* (1813), but not *T. rubra* DC. in *Prodr.* (1825); *T. mollis* var. *corallina* Spach).

Red-twigged forms of *T. platyphyllos* occur in the wild, probably throughout the range of the species, but the cultivated plants are propagated by grafting or layering, and may represent an old nursery clone. Examples are: Garnston Manor, Watford, 70 × 10¾ ft (1974); Linton Park, Kent, 80 × 9 ft (1972); Melbury, Dorset, 72 × 10½ ft (1971).

cv. 'TORTUOSA'.—Young branches curiously curled and twisted, often forming loops. This mutation was found in a batch of five hundred layers; the original plant was given to the Royal Horticultural Society and propagated by grafting at Kew, where plants were grown for some years (*T. p.* var. *tortuosa* Bean).

var. VITIFOLIA (Host) Simonkai *T. vitifolia* Host—Leaves obscurely three-lobed, rather like those of *Acer rufinerve* in shape, sparsely hairy beneath.

This is apparently little known in the wild and may be a cultivar. There is an example at Kew measuring 54 × 4½ ft (1967).

T. CAUCASICA Rupr. *T. rubra* DC. in *Prodromus* (1825), at least in part, not DC. in *Cat. Hort. Monspel.* (1813); *T. rubra* subsp. *caucasica* (Rupr.) V. Engler; *T. dasystyla sens.* Rehd., in part, not Stev.—Very near to the southeastern race of *T. platyphyllos* (subsp. *pseudo-rubra*), which it resembles in having the leaves glabrous beneath except for brownish axillary tufts, but differing in their marginal teeth, which have slender tips which are as long as, or even longer than the body of the tooth. The leaves are glossy above, rounded to slightly cordate at the base, about as wide as long. Native of the Caucasus, the Crimea and the mountains of northern Anatolia. It attains a height of over 100 ft in the wild.

The true species was distributed by Späth's nursery and is probably in cultivation as *T. dasystyla*, which indeed would be the correct name for the species if *T. dasystyla* Stev. (1831) and *T. caucasica* (1869) were considered to be conspecific.

Closely allied to *T. caucasica* are: T. BEGONIIFOLIA Stev. from the Caspian forests of Iran and bordering parts of Russia, with leaves usually longer than wide, and differently shaped fruits; and T. MULTIFLORA Ledeb., from western Transcaucasia, chiefly differing in the blunter teeth of the leaves and the many-flowered inflorescence (with up to twenty flowers). Plants distributed as *T. begoniifolia* in the last century were probably not true, but this species is now in cultivation in the Hillier Arboretum near Romsey. It was introduced by Roy Lancaster from Iran in 1972. There is no record of *T. multiflora* in cultivation.

T. DASYSTYLA Stev. *T. rubra* var. *dasystyla* (Stev.) Schneid.; *T. rubra* subsp. *caucasica* var. *typica* f. *dasystyla* (Stev.) V. Engler—An endemic of the Crimea, mainly in the southern part. The main point of difference from *T. caucasica* is that the style is densely hairy, almost woolly. There are other differences however; the leaves are downy beneath, at least on the main veins, and the tips of the leaf-teeth are shorter. It is uncertain if the true species is in cultivation. The tree originally distributed as *T. dasystyla* is *T. × euchlora* and others are probably *T. caucasica*.

T. TOMENTOSA Moench   EUROPEAN WHITE LIME

T. alba Ait., in part; *T. argentea* DC.; *T. europaea* var. *alba* (Ait.) Loud.

A tree 60 to 100 ft high, usually of broadly pyramidal habit, and with rather stiff, erect branches; young shoots woolly. Leaves 2 to 5 in. long, about as wide, roundish, heart-shaped, or nearly straight at the base, shortly and slenderly pointed, frequently with small lobes at the margins as well as the sharp, sometimes double teeth, dark green above and slightly downy at first, silvery white with a close felt beneath; stalks ¾ to 1½ in. long, felted. Flowers dull white, produced in late July and early August in five- to ten-flowered cymes, 1½ to 2½ in. long. Floral bract downy, rather longer than the cymes. Fruits $\frac{1}{3}$ to $\frac{3}{8}$ in. long, egg-shaped, with a short point, white with down and minutely warted, and faintly five-angled.

Native of the Balkans, Hungary, southwest Russia and northwest Anatolia; introduced to Britain in 1767. This tree, especially when fully grown, is handsome; in the young and intermediate states it is stiff and rather formal in habit. It thrives admirably in the south of England, and some fine examples exist there. In a breeze this tree presents a lively aspect, through the flashing of the leaves as they are turned by the wind. From *T.* 'Petiolaris' it is obviously distinct in its habit and short-stalked leaves. The flowers come out at the same time, and like those of 'Petiolaris' are toxic to bumble-bees. *T. oliveri*, the best known of the Chinese white limes, has glabrous twigs and more distantly toothed leaves.

Some of the larger specimens of *T. tomentosa* are: Kew, *pl.* 1872, 80 × 9½ ft (1974); Ockley Court, Surrey, 90 × 12 ft at 3 ft (1971); Melchet Court, Romsey, Hants (meas. by P. H. Gardner), 100 × 10½ ft (1980); Westonbirt, Glos., in The Downs, 92 × 9¾ ft (1974); Tortworth Court, Glos., 94 × 14 ft (1976).

cv. 'BRABANT'.—This name has been given to a clone propagated in Dutch nurseries and previously sent out as *T. tomentosa.* Deriving from a tree growing in the village of Hoeven in Brabant, it has a broadly conical crown with a well-developed central stem, which makes it more wind-resistant than the heavily branched form once planted in Britain. It is said to make a good specimen (*Dendroflora*, No. 7 (1970), pp. 78–9).

T. MANDSHURICA Rupr. & Maxim.   *T. argentea* var. *mandshurica* (Rupr. & Maxim.) Reg.   MANCHURIAN LIME.—This is the eastern counterpart of *T. tomentosa*, a native of N. China, the Russian Far East and Korea. It differs in the coarser toothing of its leaves, the teeth triangular, long-pointed at the apex. The first introduction to Kew, from Booth of Hamburg in 1871, was a failure, for like so many species from continental northeast Asia it started into growth early and was frequently cut by frost. The present tree thrives somewhat better, but does not flower freely.

T. TUAN Szysz.

A tree 40 to 50 ft high, young branches glabrous, or soon becoming so. Leaves thin, 2½ to 5½ in. long, 1½ to 3½ in. wide, broadly ovate with a very oblique, sometimes slightly heart-shaped base, apex slender-pointed; margins distantly and minutely toothed towards the point, but quite entire at the lower half, upper surface nearly glabrous, lower one covered with a close grey felt, and with small tufts in the vein-axils; stalks slender, downy, 1 to 2½ in. long; floral bract 3 to 5 in. long, ½ to ¾ in. wide, stellately downy.

Native of Central China; described from a specimen collected by Henry in 1888. It is a variable species. Typically, as described above, the leaves are finely and obscurely toothed, partly entire, but such forms are an extreme; more commonly they are distinctly though finely mucronate-serrate throughout. Again, though the young growths and buds are virtually glabrous in the typical state, they are markedly woolly in var. CHINENSIS (Szysz.) Rehd. &

Wils. This variety, originally described as *T. miqueliana* var. *chinensis* (also from a specimen collected by Henry), has quite strongly toothed leaves, almost glabrous beneath. There is also variation in the length of the floral-bracts, and in the length of the free part of the peduncle, which is sometimes very short, so that the cymes appear to be almost sessile and greatly overtopped by the bract. According to Wilson, both *T. tuan* and the var. *chinensis* are common in Hupeh.

Both *T. tuan* and the var. *chinensis* are said to have been introduced to this country by Wilson when collecting for Veitch, but there is no record of plants having been distributed. A tree at Borde Hill, Sussex, was bought at the Aldenham sale as *T. tuan* var. *chinensis* and planted in 1933. Having glabrous buds and young shoots it is not the variety, and how it came to Aldenham is unknown. The leaves are paper-thin, pale matt-green, finely toothed throughout, thinly stellate-downy and whitish beneath. The violet-scented flowers are freely borne in late summer, about nine to thirteen in each long-stalked cyme. It has made an open-crowned tree measuring 46 × 3¼ ft (1976).

TORREYA TAXACEAE

A genus of four or five species of evergreen trees, named in honour of Dr John Torrey, a famous American botanist. It is allied to *Taxus*. The branchlets are opposite and the linear, firm, sharp-pointed leaves are terminated by a fine hard point and are arranged in opposite spreading ranks. Flowers unisexual, the sexes either on the same or separate trees (solitary examples have borne fertile seed in this country). The male flowers are solitary in the leaf-axils and are composed of six to eight whorls of stamens. Fruit egg-shaped, consisting of a large bony seed enclosed entirely in a tough, fleshy coating.

Although not closely related, and placed in different families, there is a similarity in foliage between *Torreya* and *Cephalotaxus*. A difference can, however, be found in the undersides of the leaves: in *Torreya* the stomata are confined to two narrow longitudinal shallow grooves, while in *Cephalotaxus* the stomata are arranged in numerous lines, which occupy the greater part of the undersurface of the leaf.

Two species are American, and of these *T. californica* is the most widely planted and successful species in Britain; the other, *T.* TAXIFOLIA Arn. of Florida, is very rare in cultivation and unlikely to thrive in our climate. It is also very rare in the wild.

The torreyas are usually raised from imported seed, which quickly deteriorates; or by grafting on the common yew. Cuttings can be struck, but plants raised this way are said to grow poorly.

T. CALIFORNICA Torr.   CALIFORNIA NUTMEG
T. *myristica* Hook.   [PLATE 85

A tree 50 to 70 ft high in California (rarely 100 ft), with a straight, erect trunk and whorled branches; branches horizontal; branchlets pendulous, bearing the leaves in two flattish ranks. Leaves spreading at angles of 45° to 70° to the twig, 1½ to 3 in. long, about ⅛ in. wide, slightly convex, linear, with a slender spine-tipped point, dark glossy green above, yellowish green with a glaucous band of stomata each side the midrib beneath. The foliage as a whole is hard, stiff, and well armed by the needle-like points. Male flowers egg-shaped, ⅓ in. long, pale yellow. Fruit olive-like green, ultimately streaked with purple, about 1½ in. long by 1 in. wide, a thin, resinous flesh covering the grooved shell of the seed. *Bot. Mag.*, t. 4780.

Native of California, where it is widely spread in the forest region but nowhere common; introduced in 1851. This interesting and handsome tree is hardy and has borne fertile seed as far to the north-east as Peterborough, but the majority of the largest specimens are in the Atlantic zone. Among those measured recently by Alan Mitchell are: Mells Park, Dorset, 72 × 7¾ ft (1975); Tortworth, Glos., *pl.* 1861, 39 × 9¼ ft (1973); Tregothnan, Cornwall, 69 × 10¼ ft (1971); Benmore, Argyll, 40 × 8½ ft (1976); Castlewellan, Co. Down, 56 × 10¼ ft (1976).

T. NUCIFERA (L.) Sieb. & Zucc.
Taxus nucifera L.

A tree in Japan occasionally 80 ft high, oftener a shrub or small tree 20 to 30 ft high; in cultivation, so far as I have seen, always of a shrubby character, and not more than 10 or 12 ft high. Young shoots green, becoming in succeeding years purplish and shining. Leaves linear, ¾ to 1¼ in. long, ⅛ to 3/16 in. wide, tapered at the upper part to a slender, stiff point, very dark glossy green above, and with two glaucous stomatic strips beneath. The leaves (somewhat convex on the upper surface, stiff and hard in texture) are borne in two spreading ranks, which form a broad V-shaped channel. Fruits green, elliptical, 1 to 1⅛ in. long, ¾ in. wide. They are occasionally borne in abundance at Kew.

Native of Japan, where, like the Californian species, it is nowhere common, preferring moist heavily shaded positions. According to the younger Aiton it was cultivated by a Captain Cornwall as early as 1764, and was listed by Weston in 1775 as a plant then in commerce. *T. nucifera* is much less common in Britain than *T. californica*, from which it differs in its shorter more glossy leaves, directed downwards, and its orange-red shoots. There is an example at Wakehurst Place in Sussex, 38 × 2½ ft (1970) and another at Scorrier in Cornwall, 40 × 2¾ ft (1973).

The kernels of the nuts have an agreeable, slightly resinous flavour, and yield an oil once used in Japan for cooking.

Allied to the above is T. GRANDIS Fort. of China, of which a few plants are in cultivation. Its leaves are shorter and thinner in texture than those of

T. nucifera, and when crushed do not emit the pungent aromatic odour of that species; the bark of the young shoots, at first green, is later greyish. This torreya was discovered by Fortune in Chekiang in 1855 and introduced by him; it is a native of eastern and central China. The only sizeable examples known are: Kew, 25 × 2 ft (1970) and Borde Hill, Sussex, 29 × 1¼ ft (1968).

TRACHELOSPERMUM   APOCYANACEAE

A genus of about twenty species (perhaps fewer), mostly in the warmer parts of E. Asia and Malaysia, but the almost glabrous *T. difforme* is N. American. Leaves opposite, entire, they and the stems exuding a milky juice when cut. Flowers in the cultivated species white or yellowish, fragrant, borne in terminal cymes. Calyx small, five-lobed. Corolla with a slender tube and five spreading lobes overlapping each other to the right and convolute in the bud. Stamens inserted on the tube, their anthers united and attached to the stigma, included or slightly exserted. Ovary consisting of two carpels, each developing into a slender follicle.

Both species need the protection of a wall, and a good soil. The addition of peat to the soil when planting is an advantage. Increased by cuttings taken in July or August.

T. ASIATICUM (Sieb. & Zucc.) Nakai

Malouetia asiatica Sieb. & Zucc.; *T. crocostomum* Stapf; *T. divaricatum* Kanitz; *T. majus* Nakai *sec*. Ohwi (but not of gardens)

An evergreen climber at least 15 ft high, of dense, much branched habit; the young shoots very hairy, and the hairs persisting for several years, but almost glabrous in some forms. Leaves leathery, opposite, oval or slightly ovate, ¾ to 2 in. long, ⅜ to ¾ in. wide, mostly blunt at the apex, dark glossy green, glabrous; stalk ⅛ in. or less long. Flowers yellowish white, fragrant, produced in July and August in slender terminal cymes 2 to 2½ in. long. Calyx-lobes erect, narrow, pointed, not reaching to the top of the narrow basal part of the corolla-tube. Corolla with a tube about ⅜ in. long, and with five spreading obovate lobes, giving it a diameter of ¾ in.; lower part of corolla-tube narrow, and about twice as long as the widened upper part. Stamens inserted on the upper part of the tube and slightly protruding.

Native of Japan and of southern and central Korea. All the plants originally grown in gardens derived from one which had grown for many years on a garden wall at Kew, where it was hardy and flowered with great profusion in some years. But there have probably been other introductions since then, some perhaps less hardy. The Kew plant was originally considered to represent a distinct species, named *T. crocostomum* by Stapf, and it has been suggested that matching plants occur in China. But recent works include it in *T. asiaticum*, and it is probable that the plant came from Japan or Korea.

T. asiaticum has smaller leaves and yellower flowers than *T. jasminoides*, and is readily distinguished when in flower by the erect calyx-lobes, which in *T. jasminoides* are larger and distinctly turned back.

T. GRACILIPES Hook. f.—*T. divaricatum* var. *brevisepalum* Schneid., in part— Allied to *T. asiaticum* but the calyx minute and the inflorescences on usually longer peduncles. Native of N.E. India, S. China, Formosa and the former Indochina. Not known to be in cultivation.

T. JASMINOIDES (Lindl.) Lem.
Rhyncospermum jasminoides Lindl.

An evergreen twiner, growing 10 or 12 ft high, young shoots hairy. Leaves oval-lanceolate, 1½ to 3½ in. long, ½ to 1 in. wide, tapering at both ends, the tip blunt, downy beneath when young, becoming glabrous, dark glossy green above; stalk about ⅛ in. long. Flowers very fragrant, produced in July and August on glabrous, slender-stalked cymes, 1½ to 2 in. long, usually on short lateral twigs. Calyx-lobes spathulate, reflexed at the tips, as long as the narrow basal part of the corolla and about half as long as the total length of the tube. Corolla pure white, scarcely 1 in. across, the tube about ¼ in. long. *Bot. Mag.*, t. 4737.

Native of China and Japan (see var. *pubescens*); introduced by Fortune from Shanghai in 1844. It was long grown in greenhouses, where its flowers were prized for their fragrance, but it is also now grown on walls in the south and west. It is rather slow-growing when young, and then more vulnerable to frost than when it is established.

var. PUBESCENS Makino—Young stems downy. Leaves to about 3 in. long and 2 in. wide, downy beneath. Inflorescence axes downy. To this variety probably belongs *T. jasminoides* 'JAPONICUM', which is said to climb to the top of high walls in the South of France and in Italy. The leaves usually become bronze-tinted in autumn. Flowers white. This trachelospermum was originally known in gardens as *T. japonicum* and more recently has been identified as *T. majus* Nakai, but according to Ohwi, in *Flora of Japan*, this is a synonym of *T. asiaticum*.

cv. 'VARIEGATUM'.—Leaves shorter and broader, bordered and blotched with creamy white.

Trachelospermum Wilson 776 was raised from seeds collected in 1907 in Hupeh. There is no corresponding field-specimen, and no flowering specimen has been seen, but the plants probably belong to *T. jasminoides*. They are valued chiefly for their foliage, the leaves having conspicuously light green veins and a bronzy tint. They are sometimes called *T. jasminoides* 'Wilsonii'.

TRACHYCARPUS   PALMAE

A genus of not more than six species of fan-leaved palms, with stout, single or sometimes clustered trunks. Divisions of leaf numerous, more or less bifid at the apex; petiole not prolonged into the leaf, slender, edged with fine, sharp teeth. Inflorescences borne among the foliage, paniculate, with numerous large bracts (spathes) at the base. Flowers small, unisexual. Calyx and corolla each with three segments. Stamens six. Ovaries three, connate at the base, each (or sometimes only two or one) developing into a small one-seeded fruit.

The most recent study of the genus is: Myron Kimnach, 'The Species of *Trachycarpus*', in *Principes*, Vol. 21 (1977), pp. 155–60.

T. FORTUNEI (Hook.) H. Wendl.   CHUSAN PALM

T. excelsus H. Wendl.; *Chamaerops fortunei* Hook.; *C. excelsus* Mart., not Thunb.*

This palm, which is the only species that can be termed really hardy in this country, varies in height according to the circumstances under which it is grown. In the Temperate House at Kew there is an example about 50 ft high, but in the open air plants at least seventy years old are only about 12 or 15 ft high. In the warmest counties, however, it is 25 to 30 ft high. The stem is erect, cylindrical, clothed with coarse, dark, stiff fibres, which are really the disintegrated sheathing bases of the leaves. These fibres are employed by the Japanese and Chinese to make ropes and coarse garments. The leaves, which persist many years, are fan-shaped, 1½ to 2½ ft long, 2½ to 4 ft wide, divided at the outside into numerous deep, narrow, folded segments, 2 in. wide, tapering to a ragged point. The stalk is two-edged, and varies in length according to the age of the specimen and the conditions under which it is grown; it is usually between 2 and 3 ft long, and ½ to 1 in. wide, with small jagged teeth on the margins. The flowers are borne in a large, decurved, handsome panicle from near the top of the stem among the younger leaves; they are yellow, small, but very numerous. The panicles bear flowers usually of one sex only, the female ones being the smaller and less ornamental. Fruit a blue-black drupe about the size of a boy's marble. *Bot. Mag.*, t. 5221.   [PLATE 86

T. fortunei is widespread in central and southern China, but has so long been cultivated for its fibre, and naturalises itself so readily, that its original habitat is uncertain. It is not a native of Japan, but there too it was an important economic plant, and the first seeds to reach Europe were sent from that country by Siebold to Leyden in 1830. Not many germinated, but of the few plants raised one was sent to Kew in 1836. In 1860 this plant was 28 ft high, but no one suspecting its hardiness it was grown in the tropical palmhouse and was dead

---

* The palm described by Thunberg under the name *Chamaerops excelsus* is *Raphis excelsa* (Thunb.) Henry ex Rehd. (*R. flabelliformis* Ait.), a dwarf palm with cane-like stems. The confusion arose from the fact that Thunberg cited two Japanese names which properly belong to *Trachycarpus fortunei*.

by the end of the century. Fortune first saw this palm in China on the Island of Chusan in 1843—whence the vernacular name he chose for this species. Five years later he met it again in Chekiang and in 1849 sent a case of plants to Kew. One of these, no doubt on his recommendation, was immediately planted in the open and has survived every winter since then. The first general distribution of the Chusan palm took place in 1860, when Glendinning's nursery auctioned plants raised from seed sent by Fortune from the Ningpo area of Chekiang.

Although the Chusan palm is perfectly hardy in the south and west of Britain, in so far as it will, when properly established, withstand a temperature of 32° F or more of frost, it likes a spot screened from the north and east winds. Exposed to blasts from those quarters it will live, but has usually a miserable, battered appearance. When it was first experimented with in the open air it was usual to cover it in winter with mats or branches, but this has been found to be un-necessary. It is, however, advisable to give some protection to very young plants in severe weather. The perfect acclimatisation of this palm in the south and west is shown by the self-sown seedlings it produces as far east as Sussex. Since it is normally dioecious, both sexes have to be grown if fruits are to be obtained, but a male plant grown by E. A. Bowles at Myddelton House, Enfield, bore some female flowers one year, which fruited, and no doubt the converse also happens.

The example at Kew mentioned above grows near the main entrance and measures 25 × 1¾ ft (1972). Another of the plants sent by Fortune was, at his request, presented to the Prince Consort and still thrives at Osborne House in the Isle of Wight. It is about 40 ft high (1979), and other examples in southern and western England are of that height or near it.

Some plants cultivated in Japan and China have smaller leaves than is normal in T. *fortunei*, with stiffer segments, and have been distinguished as T. WAGNERIANUS Beccari. Similar in foliage to T. *wagnerianus* but developing several trunks from the base, is T. CAESPITOSUS Beccari, described in 1915 from a plant introduced to a California garden from Japan. A similar palm at Leonardslee in Sussex was acquired from Japan early this century and still exists; this was named T. *fortunei* var. *surculosa* by Henry in 1913.

T. MARTIANUS (Wall.) H. Wendl. *Chamaerops martianus* Wall.; T. *khasyanus* (Griff.) H. Wendl.; C. *khasyanus* Griff.; C. *griffithii* Verlot—Trunk smooth, to about 50 ft high, fibred only immediately under the leaves. Leaves divided to about half way, glaucous beneath; segments drooping at the tips. Native of the Himalaya from Nepal eastward, at low elevations, N.E. India, Burma, and probably S.W. China; introduced by Wallich around 1817, but scarcely tried out-of-doors in this country. According to Kimnach, some plants grown as T. *martianus* are really a form of T. *fortunei* but the true species is cultivated in California, from seeds obtained from the well known Indian seedsman Ghose of Darjeeling. A palm growing at Pallanza on Lake Maggiore, in the former Rovelli nursery, is almost certainly this species; see the note by Derek Fox in *Journ. R.H.S.*, Vol. 98 (1973), pp. 502–3 and fig. 251.

T. NANUS Beccari—Of dwarf habit, with a creeping trunk. Flowers in elongated panicles, springing direct from the soil. Native of Yunnan to about 8,000 ft, described from a specimen collected by the Abbé Delavay in 1887. It is probably not in cultivation.

T. TAKIL Beccari—This interesting palm was described by Beccari from a plant in his own garden at Florence, raised from seeds sent by Brandis from Kumaon in 1884. In this region of the Himalaya, to the immediate west of Nepal, *T. takil* occurs in several localities at an altitude of 6–8,000 ft. It is more closely allied to *T. fortunei* than to *T. martianus*, with which it was at first confused, differing in the closer fibres of the trunk, the less deeply divided leaves and, at least in the type, the more elegant crown. It should be hardy in Britain, as it was in Professor Beccari's garden; his plant is portrayed in *Kew Bulletin* 1912, p. 291.

TRICUSPIDARIA *see* CRINODENDRON, Vol. I

TRIPETALEIA ERICACEAE

A genus of a single species, belonging to the same group of the Ericaceae as *Cladothamnus*, *Elliottia* and *Botryostege*. The leading characters are the thin deciduous leaves; inflorescence terminal on the season's growths, paniculate; flowers with three free petals; anthers opening by slits. The present species, and *Botryostege bracteata* (originally described in *Tripetaleia*), were transferred to *Elliottia* by Bentham and Hooker in *Genera Plantarum* (1876). This judgement is upheld by Bruce A. Bohm *et al.* in 'Generic Limits in the Tribe Cladothamnae . . .', *Journ. Arn. Arb.*, Vol. 59 (1978), pp. 311-41. In this paper *Cladothamnus* is also transferred to *Elliottia*, its one species becoming *E. pyroliflora* (Bong.) Brim & P. F. Stevens.

T. PANICULATA Sieb. & Zucc.
Elliottia paniculata (Sieb. & Zucc.) Benth. & Hook. f.

A deciduous shrub 3 to 6 ft high, with angled young shoots. Leaves lanceolate to narrowly ovate or obovate, entire, pointed, tapered at the base to a short stalk, upper surface dark dull green and glabrous except along the midrib, lower surface pale green and glabrous except for copious white hairs on the midrib and base of the veins, chief veins in two or three pairs. Flowers borne on erect panicles 2 to 6 in. high, starting to open in July and continuing until September. They are terminal on the young leafy shoots. The flowers are white, tinged with pink, and have three (occasionally four or five) linear-oblong petals ⅜ in. long; calyx cup-shaped; style slender, ⅜ in. long, glabrous, standing out horizontally well beyond the petals. The bracts on the flower-stalks are small and linear.

Native of the mountains of Japan from Hokkaido southward; described in 1843 and introduced to Britain towards the end of the last century. It is a perfectly hardy shrub, growing well in the sort of soil that suits rhododendrons and a lightly shaded, protected position. In the form mentioned in previous editions the flowers apparently only had a pink flush, but in a later introduction they are a uniform light pink and very pretty. The three strap-shaped rather raggedly arranged petals give to the flowers of this species an informal appearance unusual in the Ericaceae and is in sharp contrast to the conventional urn-shaped, tubular or campanulate corolla of so many members of the family.

TRIPTERYGIUM   CELASTRACEAE

A genus of two or three species of shrubs, deciduous and more or less scandent. Leaves large, alternate; flowers small in terminal panicles; petals and stamens five; calyx five-lobed. The name refers to the three-winged fruit. Both species are easily cultivated in good loamy soil.

T. REGELII Sprague & Takeda
T. wilfordii sens. Reg., not Hook. f.

A deciduous shrub of rambling or climbing habit, with angular, warted stems. Leaves alternate, oval or ovate, broadly wedge-shaped to rounded at the base, tapered at the apex to a long and slender point, 2½ to 6 in. long, 1½ to 4 in. wide, the margin set with rounded, blunt, incurved teeth, dark green above, and except for minute down on the midrib when quite young; stalk ¼ to ¾ in. long. Flowers yellowish white, about ⅓ in. wide, produced in a panicle at the end of the shoot, supplemented by clusters in the axils of the terminal leaves, the whole forming an inflorescence up to 8 or 9 in. long and 2 or 3 in. wide, petals five, roundish obovate; calyx small, with five rounded lobes; stamens five. Fruits three-angled, each angle conspicuously winged; the wings erect, about ⅝ in. long, ¼ in. wide, membranous. The whole inflorescence is covered with short brown felt.

Native of the main and southern islands of Japan, and of Korea and Manchuria; introduced to the Arnold Arboretum by J. G. Jack in 1905 and thence to Kew. It is hardy, but not so fine a species as *T. wilfordii*. The fruits are greenish and rather like those of a wych elm with an extra wing.

T. WILFORDII Hook. f.

T. forrestii Loes.

A deciduous scandent shrub from 10 to 40 ft high in the wild; young shoots often long and unbranched, angular, downy. Leaves alternate, oval or ovate, mostly rounded at the base, contracted at the apex to a short, slender point, finely toothed, 2 to 5 in. long, about half as much wide, glabrous above, sometimes downy beneath on part of the midrib and stalk; the stalk ½ in. or less in length. Flowers very small, whitish, produced in a terminal panicle 6 in. long, flower-stalks thickly covered with pale brown down. Fruits three-winged, ⅝ in. long and wide, the wings thin and membranous like the wings of an elm fruit, chocolate brown and yellowish. *Bot. Mag.*, t. 9488.

Native of S. and E. China, Burma and Formosa; discovered by Wilford, the Kew collector, in 1858. Forrest first met with it in Yunnan in 1906 and from his specimen a new species was described—*T. forrestii*—by which name the Forrest introductions were at one time known in gardens. It has, however, since been shown that *T. forrestii* is only a state of the somewhat variable *T. wilfordii* (see further in the *Botanical Magazine*).

Forrest sent seeds in 1913, but his F. 18255 from the 1917–19 expedition was more widely distributed; this was from a semi-scandent shrub only 3 to 6 ft high with purplish crimson fruits, found on the Nmai-Salween divide. He also sent seed in 1924 from plants climbing to 20 or 30 ft. In F. 24319 from the Shweli-Salween divide the flowers were 'grey yellow' and the fruits dull crimson; in F. 24753 from N.W. of Tengyueh the flowers were creamy white and the fruits brilliant brownish crimson.

The flowers and fruits portrayed in the *Botanical Magazine* were from a plant at Caerhays in Cornwall. But the species is hardy in S. England and received an Award of Merit in August 1952 for its bronzy crimson fruits when shown by Sir Frederick Stern from his garden at Highdown near Worthing.

TROCHOCARPA   EPACRIDACEAE

A genus of a few species of evergreen trees and shrubs, ranging from N. Borneo to Australia, with four species endemic to Tasmania. Leaves leathery, short-stalked, mostly alternate. Flowers sessile in short terminal or axillary spikes, each subtended by a bract and two bracteoles. Corolla cylindric or campanulate, with spreading lobes which are usually shorter than the tube. Stamens five, inserted on the corolla-tube and alternating with the lobes. Ovary ten-chambered; style one, with a small stigma. Fruit fleshy, containing an apparently single stone, which eventually separates into ten one-seeded nutlets (pyrenes).

The best known species and the type of the genus is T. LAURINA R. Br., a small tree sometimes cultivated in the temperate house and hardy in the Scilly Islands.

T. THYMIFOLIA (R. Br.) Spreng.

Decaspora thymifolia R. Br

An erect evergreen shrub to about 3 ft high in the wild; branchlets downy. Leaves stalked, closely set, alternate, ovate or elliptic, bluntly pointed, about ⅓ in. long, thick and slightly concave. Flowers in spring and often again in the autumn, borne in small, pendent terminal spikes. Corolla red, about ⅓ in. long, with five lobes equalling the tube in length. Filaments of stamens included; anthers yellow, slightly exserted. Fruits globose, fleshy, bluish. Curtis, *End. Fl. Tasmania*, Vol. II, No. 49.

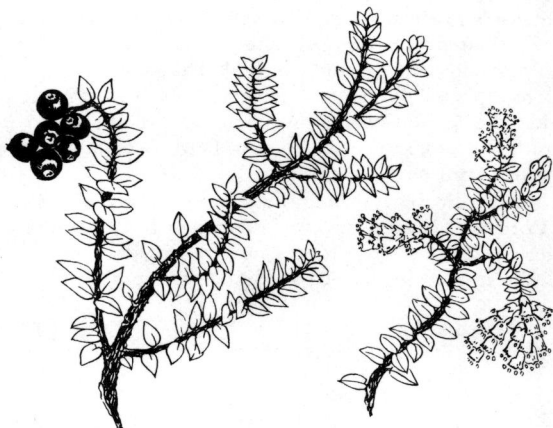

TROCHOCARPA THYMIFOLIA

Native of the mountains of Tasmania; in cultivation 1940. This attractive little shrub is proving quite hardy in a sheltered place, though late frost may damage it. It flowers well, often twice in the season, and needs a moist, acid gritty soil with leaf-mould and peat added.

TROCHODENDRON   TROCHODENDRACEAE

This is the sole member of the family to which it gives its name, and with the family Tetracentraceae constitutes the Order Trochodendrales. As in *Tetracentron* (q.v.) the structure of the wood is primitive, resembling that of the conifers. Flowers in an erect raceme-like terminal inflorescence, without sepals or petals (though the minute bracteoles at the apex of the pedicels may represent a much reduced perianth). Stamens numerous, their long filaments radiating from the base of a torus (expanded apex of the pedicel). Pistils sessile, in a single whorl, connate laterally, each with a sessile stigma and a single chamber containing numerous ovules, each pistil developing into a follicle dehiscing along the upper surface. The generic name literally means 'wheel-tree', from the manner in which the stamens radiate like the spokes of a wheel.

Euptelea, included in the Trochodendraceae by Dr Hutchinson, is by most botanists placed in a separate, also monotypic, family. On the other hand the genus *Paracryphia*, endemic to New Caledonia, is near to the Trochodendraceae.

T. ARALIOIDES Sieb. & Zucc.

An evergreen glabrous tree attaining a height of 60 to 80 ft in Japan. Leaves 3 to 5 in. long, narrowly oval or lanceolate, leathery, shallowly toothed at the upper end, lustrous green; leaf-stalks half the length of the blade. Flowers produced from April to June in erect, terminal racemes, each flower on a slender stalk 1 to 1½ in. long. There are no sepals or petals, and the numerous stamens are set round the edge of a green hemispherical disk, which is really the calyx-tube. Across the stamens the flower is ¾ in. in diameter. Carpels about ten, arranged in a ring within the stamens. *Bot. Mag.*, t. 7375.

Native of Japan from N. Honshu southwards, the Ryukyus and Formosa; also of Korea on Quelpaert Island (Cheju Do). It is a shrub or small tree in Britain, quite hardy if sheltered from cold winds, with handsome foliage recalling that of a tree-ivy. It is interesting when in bloom, the flowers being a vivid green. It was introduced from Japan by Messrs Veitch, in whose nursery at Coombe Wood it first flowered in 1894. It will grow in any good soil that is not excessively chalky.

Examples of this tree, which deserves to be more widely planted, are: East Bergholt Place, Suffolk, 30 × 2¾ ft at 2 ft (1972); Embley Park, Hants, 25 × 1½ ft (1971); Exbury, Hants, 25 × 1¼ ft (1970); Westonbirt, Glos., 26 × 2 ft (1971); Caerhays, Cornwall, 25 × 2¼ ft (1971).

TSUGA   Hemlock   pinaceae

A group of eight or ten evergreen trees of great beauty and elegance, represented on both sides of N. America, in China, Japan, and the Himalaya. They have very slender twigs, and short linear leaves, arranged, except in one species (*T. mertensiana*), mainly in two opposite ranks, each leaf seated on a cushion-like projection (as in *Picea*), and closely set on the twigs—twelve to twenty-four to the inch. They differ, however, from those of *Picea* in being always borne on a short but distinct petiole; in *Picea* they are sessile on the cushions. Cones solitary, rarely more than 1 in. long, and usually pendulous at the end of the twigs. Seeds winged. In places where they thrive, which is where the rainfall is abundant and the soil is deep and well-drained, they are not exceeded in beauty of form by any other evergreen trees. They are best propagated by means of seed; but the Japanese and Chinese species, perhaps the others also, can be propagated by cuttings.

T. canadensis (L.) Carr.   Eastern or Canada Hemlock
Pinus canadensis L.; *Abies americana* Mill.

A tree 70 to 100 ft high, with a trunk 6 to 10 ft in girth, and a head of often rounded form; bark reddish brown; young shoots bright grey, minutely hairy. Leaves very shortly stalked, $\frac{1}{4}$ to $\frac{2}{3}$ in. long, linear, but often broadest ($\frac{1}{16}$ to $\frac{1}{12}$ in.) near the rounded base, tapering thence to a bluntish point, margins toothed, dark green above, with a clear, well-defined band of stomata each side of the midrib beneath. The leaves are mainly in two opposite, spreading ranks, but there are also smaller leaves on the upper side of the branchlet pointing forward, flattened to the branchlet, often inverted and showing the white-lined lower surface. Cones $\frac{1}{2}$ to $\frac{7}{8}$ in. long, oval, borne on a short downy stalk; the scales broadly obovate, about as wide as long, minutely downy except on the exposed part.

Native of eastern North America from Nova Scotia south-west to the region of the Great Lakes, south through the eastern States to Alabama and Georgia, but confined to the mountains in the southern part of its range; introduced early in the 18th century. This beautiful tree thrives very well in the moister parts of our islands, especially where the soil is good and retentive of moisture. There are good specimens in the west of England, and I have also seen excellent ones at Murthly, and elsewhere in Perthshire. The tree as grown in this country has a strong propensity to branch into several stems near the ground, and to form a large rounded head of branches very distinct from the slenderly tapered form of *T. heterophylla*. From that species it is also distinct in the usually (not invariably) tapered leaf, in the much more clearly defined, whiter lines beneath, and in the cones being shortly stalked (stalkless in *T. heterophylla*). Visitors to Boston, Mass., have an opportunity of conveniently inspecting a primaeval wood of hemlock, which covers one of the hills in the beautiful Arnold Arboretum near that city. Here they have formed clean straight trunks, one of which (it may not have been the largest) I found in 1910 to be over 9 ft in girth.

The eastern hemlock grows more slowly in Britain than its western ally, *T. heterophylla*, and good specimens with a single bole are rare, the best being in Shropshire: Walcot Hall, 70 × 12 ft and 80 × 11¾ ft (1975); Oakley Park, 72 × 13½ ft (1971).

T. canadensis is lime-tolerant—more so than any other tsuga according to Messrs Hillier. For its use as a hedging plant see Donald Wyman, *Trees for American Gardens*, Ed. 2 (1965), pp. 458–9.

cv. 'ALBO-SPICA'.—Tips of the young shoots white. As usually seen in cultivation this is of compact habit and is quite striking when in full growth. According to van Ouden and Boom there may be more than one clone under this name.

cv. 'COLE' ('Cole's Prostrate').—A ground-hugging plant, attaining a spread of about 3 ft. Found in the wild in New Hampshire in 1931 by H. R. Cole.

cv. 'EVERITT GOLDEN'.—A slow-growing small tree of narrow habit and rugged outline. Young leaves golden-yellow, later yellowish green. Like 'Cole' this was originally found wild in New Hampshire. It is possible that plants now grown as 'Aurea' are this clone.

cv. 'FREMDII'.—A compact small tree of pyramidal habit, but spreading when old. Leaves crowded, some pointing out of the general plane. Found among seedlings by the American nurseryman C. Fremd and distributed in Europe by P. Koster of Boskoop. An example at Westonbirt, Glos., *pl.* 1939, measures 34 × 2¾ ft (1967).

cv. 'HUSSII'.—A slow-growing dwarf of irregularly branched, rugged habit. Branchlets congested, with very short, dark green leaves. Found in the USA about 1900.

f. MICROPHYLLA (Lindl.) Beissn. *Abies canadensis microphylla* Lindl.—Leaves shorter than normal. A common seed-bed variant. The plant named by Lindley was raised by Fisher and Holmes of Sheffield from American seed; it had minute leaves resembling those of *Daboecia cantabrica*.

f. MINUTA (Teuscher) Rehd. *T. canadensis* var. *minuta* Teuscher in *New Fl. and Sylv.*, Vol. 7 (1935), pp. 274–5 and fig. lxxxix—A minute form found in the Green Mountains of Vermont in 1927. Despite its small size it was fertile and surrounded by seedlings of the same habit, many of which were dug up and distributed. A named clone of this form is 'JERVIS', probably of independent origin.

cv. 'NANA'.—Of low, spreading habit, to 2 or 3 ft high, more in width. Leaves closely set. This is probably the *T. canadensis* var. *nana* of Carriere (1855).

cv. 'PROSTRATA'.—Described in the present work in Vol. III (1933), p. 482, as a perfectly prostrate plant, seen growing in the garden of Mr Renton of Branklyn, Perth, in 1931 (*T. canadensis* var. *prostrata* Bean). The origin of this plant was not stated.

cv. 'PENDULA'.—A name of uncertain application. Probably the plants now grown under this name are mostly 'Sargentii Pendula' or the other possibly

624   TSUGA

distinct pendulous form mentioned under it. But the epithet *pendula* was also
applied to trees with long pendulous ends to the branches (cf., A. D. Webster,
Hardy Coniferous Trees (1896), p. 132.

cv. 'SARGENTII PENDULA'.—A pendulously branched variant, lacking a
leading stem and forming a beehive- or dome-shaped bush, but procumbent
and wide-spreading if not trained up early in life. It was put into commerce by
Samuel Parsons of Flushing, New York, in 1879 as *Abies canadensis Sargenti
pendula*.

This hemlock is one of four found by General Howland in the Fishkill
Mountains above the Hudson River in the late 1860s. One he kept for himself;
the other three he gave to his neighbour Henry Winthrop Sargent, who in turn
planted one in his arboretum at Wodenethe, giving one to his cousin by marriage
H. H. Hunnewell of Wellesley, Mass., and the other to his young cousin
Charles Sprague Sargent, the future Director of the Arnold Arboretum, who
had taken charge of his father's garden near Boston after his return from
Europe in 1868 (see further below).

The name 'Sargentii Pendula' commemorates H. W. Sargent, from whom
Parsons is known to have obtained propagating material (Stout, *Journ. New York
Bot. Gard.*, Vol. 40 (1939), pp. 153–166). Parsons evidently thought highly of
this hemlock, for he wrote to *The Garden* in 1875 announcing it as a forthcoming
novelty, and in 1887 sent a photograph to the same periodical, from which an
engraving was made and published (*Garden*, Vol. 32 (1887), p. 363).

The plant given to C. S. Sargent was placed in the family garden at Holm Lea,
Brookline, and by about 1923 had made a flat-topped bush 6 ft high and 23 ft
across (E. H. Wilson, *Gard. Chron.*, Vol. 75 (1924), p. 107 and fig. 43). The Holm
Lea estate was broken up after Prof. Sargent's death in 1927, but the plant still
exists in a private garden. He gave an account of the four Howland hemlocks in
Garden and Forest, Vol. 10 (1897), p. 491, and Kent's description of var. *sargentiana*
is taken from this article, where it appears to be the Holm Lea plant that Sargent
describes (*Veitch's Manual of the Coniferae*, Ed. 2, pp. 465 and 466). But Kent
clearly intended that the epithet *Sargentiana* should commemorate H. W. Sargent,
who distributed General Howland's discoveries, and to use the cultivar name
'Sargentiana' for the Holm Lea plant would cause confusion. The name
'BROOKLINE' has been proposed for it (H. J. Welch, *Dwarf Conifers* (1966),
p. 323). General Howland's own plant survives, and there is still a large specimen
in the remains of H. W. Sargent's once famous arboretum (*Gard. Chron.*, Vol. 141
(1957), p. 188).

Although most plants in Europe are probably the true clone 'Sargentii
Pendula', seedlings of it have also been distributed. Some cultivated plants are
narrow-topped, with steeply pendulous branches, and it is possible that these
are a different clone, of European origin.

T. CAROLINIANA Engelm.   CAROLINA HEMLOCK

A tree usually 50 to 80 ft high; young shoots glossy, pale brown, downy
on the upper surface. Leaves linear, ¼ to ¾ in. long, mostly of uniform width,

rounded, and sometimes slightly notched at the apex, margins not toothed, dark green above, with two bands of stomata beneath. The lower leaves spread in two opposite ranks, but the shorter ones on the upper side of the branchlets are more or less erect. Cones shortly stalked, 1 to 1½ in. long; scales oblong, considerably longer than wide.

Native of the south-eastern USA, where it has a limited range in the Appalachians from Virginia to Georgia. It is easily distinguished from both *T. canadensis* and *T. heterophylla* by its entire leaf-margins and reddish brown shoots, and from the former, with which it is more likely to be confused, by the elongate cone-scales, spreading widely when the cone is ripe. See also *T. diversifolia* and *T. sieboldii*.

T. caroliniana was introduced to Kew from the Arnold Arboretum in 1886. It has never succeeded really well in Britain, although capable of withstanding severe cold. In the United States it is valued almost as much as the eastern hemlock, both as a specimen and for hedges and topiary work, and is said to be more tolerant of a polluted atmosphere.

Examples of *T. caroliniana* are: National Pinetum, Bedgebury, Kent, *pl.* 1925, 32 ft (1970); Wakehurst Place, Sussex, south of Westwood Lake, 42 × 4 ft (1973); R.H.S. Garden, Wisley, 30 × 2¼ ft (1969); Hollycombe, nr Liphook, Hants, 42 × 4 ft (1974).

T. CHINENSIS (Franch.) Pritzel

Abies chinensis Franch.; *Tsuga brunoniana* var. *chinensis* (Franch.) Mast.

A tree up to 80 ft high in China, occasionally over 100 ft high; young shoots light brown, finely downy, especially in the grooves. Leaves pectinately arranged, up to about ¾ in. long, entire (minutely toothed on seedling plants), parallel-sided, slightly notched at the apex, with two usually green stomatic bands beneath. Cones broadly ovoid, with light brown, lustrous, almost orbicular scales. *Bot. Mag.*, t. 9193.

Native of central and western China (a variety in Formosa); first seen by Père David near Mupin in W. Szechwan; described from specimens collected by Père Farges in N.E. Szechwan; introduced by Wilson for Messrs Veitch in 1902, from Hupeh. It was at first confused with *T. sieboldii*, from which it differs in its downy stems; and also with *T. yunnanensis*, see below.

T. chinensis is not in general cultivation. The best examples are: Bodnant, Denbigh, 48 × 5 ft (1974) and National Botanic Garden, Glasnevin, Eire, 33 × 3¾ ft (1974).

T. FORRESTII Downie—Although included in *T. chinensis* by Wilson this species appears to be distinct enough in its brown or reddish brown branchlets, longer leaves (to 1 in. long) and tapered cones. It was described from a specimen collected by Forrest in the Lichiang range in 1918 and was subsequently found by him in other parts of N.W. Yunnan and in bordering S.W. Szechwan (*Notes Roy. Bot. Gard. Edin.*, Vol. 14 (1923), p. 18 and Vol. 18 (1933), p. 136). A plant at Borde Hill in Sussex may belong here, and those propagated by Messrs Hillier derive from it.

T. DIVERSIFOLIA (Maxim.) Mast.   NORTHERN JAPANESE HEMLOCK

Abies diversifolia Maxim.; *T. sieboldii* var. *nana* Carr.; *Abies tsuga* var. *nana* Endl.

A tree 70 to 80 ft high in Japan, with a trunk up to 6 or 7 ft in girth; young shoots downy, becoming reddish brown. Leaves ¼ to ⅝ in. long, linear, of uniform width, margins not toothed, distinctly notched at the apex, dark glossy green above, with two clearly defined white lines of stomata beneath. Cones egg-shaped, ½ to ¾ in. long.

A native of Japan, where it attains a large size in the mountains of central and northern Honshu; introduced by John Gould Veitch and subsequently distributed by his firm as "*Abies Tsuga nana*", a name by which it was long known in gardens. It is at once distinguished from *T. sieboldii* by its closer habit and downy more highly coloured shoots, its shorter leaves, and by always starting into growth earlier in the spring. Its cones are somewhat smaller and egg-shaped (elongate and more tapered in *T. sieboldii*). In both species the leaves are notched at the apex, which helps to distinguish them from *T. caroliniana*.

With us *T. diversifolia* grows slowly and remains more of a shrub than a tree for many years. Because of its dainty habit it makes a very pleasing lawn plant, especially in spring, while the young shoots are still bright yellow-green. It has, however, remained rare and is mostly to be seen in collections. Examples of known planting date are: National Pinetum, Bedgebury, Kent, *pl.* 1926, 26 × 2¼ ft (1969); Little Hall, Canterbury, Kent, *pl.* 1906, 40 × 3¼ ft (1961); Hergest Croft, Heref., *pl.* 1916, 30 × 3 ft (1961). Other examples are: Kew, 34 × 3 ft (1973); Wakehurst Place, Sussex, 34 × 3½ ft (1966).

T. DUMOSA (D. Don) Eichl.   HIMALAYAN HEMLOCK

Pinus dumosa D. Don; *T. brunoniana* (Wall.) Carr.; *Pinus brunoniana* Wall.

A tree 120 ft high in the wild, of cedar-like habit, with spreading branches pendulous at their extremities; young shoots downy on the upper side. Leaves linear, ½ to 1⅓ in. long, 1/16 to 1/12 in. wide, tapered at the apex, shortly stalked, minutely toothed, dark green above, with the midrib deeply sunk, the lower surface silvery white, being almost entirely covered with stomata. Cones not stalked, egg-shaped, ¾ to 1 in. long; scales roundish, downy at the base outside.

Native of the Himalaya from Kumaon eastward, extending into S.E. Tibet, Burma and probably China; according to Loudon it was introduced in 1838 but there were many other sendings in the next few decades. In the interior of Sikkim and bordering parts of Nepal it reaches a huge size in the moister type of forest—Hooker found it 120 ft high and 28 ft in girth. Needing a long frost-free growing season, it is only adapted to the milder parts of the British Isles, and even there large trees are rare. The specimen at Boconnoc in Cornwall, mentioned by Elwes and Henry in 1906, measured 77 × 15¾ ft in 1957 but has since died. Of the other examples mentioned in previous editions the tree at Dropmore in Buckinghamshire is dead, but the specimen at Fota, Co. Cork, still exists; planted in 1855 it measures 62 × 9½ ft (1966). Other trees measured recently are: Hergest Croft, Heref., 38 × 3 ft (1961); Rowallane, Co. Down, 48 × 6¾ ft (1976); Powerscourt, Co. Wicklow, Eire, 52 × 5¼ ft (1975); Mount Usher, Co. Wicklow, Eire, 51 × 6¼ ft (1975).

TSUGA DUMOSA

Near London *T. dumosa* rarely reaches tree size, but is sometimes seen as an attractive shrub. A two-stemmed tree at Borde Hill in Sussex measures 30 × 2½ ft (1958).

T. YUNNANENSIS (Franch.) Mast. *Abies yunnanensis* Franch.—Little is known in cultivation of this ally of *T. dumosa*, described from specimens collected in Yunnan by the French missionary Delavay and also occurring in W. Szechwan, where Wilson saw it but did not send seed, so far as is known. A plant at Borde Hill, Sussex, in Warren Wood, was identified by A. B. Jackson as belonging to this species. It was raised from Forrest 10293 and measures 35 × 3½ ft (1974). Forrest's field-specimen under 10293 is *T. chinensis*, but the plant in question is definitely not that species.

T. HETEROPHYLLA (Raf.) Sarg.   WESTERN HEMLOCK

Abies heterophylla Raf.; *T. albertiana* (A. Murr.) Sénécl.; *Abies albertiana* A. Murr.; *T. mertensiana* Carr., in part; *Pinus mertensiana sens.* Parl., not Bong.

A tree up to 230 ft high in its native forests, with a slender, lightly branched crown; trunk cylindrical, rarely exceeding 15 ft in girth; bark furrowed, reddish brown; young shoots downy. Leaves shortly stalked, set all round the branchlet, although more crowded on the upper side, linear, curved, $\frac{1}{2}$ to 1 in. long, $\frac{1}{20}$ to $\frac{1}{16}$ in. wide, rounded at the apex, margins not toothed, sometimes grey-green, sometimes conspicuously blue-green. There are inconspicuous lines of stomata on both surfaces. Cones without stalks, rich purple when young, becoming red-brown; oval-cylindric, $1\frac{1}{2}$ to 3 in. long, $\frac{1}{2}$ to $\frac{3}{4}$ in. thick.

Native of western N. America, with its main area on or near the Pacific from Alaska to northernmost California, but also occurring in the Rocky Mountains from S.E. British Columbia to N. Idaho and N.W. Montana. It was described in 1832 from a specimen collected during the pioneering trans-continental journey of Lewis and Clark; and introduced to Scotland by John Jeffrey in 1852 for the Oregon Association. He sent only a small bag of seed, received in August 1852, but larger quantities were sent by the British Columbia Expedition in 1861–2.

Few conifers combine elegance and usefulness to such a degree as the western hemlock. In the right conditions it gains height at the rate of 2 to 3 ft a year and quickly makes a beautiful specimen when not crowded, but one needing plenty of room owing to the wide spread of its lower branches. It also yields timber of good quality and has been used in Britain to some extent as a plantation tree, especially since the Second World War. Like its companions in the forests of western N. America it likes a humid climate, and most of the finest specimens in the British Isles are to be found in the Atlantic zone, but it succeeds wherever the soil is not chalky nor inordinately heavy. The Lower Greensand seems to suit it very well, and there are three notable specimens on this formation in the Haslemere district, where the rainfall is higher than the average for S. England. It is tolerant of shade, and was used for under-planting of decrepit deciduous woodland until it was found to be subject to the butt-rot *Fomes annosus* when grown in these conditions.   [PLATE 88

The following are some of the many fine specimens measured recently by Alan Mitchell: Scone Palace, Perth, *pl.* 1866, 126 × 18¼ ft (1974); Dupplin Castle, Perths., *pl.* 1859, 134 × 12½ ft (1970); Murthly Castle, Perths., *pl.* 1860, 139 × 12¼ ft (1970); Abercairney, Perths., *pl.* 1864, 120 × 11¼ ft (1974); Dawyck, Peebl., *pl.* 1860, 120 × 11½ ft (1970); Strone, Argyll, *pl.* 1875, 124 × 14¾ ft (1976); Benmore, Argyll, 152 × 13¼ ft (1976); Cragside, Northumb., 138 × 12¾ ft + 12¼ ft (1974); Bodnant, Denb., *pl.* 1887, 131 × 12¾ ft (1974); Hafodunos, Denb., *pl. c.* 1856, 120 × 14¼ ft (1978); Drum Manor, Co. Tyrone, 108 × 18¾ ft (1976); Stourhead, Wilts, *pl.* 1871, 130 × 12¾ ft (1976); Knightshayes, Devon, 102 × 12¼ ft (1970); Honeyhanger, Shottermill, Haslemere, 92 × 13 ft (1971); Grayswood Hill, Haslemere, *pl.* 1881, 85 × 14 ft (1970); Lythe Hill, Haslemere, 112 × 12½ ft (1969).

T. MERTENSIANA (Bong.) Carr.   MOUNTAIN HEMLOCK

Pinus mertensiana Bong.; *Abies pattonii* Jeffrey, *nom. inedit.*; *A. pattoniana* A. Murr.;
A. hookeriana A. Murr.; *Tsuga pattoniana* (A. Murr.) Engelm.; *T. hookeriana* (A. Murr.)
Carr.; *Hesperopeuce mertensiana* (Bong.) Rydb.; *T. crassifolia* Flous; × *Tsugo-Picea
hookeriana* (A. Murr.) Campo-Duplan & Gaussen; × *Tsugo-Picea crassifolia* (Flous)
Campo-Duplan & Gaussen   [PLATE 87

A tree 70 to over 100 ft high, the trunk 12 ft or more in girth; bark red-
brown; young shoots downy. Leaves shortly stalked, set all round the branchlet,
although more crowded on the upper side, linear, curved, $\frac{1}{2}$ to 1 in. long,
$\frac{1}{20}$ to $\frac{1}{16}$ in. wide, rounded at the apex, margins not toothed; sometimes grey-
green, sometimes conspicuously blue-green. There are inconspicuous lines of
stomata on both surfaces. Cones without stalks, rich purple when young,
becoming red-brown, oval-cylindric, $1\frac{1}{2}$ to 3 in. long, $\frac{1}{2}$ to $\frac{3}{4}$ in. thick.

Native of western N. America from S. Alaska south through the coastal ranges
to N. California, where it is confined to the Sierra Nevada; there is also a less
extensive area in the northern Rocky Mountains. The western hemlock,
T. heterophylla, has a similar distribution but reaches its maximum development
in the coastal ranges at altitudes of up to 2,000 ft, while the mountain hemlock
does not descend below 3,000 ft except in Alaska and ascends into the subalpine
zone, where it is sometimes a stunted shrub. It was discovered in 1827 on
Baranof Island, Alaska, by the German botanist K. Mertens during the Russian
circumnavigation of the globe led by Capt. Luetke. In 1851 John Jeffrey sent
seeds from Mt Baker in Washington to the Oregon Association but they
germinated poorly and the only recorded tree from this sending was a hybrid
(see *T. × jeffreyi* below); he sent further seed in the following year from the
Cascades around 42°, and it was from a coning specimen in this consignment
that Andrew Murray described *Abies pattoniana*, naming it in honour of George
Patton of The Cairnies (later Lord Glenalmond), who was the originator of the
Oregon Association—a group 'of Gentlemen interested in the promotion of the
Arboriculture and Horticulture of Scotland' (see *Notes Roy. Bot. Gard. Edin.*,
Vol. 20 (1939), pp. 1–53). But apparently no trees from the 1852 seed have
survived and the introduction of the species is usually attributed to William
Murray, who sent seed from Mt Scott in Oregon in 1854, while collecting for
Messrs Lawson of Edinburgh. Confusingly, his brother Andrew Murray made a
new species of this collection, calling it *Abies hookeriana*.

T. mertensiana is remarkable for its radially arranged leaves with stomata on
both sides and its large cones. Engelmann gave it generic rank and more recently
some French investigators have advanced the theory that it is the result of
hybridisation between *T. heterophylla* and *Picea sitchensis*, and that the minor
variant named *T. crassifolia* by Mme Flous is a hybrid between the mountain
hemlock and *Picea engelmannii*, thus having the western hemlock and two spruces
in its genetic make-up.

The colour of the foliage of *T. mertensiana* is usually of a glaucous hue, but
of varying shades, which it is pointless to distinguish varietally. But plants
in which the leaves are silvery glaucous could take the name f. ARGENTEA
(Beissn.) Rehd.

In all its forms the mountain hemlock is remarkably beautiful. It needs a moist climate and a pure atmosphere. At Murthly Castle, near Perth, there is a group of several trees (one with pendulous branchlets) which makes one of the most beautiful garden pictures one can imagine, produced by foliage alone. The largest of these trees are: *pl.* 1862, 94 × 8¼ ft; *pl.* 1863, 85 × 8½ ft and, date unrecorded, 84 × 10¾ ft (1970). Among other Scottish trees are: Blair Atholl, Perths., *pl.* 1872, 102 × 7½ ft (1970); Fairburn, Ross and Cromarty, 102 × 7½ ft (1970). At Patterdale Hall, Westmorland, there is a broad, silvery grey bush measuring 42 × 7¾ ft (1976). Examples in eastern and southern England are: Fulmodestone, Norfolk, 72 × 5¼ ft (1969); Tilgate, Sussex, *pl.* 1905, 45 × 6½ ft (1964); National Pinetum, Bedgebury, Kent, an example of f. *argentea*, *pl.* 1925, 24 × 2¼ ft (1969).

T. × JEFFREYI (Henry) Henry *T. pattoniana* var. *jeffreyi* Henry; × *Tsugo-Piceo-Tsuga jeffreyi* (Henry) Campo-Duplan & Gaussen—As noted above, Jeffrey sent seed of *T. mertensiana* in 1851 from Mt Baker, where *T. heterophylla* also occurs. A tree in the Edinburgh Botanic Garden, raised from this batch was anomalous in having flatter, shorter, relatively broader, finely serrated leaves, more pectinately arranged, the exposed side lighter green, with fewer stomata. Henry at first gave this tsuga varietal rank but later concluded that it was a hybrid between *T. mertensiana* and *T. heterophylla*. Trees under the name *T.* × *jeffreyi* are in cultivation, but their provenance is unknown. They are very near to *T. mertensiana* but with greener leaves more pectinately arranged.

Intermediates or hybrids between the western and mountain hemlocks have been observed in the wild where the two are in contact, and are said to be common in parts of Washington, east of the Cascades (Hitchcock et al., *Vasc. Pl. Pacif. Northwest*, Part I (1969), p. 133). They have also been observed on Mt Baker, whence came the type of *T.* × *jeffreyi*.

T. SIEBOLDII   TSUGA, JAPANESE HEMLOCK
Abies tsuga Sieb. & Zucc.; *T. araragi* Koehne

A tree up to 100 ft high in Japan, with a trunk 9 ft or more in girth; but only a small bushy tree with us, although a very elegant one; young shoots perfectly glabrous. Leaves linear, of uniform width, ⅓ to 1 in. long, 1/16 to 1/10 in. wide, rounded and distinctly notched at the apex, not toothed, abruptly narrowed at the base to a short stalk; rich glossy green above, with two clearly defined white lines of stomata beneath. Cones ¾ to 1 in. long, egg-shaped; scales rounded.

Native of Japan and of one island off the coast of Korea; introduced to Europe by Siebold about 1853. J. G. Veitch brought back cones in 1861, mixed with those of *T. diversifolia*, but it is uncertain whether his firm ever distributed plants from that source. It is an important constituent of the coniferous forests of Japan from central Honshu southwards, while its ally *T. diversifolia* has a more northern distribution and ascends to higher altitudes; see that species for the points of difference. Although slow-growing, and not making a large tree in this country, the grace and beauty of Siebold's hemlock makes it well worth growing. For all that, it is rarely seen outside collections. Examples measured recently are: National Pinetum, Bedgebury, Kent, *pl.* 1926, 40 × 3 ft and 38 × 3½ ft (1969); Wakehurst Place, Sussex, 45 × 3 ft (1964); Leonardslee, Sussex, 43 × 2 ft (1969); Lydhurst, Sussex, 54 × 4½ ft (1971); Bodnant, Denb., 35 × 3½ ft (1959).

TSUSIOPHYLLUM ERICACEAE

A genus of a single species allied to *Rhododendron* but with the anthers opening by slits instead of pores and the ovary three-celled (five or more in *Rhododendron*, very rarely four). It is even closer to *Menziesia* but the three-celled ovary and the hairy outer surface of the corolla distinguish it. The generic name refers to the appressed strigose hairs on the leaves, resembling those of *Rhododendron* section *Tsusia* Planch. *emend.* Maxim. (series *Azalea* subseries *Obtusum*).

T. TANAKAE Maxim.

Rhododendron tanakae (Maxim.) Ohwi, not R. *tanakai* Hayata; R. *tsusiophyllum* Sugimoto

A spreading semi-evergreen shrub to about 1½ ft high, its short, whorled branchlets clad with appressed bristly hairs. Leaves ovate to lanceolate or oblanceolate, up to ¾ in. long and ¼ in. or slightly more wide, clustered at the ends of the branchlets, appressed-hairy on both sides, more so above. Flowers white, shortly stalked, borne in June in terminal umbellate clusters or sometimes solitary; bud-scales brown and papery. Calyx very small. Corolla tubular, about ⅜ in. long, and about ¼ in. wide across the spreading five-lobed limb, downy on the outside, silky-hairy within. Stamens five, included. Fruit a capsule, resembling that of a rhododendron but with three-chambers only.

Native of Japan, confined to a few localities in Honshu; described by Maximowicz shortly after its discovery; introduced by Wilson in 1915. It is hardy and needs the same conditions as other ericaceous shrublets, but is one of the least interesting of these. It received an Award of Merit in 1965. Propagated by cuttings.

ULEX LEGUMINOSAE

A genus of very spiny shrubs allied to the brooms, but differing in having the calyx as well as the petals yellow. Only three species, all natives of Britain, are worth cultivating; several others, mostly found in Spain and Portugal, are too tender to be of any value. The leaves are small and spine-tipped, often reduced to mere prickles; and all the species have the quality of evergreens, from the dark green of their spines and branches.

In gardens they are often useful for covering dry sunny banks or breadths of poor gravelly soil, where most shrubs would not thrive. In such places the double-flowered variety of *U. europaeus* is particularly effective in spring. The two other species have a value in flowering in late summer and autumn. None of them will thrive in shade, and they are never satisfactory in rich soil; in either case flowers will be sparsely borne, and the plants apt to get lank and ungainly. Where the soil is of good quality it is advisable not to dig it over when planting, with the view of keeping it as hard as possible. Propagation by cuttings is referred to under the notice of *U. europaeus* 'Plena'. Seeds should be sown singly in small pots and the plants put in their permanent places at their first planting, for they transplant badly. The common gorse should be sown *in situ*.

U. EUROPAEUS L.   GORSE, FURZE, WHIN

A shrub usually 2 to 4 ft high as seen wild, but occasionally 6 ft or even more high; excessively spiny. The main branches are hairy, and from them spring numerous short side branches which grow horizontally and always end in a stout, sharp spine, the whole forming an intricate formidable mass. Leaves simple, $\frac{1}{4}$ to $\frac{1}{2}$ in. long, linear, sharply pointed or reduced to mere spines. Flowers produced singly from the leaf-axils of the previous year's shoots, on hairy stalks $\frac{1}{4}$ in. long, transforming the end of the branch into a brilliant raceme of gold. Calyx large, hairy, yellow like the petals, persistent. Pods $\frac{5}{8}$ in. long, covered with brown hairs, two- or three-seeded.

Native of W. and Central Europe, and abundant in the British Isles, where it covers thousands of acres of moor, common, and heath. Whilst April and early May is the time when the gorse is in its full beauty, it starts flowering in February, and odd flowers may be found at almost all times—a characteristic on which is based the country saying, 'When furze is out of bloom, then is kissing out of fashion.' There is but little use for gorse in gardens. It may be employed for covering dry banks, but even there the double-flowered variety described below is much to be preferred.

Dried furze was at one time much used in country places for heating bakers' ovens, and is still often woven into hurdles for sheltering cattle, forming a good wind screen, not rubbed against or soon knocked down. Owing to the amount of dead twigs and spines inside the outer living layer, gorse plants are very inflammable during hot summer spells. Gorse, therefore, even the double-flowered variety, should not be planted where its firing would be a source of danger to buildings or even to valuable trees.

The common furze has also been grown to some extent as a field-crop to provide winter fodder for cattle and horses, and was then known as 'meadow furze'. The seed was undersown to a cereal crop and cut in the autumn of the following year and usually every second year thereafter. In the last century its use seems to have been largely confined to Ireland and the hill-farms of the west, but production was large enough for at least two firms to manufacture special crushers; to prepare the gorse for horses an ordinary chaff-cutter sufficed.

cv. 'FLORE PLENO' ('Plenus'). DOUBLE-FLOWERED GORSE.—In this form the stamens disappear from the flower, either partially or entirely, and are replaced by petals of varying size (*U. europaeus* var. *flore-pleno* G. Don; *U. e.* f. *plenus* Schneid.). The variety does not produce seed, and must be propagated by cuttings, which should be made of the current season's wood in August, 3 or 4 in. long, and placed in a cold frame in very sandy soil, and kept close. They will root and grow the following spring. When the roots are 1 in. long, the young plants should be potted in 3-in. pots, ready for planting out whenever required the following winter. It is slower-growing and more compact in habit than the type, and is in every way superior to it as an ornamental shrub for gardens, lasting longer in flower. Like common gorse, it needs a dry, hungry soil and a sunny position to develop its full beauty. In rich soil it grows rank and does not flower so freely. This variety first appeared in the nursery of John Miller of Bristol, about 1828. [PLATE 92

cv. 'STRICTUS'. IRISH GORSE.—Of erect, rather columnar growth, growing 6 to 10 ft high in its original form, mildly armed (*U. strictus* Mackay; *U. europ.* var. *strictus* (Mackay) D. A. Webb; *U. hibernicus* Hort.; *U. fastigiatus* Hort.). This was found early in the 19th century on the estate of the Marquess of Londonderry in Co. Down; similar plants have been noticed in Spain. Flowering sparsely, it is of little value in gardens, though it was used to some extent as a hedge. Being scarcely armed it would have been more suitable as a forage crop than the normal form, were it not that it had to be raised from cuttings. Seed sold as 'Irish Furze' was not 'Strictus' but common gorse, imported from Ireland.

U. GALLII Planch.

A dwarf, sturdy bush allied to *U. minor*, and by some writers made a variety of it. In general appearance, however, it more resembles *U. europaeus*, having the same hairy branches and stout spiny branchlets, but it is much dwarfer, usually under 2 ft. The flowers, each $\frac{5}{8}$ in. long and bright yellow, are borne from August to October; the wing-petals are curved and longer than the keel, the calyx finely downy. Pods $\frac{1}{3}$ to $\frac{1}{2}$ in. long, one- or two-seeded.

Native of W. Europe, and abundant in the southwest of England, where it makes (often in company with *Erica cinerea* and *E. tetralix*) most charming displays in autumn. The heaths of Dorset, covered with these plants, will be familiar to many. It is in some respects intermediate between *U. europaeus* and *U. minor*, resembling the former in its branches, but the latter in time of flowering and in the absence of hairs from the calyx. It is not so hardy as either of them, but apparently withstands all except the very hardest winters at Kew, especially when the plants are a few years old. In gardens it is scarcely known—being confused with *U. minor*—but is very pretty planted in poor soil, especially if associated with the two heaths just named.

U. MINOR Roth   DWARF GORSE
U. nanus T. F. Forster ex Symons

A dwarf shrub of dense, close habit, sometimes procumbent in the wild, but changing its character when introduced to the garden, and sending up slender, erect branches, 1 to 2 ft long in a single season; branchlets hairy. Leaves and branchlets as in *U. europaeus*, only smaller and less rigid. Flowers golden yellow, about half the size of common gorse, the calyx not hairy but slightly downy; wing-petals straight, shorter than the keel. Pods ½ in. long, hairy.

Native of W. Europe, and abundant in many parts of Great Britain. Although some botanists profess to find it a variety of *U. europaeus*, it is really very distinct. It is autumn-flowering, being at its best in September, when the ordinary gorse is in seed. This is its most valuable characteristic in gardens, for its long, slender stems set with flowers are often very pretty when few other shrubs are in blossom. But it needs a poor dry soil to develop its greatest beauty. In rich garden soil it gets to be 6 ft high, and very lanky.

ULMUS   ELM   ULMACEAE

A genus of between twenty and thirty species of deciduous (rarely subevergreen) trees, some of the largest size, natives of the northern hemisphere in both the Old and New Worlds, but not occurring in N. America west of the Rocky Mountains. Leaves alternate, toothed, usually oblique and sometimes very unequal at the base. Flowers produced in clusters or short racemes from axillary buds, either on the naked shoots in early spring, or on the leafy ones in autumn. They are very small, consisting of a green or red-tinged perianth with four to nine (usually about five) lobes. Stamens usually equal in number to the perianth-lobes, their purplish or reddish anthers being usually the most conspicuous feature of the flower. The fruit is most characteristic, being a flat, membranous, semi-transparent disk called a samara, enclosing the single seed in a cavity at the centre or towards the apex, where it is slightly or deeply notched.

On a healthy, unlopped adult tree the greater part of the foliage is borne on short-shoots of determinate growth, and the leaves on these, in particular the upper ones (distal and subdistal leaves) characterise the species, variety or hybrid. Leaves on suckers and water-sprouts (epicormic shoots) are of a juvenile type, and often very different from the adult ones, not only in shape and size but also in indumentum, being often

hairy even when the adult leaves are glabrous. Also uncharacteristic are
the leaves of lammas-shoots and on 'proliferating' shoots.

CLASSIFICATION

The genus has been grouped into sections largely on the basis of
flowering-time, and on characters of the perianths, inflorescence and
samaras. Foliage does not serve to distinguish the sections reliably.

sect. ULMUS (Madocarpus).—Flowers short-stalked, borne in clusters
before the leaves expand. Perianth with four to seven equal lobes.
Samaras rarely ciliate. The most important group. Apart from the
N. American *U. rubra* all are natives of the Old World. Those treated
here are:
Europe, S.W. and C. Asia: *U. carpinifolia* (also in N. Africa) and
U. angustifolia, *U. coritana*, *U. plotii*, included in *U. carpinifolia* by some
authorities; *U. elliptica*; *U. glabra*; *U. procera*; *U. pumila* (also in N.E. Asia
and the inner Himalaya).
Himalaya: *U. villosa*; *U. wallichiana* (for other Himalayan elms see
Melville and Heybroek, op. cit. in Bibliography).
E. Asia: *U. davidiana* and the closely allied *U. japonica*; *U. laciniata*;
U. macrocarpa; *U. wilsoniana*.

sect. BLEPHAROCALYX.—Flowers on slender stalks of unequal length,
opening before the leaves expand. Perianth deeply divided. Samaras
ciliate.
U. americana (N. America); *U. laevis* (W. Eurasia).

sect. CHAETOPTELEA.—Flowers before the leaves, on slender stalks,
arranged in racemes. Samaras downy and ciliate.
Two N. American species: *U. alata*; *U. thomasii*.

sect. MICROPTELEA.—Flowers opening in autumn. Perianth deeply
divided. Leaves rather leathery, falling late. Samaras glabrous.
U. parvifolia (E. Asia); *U. crassifolia* (N. America).

sect. TRICHOPTELEA.—Flowers opening in autumn as in the preceding section, but borne in pendulous racemes. Samaras hairy.
U. serotina (N. America).

The present revision has been prepared under the shadow of the new
and lethal outbreak of the Dutch elm disease, which has killed many of
the elms that ornamented the parks and gardens of this country and, far
worse, destroyed or defaced some of our loveliest rural landscapes. The
contents of the present treatment are substantially as they were in previous
editions, though it must be sadly acknowledged that many of the elms
described, including some of the largest and finest, will never be planted
again, and it is only to be hoped that resistant substitutes will eventually
become available. The taxonomy of the British elms of the *U. carpinifolia*
complex is a controversial matter, but it is one that in the main concerns
countryside elms rather than those planted for ornament. The taxonomic

position of a cultivar is also a matter of dispute in many cases, but controversy can here usually be avoided by placing the cultivar-name directly under *Ulmus* (cf. P. S. Green, op. cit., in Bibliography). For example, the name of the Jersey elm—'Sarniensis'—could be placed under *U. minor* (if that is to be taken as the earliest name for the field elm), *U. carpinifolia*, *U. angustifolia* or *U.* × *sarniensis*. This instability of nomenclature is avoided if it is called *Ulmus* 'Sarniensis'. The same device can be used in presenting the names of the many probably hybrid cultivars that have in the past been placed under *U. procera* or *U. carpinifolia*, or confusingly shuttlecocked between them. See also *U.* × *hollandica*.

SELECT BIBLIOGRAPHY

BANCROFT, HELEN.—'Notes on the Status and Nomenclature of British Elms', *Gard. Chron.*, Vol. 96 (1934), pp. 122, 139, 208, 244, 298, 334, 372.
CLAPHAM, A. R., TUTIN, T. G., and WARBURG, E. F.—*Flora of the British Isles*, ed. 1 (1952), pp. 715–24.
ELWES, H. J., and HENRY, A.—*Trees of Great Britain and Ireland*, Vol. VII (1913), pp. 1847–1929.
FONTAINE, F. J.—*Dendroflora* No. 5 (1968), pp. 37–55.
GREEN, P. S.—'Registration of Cultivar Names in *Ulmus*', *Arnoldia*, Vol. 24 (1964), pp. 41–80.
HADFIELD, MILES.—*British Trees* (1957), pp. 226–47.
HILLIER AND SONS.—Hillier's Manual of Trees and Shrubs, Ed. 3 (1973), pp. 398–402.
JACKSON, A. B.—'The British Elms', *New Flora and Sylva*, Vol. 2 (1930), pp. 219–29.
JOBLING, J., and MITCHELL, A. F.—*Field Recognition of British Elms*. Forestry Commission Booklet 42 (1974).
MEIKLE, R. D.—*British Trees and Shrubs* (1958), pp. 148–55.
MELVILLE, R.—*Ulmus*, in Stace, ed., *Hybridisation and the Flora of the British Isles* (1975), pp. 292–9.
Other studies by Dr Melville are cited in the text.
MELVILLE, R., and HEYBROEK, H. M.—'The Elms of the Himalaya', *Kew Bulletin*, Vol. 26 (1971), pp. 5–28.
MITCHELL, A. F.—*A Field Guide to the Trees of Great Britain and Northern Europe* (1974), pp. 247–54 and pl. 22.
MOSS, C. E.—'British Elms', *Gard. Chron.*, Vol. 51 (1912), pp. 199, 216, 234.
—— *The Cambridge British Flora* (1914), Vol. II, pp. 88–96.
RICHENS, R. H.—'Studies on *Ulmus*'. A series of regional studies: Part I (E. Anglia), *Watsonia*, Vol. 3 (1958), pp. 138–53; Part II (S. Cambs), *Forestry*, Vol. 31 (1959), pp. 132–46; Part III (Herts), ibid., Vol. 32 (1959), pp. 138–54; Part IV (Hunts), ibid., Vol. 34 (1961), pp. 47–64; Part V (Beds), ibid., Vol. 34 (1961), pp. 181–200; Part VI (Fenland), ibid., Vol. 38 (1965), pp. 225–35; Part VII (Essex), ibid.,

ULMUS 637

Vol. 40 (1967), pp. 185–206.
—— 'Variation, Cytogenetics and Breeding of the European Field Elm',
Annales Forestales (Anali za Sumarstvo), Vol. 7 (1976), pp. 107–41.
An invaluable survey, with an extensive bibliography.
ROSS-CRAIG, STELLA.—*Drawings of British Plants*, Part XXVII (1970),
Ulmaceae, plates 1–6.
SCHNEIDER, C.—*Plantae Wilsonianae*, Vol. III (1916), pp. 238–65.
WYMAN, D.—*Trees for American Gardens*, Ed. 2 (1965), pp. 461–73.
WILKINSON, G.—*Epitaph for the Elm*. 1978.

DUTCH ELM DISEASE*

One of the most important factors influencing the elms in Britain in the
20th century has been Dutch elm disease. This disease spreads rapidly
through elm populations causing widespread wilting and death of trees.
It was initially found in Western Europe soon after the first World War
and was given its name following the invaluable early research carried out
in Holland. It was first positively identified in Britain at Totteridge in
Hertfordshire in 1927 but was almost certainly present some years earlier.
The disease rapidly reached epidemic proportions in southern England
and during the 1930s caused widespread deaths of most species of elm
growing in the region. By 1936 the epidemic had reached its peak and
thereafter the disease declined both in numbers of trees affected and in the
severity of the symptoms on individual trees. During the subsequent two
decades local flare-ups occurred but the disease came to be regarded as an
endemic problem of no great consequence.

However in the late 1960s fresh serious outbreaks of Dutch elm disease
were reported from various localities in southern England. It soon
became clear that a second epidemic had started on a greater scale than in
the earlier outbreak. Surveys of disease incidence were instituted in 1971
in southern Britain and by 1977 about 50 per cent of the total elm population of 23 million elms present at the beginning of the survey period had
been killed.

When the disease was first identified in Europe there was considerable
argument as to its cause; however it soon became clear that a fungus, now
called *Ceratocystis ulmi* (Buism.) C. Moreau, was in fact responsible. It took
several years before further research demonstrated that the elm bark
beetles *Scolytus scolytus* (Fab.) and *Scolytus multistriatus* (Marsh.) were the
vectors which transmitted the disease from dead or dying elms to healthy
trees. These bark beetles emerge from the bark of diseased elms carrying
spores of *C. ulmi* on their bodies and often feed in the crotches of twigs
on healthy elms. The feeding wounds can thus become infected with the
disease and it subsequently spreads through the vascular system of the
tree. The characteristic symptoms of the disease, including wilting,
yellowing or browning of leaves and dieback of parts or the whole of the

* Contributed by Dr D. A. Burdekin, Forest Research Station, Forestry Commission, Alice Holt.

crown then appear. At the same time, dark brown streaks often develop in the outermost annual ring of the stem and these provide a useful diagnostic feature which can be seen when the bark is stripped away or the stem cut across.

The disease cycle is completed when the mature bark beetles invade the bark of diseased elms in order to breed. Egg-laying female beetles, together with the larvae which subsequently develop, tunnel out characteristic insect galleries within the bark. The disease can also be transmitted from one tree to its neighbour through connecting roots and this method of spread is particularly important in hedgerow trees.

Research undertaken during the course of the second epidemic in the 1970s has revealed much information about the origin of this outbreak and of the variability of the causal fungus. It was discovered that a very pathogenic, aggressive strain was responsible for the new outbreaks whereas a different, non-aggressive strain occurred in locations where infection had survived from the earlier epidemic. Detailed studies have demonstrated that the aggressive strain was probably introduced into Britain on elm logs imported from Canada and that this strain was responsible for the second and more devastating epidemic.

Much research has been directed towards finding methods of controlling the disease including trials with insecticides to kill the bark beetles and with fungicides to protect the trees from fungal invasion. None of these materials have so far proved particularly effective and the only proven method of slowing down the spread of the disease has been by sanitation felling. Experience both in the United States and in Britain has shown that an intensive programme of felling diseased trees and subsequent destruction of the bark in which the bark beetles breed can keep the disease at a relatively low level. Such sanitation felling programmes must however be initiated at an early stage in the epidemic for the task can soon become too great for the available resources.

In the longer term it may be possible to select or breed varieties of elm which are resistant to the disease and which are also suitable for planting as landscape or urban features. Resistant elms have been recognized, particularly among the Asiatic species, and breeding programmes are now in progress in Holland and USA. However, it may be some years before elm hybrids or cultivars, resistant to Dutch elm disease in its aggressive form, are available for widescale commercial planting.

U. AMERICANA L.   AMERICAN OR WHITE ELM

A tree up to 120 ft high, with a trunk 6 ft or more in diameter, forming in isolated positions a wide-spreading head of branches gracefully pendulous at the ends, the whole as much in diameter as the tree is high; bark ashy grey, furrowed; winter buds ovoid, acute or bluntish; young shoots slender, at first downy. Leaves mostly oblong-ovate to elliptic, broadest about the middle, contracted at the apex to a long, slender point, unequal at the base, which is tapered on one side of the midrib, rounded on the other, doubly toothed, 4 to 6 in. long,

1 to 3 in. wide, dark green and glabrous or scabrid above, downy to almost glabrous beneath, lateral veins crowded, straight, running out to teeth; petiole about ¼ in. long. Flowers produced before the leaves. Fruits ripening in the following spring as the leaves unfold, oval or obovate, nearly ½ in. long, beautifully fringed with pale hairs, the two incurved horns at the apex meeting and forming a small aperture, each fruit on a slender, pendulous stalk about 1 in. long.

Native of eastern and central N. America; introduced in 1752. The American elm is one of the finest and most picturesque trees of its native country, always marked by its beauty and grace of branching, and has been much planted as a shade tree since colonial times, but both in gardens and the wild it has suffered severely from Dutch Elm disease. It is closely allied to the Old World *U. laevis*.

U. ANGUSTIFOLIA (West.) West.   GOODYER'S ELM
U. campestris var. *angustifolia* West.; *U. stricta* var. *goodyeri* Melville

This elm has a limited distribution in England near the Hampshire coast south of the New Forest, and has also been reported from Brittany. From the Cornish elm (see below) it differs in the rounded crown, in the leaves of the short shoots having a broader base, and serrations with up to three secondary teeth; the petiole is slightly longer in relation to the length of the blade. For Dr Melville's interesting account of Goodyer's elm see *Journ. Bot.* (Lond.), Vol. 76 (1938), pp. 185–92. It was originally described by Thomas Goodyer in Johnson's edition of Gerard's *Herball* (1636) from plants that he had collected in September 1624 between Lymington and Christchurch and transferred to his garden at Mapledurham House near Petersfield. His phrase-name *Ulmus minor folio angusto scabro* does not at first sight agree with the Hampshire elm, but Dr Melville has pointed out that the leaves on sucker-shoots and saplings are indeed narrow and rough to the touch above and only slowly give way to the smooth and broader adult type of leaf. Goodyer's trees, about 1 ft high when he collected them, were 10 to 12 ft high in 1633 when he made his observations and would still be bearing the juvenile type of foliage.

For the name *U. angustifolia*, which is founded on Goodyer's phrase-name, see R. Melville, 'The Names of the Cornish and the Jersey Elms', *Kew Bulletin*, Vol. 14 (1960), pp. 216–8.

var. CORNUBIENSIS (West.) Melville   *U. campestris* var. *cornubiensis* West.; *U. campestris* var. *stricta* Ait.; *U. stricta* (Ait.) Lindl.; *U. nitens* var. *stricta* (Ait.) Henry; *U. carpinifolia* var. *cornubiensis* (West.) Rehd. CORNISH ELM.—A tree 80 to 100 ft high, of slender form, the upper branches ascending, the lower spreading; young shoots more or less hairy. Leaves of firm texture, somewhat concave above, narrowly to broadly obovate, cuneate and not markedly unequal at the base, 2 to 2½ in. long, 1 to 1½ in. wide, dark green, smooth and glossy above, paler beneath, with conspicuous tufts of down in the axils of the veins, marginal teeth rather blunt, entire or with up to two secondary teeth; veins in ten to twelve pairs; petiole about ⅜ in. long. Samaras obovate to orbicular, up to ⅝ in. long and ½ in. wide, with the seed situated near the notch.

A native mainly of Cornwall, but extending into Devon and south Dorset. It suckers freely, but does not often bear perfect fruit.

U. ANGUSTIFOLIA × U. GLABRA.—This hybrid occurs within the area of *U. angustifolia*. It resembles the wych elm (*U. glabra*) in habit, and often in size, but the shoots are more slender and less hairy, and the leaves are intermediate in size and shape (Melville in Stace, ed., op. cit., p. 293).

U. 'SARNIENSIS'. JERSEY or WHEATLEY ELM.—This is similar in many respects to the Cornish elm, but its branches are more stiffly erect and the tree more tapered; the leaves are proportionately broader, their serrations have one to three secondary teeth, and the axil-tufts are smaller. (*U. sarniensis* Lodd., nom. nud.; *U. campestris* var. *sarniensis* (Lodd.) Loud.; *U. campestris wheatleyi* Simon-Louis Cat. 1869; *U. stricta* var. *sarniensis* (Lodd.) Moss; *U. nitens* var. *wheatleyi* (Simon-Louis) Henry; *U. carpinifolia* var. *wheatleyi* (Simon-Louis) Bean; *U. carpinifolia* f. *sarniensis* (Lodd.) Rehd.). [PLATE 89
The Jersey elm occurs in the Channel Islands, and has been cultivated on the mainland of Britain since the early 19th century, if not earlier. Dr Moss and Helen Bancroft both suggested that it was a hybrid, though they did not agree as to the parentage. In Dr Melville's view it is a pyramidal form of a variable hybrid (*U.* × *sarniensis*) between *U. angustifolia* and *U.* × *hollandica*, occurring not only in the Channel Islands but also in adjacent parts of France, and in Cornwall, south Devon and Kent. Since in his view *U.* × *hollandica* is a triple hybrid (*U. carpinifolia* × *U. plotii* × *U. glabra*) the Jersey elm would have four species in its parentage. Other authorities, however, consider it to be no more than a pyramidal form of the common field elm. Indeed, if all taxonomic views are taken into account the cultivar name 'Sarniensis' could appear in four different combinations, each nomenclaturally correct, and it therefore seems best to place the name directly under *Ulmus*.

U. 'DICKSONII' ('Wheatleyi Aurea').—A compact, slow-growing tree with golden-yellow leaves, retaining their colour well. Raised by Messrs Dickson of Chester in 1900 and distributed by them as 'Golden Cornish Elm' (*U.* × *sarniensis* 'Dicksonii' Hillier's *Manual*).

U. 'PURPUREA'.—A medium-sized tree with a dense crown. Leaves tinged with purple when young, mostly upfolded along the midrib, to 2½ in. long on the short-shoots, larger and somewhat scabrid on extension shoots, irregularly toothed. An elm of uncertain status. It was received at Kew as *U. montana purpurea*. Henry judged it to be a hybrid, but rather confusingly transferred it to *U. campestris* (*procera*). More recently the Dutch authority F. J. Fontaine has suggested that it is a hybrid between *U. glabra* and the Cornish elm, a view which approaches that of Messrs Hillier, who place it under *U.* × *sarniensis* in their *Manual* (*U. montana purpurea* Hort. ex Henry; *U. campestris purpurea* Kirchn. sec. Henry; ? *U. scabra* [f]. *U. purpurea* K. Koch; ? *U.* × *dippeliana* f. *purpurea* (Kirchn.) Schneid.; *U.* × *hollandica* 'Purpurascens' Fontaine, not *U. procera* f. *purpurascens* (Schneid.) Rehd.; *U.* × *sarniensis* 'Purpurea' Hillier's *Manual*).

U. CAMPESTRIS L., *nom. ambig.*

It is now almost universally accepted that the name *U. campestris* L. should be abandoned as a source of confusion, but it has been so widely used until recently, and in so many different senses, that a note on it may not be out of place. In *Species Plantarum* (1753) Linnaeus describes three species of *Ulmus*, one being *U. campestris*. The other two, *U. americana* and *U. pumila*, are non-European, and there must therefore be a presumption that Linnaeus considered all the European elms to belong to a single species. If an attempt is made to ascertain which European species has the best title to the name, the evidence, such as it is, points in different ways. Under *U. campestris*, Linnaeus gives references to the works of his predecessors, and also to his *Hortus Cliffortianus*, where further references are given. These references on the whole support the argument that *U. campestris* is the proper name for *U. carpinifolia*, the common field elm of Europe, and the name has been used by most continental botanists in this sense. But in Britain the name has in the past been applied consistently to the English elm, *U. procera*, and with good reason. In *Flora Anglica*, a work written by Linnaeus and published only one year after the *Species Plantarum*, *U. campestris* L. is identified with the *Ulmus vulgatissima folio lato scabro* of Ray's *Synopsis* (1724), and this is undoubtedly the English elm. Only a few authorities have used *U. campestris* for the wych elm, *U. glabra*, yet this is the species that perhaps has the best title to the name, for it is the species with which Linnaeus was acquainted in his native Sweden, and is the only elm species represented in his herbarium.

Thus three European species—*U. glabra*, *U. carpinifolia* and *U. procera*—have some title to the name *U. campestris* L., and all have been so called by one authority or another. The name is therefore ambiguous and a permanent source of confusion. See further in Dr Melville's paper published in *Journ. Bot.* (Lond.), Vol. 76 (1938), pp. 261–5.

U. CANESCENS Melville

This species is closely allied to *U. carpinifolia*, from which it differs in the young branchlets and the leaves being densely downy. The leaves are elliptic-ovate, more bluntly toothed than in *U. carpinifolia*, with twelve to sixteen pairs of veins. It is a native of the eastern and central Mediterranean region as far west as Italy, first distinguished by Dr Melville in *Kew Bulletin*, Vol. 12 (1957), pp. 499–502.

It is possible that some trees in Italy and southeast Europe, identified as *U. procera*, belong to this species.

U. CARPINIFOLIA Gleditsch   EUROPEAN FIELD ELM, SMOOTH ELM

U. campestris L., in part, *nom. ambig.*; *U. minor* Mill. *sens.* Tutin in *Fl. Europ.*, not Mill. *sens. strict.**, *nom. ambig.*; *U. glabra* Mill., not Huds.; *U. foliacea* Gilib., *nom. pre-Linn.*; *U. nitens* Moench

A tree to about 100 ft high; young shoots almost or quite without down in the adult tree, slender. Leaves obliquely oval or ovate, doubly toothed, narrowing at the apex to a shortish point, very unequal at the base (one side of the blade being tapered, the other rounded or semi-cordate), 1½ to 4 in. long, 1 to 2 in. wide (on vigorous shoots considerably larger), upper surface glossy green and glabrous, lower surface downy only in the vein-axils or along the midrib; stalk ¼ to ½ in. long; veins in ten to thirteen pairs. Flowers crowded in dense clusters close to the leafless shoot. Samaras oval or obovate, glabrous, ½ to ⅝ in. long, notched at the top, with the seed close to the notch.

A native of Europe, N. Africa and parts of S.W. Asia. In Britain (if interpreted in a narrow sense) it is confined to eastern England from Norfolk to the Thames, and perhaps beyond the Thames to Kent, elsewhere being replaced by closely related species (*U. angustifolia*, *U. coritana* and *U. plotii*), by *U. procera* or by hybrids. It is one of the two undisputed species of British elm. The other, the wych elm, *U. glabra*, is amply distinguished by the seed being in the middle of the samara, by the very downy shoots, the much larger downy or scabrid leaves and the very short petiole, almost concealed by the lobe-like base of the longer side of the blade. The common English elm, *U. procera*, differs in its rounder leaf, downy all over beneath and rough above.

U. carpinifolia var. *italica* (Henry) Rehd. (*U. nitens* var. *italica* Henry) is described as a Mediterranean variety, differing in its leathery leaves with more numerous pairs of lateral veins. This variety is of rather dubious standing, and it may be that some of the trees mentioned by Henry were hybrids of the Hollandica group.

cv. 'BERARDII' (*U. nitens* var. *berardii* (Simon-Louis) Bean).—See *U.* 'Berardii' below.

var. CORNUBIENSIS (West.) Rehd.—See *U. angustifolia* var. *cornubiensis*.

var. DAMPIERI (Wesm.) Rehd. (*U. carpinifolia* 'Dampieri')—See *U.* 'Dampieri', under *U.* × *hollandica*.

---

* *U. minor* Mill., in *Gardeners Dictionary* (1768) is inadequately described, and there is no type-specimen. There can, however, be little doubt that it was some sort of field elm, and if all the field elms of Britain and western Europe are considered to constitute a single variable species, the correct name for that species would be *U. minor*. If, on the other hand, two or more species are to be recognised, the ambiguity of *U. minor* Mill. would exclude its use, since it is impossible to know to which element in the complex the name should be applied. The name *U. minor* Mill. is discussed by Dr R. Melville in *Journ. Bot.* (Lond.), Vol. 77 (1939), pp. 266–70, and rejected as ambiguous. The argument for the adoption of this name in place of *U. carpinifolia* is put forward by J. H. Richens in *Fedde's Repertorium*, Vol. 79 (1968), pp. 1–2. For the typification of *U. carpinifolia*, see Rehder in *Journ. Arn. Arb.*, Vol. 19 (1938), pp. 266–8, and Melville, in *Journ. Linn. Soc.*, Vol. 53 (1946), pp. 83–90.

cv. 'HOERSHOLMENSIS'.—An elegant tree with a short trunk and straight, ascending branches, forming a narrowly ovoid crown. Leaves dark green, rather narrow, cuneate at the base. Raised in 1885 by Nielsen of Hoersholm, Denmark, where it has been much planted.

cv. 'PENDULA'.—A tree with pendulous branches, very vigorous and large-leaved.

f. SARNIENSIS (Loud.) Rehd. (*U. carpinifolia* 'Sarniensis').—See *U.* 'Sarniensis', under *U. angustifolia.*

var. SUBEROSA (Moench) Rehd. *U. suberosa* Moench—Branchlets developing corky wings when two years old or later. This variant occurs occasionally in the wild, and is said to be often rather dwarf and to occur in dry habitats.

cv. 'UMBRACULIFERA'.—A tree with a dense, rounded crown, much planted in southwest and central Asia. It was introduced to Europe from Iran in the last century. It is little known in this country, but succeeds well on the continent and in eastern N. America.

cv. 'VARIEGATA'.—Leaves marked more or less copiously with white, especially on the margin.

U. 'BERARDII'.—A very interesting little elm raised in the nursery of Messrs Simon-Louis, near Metz, in 1863. Its leaves are oval, $\frac{1}{2}$ to $1\frac{1}{2}$ in. long, $\frac{1}{4}$ to $\frac{5}{8}$ in. wide, with four to seven coarse triangular teeth down each side; glabrous on both surfaces. The slender twigs and leaf-stalks are downy. The parent tree was said to be a large elm on the ramparts of Metz, and judging by the glabrous leaves of this variety it would appear to have been *U. carpinifolia.* It may, however, have been a hybrid. (*U. berardii* Simon-Louis; *U. procera* var. *berardii* (Simon-Louis) Rehd.; *U. nitens* var. *berardii* (Simon-Louis) Bean.)

U. CORITANA Melville   CORITANIAN ELM
U. coritana var. *media* Melville

A tree up to 65 ft in height with an open head of spreading and ascending branches. Trunk up to 4 ft in diameter, with coarsely furrowed bark. First-year branchlets $\frac{1}{12}$ in. in diameter, smooth, shining brown. Spur shoots with 3 to 5 leaves. Leaves leathery, glabrous and shining above, paler below, glabrous except for axillary tufts and numerous glands, narrowly to broadly ovate, acute, unequal at the base, with 8 to 12 pairs of veins, margins doubly-serrate; petiole $\frac{1}{12}$ to $\frac{5}{8}$ in. long, pubescent above, glabrous below. Flowers and fruit similar to those of *U. carpinifolia.*

var. ANGUSTIFOLIA Melville—Leaves of the short shoots narrowly ovate to ovate lanceolate $1\frac{1}{2}$ to $3\frac{1}{2}$ in. long, very unequal at the base, the long side meeting the petiole $\frac{1}{24}$ to $\frac{1}{8}$ in. below the short side; petiole $\frac{1}{4}$ to $\frac{5}{8}$ in. long.

var. ROTUNDIFOLIA Melville—Leaves of the short shoots broadly ovate, acute, 1 to $2\frac{1}{8}$ in. long, subcordate at the base, the long side meeting the petiole $\frac{1}{24}$ to $\frac{1}{8}$ in. below the short side; petiole $\frac{1}{12}$ to $\frac{2}{5}$ in. long.

This controversial species was described by Dr Melville in 1949 (*Journ. Linn. Soc. (Bot.)*, Vol. 53, pp. 263–71). As interpreted by him it has a wide range in central England, from the Thames to the Trent and from Warwick to E. Anglia, the var. *rotundifolia* being more southern in distribution and grading northward into var. *angustifolia*. It is not recognised as distinct from *U. carpinifolia* in the second edition of Clapham, Tutin and Warburg, *Flora of the British Isles*, and is considered by R. H. Richens to be 'an artificial aggregate'.

Part of its area lies within the territory of the British tribe known to the Romans as the Coritani, whence the epithet *coritana*.

U. CRASSIFOLIA Nutt. CEDAR ELM

A tree up to 80 ft high in nature, but in cultivation in England very slow-growing and forming a round-headed small tree; young shoots clothed with fine, soft, very short down; winter buds often in pairs. Leaves ovate to oblong, obliquely rounded or slightly heart-shaped at the base, bluntish or rounded at the apex, ¾ to 2 in. long, ½ to 1¼ in. wide, toothed (sometimes doubly), of firm rather hard texture, very harsh to the touch above, more or less downy beneath; stalk $\frac{1}{12}$ to ⅛ in. long. Flowers produced in clusters in the leaf-axils in August and later. Samaras ⅜ in. long, oval, tapered at both ends, deeply notched at the top, downy all over, especially on the margin.

A native of the southern USA and northern Mexico, and said by Sargent to be the common elm tree of Texas. It was introduced to Kew through him in 1876, but grew slowly and sometimes died back in winter from the failure of the new growths to ripen. It evidently needs hotter summers than ours.

U. ELLIPTICA K. Koch

Allied to *U. glabra*, which it is said to resemble in habit, but differing in the rusty hairs on the buds, the ciliate leaves, and in having samaras which are downy at the centre on both sides. Native of southwest Russia, including the Crimea and the Caucasus. It is remarkable that the characters by which *U. elliptica* differs from the common wych elm are shared by the American *U. rubra*. But the bark is less fissured than in the American tree, the leaves are thinner, more sharply toothed, and the samaras are more elongate. Henry held that the tree distributed by Späth as *U. elliptica*, and as '*U. Heyderi*' were *U. rubra*, despite the fact that both were said to have come originally from the area of *U. elliptica*.

U. GLABRA Huds. WYCH OR SCOTCH ELM

U. campestris L., in part, *nom. ambig.*; *U. montana* Stokes in Withering; *U. scabra* Mill.

A tree from 100 to 125 ft high, with a trunk sometimes 6 ft in diameter; head of branches wide-spreading, rather open; young shoots stout, downy; bud scales hairy. Leaves usually 3 to 7 in. long (sometimes more on young trees) 1½ to 4 in. wide, oval to obovate, slender-pointed, sometimes three-lobed towards the top, very unequal-sided at the base, coarsely and doubly

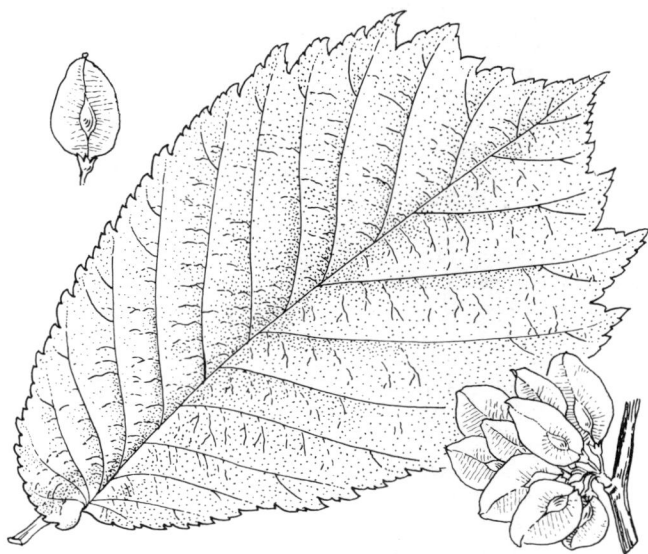

ULMUS GLABRA

toothed, upper surface dull green, very rough, lower one downy; stalk very short, never more than ¼ in. long, often quite hidden by the rounded basal half of the blade, downy; veins in fourteen to twenty pairs. Flowers in dense, stalkless clusters. Samaras oval, ¾ to 1 in. long, downy only at the slightly notched apex, the seed situated in the middle.   [PLATE 90

Native of Europe from the mountains of northern Spain to western Russia, north to the British Isles and western Scandinavia, south to the Italian Appenines and southeast Europe. It is rare in the south and east of England, but common in Scotland and Ireland, where it is one of the noblest of native trees. In the open it forms a stout, shortish trunk of great thickness, and a head of branches often pendulous at the ends. It has been confused with some of the hybrids that have arisen between it and *U. carpinifolia* or related species, but can be distinguished from these and other native British elms by the combination of the short leaf-stalks, the very rough leaves, the absence of corkiness from the two-year old shoots, and by the samaras having the seed in the middle. Moreover, the tree does not produce suckers as *U. carpinifolia* and *U. procera* do, and on that account is valued as a stock for other elms.

U. glabra is variable in the shape of its leaves. In what is regarded as the typical subspecies they are roundish or broadly obovate, with a well-developed lobe at the base on the longer side; this is southern in distribution. Trees in northern England, Scotland and most of Scandinavia have narrower leaves and a less developed basal lobe, and have been distinguished as subsp. MONTANA (Stokes) Lindqvist.

The specific epithet *glabra* refers to the bark, which is at first downy but remains smooth for many years. It is a pity that such an inept name as *U. glabra* should have to be used for this species, but it is the earliest one. The *U. glabra* of Miller is *U. carpinifolia*.

The names 'wych' or 'witch', of Anglo-Saxon derivation, were originally applied to other elms also, and to other trees with flexible branches, *U. glabra* being known as 'witch hazel'. According to Hegi, the equivalent words 'wieke' and 'wietschke' are still in use in some parts of Germany. As a result of false etymology the wych elm came to be associated with witches in popular superstition.

f. CORNUTA (David) Rehd.   *U. campestris cornuta* David; *U. tridens* Hort.; *U. tricuspis* Hort.—Leaves with one or two cusp-like lobes on each side of the apex. Such leaves occur fairly frequently on strong shoots. See also *U. laciniata* below.

cv. 'CRISPA'.—Leaves narrowly oblong-oval, 1½ to 3½ in. long, ¾ to 1½ in. wide, rather infolded, the margins very deeply and jaggedly cut into slender, often double, teeth. A curious, slow-growing form (?*U. crispa* Willd.; *U. montana* var. *crispa* Loud.).

cv. 'EXONIENSIS'.   EXETER ELM.—Branches and branchlets erect. Leaves clustered, dark green, upfolded along the midrib. Although of narrow habit when young it broadens with age. It was found at the Exeter Nursery in 1826 (*U. montana* var. *fastigiata* Loud., not *U. glabra fastigiata* Kirchn.; *U. Fordii* Hort.; *U. exoniensis* Hort.).

cv. 'HORIZONTALIS' ('Pendula').—Like 'Camperdownii', this arose as a seedling of procumbent habit, but top-grafted it becomes a flat-topped tree with horizontal or low-arching branches and pendulous branchlets and eventually rises considerably above the point of grafting. It flowers and fruits very freely. It originated in 1816 in a nursery at Perth but was first distributed by Booth of Hamburg, who is said to have bought the entire stock (*U. montana* var. *pendula* Loud.; *U. horizontalis* Hort. ex Loud.; *U. montana horizontalis* Kirchn.). See also *U.* 'Camperdownii' below.

cv. 'LUTESCENS'.—Leaves yellowish (*U. scabra* var. *lutescens* Dipp.).

U. 'CAMPERDOWNII'.—When top-grafted this makes a pendulously branched, small arbour-like tree, with a rounded crown, sinuously branched. It does not grow so tall as *U. glabra* 'Horizontalis'. It arose as a seedling at Camperdown House, Dundee, in the first half of the 19th century (*U. montana pendula Camperdownii* Hort. ex Henry; *U. montana pendula* Kirchn. *sec.* Rehd., not Loud.). Although usually placed under *U. glabra* it shows the influence of *U. carpinifolia*.

Similar to 'Camperdownii', and of the same hybrid parentage, is 'SERPENTINA', which differs in its more contorted branches. According to Henry, it was originally received at Kew under the name *U. montana pendula nova*, a name said to have been used for 'Camperdownii' also.

U. LACINIATA (Trautv.) Mayr   *U. montana* var. *laciniata* Trautv.—An ally of *U. glabra*, widely distributed in N.E. Asia, including Japan. Young branchlets glabrous or slightly downy. Leaves with one or two cusp-like lobes on each side of the apex, though some leaves lack them. Similar leaves sometimes occur in *U. glabra*, especially on sprouts.

U. × HOLLANDICA Mill.

U. × dippeliana Schneid.

In their original wild state the wych elm, *U. glabra*, and the field or smooth elm, *U. carpinifolia*, occupied largely distinct territories, and despite their affinity and overlapping flowering-times, can have hybridised but rarely. But with deliberate planting, and the destruction of natural barriers, the two elms were brought into closer contact, and the results of their crossing, showing hybrid vigour, having been selected by man at the expense of both parents. And these hybrids have in turn interacted with the original wild species, so that a very complex elm population eventually emerged, which has been the despair of taxonomists. The hybridity of some elms was first recognised only early this century and even then, and much later, many cultivars were treated as 'varieties' of one or the other parental species. Intercrossing between hybrids, and back-crossing with the original species, no doubt explains the variation that exists in this group, and even the various stunted and small-leaved elms that have on occasion arisen in seed-beds and been put into commerce are precisely what geneticists would expect to occur in the later generations of hybrid-with-hybrid crosses. The fact that self-sterility is prevalent in the elms makes suspect the status of the great majority of elm cultivars originally raised from seed.

The taxonomic treatment of the European hybrid elms would be simple if the field or smooth elm could be regarded as a single but variable species. All the hybrids between it and *U. glabra* would then fall under *U. × hollandica*. It is, however, Dr Melville's view that *U. × hollandica* is a triple hybrid—*U. glabra* × *U. carpinifolia* × *U. plotii*—and that the elm populations of Belgium, Holland, France and Germany consist 'almost exclusively' of trees of this parentage (Stace, ed., op. cit., pp. 295–6). The simple hybrid between *U. glabra* and *U. carpinifolia* is in his view typified by the Huntingdon elm, *U. vegeta*, and elms of this parentage would accordingly take the name *U. × vegeta*. The Huntingdon elm itself, as the typical nothomorph (form) of this group, would be *U. × vegeta* 'Vegeta', not *U. × hollandica* 'Vegeta', which would, however, be its correct name if *U. carpinifolia* is interpreted in a broad sense. In the following account the cultivar name of the hybrid is placed directly under *Ulmus*.

U. 'BELGICA'. BELGIAN or HOLLAND ELM.—A massive tree to about 120 ft high, broad-crowned when old. It is nearer to *U. glabra* than it is to *U. carpinifolia*, but the young shoots are more slender and become glabrous by autumn, and the leaves are narrower with a coarsely toothed, more elongated apex. It was once a common tree in the Low Countries, prized for its rapid growth even on poor soils and its good resistance to wind and atmospheric pollution. But it was one of the first elms to fall victim to the outbreak of Dutch elm disease in the 1930s (*U. belgica* Burgsd.; *U. × hollandica* var. *belgica* (Burgsd.) Rehd.; *U. hollandica* Späth, not Mill.).

U. 'DAMPIERI'.—A medium-sized tree of narrowly pyramidal habit, broadening with age. Branchlets short, slender, glabrous. Leaves bright green, of firm texture, broadly ovate to elliptic, slightly oblique at the base, to 2½ in. long, shortly stalked, deeply toothed, almost glabrous on both sides, except for axillary tufts. Little known in this country, it was used as a street-tree in Holland and in Belgium, where it originated before 1862. It proved very resistant to the first outbreak of Dutch elm disease. It has been placed under both *U. carpinifolia* and *U. glabra*, but was considered by Schneider to be a hybrid between them (*U. campestris* var. *nuda* subvar. *fastigiata Dampieri* Wesm.; *U. montana Dampieri* Kirchn.; *U. × dippeliana* f. *dampieri* Schneid.; *U. nitens* var. *Dampieri* Henry).

There is a golden-leaved branch-sport of this—U. 'DAMPIERI AUREA' ('Wredei').

U. 'HOLLANDICA'. DUTCH ELM.—A large tree, up to 110 ft in height, of somewhat open, sparse branching; trunk rather short, up to 6 ft in diameter; young shoots slightly hairy. Leaves oval or ovate, 2½ to 5 in. long, 1½ to 3 in. wide, taper-pointed, one side cordate at the base, and developed farther down the stalk than the other side, which is tapered, upper surface dark shining green, glabrous or nearly so, the lower one also bright green, downy in the vein-axils and along the midrib, nearly glabrous elsewhere; veins in ten to fourteen pairs; stalk downy, ¼ to ⅜ in. long. Samaras between ovate and obovate, ¾ to 1 in. long, with the seed close to the terminal notch. (*U. hollandica* Mill., *sens. strict.*; *U. major* Sm., *nom. illegit.*).

In its botanical characters this elm most resembles *U. carpinifolia*, but the twigs are stouter, the leaves and fruits much larger, and the habit more open. It is quick-growing and several fine specimens existed in the older parts of Kew Gardens, but all these are now gone, having become decrepit, or been killed by the Dutch elm disease, to which this clone is very susceptible. One, felled about 1910, was estimated to have been some 150 years old and measured 92 × 13½ ft; another, felled 1967, was 88 × 14¾ ft.

According to Miller, 'Hollandica' was introduced to Britain from Holland at the end of the 17th century—an origin rejected by Elwes and Henry on the grounds that there is no record of the Dutch elm, as we call it, ever having been planted in Holland, where the common clone is, or rather was, 'Belgica'. Hybrids similar to the Dutch elm have no doubt arisen spontaneously in Britain, and the clone 'Hollandica' may have done so originally, if Miller was indeed wrong in attributing to it a Dutch provenance. It is possible, however, that the Dutch elm was grown in Holland in the 17th century, later being replaced by 'Belgica', which is believed to have originated at the end of the century. In any case, the name *U. major* Sm. preferred by Elwes and Henry is illegitimate, since it is a re-naming of *U. hollandica* Mill., which Smith cited as a synonym.

U. 'SMITHII'. DOWNTON ELM.—Branches ascending; branchlets pendulous, hairy when young, developing corky ridges. Leaves rather thick, glabrous and glossy above, downy beneath, coarsely double-toothed; lateral veins in fourteen to sixteen pairs; petiole about ¾ in. long (*U. × smithii* Henry; *U. glabra* Mill. var. *pendula* Loud.; *U. pendula* W. Masters).

This elm was found at Downton Castle early in the last century among seedlings bought from Smith of Worcester. It is related to the Huntingdon elm ('Vegeta').

U. 'VEGETA'. HUNTINGDON ELM.—This fine elm, according to information given to Loudon by John Wood of Huntingdon, in 1836, was raised in the nursery of his firm about the middle of the 18th century from seed gathered in Hinchingbrook Park. It is, no doubt, a hybrid between *U. glabra* and *U. carpinifolia*, and like many hybrid trees, is of remarkably vigorous growth. One of the largest of all elms, it reaches 140 ft in height, forming a thick short trunk 5 or 6 ft in diameter with ascending branches. Leaves up to 5 or 6 in. long, more than half as wide, glabrous above and downy beneath only in the leaf-axils. Samaras oval, up to $\frac{7}{8}$ in. long, the seed not reaching to the notch at the top. This last character and its less downy leaves distinguish it from *U.* 'Hollandica'. The veins, too, are more numerous (fourteen to eighteen pairs) than in 'Hollandica'. According to Elwes it has the defect of splitting in the trunk due to its habit of forking low down. This, however, can be prevented by timely pruning. The tree produces suckers. (*U. glabra* var. *vegeta* (Lindl.) Loud.; *U. vegeta* Lindl. in *Hort. Cantab.*, *nom. nud.*; *U.* × *dippeliana* f. *vegeta* (Loud.) Schneid.; *U. Huntingdonii* Hort.).

Elms similar to the Huntingdon occur, apparently naturally, here and there in central and southern England, and other trees of the same parentage may have been raised in nurseries and distributed. According to Loudon, several elms of the Huntingdon type were grown by the nurseryman Masters of Canterbury (whose son, Dr Maxwell Masters, was editor of the *Gardeners' Chronicle*, 1865–1907). The elm raised by the nurseryman Gill of Blandford, and distributed as *U. montana superba*, may also have been of the same parentage as the Huntingdon (*Gard. Chron.*, 1845, advertisement p. 653), and may be the hybrid elm 'Superba' later cultivated to some extent in the Low Countries and in Germany (cf. Elwes and Henry, op. cit., p. 1873, and footnote 5). A great elm at Magdalen College, Oxford, when blown down in 1911 was found to contain 2,787 cubic feet of timber. It had been identified as *U.* × *vegeta* by H. J. Elwes only six years earlier, having previously been taken for a wych elm. Almost certainly it antedated the true Huntingdon. Other elms resembling the Huntingdon may be seedlings of it, as it is fairly fertile. The sale of its seedlings by the nurseryman Rivers as 'Huntingdon elm' was the subject of a court case which attracted much interest in gardening circles at the time and will no doubt feature in a history of the nomenclature of cultivated plants, if this is ever written; see *Gard. Chron.*, 1847, pp. 507, 526.

Following the earlier outbreak of Dutch elm disease a breeding programme was started in Holland with the objective of raising cultivars resistant to the disease. Some have been distributed, and introduced to Britain, but they were tested against the earlier and milder strain of the disease and have either failed to resist the new and more virulent strain or have yet to prove themselves in the new circumstances. Some have other defects. In order of release these clones are: 'CHRISTINE BUISMAN', 'BEA SCHWARZ', 'COMMELIN' and 'GROENEVELD'. Of these 'Commelin' is certainly very susceptible to the new strain of Dutch elm disease. It, and 'Groeneveld', are the results of deliberate crosses between selected trees of *U. carpinifolia* and *U. glabra*. 'Christine Buis-man' was raised from Spanish seed and may be a clone of *U. carpinifolia*. In case 'Groeneveld' turns out to be resistant to the new strain, it should be noted that it is likely to make a small tree, with a well-developed central stem but no more than 50 ft high (Fontaine, op. cit., p. 49).

U. JAPONICA (Rehd.) Sarg.

U. campestris var. *japonica* Sarg. ex Rehd.; *U. davidiana* var. *japonica* (Rehd.) Nakai; *U. propinqua* Koidz.

A tree up to 110 ft in Japan, forming broad heads of graceful, pendent branches; young shoots very downy, fawn-coloured when mature, sometimes developing corky wings. Leaves oval, inclined to obovate, 3 to 4½ in. long, 1½ to 2½ in. wide, unequal at the broadly tapered base, abruptly narrowed at the apex to a slender point, rather coarsely toothed, furnished with stiff hairs above at first, afterwards very harsh to the touch, lower surface clothed with pale down, especially on the veins and midrib; veins in seven to thirteen pairs; stalk about one-sixth of an inch long. Samaras obovate, ⅝ in. long, nearly ½ in. wide, tapered at the base, the seed being situated near to the notch, the inner edges of which are ciliate; elsewhere the fruit is glabrous.

Native of Japan and continental N.E. Asia; introduced to the Arnold Arboretum in 1895 and thence to Kew. It is closely allied to *U. carpinifolia*, which has reddish brown young bark on the twigs and leaves more unequal at the base.

By some authorities this species is considered to be only varietally distinct from U. DAVIDIANA Planch., of which little is known in cultivation. It is a native of N.E. Asia, including Sakhalin and the Kuriles.

U. LAEVIS Pall.

U. effusa Willd.; *U. pedunculata* Foug.

A tree over 100 ft high, with a trunk up to 6 ft in diameter, supporting a wide-spreading, rather open head of branches; bark brownish grey; young shoots clothed with grey down, at least at first. Leaves obliquely obovate, 2½ to 5 in. long, rather more than half as wide; with double incurved teeth at the margins, the base very unequal, being rounded at one side of the midrib, abbreviated and tapered at the other, the apex narrowed abruptly to a slender point; bright green, glabrous, or slightly harsh above, usually clothed beneath with a dense grey down; side veins up to eighteen pairs; stalk ⅛ to ¼ in. long. Samaras oval, about ½ in. long, fringed with pale hairs, and having two incurved horns at the apex, borne on slender pendulous stalks in crowded clusters.

Native of Europe from the Volga region west to eastern France, south to the Caucasus and southeast Europe; rare in the region of the Alps. It is very closely allied in botanical characters to *U. americana*, from which it is indeed difficult to distinguish it. It thrives better in this country, and the leaves seem more uniformly downy, more unequal at the base, and more frequently broadest above the middle; the winter buds are more elongated and sharply pointed. Of European elms it most resembles *U. glabra*, but is easily distinguished by the smoother, smaller leaves, and especially by the fringed samaras.

U. MACROCARPA Hance
? *U. rotundifolia* Carr.

A deciduous tree sometimes bushy but occasionally 50 ft high; young shoots hairy, often becoming furnished with two corky wings after the second year. Leaves broadly obovate to oval, obliquely rounded at the base, narrowed abruptly to a short slender apex, doubly toothed, 2 to 4 in. long, $1\frac{1}{4}$ to $2\frac{1}{2}$ in. wide, very rough with short bristles on both surfaces and with axil-tufts of down beneath; veins in ten to fourteen pairs; stalk $\frac{1}{4}$ in. or less long. Samaras flat, winged, orbicular or broadly oval inclined to obovate, slightly notched at the top, $\frac{3}{4}$ to $1\frac{1}{4}$ in. long, harshly bristly like the leaves, ciliate, with the seed in the centre.

Native of continental N.E. Asia, where it is widely distributed; described in 1868 from specimens collected by David in N. China; introduced to the Arnold Arboretum in 1908 by F. N. Meyer from N. China. It was cultivated by Vicary Gibbs at Aldenham, but beyond that little is known of it in this country. It belongs to the same section as the field and wych elms, within which Schneider grouped it with the Himalayan *U. villosa* and *U. wallichiana*.

U. PARVIFOLIA Jacq.   [PLATE 91
U. chinensis Pers.

A small tree up to 60 ft high, with a slender trunk supporting a rounded head of branches; branchlets very slender, clothed with a close, minute, grey down; winter buds small, conical. Leaves leathery, $\frac{3}{4}$ to $2\frac{1}{2}$ in. long, $\frac{1}{2}$ to $1\frac{1}{3}$ in. wide, oval, ovate or obovate, unequally rounded at the base (or one side of the midrib tapered), pointed, the margins rather evenly toothed, the teeth triangular, often blunt, upper surface lustrous green, and smooth on the smaller twigs, rather rough on vigorous shoots, lower surface pale, bright green, with tufts of down in the vein-axils, or smooth; stalk $\frac{1}{16}$ to $\frac{1}{4}$ in. long, downy, veins in ten to twelve pairs. Flowers produced in September and October in the leaf-axils. Samaras ovate-oval, $\frac{1}{3}$ in. long, not downy; seed in the centre.

Native of E. Asia, including Japan and much of China; introduced towards the end of the 18th century. This tree retains its leaves until the New Year quite fresh and green, and is well worth growing for its elegance. It is sometimes confused with *U. pumila* in gardens, but that species flowers in spring. From the other autumn-flowering elms, *U. crassifolia* and *U. serotina*, it is distinct in retaining its leaves so late, in the almost complete absence of down from beneath the leaves, and in their brighter smoother surfaces. Introduced in 1794.

cv. 'FROSTY'.—A dwarf, bushy shrub. Leaves small, densely set on the shoot, white-margined at first, finally white only on the teeth.

U. PLOTII Druce   PLOT ELM

U. sativa Mill. *sens.* Moss, not Mill.; *U. minor* Mill. *sens.* Henry, ?not Mill.

A tree to about 90 ft high; leader arching on younger trees; branches few, short, spreading; branchlets pendulous. Leaves elliptic, acute, up to 2¾ in. long and 1½ in. wide, obliquely truncate to slightly cordate at the base, dull dark green and almost glabrous above, lower surface glandular at first, glabrous except for axillary tufts, bluntly toothed; lateral veins in seven to ten pairs. Proliferating short shoots (of indefinite growth) are frequent in this species, and have smaller, rounder leaves. Samaras elliptic, about ½ in. long, bearing the seed in the upper part.

U. plotii was first distinguished as a species by Druce in 1911. He did not study it in any detail, and the first full account of it was published by Dr Melville some thirty years later (*Journ. Bot.* (Lond.), Vol. 78 (1940), pp. 181–92 and figs. 1–3). According to him, the main area of *U. plotii* is in the lower valley of the Trent around Newark-on-Trent in Nottinghamshire, below 400 ft, and on the other side of the border with Lincolnshire in the Witham valley. There are more scattered occurrences farther west, as far as Monmouthshire and Shropshire.

Druce believed that this elm was one described by Plot in his *Natural History of Oxfordshire* (1677). It is now generally accepted that this identification was incorrect—an error that makes 'Plot's elm' an inappropriate vernacular name for this species, though it does not affect the validity of its botanical name. Dr Melville suggested that it should be called Plot elm.

Elwes and Henry identified *U. plotii* with Miller's obscure *U. minor* and also confused it with Goodyer's elm, for which see *U. angustifolia*.

U. × DIVERSIFOLIA Melville   *U. minor* Mill. *sens.* some authors—A tree up to about 65 ft in height with spreading branches and slender wiry branchlets. First-year branchlets 1/24 to 1/12 in diameter, hairy at first becoming nearly glabrous by the autumn. Spur shoots of three kinds, the majority having leaves with unequal bases, a lesser number with bases equal or nearly so, and a few with both types of leaf together. Leaves with unequal bases, elliptical to obovate acute, veins eight to eleven pairs. Leaves with equal bases, elliptical acute, veins five to nine pairs. Blade of leaf 2 to 3½ in. long, slightly rough above, downy below, margin doubly-serrate. Stalk 3/16 to 3/8 in. long, downy above. Sucker shoots hairy, with leaves 2/5 to 2¾ in. long. Flowers ten to twenty-five in a cluster. Samaras ½ to ¾ in. long, ⅓ to ½ in. broad, ovate to obovate, seed near the notch.

Originally described by Dr Melville as a species, this elm is now considered by him to be a triple hybrid of the parentage *U. coritana* × *U. plotii* × *U. glabra*, the leaves being intermediate between the first two species in shape and texture, but deriving the downiness of the undersides from *U. glabra*. Its main area is in Hertfordshire, Cambridge and Suffolk. It was originally described in *Journ. Bot.* (Lond.), Vol. 77 (1939), pp. 138–45; for Dr Melville's later view see Stace, op. cit., pp. 293–4.

U. × ELEGANTISSIMA Horwood—This was described, as a species, in
Flora of Leicestershire and Rutland (1933) and is considered by Dr Melville to be
a natural hybrid between *U. plotii* and the wych elm, *U. glabra,* occurring fairly
frequently from the Trent valley southwards to Hertfordshire and Essex. It is
variable, combining in many different ways the characters of the two parents.
See further in Stace, op. cit., p. 297.

The elm named 'JACQUELINE HILLIER' is considered to belong to U. ×
elegantissima. It is shrubby, of dense habit, with small, scabrid leaves, originally
found in a garden in the Midlands, and named by Messrs Hillier in 1967.

U. PROCERA Salisb.   ENGLISH ELM

U. campestris L., in part, *nom. ambig.*; *U. surculosa* Stokes; *U. glabra* Mill. var. *pubescens*
Schneid.

A tree up to 120, or even 150 ft high, with a trunk 6 ft or more in diameter;
young shoots hairy. Leaves roundish ovate, to broadly oval, very unequal at
the base, terminated by a short, abrupt point, coarsely and doubly toothed,
2 to 3½ in. long, about two-thirds as wide, dark green and very harsh to the
touch above, paler beneath and downy all over, with conspicuous tufts of
white down in the vein-axils, along the midrib, and at the base of the cheif
veins—of which there are ten to twelve pairs. Flowers clustered closely to the
branchlet, opening early in the year, reddish. Samaras roundish, ½ in. across,
not downy, bearing the seed close to the notch at the top.

In old books on trees the English elm is usually said to be a native of Europe
introduced to Britain at the time of the Romans. It is now fairly certain that
it is genuinely wild nowhere but in southern and central England. The English
elm produces fertile seed extremely rarely. I have not myself seen a genuine
seedling, but Henry states that he raised four plants out of twenty batches
of seed sown in 1909. In many parts of southern England the elm is the dominant
tree, especially in hedgerows; all these trees, however, have sprung from root
suckers, which the elm produces freely, and which afford the best means of
propagation.

The origin of the English elm still remains a mystery. It occurs in some
parks and gardens of Spain, but apparently always planted, and there is no
evidence that it was ever introduced to England from Spain. On the other
hand, there is a tradition that the Spanish trees originally were sent from
England. Against its being a genuine native of Britain, there is the curious
fact that it is almost invariably infertile.

As a tree in the English landscape the elm impresses one by its noble stature
and bulk, its rich leafiness, and its singular beauty in winter when the finely
fretted outline of its naked branches shows in delicate tracery against the sky.
In the autumn the foliage dies off rich yellow, and lingers on the branches
longer perhaps than that of any of our native trees. This elm has an unfortunrte
propensity in age of dropping its limbs, which snap off without any warning.
This usually happens on still evenings in late summer and early autumn when

the trees are still in full leaf. It is also liable to occur during a heavy rain following a period of heat and drought. This habit makes the elm a very unsuitable tree to plant in crowded thoroughfares.

The timber of elm is valuable for its toughness and the absence of any tendency to split. It has also considerable beauty of graining and colour. Kept permanently dry or permanently wet, it is very durable. At one time, before the introduction of iron pipes, hollowed-out trunks of elm were used as water-pipes.

To the above account, taken from previous editions, there is little to add, except that it would now be more appropriate to put the whole passage in the past tense so far as much of England is concerned. Most authorities now accept that the English elm is endemic to England, occurring on the continent only as a planted tree. Reports of its occurrence on the continent were in part the result of nomenclatural confusion. To British botanists *U. campestris* has always meant the English elm, and it may have been assumed that continental botanists were using the name in the same sense, when in fact *U. campestris* as understood by them was *U. carpinifolia* or sometimes a hybrid of it. There may also have been confusion between *U. procera* and hybrids of the Hollandica group with leaves downy beneath as in our species, or even with *U. canescens*. But the origin of the English elm remains a mystery. Helen Bancroft suggested that it may have arisen in this country from hybridisation between *U. carpinifolia* and *U. glabra*. There is also the possibility that it may have originated somewhere in the western parts of continental Europe and subsequently become extinct there.

It was suggested in previous editions that the infertility of the English elm is actually a consequence of its abundant production of suckers from the roots, but this theory is disputable, to say the least of it. Many species of trees and shrubs have the ability to spread by this means without thereby losing the ability to reproduce themselves sexually. A more conventional explanation is that owing to its suckering habit it has, with the help of man, been able to maintain itself, becoming widely spread despite its infertility. It has been found that *U. carpinifolia* and at least some of its hybrids are as a rule self-sterile and therefore would not set fertile seed when grown in isolation or in the company only of other members of the same clone. But in the case of the English elm a further explanation for its infertility has been advanced, namely that the flowers are frost-tender. This might help to explain why no proven hybrids between *U. procera* and other elms exist, though it has been suggested that certain anomalous elms are the result of crossing between it and *U. carpinifolia*.

For *U. procera* var. *australis* (Henry) Rehd. (*U. campestris* var. *australis* Henry) see Elwes and Henry, op. cit., p. 1904. The elm described by Henry is really rather remote from *U. procera* and probably a hybrid, planted in S. France, Italy, etc.

cv. 'ARGENTEO-VARIEGATA' ('Foliis Variegatis').—Leaves conspicuously blotched, striped and margined with creamy white (*U. campestris argenteo-variegata* West.; *U. c. foliis variegatis* Lodd.). This sport has been in commerce since the 18th century. Its suckers are also sometimes variegated.

So far as is known, there are no other cultivars that can with any certainty be referred to *U. procera*. Those placed under it in Rehder's *Bibliography* are disposed of as follows:

f. *aurea* (Morren) Rehd.—See under *U.* 'Viminalis'.

var. *berardii* (Simon-Louis) Rehd.—See *U.* 'Berardii' under *U. carpinifolia*.

f. *marginata* (Kirchn.) Rehd.—See under *U.* 'Viminalis'.

f. *purpurascens* (Schneid.) Rehd.—This name is founded on a small-leaved elm put into commerce under the name *U. campestris myrtifolia purpurea*; it had purple-tinged leaves about 1 in. long, downy beneath.

f. *purpurea* (Wesm.) Rehd.—This appears to be the purple-leaved hybrid elm mentioned under *U. angustifolia*.

f. *vanhouttei* (Schelle) Rehd.—See *U.* 'Louis van Houtte' below.

U. 'LOUIS VAN HOUTTE' ('Vanhouttei').—Leaves entirely yellow, retaining their colour throughout the summer, scabrid on both sides; branchlets hairy (*U. campestris* 'Louis van Houtte' Deegen; *U. montana* f. *lutescens van Houttei* Hort. ex Schelle; *U. procera* f. *vanhouttei* (Schelle) Rehd.).

Although usually placed under *U. procera* this elm is of uncertain status. It is very susceptible to Dutch elm disease and it was recorded in the last edition of this work that two healthy trees at Kew in the Pagoda Vista succumbed in 1931 only a few months after the first visible evidence of the disease.

U. PUMILA L.

U. humilis Gmel.; *U. microphylla* Pers.; *U. campestris* var. *parvifolia* Loud., not *U. parvifolia* Jacq.

A small tree, 10 to 30 ft high, sometimes a shrub. Leaves oval or ovate-lanceolate, acute to acuminate at the apex, tapered or rounded at the base, and not unequal-sided there as elms usually are, rather coarsely toothed except at the base, $\frac{3}{4}$ to $2\frac{1}{4}$ in. long, $\frac{1}{3}$ to 1 in. wide, dark green and quite glabrous (or with minute tufts of down in the vein-axils) beneath; stalk downy, $\frac{1}{12}$ to $\frac{1}{6}$ in. long. Flowers borne on the naked shoots in spring, on very short stalks, and in clusters. Samaras circular or rather obovate, deeply notched at the top, $\frac{1}{2}$ in. across, the seed about the middle.

Native of Central Asia (with a southern limit in south Tibet and the inner parts of the northwest Himalaya), E. Siberia, Mongolia, N. China and Korea. It is allied to *U. carpinifolia*, differing in the smaller leaves with fewer lateral veins and the smaller samaras. It is shrubby only in very cold or exposed habitats, and becomes a tall tree in northern China, whence derive most if not all of the examples now in collections. It was introduced by Purdom for Messrs Veitch from the Peking region, while the two trees at Colesbourne were, most remarkably, raised by H. J. Elwes from a branch 2 in. thick, received from Peking, which grew when potted up (Elwes and Henry, op. cit., p. 1927). These measure 59 × 6 ft and 65 × 6¾ ft (1971). There is an example at Wakehurst Place in Sussex of about the same size, which has so far (1979) been unaffected by the Dutch elm disease (another healthy tree was blown down in the winter of 1978/9). A smaller example at Kew is one of the few remaining elms in the collection.

In the United States *U. pumila* is much planted, especially in the Middle West, being both hardy and drought-resistant, besides being so far little affected by the Dutch elm disease. It has been crossed there with *U. rubra*, one clone of this parentage being 'COOLSHADE', introduced in 1946.

U. 'PINNATO-RAMOSA'.—A very elegant, vigorous-growing, small-leaved elm, sent out by Dieck from the Zöschen Arboretum at the end of the last century as *U. pinnato-ramosa*. The branchlets, more downy than those of *U. pumila*, are arranged in two opposite rows (distichously) and the leaves are longer pointed, but otherwise similar to those of *U. pumila*. The Dieck plants were said to have come from western Siberia but similar plants have been found in Turkestan, or been raised from seed collected in that area. This elm has been identified with *U. pumila* var. *arborescens* Litvinov.

U. RUBRA Muhl.   SLIPPERY OR RED ELM
U. fulva Michx.

A tree 60 to 70 ft high, with a trunk up to 2 ft thick, supporting a spreading head of branches; young shoots very downy; winter buds $\frac{1}{4}$ in. long, covered with brown hairs. Leaves oblong-ovate, 3 to 8 in. (sometimes in young trees 10 in.) long, about half as wide, abruptly tapered to a long, slender point, obliquely rounded at the base, jaggedly or doubly toothed, upper surface very harsh to the touch through minute excrescences, lower surface downy; stalk $\frac{1}{4}$ to $\frac{1}{3}$ in. long. Flowers very short-stalked and crowded in clusters. Samaras orbicular or obovate, $\frac{1}{2}$ to $\frac{3}{4}$ in. long, slightly notched at the top, the part covering the seed (which is in the centre) coated with red-brown hairs, naked elsewhere.

A native of N. America, with a similar distribution to that of *U. americana*. It does not thrive in this country. Elwes and Henry knew of no good specimen, and the last tree to be planted at Kew attained only 38 × 2$\frac{1}{4}$ ft in sixty years. It is not planted for ornament in its native country, and is very subject there to Dutch elm disease. The popular name 'slippery elm' refers to the mucilaginous inner bark, which was used by the early settlers as a thirst quencher, and as a remedy for throat inflammations.

U. SEROTINA Sarg.   RED OR SEPTEMBER ELM

A tree 60 to 70 ft high of spreading habit, with slender, glabrous or nearly glabrous young shoots, often becoming more or less corky-winged. Leaves oblong, often inclined to obovate, 2 to 3$\frac{1}{2}$ in. long, slender-pointed, very unequal at the base, doubly toothed, with about twenty pairs of veins, glabrous and glossy above, downy on the veins beneath; stalks about $\frac{1}{4}$ in. long. Flowers in pendulous racemes 1$\frac{1}{4}$ in. long, opening in September. Samaras $\frac{1}{2}$ in. long, oblong-elliptical, deeply divided at the apex, fringed along the margins with silvery white hairs; they are ripe in November.

A native of the USA, where it is rare and local in Georgia and Alabama, north to south Kentucky and south Illinois, usually on limestone. From the other autumn-flowering American species, *U. crassifolia*, it differs in its racemose inflorescence and in having the perianth divided to the base. It has been planted to some extent as a shade-tree in Georgia and Alabama, but is of no economic importance.

U. THOMASII Sarg.   ROCK ELM
U. racemosa Thomas, not Borkh.

A tree 80 to 100 ft high, with a trunk up to 3 ft in diameter, supporting a narrow roundish head of branches. Branchlets often becoming corky. In a young state the trees are pyramidal; winter buds and young shoots downy. Leaves oval to obovate, with an abrupt, slender point and an unequal, oblique base, 2 to 4½ in. long, 1¼ to 2¾ in. wide, doubly toothed, glabrous, dark glossy green above, downy beneath; side veins in often over twenty pairs; stalk up to ¼ in. long, sometimes partially covered by the overlapping bases of the blade. Flowers in racemes 1 to 2 in. long. Samaras oval, ½ to ¾ in. long, downy all over as well as on the thickened margins, with a slight open notch at the apex, the two points erect.

A native of N. America from southeast Canada and New England west to Minnesota and Iowa, south to Tennessee. It produces a hard timber, whence 'rock elm', but is of no economic importance and grows slowly even in its native country. An example at Kew, although almost half-a-century old, measures only 38 × 1¾ ft (1970). The distinctive points of this species are the large, downy winter-buds, ¼ or ⅜ in. long, the racemose inflorescence and the shallowly notched hairy fruits.

U. ALATA Michx. WAHOO or WINGED ELM.—Allied to *U. thomasii*, differing in its smaller stature, shorter almost sessile leaves (to 2½ in. long) and the narrower ovate-lanceolate samaras. A native of the southern and southern-central USA, attaining its best development in the Mississippi delta. The twigs of this species soon develop a pair of opposite corky wings, though these are sometimes absent; in *U. thomasii* the cork appears later and is less regular. As in *U. thomasii* the leaves are downy beneath and the samaras downy on the surface and edge.

This elm is rare in Britain, where it has been confused with the corky-barked form of *U. carpinifolia*, which differs obviously in its inflorescences and samaras, and in having longer-stalked leaves glabrous beneath.

U. VILLOSA Brandis ex Gamble   CHERRY-BARK OR MARN ELM
U. laevigata Royle, *nom. nud.*

A tree to 80 ft high in its native country, with rather pendulous branches; bark at first grey and smooth, with horizontal bands of lenticels, becoming coarsely furrowed with age; growth-buds very small, obtuse; branchlets densely downy. Short shoots with up to ten leaves. Upper leaves oblong-elliptic, 2¼ to 4⅜ in.

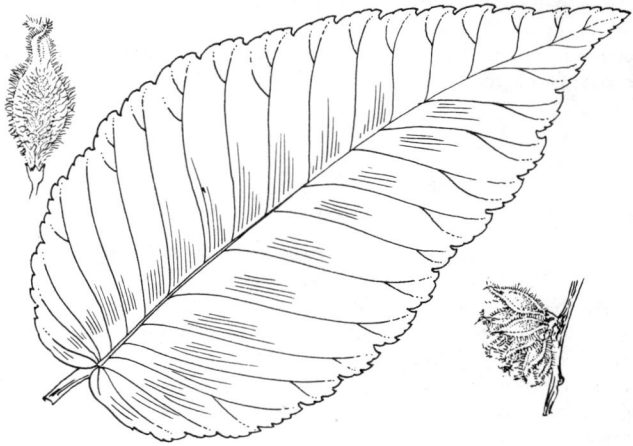

ULMUS VILLOSA

long and up to 2 in. wide, acute at the apex, rounded to truncate and only
slightly oblique at the base, the lower leaves of the shoot more ovate and
slightly cordate at the base, upper surface glabrous except on the midrib,
lower surface at first sprinkled with reddish glandular hairs, later glabrous
or slightly downy on the blade, and usually with white tufts in the lower
vein-axils, margins uniformly biserrate, each major serration with up to seven
minor teeth; petioles $\frac{3}{16}$ to $\frac{3}{8}$ in. long. Flowers in spring on the naked wood,
in dense clusters which appear whitish from the long hairs on the perianth-
segments and ovaries; anthers purplish red. Samaras elliptic, about $\frac{1}{2}$ in. long,
hairy on both sides and ciliate at the margin; seed slightly above the middle.
Bot. Mag., n.s., t. 742.

A native of the western Himalaya. It is now mainly confined to Kashmir, and
even there has become rare in the wild. A small stand exists in the Dachigam
Game Reserve near Srinagar, but elsewhere it has been reduced to the verge of
extinction in the wild by the lopping of its branches for cattle-fodder. But fine
specimens are still to be seen by temples and shrines.

U. *villosa* was first recognised as a distinct species by Royle in 1839, but
he provided no description. Dietrich Brandis gave an account of it in 1899
under the provisional name U. *villosa*, which was taken up and validated by
Gamble three years later. It was introduced to Kew in 1935 but seems to have
gone unnoticed outside Kew until Nigel Muir gave a good account of it in 1969
and called attention to its merits as an ornamental (*Gard. Chron.*, Vol. 166,
pp. 7–8).

U. *villosa* had attained a height of 72 ft at Kew by 1972, and made an elegant,
lightly branched tree, flowering very freely. Even when in good health it
sometimes had a gaunt look owing to its habit of shedding the short shoots
after bearing a heavy crop of flowers, but is now (1979) showing signs of attack

by Dutch elm disease.

The vernacular names given in the heading are those suggested by Dr Melville in his article in the *Botanical Magazine*, the first alluding to the lenticellate bar, the second being taken from one of the native names for this elm.

U. 'VIMINALIS'

U. viminalis Lodd. ex Bean; *U. campestris* var. *viminalis* (Lodd.) Loud.; *U. procera* var. *viminalis* (Loud.) Rehd.; *U. antarctica* Kirchn.

A narrow-headed, rather slender tree with drooping branches; young shoots slightly downy, slender. Leaves oblanceolate or narrowly oval, nearly always tapered at the base, terminated by a long slender point, 1 to 2 in. long, ⅓ to ¾ in. wide, very deeply toothed, the teeth narrow, often blunt, upper surface very rough, lower one downy especially in the vein-axils and on the veins.

The cultivated elm described above is considered by Dr Melville to be an extreme form, and also the type, of a natural hybrid, U. × VIMINALIS, occurring from Essex to Oxfordshire, the parents of which are *U. carpinifolia* and *U. plotii* (Stace, op. cit., p. 298).

U. 'VIMINALIS AUREA'.—Resembling the above, but leaves yellow (*U. campestris* var. *aurea* Morren; *U. campestris Rosseelsii* Hort.; *U. campestris* var. *viminalis aurea* Henry). Raised by Rosseel, Louvain, before 1866.

U. 'VIMINALIS MARGINATA'.—Leaf margins variegated with creamy white (*U. campestris viminalis marginata* Kirchn.; *U. c.* var. *viminalis variegata* Nichols.).

U. 'BETULIFOLIA'.—This appears to be allied to 'Viminalis', and is of uncertain origin, possibly a hybrid in whose origin *U. carpinifolia* has shared. The leaves are narrowly obovate, up to 2½ in. long by 1½ in. wide, the margins deeply toothed, the teeth narrow, incurved, often again toothed, very harsh to the touch above, downy in the vein-axils beneath. The habit is elegant on account of the pendulous young branchlets.

U. WALLICHIANA Planch.

U. erosa Wall. Cat., not Roth

A tree to 100 ft in the wild; branchlets slender, downy at first, becoming glabrous and dull brown or yellowish brown; buds narrowly ovate, acute, the scales ciliate and with some down on the back. Leaves three to five on the short shoots, more or less elliptic, acuminate at the apex, slightly unequal at the base, the larger leaves 2⅜ to 5 in. long, 1 to 2⅜ in. wide, slightly scabrid above, sparsely to densely downy beneath, with small axillary tufts, margins doubly serrate, the teeth convex on the back, with one to three secondary teeth; lateral veins in fifteen to seventeen pairs on the larger leaves; petiole up to ⅜ in. long, downy. Flowers before the leaves; inflorescence with a distinct central axis up to about ½ in. long, which, like the pedicels, is densely hairy; perianth-segments five or six, hairy on the back and at the edge. Samaras orbicular to obovate, slightly hairy to almost glabrous; seed at the centre.

Native of the Himalaya, from Kashmir to Nepal. In Kashmir it overlaps with *U. villosa* (q.v.), which has short shoots with more numerous leaves, these not much more than 4 in. long, acute or gradually acuminate and more evenly biserrate, the serrations not convex on the outside and with three to seven secondary teeth. In var. TOMENTOSA Melville and Heybroek the buds, young branchlets and leaf-undersides are more densely hairy and the samaras uniformly hairy on the surface and at the margins.

subsp. XANTHODERMA Melville & Heybroek—Branchlets becoming orange- or yellow-brown, glandular at first, not hairy. Inflorescence slightly glandular, almost glabrous. Samaras with a few glandular hairs. Of more western distribution than subsp. *wallichiana,* from Afghanistan to Kashmir. For a comment on this variety, see *Flora Iranica*, Part 142 (1979), p. 4.

U. wallichiana is in cultivation in Britain, but little known. Its chief importance would appear to be as a possible source of genetic factors for resistance to Dutch elm disease. The clone introduced to Holland from the Arnold Arboretum in 1930 proved tender, but its offspring showed promising characters and in 1960 Dr H. M. Heybroek of the Forest Research Station 'De Dorschkamp' at Wageningen visited the Himalaya to collect hardier provenances. See further in Melville and Heybroek, op. cit., pp. 5, 8–14, on which this account is based.

U. WILSONIANA Schneid.

A deciduous tree up to 80 ft high, with a trunk occasionally a yard in diameter; young shoots downy. Leaves ovate to oval, often with a short, slender, drawn-out apex, obliquely unsymmetrical at the base, rather evenly doubly-toothed, 2 to 4½ in. long, 1¼ to 2½ in. wide, veins in sixteen to twenty-two pairs, parallel, those in the centre of the leaf forking towards the margin, dark green and very harsh to the touch above, paler and downy beneath, especially on the midrib and veins beneath; stalk stout, downy, ⅛ to ¼ in. long. Samaras smooth, obovate, notched at the rounded apex of the wing, the seed situated just below the notch.

Native of W. China; discovered by Wilson in 1900; introduced by him to the Arnold Arboretum in 1910 by means of graftwood. It is related to *U. japonica*, but that species has leaves with usually not more than sixteen pairs of veins, also the inner margins of the notch of its samaras are edged with down. Wilson's elm often develops corky bark after the fashion of our native *U. carpinifolia*.

UMBELLULARIA LAURACEAE

A genus of two species, belonging to the same tribe of the Laurel family as the common bay laurel, and distinguished from it by technical characters of the flower. An obvious point of distinction, however, so far as the present species is concerned, is that the leaves are longer and have a pungent aroma when crushed.

U. CALIFORNICA (Hook. & Arn.) Nutt.   CALIFORNIA LAUREL

Tetranthera californica Hook. & Arn.; *Oreodaphne californica* Nees; *Laurus regia* Dougl., *nom. nud.*; *L. regalis* Hort.

An evergreen tree, 80 to 100 ft high in favourable situations in California, with a dense head of very leafy branches; young shoots at first minutely downy. Leaves alternate, leathery, with a pungent aromatic odour when crushed, narrowly oval or oblong, but tapered at both ends, 2 to 5 in. long, ¾ to 1½ in. wide, not toothed, dark green and glossy above, paler beneath, almost glabrous on both surfaces except when just unfolding. Flowers ¼ in. across, yellowish green, produced during April in terminal and axillary umbels ¾ in. wide, on a common stalk 1 in. long. Fruit roundish pear-shaped, 1 in. long, ¾ in. wide; green changing to purplish. *Bot. Mag.*, t. 5320.

Native of Oregon, from Douglas Co. southward, and of California, where it occurs throughout the coastal range almost to the Mexican border and is also found in the Sierra Nevada. The best stands occur in moist valley bottoms, but most of these have given way to cultivation. At the other extreme, it occurs at high elevations as a stunted, even prostrate shrub. All parts of the plant are aromatic, and the leaves give off a volatile oil which can cause sneezing and headache—even, it is said, unconsciousness—if sniffed too long and deeply. It may also cause skin irritation in some persons. The timber, which is heavy and often beautifully figured, was used in cabinet work and for panelling.

U. californica was discovered by Menzies in 1792, and introduced by David Douglas, who met with it in a deep and shady side-valley of the Umpqua river of Oregon in October 1826, and felled a tree to obtain the seeds. This fine tree is hardy in the open at Kew, being only occasionally injured by severe frost. On a wall it has flowered there, and borne fruit. It likes a sheltered spot, and is about equal to the bay laurel as an ornamental evergreen. The tree at Kew, near the Temperate House, mentioned in previous editions as 32 ft high, measures 52 × 7¾ ft (1967). Other notable specimens are: Warnham Court, Horsham, Sussex, 55 × 6 ft + 5¼ ft (1969); Nymans, Sussex, 40 × 7¼ ft (1966); Borde Hill, Sussex, 51 × 8¼ ft (1973, cf. 25 × 2½ ft in 1934); Langley Park, Slough, 41 × 4¼ ft (1974); Mount Usher, Co. Wicklow, Eire, 42 × 7¾ ft (1975).

UNGNADIA   SAPINDACEAE

A genus of a single species, which, like the related *Xanthoceras*, bears some resemblance to *Aesculus* in the neighbouring family Hippocastanaceae. In this instance the similarity is even reflected in the vernacular name 'Mexican Buckeye', all the other American buckeyes being species of *Aesculus*. There is, however, the obvious difference that the leaves in *Ungnadia*, as in *Xanthoceras*, are odd-pinnate and that the flowers are borne before or with the leaves, in short dense clusters.

Endlicher named the genus in honour of Baron Ungnad, Austrian envoy at Constantinople 1576–82, who sent seeds of the horse-chestnut to Clusius.

U. speciosa Endl.   Mexican or Spanish Buckeye

A deciduous shrub, or sometimes a small tree to about 35 ft high in the wild; winter-buds globose, many-scaled. Leaves alternate, odd-pinnate, on petioles up to 6 in. long; leaflets mostly five or seven, sometimes three, ovate-lanceolate, acuminate, 3 to 5 in. long, almost glabrous when mature. Flowers fragrant, about 1 in. across, borne in spring before or with the leaves in small clusters from the previous season's wood. Calyx five-lobed. Petals four or five, obovate, clawed, spreading, bright rosy pink, crested with fleshy hairs at the apex. Stamens seven to ten, long-exserted. Style one, slender, with a minute stigma. Fruit a three-valved pear-shaped capsule, roughened but not prickly. Seeds globose, dark brown or black, about ½ in. wide.

Native of northern Mexico, extending into New Mexico and Texas, where, according to Sargent, it is commonest and attains its greatest size fifty miles from the coast west of the Colorado River. Although introduced in 1850, the Mexican buckeye is little known in this country. In old horticultural works it is mentioned as a shrub to be stood outside during the summer and over-wintered under glass, but is probably hardy enough to survive most winters in Britain. There is, however, little hope of seeing unprotected plants in flower unless the wood is well-ripened before winter. It is certainly not a species for cool, rainy localities, and is most likely to succeed in eastern or south-eastern England, against a sunny wall.

VACCINIUM   ERICACEAE

A large genus of shrubs, or occasionally trees, with over 100 species, mostly in the northern hemisphere, but extending into the mountains of S. America. They are deciduous or evergreen, with alternate leaves. Flowers in terminal or axillary racemes, or sometimes solitary. Corolla four- or five-lobed, urn-shaped, campanulate or cylindric (but four-parted in *V. erythrocarpum* and *V. japonicum*). Stamens four or five; anthers awned or not, with a tubular appendage at the apex. Ovary inferior, developing into a berry crowned by the persistent calyx.

As garden shrubs, the vacciniums are chiefly valued for their fruits and the autumnal colour of their foliage. Many are pretty in flower, but none makes the fine display provided by so many of the heath family. In nature they are nearly always found on mountain and moorland, and the genus is one of the most characteristic of the lonely parts of the northern hemisphere. Many produce very palatable fruits.

Under cultivation they prefer a peaty soil, or a light loamy one devoid of lime, and improved by adding decayed leaves. They are all moisture-loving plants. All the hardier ones produce seed which should be treated as advised for rhododendrons, and the others can be propagated by cuttings made of half-ripened wood in July, and placed in sandy peaty soil in gentle bottom heat.

There is considerable confusion in the identification of the N. American species of the *V. corymbosum* group, and American authors are by no means unanimous in their estimates of the number it contains; but they are all hardy, free-growing, and handsome shrubs, the leaves turning red before they fall, but frequently persisting well into the winter.

V. ANGUSTIFOLIUM Ait. LOW BLUEBERRY
V. pensylvanicum var. *angustifolium* (Ait.) A. Gr.

V. angustifolium in its typical state is not so well known in gardens as the following variety, better known as "*V. pennsylvanicum*". The two are the same in their leading botanical characters, but the typical state is dwarfer and intricately branched, with smaller, relatively narrower leaves. Also, it is diploid, whereas the var. *laevifolium* is tetraploid, and is maintained as a separate species by some authorities under the name *V. lamarckii*.

var. LAEVIFOLIUM House *V. pensilvanicum* Lam., not Mill.; *V. lamarckii* Camp—A low, deciduous shrub, usually under 2 ft high; young shoots warted and more or less downy. Leaves nearly stalkless, lance-shaped to narrowly oval or oblong, $\frac{3}{4}$ to $1\frac{1}{2}$ in. long, $\frac{1}{8}$ to $\frac{1}{2}$ in. wide, minutely toothed, glandular at the edge when young, pointed, glabrous and bright green, the midrib downy on one or both sides. Flowers produced in April and May in short dense clusters. Corolla white tinged with red, cylindric to bell-shaped, $\frac{1}{4}$ in. or rather more long. Berries round, $\frac{1}{4}$ to $\frac{3}{8}$ in. wide, normally black, covered with a blue bloom, but white in f. LEUCOCARPUM (Deane) Rehd.; very sweet.

Native of the eastern United States and Canada; introduced in 1772. The berry has a pleasant flavour, and is one of the most valuable wild fruits of N. America, ripening earlier than those of any other species. It covers large areas of poor sandy soil and the stands are periodically burned in some areas to improve the yield. In this country it has little or no value as a fruit-bearer, but makes a pleasing low cover on peaty or light sandy soils. It attains a height of about 3 ft in gardens.

var. NIGRUM (Wood.) Dole *V. pennsylvanicum* var. *nigrum* Wood; *V. brittonii* Porter—Young stems and undersides of leaves glaucous. Berries black.

V. PALLIDUM Ait.—Similar to *V. angustifolium* var. *laevifolium* in habit and leaf-colour, but the leaves up to 2 in. long and 1 in. wide and the corollas cylindric or nearly so.

V. ARBOREUM Marsh.　　FARKLEBERRY

A shrub, or small tree, up to 30 ft high in some of its native localities, and varying also from deciduous to evergreen, according to locality; young twigs downy. Leaves ovate, obovate or oval, ½ in. to 2 in. long, half as wide, very shortly stalked, mostly pointed, minutely and sparsely glandular-toothed, the margins slightly recurved; of leathery texture, glabrous and of a very glossy dark green above, slightly downy beneath. Flowers produced during July and August singly in the axils of the leaves or in the axils of bracts on terminal racemes 1 to 2 in. long, each on a slender stalk ¼ to ⅜ in. long, with two minute bracts about the middle. Corolla white, bell-shaped, ¼ in. long, five-lobed, the lobes reflexed. Calyx small, the five lobes triangular. Fruits ¼ in. wide, black, roundish. The flower is jointed to the stalk. *Bot. Mag.*, t. 1607.

Native of the south and east United States, as far north as Virginia and Missouri; introduced to Kew by John Cree in 1765. In the British Isles it is a deciduous shrub, said by Loudon in 1837 to have been 10 ft high in the walled garden at White Knights, near Reading. It was quite hardy when grown at Kew, pretty and free-flowering, but slow in growth. The form in cultivation is, no doubt, from the northern limits of its distribution, but the evergreen tree form ought to be tried in the mildest counties.

V. ARCTOSTAPHYLOS L.　　CAUCASIAN WHORTLEBERRY

A deciduous shrub, probably 10 ft high ultimately; young wood glabrous or slightly downy. Leaves ovate-oblong, pointed, finely toothed, 1½ to 4 in. long, ¾ to 1½ in. wide, dark dull green and downy on the veins above, paler and more downy beneath; stalk $\frac{1}{12}$ in. long. Flowers produced during June, each in the axil of a bract on slightly downy racemes 1 to 2 in. long from the previous year's wood; corolla greenish white tinged with purple, bell-shaped, ⅓ in. long and wide; stamens ten, hairy; calyx with five shallow triangular lobes. The flower is distinctly jointed to the stalk just below the ovary. Berries globose, purple, ¼ to ⅓ in. across. *Bot. Mag.*, t. 974.

Native of the Black Sea region, from the Caucasus through the mountains of northern Anatolia to European Turkey and southeast Bulgaria. It was discovered about 1701 by the French botanist Tournefort, who identified it with the 'arctostaphylos' or bear's grape of the Greek physician Galen. It was introduced to Britain in 1800, and reintroduced by E. K. Balls in 1934. It is a remarkable fact of plant geography that this species has its nearest relatives in Madeira and the Azores, and is so nearly akin to the Madeiran species,

VACCINIUM ARCTOSTAPHYLOS

V. padifolium, that they had become confused in gardens by the 1830s and remained so for the next hundred years, plants sold as *V. arctostaphylos* often being really *V. padifolium*. But it is a hardier shrub, and in cultivation at least it has larger leaves, and there is the further difference that the filaments of the stamens are hairy.

The leaves, which are the largest among hardy vacciniums, die off a pretty purplish red and fall late in the autumn. It occasionally bears a second crop of flowers in September in the leaf-axils of the current year's shoots.

A decoction of the leaves was used in the Caucasus as a tea, especially by the Circassians. When dried they have the appearance and aroma of black tea, though the brew is very different and inferior in flavour. It was known as 'Broussa tea'.

V. BRACTEATUM Thunb.
Andromeda chinensis Lodd.

An evergreen shrub 3 to 6 ft high; young shoots glabrous or nearly so. Leaves narrowly oval, tapered at both ends, of thin firm texture, distantly or scarcely toothed at all, 1 to 3 in. long, ½ to 1 in. wide, dark green, glabrous; stalk ⅛ in. or less long. Racemes 1 to 2 in. long, minutely downy, carrying a dozen or more flowers, sometimes forming a kind of panicle of short flowering twigs; each flower is in the axil of a small, persistent, leafy bract of linear-lanceolate shape and ⅛ to ⅜ in. long. Corolla white, slender, cylindrical, ¼ in. long, tapering slightly to the mouth which has tiny triangular lobes, minutely downy outside; calyx lobes triangular, downy; stamens downy. Fruits globose,

$\frac{1}{4}$ in. wide, red, downy.

Native of Japan, Korea, and China; apparently first introduced by John Reeves of Canton to Loddiges' nursery at Hackney in 1829; it was figured in their *Botanical Cabinet* (t. 1648) the following year. Reintroduced from China by the late Maurice de Vilmorin in 1914. Its distinctive characters are its downy, slender corolla and conspicuous bracts like tiny leaves which are borne on the main flower-stalk. Although perhaps best suited for the south-western counties and similarly mild localities, where it should make a cheerful evergreen, it is hardy enough and flowered freely in August or September when grown at Kew, reaching 5 ft in height there.

V. CAESPITOSUM Michx.   DWARF BILBERRY

A dwarf deciduous shrub of tufted habit, 4 to 10 in., sometimes only 2 or 3 in. high; branches round, minutely downy or glabrous. Leaves obovate to narrowly wedge-shaped, tapered towards the base, toothed, usually $\frac{1}{4}$ to $1\frac{1}{2}$ in. long, about half as wide, glabrous and shining. Flowers appearing in May with the young shoots, and produced singly on decurved stalks $\frac{1}{8}$ in. long. Corolla pitcher-shaped, $\frac{1}{5}$ in. long, pale pink, five-toothed at the much contracted mouth. Berry globose, about $\frac{1}{4}$ in. wide, black with a blue bloom, sweet. *Bot. Mag.*, t. 3429.

Native of N. America, spreading across the continent from Labrador to Alaska and southwards to New York on the east, to California on the west, inhabiting mountain summits at its more southerly limits. Introduced in 1823. It is a neat little shrub, very suitable for the rock garden.

V. DELICIOSUM Piper—It was in 1915 that this species was first distinguished and named. Previous to that date it was probably confused with *V. caespitosum*, to which it is closely related. It grows 4 to 12 in. high, has the same tufted habit and the same round, glabrous young shoots, but the leaves are of thicker texture and instead of being bright green on both sides are pale and rather glaucous beneath. In shape they are obovate to oval, pointed, tapered at the base, roundish-toothed, $\frac{1}{2}$ to $1\frac{1}{4}$ in. long. Flowers solitary, drooping from the leaf-axils; corolla rather globose, pinkish, $\frac{1}{5}$ in. long. Fruit globose, black with a blue bloom, sweet.

Whilst *V. caespitosum* is distributed right across N. America, *V. deliciosum* appears to be confined to the northwest, more especially to Washington. Piper, the author of the name, describes it as abundant in the alpine meadows at about the limit of trees in the Cascade and Olympic Mountains. It is a neat, deciduous shrub suitable for a moist spot in the rock garden and was introduced some years ago to this country by F. R. S. Balfour of Dawyck.

V. CANADENSE Richards.   SOUR-TOP, VELVET LEAF

A low, much-branched deciduous shrub usually under 1 ft high; shoots very downy, even bristly. Leaves $\frac{3}{4}$ to $1\frac{1}{2}$ in. long, $\frac{1}{8}$ to $\frac{1}{2}$ in. wide, narrowly oval, pointed, not toothed, downy on both sides. Flowers produced during

May along with the young leaves in short dense clusters. Corolla bell-shaped, ¼ in. or less long, white tinged with red. Berries blue-black, ¼ in. or more wide, very agreeably flavoured. *Bot. Mag.*, t. 3446.

Native of eastern N. America; introduced in 1834. It has been much confused in gardens with the various forms of *V. angustifolium*, but is readily distinguished by its very downy entire leaves. Like that species, it gives a valuable wild fruit, its berries ripening later, and forming a useful succession to the other in N. America.

It is a matter of controversy whether this is the species described by Michaux as *V. myrtilloides*. Some authorities have adopted this name (1803) and reduced *V. canadense* Richards. (1823) to synonymy under it. Others hold that *V. myrtilloides* was simply a downy variety of *V. angustifolium*, var. *myrtilloides* (Michx.) House.

V. CORYMBOSUM L.   HIGHBUSH BLUEBERRY

A deciduous shrub, 4 to 12 ft high, forming a dense thicket of erect, much-branched stems; young shoots downy to nearly glabrous. Leaves ovate to oval-lanceshaped, 1 to 3½ in. long, half as wide; tapering at both ends, very shortly stalked, downy beneath on the midrib and veins, not toothed. Flowers

VACCINIUM CORYMBOSUM

produced during May in a series of short, few-flowered clusters near and at the leafless ends of the previous season's twigs. Corolla cylindrical, but narrowed near the mouth, ¼ to ½ in. long, white or pale pink. Berries black, covered with a blue bloom, and from ¼ to ½ in. wide, variable in size, colour, and flavour.

Native of eastern N. America; introduced in 1765. In British gardens this is the commonest, often the only N. American vaccinium. It not only grows well and blossoms freely, but its leaves turn to beautiful shades of red before falling in the autumn.

V. corymbosum is a very variable species, which has been shown to be polyploid and may be the result of hybridisation in post-glacial times between species that survived the Ice Age in southern habitats. Numerous variants have been named, and the following are still recognised in recent works of American botany:

f. ALBIFLORUM (Hook.) Camp *V. albiflorum* Hook.; *V. corymbosum* var. *amoenum* A. Gr., not *V. amoenum* Ait.; *V. c.* var. *albiflorum* (Hook.) Fern.—Like the type in habit, but with the leaves minutely toothed, and hairy on the margins when young. *Bot. Mag.*, t. 3428.

f. GLABRUM (A. Gr.) Camp *V. corymbosum* var. *pallidum* A. Gr., not *V. pallidum* Ait.—Leaves bluish beneath, and glabrous except for the minutely toothed ciliate margins. Berries deep blue.

V. ATROCOCCUM (A. Gr.) Heller *V. corymbosum* var. *atrococcum* A. Gr.— Near to *V. corymbosum* but with much more downy leaves, and often broader and shorter corollas, and black fruits. It is almost as widely distributed in the wild as *V. corymbosum*.

The highbush blueberries ·are now an important soft fruit in eastern N. America, where some 20,000 acres are devoted to their cultivation. The first of the named commercial clones were selected in the wild, or bred, by Frederick V. Coville early this century, and released to the trade from the 1920s onwards. They derive partly from *V. corymbosum* and partly from the closely related but more southern V. ASHEI, not treated here. Many of these, and others of more recent raising, are available in Britain, and could well be planted in place of unselected forms, for the autumn colour is usually as good and the fruits both larger and more abundant. But two or more clones should be grown together, as cross-pollination is necessary to ensure a good set of full-sized fruits. Netting against birds is essential and the older wood should be cut out periodically, since the quality and quantity of the fruit falls off once the laterals become twiggy. See further in *Journ. R.H.S.*, Vol. 98 (1973), pp. 401–4.

V. CRASSIFOLIUM Andr.

An evergreen shrub of more or less procumbent habit; young wood covered with fine down. Leaves set about ¼ in. apart on the twigs, oval; ⅓ to ¾ in. long, ⅛ to ⅜ in. wide, slightly toothed, quite glabrous, shining green, and of leathery texture; stalk 1/16 in. long, reddish like the young twigs. Flowers produced in May and June in short lateral and terminal racemes; corolla bell-shaped, rosy red, ¼ in. long. Berries black. *Bot. Mag.*, t. 1152.

Native of the south-eastern United States; introduced in 1787. It is not very hardy in the London district, and is better adapted to the milder parts. The only other evergreen vaccinium in cultivation with which it might be confused is *V. vitis-idaea*, which is a much sturdier shrub with larger leaves, speckled beneath with black dots.

V. DELAVAYI Franch.

A compact evergreen shrub, 1 to 3 ft high; young shoots angled, bristly. Leaves much crowded on the shoots (ten or twelve to the inch) obovate, rounded or notched at the end, tapered to the base, not toothed, margins decurved, ⅜ to ½ in. long, half as much wide, glabrous, rather leathery; stalk very short. Racemes ½ to 1 in. long, usually terminating the shoot, main flower-stalk bristly. Corolla roundish urn-shaped, ⅙ in. long, described by Forrest as creamy yellow, tinged with rose outside, also as white flushed with rose, and bright rose; calyx lobes triangular, ciliate. Fruits globose, ⅙ in. wide, described by Forrest as 'dark crimson' and 'purplish red.'

Native of S.W. China and upper Burma; discovered by Père Delavay in Yunnan and introduced by Forrest. In general appearance it much resembles *V. moupinense* and, like that species, grows in the wild on cliffs, rocks, and as an epiphyte on trees. It differs from it in the leaves having a notch at one end and in the bristly flower-stalks. But both species are closely allied to *V. nummularia*.

V. delavayi makes a neat little evergreen with small, box-like foliage, and is hardy in a sheltered position, but is not suitable for gardens subject to late spring-frost. The flowers on cultivated plants are white or cream-coloured, sometimes flushed with pink, and the young growths are tinted with red. But many years may pass before it flowers freely. The fruits on cultivated plants are bluish purple. *V. delavayi* received an Award of Merit in 1950.

V. DUCLOUXII (Lévl.) Hand.-Mazz.

Pieris duclouxii Lévl.; *V. forrestii* Diels

A deciduous shrub up to 10 ft high; young shoots round, mostly glabrous. Leaves 1½ to 3½ in. long, ½ to 1¼ in. wide; ovate-lanceolate, slender-pointed, tapered at the base, glabrous; stalk ⅛ in. long. Flowers produced in May on the lower side of axillary racemes 1 to 3 in. long; corolla white (either pure or pink tinted), cylindrical, ¼ in. long, with five teeth at the narrow mouth; glabrous outside, downy within; calyx-lobes triangular; filaments of stamens hairy. Fruits ⅕ in. wide, black-purple. *Bot. Mag.*, t. 9658.

Native of W. China, mainly in Yunnan. It was raised by J. C. Williams of Caerhays in Cornwall from seeds collected by Forrest in 1913–14, and the flowering and fruiting sprays portrayed in the *Botanical Magazine* were sent by him in 1925 and 1932 respectively. But the species has never become established in cultivation and is probably only suited to the milder parts.

V. duclouxii is one of five Chinese species all closely allied to V. DONNIANUM Wight, described in 1848 from the Khasi Hills of Assam. See further in the note by J. R. Sealy accompanying the plate in the *Botanical Magazine*. *V. donnianum* is also known as *V. sprengelii* (Don) Sleum., based on *Agapetes sprengelii* G. Don (1834), but the identity of the plant described by Don under this name is uncertain.

V. DUNALIANUM Wight

An evergreen shrub up to 20 ft high in its wild state, or almost a tree; quite devoid of down in all its parts; young shoots lightly ribbed. Leaves oval or oval-lanceolate, with a long, slender, often tail-like end, and a broadly tapered base, not toothed, 3 to 5 in. long, 1 to 1¾ in. wide, dark green, of leathery texture; stalk ⅛ to ¼ in. long. Racemes axillary, 1½ to 3 in. long; corolla waxy white, bell-shaped, ⅕ in. long, with five triangular teeth; fruits black, globose, ¼ in. wide.

Native of the eastern Himalaya, and of the mountains of N.E. India, extending into W. China; introduced in the last century from India. Wilson found it on Mt Omei in 1904 and it had previously been collected there by Henry. It is one of the strongest growing of vacciniums and is worth growing as a handsome evergreen in the south-western counties. Near London it needs cool greenhouse treatment. The slender tail-like apex of the leaf, often 1 to 1½ in. long, is distinctive.

V. ERYTHROCARPUM Michx.

Oxycoccus erythrocarpus (Michx.) Pers.

A deciduous shrub, from 3 to 6 ft high, with downy young branches. Leaves short-stalked, ovate or ovate lance-shaped, taper-pointed, 1 to 3 in. long, scarcely half as wide, bristle-toothed, tinged with red and slightly hairy when young. Flowers produced in June singly on slender pendulous stalks, about ½ in. long, from the axils of the young leaves. Corolla pale red, deeply four-lobed; the lobes narrow, ⅓ in. long, and curled back, leaving the long anthers exposed and standing close together in a sort of column. Berries acid, roundish, ¼ in. wide, turning red, then purplish black. *Bot. Mag.*, t. 7413.

Native of the mountains of the southeastern United States; introduced in 1806 by Loddiges of Hackney, but never common. It is a pretty shrub and of peculiar interest in forming a connecting link between *Vaccinium* and *Oxyoccus* (the true cranberries). It has the shrubby habit of the former, but the flower structure and arrangement of the latter. When first introduced it was hoped that it might prove of value as a fruiting bush, but like the rest of the imported species, it has never borne fruit freely enough to count for much.

V. FLORIBUNDUM H.B.K.

V. mortinia Benth.

An evergreen shrub, 2 to 4 ft high, the young shoots covered with a dark minute down. Leaves densely set on the twigs (about ⅛ in. apart), ovate, minutely toothed, ⅓ to ½ in. long, very uniform, dark green above and glabrous except for a little dark down on the midrib, paler and minutely pitted beneath with a tiny bristle in each cavity; stalk downy, ¹⁄₁₂ in. long. Flowers produced during June in short, dense racemes from the leaf-axils, and on the lower side of the twigs. Corolla rosy pink, cylindrical, about ¼ in. long; stamens hairy; calyx with five triangular lobes; flower-stalks downy. Berries red, $\frac{1}{5}$ in. in diameter. *Bot. Mag.*, t. 6872.

Native of the northern and central Andes; described from Peru; introduced by Hartweg about 1840 from the slopes of Mt Pichincha in Ecuador. It is damaged only in severe winters and succeeds admirably in the Heath Garden at Wakehurst Place in Sussex. It was also grown for over twenty-five years by R. B. Cooke in his garden in Northumberland. It is a particularly neat and pleasing shrub, although its flowers and fruits are hidden by the foliage. Its chief attraction is the beautifully red-tinted young growths, for which alone it deserves a place in any collection of ericaceous shrubs.

V. floribundum is of peculiar interest as affording one of the very few instances of a shrub found wild within a few miles of the equator, yet hardy enough to grow and flower in Britain. The fruits are sold in the market of Quito, and from the name 'mortina' by which they are known there the specific epithet used by Bentham was adapted.

V. FRAGILE Franch.

V. setosum C. H. Wright

An evergreen shrub 1 to 3 ft high; young shoots round, densely bristly. Leaves ovate to oval, tapered to both ends; finely and regularly toothed, $\frac{1}{2}$ to 1 in. long, $\frac{1}{4}$ to $\frac{3}{8}$ in. wide, nearly or quite glabrous above, downy beneath especially on the midrib. Flowers produced in a cluster of racemes during May and June from the terminal leaf-axils; racemes 1 to 2 in. long, downy; corolla urn-shaped, white to rosy red, $\frac{1}{6}$ to $\frac{1}{4}$ in. long, with five small reflexed teeth at the mouth; calyx-lobes ciliate. Fruits black, globose, $\frac{3}{16}$ in. wide.

Native of W. China; described from specimens collected by Père Delavay in Yunnan, where it reaches to above the tree-line. Wilson collected it in fruit in W. Szechwan during his first expedition for the Arnold Arboretum, but apparently it was first introduced to this country by Forrest, according to whom the fruits in their season are the principal food of the common pheasant of Yunnan. According to the field notes of its various collectors the colour of the flowers in wild plants varies from white to salmon red. The beauty of the inflorescence is heightened by the red bracts on the raceme. Lionel de Rothschild grew this species at Exbury and thought highly of it as a dwarf evergreen, but it never became fully established in gardens and is now rare.

V. GLAUCO-ALBUM C.B.Cl.

An evergreen shrub, 2 to 4 ft high; young stems soon glabrous. Leaves stiff and hard in texture, oval or ovate, $1\frac{1}{2}$ to $2\frac{1}{2}$ in. long, $\frac{5}{8}$ to $1\frac{1}{4}$ in. wide, pointed, with bristle-like teeth on the margins, green and glabrous above, of a vivid blue-white and slightly bristly on the midrib beneath. Racemes slightly downy, 2 to 3 in. long, produced from the leaf-axils, and conspicuous for their large, persistent, blue-white bracts, edged with bristles. Corolla pinkish white, $\frac{1}{4}$ in. long, cylindrical, contracted at the mouth; calyx glabrous, shallowly lobed. Berries $\frac{1}{3}$ in. in diameter, globose, black, covered with blue-white bloom.

Native of the Himalaya from E. Nepal eastwards, thence into northernmost Burma and S.E. Tibet; in cultivation by about 1900. Many of the vacciniums of the eastern Himalaya and S.E. Asia are epiphytes, but this often forms extensive thickets near the tree-line in some localities, thriving in places too dry for most rhododendrons. It is one of the finest of the hardier vacciniums, remarkable for the vivid blue-white bloom on the fruits, bracts and the under-surface of the leaves. The fruits are very freely borne in this country and are often untouched by birds until late winter.

V. glauco-album seems to have been little known in gardens outside a few collections until it received an Award of Merit when shown from Bodnant in 1931. The Bodnant form represents an old introduction and is figured in *Bot. Mag.*, t. 9536. But many of the plants now in gardens derive from the seed collected by Kingdon Ward in late 1924 on the Doshong La at the eastern end of the Himalaya (*Bot. Mag.*, t. 8924). Since then there have been several other introductions—by Ludlow and Sherriff and their companions from the Himalaya and S.E. Tibet, and by Kingdon Ward from the mountains beyond the Tsangpo and from upper Burma.

In the form originally introduced, this species was hopelessly tender at Kew, but later introductions have proved hardy in gardens south of London, in a sheltered position.

V. GAULTHERIIFOLIUM (Griff.) C.B.Cl. *Thibaudia gaultheriifolia* Griff.— Allied to the above, and with the same white undersides to the leaves and large wax-coated berries, but a taller shrub to about 8 ft high and the leaves up to almost 5 in. long. Native of the Himalaya from E. Nepal eastwards, extending through northernmost Burma into S.W. China. The date of its first introduction is uncertain, but this is Kingdon Ward's 'big-leaved vaccinium' of which he collected seed in 1931 in the Adung Valley, Burma, under KW 9197. The seed was distributed as *V. glauco-album*, though with a query. It was reintroduced by the University of North Wales expedition to eastern Nepal in 1971.

V. HIRSUTUM Buckl.

A low, deciduous shrub, 2 to 3 ft high, spreading by underground rhizomes; young shoots very downy, and remaining so the second year. Leaves ovate to oval, 1 to $2\frac{1}{2}$ in. long, $\frac{1}{2}$ to $1\frac{1}{4}$ in. wide, shortly stalked, pointed, deep green and slightly downy above, paler and more downy beneath, not toothed. Flowers in broad, short racemes, produced towards the end of May. Corolla cylindrical, narrowed towards the mouth, $\frac{3}{8}$ in. long, white tinged with pink, hairy; calyx-lobes pointed, and like the flower-stalks, very hairy. Berries $\frac{1}{4}$ in. in diameter, nearly globular, blue-black covered with gland-tipped hairs.

Native of the mountains of N. Carolina and southwards, discovered by B. S. Buckley about 1836, but lost sight of until rediscovered and brought into cultivation by Prof. Sargent in 1887. Given a position that is moist and not too sunny, it spreads rapidly by underground suckers. It is rendered very distinct by the hairiness of all its parts, more especially of its fruits, which have a sweet, pleasant, but not very pronounced flavour.

V. × INTERMEDIUM Ruthe

A natural hybrid between *V. myrtillus* and *V. vitis-idaea* that occurs occasionally where the two species are in contact, usually in disturbed habitats. It is fairly intermediate between its parents, but resembles *V. myrtillus* more closely in habit. The stems are not so markedly angular as in that species, and it inherits from *V. vitis-idaea* an evergreen or almost evergreen character. The leaf-margins are toothed, but the under-surface is not dotted. Berries dark violet. It was described from Germany in 1834, and originally discovered in this country in Maer Woods, Staffordshire, by Robert Garner in 1870. Prof. Bonney found it in 1886 on Cannock Chase, where it still occurs. It is commonest, however, on the moors of north Derbyshire.

V. JAPONICUM Miq.

A deciduous shrub 2 to 3 ft high, with angular, glabrous young shoots. Leaves ovate to ovate-oblong, pointed, heart-shaped, rounded or tapered at the base, very finely toothed, 1 to 2¼ in. long, ½ to 1¼ in. wide, glabrous, very shortly stalked. Flowers solitary in the leaf-axils, each borne in June and July on a very slender stalk ½ to ¾ in. long. Corolla pink, deeply divided into four narrow lobes which are curled back so as to leave ⅜ in. of the long anthers exposed and clustered together in a sort of column; calyx with triangular lobes. Fruits globose, ¼ in. wide, bright red, pendulous.

Native of Japan and Korea; introduced about 1893. It belongs to the same section of *Vaccinium* as the American *V. erythrocarpum*, and with it forms a connecting link with the cranberries (*Oxycoccus*). Both have the shrubby habit of *Vaccinium* but the corolla is deeply split as in *Oxycoccus*, the lobes similarly recurved.

var. SINICUM (Nakai) Rehd. *Oxycoccoides japonica* var. *sinica* Nakai—This Chinese variety has proportionately narrower leaves with down on the midrib beneath. Introduced by Wilson in 1907. *V. erythrocarpum* differs from both species and this variety in its downy shoots and more downy leaves; the flower-stalks are all very slender.

V. MACROCARPUM Ait. *see* OXYCOCCUS MACROCARPUS

V. MEMBRANACEUM Dougl. ex Torr.
V. myrtilloides Hook., not Michx.

A deciduous shrub from 1 to 5 ft high, erect-growing; branchlets angular, glabrous. Leaves ovate to oblong, pointed, rounded or tapered at the base, minutely toothed, $\frac{3}{4}$ to $2\frac{1}{2}$ in. long, $\frac{1}{3}$ to 1 in. wide, bright green and glabrous on both surfaces; stalk $\frac{1}{16}$ in. or less long. Flowers solitary in the leaf-axils on stalks $\frac{1}{4}$ to $\frac{1}{3}$ in. long. Corolla between globose and urn-shaped, $\frac{1}{4}$ in. across, greenish or pinkish white; calyx entire. Berries $\frac{1}{4}$ to $\frac{1}{3}$ in. in diameter, purplish black, sweet, but rather acid. *Bot. Mag.*, t. 3447.

Native of N. America from the region of the Great Lakes westward to British Columbia, south to northern California; discovered by Douglas about 1828. It belongs to the same group as *V. myrtillus* and *V. uliginosum*. Of little garden value.

V. MOUPINENSE Franch.

A low evergreen shrub of close, dense habit, 1 to 2 ft high; young shoots grooved, downy. Leaves leathery, entire, crowded (ten or twelve to the inch), narrowly obovate or oval, usually tapered more abruptly to the bluntish or rounded apex, $\frac{1}{3}$ to $\frac{1}{2}$ in. long, $\frac{1}{8}$ to $\frac{1}{16}$ in. wide, dark glossy green, quite glabrous except for the short stalk which is downy. Racemes $\frac{3}{4}$ to 1 in. long, carrying nine to fifteen nodding flowers, the main and individual flower-stalks quite glabrous, chocolate-red. The racemes spring mainly from the axils of the terminal leaves. Corolla $\frac{3}{16}$ in. long, urn-shaped, five-angled, contracted at the top to a small orifice where are five tiny triangular lobes; deep shining chocolate-red. Calyx glabrous, coloured like the corolla, its lobes triangular. Stamens included in the corolla, their stalks hairy. Fruits globose, $\frac{1}{4}$ in. wide, purple-black.

Native of W. Szechwan, China; discovered by David in 1869; introduced by Wilson in 1909. In the original description by Franchet the flowers are described as white and Wilson describes them as rose-pink, but on the plants which have flowered at Kew they are dark red or chocolate crimson. Wilson observes that, in its wild state, it often occurs as an epiphyte on old trees. It is a pleasing little evergreen, suitable for the rock garden and growing well in peaty or light loamy soil. It blooms in May and June. Very similar to *V. delavayi*, it can be distinguished by its leaves not being notched at the apex and by its glabrous inflorescence.

V. MYRSINITES Lam.

V. nitidum Andr.

An evergreen shrub, often low and spreading, sometimes up to 2 ft high; young shoots slender, mostly downy, minutely warty. Leaves narrowly oval or obovate, pointed, tapered at the base, indistinctly toothed, scarcely stalked; $\frac{1}{4}$ to 1 in. long, $\frac{1}{8}$ to $\frac{1}{4}$ in. wide (sometimes larger), glossy green above, paler, conspicuously veined and sometimes bristly beneath. Flowers produced in April and May in terminal and axillary clusters towards the end of the shoots. Corolla tubular, $\frac{1}{4}$ in. long, white or tinged with pink, with five small teeth; calyx triangularly toothed, glabrous like the flower-stalk. Fruits $\frac{1}{4}$ in. wide, blue-black. *Bot. Mag.*, t. 1550.

Native of S.E. United States from Virginia to Florida; cultivated by Loddiges of Hackney in 1813. It is not very hardy near London, but should succeed farther south and west. The upright form suggests *Pernettya mucronata* by its small leaves closely set on the branchlets.

V. MYRTILLUS L.   WHORTLEBERRY, BILBERRY

A deciduous shrub usually 6 to 12 in. high, sometimes more; young branches glabrous, distinctly angled, remaining green for several years. Leaves ovate, ½ to 1 in. long, often somewhat heart-shaped, regularly and bluntly toothed, bright green and quite glabrous, scarcely stalked. Flowers produced in May usually singly on drooping stalks from the leaf-axils. Corolla nearly globular, pale pink, ¼ in. long. Berries black, with a blue bloom, ⅓ in. in diameter, globular.

Native of Britain, where it is one of the commonest of mountain and moorland shrubs, also of N. and Central Europe. The bilberry is one of the most valuable wild fruits of Britain, and is frequently offered in considerable quantities in the markets of north country towns. They are used for making tarts, jelly, and are especially delicious eaten with cream and sugar. A very hardy plant, it manages to survive on the summits of our loftiest mountains. It is scarcely of sufficient interest for the garden, and does not always thrive well translated to low-level gardens, in the south at any rate. Its angled stems distinguish it from the other British species.

f. LEUCOCARPUM (Dumort.) E. Busch—Fruits white. This occurs quite frequently in the Alps; the fruits are said to be sweeter and better flavoured than the normal kind.

V. SCOPARIUM Leiberg   *V. myrtillus* var. *microphyllum* Hook.; *V. microphyllum* (Hook.) Rydb., not Bl.; *V. erythrococcum* Rydb. GROUSEBERRY.— Native of the inner ranges of western North America as far north as British Columbia. It is about half the size of *V. myrtillus* in all its parts and its berries are red.

V. NUMMULARIA C.B.Cl.

A low evergreen shrub often found growing wild in the forks of trees and having pendulous branches there. Young shoots rather slender, but made to look thicker by their dense covering of pale brown bristles, which give them an almost mossy appearance. Leaves scarcely stalked, of firm, even hard texture, bright green, wrinkled but not downy above, conspicuously veined and glabrous beneath; broadly oval to ovate, rounded or bluntish at the apex, rounded at the base, ½ to 1 in. long, ⅜ to ⅝ in. wide, margins recurved, sparingly set with bristles. The leaves are set on the twigs six or eight to the inch. Flowers opening in May and June, crowded on several racemes, each ½ to ¾ in. long, clustered at the end of the shoots. Corolla rose-red to pink or pinkish white, ⅕ in. long, 1/10 in. wide, tapering from the base to the narrow mouth; stamens hairy; calyx shallowly lobed, ciliate. Fruits globose, ⅕ in. wide, each on a stalk ¼ to ½ in. long, black, said to be edible. *Bot. Mag.*, n.s., t. 470.

Native of the eastern Himalaya and some of the lower ranges of northeastern

India. **The** younger Hooker collected it in Sikkim in 1850 and may have sent home seeds, but it first came to notice in gardens around 1930. It occurs in the Himalaya at altitudes of up to 12,000 ft and has been successfully grown in several gardens of the south and west and flowers well when once established, but it is not reliably hardy and the bronze-coloured growths may be cut by late frost.

V. nummularia is replaced in northern Burma and western China by related species, of which *V. delavayi* is the nearest allied.

V. OLDHAMII Miq.
V. ciliatum sens. G. Don, not Thunb.

A deciduous shrub up to 12 ft high in Japan, of bushy habit; young shoots glandular-downy. Leaves ovate, oval or obovate, pointed, tapered or sometimes rounded at the base; edged with fine, bristle-like teeth, 1 to 3 in. long, $\frac{1}{2}$ to $1\frac{1}{2}$ in. wide, green on both surfaces, sprinkled with bristles above, bristly on the midrib and veins beneath; stalk $\frac{1}{8}$ in. or less long. Racemes $1\frac{1}{2}$ to $2\frac{1}{2}$ in. long, carrying eight to twenty nodding flowers; flower-stalks glandular-downy with leaf-like bracts at the base. Corolla bell-shaped, $\frac{1}{6}$ in. long, reddish; stamens downy, somewhat shorter than the corolla; calyx-lobes triangular. Blooms in June. Fruits globose, $\frac{1}{3}$ in. wide, black, edible.

Native of Japan and Korea; described from a specimen collected by Richard Oldham near Nagasaki in Japan in 1863, but found earlier in Korea by Charles Wilford; introduced about 1892. Although quite hardy, it is not a particularly attractive species, but the leaves sometimes turn a good red before falling.

V. OVALIFOLIUM Sm.

A deciduous shrub of slender shape from 4 to 12 ft high; young shoots angular, not downy. Leaves oval, ovate, not toothed, blunt at the apex, rounded at the base, 1 to $2\frac{1}{2}$ in. long, $\frac{5}{8}$ to $1\frac{1}{4}$ in. wide, pale green above, rather glaucous beneath, quite glabrous; stalk $\frac{1}{16}$ in. long. Flowers solitary on a glabrous drooping stalk about $\frac{1}{4}$ in. long. Corolla egg-shaped, much narrowed at the mouth, $\frac{3}{8}$ in. long, pinkish; calyx shallowly ten-toothed. Fruits bluish purple, $\frac{2}{5}$ in. wide, acid.

A native of the northern USA and Canada from Labrador to Alaska, and of northeast Asia, including Japan. It was first collected by Archibald Menzies on the northwest coast of N. America during Vancouver's great voyage of survey in 1790–5. He found it in 'shady Alpine woods' and alludes to its 'very useful fruit'. It was probably not introduced to this country until the early years of this century and does not have sufficient beauty of flower to have secured for itself a permanent place in gardens, its chief value being the occasional rich colouring of its foliage in autumn.

V. OVATUM Pursh   BOX BLUEBERRY

An evergreen shrub of bushy habit, 10 to 12 ft high in this country; young wood purple, covered with short, dense down. Leaves ¼ to ½ in. apart, sometimes slightly heart-shaped at the base, but usually rounded or tapering; ½ to 1½ in. long, ⅓ to ¾ in. wide; of firm leathery texture, finely and regularly toothed, dark glossy green above, paler beneath and glabrous except for some short, scattered bristles beneath and some down on the midrib above. Flowers produced in May and June four to six together in short, nodding racemes from the leaf-axils, white, roundish, bell-shaped, with five small, recurved, triangular lobes. Berries black, round, ⅓ in. in diameter. *Bot. Mag.*, t. 4732.

Native of the coastal regions of western N. America from Alaska to California; discovered by Menzies during Vancouver's voyage and introduced by Douglas in 1826. Although hardy enough to withstand the hardest winters experienced at Kew it may suffer in severe frost through the cutting back of the younger growth, and prefers woodland conditions. It is a handsome shrub when seen at its best, with its densely set glossy leaves, which are pinkish brown when unfolding and often take on a purple tone in winter. It grows well in deep shade and will even give a good account of itself where the soil is rooty, but needs lighter conditions if it is to produce its flowers, which in some forms do not open until August. The fruits, not borne freely with us, are variable in quality; the finest form, with large juicy fruits, was named var. SAPOROSUM by Jepson. The foliage is useful for floral decoration, and is cut in such quantity from wild stands that the species is becoming scarce in some areas.

V. OXYCOCCUS L. *see* OXYCOCCUS PALUSTRIS

V. PADIFOLIUM Sm.   MADEIRA WHORTLEBERRY

V. maderense Link; *V. arctostaphylos sens.* Ait., in part, not L.

A deciduous shrub, 6 to 8 ft high in this country, but becoming a small tree in Madeira; young wood downy except for glabrous strips extending from the base of one leaf to the axil of the next below. Leaves ovate to oval, 1 to 2¼ in. long, ½ to 1 in. wide, rounded or tapering at the base, pointed, finely toothed, dark green and downy on the midrib above, paler and downy at the base of the midrib below; stalk 1/12 in. long. Flowers produced in June in racemes 1 to 2 in. long, from the wood of the previous year, each flower drooping and jointed at the base of the ovary to a short stalk springing from the axil of a bract about ⅓ in. long. Corolla bell-shaped, ⅓ in. long, dull yellow tinged with purple, the five lobes triangular; stamens ten, glabrous. Berries blue, globose, ⅓ to ½ in. across. *Bot. Mag.*, t. 7305.

Native of the mountains of Madeira at altitudes of 3,000 to 5,000 ft; introduced to Kew by Francis Masson on his return in 1777 from his famous collecting expedition to the Cape of Good Hope. What was believed to be one of his plants was still growing at Kew near the Main Gate in 1909; it had, at any rate, stood in that position for at least sixty years. But whilst it may thus be considered hardy, it thrives better where the climate is warmer and moister. There has been much confusion between this species and the Pontic *V. arctostaphylos*, but seen together they are quite distinct. The latter has larger leaves, is of more open growth, the stamens are hairy, and it is quite deciduous. The confusion no doubt arose because *V. padifolium* was at first identified as *V. arctostaphylos* and was well established in gardens under that name when Smith distinguished it as a separate species in 1817.

V. CYLINDRACEUM Sm. *V. longiflorum* Wikstr.—Allied to *V. padifolium* (and to *V. arctostaphylos*) but easily distinguished by the cylindrical corollas, which are up to ¾ in. long, greenish yellow tinged with purple. The flowers are borne in dense racemes up to 2 in. long, up to twenty in each, during late summer and autumn. The narrowly elliptic or lanceolate leaves do not fall until shortly before the new foliage appears in spring. Native of the Azores, not cultivated in the open air until the late 1930s, and even now uncommon in gardens, but hardy in woodland south of London. It makes a tall shrub of narrow habit.

V. PARVIFOLIUM Sm.

A deciduous shrub, varying in height from 1 to 6 ft; the stems and twigs slender, sharply angled (like *V. myrtillus*) when young, glabrous. Leaves oval, obovate, or nearly round, thin, ¼ to ½ in. long, not toothed. Flowers solitary in the leaf-axils, nodding; corolla globular, pinkish white. Berries bright red, acid but very palatable, ¼ in. across.

Native of western N. America from California to Alaska, in coastal forests. It is uncommon in gardens and its chief value as an ornament is in the beautiful red autumn colouring of its leaves.

V. PRAESTANS Lamb.

A low deciduous shrub with a creeping root-stock; its glabrous or downy mostly unbranched shoots growing 3 to 6 in. high. Leaves obovate, sometimes broadly oval, usually rounded at the apex except for a small mucro, or broadly tapered to a point; always slenderly tapered at the base to a short stalk, indistinctly toothed, 1 to 2¼ in. long, ½ to 1½ in. wide, both surfaces green, glabrous above, sparsely hairy on the veins beneath. Flowers white, tinged with pink, produced in June two or three together or solitary at the base of the leafy part of the stem, each on a short downy stalk that is furnished with two narrow, leaf-like bracts. Corolla bell-shaped, ¼ in. or less long, with erect lobes at the mouth; stamens downy; calyx-lobes rounded, ciliate. Fruits ⅖ to ½ in. wide, globose, bright glossy red, sweet and fragrant.

Native of N.E. Asia, including northern Japan, also of the Aleutians and Alaska. It was shortly described in 1810 from a specimen that had been collected in Kamchatka, but does not seem to have reached cultivation until Wilson introduced it from Sakhalin in 1914. It is remarkably distinct for a vaccinium in its low, creeping habit, in the shape and size of its leaves, and in its large fruits, which some say have a fragrance like strawberries. Inured as it is to severe cold

VACCINIUM PRAESTANS

in winter it succeeds best in the north and in Scotland. In southern England it needs the same conditions as *Therorhodion kamtschaticum*—a damp, peaty soil and a position where it gets abundant light but little direct sun.

V. STAMINEUM L.   DEERBERRY

V. candicans Mohr ex Bean; *Polycodium stamineum* (L.) Greene

A deciduous shrub, 2 to 4 ft high, of bushy, much-branched habit; twigs downy. Leaves oval or ovate, ¾ to 3 in. long, about half as wide, pointed, dark dull green above, and downy on the midrib, paler or more or less glaucous and downy beneath; leaf-stalk ⅛ in. or less long. Flowers white, with bright yellow projecting anthers, produced during May and June in downy racemes, 1 to 2 in. long, furnished with leaf-like bracts ¼ to ¾ in. long. Corolla open, bell-shaped, ¼ to ⅓ in. wide; calyx glabrous except on the ciliate margins, flower-stalk slender, downy, ⅛ to ½ in. long. Fruits greenish or yellowish, round or pear-shaped, ⅓ in. wide, not edible.

Native of eastern N. America; introduced in 1772. This is one of the prettiest vacciniums in its blossoms, which are freely borne on short, broad racemes, springing from the joints of the previous year's wood. It is distinct among cultivated vacciniums in its open corollas with protruding stamens, likened by William Marchant to those of *Solanum jasminoides* in their shape. The leafy bracts are also a distinctive feature. It has nothing to offer out of flower, as the leaves do not colour in autumn and the fruits are of no interest.

var. MELANOCARPUM Mohr *V. melanocarpum* (Mohr) Mohr—Leaves downy beneath. Calyx densely woolly. Fruits dark purple.

var. NEGLECTUM (Small) Deam *Polycodium neglectum* Small; *V. neglectum* (Small) Fern.—Young shoots, leaves and flower-stalks glabrous.

V. ULIGINOSUM L.   BOG BILBERRY

A deciduous shrub, 1 to 2 ft high, with very minutely downy or glabrous round branchlets. Leaves obovate, or almost round, not toothed, glabrous or finely downy beneath, dull glaucous green, ½ to 1 in. long, with scarcely any stalk. Flowers produced during May singly or in pairs or threes from the uppermost joints of the previous year's wood, each on a drooping stalk about ¼ in. long. Corolla pale red or white, bell-shaped, ⅙ in. long, with usually four teeth. Berries black with a blue bloom, sweet.

Native of the mountain heaths and bogs of the Northern Hemisphere and common in the north of Britain. The fruit is edible, but is said to produce headache and giddiness if eaten in quantity. It furnishes a valuable food for mountain game, but is scarcely worth cultivating in gardens. From its companion deciduous species in Britain (*V. myrtillus*), it is easily distinguished by its round stems, entire leaves, and in the parts of the flower being mostly in fours.

V. URCEOLATUM Hemsl.

An evergreen bush up to 6 ft high in its wild state; young shoots at first covered thickly with fine down. Leaves of firm, leathery texture, ovate-oblong, narrowed at the apex to a long fine point, rounded at the base, not toothed, 2 to 4 in. long, 1 to 2½ in. wide, dark green, downy only when quite young; veins very deeply impressed on the upper surface; stalk very short, about ¹⁄₁₂ in. long. Flowers in racemes 1 to 1½ in. long, springing from the leaf-axils in June. Corolla urn-shaped, ¼ in. long, pink; calyx-lobes triangular; stamens with slightly exposed anthers, their stalks downy. Fruits black, globose, ¼ in. wide.

Native of W. China; first discovered about 1887 on Mt Omei, in Szechwan by the Rev. E. Faber. Wilson found it on the same mountain in 1904 when collecting for Messrs Veitch and on Wa-shan in 1908. Introduced to Kew from Messrs Vilmorin of Paris in 1923. It is scarcely hardy enough to grow really well at Kew, but it succeeds well in Cornwall. Wilson observes that it is partial to sandstone boulders. It is more notable perhaps for its striking evergreen foliage than for any beauty of blossom.

V. VACILLANS Kalm *sec.* Torr.
V. torreyanum Camp

A deciduous shrub, 1 to 4 ft high, with mostly glabrous, yellowish green, warted branchlets. Leaves mostly oval to obovate, 1 to 2 in. long, about half as wide, nearly entire, or minutely toothed except towards the base, very shortly stalked, minutely pointed, glabrous, firm, pale or glaucous beneath.

Flowers produced during May in short clusters on the leafless tips of the previous year's shoots. Corolla greenish, pink or purplish, about ¼ in. long; calyx often reddish. Berries roundish, ¼ to ⅓ in. wide, black, usually covered with a blue bloom, very sweet.

Native of the eastern United States from Maine southwards to Georgia. This is a stiffly branched species with firm textured leaves, and is one of the most ornamental in its flowers, which, like the fruits, cover the terminal (and naked) 2 or 3 in. of the twigs. Said to favour dryish situations in the wild.

V. VIRGATUM Ait.

A deciduous shrub of erect habit, 4 to 10 ft high; young shoots minutely downy. Leaves ovate-lanceolate to oval-oblong, 1 to 3 in. long, 1 to 1½ in. wide, tapered to both ends, finely toothed or entire, bright green and glabrous above, pale or glaucous beneath; shortly stalked. Flowers white or pink, in short axillary clusters of six to ten; corolla ⅓ in. long, cylindrical but slightly tapered towards the mouth, where are five tiny, reflexed teeth; calyx five-lobed, lobes triangular. Fruits globose, black, ¼ in. wide, sometimes slightly covered with bloom. *Bot. Mag.*, t. 3522.

Native of eastern N. America from Southern Virginia southwards, often in swamps. Much confused in gardens with *V. corymbosum* which has a more urceolate, less cylindrical corolla. Probably some of the plants called *V. virgatum* in gardens and valued for their autumn tints are really *V. corymbosum*.

V. VITIS-IDAEA L.   COWBERRY

A low, evergreen, creeping shrub, 6 to 10 in. high, with round, wiry, few-branched stems, covered when young with short, black down. Leaves dark lustrous green, box-like, obovate, often notched at the apex, shortly stalked, ⅜ to 1 in. long, about half as wide, the lower surface sprinkled with black dots. Flowers produced during May and June, five to twelve together in terminal racemes less than 1 in. long. Corolla white or pinkish, bell-shaped, rather deeply four-lobed, ¼ in. long. Berries dark red, globular, acid and harsh in flavour, ⅖ in. wide.

Considered in a wide sense, this species girdles the globe in high latitudes of the northern hemisphere, and also occurs in many mountain ranges farther south. In the British Isles it is common in Scotland and northern England, and extends through Wales as far south as Somerset and Devon; it also occurs in northern and eastern Ireland. The cowberry is the handsomest of the native vacciniums, the dark glossy foliage making neat, dense tufts. In suitable positions it spreads quickly by means of its creeping root-stock, and makes a useful ground-cover on acid soils. The fruits are palatable when cooked, and put to the same culinary use as cranberries or red currants.

f. MAJUS (Lodd.) Rehd.   *V. vitis-idaea major* Lodd.—Leaves larger than in the typical state, to 1⅜ in. long. This occurs occasionally in Europe.

subsp. MINUS (Lodd.) Hultén *V. vitis-idaea minor* Lodd.; *V. v.* var. *pumilum* Hornemann—Smaller in all its parts, the leaves being up to ⅝ or ¾ in. long. This replaces the typical state in Arctic regions and in the mountains of N. America.

The specific epithet is an old generic name, *Vitis Idaea*, meaning 'Grape of Mt Ida', and was used by many of the pre-Linnaean botanists for the European vacciniums, but who coined it and why is not recorded.

VELLA   CRUCIFERAE

A genus of three shrubby species in Spain and N. Africa. Flowers yellow, with clawed petals, borne in racemes. Filaments of inner stamens paired and connate. The elongated pod-like fruit is of distinctive shape: it is constricted around the middle and only the lower portion is fertile, the upper sterile part forming a beak.

V. PSEUDOCYTISUS L.
Pseudocytisus integrifolius (Salisb.) Rehd.; *Vella integrifolia* Salisb.

A low, evergreen shrub, usually less than 2 ft high near London, but larger in milder localities; branches erect, covered the first two or three years with spiny bristles, ultimately glabrous. Leaves obovate, ½ to ¾ in. long, rounded at the apex, tapering to a short stalk at the base, covered on both surfaces and at the margin with stiff bristly hairs. Flowers in an erect, elongated, terminal raceme, 4 to 8 in. long, more crowded towards the top, the calyx erect, green, hairy; petals somewhat spoon-shaped, the terminal part yellow, and roundish; the lower part contracted into a long, slender, purplish claw; each petal about ⅓ in. long; flower-stalk 1/16 in. long.

This curious shrub is a native of the mountains of central Spain. It is not really hardy, but has stood unprotected on the rock garden at Kew for several years at a time. Our hardest winters kill it. A sunny, rather dry position should be given it. It was cultivated by Miller at Chelsea, in 1759. Propagated easily by cuttings of half-ripened wood in gentle heat. It flowers from the end of May to July. Very suitable for the Isle of Wight and similar climates.

V. SPINOSA Boiss.
Pseudocytisus spinosus (Boiss.) Rehd.

A dwarf, deciduous shrub of dense, compact habit about 1 ft high, with rigid, erect stems, the upper branchlets of which become spine-tipped; young shoots glabrous except for a few pale bristles at first. Leaves dull greyish green, linear, ½ to ¾ in. long, $\frac{1}{20}$ in. wide, fleshy, often showing a tendency to become pinnate; glabrous except for an occasional bristle like those on the young shoots. Flowers few in terminal corymbs; each flower about ⅝ in. across; petals four, yellow with brown veins, roundish obovate, narrowed at the base to a slender claw about as long as the blade. Calyx tubular, ¼ in. long, green, with four erect pointed teeth. Seed-vessel a dry two-celled pod, erect, ¼ in. long, somewhat heart-shaped, terminated by a flat, pointed beak ⅓ in. long.

Native of Spain, it was quite hardy in the rock garden at Kew, where a plant scarcely 1 ft high grew for twenty years, flowering in June. It is an interesting, but not very showy shrub. Propagated by cuttings of young wood. The bristles and spines on the stems and leaves are much more numerous and conspicuous in wild plants.

VERBENA   VERBENACEAE

The genus *Verbena* is best known in gardens by its ornamental annuals and herbaceous perennials. There is however a large group of shrubs in this genus, mostly dwarf and often with much reduced leaves, all confined to S. America (the genus itself is largely confined to the New World). The species described below is the best known of these in gardens, but others, mostly new to science, were introduced by Comber from Argentina in 1925–7, notably V. CEDROIDES Sandwith, closely allied to *V. tridens*, and V. SCABRIDO-GLANDULOSA Turrill (*Bot. Mag.*, n.s., t. 98).

V. TRIDENS Lag.   MATA NEGRA
V. carroo Speg.; *Junellia tridens* (Lag.) Moldenke

An evergreen shrub of virgate habit, 3 to 6 ft high; young shoots slender, stiffly erect, downy. The leaves on the first year shoots are arranged oppositely, seven to fourteen pairs to the inch, and are downy, from $\frac{1}{12}$ to ⅙ in. long, stalkless, consisting of three stout, sharply-pointed lobes, each lobe grooved beneath on either side of its prominent midrib. From the axils of these leaves, during the second year and afterwards, proceed short branches on which the short, simple, thick, blunt leaves are packed closely together decussately (*i.e.*, in four superposed rows), the whole making a quadrangular arrangement of leaf and stem ¼ in. wide. The second type of leaf is about $\frac{1}{12}$ in. long. On the older wood the leaf-clusters form curiously contorted masses. Flowers white to rosy-lilac, sweetly and powerfully scented, produced in terminal spikes, often one spike on each of a cluster of short twigs near the apex of the shoot; a spike carries six to twelve flowers. Corolla ¼ to ⅓ in. long, ¼ in. wide, tubular, five-lobed, downy; calyx tubular, grooved, downy, with jagged margins; stamens four, attached to and hidden in the upper half of the corolla-tube, anthers yellow; bracts three-lobed, resembling the leaves, one flower in the axil of each.

Native of Chile and Argentina in the Patagonian steppe; described in 1816. It is a noxious weed of sheep pastures in the Patagonian region of southern Argentina and in places is sufficiently abundant to provide fuel. We owe the introduction of this extraordinary shrub to Clarence Elliott, who saw it in flower in February 1928 in the neighbourhood of Last Hope Bay—'a strange country of mountain, lake and moorland, eternal devastating wind, intense cold, and rain' (C. Elliott, *Gard. Chron.*, Vol. 89 (1931), p. 378). He later procured seeds and distributed plants from his famous Six Hills Nursery at Stevenage, where it first flowered in 1932. It is perfectly hardy, but needs a sunny position. The vanilla-like fragrance of the blossom is very strong and is perceptible several yards away from the plant.

VESTIA   SOLANACEAE

A monotypic genus allied to *Cestrum*, but having flowers with exserted stamens and a capsular fruit. It was named by Willdenow in honour of Prof. Vest of Klagenfurt.

V. FOETIDA (Ruiz & Pavon) Hoffmannsegg   HUEVIL

Periphragmos foetidus Ruiz & Pavon; *Vestia lycioides* Willd.; *Cantua ligustrifolia* Juss.; *Cantua foetida* (Ruiz & Pavon) Pers.

An erect, evergreen shrub to about 6 ft high, its stems and leaves unpleasantly scented when bruised; young stems soon glabrous. Leaves alternate, bright green, glossy, slightly fleshy, obovate to oblong-elliptic, mostly 1 to 2 in. long and up to ¾ in. wide, cuneate at the base, shortly stalked, glabrous, entire. Flowers in late spring or early summer on laterals from the upper leaf-axils, nodding. Calyx campanulate, shortly five-toothed. Corolla pale yellow, funnel-campanulate, 1 to 1¼ in. long, with a spreading five-lobed limb about ¾ in. wide. Stamens five, exserted. Style one, exserted. Fruit a roundish-ovoid capsule, enclosed in the lower half by the persistent appressed calyx, two-valved, containing numerous small seeds. *Bot. Mag.*, t. 2412.

Native of Central Chile; introduced before 1809 to the Berlin Botanic Garden and thence to Britain about 1815. Although usually regarded as a greenhouse plant, it is nearly hardy in the milder parts and should survive most winters near London with the protection of a wall. It produces fertile seeds which, sown under glass in late winter, will produce plants 2 ft high by autumn. But cuttings are the usual means of increase.

This species is usually known as *Vestia lycioides*, the name given to it by Willdenow when setting up the genus. But there is no doubt that it is the plant described and figured by Ruiz and Pavon in 1799 in their genus *Periphragmos* as *P. foetidus* and must therefore take their epithet. The other species placed by them in *Periphragmos* belong to *Cantua*.

VIBURNUM   CAPRIFOLIACEAE

A genus of well over 100 species, mainly in the temperate regions of the northern hemisphere, but extending into Malaysia and S. America. They are shrubs or small trees, deciduous or evergreen. Winter-buds naked or few-scaled. Leaves opposite, simple, toothed, sometimes lobed; stipules sometimes present, small, adnate to the petiole, but more commonly lacking. Inflorescence compound, consisting of cymes arranged in the form of a flat or dome-shaped umbel, or more rarely of a panicle; it is usually terminal on short growths of the season, sometimes almost sessile on the previous season's wood. Ovary inferior. Calyx small, shortly five-toothed. Corolla white or cream-coloured, sometimes flushed with pink or almost wholly pink, rotate, campanulate or more rarely cylindric, five-lobed. Stamens five, inserted on the tube. Style sessile on the free upper part of the ovary (i.e., the portion above the insertion of the corolla and calyx). Fruit a red, blue or black one-seeded drupe.

A curious feature of several species of *Viburnum* is the presence of two distinct types of flower in one inflorescence—the one sterile and showy, consisting of a corolla without stamens or pistil, the other much smaller but perfect and fertile. The function of the large sterile ray-flowers is that of advertisement, to attract insects to the inflorescence. This really represents an interesting and unusual division of labour, for most insect-fertilised do their own advertising by means of the petals attached to the individual flower. In three species—*V. opulus*, *V. macrocephalum* and *V. plicatum*—mutations have occurred in which the inflorescence is entirely made up of sterile ray-flowers, which represents a striking increase in flower beauty. These phenomena are also exhibited by several species of *Hydrangea* (according to one controversial theory *Viburnum* actually derives from the Hydrangaceae).

The viburnums that have spread most widely in gardens are grown for their flowers. But many species have beautiful fruits, and some would be as common as the cotoneasters and pyracanthas were it not that most of these show the phenomenon of self-incompatibility. Like so many of the orchard apples, pears, plums and cherries they do not as a rule set fruit unless pollinated by another seedling or by a plant belonging to a different clone. There may be exceptions (*V. opulus* and its allies are said to be self-fertile), but cross-pollination appears to be necessary precisely in those groups that have the most ornamental fruits—the blue-fruited *V. davidii* and its allies, and the red-fruited members of the section *Odontotinus* such as *V. betulifolium* and *V. dilatatum*. Ideally, two or more compatible clones should be available, as is already the case here with *V. davidii*. A possible alternative would be to plant together different but related species that flower at the same time. But either way, insect visitors are still necessary to do the work of cross-pollination.

Viburnums are as a rule of easy cultivation, but there are some exceptions and some are not very hardy. They love moist conditions and a deep, rich loamy soil.

Most viburnums can be increased by cuttings taken in late summer and placed in gentle bottom heat, and this is the usual way of propagating the evergreen species. Seed is often slow to germinate and may not come true if taken from plants in a collection. Layering is another means, and is often used for *V. plicatum*, *V. lantanoides* and *V. furcatum*.

CLASSIFICATION

The following is based on Rehder's standard classification of *Viburnum* in Sargent, *Trees and Shrubs*, Vol. II (1908). Characters based on the stone of the fruit are omitted, as the fruits of *Viburnum* are too rarely seen on some species to make them of much value. Only species given a full treatment are listed; those given shorter descriptions belong to the same group as the one under which they are mentioned.

sect. VIBURNUM (*Lantana*).—Indumentum of stellate hairs. Winter-buds naked. Leaves usually deciduous but evergreen in some species, toothed (sometimes slightly so). Inflorescence corymbose, without ray-flowers except in *V. macrocephalum* (wild form). Fruits black or bluish black, but at first red in most species. An Old World group.

V. buddleifolium, *V. burejaeticum*, *V. carlesii*, *V. cotinifolium*, *V. lantana*, *V. macrocephalum*, *V. rhytidophyllum*, *V. utile*, *V. veitchii*.

Of the several garden-raised hybrids in this group, *V.* × *carlcephalum* and *V.* × *burkwoodii* are of particular interest as confirming the relationship of *V. carlesii* to the outwardly dissimilar *V. macrocephalum* and *V. utile* respectively.

The section PSEUDOTINUS is in most respects similar to the sect. *Viburnum*, but differs in technical characters of the fruit-stone. See *V. lantanoides*.

sect. PSEUDOPULUS.—Winter-buds with one pair of scales. Indumentum stellate. Leaves with straight veins, running to teeth. Flowers in cymes, with ray-flowers, terminating short lateral branches.

V. plicatum.

sect. LENTAGO.—Winter-buds with one pair of outer scales. Leaves deciduous, finely serrate or entire; lateral veins anastomosing before reaching the margin. Flowers in cymes, all fertile. Fruits blue-black. All species American.

V. cassinoides, *V. lentago*, *V. nudum*, *V. prunifolium*, *V. rufidulum*.

sect. ODONTOTINUS.—Winter-buds with two pairs of outer scales. Hairs when present fascicled. Leaves usually deciduous, toothed (sometimes obscurely lobed), with straight lateral veins running out to teeth. Flowers in cymes. Corolla rotate. Old and New Worlds. The fruits are blue-black in the American species *V. dentatum* and *V. molle* and in *V. acerifolium* of southwest Asia, red in the following species, all natives of E. Asia:

V. betulifolium, *V. dilatatum*, *V. erosum*, *V. foetidum* (semi-evergreen, leaves often lobed), *V. hupehensis*, *V. japonicum* (evergreen, leaves obscurely toothed), *V. phlebotrichum*, *V. setigerum*, *V. wilsonii*, *V. wrightii*.

sect. THYRSOSMA (*Solenotinus*).—Leaves deciduous or persistent. Flowers in panicles (often somewhat condensed in the winter-flowering species). Corolla rotate to cylindric. Fruits blue-black or purple. Himalaya and E. Asia.

V. farreri, *V. grandiflorum*, *V. henryi*, *V. odoratissimum*, *V. sieboldii*, *V. suspensum*.

sect. TINUS.—Leaves persistent, entire or almost so, sometimes three-veined from the base. Fruits blue or blue-black. An Old World group, from the Atlantic islands to E. Asia.

V. davidii, *V. harryanum*, *V. propinquum*, *V. rigidum*, *V. tinus*.

sect. OPULUS.—Buds with one pair of outer scales, connate at the base. Leaves deciduous, three- or five-veined from the base and usually lobed. Inflorescence with ray-flowers in some species. Fruits red or scarlet. Old and New Worlds.

V. opulus, *V. kansuense*.

Similar to this section in general appearance is *V. cylindricum* of the section MEGALOTINUS; it has a cylindric corolla with short lobes.

V. ACERIFOLIUM L.   DOCKMACKIE

A deciduous bush 3 to 6 ft high; young branches at first softly downy, becoming glabrous. Leaves maple-like, three-lobed, the side lobes with divergent, slender points, all coarsely toothed, 1½ to 4 in. long and about the same wide; rounded or heart-shaped at the base, with scattered down above, softly downy (especially at first) and covered with black dots beneath; stalk ½ to 1 in. long, downy. Flowers white, ⅕ in. in diameter, uniform and all fertile, produced during June in terminal, long-stalked cymes 2 to 3 in. across. Fruits first red, then purple-black, oval, ⅓ in. long.

Native of eastern N. America; introduced in 1736. Although one of the earliest introduced of American viburnums this is now very scarce in gardens; it has little beauty of flower, but is attractive in autumn for its crimson foliage. I have seen it growing along the roadsides in New Hampshire just as *V. opulus* does at home, but never so vigorous a shrub.·

V. ORIENTALE Pall.—A native of the western Caucasus and Asia Minor, this is closely allied to *V. acerifolium*, but can always be distinguished by the absence of the minute black dots beneath the leaf so characteristic of the American species. It is not so downy, the hairs beneath being almost confined to the vein-axils; otherwise very similar. Rare in gardens.

V. BETULIFOLIUM Batal.

A deciduous shrub up to 10 or 12 ft, branchlets glabrous, becoming brown or purplish brown. Leaves ovate to diamond-shaped, broadly wedge-shaped at the base, and often entire there, the terminal part more gradually tapered and coarsely toothed; 2 to 4 in. long, $1\frac{1}{4}$ to 3 in. wide; dark green and glabrous above, paler and also glabrous beneath, except for a few simple hairs on the veins, and sometimes tufts in the vein-axils; veins in four to six pairs; leaf-stalk $\frac{1}{2}$ to $\frac{3}{4}$ in. long, usually slightly hairy. Cymes $2\frac{1}{4}$ to 4 in. across, the main, and especially the secondary, flower-stalks usually covered with a close, pale brown, stellate down; main branches of corymb seven. Flowers white, $\frac{1}{5}$ in. across, all perfect; stamens protruded, anthers yellow. Fruits red, roundish, $\frac{1}{4}$ in. long. *Bot. Mag.*, t. 8672.

Native of western and central China; introduced by Wilson, who sent seeds in 1901, 1907 and 1910; it may also have been raised from the seeds collected by Farrer in Kansu in 1914. In its best forms *V. betulifolium* is a fine sight in autumn, when its slender branches are weighed down with the heavy trusses of bright red, translucent berries. Some plants have been noticed bearing good crops of fruit even though there is allegedly no other viburnum in the vicinity. If these are really self-fertile they would be worth propagating, but in the meantime it is safer to assume that cross-fertilisation is necessary if fruit is to be set, which means that two or more seedlings must be planted near together, or better, plants of two selected and different clones.

V. betulifolium received an Award of Merit in 1936, when shown from Trewithen in Cornwall, where there are two fine plants growing together, and a First Class Certificate in 1957 (exhibited by Sir Frederick Stern, Highdown, Sussex).

V. DASYANTHUM Rehd.—Closely allied to *V. betulifolium* and occurring in the same areas. It has similar foliage but is distinguished by having the corolla downy on the outside. Introduced by Wilson.

V. LOBOPHYLLUM Graebn.—Another close ally of *V. betulifolium* and also closely associated with it in the wild. The differences, such as they are, are that the leaves instead of being ovate to rhombic-ovate tend to be broadly ovate to obovate, with somewhat shallower teeth (but not lobed as the specific epithet would imply); the longer peduncle of the inflorescence (to 1 in. long); and the slightly larger, rounder fruits. It is interesting that wherever Wilson found *V. betulifolium* he also found this species. *Bot. Mag.*, n.s., t. 164. Plants in cultivation under this name fruit well, though whether they would do so in isolation it is impossible to say. In some the fruits are larger than in *V. betulifolium*, in others smaller. The leaves usually colour well in the autumn.

V. OVATIFOLIUM Rehd.—The plants in cultivation were apparently raised from Wilson 240, collected in Hupeh. The leaves are broad-ovate, deltoid-elliptic or roundish elliptic, colouring in the autumn. Flowers in rather small trusses. In typical *V. ovatifolium*, described from Yunnan, the leaves are relatively narrower.

V. × BODNANTENSE Stearn   [PLATE 93

A group of hybrids between *V. farreri* (*V. fragrans*) and *V. grandiflorum*. The cross was first made by Charles Lamont (*d.* 1949), Assistant Curator at the Royal Botanic Garden, Edinburgh, in 1933, but he considered the four plants he raised to be no improvement on either parent, and did not propagate them. In the following year and in 1935 the same cross was made at Bodnant, and ten plants were raised, all from seed of *V. farreri*. One of them received an Award of Merit in 1947; this clone has been named 'DAWN', and is also the type of *V. × bodnantense*, described in 1950 (*Bot. Mag.*, n.s., t. 113). The flowers of 'Dawn' are rich rose-red in bud, but the limb eventually becomes white with a strong flush of pink. In size of flower and in other botanical characters it is more or less intermediate between the parents. The flowers are more frost-resistant than in *V. grandiflorum* and more freely borne. There is a fine group at Kew by the Orangery, the tallest about 10 ft high.

Also belonging to this group is 'DEBEN', raised by Messrs Notcutt (A.M. 1962; F.C.C. 1965). It is a vigorous shrub to about 10 ft high, whose flowers are delicate shell-pink in bud, opening white with a flush of pink. The tube is a trifle longer and more slender than in 'Dawn'. They are perhaps rather less weather-proof, but lovely in a mild, dry spell in winter.

One of Mr Lamont's seedlings (see above) has been named 'CHARLES LAMONT'. The late Rowland Jackman considered it to be finer than 'Dawn', with brighter pink flowers, more freely borne (*Journ. R.H.S.*, Vol. 93 (1968), p. 419). It is still scarce in the trade.

V. BUDDLEIFOLIUM C. H. Wright

A deciduous shrub about 6 ft high; the young shoots densely covered with pale, star-like down. Leaves oblong-lanceolate, 3 to 5 in. long, 1 to 2 in. wide, pointed, rounded or slightly heart-shaped at the base, shallowly toothed; upper surface furnished with simple or forked hairs, the lower one felted with pale, stellate down; stalk $\frac{1}{4}$ to $\frac{1}{2}$ in. long. Flowers white, funnel-shaped, $\frac{1}{3}$ in. across, all perfect, produced on a short-stalked, numerously branched cyme, 3 in. across. Fruits oval, $\frac{1}{3}$ in. long, black.

Native of Central China; discovered and introduced by Wilson in 1900. It belongs to the Lantana group, differing from *V. lantana* in its narrow, oblong leaves. It has no ornamental value.

V. BUREJAETICUM Reg. & Herder

V. burejanum Herder

A deciduous shrub whose young shoots are covered at first with a dense, stellate down, becoming almost white and glabrous the second year. Leaves ovate, oval or slightly obovate, tapered, rounded, or slightly heart-shaped at the base, tapered and often blunt at the apex, 2 to 4 in. long, 1 to 2 in. wide, evenly and angularly toothed, with scattered, mostly simple hairs above, and scattered stellate ones beneath, chiefly on the veins, becoming almost glabrous; stalk ¼ to ½ in. long, scurfy. Flowers white, uniform and perfect, ¼ in. wide, produced in stalked usually five-branched cymes, 2 in. across; the stalks covered with stellate scurfy down.

Native of N. China, Korea, and the Ussuri region of Russia, described from the Bureia Mountains. Rare in cultivation.

V. × BURKWOODII Burkwood & Skipwith ex Anon. in *Gard. Chron.*, Vol. 85 (1929), p. 285

This charming hybrid was raised by Messrs Burkwood and Skipwith in their nursery at Kingston-on-Thames in 1924. It was raised from *V. utile* pollinated by *V. carlesii* and has inherited the evergreen character of the seed parent. Its ovate, pointed leaves are 1½ to 4 in. long, ¾ to 1¾ in. wide, rounded or slightly heart-shaped at the base, indistinctly toothed, dark, slightly burnished green above, thickly covered beneath with pale brown stellate down; leaf-stalks ¼ in. or less long, covered (like the young shoots) with the same kind of down as the leaves. Flowers charmingly fragrant, produced in late April and May in rounded, five-rayed, terminal, well-filled clusters 2½ to 3½ in. across. Each flower is about ½ in. wide, the corolla having five spreading, rounded lobes, pinkish when quite young, afterwards pure white; anthers pale yellow.

This viburnum is very hardy, and with the beauty of its fragrant flowers, and its easy propagation from cuttings, it has become one of the most widely planted of viburnums, thriving even in a smoky environment. It grows to about 8 ft high and slightly more in width. It is not completely evergreen. Although the main flowering season is spring, some trusses may open in early winter if the weather is mild.

This hybrid is mainly represented in cultivation by the original clone 'Burkwoodii'. Believed to be of the same parentage are 'CHENAULTII', of more compact habit and more moderate growth, with slightly duller more persistent leaves; and 'PARK FARM', which grows as tall as 'Burkwoodii', of which it is a sister-seedling, and differs in its slightly larger clusters of flowers pinker in the bud. Its older leaves colour in the autumn.

V. 'ANNE RUSSELL'.—A backcross of *V. carlesii* with 'Burkwoodii', raised by Messrs L. R. Russell of Windlesham and named in 1951. Leaves slightly glossy, broad-elliptic. Flowers about ⅝ in. wide, pink in the bud, opening white, in perfectly shaped, not too dense trusses about 3 in. wide, very fragrant.

Of compact habit. This fine viburnum is 6 ft high and 8 ft across at Wakehurst Place in Sussex. Award of Merit 1957.

V. 'FULBROOK'.—When this viburnum received an Award of Merit in 1957 the parentage was stated to be *V. carlesii* crossed with *V. × burkwoodii*, but it shows no influence of the latter and is quite unlike 'Anne Russell'. It is of rather open, graceful habit. Flowers in late April or May, white when expanded, in broad, rather lax trusses about 4 in. wide. Raised by Miss Florence Paget and distributed by Mrs Douglas Gordon of Fulbrook House, Elstead, Surrey.

V. × CARLCEPHALUM Burkwood & Skipwith ex A. V. Pike

A hybrid between *V. carlesii* (seed-parent) and *V. macrocephalum* f. *keteleeri*, raised by Messrs Burkwood and Skipwith about 1932 but scarcely known until it received an Award of Merit in 1946. It is a shrub to about 8 ft high and wide, with leaves resembling those of *V. carlesii* but larger and with a slight gloss, sometimes colouring in the autumn. Flowers in May in dense trusses up to 6 in. wide, up to 100 in each, fragrant, pink in the bud, white with a flush of pink when expanded. A fast-growing shrub, thriving and flowering well even in cold localities, but lacking in charm, with its stiff branching and rather lumpy trusses.   [PLATE 94

V. CARLESII Hemsl.

A deciduous shrub of rounded habit, 4 to 8 ft high; young shoots densely clothed with starry down. Leaves broadly ovate, with often a slightly heart-shaped base, pointed, irregularly toothed, 1 to 3½ in. long, ¾ to 2½ in. wide, dull green above, greyish below, both surfaces soft with starry down; stalk about ¼ in. long. Inflorescence a terminal, rounded cluster 2 to 3 in. across, composed of very fragrant flowers, all fertile. It reaches the bud state in autumn, and remains exposed through the winter, the flowers expanding in April and May. Corolla ½ in. across, at first pink then white, with a slender tube ⅓ in. long. Fruits jet-black, ¼ in. long, egg-shaped but flattened. *Bot. Mag.*, t. 8114.

Native of Korea and of Tsushima Island, Japan; described in 1885 from a specimen collected by W. R. Carles of the British Consular Service, who explored in Korea 1883–5. It was introduced from Korea to Japan around 1897 by Alfred Unger of L. Boehmer and Co. of Yokohama. A single plant was sent to Kew by this firm in 1902, which represented its first introduction to Europe; it flowered in the open ground at Kew in 1906. But the species was apparently first distributed in Europe by Messrs Lemoine, who announced in 1905 that they had bought Messrs Boehmer's entire stock.

V. carlesii is one of the most delightful of the viburnums, not only for the beauty of its flowers, but for a fragrance unrivalled for sweetness in the genus. But it is now overshadowed by its hybrids; see below and *V. × burkwoodii*.

The following selections of *V. carlesii* were raised from Korean seed at the Slieve Donard Nursery, Co. Down:

VIBURNUM CARLESII

cv. 'AURORA'.—Flowers red in the bud, opening pure light pink. Young leaves light green, some flushed with copper.

cv. 'CHARIS'.—Rather more vigorous than the older forms of *V. carlesii*. Flowers white.

cv. 'DIANA'.—Open flowers pink, with a slightly more purplish tone than in 'Aurora'. It is also a little more vigorous. Young leave slight chocolate-coloured.

V. BITCHIUENSE Makino *V. carlesii* var. *bitchiuense* (Makino) Nakai; *V. carlesii* var. *syringiflorum* Hutch.—Very closely allied to *V. carlesii*, but with relatively narrower, ovate or oblong, obtuse leaves, sometimes slightly cordate at the base. A further botanical difference has been given, namely that the stamens are inserted on the lower one-quarter to one-third of the tube and have filaments twice as long as the anthers, while in *V. carlesii* the stamens are inserted around the midpoint of the tube and the filaments are shorter than the anthers. As the two species are usually seen in gardens, *V. bitchiuense* is a taller shrub than *V. carlesii*, of more open habit, to about 8 ft high and as much in width, its laxer inflorescences are slightly smaller and the corollas narrower across the limb with a longer tube.

Native of southern Japan and Korea; described in 1902. It was introduced to Britain about 1911 as "*V. carlesii*" and was for a time regarded as an inferior form of that species. It was at one time used by Japanese export nurseries as a stock for *V. carlesii* and often the scion was dead by the time the plants arrived in this country. In this way poor forms were introduced to gardens. Eventually selections arrived, which proved to be almost as fine as *V. carlesii* itself.

V. × JUDDII Rehd.—A hybrid between *V. carlesii* (seed-parent) and *V. bitchiuense*, raised by William H. Judd at the Arnold Arboretum in 1920, though not described and named until 1935. In its botanical characters it is intermediate between the parents. The corymbs are rather laxer than in *V. carlesii* and a trifle wider with more numerous not quite so sweetly scented flowers. It is a plant of good constitution, growing to about 5 ft high. Award of Garden Merit 1960.

V. CASSINOIDES L.   WITHE-ROD

A deciduous shapely bush of rounded form, rarely more than 6 to 8 ft high in Britain, but said to be occasionally a small tree in the southern United States; young wood scurfy. Leaves ovate to oval with a short, slender, often bluntish apex, rounded or wedge-shaped at the base, 1½ to 4½ in. long, ¾ to 2¼ in. wide, irregularly and shallowly round-toothed, or merely wavy at the margin, thick and firm in texture, dull dark green and glabrous or nearly so above, somewhat scurfy beneath; stalk scurfy, ¼ to ¾ in. long. Flowers all uniform and perfect, yellowish white, ε in. wide, produced in early June in cymes 2 to 4 in. across, the main-stalk of which is shorter than the branching portion. Fruits blue-black when ripe.

Native of eastern N. America; introduced, according to Aiton, in 1761. There is much confusion between this species and *V. nudum* (q.v.), but *V. cassinoides* has dull green leaves and very scurfy young shoots, leaf-stalks, and flower-stalks, and a short-stalked inflorescence. In *V. nudum* the leaves are glossy, the shoots, etc., comparatively free from scurf, and the inflorescence usually long-stalked.

V. cassinoides is one of the finest American viburnums. The leaves, chocolate- or bronze-tinted when young turn bright red before they fall, and the fruits as they mature pass from green to pink or red and finally to dark blue.

V. COTINIFOLIUM D. Don
V. polycarpum Wall. ex DC.

A deciduous shrub 6 to 12 ft high, whose young branchlets, under-surface of leaves (upper surface to a less extent) and the flower-stalks are clothed with a dense, grey, stellate down. Leaves ovate, oval or nearly round; the base rounded, the apex shortly pointed or rounded, 2 to 5 in. long, two-thirds to nearly as wide, finely toothed. Flowers white, tinged with pink, widely funnel-shaped, ¼ in. long, produced during May in rounded usually five-branched cymes 2 to 3 in. across. Fruits ovoid, red, ultimately black, ⅓ to ½ in. long.

Native of the Himalaya from Bhutan to Baluchistan; introduced about 1830. This species is closely allied to *V. lantana*, and is very similar in foliage and general appearance, but differs in the following respects: cymes more often five-rayed than seven-rayed, corolla tinged with pink, and distinctly funnel-shaped, the corolla-tube longer than the lobes. The true plant is rare in gardens, and not so hardy as *V. lantana*.

V. CYLINDRICUM Buch.-Ham. ex D. Don (1825)

V. coriaceum Bl. (1826)

An evergreen shrub (in some of its native habitats a tree 40 to 50 ft high), branchlets warted, otherwise glabrous. Leaves oval, oblong, or somewhat obovate, 3 to 8 in. long, 1½ to 4 in. wide, wedge-shaped or sometimes rounded at the base, slender-pointed at the apex, the terminal half usually remotely toothed, upper surface dark dull green and covered with a thin, waxy layer, which cracks and turns grey when the leaf is rubbed or bent, both surfaces quite glabrous; stalk ½ to 1½ in. long. Flowers white, quite tubular, about ⅕ in. long, produced from July to September in usually seven-rayed cymes 3 to 5 in. across. The cymes are rendered pretty by the protruded bunch of lilac-coloured anthers. Fruits egg-shaped, ⅙ in. long, black.

Native of the Himalaya and China; introduced to Kew from India in 1881, and later from Yunnan through the Jardin des Plantes, Paris, in 1892. Most of the plants now in cultivation are Chinese, and these are probably hardier than the Indian ones. They have at any rate succeeded very well. Two characters make this species very distinct, the tubular corolla with erect, not spreading lobes, and the curious waxy covering of the leaves; the latter only shows itself when the leaf is touched or bent; ordinarily they are of a dingy dark green.

V. DAVIDII Franch.

An evergreen shrub of apparently low, compact habit, and about 3 to 5 ft high; young branches warted. Leaves leathery, narrowly oval or slightly obovate, tapered at the base, more slenderly so at the apex; 2 to 6 in. long, 1 to 2½ in. wide; strongly and conspicuously three-veined, often obscurely or shallowly toothed near the apex, dark green above, pale below, glabrous on both surfaces except for small tufts of down in the vein-axils beneath; stalk ¼ to 1 in. long. Flowers dull white, ⅛ in. wide, densely crowded in stalked stiff cymes, 2 to 3 in. across. Fruits blue, ¼ in. long, narrow oval. *Bot. Mag.*, t. 8980.

Whether or not *V. davidii* is actually dioecious, it is so in effect, but many nurserymen now offer both 'male' and 'female' plants, and both must be grown if the beautiful fruits are to be seen. It received an Award of Merit for these in 1971.

V. 'JERMYN'S GLOBE'.—A seedling of *V. davidii* of which the other parent was probably *V. calvum*. Although the original plant is not as globular in habit now as when it was first named, it is of low, compact and roundish habit. The foliage is not unlike that of *V. tinus*.

VIBURNUM DAVIDII

V. CINNAMOMIFOLIUM Rehd.—This is very similar to *V. davidii*, with the same conspicuously three-veined leaves, but it is a bigger shrub, or sometimes a tree 20 ft high; its inflorescence is much larger and more lax, its almost entire leaves are not so thick, and its fruits smaller. There is a plant about 15 ft high and wide on a garden wall at Borde Hill in Sussex, which is perfectly hardy. Wilson discovered this species on Mt Omei in W. Szechwan.

V. DENTATUM L.   SOUTHERN ARROW-WOOD

V. venosum var. *canbyi* Rehd.; *V. pubescens* var. *canbyi* (Rehd.) Blake

A shrub to about 10 or 15 ft in the wild, with a close grey or brownish bark; branchlets usually downy, sometimes glabrous or almost so. Leaves of thin texture, ovate to broadly so or roundish, 2 to 4½ in. long, 1 to 4 in. wide, sometimes even broader than long, shortly acuminate at the apex, rounded or slightly cordate at the base, almost glabrous above, sparsely stellate-downy beneath; veins five to eleven on each side, straight, impressed above and prominent beneath; marginal teeth usually large and triangular; petiole slender, to 1 in. long, clad with stiff down, usually without stipules. Flowers white, perfect and regular, about ⅙ in. wide, borne around midsummer. Corymbs up to 4½ in. wide, on peduncles 1½ to 2½ in. long, its branches downy or Corymbs up to 4½ in. wide, on peduncles 1½ to 2½ in. long, its branches downy or sometimes glabrous. Fruits blue-black, roundish oval, up to ⅜ in. long; stone with a narrow and deep groove on one side.

Native of eastern N. America, mostly south of New York; described by

Linnaeus from a specimen collected in Virginia and cultivated since the 18th century.

var. DEAMII (Rehd.) Fern. *V. pubescens* var. *deamii* Rehd.; *V. pubescens* var. *indianense* Rehd.—Leaves glabrous or almost so. Petioles usually with a pair of narrow stipules at the base. Of more western distribution than typical *V. dentatum*.

var. PUBESCENS Ait. *V. pubescens* (Ait.) Pursh; *V. nervosum* Britt.—Branchlets and undersides of leaves densely downy; leaves of thicker texture. *V. dentatum* 'LONGIFOLIUM' is a cultivated form of this with leaves longer than wide (*V. longifolium* Lodd. ex Loud.).

V. RECOGNITUM Fern. ARROW-WOOD. *V. dentatum* of many authors, not L.; *V. dentatum* var. *lucidum* Ait., not *V. lucidum* Mill.—Very closely allied to the preceding, but essentially glabrous. Stones of fruits, according to Fernald, globose-ovoid, with a shallow furrow. It is of more northern distribution than *V. dentatum*. Introduced in the 18th century.

None of the species and varieties in this complex group are of much value in British gardens and now rarely seen outside collections. The young shoots that spring from the base are straight and erect, and it was their use by the Indians as arrows that gave rise to the popular name.

V. DILATATUM Thunb.   [PLATE 95

A deciduous shrub, 6 to 10 ft high (sometimes taller), with erect stems; young branchlets very downy. Leaves broadly ovate, roundish or obovate; 2 to 5 in. long, and from half to about as much wide, widely toothed, pointed, tapering, rounded or heart-shaped at the base, hairy on both sides; stalk $\frac{1}{4}$ to $\frac{3}{4}$ in. long; veins in five to eight pairs. Flowers pure white, all fertile, $\frac{1}{4}$ in. across, produced in June in hairy, stalked, mostly five-rayed cymes, 3 to 5 in. across. Fruits bright red, roundish ovoid, $\frac{1}{3}$ in. long. *Bot. Mag.*, t. 6215.

Native of Japan and China; it first flowered with Messrs Veitch in 1875 and had attained a height of 20 ft in their nursery by 1900. This fine viburnum is remarkably profuse in its flowering, the trusses being produced not only at the top of the branch but from short twigs down the sides as well. It is even more beautiful in its fruits but does not set them freely unless at least two seedlings, or plants of different clones, are grown. Obviously clones of known character are preferable and with this point in mind two were selected and named by Donald Egolf at the United States National Arboretum in 1958—'CATSKILL' and 'IROQUOIS' (*Baileya*, Vol. 14 (1966), pp. 109–12).

f. XANTHOCARPUM Rehd.—Fruits yellow. *Bot. Mag.*, n.s., t. 103. Originally described from a plant cultivated in the USA but also known in Japan. Award of Merit 1936.

V. 'ONEIDA'.—A hybrid between *V. dilatatum* and *V. lobophyllum*, raised by Donald Egolf (see above) and selected in 1961. A shrub to about 10 ft high and almost as wide, of tiered habit. Leaves dark green, glossy, variable in shape, up to 4 in. long. Flowers creamy white in trusses up to almost 6 in. wide, mostly in May but then intermittently, so that some flowers are still borne when the dark red glossy fruits are ripe (*Baileya*, loc. cit., pp. 115–17).

V. PARVIFOLIUM Hayata—A sparsely branched, slow-growing shrub, making short annual growths. Leaves ovate or broad-elliptic, to about 1 in. long, glossy on both sides, hairy on the veins beneath, jaggedly toothed; lateral veins in four to six pairs, impressed above and prominent beneath. Flowers in hairy cymes about 1 in. wide. Fruits described as oblong to globose, red, about ⅜ in. long. Native of the mountains of Formosa. Perfectly hardy, but rare in cultivation.

V. EROSUM Thunb.

A deciduous shrub of erect habit up to 6 ft high; branches slender, covered with pale brown down when young. Leaves oval-ovate or somewhat obovate, wedge-shaped or rounded at the base, pointed; 1½ to 3½ in. long, 1 to 2 in. wide; sharply toothed, stellately downy on both surfaces, especially beneath; stalks ¼ in. or less long. Flowers white, ⅙ in. across, produced in May in rather loose, slender, scurfy-stalked, usually five-branched cymes, 2 to 3½ in. across; stamens rather longer than the corolla. Fruits red, roundish-ovoid, ¼ in. long.

Native of Japan and China; introduced by Fortune from China in 1844, later by Maries and by Sargent from Japan. It was cultivated for some years in the Royal Horticultural Society's garden at Chiswick, but never seems to have secured a permanent place in gardens. It is, perhaps, not perfectly hardy. Among the red-fruited viburnums this species is marked by the stalks of the leaves being so short.

V. ICHANGENSE (Hemsl.) Rehd. *V. erosum* var. *ichangense* Hemsl.—This close ally of *V. erosum* was discovered in Hupeh by Henry, and introduced by Wilson in 1901, and several times since. It flowered at Coombe Wood in 1906. The leaf-stalks are very short, as in *V. erosum*, but the blades are smaller, ovate-lanceolate, and slender-pointed. The flowers are in smaller cymes, 1 to 1½ in. wide, the stamens are shorter than the corolla; the calyx-tube is conspicuously and densely woolly. Fruits red, as in *V. erosum*.

V. FARRERI Stearn

V. fragrans Bge., not Loisel.

A deciduous shrub 10 ft or more high, and as much in diameter, with nearly glabrous young shoots. Leaves obovate or oval, pointed, much tapered towards the base, strongly toothed, with about six pairs of parallel veins, 1½ to 4 in. long, 1 to 2¾ in. wide, glabrous except for tufts of down in the vein-axils beneath, stalk ⅓ to ¾ in. long. Flower-clusters terminal and lateral, 1½ to 2 in. wide; flower-stalks slightly hairy or glabrous. Each flower is ⅓ to ⅝ in. wide, white, or tinged with pink on opening, the corolla-lobes rounded and spreading, the tube slender and ⅜ in. long. Calyx ⅛ in. long, with five small, rounded, membranous lobes. Stamens short and inserted about midway on the tube. Fruits said to be brilliant red and edible. *Bot. Mag.*, t. 8887.

Native of Kansu, China; first introduced by W. Purdom for Messrs Veitch under his numbers 689 ('white flowers') and 690 ('pink'). The first flowers I saw were sent to me from Wakehurst, Sussex, in March 1920, which had been gathered from one of Purdom's plants obtained from the Coombe Wood Nursery some years previously. As a rule it starts flowering in November and continues through the winter, the flowers being able to bear ten or twelve degrees of frost without injury. They have a very charming heliotrope-like fragrance. According to Farrer, this is the best beloved and most universal of garden plants all over N. China, and it is curious that so popular a shrub should have been so long in reaching this country. A dried specimen at Kew is dated from St Petersburg as long ago as 1835; and the Russian traveller, Potanin, found it wild and cultivated in Kansu in 1885. But nothing was heard of it until Farrer found it, wrote about it, and sent home seeds. In his *On the Eaves of the World*, Vol. I, pp. 96, 97, he alludes to it as 'this most glorious of shrubs'. This is a courageous statement to make, but, seen at its best on a winter's day well in bloom and filling the air with its fragrance, it is a singularly delightful shrub. Farrer found it 10 ft high and more in diameter, growing wild in cold bleak regions, so it is absolutely hardy. He describes the fruit as well-flavoured and of a 'glossy scarlet.' There are two forms in cultivation, one with bronzy young leaves and shoots and flowers that are pink in bud; and another with green shoots and foliage and pure white flowers. They vary also in mode of growth. Some are stiffly erect whilst others are more sprawling, their lower branches layering themselves freely. This is certainly the best midwinter blossoming shrub introduced since the advent of *Hamamelis mollis*.

To the above account, taken almost unchanged from previous editions, there is little to add, except to stress the variability of this species in its garden value. The Purdom introduction first reached gardens through the winding-up sale of the Coombe Wood Nursery in 1913–4. The seeds gathered by Farrer were distributed by the Royal Horticultural Society, excepted for those kept by Farrer himself for the nursery he had founded. It has been said that the Purdom introduction was the finer of the two, but in fact the plants from the seeds collected on Farrer's expedition (in which, be it said, Purdom too took part) proved to be very variable and it is doubtful whether the best of these were inferior to those raised from the Purdom introduction. The form with pure white flowers and light green leaves mentioned above has been named 'CANDIDISSIMUM'.

The name *V. fragrans*, long used for this plant, is invalid, having been used earlier for another species.

V. FOETIDUM Wall.

V. ceanothoides C. H. Wright; *V. foetidum* var. *ceanothoides* (C. H. Wright) Hand.-Mazz.; *V. rectangulatum* Graebn.; *V. foetidum* var. *rectangulatum* (Graebn.) Rehd.

A semi-evergreen shrub up to 10 ft high; young shoots angular and reddish, downy, the hairs either simple or clustered. Leaves 1 to 3 in. long, about half as wide, either broadly ovate with a rounded base and more or less trilobed towards the apex, or, on the older shoots, varying to broadly lanceolate and obovate, coarsely toothed, more or less finely downy, especially on the three or four pairs of veins and the reddish stalks. Flowers individually stalkless, in rounded branched clusters 2 in. wide, opening in July, each about ¼ in. wide; petals white, anthers violet. Fruits closely packed, scarlet-crimson, broadly oval to orbicular, ¼ in. wide. *Bot. Mag.*, t. 9509.

Native of northeast India and Bhutan, northern Burma, and of western and central China. Probably the first introduction was by Wilson from W. Szechwan, but seed was later sent by Forrest from Yunnan. The fine form that received an Award of Merit for its fruits when shown from Exbury in 1934 was raised from Forrest 27410, but this is not altogether hardy. There is, however, a compact and twiggy form in gardens which is hardy and was probably raised from seeds collected by Farrer. The specific epithet refers to the odour of the crushed leaves, still evident in century-old herbarium specimens. Being of wide range *V. foetidum* exhibits minor variations which it is pointless to categorise. Even on one and the same plant the foliage varies, leaves on strong shoots being often definitely three-lobed.

V. GRANDIFLORUM Wall.
V. nervosum sens. Hook. f. & Thoms., not D. Don

A deciduous shrub of stiff habit, or, in the wild, sometimes a small tree; young shoots softly downy at first, becoming dark brown by winter. Leaves of firm texture, dullish green, narrowly oval, tapered towards both ends, pointed, finely and regularly toothed, 3 to 4 in. long, half as much wide; veins parallel in six to ten pairs, very downy beneath; stalk ¾ to 1 in. long, purplish. Flowers fragrant, produced in February and March (sometimes earlier) in a cluster of stalked corymbs at the end of the preceding summer's growth, the whole making a many-flowered inflorescence 2 to 3 in. across. The tube of the corolla is slenderly cylindrical, ½ in. long, spreading at the mouth into five roundish ovate lobes and measuring there ½ to ¾ in. wide. On first opening the corolla is flushed with pale rose, afterwards it is almost pure white; anthers pale yellow. Calyx reddish, with five minute, pointed lobes. Bracts linear, ¼ to ½ in. long and, like the main and secondary flower-stalks, downy. Fruits oval, ½ to ¾ in. long, ultimately blackish purple, said to be edible. *Bot. Mag.*, t. 9063.

Native of the Himalaya from Chamba eastward, possibly extending into parts of western China; introduced from Bhutan by Roland Cooper for A. K. Bulley in 1914. In its wood it is quite as hardy as *V. farreri*, but the flowers are more likely to be damaged or destroyed by frost. The two are closely allied, but *V. grandiflorum* differs in having the undersides of the leaves, the inflorescence-axes and inner bud-scales densely hairy, and in the longer corolla-tubes of its flowers. As seen in gardens it is of different aspect, with its rather gaunt habit. Its flowers vary in colour as they do in *V. farreri*. The flowers are usually bright rosy pink and fade only gradually to white as they age, but some plants of the

Cooper introduction had almost white flowers. In 'SNOW WHITE' (A.M. 1970) the limb is white, flushed with pink; this was raised from seeds collected in Nepal by Col. Donald Lowndes in 1950 (*Journ. R.H.S.*, Vol. 95 (1970), fig. 172).

V. FOETENS Decne.—Near to *V. grandiflorum* both botanically and geographically, and often included in it without distinction. The only essential difference is that the leaves are glabrous or almost so beneath and that the inflorescence too is glabrous, except at the nodes. The corollas are almost as long as in *V. grandiflorum* and longer than those of *V. farreri*. In both species there is variation in the inflorescence, which may be congested with the outer bud-scales persisting beneath, or elongate, with the scales fallen away. Nevertheless, some garden plants of *V. foetens* are very distinct from *V. grandiflorum* in their roundish, stiff habit, in their stout branchlets, and in not opening their flowers until late in the winter.

V. foetens has a more western distribution than *V. grandiflorum*, from Chamba, where it overlaps with that species, to Chitral and Hazara. It was described in 1844 from a specimen collected by Jacquemont, the French traveller. So far as is known it did not reach commerce until Messrs Marchant distributed plants, raised from seeds which according to their catalogue of 1937 had been collected by Capt. Simpson-Hayward. In 1934 George M. Taylor, the well-known plantsman, of Longniddry, East Lothian, received seeds from which he raised two plants. The specimen he sent to Kew from one of these was certainly *V. foetens*, except that the corolla-tubes were unusually short. He also gave to Kew some of the seeds he received in 1934, and a plant raised from them had glabrous leaves as in *V. foetens* but hairy inflorescence-axes as in *V. grandiflorum*. When enquiries were made early in the 1950s about the provenance of the seeds, Mr Taylor said they had been collected in Korea, where neither *V. grandiflorum* nor *V. foetens* occur, and *V. farreri* only as a cultivated plant. On the whole it seems likely that Mr Taylor's memory betrayed him, and that the seed was in fact collected in the western Himalaya.

V. HARRYANUM Rehd.

An evergreen shrub ultimately 6 to 8 ft high, of bushy habit, sometimes taller; young shoots clothed with a minute, dark down. Leaves orbicular to obovate or broadly ovate, tapered at the base, rounded at the apex except for a small mucro, margins entire, or with a few obscure teeth; $\frac{1}{4}$ to 1 in. long, from two-thirds to nearly as wide, dark dull green above, paler beneath, quite glabrous on both surfaces; leaf-stalk about $\frac{1}{12}$ in. long, reddish. Inflorescence a terminal, compound umbel, $1\frac{1}{2}$ in. across. Flowers pure white, $\frac{1}{8}$ in. across. Fruits ovoid, pointed, E in. long, shining, black.

Native of W. China; discovered and introduced in 1904 by Wilson, who remarks that it is rare on mountains at 9,000 ft. It is quite distinct from any other cultivated evergreen viburnum in its small privet-like leaves. It appears to be fairly hardy, and flowered for the first time in cultivation in 1914. It was named in compliment to Sir Harry Veitch.

V. HENRYI Hemsl.

An erect, evergreen shrub becoming 10 ft high, and having a tree-like habit; branchlets stiff, glabrous. Leaves narrowly oval, oblong or obovate, 2 to 5 in. long, 1 to 1¾ in. wide, shortly pointed, wedge-shaped or rounded at the base, shallowly toothed, dark shining green above, paler beneath, glabrous on both sides or slightly furnished with stellate down on the stalk and midrib; stalk slightly winged, ½ to ¾ in. long. Panicles stiff, pyramidal, 2 to 4 in. wide at the base, and about as long; flowers perfect and uniform, fragrant, white, ¼ in. across, opening about midsummer. Fruits oval, ⅓ in. long, at first red, then black. *Bot. Mag.*, t. 8393.

Native of Central China; discovered by Henry in 1887 in the Patung district of Hupeh; introduced by Wilson in 1901. It is distinct among hardy viburnums in its long, narrowish, nearly or quite glabrous leaves, and its stiff, thin, erect, formally branched habit; also in its pyramidal panicles (but see *V. erubescens* below). It was given a First Class Certificate in September 1910 for its beauty in fruit, and an Award of Garden Merit in 1936.

V. ERUBESCENS Wall. *V. pubigerum* Wight & Arn.—This variable species is of wide distribution, from Ceylon and peninsular India to the Himalaya from Kumaon eastward and the hills of Assam; also of northern Burma and China. It is a shrub or small tree with deciduous or sub-evergreen foliage. Leaves elliptic, elliptic-ovate or oblong, acuminate, to about 4 in. long, downy or glabrous beneath, serrate, veins in five to seven pairs, prominent beneath; petioles reddish, about 1 in. long. Flowers white flushed with pink, borne around midsummer in lax, pendent panicles up to 3 or 4 in. long. Corolla with a slender tube about ⅜ in. long. Anthers almost sessile. Fruits black.

The cultivated plants probably derive mostly from the seeds collected by Wilson in 1910 in W. Szechwan. This is a hardy form with longer panicles than normal and was referred by Rehder to his var. GRACILIPES. It was probably reintroduced by Forrest from Yunnan and certainly more recently by Ludlow, Sherriff and Hicks from Bhutan.

V. erubescens has in common with *V. henryi* its usually long panicles, but differs among other characters in the long tubes of the corollas. It is really quite near to *V. grandiflorum* and its allies, but the flowers are borne after the leaves.

V. × HILLIERI Stearn—A seedling of *V. henryi*, which must have been pollinated by a plant of *V. erubescens* growing nearby. It was raised by Messrs Hillier and described in 1956. The leaves are broader than in *V. henryi* and the habit more lax. The flowers have a longer tube than in *V. henryi*, but shorter than in *V. erubescens*, and are also intermediate in the length of the stamens, which as in *V. henryi* have distinct filaments, though shorter (in *V. erubescens* the anthers are almost sessile). See further in *Journ. R.H.S.*, Vol. 81 (1956), pp. 538–40, with figs., and *Bot. Mag.*, n.s., t. 680. The original clone of *V.* × *hillieri* has been named 'WINTON'. It received an Award of Merit in 1956.

V. HUPEHENSE Rehd.

A deciduous shrub, the young shoots stellately hairy the first year, purplish brown the second. Leaves broad-elliptic or roundish ovate, long-pointed, truncate or slightly heart-shaped at the base, coarsely toothed, dark green and covered with loose stellate down above, paler and more downy beneath, 2 to 3 in. long, 1¼ to 2¼ in. wide, veins in six to eight pairs; leaf-stalk grooved, ½ to ¾ in. long, densely downy; stipules narrowly lanceolate, downy. Corymbs about 2 in. wide, the main and secondary flower-stalks covered densely with stellate down; branches of the corymb usually five. Fruits egg-shaped, red, ⅓ to ⅖ in. long. *Bot. Mag.*, n.s., t. 41.

Native of Hupeh and Szechwan, China; discovered by Henry; introduced by Wilson in 1908. It is nearly related to *V. dilatatum* (from which it differs in its orbicular-ovate leaves, and stipuled leaf-stalks), and to *V. betulifolium*, from which it is distinct in being downy on both leaf surfaces. As a fruiting shrub it is scarcely inferior to either. Award of Merit 1952.

V. JAPONICUM (Thunb.) Spreng.

Cornus japonicus Thunb.; *V. macrophyllum* Bl.

A sturdy, evergreen bush up to 6 ft high in this country, with thick, glabrous young shoots. Leaves leathery, usually ovate (sometimes very broadly so), but also roundish, oval or obovate, 3 to 6 in. long, half to nearly as much wide, abruptly pointed or with a short, slender apex, the base entire and rounded or tapering, the terminal part remotely and shallowly toothed or merely wavy; both surfaces quite glabrous, the upper one dark glossy green, the lower one paler but with innumerable tiny dark dots; stalk ½ to 1¼ in. long. Flowers uniformly perfect, ⅜ in. wide, white, very fragrant, produced in rounded short-stalked, often seven-rayed cymes 3 to 4½ in. across. Fruits round-oval, ⅓ in. long, red.

Native of Japan; probably first introduced by Maries in 1879. Richard Oldham, who collected it in Nagasaki in 1862, describes it as 'a small tree on the hills,' but it does not make more than a sturdy bush with us. It appears to be quite hardy at Kew, but grows slowly in the open, and is no doubt happier in a warmer climate. On a wall it makes a pleasing and striking evergreen. This species has been much confused in gardens with *V. odoratissimum*, but it may be distinguished in the following respects: The young wood is not so warted as in *V. odoratissimum*; the secondary veins run out to the margin of the leaf; the inflorescence is rounded and umbel-like rather than paniculate.

V. KANSUENSE Batal.

A deciduous shrub, 4 to 8 ft high, with glabrous, ultimately greyish branchlets. Leaves ovate to roundish in main outline, but deeply three- or five-lobed, the lobes coarsely toothed and taper-pointed, the base wedge-shaped, rounded or slightly heart-shaped, 1 to 2 in. long, and from two-thirds to fully as much in width, dark green, and with appressed hairs above, especially on the veins; much paler beneath, with conspicuous tufts of pale down in the vein-axils, and with hairs along the midrib and veins; leaf-stalk ½ to 1 in. long, slender, glabrous; three or five main veins radiate from the top of the leaf-stalk. Corymbs without sterile flowers, 1 to 1½ in. across, often seven-rayed. Flowers pinkish white, ¼ in. wide; calyx glabrous. Fruits red, ⅓ to ½ in. long, oval to roundish.

Native of China, where it is apparently widely spread, being found in Kansu, Szechwan, and Yunnan; introduced by Wilson in 1908. It belongs to the Opulus group, but is distinct in having no marginal showy sterile flowers, which the other Chinese species (*V. sargentii*) has. The leaves also are very distinct in their frequently small size and deep lobing, some suggesting a small maple leaf. An elegant shrub needing a shaded position. The leaves turn dull red in autumn.

V. LANTANA L.   WAYFARING TREE

A vigorous deciduous bush, sometimes almost tree-like, 12 to 15 ft high; young shoots, buds, lower surface of leaves and flower-stalks all covered with a dense coat of pale, minute, starry down. Leaves broadly ovate or inclined to oblong, the base heart-shaped, the apex pointed or bluntish, minutely toothed, 2 to 5 in. long, 1½ to 4 in. wide, upper surface velvety with stellate down, at least at first; stalk ½ to 1¼ in. long. Flowers white, ¼ in. across, uniform and perfect, produced in May and June in stalked, usually seven-rayed cymes, 2 to 4 in. wide. Fruits oblong, ⅓ in. long, at first red, ultimately black.

Native of Europe, including England as far north as Yorkshire; also of N. Africa, Asia Minor, the Caucasus and northwest Iran. It is the type species of the Lantana group of viburnums, characterised by naked winter buds, deciduous foliage, a scurfy stellate down, and fruits at first red, then black. *V. lantana* is itself an ornamental shrub, pretty in flower, in fruit, and sometimes in its red autumn tints; useful for planting in tall shrubberies or in thin woodland. There is a variety whose leaves are blotched and spotted with yellow, but I have never seen it in a condition that would justify one in planting it.

A variant with leaves even more wrinkled than in the normal form was named var. RUGOSUM by Lange, while var. DISCOLOR Huter is described as having the leaves smaller and firmer, white-tomentose beneath.

V. LANTANOIDES Michx.   HOBBLE BUSH
V. alnifolium Marsh., *nom. confus.*

A strong-growing, rather coarse-habited, deciduous shrub, 6 to 10 ft high; the central shoots erect, the lower ones spreading, often prostrate; young bark covered with a thick scurfy down. Leaves in distant pairs, broadly ovate to roundish, the points short and abrupt, the base heart-shaped, margins irregularly toothed, 4 to 8 in. long, nearly as broad, upper surface dark green, at first downy, but becoming glabrous; lower surface with much stellate down on the midrib and veins, especially when young; stalk 1 to 2½ in. long, scurfy downy. Flowers white, produced in stalkless cymes with usually five divisions, and 3 to 5 in. across; marginal flowers sterile, and ¾ to 1 in. across; central ones perfect and much smaller. Fruits red, turning black-purple, ⅓ in. long, broadly oval. *Bot. Mag.*, t. 9373.

Native of eastern N. America; introduced in 1820. It is not an easy plant to suit, needing woodland conditions and there only thriving away from the root-run of large trees, as it needs abundant moisture to be seen at its best. It is very distinct in its large leaves, which turn deep claret-red in the autumn, and from our native *V. lantana* is well distinguished in having large, sterile marginal flowers. The popular name refers to its prostrate lower branches, which often take root and trip up the unwary traveller through its native haunts. The venation of the leaves is handsome; the primary veins branch on the lower side only, and are connected by thin parallel nerves almost at right angles.

V. lantanoides received an Award of Merit in 1952.

V. FURCATUM Bl.—The Japanese ally of *V. lantanoides* (also found in Formosa), needing the same conditions. This also has the showy sterile marginal flowers, but its stems are more uniformly erect. It differs also in the shorter stamens, which are only half the length of the corolla, and in the shape of the furrow in the seed. The foliage turns brilliant scarlet to reddish purple in autumn. It is a bush 12 ft or more high in the wild. Introduced in 1892.

V. SYMPODIALE Graebn.—This is closely allied to both the preceding, especially to *V. furcatum*, but differs in having stipules on the leaf-stalks, and in its smaller, ovate, more finely toothed leaves. It was collected in Central China by Wilson in 1900, and may be in cultivation.

Another member of the group, rare in cultivation, is V. CORDIFOLIUM Wall. ex DC., a native of the eastern Himalaya, northern Burma and S.W. China. It differs from the preceding in having no sterile flowers in the inflorescence.

V. LENTAGO L.   SHEEPBERRY

A robust deciduous shrub or small tree up to 20 or 30 ft high; young wood with a slight reddish scurf; winter buds grey. Leaves ovate to obovate, wedge-shaped or rounded at the base, the apex as a rule long and taper-pointed, finely, sharply and regularly toothed; dark, shining green above, smooth on both sides except for a short, scurfy down on the midrib and veins, 2 to 4 in. long, half as wide; stalks mostly winged, ½ to 1 in. long. Flowers creamy white, ¼ in. across, agreeably fragrant, all perfect, produced in May and June in a terminal stalkless cyme, 3 to 4½ in. across. Fruits oval, blue-black, ½ to ⅝ in. long, covered with bloom.

Native of eastern N. America from Canada to Georgia, west to Missouri; introduced in 1761. Although this species does not bear fruit freely in this country, it is well worth growing for its flowers and autumn colour, and as a small, handsome tree. It has been confused with *V. prunifolium* and *V. rufidulum*, but differs in the leaves being long and taper-pointed, with broadly winged stalks. It is the type of the small section *Lentago* (see p. ooo).

V. MACROCEPHALUM Fort.

A deciduous or partially evergreen shrub up to 12 or 20 ft high, forming a large rounded bush, the young shoots covered with a close scurf which, seen under the lens, is found to be minute stellate down. Leaves ovate, occasionally oval or oblong, rounded at the base, rounded or pointed at the apex, 2 to 4 in. long, $1\frac{1}{4}$ to $2\frac{1}{2}$ in. wide, dull green, and with scattered hairs above, covered with stellate down beneath; stalk $\frac{1}{3}$ to $\frac{3}{4}$ in. long. Flowers pure white, all sterile, 1 to $1\frac{1}{4}$ in. across, forming a huge, globular truss 3 to 6 in. wide opening in May.

x $\frac{2}{3}$

VIBURNUM MACROCEPHALUM

This is Fortune's type and was introduced by him from China in 1844. Being perfectly sterile, it has, of course, no place in nature, and is a purely garden plant, once distinguished as *V. macrocephalum sterile* but under modern rules should strictly be f. *macrocephalum*. It is the most striking, if not the most beautiful of viburnums, its truss exceeding in bulk that of any other species. Near London, it lives in a sheltered spot in the open, but is better on a wall, where a well grown plant makes a very fine display in May. Fortune saw it 20 ft high on the island of Chusan.

f. KETELEERI (Carr.) Rehd. *V. keteleeri* Carr.; *V. arborescens* Hemsl.— This is the normal wild form, a native of China, and has only the marginal flowers of the showy, sterile kind, the small perfect ones filling the centre of the cyme, which is 3 to 5 in. across, and comparatively flat. It is somewhat hardier than the wholly sterile plant, but is now uncommon in gardens.

V. MOLLE Michx.
V. demetrionis Deane

A deciduous shrub of bushy habit, 6 to 12 ft high; young shoots glabrous and bright green at first, soon turning grey; older bark peeling. Leaves broadly ovate to roundish, 2 to 5 in. long, $1\frac{3}{4}$ to $3\frac{3}{4}$ in. wide, mostly heart-shaped at the base, slender-pointed, coarsely triangular toothed, the teeth twenty to thirty on each side, upper surface dark green and glabrous, paler and more or less downy beneath; stalk $\frac{1}{2}$ to almost 2 in. long. Flowers white, all perfect, $\frac{1}{4}$ in. across, produced in long-stalked cymes 2 to 4 in. wide. Fruits scarcely $\frac{1}{2}$ in. long, oval, much compressed, blue-black.

Native of eastern-central North America, rare in gardens (the plant once grown as *V. molle* was a form of the *V. dentatum* complex). *V. molle* is very distinct from other American viburnums with blue-black fruits in the combination of the loose peeling bark of the older branches, the long-stalked leaves, and the presence of a pair of glandular-downy stipules on the petiole.

V. RAFINESQUIANUM Schultes   *V. villosum* Raf., not Swartz; *V. affine* var. *hypomalacum* Blake—A shrub to about 8 ft high; bark not peeling. Leaves from narrow- to broad-ovate, acuminate or acute, truncate to slightly cordate at the base, with up to ten rather coarse teeth on each side, softly downy beneath. Petioles very short (rarely more than $\frac{1}{4}$ in. long), usually furnished with stipules. Cymes to about 3 in. wide. Fruits blue-black, ellipsoid; stone flattened, shallowly grooved. Native of eastern N. America, sometimes cultivated for its scarlet autumn colouring. It has been confused with the downy form of *V. dentatum* (var. *pubescens*) but differs, among other characters, in the very shortly stalked leaves.

var. AFFINE (Blake) House—Leaves more or less glabrous beneath.

V. NUDUM L.

A deciduous shrub up to 10 ft high; young shoots slightly scurfy and downy. Leaves oval, ovate or lance-shaped, 2 to $4\frac{1}{2}$ in. long, 1 to $2\frac{1}{4}$ in. wide, minutely and irregularly toothed to almost entire, dark glossy green and glabrous above, paler, somewhat scurfy or glabrous beneath; stalk $\frac{1}{4}$ to $\frac{5}{8}$ in. long. Flowers yellowish white, uniform and perfect, $\frac{1}{5}$ in. across, produced in early June on cymes 2 to 4 in. wide; the main-stalk as long or longer than the branched flowering portion. Fruits $\frac{1}{3}$ in. long, oval, blue-black.

Native of eastern N. America; introduced in 1752. This viburnum is closely akin to *V. cassinoides*, under which species the distinctions between the two are explained. It is a handsome, shiny-leaved shrub which flowers freely. It has a more southern distribution than *V. cassinoides*, and does not, apparently, reach into Canada.

V. odoratissimum Ker-Gawler

V. awabuki K. Koch; *V. awafuki* Hort.

An evergreen shrub, 10 to 25 ft high, with warted bark, free from down. Leaves leathery, oval to obovate, 3 to 8 in. long, 1½ to 4 in. wide, wedge-shaped at the base, rounded or with a short, blunt tip at the apex, entire or with a few obscure teeth towards the end, glossy green and glabrous above, paler beneath and glabrous except for tufts of down in the vein-axils; stalk ½ to 1¼ in. long. Flowers pure white, fragrant, all perfect, produced in stalked, broadly pyramidal panicles, 3 to 6 in. high, 2½ to 5 in. wide at the base. Fruits red at first, ultimately black.

Native of northeastern India, southeast continental Asia, Japan, Formosa, the Philippines and the Celebes; introduced about 1818. This shrub grows well and makes a handsome bush in the southwestern counties, but is not very hardy near London—not so hardy even as *V. japonicum*, with which it was much confused. Its pyramidal inflorescence best distinguishes it from that species, but the venation of the leaf also is different in the veins splitting up and not running out to the margin, a character which enables it to be recognised when out of bloom.

The Japanese race of *V. odoratissimum* is sometimes separated as a distinct species—*V. awabuki*—though Rehder considered that it did not even merit varietal status. In *V. odoratissimum sens. strict.* the corolla-lobes are large, and longer than the tube; in the Japanese plants the corolla-tube is longer, and the lobes relatively shorter.

V. opulus L.   Guelder Rose

A deciduous shrub forming a thicket of erect, grey stems, 10 to 15 ft high; young wood glabrous, ribbed. Leaves three- (sometimes four- or five-) lobed, maple-like, 2 to 4 in. long, often as much or more wide, the base truncate, the lobes pointed, coarsely and irregularly toothed, dark green and glabrous above, more or less downy beneath; stalk ½ to 1 in. long, with two thin linear stipules at the base and large, concave glands near the leaf-blade. Cymes 2 to 3 in. across, with a border of sterile, showy white flowers, ¾ in. in diameter, the centre composed of small fertile flowers; anthers yellow. It blossoms in early June. Fruits bright red, globose, ⅓ in. wide.

Native of Europe (including Britain), northwest Africa, Asia Minor, the Caucasus, and parts of Central Asia. Whilst in beauty of flower the Guelder rose is inferior to many viburnums, it is inferior to none in this country in its fruits, or in the rich hues of its decaying foliage. Many other species, no doubt, have fruits as beautiful, but they do not set them in our gardens with the certainty of this. Its only fault is that black fly (bean aphis) also like it.

cv. 'Aureum'.—Leaves bright yellow. This needs some shade.

var. americanum Ait. *V. trilobum* Marsh.—This is the American race of *V. opulus*, differing from it in minor characters such as the more glabrous undersides of the leaves, the shallower, broader channel of the petiole, and the smaller, convex or club-shaped petiolar glands. It is of wide distribution in Canada and the northern USA, as far west as British Columbia and Washington. It is known in N. America as 'cranberry'; the fruits are pleasanter to the taste than those of its Old World ally and are used to make jam.

cv. 'COMPACTUM'.—Of lower and denser habit. Free fruiting. Award of Merit 1962. Not to be confused with 'Nanum'.

cv. 'FRUCTU-LUTEO'.—Fruits translucent yellow. In cultivation 1901 and probably much earlier (*V. o. fructu-luteo* Hort. ex Kew Hand-list; *V. o.* var. *luteum* Bean). Edwin Beckett, the garden manager at Aldenham, suggested that the name *fructu-luteo* should be used for a form with fruits pink at first, becoming dark chrome yellow, slightly tinged with pink (*Gard. Chron.*, Vol. 88 (1930), p. 281.

cv. 'NANUM'.—A curious dwarf form of tufted habit, growing 1 to 3 ft high. Its leaves are ¾ to 1½ in. wide. It rarely or never flowers.

cv. 'NOTCUTT'S VARIETY'.—A vigorous bush to about 12 ft high, with larger fruits than usual and fine autumn colour.

cv. 'ROSEUM' ('Sterile'). SNOWBALL TREE.—In this form all the flowers are of the sterile kind, and the cyme in consequence becomes transferred into a globose head of white, closely packed blossoms, 2 to 2½ in. across. This is one of the most beautiful of hardy shrubs, but of course the fruiting beauty of the common Guelder rose is sacrificed.

'Roseum' has been known in European gardens since the 16th century and was usually called '*Sambucus rosea*' or Rose Elder, probably from a fancied resemblance of the florets to single roses. The name 'Guelder Rose', now used for the species as a whole, originally belonged exclusively to this double form and suggests that it originated in the region of the Netherlands known as Gelderland. Miller called the wild form 'Marsh Elder' or 'Guelder Rose with flat flowers'.

cv. 'XANTHOCARPUM'.—Plants under the name *V. opulus xanthocarpum* were distributed by Späth's nursery, Berlin, from 1910 onward, and were described as having golden-yellow fruits. Two plants have received awards in Britain under this name. One was shown by Sir William Lawrence in 1932; it came from Aldenham, and was 'Xanthocarpum' as understood by Edwin Beckett (see his article under 'Fructu-luteo'), in which the fruits are golden-yellow from the start. This appears to have been a rather weak-growing plant; it received an Award of Merit, however. The plant that received a First Class Certificate when shown by the Crown Estate Commissioners, Windsor, in 1966, had fruits starting apricot yellow and becoming translucent saffron-yellow (*Journ. R.H.S.*, Vol. 91 (1966), p. 519 and fig. 251).

V. EDULE (Michx.) Raf. *V. opulus* var. *edule* Michx.; *V. o. pauciflorum* Raf.; *V. pauciflorum* (Raf.) Torr. & Gr.—Although allied to *V. opulus* and with similar foliage, this differs in the absence of sterile flowers from the inflorescence. It could be confused with *V. acerifolium*, but differs in its red fruits and in having winter-buds with a single pair of outer scales. It is of wide distribution in N. America, ranging across the continent in high latitudes, south in the mountains. The fruits are used for making jam.

V. SARGENTII Koehne—An ally of *V. opulus* in N.E. Asia, introduced to Europe through Prof. Sargent in 1892. It is a coarser growing shrub with often larger leaves, a corky bark, and smaller fruits. It is not so useful and well-doing a shrub as *V. opulus* in Britain, starting earlier into growth, and being subject to injury by spring frosts. There is a yellow-fruited form—f. FLAVUM Rehd.— which comes partly true when raised from seeds (see D. Wyman, *Shrubs and Vines* (1969), p. 494). This is in cultivation in Britain and received an Award of Merit in 1967. In 'ONONDAGA' the young leaves are coloured deep bronze and retain a tinge of that colour when older. Flower buds red. This very ornamental viburnum was raised by Donald Egolf at the United States National Arboretum and is cultivated in the Hillier Arboretum near Romsey.

V. PHLEBOTRICHUM Sieb. & Zucc.

A shrub to about 8 ft high, slenderly branched, the branchlets glabrous or slightly downy. Leaves thin, ovate or elliptic, up to $3\frac{1}{2}$ in. long (longer on extension growths), glabrous or slightly downy above, appressed-hairy on the veins beneath, coarsely and sharply serrate, the teeth tipped with horny mucros; petioles to about $\frac{5}{8}$ in. long. Inflorescences terminating short two-leaved laterals, rather few-flowered, about 2 in. wide, more or less pendulous. Flowers white, small. Stamens shorter than the corolla-lobes. Fruits red, ovoid, about $\frac{3}{8}$ in. long.

Native of Japan; plants originally grown under this name were *V. setigerum*, and the true species was little known until the 1930s. From that species it differs in the smaller and more shortly stalked leaves, shorter stamens and drooping inflorescences. The leaves colour crimson in the autumn.

V. PLICATUM Thunb. [PLATE 96

Thunberg described two forms of this species: *V. tomentosum* was the normal wild form, and the name *V. plicatum* was given by him to a form with 'snowball' inflorescences cultivated in the gardens of Japan. Since the former name had priority, the garden plants with the snowball type of inflorescence took their natural place as a variety or botanical form of the wild *V. tomentosum*. Unfortunately, as Dr Rehder pointed out in 1945, the name *V. tomentosum* Thunb. (1784) is antedated by *V. tomentosum* Lam. (1778), which is *V. lantana*. It is therefore illegitimate, and the order of nature must be reversed: *V. plicatum* becomes the correct name for the species as a whole; the snowball forms are collectively *V. plicatum* f. *plicatum*, and the wild plants *V. plicatum* f. *tomentosum*. In the following account the wild form and the cultivars with normal inflorescences are treated first.

f. TOMENTOSUM (Thunb.) Rehd. *V. tomentosum* Thunb., not Lam.; *V. plicatum* var. *tomentosum* (Thunb.) Miq.—A deciduous shrub of bushy habit, 6 to 10 ft high, the branches mostly horizontal, covered when young with a minute, starry down. Leaves ovate or oval, tapered to a point, rounded or wedge-shaped at the base, 2 to 4 in. long, 1 to 2½ in. wide, toothed except at the base, dull dark green above with scattered hairs at first, pale, greyish, and stellately

VIBURNUM PLICATUM f. TOMENTOSUM

downy beneath; stalk ½ to ¾ in. long. Inflorescence a flat umbel 2½ to 4 in. across, borne at the end of a short, usually two-leaved twig; the centre is filled with the small perfect flowers, surrounded by a few large, white sterile ones, 1 to 1½ in. across. Blossoms in early June. Fruits roundish egg-shaped, at first coral red, finally blue-black.

Native of Japan and China; introduced from Japan about 1865; reintroduced by Maries (see 'Mariesii') and later by Wilson. It seems to have been over-shadowed by typical *V. plicatum*, i.e., the snowball forms, until the selection 'Mariesii' began to spread into gardens early this century. It is now almost wholly represented in gardens by named selections.

cv. 'CASCADE'.—See under 'Rowallane'.

cv. 'MARIESII'.—A selection with larger inflorescences than normal and ray-flowers up to about 1¾ in. wide, on common stalks to about 2½ in. long. A splendid shrub of tabular habit, growing best in a moist, light soil, where it will spread indefinitely by self-layering if permitted to do so. In the drier parts of the country it may need some shade, but elsewhere is quite happy in full sun. A large plant is a fine sight in early summer, each branch set from end to end with pure white corymbs. As always in the wild form, the ray-flowers are very asymmetrical and irregular—more so than in *V. lantanoides* and *V. furcatum* and much more so than in *V. opulus*, where the ray-flowers are fairly regular. Most flowers are slightly rotated on their axis, so that one of the two innermost segments appears to be basal, and is very small and rounded. The leaves turn dull crimson or purplish red in autumn, but fruit is not freely set. 'Mariesii' was introduced by Charles Maries, who collected for Messrs Veitch in Japan and China 1877–9. It was named by Veitch in 1902 and received an Award of Garden Merit in 1929.

The similar 'LANARTH' was raised at Lanarth and received an Award of Merit when shown from Exbury in 1930. It has become confused with 'Mariesii', but so far as can be ascertained the true 'Lanarth' has ray-flowers to well over 2 in. wide; the common-stalk of the inflorescence is about ½ in. longer than in 'Mariesii' and rather stout; and the fertile flowers are not so densely arranged. It is really a rather coarser plant, and certainly no more effective in the garden. It is possible that some plants grown as 'Lanarth' are really 'Mariesii' and *vice versa*, while plants imported from the continent under one or the other name are said in some cases to be the normal wild form (these may be seedlings).

cv. 'NANUM SEMPERFLORENS' ('Watanabei')—An anomalous form of very dense habit, producing rather small inflorescences throughout the summer, but giving a good display at the normal season once it is established. Despite the epithet *nanum* it has attained a height of 6 ft in the Hillier Arboretum. It can be kept dwarf by pruning in early spring; by this treatment flowering is delayed until late summer and autumn, when a splash of white in the garden is more appreciated. This viburnum was introduced to Britain from Wada's nursery; the original plant was found wild in Japan.

cv. 'PINK BEAUTY'.—Ray-florets becoming pink as they age; raised in the USA. It is the counterpart among viburnums of *Hydrangea serrata* 'Rosalba'.

cv. 'ROWALLANE'.—Of much more moderate growth than 'Lanarth' or 'Mariesii', with smaller, broadly ovate, short-acuminate leaves. The flower-trusses are not so wide as in those plants, but the ray-florets are still large, and form a neat circle round the rather few fertile flowers. This exquisite viburnum apparently originated at Rostrevor in N. Ireland, where it may have been raised from seeds sent by Wilson, but was first distributed from Rowallane. It received an Award of Merit in 1942, and a First Class Certificate in 1956, on both occasions when exhibited by Collingwood Ingram from his garden in Kent (*Journ. R.H.S.*, Vol. 67 (1942), fig. 110; ibid., Vol. 81 (1956), fig. 135). It fruits freely in some gardens. A selection from its seedlings, raised in Holland, is 'CASCADE' (*Dendroflora*, No. 8, p. 69, fig.).

f. PLICATUM *V. dentatum* L. *sens*. Thunb., not L.; *V. plicatum* Thunb.; *V. tomentosum* var. *plicatum* (Thunb.) Maxim.; *V. t.* var. *sterile* K. Koch— This stands in the same relation to *V. plicatum* f. *tomentosum*, i.e., to the normal wild state of the species, as the common Snowball tree does to *V. opulus*, the whole of its flowers being transformed into the showy sterile kind, and the inflorescence from a flat umbel to an almost globose one. It has long been cultivated in the gardens of Japan and China, and was first mentioned under its Japanese name by Kaempfer in 1712. It was introduced by Fortune from China in 1846 to the garden of the Horticultural Society, but seems to have been overshadowed by *V. macrocephalum*, which came at the same time. But around 1860 Fortune reintroduced it from Japan for Standish, and by the 1870s it was being grown in the open in many gardens and found to be hardy. It grows to about 8 ft high and almost as much wide. Towards the end of the last century the German firm of Hesse distributed a form they called *V. plicatum grandiflorum*, which, judging from a plant received from them at Kew in 1898 had flower-heads no larger than in the second Fortune introduction, but densely set on the branch and with broadish, neatly toothed leaves. This seems very similar to plants distributed in the USA at about the same time by Meehan's and Parsons' nurseries as *V. tomentosum rotundifolium*. But there is another clone under the name *V. plicatum grandiflorum* with larger trusses and a more tiered habit of growth than the old form.

V. PROPINQUUM Hemsl.

An evergreen shrub of bushy habit, with glabrous, shining, angular young shoots. Leaves three-veined, ovate-lanceolate to oval, wedge-shaped or rounded at the base, pointed, shallowly and sparsely toothed, 2 to $3\frac{1}{2}$ in. long, $\frac{3}{4}$ to $1\frac{1}{4}$ in. wide, dark glossy green and glabrous; stalk $\frac{1}{4}$ to $\frac{5}{8}$ in. long. Flowers greenish white, $\frac{1}{6}$ in. across, all perfect, produced in usually seven-branched cymes $1\frac{1}{2}$ to 3 in. wide. Fruits blue-black, egg-shaped, $\frac{1}{5}$ in. long.

Native of Central and Western China, Formosa and the Philippines; discovered by Henry and introduced by Wilson for Messrs Veitch in 1901, and again later. It is hardy at Wakehurst Place in Sussex and at Exbury on the Solent, but needs a sheltered position. Even on one plant the leaves vary in shape; forms with consistently narrow leaves have been distinguished in gardens by the epithets *angustifolium* and *lanceolatum*. It is distinct from all other cultivated viburnums except *V. davidii* and *V. cinnamomifolium*, but the former is dwarf and the latter has uniform, broader, scarcely toothed leaves.

The following belong, like *V. propinquum*, to the section *Tinus*, but neither is common in cultivation:

V. ATROCYANEUM C.B.Cl. *V. wardii* Hort., not W.W.Sm.—Leaves oblong, entire or indistinctly toothed, pinnately veined, about 3 in. long. Inflorescence almost sessile. Fruits about $\frac{3}{16}$ in. long, hard, dark blue with a metallic lustre. Described from a specimen collected by Griffith in the Mishmi Hills of Assam. Most plants in cultivation derive from Kingdon Ward's 9198, collected in 1931 a short distance farther east, in the Adung valley of northwest Burma. He wrote in his field note: 'Berries small, steely blue, with a curious lustre, giving an effect of lalique. The winter contrast of red stems, dark green leaves, and bunches of gleaming opalescent fruits is charming. Found up to 7,000 ft, usually on open cliffs and ridges in full sun.'

V. CALVUM Rehd. *V. schneiderianum* Hand.-Mazz.—Leaves shortly stalked, narrowly ovate to rhombic-elliptic, to about 3 in. long and $1\frac{3}{4}$ in. wide, entire or with a few callous teeth, glabrous, glossy, veins five to eight on each side, deeply impressed above. Flowers white or cream-coloured, in rounded terminal corymbs about 3 in. wide, on peduncles 1 to 1¾ in. long. Fruits blue-black, about ¼ in. long. A native of southwest China, described from a specimen collected by Henry in southern Yunnan; introduced by Forrest. It has long been grown at Exbury and has been distributed under its synonymous name. It is the probable pollen-parent of 'Jermyns Globe' (see under *V. davidii*).

V. PRUNIFOLIUM L.   BLACK HAW

A deciduous, tall shrub or sometimes a small tree, 20 to 30 ft high; branchlets rigid, glabrous and reddish when young. Leaves glabrous, ovate, oval or obovate, sometimes roundish, 1½ to 3½ in. long, 1 to 2 in. wide, rounded or wedge-shaped at the base, blunt or short-pointed at the apex, dull green above, pale below; stalks not or slightly winged, reddish, ⅓ to ¾ in. long. Flowers white, ¼ in. across, uniformly perfect, produced during June in scarcely stalked cymes 2 to 4 in. across. Fruits dark blue, oval, ½ to ⅔ in. long, sweet and edible.

Native of eastern and eastern-central N. America, ranging in the south as far west as Texas: introduced early in the 18th century. This makes a very handsome small tree, especially if kept to a single stem when young, forming a shapely rounded head of branches. The leaves colour red and yellow in the autumn. It is allied to *V. lentago* and *V. rufidulum*.

V. RHYTIDOPHYLLUM Hemsl.

An evergreen shrub 10 to 20 ft high, and more through; the stout branches thickly covered with starry down. Leaves ovate-oblong, 3 to 7½ in. long, 1 to 2½ in. wide, pointed or blunt at the apex, rounded or slightly heart-shaped at the base, upper surface glossy, not downy, but deeply and conspicuously wrinkled, lower one grey with a thick felt of starry down; stalk ½ to 1¼ in. long. Flowers produced on large terminal umbel-like trusses 4 to 8 in. across, which form into bud in the autumn and remain exposed all through the winter, and until the blossoms expand the following May or June. They are a dull yellowish white, about ¼ in. diameter. Fruits oval, ⅓ in. long, at first red, then shining black. *Bot. Mag.*, t. 8382.

Native of central and western China; introduced by Wilson for Messrs Veitch in 1900. It is hardy, and flowers well in spite of forming its inflorescences and partially developing them in autumn. Its beauty is in its bold, wrinkled, shining leaves; the flowers are dull and not particularly attractive. In previous editions it was called 'one of the most distinct and striking, not only of viburnums but of all the newer Chinese shrubs.' But this praise only applies to plants grown in a good soil and a sheltered position. Wind-tattered or starved plants are downright ugly, and are responsible for the execration in which this species is held by some gardeners. The plant at Borde Hill in Sussex, mentioned in previous editions as 20 ft high and 30 ft across, enjoyed an exceptionally good soil as it grew in the former kitchen garden.

VIBURNUM RHYTIDOPHYLLUM

f. ROSEUM Hort.—Flowers pink in the bud, becoming white flushed with pink. This has probably been raised in several gardens. The name was originally given to a plant grown by Mrs Sebag-Montefiore in her garden near Plymouth (*Gard. Chron.*, Vol. 104 (1938), p. 171 and supplementary illustration).

V. × RHYTIDOCARPUM Lemoine—A hybrid between *V. rhytidophyllum* and *V. buddleifolium* raised by Lemoine and put into commerce in 1936. It is of no horticultural interest.

V. × RHYTIDOPHYLLOIDES Suring. *V. lantanophyllum* Lemoine ex Rehder's *Manual*—A hybrid between *V. rhytidophyllum* and *V. lantana*, said to occur fairly frequently among seedlings of the former when *V. lantana* is growing in the vicinity. The original clone, named 'HOLLAND' was raised about 1925. It resembles *V. rhytidophyllum* but the leaves are larger and less wrinkled, almost deciduous.

V. rhytidophyllum × *V. utile* 'PRAGENSE'.—Leaves shorter than in the former, which was the seed-parent, elliptic or elliptic lanceolate, 2 to 4 in. long, glossy and slightly wrinkled above, felted beneath. The branchlets are more slender than in that species and the habit laxer. Raised by Josef Vik in the

Municipal Nurseries, Prague, and named in 1959. Like the seed-parent it is hardy in Czechoslovakia, while *V. utile* is tender (*Gard. Chron.*, Vol. 147 (1960), p. 317).

V. RIGIDUM Vent.

V. rugosum Pers.; *V. tinus* var. *strictum* Ait. f.; *V. tinus* subsp. *rigidum* (Vent.) P. Silva

An evergreen shrub of bushy rounded habit and rather open branching, 6 to 10 ft high and as much wide; young shoots covered with hairs. Leaves ovate or oval, toothless, pointed to rounded at the apex, wedge-shaped at the base; 2 to 6 in. long, 1 to 3 in. wide; dark dull green and roughish with appressed hairs above, paler and furnished beneath with soft grey hairs, especially on the prominent midrib and veins; margins ciliate; stalk up to ¾ in. long. The inflorescence is a flattish corymb 3 to 4½ in. wide, carrying numerous white flowers, each about ⅕ in. wide; stigma rose-coloured; main and secondary flower-stalks hairy. Fruits egg-shaped, ¼ to ⅓ in. long, blue, finally black. *Bot. Mag.*, t. 2082.

Native of the Canary Islands; introduced by Masson, the Kew collector, on his way home from S. Africa in 1778. This evergreen is not hardy at Kew except against a wall, but succeeds well in the southern and western maritime counties. In Cornwall it blossoms from February to April. Most nearly akin to *V. tinus*, it is well distinguished by its much larger, dull, very hairy leaves; nor is it so densely leafy in habit.

V. RUFIDULUM Raf.   SOUTHERN BLACK HAW

V. rufotomentosum Small; *V. prunifolium* var. *ferrugineum* Torr. & Gr.

A deciduous shrub of very rigid, thin habit, described as becoming a tree often 40 ft high in the wild; young shoots more or less covered with a rust-coloured down; winter buds reddish brown. Leaves stiff and leathery, oval, ovate, or obovate; rounded, blunt, or shortly pointed at the apex, wedge-shaped or rounded at the base, toothed, 2 to 4 in. long, 1 to 1½ in. wide, dark shining green above, covered beneath when young with a reddish short down, much of which falls away before the leaf drops; stalks ¼ to ½ in. long, stout, more or less winged, and densely covered with rusty coloured down. Flowers white, all perfect, ⅓ in. across, borne on cymes 3 to 5 in. across. Fruits blue, ½ to ⅔ in. long.

Native of the southern United States; introduced to Kew in 1902. It belongs to the same group as *V. prunifolium* and *V. lentago*, from both of which it differs in its dense covering of rusty down especially on the leaf-stalk and midrib. Its habit, too, as a young shrub, is curiously rigid and its foliage narrower. It does not flower very freely in Britain but the leaves sometimes colour red in the autumn.

V. SCHENSIANUM Maxim.
V. giraldii Graebn.

A deciduous shrub with slender branches; young shoots, undersurface of leaves, leaf-stalks and flower-stalks clothed with starry down. Leaves oval or ovate, often blunt or rounded at the end, toothed; 1 to 2¾ in. long, ¾ to 1½ in. wide, veins in five or six pairs; stalks ⅙ to ⅓ in. long. Flowers dullish white, ¼ in. wide, borne in May and June on a five-branched cyme, 1½ to 3½ in. across. Fruits egg-shaped, ⅓ to ½ in. long, turning red, finally black. Ovary glabrous.

Native of northwest and central China; introduced about 1910. It belongs to the same group in the genus as the well-known *V. lantana*, in which it is distinguished by its comparatively small leaves, whose veins do not run out fully to the margins, but subdivide and die out before reaching them. Its nearest ally is *V. burejaeticum*, which is distinguished by its downy ovary and more generally pointed leaves. It does not appear to have any great garden value.

V. SETIGERUM Hance
V. theiferum Rehd.

A deciduous shrub of erect habit, up to 12 ft high, with glabrous grey stems. Leaves ovate-lanceolate, rounded at the base, long, and taper-pointed, widely and sharply toothed, 3 to 6 in. long, 1¼ to 2½ in. wide, dark green above, and glabrous on both surfaces, with the exception of long hairs on the midrib and on the parallel veins beneath, which mostly fall away by autumn; veins in six to nine pairs, running out to the teeth; stalk ½ to 1 in. long, hairy like the midrib. Cymes 1½ to 2 in. across, five-branched, terminal on short, lateral, two-leaved twigs. Flowers white, ¼ in. wide, all perfect. Stamens included. Fruits red, egg-shaped, nearly ½ in. long.

Native of Central and W. China; introduced in 1901 by Wilson. It is allied to *V. phlebotrichum*, but has larger, longer stalked leaves. Rehder's specific name refers to the use of the leaves by the monks of Mount Omei as a kind of tea.

cv. 'AURANTIACUM'.—Fruits bright orange. Raised at the Arnold Arboretum from seeds collected by Wilson in China in 1907.

V. SIEBOLDII Miq.

A deciduous, strong-growing shrub 6 to 10 ft high, or a small tree with stiff, spreading branches, stellately downy and grey when young. Leaves mostly obovate or approaching oblong, pointed or rounded at the apex, and tapered at the base, prominently parallel-nerved, coarsely toothed except towards the stalk; 2 to 5 in. long, 1½ to 3 in. wide, dark glossy green and glabrous above, glabrous beneath or downy, chiefly on the veins; stalk ¼ to ¾ in. long. Flowers creamy white, ⅓ in. across, all perfect, produced in long-stalked cymes 3 to 4 in. across. Fruits oval, about ½ in. long, at first pink then blue-black.

Native of Japan; cultivated in Britain since the end of the 19th century. This is a vigorous and handsome shrub usually more in spread than it is high, distinguished by its large, strongly veined, often obovate leaves, which have a disagreeable scent when crushed. It does not flower and fruit so well in Britain as it does in climates with warmer summers and colder winters than ours. It is considered to be one of the finest of all viburnums in the northeastern United States.

cv. 'Reticulatum'.—With smaller leaves and inflorescences than normal, the former quite glabrous (*V. reticulatum* Hort.).

V. suspensum Lindl.

V. sandankwa Hassk.

An evergreen shrub 6 to 12 ft high; branchlets warted and furnished with starry down when quite young only. Leaves leathery, ovate or inclined to oval, pointed, rounded or broadly wedge-shaped at the base, toothed at the terminal two-thirds or scarcely toothed at all; 2 to 5 in. long, 1½ to 3 in. wide; glossy green and quite glabrous; chief veins in four or five pairs; stalk ¼ to ½ in. long. Inflorescence a corymbose panicle 2½ to 4 in. long and nearly as wide. Flowers fragrant, white, faintly tinted with rose; corolla-tube cylindrical, spreading at the top into five rounded lobes and measuring ⅓ in. in diameter. Calyx five-toothed, the teeth triangular, pointed, ciliate; bracts awl-shaped, ⅛ in. long; flower-stalks minutely downy. Fruits globose, red, crowned with the persisting style. *Bot. Mag.*, t. 6172.

Native of the Ryukyus, cultivated in Japan; introduced to Belgium about 1850. This has been tried out-of-doors at Kew with indifferent success even on a wall, but it succeeds very well in the Scilly Isles and Cornwall and it is occasionally sent to Kew to be named from other mild parts of Britain. This shrub does not flower freely in this country, probably for lack of sufficient sunshine, for it flowered well in several gardens in March 1922, owing no doubt to the phenomenal heat and dryness of the previous summer.

V. tinus L.   Laurustinus   [Plate 97

V. tinus var. *hirsutum* West. (and of Ait.); *V. tinus* var. *virgatum* Ait.

A dense-habited, much-branched evergreen shrub of rounded form, 6 to 12 ft high, often more in diameter, and furnished to the ground; young shoots smooth, or slightly hairy. Leaves not toothed, narrowly ovate, approaching oblong, tapered at both ends, 1½ to 4 in. long, ¾ to 1½ in. wide, dark glossy green above, paler beneath, and with tufts of down in the lower vein-axils; stalk ⅓ to ¾ in. long, often more or less hairy. Flowers white, about ¼ in. across, uniform and perfect, produced in winter and spring in terminal cymes 2 to 4 in. across. Fruits ovoid, tapering towards the top, ¼ in. long, deep blue, finally black.

Native of S. Europe, mainly in the Mediterranean region, and of N. Africa, occurring in the more luxuriant type of macchia vegetation with bay laurel, myrtle, phillyrea, *Rhamnus alaternus*, etc., or as undergrowth in woodland. It has been cultivated in Britain since the 16th century. In southern gardens the laurustinus is one of the most useful of evergreen shrubs, forming rich masses of greenery and opening its flowers any time between November and April, according to the weather. It will thrive in moderate shade, but flowers more freely in full sun. The fruits, indigo-blue, ultimately black, are not frequently seen with us, though often enough for it to escape from gardens in some places. From all other cultivated hardy viburnums this is distinguished by its luxuriant masses of entire, evergreen leaves.

V. tinus is a very variable species in the size and shape of its leaves, in the presence or absence of hairs from the young growths, petioles and the undersides and margins of the leaves, as well as in the size and density of the inflorescence. But the variations do not seem to be well correlated on wild plants, which may explain why in recent times botanists have not troubled to inventorise all the character-combinations that occur. However, forms with large leaves and inflorescences appear to be commonest in N. Africa. Some are very striking, judging from herbarium specimens, and seem to be really nearer to *V. rigidum* of the Canary Islands or the Azores subspecies of *V. tinus* than to *V. tinus* as it exists on the northern shores of the Mediterranean or in the Adriatic. No doubt there are also physiological variations, some races being adapted to the Mediterranean type of climate, others needing moister conditions and perhaps nearer to *V. tinus* as it existed in earlier epochs, when the climate of the Mediterranean region was rainier than it has been since the Ice Age.

Many forms of *V. tinus* have been selected in gardens, of which the following are the most important:

cv. 'CLYNE CASTLE'.—A hardy selection with large glossy leaves.

cv. 'EVE PRICE'.—Of compact habit, with smaller leaves than normal. Flowers bright pink in the bud. Award of Merit 1961. The original plant grows at Wakehurst Place in Sussex, and was bought by Gerald Loder from Messrs Dickson of Chester.

cv. 'FRENCH WHITE'.—A vigorous form, often sold for hedging as *V. tinus* simply. Young stems hairy; leaves moderately glossy. Flowers white when fully open.

cv. 'GWENLLIAN'.—Leaves dull green, elliptic-oblong, shorter than in most forms of the species. Flowers in small but very numerous trusses, rich pink in the bud, blush when expanded. Fruits freely borne. The original plant of this very attractive viburnum grows at Kew, where it was selected and named by the late Sydney Pearce. It is about 10 ft high and 15 ft wide.

f. HIRTUM Hort.—Shoots, leaf-stalks and bases of the leaves clothed with bristly hairs. Leaves larger than normal and of a different shape, being as much as 3 to 4 in. long and 2 in. wide, rounded or even slightly heart-shaped at the base, and once grown for early flowering in cool greenhouses. The provenance of this form, described in previous editions, is unknown, but a similar plant received an Award of Merit when shown from Highdown in 1939; this was raised from seeds collected by E. A. Bowles in Algeria.

The laurustinus grown in gardens early in the 19th century as var. *hirtum*, which was probably the var. *hirtum* of Aiton and the var. *hirsutum* of Weston, had hairy foliage but the leaves were of normal size and it was hardy.

cv. 'LUCIDUM'.—In habit this is more open, and less compact than the type, and altogether a stronger grower. It also bears larger leaves and trusses, and the individual flower is nearly ½ in. across, sometimes pinkish. The largest leaves are 4 in. long, and 2½ in. wide. Very useful and effective in the milder counties, it is not so hardy as the type. The varietal name refers to the glabrous, burnished young shoots, and to the glossy surfaces of the leaf, the lower one with only a few tufts of hairs in the vein-axils. (*V. lucidum* Mill.; *V. tinus* var. *lucidum* (Mill.) Ait.; *V. tinus splendens* West.).

Whether or not the plant described above is clonally the same as the one known to Miller it is impossible to say, but everything that Miller said about his plant agrees with the above. The origin of neither is known, but Martyn, in his edition of Miller's *Dictionary* (1807) gives 'Mount Atlas and Barbary' as the habitat of var. *lucidum*. It is likely, however, that there are several clones or seedlings in gardens. The plant that received an Award of Merit in 1972 was shown by Miss Godman from her garden at South Lodge near Horsham. In this the flowers are almost ½ in. across.

A clone in commerce as 'Lucidum' has leaves no more than 2 in. long on the flowering shoots, though to almost twice that length on strong growths. Correctly named or not, it is one of the finest and hardiest of the cultivars, with bright green, very glossy leaves, and one of the last to open its flowers.

cv. 'PURPUREUM'.—Leaves dark green, tinged with purple.

cv. 'PYRAMIDALE' ('Strictum').—Of erect habit, suitable for hedges.

subsp. RIGIDUM (Vent.) P. Silva—See *V. rigidum*, but the subspecific status for this is probably more appropriate.

subsp. SUBCORDATUM (Trel.) P. Silva—Leaves broadly elliptic, truncate to slightly cordate at the base. There is variation in the amount of hairs on the vegetative parts, as in Mediterranean plants. The inflorescence is up to 4 in. wide, rather dense in some plants. Native of the Azores, but forms similar in most respects occur in N. Africa.

cv. 'VARIEGATUM'.—A portion of the leaf, sometimes all one side, yellow. This is tender.

V. UTILE Hemsl.

An evergreen shrub of rather thin, open habit, 5 or 6 ft high, the slender branches clothed at first with a pale, starry down. Leaves usually narrowly ovate or nearly oblong, 1 to 3 in. long, ¼ to 1¼ in. wide, of firm texture, glabrous and dark glossy green above, prominently veined and white beneath, with a dense covering of starry down, margins entire, apex tapered but bluntish; base rounded to wedge-shaped; stalk ⅙ to ⅓ in. long. Flowers all fertile, produced during May densely packed in terminal, rounded trusses, 3 in. across, the branches of the inflorescence stellately downy. Each flower is ⅓ in. wide, white. Calyx glabrous, with shallow, rounded lobes. Fruits blue-black, oval, ¼ in. long. *Bot. Mag.*, t. 8174.

Native of China; introduced in 1901 by Wilson. It has proved quite hardy since its introduction, and is a pretty, graceful shrub. According to Wilson, it grows on limestone. Award of Merit 1926.

V. VEITCHII C. H. Wright

A deciduous shrub about 5 ft high; young branches, leaf-stalks, and under-surface of leaves densely clothed with stellate down. Leaves ovate, pointed, heart-shaped at the base, 3 to 5 in. long, 2 to 3 in. wide, sharply and widely toothed; upper surface with scattered stellate down. Flowers white, uniform and perfect, $\frac{1}{4}$ in. across; produced on a stoutly stalked, very scurfy-downy cyme, that is 4 or 5 in. across. Fruits red, then black.

Native of Central China, discovered and introduced in 1901 by Wilson, for Messrs Veitch. It is one of the Lantana group, differing from *V. lantana* itself in the more remote marginal teeth, and in the calyx being felted with starlike down. Wilson found it as a bush about 5 ft high, but rare; he considers it to be about the most ornamental of the Lantana group. But it makes a straggly shrub in gardens.

V. WILSONII Rehd.

A deciduous shrub 6 to 10 ft high with very downy young shoots. Leaves ovate to roundish oval, rounded or broadly tapered at the base, the apex slender or even tail-like, toothed, $1\frac{1}{2}$ to $3\frac{1}{2}$ in. long, half as much wide, dark green and with usually some hairs above, at least on the veins; clothed beneath either on the veins and midrib with mostly simple hairs, or all over the lower surface with star-shaped hairs and some long simple ones; veins in six to nine pairs; stalk $\frac{1}{4}$ to $\frac{3}{8}$ in. long, hairy and starry-downy. Flowers white, all fertile, $\frac{1}{4}$ in. wide, opening in June in a terminal five- or six-branched corymb 2 to 3 in. wide; main and secondary flower-stalks velvety with down. Corolla $\frac{1}{5}$ in. wide, the lobes roundish ovate; calyx downy. Fruits bright red, egg-shaped, $\frac{1}{3}$ in. long, slightly hairy.

Native of Szechwan, China; discovered by Wilson in 1904 and introduced by him in 1908 to the Arnold Arboretum, Mass., whence it was obtained for Kew the following year. The plants which were raised from Wilson's No. 1120 flower and bear fruit regularly at Kew. Rehder compared it with *V. hupehense*, but that species has stipules attached towards the base of the leaf-stalks which are absent in *V. wilsonii*. In the downiness of leaf and inflorescence they are very similar.

V. MULLAHA Buch.-Ham. ex D. Don   *V. stellulatum* Wall. ex DC.—*V. wilsonii* is closely related to this species, described in 1825. Rehder acknow-ledged the resemblance, giving as the differential character that in *V. wilsonii* the inflorescence-axes and the outside of the corolla are velvety; in *V. mullaha* the inflorescence is merely downy and the corollas glabrous. This species is a native of the Himalaya from Kashmir eastwards, and of parts of S.E. Asia. It is in cultivation from Ludlow, Sherriff and Hicks 19820, collected in Bhutan.

V. WRIGHTII Miq.

A deciduous shrub, 6 to 10 ft high, with erect stems; young branches glabrous. Leaves 2 to 5 in. long, 1 to 2½ in. wide, mostly ovate and rounded at the base, but sometimes obovate and tapered at the base, slenderly and often abruptly pointed, somewhat distantly toothed; bright green and almost glabrous above, paler beneath with tufts of hairs in the vein-axils; veins in six to ten pairs; stalk ¼ to ¾ in. long. Flowers all perfect, produced in May on glabrous or downy-stalked, five-rayed cymes, 2 to 4 in. across, the flowers themselves scarcely stalked, white. Fruits round-ovoid, red, ⅓ in. long.

Native of Japan. This handsome-fruited species is closely related to *V. dilatatum*, but that species is at once distinguished by the extremely downy character of its leaves, young branches and inflorescence. Another ally is *V. phlebotrichum*, but that is very distinct in its smaller shorter-stalked leaves, the more numerous silky, whitish hairs on the veins beneath, the quite glabrous and slender-stalked, pendulous cymes, and the very short stamens.

The fruits of *V. wrightii*, if formed, ripen in August and the leaves colour red in the autumn.

cv. 'HESSEI'.—Dwarfer, with smaller inflorescences. This and similar plants were raised in Germany from seeds collected by Prof. Sargent in Japan in the 1890s (*V. hessei* Koehne; *V. wrightii* var. *hessei* (Koehne) Rehd.).

VILLARESIA (CITRONELLA)   ICACINACEAE

A genus of about thirty species, mostly tropical and subtropical, in Central and South America, Australia, Malaysia and the islands of the S.W. Pacific. The family Icacinaceae is quite closely allied to the Holly family (Aquifoliaceae), so the holly-like appearance of the species here described is not altogether fortuitous. It can, however, be distinguished from any member of the genus *Ilex* by the paniculate inflorescence, the valvate not imbricate petals, and the characteristic pits on the underside of the leaves.

It must be added that the generic name *Villaresia*, in the present sense, is invalid and should give way to CITRONELLA (see the article accompanying the plate in the *Botanical Magazine* and *Journ. Arn. Arb.*, Vol. 21, p. 473).

V. MUCRONATA Ruiz & Pavon   NARANJILLO

CITRONELLA MUCRONATA (R. & P.) D. Don; *Citronella chilensis* (Mol.) Munz; *Villaresia chilensis* (Mol.) Stuntz; *Citrus chilensis* Mol. *sec.* Miers, (?) not Mol.

An evergreen tree up to 60 ft high; young shoots downy, ribbed. Leaves alternate, of hard leathery texture like those of a holly, ovate or oblong, pointed,

1½ to 3½ in. long, ¾ to 2 in. wide, entire on the flowering shoots of adult trees, spiny, much larger, and more rounded at the base on young ones, glabrous and dark glossy green; stalk ⅛ to ¼ in. long, downy. Flowers fragrant, ⅜ in. wide, yellowish white and densely crowded in a cluster of panicles, each 1 to 2 in. long and produced in the terminal leaf-axils and at the end of the shoot in June. The individual flower, which has its various parts in fives, is almost stalkless, but the main and secondary flower-stalks are clothed with brown down. Fruit an egg-shaped drupe ⅔ in. long, containing one fleshy seed surrounded by a hard shell. *Bot. Mag.*, t. 8376.

A native of central Chile, rare in the wild. It was introduced by the Hon. W. Fox-Strangways about 1840 to the garden at Abbotsbury in Dorset, where there is still a tree of this species. The only other sizeable example in the open air grows in the National Botanic Garden at Glasnevin, Eire. It measures 25 × 2 ft (1974).

VINCA   Periwinkle   apocynaceae

A genus of seven species of subshrubs and herbaceous perennials, natives of the Old World from Europe and North Africa to Southwest and Central Asia (the tropical plant well-known as *V. rosea* is now usually placed in a separate genus as *Catharanthus roseus*). Leaves opposite, simple. Flowers solitary in the leaf-axils, stalked. Sepals five. Corolla with a narrowly funnel-shaped tube, which is hairy above the insertion of the stamens, and a spreading, five-lobed limb. Stamens five, inserted about the middle of the tube. Pistil of two united carpels and a single style, the latter widening upward into a disk-like structure known as a clavuncle, which bears a conical stigma surrounded by conspicuous tufts of hair. The fruit consists of two slender many-seeded follicles, rarely produced in Britain on open-ground plants.

The three species described are evergreen trailing shrubs, propagated with the greatest of ease by means of cuttings a few inches long, or by taking up old patches and dividing them. *V. major* is sometimes attacked by a parasitic fungus which turns the leaves yellow and ultimately kills them.

The most valuable work on the genus is: W. T. Stearn, 'A Synopsis of the Genus *Vinca* . . .' in Taylor, W. I., and Farnsworth, N. R. (eds), *The Vinca Alkaloids* (1973), pp. 19–94. There is a copy of this work in the Library of the Royal Horticultural Society. Dr Stearn's contribution also comprises the cultivars of *V. major* and *V. minor*.

V. DIFFORMIS Pourr.

V. media Hoffmanns. & Link; *V. acutiflora* Bertol.

A trailing sub-shrubby plant in Britain usually dying back in winter, probably evergreen in S. Europe, of spreading growth, quite glabrous in leaf and stem. Leaves ovate, broadly wedge-shaped or rounded at the base, more tapered towards the apex, 1½ to 3 in. long, ¾ to 2 in. wide, entire, rich green on both surfaces, but rather paler beneath; stalk E to ⅓ in. long. Flowers solitary in the leaf-axils, produced in November and December on stalks 1 to 1½ in. long. Corolla 1½ in. across, very pale lilac-blue, the lobes obovate or rather rhomboidal, pointed; calyx-lobes awl-shaped, ¼ in. long. Fruit awl-shaped, 1½ in. long. *Bot. Mag.*, t. 8506.

Native of southwest Europe as far east as Italy, and of northwest Africa. It resembles *V. major* in general appearance, but is easily distinguished by the absence of hairs on stem and leaf-margin and by the non-ciliate sepals. It is not so hardy as *V. major*, and at Kew it flowers too late to expand properly out-of-doors, but taken up and put under glass provides a continuous display during the darkest months of the year.

V. MAJOR L.   GREATER PERIWINKLE   [PLATE 98

An evergreen shrub whose barren stems are long and trailing, its flowering ones erect and 1 to 2 ft high, glabrous except for a few dark bristles at the joints. Leaves opposite, ovate, 1 to 3 in. long, half to two-thirds as wide, pointed, dark green, glossy on both surfaces, glabrous, but edged with minute hairs; stalk ⅓ to ½ in. long. Flowers bright blue, solitary in the leaf-axils on a slender stalk 1 to 2 in. long; corolla 1½ in. across, the base a funnel-shaped tube spreading at the mouth into five deep, broadly obovate lobes; calyx-lobes five, narrowly linear, nearly ½ in. long, with hairs on the margin. Fruits glabrous, awl-shaped, long-pointed, 1½ to 2 in. long.

Native of the western and central Mediterranean region, with a subspecies ranging as far east as the Caucasus; long cultivated and seemingly wild in many localities outside its natural range, including Britain. It is useful for growing in semi-shaded positions where it makes pleasant ground cover, but not flowering so well there as in the full sun. The first flowers appear in May and continue until September. It should be trimmed over annually in spring, cutting away the old growths. Distinct from *V. minor* in its large, broad-based, often heart-shaped leaves, and from *V. difformis* in its ciliate leaves and calyx-lobes.

cv. 'OXYLOBA'.—Corolla segments narrower and more acute than normal, deep violet-blue. Leaves oblong-lanceolate, conspicuously hairy at the edge. Originally described in 1930 from a plant cultivated by E. A. Bowles at Myddelton House, Enfield. Possibly a geographical variant, since it is matched by plants growing wild in southern Italy (Stearn, *Gard. Chron.*, Vol. 88 (1930), p. 156, and in *Vinca Alkaloids*, pp. 85–6). It has been confused with subsp. *hirsuta*.

subsp. HIRSUTA (Boiss.) Stearn   *V. major* var. *hirsuta* Boiss.; *V. pubescens* D'Urville—Hairs on the petiole, leaf-margin and calyx longer and denser than in the typical state. Native of the western Caucasus and northern Asia Minor.

cv. 'RETICULATA'.—Young leaves golden-veined, later normally coloured and dark green. Flowers with broad segments, rich blue.

cv. 'VARIEGATA' ('Elegantissima').—Leaves boldly margined with creamy white, the margination yellower on the young foliage. Flowers rich lavender-blue.

V. MINOR L.   LESSER PERIWINKLE

An evergreen trailing shrub rarely more than 6 in. above the ground, forming in time a dense mat; stems glabrous, wiry. Leaves oval, or slightly obovate, always tapered at the base; ¾ to 2 in. long, ½ to ¾ in. wide, quite glabrous and of a deep glossy green on both sides. Flowers 1 in. across, bright blue, produced from April until autumn in the leaf-axils of the young growths. Corolla-lobes obovate; calyx glabrous, its lobes about ⅛ in. long.

Native of southwest and central Europe, Asia Minor and the Caucasus. Like *V. major*, it is found apparently wild in England, but is doubtless an escape from cultivation. It is, of course, easily distinguished from that species by the smaller flowers, whose calyx-lobes are shorter and broader, and by the smaller narrow-based leaves.

V. minor makes a neater and closer ground-cover than *V. major* and is very useful for steep banks, so long as they are not exposed to the hottest sun. In deep shade it will form dense weed-proof mats of indefinite extent, but does not flower much under such conditions. Even in lighter conditions its flowers tend to be concealed by the old foliage, except at the edge of the colony, so some trimming is desirable in early spring, if plants are grown primarily for their flowers. It is an interesting feature of both this species and *V. major* that the growths produce flowers only in their first year, though they may live and elongate for more than one, eventually reaching the ground and rooting, to give a flowering-shoot the next year. But much of the work of extending the colony is done by horizontal shoots, which may run along the ground for a considerable distance before rooting.

f. ALBA Dipp.   *V. minor alba* West.—Flowers white. Selected forms are: 'BOWLES' WHITE', with rather large flowers, flushed with pink in the bud; and 'GERTRUDE JEKYLL', a neat grower with rather smaller flowers and leaves than normal, but the flowers freely borne above the foliage.

cv. 'ALBA VARIEGATA'.—Leaves edged with light yellow. Flowers white.

cv. 'ARGENTEO-VARIEGATA'.—Leaves shorter, proportionately broader, blotched with white. Flowers blue.

f. ATROPURPUREA (Sw.) Rehd.   *V. minor atropurpurea* Sw.; *V. m. flore puniceo* Lodd. ex Loud.; *V. m.* var. *punicea* Bean; *V. m. purpurea* Hort.—Flower colour inclining to plum-purple or vinous purple. The clone usually sold as 'Atropurpurea' has flowers near to the shade known as Violet Purple, with a curious black sheen when looked at close to. This is scarcely the same as 'BURGUNDY', which has not been seen.   [PLATE 99 cv. 'Aureo-variegata'.—Leaves striped and margined with yellow. Flowers blue. 'Alba Variegata' is sometimes sold under this name.

cv. 'Bowles' Variety'.—Flowers on the small side, but well-shaped and a pure light blue. Some plants sold under this name may be 'La Grave'.

cv. 'La Grave'.—Flowers lavender-blue, large; leaves broader than normal. Collected by E. A. Bowles in the 1920s in the churchyard of La Grave in the Dauphine (Stearn, op. cit., p. 52). Also known as 'Bowles' Variety'.

Most of the colour variants have double forms in which the stamens have been converted into petaloids.

VISCUM   Mistletoe   loranthaceae

A genus of about sixty-five species of hemiparasitic shrubs, mostly in the African tropics and subtropics, a few extending into Temperate Asia and Australia, only two into Europe. They draw most of their nourishment from the host plant by means of suckers (haustoria) inserted in its conductive tissues, but are not true parasites, the leaves being green and photosynthetic. Annual growths dichotomous, short, each with a pair of opposite, leathery, persistent, parallel-veined leaves. Flowers inconspicuous, cymose, unisexual. Perianth with two or four segments. Anthers sessile, adnate to the perianth segments and containing numerous pollensacs. Ovary inferior, with an almost sessile stigma. Fruit berry-like, containing one to three embryos embedded in the endosperm (neither ovules nor true seeds are developed in the Loranthaceae, but the contents of the mature ovary are usually termed a single seed lacking a testa).

The common mistletoe is the only representative of the Loranthaceae in Britain, but Loranthus europaeus L. is found in southern Europe as far west as Italy, in parts of Central Europe and in southwest Asia. It lives mainly on species of oak, the leaves are deciduous and the fruits are borne in spikes.

The only other European member of the Loranthaceae is Arceuthobium oxycedri (DC.) Bieb., which occurs in S. Europe as a parasite on junipers. It is a true parasite with scale-like leaves and a dry, dehiscent fruit. An interesting species in the genus *Arceuthobium* is *A. minutissimum* Hook. f. of the Himalaya, a parasite on *Pinus wallichiana* of which nothing shows save a pair of green scales embracing the solitary flower—'The most minute dicotyledonous plant that I can call to mind.' (Hooker).

V. album   Mistletoe

An evergreen shrub of tawny, yellowish aspect, parasitic on various trees, usually in the form of a rounded, pendulous bush; branches glabrous, bifurcating at each joint. Leaves opposite, narrowly oblong or obovate, tapering at the base, rounded at the apex, 1½ to 4 in. long, ¼ to 1 in. wide, not stalked. Flowers inconspicuous, almost stalkless, and produced in the forks of the branches, the sexes often on separate plants. Fruit a white, translucent berry ⅓ in. wide, whose single seed is embedded in a very viscid pulp; ripe in midwinter.

Native of Europe, where it is widespread, and of temperate Asia and N. Africa. It is found in Britain as far north as N. Wales and Yorkshire. The mistletoe is frequently cultivated in gardens for its interest and associations, and nurserymen supply it growing on apple trees. The two sexes should be obtained if possible on the same or separate host plants. It must be propagated by seed, and this is best done by bursting the berry on the youngish bark of the host plant. The glutinous substance in which the seed is embedded soon hardens and attaches the seed securely. It is not necessary to make a slit in the bark for the seed.

In nature the mistletoe is spread by birds. They eat the fruits, but a thick coating of gluten remains on the seeds which is made even more adhesive by partial digestion. It grows most commonly perhaps on apple trees, so much so as to be a pest in some of the west country orchards. Although, because of its association with the rites of the ancient Druids, its most famous host plant is the oak, it is in reality very rarely seen on that tree. But there appear to be few of our native trees on which it will not grow.

The ancient belief was that the seed would only germinate after it had passed through the digestive tract of a bird. But the naturalist John Ray doubted this, and at his suggestion 'Mr Doody, an Apothecary of London, inserted a seed of the mistletoe into the bark of a White Poplar tree . . . with complete success . . . . It is wonderful, as Ray justly observes, that Botanists and Philosophers should choose to argue pro and con so many years, not to say centuries, rather than take the pains to consult nature, and determine the matter by an easy experiment.' (Thomas Martyn, in his edition of Miller's *Dictionary*, Vol. II (1809)).

On the continent of Europe two races of mistletoe occur in addition to the typical one, both confined to conifers: subsp. ABIETIS (Wiesb.) Abromeit, found on *Abies*, and subsp. AUSTRIACUM (Wiesb.) Vollmann, living mainly on pines and occasionally on spruce (and artificially established on other conifers, including *Cedrus atlantica*).

VITEX   VERBENACEAE

The two species described here are almost hardy representatives of a mainly tropical and subtropical genus, which contains tall timber trees such as *V. altissima* of southern India and *V. lignum-vitae* of southeast Australia. Leaves deciduous or persistent, palmately compound, usually opposite. Flowers stalked, in cymose clusters, which are often arranged in the form of panicles. Calyx campanulate. Corolla tubular or funnel-shaped, with a five-lobed limb. Stamens four. Fruit a drupe, with a four-celled stone.

The two species treated here are propagated by cuttings of half-ripened wood. Flowering as they do on the wood of the current season they should be pruned in much the same way as the late-flowering buddleias, once a framework of branches has been formed.

V. AGNUS-CASTUS L.   CHASTE-TREE   [PLATE 100

A deciduous shrub of free, spreading habit; young shoots covered with a minute grey down. Leaves opposite, composed of five to seven radiating leaflets borne on a main-stalk 1 to 2½ in. long, leaflets linear lance-shaped, 2 to 6 in. long, ¼ to ¾ in. wide, tapering gradually towards both ends, not toothed, dark green above, grey beneath with a very close felt; stalks of leaflets ¼ in. or less long. Flowers fragrant, produced during September and October in whorls on slender racemes which are 3 to 6 in. long, sometimes branched, and borne in numbers on the terminal part of the current season's growth, at the end and in the leaf-axils, the whole forming a large panicle. Corolla violet, tubular, ⅓ in. long, with five expanding lobes; stamens four, protruded; calyx funnel-shaped, downy, shallowly lobed. *Bot. Mag.*, n.s., t. 400.

A native of the Mediterranean region, where it grows in river-beds, often with tamarisk and oleander; and of Southwest and Central Asia; in cultivation in Britain by the 16th century. Near London it needs the protection of a wall, given which it is quite safe; a plant lived at Kew on a west wall for over sixty years. It flowers freely in warm seasons, and its crowd of panicles is sometimes very effective. The entire plant has an aromatic, pungent odour.

f. ALBA (West.) Rehd.—Flowers white. In cultivation since the 18th century. Award of Merit when exhibited from Kew on September 1, 1959.

f. LATIFOLIA (Mill.) Rehd.   *V. agnus castus* var. *latifolia* (Mill.) Loud.; *V. latifolia* Mill.; *V. macrophylla* Hort.—Leaflets relatively broader than normal, to about 1 in. wide. First recorded in the 16th century. Cultivated plants are vigorous and slightly hardier. Award of Merit 1964, when exhibited on September 15 by the Crown Estate Commissioners, Windsor Great Park.

Agnus Castus is an old generic name, used for the present species in pre-Linnaean times. The Greek word for the plant was 'agnos' and is of unknown, perhaps pre-Hellenic derivation. But the same word as an adjective meant holy or chaste in Greek, and it is probably from the apparent identity of these two words that the plant became a symbol of chastity in ancient Greece. In later times, no doubt for the same reason, the seeds were regarded as antaphrodisiac, though Philip Miller thought that judging by the taste and smell the plant was more likely to provoke lust than to allay it. Pomet, in his *Histoire des Drogues* (1694), was also sceptical. To quote from the English translation of 1712: 'This plant bears . . . the name of *Agnus Castus*, because the *Athenian*

Ladies who were willing to preserve their Chastity, when there were places consecrated to the Goddess *Ceres*, made their Beds of the Leaves of this Shrub, on which they lay: But it is by way of Ridicule that the Name of *Agnus Castus* is now given to this Seed, since it is commonly made use of in the Cure of Venereal Cases, or to assist those who have violated, instead of preserv'd their Chastity.' The fact that in Latin 'agnus' means lamb made the plant sound even more innocent, and explains the name 'Chaste Lamb tree', also used for it.

V. NEGUNDO L.

A deciduous shrub up to 10 ft high. Leaves digitate, with three to seven, usually five, stalked leaflets, which are elliptic or narrow-elliptic or oblong to lanceolate, acute to acuminate at the apex, cuneate at the base, $2\frac{5}{8}$ to 4 in. long, $\frac{5}{8}$ to $\frac{13}{16}$ in. wide, margin entire or with a few large blunt lobule-like teeth. Inflorescence a loose panicle terminating the shoots of the year and made up of several spikes 6 to 9 in. long, along which the flowers are arranged in rather distant clusters. Calyx downy, deeply and sharply lobed. Corolla violet-blue, rather smaller than in *V. agnus-castus*.

VITEX NEGUNDO

V. negundo is widely distributed from India and Ceylon to China and Formosa, commonly in waste places near villages. Although cultivated in Europe since the 17th century, it has always been rare in Britain, where it is usually represented by the following variety:

var. HETEROPHYLLA (Franch.) Rehd. *V. incisa* Lam.; *V. negundo* var. *incisa* (Lam.) C.B.Cl.; *V. chinensis* Mill.; *V. incisa* var. *heterophylla* Franch.—This differs by its leaflets being smaller, 1 to 2 in. (rarely to 3½ in.) long, ½ to ¾ in. wide, coarsely and deeply toothed or lobulate, the divisions often reaching to the midrib or nearly so. *Bot. Mag.*, t. 364. It is a native of the northern provinces of China. It was introduced to Paris around the middle of the 18th century by one of the French Jesuit missionaries, and to Britain (*c.* 1758) by Philip Miller, who 'was favoured with some young plants by Monsieur Richard, gardener to the King at Versailles.'

Although not truly tender, this is a shrub that needs abundant summer heat if it is to flower and ripen its wood, and succeeds with the protection of a wall. But being in competition with the many finer plants adapted for wall cultivation, it scarcely keeps its place in gardens.

An extreme form of the var. *heterophylla* is 'MULTIFIDA', dwarf and free flowering, at least in France, with very deeply dissected leaves (*Rev. Hort.*, 1870, p. 415).

VITIS   VINE   VITACEAE

A genus of tendril-climbers, rarely shrubs; tendrils without adhesive tips, borne opposite the leaves, but often lacking at every third node. Bark on older branches dividing into long fibrous strips. Leaves mostly simple, toothed and more or less lobed (partly compound in *V. piasezkii*). Flowers in panicles, polygamo-dioecious (flowers male on some plants, bisexual on others). Calyx minute. Petals five, united at their ends into a sort of cap, which is shed as the flower expands. Stamens five, alternating with five nectar-bearing glands. Style short. Fruit a berry; seeds two to four, pear-shaped, pointed at the base.

There are about fifty species of *Vitis*, confined to the northern hemisphere, with a maximum development in N. America. Some of the American species, and even more their hybrids, have acquired immense economic importance in wine-growing areas as root-stocks for the grape-vine, being resistant to the 'phylloxera' root-louse, which made its appearance in France about 1867 and at one time threatened the very existence of the European wine-growing industry. Some of the American species, notably *V. labrusca*, have parented vineyard varieties widely cultivated in N. America outside California. But as ornamentals the American species are rare in gardens. Better known are those from China and Japan grown for their fine foliage and autumn colour, such as *V. coignetiae* and *V. davidii*. The most important species in the genus *Vitis* is of course the European grape vine *V. vinifera*. This species would hardly fall within the compass of this work were it not that it has produced a few varieties grown chiefly for ornament.

The chief difficulty with vines is the provision of suitable support. Best of all, perhaps, is a pergola on which the shoots can be trained and pruned back annually as much as is necessary. They can also be trained up posts, when, if the shoots are allowed to hang loosely, they are very elegant. Pruning is much the same as that given to grape vines under glass. The young shoots are shortened back once or twice during the summer, and then cut hard back in December or January to one or two buds (later pruning causes bleeding). The same pruning should be given to plants grown on walls. Whole trees or large shrubs can be given up to them, over which they can ramble at will, and this, approaching nature as it does, shows the more vigorous ones at their best. *Vitis coignetiae* is often grown in this way, but needs a large tree for its support.

Some species such as *V. coignetiae* are difficult to increase except by layers or seeds, but most of the true vines can be propagated by cuttings, or preferably by 'eyes'. An eye consists of a single bud of the previous summer's shoot, with about half an inch of wood at each side, cut slanting fashion, so that the cut surfaces almost meet beneath the bud. These are made in early spring, each one placed on the surface of a small pot of sandy soil, the bud only uncovered, then put in gentle bottom heat. Cuttings are made one or two joints long, at the time the leaves are falling in autumn, and put under a handlight or in a cool frame.

For other species treated under *Vitis* in previous editions, see *Ampelopsis*, *Cissus* and *Parthenocissus*.

V. aestivalis Michx.   Summer Grape

A very vigorous deciduous climber, growing to a great height when support is available; branchlets round, glabrous or loosely downy. Leaves very large, 4 to 12 in. across, about as much long, varying from deeply three- or five-lobed to scarcely lobed at all, teeth shallow and broad, pointed at the apex, deeply heart-shaped at the base, dull green, ultimately glabrous above, covered beneath with more or less persistent floss which is rusty red at first, changing to brown with age. Flowers in panicles up to 8 or 10 in. long. Berries globose, $\frac{1}{3}$ in. in diameter, black with a blue bloom, agreeably flavoured.

Native of the eastern and central United States; introduced in the 17th century. On the young stems there is a tendril missing from every third joint, and in its large-leaved state it can thus be distinguished from *V. labrusca*, which has a tendril or panicle opposite every leaf. The viticultural variety 'Norton's Virginia' is near to *V. aestivalis*.

var. bourquiniana (Munson) Bailey *V. bourquiniana* Munson—A group of viticultural varieties, or more probably hybrids, of *V. aestivalis*, differing from it in the larger, juicy berries, the thinner leaves only slightly downy beneath, the down grey or dun-coloured. Originally grown by the Bourquin family in Georgia. Here belong 'Herbemont' and several other sorts.

V. CINEREA Engelm ex Millardet SWEET WINTER GRAPE.—This vine, a native of the central and southern USA, is allied to, and was at one time regarded as a variety of *V. aestivalis*. It has angular, downy branchlets (as contrasted with the round, almost glabrous ones of *V. aestivalis*); the down beneath the leaf is grey or whitish, and the berries have little or no bloom.

V. AMURENSIS Rupr.

A strong-growing, deciduous vine of somewhat similar character to, but quite distinct from, *V. vinifera*, with reddish young shoots, flossy when young, a thick, hard disk of wood dividing the pith at the joints. Leaves 4 to 10 in. wide, somewhat longer, five-lobed, often deeply so, the middle lobe then of broadly ovate form, with a slender abrupt point; the base has a deep, round, broad sinus; under-surface somewhat downy.

Native of N. China, Korea and the Russian Far East, with a variety in Japan; introduced to Europe about the middle of the last century. It is worth growing for the usually fine crimson and purple autumn hues of its noble foliage.

V. ARGENTIFOLIA Munson

?*V. bicolor* Leconte, not Raf.; *V. aestivalis* var. *argentifolia* (Munson) Fern.

A vigorous deciduous climber, with round shoots free from down, but usually very glaucous, a tendril missing from every third joint. Leaves 4 to 12 in. wide and long, three- or five-lobed, irregularly and shallowly toothed, usually glabrous on both surfaces, and vividly blue-white beneath. In other respects this vine is similar to *V. aestivalis*, to which it is most nearly allied. It is a native of the eastern and central United States, and in cultivation in Britain makes a luxuriant climber. Visitors to Goat Island, Niagara Falls, will have noticed its abundance there, associating with *Celastrus scandens* and other climbers in the production of a beautiful and luxuriant effect; this vine is conspicuous in the blue-white young shoots and under-surface of the leaves, to which the popular name refers.

V. CALIFORNICA Benth.

A deciduous climber, reaching 20 to 30 ft in height, the young shoots covered at first with grey cobwebby down, nearly glabrous later. Leaves roundish cordate or kidney-shaped, occasionally three-lobed, 2 to 4 in. wide, and about as long, rounded at the apex, the sinus at the base often wide and rounded, the margins set with fairly even, broadly triangular teeth, scarcely

$\frac{1}{8}$ in. deep, upper surface glabrous, lower one usually grey with down; stalk 1 to 2 in. long, grey downy like the young shoot. Berries $\frac{1}{3}$ in. in diameter, black, covered with purple bloom.

Native of western N. America from Oregon to California. This is a very well-marked vine in the round-ended, shallowly and evenly toothed leaves. It has no value, even in its own home, as a fruit-bearer, but is certainly very handsome in autumn, its leaves turning a deep crimson before they fall.

V. CANDICANS Engelm. ex A. Gr.   MUSTANG GRAPE

A vigorous deciduous climber, shoots covered with a dense white wool, a thick disk interrupting the pith at the joints. Leaves 2 to $4\frac{1}{2}$ in. wide, broadly heart-shaped to kidney-shaped, sometimes entire or with only a wavy outline, sometimes obscurely three-lobed; on young plants or strong sucker shoots the leaves are sometimes deeply three-, five-, or seven-lobed, but even then scarcely or very shallowly toothed. On first expanding the upper surface is woolly, but the wool soon falls away, leaving it a dull, dark green, whilst the under-surface remains covered with a thick white felt. The stalk is one-fourth to half as long as the blade, and white-woolly. Berries globose, about $\frac{2}{3}$ in. wide, purplish, and unpleasantly flavoured.

Native of the USA from Oklahoma and Arkansas to Texas, often found on limestone. It is one of the most distinct of American grape-vines in the broad, almost entire leaves and vivid white wool beneath, suggesting a white poplar leaf. It was quite hardy when grown at Kew. Allied to it, and perhaps a hybrid from it is:

V. DOANIANA Munson—It is distinct from *V. candicans* in the always three-lobed and coarsely toothed leaf, the upper surface of which is bluish green strewn with patches of white wool. The young shoots and leaves are quite white all over at first, and much of the wool persists beneath. It was originally found wild in Texas, but is commonest in one area of Oklahoma. It was introduced to Kew in 1892 and was quite hardy there.

V. COIGNETIAE Pulliat ex Planch.   [PLATE 101
V. thunbergii Hort., not Sieb.

A very vigorous deciduous climber, reaching the tops of the highest trees; young shoots round, ribbed, and at first covered with a loose greyish floss; there is a tendril missing at every third joint. Leaves perhaps the largest among vines, being sometimes 12 in. long and 10 in. broad, ordinarily 4 to 8 in. wide; they are roundish in the main, rather obscurely three- or five-lobed, the lobes and apex pointed, the base deeply heart-shaped, shallowly to coarsely toothed, dark green and glabrous above, covered beneath with a thick rusty brown felt; stalk from 2 to 6 in. long, somewhat woolly. Berries black with a purple bloom, $\frac{1}{2}$ in. wide.

Native of Japan; described in 1887 from plants which had been raised from seeds collected in 1875 by Mme Coignet, daughter of the French rosarian Jean Sisley, who was travelling in Japan with her husband. It was apparently first introduced, however, to Anthony Waterer's Knap Hill nursery through Messrs Jardine and Matheson, East India merchants. The original plant there grew up some trees by the road and made a glorious display of crimson every autumn. Owing to difficulty in propagation it spread very little in cultivation, only a few layers being produced each year. No name could be found for the Knap Hill plant until about 1894, when it was identified as *V. coignetiae* at Kew. By that time the species was becoming more widely available. A large quantity of seed was sent to France in 1884 by Degron, collecting for the French government, and Messrs Späth of Berlin were offering plants from imported seeds by 1893–4. There was a second introduction to Britain by J. G. Veitch in 1892 but his firm appears to have distributed the species as *V. thunbergii*, the plants they sold as *V. coignetiae* probably being the true *V. thunbergii*.

In the forests of Hokkaido, according to Sargent, *V. coignetiae* 'climbs into the tops of the largest trees, filling them with its enormous leaves, which in autumn assume the most brilliant hues of scarlet.' It was in this way that it grew at Knap Hill and now does in many gardens. In the size of its leaves and in the richness of its colour in autumn, it is undoubtedly the finest of the true vines. Seedling plants have leaves very much more deeply lobed than fully grown ones.

V. 'PULCHRA'.—This vine was originally sent out by Veitch's nursery as "*V. flexuosa major*". It does not belong to *V. flexuosa* and was named *V. pulchra* by Dr Rehder in 1913. It is a vigorous deciduous climber with reddish, soon quite smooth shoots. Leaves roundish ovate, 3 to 6 in. wide, often as broad as, or broader than long, sometimes slightly three-lobed, oftener coarsely toothed, the base widely and shallowly heart-shaped, the apex shortly or slenderly pointed, glabrous or nearly so above, covered with grey down beneath; stalk 1½ to 4 in. long. Flowers produced in June on slender panicles up to 4 in. long.

The origin of this vine is not known, but it probably came from N.E. Asia and appeared in cultivation about 1880. It is near to *V. coignetiae* and Rehder suggested that it might be a hybrid between it and *V. amurensis*. Its chief value is in the rich purple and blood-red tints of its dying leaves, which are also coloured when unfolding.

V. CORDIFOLIA Michx.   FROST OR CHICKEN GRAPE
V. vulpina L., in part, *nom. confus.*

A very vigorous vine, whose main stem in the wild is sometimes from 1½ to 2 ft thick; young shoots smooth or only slightly hairy, a tendril missing from every third joint. Leaves thin, roundish ovate, with a heart-shaped base (the sinus pointed and narrow), 3 to 5 in. wide, rather more in length, slenderly pointed, coarsely and irregularly toothed, unlobed or sometimes obscurely three-lobed, glossy and glabrous above, glabrous or downy on the veins beneath; stalk often as long as the blade. Flowers in drooping panicles, 4 to 12 in. long. Berries globose, ⅓ to ½ in. in diameter, black.

Native of the eastern and southern United States; introduced in 1806. The berries are moderately well-flavoured after they have been touched with frost in America, harsh and acid before; in var. FOETIDA Engelm., described from the Mississippi basin, they have a pungent, foetid odour.

This is probably the species to which the name *V. vulpina* L. should be applied, but it has so often been used for *V. riparia* as to be a source of confusion.

V. DAVIDII (Carr.) Foëx
Spinovitis davidii Carr.; *Vitis armata* Diels & Gilg

A luxuriant, deciduous climber, the young shoots not downy, but covered with spiny, gland-tipped, somewhat hooked bristles, which give them a very rough appearance. Leaves heart-shaped, slender-pointed, toothed, 4 to 10 in. long, 2½ to 8 in. wide, shining dark green and glabrous above, bluish or greyish green beneath, and downy only in the vein-axils, but more or less glandular-bristly, as is also the leaf-stalk, which is from half to nearly as long as the blade. Fruits about ⅔ in. in diameter, black, and of a pleasant flavour.

VITIS DAVIDII

Native of China; discovered by Père David in Shensi in 1872 and introduced by him to the garden of the Paris Museum at the same time. It had reached Kew by 1885, but the plants now in cultivation probably all derive from a re-introduction by Wilson for Messrs Veitch in 1900 (*Journ. R.H.S.*, Vol. 28 (1903–4), p. 393 and figs 83, 88, 104). The seedlings grew very fast, and were

6 to 10 ft high when only two or three years old. The species is variable in its foliage and in the length and density of the prickles on its stems, but as now seen in gardens it is one of the most handsome of the vines and, with its armed stems, one of the most distinct. The leaves usually colour brilliant red in the autumn.

cv. 'VEITCHII'.—A selection by Messrs Veitch from the seedlings raised from the Wilson introduction. 'This form is unusually bold and handsome, more vigorous than the type, producing a shining, bronzy green colour all through the summer, and in autumn assumes the richest hues.' (J. Veitch, *Journ. R.H.S.*, Vol. 28 (1903–4), p. 393 and figs 84, 89, as *V. armata Veitchii*). Gagnepain, in *Plantae Wilsonianae*, Vol. 1, p. 104, suggested that this might be referable to his *V. armata* var. *cyanocarpa* (*V. davidii* var. *cyanocarpa* (Gagnep.) Sarg.). This was defined as having sparser prickles, and was described from specimens collected by Wilson in W. Hupeh in 1907 for the Arnold Arboretum.

V. FLEXUOSA Thunb.

A slender-stemmed, elegant climber; shoots glabrous, or downy only when quite young. Leaves roundish ovate and heart-shaped at the base, or triangular and truncate at the base, often contracted at the apex to a slender point; amongst the smallest in the genus, being ordinarily 2 to 3½ in. across, of thin firm texture, glabrous and glossy above, downy on the veins and in the vein-axils beneath. Inflorescence slender, 2 to 6 in. long. Fruits about the size of a pea, black.

Native of Japan, Korea and China. Long cultivated but now represented in British gardens mainly by the following:

var. PARVIFOLIA (Roxb.) Gagnep. *V. flexuosa* f. *parvifolia* (Roxb.) Planch.; *V. parvifolia* Roxb.—As introduced by Wilson from W. Hupeh around 1900 this differs in the smaller leaves, shining bronzy green above, purple beneath when young, and is one of the daintiest in leaf of the true vines. It received an Award of Merit in 1903 when exhibited by Messrs Veitch under the name *V. flexuosa Wilsonii*, though some of the plants they distributed under this name were apparently *Ampelopsis bodinieri*.

The var. *parvifolia* is widely distributed from the Himalaya to southern and central China and Formosa, and is regarded as a distinct species by many authorities.

V. LABRUSCA L.   NORTHERN FOX GRAPE

A vigorous deciduous climber, with very woolly young shoots carrying a tendril or an inflorescence at every joint. Leaves thick-textured, unlobed, or three-lobed (sometimes deeply) towards the top; shallowly and irregularly toothed, broadly ovate or roundish, 3 to 7 in. wide and long, the base heart-shaped, upper surface dark green, becoming glabrous, lower one covered with rusty-coloured (at first whitish) wool; stalk more than half as long as the blade. Panicles 2 to 4 in. long. Berries globose, ⅔ in. in diameter, thick-skinned, dark purple with a musky or foxy aroma.

Native of eastern N. America from New England southwards; introduced in 1656. Of the wild grape vines of N. America this is the most important in an economic sense, and has produced more varieties cultivated for their fruit than any other. It is a vigorous species, and although it has not the least value as a fruiting vine in this country, it is worth growing for its fine foliage and luxuriant growth. It is distinguished by having a tendril or an inflorescence opposite each leaf.

The Labruscan viticultural varieties have been given the collective name V. LABRUSCANA Bailey. Mostly they derive from *V. labrusca* crossed with *V. vinifera*, though other N. American species may enter into the parentage of some. They are the varieties mainly used in eastern N. America for wine-making, but the wines of California are made from *V. vinifera*, as in Europe. The Labruscan wines have what is usually termed a 'foxy' flavour, and only a few British wine merchants list any.

V. MONTICOLA Buckl.

A deciduous climber up to 30 ft high, with slender, angled, slightly downy branchlets. Leaves 2 to 4 in. across, about the same in length, heart-shaped at the base, the sinus broad and rounded, sharply, sometimes slenderly pointed, coarsely triangular-toothed, and slightly three-lobed, of thinnish texture, dark green above, greyish beneath, both surfaces shining, woolly on the veins beneath when young; stalk about half the length of the blade. Berries globose, $\frac{1}{2}$ in. wide, black and sweet.

Native of southwest Texas; introduced in 1898. There is a thin diaphragm interrupting the pith at the joints. V. CHAMPINII Planch., from the same region, and introduced at the same time, is similar but has larger berries, also black. It is possibly a hybrid of *V. candicans*.

V. PIASEZKII Maxim.

Parthenocissus sinensis Diels & Gilg; *Psedera sinensis* (Diels & Gilg) Schneid.; *Vitis sinensis* Veitch

A vigorous, deciduous climber, with rarely branching tendrils, young shoots at first flossy, then glabrous. Leaves very variable, 3 to 6 in. long, 2½ to 5 in. wide, three-lobed with a heart-shaped base, or composed of three or five taper-based leaflets, the middle one of which is stalked and oval or obovate, the side ones or at least the lower pair obliquely ovate and stalkless. The merely lobed leaves differ much in the depth of the lobes, which are sometimes little more than large triangular teeth, but showing every intermediate condition between that and the tri- or quinque-foliolate ones. The margins are sharply toothed, the upper surface dark green, downy on the veins, the lower surface more or less brown-felted; stalks purplish, half to two-thirds as long as the blade. Fruits black-purple, globose, $\frac{1}{3}$ in. wide, in slender, sometimes forked branches 4 or 5 in. long. *Bot. Mag.*, t. 9565.

Native of western and central China; introduced by Wilson in 1900. It was at first known as *V. sinensis* and received an Award of Merit when exhibited under that name by Messrs Veitch in 1903. It is remarkably variable in the shape of its leaves. Those at the base of the shoot are simple but higher they become progressively more lobed, eventually becoming palmately compound at the apex of the growth. The leaves are tinted when young and colour red or bronze before falling.

var. PAGNUCCII (Romanet du Caillaud) Rehd. *V. pagnuccii* R. du Caillaud— This differs only in having more glabrous shoots and leaves. It was introduced to Kew in 1899, but it had previously been cultivated in France. It has the same garden value as the type.

V. BETULIFOLIA Diels & Gilg—Similar to *V. piasezkii,* but with always simple leaves (sometimes slightly lobed), mostly ovate and up to 4 in. long, almost glabrous beneath when mature. Fruits blue-black, with a slight bloom. Native of Central and Western China. Although Wilson collected specimens during his expeditions for Messrs Veitch he apparently first sent seeds to the Arnold Arboretum, in 1907–8 and again in 1911. It received an Award of Merit when exhibited in 1917 by Mrs Berkeley of Spetchley Park, both for its fruits and for its autumn colour. Mrs Berkeley was the sister of the more renowned Miss Ellen Willmott, but was herself a talented gardener.

V. QUINQUANGULARIS Rehd.

V. pentagona Diels & Gilg, not Voigt; *V. ficifolia* var. *pentagona* Pamp.

A deciduous climber; young shoots clothed with a whitish felt, and attaching themselves to their supports by twining tendrils, which are also felted. Leaves ovate with a heart-shaped or truncate base and pointed apex, usually but not always shallowly three- or five-lobed, unevenly and shallowly toothed, 3 to 6 in. long, three-fourths as wide, dark green above and at first downy, becoming glabrous, clothed beneath with a vividly white, close felt which remains until the leaves fall; stalk 1 to 3 in. long; veins in six to nine pairs. Berries globose, $\frac{1}{3}$ in. wide, blue-black, borne in slender bunches 4 to 6 in. long.

Native of Western and Central China; introduced by Wilson in 1907, in the autumn of which year seeds were sent to Kew from the Arnold Arboretum (W.134). This is a very distinct and ornamental species on account of the white felt that covers the under-surface of the leaf. No cultivated species of *Vitis* has this character more marked. It is a vigorous grower and very hardy.

var. BELLULA (Rehd.) Rehd. *V. pentagona* var. *bellula* Rehd.—This differs from the type chiefly in its much smaller leaves, $1\frac{1}{2}$ to $2\frac{1}{2}$ in. long.

V. RIPARIA Michx.   RIVERBANK GRAPE

V. vulpina L., in part; *V. odoratissima* Donn

A vigorous, deciduous, scrambling bush or climber with glabrous young shoots. Leaves thin, 3 to 8 in. wide, usually somewhat longer, broadly heart-shaped, with a finely tapered point and coarse, triangular, unequal teeth, usually more or less three-lobed, shining green on both surfaces, downy on the veins beneath; stalk from half to quite as long as the blade. Flowers sweetly scented like mignonette, produced in panicles 3 to 8 in. long. Berries globose, ⅓ in. in diameter, black-purple, covered thickly with blue bloom.

Native of eastern and central North America; introduced in 1806. It is worth growing for its vigorous, leafy habit and sweet-scented flowers. It strikes very readily from cuttings and has in consequence been much used as a phylloxera-proof stock on which the wine-producing vines of France have been grafted.

The confused name *V. vulpina* L. was applied to this species in previous editions, as by other authorities, but probably belongs properly to *V. cordifolia*. The two are allied, and both have a tendril missing from every third joint, but the present species differs in its more commonly three-lobed leaves with larger more persistent stipules, and in its blue-bloomed fruits. The name *V. vulpina* has also been used for *V. rotundifolia*.

V. 'BRANT'.—A hybrid of 'Clinton', which was one of the earliest selections of *V. riparia* or possibly a hybrid between it and *V. labrusca*. The other parent, which contributed the pollen, was *V. vinifera* 'Black St Peters'. It was raised by Charles Arnold of Paris, Canada, in the 1860s and was introduced to Britain in 1886. The rather small, black fruits frequently ripen with us, at least in southeastern England, and if not used to make wine the juice is excellent when drunk fresh. One grower at least has dried the fruits, which contain few or no seeds, and used them in chutney. But this vine is chiefly grown for its remarkable autumn colour: the leaves turn deep bronzy red in autumn, except for the veins, which remain green. The deeply lobed leaves are an inheritance from 'Black St Peters'. Award of Merit 1970, when exhibited by W. J. Tjaden of Welling, Kent, whose note on 'Brant' will be found in *Journ. R.H.S.*, Vol. 96 (1971), pp. 133-4.

V. PALMATA Vahl   *V. rubra* Michx. RED GRAPE.—A native of the southern central USA, allied to *V. riparia*. It has glabrous, bright red young branches and leaf-stalks; leaves three- or five-lobed, the lobes long and slenderly pointed. Berries black, without bloom.

V. ROMANETII A. David ex Foëx

A vigorous, deciduous climber; young shoots downy, and mixed with the down are numerous gland-tipped, stiff, erect bristles. Leaves of firm texture, three-lobed with a deep, narrow opening where the stalk is attached, shallowly toothed, each tooth ending in a bristle-like tip, 6 to 10 in. long, 4 to 7 in. wide, upper surface slightly downy on the nerves, or almost smooth and dark green, lower surface covered with a dense grey felt, with the midrib and veins hairy, a few large gland-tipped bristles mixed with the hairs. Stalk one-third to one-half as long as the blade, with a mixture of down and glandular bristles as on the shoot, but with the bristles more numerous. Fruits black, ⅓ to ½ in. in diameter.

Native of China; discovered and introduced by Père David in 1872-3; it was reintroduced in 1880, on both occasions from Shensi. It is one of the finest of the true vines, though rather tender in the young state, and until the main stem becomes quite woody. From *V. davidii* it is very distinct in the felted undersurface of the leaves, and it is not possible to confuse it with any other species.

V. ROTUNDIFOLIA Michx.   MUSCADINE

V. vulpina sens. Torr. & Gr., not L.; *Muscadinia rotundifolia* (Michx.) Small

A vigorous, deciduous climber, with stems up to 90 ft in length, in the wild, the bark adhering (not shredding); young shoots warted; tendrils unbranched. Leaves broadly ovate or roundish, always broadly heart-shaped at the base, pointed, 2 to 4½ in. long and wide, of firm texture, seldom lobed, but with the marginal teeth large, irregular, triangular, upper surface glossy dark green, glabrous, the lower one yellowish green, also glossy, with down in the vein-axils and sometimes on the veins; stalk usually shorter than the blade. Berries roundish, ⅔ to 1 in. in diameter, dull purple without bloom, skin thick and tough; flavour musky.

Native of the southern USA and the source of some American vineyard varieties, such as 'Scuppernong'. Nearly allied to it is V. MUNSONIANA Simpson ex Munson, a native of Florida and Georgia, which has been in cultivation here but is not very hardy. It has smaller fruits with a tender skin and acid flesh. These two species are distinguished from all other members of the genus by the pith running uninterrupted through the joints of the stem, and by the non-shredding bark and unforked tendrils.

V. RUPESTRIS Scheele   BUSH GRAPE

A deciduous bush up to 6 or 7 ft high, usually without tendrils; young shoots glabrous or nearly so. Leaves kidney-shaped or broadly heart-shaped, 2 to 4½ in. wide, scarcely so long, abruptly and slenderly pointed, coarsely, irregularly, and sharply toothed, but not or only slightly lobed, glossy, bluish green on both surfaces and glabrous, except that the veins beneath are sometimes downy; stalk rather shorter than the blade. Berries roundish, about ½ in. wide, purple-black with a slight bloom, agreeably flavoured.

Native of the southern USA. Interesting as a bushy, not climbing vine.

V. THUNBERGII Sieb. & Zucc.

? *V. ficifolia* Bge.; *V. labrusca sens.* Thunb., not L.; *V. sieboldii* Hort.

A slender-stemmed, only moderately vigorous, deciduous climber, the young shoots angled, more or less woolly. Leaves variable, but deeply three- or five-lobed, usually 2½ to 4, sometimes 6 in. across, heart-shaped at the base. Lobes ovate, often penetrating half or more than half the depth of the blade, the space (or sinus) between the lobes often expanding and rounded at the bottom, sharply, shallowly, and irregularly toothed, dark dull green and glabrous above, covered with a rusty brown felt beneath; leaf-stalk about half the length of the blade. Berries in bunches 2 or 3 in. long, black with a purple bloom, ⅜ in. or less in diameter. *Bot. Mag.*, t. 8558.

A native of Japan and Korea; introduced probably by Siebold. It is uncommon in gardens and a rather weak grower at Kew, the ends of the shoots dying back considerable every winter. Canon Ellacombe grew it at Bitton in Gloucestershire, where the leaves turned rich crimson in autumn and fruit was occasionally borne. It is chiefly remarkable for its fig-like leaves. It was originally confused with the American *V. labrusca*, which has a tendril at every node on the young shoot, while in *V. thunbergii* a tendril is missing from every third one. The name *V. thunbergii* was at one time used in gardens for *V. coignetiae*, from which the present species differs obviously in its smaller, lobed leaves and its slender, less woody, five-angled young shoots.

V. VINIFERA L.   COMMON GRAPE VINE

A deciduous climber growing to a considerable height if given support, its young shoots sometimes glabrous, sometimes cobwebby. Leaves mostly 3 to 6 in. wide and long, three- or five-lobed, variable in toothing, widely to narrowly cordate at the base (in some deeply lobed forms the sinus between the lobes is almost closed by the overlapping of their lower margins), upper surface almost glabrous at maturity, the underside hairy or woolly, rarely becoming quite glabrous. Further description is unnecessary, for *V. vinifera* is the common grape vine, with all its variations in size of truss and size and colouring of the grapes. The seeds are usually not more than two in number (none in some seedless sorts such as the currant grape), with a definite beak at one end.

The grape vine has been cultivated in all the warm temperate parts of Europe, and in parts of Asia back into the unrecorded past, and its really native country is now a matter of conjecture. But the general opinion is that it originated in Asia Minor and the Caucasian region. It is perfectly hardy in most parts of Britain, but needs a certain minimum of summer heat if it is to ripen its grapes. In the early middle ages the summers in northern Europe were warmer than today and the vine was widely cultivated for the making of wine. Then the summers deteriorated and the northern limit of wine-growing moved southward (and in mountainous regions downwards). Parkinson, early in the 17th century, remarked that less wine was produced than in times past and suggested 'our unkindly yeares' as one of the reasons. At the present time there are probably more true vineyards in England (as distinct from plantings on or with the protection of walls) than at any time since the middle ages. But the revival of English viticulture is due partly to economic factors and partly to the availability of early ripening clones.

subsp. SYLVESTRIS (Gmel.) Berger & Hegi *V. sylvestris* Gmel.—This is the wild grapevine, from which the cultivated sorts have evolved over the past five millennia or so, though the first stages of domestication probably took place in the eastern part of its range. It is dioecious, and generally the male plants have the leaves more deeply lobed than the female. The berries are sour, about ¼ in. wide. The seeds are roundish, with a short beak, and usually three in number (in the cultivated grapevine the seeds are usually two, ovate, tapered into a longer beak). Native of southern Central and S.E. Europe, east through Asia Minor to the Caucasus, N. Iran and Turkestan. It is uncommon in Central Europe, but occurs in the valleys of the upper Rhine and upper Rhone and is of interest as one of the few lianes in the European flora, climbing high into alders, elms, willows and poplars (Hegi, *Fl. Mitteleuropa*, Vol. VI, pp. 364–5). It is claimed that from a study of the seeds found in prehistoric sites it is possible to determine whether the fruits were originally gathered from wild grapevines or early cultivars.

The following grapevines are cultivated for ornament:

cv. 'APIIFOLIA' ('Laciniosa'). PARSLEY VINE, CIOTAT.—The striking and handsomely cut leaves of this variety consist of three to five main divisions, sometimes stalked. These are again variously cut into deep narrow lobes, pointed or bluntish. It makes a very effective climber, although the leaves are certainly not subdivided enough to suggest parsley. Cultivated in Britain since the middle of the 17th century.

cv. 'BRANT' ("Brandt").—See *V.* 'Brant', under *V. riparia.*

cv. 'INCANA'.—Leaves three-lobed or unlobed, the upper surface as well as the lower one covered with a grey, cobweb-like down, giving them, especially when young, a whitish appearance, as if dusted with flour. Fruits black. It is uncertain whether this is the same as the old variety variously known as the Miller, Meunier or Black Cluster grape, or as Miller's Burgundy, though this certainly had hoary leaves. This produced a moderately good fruit, which usually ripened in this country.

cv. 'PURPUREA'.—Youngest leaves whitish with down, becoming plum-purple and darkening to rich purple in autumn. Grapes dark purple, with a harsh juice (*V. vinifera purpurea* Veitch, *Journ. R.H.S.*, Vol. 28 (1903–4). This is probably the same as, or a derivative from, the old so-called Claret grape, and not the Teinturier grape, but the two seem to have been long confused. Both yielded a dark juice, that of the Teinturier being so dark that it was used to colour table wines. But the leaves of this did not colour until the fruit was ripe.

V. WILSONIAE Veitch ex *Gard. Chron.*

V. reticulata Pamp., not M. A. Laws.

A very vigorous deciduous climber with woolly young shoots. Leaves roundish ovate, more or less heart-shaped at the base, short-pointed, wavy-toothed, 3 to 6 in. wide, woolly all over when young, becoming glabrous above but remaining cobwebby beneath especially on the veins. The fruits are black covered with a purplish bloom and scarcely ½ in. long, borne in slender bunches 3 to 5 in. long, often unbranched.

Native of Central China; discovered by Wilson in 1902 and introduced then or in 1904 when collecting for Messrs Veitch. They exhibited it in October

1909, as *V. Wilsonae,* and it was then given an Award of Merit for its fine deep red autumn colour. In recording this, the *Gardeners' Chronicle* published a short description of the plant, and a figure of one leaf. It is a vigorous vine, but now uncommon.

WATTAKAKA   ASCLEPIADACEAE

The species described here, often known as *Dregea sinensis,* is of interest as an almost hardy relative of *Hoya,* many species of which are cultivated for ornament in greenhouses. For a discussion of the taxonomic position of *Wattakaka,* and details of flower-structure, see the article in the *Botanical Magazine* cited below. The generic name comes from 'Wattakaka-kodi', the native name on the Malabar Coast for the type-species of the genus, *W. volubilis* (L.f.) Stapf (*Asclepias volubilis* L.f.).

W. SINENSIS (Hemsl.) Stapf
Dregea sinensis Hemsl.

A climbing or creeping evergreen shrub with densely downy young stems and a dark, warty bark. Leaves opposite, ovate, cordate at the base, tapered at the apex to a sharp point, $1\frac{3}{4}$ to 4 in. long, up to 3 in. wide, slightly hairy above, densely clad beneath with a velvety down; petioles $\frac{5}{8}$ to $1\frac{5}{8}$ in. long. Flowers fragrant, slenderly stalked, produced in June or July up to twenty-five together in nodding, downy, umbel-like inflorescences, which spring from the stem near the leaf-insertions; common stalk $1\frac{1}{4}$ to $2\frac{1}{4}$ in. long. Calyx downy, deeply divided into five ovate or oblong-ovate segments. Corolla about $\frac{5}{8}$ in. wide; segments five, broadly ovate-elliptic, ciliate, white or cream-coloured, speckled and streaked with red. The centre of the flower is almost filled by the dome-shaped stigmatic head, surrounded by the five knob-shaped append-ages of the corona and the membranous tips of the anthers. Fruit of two spindle-shaped follicles 2 to $2\frac{3}{4}$ in. long; seeds with a tuft of hairs at one end. *Bot. Mag.,* t. 8976.

Native of China; discovered by Henry in 1887 growing near Ichang in Hupeh, and introduced by Wilson from the same locality in 1907. The first recorded flowering in Britain was in 1922 at Aldenham, where it had grown on a wall unharmed by frost for many years. Although an interesting and attractive species, it is not showy, and has never become frequent in gardens.

WEIGELA   CAPRIFOLIACEAE

A genus of deciduous shrubs, closely allied to the honeysuckles, with about twelve species in temperate E. Asia. Leaves opposite, serrate. Flowers borne on short lateral twigs on the year-old branches. Calyx five-lobed, persistent. Corolla tubular-campanulate or funnel-shaped, more or less regular (not two-lipped). Stamens five, shorter than the corolla. Ovary inferior, slender and elongate. Style one, with a capitate or hood-like stigma. Fruit a capsule, opening by two valves. Seeds numerous, very small. The genus was named by Thunberg after the German botanist C. E. von Weigel (*d.* 1831). The spelling 'Weigelia' is incorrect.

The genus *Diervilla* is closely allied, but its flowers are borne at the ends of the current season's growths and the corolla is two-lipped; it is confined to N. America. *Weigela* was at one time included in *Diervilla*, but it is now usual to keep them separate.

There are few more beautiful summer-flowering shrubs than the weigelas. The first of them, *W. florida*, was introduced in 1845. Afterwards other species arrived and hybridising was commenced, with the result that a very fine race of garden varieties has been produced, showing great variety in colour. The best known of those now in cultivation are described after the species. These now surpass the original species in effectiveness, and the latter are becoming scarce.

The weigelas are easily cultivated, and there are only two, *W. hortensis* and *W. middendorfiana*, that show any sign of tenderness. Being gross feeders, they need a rich soil. They should be pruned as soon as the flowers are past, by entirely removing the old shoots that have flowered, leaving the young shoots of the year untouched, to produce their crop the following year. They are easily propagated by cuttings of half-ripened growths.

Apart from the anomalous *W. middendorffiana* and *W. maximowiczii*, the species here described belong to the following two sections:

sect. WEIGELA (*Utsugia*).—Calyx divided to the base into linear segments. Seeds narrowly winged. Here belong: *W. coraeensis, W. floribunda, W. hortensis, W. japonica.*

sect. CALYSPHYRUM.—Calyx divided to about the middle; lobes lanceolate or deltoid. Seeds unwinged. Here belong only *W. florida* and *W. praecox.*

W. CORAEENSIS Thunb.

Diervilla coraeensis (Thunb.) DC.; *W. grandiflora* (S. & Z.) K. Koch;
Diervilla grandiflora Sieb. & Zucc.;

A deciduous shrub, 6 to 12 ft high; young branchlets glabrous. Leaves 3 to 5 in. long, 2 to 3 in. wide, oval or obovate, with a long, abrupt point, nearly or quite glabrous above, hairy on the midrib and chief veins below; stalks ¼ to ¾ in. long, bristly. Flowers produced during June in sessile corymbs usually of threes, terminating short lateral twigs. Corolla bell-shaped, abruptly narrowed near the base, 1 to 1¼ in. long, ¾ in. across at the five-lobed mouth, not downy, pale rose at first, changing to carmine. Calyx with five narrow, linear lobes ⅓ in. long; ovary glabrous.

Native of Japan, but not of Korea, as the specific epithet implies. A very handsome, free-flowering shrub, introduced to Europe about the middle of the 19th century, and the parent of many modern garden hybrids. Its distinguishing characters from *D. florida* are the linear calyx-lobes, reaching to the base, the longer-stalked leaves, and the glabrous ovary.

cv. 'ALBA'.—Flowers greenish yellow, aging to pink. A remarkable plant, possibly a hybrid (*Diervilla grandiflora alba* Dipp.; *D. g.* var. *arborea* Rehd.).

W. AMABILIS (Carr.) Planch. *Diervilla amabilis* Carr.—Although considered by Rehder to be synonymous with *W. coraeensis*, it is possible that the plant in Van Houtte's nursery, described by Planchon in 1853 as *W. amabilis*, was in fact a hybrid. Certainly the plant sent under this name by Van Houtte to Vilmorin was not pure *W. coraeensis*, judging from a specimen collected from it in 1861 and preserved in the Kew Herbarium, the undersides of the leaves and the inflorescence being much hairier than is normal in that species. A weigela imported into this country from the continent under the name *W.* 'Rosea' seems to be very similar to *W. amabilis*, and is certainly not *W. florida*, of which *W. rosea* is a synonym. The flowers are blush coloured with deeper lilac-pink shadings, abruptly narrowed to a short tube; the veins of the leaf-undersides are hairy, as are the ovaries and the outside of the corolla; the calyx is divided to the base into narrow segments. [PLATE 103

According to Van Houtte, *W. amabilis* produced variegated seedlings. It is possible that the variegated clones now in commerce under the erroneous names *W. florida* 'Variegata' and *W. praecox* 'Variegata' derive from *W. amabilis*. See further under *W. florida*.

W. FLORIBUNDA (Sieb. & Zucc.) K. Koch
Diervilla floribunda Sieb. & Zucc.

A shrub 4 to 8 ft high, with slender, supple branches clothed with soft hairs. Leaves ovate or oval with long, tapering points, wedge-shaped at the base, 3 or 4 in. long, and about half as wide on the long, barren, first-year shoots, considerably smaller on the lateral flowering twigs, toothed, downy on both surfaces, especially beneath. Flowers produced during June in corymbs terminating, and in the leaf-axils of, the short side twigs. Corolla funnel-shaped, 1 in. long, with five spreading lobes at the mouth, where it is ⅝ in. across, downy outside, of a dark, almost blood-red. Calyx ½ in. long, consisting of a tube and five narrow linear lobes, hairy. Seed-vessel cylindrical, narrow, downy.

Native of the mountains of Japan; introduced to Europe about 1860. The typical *W. floribunda* is now rare in cultivation, but it is the species whose characteristics and colour of flower are dominant in the crimson-flowered hybrids such as 'Eva Rathke' and 'Styriaca'. It is allied to *W. coraeensis*, but that is an almost glabrous species, with lighter coloured flowers.

W. FLORIDA (Bge.) A.DC.

Calysphyrum floridum Bge.; *Diervilla florida* (Bge.) Sieb. & Zucc.; *Weigela rosea* Lindl.

A shrub 6 to 9 ft high, of spreading habit and arching branches; young shoots with two lines of short hairs. Leaves oval or oval-lanceolate, long-pointed, toothed except at the base, felted on the midrib beneath, 2 to 4½ in. long, ¾ to 1½ in. wide, very shortly stalked. Flowers often in terminal threes or fours on short lateral twigs. Corolla funnel-shaped, 1¼ in. long, with five spreading rounded lobes at the mouth, where it is as much in diameter, deep rose on the outside, paler and becoming almost white within; stigma bilobed. Calyx divided to about the middle into five narrowly triangular lobes. Ovary slightly downy. *Bot. Mag.*, t. 4396.

Native of N. China and Korea; described by Bunge from plants cultivated in Peking gardens; Fortune first saw it in the garden of a mandarin on the island of Chusan, and purchased a plant from a Shanghai nursery which he sent to the Horticultural Society in 1844. Revisiting the nursery some years later he was asked by the nurseryman and his sons how this and the other plants he bought had thrived. 'I told them that most of the plants had arrived safely in England, that they had been greatly admired, and that the beautiful Weigela had even attracted the notice of her Majesty the Queen. All these statements, more particularly the last, seemed to give them great pleasure; and they doubt-less fancied the Weigela of more value ever afterwards.' (Fortune, *Gard. Chron.*, 1850, p. 757).

For half a century or more after its introduction *W. florida* was the commonest of the weigelas in gardens, but has now largely given way to its cultivars and hybrids. Most of the latter with pink or carmine flowers derive from it.

cv. 'FOLIIS PURPUREIS'.—A low-growing shrub. Leaves dark green with a metallic lustre, flushed with purple. Flowers pink; ovaries dark purple. Raised in France from Chinese seed (*Rev. Hort.*, 1921, pp. 278–9).

var. VENUSTA (Rehd.) Nakai *Diervilla florida* var. *venusta* Rehd.; *W. venusta* (Rehd.) Stapf.—A graceful shrub to 6 or 9 ft high, differing from the typical state in its smaller, obovate leaves. Corolla narrowing rather more gradually towards the base. Introduced to the Arnold Arboretum by J. G. Jack in 1905 from Korea, of which it is a native. Plants cultivated in Britain probably all derive from the seeds collected by Wilson in Korea in 1918. *Bot. Mag.*, t. 9080.

W. PRAECOX (Lemoine) Bailey *Diervilla praecox* Lemoine—Closely allied to the preceding but with the leaves hairy all over the undersurface and also hairy, though more sparsely, above. Flowers fragrant, 1½ in. long, rose-coloured, yellow in the throat, produced in clusters of three or five in May, some three or four weeks before the other cultivated species. The calyx resembles that of *W. florida* in lobing but it and the ovary are more downy. Native mainly of Korea and the Ussuri region of Russia, but also reported from the Japanese island of Kyushu. It was put into commerce by Lemoine in 1894 and named by him. He used it to produce such hybrids as 'Avalanche', 'Espérance', 'Fleur de Mai' and 'Gracieux'.

Under the names *W. florida* 'Variegata' and *W. praecox* 'Variegata' two rather similar hybrids are in cultivation, both erect growing and with creamy yellow or ivory-white margination to the leaves. In one, which seems to be the commoner, the leaves are mostly rather narrow, with an abnormal, irregular margin, and are often pink-tinged at the edge. Despite its virus-ridden look it is free-flowering, bearing blush flowers with pink shading in large clusters. The other has generally broader, even obovate leaves, the margins of which are mostly regular, except near the flowers. Neither belongs to *W. florida* nor to *W. praecox*, though the influence of the former is shown in both clones, which are probably seedlings of *W. amabilis*, (see *W. coraeensis*).   [PLATE 102

There is another variegated weigela in gardens quite different from the above. It is of low, spreading habit; the leaves have a pure and vivid white margination and the flowers are dull reddish pink from dark red buds. This is near to *W. florida*, but probably a hybrid. Its correct cultivar-name is uncertain.

W. JAPONICA Thunb.
Diervilla japonica (Thunb.) DC.

A deciduous shrub, 6 to 9 ft high; young shoots nearly glabrous. Leaves oval or ovate, 2 to 4 in. long, about half as wide, toothed, long and taper-pointed, densely felted with pale down beneath, slightly hairy above; stalk ¼ in. or less long, bristly on the edges. Flowers mostly in threes, terminal and in the leaf-axils of short side twigs, forming a leafy panicle 3 to 5 in. long. Corolla between funnel- and bell-shaped, 1 to 1¼ in. long, less in width, rather downy outside; pale rose at first, changing to carmine. Calyx-lobes linear, more or less downy.

Native of Japan; introduced to the USA in 1892 and to Britain soon afterwards. It is allied to *W. floribunda*, which it resembles in having the calyx divided to the base into linear lobes and in the often downy outer surface of the corolla, but the down on the undersurface of the leaves is more velvety and the flowers differ much in colour from the almost blood-red ones of *W. floribunda*.

var. SINICA (Rehd.) Bailey *Diervilla japonica* var. *sinica* Rehd.—Taller, to 20 ft. Leaves downy beneath all over the blade, longer-stalked. Corolla tubular at the base, pale pink. Ovary densely downy. Introduced by Wilson from W. Hupeh, China, in 1908, but uncommon in this country.

W. HORTENSIS (Sieb. & Zucc.) K. Koch *Diervilla hortensis* Sieb. & Zucc.; *Diervilla japonica* var. *hortensis* (Sieb. & Zucc.) Maxim.; *W. japonica* sens. Miq., not Thunb.—Allied to *W. japonica*, but with the leaves velvety and almost white beneath, the corolla glabrous outside. Described by Siebold and Zuccarini from Japanese garden plants with deep red (typically) or white flowers (*W. hortensis* f. ALBIFLORA (Sieb. & Zucc.) Rehd.). They surmised that it was an importation from China or Korea, though it is in fact a Japanese native.

This species is now mainly represented in cultivation by 'NIVEA', with pure white flowers, introduced to Europe from Japan early in the 1860s and to Britain well before 1878 (*W. hortensis nivea* Bonard; *W. nivea* Carr.; *W. h.* f. *albiflora* (Sieb. & Zucc.) Rehd., at least in part). 'Nivea' is one of the finest of the weigelas. It received a First Class Certificate in 1891. The typical red-flowered state of *W. hortensis* is now uncommon but had reached Europe by the 1870s. Both this and 'Nivea' were used in the breeding of the present garden race of weigelas.

W. MIDDENDORFFIANA (Trautv. & Mey.) K. Koch

Calyptrostigma middendorffianum Ttautv. & Mey.;
Diervilla middendorffiana (Trautv. & Mey.) Carr.

A shrub 3 to 5 ft high; young shoots glabrous except for two downy ridges. Leaves 2 or 3 in. long, 1 to 1½ in. wide, ovate-lanceolate, toothed, wrinkled, slightly hairy on the margins and on the chief veins when young only; stalk ⅙ in. or less long. Flowers in terminal cymose clusters. Corolla bell-shaped, sulphur-yellow, stained with orange on the lower lobes, 1¼ in. long, 1 in. wide across the mouth, where are five spreading lobes. Calyx two-lipped, the upper lip with three narrow lobes, the lower one with two broader, deeper ones, all fringed with short bristles. Flowers in April and May. *Bot. Mag.*, t. 7876.

Native of Japan, N. China, Korea and the Russian Far East; introduced to Europe in 1850. A beautiful shrub, very distinct in its yellow blossoms. But it suffers much from late spring frosts in some gardens.

It is the only species of the section *Calyptrostigma*. Apart from its yellow flowers, its distinctive characters are the two-lipped calyx, the hooded (calyptrate) stigma and the connate anthers.

Another yellow-flowered, anomalous species is W. MAXIMOWICZII (S. Moore) Rehd., which is similar to *W. middendorffiana* in most of its botanical characters, but inferior to it as an ornamental. Native of Japan.

THE HYBRID WEIGELAS

The species of *Weigela* (apart from the two with yellow flowers) have been largely replaced in gardens by nursery-bred hybrids. This was true even in the early years of this century, when many hybrids were already available, some dating back to the 1860s. The species with the shortest life in gardens was *W. praecox*, which was thrown into the melting-pot as soon as it reached Lemoine's nursery and produced its first hybrids only five years after it was named. This species, and the related *W. florida*, belong to the section *Calysphyrum*, in which the calyx is divided only in the upper half. The influence of one or the other shows in the incompletely divided calyx of so many hybrids. A completely divided calyx is shown only by hybrids in which the influence of *W. coraeensis*, *W. floribunda* or *W. hortensis* is dominant.

The dark red colouring of 'Eva Rathke' derives from *W. floribunda* and probably the later sorts in the same style were bred from it. These dark-coloured hybrids, and the reddish pink 'Abel Carrière' are the only ones to have received awards in this country, but they are much less effective in the garden than those with lighter coloured flowers. *W. coraeensis* has been an important parent, perhaps through its probable hybrid *W. amabilis*, contributing a many-flowered inflorescence and a tendency to glabrousness; its influence is often to be seen in a short corolla-tube, abruptly widening. Hybrids with the leaves densely and softly hairy beneath have *W. hortensis* in their parentage.

Most of the cultivated weigelas flower in early June. Not many of Lemoine's May-flowering hybrids of *W. praecox* have found their way into British gardens, though some of the finest are among them, nearly all in light shades.

The hybrids need a good garden soil and are moisture-loving. For pruning, see the generic introduction. The following selection is confined to those that are in general commerce in this country, and a few others, old and new, that seem to deserve wider cultivation. The parentage, when given, is that suggested by Rehder. For three variegated hybrids, see under *W. florida*.

'ABEL CARRIÈRE'. Lemoine, 1876; *W. florida* × *W. hortensis*.—Flowers deep rosy carmine, opening from darker coloured buds, with a faint yellowish flare in the lower part of the throat. Style long-exserted. A.M. 1939.

'AVALANCHE'.—See 'CANDIDA'.

'BRISTOL RUBY'.—Flowers red from almost black buds. Vigorous and tall-growing. Raised in the USA, probably from 'Eva Rathke'. A.M. 1954.

'CANDIDA'.—Flowers pure white even in the bud-stage. Leaves bright green, almost glabrous beneath, long-acuminate at the apex. Very near to *W. coraeensis*. It is also sold as 'Avalanche', but the Lemoine hybrid of that name had white flowers with a pink tinge; it was a hybrid of *W. praecox*.

A newer hybrid with almost pure white flowers is 'BRISTOL SNOWFLAKE', raised in the USA around 1955. It is more vigorous than 'Candida'.

'CONQUÊTE'. Lemoine, 1907.—Flowers very large, almost 2 in. wide at the mouth, deep rosy pink, slightly darker on the reverse. Sparse, spreading habit.

'DAME BLANCHE'. Lemoine, 1900; *W. coraeensis* × *W. hortensis*.—Flowers white, tinged with pink on the outside. The buds are creamy white with a lilac flush.

'ESPÉRANCE'. Lemoine, 1906; *W. praecox* hybrid.—Flowers large, white flushed with salmon pink.

'EVA RATHKE'.—Flowers vivid crimson red from dark buds. Free-flowering to excess and weak-growing in consequence. An old hybrid of *W. floribunda* crossed with *W. coraeensis*, raised by Rathke of Praust near Danzig, in the 1880s. F.C.C. 1893. 'EVA SUPREME' is the result of a cross between 'Eva Rathke' and 'Newport Red', raised at the Boskoop Experimental Station and released to the trade in 1960. It is similar in flower-colour but much more vigorous (*Dendroflora*, No. 1 (1964), p. 36).

'FÉERIE' ('Fairy'). Lemoine, 1925.—Flowers uniform bright rosy pink, funnel-shaped, with a slender tube, in large trusses.

'FLORÉAL'. Lemoine, 1901.—Flowers clear rosy pink on the outside, with a deep, carmine pink throat. A hybrid of *W. praecox* flowering about mid-May.

Other pink-flowered hybrids of *W. praecox* raised by Lemoine, and flowering in May are: 'BOUQUET ROSE', 'FLEUR DE MAI' (one of the earliest), 'IDÉAL', 'LE PRINTEMPS', 'MAJESTUEUX', 'SÉDUCTION'.

'GUSTAVE MALLET'. Billiard; *W. coraeensis* × *W. florida*.—Flowers large, rosy pink with a paler limb, deep pink in the bud.

'LOOYMANSII AUREA'.—Leaves light golden-yellow in spring and early summer, with a very narrow red rim. Flowers light pink, with a slender tube. It was found in 1873 among seedlings of "*W. amabilis*" (q.v. under *W. coraeensis*) and shows the influence of *W. florida* in its calyx. Not vigorous, and best in light shade.

'NEWPORT RED'.—Similar to 'Bristol Ruby' but the flowers slightly lighter in colour; equally vigorous. Often known as 'Vanicek', it was raised by V. A. Vanicek, Newport, Rhode Island, USA in the 1930s.

'ROSABELLA'.—Although as yet scarcely known in this country, this is stated to be a fine hybrid, with rose-coloured flowers, lighter on the limb, with flowers almost as large as in 'Conquête' and a stronger grower. It is the product of the cross between 'Eva Rathke' and 'Newport Red' made in Holland which also produced 'Eva Supreme' (see 'Eva Rathke').

'ROSEA'.—See under *W. coraeensis*.

'STYRIACA'.—Flowers small, carmine-rose, deeper in the bud, borne very freely on arching branches. Raised by W. Klenert, nurseryman of Graz, Austria, and put into commerce in 1908. It was a selection from several hundred seedlings, and was considered by Rehder to be *W. coraeensis* × *W. floribunda*.

'VAN HOUTTEI'.—This may be the correct name for a hybrid between *W. florida* and *W. hortensis*, with light green leaves softly downy beneath. Flowers red in the bud, opening white flushed with apple-blossom pink inside, marbled with a slightly darker pink outside.

WEINMANNIA   CUNONIACEAE

A genus of about 150 species of trees and shrubs, mainly tropical and largely confined to the southern hemisphere. Leaves pinnate or trifoliolate. Flowers small, in clusters or racemes. Calyx-lobed and petals four or five, imbricated. Stamens eight to ten. Ovary two-chambered, with two styles. Fruit a leathery capsule, opening by two valves and containing numerous, usually hairy seeds. The genus was named by Linnaeus in honour of J. W. Weinmann (*d.* 1745), an apothecary of Regensburg (Ratisbon).

W. TRICHOSPERMA Cav.

A tree up to 100 ft high in its native forests, though usually not taller than about 40 ft; young shoots and flower-stalks furnished with brown down. Leaves opposite, pinnate, 1½ to 3 in. long, composed of nine to nineteen leaflets set ¼ to ½ in. apart, the space between each pair filled on each side of the main-stalk with a triangular wing. Leaflets stalkless, oval or obovate, ⅜ to ¾ in. long, with two to six conspicuous teeth on each side, dark lustrous green and glabrous except for a few bristles at the base of each pair of leaves. Flowers fragrant, quite small, white, packed closely on a cylindrical mignonette-like raceme 1½ to 2 in. long, opening in May, the stamens with their pink-tipped anthers being the most conspicuous feature. Fruit a dry, two-celled capsule, ⅛ in. long, red when young.

Native of the Chilean Andes in the region of *Nothofagus* forests, from about 35° to 45° S. This is one of the handsomest evergreens brought from Chile. Its leaves, rather fern-like in character, are distinct among evergreen species in

WEINMANNIA TRICHOSPERMA

the conspicuously winged rachis, and the white flowers, freely produced on cultivated plants, are followed by bronzy immature capsules, which in the Chilean forests give an autumnal tinge to the slopes when it occurs in large stands. It is not quite hardy near London, though it can be grown on a wall there. At Wakehurst Place in Sussex an example about 12 ft high has lived in the open for many years and flowers freely. At Tregothnan in Cornwall there is a tree measuring 59 × 3¾ ft (1971).

W. RACEMOSA L.f. KAMAHI.—This is the counterpart of *W. trichosperma* in New Zealand, where it ranges from North Island to Stewart Island. It is probably no more tender than the Chilean species, but lacks its distinctive foliage, as the leaves of adult plants are reduced to a single leaflet and are thus in effect simple. In juvenile plants many of the leaves are divided into three leaflets, or three-lobed. It is in cultivation, but no sizeable specimen has been recorded.

WISTERIA   LEGUMINOSAE

A small genus of exceedingly ornamental climbers, represented in the eastern United States and in N.E. Asia. The leaves are alternate, deciduous, and unequally pinnate. Flowers in handsome axillary or terminal racemes, and mostly of a pale bluish lilac or white. Pods rather resembling those of kidney beans in shape. The genus was named in honour of Caspar Wistar, a professor of anatomy in the University of Pennsylvania in the early part of the 19th century. Nuttall, however, spelt the generic name 'Wisteria'. This may have been a slip of the pen or printer's error, but it is equally possible, indeed more likely, that he chose deliberately to spell it thus, on the grounds that this rendering accorded better with the way the Professor actually pronounced his surname, with a mute 'a'. In any case, the name *Wisteria*, thus spelt, is conserved under the *International Code of Botanical Nomenclature*. It should be pronounced with a short 'e'.

Wisterias are of easy cultivation, and quite hardy in the southern half of England. In some places to the north-east they should be grown on walls. They like a good loamy soil, but are not fastidious at the root; a sunny position, however, is essential. The only problem in connection with their culture is the provision of suitable support. For *W. sinensis*, walls and pergolas afford the best means of displaying its attractions, and for the display of the long racemes of *W. floribunda* 'Multijuga' an overhead trellis work is desirable. An arbour framed in iron will quickly be covered by either of these species, and as the branches twist round the rods any unsightliness is soon hidden.

Seeds are not frequently produced, and do not afford a reliable means of increase. But layering may be adopted, and shoots grafted in spring on pieces of root from the same species, or *W. sinensis*, unite readily under glass. Cuttings of August wood made of the lower part of the season's shoots may also be tried.

W. FLORIBUNDA (Willd.) DC.   [PLATES 105, 106

Glycine floribunda Willd., a name founded on *Dolichos polystachyos sens.* Thunb., not L.f.

A deciduous climber to 30 ft or more high, its stems twining clockwise.

Leaves 10 to 15 in. long, consisting usually of eleven to nineteen leaflets which are downy when young, soon almost or quite glabrous and of a glossy dark green, of ovate shape and 1½ to 2½ in. long. Racemes borne on short leafy shoots, normally 5 to 10 in. long, slender, the flowers opening successively from the base, each on a slightly downy stalk ½ to 1 in. long; they are (normally) violet- or purplish-blue and fragrant; standard petal ¾ in. wide; calyx ⅓ in. long, bell-shaped, with triangular teeth. Pods 3 to 6 in. long, velvety.

Native of Japan, where it is common in the wild and known as 'Fuji' (sometimes transliterated as 'Fudsi'). It is really very near to W. sinensis but with more numerous leaflets and with stems that twine clockwise, i.e., ascend from left to right towards the growing point. As seen in cultivation it also differs in the generally longer racemes and in flowering two or three weeks later. These cultivated forms are the product of many centuries of selection in the gardens of Japan and differ among themselves in length of raceme and colour of flower. Collectively they take the name W. floribunda f. MACRO-BOTRYS (Neubert) Rehd. & Wils. There was a famous plant at the Kameido Shrine in Tokyo, illustrated and described by many travellers. It formed a huge arbour, extending partly over a piece of water spanned by a semicircular Japanese bridge. With its thousands of slender, pendulous racemes 3 to 4 ft long, crowded with lilac blossoms 'odorous of honey and buzzing with bees', it made, no doubt, one of the most famous floral exhibitions on the globe. Another remarkable plant grew at Ushijima, near Kasukabe, in the Tokyo area, where Wilson measured trusses over 5 ft long.

W. floribunda (in one or more of its cultivated forms) was introduced to Holland by Siebold in 1856, but seems to have attracted little notice at first, perhaps because it was slow to flower, or because it was wrongly pruned. The selection 'Multijuga', sent out by Van Houtte in the 1870s was the first to reach this country, but others came later in the century, most of them probably distributed by the Yokohama Nursery Company, through which so many Japanese garden plants reached the European trade.

W. floribunda, in its long-racemed garden forms, is not so well adapted for walls as W. sinensis; it should be trained in such a way as to allow the racemes to hang freely, as on overhead trellises. Where climbing space is not available, it can be treated as a bush. Two plants in the Kew collection have been treated like this for a century or so, and are still only about 8 ft high. The branches are spurred back every year, and produce an amazing profusion of racemes.

All forms of W. floribunda can be increased by layers, and by grafting twigs on pieces of its own roots in spring. Unlike W. sinensis, it produces fertile seed fairly regularly in this country. James Comber raised many seedlings at Nymans in Sussex, the best of which were the equal of the named sorts, and E. A. Bowles had a home-raised seedling in his garden at Myddelton House.

cv. 'ALBA'.—Flowers white, sometimes tinged with lilac; racemes rather shorter than in 'Multijuga'. Very beautiful. There is also a white-flowered form with very long racemes, not free-flowering. [PLATE 104

cv. 'COELESTINA'.—Flowers lavender-blue. Named by Sprenger, the Naples nurseryman, in 1911. This may be the form distributed by the Yokohama Nursery Company; see the note by Boehmer, the nursery manager, in *Gard.*

Chron., Vol. 33 (1903), p. 347. Also with flowers of this shade is 'GEISHA', mported from Japan by James Russell.

cv. 'MULTIJUGA'.—Standard pale violet with a yellow mark at the base; wings and keel violet blue. Racemes to about 2 ft long (*W. multijuga* Van Houtte, *Flore des Serres*, Vol. 19 (1873), t. 2002; *W. chinensis* var. *multijuga* (Van Houtte) Hook. f., *Bot. Mag.*, t. 7522; *W. floribunda* f. *macrobotrys* (Neubert) Rehd. & Wils., in part only). Introduced to Kew from Van Houtte in 1874. The original plants still exist (see above). The name *W. multijuga* was at one time used in a general sense for all the known forms of *W. floribunda*.

cv. 'ROSEA'.—Standards pale rose, wings and keel purple. Racemes to about 18 in. long. In cultivation 1903.

cv. 'RUSSELLIANA'.—Flowers darker than in 'Multijuga', marked with creamy blotches. Raised by Messrs L. R. Russell, then of Richmond, before 1904.

cv. 'VIOLACEA PLENA'.—In this the flowers are lilac, but owing to the stamens becoming transformed into petals, they lose their pea-flower shape and become rosettes. This spoils rather than improves the flower, and the plant does not blossom so freely. Introduced from Japan to the USA in the 1860s and thence to Britain about 1870 (*W. sinensis flore pleno* Carr.; *W. chinensis* var. *flore-pleno* Bean; *W. sinensis* var. *violaceo-plena* Schneid.).

Under the name 'ISSAI', and also as *W. floribunda praecox*, Mr K. Wada, the Japanese plantsman, distributed several clones which are thought in all probability to be hybrids between this species and *W. sinensis*, in which case they would belong to *W.* × *formosa*. Two of the Wada clones have been further distributed by Dutch nurseries and are described by Mr H. J. Grootendorst in *Dendroflora*, No. 5 (1968). The clone described as 'Issai' simply has short racemes of lilac blue flowers very freely borne even on young plants; leaflets mostly thirteen. In 'ISSAI PERFECT', also free-flowering, the trusses are somewhat longer (to 15 in. against 10 in. in 'Issai'), more pointed at the end; leaflets fifteen. Both flower at the same time as *W. sinensis*. Both these clones twine clockwise as in *W. floribunda*. But some plants in the Arnold Arboretum received as 'Issai' twine anti-clockwise as in *W. sinensis* (D. Wyman, *Shrubs and Vines for American Gardens* (1969), p. 568).

W. × FORMOSA Rehd.

A hybrid between *W. sinensis* and *W. floribunda* 'Alba', raised in Professor C. S. Sargent's garden at Holm Lea, Brookline, Massachusetts, in 1905. Leaflets nine to fifteen to a leaf; racemes about 10 in. long; flowers E in. long, each on a stalk $\frac{1}{5}$ to $\frac{1}{2}$ in. long, pale violet, produced in May and June. Young shoots silky downy as are the leaves at first, afterwards becoming bright green on both surfaces and glabrous above. Rehder observes that all the flowers of one raceme open at nearly the same time, unlike those of *W. floribunda* which open successively from the base upwards; also that this hybrid is superior in beauty to both parents. It was introduced to Kew in 1922.

It is possible that some reputedly inferior forms of *W. sinensis* are really hybrids of this parentage. See also 'Issai' under *W. floribunda*.

W. FRUTESCENS (L.) Poir.

Glycine frutescens L.

A deciduous climber, spreading 30 to 40 ft from its base, and enveloping trees and shrubs in its wild state; young shoots yellowish. Leaves pinnate, 7 to 12 in. long, with four and a half to seven and a half pairs of leaflets of nearly uniform size, ovate, 1½ to 2½ in. long, up to 1⅛ in. wide, slightly downy only when young. Racemes terminal on the shoots of the year, very downy, 4 to 6 in. long, the shorter ones erect. Flowers much crowded, fragrant, each about ¾ in. long, pale lilac-purple, with a yellow spot; calyx slenderly bell-shaped, ¼ in. long, downy, with five short triangular teeth, and like the flower-stalk, downy. Pods glabrous, much more cylindrical and swollen where the seeds are fixed, than in the Asiatic species. Seeds also rounder. *Bot. Mag.*, t. 2103.

Native of the southern USA as far west as Texas; introduced in 1724. It is not so strong a grower as *W. sinensis* or *W. floribunda*, nor does it ever produce so fine a display. It blooms from the latter half of June until the end of August.

cv. 'NIVEA'.—Flowers white (*W. frutescens nivea* Lescuyer; *W. frut. alba* T. Moore). In cultivation in Europe by 1854.

W. MACROSTACHYS (Torr. & Gr.) Robins. & Fern. *W. frutescens* var. *macrostachys* Torr. & Gr.; *Glycine frutescens* var. *magnifica* Herincq; *W. frut. magnifica* André.—In habit, foliage and flower characters this is very like the preceding, but is a handsomer plant and differs in the following respects: Leaflets somewhat larger; racemes 8 to 12 in. or even more long, with up to ninety flowers on each; calyx teeth longer in proportion to the tube; flower-stalk and calyx very glandular as well as hairy. Well worthy of cultivation. Native of the southern USA, extending north in the Mississippi basin to southern Illinois.

W. JAPONICA Sieb. & Zucc.

Milletia japonica (Sieb. & Zucc.) A. Gr.

A deciduous climber with slender twining stems. In the wild it climbs up bushes and small trees, which eventually become almost entirely enveloped by it. Leaves 6 to 9 in. long, composed of nine to thirteen leaflets, which are ovate, rounded, or slightly heart-shaped at the base, 1½ to 2½ in. long, ½ to ¾ in. wide, bright glossy green and glabrous below. Racemes axillary, often branched, very slender, many-flowered, 6 to 12 in. long. Flowers white or pale yellow, ½ in. or so long (the smallest of wisterias), each produced on a stalk E in. long. Calyx bell-shaped, 3/16 in. long, glabrous except for the ciliate margins, five-toothed. Pods 3 to 4 in. long, ⅓ in. wide, quite glabrous, six- to seven-seeded.

Native of Japan; introduced for Messrs Veitch by Maries, in 1878. It first flowered in August 1884 at the Coombe Wood nursery. One of the most distinct of wisterias, belonging, perhaps, to another genus, this species never appears to have had full justice done to it in this country. It is worth growing if only for the lateness of its flowers (July and August). The often branching racemes, small flowers, and almost entire absence of down, distinguish it clearly. According to Siebold, a tree enveloped by this wisteria in full flower forms a 'magnificent *coup d'oeil*, giving to vegetation an aspect of wild beauty.' The taxonomic status of this species is controversial. In some respects it approaches *Milletia*, a genus of mainly tropical climbers.

W. SINENSIS (Sims) Sw.

Glycine sinensis Sims; *W. chinensis* DC.; *W. consequana* Loud.

A strong-growing, deciduous climber, capable of covering lofty trees. The trunks of old specimens, although often decayed and hollow, attain great dimensions for a climber, and on some of the older plants in this country are over 5 ft in circumference. The branches, which are covered with silky down when young, support themselves by twining round whatever support is available. Leaves pinnate, 10 to 12 in. long, consisting usually of eleven leaflets, which are elliptical or ovate, deep rich green and glabrous above, somewhat hairy beneath, especially on the midrib, 1½ to 4 in. long, ½ to 1½ in. wide, increasing in size towards the end of the leaf. Racemes 8 to 12 in. long, produced in May from the buds of the previous season's growth. Flowers mauve or lilac-coloured, borne singly on stalks ½ to ¾ in. long, each flower about 1 in. long, pea-flower shaped, with a fine, rounded standard petal ¾ in. wide. Pod rather like that of a kidney bean, 5 or 6 in. long, club-shaped, ¾ in. wide towards the end, tapering gradually towards the base, covered with a velvety pile, and containing two or three seeds. *Bot. Mag.*, t. 2083.

Native of Central China. It is probable that even now most plants of the true *W. sinensis* descend by vegetative propagation from one that grew at Canton in the garden of a Chinese merchant whose name was rendered by European residents as 'Consequa'. He seems to have been a popular figure, and famous enough to be honoured by an obituary in *The Times*, when he died in 1833. The plant had been given to him by his nephew, who brought it from Changchow in Fukien province. John Reeves, Chief Inspector of Tea at Canton, is usually credited with its introduction, and this is true in the sense that he persuaded Consequa to propagate his plant. This he did, and two of the propagations reached Britain in May 1816. One was brought by Capt. Robert Welbank, commander of the EIC merchantman *Cuffnels*, who gave it to his brother-in-law C. H. Turner of Rooksnest Park, Godstone, Surrey, where it flowered in 1819. At that time it had already been propagated by the gardener McLeay, and plants were given by Turner to the nurseryman Loddiges and to the Horticultural Society. The other plant was brought by Capt. Richard Rawes of camellia fame, on the *Warren Hastings*; this was given to T. C. Palmer of Bromley in Kent, from whom the nurseryman Lee of Hammersmith received his stock. The retail price of the first plants to be sold was six guineas, an enormous sum for those days, but by 1835 it had dropped to a mere 1/6 to 2/6 a plant.

No climber ever brought to this country has added more to the beauty of gardens. It flowers towards the end of May, and there is frequently a second smaller crop in August. It is as remarkable for its rapid growth as for its wealth of blossom. Where wall space is available, it will extend forty yards or more from each side of the stem. In full blossom, when every twig is garnished with pale lilac flowers, few plants are so lovely. It may be used in several ways; the commonest is as a wall plant on houses, also as a pergola plant, and for covering arches. At Kew, an old specimen which up to 1860 grew on a house there, was trained over a large iron cage erected for it when the house was demolished; it was old then, but is still a fine feature. On the continent, especially in Italy, it is frequently planted so as to overrun large trees; in such a way it makes gorgeous displays there in April. When the plant has filled its destined space, it becomes necessary to prune the long, slender shoots back to within an inch or two of the older wood; otherwise it soon becomes an inextricable tangle. This is done in late summer. This wisteria may also be treated as a shrub, 5 to 8 ft high, by an annual hard pruning. Seed is only ripened in unusually hot years.

cv. 'ALBA'.—Flowers white. Introduced by Fortune for the Horticultural Society in 1846. (*W. sinensis alba* Lindl.). The wisteria once known in gardens as "*W. alba plena*" is *W. venusta*.

cv. 'PROLIFIC'.—Near to the common form, but even freer in flower and with a longer, more pointed truss. It was probably selected by P. J. C. Oosthoek and Co. of Boskoop, Holland, and distributed by the firm's successor W. J. Hooftman and by other Dutch nurseries (H. J. Grootendorst, *Dendroflora*, No. 5 (1968), pp. 63–4). It is in the British trade as 'Oosthoek's Variety'.

There seem to have been very few reintroductions of *W. sinensis* since it first reached Europe. Wilson knew of only one plant not of the Consequa provenance; this had been raised at the Arnold Arboretum from seeds received from Shanghai in 1887. He himself saw the species growing wild in W. Hupeh, but did not collect seeds. Yet inferior forms have been distributed. On the continent *W. sinensis* sets seed fairly frequently, and though it is unlikely that any reputable nurseryman ever relied on this means of propagation it may be that inferior seedlings have inadvertently been used as stock plants. An accusing finger has also been pointed at Japanese nurserymen. These rogue plants, some of which may be hybrids, either do not flower, or do so when the leaves are already fully expanded.

W. VENUSTA Rehd. & Wils.

W. brachybotrys cv. 'Alba' Ohwi; *W. brachbotrys* var. *alba* W. Mill.; *W. brachybotrys* f. *alba* (W. Mill.) Hurusawa

A deciduous climber growing 30 ft and upwards high; young shoots softly downy. Leaves pinnate, 8 to 14 in. long, composed usually of eleven leaflets, sometimes nine or thirteen; main-stalk downy. Leaflets oval to ovate, with a tapered apex and usually rounded base, $1\frac{1}{2}$ to $3\frac{1}{2}$ in. long, $\frac{1}{2}$ to $1\frac{1}{2}$ in. wide, both surfaces, but especially the lower one, softly downy. Racemes pendulous 4 to 6 in. long, 3 to 4 in. wide, the stalks densely downy. Flowers white, opening in May and June, slightly fragrant, 1 to $1\frac{1}{4}$ in. long; standard petal roundish, 1 in. wide, stained with yellow at the base. Calyx downy, cup-shaped, about $\frac{1}{2}$ in. wide, the lobes triangular or awl-shaped; flower-stalks about

1½ in. long at the base of the raceme, becoming shorter towards the end.
Pods 6 to 8 in. long, velvety. *Bot. Mag.*, t. 8811.

This wisteria, long cultivated in Japan, where it is known as 'Shira Fuji',
is a white-flowered form of a species native to that country (see f. *violacea*).
It was described in 1916 but was first seen in Britain in May 1912, when it was
shown in bloom in the Japanese section of the International Horticultural
Exhibition at Chelsea, under the probably correct name of *W. brachybotrys*, and
was introduced to Kew in the following year from the Yokohama Nursery
Company.

W. venusta used to grow luxuriantly in the garden of Hugh Wormald at
Heathfield, East Dereham, Norfolk (*Journ. R.H.S.*, Vol. 59 (1934), p. 280 and
fig. 99); ibid., Vol. 73 (1948), p. 331 and fig. 137). It received an Award of
Merit in 1945 when exhibited by Messrs Notcutt, who acquired their original
stock from Mr Wormald, and a First Class Certificate in 1948. Earlier it had been
largely imported from Japan but the plants were often badly grafted and short-
lived. It can be kept permanently in a shrubby state by shortening the long
shoots once or twice in summer and then pruning them to within an inch or two
of the base in winter. It is quite hardy and a beautiful wisteria. Besides being
larger than those of the white-flowered forms of *W. sinensis* and *W. floribunda*
its flowers are of greater substance and all open more or less simultaneously.

cv. 'ALBA PLENA'.—This has more or less double white flowers. It appears
to have been in cultivation previous to the introduction of the normal form,
as "*W. sinensis alba plena*".

f. VIOLACEA Rehd. *W. brachybotrys* Sieb. & Zucc., *nom. confus.*—This is the
normal wild form with purplish flowers, a native of Japan in the western part
of the main island and in the southern islands (Shikoku and Kyushu). The
name *W. venusta* f. *violacea* is founded on a specimen collected by Wilson in
Kyushu on Mt Kirishima in 1914, but this wild form had been found earlier by
Richard Oldham, who collected it near Nagasaki in 1863.

There seems to be really little doubt that it was the wild form of *W. venusta*
that Siebold and Zuccarini described and figured in *Flora Japonica*, Vol. I,
p. 92 and t. 45 (1839) under the name *W. brachybotrys*, from a specimen collected
near Nagasaki. This is at any rate the view of at least two leading Japanese
botanists. The identity of this wisteria was long in doubt, and since the publication of the second volume of *Plantae Wilsonianae* in 1916, Wilson's view has
been generally accepted that *W. brachybotrys* was synonymous with *W. floribunda*
and simply the wild, short-racemed form of that species. However at that time
Rehder and Wilson were apparently unaware that a wild counterpart of *W. venusta*
existed in Japan and suggested wrongly that *W. venusta* was a garden variety
of a species occurring in N. China (this is actually *W. villosa* Rehd.). It was not
until 1926 that Rehder acknowledged the existence of a wild form of *W. venusta*
in Japan, naming it *W. venusta* f. *violacea*.

If the view of Japanese botanists such as Ohwi and Makino were to be
accepted, *W. venusta* would take the name *W. brachybotrys* 'Alba' and *W. venusta*
f. *violacea* Rehd. would become *W. brachybotrys* simply. It is probably the latter
that was introduced to the Ghent Botanic Garden by Siebold in 1830, actually as

W. brachybotrys. It has generally been assumed, however, that this was *W. floribunda.*

XANTHOCERAS   SAPINDACEAE

The beautiful and interesting species described below is the only member of its genus, and one of the few hardy members of the Sapindaceae. Leaves odd-pinnate, alternate. Flowers regular, unusually large for this family. Sepals and petals five. Stamens eight. Stigma three-lobed. Fruit a leathery, thick-walled capsule, containing several large seeds. The generic name (from the Greek words for yellow and horn) refers to the long horn-like appendages of the floral disk.

X. SORBIFOLIUM Bunge   [PLATE 107

A deciduous shrub or small tree up to 20 ft high, of rather stiff, erect habit, with a large quantity of pith in the young branches. Leaves pinnate, alternate, 5 to 8 in. long, glabrous; leaflets nine to seventeen, borne on the upper two-thirds of the main-stalk, 1½ to 2½ in. long, lanceolate, deeply and sharply toothed. Flowers produced during May in erect panicles at the end of the shoots of the previous year, and from the side buds; the terminal panicle is considerably the largest, and up to 8 in. long; the side ones about half as large. Each flower is 1 to 1¼ in. across, the five white petals having a carmine stain at the base. Fruit a top-shaped capsule 2 in. wide at the top, tapering to a stout stalk at the base; it is three-valved, and as the valves open they release the rather numerous seeds, which resemble chestnuts, but are only ⅓ to ½ in. wide. *Bot. Mag.*, t. 6923.

Native of N. China; introduced by Père David, who sent a seedling to the Paris Museum in 1866, which was successfully established and had fruited by 1873. Although hardy, it is a heat-loving species which is really most at home in southeastern England, East Anglia and the E. Midlands. Where the summers are cooler than average it needs the additional heat provided by a wall to ripen its wood and to set flower-bud. Grown in the open it needs a position not subject to late frost, as the flowers come out with the young leaves. Still, it is one of the most beautiful of small trees when seen at its best. An interesting feature of the flower is the change of colour of the flare as the flower ages— at first yellow, later pink. There is a fine specimen of this species at the Garden House, Saltwood, Kent.

X. sorbifolium is best propagated by seeds, but, failing these, cuttings of roots may be used, placing them in gentle heat in April. Like other pithy stemmed shrubs and trees, the plant is rather subject to the attacks of the coral-spot fungus. Branches attacked should be cut off as soon as noticed, and burnt, the wounds being coated over with a proprietary dressing.

XANTHOCERAS SORBIFOLIUM

XANTHORHIZA   RANUNCULACEAE

A genus of a single species in Atlantic North America. Flowers in compound racemes. Sepals five, deciduous. Petals five. Pistils five to fifteen, developing into follicles, each containing a single seed. The generic name refers to the yellow roots of the species.

X. SIMPLICISSIMA Marsh.   YELLOWROOT
Zanthorhiza apiifolia L'Hérit.

A deciduous shrub with creeping roots and erect stems from 1 to 2 ft high. The handsome leaves are pinnate, consisting of three or five stalkless leaflets which are themselves deeply and irregularly toothed, 1 to 3 in. long, the basal pair two- or three-lobed. The naked base of the main leaf-stalk varies from 3 to 6 in. in length. Flowers produced in March and April, along with the young leaves in a cluster of more or less drooping panicles 3 to 5 in. long; individually the flowers are very small ($\frac{1}{8}$ to $\frac{1}{4}$ in. wide), lurid purple, petals five, triangular, pointed. *Bot. Mag.*, t. 1736.

XANTHORHIZA SIMPLICISSIMA

This interesting little shrub is a native of the eastern United States, where it extends from Pennsylvania to Florida, being most abundant in Virginia and N. Carolina; introduced to England about 1776, but, on account of its lack of any striking beauty of flower, has never become common. The foliage, however, is attractive, and the flowers are amongst the first to appear in spring. Easily increased by division in February. Spreading freely by suckers it makes a good ground cover in damp, semi-shaded places. It has also been used as a game-covert.

XYLOSMA   FLACOURTIACEAE

A genus of predominantly tropical trees and shrubs. Leaves leathery, toothed. Flowers small, unisexual, in axillary clusters. Sepals imbricated. Petals none. Stamens numerous. Fruit a few-seeded berry.

X. JAPONICA [Thunb.] A. Gr.

Apactis japonica Thunb.; *Hisingera japonica* Sieb. & Zucc.; *H. racemosa* Sieb. & Zucc., not Presl; *Flacourtia japonica* Walp.; *Xylosma racemosa* (Sieb. & Zucc.) Miq.; *X. congesta* (Lour.) Merr.; *Croton congestum* Lour.

The typical state of this species occurs in Japan, Korea and E. China. The plants raised from the seeds collected by Wilson in W. Hupeh and W. Szechwan belong to the following variety:

var. PUBESCENS Rehd. & Wils. (under *X. racemosum*).—An evergreen tree up to 80 ft high in the wild, but a small, bushy tree in cultivation; young shoots covered with short pale hairs. Young plants are armed with straight, sharp, axillary spines ½ to over 1 in. long. Leaves alternate, ovate to roundish ovate, toothed except near the base, which is rounded or broadly wedge-shaped, apex acuminate, ¾ to 3 in. long, ½ to 1½ in. wide, firm in texture, dark glossy green, glabrous on both surfaces; leaf-stalk ⅛ to ¼ in. long, downy like the shoots. Flowers unisexual, small, yellow, fragrant, produced from the leaf-axils in short racemes ¼ to 1 in. long. Fruits about the size of small peas, black-purple, the style adhering at the top.

Wilson considered this to be one of the finest evergreen trees of China, but it grows at lower elevations than most of his introductions and remains a small tree in Britain, where it is very rare and really of little worth as an ornamental.

YUCCA   AGAVACEAE

A genus of about thirty species in the New World, ranging from Central America and the West Indies northward through Mexico to the USA, as far North as California and N. Dakota in the west, and New Jersey in the east. The leaves are borne in rosettes at the apex of a woody stem which in the acaulescent species does not rise much above ground-level, in others attaining a height of 30 ft or even more. Leaves long and relatively narrow, pointed, crowded in a spherical or hemispherical head. Flowers white, creamy white or greenish, sometimes stained with pink or purple on the outside, drooping or sometimes horizontal, produced in erect panicles or racemes. They are composed of three outer segments and three inner ones, free except at the base. Stamens six, with short filaments. Ovary sessile or shortly stalked, three-chambered. Style short and thick, with three erect lobes, the stigmatic surfaces facing inward (but see *Y. whipplei*). Fruits dry or sometimes fleshy, ovoid or oblong, up to 2 or 3 in. long. Seeds dark brown or black.

We owe the generic name to John Gerard, who believed that his plant (see *Y. gloriosa*) was the 'yuca' of the Caribbean. In fact the plant so called was the common manihot or cassava. Gerard's error was pointed out by Parkinson in the *Paradisus* (1629), but the name persisted and was legitimised by Linnaeus in its present sense.

In their general aspect the yuccas are quite distinct from any other group of hardy shrubs. Their foliage is essentially of a tropical or subtropical character, which, combined with a peculiar stateliness and beauty of flower, gives the genus a unique value in gardens. They are especially suitable in formal arrangements, either isolated or in groups, and their effectiveness in flower is enhanced if a dark background can be given them.

Considering the regions of which the species described below are native, it is remarkable that they are so hardy and adaptable to our climate. The commonest species come from the coast regions of the south-eastern United States, yet they withstand 30° or 32° F of frost uninjured. Compared with wild plants, our garden ones have longer, larger leaves, but smaller inflorescences. They appear to thrive in any soil, but prefer a sandy loam in a position fully exposed to the south. In such a position they never suffer from drought, nor do they, except for a diminished crop of blossom, appear to be affected by cold, wet seasons. As the stems lengthen, they ultimately decay at the older part, and fall over by their own weight. The tops can be made to strike root by trimming off half the leaves, and placing the stem in a pot of sandy soil, giving it a place in a greenhouse until rooted. The dwarf species like *Y. flaccida* can be increased by division, and most of the species produce rhizomatous underground stems which make plants when cut off and potted.

In the wild, most species of yucca are dependent for pollination and seed-setting on the activities of yucca moths, allies of the gipsy moths. The moths emerge from the pupae some days before the flowering of the species with which they are associated. With the aid of special tentacles on its maxillae, the female gathers pollen from one flower and rolls it into a ball. It then lays its eggs in the ovary of another flower and with an extraordinary deliberateness rams the pollen-ball into the stigmatic chamber of the ovary. The grubs when they emerge feed on the ovules, but leave enough to provide an adequate supply of seeds. The yucca is entirely dependent on the services of the moth, and equally the moth could not exist without the yucca. But two species, *Y. aloifolia* and *Y. whipplei*, do not need the moth for pollination, and some species supposed to do so have set seed in European gardens. In any case, the yuccas hybridise readily when artificially cross-pollinated. It should be added that, according to Trelease, *Y. gloriosa* and *Y. recurvifolia* flower too late even in the wild to receive the services of any yucca moth, and are consequently sterile. He suspected that both might be the result of past hybridisation between fertile species, presumably *Y. aloifolia* and *Y. filamentosa* (in which Trelease included the species now known as *Y. smalliana*), and owe their perpetuation to vegetative means.

SELECT BIBLIOGRAPHY

BOWLES, E. A.—'Yuccas for English Gardens', *Journ. R.H.S.*, Vol. 47 (1922), pp. 105–9.
McKELVEY, SUSAN.—*Yuccas of the Southwestern USA*. Part 1 (1938), Part 2 (1947).
MOLON, G.—*Le Yucche*, 1914.
RUSSELL, J.—'Yuccas in Britain', *Journ. R.H.S.*, Vol. 96 (1971), pp. 491–5.
TRELEASE, W.—'The Yucceae', *Missouri Bot. Gard. 13th Report* (1902), pp. 27–133.
WEBBER, J. M.—*Yuccas of the Southwest*. Agricultural Monograph No. 17 (1953).

Y. FILAMENTOSA L. SPOONLEAF YUCCA [PLATE 109

Y. concava Haw.; *Y. filamentosa* var. *concava* (Haw.) Bak.

A low evergreen shrub, the stem of which does not rise above ground-level, and which increases and spreads by means of side-growths from the base. Leaves stiffly erect or spreading, 1 to 2½ ft long, 1½ to 4 in. wide, rather abruptly narrowed at the apex, where the margins are usually infolded. From the margins of the leaves, curly thread-like filaments 2 to 3 in. long break away, and are especially numerous towards the base. Flowers pendulous, yellowish white, 2 to 3 in. across, produced during July and August in erect, conical, glabrous panicles 3 to 6 ft high, looser and broader than in either *Y. gloriosa* or *Y. recurvifolia*. Petals rounded at the apex, then abruptly narrowed to a short tip. Style about ⅜ in. long.

Native of the coastal plain of the southeastern USA as far north as southern New Jersey, in sand-dunes, waste ground and pine woodland; in cultivation by the second half of the 18th century and probably earlier, but the splitting of *Y. filamentosa* as once understood into two species makes its early history in gardens uncertain. This is a very hardy and beautiful yucca, forming low tufts from which the stately panicles spring in profusion. It should be planted in broad masses with, if possible, a dark, evergreen background. It flowers in a small state. Easily propagated by division. Some plants grown as *Y. filamentosa* are *Y. smalliana* (see below) and in the past there has been confusion between *Y. filamentosa* and *Y. flaccida*.

cv. 'VARIEGATA'.—Leaves with a well defined margin of white, becoming duller and pink-tinged. Centre of leaf dark glaucous green, streaked and edged with lines of paler green. In cultivation by the 1860s. It is more tender than the normal form and in Victorian times was often grown under glass, where the variegation was purer, and the leaves less rigid.

Y. SMALLIANA Fern. *Y. filamentosa sens.* Small (and of other authors, in part), not L.—Allied to the preceding, but with thinner, flatter, narrower leaves $\frac{7}{8}$ to $1\frac{3}{4}$ in. wide, long-tapered at the apex. Main axis of panicle downy. Flowers to 2 in. long; petals gradually acuminate at the apex. Style very short or none. Native of the southeastern USA as far north as N. Carolina, west to Louisiana and Tennessee. Some plants grown as *Y. filamentosa* belong here, but so far as is known there is no difference in hardiness, nor much in horticultural value.

The American botanist J. K. Small was the first to recognise that *Y. filamentosa* as understood by earlier authorities really comprised two species. But he assumed that the true *Y. filamentosa* of Linnaeus was the species described above, and took up the name *Y. concava* Haworth for the other. However, an examination of the type-specimen of *Y. filamentosa*, preserved in the herbarium of the British Museum of Natural History showed that *Y. concava* Haw. was the true *Y. filamentosa*. The other species therefore needed a new name and this was provided by Fernald in 1944 (*Rhodora*, Vol. 46, pp. 5–8). The specimen was collected by Clayton in eastern Virginia and was described by Gronovius in *Flora Virginica* (1739) under a phrase-name. The name *Y. filamentosa* L. is based on this.

Y. FLACCIDA Haw.

Y. filamentosa var. *flaccida* (Haw.) Engelm.; *Y. puberula* Haw.; *Y. meldensis* Engelm.

A low evergreen shrub, whose stem, like that of *Y. filamentosa*, does not arise above ground-level, spreading by sucker growths. Leaves 1 to $1\frac{1}{4}$ ft long, 1 to $1\frac{1}{2}$ in. wide, green or glaucous, and bent downwards above the middle, long-pointed with straightish, thread-like fibres separating from the margin, and 2 in. or more long. Flowers as in *Y. filamentosa*, but borne on a downy, shorter panicle. Seeds dull, $\frac{1}{3}$ in. long, produced in a capsule 2 to 3 in. long.

Native of the southeastern USA from N. Carolina to Alabama, with a more inland distribution than *Y. filamentosa* and *Y. smalliana*; introduced from Georgia and described by Haworth in 1819 from plants cultivated in his garden at Chelsea. The bent back apices of the leaves and the straighter marginal threads distinguish it from the two species mentioned, but its affinity is with *Y. smalliana* rather than with *Y. filamentosa*, having the leaves gradually tapered to the apex as in that species, and a usually downy panicle. It is rather more vigorous than either of its allies, and perhaps freer flowering, but its lax leaves can be damaged by wind.

cv. 'GOLDEN SWORD'.—Leaves banded with creamy yellow at the centre.

f. INTEGRA Trel. *Y. glauca* Sims, *Bot. Mag.*, t. 2662, not Nutt.—Leaf margins without fibres; leaves smaller. Flower-stalks glabrous.

cv. 'IVORY'.—A very free-flowering clone, selected by the late Rowland Jackman from a batch of seedlings. Unlike some forms of *Y. flaccida* the flowers are poised horizontally, as they were in the original introduction. It received an Award of Merit in 1966 when shown by S. M. Gault, who made this yucca a feature of the Queen Mary's Garden in Regent's Park. It received a First Class Certificate two years later.

var. MAJOR (Bak.) Rehd. *Y. orchioides major* Bak.; *Y. flaccida* var. *glaucescens* (Haw.) Trel.; *Y. glaucescens* Haw.—Leaves glaucous, broader, rather more erect. Panicle very downy. Petals more attenuate. According to Trelease, this is the form of *Y. flaccida* commonly cultivated in American gardens. It was introduced to Britain in 1816.

cv. 'ORCHIOIDES'.—A depauperate form with an unbranched inflorescence; leaves stiffer and more erect than normal. A seedling raised in France (*Y. orchioides* Carr.; *Y. flaccida* f. *orchioides* (Carr.) Trel.).

Y. GLAUCA Nutt.
Y. angustifolia Pursh

An evergreen shrub with a low, often prostrate stem carrying a hemispherical head of leaves 3 to 4 ft across. Leaves narrow linear, 1 to 2½ ft long, ½ to ¾ in. wide, tapering to a long fine point, of a glaucous green, the margins white, and beset with a few threads. Raceme, erect, 3 to 4½ ft high, rarely branched. Flower dull greenish white, 2½ to 3 in. long, pendulous. *Bot. Mag.*, t. 2236.

Native of the western central USA from Texas and New Mexico north to Kansas, Nebraska and southern Wyoming; it was discovered by Thomas Nuttall, and introduced by him in 1811, when he visited England, bringing with him specimens, seeds and bulbs of some of the plants he had found during his journey from the Great Lakes to the Missouri and then south to New Orleans. His description was published in 1813 in a catalogue of Fraser's nursery.

Y. glauca is quite hardy at Kew and Edinburgh, but does not flower with the freedom and regularity of *Y. gloriosa* and *Y. recurvifolia*. Neither is it so striking, being of a pale green, rather than truly white. Still, it is quite handsome.

var. STRICTA (Sims) Trel. *Y. stricta* Sims *sec.* Trel.; *Y. glauca* var. *gurneyi* McKelvey—A robust variety with a more branched inflorescence, described from plants collected by James Gurney in Seward County, Kansas, and brought into cultivation in the Missouri Botanic Garden. Trelease identified this variety with the yucca named *Y. stricta* in *Bot. Mag.*, t. 2222 (1821). This was said to have been raised from seeds sent by John Lyon in 1816 from the Carolinas, where *Y. glauca* does not occur. Either Lyon obtained the seeds from a collector who had visited the area of *Y. glauca*, or Sims was mistaken and the plant had in fact been raised from the seeds of *Y. glauca* brought to England by Nuttall.

Y. × KARLSRUHENSIS Graebener—A hybrid of *Y. glauca* (seed-parent), raised by Graebener, head gardener to the Archduke of Baden at Karlsruhe and described in 1903. *Y. filamentosa* was stated to be the other parent, but the rather lax foliage suggests that it may have been *Y. flaccida*. It is little known in this country, but is valued in Germany and Central Europe. According to the original description the leaves are glaucous, about ⅝ in. wide; flowers white with a reddish stain on the outside. For an illustrated note on this hybrid by Camillo Schneider, see *New Fl. and Sylv.*, Vol. 1 (1928), p. 34 and fig. x.

A cross between *Y. glauca* (seed-parent) and *Y. filamentosa* was made by

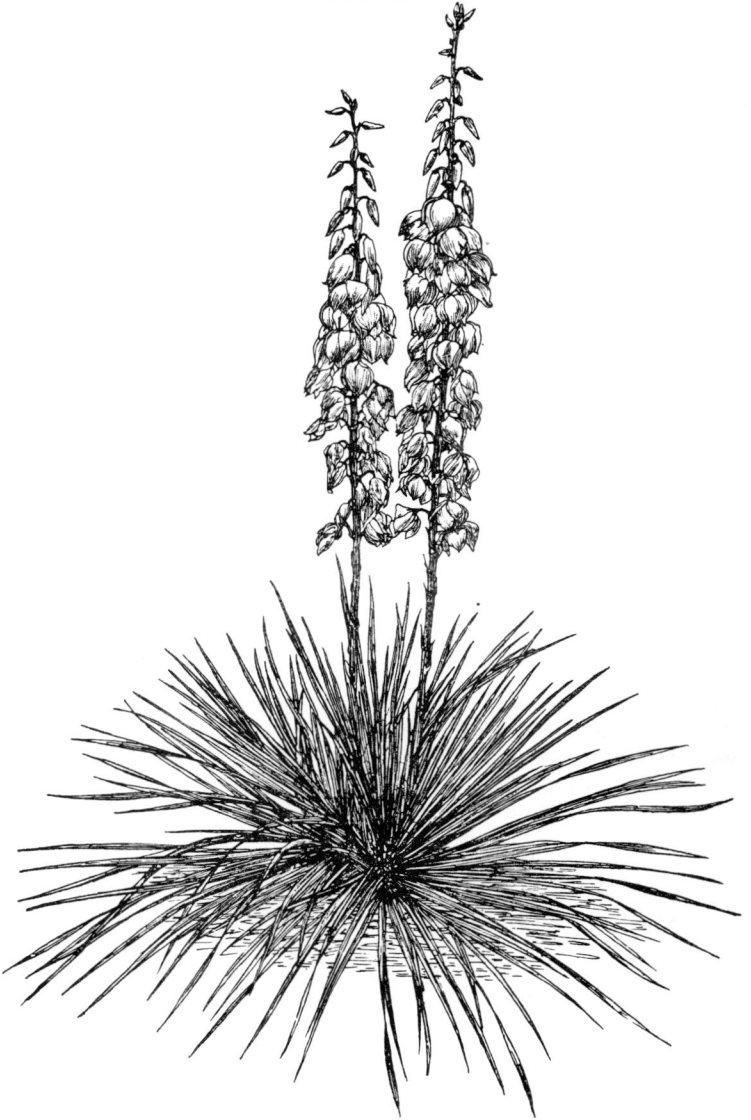

Yucca glauca

Thomas Javit at St Etienne, France, in 1921. The plants first flowered with him in 1927 and two years later in the Vilmorin nurseries at Verrières, to which he had given some of the seed. A merit of this cross was that the inflorescences were almost as wide as in *Y. filamentosa*, but produced earlier (*Gard. Chron.*, Vol. 87 (1934), p. 66 and fig. 27).

Y. ANGUSTISSIMA Engelm. ex Trel.—Near to *Y. glauca* but with more slender leaves, up to barely ¼ in. wide. The flower-spike is racemose as in that species, but shorter. Flowers white, with a pale green or white, oblong style (dark green and swollen in *Y. glauca*). Native of the southwestern USA, common in northwestern Arizona. It is hardy.

Y. GLORIOSA L. [PLATE 108

An evergreen shrub up to 6 or 8 ft high in this country, sometimes branched, but oftener consisting of a single, thick, fleshy stem, crowned with a cluster of numerous stiff, straight, spine-tipped leaves 1½ to 2 ft long, by 2 to 3 in. wide, glaucous green when young, quite glabrous. Flowers produced from July to September, crowded on an erect, narrowly conical panicle, 3 to 8 ft high, and 1 ft wide. The flowers are pendulous, creamy white, sometimes tinged with red or purple outside, the six parts of the perianth oblong-lanceolate and pointed. The fruit is an oblong capsule 2 to 2½ in. long, six-ribbed; seeds glossy, ¼ in. long.

Native of the coast region of eastern N. America, from S. Carolina to N.E. Florida, often on sand-dunes. It was cultivated by Gerard in his garden at Holborn late in the 16th century*, and has long been a favourite in the gardens of south and western Britain. Even now, in the 20th century, gardens can show no more striking a feature than a group of plants in flower. It is closely allied to *Y. recurvifolia*, but easily distinguished by the straight, rigid leaves. It is not so common as that species, and although quite hardy in not being affected by frost, is apparently more subject to decay and injury by winter damp and snow. Also, as Parkinson put it, it flowers 'now and then, but not every yeare' and is inclined to develop its spikes so late in some seasons that the flowers fail to open.

cv. 'MEDIO-STRIATA'.—With a whitish stripe down the centre of each leaf.

cv. 'NOBILIS' ('Ellacombei').—Leaves persistently glaucous, the outer ones recurving and sometimes twisted on one side, not or scarcely ribbed. Petals red

---

* Gerard obtained his plant from Thomas Edwards, an apothecary of Exeter. Its original provenance is unknown, but the likeliest guess is that it came from Roanoke Island, which had been explored in the 1580s and was the site of the first but short-lived English settlement in what is now the United States. Gerard's plant had not flowered by the time he published his *Herball*, but it did so later. It died when an attempt was made to move it to another garden after his death. But he had given a propagation to Robin, the French king's gardener. Parkinson had a plant, and gives a good account of the species in his *Paradisus*.

on the back. It approaches *Y. recurvifolia* in habit. The original plant was said by Canon Ellacombe to have been obtained from Loddiges' nursery at Hackney by his father. It is a fine yucca, perhaps from *Y. gloriosa* crossed with *Y. recurvifolia* (*Y. ellacombei* Bak.; *Y. gloriosa nobilis* Carr.).

cv. 'SUPERBA'.—Leaves very rigid, glaucous, 2 to 2¼ in. wide. Panicle denser and shorter than normal. Flowers white. *Bot. Reg.*, t. 1698 (*Y. superba* Haw.; *Y. gloriosa* var. *superba* (Haw.) Bak.

var. PLICATA (Carr.) Engelm.   *Y. gloriosa plicata* Carr.—Leaves glaucous, very concave, infolded at the tips. This variety occurs in the wild.

cv. 'VARIEGATA'.—Leaves striped with dull yellow.

Y. × VOMERENSIS Sprenger—A hybrid between *Y. gloriosa* (pollen-parent) (not *Y. recurvifolia* as stated by Molon) and *Y. aloifolia*, raised by Karl Sprenger at Vomero near Naples early this century. Trunk as in *Y. gloriosa*. Leaves 2 to 2½ ft long, about 1¾ in. wide, of thicker texture than in *Y. gloriosa*. Flowers pendulous, greenish white with a pink flush on the outside, borne throughout most of the length of a stem 6 to 8 ft high; panicle broad. According to Sprenger, a beautiful and vigorous yucca, flowering several times during the summer (at Vomero). It is not certain whether this hybrid is now in cultivation in Britain. No confirmation has been found for the statement that Sprenger distributed several of his hybrids as *Y. × vomerensis*, though it is very likely that the name was misused in other gardens for Sprenger's productions in general.

A plant under the tentative name of *Y. × vomerensis* 'East Lodge Variety' received an Award of Merit on 15 July 1941 when exhibited by W. B. Cranfield of East Lodge, Enfield Chase.

Y. ALOIFOLIA L.—Mentioned above as a parent of *Y. × vomerensis*, this is an arborescent yucca with a slender trunk up to 25 ft high, though usually much shorter. Leaves rigid, flat, spine-tipped, edged with horny teeth, 1 to 1¾ in. wide. Flowers creamy white, usually tinged with green or purple outside, borne in broad, compact panicles near to the leaves. Unlike most yuccas it does not need the pronuba moth for fertilisation of its flowers, and fruits freely in the Mediterranean region, where it is naturalised in some areas. Native of the West Indies and the southeastern USA; introduced to Europe in the 17th century. It is hardy only in the mildest parts. The var. DRACONIS (L.) Engelm. (*Y. draconis* L.) is a variant with laxer, rather broad leaves. The epithet *aloifolia* refers to the stoutness of the leaves, resembling those of an aloe.

Y. RECURVIFOLIA Salisb.   [PLATE 110

Y. recurva Haw.; *Y. gloriosa* var. *recurvifolia* (Salisb.) Engelm.; *Y. pendula* Groenland

An evergreen shrub up to 6 or 8 ft high, more or less branched. Leaves at first glaucous, 2 to 3 ft or even more long, 1½ to 2¼ in. wide, tapering to a fine stiff point, all but the upper leaves much recurved. Flowers creamy-white, 2 to 3 in. across, in an erect panicle 2 to 3 ft high, not so tall nor standing so clear of the leaves as in *Y. gloriosa*, and with the flowers more loosely arranged; the parts of the flower as in *Y. gloriosa.* Fruit 2 to 2½ in. long; seeds not glossy, about ⅓ in. long.

Native of the coast region of the south-eastern United States, especially of Georgia; introduced in 1794. This is the commonest and most easily cultivated of yuccas, and although not so striking as *Y. gloriosa* in flower, is a more graceful plant and hardier—or, at any rate, resists snow and damp better. It flowers in late summer, and withstands the smoke of London admirably. It associates well with a formal arrangement of paths and lawns, and gives a very pleasing exotic effect.

cv. 'ELEGANS MARGINATA'.—Leaves bordered with pale yellow. This came into cultivation early in the 1870s and was considered to be one of the finest of the variegated yuccas, but is now uncommon.

cv. 'VARIEGATA'.—Leaves with dull yellow striations down the centre.

Y. 'VITTORIO EMMANUELE II'.—Leaves bluish green, 2 to 2½ ft long and about 2¼ in. wide, tapered at the apex, rigid, borne on an eventually tall, branched trunk. Flowers in July and August, red in the bud, white inside, campanulate, borne in broad panicles up to 6 ft long, the lowermost flowers near the leaves. Petals acuminate. A hybrid between *Y. recurvifolia* and *Y. aloifolia* 'Purpurea', raised by Karl Sprenger at Vomero, where it first flowered in 1901. He considered it to be the finest of his hybrids. It is figured in Molon, *Le Yucche,* frontispiece.

Y. RUPICOLA Scheele

A nearly stemless plant, consisting above ground mainly of a dense rosette of leaves, which are 2 to 2½ ft long, 1 to 1½ in. wide, pale glaucous green, the margins finely toothed, cartilaginous, and yellowish. Flowers in a much-branched, glabrous panicle 4 to 6 ft high, the branches slender, semi-erect. Flowers pendulous, somewhat bell-shaped, milky white; the three outer parts of the perianth oblong, ¾ in. wide; the inner ones broader (1 in. wide); all 2¼ in. long, and pointed. *Bot. Mag.,* t. 7172.

Native of the southwestern USA; introduced about 1850. It flowered with Canon Ellacombe at Bitton in 1890 and was hardy with E. A. Bowles at Myddel-ton House, Enfield, but is now rare in cultivation. It grows outside at Edinburgh and flowers regularly in several East Lothian gardens. It is allied to *Y. glauca.*

Y. WHIPPLEI Torr.   OUR LORD'S CANDLE
Hesperoyucca whipplei (Torr.) Trel.

An evergreen, mostly stemless shrub, producing from a rootstock a hemispherical rosette of much crowded leaves up to 6 ft in diameter. Leaves 1 to 3 ft long, ½ to 1¼ in. wide at the base, terminated by a sharp slender spine, margins very finely toothed; the whole leaf is rather glaucous. Flowers pendent, fragrant, closely packed on the upper part of a perfectly erect, stout stem 8 to 15 ft high and 3 to 5 in. in diameter at the base; the inflorescence itself as much as 7 ft long and 1 to 2 ft wide. The six segments of the perianth are ovate-lanceolate, pointed, greenish-white tipped and edged with purple, more or less incurved, and give the flower a diameter of $2\frac{1}{2}$ to 3 in. *Bot. Mag.*, t. 7662.

A native mainly of southern California, in the coastal ranges south of San Francisco, extending inland to the borders of the Mojave desert; also reported from the Mexican State of Baja California and from Arizona. It was known to the Spanish missionaries in California, but was first described from specimens collected during Lt Whipple's exploration for a railway route from the Mississippi to the Pacific in 1853–4. *Y. whipplei* is a splendid species, surpassing all other yuccas, Sargent remarks, in the height and beauty of its panicles. 'From day to day the waxen tapers on the distant slopes increase in height as the white bells climb the slender shafts. At length each cluster reaches its perfection, and becomes a solid distaff of sometimes two—yes, even six—thousand of the waxen blossoms' (Mary Parsons, *The Wild Flowers of California* (1904), p. 70).

Y. whipplei first flowered in Britain, under glass, with Mr Peacock at Hammersmith, in 1876. So far as is known, its first flowering in the open air in this country was with Mr Fletcher at Aldwick Manor near Bognor in 1910 (*Gard. Chron.*, Vol. 51 (1912), Feb. 17, supplementary illustration and p. 106); the plant had been received some six or eight years earlier. A very magnificent example flowered with W. M. Christie at Watergate near Chichester in 1921 (*Journ. R.H.S.*, Vol. 47 (1922), fig. 23).

Y. whipplei is nearly hardy, and probably more plants have been killed by winter wet than by frost. At Bodnant in North Wales, where it has flowered several times, the plants are given no more than overhead protection in winter, to keep rain out of the crown. But all the recorded flowerings of *Y. whipplei* in Britain are from gardens near the south or west coasts. Yuccas need warmth in late summer and autumn if they are to form their embryonic flower-spikes and it may be that *Y. whipplei* needs more warmth than most other hardy species at that time. A plant at Borde Hill in Sussex vegetated for thirty years before flowering, which at least says something for its hardiness. At Bodnant, however, a plant from home-raised seeds set in 1944 flowered in 1951.

In its typical state *Y. whipplei* is monocarpic, i.e., the whole plant dies after flowering, and it seems that most plants grown in Britain are of this nature. Fortunately this species is self-fertile and produces good seed in this country, though artificial pollination is advisable, to secure a good set. But some wild plants are perennial and the following varieties or subspecies of these have been distinguished:

var. CAESPITOSA M. Jones *Y. w.* subsp. *caespitosa* (M. Jones) Haines—This forms large clumps and several rosettes may bear flowers simultaneously. This variety is confined to hot and dry localities bordering the Mojave desert and would probably be unsuitable for British conditions.

var. INTERMEDIA (Haines) Webber *Y. w.* subsp. *intermedia* Haines—In this variety only one rosette flowers in each season, but flowering stimulates the formation of adventitious growth-buds beneath the spike, from which new rosettes emerge. Leaves to 3 ft long. This occurs in Los Angeles and Ventura Counties.

var. PERCURSA (Haines) Webber *Y. w.* subsp. *percursa* Haines—This spreads by rhizomes, forming a lax clump. Only one rosette produces a spike, but the younger rosettes perpetuate the plant. Leaves about 1¾ ft long. Monterey and Sta Barbara Counties.

Y. whipplei differs from other yuccas in having a slender, conic-cylindric style rising abruptly from the top of the ovary and enlarging into a capitate, papillate stigma—or as J. G. Baker put it: into 'a stigma shaped like the top of a music-stool, and encrusted all over with white crystalline papillae, like those of an Ice-plant.' Another distinction lies in the glutinous pollen, and there are certain differences in the structure of the capsule. The anomalous characters of this species were recognised by Dr Engelmann, who proposed for it the rank of subgenus, named *Hesperoyucca*. This was raised to generic rank by Trelease in 1893. But other botanists have continued to include it in *Yucca* as a monotypic section or subgenus.

ZANTHOXYLUM RUTACEAE

A widely spread genus of shrubs and trees belonging to the Rue family, of which some half a dozen hardy species are in cultivation. Their leading characteristics are the strong, aromatic, sometimes unpleasant odour of the crushed leaves, the spiny young branches and leaf-stalks, the trifoliolate or pinnate, alternate leaves, the small, mostly unisexual flowers, the two-valved roundish capsules which split downward, and the shining black or blue seeds which, after the bursting of the capsules, often remain for some time attached by a short thread. For the two sections of the genus see under *Z. schinifolium*.

These species are not in the first rank of ornamental shrubs, but well-grown specimens are handsome in foliage. They like a good deep soil, and are best propagated by seeds; when these are not available they may be increased by cuttings made of the young wood in July, or of the roots in spring. The fruits and seeds of some species have a pungent pepper-like taste and are used as a condiment, and the bark contains a powerful stimulant and tonic principle sometimes employed in medicine. The generic name, sometimes spelt *Xanthoxylum*, refers to the yellowness of the wood of some species.

Z. AILANTHOIDES Sieb. & Zucc.
Fagara ailanthoides (Sieb. & Zucc.) Engl.

A deciduous, dioecious tree 50 to 60 ft high in Japan, the branchlets very stout, glabrous, densely set with short, stiff spines. Leaves pinnate, variable in size, normally to about 18 in. long on adult trees, but up to 3 ft long in the juvenile state; lateral leaflets in five to eleven pairs, 2 to 5 in. long, ovate or ovate-lanceolate, finely toothed, smooth, dark green. Flowers in terminal cymes up to 5 in. across; sepals and petals minute, the latter greenish yellow. Stamens yellowish. Seeds black, compressed and tapering at one end.

A species of wide distribution in E. Asia, from Formosa through Japan to S. Korea, and eastern China. Often introduced, it is tender, but very handsome and vigorous when young.

Z. AMERICANUM Mill.   PRICKLY ASH, TOOTHACHE TREE
Z. fraxineum Willd.

A spreading, round-headed, deciduous shrub, usually 6 to 10 ft high in this country, but capable of growing twice as high; young shoots brown, downy, becoming smooth and grey with age; armed with stiff spines ½ in. or less long, in pairs. Leaves pinnate, 6 to 8 in. long with usually five to eleven, but sometimes thirteen, leaflets, often with one or two spines on the main-stalk where the leaflets are attached. Leaflets 1½ to 2½ in. long, ovate or oval, downy beneath especially on the midrib, minutely or not at all toothed. Flowers crowded at the joints of the previous season's shoots, very small, yellowish green. Fruit a blackish, fragrant, two-valved capsule; seeds black and shining.

Native of the eastern United States; introduced during the middle years of the 18th century. This shrub is said to have been at one time common in gardens; it is no longer so. The bark and capsules have a pungent, acrid taste, and one of the popular names is given because they have been chewed to alleviate toothache. It is very easily distinguished from the other species here included by the very downy under-surface of the leaves.

Z. PIPERITUM (L.) DC.   JAPAN PEPPER
Fagara piperita L.

A compact, rounded, deciduous shrub; young shoots more or less downy when young, armed with flattish spines ½ in. long arranged in pairs at each node. Leaves pinnate, from 3 to 6 in. long, with eleven to twenty-three leaflets, the main-stalk downy, having a few small spines on the lower side, and slightly winged. Leaflets ¾ to 1½ in. long, ovate, stalkless, toothed, with an occasional prickle on the midrib which is also downy above, dark green, but often yellow in the centre when young. Flowers in panicles 2 in. long at the end of short axillary twigs, small, green. Fruits reddish, dotted with glands. Seeds black, about the size of large shot.

Native of China and Japan, this shrub is, on the whole, the prettiest of these hardy species. Its neat, bushy habit and graceful foliage consisting of numerous small leaflets render it quite distinct among hardy shrubs. It most nearly resembles *Z. schinifolium*, but is easily distinguished by having its spines in pairs and flowers with a single whorl of segments (in *Z. schinifolium* the perianth is differentiated into sepals and petals). The seeds when ground are used by the Japanese as pepper.

ZANTHOXYLUM PIPERITUM

Z. PLANISPINUM Sieb. & Zucc.

Z. alatum var. *planispinum* (Sieb. & Zucc.) Rehd. & Wils.; *Z. alatum* f. *subtrifoliolatum* Franch.

A deciduous shrub up to 12 ft high, with glabrous, spiny branches; spines in pairs, thin, broad and flat at the base, $\frac{1}{4}$ to $\frac{3}{4}$ in. long, shining. Leaves 5 to 10 in. long, trifoliolate or pinnate, with usually three or five, rarely seven stalkless leaflets, the main-stalk distinctly winged, often $\frac{3}{8}$ in. wide. Leaflets increasing in size towards the end of the leaf, the terminal one largest and as much as 5 in. long; others are only half as long; ovate or lanceolate, finely toothed, acuminate. Flowers yellowish, in small panicles $\frac{1}{2}$ to $1\frac{1}{2}$ in. long produced from the leaf-axils in spring. Fruit red, warted; seeds black, shining, about the size of large shot. *Bot. Mag.*, t. 8754.

Native of Japan, Korea, Formosa and China; in cultivation by the 1870s. It is easily recognised among the other hardy species by its very distinctly winged rachis and broad spines. Although deciduous, it will in mild seasons retain its leaves up to Christmas, fresh and green. After a hot summer it bears the red fruits freely, and is then very handsome. The fruiting spray depicted in the *Botanical Magazine* was sent by Canon Ellacombe from his garden in Gloucestershire in December 1914 and shows how the leaflets of this species roll their margins inwards during cold weather. It is almost hardy, suffering only in severe winters.

Z. SCHINIFOLIUM Sieb. & Zucc.

Fagara schinifolia (Sieb. & Zucc.) Engl.; *F. mantchurica* (Benn.) Honda; *Z. mantchuricum* Benn.

A deciduous shrub, whose glabrous branches are armed with solitary spines up to ½ in. long. Leaves pinnate, 3 to 7 in. long, spiny on the main-stalk, and composed of eleven to twenty-one leaflets, which are ¾ to 1½ in. long, lanceolate, shallowly toothed, nearly or quite glabrous, deep green above, paler beneath. Flowers in a terminal flattish cluster, 2 to 4 in. across; each flower about ⅛ in. across. Fruit green; seeds blue.

Native of China, Korea and Japan. It very much resembles *Z. piperitum* in leaf, but differs in its spines being solitary (not in pairs) and in bearing its flowers in wider inflorescences in August at the ends of long branchlets. For the difference in flowers, see *Z. piperitum*. The two species belong in fact to different sections of the genus, *Z. schinifolium* belonging, like *Z. ailanthoides*, to the section *Fagara*, the other species treated all belonging to the section *Zanthoxylum*.

Z. SIMULANS Hance

Z. bungei Planch., *nom. nud.*

A deciduous bush of graceful, spreading habit 10 ft or more high; said sometimes to be a small tree over 20 ft high. Branchlets downy or glabrous, armed with broad, flat spines ¼ to ¾ in. long. Leaves pinnate, 3 to 5, sometimes 9 in. long, aromatic; leaflets seven to eleven, broadly ovate, ½ to 2 in. long, slightly toothed; there are often a few spiny bristles on the upper surface, also on the midrib below; the main-stalk is armed beneath with short spines, also above, where the leaflets are attached. The inflorescence is a small panicle produced at the end of short, axillary twigs. Fruits reddish, with dark dots.

Native of China; introduced to Kew in 1869. One of the hardiest of the genus.

ZAUSCHNERIA   CALIFORNIA FUCHSIA
ONAGRACEAE

A genus of four species in western North America (mainly in California) and northwest Mexico; they are herbaceous perennials but two become woody at the base and are usually classified as shrubs. Leaves more or less sessile, the lower ones opposite, the upper alternate. Flowers similar in most characters to those of the allied *Fuchsia*. Ovary globose, enclosed in the base of the floral tube (receptacle), which narrows above the ovary and then expands funnelwise. Sepals and petals four, spreading. Stamens eight. Fruit a capsule, resembling that of *Epilobium*. Seeds numerous, with a tuft of hairs at one end. The generic name commemorates Dr M. Zauschner, Professor of Natural History in the University of Prague (*d.* 1799).

The most recent study of the genus is: Clausen, Keck and Hiesey, 'The Genus *Zauschneria*', Carnegie Inst. Washington Publication 520, pp. 213–59 (1940), a work based on experimental cultivation and cytological study (botanical treatment and key by D. D. Keck.) See also: Munz, *A California Flora* (1959), pp. 926–7.

Z. CALIFORNICA Presl

?*Z. mexicana* Presl; *Z. californica* subsp. *angustifolia* Keck

A subshrub to about 3 ft high. Leaves opposite to alternate, sessile, lanceolate, narrow-lanceolate, oblong or linear, $\frac{3}{16}$ to $1\frac{5}{8}$ in. long, to about $\frac{3}{16}$ in. wide, downy, sometimes densely so, on both sides, or the underside woolly, entire or distantly toothed. Flowers in terminal, bracted spikes. Receptacle scarlet, up to $1\frac{5}{8}$ in. long, funnel-shaped above the ovary, with spreading, triangular lobes, which are up to $\frac{1}{2}$ in. long. Petals coloured like the receptacle, inserted at the mouth of the tube, deeply notched, spreading. For other characters, see generic introduction. *Bot. Mag.*, n.s., t. 19.

Native of California and probably of Mexico; introduced in the 1840s. Clausen and his co-workers (see reference above) have shown that *Z. californica*, as usually understood, is a tetraploid species of hybrid origin, deriving from the two diploid species *Z. cana* and *Z. septentrionalis* (see below). *Z. californica* is variable, and some forms so closely approach one or other of the parents that they cannot be named with any certainty without knowledge of their chromosome-number. It is probable, however, that most of the plants cultivated as *Z. californica* do belong to that species, and those grown as *Z. mexicana* also, but there is need for further investigation. Numerous zauschnerias are or have been in commerce, differing little in colour of flower, but varying in habit, and in the relative width and the indumentum of the leaves; also—an important consideration—in flowering-time, for some forms of *Z. californica* produce their inflorescences so late that the flowers do not open in a dull, wet autumn.

Z. californica, in its dwarfer forms, is suitable for the rock garden, and is nearly hardy in a well-drained soil. Cuttings taken in late summer and over-wintered in a frost-free house or frame will flower the following season.

subsp. LATIFOLIA (Hook.) Keck *Z. californica* var. *latifolia* Hook.; *Z. latifolia* (Hook.) Greene—Not very clearly demarcated from the typical subspecies, but said to be herbaceous, to about 2 ft high, with elliptic or broadly ovate leaves up to $\frac{5}{8}$ in. or slightly more wide, grey or green. To this subspecies probably belonged the plant that received an Award of Merit when shown by Tom Hay from Hyde Park in 1928, though this grew to 4 ft high.

Z. CANA Greene *Z. californica* var. *microphylla* A. Gr.; *Z. microphylla* (A. Gr.) Moxley—A subshrub to about 2 ft high. Leaves grey and densely indumented, linear or almost thread-like, to $\frac{1}{12}$ in. wide. Flowers smaller than in *Z. californica*, to $1\frac{1}{2}$ in. long. The plant that received an Award of Merit in 1928 when shown by Tom Hay as *Z. microphylla* was probably the true *Z. cana*.

Z. cana is one of the diploid parents of *Z. californica* (see above) and is difficult to distinguish from narrow-leaved forms of that species. The other parent of *Z. californica*—*Z. septentrionalis* Keck—is herbaceous, and not treated here. It occurs in the Redwood region north of San Francisco.

ZELKOVA   ULMACEAE

Nearly allied to the elms, the four species of *Zelkova* in cultivation are amongst the most interesting and handsome of hardy trees. They have smooth, beech-like trunks with a scaling bark, and deciduous, alternate, coarsely toothed, feather-nerved leaves, usually harsh to the touch like those of elm. Flowers unisexual; both sexes produced on the same twig, the males at the base, the females solitary or few in the leaf-axils above them; both sexes small, green, and of no beauty. Seed-vessel roundish, $\frac{1}{6}$ to $\frac{1}{4}$ in. long, with the calyx adhering at the base, slightly horned at the top.

The zelkovas should be grown in deep, moist, loamy soil where the position is moderately sheltered. *Z. carpinifolia* is the best known of them, and appears to be adapted to all but the most inclement parts of Britain. Both it and *Z. serrata* should be raised from imported seed, although they can probably be grafted on elm, as are the other two species.

The generic name derives from the local word for *Z. carpinifolia*. An earlier name for the genus is *Abelicea*, for the origin of which see *Z. abelicea*, but *Zelkova* has been conserved.

Z. ABELICEA (Lam.) Boiss.

Quercus abelicea Lam.; *Ulmus abelicea* Sm.; *Planera abelicea* (Sm.) Schultes; *Z. crenata* var. *cretica* Spach; *Z. cretica* Spach

A small, much branched tree or tall shrub, mostly 15 to 20 ft high, rarely attaining 50 ft in the wild; young branchlets densely white-downy or white-tomentose. Leaves crowded on the short lateral branches, more widely spread on the longer shoots, very shortly stalked, ovate or oblong-ovate, cordate, truncate or rounded at the base, obtuse at the apex, coarsely toothed, with four or five (rarely six) teeth on each side, mostly $\frac{3}{8}$ to 1 in. long and $\frac{5}{16}$ to $\frac{3}{4}$ in. wide, sometimes to $1\frac{3}{8}$ in. long and $\frac{3}{4}$ in. wide, dark green and with scattered hairs above, lower surface whitish or light green, downy (densely so along the midrib and main veins). Fruits clustered, rounded, or deeply two- or three-lobed, about $\frac{3}{16}$ in. long, brown, downy.

Native of Crete (there is one unsubstantiated record from Cyprus). It is allied to *Z. carpinifolia*, but with much smaller leaves edged with fewer teeth. Some specimens of *Z. carpinifolia* from Iran have small leaves and hairy young shoots and leaves, as in *Z. abelicea*, but the number of marginal teeth is greater (nine to eleven).

The first account of this species was contained in a letter to Clusius (Charles d'Escluse) from Honorius Bellus (Onorio Belli), written from Cydonia (Kandia) in October 1594, and published in Clusius' *Rariorum Plantarum Historia* (1601). Bellus gave a brief description of the plant, which he called '*Abelicea*' from its Greek name *apelikea*. He mentioned that the scent of the wood and sawdust resembled that of sandalwood, 'whence it may be called Pseudosantalus Creticus.' This name was taken up by Caspar Bauhin in his *Pinax* (1623) and was the basis on which Spach named the plant *Zelkova cretica* in 1842. But Lamarck had named it *Quercus abelicea* in 1785. The epithet *abelicea* was the first to be applied to the plant under the Linnean system of nomenclature and is the one that must be used.

In his letter (see above) Bellus said he was sending seeds, but Clusius makes no reference to them. The first recorded introduction to Britain was in 1929, when Major Bonakis collected plants in the Levka Mountains, which were flown to England from Soudhas Bay by a flying-boat of Imperial Airways. The introduction was arranged by G. P. Baker; see his article in *Journ. R.H.S.*, Vol. 54 (1929), pp. 389, 392–3 and figs 141, 142.

There are two thriving examples of *Z. abelicea* at Kew, one with a single trunk and about 20 ft high, the other with several stems and about 15 ft high. The species is also represented in the R.H.S. Garden at Wisley.

Z. CARPINIFOLIA (Pall.) K. Koch [PLATE 111

Rhamnus carpinifolius Pall.; *Zelkova crenata* (Michx. f.) Spach; *Planera crenata* Michx. f.; *Zelkova ulmoides* (Güldenstädt) Schneid.; *Rhamnus ulmoides* Güldenstädt; *Abelicea ulmoides* (Güldenstädt) Kuntze; *Planera richardii* Michx.

A tree 100 ft high, with a smooth, beech-like trunk, usually comparatively short (10 to 20 ft high), dividing into a great number of erect, crowded branches; bark peeling off in flakes; young twigs very downy. Leaves 1½ to 3 in. long, ¾ to 1¾ in. wide, ovate or oval, rounded or slightly heart-shaped at the base, with seven to eleven coarse sharp teeth down each side, dark green and with scattered hairs above, paler and more downy beneath; stalk about ⅛ in. long. Flowers on short twigs, the males at the naked base of the twigs, the females in the leaf-axils above them. Fruits about the size of a small pea, distinctly ridged above.

Native of the Transcaucasian forests of Russia, and of bordering Iran and northeast Anatolia; introduced to France in 1760 and to Britain probably at the same time, certainly by the 1780s. This remarkable tree is undoubtedly one of the most picturesque and distinct of any that can be grown in this country. It is slow-growing and long-lived, and might well be used as a commemorative tree. The densely clustered branches, much divided at their extremities, suggest a monstrous besom. The timber is of good quality, being tough and durable.

Early this century the largest tree in the country grew at Wardour Castle in Wiltshire, but this was felled about 1936; it was about 100 ft high and had no distinct trunk, but a clustered group of more than a dozen stems (Elwes and Henry, *Tr. Gt. Brit. and Irel.*, Vol. IV (1906), pl. 248).

At Kew there are three large specimens of this tree: in front of the Herbarium, 86 × 13 ft (1972; cf. 60 × 9¼ ft in 1906); behind the Herbarium, 84 × 12½ ft (1972); near the Main Gate, 77 × 11½ ft (1963). At Syon House, on the other side of the Thames from Kew, there are two old trees: 108 × 16¾ ft (1976) and 87 × 15¾ ft at 3 ft (1967).

ZELKOVA CARPINIFOLIA

Others recorded recently by Alan Mitchell are: Capel House, Enfield, Middx., 105 × 14½ ft (1976); Albury House, Surrey, 95 × 15¼ ft at 2 ft (1973); Worlingham, Suff., 92 × 21¼ ft at 2 ft (1968); University Parks, Oxford, 102 × 9¾ ft and 90 × 14½ ft (1975); Croome Court, Worcs., 82 × 23 ft at 3 ft (1964); Pitt House, Chudleigh, Devon, 82 × 19½ ft and 98 × 19¼ ft (1975); Bicton, Devon, 105 × 19 ft at 3 ft (1967); National Botanic Garden, Glasnevin, Eire, 70 × 12¼ ft (1974).

Z. SERRATA (Thunb.) Makino

Corchorus serratus Thunb.; *Corchorus hirtus* Thunb., not L.; *Ulmus keaki* Sieb.; *Planera acuminata* Lindl.; *Zelkova acuminata* (Lindl.) Planch.; *Z. keaki* (Sieb.) Maxim.; *Abelicea hirta* (Thunb.) Schneid.; *Zelkova hirta* (Thunb.) Schneid.; *Z. formosana* Hayata

A tree 100 or even 120 ft high in Japan, with a tall, smooth, grey trunk,

5 to 10 ft in diameter; young shoots at first slightly downy, soon becoming almost glabrous. Leaves ovate or ovate-lanceolate, 2 to 4½ in. long, ¾ to 2 in. wide, long and taper-pointed, rounded or slightly heart-shaped at the base, with six to thirteen coarse teeth at each side, each tooth with a short, slender point, dark green and furnished with short, scattered hairs above, paler and glabrous beneath; stalk ⅛ to ¼ in. long. Flowers produced in April and May on short twigs, the males being borne two or more together at each joint of the leafless bases of the twigs, the females solitary in the axils of the leaves at the end; both small, green, and of no beauty. Fruits roundish, about ⅛ in. in diameter. Fading leaves often good red and orange.

Native of Japan, Formosa and probably of continental E. Asia; introduced from Japan by J. G. Veitch in 1861. Although this distinctive species is one of the most important forest trees of Japan it has not succeeded so well in this country as *Z. carpinifolia*. With more spreading branches than in that species it makes a less striking tree, but still elegant and interesting. In a young state it is sometimes injured by spring frost. From *Z. carpinifolia* it is distinguished by the taper-pointed, thinner leaves with narrower, longer-pointed teeth. It is proving susceptible to dutch elm disease.

The following examples have been recorded: Kew, 56 × 6 ft (1974); Tilgate Park, Crawley, Sussex, 60 × 6¾ ft (1974); Lower Sheriffs Farm, West Hoathly, Sussex, from seeds brought back from Japan in 1890, 62 × 9 ft (1976); Whitfield House, Heref., 50 × 7½ ft (1973); Highnam Court, Glos., 58 × 6¾ ft (1970); Kilmacurragh, Co. Wicklow, Eire, 48 × 9¼ ft (1966).

Z. SINICA Schneid.

A deciduous tree 50 to 60 ft high, with a trunk up to 6 ft in girth; young shoots greyish woolly in the early part of the season, becoming bright brown and glabrous by autumn. Leaves firm in texture, alternate, ovate to ovate-lanceolate, rounded at the base, the apex mostly slenderly pointed, the margins coarsely toothed, ciliate; veins seven to twelve each side of the midrib, each one running out to the point of a marginal tooth, 1 to 2½ in. long, ⅔ to 1⅛ in. wide, dark dull green and harsh to the touch above, greyish beneath and downy especially on the midrib and veins; stalk very short, downy. Fruits solitary as a rule, produced from the under side of the leaf-axils, veined, very shortly stalked, roughly obovoid, ⅕ to ¼ in. wide.

Native of central and eastern China; introduced by Wilson to the Arnold Arboretum in 1908 and to Kew in 1920, when seeds were received from Messrs Vilmorin. Although known since the last century it was originally confused with *Z. serrata*, to which it is indeed closely related, but has leaves with fewer lateral veins and coarser teeth. It was first described in 1916.

A plant at Kew from the introduction of 1920 measures 32 × 3¼ ft (1974) and there are examples of about the same size in other collections.

Z. 'VERSCHAFFELTII'

Z. verschaffeltii (Dipp.) Nichols.; *Z. japonica* var. *verschaffeltii* Dipp.; *Ulmus campestris* var. *verschaffeltii* Lav.; *Ulmus pendula laciniata Pitteursii* Hort.

A small tree, or unless trained to a single stem, often a bush, with slender spreading shoots, slightly hairy when quite young; winter buds often in pairs. Leaves $1\frac{1}{2}$ to $2\frac{1}{2}$ in. long, $\frac{3}{4}$ to $1\frac{3}{4}$ in. wide, oval or ovate, with usually six to nine coarse, triangular teeth at the sides (usually fewer on one side than on the other), the larger teeth $\frac{1}{4}$ in. deep, upper surface dark green and with stiff, short hairs, lower surface with more numerous, softer hairs; stalk $\frac{1}{8}$ to $\frac{3}{16}$ in. long. Fruits (according to Henry) like that of *Z. carpinifolia*, but somewhat smaller.

There is no wild-collected specimen of this tree in the Kew Herbarium, and its origin is not definitely known. Dippel, who first distinguished it as a zelkova, and figured it in his *Handbuch der Laubholzkunde*, Vol. II, fig. 14, in 1892, suggested an Eastern Asiatic origin for it. It appears, however, to have considerable affinity with *Z. carpinifolia*, and is more likely to be of Caucasian origin. Henry suggests it may be a hybrid between *Z. abelicea* and *Z. carpinifolia*, but does not explain how such a cross can have been effected. It is a pretty and distinct shrub or tree, well marked by the deep angular cutting of the leaf-margins. It has been cultivated at Kew since 1886, and is perfectly hardy, slow growing, and forming a bushy head. It was long thought to be an elm, but it fruited at Paris in 1908, and was conclusively shown to be a zelkova.

ZENOBIA   ERICACEAE

A genus of a single species, having much in common with *Leucothoë* but distinguished from it by its campanulate flowers. It is part of *Andromeda* as understood by early botanists, and David Don, in splitting the genus, gave most of the segregate genera names that, like *Andromeda*, were drawn from Greek mythology. But *Zenobia* commemorates an historic personage—the self-styled Queen of Palmyra, who led a massive revolt in the eastern provinces of Rome in the third century A.D.

Z. PULVERULENTA (Bartr. ex Willd.) Pollard

Andromeda pulverulenta Bartram ex Willd.; *A. speciosa* var. *pulverulenta* Michx.; *Zenobia speciosa* var. *pulverulenta* (Michx.) DC.

A deciduous or sub-evergreen shrub of somewhat irregular, thin habit, 4 to 6 ft high, glabrous. Leaves alternate, oblong-ovate or oblong-elliptic, tapering at the base, pointed or rounded at the apex, shallowly toothed, they and the young shoots coated with a glaucous-white bloom. Flowers fragrant, pendent, borne in June and July, in axillary clusters on the terminal portion of the shoots of the previous year, each on a stalk ½ to ¾ in. long, forming in effect leafy or naked racemes 4 to 8 in. long. Corolla pure white, broadly bell-shaped, about ⅜ in. wide, with five shallow lobes. Calyx-lobes five, triangular, persisting at the base of the dry, flattish-globose (or orange-shaped) capsule.

ZENOBIA PULVERULENTA

Native of the eastern USA from southeast Virginia to S. Carolina; introduced around 1800. With fragrant flowers like large lilies-of-the-valley and glaucous young stems and leaves this is one of the loveliest of ericaceous shrubs. It is perfectly hardy and tolerates almost full sun, but suffers in droughty summers, especially when planted in the root-run of large trees, and the young growths may be cut by late frost. It needs a moist, peaty or leafy soil, and may be propagated by cuttings of half-ripened wood placed in gentle heat about July. This method is preferable to raising from seed, as there is some variation in the whiteness of the foliage and the size of flower, and the best forms cannot be relied upon to come true from seed. The flowering part of the shoot, from beneath which the young shoots spring, should be cut off as soon as the flowers have faded, if seed is not required.

Z. *pulverulenta* received a First Class Certificate in 1934.

f. NITIDA (Michx.) Fern. *Andromeda speciosa* var. *nitida* Michx.; *A. cassinefolia* Vent.; *A. cassinefolia* var. *nuda* Vent.; *Zenobia speciosa* (Michx.) D. Don; *Z. pulverulenta* var. *nuda* (Vent.) Rehd.; *Z. cassinefolia* (Vent.) Pollard—In all essential characters this is identical with typical *Z. pulverulenta*, but the leaves are green on both sides and the stems are scarcely glaucous. *Bot. Mag.*, t. 970.

This green-leaved form is less striking than the typical glaucous form, but still charming. It received an Award of Merit in 1934 as *Z. speciosa* and in 1965 as *Z. pulverulenta* var. *nuda*. It is still usually known in gardens as *Z. speciosa*, but is certainly not specifically distinct from *Z. pulverulenta* and, if it were, it would take the name *Z. cassineifolia*.

cv. 'QUERCIFOLIA'.—This has the leaf-margins set with shallow wavy lobes. A curiosity.

ZIZIPHUS   RHAMNACEAE

Ziziphus has some forty species of deciduous or evergreen trees and shrubs, natives mainly of tropical and warm temperate regions. Leaves alternate, three- or five-nerved from the base, stipules present, usually converted into spines. Flowers small, perfect, in axillary cymes. Petals five. Stamens five, opposite the petals. Fruit an edible, fleshy drupe; stone with usually two seeds.

The generic name derives from the ancient Greek word for *Z. jujuba*. It is sometimes wrongly spelt "*Zizyphus*".

Z. JUJUBA Mill.   JUJUBE
Rhamnus Zizyphus L.; *Z. sativa* Gaertn.; *Z. vulgaris* Lam.

A small deciduous tree up to 30 ft high, with glabrous, spiny branches; spines in pairs, the longer one up to $1\frac{1}{4}$ in. long, straight, the shorter one decurved. Leaves alternate, oval, ovate to ovate-lanceolate, shallowly round-toothed, blunt or rounded at the apex, glabrous, or downy only on the veins beneath; three-veined at the base. Flowers less than $\frac{1}{4}$ in. across, yellowish, borne two or three together on short stalks in the leaf-axils. Fruits fleshy and rather like small plums, roundish egg shaped, $\frac{1}{2}$ to 1 in. long, dark red, or almost black when ripe.

Native originally of temperate Asia as far east as China, but now widely cultivated and naturalised outside its natural range, notably in the Mediterranean region, to which it was introduced in Roman times. Although cultivated off and on in this country since the 17th century it does not succeed in our climate— perhaps more from lack of adequate summer heat than actual tenderness. The fruits have a pleasant acid taste when fresh, but are more palatable when dried; they are commonly eaten in both states in the Mediterranean. Some Chinese cultivated varieties were introduced to the USA in 1906 for trial and the tree has been grown there commercially, though only to a limited extent. In earlier centuries the fruits were used medicinally to alleviate coughs and sore throat and were imported into this country in large quantities from Provence and the Iles d'Hyères.

var. INERMIS (Bunge) Rehd. *Z. vulgaris* var. *inermis* Bunge—Branches unarmed. Originally described from N. China.

Z. LOTUS (L.) Lam. *Rhamnus lotus* L.—Allied to the preceding but always an intricately branched shrub, with greyish or whitish twigs and shallowly crenate leaves. The fruits are edible but less palatable than those of the common jujube. This is believed to be genuinely native in the warmer parts of the Mediterranean region, but its main distribution is in N. Africa and the Near East. It is tender.

Other species occurring in the lands bordering the Mediterranean region are Z. MAURITIANA Lam., a mainly tropical, wide-ranging species; and Z. SPINA-CHRISTI (L.) Desf. (*Rhamnus spina-christi* L.), valued both for its fruits and as a shade-tree in northern Africa, Arabia and the Near East.

INDEX

As the general arrangement of this work is alphabetical it has not been considered necessary to index names which appear in their proper sequence. The following is an index of 'popular' or English names; of the more important synonyms which, in accordance with the usual practice, are given in italics; and of a number of trees and shrubs which are not described in their alphabetical order but under related plants.

The attention of the reader is called to the glossary of botanical terms on p. 112, and to the glossary of nursery terms on p. 54 of Vol. I of this edition. The plates are listed on pp. xiii–xv; the line drawings on pp. xi–xii.

INDEX